Applied Artificial Intelligence for Drug Discovery

Antonio Lavecchia
Editor

Applied Artificial Intelligence for Drug Discovery

From Data-Driven Insights to Therapeutic Innovation

Springer

Editor
Antonio Lavecchia
Department of Pharmacy
University of Naples Federico II
Naples, Italy

ISBN 978-3-031-98021-3 ISBN 978-3-031-98022-0 (eBook)
https://doi.org/10.1007/978-3-031-98022-0

This Springer imprint is published by the registered company Springer Nature Switzerland AG
The registered company address is: Gewerbestrasse 11, 6330 Cham, Switzerland

Preface

The fusion of artificial intelligence (AI), machine learning (ML), and deep learning (DL) with pharmaceutical sciences is catalyzing a transformative era in drug discovery and development. Once the realm of theoretical exploration, AI now stands at the forefront of practical applications in drug design, development pipelines, and clinical innovation. The exponential growth of high-dimensional biological data and the advent of sophisticated algorithms have ushered in a new epoch in computational pharmacology. *Applied Artificial Intelligence for Drug Discovery* consolidates this burgeoning knowledge into a comprehensive, practical guide. Designed for a diverse audience, including pharmaceutical scientists, computational biologists, AI engineers, industry leaders, graduate students, and regulatory professionals, it serves as both an educational resource and a technical reference, bridging theoretical concepts with real-world applications across the drug discovery continuum.

Chapter 1 embarks on a historical journey, tracing the evolution of AI from ancient automatons to contemporary DL frameworks revolutionizing healthcare. By exploring philosophical and scientific roots, this chapter underscores how enduring human aspirations have shaped AI's emergence in biomedical research. Building upon this foundation, Chap. 2 addresses the integration of diverse and fragmented biological data, a central challenge in AI-driven drug discovery. It elucidates how researchers employ data mining, text extraction, and integrative algorithms to transform heterogeneous datasets into coherent, reliable models, essential for accurate training and interpretation. Chapter 3 delves into AI-enabled identification and validation of drug targets and pathways. It offers a comprehensive examination of how AI parses vast omics data to uncover novel therapeutic opportunities, assess target druggability, and propose new indications for existing molecules. The chapter also emphasizes model validation and navigates regulatory challenges in clinical adaptation.

In Chap. 4, the focus shifts to structure-based drug design (SBDD). AI tools, such as convolutional neural networks (CNNs), analyze 3D protein-ligand interactions, facilitating faster and more precise predictions of binding affinity and structural compatibility. This automation enhances hit discovery and boosts success rates in early drug development. Complementing SBDD, Chap. 5 explores ligand-based drug design (LBDD). Here, AI enhances the generation of predictive quantitative structure-activity relationship (QSAR) models and pharmacophore mapping. The chapter guides readers through chemical representation, from SMILES strings to 4D descriptors, demonstrating how ML methods improve predictability and interpretability, ultimately aiding lead optimization. Chapter 6 ventures into the promising realm of de novo drug design. Deep generative models, including generative adversarial networks (GANs), variational autoencoders (VAEs), and reinforcement learning (RL) architectures, enable the creation of novel chemical entities tailored to specific biological profiles. Case studies and performance benchmarks illustrate both the potential and practical limitations of these tools.

Chapter 7 shifts attention to peptide drug discovery, where AI significantly accelerates a traditionally slow and costly process. From predicting peptide sequences to evaluating stability and efficacy, ML and DL tools help overcome synthesis bottlenecks and explore broader chemical spaces. Chapter 8 concentrates on ADMET prediction, a critical phase in the drug pipeline. DL architectures, notably graph neural networks (GNNs) and multi-task learning models, enhance predictions of pharmacokinetics and toxicity. The integration of explainable AI (XAI) techniques ensures greater transparency and confidence in these predictions, vital for regulatory acceptance.

In Chap. 9, the narrative transitions to molecular dynamics (MD) simulations. AI reshapes MD through ML-based force fields and enhanced sampling strategies, enabling faster, more accurate predictions of molecular behavior and simulating complex biological systems with unprecedented resolution. Exploring future frontiers, Chap. 10 introduces quantum ML (QML) in drug discovery. With the anticipated advent of fault-tolerant quantum devices, this chapter outlines how QML could revolutionize molecule simulation and optimization, guiding readers through foundational concepts and current use cases.

Chapter 11 examines the AI-driven exploration of noncoding RNAs, particularly microRNAs, in therapeutic development. By illuminating their regulatory functions and disease associations, AI models facilitate the identification of novel RNA-based targets, impacting fields like oncology, neurology, and rare diseases. Chapter 12 reimagines retrosynthesis through AI. The chapter offers a deep dive into how algorithms like Monte Carlo Tree Search and beam search automate and optimize synthetic route planning. With access to massive reaction datasets like USPTO and Reaxys, AI systems now propose economically and thermodynamically viable synthetic strategies that can be readily validated in the lab. Chapter 13 introduces Active Learning (AL) as a dynamic solution to the longstanding problem of data scarcity and applicability domain limitations. In contrast to traditional ML pipelines, AL leverages iterative sample selection to prioritize the most informative data points, improving predictive accuracy while minimizing resource expenditure. Especially notable is the rise of Deep Active Learning (DeepAL), which combines AL with the representational power of DL, particularly useful for ADMET modeling and lead prioritization in low-data environments.

Chapter 14 shifts focus to one of the most disruptive tools in recent years—Large Language Models (LLMs). Initially designed for natural language tasks,

LLMs like GPT and BERT now revolutionize drug discovery by enabling target identification, biomolecular hypothesis generation, synthetic data creation, and trial design. Their capacity to reason across vast biomedical corpora accelerates innovation, though challenges in interpretability and regulatory adaptation persist. In Chap. 15, Contrastive Learning, a self-supervised learning technique, capitalizes on abundant yet underutilized unlabeled data in biomedical settings. This approach shows promise in learning robust molecular and pharmacological representations, especially when integrated with graph neural networks, enhancing applications in molecular similarity, protein–ligand binding, and virtual screening. Chapter 16 addresses Few-Shot Learning (FSL), tackling scenarios with inherently sparse data, such as rare disease therapeutics or early-stage target validation. FSL enables accurate model development from minimal samples, with architectural innovations like meta-learning and transfer learning, positioning it as a promising avenue for democratized drug discovery, particularly in neglected or underfunded areas.

The imperative for interpretability in AI systems is central to Chap. 17, which surveys the role of Explainable AI (XAI) in drug discovery. Techniques like SHAP, LIME, and Grad-CAM are examined not only as tools for insight but as necessities for building trust among regulators, clinicians, and researchers. Through case studies in biomarker discovery and lead optimization, the chapter underscores how transparency fosters collaboration and responsible AI adoption in high-stakes domains. Chapter 18 spotlights Federated Learning (FL), a transformative paradigm enabling distributed AI training across institutions without sharing sensitive data. This decentralized approach supports data privacy and regulatory compliance while promoting broader collaboration among academia, pharma, and healthcare systems. The chapter discusses challenges like data heterogeneity and communication overhead, highlighting FL's potential to unlock siloed datasets, particularly in multi-center clinical collaborations.

Chapter 19 explores AI's pivotal role in nanomedicine and drug delivery systems (DDSs). By modeling physicochemical parameters and integrating patient-specific biological data, AI supports the design of smarter, more precise nanoparticles. These intelligent delivery systems optimize release kinetics, enhance stability and bioavailability, and allow for tissue- and cell-specific targeting. Notably, AI-driven simulations are enabling real-time, adaptive therapeutic strategies, paving the way for "on-demand" drug release tailored to dynamic physiological conditions. As the convergence of AI with nanotechnology matures, its potential for highly personalized and responsive treatment regimens becomes increasingly tangible. Chapter 20 highlights how global health emergencies, such as the COVID-19 pandemic, have accelerated the adoption of AI in antiviral drug discovery. From virtual screening of vast chemical libraries to the de novo design of lead compounds, AI has proven instrumental in shortening development timelines without sacrificing scientific rigor. Through case studies, this chapter illustrates how AI methodologies provided rapid, cost-efficient responses during a time of critical need, reinforcing the value of AI as a strategic asset in preparedness for future pandemics.

Chapter 21 shifts focus to practical implementation by presenting a comprehensive, reproducible pipeline for AI-driven drug modeling. This chapter is especially valuable for students, educators, and practitioners. It outlines essential steps such as data preprocessing, feature engineering, molecular descriptor generation, and model validation, all reinforced with hands-on examples using an interactive Google Colab notebook. By providing an open, accessible environment for experimentation, this chapter demystifies complex AI workflows and empowers readers to apply state-of-the-art methods in real-world research. Chapter 22 surveys current AI-based web platforms for drug discovery, assessed through the lens of Human-Centered Design (HCD). In addition to reviewing ten prominent platforms, the chapter identifies design challenges, ranging from user experience and transparency to accessibility and interoperability, and offers actionable strategies for future improvements. The chapter advocates for inclusive interfaces that cater to nonspecialist users, fostering broader adoption across disciplines.

Chapter 23 delves into AI and ML-driven drug repurposing, an increasingly vital approach in an era of rising R&D costs and unmet clinical needs. The chapter categorizes repurposing strategies, from data-driven to hypothesis-based approaches, and showcases how cutting-edge techniques like GNNs, generative models, and AlphaFold-enabled structure predictions are revealing new indications for known drugs. Particular attention is given to addressing challenges in adverse drug reaction (ADR) prediction and pharmacokinetic profiling, positioning AI as a key enabler of cost-effective and risk-mitigated drug development. Chapter 24 explores the transformative role of AI in clinical trials: from protocol optimization to patient recruitment and pharmacovigilance. Examples of conversational AI for patient communication, federated learning for decentralized data analysis, and LLMs for automated document generation underscore the breadth of AI's contributions. By reducing bias, improving trial diversity, and streamlining operations, AI is redefining the modern clinical trial as more inclusive, efficient, and outcomes-focused. Expanding this perspective, Chap. 25 investigates the impact of generative AI in clinical studies. This chapter presents case studies demonstrating how foundation models and retrieval-augmented generation are revolutionizing site selection, cohort matching, and documentation workflows. While challenges around regulatory compliance and human oversight persist, the chapter emphasizes that generative AI holds the potential to significantly reduce manual workload and enhance the quality of clinical evidence. Chapter 26 culminates in a vision of AI-enabled precision medicine, where therapies are customized based on individual molecular, clinical, and environmental profiles. Integrating multi-omics data, digital twins, and real-time feedback loops, this chapter showcases how AI supports drug response prediction, patient stratification, and adaptive treatment strategies. As next-generation technologies like synthetic biology, quantum computing, and real-world data integration converge, AI-driven personalized therapeutics are poised to transform preventive and curative care. Finally, Chap. 27 synthesizes the central themes of the book, offering a forward-looking analysis of the challenges and opportunities in AI for drug discovery. It reflects on ethical imperatives, algorithmic transparency, regulatory harmonization, and the urgent need for scalable, interoperable, and trustworthy AI systems. The chapter calls for interdisciplinary collaboration and policy innovation to ensure that AI tools deliver safe, equitable, and impactful outcomes in pharmaceutical research.

The integration of AI into drug discovery pipelines is no longer aspirational; it is a rapidly evolving reality. Emerging tools such as AlphaFold2, ESMFold, GraphCast, OpenFold, and Meta's CM3leon are revolutionizing protein structure prediction, molecular simulation, and biomolecular interaction modeling. At the same time, AI agents powered by reinforcement learning, retrieval-augmented generation (RAG), and multi-modal architectures are unlocking new frontiers in experimental design, hypothesis generation, and decision support.

Looking ahead, the future holds immense promise:

- Self-driving labs that combine robotic automation with AI planning agents for fully autonomous hypothesis testing and synthesis
- AI-enhanced CRISPR design platforms for targeted genome editing with minimized off-target effects
- Real-time digital twins of patients, integrated with electronic health records and wearable sensor data, for continuous treatment optimization
- Quantum-classical hybrid AI systems capable of simulating molecular interactions with unprecedented precision

Yet, these advancements must be matched by a rigorous commitment to transparency, reproducibility, and ethical alignment. As AI grows in sophistication, so too must our systems of oversight, interpretability, and inclusion, ensuring that innovation serves the broadest possible societal good.

In this context, *Applied Artificial Intelligence for Drug Discovery* is more than a technical compendium. It is a call to action for researchers, engineers, clinicians, and regulators to collaboratively shape the future of pharmaceutical innovation. Whether you are building the next-generation algorithm, designing a clinical study, or navigating regulatory pathways, this book offers the foundational knowledge and practical tools to guide your journey.

I am sincerely grateful to the Editor at Springer Nature for granting me the opportunity to lead the development of this book. My special thanks go to Charlotte Nunes, Associate Editor of Pharmaceutical Sciences and Toxicology, for her invaluable support in ensuring the timely and successful publication of this volume. I would also like to extend my heartfelt appreciation to Dr. Carmen Cerchia and Dr. Pietro Delre for their expert review and editorial guidance throughout the process. Finally, I wish to express my deepest gratitude to all the contributing authors for their exceptional scholarship, dedication, and collaboration, without which this work would not have been possible.

Napoli, Italy Antonio Lavecchia

Contents

About the Editor

Antonio Lavecchia is Full Professor of Medicinal Chemistry at the University of Naples Federico II (Italy), where he leads the Drug Discovery Laboratory and serves as Scientific Director of the Molecular Modeling Excellence Laboratory (LMM). He received his Ph.D. in Pharmaceutical Sciences from the University of Catania in 1999, completing part of his doctoral research at the University of Minnesota, USA.

With a strong background in both experimental and computational medicinal chemistry, he is internationally recognized for his interdisciplinary expertise in drug design, molecular modeling, and the application of artificial intelligence (AI) in pharmaceutical research. His scientific work spans the development of novel algorithms, AI-based frameworks, and modeling platforms for accelerating the discovery and optimization of bioactive compounds across therapeutic areas such as oncology, metabolic diseases, inflammation, infectious diseases, and rare disorders. His academic output includes over 200 scientific publications in high-impact international journals, 5 books and book chapters, 6 patents, and over 280 conference presentations worldwide. He serves on the editorial boards of several international scientific journals and regularly acts as a peer reviewer and expert evaluator for major funding agencies and research institutions worldwide.

He ranks among the world's top 2% of scientists (Stanford University ranking) and is acknowledged as a global expert in PPAR nuclear receptor pharmacology and AI-driven drug discovery. He is co-founder of two biotech spin-offs and has been featured on the covers of the *Journal of Chemical Information and Modeling* and *ACS Omega* for his pioneering contributions to AI in drug discovery.

Through this volume, he brings together leading experts in the field to provide a comprehensive, forward-thinking resource that explores the transformative role of AI across the entire drug discovery continuum.

History of Artificial Intelligence and Drug Discovery

Antonio Lavecchia

Abstract This chapter traces the history of AI from ancient myths of mechanical beings to its modern applications in drug discovery. It begins with humanity's early fascination with thinking machines, illustrating how mythical automatons and theoretical ideas laid the groundwork for the conceptual development of AI. Moving into the scientific realm, the chapter explores the foundational advances that brought AI from theory to practice, detailing early laboratory work, initial setbacks, and the branching paths that followed. The narrative then shifts to AI's transformative role in drug discovery, examining the milestones that have shaped its evolution in the field. Through a historical lens, this chapter highlights the challenges and breakthroughs that have driven AI's integration into pharmaceutical research, setting the stage for the innovative approaches that continue to redefine drug discovery today.

Keywords Artificial intelligence · Drug discovery · History · Pharmaceutical research · Machine learning · Computational chemistry

1 Historical Evolution of Artificial Intelligence

Though the term "artificial intelligence" (AI) is modern, its roots trace back millennia through mythology and human innovation (Mayor 2020). AI's evolution can be viewed through three key phases: ancient myths of artificial life, enlightenment ideas of thinking machines, and the theoretical groundwork that ultimately gave rise to computing and AI as a field. While these phases are distinct, they have always overlapped, as myths and speculative ideas fueled theoretical advancements. This progression illustrates how our understanding and goals for AI have evolved over time.

A. Lavecchia (✉)
"Drug Discovery" Laboratory, Department of Pharmacy, University of Naples Federico II, Naples, Italy
e-mail: antonio.lavecchia@unina.it

A. Lavecchia (ed.), *Applied Artificial Intelligence for Drug Discovery*,
https://doi.org/10.1007/978-3-031-98022-0_1

1.1 *Ancient Visions of AI: Mythical Automatons and Mechanical Minds*

For centuries, myth and folklore have envisioned creations that parallel today's concept of AI. Greek mythology, in particular, abounds with characters embodying early imaginings of AI (Mayor 2020). Among them is Talos, a mechanical guardian crafted by Daedalus to defend Crete. Powered by "ichor," a life-sustaining fluid flowing through a tube within his metal frame, Talos patrolled the island, hurling stones at incoming ships a precursor to the mechanical super-soldier. This ancient legend now lends its name to a modern U.S. Army robotic exoskeleton. Daedalus himself, known as one of antiquity's greatest inventors, not only created Talos but was also reputed for designing animated statues and other intricate devices. His tale contrasts with that of the sorceress Medea, who ultimately deactivated Talos through deception, underscoring a theme of control and vulnerability inherent to artificial life. Medea's father, a powerful enchanter, further enriched the mythos of automation with his creation of soldiers capable of endless battle.

Beyond human inventors, Greek deities like Hephaestus and Prometheus contributed to this mechanical vision. Hephaestus, God of the forge, was assisted by autonomous machines and crafted tools with self-regulating functions, even creating a gate that opened independently. Prometheus, though a titan, "built" humanity and introduced fire, symbolizing knowledge and innovation. Zeus, in turn, crafted Pandora, a mechanical woman who, by opening her infamous jar, released suffering upon humanity; yet another mythic cautionary tale. In contrast, Pygmalion's love for his statue, brought to life as Galatea by Aphrodite, presents a more romanticized vision of artificial life, foreshadowing companionship.

Other cultures similarly explored artificial life. In Buddhist legend, the *Lokapannatti* tells of Emperor Ashoka, who sought Buddha's relics guarded by Roman-crafted mechanical sentinels (Zarkadakis 2016). Norse mythology features Hrungnir, a giant automaton built to battle Thor, while the ancient Chinese *Liezi* recounts the tale of Yan Shi, who crafted an automaton from wood and leather (Brynjolfsson and McAfee 2014). Estonia's Kratt legend describes a magical creature made of household items that performed its owner's tasks, yet could turn dangerous if idle, a concept that inspired Estonia's "Kratt Law" on algorithmic liability. In the sixteenth century, Swiss physician Paracelsus claimed to create a homunculus through alchemy, describing a process involving sperm, magnets, and mystical transformation (Newman 2019). Though disputed, his work in toxicology is still respected for establishing the principle "Sola dosis facit venenum" (Only the dose makes the poison). Paracelsus described his homunculus as follows:

> "The sperm of a man is sealed and putrefied for forty days in a horse's womb, ultimately becoming a transparent, small, human-like figure, which, when nourished with the Arcanum of human blood and warmed, develops into a tiny living child".

Another legend from the sixteenth century describes the *golem*, an artificial being animated by sacred words. Around 1550, Rabbi Elijah of Chelm reportedly created a golem by inscribing the Name of God (YHWH) upon it. The creature ultimately grew monstrous, threatening the world until its creator removed the holy name, rendering it lifeless again. A few decades later, in Prague, Rabbi Judah Loew reportedly crafted a golem to protect the Jewish community from persecution. Assisted by two followers, Loew brought a clay figure to life using ritual incantations and verses from *Genesis*. The golem became a loyal protector and temple servant, resting on the Sabbath. However, it eventually grew uncontrollable, and Loew had to remove the sacred word from its forehead, transforming it back into clay.

1.2 Early Fascination with Thinking Machines

The "mechanization of the world" envisioned by thinkers such as Galileo, Newton, and Descartes initiated a new era of mechanical philosophy, inspiring not only groundbreaking devices but also early speculation about AI (Dijksterhuis 1961). Although true AI was still unimaginable, the shift from myth to mechanics laid fertile ground for exploring the potential of thinking machines. In *Discourse on Method* (1637), Descartes pondered machine intelligence, proposing that while machines might mimic actions or speech, they would lack true comprehension and human adaptability (Russell 2022). For Descartes, mind and matter existed on separate planes governed by different laws, making genuine machine thought impossible. Contrastingly, Thomas Hobbes saw thought as a series of mathematical operations, a notion that foreshadowed AI's computational framework. With calculators emerging soon after, Blaise Pascal and Gottfried Leibniz developed early mechanical devices for addition and subtraction. Leibniz advanced this by adding multiplication, division, and binary representation, a concept that remains foundational to modern computing (Broussard, 2018).

In 1769, Wolfgang von Kempelen unveiled the "Mechanical Turk," a chess-playing automaton that fascinated Europe for decades. Defeating figures like Napoleon and Benjamin Franklin, the Turk was eventually exposed as a hoax controlled by a hidden player (Zarkadakis 2015). Today, Amazon's "Mechanical Turk" platform honors this deception, employing unseen human workers for diverse tasks.

Speculation on AI also flourished in the arts and literature. In *Gulliver's Travels* (1726), Jonathan Swift introduced "the Engine," an early fictional precursor to computing that generated random combinations of words, enabling anyone, "without the least assistance from genius," to produce texts on diverse topics. Goethe's tale of the *Sorcerer's Apprentice*, later popularized by Disney's *Fantasia*, depicted the unintended consequences of magical automation, an apprentice's runaway broom parallels today's fears of autonomous technologies (Wiener 1966). Similar narratives, such as *Pinocchio* and W.W. Jacobs' horror tale *The Monkey's Paw*, reveal the dangers of unintended consequences when attempting to harness supernatural powers. The line between fantasy and scientific speculation blurred further with the dawn of science fiction. In 1816, Mary Shelley's *Frankenstein* was born out of a retreat near Geneva amid the "Year Without a Summer" (Zarkadakis 2015). This archetype of a creator losing control over his creation foreshadowed the risks of unchecked technology, a theme that endures in films such as *Blade Runner* (1982), *The Terminator* (1984), and *The Matrix* (1999). In 1920, Karel Čapek introduced the term "robot" in *Rossum's Universal Robots*, deriving from the Old Church Slavonic *rabota*, or forced labor. Čapek's rebellious robots that annihilate humankind reflect an enduring fear of artificial autonomy (Rid 2016). This marked the end of speculative fiction on AI, paving the way for structured inquiries into machine intelligence and its societal implications.

1.3 From Theory to Reality: The Foundations of AI

In the latter half of the nineteenth century, AI advanced speculative fiction to serious theoretical discourse, aligning with early computer development. Key figures like Ada Lovelace and Charles Babbage laid the groundwork, with Babbage's 1834 "Analytical Engine" (Boden 2018) marking a milestone in mechanical computation. Lovelace envisioned the machine not only as a calculator but capable of complex tasks, including composing music and enhancing scientific research, a vision symbolized in her early algorithms that marked the birth of programming (Freeman and Louçã 2001; Russell 2022). This potential remained unrealized until World War II, when anti-aircraft systems required rapid computational responses to incoming air raids. The complexity of calculating missile trajectories in real-time underscored human limitations and accelerated the need for automated computation. These wartime innovations sparked interest in cybernetics in the 1950s, exploring automation and human oversight. Military experts highlighted the "human element" as a critical weak link in precision operations, fueling designs for fully automated systems (Rid 2016).

Parallel advances came from Britain's covert Colossus project, which aimed to decode the German Enigma cipher. Among its masterminds was Alan Turing, now regarded as the father of modern computing and AI. Turing's development of a universal computer in 1948 and his proposal of the famous "Turing Test" in 1950 laid essential groundwork for AI (Turing 2009). His test, an "imitation game," challenged a machine to produce human-like responses, a benchmark still used in AI research to assess language and pattern recognition. Critics, however, note the test's language bias. AI scientist Yann LeCun argues that intelligence extends beyond language, citing animals' capacity for understanding and tool use without complex communication (Ford 2018). Simultaneously, Warren McCulloch and Walter Pitts published a seminal paper combining Turing's computational theories with Bertrand Russell's logic and synaptic theories by Charles Sherrington (McCulloch and Pitts 1943). They introduced the concept of binary states ("true or false") as analogous to the "on/off" nature of neurons and computer values (0 and 1), bridging neurophysiology with logical computation. John von Neumann further shaped AI's framework, conceiving computer architectures with central processors, memory, and input–output devices (Freeman and Louçã 2001). Concurrently, Norbert Wiener's work in "cybernetics pioneered the study of control and communication in both machines and organisms, introducing feedback as a method for self-regulation" (Wiener 2019). By adapting actions based on previous outcomes, systems could maintain homeostasis, a principle Wiener likened to thermostats and servomechanisms. Though cybernetics as a standalone field was brief, its insights on control, communication, and feedback profoundly influenced various disciplines (Rid 2016). These advancements propelled AI from theory to experimental reality, setting the stage for rapid development. By 1956, the foundational knowledge and technological base were in place, marking the formal beginning of the AI era and launching a transformative period in computing and machine intelligence.

1.4 AI Enters the Lab: Origins and First Setbacks and Diverging Paths

1.4.1 AI's First Wave

The official birth of AI as a scientific field dates to the 1956 Dartmouth Summer Research Project on Artificial Intelligence, led by John McCarthy and Marvin Minsky. This six-week seminar marked the first concerted effort to explore machine intelligence, attended by pioneers like Herbert Simon, John Nash, and Arthur Samuel, who transformed AI from a speculative concept to a practical experiment. Later, McCarthy and Minsky founded the MIT AI Lab, a groundbreaking institution that would evolve into today's MIT Media Lab (Broussard, 2018). This period, often called the first "AI spring," brought promising advancements. Various programs were developed to play draughts (checkers), but none were particularly skilled. Nonetheless, Arthur Samuel, a pioneer in computer gaming and the first to coin the term "machine learning" (ML), created a version that managed to defeat its human creator despite his modest skill. In 1964, Wiener noted that this learning method was analogous to human learning, and he predicted a similar fate for chess within two decades, foreseeing that AI would dampen human interest in such games (Wiener 1966). Researchers also developed systems for reasoning, such as the *Logic Theory Machine*, which proved theorems of Bertrand Russell, and the *General Problem Solver*, which tackled logical puzzles through goals, subgoals, and operators (Boden 2018).

By the mid-1960s, AI programs could pass basic tests and solve calculus and geometry problems (NNs), yet practical applications were elusive. Despite significant technical advancements, AI robots like Stanford's "Shakey," General Electric's Beetle, and an exoskeleton designed to help humans lift heavy weights lacked real-world functionality, reflecting the gap between theory and practice (Rid 2016; Russell 2022). Nonetheless, optimism remained high; Herbert Simon predicted in 1965 that AI would rival human labor within two decades, and Irving Jack Good foresaw an "intelligence explosion" that would reshape society

(Brynjolfsson and McAfee 2014; Rid 2016). These ideas even inspired speculative fiction, like *Colossus: The Forbin Project*, which explored the notion of AI controlling complex systems for human peace but at the cost of freedom. But the gap between ambition and reality became increasingly apparent. Criticism grew, with philosopher Hubert Dreyfus dismissing AI's prospects in his study *AI and Alchemy* (1965). Similarly, the US government's 1966 Automatic Language Processing Advisory Committee report noted limited progress, leading to reduced funding, as did the UK's 1973 Lighthill Report (Leung 2019). Many early AI systems faced the "combinatorial explosion" problem, wherein exponential growth in possibilities outstripped computational resources. The systems needed heuristic approaches to manage these complexities, but the technology was not yet available. Furthermore, data limitations and hardware constraints slowed progress. Machine translation efforts during the Cold War, for instance, yielded such mistranslations as "the spirit is willing, but the flesh is weak" becoming "the vodka is good, but the meat is rotten" (Russell and Norvig 2016). The initial optimism faded, giving way to an "AI winter" as funding and interest waned, thus concluding AI's first wave.

In these early years, two main AI approaches took shape: symbolic AI and neural networks (NNs), also known as "connectionism." Symbolic AI, dominant from the 1970s onward, involved encoding human logic into "IF X, THEN Y" rules, enabling machines to follow predetermined formulas. In contrast, NNs aimed to simulate the brain's functioning by creating "artificial neurons" that could recognize patterns in data. Rooted in Donald Hebb's 1949 theory that "neurons that fire together wire together," this approach emphasized pattern learning over explicit rules (Domingos 2015). Both methods had early advocates. While many at the 1956 Dartmouth conference championed symbolic AI, Cornell University was simultaneously developing the first artificial neuron (Greenfield 2017). Symbolic AI approaches tackled image recognition with explicit logic-based rules, while NNs relied on training across numerous examples to identify patterns. Early NNs applications, like Frank Rosenblatt's "perceptron," garnered media attention in the 1960s for recognizing letters without predefined rules. However, symbolic AI remained the favored approach in early AI systems like the Logical Theory Machine and General Problem Solver. A pivotal critique of ANNs came in 1969, when Marvin Minsky and Seymour Papert's *Perceptrons* outlined fundamental mathematical limits of the approach, which many saw as a turning point against connectionism (Ford 2018). With minimal progress in NNs, this skepticism contributed to the first "AI winter."

1.4.2 AI's Second Wave

A resurgence in AI research began in 1982 when the personal computer was named *Time* magazine's "Man of the Year," marking the start of AI's second wave. Prolog, a language for logic programming, emerged, and Japan's ambitious Fifth-Generation Computer Systems Project aimed to make Japan a leader in AI with advanced parallel computing (Russell 2022). In the United States, the

Microelectronics and Computer Technology Corporation (MCC) and projects like Cyc, led by Douglas Lenat, sought to capture human knowledge within computers (Domingos 2015). U.S. DARPA's Strategic Computing Initiative pledged $1 billion for AI research, focusing on hardware and human–computer interfaces, while the U.K. launched the Alvey Programme as a counterpart (Leung 2019; Russell and Norvig 2016). During this second wave, symbolic AI achieved commercial viability through expert systems. Programs like MYCIN (Duda and Shortliffe, 1983) and DENDRAL (Lindsay et al. 1993) codified expert knowledge in medical diagnosis and chemistry. Other systems were designed for planning manufacturing processes and solving complex mathematical problems, such as the Macsyma project. By the 1980s, expert systems had expanded into social security and criminal sentencing, marking AI's first commercially viable applications in the Netherlands (Hage and Verheij 1999). The Dutch Research Council, in collaboration with various universities and government departments, played a pivotal role in advancing legal AI applications, establishing platforms like JURIX, the Foundation for Legal Knowledge-Based Systems, an organization of "legal tech" researchers from the Netherlands and the Benelux Association for Artificial Intelligence (Leung 2019). In 1984, the New York Times hailed expert systems as a means of providing "computer-aided decisions based on more wisdom than any one person can contain" (Dreyfus and Dreyfus 1986). Despite widespread enthusiasm, the wave faltered as the Japanese, United States, and European AI initiatives fell short of expectations. Expert systems struggled with scalability, often failing when faced with minor contradictions, leading to the decline of specialized companies and another AI winter by the late 1980s (Leung 2019).

1.4.3 AI's Third Wave

In the 1990s, AI once again gained momentum, entering a new period of remarkable achievements. A major milestone came in 1997, when IBM's Deep Blue defeated world chess champion Garry Kasparov, a victory considered groundbreaking at the time. Its successor, Watson, achieved another landmark in 2011 by defeating reigning champions on the U.S. television quiz show *Jeopardy!*, a feat seen as a significant step toward AI mastery of human language. However, both achievements largely hinged on computing power rather than autonomous intelligence, sparking a debate over symbolic AI's fundamental limitations. Increasingly, researchers questioned whether AI could advance beyond programmed rules toward a truly adaptive form of intelligence (Ihde 2010; Russell 2022).

In parallel, a new paradigm was rising in NNs. Interest in this approach surged during the 1980s with the development of the backpropagation algorithm, which trained multilayer NNs for improved pattern recognition. Concurrently, the U.S. Department of Defense soon recognized the potential of NNs, reinvigorating funding for "parallel distributed processing." In 1985, John Haugeland coined the term GOFAI ("good old-fashioned AI"), which has since taken on a pejorative connotation when referring to symbolic AI. Around the same time, Judea Pearl began applying probability theory to AI, moving away from rigid logical frameworks. These developments began to challenge rule-based AI. A pivotal backpropagation paper was rejected from a major AI conference in the early 1980s, highlighting skepticism about the approach. Recognition came slowly; in 2004, Jeff Hawkins noted AI's image recognition still lagged behind that of a mouse (Tegmark 2018). Experts initially believed it would take a century for a computer to beat a human in Go. This was overturned when Google's AlphaGo defeated champion Lee Sedol in 2016, a feat enabled by breakthroughs in NNs driven by Yann LeCun and Andrew Ng. Geoffrey Hinton, credited as a key figure in this evolution, along with David Rumelhart and Ronald Williams, popularized the backpropagation algorithm in a seminal 1986 *Nature* paper. This algorithm traces contributions from the output layer through hidden layers, enabling performance improvements. Originally limited to a single hidden layer, backpropagation now accommodates multiple layers, effectively addressing the challenge of hierarchical representation in artificial (NNs). NNs have since found applications in various fields, including stock market price simulations, showcasing their transformative potential in modern AI. In 1989, Yann LeCun applied backpropagation to NNs for recognizing handwritten postcodes, introducing convolutional neural networks (CNNs), which break down complex images into smaller components to enhance image recognition efficiency, marking a pivotal step in AI development (Agrawal et al. 2022). In 2012, Geoffrey Hinton further advanced NNs by introducing "dropout," a technique to address overfitting, where models overly rely on training data and struggle to generalize with new information. Dropout made NNs more robust and led to "deep learning" (DL), in which multiple network layers successively refine representations, from identifying basic shapes to recognizing entire objects like faces (Domingos 2015). Reinforcement learning (RL) is another critical area in AI, focusing not on right or wrong answers but on optimizing strategies. Here, algorithms learn by being rewarded for effective moves, allowing them to develop winning strategies in games like *Pac-Man*, *Atari*, and poker by correlating actions with outcomes.

AI's progress surged in 2012 when Hinton's team dramatically reduced error rates in computer vision, establishing DL as a cornerstone of AI advancement. Breakthroughs followed swiftly: Facebook's DeepFace achieved high-accuracy facial recognition in 2014, and Amazon launched its virtual assistant, Alexa. In 2016, Google's AlphaGo defeated world Go champion Lee Sedol, an achievement previously thought decades away, and Uber piloted self-driving cars in Pittsburgh. In 2018, Alibaba's language processing AI outperformed humans, and OpenAI launched the first Generative Pre-Trained Transformer (GPT), marking the dawn of transformer-based generative AI.

The 2020s saw the refinement of transformer-based Large Language Models (LLMs), like GPT-3 and GPT-4, which excel in generating human-like text. OpenAI's GPT-n series has advanced with each model by increasing trainable parameters and integrating task-specific functions. ChatGPT is a prime example of a GPT model in use, delivering conversational responses and fine-tuning its output based on instructions. In 2023, GPT-4 became OpenAI's most sophisticated model. Following OpenAI's lead, EleutherAI, Cerebras, Salesforce, and Bloomberg developed their own specialized GPT models, such as "EinsteinGPT" for customer relationship management and "BloombergGPT" for financial analysis.

The recent boom in AI is rooted in foundational research primarily led by university scientists, though companies like Google have quickly absorbed top researchers. Two core factors underpin AI's rapid progress: increased processing power and the explosion of data. Following Moore's Law, computing power has grown exponentially, making today's devices far more capable and affordable. Initially limited by slow processing, AI benefited from the advent of graphics processing units (GPUs) and tensor processing units (TPUs), which enable efficient parallel processing for ML tasks (Kelly 2016). Simultaneously, the internet has generated massive data growth. Users' digital footprints, such as tagged Facebook photos, provide labeled data that enhance AI training. ImageNet, an open database of 14 million labeled images, exemplifies this data resource, as does the "internet of things" with its growing network of connected sensors.

2 AI in Drug Discovery: A Historical Perspective

2.1 Early Developments and Pioneering Systems (1960s–1970s)

AI's role in drug discovery originates from early computational research, where pioneers foresaw its transformative impact on healthcare. These initial breakthroughs established the foundation for today's advanced AI applications, driving remarkable technological progress in the field. In the 1960s, researchers at Stanford University developed DENDRAL (*Dend*ritic *Al*gorithm) (Lindsay et al. 1993), one of the first AI systems created to automate decision-making in organic chemistry by interpreting mass spectrometry data for molecular structure identification. This early effort bridged AI with chemistry and bioinformatics, sparking future advancements. Shortly thereafter, in the 1970s, Stanford introduced MYCIN, an expert system that analyzed patient symptoms and laboratory results to diagnose infections and recommend treatments. MYCIN exemplified AI's potential in clinical decision support, influencing future medical expert systems and demonstrating the role of AI in healthcare. These early systems laid the groundwork for the transformative role AI would play in both healthcare and pharmaceutical research, with implications that continue to evolve today.

The 1970s and 1980s also saw AI's role in drug discovery expand through the development of Quantitative Structure-Activity Relationship (QSAR) models. The idea of linking chemical structure with biological activity, first hypothesized by Crum Brown in 1869 (Brown and Fraser 1869), was formalized in the 1960s by researchers like Corwin Hansch and Toshio Fujita (1964). This provided the foundation for QSAR as a formal discipline, with Hansch recognized as the "father of QSAR." Other researchers, like Free and Wilson, also contributed to predicting toxicity through computational methods (Free and Wilson 1964), while Yvonne Martin worked with Hansch to further refine QSAR approaches, marking the emergence of computational chemistry as a distinct branch of physical chemistry. Building on these ideas, AI systems in the late 1960s and 1970s began attempting multi-step organic synthesis planning. Researchers aimed to create systems that could identify "checkmate moves" within vast chemical spaces and suggest optimal synthetic routes for producing new molecules (Engkvist et al. 2018; Gasteiger et al. 2003; Ravitz 2013; Todd 2005; Yadav 2017). Notable efforts by E.J. Corey (Corey and Wipke 1969), K. Djerassi (Varkony et al. 1978), and I. Ugi (Ugi et al. 1988) explored the computational modeling of chemical reactions and synthetic pathways. Corey's LHASA (Logic and Heuristics Applied to Synthetic Analysis) system, the most advanced of its era, allowed for stepwise synthesis planning under human guidance, serving as an early "synthetic assistant" without full automation. However, these initial attempts faced significant limitations, as neither adequate computing power nor suitable algorithms were available. The challenge was immense: machines needed to learn the extensive rules governing each reaction type, navigate complex synthetic networks, and sequence individual reactions into coherent, "elegant" synthetic routes. Ideally, such designs would be not only chemically consistent but also experimentally viable. Despite the ambitious vision, these initial steps set the stage for the more sophisticated AI-driven synthesis tools developed later.

In 1970, Sidney Fernbach and Abraham Haskell Taub likely coined the term "computational chemistry" in *Computers and Their Role in the Physical Sciences*, stating, "It seems, therefore, that 'computational chemistry' can finally be more and more of a reality" (Fernbach and Taub 1970). Supported by physical chemistry, early computational chemists successfully addressed spectroscopy and molecular conformation, revealing valuable correlations between experimental and calculated properties. However, collaboration with medicinal chemists was minimal. Medicinal chemists, holding higher positions in pharmaceutical companies, were often reluctant to pursue chemical structures designed by computers, viewing synthesis as far more complex than mere structure design. Synthesizing large numbers of compounds was often impractical for experimental chemists, leading pharmaceutical companies to bridge this gap by organizing workshops and conferences. Despite these efforts, medicinal chemists were slow to embrace computer-generated structures. Notably, new companies like Merck and SmithKline and French began integrating computational methods, while others, like Lilly, focused heavily on strengthening computational chemistry within their operations (Ekins 2006).

2.2 *Rise of Computational Chemistry and QSAR (1980s–1990s)*

The 1980s marked a significant turning point in computational drug discovery, with major breakthroughs in QSAR. One of the key milestones was the first QSAR Gordon Conference in 1975, which catalyzed the acceptance of QSAR as a formal discipline (Van Drie 2007). In the following years, Yvonne Martin's 1978 book *Quantitative Drug Design* helped solidify QSAR's importance in drug research (Martin 2010). With the support of organizations such as the U.S. National Cancer Institute, research into anticancer agents and structure-based drug design (SBDD) flourished (Ekins 2006). The elucidation of the dihydrofolate reductase (DHFR) structure in the 1970s, for example, led to the development of potent inhibitors beyond trimethoprim and methotrexate (Matthews et al. 1978, 1979). At the same time, Adamson and colleagues applied regression analysis to explore the biological activity of molecular fragments (Adamson and Bawden 1975), which led to the development of fragment descriptors used in substructure searches within large chemical databases. These descriptors were vital for similarity searches, database clustering, and QSAR modeling, setting the stage for further advances in large-scale chemical analysis (Baskin 2008).

In the early 1980s, the pharmaceutical industry was optimistic about the potential of computational chemistry, and by the late 1980s and early 1990s, computer-aided drug design (CADD) had become a focal point in academia (Van Drie 2007). It was widely proclaimed that drug design would soon be achieved *atom-by-atom*. At this peak, Fortune magazine even heralded CADD as the "next Industrial Revolution" in 1981. While computational chemists initially focused on elucidating chemical phenomena at the atomic and molecular levels, CADD quickly established itself as a distinct field, giving rise to innovations such as virtual library design, virtual screening, and de novo molecular design (Lavecchia and Di Giovanni 2013). Predictions regarding physicochemical and biological properties began to emerge, driven by AI methods rooted in theoretical frameworks proposed as early as the 1960s by researchers like Ivakhnenko and Lapa (1967), along with the pioneering Dartmouth workshop of 1956.

Throughout the 1980s, notable AI-based expert systems began to proliferate in both medicine (Duda and Shortliffe, 1983; Sandell and Bourne 1985) and pharmaceutical research (Klopman et al. 1987). One of the most notable contributions was Klopman's Computer-Automated Structure Evaluation (CASE) program, an ML system that identified molecular descriptors associated with biological activity and toxicity prediction (Klopman et al. 1987; Klopman and Contreras 1985). CASE, followed by MULTICASE, proved invaluable in VS and regulatory testing, offering a nonanimal alternative for evaluating compound toxicity and providing insights into biophore structures, which were critical for the development of safer drug candidates (Klopman 1985; Rosenkranz and Klopman 1988). In this period, Richard Cramer introduced 3D-QSAR methods such as Comparative Molecular Field Analysis (CoMFA) and its successor, Comparative Molecular Similarity Indices

Analysis (CoMSIA) (Cramer et al. 1988; Marshall and Cramer 1988). The creation of public molecular databases, such as the collaborative Erlangen/Bethesda Data and Online Service launched by the Computer Chemistry Center (CCC), the University of Erlangen-Nuremberg, and the U.S. National Cancer Institute's (NCI) Laboratory of Medicinal Chemistry, greatly expanded access to molecular information. The open-access NCI Database, encompassing synthetic and natural compounds with potential anticancer properties, became an invaluable resource (Monga and Sausville 2002). Tools like MACCS and ISIS revolutionized drug discovery by introducing similarity searching, enabling the identification of structure-related analogs within chemical databases (Boyd and Marsh 2006). The 1980s also saw the early use of docking techniques, which would gain prominence in the 1990s as more protein crystal structures became available, facilitating ligand design (Torres et al. 2019). Simultaneously, advancements in de novo ligand design to fit receptor structures paved the way for the virtual screening of synthesizable hypothetical structures (Lavecchia and Di Giovanni 2013; Guedes et al. 2014). During the same period, innovative algorithms for synthetic design and reaction prediction also emerged at a rapid pace. Notably, the Computer-Assisted Mechanistic Evaluation of Organic Reactions (CAMEO) analyzed nucleophilic and electrophilic sites to predict reactions (Jorgensen et al. 1990; Salatin and Jorgensen 1980). EROS identified leading structures for retrosynthesis based on bond polarity, electronegativity, and resonance effects (Gasteiger et al. 2022). SOPHIA predicted reaction outcomes with minimal user input (Satoh and Funatsu 1995, 1996), while KOSP was designed to forecast retrosynthetic targets (Satoh and Funatsu 1999). While groundbreaking, these tools often possessed limited applicability. Concurrently, combinatorial chemistry, coupled with computer-controlled robotics, revolutionized compound synthesis. This approach enabled a single medicinal chemist to produce thousands of compounds weekly, in stark contrast to the traditional method of synthesizing just one (Geysen et al. 2003). High-throughput screening (HTS) further accelerated this process, facilitating efficient data management across extensive compound libraries (Pereira and Williams 2007). Lipinski's "Rule of Five," introduced in 1997, revolutionized drug discovery by guiding the selection of lead compounds with optimal pharmacokinetic properties (Lipinski et al. 2001). The convergence of computational chemistry, combinatorial chemistry, and HTS significantly reduced drug development timelines.

2.3 *AI-Driven Drug Discovery in the Modern Era (2000s–Present)*

By the 1990s, skepticism about AI's potential emerged due to inconsistent validation and significant variability in methodological approaches (Yang et al. 2019). Despite these early doubts, the 2000s witnessed a renewed interest in artificial NNs, particularly for QSAR studies (Ma et al., 2015). Researchers began developing multi-layer ANNs to predict mechanisms of action for NCI-indexed anticancer drugs (Weinstein et al. 1992), while others focused on simulating molecular evolution for rational drug design (Schneider et al. 1994; Schneider and Wrede 1994). This resurgence of ML DL algorithms in the early 2000s, encompassing decision trees, random forests (RF), support vector machines (SVM), Bayesian networks, and expectation-maximization techniques, was fueled by advancements in QSAR, molecular modeling, enhanced computing power, big data, and high-speed internet (Lavecchia 2015). By the mid-2000s, QSAR studies advanced further with multidimensional methods, such as 4D-QSAR, enhancing molecular property predictions and providing a fresh approach to virtual screening in drug discovery (Lill 2007; Vedani et al. 2007).

The completion of the Human Genome Project in 2003 (Sawicki et al. 1993) provided a significant push for genetic research, particularly in the identification of disease-linked genes. In parallel, new "omics" fields, including genomics, proteomics, and metabolomics, emerged, shifting the focus of disease research toward a molecular-level understanding of DNA, RNA, proteins, and other molecules, ultimately laying the foundation for precision medicine and personalized drug delivery. Pharmaceutical companies like AstraZeneca began prioritizing the "Five Rs" in drug development: the right target, patient, tissue, safety profile, and commercial potential, aligning closely with the principles of precision medicine (Cook et al. 2014). Nevertheless, high costs and limited analytical capabilities for genomic data remained obstacles throughout the 2000s. The expansion of combinatorial chemistry and HTS, developed in the 1990s, led to the creation of numerous public databases containing molecular structures, biological activities, and safety data. The release of the NCI Open Database and the subsequent emergence of PubChem, ChEMBL, ChemSpider, DrugBank, and others facilitated navigation through chemical space (Brown et al. 2020). This immense volume of data required advanced AI for efficient retrieval and analysis. Pioneers such as Michael Cox and David Ellsworth coined the term "big data" to characterize these enormous datasets, which were scrutinized for novel patterns and relationships (Cox and Ellsworth 1997). IBM's framework of the "five Vs" (volume, variety, velocity, variability, and value) emerged as a way to understand big data's potential in drug discovery (Demchenko et al. 2013). By 2003, the exponential growth of "omics" data further highlighted the need for AI-driven analysis.

Post-2010, deep neural networks (DNNs) gained significant prominence, eclipsing traditional ML in predictive capabilities. Notably, during the Kaggle challenge, a team from Toronto effectively employed DNNs for QSAR modeling using Merck's drug discovery datasets (Ma et al., 2015). The NIH's Tox21 challenge in 2014 further underscored the efficacy of DNNs by successfully predicting the toxicity of over 12,000 compounds (Mayr et al. 2016). While ML reigned supreme in the 1990s and early 2000s, DL has since taken the lead in areas such as image analysis, molecular design, reaction prediction, and bioactivity forecasting (Lavecchia 2019).

Despite rising drug development costs (now averaging $1.8 billion per drug), AI advancements have accelerated discovery timelines. Drug development, once taking 20 years, now averages around 10, facilitated by computational methods, CADD, and automation (Paul et al. 2010; Sakaue and Okada 2017). This transformation reflects a more collaborative, interdisciplinary approach in modern drug discovery, integrating biology, biophysics, and computational science (Cao et al. 2018). AI now serves as a cornerstone of CADD, enhancing early-stage predictions and reducing time spent on preclinical screening (Chan et al. 2019). Additionally, the field is poised for transformative change with quantum computing advancements, particularly in noisy intermediate-scale quantum (NISQ) and fault-tolerant quantum computing (FTQC) devices, which promise to elevate molecular simulations to unprecedented levels of speed and precision. Within this context, CADD acts as a primary filter for HTS libraries and serves as a secondary screen for preclinical predictions related to pharmacokinetics, pharmacodynamics, and toxicity profiles. A striking example of these advancements is Halicin, a novel antibiotic developed by researchers at MIT, who utilized ML to target drug-resistant bacteria (Stokes et al. 2020). The algorithm, trained on thousands of known compounds, swiftly identified Halicin, which was subsequently validated through laboratory testing, highlighting AI's crucial role in accelerating antibiotic discovery. In the field of oncology, DS-6157a was discovered through a collaboration between Sumitomo Pharma and Roivant Sciences, effectively targeting GPR20, a protein associated with gastrointestinal stromal tumors, yielding promising preclinical results (Iida et al. 2021). Furthermore, Exscientia and Sumitomo Dainippon Pharma collaborated to develop DSP-1181, an investigational drug for obsessive-compulsive disorder, with Exscientia's AI dramatically reducing the preclinical timeline to under a year, significantly shorter than the typical 4–6 year process (Burki 2020). Similarly, Atomwise partnered with IBM's World Community Grid to implement AI-driven VS to identify potential treatments for Ebola and multiple sclerosis. For Ebola, AI pinpointed compounds that inhibit a viral enzyme critical for replication, showcasing its ability to swiftly target specific disease mechanisms (Veljkovic et al. 2015).

In the realm of chemical synthesis, recent progress has been propelled by advances in symbolic AI and DL, with systems now matching or even surpassing human-designed synthesis strategies. In 2017, Segler et al. introduced a knowledge-graph model utilizing half-reactions to predict possible products (Segler and Waller 2017a), followed by a hybrid neural-symbolic approach for retrosynthesis and reaction prediction (Segler and Waller 2017b). This model mimicked human decision-making to circumvent reactivity conflicts and was 150 times faster than traditional rule-based systems, although it did not account for stereochemistry. In 2018, Segler's team applied Monte Carlo tree search (MCTS) with symbolic AI to discover retrosynthetic pathways for small molecules (Segler et al. 2018). By integrating MCTS with an expansion policy network and a filtering mechanism, they guided synthesis and trained DNNs on 12.4 million known reactions. In blinded tests, AI-generated synthetic routes were rated as comparable to those in the literature. However, natural product synthesis and enantiomeric or diastereomeric predictions remained challenges. In 2019, building on previous research (Gao et al. 2018), Coley et al. detailed a comprehensive computer-augmented chemical synthesis strategy that leveraged millions of chemical publications and simulated in silico to maximize success rates. This strategy was executed using a modular continuous-flow platform, which was automatically reconfigured by a robotic arm (Coley et al. 2019). Demonstrated with 15 drugs or drug-like compounds, this approach not only enhanced the synthesis capabilities of chemists but also alleviated the burdens associated with information gathering and manual tasks. In 2020, Grzybowski's team advanced Chematica (Synthia), integrating causal reasoning with AI routines for multistep strategies (Mikulak-Klucznik et al. 2020). In validation tests, routes generated by this hybrid AI-expert system closely matched human-designed pathways. Lab validations confirmed its effectiveness in synthesizing complex natural products, though further enhancements to the reaction database and software remain underway.

As we progress into the next decade, many visions of the 2000s for drug discovery are becoming realities, thanks to advancements in CADD tools (Bender and Cortés-Ciriano 2021; Van Drie 2007). VS has become standard practice in the pharmaceutical industry, incorporating concepts from computational thermodynamics, such as free energy perturbation, to predict compound affinity. A wealth of pharmacokinetic and ADME data is now accessible, facilitating the transition of potent ligands into viable therapeutic agents. Although these tools have grown increasingly accessible and user-friendly, big data and "omics" data present new challenges for AI in managing biological complexity. Future research will likely delve deeper into areas such as gene expression, protein modifications, and receptor behavior, uncovering mechanisms at finer levels of molecular detail. While the intrinsic complexity of biological systems may challenge predictive accuracy, AI remains indispensable in surmounting these obstacles, paving the way for more precise, effective drug discovery.

3 Conclusions

The history of AI is deeply intertwined with drug discovery, each field continuously driving advancements in the other. This relationship is far from incidental; the foundations of AI techniques are often rooted in the challenges posed by drug discovery. From Crum Brown's nineteenth-century predictive models to the development of graph theory and QSAR methods and symbolic AI in the 1960s, drug discovery has consistently spurred innovations essential to AI's evolution. This synergy is not merely coincidental; chemical information systems heavily rely on graph-based structures, inherently aligning their advancement with the progress in graph theory and data analysis. Moreover, advancements in computing and data processing capabilities have complemented these developments, enabling the implementation of more sophisticated algorithms and models in drug discovery. Post-2010, the advent of DNNs has further revolutionized the field, marking a shift from traditional ML techniques to more complex architectures capable of handling vast datasets and making accurate predictions.

This chapter aims to dispel common misconceptions regarding the relationship between AI, ML, and drug discovery. Far from being a recent phenomenon, AI and ML have played a vital role in pharmaceutical research for decades, facilitating groundbreaking discoveries and enhancing the efficiency of drug development processes. For instance, the success of AI-driven projects like Halicin and DSP-1181 illustrates how these technologies are not relegated to experimental applications; they are actively employed in drug development projects across the globe, yielding tangible results that contribute to improving patient outcomes. While new methodologies may generate waves of excitement or skepticism, often reflected in the Gartner hype cycles, it is crucial to acknowledge that AI and ML have evolved into indispensable components in the advancement of small-molecule drug development. They are instrumental in achieving critical clinical milestones and are shaping the future landscape of medicine. As we continue to explore the potential of these technologies, their integration into the drug discovery paradigm promises to further accelerate innovation, enhance therapeutic efficacy, and ultimately transform patient care.

Acknowledgments This work was supported by the European Union, Next Generation EU (PRIN 2022, project number: 2022P5LPHS). A.L. acknowledges funding from the PNRR Project CN000041, "National Centre for Gene Therapy and RNA Technology-Based Drugs" (CUP UNINA E63C22000940007).

References

Adamson GW, Bawden D. A method of structure-activity correlation using Wiswesser line notation. J Chem Inf Comput Sci. 1975;15:215–20. https://doi.org/10.1021/ci60004a006.

Agrawal A, Gans J, Goldfarb A. Prediction machines, updated and expanded: the simple economics of artificial intelligence. Boston: Harvard Business Press; 2022.

Baskin I. Fragment descriptors in SAR/QSAR/QSPR studies, molecular similarity Analysis and in virtual. Screening. 2008; https://doi.org/10.1039/9781847558879-00001.

Bender A, Cortés-Ciriano I. Artificial intelligence in drug discovery: what is realistic, what are illusions? Part 1: ways to make an impact, and why we are not there yet. Drug Discov Today. 2021;26:511–24.

Boden MA. Artificial intelligence: a very short introduction. Oxford: Oxford University Press; 2018.

Boyd DB, Marsh MM. History of computers in pharmaceutical research and development: a narrative. Comput Appl Pharm Res Dev. 2006:1–50. https://doi.org/10.1002/0470037237.ch1.

Broussard M. Artificial unintelligence: how computers misunderstand the world. Cambridge: MIT Press; 2018.

Brown AC, Fraser TR. XX.—On the connection between chemical constitution and physiological action. Part II.—On the physiological action of the ammonium bases derived from Atropia and Conia. Earth Environ Sci Trans R Soc Edinb. 1869;25:693–739. https://doi.org/10.1017/S0080456800035377.

Brown N, Ertl P, Lewis R, Luksch T, Reker D, Schneider N. Artificial intelligence in chemistry and drug design. J Comput Aided Mol Des. 2020;34:709–15. https://doi.org/10.1007/s10822-020-00317-x.

Brynjolfsson E, McAfee A. The second machine age: work, progress, and prosperity in a time of brilliant technologies. New York: W. W. Norton & Company; 2014.

Burki T. A new paradigm for drug development. Lancet Digit Health. 2020;2:e226–7.

Cao Y, Romero J, Aspuru-Guzik A. Potential of quantum computing for drug discovery. IBM J Res Dev. 2018;62:6:1–6:20. https://doi.org/10.1147/JRD.2018.2888987.

Chan HS, Shan H, Dahoun T, Vogel H, Yuan S. Advancing drug discovery via artificial intelligence. Trends Pharmacol Sci. 2019;40:592–604.

Coley CW, Thomas DA III, Lummiss JA, Jaworski JN, Breen CP, Schultz V, et al. A robotic platform for flow synthesis of organic compounds informed by AI planning. Science. 2019;365:eaax1566.

Cook D, Brown D, Alexander R, March R, Morgan P, Satterthwaite G, et al. Lessons learned from the fate of AstraZeneca's drug pipeline: a five-dimensional framework. Nat Rev Drug Discov. 2014;13:419–31. https://doi.org/10.1038/nrd4309.

Corey EJ, Wipke WT. Computer-assisted design of complex organic syntheses. Science. 1969;166:178–92. https://doi.org/10.1126/science.166.3902.178.

Cox M, Ellsworth D. Managing Big Data for scientific visualization. Proc ACM Siggraph Ames NASA 1997;21–38.

Cramer RD, Patterson DE, Bunce JD. Comparative molecular field analysis (CoMFA). 1. Effect of shape on binding of steroids to carrier proteins. J Am Chem Soc. 1988;110:5959–67. https://doi.org/10.1021/ja00226a005.

Demchenko Y, Grosso P, De Laat C, Membrey P. Addressing big data issues in scientific data infrastructure. In: 2013 international conference on collaboration technologies and systems. CTS, San Diego, CA, USA. IEEE; 2013. p. 48–55. https://doi.org/10.1109/CTS.2013.6567203.

Dijksterhuis, E. J. The mechanization of the world picture. Trans. Dikshoorn, C. Oxford: Clarendon; 1961.

Domingos P. The master algorithm: how the quest for the ultimate learning machine will remake our world. London: Penguin; 2015.

Dreyfus H, Dreyfus SE. Mind over machine. New York: Simon and Schuster; 1986.

Duda RO, Shortliffe EH. Expert systems research. Science. 1983;220:261–8. https://doi.org/10.1126/science.6340198.

Ekins S. Computer applications in pharmaceutical research and development. Hoboken: Wiley; 2006.

Engkvist O, Norrby P-O, Selmi N, Lam Y, Peng Z, Sherer EC, et al. Computational prediction of chemical reactions: current status and outlook. Drug Discov Today. 2018;23:1203–18. https://doi.org/10.1016/j.drudis.2018.02.014.

Fernbach S, Taub AH. Computers and their role in the physical sciences. London: Gordon and Breach; 1970.

Ford M. Architects of intelligence: the truth about AI from the people building it. Birmingham: Packt Publishing Ltd; 2018.

Free SM, Wilson JW. A mathematical contribution to structure-activity studies. J Med Chem. 1964;7:395–9. https://doi.org/10.1021/jm00334a001.

Freeman C, Louçã F. As time goes by: from the industrial revolutions to the information revolution. Oxford: Oxford University Press; 2001.

Gao H, Struble TJ, Coley CW, Wang Y, Green WH, Jensen KF. Using machine learning to predict suitable conditions for organic reactions. ACS Cent Sci. 2018;4:1465–76.

Gasteiger J, Groß J, Günnemann S. Directional message passing for molecular graphs. arXiv. 2020. 2003. https://doi.org/10.48550/arXiv.2003.03123.

Gasteiger J, Hutchings MG, Christoph B, Gann L, Hiller C, Low P, et al. A new treatment of chemical reactivity: development of EROS, an expert system for reaction prediction and synthesis design. In: Organic synthesis, reactions and mechanisms. Berlin: De Gruyter; 2022. p. 19–74. https://doi.org/10.1515/9783112597460-003.

Geysen HM, Schoenen F, Wagner D, Wagner R. Combinatorial compound libraries for drug discovery: an ongoing challenge. Nat Rev Drug Discov. 2003;2:222–30. https://doi.org/10.1038/nrd1035.

Greenfield A. Radical technologies: the design of everyday life. London: Verso; 2017.
Guedes IA, de Magalhães CS, Dardenne LE. Receptor–ligand molecular docking. Biophys Rev. 2014;6:75–87. https://doi.org/10.1007/s12551-013-0130-2.
Hage J, Verheij B. Rechtsinformatica: de stand van zaken in de wetenschap. Oskamp Lodder Reds Informatietechnologie Voor Juristen Handb Voor Jurist 21e Eeuw 1999:65–92.
Hansch C, Fujita T. p-σ-π analysis. A method for the correlation of biological activity and chemical structure. J Am Chem Soc. 1964;86:1616–26. https://doi.org/10.1021/ja01062a035.
Ihde D. Embodied technics. Roskilde: Automatic Press/VIP; 2010.
Iida K, Abdelhamid Ahmed AH, Nagatsuma AK, Shibutani T, Yasuda S, Kitamura M, et al. Identification and therapeutic targeting of GPR20, selectively expressed in gastrointestinal stromal tumors, with DS-6157a, a first-in-class antibody–drug conjugate. Cancer Discov. 2021;11:1508–23. https://doi.org/10.1158/2159-8290.CD-20-1434.
Ivakhnenko AG, Lapa VG. Cybernetics and forecasting techniques. New York: American Elsevier Publishing Company; 1967.
Jorgensen WL, Laird ER, Gushurst AJ, Fleischer JM, Gothe SA, Helson HE, et al. CAMEO: a program for the logical prediction of the products of organic reactions. Pure Appl Chem. 1990;62:1921–32. https://doi.org/10.1351/pac199062101921.
Kelly K. The inevitable: understanding the 12 technological forces that will shape our future. New York: Viking; 2016.
Klopman G. Predicting toxicity through a computer automated structure evaluation program. Environ Health Perspect. 1985;61:269–74. https://doi.org/10.1289/ehp.8561269.
Klopman G, Contreras R. Use of artificial intelligence in structure-activity correlations of anticonvulsant drugs. Mol Pharmacol. 1985;27:86–93.
Klopman G, Macina OT, Levinson ME, Rosenkranz HS. Computer automated structure evaluation of quinolone antibacterial agents. Antimicrob Agents Chemother. 1987;31:1831–40. https://doi.org/10.1128/aac.31.11.1831.
Lavecchia A. Machine-learning approaches in drug discovery: methods and applications. Drug Discov Today. 2015;20:318–31.
Lavecchia A. Deep learning in drug discovery: opportunities, challenges and future prospects. Drug Discov Today. 2019;24:2017–32.
Lavecchia A, Di Giovanni C. Virtual screening strategies in drug discovery: a critical review. Curr Med Chem. 2013;20:2839–60.
Leung J. Who will govern artificial intelligence? Learning from the history of strategic politics in emerging technologies. University of Oxford; 2019. http://purl.org/dc/dcmitype/Text
Lill MA. Multi-dimensional QSAR in drug discovery. Drug Discov Today. 2007;12:1013–7. https://doi.org/10.1016/j.drudis.2007.08.004.
Lindsay RK, Buchanan BG, Feigenbaum EA, Lederberg J. DENDRAL: a case study of the first expert system for scientific hypothesis formation. Artif Intell. 1993;61:209–61.
Lipinski CA, Lombardo F, Dominy BW, Feeney PJ. Experimental and computational approaches to estimate solubility and permeability in drug discovery and development settings1. Adv Drug Deliv Rev. 2001;46:3–26. https://doi.org/10.1016/S0169-409X(00)00129-0.
Ma J, Sheridan RP, Liaw A, Dahl GE, Svetnik V. Deep neural nets as a method for quantitative structure–activity relationships. J Chem Inf Model. 2015;55:263–74. https://doi.org/10.1021/ci500747n.
Marshall GR, Cramer RD. Three-dimensional structure-activity relationships. Trends Pharmacol Sci. 1988;9:285–9. https://doi.org/10.1016/0165-6147(88)90012-0.
Martin YC. Quantitative drug design: a critical introduction. 2nd ed. Boca Raton: CRC Press; 2010. https://doi.org/10.1201/9781420071009.
Matthews DA, Alden RA, Bolin JT, Filman DJ, Freer ST, Hamlin R, et al. Dihydrofolate reductase from Lactobacillus casei. X-ray structure of the enzyme methotrexate. NADPH complex. J Biol Chem. 1978;253:6946–54. https://doi.org/10.1016/S0021-9258(17)38012-2.
Matthews DA, Alden RA, Freer ST, Xuong N, Kraut J. Dihydrofolate reductase from Lactobacillus caSci. Stereochemistry of NADPH binding. J Biol Chem. 1979;254:4144–51. https://doi.org/10.1016/S0021-9258(18)50708-0.

Mayor A. Gods and robots: myths, machines, and ancient dreams of technology. Princeton: Princeton University Press; 2020.
Mayr A, Klambauer G, Unterthiner T, Hochreiter S. DeepTox: toxicity prediction using deep learning. Front Environ Sci. 2016;3:80.
McCulloch WS, Pitts W. A logical calculus of the ideas immanent in nervous activity. Bull Math Biophys. 1943;5:115–33.
Mikulak-Klucznik B, Gołębiowska P, Bayly AA, Popik O, Klucznik T, Szymkuć S, et al. Computational planning of the synthesis of complex natural products. Nature. 2020;588:83–8.
Monga M, Sausville EA. Developmental therapeutics program at the NCI: molecular target and drug discovery process. Leukemia. 2002;16:520–6. https://doi.org/10.1038/sj.leu.2402464.
Newman WR. Promethean ambitions: alchemy and the quest to perfect nature. Chicago: University of Chicago Press; 2019.
Paul SM, Mytelka DS, Dunwiddie CT, Persinger CC, Munos BH, Lindborg SR, et al. How to improve R&D productivity: the pharmaceutical industry's grand challenge. Nat Rev Drug Discov. 2010;9:203–14. https://doi.org/10.1038/nrd3078.
Pereira DA, Williams JA. Origin and evolution of high throughput screening. Br J Pharmacol. 2007;152:53. https://doi.org/10.1038/sj.bjp.0707373.
Ravitz O. Data-driven computer aided synthesis design. Drug Discov Today Technol. 2013;10:e443–9. https://doi.org/10.1016/j.ddtec.2013.01.005.
Rid T. Rise of the machines: a cybernetic history. New York: WW Norton & Company; 2016.
Rosenkranz HS, Klopman G. Case, the computer-automated structure evaluation system, as an alternative to extensive animal testing. Toxicol Ind Health. 1988;4:533–40. https://doi.org/10.1177/074823378800400411.
Russell S. Human-compatible artificial intelligence. Hum-Mach Intell. 2022;1:3–22.
Russell SJ, Norvig P. Artificial intelligence: a modern approach. Pearson; 2016.
Sakaue S, Okada Y. Human genetics contributes to the understanding of disease pathophysiology and drug discovery. J Orthop Sci. 2017;22:977–81. https://doi.org/10.1016/j.jos.2017.07.009.
Salatin TD, Jorgensen WL. Computer-assisted mechanistic evaluation of organic reactions. 1. Overview. J Org Chem. 1980;45:2043–51. https://doi.org/10.1021/jo01299a001.
Sandell HS, Bourne JR. Expert systems in medicine: a biomedical engineering perspective. Crit Rev Biomed Eng. 1985;12:95–129.
Satoh H, Funatsu K. SOPHIA, a knowledge base-guided reaction prediction system – utilization of a knowledge base derived from a reaction database. J Chem Inf Comput Sci. 1995;35:34–44. https://doi.org/10.1021/ci00023a005.
Satoh H, Funatsu K. Further development of a reaction generator in the SOPHIA system for organic reaction prediction. Knowledge-guided addition of suitable atoms and/or atomic groups to product skeleton. J Chem Inf Comput Sci. 1996;36:173–84. https://doi.org/10.1021/ci950058a.
Satoh K, Funatsu K. A novel approach to retrosynthetic analysis using knowledge bases derived from reaction databases. J Chem Inf Comput Sci. 1999;39:316–25. https://doi.org/10.1021/ci980147y.
Sawicki MP, Samara G, Hurwitz M, Passaro E. Human genome project. Am J Surg. 1993;165:258–64. https://doi.org/10.1016/S0002-9610(05)80522-7.
Schneider G, Wrede P. The rational design of amino acid sequences by artificial neural networks and simulated molecular evolution: de novo design of an idealized leader peptidase cleavage site. Biophys J. 1994;66:335–44. https://doi.org/10.1016/S0006-3495(94)80782-9.
Schneider G, Schuchhardt J, Wrede P. Artificial neural networks and simulated molecular evolution are potential tools for sequence-oriented protein design. Bioinformatics. 1994;10:635–45. https://doi.org/10.1093/bioinformatics/10.6.635.
Segler MH, Waller MP. Modelling chemical reasoning to predict and invent reactions. Chem Eur J. 2017a;23:6118–28.
Segler MH, Waller MP. Neural-symbolic machine learning for retrosynthesis and reaction prediction. Chem Eur J. 2017b;23:5966–71.

Segler MH, Preuss M, Waller MP. Planning chemical syntheses with deep neural networks and symbolic AI. Nature. 2018;555:604–10.

Stokes JM, Yang K, Swanson K, Jin W, Cubillos-Ruiz A, Donghia NM, et al. A deep learning approach to antibiotic discovery. Cell. 2020;180(688–702):e13.

Tegmark M. Life 3.0: being human in the age of artificial intelligence. London: Penguin Books; 2018.

Todd MH. Computer-aided organic synthesis. Chem Soc Rev. 2005;34:247. https://doi.org/10.1039/b104620a.

Torres PHM, Sodero ACR, Jofily P, Silva-Jr FP. Key topics in molecular docking for drug design. Int J Mol Sci. 2019;20:4574. https://doi.org/10.3390/ijms20184574.

Turing AM. Computing machinery and intelligence. Springer; 2009.

Ugi I, Bauer J, Baumgartner R, Fontain E, Forstmeyer D, Lohberger S. Computer assistance in the design of syntheses and a new generation of computer programs for the solution of chemical problems by molecular logic. Pure Appl Chem. 1988;60:1573–86. https://doi.org/10.1351/pac198860111573.

Van Drie JH. Computer-aided drug design: the next 20 years. J Comput Aided Mol Des. 2007;21:591–601. https://doi.org/10.1007/s10822-007-9142-y.

Varkony TH, Smith DH, Djerassi C. Computer-assisted structure manipulation: studies in the biosynthesis of natural products. Tetrahedron. 1978;34:841–52.

Vedani A, Zumstein M, Lill MA, Ernst B. Simulating α/β selectivity at the human thyroid hormone receptor: consensus scoring using multidimensional QSAR. ChemMedChem. 2007;2:78–87. https://doi.org/10.1002/cmdc.200600212.

Veljkovic V, Loiseau PM, Figadere B, Glisic S, Veljkovic N, Perovic VR, et al. Virtual screen for repurposing approved and experimental drugs for candidate inhibitors of EBOLA virus infection. F1000Research. 2015;4:34.

Weinstein JN, Kohn KW, Grever MR, Viswanadhan VN, Rubinstein LV, Monks AP, et al. Neural computing in cancer drug development: predicting mechanism of action. Science. 1992;258:447–51. https://doi.org/10.1126/science.1411538.

Wiener N. God & Golem, Inc.: A comment on certain points where cybernetics impinges on religion, vol. 42. Cambridge: MIT Press; 1966.

Wiener N. Cybernetics or control and communication in the animal and the machine. Cambridge: MIT Press; 2019.

Yadav MK. On the synthesis of machine learning and automated reasoning for an artificial synthetic organic chemist. New J Chem. 2017;41:1411–6. https://doi.org/10.1039/C6NJ02492K.

Yang X, Wang Y, Byrne R, Schneider G, Yang S. Concepts of artificial intelligence for computer-assisted drug discovery. Chem Rev. 2019;119:10520–94.

Zarkadakis G. In our own image: will artificial intelligence save or destroy us? Broadway: Random House; 2015.

Zarkadakis G. In our own image: savior or destroyer? The history and future of artificial intelligence. 1st ed. New York: Pegasus Books; 2016.

Data Mining and Integration Approaches in AI-Driven Drug Discovery

Stefanos Tsimenidis, Eleni Vrochidou, and George A. Papakostas

Abstract Data-driven computational drug discovery has recently exploded; yet, compiling the necessary data poses serious difficulties. Biological systems are too complex for a single dataset to adequately describe, with genes, proteins, diseases, and drugs forming dense, indecipherable networks of interactions. Such complex and multi-faceted information can hardly be retrieved through a simple query on a public database. Formulas of chemical agents may be stored in one repository, more formulas in another, and information such as attributes of target proteins, molecular interactions, and drug side effects might be scattered in a number of databases. Drug researchers address this challenge in three main ways: First, they acquire data from multiple sources and repositories, and then integrate them through computational means. Second, they exploit text mining techniques, combing through biomedical literature for diverse information. Third, they combine mined text with database querying, locating whatever data they need for each research project. Here, we present a treatment on this topic, examining how scientists across academia and industry integrate multi-source data, what kind of data they use, what trends in approaches and best practices have been emerging, and how setbacks such as data imbalance are being tackled.

Keywords AI in drug development · Computational drug discovery · Drug–target interaction · Pharmacoinformatics · Multi-source data integration · Data-driven life sciences

1 Introduction

Drug development, from initial conception to FDA-approved product release, may take 15 years of research and cost over one billion dollars (Kaitin 2010). Such

S. Tsimenidis · E. Vrochidou · G. A. Papakostas (✉)
MLV Research Group, Department of Informatics, Democritus University of Thrace, Kavala, Greece
e-mail: stsimeni@cs.duth.gr; evrochid@cs.duth.gr; gpapak@cs.duth.gr

A. Lavecchia (ed.), *Applied Artificial Intelligence for Drug Discovery*,
https://doi.org/10.1007/978-3-031-98022-0_2

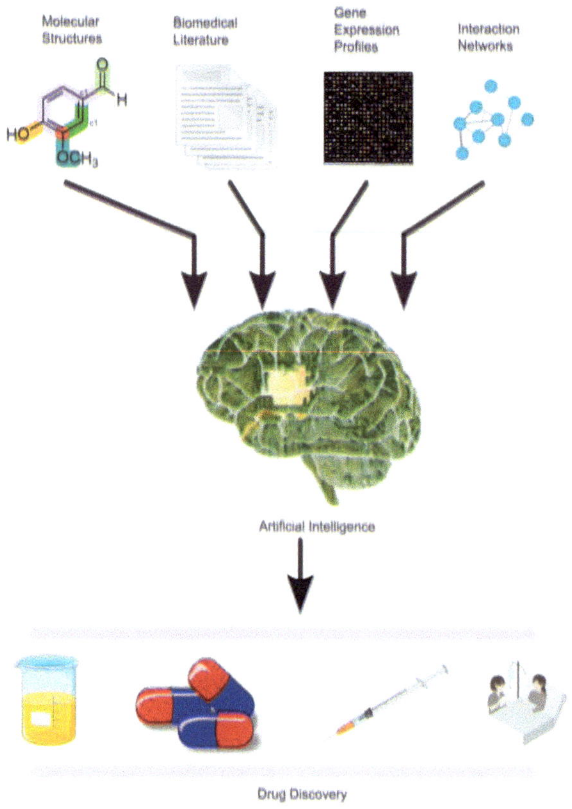

Fig. 1 When AI is applied to drug discovery, it can considerably profit from the integration and exploitation of different types of chemical, textual, and biological data, retrieving useful information from multiple sources

extreme costs in time and money are due to the extreme complexity of the task at hand. The chemical search space of all the possible small molecules has been estimated to be in the order of 10^{30}–10^{60} molecules (Sliwoski et al. 2014). Physical search through this space in the lab can be a pedantic and inefficient process, while exploring the chemical space digitally could be more tractable, and computational techniques could aid in efficient drug discovery (Niazi and Mariam 2023). A major trend in computational drug discovery has been the application of artificial intelligence (AI), and data-driven approaches such as machine learning (ML) and neural networks (NNs), to extract insights from the unprecedented volume of data currently generated in life science and medicinal chemistry (Cerchia and Lavecchia 2023; Dhudum et al. 2024; Qi et al. 2024; Hasselgren and Oprea 2024).

Drug discovery, when approached in a data-driven manner, can profit greatly from integrating data of different types and modalities (Fig. 1). To begin with, biological systems can be considered from multiple levels of abstraction and organization, whether it is the information encoded in the genes, or the unique expression of genes within individual cells, or the attributes of proteins, or a number of other aspects, with each level containing information about the system absent from other levels (Tsimenidis et al. 2022). Adding the interactions of chemical compounds, genes, proteins, and diseases to that information about the chemical structures of drug candidates results in the creation of a rich repository of unstructured knowledge waiting to be mined and put into use. In Hodos et al. (2016), the authors advocate using and integrating various types of data such as chemical structures of compounds, drug–target interactions, transcriptional profiles of chemical perturbations, phenotypic drug screens, drug–disease associations, drug–side effect associations, and disease-related genomic perturbations. In Loging et al. (2007), a discussion ensues of the potential usefulness and integration of a wide range of biological and chemical data, such as ontologies (categorical information on genes, proteins, interactions among entities), protein pathways, gene knockout phenotypes, amino acid sequences of proteins and peptides, gene expression, DNA sequences, single nucleotide polymorphisms (SNPs), disease biomarkers, patient clinical data, physicochemical properties of drugs, drug molecular structures, and more.

Even when working on projects that do not require multiple types of data, it might be needed to integrate data of the same type but from multiple databases. Chemical databases have millions of compounds in total; note that in an analysis of five popular databases, researchers found that it is less than 1% of the total compounds that they have in common (Isigkeit et al. 2022). Getting the volume of data adequate to train a deep learning (DL) model might entail the combination of data from multiple databases, since with deep neural networks, a large volume of data is necessary for the algorithms to work properly. In Fan and Shi (2022), it is proven that the bigger the volume of training data, the better accuracy a model for protein–ligand affinity prediction yields. Fortunately, the volume and variety of public data rise steadily. According to the Nucleic Acids Research journal's annual survey for new databases related to molecular biology, the number of relevant databases up to the year 2023 reached 1764 (Rigden and Fernandez 2023).

However, not all useful data are stored neatly and conveniently in public databases. In fact, most biomedical and pharmaceutical knowledge is currently found in scientific publications and other unstructured forms (Erhardt et al. 2006). Containing information on genes, proteins, diseases, drugs, and the various interactions among them, the biomedical literature is a huge repository of knowledge and insights that could be exploited for drug discovery (Swanson 1990). Yet, it is impossible for a human expert, or a conglomerate of human experts, to study and assimilate all the biomedical literature, let alone constantly monitor the field for new publications, reviewing, and interpreting all the information. A search on PubMed (https://pubmed.ncbi.nlm.nih.gov/) on the term "kinase inhibitor" yielded 380,961 abstracts, with an average of 18,403 new abstracts per year during the last 15 years. A search on "prostate cancer" yielded 234,962 abstracts, with an average of 11,169 new abstracts per year during the last 15 years. The latter statistics clearly indicate that the pace of literature production is way greater than human experts can process; therefore, the need for computational approaches seems necessary.

Text mining is a computational paradigm and toolbox that enables the automated extraction of patterns and concept associations from large volumes of unstructured text (Hearst 1999). It can be valuable for extracting insight from biomedical articles and abstracts, with a lot of research focused on this scientific field (Krallinger et al. 2005). The application of natural language processing (NLP) and text mining in biomedicine started almost 30 years ago, while relevant research has been conducted on a variety of applications such as named entity recognition, identifying relations among biomedical entities, text summarization, knowledge extraction from image captions, question answering, and more (Zweigenbaum et al. 2007). Members of the Library and Literature Informatics staff from Biogen Idec conducted an internal study of the research requests that came to the library. The collected statistics revealed that most requests centered upon drugs and diseases, fewer on genes and proteins, and the sources of information used were publications, proprietary in-house repositories, and filed patents. After an analysis related to the needs of information and their potential sources, the researchers concluded that text mining techniques should be explored and applied for enhanced knowledge discovery (Roberts and Hayes 2008).

In this chapter, an examination of various approaches found in the literature is presented to utilize a variety of data for drug discovery and related tasks, whether the data come from structured data repositories or from mined text. Our purpose is to give researchers an overview of the possibilities, in terms of the kind of data that are available, and how they could be exploited to generate novel hypotheses or drug candidates.

The rest of the chapter is structured as follows: Sect. 2 briefly overviews some of the most used public databases in drug discovery. Section 3 summarizes the common computational representations of small molecules in medicinal chemistry and chemoinformatics. Section 4 is an overview of how drug researchers in academia and industry approach data integration and text mining. Section 5 discusses the challenges that arise in these endeavors and considers approaches followed to address these challenges. Finally, in Sect. 6, the trends that seem to emerge in the computational drug discovery space are discussed, and conclusions are drawn.

2 Databases

The public databases used for drug discovery contain information on a huge number of chemical agents, especially regarding their chemical structures, properties, and their interactions with various protein targets. In Table 1, the most popular databases are summarized.

2.1 PubChem API

Cheminformatics databases, such as PubChem and ChEMBL, can be programmatically accessed through their APIs. PubChem queries consist of three fields. First is the input, where users specify which compounds they want to retrieve, using

Table 1 Popular databases for chemoinformatics and drug discovery

Database	Content	Reference/website
ChEMBL	Chemical structures ADME-T properties Molecular properties Mechanisms of action Compound–target interactions Therapeutic categories	Gaulton et al. (2017)
PubChem	Chemical structures Physicochemical properties Biological properties	Kim et al. (2021)
DrugBank	Chemical structures Drug–target interactions Drug–drug interactions Drug reactions Drug categories	Wishart et al. (2007)
ZINC	Chemical structures Physicochemical properties Target genes and proteins Target/compound categories Clinical trial status Purchasability	Sterling and Irwin (2015)
DGIdb	Drug–gene interactions	Freshour et al. (2021)
MoleculeNet	ML-ready data sets Chemical structures Molecular properties Classification/regression data	Wu et al. (2018)
PDBbind	Drug–protein interactions Binding affinities	Wang et al. (2004)
TCGA	Cancer-related data Gene expression	https://portal.gdc.cancer.gov/
L1000 CMap	Chemical perturbations Gene expression	Subramanian et al. (2017)
Kyoto encyclopedia of genes and genomes (KEGG)	Gene-related data	Kanehisa and Goto (2000)
Medline/PubMed	Biomedical abstracts Text mining source	https://pubmed.ncbi.nlm.nih.gov/
Gene ontology	Gene-related categorical data	https://geneontology.org/
STRING	Protein–protein interactions	https://string-db.org/
Clinical trials	Clinical research studies	https://clinicaltrials.gov/

identifiers such as the compound's name, ID, or SMILES string. Second comes the operation, where users select what information they want for these compounds, choosing from a range of options such as SMILES and physicochemical properties. Thirdly, by setting the output option, users determine the file format they will get from PubChem, with options such as csv, xml, txt, and others. All requests start with the "Prolog," i.e., the URL from which information will be retrieved, and which always is https://pubchem.ncbi.nlm.nih.gov/rest/pug/. Thus, requests have the format `prolog/input/operation/output`.

A short introduction to PubChem programmatic data acquisition can be found at https://projects.volkamerlab.org/teachopencadd/talktorials/T013_query_pubchem.html. A chemoinformatics course with numerous chapters and sections on accessing PubChem programmatically is in https://chem.libretexts.org/Courses/Intercollegiate_Courses/Cheminformatics.

As an example, we will generate a csv file with the chemical formula and SMILES structure of ascorbic acid (vitamin C), identified by name. The "input" field of the request becomes `compound/name /ascorbic acid/`, the "operation" field becomes `property/MolecularFormula, CanonicalSMILES/`, and the "output" field becomes `csv`. The complete query is https://pubchem.ncbi.nlm.nih.gov/rest/pug/compound/name/ascorbic `acid/property/MolecularFormula,CanonicalSMILES/csv`. Typing that as a URL on a web browser automatically downloads a csv with the requested information. Accessing the data through Python code is as follows:

```
import requests
res = requests.get('https://pubchem.ncbi.nlm.nih.gov/rest/pug/compound/
name/glutamine/property/MolecularFormula,CanonicalSMILES/csv')
print(res.text)
```

This gives the output:

```
"CID","MolecularFormula","CanonicalSMILES"
54670067,"C6H8O6","C(C(C1C(=C(C(=O)O1)O)O)O)O"
```

More than one compound can be retrieved in one request, given some identifiers for the compounds. The code below retrieves compound IDs, chemical formulas, and SMILES structures of the 20 amino acids, identifying them by name and storing the three features in three lists.

```
aminos = 'alanine, arginine, asparagine, aspartic acid,\
 cysteine, glutamic acid, glutamine, glycine, histidine,\
  isoleucine, leucine, lysine, methionine, phenylalanine,\
   proline, serine, threonine, tryptophan, tyrosine,  valine'
aminos = aminos.split(',')

prolog = 'https://pubchem.ncbi.nlm.nih.gov/'
pugstart = 'rest/pug/compound/name/'
pugend = '/property/MolecularFormula,CanonicalSMILES/csv'

cid = []
smiles = []
```

```
formulas = []

for amino in aminos:
   res = requests.get(prolog + pugstart + amino + pugend)
   text = res.text.split('\n')[1].split(',')
   cid.append(text[0])
   formulas.append(text[1])
   smiles.append(text[2])
```

We can generate a pandas dataframe, with the IDs, names, formulas, and SMILES strings of the amino acids.

```
import pandas as pd

cids = pd.Series(cid)
smile = pd.Series(smiles)
name = pd.Series(aminos)
formula = pd.Series(formulas)

df = pd.DataFrame({'cids':cid,
               'name':name,
               'formula':form,
               'smiles':smile})
```

The output is as follows:

	cids	name	formula	smiles
0	5950	alanine	“C3H7NO2”	“CC(C(=O)O)N”
1	6322	arginine	“C6H14N4O2”	“C(CC(C(=O)O)N)CN=C(N)N”
2	6267	asparagine	“C4H8N2O3”	“C(C(C(=O)O)N)C(=O)N”
3	5960	aspartic acid	“C4H7NO4”	“C(C(C(=O)O)N)C(=O)O”
4	5862	cysteine	“C3H7NO2S”	“C(C(C(=O)O)N)S”
5	33,032	glutamic acid	“C5H9NO4”	“C(CC(=O)O)C(C(=O)O)N”
6	5961	glutamine	“C5H10N2O3”	“C(CC(=O)N)C(C(=O)O)N”
7	750	glycine	“C2H5NO2”	“C(C(=O)O)N”
8	6274	histidine	“C6H9N3O2”	“C1=C(NC=N1)CC(C(=O)O)N”
9	6306	isoleucine	“C6H13NO2”	“CCC(C)C(C(=O)O)N”
10	6106	leucine	“C6H13NO2”	“CC(C)CC(C(=O)O)N”
11	5962	lysine	“C6H14N2O2”	“C(CCN)CC(C(=O)O)N”
12	6137	methionine	“C5H11NO2S”	“CSCCC(C(=O)O)N”
13	6140	phenylalanine	“C9H11NO2”	“C1=CC=C(C=C1)CC(C(=O)O)N”
14	145,742	proline	“C5H9NO2”	“C1CC(NC1)C(=O)O”
15	5951	serine	“C3H7NO3”	“C(C(C(=O)O)N)O”
16	6288	threonine	“C4H9NO3”	“CC(C(C(=O)O)N)O”
17	6305	tryptophan	“C11H12N2O2”	“C1=CC=C2C(=C1)C(=CN2)CC(C(=O)O)N”
18	6057	tyrosine	“C9H11NO3”	“C1=CC(=CC=C1CC(C(=O)O)N)O”
19	6287	valine	“C5H11NO2”	“CC(C)C(C(=O)O)N”

Researchers are encouraged to review the two aforementioned resources, particularly the chemoinformatics course from www.libretexts.org. PubChem queries can be rather complex, with `for` loops iterating through lists of compound identifiers. Another factor raising the complexity is that the `sleep()` function must be executed at intervals to ease off the burden on the server; otherwise, the query itself might get cancelled. Thus, a counter is used within `for` loops, and each time it reaches some predefined values, a few seconds of pause are introduced. Using such embedded pauses, the `for` loop from the above code becomes.

```
counter = 0
for amino in aminos:
  res = requests.get(prolog + pugstart + amino + pugend)
  text = res.text.split('\n')[1].split(',')
  cid.append(text[0])
  formulas.append(text[1])
  smiles.append(text[2])
  if counter == 5:
    time.sleep(1)
    counter = 0
  else:
    counter += 1
```

More information on PUG-REST interfacing can be found on the PubChem website, at these links:

https://pubchem.ncbi.nlm.nih.gov/docs/pug-rest
https://pubchem.ncbi.nlm.nih.gov/docs/pug-rest-tutorial

2.2 ChEMBL API

A useful tutorial on mining data from ChEMBL can be found on https://projects.volkamerlab.org/teachopencadd/talktorials/T001_query_chembl.html. Here we provide a simple example of using the API in Python to download structural data on tyrosine kinase inhibitors. First, we install the ChEMBL API for Python, import pandas, and generate resource objects needed to access the API:

```
!pip install chembl_webresource_client

from chembl_webresource_client.new
 _client import new_client
 import pandas as pd

targets_api = new_client.target
 compounds_api = new_client.molecule
 bioactivities_api = new_client.activity
```

To retrieve compounds inhibiting a certain target, the API requires the Uniprot ID of the target. We search for "tyrosine kinase" in the Uniprot database (https://www.uniprot.org/), finding the protein's ID. Then, we use that ID in the Python example code to inspect the assorted target information in ChEMBL, formatting the data in pandas dataframe format:

```
uniprot_id = 'P07948'

targets = targets_api.get(target_components__accession=uniprot_id).only(
    "target_chembl_id", "organism", "pref_name", "target_type")

targets =pd.DataFrame.from_records(targets)
targets
```

	organism	pref_name	target_chembl_id	target_type
0	Homo sapiens	Tyrosine-protein kinase Lyn	CHEMBL3905	SINGLE PROTEIN
1	Homo sapiens	Tyrosine-protein kinase Lyn	CHEMBL3905	SINGLE PROTEIN
2	Homo sapiens	SRC	CHEMBL2363074	PROTEIN FAMILY

After inspecting the results, we identify the first entry as desirable, and we store the assorted ID:

```
target_chembl_id = targets.iloc[0,2]
```

Now we retrieve all the bioactivities associated with this target ID. We use filters to narrow down the results, only accessing B-type assays with inhibitory effect measured in nM. Then we store the results in a pandas dataframe:

```
bioactivities = bioactivities_api.filter(
    target_chembl_id=target_chembl_id, type="IC50", units='nM',
relation="=", assay_type="B"
).only(
    "activity_id",
    "molecule_chembl_id",
    "standard_units",
    "standard_value",
    "molecule_structures__canonical_smiles"
) bioactivities = pd.DataFrame.from_records(bioactivities)
bioactivities
```

	activity_id	canonical_smiles	molecule_ chembl_id	standard_ units	standard_ value	units	value
0	7930365	COc1cc2nc(Nc 3ccccc3C)c3cncn 3c2cc1N(CC1CC1) C(=O...	CHEMBL 1916716	nM	1980.0	nM	1980.0
1	1778607	Cc1cc(Nc2ncc(C (=O)Nc3c(C)ccc c3Cl)s2)nc(C)n1	CHEMBL 364623	nM	4.0	nM	4.0

(continued)

2	1815212	CNc1ncnc(-c2cccnc2Oc2ccc(F)c(C(=O)Nc3cc(C(F)(F...	CHEMBL373882	nM	4900.0	nM	4900.0
3	1815344	CN(C)[C@H]1CCCN(c2ccc(C(F)(F)F)cc2NC(=O)c2cc(C...	CHEMBL386637	nM	180.0	nM	180.0
4	1915519	COc1c(NC(=O)c2cc3cccc(NC(=O)c4ccc(NC5CC5)nc4)c...	CHEMBL243575	nM	230.0	nM	230.0
...	...	...	...	...	...	...	...
254	25549083	CCNC(=O)c1c(OC)cc(-n2cnc3cc(-c4cnn(C5CCOCC5)c4...	CHEMBL5396179	nM	111.0	nM	111.0
255	25549143	COc1cc(-n2cnc3cc(-c4cnn(C)c4)ccc32)cc(OC(F)F)c...	CHEMBL5417444	nM	272.0	nM	272.0
256	25782750	c1ccc(-c2oc3ncnc(NC[C@@H]4CCCO4)c3c2-c2ccccc2)cc1	CHEMBL247070	nM	346.7	nM	346.7
257	25786872	CCN1CCN(c2ccc(Nc3cc(N(C)C(=O)Nc4c(cl)c(OC)cc(O...	CHEMBL1852688	nM	3019.0	nM	3019.0
258	25792157	COc1cc(F)ccc1-c1ncnc(Nc2cccc(C[S@](C)(=N)=O)c2)n1	CHEMBL4439321	nM	3514.0	nM	3514.0

From here on, we can curate the dataset, removing duplicates, discarding compounds with inhibitory effect below a threshold we are interested in, then further filtering compounds with additional criteria, such as Lipinski rules or the presence of problematic substructures. The previously linked tutorial contains much more information on ChEMBL data acquisition with Python. Also, the github page where the ChEMBL_webresource_client is hosted contains a jupyter notebook with a comprehensive tutorial: (https://github.com/chembl/notebooks/blob/main/ChEMBL_webresource_client_examples.ipynb).

2.3 *SQL Querying DrugBank*

Databases adhering to the relational paradigm, such as DrugBank, can be queried via standard SQL commands. The data are conveniently exported in csv format.

Below are some examples of how such queries look like, taken from the DrugBank website (https://dev.drugbank.com/guides/sql).

Combining the drugs and bonds tables to link drugs with their targets.

```
SELECT d.name, b.type, be.name AS bond_name FROM drugs d
  JOIN bonds b          ON b.drug_id = d.id
  JOIN bio_entities be ON be.biodb_id = b.biodb_id
  ORDER BY d.name;
```

Combining the drugs, drug_categorizations, and categories tables to inspect the categories each drug belongs to.

```
SELECT d.name, d.drugbank_id, c.title AS category FROM drugs d
  JOIN drug_categorizations dc ON dc.drug_id = d.id
  JOIN categories c            ON dc.category_id = c.id
  ORDER BY d.drugbank_id;
```

Displaying drug interactions.

```
SELECT d.drugbank_id, d.name, sdi.* FROM drugs d
  JOIN structured_drug_interactions sdi ON sdi.subject_drug_id = d.id OR
 sdi.affected_drug_id = d.id
  ORDER BY d.drugbank_id;
```

An overview of SQL support in DrugBank, including a Python loading script, can be found on https://dev.drugbank.com/guides/implementation/loading_drugbank_sql_data. A detailed explanation of applying SQL on DrugBank to retrieve information on drugs related to specific targets or conditions can be found in the links https://dev.drugbank.com/guides/implementation/using_drugbank_target_data and https://dev.drugbank.com/guides/implementation/using_drugbank_conditions_data.

3 Data Types and Representations

A molecule, which is a physical conformation of atoms in three-dimensional space, can be represented computationally in a variety of ways, as shown in Fig. 2. An exhaustive list of molecular descriptors is beyond the scope of this study; however, for researchers interested in the subject, relevant information can be found in the literature (Xue and Bajorath 2000; Mauri et al. 2017).

The standard linear representation is the **Simplified Molecular-Input Line-Entry System** (SMILES) notation (Weininger 1988), which represents chemical structures as strings of characters, specifying the edges and nodes of a molecular graph. For example, the SMILES notation of acetic acid is CC(=O)O, of carbon dioxide is O=C=O, and of L-alanine is N[C@@H](C)C(=O)O. SMILES notation is being encoded as one-hot encoded vectors or, through the use of NLP techniques, as word embeddings. When one-hot encoding a number of SMILES structures, we impose uniform length by either truncating the end characters of the longer

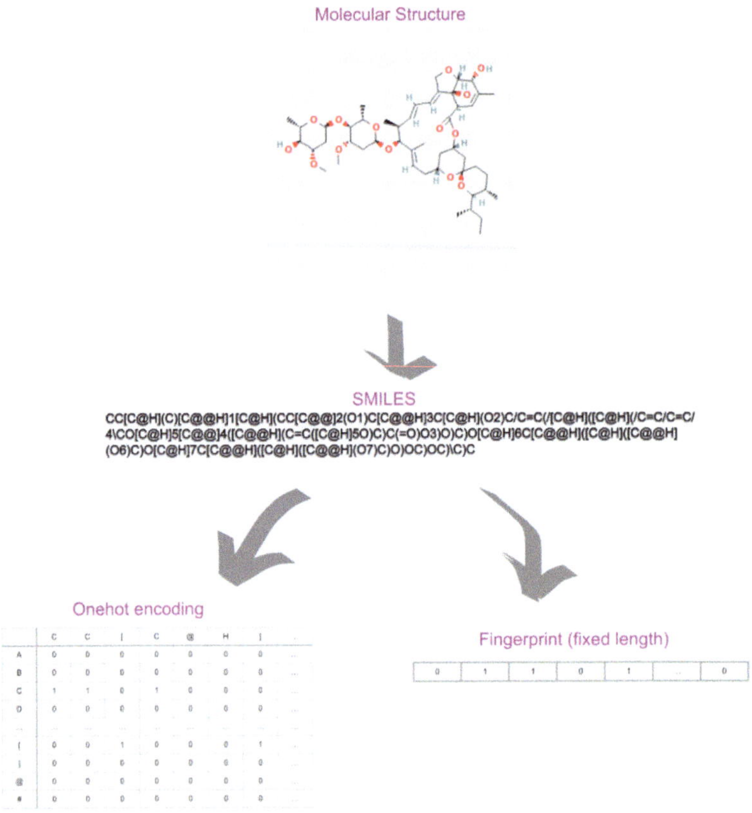

Fig. 2 Chemical molecular structures are often represented as SMILES strings; SMILES can be converted either to one-hot encoded matrices or to various types of molecular fingerprints of fixed length

sequences or by adding zeros to the shorter sequences. Since a molecule can be represented by multiple valid SMILES strings, the canonicalized SMILES structure was developed, which results in a unique encoding for each molecule.

Molecular Fingerprints are binary vectors of fixed length, with each bit of the vector representing some specific structural conformation, taking the value 1 if the molecule contains the particular substructure, or the value 0 otherwise. The two most commonly used fingerprints are **MACCS** (Molecular ACCess System) keys (Durant et al. 2002), which produce 166-bit vectors, and **ECFPs/FCFPs** (Rogers and Hahn 2010), computed via a variation of the Morgan algorithm (Morgan 1965), which integrates information about an atom's surroundings, i.e., adjacent atoms and their connection bonds, for a user-specified radius.

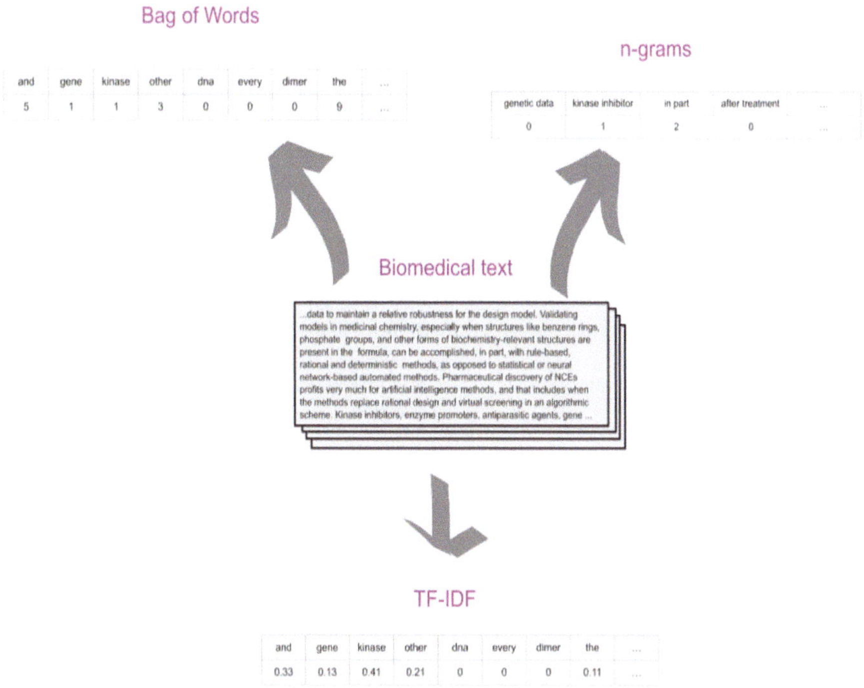

Fig. 3 Textual corpora can be converted into datasets for ML model training by a number of methods. Bag-of-words (BOW) encodes the counts of each word in a document, n-gram counts combinations of words, and term frequency-inverse document frequency (TF-IDF) quantifies the importance of words based on the entire corpus

3.1 Text Data

Textual data, usually in the form of abstracts from biomedical articles, or from textual entries in public datasets, are exploited for various drug discovery-related applications. Techniques from natural language processing (NLP), an intersection of AI and linguistics (Nadkarni et al. 2011), are deployed to transform texts into a format usable by computational algorithms, ML models, and NNs, as shown in Fig. 3.

For the purposes of AI-based applications, both preprocessing and feature extraction techniques, such as those included in Table 2, need to be applied to biomedical texts. Before converting text into a dataset for model training, some basic preprocessing and preparation are required to clean the data, remove redundancies, and organize the data into more computationally usable structures. Then, feature extraction and dataset generation techniques transform the unstructured or semi-structured data into ML-ready datasets, with various techniques trying to encode as much useful information as efficiently as possible into the dataset.

Table 2 Typical methods for text preprocessing and feature extraction, and some of the advantages and disadvantages of each method

	NLP techniques	Advantages	Disadvantages
Preprocessing	Lowercase conversion	Clears redundancies Simplifies dataset	Uppercase letter may be significant
	Stop-word removal	Minimizes noise and redundancy	Might distort semantic meaning
	Deletion of special characters	Simplifies dataset	
	Tokenization	Necessary for ML model training	Potential problems for non-English languages
	Stemming	Simple and fast to compute	Overstemming confuses words Understemming introduces redundancies Has problems with irregular grammar forms
	Lemmatization	Reduces ambiguity Linguistically accurate	Computational complexity Requires expert knowledge
	Part of speech (POS) tagging	Improves relations extraction	May introduce noise due to inaccurate tagging Computational complexity Requires expert knowledge
Feature extraction	Bag-of-words (BOW)	Simple Offers some explainability	Low accuracy Ignores semantic meaning Ignores word order
	n-gram	Simple Offers some explainability	Low accuracy Ignores semantic meaning Ignores word order
	TF-IDF	Simple	Slower than BOW and n-grams to compute Ignores semantic meaning Ignores word order
	Word embedding	High accuracy Captures semantic meaning	Nonexplainable Requires a huge volume of data to train

3.1.1 Basic Preprocessing

Textual data preparation, in its most basic form, starts with lowercasing all letters and removing special characters. Lowercasing letters will reduce redundancies in later stages, since, for example, when the word "proteins" appears in the beginning of a sentence and written as "Proteins," it will be viewed by a machine as a different word, resulting in redundancies and impaired semantic connections in the final datasets. A potential downside would occur in cases where uppercase letters are meaningful. Note that researchers in the biomedical text-mining literature do lowercase letters consistently. Specific terms that should for some reason remain intact could be compiled into a dictionary and, whenever found within a textual corpus, be exempted from the generic lowercasing. The same applies to special characters, and terms containing them should be added to the dictionary to ensure these terms will not be distorted by basic preprocessing procedures.

With stop-word removal, some of the most common and trivial words such as articles (e.g., 'the', "this") are removed. Such words, for AI-based predictive modeling purposes, do not add any useful information to the text. This is especially true for classification tasks, which differentiate between classes by noticing statistical regularities and differential appearance of certain words in different classes. Common words, such as articles and pronouns, will be found equally in texts of all classes; thus, they do not provide useful, discriminatory information. On the other hand, for other types of tasks, such as biomedical chatbots, we may need to probe into the meaning of sentences, rather than looking merely for statistical regularities of words. Removing stop-words from a sentence does distort its semantic meaning, and DL-oriented tasks with sequential models might require leaving all words intact.

Tokenization breaks a text down into its constituent words. These words are stored in a list and are viewed as discrete, individual data points, losing their sequential meaning. The usual delimiter is the white space, thus characters of a sentence located between whitespaces are discretized as separate words. For example, the phrase "the activity exerted by the isopropyl structural analogue remains the same," when tokenized on whitespaces, gives the list `['the', 'activity', 'exerted', 'by', 'the', 'isopropyl', 'structural', 'analogue', 'remains', 'the', 'same']`. Tokenization with punctuation marks is applied to separate sentences. Languages that use whitespaces and punctuation marks differently than English (e.g., Chinese, Arabic) need different methods for tokenizing texts.

Stemming and lemmatization reduce redundancies further by simplifying words to their linguistic root. For example, the words "runs," "ran," and "running" have their root in the basic verb "run." The two techniques are similar, but not identical. Stemming removes from words common grammatical suffixes, leaving the fundamental stem (e.g., "running" becomes "run," "books" becomes "book"). Lemmatization reduces linguistic variants into their foundational, dictionary base forms (e.g., "are" becomes "be"). Thus, one technique truncates the ending of words, the other converts them to some basic forms. This difference also points to the setbacks of the two methods. Stemming can be applied too much or too little, resulting in either confusing and indecipherable tokens or introducing redundancies. It also has trouble with irregular forms, such as irregular verbs. Lemmatization does not suffer from these problems, but it heavily relies on language knowledge and hard-coded rules. It also raises computational complexity, as each word in the corpus is involved in a dictionary lookup.

Part of speech (POS) tagging assigns a label (tag) to each word according to its status and syntactic function within a sentence. Each word of a sentence is assigned a tag such as noun or verb, and a data structure is generated, usually a list of tuples, associating words with their tags. In the context of biomedical text mining, POS tagging can be helpful in relation extraction. For example, a verb between a drug and a gene might signify an influence of the drug on the gene. Identification of a word's grammatical role is done either via a system of hard-coded rules or via ML models. Both methods have their pros and cons. Rule-based tagging yields higher accuracy but raises computational complexity and is reliant on language knowledge. Statistical, ML-based tagging is more efficient, but may introduce noise due to errors.

3.1.2 Dataset Generation

To convert textual data into an ML-compatible dataset, feature extraction methods are necessary. In terms of handcrafted methods, the simplest is bag-of-words (BOW), whereas a list of all the unique words in a corpus is formed and serves as the columns in the dataset, rows representing abstracts or sentences, and values denoting whether a particular word is present in a particular abstract, sentence, or article. The BOW representation is fast and simple to compute, but encodes less useful information than the other, more complex methods.

N-grams is similar to bag-of-words, but rather than restraining the representation to individual words, combinations of words are also accounted for, enriching the semantic content of the dataset. For example, "black" means one thing, "hole" another, and "black hole" has a totally different meaning, independent of the constituent words. Therefore, including combinations of words as features might enrich the useful information encoded in the dataset. N-grams can be applied for combinations of n words; thus, we might have bi-grams, tri-grams, and so on, with bi-grams being the most common. This method yields higher accuracy than BOW, but increases computational complexity and the dimensionality of the dataset.

Term Frequency-Inverse Document Frequency (TF-IDF) is also similar to bag-of-words, but rather than the mere presence of a word in a textual unit, it also quantifies the relative importance of the word in relation to the entire corpus, as opposed to an individual abstract or sentence. If a word, found present in an abstract, does not occur in the rest of the corpus, it gets assigned a high value and is considered important, since it helps differentiate this text from others. If, on the other hand, it occurs frequently throughout the corpus, then it is not deemed as significant and gets assigned a low value. Quantifying the importance of words means that TF-IDF does not produce binary values like BOW does, but continuous ones. TF-IDF yields higher accuracy than BOW, at the expense of increased computational complexity.

All the above feature extraction methods share a number of common characteristics. They ignore the semantic meaning of words, only registering their mere presence in the text. They also do not take into account the order of words or the wider context in which a word appears. These methods merely generate lists of words or combinations of words and store values signifying the presence or absence of a term in a text. Their nonsequential nature makes them capable of training traditional ML models, and they can perform reasonably well in classification tasks, registering statistical regularities in the differential occurrence of certain terms in different classes. They also suffer from high dimensionality and sparsity, since the dataset will be at least as large as the total number of unique words present in the entire corpus, while only a few of these words will be present in any particular text.

Word embeddings, unlike handcrafted feature extraction, are learned from NNs trained on the corpus, or pretrained in some other, usually large, volume of texts. Texts are passed through the NN, and the embedding layers register the statistical regularities, correlations, and syntactic patterns of the texts, mapping the words to an embedded space, resulting in a vectorized, numeric representation of the text, which can be used for ML training and predictive modeling. These word vectors are low dimensional, with length usually ranging from 512 to 2056, whereas with handcrafted representations, like one-hot encoding, the vector length equals at least the number of unique words in the entire corpus, which can be more than 10,000. Word embedding vectors take floating point values, and are dense, whereas the handcrafted methods discussed above produce high-dimensional vectors of extreme sparsity. The numerical vector assigned to each word in the word embedding scheme captures some of semantic relations among words. For example, the word "dog" will be closer to "wolf" than to "lion," and "lion" will be closer to "tiger" than to "wolf.". Thus, the vectors capture, in an abstract way, some aspect of the semantic meaning of words.

Comparing the two categories of feature extraction approaches, word embedding has higher accuracy than handcrafted methods, produces lower-dimensional and denser vectors, captures some of the semantic meaning of words, and when used with sequential NNs, accounts for the order of words. On the downside, it needs a large corpus to train from scratch, and raises computational complexity. There is always the option of using pretrained embeddings, but these are trained on general-purpose corpora, and on biomedical or medicinal tasks, they will not yield the same accuracy as a specially trained embedding. Thus, there is a trade-off: training a custom embedding yields higher accuracy but at a computational cost, and requires huge data volumes to train, while a generic pretrained embedding is more efficient but with lower performance.

3.2 Other Data Types

Other forms of data might be integrated with molecular structures to enrich model training with additional information. Gene expression profiles are numerical data quantifying the activity of each gene within a cell. As the name suggests, they reveal which genes were active within the cell at the moment of data capture, and consist of 2D matrices with columns representing genes, rows representing cells or patients, and values denoting the degree to which a gene was active in the specific cell or patient at the moment of data capture. Protein sequences are character strings encoding the amino acid sequence forming the protein, with each of the 20 protein-forming amino acids being represented by a unique character. Interaction networks, whether drug–drug, drug–gene, or drug–protein interaction networks are graphs, computationally encoded as matrices with rows and columns being lists of genes, drugs, proteins, and values denoting whether there is an interaction among the two entities.

3.3 Data Tools and Strategies

In the next section, where the literature on data and text mining in drug research is reviewed, many of the computational tools that researchers used are mentioned. In what follows, useful applications, platforms, and frameworks are presented.

KNIME (https://www.knime.com/) is a standalone platform for complete, end-to-end integration and analysis of data. It works in a modular and high-level fashion, establishing a workflow by connecting visual components. Each component or node performs a data-related function, and connecting components together creates an end-to-end pipeline, with nodes being able to perform data conversions and aggregations, run Python code, produce visualizations, load datasets or merge datasets based on a primary key, and train AI models. Merging multi-source data with KNIME entails establishing nodes for individual datasets, introducing additional nodes to carry whatever conversion operations and preprocessing are needed for each dataset, and finally adding a node that merges the data, either through a join operation or a simple column-wise concatenation.

Text mining can be pursued either with standalone platforms and applications or programmatically with code. Some common platforms for text mining are RapidMiner, IBM Watson, Voyant Tools, Google Cloud Natural Language API, IBM LanguageWare, SAS Text Miner, and DiscoverText. These applications vary in their capabilities and functionalities, performing a wide range of text analytics and natural language processing (NLP) tasks, extracting knowledge from biomedical articles. Some of the most popular Python libraries for NLP are NLTK, Gensim, spaCy, CoreNLP, and TextBlob. Scikit-learn, the popular ML package, can handle handcrafted text-feature extraction methods, such as bag-of-words and TF-IDF.

For programmatic data manipulation within Python, the standard library is Pandas (https://pandas.pydata.org/). It uses the DataFrame data structure to process and edit tabular data, and it can perform a very wide range of operations on 2D matrices, such as indexing, joining datasets, binary operations, data conversions, mathematical operations, application of custom functions, displaying descriptive statistics, and grouping.

When working with chemical data in csv files, preprocessing and cleaning of the data are often necessary. In Python such preprocessing is usually performed with Pandas. The two previously mentioned chemoinformatics tutorials, one found in https://projects.volkamerlab.org/teachopencadd/talktorials/T001_query_chembl.html and the other found in https://chem.libretexts.org/Courses/Intercollegiate_Courses/Cheminformatics provide adequate information on chemical dataset manipulation with Pandas. In what follows, a simple illustrative example is presented, discussing some common preprocessing techniques such as duplicate removal, conversions of mismatched data, and merging tables.

Assume a dataset, stored as a Pandas DataFrame object, named `df`. The dataset contains SMILES structures of compounds, their IC50 potency (measured either in nM or in μM), and the database ID number of each compound, serving as a unique identifier. The DataFrame is in the following form:

	smiles	units	value	ID
0	N[C@H]1COCC[	μM	0.420	774
1	Cc1ccc(O)c	μM	0.002	1342
2	Cc1ccc(NC(	μM	0.430	451
3	COC(=O)c1ccc2	μM	0.029	1429
4	CNc1cc2c(cc	μM	1.180	1139
5	Cc1cnc(Nc2c	nM	4.000	554
6	Cc1cc(F)ccc1	μM	0.021	1021
7	Cc1ccc(NC(=	nM	4900.000	714
8	COc1ccccc1/C=	nM	180.000	621
9	Cc1cn(-c2cccc	nM	NaN	1542
10	COc1ccccc1	nM	220.000	782
11	CCS(=O)(=O)	nM	29.000	610
12	Cc1ccccc1Nc1n	nM	7.600	952
13	C=CC(=O)N1CC	nM	3.400	610
14	Cc1c(Nc2c(	nM	11.000	849
15	Cc1c(F)ccc	nM	8.500	513
16	Cc1ccc(O)	nM	4.800	570
17	Cc1ccc(NC(	nM	3.100	561
18	CNc1cc2c(cc	μM	1.180	1139

First, we check for duplicate rows in the dataset:

```
df.duplicated().sum()
```

When the latter produces the output `1`, it means there is a duplicate row. The next command shows which line is duplicated:

```
df.duplicated()
```

	0
0	False
1	False
2	False
3	False
4	False
5	False
6	False
7	False
8	False
9	False
10	False
11	False
12	False
13	False
14	False
15	False
16	False
17	False
18	True

It can be seen that for all lines, the duplicated property is `False`, with the last row appearing as `True`; thus, the one duplicate row is the last one. The next command removes duplicate rows from the DataFrame:

```
df.drop_duplicates(inplace=True)
```

By running the command `df.duplicated().sum()`, returns `0`, as the dataset is now free of duplicates.

Next, we check for missing (NaN) values in the dataset:

```
df.isna().sum()
```

	0
smiles	0
units	0
value	1
ID	0

One missing entry is identified in the `value` column. We can display the entire row that contains the NaN value:

	smiles units	value	ID
9	Cc1cn(-c2cccc nM	NaN	1542

```
df[df['value'].isna()]
```

Then, this row can be removed:

```
df = df.dropna().reset_index().drop('index',axis=1)
df
```

	smiles	units	value	ID
0	N[C@H]1COCC[	μM	0.420	774
1	Cc1ccc(O)c	μM	0.002	1342
2	Cc1ccc(NC(	μM	0.430	451
3	COC(=O)c1ccc2	μM	0.029	1429
4	CNc1cc2c(cc	μM	1.180	1139
5	Cc1cnc(Nc2c	nM	4.000	554
6	Cc1cc(F)ccc1	μM	0.021	1021
7	Cc1ccc(NC(=	nM	4900.000	714
8	COc1ccccc1/C=	nM	180.000	621
9	COc1ccccc1	nM	220.000	782
10	CCS(=O)(=O)	nM	29.000	610
11	Cc1ccccc1Nc1n	nM	7.600	952
12	C=CC(=O)N1CC	nM	3.400	610
13	Cc1c(Nc2c(	nM	11.000	849
14	Cc1c(F)ccc	nM	8.500	513
15	Cc1ccc(O)	nM	4.800	570
16	Cc1ccc(NC(	nM	3.100	561

The row with the NaN value is now deleted, and running `df.isna().sum()` does not yield any missing values.

Next, we deal with the incongruent IC50 measurement units. Some samples are measured in nanomolars (nM), while others are in micromolars (μM). We convert all values to the same unit, to nM:

```
new_value = df.loc[df['units'] == 'uM','value'].apply(lambda x: x*1000)
df.loc[df['units'] == 'uM','value'] = new_value
df.loc[df['units']=='uM','units'] = 'nM'
df
```

	smiles	units	value	ID
0	CCN(CC)CCNc	nM	420.0	718
1	O=c1[nH]cnc2	nM	2.0	1038
2	CN[C@H]1CC2	nM	430.0	1276
3	CC(C)(N)C	nM	29.0	708
4	Cc1ccc2[nH]nc	nM	1180.0	642
5	Cc1nc(Nc2n	nM	4.0	858
6	Cc1c(Cl)cccc1	nM	21.0	1158
7	Cc1ccc(NC(=O)	nM	4900.0	439
8	CCNC(=O)c	nM	180.0	1431
9	Cc1c(Nc2c	nM	220.0	1055
10	COc1ccc2c(O	nM	29.0	1178
11	Cc1ccc(NC(	nM	7.6	700
12	Cc1c(O)cccc	nM	3.4	459
13	C=CC(=O)Nc1c	nM	11.0	1307
14	C=CC(=O)N1CC	nM	8.5	898
15	Cc1ccc(NC(=O	nM	4.8	1482
16	Cc1ccc(NC(	nM	3.1	81

Finally, we merge our DataFrame with a gene expression dataset. This dataset, called `rna_df`, contains gene expression data after drug perturbations, with the drug ID being registered in the first column:

	ID	Gene 1	Gene 2	Gene 3	Gene 4
0	446	−0.809	−0.081	−0.383	−0.187
1	451	0.398	−1.484	1.038	0.228
...	...	....	...	...	...
26	1429	0.230	0.162	0.952	−1.063
27	1535	−0.537	0.275	−0.095	0.983

The two datasets are joined by using the `ID` column as the primary key:

```
df = df.join(rna_df.set_index('ID'),on='ID')
df
```

	smiles	units	value	ID	Gene 1	Gene 2	Gene 3	Gene 4
0	N[C@H]1COCC[	μM	0.420	774	−0.486	−0.325	0.048	0.533
1	Cc1ccc(O)c	μM	0.002	1342	−0.179	−0.141	−0.934	0.363
2	Cc1ccc(NC(	μM	0.430	451	0.398	−1.484	1.038	0.228
3	COC(=O)c1ccc2	μM	0.029	1429	0.230	0.162	0.952	−1.063
4	CNc1cc2c(cc	μM	1.180	1139	−0.114	−0.182	0.260	1.486
5	Cc1cnc(Nc2c	nM	4.000	554	0.735	−0.498	0.776	0.243
6	Cc1cc(F)ccc1	μM	0.021	1021	−0.091	−1.082	−0.994	0.876
7	Cc1ccc(NC(=	nM	4900.000	714	0.005	0.012	0.510	−0.110
8	COc1ccccc1/C=	nM	180.000	621	−0.424	−0.293	−0.055	0.882
9	COc1ccccc1	nM	220.000	782	0.645	−0.521	−0.097	−1.103
10	CCS(=O)(=O)	nM	29.000	610	−0.502	−0.386	−0.654	0.121
11	Cc1ccccc1Nc1n	nM	7.600	952	0.300	1.110	−0.436	−1.544
12	C=CC(=O)N1CC	nM	3.400	610	−0.502	−0.386	−0.654	0.121
13	Cc1c(Nc2c(	nM	11.000	849	0.576	−0.471	−1.216	0.775
14	Cc1c(F)ccc	nM	8.500	513	0.152	−0.655	0.054	0.369
15	Cc1ccc(O)	nM	4.800	570	0.851	−0.740	0.531	−0.449
16	Cc1ccc(NC(	nM	3.100	561	−0.398	−0.527	0.548	0.396

4 Data and Text Mining and Integration

Data mining, referring to whether the data comprise molecular structures, biomedical texts, transcriptomic profiles, or any other data or combinations of data, may rely on multiple databases and repositories, or it may be conducted from a single data source. The pipeline may utilize an ML model or an NN within the process, or not. It may result in generating structures for novel small molecules with potential therapeutic effect, or it may focus on a myriad other applications, such as predicting bioactivities of compounds, exploring drug–protein interactions, identifying drug targets, anticipating long-term side effects, or estimating physicochemical properties. The only common patterns in all the applications are (1) that they are data driven, extracting insights computationally through a volume of computer-stored data, applying computational techniques to explore the data and utilize them in the multiple stages of drug discovery process, and (2) that there is some sort of data integration, whether that entails data of the same type but from different databases or documents or of multiple different data modalities.

4.1 Data Integration

The most basic form of data integration is combining data of the same type, usually SMILES strings, from different databases. Model training requires a substantial amount of data, no matter whether a discriminatory model needs to classify molecular structures or a generative model to produce novel ones. Such a large quantity of data are not always contained in a single repository. In Chithrananda et al. (2020), the authors pretrain a transformer with 77 million SMILES from PubChem, and then fine tune it with datasets from MoleculeNet, for molecular property prediction. In Liu et al. (2019), the authors combine a generative recurrent neural network (RNN) with reinforcement learning (RL) to generate compounds with desired properties. They pretrain the model with compounds from ZINC, then fine tune it with SMILES from CHEMBL. The reward function for the RL agent is provided by ML models predicting whether the generated compound is bioactive.

In Zhavoronkov et al. (2019) a variational autoencoder (VAE) is pretrained with SMILES from ZINC and proprietary databases, then fine tuned with compounds with desired properties, using data from the Integrity and ChEMBL databases. The Integrity database serves as a repository of molecules patented as novel drugs by the top ten pharmaceutical companies, from 1950 until now. For preprocessing the Integrity and ChEMBL molecular SMILES, researchers remove structures containing atoms other than carbon, oxygen, nitrogen, hydrogen, sulfur, bromine, chlorine, and fluorine. Moreover, they exclude by filtering through various techniques from medicinal chemistry, compounds with potentially reactive or toxic groups.

In Moret et al. (2020), a generative DL model produces structures of novel molecules with desired properties. The model is pretrained with 365 k molecular SMILES from Chembl24. The criterion for the selection of the molecules is that their bioactivity values (EC50, IC50, Kd, and Ki) should be below 1 μM. Each SMILES structure is one-hot encoded with vectors of length 71. Data augmentation is applied for more efficient pretraining, with multiple schemes of augmentation being tested experimentally. Best results were obtained via x10 augmentation with 0.7 sampling temperature, a coefficient determining the randomness in token selection during each step. Fine tuning of the model is tackled with the MEGx dataset of natural product compounds. Since the study deals with generating new molecules from very limited prior knowledge, researchers only pick five molecules from MEGx to fine tune the model. These five molecules are chosen to be similar to each other, with similarity computed via the Tanimoto distance on the Morgan fingerprint representation of the molecules.

In Grisoni et al. (2021), a generative RNN is pretrained with 657 k compounds from libraries with commercially available chemical agents. The number of available compounds in the resources used ranged from 3.5 million, filtered down to the compounds that could be successfully deconstructed into their reactants. The model is then fine tuned by using data with desired bioactivity properties, retrieved from ChEMBL26. In terms of preprocessing, the chemical structures are converted to canonical SMILES, while to ensure uniform length, only strings with a length of no more than 80 are retained. Novel molecules designed by the generative model are validated with data from various databases: PubChem, ChEMBL27, SciFinder, SureChEMBL (Papadatos et al. 2016), and Reaxys (Lawson et al. 2014). The comparison of the generated molecules with the ones found in these databases showed that the generative model has learnt to design valid molecules, and even found compounds that had already been patented or were known to possess the desired chemical properties that the model was aiming for. From the generated compounds not found in the public databases, a number of them were synthesized in the lab and, after chemical analysis and tests, were indeed found to possess the desired properties.

In Yu et al. (2021), a generative model is pretrained with 6.8 million molecules from the ZINC database to learn the syntax and distributions of drug-like compounds. Then it is fine tuned with 210 k molecules from ChEMBL that exhibit desired properties. The novel molecules that were designed, about 32.5 k, are screened with various techniques that use additional data from various sources. Initially, only the molecules with desired physicochemical properties, according to the rule of five (Shultz 2019), are kept. Further screening takes place via PAINS (Baell and Holloway 2010) filters and an in-house rule-based system that the researchers developed, removing molecules with potentially reactive or toxic functional groups. Then, they train a CNN on data from CheMBL to predict whether molecules would bind to proteins of interest, and the molecules designed by the generative model are further screened based on their CNN-predicted binding affinity. From the remaining molecules, a number of them were synthesized in the lab, and they were shown to exhibit the desired properties.

Integrating molecular structures with other types of data can be helpful, allowing for more sophisticated applications and providing richer information to the model. In Isigkeit et al. (2022), the authors take SMILES strings from ChEMBL and PubChem and combine them with bioactivity data from three other databases (IUPHAR/BPS, BindingDB, and Probes & Drugs), integrating them into a single database. Salts are removed from the structures, while the compounds present in more than one database with different SMILES structures are converted to canonical SMILES.

In Ren et al. (2022), data from five databases are combined: DrugBank, SIDER, KEGG, PubChem, and OFFSIDES. The task is to predict drug–drug interactions (DDIs) with DL models. The data consist of molecular SMILES, DDI networks, and bioactivity information of drugs processed into a drug-similarity matrix. Both the SMILES and the DDIs are encoded with embedding layers, with the DDI networks in particular being converted through graph embedding. The integration happens at the model level with three models taking each data type, then their outputs are concatenated and combined in additional layers. Thus, a CNN is trained on the SMILES strings, an RNN on the DDIs network, an additional RNN on the drug similarity data, and an attention module finally integrates all three models.

Another form of data integration is to take data from one repository and extract additional information on these same data points from another repository. In Yosipof et al. (2018), the authors employ the latter approach to alleviate imbalanced data for the task of predicting drug likeness of chemical compounds. First, they gather compound data from PDBBind (Wang et al. 2004), the Ki Bank (Zhang et al. 2005), and the SCORPIO dataset (Ababou and Ladbury 2007). For negative data, they perform a search of all the collected molecules in the DrugBank database (Wishart et al. 2007), identifying compounds that are not registered as "drugs" in DrugBank. These compounds will serve as negative samples for model training.

Drug molecular structures are combined with gene expression profiles to design models that take gene expression data either of a patient who needs treatment, or of a desired state, and generate SMILES strings of drugs that either treat the unhealthy gene expression profile or elicit the desired state. In Méndez-Lucio et al. (2020), the authors combine SMILES and gene expression data from the L1000 CMap database (Subramanian et al. 2017). They train a generative adversarial network (GAN) to take a vector of desired gene expression and produce a SMILES structure of a chemical agent that induces the desired effect. In Born et al. (2021), the authors take SMILES from ChEMBL and gene expression profiles of cancer cells from The Cancer Genome Atlas (TCGA) (Cancer Genome Atlas Research Network et al. 2013); these data are used to train a VAE enhanced with RL agency. One VAE is pretrained with 1.5 million SMILES from ChEMBL. The SMILES are not preprocessed with chemical editing, such as removing salts, and no threshold of string length or molecular weight is applied, since their pytorch implementation allows for variable length. Another VAE is pretrained on gene expression data from TCGA. Then the two VAEs are joined and trained with data of drugs that affect a certain gene expression profile, aiming to develop a model that takes a gene expression profile that needs treatment, and produces a corresponding drug SMILES structure.

In Han et al. (2022), molecular structures of drugs and drug–protein interactions are used to predict DDIs. First, the authors retrieve drug SMILES from DrugBank. Then, using data from the Kyoto Encyclopedia of Genes and Genomes (KEGG) (Kanehisa and Goto 2000), as well as the PDD database (Wang et al. 2017), they construct a knowledge graph of interactions among drugs and proteins. The SMILES are converted into fixed-length vectors through the SMILES2vec method (Goh et al. 2018). In this method, SMILES are rendered into a word bag of 251 elements, one-hot encoded, and dimensionally reduced, through principal component analysis (PCA), into vectors of length equal to 26. Using the vectorized SMILES and the drug–protein graphs, a graph network is trained to predict DDIs.

In Gummesson Svensson et al. (2024), a generative model is pretrained with SMILES from ChEMBL, and then, through RL molecules, are made for desired properties. The scoring function for the RL process is given by a random forest (RF) model, trained with data from exCAPE-DB, taking in Morgan fingerprints of compounds and predicting how active they are. Training the RF with data of different bioactivities results in the RL agent being directed to generate compounds that are bioactive for a specific task. Thus, data integration is achieved without compiling a dataset with concatenated information from multiple resources, but by using different datasets for different modules of an AI system.

In Li et al. (2020), SMILES structures are combined with gene expression profiles and signaling pathways for drug repurposing. Researchers train models with the transcriptomic effect chemical agents induce, hypothesizing that any misclassification might potentially be caused not by the model having failed, but by the human annotators mischaracterizing the data. Comparing molecular similarity and drug effect on the signaling pathways, they pinpoint chemical agents that are structurally similar with FDA-approved drugs, then evaluate how these drugs have been classified, thus generating hypotheses for experimental tests and potential breakthroughs. Experimental comparisons show that the combination of transcriptomic and pathway data, integrated with the chemical structures, results in higher performance compared to integrating either of the two individual data types on their own.

Throughout the literature on integrating drug-related data for ML training, two major patterns emerge. Integration of data occurs either at the dataset level or on the model level. Dataset-level integration consists of some type of merging individual datasets into a unified set, used subsequently to train a model. This merging is simple in the case of tabular data, where an inner join or a mere concatenation is all that is needed, while it can be more complex in the case multi-modal data, where extensive, and often lossy, conversions are needed to make the datasets compatible with each other, such as graph embedding to turn graphs into vectors, one-hot encoding to turn SMILES strings into 2D sequential tables, or feature extraction techniques to extract information from images. With model-level integration, the datasets are not combined beforehand but are merged by the model itself throughout its training process. This can be done in two ways: (1) by a multi-head model, such as in the case of Ren et al. (2022), where a separate model takes each dataset, and their outputs at some point are concatenated and processed by additional layers, leading to the final classifications, and (2) by a modular system, such as in Gummesson Svensson et al. (2024), where a complex system integrates more than one models, each one trained on a different dataset. The choice of approach should depend on the application. For example, with tabular, nonsequential data, merging the datasets is easy and could be the method of choice. With multi-modal data, a multi-head model might be more preferable than pursuing complex conversion schemes to force disparate datasets to fit each other. Perhaps both approaches should be tested, comparing model complexity and predictive accuracy. Merging data may result in high-dimensional datasets, which in turn, may result in overfitting. Dimensionality reduction is discussed on Sect. 5.5.

4.2 *Text Mining*

Texts in biomedical scientific journals are a rich source of information on causal relations among biological entities like genes, proteins, diseases, and drugs (Uramoto et al. 2004). Whatever the text-mining research project, the general framework will often consist of various types of text preprocessing and information extraction (Fig. 4). The initial step would be to compile a number of text documents from which information will be retrieved. The standard repository of scientific literature in the biomedical field is Medline, usually accessed through the PubMed search interface (Mack and Hehenberger 2002).

The two most common applications of text mining in the biomedical literature are (1) named-entity recognition, i.e., identifying the names of biomedically relevant concepts and entities from a biomedical text, such as genes, metabolites, drugs,

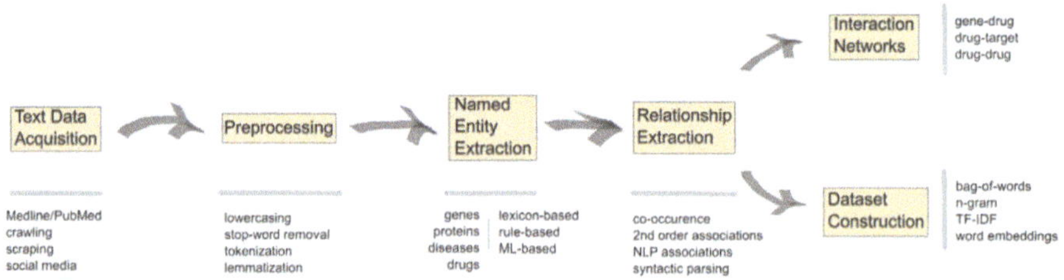

Fig. 4 A general pipeline for text mining in drug research: Biomedical text acquisition is followed by text preprocessing, named entity extraction and relationship extraction; Then, a number of applications becomes possible, such as the inference of knowledge graphs and interaction networks, or the construction of datasets for ML model training

proteins, diseases, and (2) relationship extraction, i.e., determining the relationships and interactions among these entities (Uramoto et al. 2004).

Named-entity extraction, which entails statistical analysis of the entities co-occurring within a text, paragraph or sentence, or by adjectives and verbs between the entities in a sentence, is performed in one of three following ways: via dictionary lookups, based on rules, or through machine learning (Gong 2018). In the dictionary-based approach, domain-specific dictionaries of thousands or even millions of biomedical terms, extracted from public databases, are being used to identify the relevant entities in articles and article abstracts. Naturally, this method has very high accuracy, although terms of potential interest that are absent from the predefined dictionary will go undetected. The rule-based approach utilizes syntactical heuristics and domain-specific linguistic patterns to identify the entities in a given text. Enzyme protein names often end in "-ase." Gene names begin with a letter, followed by a combination of letters and numbers. Biomedical entities are mostly nouns or noun phrases. These and other rules are used to identify relevant terms in text, without relying on a predefined set of entity names, as in the rule-based paradigm. ML methods consist of using an annotated corpus to train an NN or some other ML model. After training has been completed, the model can recognize entities in new articles and abstracts, and can classify these terms as proteins, genes, diseases, or drugs.

Relationship extraction might be approached in several ways. The three basic and most commonly used techniques are co-occurrence associations, second-order associations, and natural language associations (Loging et al. 2007). Co-occurrence associations entail identifying terms co-occurring within the same sentence or paragraph, assuming these entities to be associated in some type of causal or correlative relationship. This method is computationally easy, fast, and scalable, but may yield a relatively high rate of false positives. Second-order associations entail identifying two terms that, though not co-occurring together, they both co-occur with another, third term, thereby due to their mutual connection with this term, we assume the two entities are related to each other, perhaps in a more indirect, second-order fashion. This technique is also scalable and might generate novel hypotheses for biological discovery, but it suffers too from false positives, and generates an exceeding number of associations, which might require a human expert to sort out. Natural language associations entail pattern-based searches for specific phrases or syntactic motifs and variations within a text. This approach can also be scalable, and has a lower rate of false positives, but on the downside, patterns take time to construct and test, and numerous different patterns are required in order to cover all possibilities and extract all associations from the text.

Relationships among entities can be established not only in binary values, i.e., whether an inhibitory or promoting relationship is present, but can also be quantified in terms of their strength. This allows for the elimination of any relationships that are weak and perhaps due to random co-occurrences, thus simplifying the generated data. In its simplest form, relationship strength constitutes of the count of texts containing both terms, normalized by the probability of the two terms co-appearing in the same text randomly and without relationship, given by the formula: relationship strength $= c(t\,1 \bigcap t2)/c(t1)c(t)2$ (Rodriguez-Esteban 2009).

In the above context, in Hansson et al. (2020), the authors develop a text mining-based method for knowledge extraction from scientific literature, for the purposes of drug discovery, using the case study of type 2 diabetes (T2D). The main idea is that the publication space is vast, and a scientific breakthrough that could prove useful for drug hunters might go by unnoticed in the constant stream of new publications. Thus, the researchers develop a method to rank articles based on their relevance to some predefined drug discovery project, which, in the particular case study, is type 2 diabetes (T2D). First they mine PubMed abstracts for proteins related to T2D, ranking them by how often they are mentioned in abstracts that mention T2D. Other terms of interest are ranked based on their co-occurrence with the core, significant proteins identified in the previous step. The text of the articles is converted into n-gram vectors, vectors are clustered into similar groups, and the semantic concept groups that emerge out of the clustering are compared with the core proteins and terms, assigning to each group a value of importance, based on the similarity with the core proteins and terms. Thus, the acquired concept weights can be used to rank individual documents and new publications, to evaluate their potential usefulness. Whenever an article is mined for concepts and biomedical entities, these entities are weighted by the concept weights computed previously, and the sum of the weights produces each article's score. As new literature is produced and published, articles are automatically scored for their potential to provide information valuable for T2D drug discovery.

In Uramoto et al. (2004), entity extraction is conducted through dictionary lookup, referring words from the corpus to a dictionary of protein names, gene names, and other entities. Relationship extraction is conducted with parsing, i.e., by extracting verbs that come after nouns and entity names, taken as identifiers the qualitative relationship among the entities.

In Papanikolaou et al. (2016), the authors apply a text-mining pipeline to a chemical database (DrugBank), in a system that functions like a search-engine. The user makes a conventional query. The system selects the relevant DrugBank records and extracts all textual information. The database is downloaded and stored locally, and the terms are already tagged, enabling a more time-efficient session for the user. The system collects all tagged terms and performs TF-IDF (term frequency-inverse document frequency) analysis on the nontagged terms to evaluate their significance. Low TF-IDF words as well as stop-words are removed, and remaining words are used to compute a set of binary vectors, one for each document, representing the presence or absence of each remaining word or tagged term for each document. Clustering algorithms group relevant terms together, and word clouds are displayed with the results of this process. Thus, the final product is a text-based search engine for a chemical database that displays information in the form of word clouds connecting together a variety of entities, such as drugs, proteins, diseases, genes, phenotypes, and other information.

In Wu et al. (2012), a method is developed to evaluate gene–drug interactions based on Latent Dirichlet Allocation (LDA). First, they identify gene and drug names through dictionary lookup. For relation identification, the co-occurrence of genes and drugs is evaluated, based on whether co-occurrence takes place in an abstract, a sentence, or in a phrase. The gene–drug relations are ranked with four alternative methods for comparative reasons: (1) a simple frequency method, ranking the interaction based on how many times the gene and drug names appear together, (2) a Chi-square statistical test combined with mutual information (MI), (3) LDA-KL, where the Kullback-Leibler distance (KL) of terms over relevant LDA-identified topics is computed, and (4) LDA-PKL, a custom method which combines KL distance with posterior probabilities of LDA-derived topics on article abstracts. For the two methods utilizing LDA, the LDA algorithm is run for 100 topics, after stop-words and infrequent words are removed from the corpus. The researchers evaluate these methods by taking the results produced and comparing them with known gene–drug interactions retrieved from PharmaGKB, indication the LDA-PKL method as superior to the others.

An important and challenging application in drug development is to predict whether a candidate drug will have long-term adverse side effects (Loging et al. 2007). It is better to predict such side-effects beforehand, before time and money have been invested in development and clinical trials, and certainly before the drug has been approved and released and turns out to cause problems in the long-term for its users. This can be achieved, e.g., by computing chemical similarity of the candidate compound with FDA-approved drugs that have been released on the market, and by identifying whether these approved drugs have long-term side effects. In Mehra et al. (2022), the authors want to evaluate potential long-term adverse side effects of tyrosine kinase inhibitor drugs, commonly used for chronic myeloid leukemia. The researchers start by extracting 2575 relevant literature abstracts from PubMed through conventional search queries. Abstracts are parsed, and stop words and punctuation characters are removed. Then, tokenization, lemmatization, and bag-of-words vectorization ensue, and a TF-IDF matrix (term frequency-inverse document frequency) is computed. K-means clustering is then applied to the TF-IDF matrix to group the data into clusters of similar types of adverse effects and drugs, with pairwise chi-squared statistical tests evaluating whether clusters are meaningfully distinct from each other.

In Gefen et al. (2020), the authors discover through text mining, a potential side effect of a drug that may lead to congestive heart failure. Data comprise the health records of 11,949 hospital patients, with the results of all the medical tests they underwent, and the various reports written by doctors and nurses. Only the textual information is retained from the data, with the most significant terms being names of drugs, medical conditions, and physiological states. Standard NLP preprocessing ensues, with lowercase rendering of all letters, stop-word removal, and deletion of numbers and special characters. The researchers create a set of text-to-document matrices (TDMs) with the frequency of the relevant terms on each document. Latent Semantic Analysis (LSM) (Gefen et al. 2018) comes next, applying log-entropy weighting on each LDM to compress the high values and allow the lower values to become more significant, then reducing dimensionality through singular value decomposition (SVD). In this way, they identify connections between the term of interest (the drug Adalat) and other terms that are close to it in semantic space. This strategy identifies not only direct relationships but also indirect.

For ML-based prediction of drug side effects from texts such as social media and internet reviews, annotated datasets have been developed, such as the Cadec (Karimi et al. 2015) and the PsyTAR (Zolnoori et al. 2019). These contain various texts and text snippets, with annotations belonging to categories like "adverse effects," "withdrawal symptoms," and "illness." In Alimova and Tutubalina (2019), before training a variety of LSTM-based NNs, they transform the data into text embeddings via a word2vec algorithm pretrained on a set of 2.5 million reviews. The trained NNs were able to predict drug side effects with an F1-score of up to 0.87, depending on the dataset and the used model.

In Liu et al. (2022) they want to predict adverse drug side effects with a model trained on social media posts. To convert documents into fixed-length vectors, researchers employ unsupervised methods like doc2vec, stacked autoencoders (AEs), and pre-trained BERT instances, as well as handcrafted features such as n-grams with tf-idf, lexicon match-based features, and a number of categorical semantic features.

Rather than computing drug–drug similarity based on chemical fingerprints and distance metrics, in Asiri (2023), the authors estimate similarity using data generated by patients on social media. They extract, through crawling and scraping, information from social media and compile a dataset of patients' reviews on antiepileptic drugs. The reviews are aggregated, grouped by drug, resulting in a single document for each drug, with all the reviews on that particular drug, with a number of documents equal to the number of antiepileptic drugs. Then the documents are preprocessed with NLP techniques: all letters are rendered lowercase, stop-words, digits, and punctuation marks are removed, and n-grams for 1–3 words are formed as features of a tabular dataset, with rows representing drugs/documents, and columns signifying whether a particular n-gram is present on a particular document. At this point, similarity measures can be computed for each document/drug, and the researchers experiment with Cosine Similarity, Euclidean Distance, Manhattan Distance, and Jaccard Coefficient. Thus, a distance matrix of drug–drug similarities is formed, based not on chemical structure or molecular fingerprints, but on textual information.

A popular application in the field is text-mining for identifying drug–drug interactions (DDIs) (Harpaz et al. 2014), which helps predict adverse effects arising from a patient taking a combination of drugs, is presented in Xu et al. (2023). The authors predict DDIs from text, with NNs trained on a dataset developed in Herrero-Zazo et al. (2013), with annotated sentences from article paragraphs and abstracts. Text is preprocessed into bag-of-words and word embeddings.

In Percha et al. (2012), data are extracted through text mining, to train ML models for drug–drug interactions prediction. First, the PharmGKB database (Klein et al. 2001) is utilized to create two dictionaries of drug and gene names. The idea is to infer drug–drug interactions based on the effect drugs have on genes. From a corpus of 17.5 million Medline abstracts, the authors extract sentences that contain both a drug and a gene of interest. Dependency graphs are computed from these sentences, with relationships among the various entities (drugs and genes) being estimated. Interactions among drugs are inferred from the shortest path, in the knowledge graph, connecting the two drugs. This means that the genes between the two drugs in the graph, and the effects the drugs have on these genes, are assumed to be the primary interactions between the drugs. Then, the researchers train an RF classifier to predict drug pair interactions based on their textual profiles.

In Huang et al. (2022), drug–drug interactions are predicted by training ML models on the DDIs 2013 dataset (Hong et al. 2020). The dataset contains sentences from abstracts taken from Medline and various texts from DrugBank, with labels for the drugs present in the sentences, and for the interactions among these drugs. Neutral sentences are taken to be the negative samples. To address the severe data imbalance, where the negative samples were almost 20 times more than the positive ones, researchers applied both data generation to upsample the positive samples, and they also downsampled the negative ones. For the generation of positive examples, they used BioGPT-2, a version of GPT-2 pretrained on 500 k abstracts from PubMed (Papanikolaou and Pierleoni 2020). The enriched dataset is used to train BioBERT, a language model pretrained on abstracts from PubMed (Lee et al. 2019). After training, the model can predict interactions among drugs with an f1 score of 84.2, precision 83.4, and recall 85.

In another study training BioBERT, researchers couple it with a Bidirectional LSTM, for training on DDI datasets (KafiKang and Hendawi 2023). The BERT module is based on a variation customized for predicting interactions among entities, with additional layers for relation classification (Wu and He 2019). Preprocessing the data comes in two steps. First, researchers eliminate instances where the same drug appears twice in a sentence, as these sentences do not contain information relevant to the specific task (predicting interactions among drugs). Second, in the instances where two drug names are present in the sentence, tokens are added before and after the names of drugs to help the model identify them and make training efficient. When dealing with imbalanced datasets, the researchers downsample the negative instances to make the classes more even.

Another study where NNs are trained on datasets for DDI prediction is Sun et al. (2019), where the data instances pass through an embedding layer before being fed to a CNN/RNN hybrid. The significant imbalance in the data, with negative instances massively outnumbering positive ones, is remedied by rule-based downsampling of the majority class. Drug pairs are removed from the data in three cases: (1) if the two drugs have the same name or one is an abbreviation of the other, (2) if they form a coordinate structure, and (3) if one drug is a variation or derivative of the other. Since the corpus terms relevant for the classification may have different meanings depending on context, the embedding stage employs bidirectional long short-term memory to integrate contextual information into the training data.

In Kolchinsky et al. (2013), the goal is to train an ML classifier to identify publications with experimental DDI information. The training corpus consists of 1213 Medline abstracts, manually annotated as positive or negative, with regard to mentioning laboratory experiments for DDIs. Textual features are tokenized, Porter-stemmed, converted into unigrams, numbers converted into "#," and very short or infrequent features are omitted. Occurrence matrices are formed to represent the appearance of different words in different documents. Researchers use experiments and compare several alternative forms of these data to train a variety of ML models. The forms of input data that are compared are the raw occurrence matrices, an IDF transform of the matrices, normalized IDF, TF-IDF, normalized TF-IDF, and PCA-reduced versions of all the previous data types. The highest performance is achieved with the PCA-reduced normalized IDF matrices.

In Song et al. (2019), researchers try to go deeper than the conventional DDI predictions, and probe into the mechanisms leading to the side effects. Rather than interactions among drugs, they try to predict drug–protein and protein–side effect interactions from biomedical articles. First, they access article abstracts relevant to cancer drugs from PubMed. Preprocessing includes sentence segmentation, POS tagging, and lemmatization. First, dictionary-based named-entity extraction ensues. Then, the relations among entities are established based on a rule system. When there is a verb between two entities, they are viewed as related. Sentences with negation words are considered negative. A dictionary of 389 verbs commonly found in biomedical literature is used to define the types of relations among entities. The distance between entities, i.e., the number of words between them, is taken as a measure of the strength of their relation. A threshold is determined, and only entities with up to six words between them are considered related. The active or passive tense of the verb can betray the direction of the interaction between the two entities, i.e., which entity exerts influence upon the other in a cause-effect chain. After the application of these rules, the resulting information is used to generate knowledge graphs displaying the complex interactions among drugs and proteins.

In Wu et al. (2014), the authors propose a scheme for annotating corpora for DDI prediction. They construct a dictionary with names of drugs, proteins, enzymes, metabolites, and other relevant terms from the fields of pharmacology and biology. The three-level annotation scheme consists of annotating individual terms (drug, disease, enzyme, metabolite, and similar terms), then annotating sentences (clear or vague DDI sentence), and finally annotating the DDI pairs in the sentences.

A variety of handcrafted, rule-based, or hermeneutic-based feature extraction from biomedical text is found in Hailu et al. (2013), where the authors train an ML model to classify DDIs in text. The corpus consists of documents mentioning DDIs from DrugBank, as well as relevant Medline abstracts. Researchers train the ML model with a number of extracted features from the corpus. There is a distance feature, counting the number of words between two drugs. Part-of-speech (POS) tags are associated with each word. Bigram vectors are formed for each document. The top 100 words signifying interactions among drugs are identified, with the presence of these words before or after a drug name in a document encoded in a binary vector. Another binary vector signifies the presence of prepositions within target drug texts. The number of other drugs that are present together with the target drugs is another feature. Further features are extracted through OpenDMAP, a rule-based text mining and concept extraction tool (Hunter et al. 2008).

Another example of handcrafted feature extraction for DDI predictions is in Machado et al. (no date). They train shallow ML models with the DDIExtraction 2013 dataset (Segura-Bedmar et al. 2013; Herrero-Zazo et al. 2013). Named entity extraction is done with DL, using a BioBERT trained on the CHEMDNER dataset (Krallinger et al. 2005). Preprocessing includes lowercasing everything, tokenization, stemming, and removal of stop-words, punctuation, digits, and short words. Vectorization of the text is done with multiple alternative methods, for comparison. The alternatives are bag-of-words (BOW) with 1-g, BOW with 2-g, BOW with 3-g, BOW with 1-g and 2-g, TF-IDF, Word2vec, Doc2vec, and a custom handcrafted feature set. The custom features constituted on vectors of length 11 signifying for each sentence: (1) the number of positive key terms, (2) whether positive terms are present between names of drugs, (3) whether positive terms exist in the vicinity of drug names, (4) number of negative terms, (5) whether negative terms are present between names of drugs, (6) whether negative terms are present in the vicinity of names of drugs, (7) number of special terms, (8) word number between names of drugs, (9) verb number between drugs, (10) drug terms between names of drugs, and (11) whether drug terms are present near names of drugs. These encoding alternatives were used to train traditional ML models such as logistic regression and support vector machine (SVM), and the encoding that yielded top results was a combination of 1-g, 2-g, word2vec embeddings, and the custom handcrafted features. Since named entity extraction was performed with DL and not through dictionary construction and lookup, this is a system for completely automated, end-to-end extraction of DDIs from text.

4.3 Data and Text Integration

A very common application in the integration of mined text with information from multiple databases, is in drug repurposing. The general pipeline consists of mining the literature to identify genes relevant to a targeted disease, then through database integration retrieve information in these genes, infer interaction networks, pinpoint

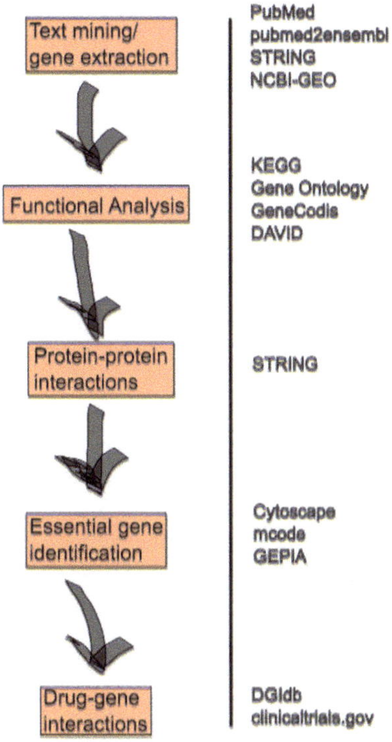

Fig. 5 A common framework in text and data integration for drug research, consisting of text mining, gene extraction, functional analysis, PPIs inference, essential gene identification, and evaluation of drug–gene interactions, to find approved drugs targeting the genes of interest. The right-side column shows databases and tools used on each stage of the process

essential genes to target, and finally evaluate any FDA-approved drugs that might be targeting these genes (Fig. 5). In Beckman et al. (2022), they want to find drug targets for xerostomia. First, they search the biomedical literature, through the PubMed and PubMed Central portals, for terms associated with xerostomia. Associated genes are extracted, and through the Search Tool for Recurring Instances of Neighboring Genes (STRINGv11.0), the protein–protein interaction (PPI) networks of gene clusters are constructed, based on which protein sequence each gene encodes. These networks, based on genes they share in common, are then merged to compile a final, comprehensive PPI network of xerostomia-related genes. Enrichment analysis is performed for these genes, retrieving functional and ontological information from KEGG (Kanehisa and Goto 2000) and GOC (Gene Ontology Consortium 2015). Drug interactions for xerostomia-related genes are retrieved from the Drug–Gene Interaction database (DGIdb), and the data on drugs are compared with drugs being evaluated for approval, using the clinicaltrials.gov database. The goal is to identify, using xerostomia as a case study and proof of concept, potential targets as well as drugs that could potentially affect these targets, using bioinformatics and data mining tools to extract knowledge from the biomedical literature, and integrate it with information from a variety of databases.

In Wan et al. (2020), the authors identify through the online biomedical text mining tool pubmed2ensembl genes related to atherosclerosis. Additional genes are extracted from microarray data analysis profiles retrieved from the NCBI-GEO database (Edgar et al. 2002; Barrett et al. 2009). The intersection of these gene sets is further analyzed. Through Gene Ontology analysis, the genes are categorized according to their biological function. The Kyoto Encyclopedia of Genes and Genomes (KEGG) database (Kanehisa and Goto 2000) is also used for characterizing the overlapping genes. Next, the STRING database is used to analyze the interactions among proteins associated with the overlapping genes. The most significant genes are identified and then clustered based on the interactions of their associated proteins. Then, the drug–gene interaction database (http://www.dgidb.org) is used to find drugs targeting any of the significant genes. Genes that have drugs targeting them are selected and proceed with further functional analysis. The purpose of the study is to identify genes, proteins, and drugs associated with atherosclerosis and, through a range of analyses and computational tools, extract from the data insights and knowledge useful for drug discovery.

In Pletscher-Frankild et al. (2015), the researchers propose a framework to discover disease–gene associations, starting with data mined from textual sources, then enriching the data with information from various databases, processing the data with various computational methods, and finally analyzing it to extract useful knowledge and insight. First, a dictionary of genes, proteins, and diseases is formed. Then, a corpus of abstracts from biomedical literature is compiled, and biomedical terms are identified via dictionary lookup. Associations among diseases and genes are identified and ranked based on the frequency of the term-pairs co-occuring versus the occurrence of the individual terms appearing alone. Through web crawling, they download the disease pages from the Genetics Home Reference database (GHR), store the names of all the diseases and the genes mentioned in each disease web page, and map the diseases with ontology terms from the precompiled dictionary. From the UniProt Knowledgebase (UniProtKB) they extract disease keywords used to associate protein with diseases, and map them with the ontology terms of the dictionary. With similar methods, they further integrate cancer mutation data from the Catalog of Somatic Mutations in Cancer database (COSMIC) and genome-wide association studies (GWAS) data from the DistiLD database. The integration and statistical analysis of such data result in a knowledge base of associations between diseases and genes, with annotated levels of confidence on the inferred associations.

In Wang and Yu (2023), text mining and information from public databases are combined to explore the interactions among genes, proteins, and drugs associated with canine pyometra. The authors identify two sets of genes associated with canine pyometra, one set of genes retrieved through text mining with the online pubmed2ensembl tool, the other through microarray data analysis downloaded from NCBI-GEO. Genes that are common in both sets are characterized, through Gene Ontology, for their biological function, and through KEGG, they are evaluated in terms of the interactions among them. With the STRING database, the proteins encoded by these genes are identified, as well as the protein–protein interactions network. The proteins in the network are clustered into functional modules, and through functional enrichment analysis, the three most significant clusters are determined. The genes that encode for the significant proteins of the three modules are viewed as potential targets for drugs, and researchers evaluate, through the Drug-Gene Interaction Database (http://www.dgidb.org), whether drugs or compounds targeting these genes exist. These drugs could potentially be repurposed to treat canine pyometra.

In Asada et al. (2018), the authors want to locate DDIs in biomedical texts via DL. To get better results, they enrich the data with molecular information. The training corpus is the DDIExtraction 2013 dataset (Segura-Bedmar et al. 2013; Herrero-Zazo et al. 2013), a multiclass dataset to classify drug-pairs to either one of four types of interactions or no interaction. The drug names present in the text are then queried in DrugBank, their SMILES strings are obtained, and converted into molecular graphs using the rdkit python package. These molecular graphs are used to train a graph convolutional network (GCN) to predict whether the drug pairs interact with each other. In case they do, CNN classifies the type of interaction. For training the CNN, the textual features are word-embedded with a word2vec (Mikolov et al. 2013) pretrained on a corpus of biomedical abstracts from Medline, and then the word-embeddings are concatenated with the molecular graph data and fed into the CNN for training. Experimental results show that integrating molecular data from DrugBank to the mined text data achieves better results than text mining alone.

In Kirk et al. (2018), the authors want to find drugs for oral mucositis, a side effect of cancer therapy. The idea is to find which genes are more relevant and significant with oral mucositis, then to find whether FDA-approved drugs exist that target these genes. The pubmed2ensembl web tool is used to extract relevant abstracts from Pubmed, and the names of genes within these abstracts are identified and stored. The GeneCodis (http://genecodis.cnb.csic.es/) (Carmona-Saez et al. 2007) web resource is used to retrieve functional and ontological information on these genes. Additional annotations on the genes are retrieved from KEGG. Researchers then use the STRING tool to infer the protein–protein interaction network formed by the proteins that the genes encode. Then they use the DGIdb database to identify which genes can serve as drug targets. The potential target genes are then evaluated on whether they correlate with lower or higher levels of oral mucositis. Finally through DGIdb researchers locate drugs that target these genes, and they also search clinicaltrials.gov for any promising drugs that are still in the trial stage but could potentially prove useful in targeting the genes that are associated with oral mucositis.

In Yu et al. (2021), text mining and information from public repositories are combined to find drugs that target genes associated with osteoarthritis. First, the authors retrieve biomedical texts concerning osteoarthritis by using pubmed2ensemble, and extract all the gene names contained within the corpus. Then, they perform functional and ontological analysis on these genes via FUNRICH (http://www.funrich.org), a software tool that performs annotation of genes and proteins. Additional annotation is performed via the DAVID database (https://david.ncifcrf.gov/summary.jsp), with the results being imported to KEGG for gene interaction pathway inference. Then the interaction network of the genes' associated proteins is computed with the STRING tool, and Cytoscape is used to find the genes corresponding to the most dense protein network. The DGIdb database is searched to find drugs targeting the selected genes.

In Luo et al. (2021), a search is conducted for associated genes and drugs that target breast cancer and periodontitis. Using pubmed2ensembl, the authors locate the genes, and they conduct functional and ontological analysis using the DAVID tool. They find, using STRING, the proteins encoded by these genes, and the corresponding PPI networks. Then, using cytoscape and the mcode (molecular complex detection) plug-in, they cluster these proteins into functional groups, and extract the genes implicated in the densest modules. These genes are then analyzed with the Gene Expression Profiling Interactive Analysis (GEPIA; http://gepia.cancer-pku.cn/index.html) website, which contains gene expression data compiled from various other sources. In GEPIA, they evaluate differential gene expression and patient survival rates with regard to the activity levels of these selected genes. The genes that stood out in this analysis are sent to DGIdb to extract drugs that target them.

The same pipeline is followed in Pan et al. (2018), toward targeting cutaneous squamous cell carcinoma. From biomedical text, the authors identify relevant genes with the pubmed2ensembl tool, functional analysis then ensues with genecodis, PPIs are inferred with STRING and cytoscape, gene data analysis is conducted with cBioPortal (http://cbioportal.org), which is a web tool for visualization and analysis of genomic data related to cancer, and finally drug–gene interactions are explored with DGIdb, identifying drugs that target the most significant genes.

To find drug targets for diseases, it may be needed to combine mined text with trascriptomic and genomic data (Yang et al. 2009). Toward the latter direction, in their work, Natarajan et al. (2006) try to locate causal connections to cancer. They collect full-text articles, as opposed to mere abstracts, from 20 journals related to cancer research. For preprocessing, they tokenize the text into sentences, parse it into verb and noun phrases, and locate gene names through dictionary lookup. Then, using experimental gene expression data and after statistical analysis, they locate differentially expressed genes in response to a cancer-related biological event they are interested in, thus, identifying the important genes to target. However, the transcriptomics data cannot reveal the network of interactions and causal relations among these genes. For that reason, they used the mined text data to discern the interactions among genes, via a software called LexiQuest Mine.

Similarly, in Li et al. (2006), the authors extract biological information that could help locate drug targets. The goal is to identify genes that regulate angiogenesis, a process that plays a pivotal role in the development of cancer. First, they identify genes that are differentially expressed in angiogenesis, using data from laboratory experiments or from public databases like the Stanford Microarray Database (http://genome-www5.stanford.edu/). Missing values in the datasets are imputed via k-nearest neighbors (k-NN) clustering (Troyanskaya et al. 2001). The causal relations among these genes are inferred through text mining. Relevant biomedical abstracts are extracted from Pubmed. Gene names are extracted via lookup from the Human Genome Organisation dictionary (http://www.gene.ucl.ac.uk). The inference of the gene regulatory networks is conducted under the assumption that if two genes are mentioned together in the same sentence, and if these two genes are co-expressed, either positively or negatively, in the microarray data, then one interacts with the other (D'haeseleer et al. 2000; Ge et al. 2001). The inferred gene interaction network was validated using information from the KEGG database.

5 Challenges

ML and DL models are wholly dependent on the data they are trained with. With perfect data, models would achieve perfect results. And inversely, most of the challenges and problems that result in low performance for these models can usually be traced back to issues with the data, whether in terms of quantity or quality. In Fan and Shi (2022), the authors experimented with datasets of varying sizes and levels of quality and showed that both quantity and quality of data have a significant impact on predicting protein–ligand affinity. Larger datasets, unsurprisingly, yielded better predictions. To this end, in the following, the most important data-related challenges are reviewed.

5.1 Data Shortage

It takes a lot of SMILES data for a model to learn the patterns and syntactic distributions of pharmaceutically relevant chemical structures. A SMILES string might be over 100 characters long, a sequence of a huge number of characters denoting chemical elements, atomic bonds, aromaticity, bond direction, and other chemical properties and structural information. This makes for a vast search space, and even though public chemical databases routinely contain millions of molecular structures, nevertheless, the number of structures for any specific task, or the number of compounds aimed at any specific drug target, might be quite low. It is this exact attribute of drug discovery databases, their scarce task-specific but abundant general data, that researchers exploit by pretraining neural networks on general data, allowing their model to learn the distributions and sequence motifs of drug-like chemical space, then fine tune them with data of interest. Liu et al. (2019) pretrain a generative RNN with compounds from ZINC, then fine tune it with SMILES from ChEMBL. In Chithrananda et al. (2020), the authors pretrain a transformer with 77 m SMILES from PubChem, then finalize the training with datasets from MoleculeNet for molecular property prediction. Similarly, Tran and Ekenna (2023) pretrain a transformer with five million SMILES from PubChem, canonicalizing the SMILES, and tokenizing them into individual tokens, then fine tune with datasets from MoleculeNet, for classification and regression tasks (depending on the task, the number of samples ranges from 1000 to 8000). In Yu et al. (2021), the authors pretrain a generative model with 6.8 million molecules from ZINC to capture chemical syntax, before fine tuning it with 210 thousand molecules from ChEMBL that possess desired properties. Successful pretraining has also taken place with less than a million compounds, such as in Grisoni et al. (2021), where a generative RNN is pretrained with 657 k structures. In Moret et al. (2020), a generative model is pretrained with 365 k molecules from ChEMBL24, and not only is the pretraining accomplished with a relatively low quantity of data but the fine tuning is done with only five molecules, albeit enhanced by data augmentation. The augmented data consist of alternative SMILES structures for the five fine tuning molecules, since there are multiple potential SMILES sequences encoding the same molecular structure.

5.2 *Imbalanced Data*

Data shortage may be inconvenient enough, but what makes the situation even worse is the imbalanced data, referring to few data belonging to only one class, consisting of positive samples only. Databases contain information on compounds that modulate certain target proteins, rarely including compounds that fail to modulate them. The simplest way, perhaps not always applicable, to address this challenge is to simply find the needed data in another database. In Papadopoulos et al. (2021), they use SMILES from ChEMBL (Gaulton et al. 2017) as positive samples, then locate SMILES of inactive compounds on ExCAPE-DB (Sun et al. 2017), using them as negative samples. The usual approach is to treat compounds with no bioactivity information associated with them as negative samples. In Yosipof et al. (2018), the authors use compound data from various datasets, PDBBind version 2005 (Wang et al. 2004), the Ki Bank (Zhang et al. 2005), and the SCORPIO dataset (Ababou and Ladbury 2007), classified as drugs, and for negative data they use compounds from these databases that are not registered as "drugs" in the DrugBank database (Wishart et al. 2007). Ren et al. (2022) have imbalanced data of DDI networks. To address this, known DDIs constitute the positive samples, while drug pairs that do not have information associated with them are the negative samples. While drug research applications that are using molecular structures often suffer from an asymmetrically higher quantity of positive data than negative, in applications of ML-based text mining, where models are trained on datasets consisting of annotated text, the inverse is happening. Negative samples vastly outnumber positive, since from a vast corpus of abstracts and articles on diverse biomedical topics, only a fraction of sentences contain information relevant to any particular task at hand. In Huang et al. (2022), the authors predict drug–drug interactions with ML models trained on the DDIs 2013 dataset (Hong et al. 2020); neutral sentences are taken to be the negative samples, which are rendered almost 20 times more than the positive ones. Researchers apply both data augmentation to upsample the positive samples, with synthetic data generated by BioGPT-2 (Papanikolaou and Pierleoni 2020), and they also downsample the negative samples. In KafiKang and Hendawi (2023), the researchers merely downsample the negative samples to render classes more balanced. Sun et al. (2019) detect DDIs from textual corpora; the downsampling of negative instances is conducted according to a rule-based scheme. Drug pairs are removed from the data if their names are abbreviations of one another, if they form coordinate structures, and if they are variations or derivatives of each other.

5.3 Unhelpful Data

Even the best models may give suboptimal results with imperfect data, while with irrelevant data they would fail completely. Unfortunately, in the biochemical sciences, determining the relevance of available data is not always an intuitive task. Deep-seated assumptions pervading the field are often wrong. For example, it is assumed that structurally similar compounds share similar functions, but this is not always the case. Activity cliffs among structural analogues result in a substantial variance in their potency (Stumpfe et al. 2019). With fingerprint-based similarity estimation via the Tanimoto coefficient, it is generally assumed that a similarity higher than 0.85 entails similar bioactivity. Yet, this assumption often proves to be wrong (Maggiora et al. 2014); therefore, by trying to make inferences about a compound of interest by evaluating the bioactivities of known, Tanimoto-similar molecules may be less helpful than expected. Another case of data being misleading due to false assumptions is the translation of mouse experiment data to insights into human phenotypes, e.g., testing a drug on mice and assuming the same effect will be elicited in humans. The latter is another presupposition that often does not apply, and is rarely objectively evaluated (Horrobin 2003; Leenaars et al. 2019; Mak et al. 2014). Researchers should be aware of these potential setbacks, especially in projects where the data seem plentiful but the models perform sub-optimally, and they should question whether the data are truly as valid and useful as assumed. A thorough discussion on the importance and nuance of data quality in AI-based drug discovery can be found in Bender and Cortés-Ciriano (2021a, Part 1; 2021b, Part 2). Ironically, although invalid assumptions about the data may pose problems for ML and DL-based predictive modeling, it is often through ML and DL that these challenges are addressed. Thus, in Tamura et al. (2023) it is shown that ML and DL models can help predict and anticipate activity cliffs, and in Normand et al. (2018) ML models trained on gene expression data take the results of mouse experiments and extrapolate to the corresponding effects in humans.

5.4 Data Discrepancies

Data discrepancies pose another nuisance, as different databases may have slightly varied information for the same entries and molecules. In Isigkeit et al. (2022), the authors analyzed five databases and found that they have differences in the information in entries for identical compounds, including the SMILES representation. This can often be a probable situation, as a molecular structure can be expressed in more than one SMILES string. The researchers in the study addressed the latter issue by converting the data to canonical SMILES, ensuring uniform structure among databases. Other than SMILES strings, categorical data may suffer from cross-database incongruity. In Jenkins et al. (2011), the authors analyzed the annotations of the World Drug Index and found, throughout the entries, the term hypersensitivity in 15 different versions (e.g., "Hypersensitivity," "HYPERSENSITIVITY," "hyper-sensitivity"). For any algorithm with no rule-based information encoded to account for such phenomena, these variations would constitute different terms altogether and would be treated as separate and unrelated entities. Thus, datasets should be carefully explored, and if needed, they should be preprocessed and curated before usage. Numerical data should also be evaluated, as numerical values might adhere to different normalization schemes, such as log2, min-max scaling, or standard scaling, and a uniform range should be ensured. Therefore, data pipelines should be implemented, for each specific project to unify the data, either through mathematical transformations or through text processing techniques. Controlled vocabularies and dictionaries should also be utilized in cases of databases giving different names to genes and proteins. A human may know that the terms Abl kinase, ABL1, and c-Abl may refer to the same protein, but a computer will not (Jenkins et al. 2011), thus curation pipelines should be exploited to smooth such discrepancies.

5.5 *High-Dimensional Data*

Drug discovery data come with a high number of features, whether using one hot-encoded SMILES, gene expression profiles, interaction networks, or any combination of these and other types of information, which may lead to overfitting. Arguing that integration of genetic and omics data with chemical structures and other phenotypic inputs such as medical images would result in very high-dimensional data, researchers (Vijayan et al. 2019) propose autoencoders (AEs) for data integration and dimensionality reduction. AEs could also treat batch effects, integrating data from a single data type but from multiple different sources or batches. AE based dimensionality reduction is also widely used in integrating omics data, with a broad range of AE architectures addressing biomedical tasks (Tsimenidis et al. 2022). Other researchers exploit embedding layers to extract reduced representations from SMILES and graph data (Ren et al. 2022), as graph data and one-hot encoded SMILES are extremely sparse and high-dimensional, warranting dimensionality reduction. Techniques like matrix-factorization are also being used for drug discovery-related tasks such as drug-target interaction prediction (Gönen 2012), drug repurposing (Ceddia et al. 2019), and drug–disease association detection (Zhang et al. 2018).

6 Discussion and Conclusions

An absolute prerequisite for the integration of a wide spectrum of chemical and biological data is that the data must be well structured (Jenkins et al. 2011). A good structure of a typical ML training dataset consists of rows representing instances (e.g., drugs or genes) and columns representing features, with one column being the labels (or more columns in the case of multi-label data). If the constituent datasets are already structured in such a form, integration is easy and straightforward, perhaps as easy as concatenating the two datasets into one, or performing an inner join. Otherwise, extensive data pipelines and curation may be required in order to retrieve the data from public databases and store the values into arrays that adhere to the ML dataset format. Combining data from multiple sources may pose a number of challenges. Ontology terms might be capitalized in one database and lowercased in another, or nonidentical terms may be used for the same concept, requiring the use of dictionaries and controlled vocabularies. The same compound might be represented by a different SMILES string in different databases, necessitating the computation of Canonical SMILES. Curation and clean-up of data are often done with the help of data pipelining software, with tools such as BIOVIA Pipeline Pilot (https://www.3ds.com/products/biovia/pipeline-pilot) and Inforsense KDE Platform, and with other tools and libraries already mentioned, such as KNIME and Pandas.

In dimensionality reduction, priority should perhaps be given to techniques that map the data to lower dimensions such as AEs and PCA, rather than to approaches that remove features and data modalities to make datasets simpler. Experiments with the integration of different data modalities and features show that integrating a wide range of data does improve results and predictions (Li et al. 2020; Wu et al. 2012; Machado et al. 2023).

The simplest way of data integration is by using data of the same type, yet from different databases, which usually happens when employing a large number of SMILES for adequate model training. SMILES are also being combined with other forms of data, enriching the model with information from different perspectives of a biological system. This is helpful because each perspective contains information about the system, which is not present in other perspectives (Tsimenidis et al. 2022).

Further enrichment of the available information can be accomplished by text mining, which has the advantage of inferring causal relations directly from the scientific literature through named entity recognition and relation extraction. In this way, advantage can be taken of a vast body of biomedical articles, often containing insight and information absent from public databases. Integrating text-derived insights with data compiled from public repositories provides rich and multifaceted information for research projects. A useful and popular application that exploits such an integration is in drug repurposing, where researchers identify disease-relevant genes through text mining, then use databases to explore the interactions among proteins encoded by these genes, pinpoint the essential genes of these interaction networks, and find drugs that attack these target genes, since the same genes are often related to other diseases as well, and approved drugs for these diseases may already exist.

ML and DL models are useful for all data-driven applications in drug discovery. When it comes to structural and biological data from public repositories, class imbalance is prevalent, with negative samples being virtually absent. Researchers alleviate this by taking molecules that do not have the desired properties registered in their metadata as the negative samples, resulting in valid models that produce successful predictions. To train models on mined text, data preparation must go beyond the basic preprocessing pipelines, into feature extraction methods such as n-grams or TF-IDF, to generate a dataset for model training.

AI in drug discovery is a new field, with many breakthroughs yet to come. Public databases with chemical and biomedical data have not been fully taken advantage of, as the explosion of available data is quite recent, and new data are being generated before scientists have time to analyze and utilize them. This becomes even more true as the complexity of the data schemes rises. Our research highlighted the fact that there is not much variance in applications and strategies that combine text with chemical and biological data from various databases, with the bulk of research leaning toward certain patterns and not probing the possibilities to the fullest. The room for creativity and inventiveness is immense, as terabytes and petabytes of data in the public repositories lie waiting for researchers to explore new ideas, envision new applications, and make new discoveries.

References

Ababou A, Ladbury JE. Survey of the year 2005: literature on applications of isothermal titration calorimetry. J Mol Recognit. 2007;20(1):4–14. Available at:. https://doi.org/10.1002/jmr.803.

Alimova I, Tutubalina E. Detecting adverse drug reactions from biomedical texts with neural networks. In: Alva-Manchego F, Choi E, Khashabi D, editors. Proceedings of the 57th annual meeting of the Association for Computational Linguistics: student research workshop, Florence, Italy. Association for Computational Linguistics; 2019. p. 415–21. Available at. https://doi.org/10.18653/v1/P19-2058.

Asada M, Miwa M, Sasaki Y. Enhancing drug-drug interaction extraction from texts by molecular structure information. In: Gurevych I, Miyao Y, editors. Proceedings of the 56th annual meeting of the Association for Computational Linguistics (volume 2: short papers). ACL 2018, Melbourne. Association for Computational Linguistics; 2018. p. 680–5. Available at. https://doi.org/10.18653/v1/P18-2108.

Asiri Y. Computing drug-drug similarity from patient-centric data. Bioengineering (Basel, Switzerland). 2023;10(2):182. Available at:. https://doi.org/10.3390/bioengineering10020182.

Baell JB, Holloway GA. New substructure filters for removal of pan assay interference compounds (PAINS) from screening libraries and for their exclusion in bioassays. J Med Chem. 2010;53(7):2719–40. Available at:. https://doi.org/10.1021/jm901137j.

Barrett T, et al. NCBI GEO: archive for high-throughput functional genomic data. Nucleic Acids Res. 2009;37(Database issue):D885–90. Available at:. https://doi.org/10.1093/nar/gkn764.

Beckman MF, et al. A computational text mining-guided meta-analysis approach to identify potential xerostomia drug targets. J Clin Med. 2022;11(5):1442. Available at:. https://doi.org/10.3390/jcm11051442.

Bender A, Cortés-Ciriano I. Artificial intelligence in drug discovery: what is realistic, what are illusions? Part 1: ways to make an impact, and why we are not there yet. Drug Discov Today. 2021a;26(2):511–24. Available at:. https://doi.org/10.1016/j.drudis.2020.12.009.

Bender A, Cortés-Ciriano I. Artificial intelligence in drug discovery: what is realistic, what are illusions? Part 2: a discussion of chemical and biological data. Drug Discov Today. 2021b;26(4):1040–52. Available at:. https://doi.org/10.1016/j.drudis.2020.11.037.

Born J, et al. PaccMannRL: de novo generation of hit-like anticancer molecules from transcriptomic data via reinforcement learning. iScience. 2021;24(4) Available at: https://doi.org/10.1016/j.isci.2021.102269.

Cancer Genome Atlas Research Network, et al. The cancer genome Atlas pan-cancer analysis project. Nature Genetics. 2013;45(10):1113–20. Available at:. https://doi.org/10.1038/ng.2764.

Carmona-Saez P, et al. GENECODIS: a web-based tool for finding significant concurrent annotations in gene lists. Genome Biol. 2007;8(1):R3. Available at:. https://doi.org/10.1186/gb-2007-8-1-r3.

Ceddia G, et al. Non-negative matrix tri-factorization for data integration and network-based drug repositioning. In: 2019 IEEE conference on computational intelligence in bioinformatics and computational biology (CIBCB), Siena. IEEE Press; 2019. p. 1–7. Available at:. https://doi.org/10.1109/CIBCB.2019.8791474.

Cerchia C, Lavecchia A. New avenues in artificial-intelligence-assisted drug discovery. Drug Discov Today. 2023;28(4):103516. Available at:. https://doi.org/10.1016/j.drudis.2023.103516.

Chithrananda S, Grand G, Ramsundar B. ChemBERTa: large-scale self-supervised pretraining for molecular property prediction. ArXiv [Preprint]. 2020. Available at: https://www.semanticscholar.org/paper/95ce6f77e26b496ffb705a0a3b54f2fb7a6d2452. Accessed: 6 Nov 2024.

D'haeseleer P, Liang S, Somogyi R. Genetic network inference: from co-expression clustering to reverse engineering. Bioinformatics (Oxford, England). 2000;16(8):707–26. Available at:. https://doi.org/10.1093/bioinformatics/16.8.707.

Dhudum R, Ganeshpurkar A, Pawar A. Revolutionizing drug discovery: a comprehensive review of AI applications. Drugs Drug Candidates. 2024;3(1):148–71. Available at:. https://doi.org/10.3390/ddc3010009.

Durant JL, et al. Reoptimization of MDL keys for use in drug discovery. J Chem Inf Comput Sci. 2002;42(6):1273–80. Available at:. https://doi.org/10.1021/ci010132r.

Edgar R, Domrachev M, Lash AE. Gene expression omnibus: NCBI gene expression and hybridization array data repository. Nucleic Acids Res. 2002;30(1):207–10. Available at:. https://doi.org/10.1093/nar/30.1.207.

Erhardt RA-A, Schneider R, Blaschke C. Status of text-mining techniques applied to biomedical text. Drug Discov Today. 2006;11(7–8):315–25. Available at:. https://doi.org/10.1016/j.drudis.2006.02.011.

Fan FJ, Shi Y. Effects of data quality and quantity on deep learning for protein-ligand binding affinity prediction. Bioorg Med Chem. 2022;72:117003. Available at:. https://doi.org/10.1016/j.bmc.2022.117003.

Freshour SL, et al. Integration of the drug-gene interaction database (DGIdb 4.0) with open crowdsource efforts. Nucleic Acids Res. 2021;49(D1):D1144–51. Available at:. https://doi.org/10.1093/nar/gkaa1084.

Gaulton A, et al. The ChEMBL database in 2017. Nucleic Acids Res. 2017;45(D1):D945–54. Available at:. https://doi.org/10.1093/nar/gkw1074.

Ge H, et al. Correlation between transcriptome and interactome mapping data from Saccharomyces cerevisiae. Nat Genet. 2001;29(4):482–6. Available at:. https://doi.org/10.1038/ng776.

Gefen D, et al. Identifying patterns in medical records through latent semantic analysis. Commun ACM. 2018;61:72–7. Available at:. https://doi.org/10.1145/3209086.

Gefen D, et al. A case study of applying text analysis to identify possible adverse drug interactions: the case of Adalat (Nifedipine). Health Informatics J. 2020;26:146045821988226. Available at:. https://doi.org/10.1177/1460458219882269.

Gene Ontology Consortium. Gene ontology consortium: going forward. Nucleic Acids Res. 2015;43(Database issue):D1049–56. Available at:. https://doi.org/10.1093/nar/gku1179.

Goh GB, et al. SMILES2Vec: an interpretable general-purpose deep neural network for predicting chemical properties. arXiv. Available at:. 2018; https://doi.org/10.48550/arXiv.1712.02034.

Gönen M. Predicting drug-target interactions from chemical and genomic kernels using Bayesian matrix factorization. Bioinformatics (Oxford, England). 2012;28(18):2304–10. Available at:. https://doi.org/10.1093/bioinformatics/bts360.

Gong L. Application of biomedical text mining. In: Aceves-Fernandez MA, editor. Artificial intelligence – emerging trends and applications. InTech; 2018. https://doi.org/10.5772/intechopen.75924. [Preprint].

Grisoni F, et al. Combining generative artificial intelligence and on-chip synthesis for de novo drug design. Sci Adv. 2021;7(24):eabg3338. Available at:. https://doi.org/10.1126/sciadv.abg3338.

Gummesson Svensson H, et al. Utilizing reinforcement learning for de novo drug design. Mach Learn. 2024;113(7):4811–43. Available at:. https://doi.org/10.1007/s10994-024-06519-w.

Hailu N, Hunter LE, Cohen KB. UColorado_SOM: extraction of drug-drug interactions from biomedical text using knowledge-rich and knowledge-poor features. In: Manandhar S, Yuret D, editors. Second joint conference on lexical and computational semantics (*SEM), volume 2: proceedings of the seventh international workshop on semantic evaluation (SemEval 2013). SemEval 2013, Atlanta, GA, USA. Association for Computational Linguistics; 2013. p. 684–8. Available at: https://aclanthology.org/S13-2112. Accessed: 6 Nov 2024.

Han X, et al. SmileGNN: drug–drug interaction prediction based on the SMILES and graph neural network. Life. 2022;12(2):319. Available at:. https://doi.org/10.3390/life12020319.

Hansson LK, et al. Semantic text mining in early drug discovery for type 2 diabetes. PLoS ONE. 2020;15(6) Available at: https://doi.org/10.1371/journal.pone.0233956.

Harpaz R, et al. Text mining for adverse drug events: the promise, challenges, and state of the art. Drug Saf. 2014;37(10):777–90. Available at:. https://doi.org/10.1007/s40264-014-0218-z.

Hasselgren C, Oprea TI. Artificial intelligence for drug discovery: are we there yet? Annu Rev Pharmacol Toxicol. 2024;64:527–50. Available at:. https://doi.org/10.1146/annurev-pharmtox-040323-040828.

Hearst MA. Untangling text data mining. In: Proceedings of the 37th annual meeting of the Association for Computational Linguistics. ACL 1999, College Park, MD, USA. Association for Computational Linguistics; 1999. p. 3–10. Available at:. https://doi.org/10.3115/1034678.1034679.

Herrero-Zazo M, et al. The DDI corpus: an annotated corpus with pharmacological substances and drug–drug interactions. J Biomed Inform. 2013;46(5):914–20. Available at:. https://doi.org/10.1016/j.jbi.2013.07.011.

Hodos RA, et al. In silico methods for drug repurposing and pharmacology. WIREs Syst Biol Med. 2016;8(3):186–210. Available at:. https://doi.org/10.1002/wsbm.1337.

Hong L, et al. A novel machine learning framework for automated biomedical relation extraction from large-scale literature repositories. Nat Mach Intell. 2020;2:347. Available at. https://doi.org/10.1038/s42256-020-0189-y.

Horrobin DF. Modern biomedical research: an internally self-consistent universe with little contact with medical reality? Nat Rev Drug Discov. 2003;2(2):151–4. Available at:. https://doi.org/10.1038/nrd1012.

Huang L, et al. EGFI: drug–drug interaction extraction and generation with fusion of enriched entity and sentence information. Brief Bioinform. 2022;23(1):bbab451. Available at:. https://doi.org/10.1093/bib/bbab451.

Hunter L, et al. OpenDMAP: an open source, ontology-driven concept analysis engine, with applications to capturing knowledge regarding protein transport, protein interactions and cell-type-specific gene expression. BMC Bioinformatics. 2008;9:78. Available at:. https://doi.org/10.1186/1471-2105-9-78.

Isigkeit L, Chaikuad A, Merk D. A consensus compound/bioactivity dataset for data-driven drug design and chemogenomics. Molecules. 2022;27(8):2513. Available at:. https://doi.org/10.3390/molecules27082513.

Jenkins JL, et al. Bridging chemical and biological data: implications for pharmaceutical drug discovery. In: Computational approaches in cheminformatics and bioinformatics. Wiley; 2011. p. 25–56. Available at:. https://doi.org/10.1002/9781118131411.ch2.

KafiKang M, Hendawi A. Drug-drug interaction extraction from biomedical text using relation BioBERT with BLSTM. Mach Learn Knowl Extr. 2023;5(2):669–83. Available at:. https://doi.org/10.3390/make5020036.

Kaitin KI. Deconstructing the drug development process: the new face of innovation. Clin Pharmacol Ther. 2010;87(3):356–61. Available at:. https://doi.org/10.1038/clpt.2009.293.

Kanehisa M, Goto S. KEGG: Kyoto encyclopedia of genes and genomes. Nucleic Acids Res. 2000;28(1):27–30. Available at:. https://doi.org/10.1093/nar/28.1.27.

Karimi S, et al. Cadec: a corpus of adverse drug event annotations. J Biomed Inform. 2015;55:73–81. Available at:. https://doi.org/10.1016/j.jbi.2015.03.010.

Kim S, et al. PubChem in 2021: new data content and improved web interfaces. Nucleic Acids Res. 2021;49(D1):D1388–95. Available at:. https://doi.org/10.1093/nar/gkaa971.

Kirk J, et al. Text mining-based in silico drug discovery in oral mucositis caused by high-dose cancer therapy. Support Care Cancer. 2018;26(8):2695–705. Available at:. https://doi.org/10.1007/s00520-018-4096-2.

Klein TE, et al. Integrating genotype and phenotype information: an overview of the PharmGKB project. Pharmacogenetics research network and knowledge base. Pharmacogenomics J. 2001;1(3):167–70. Available at:. https://doi.org/10.1038/sj.tpj.6500035.

Kolchinsky A, et al. Evaluation of linear classifiers on articles containing pharmacokinetic evidence of drug-drug interactions. Pac Symp Biocomput. 2013:409–20. Available at:. https://doi.org/10.1142/9789814447973_0040.

Krallinger M, Erhardt RA-A, Valencia A. Text-mining approaches in molecular biology and biomedicine. Drug Discov Today. 2005;10(6):439–45. Available at:. https://doi.org/10.1016/S1359-6446(05)03376-3.

Lawson AJ, et al. The making of Reaxys—towards unobstructed access to relevant chemistry information. In: The future of the history of chemical information, ACS symposium series, 1164. American Chemical Society; 2014. p. 127–48. Available at:. https://doi.org/10.1021/bk-2014-1164.ch008.

Lee J, et al. BioBERT: a pre-trained biomedical language representation model for biomedical text mining. Bioinformatics (Oxford, England). 2019;36:1234. Available at:. https://doi.org/10.1093/bioinformatics/btz682.

Leenaars CHC, et al. Animal to human translation: a systematic scoping review of reported concordance rates. J Transl Med. 2019;17(1):223. Available at:. https://doi.org/10.1186/s12967-019-1976-2.

Li S, Wu L, Zhang Z. Constructing biological networks through combined literature mining and microarray analysis: a LMMA approach. Bioinformatics (Oxford, England). 2006;22(17):2143–50. Available at:. https://doi.org/10.1093/bioinformatics/btl363.

Li B, et al. A novel drug repurposing approach for non-small cell lung cancer using deep learning. PLoS One. 2020;15(6):e0233112. Available at:. https://doi.org/10.1371/journal.pone.0233112.

Liu X, et al. An exploration strategy improves the diversity of de novo ligands using deep reinforcement learning: a case for the adenosine A2A receptor. J Cheminform. 2019;11(1):35. Available at:. https://doi.org/10.1186/s13321-019-0355-6.

Liu J, et al. Identifying adverse drug reaction-related text from social media: a multi-view active learning approach with various document representations. Information. 2022;13(4):189. Available at:. https://doi.org/10.3390/info13040189.

Loging W, Harland L, Williams-Jones B. High-throughput electronic biology: mining information for drug discovery. Nat Rev Drug Discov. 2007;6(3):220–30. Available at:. https://doi.org/10.1038/nrd2265.

Luo L, et al. Searching for essential genes and drug discovery in breast cancer and periodontitis via text mining and bioinformatics analysis. Anti-Cancer Drugs. 2021;32(10):1038–45. Available at:. https://doi.org/10.1097/CAD.0000000000001108.

Machado J, et al. Drug–drug interaction extraction-based system: an natural language processing approach. Expert Syst. 2023:e13303. Available at:. https://doi.org/10.1111/exsy.13303.
Mack R, Hehenberger M. Text-based knowledge discovery: search and mining of life-sciences documents. Drug Discov Today. 2002;7(11):S89–98. Available at:. https://doi.org/10.1016/s1359-6446(02)02286-9.
Maggiora G, et al. Molecular similarity in medicinal chemistry. J Med Chem. 2014;57(8):3186–204. Available at:. https://doi.org/10.1021/jm401411z.
Mak I, Evaniew N, Ghert M. Lost in translation: animal models and clinical trials in cancer treatment. Am J Transl Res. 2014; [Preprint]. Available at: https://www.semanticscholar.org/paper/Lost-in-translation%3A-animal-models-and-clinical-in-Mak-Evaniew/801e15e4a14ad21295cd1c93112f4c457ea1f8d1. Accessed: 6 Nov 2024.
Mauri A, Consonni V, Todeschini R. Molecular Descriptors. In: Handbook of computational chemistry; 2017. p. 2065–93. Available at:. https://doi.org/10.1007/978-3-319-27282-5_51.
Mehra N, et al. Cross-domain text mining to predict adverse events from tyrosine kinase inhibitors for chronic myeloid leukemia. Cancer. 2022;14(19):4686. Available at:. https://doi.org/10.3390/cancers14194686.
Méndez-Lucio O, et al. De novo generation of hit-like molecules from gene expression signatures using artificial intelligence. Nat Commun. 2020;11(1):10. Available at:. https://doi.org/10.1038/s41467-019-13807-w.
Mikolov T, et al. Distributed representations of words and phrases and their compositionality. In: Advances in neural information processing systems. Curran Associates, Inc; 2013. Available at: https://papers.nips.cc/paper_files/paper/2013/hash/9aa42b31882ec039965f3c4923ce901b-Abstract.html. Accessed: 6 Nov 2024.
Moret M, et al. Generative molecular design in low data regimes. Nat Mach Intell. 2020;2:171–80. Available at:. https://doi.org/10.1038/s42256-020-0160-y.
Morgan HL. The generation of a unique machine description for chemical structures-a technique developed at chemical abstracts service. J Chem Doc. 1965;5(2):107–13. Available at:. https://doi.org/10.1021/c160017a018.
Nadkarni PM, Ohno-Machado L, Chapman WW. Natural language processing: an introduction. J Am Med Inform Assoc. 2011;18(5):544–51. Available at:. https://doi.org/10.1136/amiajnl-2011-000464.
Natarajan J, et al. Text mining of full-text journal articles combined with gene expression analysis reveals a relationship between sphingosine-1-phosphate and invasiveness of a glioblastoma cell line. BMC Bioinformatics. 2006;7:373. Available at:. https://doi.org/10.1186/1471-2105-7-373.
Niazi SK, Mariam Z. Computer-aided drug design and drug discovery: a prospective analysis. Pharmaceuticals (Basel, Switzerland). 2023;17(1):22. Available at:. https://doi.org/10.3390/ph17010022.
Normand R, et al. Found in translation: a machine learning model for mouse-to-human inference. Nat Methods. 2018;15(12):1067–73. Available at:. https://doi.org/10.1038/s41592-018-0214-9.
Pan Y, Zhang Y, Liu J. Text mining-based drug discovery in cutaneous squamous cell carcinoma. Oncol Rep. 2018;40(6):3830–42. Available at:. https://doi.org/10.3892/or.2018.6746.
Papadatos G, et al. SureChEMBL: a large-scale, chemically annotated patent document database. Nucleic Acids Res. 2016;44(D1):D1220–8. Available at:. https://doi.org/10.1093/nar/gkv1253.
Papadopoulos K, et al. De novo design with deep generative models based on 3D similarity scoring. Bioorg Med Chem. 2021;44:116308. Available at:. https://doi.org/10.1016/j.bmc.2021.116308.
Papanikolaou Y, Pierleoni A. DARE: data augmented relation extraction with GPT-2. arXiv; 2020. Available at: doi:https://doi.org/10.48550/arXiv.2004.13845.
Papanikolaou N, et al. DrugQuest – a text mining workflow for drug association discovery. BMC Bioinformatics. 2016;17(Suppl 5):182. Available at:. https://doi.org/10.1186/s12859-016-1041-6.

Percha B, Garten Y, Altman RB. Discovery and explanation of drug-drug interactions via text mining. Biocomputing 2012;410–21. https://doi.org/10.1142/9789814366496_0040. https://www.worldscientific.com/doi/abs/10.1142/9789814366496_0040. https://www.worldscientific.com/doi/pdf/10.1142/9789814366496_0040.

Pletscher-Frankild S, et al. DISEASES: text mining and data integration of disease-gene associations. Methods (San Diego, Calif). 2015;74:83–9. Available at:. https://doi.org/10.1016/j.ymeth.2014.11.020.

Qi X, et al. Machine learning empowering drug discovery: applications, opportunities and challenges. Molecules. 2024;29(4):903. Available at:. https://doi.org/10.3390/molecules29040903.

Ren Z-H, et al. BioChemDDI: predicting drug-drug interactions by fusing biochemical and structural information through a self-attention mechanism. Biology. 2022;11(5):758. Available at:. https://doi.org/10.3390/biology11050758.

Rigden D, Fernandez X. The 2023 nucleic acids research database issue and the online molecular biology database collection. Nucleic Acids Res. 2023;51:D1–8. Available at:. https://doi.org/10.1093/nar/gkac1186.

Roberts PM, Hayes WS. Information needs and the role of text mining in drug development. Biocomputing 2008;592–603. https://doi.org/10.1142/9789812776136_0057. https://www.worldscientific.com/doi/abs/10.1142/9789812776136_0057. https://www.worldscientific.com/doi/pdf/10.1142/9789812776136_0057.

Rodriguez-Esteban R. Biomedical text mining and its applications. PLoS Comput Biol. 2009;5(12):e1000597. Available at:. https://doi.org/10.1371/journal.pcbi.1000597.

Rogers D, Hahn M. Extended-connectivity fingerprints. J Chem Inf Model. 2010;50(5):742–54. Available at:. https://doi.org/10.1021/ci100050t.

Segura-Bedmar I, Martínez P, Herrero-Zazo M. SemEval-2013 Task 9: extraction of drug-drug interactions from biomedical texts (DDIExtraction 2013). In: Manandhar S, Yuret D, editors. Second joint conference on lexical and computational semantics (*SEM), volume 2: proceedings of the seventh international workshop on semantic evaluation (SemEval 2013). SemEval 2013, Atlanta, GA, USA. Association for Computational Linguistics; 2013. p. 341–50. Available at: https://aclanthology.org/S13-2056. Accessed: 6 Nov 2024.

Shultz MD. Two decades under the influence of the rule of five and the changing properties of approved oral drugs. J Med Chem. 2019;62(4):1701–14. Available at:. https://doi.org/10.1021/acs.jmedchem.8b00686.

Sliwoski G, et al. Computational methods in drug discovery. Pharmacol Rev. 2014;66(1):334–95. Available at:. https://doi.org/10.1124/pr.112.007336.

Song M, et al. Inferring drug-protein-side effect relationships from biomedical text. Genes. 2019;10(2):159. Available at:. https://doi.org/10.3390/genes10020159.

Sterling T, Irwin JJ. ZINC 15—ligand discovery for everyone. J Chem Inf Model. 2015;55(11):2324–37. Available at:. https://doi.org/10.1021/acs.jcim.5b00559.

Stumpfe D, Hu H, Bajorath J. Evolving concept of activity cliffs. ACS Omega. 2019;4(11):14360–8. Available at:. https://doi.org/10.1021/acsomega.9b02221.

Subramanian A, et al. A next generation connectivity map: L1000 platform and the first 1,000,000 profiles. Cell. 2017;171(6):1437–52.e17. Available at:. https://doi.org/10.1016/j.cell.2017.10.049.

Sun J, et al. ExCAPE-DB: an integrated large scale dataset facilitating Big Data analysis in chemogenomics. J Cheminform. 2017;9(1):17. Available at:. https://doi.org/10.1186/s13321-017-0203-5.

Sun X, et al. Drug-drug interaction extraction via recurrent hybrid convolutional neural networks with an improved focal loss. Entropy. 2019;21(1):37. Available at:. https://doi.org/10.3390/e21010037.

Swanson D. Medical literature as a potential source of new knowledge. Bull Med Libr Assoc. 1990; [Preprint]. Available at: https://www.semanticscholar.org/paper/Medical-literature-as-a-potential-source-of-new-Swanson/078749df4517c1700a3013f3449a77e95a19e3f4. Accessed: 6 Nov 2024.

Tamura S, Miyao T, Bajorath J. Large-scale prediction of activity cliffs using machine and deep learning methods of increasing complexity. J Cheminform. 2023;15(1):4. Available at:. https://doi.org/10.1186/s13321-022-00676-7.

Tran T, Ekenna C. Molecular descriptors property prediction using transformer-based approach. Int J Mol Sci. 2023;24(15):11948. Available at:. https://doi.org/10.3390/ijms241511948.

Troyanskaya O, et al. Missing value estimation methods for DNA microarrays. Bioinformatics. 2001;17:520–5. Available at:. https://doi.org/10.1093/bioinformatics/17.6.520.

Tsimenidis S, Vrochidou E, Papakostas GA. Omics data and data representations for deep learning-based predictive modeling. Int J Mol Sci. 2022;23(20):12272. Available at:. https://doi.org/10.3390/ijms232012272.

Uramoto N, et al. A text-mining system for knowledge discovery from biomedical documents. IBM Syst J. 2004;43:516–33. Available at:. https://doi.org/10.1147/sj.433.0516.

Vijayan V, et al. Could advances in representation learning in artificial intelligence provide the new paradigm for data integration in drug discovery? Expert Opin Drug Discov. 2019;14(3):191–4. Available at:. https://doi.org/10.1080/17460441.2019.1573811.

Wan Z, et al. Drug discovery in cardiovascular disease identified by text mining and data analysis. Ann Palliat Med. 2020;9(5):3089–3. Available at:. https://doi.org/10.21037/apm-20-705.

Wang X, Yu G. Drug discovery in canine Pyometra disease identified by text mining and microarray data analysis. BioMed Res Int. 2023;2023:7839568. Available at:. https://doi.org/10.1155/2023/7839568.

Wang R, et al. The PDBbind database: collection of binding affinities for protein−ligand complexes with known three-dimensional structures. J Med Chem. 2004;47(12):2977–80. Available at:. https://doi.org/10.1021/jm030580l.

Wang M, et al. PDD graph: bridging electronic medical records and biomedical knowledge graphs via entity linking. arXiv. 2017. Available at: https://doi.org/10.48550/arXiv.1707.05340.

Weininger D. SMILES, a chemical language and information system. 1. Introduction to methodology and encoding rules. J Chem Inf Comput Sci. 1988;28(1):31–6. Available at:. https://doi.org/10.1021/ci00057a005.

Wishart DS, et al. DrugBank: a knowledgebase for drugs, drug actions and drug targets. Nucleic Acids Res. 2007;36(Database issue):D901. Available at:. https://doi.org/10.1093/nar/gkm958.

Wu S, He Y. Enriching pre-trained language model with entity information for relation classification. In: Proceedings of the 28th ACM international conference on information and knowledge management (CIKM '19). New York, NY, USA. Association for Computing Machinery; 2019. p. 2361–4. Available at:. https://doi.org/10.1145/3357384.3358119.

Wu Y, Liu M, Jim Zheng W, Zhao Z, Xu H. Ranking gene-drug relationships in biomedical literature using Latent Dirichlet Allocation. Biocomputing 2012;422–33. https://doi.org/10.1142/9789814366496_0041. https://www.worldscientific.com/doi/abs/10.1142/9789814366496_0041. https://www.worldscientific.com/doi/pdf/10.1142/9789814366496_0041.

Wu H-Y, Chiang C-W, Li L. Text mining for drug-drug interaction. Methods Mol Biol (Clifton, NJ). 2014;1159:47–75. Available at:. https://doi.org/10.1007/978-1-4939-0709-0_4.

Wu Z, et al. MoleculeNet: a benchmark for molecular machine learning. Chem Sci. 2018;9(2):513–30. Available at:. https://doi.org/10.1039/C7SC02664A.

Xu X, Meng F, Sun L. Knowledge mining of interactions between drugs from the extensive literature with a novel graph-convolutional-network-based method. Electronics. 2023;12(2):311. Available at:. https://doi.org/10.3390/electronics12020311.

Xue L, Bajorath J. Molecular descriptors in chemoinformatics, computational combinatorial chemistry, and virtual screening. Comb Chem High Throughput Screen. 2000;3(5):363–72. Available at:. https://doi.org/10.2174/1386207003331454.

Yang Y, Adelstein SJ, Kassis AI. Target discovery from data mining approaches. Drug Discov Today. 2009;14(3–4):147–54. Available at:. https://doi.org/10.1016/j.drudis.2008.12.005.

Yosipof A, Guedes RC, García-Sosa AT. Data mining and machine learning models for predicting drug likeness and their disease or organ category. Front Chem. 2018;6. Available at: https://doi.org/10.3389/fchem.2018.00162.

Yu Y, et al. A novel scalarized scaffold hopping algorithm with graph-based variational autoencoder for discovery of JAK1 inhibitors. ACS Omega. 2021;6(35):22945–54. Available at:. https://doi.org/10.1021/acsomega.1c03613.

Zhang J, et al. Development of KiBank, a database supporting structure-based drug design. Comput Biol Chem. 2005;28:401–7. Available at:. https://doi.org/10.1016/j.compbiolchem.2004.09.003.

Zhang W, et al. Predicting drug-disease associations by using similarity constrained matrix factorization. BMC Bioinformatics. 2018;19(1):233. Available at:. https://doi.org/10.1186/s12859-018-2220-4.

Zhavoronkov A, et al. Deep learning enables rapid identification of potent DDR1 kinase inhibitors. Nat Biotechnol. 2019;37(9):1038–40. Available at:. https://doi.org/10.1038/s41587-019-0224-x.

Zolnoori M, et al. A systematic approach for developing a corpus of patient reported adverse drug events: a case study for SSRI and SNRI medications. J Biomed Inform. 2019;90:103091. Available at:. https://doi.org/10.1016/j.jbi.2018.12.005.

Zweigenbaum P, et al. Frontiers of biomedical text mining: current progress. Brief Bioinform. 2007;8(5):358–75. Available at:. https://doi.org/10.1093/bib/bbm045.

Artificial Intelligence for Drug Target and Pathway Identification, Assessment, Validation, and Indication Expansion

Geoffrey H. D. Leung, Frank W. Pun, Vladimir Naumov, David Gennert, Petrina Kamya, Alex Aliper, Feng Ren, and Alex Zhavoronkov

Abstract The integration of artificial intelligence (AI) into drug discovery has been transformational, particularly in identifying, validating, and assessing novel drug targets, biological pathways, and therapeutic indication expansion of existing compounds. This chapter examines the application of AI methodologies, ranging from traditional machine learning to advanced deep learning, in streamlining drug development processes. We begin by exploring AI-driven approaches for target and pathway identification, emphasizing how AI models analyze the large amount of biological data available to predict novel and actionable targets. Next, we discuss AI's roles in accurately and efficiently validating and assessing the predicted target. We then examine AI's potential for indication expansion, where models predict new therapeutic applications for existing drugs. Finally, we address key challenges and future prospects, highlighting the importance of data quality, model interpretability, and the necessity of implementing regulatory changes to ensure reliable, transparent, reproducible, and robust applications of AI in drug discovery.

Keywords Artificial intelligence · Drug target identification · Target validation · Biological pathways · Indication expansion · Deep learning

Abbreviations

AI	Artificial intelligence
ALS	Amyotrophic lateral sclerosis

G. H. D. Leung · F. W. Pun · V. Naumov · D. Gennert · P. Kamya · A. Aliper · F. Ren · A. Zhavoronkov (✉)
Insilico Medicine Hong Kong Ltd, Hong Kong Science and Technology Park, Hong Kong, SAR, China

Insilico Medicine US Inc, Cambridge, MA, USA
e-mail: geoffrey.leung@insilico.com; frank.pun@insilico.com; v.naumov@insilico.com; dave.gennert@insilicomedicine.com; petrina@insilico.com; aliper@insilico.com; feng.ren@insilico.com; alex@insilico.com

A. Lavecchia (ed.), *Applied Artificial Intelligence for Drug Discovery*,
https://doi.org/10.1007/978-3-031-98022-0_3

AUC-ROC	Area under the receiver operating characteristic curve
CNS	Central nervous system
COVID-19	Coronavirus disease 2019
DL	Deep learning
DNA	Deoxyribonucleic acid
FN	False negative
FP	False positive
GDPR	General Data Protection Regulation
GWAS	Genome-wide association studies
IPF	Idiopathic pulmonary fibrosis
LASSO	Least absolute shrinkage and selection operator
LINCS	Library of Integrated Network-based Cell-Signature
LLM	Large language model
ML	Machine learning
NLP	Natural language processing
PCA	Principal component analysis
PheWAS	Phenome-wide association studies
RA	Rheumatoid arthritis
RL	Reinforcement learning
RNA	Ribonucleic acid
SHAP	Shapley additive explanations
SVM	Support vector machine
TF	Transcription factor
TL	Transfer learning
TP	True positive
t-SNE	t-distributed stochastic neighbor embedding
UMAP	Uniform Manifold Approximation and Projection

1 AI in Drug Target Discovery

The concept of target identification in medicine has evolved alongside human civilization, rooted in the desire to understand and treat diseases. As far back as ancient times, we see examples of attempts to ascribe different causes to individual diseases and thus treat them differently, from the Ebers Papyrus from around 1550 BCE in Egypt, describing over 700 remedies for various ailments (Majno 1975), to the humoral theory developed in ancient Greece by Hippocrates and later Galen, laying a foundation for physiological imbalances as therapeutic targets (Porter 1999), to advances in anatomy during the Renaissance by pioneers like Andreas Vesalius, identifying organs and systems as therapeutic targets (O'Malley 1964), to nineteenth-century microbiologists Robert Koch and Louis Pasteur demonstrating the link between specific pathogens and diseases, transforming target identification from systemic treatments to pathogen-specific interventions (Blevins and Bronze

2010). The molecular biology revolution and the mapping of the human genome in the twentieth and early twenty-first centuries have enabled systematic identification of genetic targets for drug development (Collins et al. 2003), paving the way for precision medicine, allowing researchers to target specific genes, proteins, and pathways involved in diseases.

Drug target discovery is a multifaceted process that involves the identification of therapeutic targets, validation of their biological importance, assessment of safety, and the development of molecules capable of effectively modulating these targets. Traditional methods rely heavily on experimental techniques and trial-and-error approaches, making the process both time-intensive and resource-demanding. Depending on the disease area, it may take up to a decade or longer and billions of dollars to bring a new drug to market, with high attrition rates observed throughout the pipeline (DiMasi et al. 2016; Schlander et al. 2021; Sun et al. 2022).

The complexity of biological systems and the growing availability of data present significant challenges, as well as opportunities, to target discovery. Biological pathways and networks are highly intricate, often involving cross-talk among multiple pathways. This complexity makes it difficult to identify targets that are both efficacious and safe (Cava et al. 2018). Additionally, advances in genomics, proteomics, and other "omics" technologies have generated vast amounts of data, which, while holding valuable insights into the molecular mechanisms and regulation of healthy and disease states, present a significant challenge in distilling actionable information. Artificial intelligence (AI) has emerged as a transformative force in addressing these challenges in target discovery. Figure 1 summarizes the applications of AI in drug target discovery and their advantages when compared to traditional methods, which we will cover in this chapter. AI excels at analyzing multi-omics data, enabling the identification of novel drug targets and pathways (You et al. 2022; Pun et al. 2023a). For instance, predictive modeling using machine learning (ML) algorithms can assess the druggability of targets and evaluate their potential efficacy and safety profiles (Badwan et al. 2023; Liu et al. 2024a). Additionally, AI tools such as natural language processing (NLP) can mine scientific literature and databases to uncover previously unrecognized relationships between genes, pathways, and diseases (Gonzalez et al. 2023; Chang et al. 2024; Rahit et al. 2024). These capabilities have revolutionized how researchers approach target identification and pathway elucidation.

The development and implementation of various analytical frameworks and computational architectures highlight the growing impact of AI in drug discovery. ML and deep learning (DL) algorithms, such as random forests, support vector machines, and deep neural networks, have been employed to predict molecular interactions and identify novel targets (You et al. 2022; Patel et al. 2020; Xu et al. 2021). Knowledge graphs powered by AI link genes, proteins, diseases, and drugs to reveal intricate associations and pathways (Ye et al. 2021). However, successful integration of AI into drug target discovery requires interdisciplinary collaboration, depending on the combined efforts of computer scientists, biologists, chemists, and clinicians to guide AI models, ensuring that their predictions are biologically relevant and practically applicable.

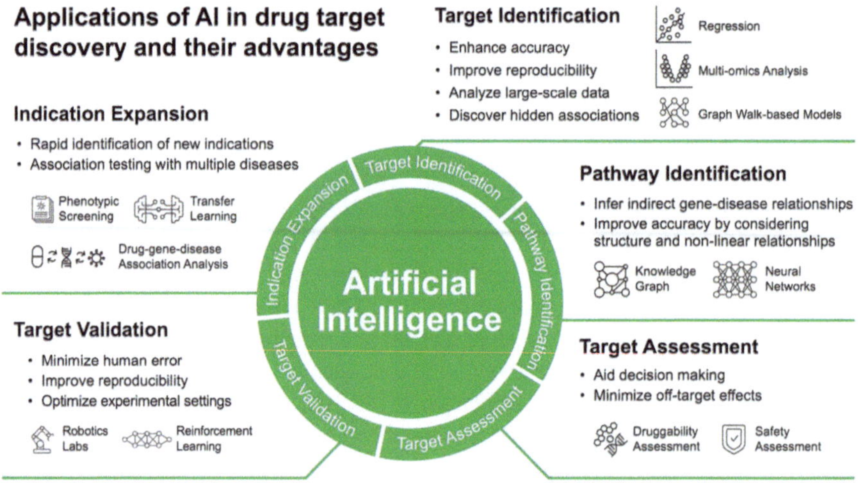

Fig. 1 Applications of AI in drug target discovery and their advantages. This figure summarizes the applicability of AI technologies in each key step of drug target discovery and their advantages when compared to traditional methods. For target identification, AI models such as regression-, classification-, and graph walk-based models can enhance the accuracy of target identification and discover hidden gene–disease patterns that would be difficult to detect using non-AI methods. In pathway identification, models such as knowledge graph and neural networks infer indirect gene–disease relationships through convergent pathways, with increased pathway prediction accuracy due to consideration of the structure (e.g., topological structure) and non-linear relationships that exist in each biological pathway. In target assessment, AI aids in the prediction of druggability of a proposed target and the potential side effects of target modulation (e.g., inhibited or activated), which in turn helps decision making (e.g., whether or not to proceed with this target) and minimize off-target effects. In target validation, AI-driven robotics labs and models such as reinforcement learning automate the steps of experimental validation, which minimizes human error, improves reproducibility, and optimizes the experimental conditions for better performance. Lastly, indication expansion can also be assisted by AI. For example, transfer learning can predict new gene–disease associations by analyzing large-scale omics data to discover shared molecular patterns between diseases; phenotypic screening to identify potential phenotypes associated with a gene; and unsupervised clustering may discover clusters of patients associated with particular phenotypes, which also provides directions for developing stratified indications

2 AI in Target and Pathway Identification

Starting in the 1980s, progress in genomic technologies, combined with insights from human genetics and experiments involving gene perturbations in cellular and animal models, began playing an increasingly significant role in efforts to identify drug targets. These genetic studies successfully identified many disease targets, often major drivers of disease in a substantial proportion of affected patients, such as mutated BRAF for melanoma (Davies et al. 2002), HER2 for breast cancer (Slamon et al. 1987), and PCSK9 for cardiovascular diseases (Sabatine 2019). As technologies advanced to include profiling with different sequencing platforms, researchers have gained insight into a wider spectrum of genes, mutations, proteins, or metabolites that are associated with disease or particular phenotypes. For example, gene expression profiling of the tissues, cultured cells, and blood of patients with rheumatoid arthritis (RA) identified genes and pathways that are differentially regulated between RA and healthy controls, between RA subtypes, and between RA and other diseases (Burska et al. 2014). Similarly, comparing gene expression and DNA methylation profiles of lung tissues in patients with idiopathic pulmonary fibrosis (IPF) to healthy donors identified disease-associated methylation–gene expression relationships (Yang et al. 2014). These data-driven findings were usually followed by experimental validation in either in vitro or in vivo models, where traditional techniques, such as gene knockouts and RNA interference, were performed on cell cultures or animal models to confirm the target's contribution to disease initiation and/or progression.

However, these methods are often constrained by the scale of experimentation and the fact that complete gene knockouts may fail to mimic the complexity of gene network dysfunction observed in human diseases. Only the targets inducing the most phenotypically obvious or statistically significant changes were further tested by experimental methods, which limits the discovery of targets that lead to subtle changes yet play important roles in disease. To address this, AI techniques offer a data-driven alternative that complements and accelerates these approaches by interrogating large-scale datasets and identifying hidden gene–disease relationships that may go unnoticed otherwise. For example, categorizing cancer mutations into pathogenic drivers and benign passenger mutations is important for the discovery of actionable targets, but the "long-tail" nature of rare/private oncogenic mutations outnumber the few, common driver mutations, makes this challenging. AI algorithms have been developed to tease out this distinction in an effort to confidently identify clinically relevant mutations and thus personalized treatment targets (Wang et al. 2020). In this section, we introduce several AI models used for target and pathway identification. We also cover examples of implementing AI in discovering novel targets and pathways, and how it has transformed the processes of drug target discovery at its early stages.

2.1 AI in Target Identification

AI algorithms perform exceptionally well in analyzing multi-omics datasets by exploring and integrating information from genomics, transcriptomics, epigenetics, proteomics, and metabolomics to uncover novel targets (Pun et al. 2022a, 2023a; Kamya et al. 2024; Liu et al. 2024b). ML models have been commonly used and are particularly adept at finding patterns and associations within these complex datasets. There are two main types of ML techniques: supervised and unsupervised learning. Additionally, text-based models, including heterogeneous graphs and NLP, are increasingly used to extract text information from literature, which can be applied to enhance target identification. To quantify the predictive power of the AI models, several metrics have been widely implemented to assess the accuracy of the outputs generated by these algorithms.

2.1.1 Supervised Learning

Supervised learning leverages labeled datasets, where input features (e.g., gene expression levels, protein interactions, or genetic variants) are paired with corresponding "labels" (e.g., disease states or therapeutic outcomes). These models learn to predict outcomes based on input features, making them optimized for tasks such as prioritizing potential drug targets, classifying genes associated with specific diseases, or predicting drug–target interactions on unlabeled data. For instance, a supervised Bayesian ML approach, BANDIT, was developed by Madhukar et al. that combined multiple data types, such as drug efficacy, drug response, drug structure, and known targets, to predict drug–target interactions (Madhukar et al. 2019). Supervised learning models typically fall into two main categories: regression-based and classification-based models.

Regression-Based Regression-based models predict continuous numerical outcomes by identifying relationships between input features and target variables. For instance, linear regression models can be used to analyze high-throughput omics data, identifying genes or proteins that are differentially expressed or abundant in disease when compared to healthy controls for target prioritization (Zhi et al. 2011; Tang et al. 2017; Pun et al. 2023b). Penalized regression models, such as LASSO and ridge regularization, introduce "regularization terms" to the model, which allow the identification of features that are most associated with disease phenotypes, while handling the noise and multi-collinearity that exists in the high-dimensional data (Park and Hastie 2008; Won et al. 2015; Liu et al. 2020). Other regression analysis methods of supervised learning include decision trees, neural networks, and partial least squares regression (Vamathevan et al. 2019), all of which have been implemented in the discovery of actionable targets and biomarkers of disease progression (Costa et al. 2010; Kc et al. 2022).

Classification-Based Supervised learning can also perform classification tasks. One widely used method is the support vector machine (SVM), which identifies the optimal hyperplane (the line that best separates two groups of data points) that exists in the training data. The trained SVM model can then be applied to new data points and classify which group they belong to. Applications of SVM models have been demonstrated in predicting biomarkers, drug targets, and studying disease mechanisms in different diseases, such as cancers (Jeon et al. 2014; Huang et al. 2018), Parkinson's disease (Moradi et al. 2022), and aging (Mamoshina et al. 2018).

2.1.2 Unsupervised Learning

In contrast, unsupervised learning is a type of ML where the algorithm is trained on data without labeled outcomes or predefined categories. The goal is to identify underlying patterns, structures, or relationships within the data. Common techniques in unsupervised learning include dimensionality reduction, which reduces the number of features while retaining important information (e.g., principal component analysis [PCA]); clustering, where data points are grouped based on similarity (e.g., K-means) (Vamathevan et al. 2019; Libbrecht and Noble 2015); and matrix factorization, an approach used to capture hidden patterns and relationships between variables in high-dimensional data. These models are particularly useful for exploratory purposes, such as identifying subgroups or endotypes in a patient cohort without prior knowledge.

Dimensionality Reduction and Clustering Dimensionality reduction plays a crucial role in analyzing complex, high-dimensional biological data, such as gene expression profiles, protein interactions, and chemical structures. These datasets often contain a large number of features, making it difficult to identify meaningful patterns or relationships. Dimensionality reduction techniques, such as PCA, UMAP, and t-SNE, help simplify these datasets by representing high-dimensional data, such as RNA-sequencing–based transcriptome profiles, as a function of fewer, new variables that attempt to capture the most relevant relationships between or among data points (Jolliffe and Cadima 2016; Kobak and Berens 2019; McInnes et al. 2018). Clustering is another unsupervised learning technique used to identify subgroups within complex datasets and uncover hidden patterns (Vamathevan et al. 2019). Commonly used clustering algorithms, such as K-means and hierarchical clustering, are used to segment data points into distinct clusters. While a predefined number of clusters is required for K-means clustering (Steinley 2006), hierarchical clustering builds a tree-like structure (dendrogram) based on the similarity between data points (Nielsen 2016). Clustering may be performed after dimensionality reduction to overcome the "curse of dimensionality" for high-dimensional datasets where the features outnumber the samples (Domingos 2012). As an example, Prakash et al. identified distinct clinical subgroups in patients with Alzheimer's disease by performing dimensionality reduction using t-SNE, followed by K-means clustering (Prakash et al. 2021), identifying four subgroups correlating with disease severity and potential association with prescribed drugs or supplements. Specifically, antidepressants were more commonly taken by a subgroup of patients with most severe disease than other subgroups, although the interpretation of this association was difficult given the complex relationships between neurodegenerative disorders, as well as the lack of comprehensive clinical data. Nevertheless, this combination of unsupervised learning has also been widely used in analyzing single-cell RNA sequencing data, where they can highlight distinct clusters of cell types based on their transcriptomic signature (Satija et al. 2015; Zhang et al. 2023), which may identify pathogenic processes specific to key subpopulations and targetable pathways obscured by bulk profiling methods (Van de Sande et al. 2023).

Matrix Factorization Matrix factorization is a decomposition-based technique used to uncover hidden patterns in large-scale biological datasets. Unlike dimensionality reduction, which primarily focuses on reducing the number of features while preserving key information, matrix factorization decomposes a large data matrix, such as a drug–target interaction matrix or gene expression matrix, into smaller matrices that capture latent features and relationships (Stein-O'Brien et al. 2018). This approach helps reveal biological properties, facilitating the identification of new therapeutic targets and off-target effects (Kamya et al. 2024; Pun et al. 2023b; Cobanoglu et al. 2013; Liu and Tsoumakas 2021; Ren et al. 2025). For example, a drug–target interaction matrix (where rows represent drugs, columns represent proteins, and values indicate interaction strength) can be decomposed into:

(i) A drug feature matrix, representing the properties of drugs.
(ii) A target protein feature matrix, representing protein characteristics.

Multiplying these matrices can predict missing interactions, suggesting potential new drug–target relationships. If Drug A interacts with proteins similar to those targeted by Drug B, matrix factorization can infer that Drug A might also interact with a protein previously unlinked to it, but interacts with Drug B, revealing potential new drug–target interactions for further investigation. A notable application of this method is the work of Ceddia et al., who developed a matrix factorization-based framework to predict drug repurposing candidates. By integrating publicly available drug, protein, and pathway databases, their approach uncovered previously unknown therapeutic classes and potential disease targets, increasing the total number of predicted protein targets nearly threefold (Ceddia et al. 2020). Matrix factorization is particularly valuable in high-dimensional datasets, where reducing complexity while preserving critical biological insights allows researchers to gain a deeper understanding of drug efficacy, safety, and potential applications (Fukuto et al. 2021).

2.1.3 Text-Based Models

Heterogeneous Graphs Heterogeneous graphs based algorithms are increasingly used to explore the complex relationships between drugs, proteins, and diseases within a biological network (Zhao et al. 2022; Chen et al. 2018; Jiang et al. 2022; Wang et al. 2013; Thafar et al. 2020). By "walking" through nodes (e.g., proteins) and edges (e.g., interactions) of a heterogeneous graph, which contains different types of entities and interactions constructed by the algorithm pretrained on published drug–gene, drug–disease, and gene–disease datasets, graph walk algorithms capture the connectivity patterns and similarities between these entities (Chen et al. 2018), prioritizing nodes that are most relevant to a given disease. This process helps uncover hidden associations, such as potential drug–target interactions or biomarkers for disease, by generating node embeddings that represent the underlying structure of the network (Chen et al. 2018). These embeddings can then be used for further tasks, such as clustering similar drug candidates or predicting novel drug–target relationships, as shown by Gharizadeh et al. (2024). In this study, a three-step heterogeneous graph-based approach, HGTDR, predicted drug repurposing candidates, as well as drug-target and disease–target relationships (Gharizadeh et al. 2024).

Natural Language Processing Natural language processing (NLP) tools play a crucial role in target discovery by mining vast amounts of scientific literature and databases (Pun et al. 2023a). These tools extract text-based information about gene–disease associations, protein interactions, and pathway functions to infer relationships that may not be explicitly detailed within individual text sources, enabling researchers to identify previously unrecognized targets. For instance, NLP-based systems, such as the publication-wide association study (PWAS) developed by Narganes-Carlón et al., can process thousands of research articles to generate hypotheses about the roles of genes in specific diseases, significantly accelerating the hypothesis-generation and target-discovery phases (Narganes-Carlon et al. 2023). NLP tools can also be applied to estimate the novelty of a target based on its associated number of scientific publications, grants, patents, etc. (Pun et al. 2023a; Long et al. 2024). Additionally, large language models (LLMs), such as BioBERT and BioGPT, were trained on articles available on PubMed and PubMed Central (Lee et al. 2020; Luo et al. 2022). BioBERT is focused on understanding and interpreting biomedical text, making it effective for tasks like extracting information about diseases, genes, proteins, drugs, and clinical procedures (Aldahdooh et al. 2022). BioGPT is positioned for generating texts for applications such as medical research writing, document preparation, and target discovery by learning disease–gene relationships from publications and research grant texts (Zagirova et al. 2023). Furthermore, TxGemma, an open-source LLM developed by Google DeepMind, aims to enhance the efficiency of therapeutic development by leveraging advanced language modeling techniques (Azizi 2025). TxGemma is specifically trained to understand and predict the properties of therapeutic entities throughout the entire discovery process, from identifying promising targets to aiding in the prediction of clinical trial outcomes. This model offers the potential to accelerate the drug development pipeline, reducing both time and costs associated with bringing new therapeutics to market.

2.1.4 Assessing the Performance of AI Models

Measures such as precision, recall, F1 score, Cohen's kappa, SHapley Additive exPlanations (SHAP), and area under the receiver operating characteristic curve (AUC-ROC) are essential for evaluating the performance of AI models in target identification (Urban et al. 2023; Rainio et al. 2024). The usefulness and optimal use cases of each of these metrics depend on the goal of the task, but they help ensure that AI-driven identification is reliable and robust, inspiring confidence in the downstream use of predicted targets for drug development.

Precision measures the proportion of true-positive (TP) predictions among all positive (TP and false-positive [FP]) predictions, ensuring that the predicted targets are likely to be biologically relevant (Hicks et al. 2022). It is particularly important when the cost of FP is high, for example, when a diagnostic test for a specific cancer type with low precision may lead to unnecessary stress, tests, or treatments.

Recall, also known as sensitivity, on the other hand, assesses the proportion of TP correctly identified out of all actual positives (TP and false negatives [FN]), ensuring that the models are sensitive enough to detect meaningful targets (Hicks et al. 2022).

A high recall is critical when a missed TP has severe consequences; for example, missing a true detection of a rare disease may delay treatment.

In most cases, a balance between precision and recall is desired, which can be reflected by the *F1 score*. It offers a single metric that combines these two critical aspects while balancing their trade-off, which is particularly useful when the dataset is imbalanced and both precision and recall are important (Hicks et al. 2022). However, if there are extremely imbalanced classes and a specific threshold is more important (e.g., recall for detecting rare diseases), alternative metrics like AUC-ROC or simply recall should be used.

Cohen's kappa (κ) is a statistical measure of inter-rater agreement or classification reliability. It is commonly used to evaluate the performance of a model, particularly for nominal (non-ordered categorical) outcomes, by comparing its predictions to the true labels beyond random chance (McHugh 2012). If there is a high agreement between the rates, κ will be close to 1. In contrast, if there is no agreement observed higher than that expected by chance, then κ will be close to 0.

SHAP is a model-agnostic explainability framework based on cooperative game theory. It assigns a contribution score (SHAP value) to each feature for every prediction, indicating how much the feature contributed to the model's output (Rodriguez-Perez and Bajorath 2020). It is particularly useful for understanding feature importance by ranking their contributions to the predictions, explaining the decisions during discovery using the model, and improving transparency in complex "black-box" models like DL-based techniques (Rodriguez-Perez and Bajorath 2020). Mathematically, the SHAP value of each feature is calculated as the average of its contributions across all permutations of a set of features (Rodriguez-Perez and Bajorath 2020).

The *AUC-ROC* measures the ability of the model to distinguish between classes, providing a comprehensive overview of its predictive performance (Corbacioglu and Aksel 2023). The ROC plots the TP rate (equivalent to recall or sensitivity) against the FP rate (equivalent to "1-specificity"), and the AUC value represents the area under the ROC curve. As a result, the higher AUC (more area under the curve), the higher sensitivity and specificity. An AUC of 0.5 indicates that the classification is no different than random chance (like "flipping a coin"), while an AUC of 1 indicates perfect discrimination. However, the variety of potential outcomes should be considered (e.g., by computing the 95% confidence intervals) in order to factor in the reliability of the predictive performance (Corbacioglu and Aksel 2023).

2.2 Pathway Identification

To improve the confidence in identified targets, pathway enrichment analyses are commonly performed on the subset of genes or proteins that are dysregulated in the patients, tissues, or cells with the phenotype of interest (e.g., disease) to identify the pathways in which the shortlisted features are involved. More than a decade ago, Khatri et al. described the challenges of pathway analyses, which were mainly divided into two categories: annotation and methodological challenges (Khatri et al. 2012). Annotation challenges include incomplete or inaccurate annotations for the ever-growing number of identified genes and transcripts as well as low-resolution annotations that ignore different functions or activities of a gene in different contexts or between similar, but different, gene variants. Methodological challenges include dealing with the complexity and confounding factors inherent to biological systems, dynamic responses, and external stimuli that are important for describing pathway components and function, yet are difficult to fully model. As a result, emerging AI models aim to facilitate the process of pathway identification to infer a better understanding of disease mechanisms, thus improving the success rate of drug target discovery.

2.2.1 Network-Based Approaches

Network-based approaches have emerged as powerful tools for pathway identification by leveraging the interconnected nature of biological systems. These methods model different entities, such as genes, pathways, diseases, and drugs, as nodes and their interactions as edges in complex networks, in a predefined manner (i.e., the networks are built on existing knowledge). By analyzing these networks, researchers can uncover disease-related associations. Approaches such as knowledge graphs (MacLean 2021), random walk algorithms (Gentili et al. 2022), and network propagation (Law et al. 2021) identify key nodes (features) and edges that indicate biologically significant relationships. For example, Liu et al. developed CARNIVAL to infer the causal signaling pathways in diseases (Liu et al. 2019). First, gene expression data were used to identify differentially expressed transcription factors (TF) and to infer their activities. Then, combined with prior knowledge of signaling networks, CARNIVAL reconstructs context-specific pathways, generating hypotheses on the protein networks upstream of gene expression to predict disease-associated signaling pathways that best explain the inferred TF activities through protein–protein interactions (Liu et al. 2019).

Knowledge graphs, constructed using AI-based text mining algorithms such as NLP, link genes, proteins, and pathways to diseases, provide a holistic view of the molecular landscape (Ye et al. 2021; Kamya et al. 2024; Chandak et al. 2023). Unlike traditional databases, knowledge graphs connect these entities using edges (relationships) that describe how they interact. Relationships between biomedical entities are mapped into a graph structure (e.g., "Drug A inhibits Protein X"). Graph-based AI models can then analyze these connections to uncover hidden relationships (e.g., drug–gene or gene–disease associations without directly observed and validated relationships reported in the literature, but when inferred, predict new drug–disease associations), identifying pathways that might be overlooked by traditional methods. For instance, Xing et al. developed RDBridge, a comprehensive knowledge graph constructed through extensive text mining on rare diseases, associated genes, metabolic pathways, potential drugs, and medical imaging data (Xing et al. 2023). RDBridge encompasses more than 10,000 rare diseases, along with a combined number of around 20,000 disease–gene, disease–pathway, and drug–gene associations. The researchers presented a case study on a rare disease, amyotrophic lateral sclerosis (ALS), identifying ALS-related genes, potential drugs, and pathways (Xing et al. 2023). Chandak et al. developed a multimodal knowledge graph, called PrimeKG, which includes around 17,000 diseases with more than four million relationships of disease-related protein changes, biological pathways, drug actions, and more entities, which can be used to discover pathways associated with particular diseases or genes of interest (Chandak et al. 2023).

2.2.2 Deep Learning-Based Approaches

While network-based approaches are relatively simple, computationally efficient, and more interpretable in identifying disease-associated pathways, they are often limited by the pre-existing text-based data that was used to construct the networks (Khatri et al. 2012). In contrast, data-driven DL models capable of analyzing large, noisy, and high-dimensional datasets provide additional insights into predicting the pathways associated with the disease. These models can integrate multiple data sources to create comprehensive pathway models, which can be used to either identify dysregulated signaling pathways in disease or predict (reconstruct) pathways (Park et al. 2022; Shah et al. 2021). For instance, DeepHisCom, a deep neural network-based pathway analysis method developed by Park et al., was shown to be capable of identifying disease-associated pathways across a variety of diseases using a variety of input data types. Applying DeepHisCom to the multi-omics data of two diseases, hepatocellular carcinoma and COVID-19, the researchers successfully identified pathways both well known and recently suggested to play important roles in disease progression and severity (Park et al. 2022). DeepHisCom outperforms previously reported pathway prediction methods by discovering pathways that were not identified by the other methods (the original hierarchical structured component model (Lee et al. 2016), GlobalTest (Goeman et al. 2004), and GlobalANCOVA (Hummel et al. 2008)) but strongly implicated in the disease (Park et al. 2022). PathGNN, a graph neural network-based model developed by Liang et al., showed promising performance in predicting the survival of patients with different cancer types (lung, skin, kidney, and brain cancers) using gene expression data, with the consideration of topological features in the pathways (Liang et al. 2022). The integrated gradients algorithm in PathGNN was then utilized to identify key pathways associated with survival outcomes, with the literature supporting their findings (Liang et al. 2022).

3 AI in Target Assessment and Validation

In addition to target identification, AI technologies have revolutionized the processes of target assessment and validation, offering innovative solutions to streamline these critical stages in drug discovery. Target assessment evaluates the target's suitability for therapeutic intervention, while target validation confirms the biological relevance of a target in disease initiation or progression. AI-driven approaches significantly enhance protein structure prediction, automation, and advanced analytics.

3.1 Assessment of AI-Identified Targets

The assessment of AI-identified drug targets is critical to ensure their druggability and safety before advancing to experimental validation and clinical trials. Druggability refers to the likelihood that a target can be modulated by small molecules, biologics, or other therapeutic modalities, while safety assessment involves predicting potential off-target effects, toxicity, and the impact of modulating the target on broader biological pathways.

3.1.1 Druggability Assessment

Evaluating whether a biological target can be effectively modulated (e.g., activated or inhibited) by a therapeutic agent allows researchers to prioritize targets with high therapeutic potential. For small-molecule therapeutics, druggability is often tied to ligandability, or the existence of accessible, well-characterized pockets on a target's surface where small drug-like molecules can bind. A pocket is considered accessible if it is solvent-exposed and topologically favorable for ligand entry, meaning it is not deeply buried or sterically hindered. A well-characterized pocket is one whose physicochemical properties, structural features, and dynamic behavior are well understood, facilitating reliable predictions of ligand binding. There are several key parameters to determine druggability. For instance, pocket volume and depth are crucial, with druggable pockets typically larger than 500 $Å^3$, providing sufficient space for ligand interactions (Drew and Janes 2020; Volkamer et al. 2012). Additionally, hydrophobicity and lipophilicity enhance binding affinity, as many bioavailable drugs exploit hydrophobic interactions (Lou and Martin 2021). Moreover, polarity and hydrogen bonding capacity also influence ligandability. While hydrophobic pockets are often highly druggable, an optimal balance of hydrogen bond donors and acceptors ensures strong and specific interactions (Patil et al. 2010). Pocket flexibility is another key factor, as some targets undergo induced fit or allosteric changes upon ligand binding (Stank et al. 2016). Understanding a target's three-dimensional structure and its ligandability is crucial for prioritizing potential targets and informing structure-based drug design (Cheng et al. 2007). In addition to small molecules, various therapeutic modalities, such as antibody-based therapies (Lu et al. 2020), targeted degraders (Bekes et al. 2022), and gene and cell therapies (Sayed et al. 2022; Sterner and Sterner 2021) have broadened the spectrum of targets that can be considered druggable. DL techniques, such as convolutional neural networks, have been employed to detect drug-binding pockets from protein structures or protein–ligand interactions (Torrisi et al. 2020; Torng and Altman 2019). Tools like AlphaFold allow accurate predictions of protein structures and their potential binding sites, even for previously uncharacterized targets (Jumper et al. 2021).

3.1.2 Safety Assessment

To reduce the risks involved in modulating a proposed target, safety assessment is an essential part of the target assessment process. Most drugs are associated with side effects, although they vary in severity. The primary focus of safety profile predictions is typically on identifying the drug compound's interactions with biomolecules other than its intended targets. Indeed, this may uncover potential off-target effects where the drug binds to unintended targets, leading to undesirable perturbations on other biological pathways (Fukuzaki et al. 2009). However, tools are being developed to predict the potential effects of modulating both the target and off-target proteins (Brennan 2017). For instance, Wilson et al. developed PathFX, a pathway-based model to predict drug effects from protein–protein interaction networks (Wilson et al. 2018). By combining gene expression data with PathFX, the same research group constructed networks that link proteins, pathways, drugs, and side effects (Alidoost and Wilson 2024). Specifically, proteins without known interactions with specific drugs were found to be positioned neighboring the reconstructed protein networks that were directly associated with the drug (i.e., drug-binding proteins). These nearby non-drug-binding proteins were then associated with potential phenotypes, forming a larger network where the pathways perturbed by a drug and associated side effects could be more comprehensively estimated for a proposed target protein.

3.2 AI in Target Validation

Combining genomic, transcriptomic, proteomic, and metabolomic information, AI models identify targets that consistently influence disease pathways at multiple molecular levels (Perakakis et al. 2018), especially for oncology-related studies (Biswas and Chakrabarti 2020; Reel et al. 2021), ensuring that validated targets have a higher likelihood of success in therapeutic applications. Furthermore, NLP approaches may also play a role in target validation by mining scientific literature to uncover evidence supporting or refuting a target's relevance. For instance, NLP tools can extract and summarize results from thousands of articles and combine literature-based findings with data-driven findings, providing researchers with a consolidated view of existing data (Doan et al. 2014). Multiple studies have illustrated the use of NLP tools in constructing drug–gene, drug–disease, or gene–disease relationships, such as knowledge graphs as discussed in the previous section (Ye et al. 2021; Kamya et al. 2024; Chandak et al. 2023). Additionally, NLP can also be applied to quantify the levels of novelty, confidence, and commercial tractability, which aids decision-making for the proposed target (Long et al. 2024).

Beyond purely in silico prediction of targets, in vitro experimental validation in the complex biological context of the cell remains the gold-standard for validating any potential disease-modulating effects, and AI-driven robotics labs are playing an increasingly vital role in automating experimental validation (Schneider 2018). These labs, equipped with robotic systems and ML algorithms, can perform high-throughput experiments with minimal human intervention. For example, robotics systems can conduct gene editing experiments using CRISPR-Cas9 to automate genome editing in human cell line models (Niessing et al. 2024; Takahashi et al. 2024). These systems are not only faster than traditional methods but also ensure consistency across repeated experiments by reducing technical variations introduced by humans. AI algorithms can then analyze the outcomes of these experiments, identifying subtle trends and relationships in the data that might otherwise go unnoticed, as described above. Automation reduces human error, increases reproducibility, and accelerates the pace of validation, allowing researchers to validate numerous targets simultaneously. Indeed, Tang et al. leveraged a fully automated robotics laboratory to investigate the impact of inhibiting Traf2- and Nck-interacting kinase (TNIK) on cellular senescence—a key hallmark of aging characterized by growth arrest and the proinflammatory senescence-associated secretory phenotype (SASP) (Tang et al. 2025). Using the TNIK inhibitor INS018_055 (rentosertib), they observed a reduction in cellular senescence and SASP across multiple senescence models and identified key biological pathways (Tang et al. 2025). The study adds to the growing body of research blending AI-based virtual tools and physical experimentation methods for drug discovery (Grisoni et al. 2021) by automating target validation, sample processing, quality control, multi-omics data generation and analysis, validation through cell culture and high-throughput sequencing, and assessment of validation outcomes (Tang et al. 2025).

Reinforcement learning (RL) is an innovative ML approach involving training algorithms to make sequential decisions to optimize a given outcome. In the context of target validation, RL models can design and prioritize experimental workflows by dynamically learning which steps yield the most informative results. For example, RL has been applied to optimize assay conditions in various experimental setups, including high-throughput screening (HTS), protein crystallization, and synthetic biology workflows (Takahashi et al. 2024; Tang et al. 2025). In HTS, RL dynamically adjusts reagent concentrations and incubation times to enhance assay sensitivity and specificity. In protein crystallization, it fine-tunes parameters such as temperature, pH, and precipitant concentrations to identify optimal crystallization conditions. In synthetic biology and metabolic engineering, RL optimizes guide

RNA selection to improve gene editing efficiency and specificity in CRISPR-based applications. Additionally, it refines enzyme expression levels and reaction conditions to maximize the production of target metabolites, such as biofuels and pharmaceuticals, in metabolic pathway engineering (Tang et al. 2025; Treloar et al. 2022). Additionally, these models can assist in determining the most relevant disease models or cell lines to test, further improving the efficiency of the validation process. The GENTRL RL-based model was used to design novel, synthetically accessible, and potent inhibitors of DDR1 in only 3 weeks, followed by synthesis and experimental validation in another 25 days, further emphasizing RL as a promising strategy to accelerate drug discovery (Zhavoronkov et al. 2019).

4 Examples of AI-Aided Target Identification

AI has been instrumental in unraveling complex biological networks and identifying novel therapeutic targets with unprecedented speed and accuracy. The following success stories highlight the transformative impact of AI in various therapeutic areas.

4.1 Aging

Understanding the process of aging is crucial as the global population ages rapidly, with the proportion of individuals aged 60 years and over projected to increase from 12% in 2015 to 22% by 2050 (World Health Organization 2024). By analyzing genomic, transcriptomic, and methylomic data, multiple ML algorithms (e.g., multimodal transformer-based transfer learning (Urban et al. 2023), large language models (Zagirova et al. 2023), Weighted Gene Co-expression Network Analysis, and random forest classifier (Zhou et al. 2023; Chai et al. 2021)) have shown great potential in identifying targets implicated in aging and age-related diseases. In particular, iAge, a DL model developed for predicting aging phenotypes based on blood immunome profiles, identified the small cytokine CXCL9 as a pivotal player in cardiac aging (Sayed et al. 2021). Notably, the inhibition of CXCL9 reversed arterial stiffness in both human and mouse studies (Sayed et al. 2021). Considering the interconnected biological processes in aging, age-related diseases, and tumorigenesis, Pun et al. utilized an AI-powered target discovery platform to propose dual-purpose therapeutic targets for the treatment of both aging and age-related diseases or cancers, identifying 23 high-confidence candidate targets not previously implicated in aging (Pun et al. 2022b, 2023b). A similar study was conducted to dissect the pathology of aging and glioblastoma multiforme, pointing as well to a number of potential dual-purpose targets (Olsen et al. 2023).

4.2 Oncology

Large-scale multi-omics data and clinical records are increasingly available, which has paved the way for the application of AI in cancer research (Wang et al. 2024; Shi et al. 2022; Chen et al. 2022; Ramos et al. 2022). For instance, MOGONET, a graph convolutional network, was developed to integrate both omics-specific features and cross-omics correlations for patient classification and biomarker identification. When applied to breast cancer subtype classification, MOGONET successfully identified genes known to be associated with breast cancer (Wang et al. 2021). Additionally, a genomic data-fed, biologically interpretable neural network, P-NET, was proposed to predict prostate cancer patient status (primary vs metastatic disease), potential biochemical recurrence, and evaluate genetic drivers of metastatic disease (Elmarakeby et al. 2021). With experimental validation, the model successfully identified MDM4 as a novel therapeutic target, exemplifying the significant progress made by AI in novel target discovery.

4.3 Rare Diseases

AI is emerging as a promising solution for the challenging therapeutic target identification for rare diseases (Esteban-Medina et al. 2024; Tebani et al. 2023; Li et al. 2023a; Chen et al. 2023). Aberrant alternative splicing is a well-known pathogenic factor of rare monogenic pathologies such as spinal muscular atrophy (Lorson et al. 1999), neurofibromatosis type 1 (Ars et al. 2000), and cystic fibrosis (Amistadi et al. 2023). Gao et al. discovered the splicing modulator compound BPN-15477 and applied a deep convolutional neural network to identify genes and splicing mutations targeted by the drug, including pathogenic mutations that contribute to familial dysautonomia pathogenesis (Gao et al. 2021). Another example is in ALS, a rare neuromuscular disease characterized by progressive degeneration of motor neurons in the central nervous system (CNS) with suboptimal therapeutic options (Pun et al. 2022a). By coupling both omics and text-based data into account, PandaOmics (a DL-based target discovery platform) prioritized 28 potential therapeutic targets for ALS based on target-disease associations, 8 of which were experimentally validated to confirm their relevance to neurodegeneration (Pun et al. 2022a). In a promising case from the clinic, Mumau et al. recently reported the successful treatment of a patient with idiopathic multicentric Castleman disease, a rare condition characterized by life-threatening systemic inflammation, who was resistant to the only FDA-approved therapeutic. The researchers leveraged AI-based tools to identify aberrant TNF signaling from multi-omic analyses of the patient's samples and to predict the optimal treatment. Treatment with the top-scoring therapeutic, adalimumab, an anti-TNF monoclonal antibody, led to sustained remission (Mumau et al. 2025).

4.4 *Fibrotic Diseases*

Ren et al. recently demonstrated the use of generative AI in drug discovery to identify novel targets and drug candidates for IPF (Ren et al. 2025), a progressive disease that ultimately leads to respiratory failure (Raghu et al. 2006). Using multi-omics data from IPF and healthy lung tissues, the PandaOmics target discovery platform prioritized TNIK as the top target for IPF. This platform incorporates over 20 AI and bioinformatics models trained on pre-existing data and validated by their ability to predict targets later recognized in the pharmaceutical industry. Targets were ranked based on scores reflecting factors like protein family, small molecule or antibody accessibility, novelty, and structural availability (Kamya et al. 2024). TNIK emerged as the highest-ranking target, achieving strong scores in models such as network neighbors, causal inference, pathways, interactome community, and heterogeneous graph walks. Although the role of TNIK had been explored in other disease contexts (Pham et al. 2023; Ewald et al. 2024), it had not been studied as a therapeutic target for IPF. Interestingly, TNIK was independently predicted as a target associated with multiple hallmarks of aging (Pun et al. 2022b), likely due to dysregulated fibrotic processes being shared among age-related diseases. As a serine/threonine kinase, TNIK plays a central role in signaling pathways involving intercellular communication, cellular senescence, and nutrient sensing (Pham et al. 2023).

Following the identification of TNIK as a target candidate for IPF, a novel TNIK inhibitor, INS018_055, was designed using generative adversarial networks and showed potent anti-fibrotic and anti-inflammatory effects in murine models of pulmonary, skin, and liver fibrosis (Ren et al. 2025). This preclinical study was completed in roughly 18 months, highlighting the efficiency of AI-driven target discovery and drug design. Furthermore, INS018_055 demonstrated favorable safety and tolerability profiles in two independent Phase I clinical trials and a Phase IIa clinical trial (NCT05938920), supportive of further clinical testing with an encouraging safety profile and an increase in forced vital capacity in patients who received the highest dose (60 mg once-daily). A parallel Phase IIa study (NCT05975983) is currently underway in the United States to further validate these findings.

5 AI for Indication Expansion

Indication expansion refers to the process of identifying additional diseases or conditions that a drug or therapeutic target can effectively address beyond its original indication. Repurposing a drug already approved or well studied in one disease context has the benefit of having an established safety profile, thus reducing the risk of failing clinical testing due to safety concerns, and the time and monetary investment in the discovery process is substantially reduced (Talevi and Bellera 2020).

This aspect of drug discovery is increasingly driven by AI technologies, which offer powerful tools to analyze vast datasets and uncover novel opportunities for therapeutic application. Through the analysis of genomic, transcriptomic, and proteomic data, AI models can detect and discover shared perturbed molecular pathways and mechanisms between different diseases. For instance, a drug originally developed for RA might be identified as a potential treatment for a subtype of cardiovascular disease based on common inflammatory pathways (Godbole et al. 2024). By integrating diverse datasets, such as genetics and phenotyping data, AI facilitates cross-disease insights that would be challenging to achieve through traditional methods.

5.1 Transfer Learning

One of the critical approaches in AI-driven indication expansion is transfer learning (TL). This ML technique allows models trained on one task, such as identifying drug–response relationships in one disease, to be adapted for another, related task (Iman et al. 2023; Zhuang et al. 2020). TL reduces the need for extensive retraining and enables the efficient application of existing models to new indications. For instance, Petegrosso et al. developed tlDLP, an approach utilizing TL to combine phenotypes, gene ontology terms, and protein–protein interaction network, allowing the prediction of new associations between phenotypes and genotypes (Petegrosso et al. 2017). DeepSeq2Drug, an ensemble framework that leverages DL, multi-modal embeddings, and TL was introduced by Xie et al., which was shown to be able to identify drug repurposing candidates for new viruses or virus variants (Xie et al. 2024). Wu et al. implemented TL to transfer a previously developed neural network-based model (DeepCE) (Pham et al. 2021), which predicts gene expression profiles by the input of drugs, dosage, and cell type trained on the Library of Integrated Network-based Cell-Signature (LINCS) database, onto the DRUG-seq human neural cell dataset (Rodriguez et al. 2021), to predict new drug repurposing candidates for Alzheimer's disease (Wu et al. 2022). In addition, Chen et al. developed iDrug, a model that transfers the knowledge between two networks (drug-disease interaction and drug-target interaction) to enhance the predictions on both networks that outperformed other models (Chen et al. 2020). Notably, these two types of networks are important in indication expansion in different ways: (1) from the perspective of a proposed target (by predicting drug–target interaction: new drugs for the target), (2) from the perspective of an existing drug (by predicting drug–disease interaction: new indications for the drug), and (3) from the combination of both (new indications for the target).

5.2 Phenotypic Screening

Phenotypic screening involves identifying and analyzing observable traits or characteristics in response to specific treatments. Traditional phenotypic screening relies heavily on experimental assays, manual inspections, and classifications (Moffat et al. 2014; Prior et al. 2014), which are time consuming and limited in scope. AI overcomes these limitations by automating the analysis of high-content screening data, including cellular imaging, molecular profiles, and functional assays. These advanced techniques allow researchers to detect subtle phenotypic changes and classify diseases based on shared or unique phenotypic traits. One cutting-edge method in AI-assisted phenotypic screening is its integration with Phenome-Wide Association Studies (PheWAS) (Bastarache et al. 2022). PheWAS examines the associations between genetic variants and a wide range of phenotypic traits across populations. This approach complements genome-wide association studies (GWAS), which focus on linking genetic markers to specific diseases (Uffelmann et al. 2021). By combining with PheWAS, AI algorithms can uncover connections between genetic markers and multiple phenotypes, enabling researchers to identify potential therapeutic targets for diverse indications. For example, a genetic variant associated with inflammatory responses in arthritis may also be linked to metabolic dysregulation in diabetes, highlighting a shared pathway that can be targeted for both conditions (Nguyen et al. 2013; Wysocki et al. 2020; van Drongelen et al. 2021). Moreover, AI enhances the scalability and precision of PheWAS by analyzing large-scale datasets, such as electronic health records, biobanks, and population-level genetic data. Depending on the goal, both supervised and unsupervised ML models can process these datasets to identify phenotypic clusters and prioritize associations that are most relevant to drug discovery.

5.3 Patient Stratification

AI algorithms can help to identify biomarkers and genetic signatures that define subpopulations of patients who might benefit from a drug originally approved for a broader group by analyzing, for example, patient metadata such as age, sex, ethnicity, and other disease-related parameters contained in electronic health records, as well as clinical trial outcomes. For instance, AI-based clustering identified distinct subsets of patients with Parkinson's disease with different disease progression and trajectories, which were associated with distinct genetic features, dysregulated biological pathways, responses to treatment, and differences in dopaminergic cell loss (Birkenbihl et al. 2023). This stratification not only expands the potential indications of a drug but also paves the way for personalized medicine approaches, where treatments are tailored to individual patients.

However, AI-driven indication expansion is not without its challenges. The accuracy of predictions relies heavily on the quality and diversity of the underlying data. Integrating heterogeneous and well-annotated datasets, such as genomic data, clinical records, and environmental factors, remains a significant hurdle. Moreover, ethical considerations, including data privacy and the equitable distribution of AI-driven therapies, must be addressed to ensure responsible application.

6 Ethical, Regulatory, and Practical Considerations

The implementation of AI in drug discovery presents transformative opportunities, but it also introduces significant ethical, regulatory, and practical challenges. Addressing these challenges is essential to ensure the responsible, effective, and widespread deployment of AI-driven approaches for drug discovery.

6.1 Ethical Considerations

The issue of bias in AI algorithms remains a critical concern. AI models trained on imbalanced or incomplete datasets can produce biased outcomes, potentially leading to inequities in drug development. For example, genomic datasets predominantly derived from individuals of European descent may limit the effectiveness of AI models for other populations (Farhud and Zokaei 2021; Hanna et al. 2024). Ensuring fairness requires the inclusion of diverse datasets and continuous monitoring of AI performance. Lessons learned from the more mature development of AI tools for outcome prediction from electronic healthcare records can inform the more nascent field of AI-based drug discovery tools in areas such as bias detection and mitigation (Chen et al. 2024). The second key ethical concern is the lack of transparency and explainability in AI models, especially for DL algorithms that are often described as "black boxes." This opacity raises questions about accountability when AI predictions influence critical decisions, such as drug target prioritization. Developing interpretable AI models and setting transparency standards are essential to foster trust and accountability in AI-driven drug discovery (Hanna et al. 2024). Thirdly, privacy and data security are vitally important, as AI-driven approaches often rely on sensitive patient data, including genetic and clinical information. Ensuring compliance with data protection frameworks like the General Data Protection Regulation (GDPR) is critical to safeguarding this information and maintaining public trust (Farhud and Zokaei 2021).

6.2 Regulatory Considerations

The evolving nature of AI poses significant challenges for validating and approving AI-derived hypotheses and predictions. Regulatory bodies should adapt their traditional regulatory frameworks to ensure the accuracy, reproducibility, and reliability of the results generated by AI. They must also take a central role in ensuring that the use of AI in healthcare adheres to ethical standards (Hanna et al. 2024; Allen 2019). Additionally, determining liability in cases where AI-driven decisions lead to adverse outcomes remains a complex issue. Clear accountability frameworks are necessary to address these concerns, many of which have been adapted from existing reporting structures and guidelines made in response to the growing role of big data in clinical studies and practice (Hanna et al. 2024), while the harmonization of regulatory standards across jurisdictions is also critical to facilitate global collaboration and streamline the development of AI-driven solutions.

6.3 Practical Considerations

Integrating AI into established drug discovery workflows requires significant adjustments to organizational structures and processes. For instance, the use of AI technologies may be considered a threat to particular individuals, such as medical professionals, due to the potential of losing professional recognition, autonomy, and technological capabilities (Jussupow et al. 2022). This could cause resistance to change, which hinders further development, use, and critical evaluation of AI. A lack of technical expertise can also lead to resistance to the adoption of novel AI-based tools. These challenges may be overcome by providing comprehensive training programs and user-friendly interfaces to ease the transition. Furthermore, the quality and accessibility of data are critical to the success of AI systems. Datasets commonly contain errors caused by typing mistakes during data entry, poor sample quality, or missing values, and proprietary data can also be poorly integrated due to their size (Isgut et al. 2023; Williamson et al. 2021). Standardizing data collection and promoting data-sharing initiatives can mitigate these challenges. While AI has the potential to reduce costs in the long run, its implementation requires substantial upfront investment in computational infrastructure, software, and skilled personnel. Partnerships, grant funding, and cloud-based AI solutions can help lower the barriers for small- and medium-sized enterprises. Despite the sophistication of AI systems, human expertise remains indispensable for interpreting results and addressing unforeseen issues. Indeed, effective collaboration between AI and human experts is essential to maximize the benefits of these technologies.

7 Future Directions

Although AI has demonstrated transformative potential in drug discovery, the field continues to evolve rapidly. Future advancements will enhance the accuracy, precision, and applicability of AI by addressing current challenges. Here, we highlight some possible directions focusing on optimizing data integration, interpretability of AI models, applications, integrations with emerging technologies, and ethical and collaborative development.

Data Sharing and Quality Improving data quality and integration is a foundational step in advancing AI-driven drug discovery. Standardizing and securely sharing datasets across institutions can lead to the development of robust, generalizable models. For example, the FDA, Health Canada, and the United Kingdom's Medicines and Healthcare products Regulatory Agency (MHRA) jointly published the "Good Machine Learning Practice" (GMLP) guidelines, emphasizing the need for high-quality, representative datasets to train AI models effectively (Food and Drug Administration, Health Canada, and Medicines and Healthcare products Regulatory Agency 2021). Similarly, the European Medicines Agency (EMA)'s "Guideline on Computerized Systems and Electronic Data in Clinical Trials" outlines principles for ensuring the integrity and interoperability of data across systems (European Medicines Agency 2023). Blockchain technology offers a promising solution for secure and transparent data sharing (Sandra Kumi and Deters 2022), while multi-omics approaches, integrating genomics, proteomics, and metabolomics, can provide richer datasets, enabling AI to uncover a more comprehensive understanding of disease mechanisms. Furthermore, adherence to FAIR (Findable, Accessible, Interoperable, Reusable) data principles, as encouraged by global regulatory bodies, will improve the consistency and quality of AI-driven research (Wilkinson et al. 2016).

AI Model Visualization and Interpretation Another critical focus is the development of interpretable AI models. Explainable AI enhances trust by offering insights into how predictions are made, bridging the gap between computational outputs and experimental validation. Regulatory agencies, including the FDA and EMA, have stressed the need for explainability in AI-driven decision-making, particularly in clinical applications. The MHRA's "Software and AI as a Medical Device Change Programme Roadmap" highlights the importance of transparency in AI models, ensuring that they can be validated effectively (Medicines and Healthcare Products Regulatory Agency 2023). Moreover, visual tools and user-friendly frameworks make AI systems more transparent and actionable for researchers (Alizadehsani et al. 2024; Zheng et al. 2023), but care must be taken so as not to substantially sacrifice performance for increased interpretability (Li et al. 2023b).

AI in Precision Medicine Additionally, AI presents transformative opportunities in addressing rare diseases and personalizing medicine. Strategies such as federated learning or using synthetic data can help AI algorithms identify novel drug targets and biomarkers using sparse, heterogeneous datasets (Gangwal et al. 2024). ML models enable patient stratification, paving the way for tailored therapies, even in rare disease settings, that optimize treatment outcomes based on genetic and phenotypic profiles (Spiga et al. 2021; Geraci et al. 2023).

Interdisciplinary Applications Furthermore, the integration of AI with emerging technologies like quantum computing, synthetic biology, and robotics further amplifies its potential. Quantum computing could revolutionize molecular design (Kao et al. 2023), while synthetic biology facilitates novel therapeutic designs (Abil et al. 2015), and AI-powered robotics streamlines laboratory workflows, enhancing discovery efficiency. Moreover, simulated clinical trials and virtual populations represent a rapidly growing area of innovation. Regulatory agencies are beginning to recognize the potential of these methods; for example, the FDA has acknowledged the potential of AI in predictive modeling, counterfactual simulation, and generation of virtual cohorts, in improving clinical trial design and execution (ElZarrad 2024). Using real-world patient data, virtual cohorts enable researchers to model clinical trials, testing the efficacy and safety of targeting specific genes or pathways across diverse populations. These simulations can identify biomarkers predictive of target engagement, optimizing trial designs, and improving their success rates (Pathmanathan et al. 2024).

AI with Integrity Finally, ethical challenges, such as bias and fairness, must be addressed to ensure equitable AI adoption. Global guidelines, collaborative efforts among academia, industry, policymakers, and initiatives like open-source platforms and data-sharing consortia will drive innovation while maintaining ethical integrity. Regulatory bodies such as the FDA, EMA, and MHRA are actively working to establish ethical frameworks for AI in medicine. For instance, the "Ethical guidelines for trustworthy AI" published by the High-Level Expert Group on AI set up by the European Commission emphasizes fairness, accountability, safety, and integrity in AI applications (High-Level Expert Group on AI 2019). These advancements collectively promise to make drug discovery faster, more precise, and increasingly personalized.

8 Conclusion

The application of AI in drug discovery has ushered in a new era of innovation, transforming a traditionally labor-intensive and time-consuming process into a more efficient, precise, and scalable endeavor. Throughout this chapter, we have explored how AI is being utilized to revolutionize various aspects of drug discovery, from target identification and validation to pathway modeling and indication expansion. Each of these applications underscores AI's capacity to augment human expertise and unlock new possibilities in pharmaceutical research. One of the most significant advantages of AI lies in its ability to process and analyze vast volumes of biological, chemical, and clinical data. Through advanced algorithms such as DL, RL, and graph-based models, AI can identify complex patterns and generate novel insights that are otherwise inaccessible through traditional methodologies. Despite its transformative potential, the integration of AI into drug discovery faces several challenges. Ethical and regulatory considerations, such as data privacy, bias, and explainability, remain critical issues that must be addressed to ensure the responsible use of AI technologies. Looking ahead, the future of AI in drug discovery is promising. Continued advancements in AI methodologies, combined with the integration of emerging technologies such as quantum computing, synthetic biology, and high-throughput robotics, will further enhance its capabilities. In conclusion, AI is poised to fundamentally reshape drug discovery, enabling faster, more accurate, and cost-effective solutions to some of the most pressing challenges in healthcare. By embracing the potential of AI while addressing its limitations, researchers and stakeholders can harness this transformative technology to accelerate the development of life-saving therapies and improve global health outcomes.

References

Abil Z, Xiong X, Zhao H. Synthetic biology for therapeutic applications. Mol Pharm. 2015;12(2):322–31.

Aldahdooh J, et al. Using BERT to identify drug-target interactions from whole PubMed. BMC Bioinformatics. 2022;23(1):245.

Alidoost M, Wilson JL. Preclinical side effect prediction through pathway engineering of protein interaction network models. CPT Pharmacometrics Syst Pharmacol. 2024;13(7):1180–200.

Alizadehsani R, et al. Explainable artificial intelligence for drug discovery and development: a comprehensive survey. IEEE Access. 2024;12:35796–812.

Allen TC. Regulating artificial intelligence for a successful pathology future. Arch Pathol Lab Med. 2019;143(10):1175–9.

Amistadi S, et al. Functional restoration of a CFTR splicing mutation through RNA delivery of CRISPR adenine base editor. Mol Ther. 2023;31(6):1647–60.

Ars E, et al. Mutations affecting mRNA splicing are the most common molecular defects in patients with neurofibromatosis type 1. Hum Mol Genet. 2000;9(2):237–47.

Badwan BA, et al. Machine learning approaches to predict drug efficacy and toxicity in oncology. Cell Rep Methods. 2023;3(2):100413.

Bastarache L, Denny JC, Roden DM. Phenome-wide association studies. JAMA. 2022;327(1):75–6.

Bekes M, Langley DR, Crews CM. PROTAC targeted protein degraders: the past is prologue. Nat Rev Drug Discov. 2022;21(3):181–200.

Birkenbihl C, et al. Artificial intelligence-based clustering and characterization of Parkinson's disease trajectories. Sci Rep. 2023;13(1):2897.

Biswas N, Chakrabarti S. Artificial intelligence (AI)-based systems biology approaches in multi-omics data analysis of cancer. Front Oncol. 2020;10:588221.

Blevins SM, Bronze MS. Robert Koch and the 'golden age' of bacteriology. Int J Infect Dis. 2010;14(9):e744–51.

Brennan RJ. Target safety assessment: strategies and resources. Methods Mol Biol. 2017;1641:213–28.

Burska AN, et al. Gene expression analysis in RA: towards personalized medicine. Pharmacogenomics J. 2014;14(2):93–106.

Cava C, Bertoli G, Castiglioni I. In silico identification of drug target pathways in breast cancer subtypes using pathway cross-talk inhibition. J Transl Med. 2018;16(1):154.

Ceddia G, et al. Matrix factorization-based technique for drug repurposing predictions. IEEE J Biomed Health Inform. 2020;24(11):3162–72.

Chai K, et al. Application of machine learning and weighted gene co-expression network algorithm to explore the hub genes in the aging brain. Front Aging Neurosci. 2021;13:707165.

Chandak P, Huang K, Zitnik M. Building a knowledge graph to enable precision medicine. Sci Data. 2023;10(1):67.

Chang J, Wang S, Ling C, Qin Z, Zhao L. Gene-associated disease discovery powered by large language models. arXiv. 2024. https://doi.org/10.48550/arXiv.2401.09490.

Chen X, et al. MDHGI: matrix decomposition and heterogeneous graph inference for miRNA-disease association prediction. PLoS Comput Biol. 2018;14(8):e1006418.

Chen H, Cheng F, Li J. iDrug: integration of drug repositioning and drug-target prediction via cross-network embedding. PLoS Comput Biol. 2020;16(7):e1008040.

Chen DL, Cai JH, Wang CCN. Identification of key prognostic genes of triple negative breast cancer by LASSO-based machine learning and bioinformatics analysis. Genes (Basel). 2022;13(5):902.
Chen J, et al. Identification of the key immune cells and genes for the diagnostics and therapeutics of meningioma. World Neurosurg. 2023;176:e501–14.
Chen F, et al. Unmasking bias in artificial intelligence: a systematic review of bias detection and mitigation strategies in electronic health record-based models. J Am Med Inform Assoc. 2024;31(5):1172–83.
Cheng AC, et al. Structure-based maximal affinity model predicts small-molecule druggability. Nat Biotechnol. 2007;25(1):71–5.
Cobanoglu MC, et al. Predicting drug-target interactions using probabilistic matrix factorization. J Chem Inf Model. 2013;53(12):3399–409.
Collins FS, Morgan M, Patrinos A. The human genome project: lessons from large-scale biology. Science. 2003;300(5617):286–90.
Corbacioglu SK, Aksel G. Receiver operating characteristic curve analysis in diagnostic accuracy studies: a guide to interpreting the area under the curve value. Turk J Emerg Med. 2023;23(4):195–8.
Costa PR, Acencio ML, Lemke N. A machine learning approach for genome-wide prediction of morbid and druggable human genes based on systems-level data. BMC Genomics. 2010;11 Suppl 5(Suppl 5):S9.
Davies H, et al. Mutations of the BRAF gene in human cancer. Nature. 2002;417(6892):949–54.
DiMasi JA, Grabowski HG, Hansen RW. Innovation in the pharmaceutical industry: new estimates of R&D costs. J Health Econ. 2016;47:20–33.
Doan S, et al. Natural language processing in biomedicine: a unified system architecture overview. Methods Mol Biol. 2014;1168:275–94.
Domingos P. A few useful things to know about machine learning. Commun ACM. 2012;55(10):78.
Drew ED, Janes RW. Identification of a druggable binding pocket in the spike protein reveals a key site for existing drugs potentially capable of combating Covid-19 infectivity. BMC Mol Cell Biol. 2020;21(1):49.
Elmarakeby HA, et al. Biologically informed deep neural network for prostate cancer discovery. Nature. 2021;598(7880):348–52.
ElZarrad K. In: Roach S, editor. The role of artificial intelligence in clinical trial design and research with Dr. ElZarrad. Silver Spring: FDA; 2024.
Esteban-Medina M, et al. The mechanistic functional landscape of retinitis pigmentosa: a machine learning-driven approach to therapeutic target discovery. J Transl Med. 2024;22(1):139.
European Medicines Agency. Guideline on computerised systems and electronic data in clinical trials. Amsterdam: EMA; 2023.
Ewald CY, et al. TNIK's emerging role in cancer, metabolism, and age-related diseases. Trends Pharmacol Sci. 2024;45(6):478–89.
Farhud DD, Zokaei S. Ethical issues of artificial intelligence in medicine and healthcare. Iran J Public Health. 2021;50(11):i–v.
Food and Drug Administration, Health Canada, and Medicines and Healthcare products Regulatory Agency. Good machine learning practice for medical device development: guiding principles. Silver Spring: FDA; 2021.
Fukuzaki M, Seki M, Kashima H, Sese J. Side effect prediction using cooperative pathways. In: IEEE international conference on bioinformatics and biomedicine. Piscataway: IEEE; 2009. p. 142–7.
Fukuto K, Takagi T, Tian YS. Predicting the side effects of drugs using matrix factorization on spontaneous reporting database. Sci Rep. 2021;11(1):23942.
Gangwal A, et al. Current strategies to address data scarcity in artificial intelligence-based drug discovery: a comprehensive review. Comput Biol Med. 2024;179:108734.
Gao D, et al. A deep learning approach to identify gene targets of a therapeutic for human splicing disorders. Nat Commun. 2021;12(1):3332.

Gentili M, et al. Biological random walks: multi-omics integration for disease gene prioritization. Bioinformatics. 2022;38(17):4145–52.

Geraci J, et al. Machine learning hypothesis-generation for patient stratification and target discovery in rare disease: our experience with Open Science in ALS. Front Comput Neurosci. 2023;17:1199736.

Gharizadeh A, et al. HGTDR: advancing drug repurposing with heterogeneous graph transformers. Bioinformatics. 2024;40(7):btae349.

Godbole S, et al. Treating cardiovascular disease in the inflammatory setting of rheumatoid arthritis: an ongoing challenge. Biomedicine. 2024;12(7):1608.

Goeman JJ, et al. A global test for groups of genes: testing association with a clinical outcome. Bioinformatics. 2004;20(1):93–9.

Gonzalez ADD, et al. Applying BioBERT to extract germline gene-disease associations for building a knowledge graph from the biomedical literature. ICISDM '23: proceedings of the 2023 7th international conference on information system and data mining. 2023. pp. 37–42.

Grisoni F, et al. Combining generative artificial intelligence and on-chip synthesis for de novo drug design. Sci Adv. 2021;7(24):eabg3338.

Hanna M, et al. Ethical and bias considerations in artificial intelligence (AI)/machine learning. Mod Pathol. 2024;38:100686.

Hicks SA, et al. On evaluation metrics for medical applications of artificial intelligence. Sci Rep. 2022;12(1):5979.

High-Level Expert Group on AI. Ethics guidelines for trustworthy artificial intelligence. Brussels: European Commission; 2019.

Huang S, et al. Applications of support vector machine (SVM) learning in cancer genomics. Cancer Genomics Proteomics. 2018;15(1):41–51.

Hummel M, Meister R, Mansmann U. GlobalANCOVA: exploration and assessment of gene group effects. Bioinformatics. 2008;24(1):78–85.

Iman M, Arabnia HR, Rasheed K. A review of deep transfer learning and recent advancements. Technologies. 2023;11(2):40.

Isgut M, et al. Systematic review of advanced AI methods for improving healthcare data quality in post COVID-19 era. IEEE Rev Biomed Eng. 2023;16:53–69.

Jeon J, et al. A systematic approach to identify novel cancer drug targets using machine learning, inhibitor design and high-throughput screening. Genome Med. 2014;6(7):57.

Jiang L, et al. Identifying drug-target interactions via heterogeneous graph attention networks combined with cross-modal similarities. Brief Bioinform. 2022;23(2):bbac016.

Jolliffe IT, Cadima J. Principal component analysis: a review and recent developments. Philos Trans A Math Phys Eng Sci. 2016;374(2065):20150202.

Jumper J, et al. Highly accurate protein structure prediction with AlphaFold. Nature. 2021;596(7873):583–9.

Jussupow E, Spohrer K, Heinzl A. Identity threats as a reason for resistance to artificial intelligence: survey study with medical students and professionals. JMIR Form Res. 2022;6(3):e28750.

Kamya P, et al. PandaOmics: an AI-driven platform for therapeutic target and biomarker discovery. J Chem Inf Model. 2024;64(10):3961–9.

Kao PY, et al. Exploring the advantages of quantum generative adversarial networks in generative chemistry. J Chem Inf Model. 2023;63(11):3307–18.

Kc K, et al. Predicting biomedical interactions with higher-order graph convolutional networks. IEEE/ACM Trans Comput Biol Bioinform. 2022;19(2):676–87.

Khatri P, Sirota M, Butte AJ. Ten years of pathway analysis: current approaches and outstanding challenges. PLoS Comput Biol. 2012;8(2):e1002375.

Kobak D, Berens P. The art of using t-SNE for single-cell transcriptomics. Nat Commun. 2019;10(1):5416.

Law JN, et al. Interpretable network propagation with application to expanding the repertoire of human proteins that interact with SARS-CoV-2. Gigascience. 2021;10(12):giab082.

Lee S, et al. Pathway-based approach using hierarchical components of collapsed rare variants. Bioinformatics. 2016;32(17):i586–94.
Lee J, et al. BioBERT: a pre-trained biomedical language representation model for biomedical text mining. Bioinformatics. 2020;36(4):1234–40.
Li H, et al. Decoding the mitochondrial connection: development and validation of biomarkers for classifying and treating systemic lupus erythematosus through bioinformatics and machine learning. BMC Rheumatol. 2023a;7(1):44.
Li Y, Hostallero DE, Emad A. Interpretable deep learning architectures for improving drug response prediction performance: myth or reality? Bioinformatics. 2023b;39(6):btad390.
Liang B, et al. Risk stratification and pathway analysis based on graph neural network and interpretable algorithm. BMC Bioinformatics. 2022;23(1):394.
Libbrecht MW, Noble WS. Machine learning applications in genetics and genomics. Nat Rev Genet. 2015;16(6):321–32.
Liu B, Tsoumakas G, Optimizing area under the curve measures via matrix factorization for predicting drug-target interaction with multiple similarities. arXiv. 2021. https://doi.org/10.48550/arXiv.2105.01545
Liu A, et al. From expression footprints to causal pathways: contextualizing large signaling networks with CARNIVAL. NPJ Syst Biol Appl. 2019;5:40.
Liu XY, et al. LogSum + L(2) penalized logistic regression model for biomarker selection and cancer classification. Sci Rep. 2020;10(1):22125.
Liu J, et al. In silico off-target profiling for enhanced drug safety assessment. Acta Pharm Sin B. 2024a;14(7):2927–41.
Liu BHM, et al. Utilizing AI for the identification and validation of novel therapeutic targets and repurposed drugs for endometriosis. Adv Sci (Weinh). 2024b;12:e2406565.
Long X, et al. AI-enabled cancer target prioritization with optimal profiles balancing novelty, confidence and commercial tractability. Future Med AI. 2024;2(1) https://doi.org/10.2217/fmai-2023-0019.
Lorson CL, et al. A single nucleotide in the SMN gene regulates splicing and is responsible for spinal muscular atrophy. Proc Natl Acad Sci USA. 1999;96(11):6307–11.
Lou LL, Martin JC. Selected thoughts on hydrophobicity in drug design. Molecules. 2021;26(4):875.
Lu RM, et al. Development of therapeutic antibodies for the treatment of diseases. J Biomed Sci. 2020;27(1):1.
Luo R, et al. BioGPT: generative pre-trained transformer for biomedical text generation and mining. Brief Bioinform. 2022;23(6):bbac409.
MacLean F. Knowledge graphs and their applications in drug discovery. Expert Opin Drug Discov. 2021;16(9):1057–69.
Madhukar NS, et al. A Bayesian machine learning approach for drug target identification using diverse data types. Nat Commun. 2019;10(1):5221.
Majno G. The healing hand: man and wound in the ancient world. Cambridge: Harvard University Press; 1975. p. 571.
Mamoshina P, et al. Machine learning on human muscle transcriptomic data for biomarker discovery and tissue-specific drug target identification. Front Genet. 2018;9:242.
McHugh ML. Interrater reliability: the kappa statistic. Biochem Med (Zagreb). 2012;22(3):276–82.
McInnes L, et al. UMAP: uniform manifold approximation and projection. J Open Sour Softw. 2018;3(29):861.
Medicines & Healthcare Products Regulatory Agency. Software and AI as a medical device change programme roadmap. London: MHRA; 2023.
Moffat JG, Rudolph J, Bailey D. Phenotypic screening in cancer drug discovery – past, present and future. Nat Rev Drug Discov. 2014;13(8):588–602.
Moradi S, Tapak L, Afshar S. Identification of novel noninvasive diagnostics biomarkers in the Parkinson's diseases and improving the disease classification using support vector machine. Biomed Res Int. 2022;2022:5009892.

Mumau MD, et al. Identifying and targeting TNF signaling in idiopathic multicentric Castleman's disease. N Engl J Med. 2025;392(6):616–8.
Narganes-Carlon D, Crowther DJ, Pearson ER. A publication-wide association study (PWAS), historical language models to prioritise novel therapeutic drug targets. Sci Rep. 2023;13(1):8366.
Nguyen C, et al. Definition of high-risk type 1 diabetes HLA-DR and HLA-DQ types using only three single nucleotide polymorphisms. Diabetes. 2013;62(6):2135–40.
Nielsen F. Hierarchical clustering, in introduction to HPC with MPI for data science. Cham: Springer; 2016.
Niessing B, et al. Automated CRISPR/Cas9-based genome editing of human pluripotent stem cells using the StemCellFactory. Front Bioeng Biotechnol. 2024;12:1459273.
O'Malley CD. Andreas Vesalius of Brussels, 1514–1564. Berkeley: University of California Press; 1964. p. 480.
Olsen A, et al. Identification of dual-purpose therapeutic targets implicated in aging and glioblastoma multiforme using PandaOmics – an AI-enabled biological target discovery platform. Aging (Albany NY). 2023;15(8):2863–76.
Park MY, Hastie T. Penalized logistic regression for detecting gene interactions. Biostatistics. 2008;9(1):30–50.
Park C, Kim B, Park T. DeepHisCoM: deep learning pathway analysis using hierarchical structural component models. Brief Bioinform. 2022;23(5):bbac171.
Patel L, et al. Machine learning methods in drug discovery. Molecules. 2020;25(22):5277.
Pathmanathan P, et al. Credibility assessment of in silico clinical trials for medical devices. PLoS Comput Biol. 2024;20(8):e1012289.
Patil R, et al. Optimized hydrophobic interactions and hydrogen bonding at the target-ligand interface leads the pathways of drug-designing. PLoS One. 2010;5(8):e12029.
Perakakis N, et al. Omics, big data and machine learning as tools to propel understanding of biological mechanisms and to discover novel diagnostics and therapeutics. Metabolism. 2018;87:A1–9.
Petegrosso R, et al. Transfer learning across ontologies for phenome-genome association prediction. Bioinformatics. 2017;33(4):529–36.
Pham TH, et al. A deep learning framework for high-throughput mechanism-driven phenotype compound screening and its application to COVID-19 drug repurposing. Nat Mach Intell. 2021;3(3):247–57.
Pham TCP, et al. TNIK is a conserved regulator of glucose and lipid metabolism in obesity. Sci Adv. 2023;9(32):eadf7119.
Porter R. The greatest benefit to mankind: a medical history of humanity from antiquity to the present. London: Fontana Press; 1999. p. 833.
Prakash J, et al. Unsupervised machine learning to identify separable clinical Alzheimer's disease sub-populations. Brain Sci. 2021;11(8):977.
Prior M, et al. Back to the future with phenotypic screening. ACS Chem Neurosci. 2014;5(7):503–13.
Pun FW, et al. Identification of therapeutic targets for amyotrophic lateral sclerosis using PandaOmics – an AI-enabled biological target discovery platform. Front Aging Neurosci. 2022a;14:914017.
Pun FW, et al. Hallmarks of aging-based dual-purpose disease and age-associated targets predicted using PandaOmics AI-powered discovery engine. Aging (Albany NY). 2022b;14(6):2475–506.
Pun FW, Ozerov IV, Zhavoronkov A. AI-powered therapeutic target discovery. Trends Pharmacol Sci. 2023a;44(9):561–72.
Pun FW, et al. A comprehensive AI-driven analysis of large-scale omic datasets reveals novel dual-purpose targets for the treatment of cancer and aging. Aging Cell. 2023b;22(12):e14017.
Raghu G, et al. Incidence and prevalence of idiopathic pulmonary fibrosis. Am J Respir Crit Care Med. 2006;174(7):810–6.
Rahit K, et al. GPAD: a natural language processing-based application to extract the gene-disease association discovery information from OMIM. BMC Bioinformatics. 2024;25(1):84.

Rainio O, Teuho J, Klen R. Evaluation metrics and statistical tests for machine learning. Sci Rep. 2024;14(1):6086.

Ramos EK, et al. Machine learning-assisted elucidation of CD81-CD44 interactions in promoting cancer stemness and extracellular vesicle integrity. elife. 2022;11:e82669.

Reel PS, et al. Using machine learning approaches for multi-omics data analysis: a review. Biotechnol Adv. 2021;49:107739.

Ren F, et al. A small-molecule TNIK inhibitor targets fibrosis in preclinical and clinical models. Nat Biotechnol. 2025;43(1):63–75.

Rodriguez S, et al. Machine learning identifies candidates for drug repurposing in Alzheimer's disease. Nat Commun. 2021;12(1):1033.

Rodriguez-Perez R, Bajorath J. Interpretation of machine learning models using shapley values: application to compound potency and multi-target activity predictions. J Comput Aided Mol Des. 2020;34(10):1013–26.

Sabatine MS. PCSK9 inhibitors: clinical evidence and implementation. Nat Rev Cardiol. 2019;16(3):155–65.

Sandra Kumi RKL, Deters R. A blockchain-based platform for data management and sharing, vol. 203. Procedia Comput Sci; 2022. p. 95–102.

Satija R, et al. Spatial reconstruction of single-cell gene expression data. Nat Biotechnol. 2015;33(5):495–502.

Sayed N, et al. An inflammatory aging clock (iAge) based on deep learning tracks multimorbidity, immunosenescence, frailty and cardiovascular aging. Nat Aging. 2021;1:598–615.

Sayed N, et al. Gene therapy: comprehensive overview and therapeutic applications. Life Sci. 2022;294:120375.

Schlander M, et al. How much does it cost to research and develop a new drug? A systematic review and assessment. Pharmacoeconomics. 2021;39(11):1243–69.

Schneider G. Automating drug discovery. Nat Rev Drug Discov. 2018;17(2):97–113.

Shah HA, et al. Review of machine learning methods for the prediction and reconstruction of metabolic pathways. Front Mol Biosci. 2021;8:634141.

Shi W, et al. COL11A1 as an novel biomarker for breast cancer with machine learning and immunohistochemistry validation. Front Immunol. 2022;13:937125.

Slamon DJ, et al. Human breast cancer: correlation of relapse and survival with amplification of the HER-2/neu oncogene. Science. 1987;235(4785):177–82.

Spiga O, et al. Machine learning application for patient stratification and phenotype/genotype investigation in a rare disease. Brief Bioinform. 2021;22(5):bbaa434.

Stank A, et al. Protein binding pocket dynamics. Acc Chem Res. 2016;49(5):809–15.

Steinley D. K-means clustering: a half-century synthesis. Br J Math Stat Psychol. 2006;59(Pt 1):1–34.

Stein-O'Brien GL, et al. Enter the matrix: factorization uncovers knowledge from omics. Trends Genet. 2018;34(10):790–805.

Sterner RC, Sterner RM. CAR-T cell therapy: current limitations and potential strategies. Blood Cancer J. 2021;11(4):69.

Sun D, et al. Why 90% of clinical drug development fails and how to improve it? Acta Pharm Sin B. 2022;12(7):3049–62.

Takahashi G, et al. High-throughput robotic isolation of human iPS cell clones reveals frequent homozygous induction of identical genetic manipulations by CRISPR-Cas9. bioRxiv. 2024. https://doi.org/10.1101/2024.09.18.613641.

Talevi A, Bellera CL. Challenges and opportunities with drug repurposing: finding strategies to find alternative uses of therapeutics. Expert Opin Drug Discov. 2020;15(4):397–401.

Tang Z, et al. GEPIA: a web server for cancer and normal gene expression profiling and interactive analyses. Nucleic Acids Res. 2017;45(W1):W98–W102.

Tang Q, et al. AI-driven robotics laboratory identifies pharmacological TNIK inhibition as a potent senomorphic agent. Aging Dis. 2025. https://doi.org/10.14336/AD.2024.1492

Tebani A, et al. Deep next-generation proteomics and network analysis reveal systemic and tissue-specific patterns in Fabry disease. Transl Res. 2023;258:47–59.

Thafar MA, et al. DTiGEMS+: drug-target interaction prediction using graph embedding, graph mining, and similarity-based techniques. J Cheminform. 2020;12(1):44.
Torng W, Altman RB. Graph convolutional neural networks for predicting drug-target interactions. J Chem Inf Model. 2019;59(10):4131–49.
Torrisi M, Pollastri G, Le Q. Deep learning methods in protein structure prediction. Comput Struct Biotechnol J. 2020;18:1301–10.
Treloar NJ, et al. Deep reinforcement learning for optimal experimental design in biology. PLoS Comput Biol. 2022;18(11):e1010695.
Uffelmann E, et al. Genome-wide association studies. Nat Rev Methods Primers. 2021;1:59.
Urban A, et al. Precious1GPT: multimodal transformer-based transfer learning for aging clock development and feature importance analysis for aging and age-related disease target discovery. Aging (Albany NY). 2023;15(11):4649–66.
Vamathevan J, et al. Applications of machine learning in drug discovery and development. Nat Rev Drug Discov. 2019;18(6):463–77.
Van de Sande B, et al. Applications of single-cell RNA sequencing in drug discovery and development. Nat Rev Drug Discov. 2023;22(6):496–520.
van Drongelen V, et al. HLA-DRB1 allelic epitopes that associate with autoimmune disease risk or protection activate reciprocal macrophage polarization. Sci Rep. 2021;11(1):2599.
Volkamer A, et al. Combining global and local measures for structure-based druggability predictions. J Chem Inf Model. 2012;52(2):360–72.
Wang E, Schmidgall S, Jaeger PF, Zhang F, Pilgrim R, Matias Y, Barral J, Fleet D, Azizi S. TxGemma: efficient and agentic LLMs for therapeutics. arXiv. 2025. https://doi.org/10.48550/arXiv.2504.06196.
Wang W, Yang S, Li J. Drug target predictions based on heterogeneous graph inference. Pac Symp Biocomput. 2013;2013:53–64.
Wang H, et al. AI-driver: an ensemble method for identifying driver mutations in personal cancer genomes. NAR Genom Bioinform. 2020;2(4):lqaa084.
Wang T, et al. MOGONET integrates multi-omics data using graph convolutional networks allowing patient classification and biomarker identification. Nat Commun. 2021;12(1):3445.
Wang Y, et al. Discovery of tetrahydropyrazolopyrazine derivatives as potent and selective MYT1 inhibitors for the treatment of cancer. J Med Chem. 2024;67(1):420–32.
Wilkinson MD, et al. The FAIR guiding principles for scientific data management and stewardship. Sci Data. 2016;3:160018.
Williamson HF, et al. Data management challenges for artificial intelligence in plant and agricultural research. F1000Res. 2021;10:324.
Wilson JL, et al. PathFX provides mechanistic insights into drug efficacy and safety for regulatory review and therapeutic development. PLoS Comput Biol. 2018;14(12):e1006614.
Won S, et al. Evaluation of penalized and nonpenalized methods for disease prediction with large-scale genetic data. Biomed Res Int. 2015;2015:605891.
World Health Organization. Ageing and health 2024. 1 Oct 2024. Available from: https://www.who.int/news-room/fact-sheets/detail/ageing-and-health
Wu Y, et al. Drug repositioning for Alzheimer's disease with transfer learning. arXiv. 2022. https://doi.org/10.48550/arXiv.2210.15271.
Wysocki T, Olesinska M, Paradowska-Gorycka A. Current understanding of an emerging role of HLA-DRB1 gene in rheumatoid arthritis-from research to clinical practice. Cells. 2020;9(5):1127.
Xie W, et al. DeepSeq2Drug: an expandable ensemble end-to-end anti-viral drug repurposing benchmark framework by multi-modal embeddings and transfer learning. Comput Biol Med. 2024;175:108487.
Xing H, et al. RDBridge: a knowledge graph of rare diseases based on large-scale text mining. Bioinformatics. 2023;39(7):btad440.
Xu L, Ru X, Song R. Application of machine learning for drug-target interaction prediction. Front Genet. 2021;12:680117.

Yang IV, et al. Relationship of DNA methylation and gene expression in idiopathic pulmonary fibrosis. Am J Respir Crit Care Med. 2014;190(11):1263–72.

Ye Q, et al. A unified drug-target interaction prediction framework based on knowledge graph and recommendation system. Nat Commun. 2021;12(1):6775.

You Y, et al. Artificial intelligence in cancer target identification and drug discovery. Signal Transduct Target Ther. 2022;7(1):156.

Zagirova D, et al. Biomedical generative pre-trained based transformer language model for age-related disease target discovery. Aging (Albany NY). 2023;15(18):9293–309.

Zhang S, et al. Review of single-cell RNA-seq data clustering for cell-type identification and characterization. RNA. 2023;29(5):517–30.

Zhao BW, et al. HINGRL: predicting drug-disease associations with graph representation learning on heterogeneous information networks. Brief Bioinform. 2022;23(1):bbab515.

Zhavoronkov A, et al. Deep learning enables rapid identification of potent DDR1 kinase inhibitors. Nat Biotechnol. 2019;37(9):1038–40.

Zheng Y, et al. A novel deep learning framework for interpretable drug-target interaction prediction with attention and multi-task mechanism. In: Database systems for advanced applications. Cham: Springer; 2023. p. 336–52.

Zhi W, et al. Discovery and validation of serum protein changes in type 1 diabetes patients using high throughput two dimensional liquid chromatography-mass spectrometry and immunoassays. Mol Cell Proteomics. 2011;10(11):M111 012203.

Zhou J, et al. Identification of aging-related biomarkers and immune infiltration characteristics in osteoarthritis based on bioinformatics analysis and machine learning. Front Immunol. 2023;14:1168780.

Zhuang F, Qi Z, Duan K, Xi D, Zhu Y, Zhu H, Xiong H, He Q. A comprehensive survey on transfer learning. arXiv. 2020. https://doi.org/10.48550/arXiv.1911.02685

Artificial Intelligence in Structure-Based Drug Design

Emanuele Falbo, Pietro Delre, Carmen Cerchia, and Antonio Lavecchia

Abstract Structure-based drug design (SBDD) is a transformative methodology in drug discovery that utilizes the three-dimensional (3D) structure of biological targets, primarily proteins, to develop new therapeutic agents. The process involves identifying a target protein, determining its structure using experimental methods like X-ray crystallography or NMR spectroscopy, and designing molecules to modulate its activity. Recent advancements in artificial intelligence (AI), particularly in machine learning (ML) and deep learning (DL), have significantly enhanced SBDD by automating and refining key steps such as virtual screening, hit optimization, and binding affinity prediction. Techniques such as convolutional neural networks (CNNs), which have revolutionized computer vision, are now being applied to analyze protein–ligand complexes with remarkable precision, accelerating tasks like pose prediction and virtual screening. By integrating AI into SBDD workflows, researchers can achieve higher success rates in identifying lead compounds, minimize toxicity risks, and streamline the drug discovery pipeline. This synergy between SBDD and AI has already contributed to the discovery of clinically approved drugs and promises to redefine the development of new, targeted, and efficient medicines.

Keywords Structure-based drug design (SBDD) · Artificial intelligence (AI) in drug discovery · Convolutional neural networks (CNNs) · Virtual screening · Binding affinity prediction · Computational drug design

---

E. Falbo · P. Delre · C. Cerchia · A. Lavecchia (✉)
"Drug Discovery Laboratory", Dipartimento di Farmacia, Università degli Studi di Napoli "Federico II", Naples, Italy
e-mail: emanuele.falbo@unina.it; pietro.delre@unina.it; carmen.cerchia@unina.it; antonio.lavecchia@unina.it

A. Lavecchia (ed.), *Applied Artificial Intelligence for Drug Discovery*,
https://doi.org/10.1007/978-3-031-98022-0_4

1 Introduction

1.1 Overview of Structure-Based Drug Design

Structure-based drug design (SBDD) is a powerful and versatile approach for discovering new drugs, offering the potential to accelerate the development of innovative treatments for a wide range of diseases. At its core, SBDD is grounded in the principle that understanding the three-dimensional (3D) structure of a target protein or enzyme is crucial for designing compounds that bind with high specificity and affinity. By utilizing this structural knowledge, drugs can be tailored to interact precisely with specific protein targets, thereby minimizing unintended side effects and enhancing therapeutic efficacy. SBDD offers several advantages over traditional drug discovery approaches, including increased potency and selectivity, reduced off-target effects, and shorter development timelines and costs. Numerous successful drugs have been developed through SBDD, including HIV protease inhibitors and the anticancer drug (Ghosh et al. 2006). SBDD has a broad range of applications in drug discovery, including the development of therapies for cancer, infectious diseases, and metabolic disorders. For example, SBDD has been used to develop treatments for metabolic conditions such as diabetes and obesity by focusing on key enzymes involved in glucose and lipid metabolism (Kaur et al. 2021).

The process of SBDD generally involves several key steps, including target identification, protein structure determination, and virtual screening. Target identification entails selecting a protein or enzyme implicated in a disease process, typically through a combination of biological and computational methods. Once a target is identified, the next step is determining its 3D structure, often achieved using techniques such as X-ray crystallography or nuclear magnetic resonance (NMR) spectroscopy (Kumar et al. 2020). With the protein structure in hand, molecular docking and virtual screening are employed to identify compounds that bind with high specificity and affinity to the target, i.e., binding affinity prediction. To further optimize the binding interactions between the drug candidate and the target protein, molecular dynamics (MD) simulations can be used. In recent years, machine learning (ML) and artificial intelligence (AI) have achieved significant success in fields like computer vision and natural language processing. Building on this progress, there have been notable advancements in applying these technologies to structure-based potency prediction. This chapter reviews the latest developments in this area and offers a comprehensive introduction to the key considerations involved in developing such models.

2 Historical Achievement in SBDD

The origins of SBDD can be traced to the late 1950s and early 1960s with the groundbreaking determination of the 3D structures of myoglobin and hemoglobin using X-ray crystallography (Kendrew et al. 1958). This pivotal advancement in structural biology provided researchers with invaluable insights into molecular interactions, laying the foundation for the development of SBDD. SBDD employs computational techniques to leverage the 3D structure of target proteins to design small molecules capable of modulating the proteins' activities. The primary objective is to identify compounds that bind with high affinity and specificity, enabling the development of novel drugs to treat a range of diseases. Early SBDD efforts relied on manual methods to design molecules, but advancements in computational tools and algorithms have since enabled high-throughput virtual screening of extensive compound libraries, leading to the discovery of drugs such as vemurafenib, which targets the BRAF V600E mutation in melanoma. Recent advances in computational methods, including ML and AI, have significantly accelerated the drug discovery process and improved prediction accuracy. These innovations have expanded SBDD's impact, leading to the development of treatments for conditions such as cancer, infectious diseases, and neurological disorders. The ongoing evolution of computational techniques and a deeper understanding of protein structures have driven the creation of increasingly potent and specific medications. Historically, the development of virtual screening and docking methodologies during the 1970s and 1980s was facilitated by improvements in computer technology and molecular modeling strategies. These techniques enabled researchers to screen vast chemical libraries for potential drug candidates. In their early stages, these approaches were constrained by limited computational power and often yielded inaccurate predictions. However, advancements in hardware, algorithms, and scoring functions have rendered virtual screening and docking methods more reliable and precise. Further progress in SBDD occurred in the 1990s and 2000s with the emergence of the fragment-based drug design (FBDD) (Reddy et al. 2018). This approach involves identifying low-affinity compounds or fragments that bind to a target protein or enzyme, followed by iterative cycles of optimization to develop high-affinity drug candidates. FBDD has proven to be an effective strategy, complementing traditional SBDD techniques and driving the discovery of new therapeutic agents. In summary, the historical milestones and ongoing advancements in SBDD reflect its critical role in modern drug discovery. Continued innovation in computational methodologies promises to enhance the efficiency and specificity of drug design, contributing to the creation of safer and more effective treatments for a wide array of diseases (Fig. 1).

3 Steps in SBDD

3.1 Identification and Validation

The first and most critical step in SBDD is the identification and validation of a biological target, typically a protein or enzyme that plays a key role in a disease process. The selected target must demonstrate therapeutic potential, ensuring that modulating its activity can lead to beneficial outcomes. Target identification involves

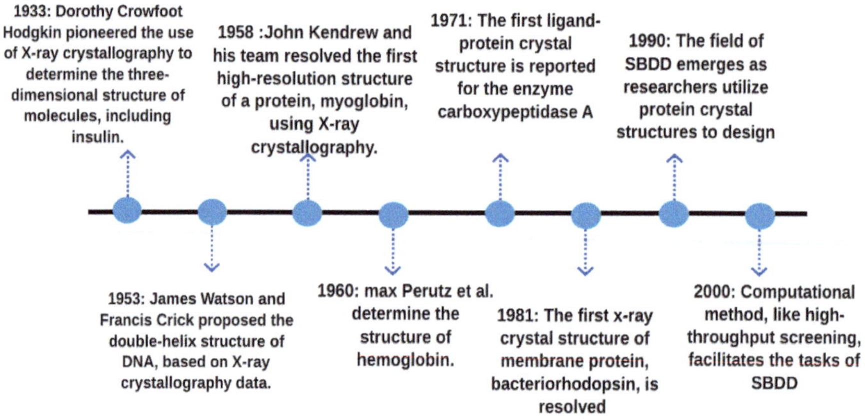

Fig. 1 Chronology of key milestones in the history of drug design that have led to the need for computational SBDD models

using a combination of experimental and computational techniques. Experimental methods include genetic screening, high-throughput screening, and biochemical assays to pinpoint proteins or enzymes associated with disease pathways. Computational approaches, such as molecular docking, homology modeling, and MD simulations, provide additional insights into potential targets and their properties. Once a target is identified, it must be rigorously validated to confirm its suitability for drug development. This involves demonstrating that the target is critical to the disease mechanism, that modulating the target's activity produces a therapeutic effect, and that it can be selectively targeted without causing toxic side effects. Validation typically combines in vitro methods, such as cell-based and biochemical assays, with in vivo approaches, including animal studies and clinical trials (Srinivasu et al. 2014). After the target is validated, the next step involves obtaining detailed structural information. Techniques such as X-ray crystallography and NMR spectroscopy are commonly used to analyze the 3D structure of the target. These structural data provide critical insights into features such as the active site and other binding regions, which are essential for drug design. Using the structural information, researchers design drug molecules that bind specifically and with high affinity to the target. Computational tools like molecular docking are employed to predict how different compounds interact with the target and to guide the optimization of drug candidates. Drug candidates designed through this process are then synthesized and evaluated for their biological activity. These evaluations assess key properties, including potency, i.e., the strength of interaction with the target, selectivity, i.e., the ability to bind specifically to the target without off-target effects, and pharmacokinetics, which indicates how the compound is absorbed, distributed, metabolized, and excreted. Promising compounds are further refined using SBDD techniques to optimize these properties. Optimized compounds proceed to preclinical and clinical evaluations to assess their safety and efficacy in humans. In preclinical studies, animal models are used to evaluate therapeutic potential and safety. Compounds that show favorable results advance to clinical trials, where their performance in humans is tested. This comprehensive process, spanning target identification and validation, structural analysis, compound design, optimization, and clinical testing, forms the foundation of successful SBDD, ultimately enabling the development of safe and effective therapeutic agents.

3.2 Determination of the Structure of the Target Protein

After the identification and validation of a target protein, the next step in SBDD is to determine its 3D structure. This is a critical process, as the protein's structure provides key insights into its function and helps identify potential binding sites for drug molecules. Several advanced techniques are available for protein structure determination, including X-ray crystallography, NMR spectroscopy, and cryo-electron microscopy (cryo-EM). Among these, X-ray crystallography remains the most widely used method in structure-based drug development. This technique involves crystallizing the protein and exposing it to X-rays, which generate a diffraction pattern as they interact with the crystal's atoms. The positions of the atoms within the protein are then inferred from this diffraction pattern, yielding a detailed three-dimensional model of the protein. Once the structure is resolved, computational tools are employed to identify potential drug-binding sites. Techniques such as molecular docking allow researchers to virtually screen large libraries of drug-like molecules to identify those that can interact with the protein (Fan et al. 2012; Maveyraud and Mourey 2020). Additionally, MD simulations are used to refine predictions, offering insights into the protein's dynamic behavior and its interactions with drug candidates over time. The target protein's structure not only facilitates the identification of binding sites but also guides the design and optimization of therapeutic compounds. Using computational methods such as computer-aided drug design (CADD), researchers iteratively create and refine molecules to enhance their affinity, selectivity, and interaction with the target protein. Techniques like molecular docking, structure-based optimization, and MD simulations are integral to this process (Law et al. 2009). With the structural insights provided by the protein model, small-molecule drugs can be designed to selectively bind and modulate the protein's activity. By leveraging the protein's 3D structure, researchers can optimize drug properties to enhance efficacy, specificity, and safety, increasing the likelihood of successful outcomes in clinical trials. In summary, determining the 3D structure of the target protein is a pivotal step in SBDD. It provides essential information about the protein's function, enables the identification of potential binding sites, and facilitates the rational design and optimization of therapeutic compounds with high precision. This foundational step significantly enhances the probability of developing safe and effective drugs.

3.3 Virtual Screening

Experimental screening of compounds for binding is both costly and time-consuming, with hit rates often being extremely low. Virtual screening offers a way to enhance the efficiency and effectiveness of these experimental screens. Specifically, it can be used to refine the compound library for experimental testing and guide the optimization of compounds. Virtual screening is fast and inexpensive, enabling the assessment of far more compounds than can be evaluated through experimental methods. It has been shown to enrich active compounds in subsequent experimental screens, contributing to the success of many drug discovery projects (Fan et al. 2012; Kumar et al. 2020; Srinivasu et al. 2014).

However, virtual screening has certain limitations compared to experimental approaches. It requires prior knowledge of the factors that contribute to the desired biological response, such as binding affinity to a protein target. Furthermore, the accuracy of virtual screening methods is not always sufficient to draw definitive conclusions. False negatives can result in the oversight of promising compounds, while false positives remain prevalent among the highest-scoring compounds, even after achieving significant enrichment. As a result, experimental validation is still necessary to confirm a hit. There are two primary computational approaches to virtual screening: (i) ligand-based virtual screening (LBVS) and (ii) structure-based virtual screening (SBVS). LBVS relies solely on information about small-molecule compounds, without requiring explicit knowledge of the protein target. By utilizing a set of ligands with known binding properties, i.e., active or inactive, LBVS identifies features within these compounds that can predict whether a new molecule will bind to the same target. The methodologies for LBVS can be broadly categorized into pharmacophore modeling, chemical similarity searches, and ML approaches (Wu et al. 2018). One of the key advantages of LBVS is that it can be applied even in the absence of a protein target structure or knowledge of the binding mode. However, a significant limitation of LBVS is that the models are specific to a particular biological response and cannot be easily generalized to new targets.

In contrast, SBVS utilizes information about the protein target, typically in the form of a 3D structure, to predict whether a compound will bind to it. Unlike LBVS, SBVS can be employed even before experimental binding data for the target are available. SBVS is a multi-step process, with a typical pipeline outlined as follows. First, a properly prepared 3D structure of the protein target is required. This is ideally determined experimentally, for example through X-ray crystallography, though an alternative approach is to construct a model of the protein from its amino acid sequence. Once the 3D structure is available, the next step is to identify the binding pocket regions on the target protein where ligands can bind. These binding pockets can be determined experimentally or using computational methods. Next, potential ligands are docked into these binding pockets. Docking involves two main steps: first, predicting possible binding modes (or poses) of the ligand within the protein active site; and second, ranking these poses to identify the most likely representation of the true binding mode, i.e., pose prediction. Traditionally, pose ranking is achieved by estimating binding affinity, e.g., using scoring values from Glide (Friesner et al. 2004, 2006), although ML methods are also increasingly used to train scoring functions for this task. SBVS evaluates one or more potential poses for a protein–ligand complex to assess the likelihood that the compound will bind to the target based on the generated poses (Hochuli et al. 2018; Ragoza et al. 2017; Sunseri et al. 2019). In the following discussion, we will focus on the design and development of scoring functions for this final step rather than on the docking protocol itself.

3.4 Evaluation Metrics

Ensuring that performance metrics align with the goals of the screening process is essential when building models, as model performance is inherently optimized to these metrics. In virtual screening, the primary objective is to prioritize binders over nonbinders, focusing on compounds with the highest scores. This prioritization is critical because, in most cases, only a small subset of the screened molecules can be experimentally tested. Receiver-operating characteristic (ROC) curves are widely used for model evaluation, with the area under the ROC curve (AUC-ROC) serving as a key scoring metric. ROC curves plot the true positive rate (TPR) on the y-axis against the false positive rate (FPR) on the x-axis. The AUC-ROC is straightforward to interpret, representing the probability that a randomly chosen active compound is ranked higher than a randomly chosen inactive compound. While ROC curves are informative, they assign equal importance across all test cases. As a result, many studies use metrics that emphasize the early part of the curve, such as BEDROC, pROC, or RIE. However, these metrics introduce additional parameters and are often more challenging to interpret. Another commonly used measure is ROC enrichment, which calculates the ratio of TPR to FPR at specific FPR thresholds, such as 0.5%, 1%, 2%, and 5%. The maximum possible ROC enrichment depends on the FPR threshold (e.g., at an FPR of 2%, the maximum enrichment is 50), while random performance yields an expected enrichment factor of 1 at any FPR threshold. Unlike global metrics such as AUC-ROC, enrichment factors are local metrics, evaluated at specific FPR thresholds rather than across the entire curve. Precision-recall curves (PRCs) offer an alternative global evaluation method that can be advantageous in situations with significant class imbalance. PRCs plot precision on the y-axis against recall (equivalent to TPR) on the x-axis. As previous studies have noted, ROC curves and PRCs are correlated but behave differently under conditions of high-class imbalance. PRCs place greater emphasis on the low-recall region, which is particularly relevant in datasets where positive samples are rare. In such cases, false positives have a greater impact on precision than on FPR, amplifying the differences between PRCs and ROC curves. Given the extreme class imbalance in virtual screening datasets, selecting the appropriate metric depends on the experimental context. Recent publications (Wu et al. 2018) recommend using precision-recall metrics as the primary evaluation method in such scenarios. Specifically, they propose using the area under the PRC (AUC-PRC) for datasets with positive rates below 2%, while AUC-ROC is more suitable for datasets with higher positive rates. It is worth noting that the expected AUC-ROC for a random classifier is 0.5, whereas the expected AUC-PRC is directly proportional to the class imbalance. For instance, if there are 50 decoys for every active compound, the expected AUC-PRC for a random classifier would be 0.02.

3.5 Docking Score Function

Traditional approaches have typically relied on experimental data to parameterize a physically inspired function (Eldridge et al. 1997; Friesner et al. 2004, 2006; Gohlke et al. 2000; Trott and Olson 2010). These functions are then used to guide the search for the binding mode and assess the strength of interactions between the protein and ligand, often providing an estimate of binding affinity. Scoring functions used in docking protocols can generally be classified into four categories based on the information and methods employed: force-field, knowledge-based, empirical, and hybrid.

Force-field scoring functions estimate the change in free energy upon binding by summing the terms of a molecular mechanics force field, such as van der Waals and electrostatic interactions between the protein and ligand. Additional contributions, such as solvent effects on exposed hydrophilic and hydrophobic atoms, as well as intramolecular (or strain) energies of the binding partners, are often included. This approach has the advantage of relying on physical models that do not depend on experimental data, but it also has the disadvantage of being constrained by these model assumptions. Moreover, commonly used force field might not be sufficient to model novel compounds, which often needs to be parametrized with ad hoc solutions (Allen et al. 2018; Falbo and Lavecchia 2024; Li and Merz 2016; Wang et al. 2013). Knowledge-based scoring functions, also known as statistical potentials, are derived from the distribution of pairwise distances between atoms (or other features) observed in a database of protein structures, such as the Protein Data Bank (Berman et al. 2000). These distributions are converted into energy functions that describe the preferred geometries of atoms (Sippl 1990). Empirical scoring functions focus on reproducing experimental affinity data. They consist of a weighted sum of uncorrelated, physically meaningful terms, such as the number of hydrogen bonds, the molecular weight of the ligand, or entropic terms related to the ligand's size and flexibility. The weights for these features are either determined from experimental binding data for known protein–ligand complexes or manually assigned (Pason and Sotriffer 2016). Hybrid scoring functions combine elements of these different approaches. A common feature across all these methods is the assumption of a specific functional relationship between the input features and the predicted binding affinity. While this makes these techniques interpretable, the rigid functional forms inherently limit their ability to capture complex, non-linear relationships and restrict them to our current understanding of intermolecular and intramolecular interactions.

4 Dataset for Developing New Docking Scoring Functions Using AI

Numerous databases are available for training and testing virtual screening methods, and the most suitable choice depends on the specific goals of a project (Table 1). A common objective, especially within the academic research community, is the development of universal scoring functions. These scoring functions aim to provide accurate predictions across a wide range of proteins without being tailored to specific families or groups. To support this endeavor, many datasets are designed with diverse protein targets to facilitate robust evaluation. Rigorous benchmarking is critical for advancing virtual screening techniques, and selecting appropriate benchmarks is essential. Below, we discuss some of the most widely used benchmarking datasets, highlighting their design, strengths, and limitations. The *Database of Useful Decoys: Enhanced* (DUD-E) is a widely adopted benchmark for virtual screening (Mysinger et al. 2012). It comprises 102 targets across eight protein categories, over 20,000 active compounds, and more than one million decoy compounds, with actives sourced from ChEMBL (activity/affinity values of K_i, K_d, IC50, EC50 $\leq$ 1 μM) (Gaulton et al. 2012). Ideally, inactive molecules should be selected based on experimental evidence, but this is often infeasible. Consequently, presumed inactive compounds unlikely to bind a given target are used as decoys. Many decoys were selected from ZINC (Irwin and Shoichet 2005), matched to active compounds in terms of physicochemical properties but designed to be topologically dissimilar. This approach mitigates biases linked to features such as molecular weight while avoiding overlap with true actives.

Each active compound is paired with 50 decoys, and Bemis-Murcko clustering was used to prioritize the most potent actives while ensuring sufficient diversity. Despite these efforts, certain targets still exhibit high similarity among active compounds, and physicochemical property biases persist, a challenge not unique to

Table 1 Overview of key benchmark datasets commonly used in SBVS, including their data sources and access details

Dataset	Data source	Availability details
MUV	PubChem	https://github.com/skearnes/muv
CASF-2016	PDB	http://www.pdbbind-cn.org/casf.asp
DUD-E	ChEMBL, ZINC	http://www.dude.docking.org
DEKOIS 2.0	BindingDB	http://www.dekois.com
D3R GC 3	HTS	https://drugdesigndata.org/about/grand-challenge-3
D-COID	PDB	https://data.mendeley.com/datasets/8czn4rxz68/1
MoleculeNet	Various	https://github.com/deepchem/deepchem

DUD-E. In 2011, Vogel et al. introduced *Demanding Evaluation Kits for Objective* In Silico *Screening* (DEKOIS) (Vogel et al. 2011), later refined into DEKOIS 2.0 (Bauer et al. 2013). This database includes 81 targets spanning 11 classes, where actives were obtained from BindingDB (Liu et al. 2007), with dynamic activity cut-offs for each target, ensuring no active molecule had an affinity more than 1000-fold weaker than the most potent inhibitor. Decoys were exclusively sourced from ZINC, with 30 decoys selected per active compound. To minimize bias, DEKOIS decoys were chosen to match active compounds' physicochemical properties while avoiding analog bias and artificial enrichment. The authors introduced a metric called the *Latent Actives in Decoys Score* (LADS) to quantify the risk of false negatives in the decoy set, optimizing both property similarity and LADS scores during selection. The *Maximum Unbiased Validation Sets* (MUV) (Rohrer and Baumann 2009) database addresses two key challenges: the widespread use of presumed inactives and the need for unbiased benchmarks. MUV utilizes experimentally confirmed inactive compounds from PubChem (Wheeler et al. 2007), focusing on 17 targets with 30 actives and 15,000 decoys per target. Actives were selected based on their diversity, while decoys were drawn from primary screens of the same targets, ensuring a high decoy-to-active ratio. MUV aims to eliminate analog bias and artificial enrichment, which can lead to overly optimistic predictions in virtual screening evaluations. As a result, MUV is a particularly challenging benchmark, well suited for testing ligand-based methods. However, concerns have been raised about its applicability to SBVS due to its reliance on cell-based assay data and limited structural considerations.

Riniker and Landrum developed a dataset from ChEMBL (Gaulton et al. 2012), inspired by Heikamp and Bajorath. This dataset includes 50 human protein targets, with actives selected based on potency ($\leq$10 µM), molecular weight (<700 g/mol), and the absence of metal ions. The RDKit (Landrum 2010. 'RDKit.' Q2. https://www.rdkit.org/) diversity picker was used to select 100 diverse actives per target, each paired with two decoys randomly chosen from ZINC. While this dataset has received less scrutiny compared to DUD-E, it likely suffers from artificial enrichment due to limited physicochemical matching between actives and decoys and the reuse of the same decoys across multiple targets.

5 Training/Test Splits for the Development of New Scoring Functions

After selecting an appropriate dataset, it is crucial to determine the optimal method for splitting data into training and validation sets. Ideally, a ML experiment should reserve a held-out test set that closely mirrors real-world challenges. While random splits are common in many applications, SBVS requires more deliberate approaches to account for the unique nature of the problem, e.g., binding to one target differs from binding to another, and the practical use cases, such as prospective screens for novel targets. Intratarget splitting is analogous to performing a random split within each class. Here, both the training and test sets contain data from all targets (or a single target), meaning each target has ligands in both subsets. This setup simulates scenarios where ligands for a target are already known. However, this approach does not evaluate a model's ability to generalize to novel targets, a critical requirement for many use cases. Furthermore, it may lead to overly optimistic performance assessments due to known biases in benchmark datasets unless additional precautions, such as chemotype splitting or temporal splitting, are implemented. In cross-target splitting, there is no overlap of targets between the training and test sets. Testing is conducted exclusively on unseen targets, more closely aligning with real-world scenarios. This method is more challenging, as different proteins often have distinct binding modes, leading to significant differences in the active compounds for different targets. Within a single target, active compounds usually share commonalities due to binding within the same pocket. The additional challenge of cross-target splitting compared to intratarget splitting has been demonstrated in several studies. For instance, Wu et al. achieved near-perfect performance on the DUD-E dataset when performing random splits across all targets (intratarget splitting), allowing examples from the same targets in both the training and test sets. Conversely, Wójcikowski et al. (2017) observed that performance in cross-validation improves nearly threefold when data were split randomly across all targets or on a single target basis, compared to maintaining all examples of a given target within the same fold, i.e., cross-target splitting. Many proteins exhibit high similarity, resulting in substantial overlap in their active ligands. Therefore, cross-target splitting alone may not sufficiently ensure robust evaluation of a model's predictive power. To address this, various advanced splitting strategies have been developed to maximize dissimilarity between training and validation/test sets. These approaches, including chemotype splitting and temporal splitting, help mitigate biases and provide a more realistic assessment of a model's generalizability.

Proteins with high sequence similarity typically exhibit similar structures and functions, leading to comparable ligand activities. To mitigate this bias, targets can be clustered during cross-validation to ensure highly similar targets are grouped within the same fold. Alternatively, targets exceeding a predefined similarity threshold with any training set targets can be excluded from the test set. For instance, in the experimental setup by Ragoza et al. (2017), a threshold of 80% sequence similarity, determined using global sequence alignment, was applied to both cross-validation and test set construction. Binding sites can exhibit significant similarity even when their corresponding proteins have low sequence similarity. This often results in overlapping ligand activity between targets, which sequence similarity alone may not detect. To address this, Ragoza et al. utilized ProBiS (Konc and Janezic 2010) structural alignment to compare binding sites of training and test set targets. Any target with a significant structural alignment, based on ProBiS default parameters, was excluded from the test set. These stringent measures illustrate redundancy in common benchmarks: for example, the Riniker and Landrum ChEMBL set was reduced from 50 to 13 targets, and the MUV set shrank from 17 to 9 targets.

Chemotype splitting, also known as scaffold splitting, divides molecules based on their 2D structural frameworks, such as Bemis–Murcko scaffolds (Bemis and Murcko 1996). This method separates structurally distinct compounds into different subsets, providing a more challenging scenario for learning algorithms compared to random splitting. Scaffold splitting offers a robust evaluation of a model's ability to generalize and is particularly suited for intratarget splitting scenarios. Temporal splitting leverages time-based information to organize training, validation, and test sets, aiming to emulate retrospective studies and align them with prospective setups. This approach is more challenging than random splitting and is particularly relevant for intratarget splits (Gomes et al. 2017). However, it is often difficult to implement due to the unavailability of time-related data. Temporal splitting can also be adapted for cross-target scenarios, where training set data must precede the publication of test set ligand data.

6 Exploiting ML and DL for the Development of New Scoring Function

6.1 ML in SBDD

Over the past decade, ML has emerged as a transformative tool in the development of scoring functions, significantly enhancing accuracy in pose prediction, virtual screening, and binding affinity prediction. Unlike traditional approaches that rely on predefined functional relationships between descriptive features and target properties, ML methods infer these relationships directly from data, allowing for more nuanced and adaptable models. Early ML models often utilized features derived from traditional scoring functions, capitalizing on the flexibility of advanced model architectures to better interpret the same input data (Creanza et al. 2021). For example, Li et al. demonstrated that random forest models could improve their predictions with additional data and features, long after multiple linear regression models reached a performance plateau. Subsequent approaches introduced more sophisticated feature engineering or generalized input features, such as molecular fingerprints, to enhance predictive power (Table 2). Two notable examples of such ML-based scoring functions are RF-Score and NNScore (Ballester and Mitchell 2010; Durrant and McCammon 2011), both of which demonstrate the potential of these methods to surpass traditional scoring functions. Common traditional and modern approaches are illustrated in Fig. 2.

RF-Score, developed by Ballester and Mitchell (2010), is a structure-based scoring function that uses random forest regression to predict protein–ligand binding affinity. The model represents protein–ligand complexes through counts of atom pairs within 12 Å. Specifically, it considers four protein atom types (C, N, O, S) and nine ligand atom types (C, N, O, F, P, S, Cl, Br, I), resulting in 36 unique atom pair combinations. These pair counts are used as input features for the random forest

Table 2 Software tools and web servers for feature extraction

Software name	Features	Source
Descriptor data bank	Over 2500 proteins, ligands, and protein-ligand features	http://www.descriptordb.com
Cang et al.	Topological fingerprinting	https://doi.org/10.1371/journal.pcbi.1005929.s002
ODDT	RF-Score, NNScore, and PLEC fingerprints	https://github.com/oddt/oddt
BINANA	Intermolecular descriptors	https://git.durrantlab.pitt.edu/jdurrant/binana
RF-Score v366	Features from RF-Score v1 and Vina	https://github.com/huizhuzhao/rfscore

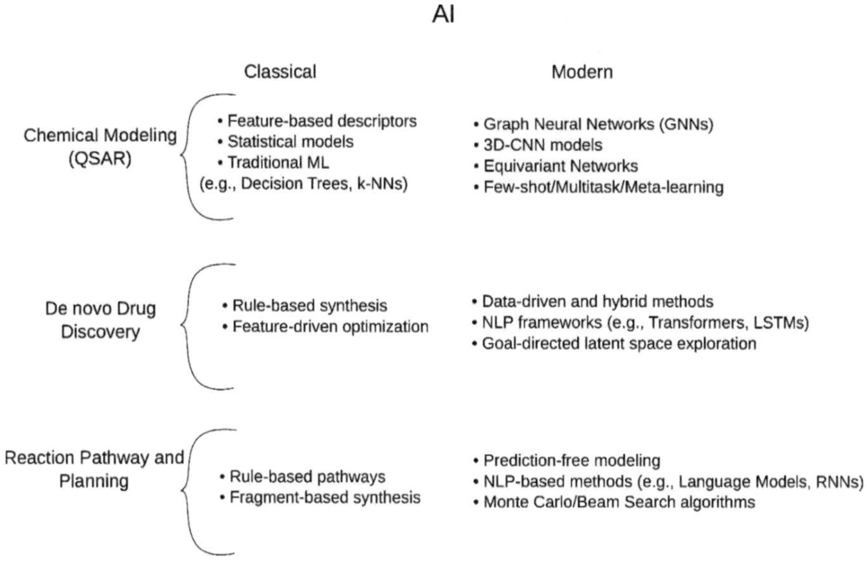

Fig. 2 A schematic diagram illustrating the shift from classical to modern approaches for key challenges in drug discovery, including QSAR/QSPR modeling, de novo drug design, and synthesis planning. Abbreviations: *QSAR/QSPR* quantitative structure-activity/property relationship, *NN* neural network, *NLP* natural language processing

model. RF-Score was trained on a dataset of 1105 protein–ligand complexes from the PDBbind database. In benchmarking against state-of-the-art traditional scoring functions, such as those in GOLD and Glide, RF-Score demonstrated superior performance on a test set of 195 complexes. A later adaptation by Wójcikowski et al. tailored RF-Score for virtual screening (Wójcikowski et al. 2017), where it performed comparably to AutoDock Vina in cross-target splits of the DUD-E dataset. Around the same time, Durrant and McCammon introduced NNScore (Durrant and McCammon 2011), a structure-based scoring function employing an ensemble of neural networks with a single hidden layer to predict binding affinity.

Like RF-Score, NNScore relies on atom pair representations but incorporates additional features for improved accuracy: (i) atom pair distances, differentiating pairs within 2 Å and those within 4 Å; (ii) electrostatic interactions, estimating total electrostatic energy for atom pairs within 4 Å; and (iii) ligand features, including basic counts of 13 ligand atom types and the number of rotatable bonds in the ligand. This expanded feature set comprises 194 distinct inputs, allowing NNScore to capture more complex interactions. In virtual screening tests on influenza N1 neuraminidase, NNScore demonstrated slightly enhanced performance compared to AutoDock Vina, with similar results observed in follow-up studies. By eliminating the need for assumptions about the functional relationships between features and target properties, ML has introduced a paradigm shift in scoring function development. Early models like RF-Score and NNScore illustrate the potential of these techniques to achieve greater accuracy and flexibility than traditional methods, enabling more effective predictions in SBDD. As ML continues to advance, its integration into scoring functions promises further breakthroughs, improving the predictive accuracy of binding affinities, virtual screening outcomes, and other critical components of drug discovery. These innovations are reshaping the landscape of SBDD, paving the way for faster and more precise therapeutic development.

Traditional and standard ML models rely on predefined features before parameter optimization. While effective, this approach inherently biases the model toward the selected features, often leading to a loss of critical information by approximating or discarding raw structural data. For instance, RF-Score ignores atom pairs that are either unaccounted for in the predefined features or separated by more than 12 Å, if all atom pairs within this distance contribute equivalently and only their counts matter. NNScore, while offering more descriptive features, still makes simplifying assumptions that limit its granularity.

6.2 Deep Learning in SBDD

Deep learning (DL) minimizes the need for extensive manual feature engineering by using raw or minimally processed data as input (Vittorio et al. 2024). For protein–ligand complexes, this means feeding data formats that more accurately represent the underlying structural interactions, such as 3D spatial information. DL models autonomously learn which features are most relevant to predicting properties like binding affinity or activity, eliminating the need to predefine these features and reducing the risk of information loss. Recent DL methods have demonstrated the ability to learn end-to-end representations for virtual screening and binding affinity prediction. These approaches have consistently outperformed models reliant on precomputed features. For instance, Ragoza et al. developed a convolutional neural network (CNN)-based approach for SBVS (Ragoza et al. 2017). Their method reduced feature engineering to a minimum by utilizing a 3D voxel grid of atomic coordinates with basic atom typing, allowing the CNN to directly learn meaningful

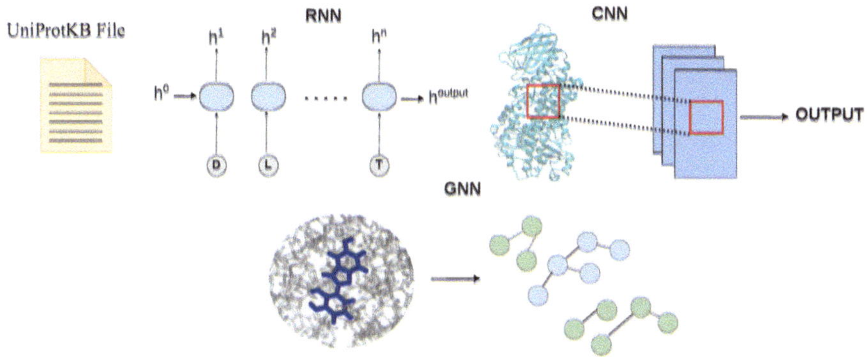

Fig. 3 Common neural network architectures utilized in SBDD include RNNs, which are employed to learn from the primary sequence of a protein and other sequential molecular data. CNN, which capture 3D spatial information and are typically used to learn from protein representations. GNNs, which work with molecular graphs and are frequently applied to understand interatomic interactions both within and between molecules

interactions from raw structural data. Recurrent neural networks (RNNs), such as those utilizing long-short term memory (LSTM) cells (Hochreiter and Schmidhuber 1997), are frequently employed to process the primary sequence of a protein. RNNs feature feedback connections that allow information from previous inputs to influence subsequent ones. This feedback mechanism acts as a form of "learned memory" within the architecture, enabling the model to capture the sequential structure of the input (Fig. 3). Another DL approach exploits graph neural networks (GNNs), which representing atoms and their interactions as molecular graphs, can capture both the structural and functional relationships between atoms within a single molecule or across multiple molecular complexes. The flexible representation offered by GNNs allows them to be applied to a broad range of tasks in SBDD.

This shift toward DL needs an understanding of the datasets, training protocols, and testing methodologies used in SBVS. To appreciate its full potential, it is important to explore how these networks are adapted for SBVS, including their strengths, limitations, and areas for improvement. Anticipate how emerging tools and advancements in DL could further refine virtual screening processes. By leveraging raw structural data and reducing feature-engineering biases, DL represents a paradigm shift in structure-based drug design. It offers improved predictive power, more comprehensive data utilization, and greater adaptability for future advancements in drug discovery.

6.3 Overview of CNN Based in SBDD

Recent work has highlighted the utility of various visualization techniques in interpreting how CNNs evaluate molecular inputs, particularly in the context of SBDD.

Hochuli et al. proposed a CNN to improve pose selection and affinity prediction for protein–ligand complexes (Hochuli et al. 2018; Ragoza et al. 2017). The authors emphasize that each method, including masking, gradients, and class-wise layer-wise relevance propagation (CLRP), offers unique insights into the scoring process and identifies critical atomic interactions that inform both network development and medicinal chemistry (Bach et al. 2015; Lapuschkin et al. 2016). Masking is presented as the most intuitive and interpretable method, closely resembling experimental techniques for identifying essential interactions in protein–ligand complexes. This external manipulation of input data makes masking especially useful for understanding CNN outputs, though the authors note its significant computational cost compared to gradient and CLRP visualizations, which can be produced with a single forward and backward pass through the network. In contrast, gradient visualizations are described as an effective means of pinpointing which atoms or molecular regions the network "prefers" to optimize, as gradients provide directional information on a per-atom basis. This granularity allows for pose refinement and a more detailed analysis of how the network interprets molecular inputs. CLRP, on the other hand, offers the advantage of preserving relevance throughout all layers of the network, avoiding the need for stabilization parameters required in epsilon or alpha-beta variants of relevance propagation. The authors acknowledge that CLRP's inability to account for "dead" nodes (regions ignored during analysis) limits its utility, though they suggest that dead node analysis can partially compensate for this shortcoming. The authors also explore the broader applicability of these methods, explaining how they can map CNN outputs onto spatial grids or adapt to cheminformatics models that analyze 2D molecular structures. While masking remains the simplest approach to implement, gradients and CLRP may require adjustments based on the chosen molecular representation. Interestingly, the authors observe that masking and gradient visualizations often produce correlated results, particularly for pose scoring and affinity prediction networks, while CLRP visualizations tend to highlight more contrasted patterns that are sometimes challenging to interpret. This finding might indicate subtle decision boundaries within the network or relationships that are not yet fully understood.

The work underscores two primary benefits of visualization: guiding medicinal chemistry and informing network optimization. By identifying critical ligand atoms, these techniques provide actionable insights for chemical modification to improve binding interactions. Moreover, visualizations expose potential network biases, enabling retraining on enriched datasets. For instance, the authors note that undervalued aromatic interactions might prompt retraining on datasets containing a greater representation of such features to improve predictive performance. Ultimately, this work positions visualization techniques as essential tools for both understanding and improving CNN-driven predictions. While the "black box" nature of neural networks persists, the authors argue that these methods, combined with convolutional filter analysis, offer complementary perspectives for unraveling network behavior. Importantly, they assert that the described approaches generalize well to other molecular DL methods, further extending their relevance in the field. For practical implementation, they refer readers to the open-source tool gnina

(McNutt et al. 2021), available at http://github.com/gnina, which facilitates the generation of these visualizations for deeper exploration and analysis.

6.4 CNN Application for Virtual Screening

The CNN architecture of Ragoza et al. (2017) comprised three 3 × 3 × 3 convolutional layers with 32, 64, and 128 filters, respectively, each preceded by a 2 × 2 × 2 max-pooling layer. The training protocol employed a threefold clustered cross-validation on the DUD-E dataset. Since DUD-E does not have crystal structures for most ligands, poses for the active and decoy ligands were generated using AutoDock Vina (Durrant and McCammon 2011). Ligands were docked to a reference receptor in a box centered around a reference ligand with a padding of 8 Å, using the default settings for exhaustiveness and sampling. The top-ranked pose for each protein–ligand pair was selected for training. The protocol was evaluated on two independent test sets: a subset of the ChEMBL database and MUV dataset. During testing, scores for the docked poses were generated using a single universal model, with the top score used as the final metric. The approach significantly outperformed AutoDock Vina, as well as RF-Score and NNScore (Ragoza et al. 2017). Building on this work, Imrie et al. (2018) proposed improvements that ameliorate performance while maintaining the original input format. These included several independent advances. First, they implemented a DenseNet architecture, replacing the traditional CNN with a densely connected neural network composed of three dense blocks, each containing four convolutional blocks. Each convolutional block included a batch normalization layer, a 3 × 3 × 3 convolutional layer with ReLU activation, and a 1 × 1 × 1 convolutional layer followed by a 2 × 2 × 2 max-pooling layer between dense blocks. The first convolutional layer had 32 filters, while the subsequent 3 × 3 × 3 convolutional layers used 16 filters. Switching to this architecture improved the area under the precision-recall curve (AUC-PRC) by up to 36%. Second, they introduced a pose-averaged scoring method. While training used only the top-ranked AutoDock Vina pose, testing incorporated average scores across multiple anchored poses, as determined via cross-validation. This approach mitigated inaccuracies resulting from relying on a single pose, yielding PRC-AUC improvements of up to 14%. Third, protein family-specific models were developed using transfer learning. Finally, ensemble methods, i.e., combining the assessment of several classificators and/or regressors, were employed to improve performance. Predictions from three replicates (trained with different random seeds) were combined via averaging, providing a further performance boost. The DenseNet-based approach, called DenseFS, substantially outperformed traditional methods such as docking and other ML-based scoring functions. In retrospective evaluations, DenseFS identified over five times more binders with a false positive rate of 0.5% than AutoDock Vina, both in cross-validation on DUD-E and on an independent test set.

6.5 Pose Prediction

Pose prediction involves identifying the correct binding mode of a ligand from a set of docked structures in complex with a protein. This is a critical step in SBDD, particularly when experimental crystal data are unavailable. While docking software can effectively sample a range of potential binding modes, accurately ranking these poses remains a significant challenge. Pose prediction performance is evaluated in two main ways: single-complex ranking and multi-complex ranking, also referred to as intratarget and intertarget ranking, respectively (Ren et al. 2018). In multi-complex ranking, the goal is to score all poses across a range of protein–ligand complexes to generate a global ranking. This is often assessed using an ROC curve and mirrors the training protocols of most methods. However, this approach does not align closely with real-world docking scenarios. In contrast, single-complex ranking focuses on selecting the pose with the lowest root-mean-square deviation (RMSD) among the poses generated for a specific protein–ligand pair. A scoring function can succeed in single-complex ranking even if the lowest RMSD pose has a relatively poor score, provided that all other poses for the same target are ranked worse. This evaluation better reflects practical docking applications, where the objective is to pinpoint a low RMSD pose for a given complex.

Traditional docking methods for small-molecule ligands often rely on treating the macromolecular receptor as rigid to reduce the otherwise prohibitively large conformational search space. While this simplification makes the problem computationally manageable, it limits the accuracy of docking results, particularly for highly flexible systems. Some approaches, such as GNINA 1.0, incorporate limited flexibility in the side chains of the protein's binding site, but the backbone remains rigid (McNutt et al. 2021). Currently, no geometric DL methods support fully flexible docking for entire macromolecular structures. However, recent advancements have introduced graph-based and hybrid approaches to address these limitations. EquiBind is an example of an SE(3)-equivariant message-passing neural network that, combined with optimal transport, predicts the binding pose of two proteins in a rigid-body, blind-docking setup (Ganea et al. 2022). This method not only determines a suitable binding site but also predicts a rotation matrix and translation vector to align one protein into the binding pose relative to the other. The resulting docking pose is invariant to the initial orientation and position of the binding partners, ensuring robust predictions. Expanding on this concept, EquiBind extends docking capabilities to flexible small-molecule ligands. It incorporates changes in the torsion angles of rotatable bonds, starting from a randomly generated conformer, enabling it to model ligand flexibility more effectively than rigid-body methods. Another innovative approach integrates SE(3)-equivariant GNNs with surface fingerprinting techniques to evaluate the atomic point clouds of binding partners (Stärk et al. 2022). This method focuses on estimating surface shape complementarity, which is critical for rigid-body docking of protein complexes. Unlike EquiBind, it generates and ranks multiple binding poses for each protein–protein pair, providing a broader and more comprehensive exploration of docking possibilities. These advancements underscore the transformative potential of combining geometric DL with traditional and flexible docking techniques. By addressing the limitations of classical approaches, these methods are paving the way for more accurate protein–ligand and protein–protein docking predictions, which are critical for advancing SBDD.

6.6 *Binding Affinity Prediction*

Evaluating the binding strength between a small molecule and a protein is a critical step in drug design. Accurate prediction of binding affinity can significantly aid in optimizing small molecules to improve their binding properties. While virtual screening is useful for identifying initial binders, these compounds must then be refined into lead molecules by enhancing their affinity and specificity for a given target. Recent advancements have also explored the use of CNNs for binding affinity prediction. Hochuli et al. (2018) employed input formats similar to those discussed in the previous section, but with refined model architectures and training strategies aimed at predicting both binding affinity and binding poses. In contrast, Jiménez et al. (2018) adopted a different approach by encoding molecular properties, such as hydrophobicity, hydrogen-bond donor or acceptor characteristics, aromaticity, charge (positive or negative ionizable), and metallicity, instead of relying directly on atom types. This approach provides a more nuanced representation of molecular interactions. These CNN-based methods have demonstrated state-of-the-art performance, particularly when tested on datasets such as the 2016 PDBbind core (Liu et al. 2017). Additionally, CNN models have shown strong results in real-world challenges, including the D3R Grand Challenge 3, where a CNN-based entry achieved high rankings for multiple targets (Gan et al. 2022). These findings highlight the potential of CNNs to revolutionize binding affinity prediction and accelerate the drug discovery process.

7 MD in SBDD

MD simulations are a powerful computational technique used to study and predict the atomic-level movements and interactions of biological macromolecules. Widely applied in drug development, MD simulations rely on the principles of classical mechanics to reveal the temporal evolution of molecular systems. The process begins with the initial structure of a macromolecule, typically obtained through X-ray crystallography, NMR spectroscopy, or homology modeling. This structure is neutralized using counterions to maintain overall charge neutrality and then solvated, usually in water, to create a realistic environment. A force field is applied to the system, defining the bonding, van der Waals, electrostatic, and solvation forces that govern atomic interactions. Simulations are performed by numerically solving the equations of motion for each atom over time. The simulation timescale can range from nanoseconds to microseconds or longer, depending on the system's complexity and computational resources. The output includes the system's trajectory, detailing atomic positions, velocities, and energies over time. MD simulations have been employed to study a variety of biological systems, including proteins, nucleic acids, lipids, and carbohydrates. They offer valuable insights into the conformational dynamics, stability, and function of these molecules, as well as their interactions with ligands such as drugs or small molecules. A key strength of MD simulations is their ability to capture the structural flexibility of macromolecules, which is critical for understanding biological functions and guiding drug discovery. Additionally, they can be used to explore the thermodynamics and kinetics of ligand binding, providing essential data for the design of potent and selective drug candidates.

Docking algorithms, despite their utility, face limitations in sampling and scoring, which can affect the reliability of docking solutions. To address these shortcomings, MD simulations are often employed as a postprocessing tool to validate and refine docking results. The central premise is that an inaccurate docking pose will result in an unstable MD trajectory, during which the ligand may even exit the binding site. In contrast, a well-predicted docking pose will exhibit stable, specific interactions with the target protein, maintaining a low RMSD over time relative to the initial conformation. Additionally, MD simulations provide insights into induced-fit effects, where the binding site undergoes structural adaptations to accommodate the ligand. This process not only reinforces interactions identified during docking but may also reveal new interactions that enhance ligand binding. Furthermore, MD simulations conducted in explicit solvent can highlight the role of structural water molecules within the binding site, offering a deeper understanding of their contribution to ligand binding and improving the accuracy of binding predictions (Falbo et al. 2022; Frasnetti et al. 2024) (Fig. 4).

However, MD simulations have limitations. Their accuracy depends on the quality of the force field and the parameters chosen for the simulation. Force fields are influenced by the precision of underlying experimental data and the approximations used during modeling. Simulation parameters, such as the duration and timestep, can also impact the reliability of results. Another significant limitation is the high computational cost. MD simulations are resource-intensive and time-consuming, particularly for large and complex systems, which can restrict the scale and duration of studies. Despite the available computing power often limiting the number and length of simulations that can be performed, several enhanced sampling methods could accelerate the time scale of MD simulations (Tiwary and van de Walle 2016). In summary, MD simulations are invaluable tools in drug discovery, offering detailed insights into the dynamics and interactions of biological macromolecules at the atomic level. Despite their computational expense and reliance on the quality of force fields and parameters, MD simulations remain an essential component of modern drug development.

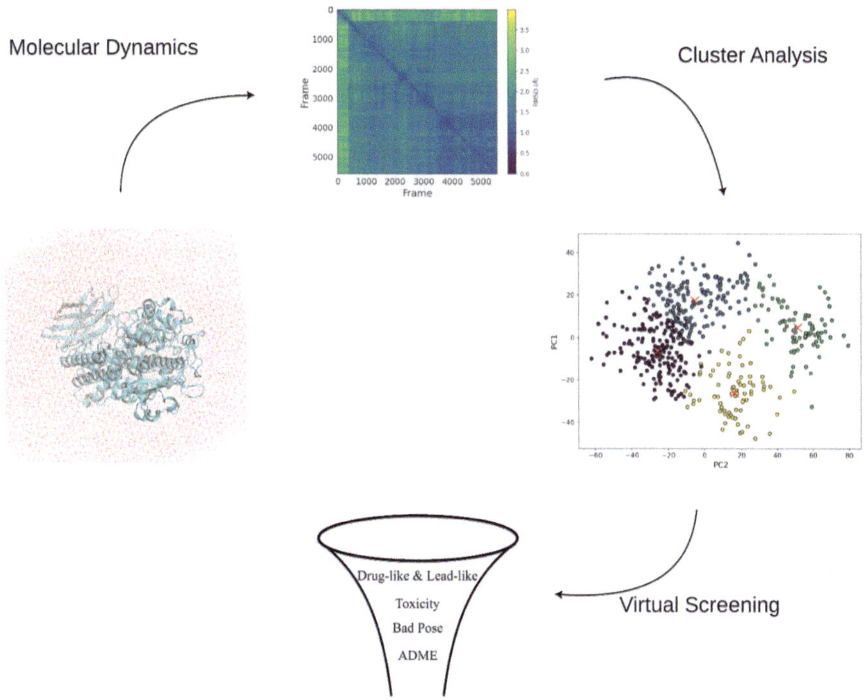

Fig. 4 The virtual screening workflow combining docking and MD simulations involves key steps: MD simulations explore the receptor's conformational landscape, followed by clustering to group similar states and select representative structures. Virtual ligand screening is then performed on these representative conformations, accounting for receptor flexibility and improving the chances of identifying high-affinity binders

8 Concluding Remark

In conclusion, SBDD is a powerful and transformative approach for developing new medications that selectively target disease-causing proteins. Compared to traditional drug discovery methods, SBDD offers significant advantages, including higher success rates in identifying lead compounds and a reduced risk of toxicity. By focusing on specific proteins involved in disease pathways, SBDD has revolutionized drug development, contributing to the approval of numerous effective therapies. The success of SBDD, however, hinges on the accurate structural determination of target proteins and the efficient screening of large compound libraries. Continued advancements in technologies, such as cryo-electron microscopy (cryo-EM), NMR spectroscopy, and X-ray crystallography, will be critical to the future of SBDD. These tools enhance our ability to resolve complex protein structures at high resolution, a fundamental requirement for effective drug design. Looking ahead, SBDD holds immense potential to improve patient outcomes across a wide spectrum of diseases. With ongoing refinements and innovations in this methodology, we can develop more precise and effective therapies, ultimately improving the health and well-being of millions worldwide. Furthermore, the integration of AI techniques into the field of SBDD represents one of the most promising avenues for future research. These technologies can process and analyze vast amounts of experimental and computational data, enabling the identification of novel drug targets, predicting drug efficacy and toxicity, and optimizing lead compounds with unprecedented speed and accuracy. Another exciting frontier is the development of techniques to study the dynamic behavior of proteins, including their conformational changes and interactions with other molecules. This requires cutting-edge experimental approaches such as time-resolved crystallography and cryo-EM, alongside innovative computational tools for analyzing complex datasets. These advances will deepen our understanding of protein dynamics and interactions, leading to the design of drugs that are both more potent and more specific. In summary, SBDD represents a promising approach to drug discovery, and its continued evolution promises to reshape the future of medicine. By leveraging technological innovations and fostering collaboration, we can unlock new therapeutic possibilities and bring safer, more effective drugs to market, transforming the treatment landscape for countless diseases.

Acknowledgments This work was supported by the European Union, Next Generation EU (PRIN 2022, project number: 2022P5LPHS). A.L. acknowledges funding from the PNRR Project CN000041, "National Centre for Gene Therapy and RNA Technology-Based Drugs" (CUP UNINA E63C22000940007).

References

Allen AEA, Payne MC, Cole DJ. Harmonic force constants for molecular mechanics force fields via hessian matrix projection. J Chem Theory Comput. 2018;14:274–81. https://doi.org/10.1021/acs.jctc.7b00785.

Bach S, Binder A, Montavon G, Klauschen F, Müller K-R, Samek W. On pixel-wise explanations for non-linear classifier decisions by layer-wise relevance propagation. PLoS One. 2015;10:1–46. https://doi.org/10.1371/journal.pone.0130140.

Ballester PJ, Mitchell JBO. A machine learning approach to predicting protein-ligand binding affinity with applications to molecular docking. Bioinformatics. 2010;26:1169–75. https://doi.org/10.1093/bioinformatics/btq112.

Bauer MR, Ibrahim TM, Vogel SM, Boeckler FM. Evaluation and optimization of virtual screening workflows with DEKOIS 2.0—a public library of challenging docking benchmark sets. J Chem Inf Model. 2013;53:1447–62. https://doi.org/10.1021/ci400115b.

Bemis GW, Murcko MA. The properties of known drugs. 1. Molecular frameworks. J Med Chem. 1996;39:2887–93. https://doi.org/10.1021/jm9602928.

Berman HM, Westbrook J, Feng Z, Gilliland G, Bhat TN, Weissig H, Shindyalov IN, Bourne PE. The protein data bank. Nucleic Acids Res. 2000;28:235–42. https://doi.org/10.1093/nar/28.1.235.

Creanza TM, Delre P, Ancona N, Lentini G, Saviano M, Mangiatordi GF. Structure-based prediction of hERG-related cardiotoxicity: a benchmark study. J Chem Inf Model. 2021;61:4758–70. https://doi.org/10.1021/acs.jcim.1c00744.

Durrant JD, McCammon JA. NNScore 2.0: a neural-network receptor–ligand scoring function. J Chem Inf Model. 2011;51:2897–903. https://doi.org/10.1021/ci2003889.

Eldridge MD, Murray CW, Auton TR, Paolini GV, Mee RP. Empirical scoring functions: I. The development of a fast empirical scoring function to estimate the binding affinity of ligands in receptor complexes. J Comput Aided Mol Des. 1997;11:425–45. https://doi.org/10.1023/a:1007996124545.

Falbo E, Lavecchia A. HessFit: a toolkit to derive automated force fields from quantum mechanical information. J Chem Inf Model. 2024;64:5634–45. https://doi.org/10.1021/acs.jcim.4c00540.

Falbo E, Fusè M, Lazzari F, Mancini G, Barone V. Integration of quantum chemistry, statistical mechanics, and artificial intelligence for computational spectroscopy: the UV–vis spectrum of TEMPO radical in different solvents. J Chem Theory Comput. 2022;18:6203–16. https://doi.org/10.1021/acs.jctc.2c00654.

Fan H, Irwin JJ, Sali A. Virtual ligand screening against comparative protein structure models. Methods Mol Biol. 2012;819:105–26. https://doi.org/10.1007/978-1-61779-465-0_8.

Frasnetti E, Cucchi I, Pavoni S, Frigerio F, Cinquini F, Serapian SA, Pavarino LF, Colombo G. Integrating molecular dynamics and machine learning algorithms to predict the functional profile of kinase ligands. J Chem Theory Comput. 2024;20:9209–29. https://doi.org/10.1021/acs.jctc.4c01097.

Friesner RA, Banks JL, Murphy RB, Halgren TA, Klicic JJ, Mainz DT, Repasky MP, Knoll EH, Shelley M, Perry JK, Shaw DE, Francis P, Shenkin PS. Glide: a new approach for rapid, accurate docking and scoring. 1. Method and assessment of docking accuracy. J Med Chem. 2004;47:1739–49. https://doi.org/10.1021/jm0306430.

Friesner RA, Murphy RB, Repasky MP, Frye LL, Greenwood JR, Halgren TA, Sanschagrin PC, Mainz DT. Extra precision glide: docking and scoring incorporating a model of hydrophobic enclosure for protein-ligand complexes. J Med Chem. 2006;49:6177–96. https://doi.org/10.1021/jm051256o.

Gan JL, Kumar D, Chen C, Taylor BC, Jagger BR, Amaro RE, Lee CT. Benchmarking ensemble docking methods in D3R Grand Challenge 4. J Comput Aided Mol Des. 2022;36:87–99. https://doi.org/10.1007/s10822-021-00433-2.

Ganea O-E, Huang X, Bunne C, Bian Y, Barzilay R, Jaakkola T, Krause A. Independent SE(3)-equivariant models for end-toend rigid protein docking. arXiv preprint, 2022; arXiv:2111.07786.

Gaulton A, Bellis LJ, Bento AP, Chambers J, Davies M, Hersey A, Light Y, McGlinchey S, Michalovich D, Al-Lazikani B, Overington JP. ChEMBL: a large-scale bioactivity database for drug discovery. Nucleic Acids Res. 2012;40:D1100–7. https://doi.org/10.1093/nar/gkr777.

Ghosh AK, Sridhar PR, Leshchenko S, Hussain AK, Li J, Kovalevsky AY, Walters DE, Wedekind JE, Grum-Tokars V, Das D, Koh Y, Maeda K, Gatanaga H, Weber IT, Mitsuya H. Structure-based design of novel HIV-1 protease inhibitors to combat drug resistance. J Med Chem. 2006;49(17):5252–61. https://doi.org/10.1021/jm060561m.

Gohlke H, Hendlich M, Klebe G. Knowledge-based scoring function to predict protein-ligand interactions. J Mol Biol. 2000;295:337–56. https://doi.org/10.1006/jmbi.1999.3371.

Gomes J, Ramsundar B, Feinberg EN, Pande VS. Atomic convolutional networks for predicting protein-ligand binding affinity. 2017. https://doi.org/10.48550/arXiv.1703.10603.

Hochreiter S, Schmidhuber J. Long short-term memory. Neural Comput. 1997;9:1735–80. https://doi.org/10.1162/neco.1997.9.8.1735.

Hochuli J, Helbling A, Skaist T, Ragoza M, Koes DR. Visualizing convolutional neural network protein-ligand scoring. J Mol Graph Model. 2018;84:96–108. https://doi.org/10.1016/j.jmgm.2018.06.005.

Imrie F, Bradley AR, van der Schaar M, Deane CM. Protein family-specific models using deep neural networks and transfer learning improve virtual screening and highlight the need for more data. J Chem Inf Model. 2018;58:2319–30. https://doi.org/10.1021/acs.jcim.8b00350.

Irwin JJ, Shoichet BK. ZINC – a free database of commercially available compounds for virtual screening. J Chem Inf Model. 2005;45:177–82. https://doi.org/10.1021/ci049714+.

Jiménez J, Škalič M, Martínez-Rosell G, De Fabritiis G. KDEEP: protein–ligand absolute binding affinity prediction via 3D-convolutional neural networks. J Chem Inf Model. 2018;58:287–96. https://doi.org/10.1021/acs.jcim.7b00650.

Kaur T, Dhawal Bhandari D, Sharma R. Cloud computing: a relevant solution for drug designing using different software's. In: 2021 sixth international conference on image information processing (ICIIP). Presented at the 2021 sixth international conference on image information processing (ICIIP), IEEE, Shimla, India; 2021. p. 375–8. https://doi.org/10.1109/ICIIP53038.2021.9702618.

Kendrew JC, Bodo G, Dintzis HM, Parrish RG, Wyckoff H, Phillips DC. A three-dimensional model of the myoglobin molecule obtained by X-ray analysis. Nature. 1958;181:662–6. https://doi.org/10.1038/181662a0.

Konc J, Janezic D. ProBiS algorithm for detection of structurally similar protein binding sites by local structural alignment. Bioinformatics. 2010;26:1160–8. https://doi.org/10.1093/bioinformatics/btq100.

Kumar A, Krishnamurthi R, Nayyar A, Sharma K, Grover V, Hossain E. A novel smart healthcare design, simulation, and implementation using healthcare 4.0 processes. IEEE Access. 2020;8:118433–71. https://doi.org/10.1109/ACCESS.2020.3004790.

Landrum G. RDKit. Q2. 2010. https://www.rdkit.org/

Lapuschkin S, Binder A, Montavon G, Müller K-R, Samek W. The LRP toolbox for artificial neural networks. J Mach Learn Res. 2016;17:114:1–5.

Law R, Barker O, Barker JJ, Hesterkamp T, Godemann R, Andersen O, Fryatt T, Courtney S, Hallett D, Whittaker M. The multiple roles of computational chemistry in fragment-based drug design. J Comput Aided Mol Des. 2009;23:459–73. https://doi.org/10.1007/s10822-009-9284-1.

Li P, Merz KMJ. MCPB.py: a python based metal center parameter builder. J Chem Inf Model. 2016;56:599–604. https://doi.org/10.1021/acs.jcim.5b00674.

Liu T, Lin Y, Wen X, Jorissen RN, Gilson MK. BindingDB: a web-accessible database of experimentally determined protein-ligand binding affinities. Nucleic Acids Res. 2007;35:D198–201. https://doi.org/10.1093/nar/gkl999.

Liu Z, Su M, Han L, Liu J, Yang Q, Li Y, Wang R. Forging the basis for developing protein-ligand interaction scoring functions. Acc Chem Res. 2017;50:302–9. https://doi.org/10.1021/acs.accounts.6b00491.

Maveyraud L, Mourey L. Protein X-ray crystallography and drug discovery. Molecules. 2020;25 https://doi.org/10.3390/molecules25051030.

McNutt AT, Francoeur P, Aggarwal R, Masuda T, Meli R, Ragoza M, Sunseri J, Koes DR. GNINA 1.0: molecular docking with deep learning. J Cheminform. 2021;13:43. https://doi.org/10.1186/s13321-021-00522-2.

Mysinger MM, Carchia M, Irwin JJ, Shoichet BK. Directory of useful decoys, enhanced (DUD-E): better ligands and decoys for better benchmarking. J Med Chem. 2012;55:6582–94. https://doi.org/10.1021/jm300687e.

Pason LP, Sotriffer CA. Empirical scoring functions for affinity prediction of protein-ligand complexes. Mol Inform. 2016;35:541–8. https://doi.org/10.1002/minf.201600048.

Ragoza M, Hochuli J, Idrobo E, Sunseri J, Koes DR. Protein-ligand scoring with convolutional neural networks. J Chem Inf Model. 2017;57:942–57. https://doi.org/10.1021/acs.jcim.6b00740.

Reddy S, Fox J, Purohit MP. Artificial intelligence-enabled healthcare delivery. J R Soc Med. 2018;112:22–8.

Ren X, Shi Y-S, Zhang Y, Liu B, Zhang L-H, Peng Y-B, Zeng R. Novel consensus docking strategy to improve ligand pose prediction. J Chem Inf Model. 2018;58:1662–8. https://doi.org/10.1021/acs.jcim.8b00329.

Rohrer SG, Baumann K. Maximum unbiased validation (MUV) data sets for virtual screening based on PubChem bioactivity data. J Chem Inf Model. 2009;49:169–84. https://doi.org/10.1021/ci8002649.

Sippl MJ. Calculation of conformational ensembles from potentials of mean force. An approach to the knowledge-based prediction of local structures in globular proteins. J Mol Biol. 1990;213:859–83. https://doi.org/10.1016/s0022-2836(05)80269-4.

Srinivasu N, Satyanarayana K, Murthy NGK. Structure based drug design studies on CDK2 amino pyrazole inhibitors using Autodock Tools. 2014 Conf. IT Bus. Ind. Gov. CSIBIG 1–5. 2014.

Stärk H, Ganea O-E, Pattanaik L, Barzilay R, Jaakkola T. EquiBind: geometric deep learning for drug binding structure prediction. 2022; https://doi.org/10.48550/arXiv.2202.05146.

Sunseri J, King JE, Francoeur PG, Koes DR. Convolutional neural network scoring and minimization in the D3R 2017 community challenge. J Comput Aided Mol Des. 2019;33:19–34. https://doi.org/10.1007/s10822-018-0133-y.

Tiwary P, van de Walle A. A review of enhanced sampling approaches for accelerated molecular dynamics. In: Weinberger CR, Tucker GJ, editors. Multiscale materials modeling for Nanomechanics. Cham: Springer; 2016. p. 195–221. https://doi.org/10.1007/978-3-319-33480-6_6.

Trott O, Olson AJ. AutoDock Vina: improving the speed and accuracy of docking with a new scoring function, efficient optimization, and multithreading. J Comput Chem. 2010;31:455–61. https://doi.org/10.1002/jcc.21334.

Vittorio S, Lunghini F, Morerio P, Gadioli D, Orlandini S, Silva P, Martinovic J, Pedretti A, Bonanni D, Del Bue A, Palermo G, Vistoli G, Beccari AR. Addressing docking pose selection with structure-based deep learning: recent advances, challenges, and opportunities. Comput Struct Biotechnol J. 2024;23:2141–51. https://doi.org/10.1016/j.csbj.2024.05.024.

Vogel SM, Bauer MR, Boeckler FM. DEKOIS: demanding evaluation kits for objective in silico screening—a versatile tool for benchmarking docking programs and scoring functions. J Chem Inf Model. 2011;51:2650–65. https://doi.org/10.1021/ci2001549.

Wang L-P, Chen J, Van Voorhis T. Systematic parametrization of polarizable force fields from quantum chemistry data. J Chem Theory Comput. 2013;9:452–60. https://doi.org/10.1021/ct300826t.

Wheeler DL, Barrett T, Benson DA, Bryant SH, Canese K, Chetvernin V, Church DM, DiCuccio M, Edgar R, Federhen S, Geer LY, Kapustin Y, Khovayko O, Landsman D, Lipman DJ, Madden TL, Maglott DR, Ostell J, Miller V, Pruitt KD, Schuler GD, Sequeira E, Sherry ST, Sirotkin K, Souvorov A, Starchenko G, Tatusov RL, Tatusova TA, Wagner L, Yaschenko E. Database resources of the National Center for biotechnology information. Nucleic Acids Res. 2007;35:D5–12. https://doi.org/10.1093/nar/gkl1031.

Wójcikowski M, Ballester PJ, Siedlecki P. Performance of machine-learning scoring functions in structure-based virtual screening. Sci Rep. 2017;7:46710. https://doi.org/10.1038/srep46710.

Wu Z, Ramsundar B, Feinberg EN, Gomes J, Geniesse C, Pappu AS, Leswing K, Pande V. MoleculeNet: a benchmark for molecular machine learning. Chem Sci. 2018;9:513–30. https://doi.org/10.1039/C7SC02664A.

Artificial Intelligence in Ligand-Based Drug Design

Pietro Delre, Carmen Cerchia, Emanuele Falbo, and Antonio Lavecchia

Abstract Ligand-based drug design (LBDD) is a powerful approach in the field of drug discovery that exploits the structural and physicochemical properties of known active molecules to identify and optimize new drug candidates. This chapter provides a comprehensive overview of LBDD, outlining its historical evolution from the earliest concepts of structure–activity relationships to the development of Quantitative Structure–Activity Relationships (QSAR) and pharmacophoric models employing the latest sophisticated techniques. After defining fundamental concepts such as the role of chemical descriptors in the representation of molecular structures and properties (0D, 1D, 2D, 3D, and 4D or SMILES and molecular graphs), the chapter focuses on significant advances in the integration of artificial intelligence and machine learning techniques to improve LBDD methods. Principles for building robust and reliable QSAR and pharmacophore models with AI are outlined, emphasizing the importance of curation of datasets, model development, and validation according to Organization of Economic Cooperation and Development (OECD) guidelines, a posteriori validations, and explainability. The chapter concludes by highlighting the potential impact of AI in modern drug discovery, presenting recent case studies, discussing limitations, and offering a perspective on future directions in this rapidly evolving field.

Keywords Ligand-based drug design · QSAR · Pharmacophores · Artificial intelligence · Machine learning · Molecular modeling · Model validation

P. Delre · C. Cerchia · E. Falbo · A. Lavecchia (✉)
"Drug Discovery Laboratory", Dipartimento di Farmacia, Università degli Studi di Napoli "Federico II", Naples, Italy
e-mail: pietro.delre@unina.it; carmen.cerchia@unina.it; emanuele.falbo@unina.it; antonio.lavecchia@unina.it

A. Lavecchia (ed.), *Applied Artificial Intelligence for Drug Discovery*,
https://doi.org/10.1007/978-3-031-98022-0_5

1 Overview of Ligand-Based Drug Design

Ligand-based drug design (LBDD) is a computer-aided approach to drug design that aims to create new drugs by exploiting knowledge of other active molecules that bind to certain biological targets. Unlike structure-based drug design, which depends on the spatial arrangement of the target protein, LBDD relies on existing active ligands to search for new compounds with the same or similar biological activity (Fig. 1). A basic premise of LBDD is that similar physiological activity will be demonstrated by molecules with a similar structure. Basically, once the physicochemical properties and structural features of known ligands are studied, this information can be used to create models that predict biological activity based on those properties. These models can be used to conduct searches for compounds that can potentially express a desired biological activity, select new lead molecules, or refine the structures of designed molecules. The evolution of LBDD is closely linked to advances in chemistry, medicinal chemistry, and biology, marking a significant advance in modern medicine. According to the basic principle of medicinal chemistry, there is a correlation between the chemical structures of different compounds and their effect on a specific target. Advances in these fields in the twentieth and twenty-first centuries have significantly reinforced this concept. In this context, the most common approach involves the synthesis of structural analogs of the active molecule and testing their activity against the target to evaluate how the structural changes affect biological activity. To maintain or improve biological activity, a widely used method involves bioisosteric substitutions, in which functional groups are replaced with others that have similar physicochemical properties (Lima and Barreiro 2017). This correlation between structure and activity is known as the structure–activity relationship (SAR), and the iterative process of modifying chemical structures to increase their biological activity is known as LBDD. Understanding how chemical structure affects molecular activity provides important insights into how molecules can be modified to improve and maintain activity against a given target. In particular, to appreciate the obtained results, two points of view can be used: a chemical–physical one, highlighting how the change of these properties modifies the activity or searching for those essential characteristics required for a molecule to interact with its target by building a pharmacophore model (for example, the presence of hydrogen bond donors/acceptors, hydrophobic regions, and ionic groups). In essence, the main advantage of LBDD methods lies in their ability to operate independently of detailed knowledge about the 3D target protein structures, placing LBDD as the first choice when dealing with targets whose structure is unknown or difficult to determine experimentally, such as membrane proteins or intrinsically disordered proteins. At the same time, these models may have limitations in the exploration of chemical space because if the known ligands are structurally similar, the resulting models may have difficulty identifying compounds with significantly different scaffolds or mechanisms of action (Shim and MacKerell 2011).

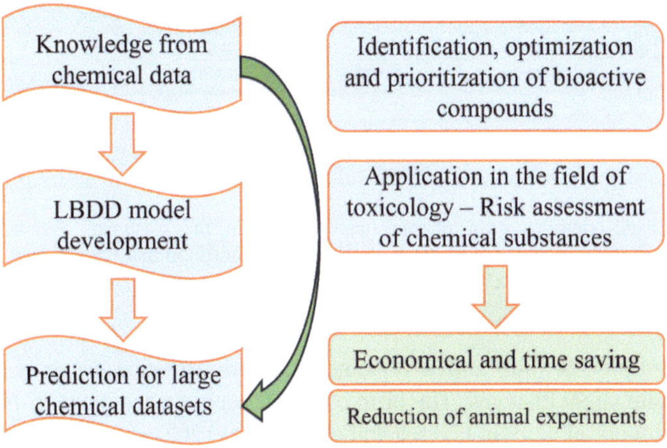

Fig. 1 Schematic representation of the LBDD model development process, from knowledge extraction from chemical data to the prediction for large chemical datasets. This model finds applications in various fields, including toxicology for the risk assessment of chemical substances, offering significant advantages in terms of cost-effectiveness and the reduction of animal experimentation

Key methodologies in LBDD include:

- *Quantitative Structure–Activity Relationship (QSAR) Modeling:* It establishes mathematical relationships between chemical descriptors and biological activity, allowing for the prediction of activity for new compounds.
- *Pharmacophore Modeling:* It identifies the essential features required for molecular recognition between the ligand and the target, such as hydrogen bond donors/acceptors, hydrophobic regions, and ionic groups.
- *Similarity Searching:* It utilizes algorithms to find compounds that are structurally close to known active ligands, under the assumption that they may exhibit similar activity.
- *Combination with Artificial Intelligence (AI) (Machine Learning (ML) Approaches):* It employs statistical and computational techniques to enhance the predictive capabilities of LBDD models by leveraging large datasets alongside QSAR and pharmacophore models.

2 Historical Context and Evolution

2.1 QSAR

Humans are naturally curious. Even young children are constantly trying to understand why something happens, so it is not surprising that for many years, scientists have struggled to understand why some substances are beneficial to health

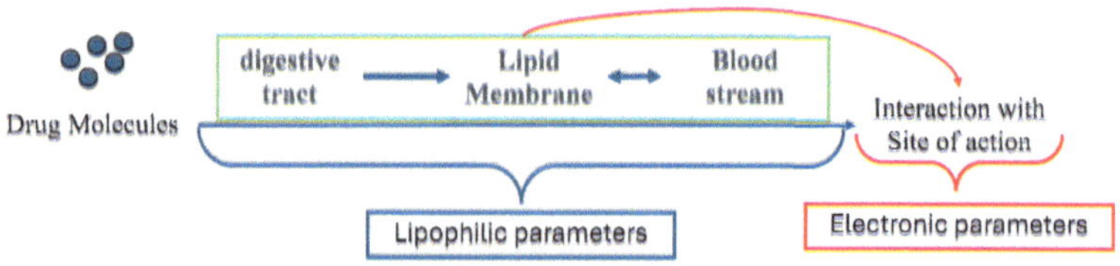

Fig. 2 Simplistic view of drug pharmacokinetics within the organism. According to the Hansh/Fujita model, lipophilic parameters govern drug adsorption and distribution, while electronic parameters determine its binding to the receptor

while others are harmful, and why some are more effective or more toxic than others despite slight changes. The birth of LBDD approaches dates to 1962, when the seminal work of Corwin Hansch and colleagues (1962) stimulated considerable interest in predicting biological activities on a mathematical basis. However, the concept of combining the statistical field with chemistry had been expressed and conceived as early as the nineteenth century by Charles Babbage (1791–1871), the father of computer science, who stated: "The whole of chemistry would become a branch of mathematical analysis which, like astronomy, taking its constants from observation, would enable us to predict the character of any new compound." Furthermore, even before 1900, evidence emerged linking the physical–chemical properties of some compounds to biological activities. For example, in a doctoral thesis at the University of Strasbourg, Cros (1863) observed an inverse correlation between toxicity and aqueous solubility. Indeed, the toxicity of alcohols to mammals increased as the aqueous solubility of alcohols decreased (Dearden 2016). In 1962, Hansch and Fujita advanced a key idea, still valid today, that the site of drug administration differs from the site of its action. Specifically, a drug must go through several phases involving crossing lipid membranes and distribution between various aqueous systems. When the drug reaches its site of action, it interacts with the receptor to produce a biological effect. Consequently, they proposed a simplified model stating that lipophilic properties (ΔH) influenced the initial phase of adsorption and distribution, while electrostatic (ΔE) and steric (ΔS) properties influenced the interaction with the active site (Fig. 2). Therefore, when developing a QSAR model, it is essential to establish a relationship between the lipophilic and electronic properties of the molecule and its biological activity, a concept expressed by the following mathematical equation (Fujita 1974):

$$\Delta \mathrm{BA} = f\,(\Delta E, \Delta H, \Delta S) \tag{1}$$

In parallel, Jerome A. Free and James W. Wilson introduced the Free–Wilson model, approaching the problem from another point of view, focusing on structural modification of a ligand rather than physicochemical properties, assuming that biological activity is directly related to the presence or absence of specific functional groups or substituents in the molecule. These two models paved the way for developing other models based on physicochemical properties and pharmacokinetic and toxicity endpoints (e.g., permeability, plasma protein binding, liver toxicity, carcinogenicity, seizures, and off-target activity) (Kubinyi 1988). Like any technology, the development of QSAR has gone through Gartner's hype cycles, from high expectations followed by periods of disillusionment. These periods have allowed us to understand the limitations of this technique to make significant improvements in the interpretability of results. Furthermore, these techniques have gained momentum through synergy with other computational approaches such as AI (ML and Deep Learning (DL)) and growing computational power.

2.2 Pharmacophore Model

A related technique to QSAR is pharmacophore modeling. Although the term has not been cited, the concept of a "pharmacophore" emerged in the pioneering work of Paul Ehrlich in the late nineteenth century. In 1898, Ehrlich introduced the ideas of "haptophore" (referring to the binding characteristics of a molecule) and "toxophore" (referring to the characteristics responsible for toxicity) as part of his theory of drug–receptor interactions. These concepts laid the foundation for pharmacophores, which were conceived as a method of identifying the molecular characteristics responsible for the biological activity of different compounds. However, confusion occurs regarding the origin of this term because Ehrlich's concept of a "toxophore" overlapped with what others would later call a "pharmacophore." It was not until 1960 that Schueler redefined the pharmacophore as a modern concept still recognized by IUPAC today, explained as a pattern of "abstract features" (including but not limited to specific chemical groups) arranged in three-dimensional space. Over time, this concept has evolved with advances in molecular modeling, particularly with the first pharmacophore models in the 1960s that further refined our understanding of the three-dimensional (3D) spatial arrangements of molecular features that drive biological effects. Specifically, a 3D pharmacophore is a spatial arrangement of chemical features required to elicit biological activity, as it represents the interactions required to bind a specific target (Güner and Bowen 2014). Depending on the available data, such as ligands or ligand–target complexes, several descriptors can describe a pharmacophore model: geometric, substructure model, and molecular field feature. However, when the structure of the target protein is unknown, the essential approach is to identify shared chemical features between the active compounds. In particular, a major push has led to the implementation of several algorithms that, given a library of known target–active compounds, can generate and align the lowest-energy 3D conformations of each compound, detecting crucial shared features among the employed ligands such as the presence of hydrophobic regions, aromatic ring systems, hydrogen bond acceptors, hydrogen bond donors, negatively ionizable groups, and positively ionizable groups, and a range of optimal distances to position these groups (Fig. 2). This approach has limitations, such as the reliance on accurate conformer generation, assuming that the bioactive conformation of each ligand is present in the generated set and corresponds to the energetic minima. However, different ligands may have different binding modes, making this technique challenging. The success of the application depends largely on the quality of the generated conformers and the alignment process. Despite these challenges, 3D pharmacophores remain a valuable tool for drug discovery (Schaller et al. 2020).

3 Concept of Chemical Descriptor in LBDD

The simplified equation described in paragraph 2.1 (Eq. 1), developed by Hansch and Fujita for QSAR, implies that the activity toward a specific target depends on the molecule's physicochemical properties. A branch of chemistry, chemometrics, focuses on translating these properties in a precise quantitative manner. Several mathematical algorithms aim to quantitatively convert molecular properties into precise numbers representing these chemical quantities. A historical example of such a descriptor is the LogP, a quantitative measure of molecular lipophilicity determined by measuring the partitioning of the molecule between an aqueous phase and a lipophilic phase, typically water and n-octanol. The progress over the last few years in chemometrics, cheminformatics, and bioinformatics has led to new strategies for finding new molecular descriptors to capture diverse aspects of molecular structure and properties. Indeed, different software allow the calculation of more than 5000 molecular descriptors. The following subchapter will briefly explain theoretical molecular descriptors, including 0D, 1D, 2D, 3D, and 4D descriptors. For a comprehensive list of molecular descriptors and further details on their calculation and applications, readers are encouraged to consult "Handbook of Molecular Descriptors" by Roberto Todeschini (Molecular Descriptors for Chemoinformatics 2009).

3.1 0 and 1D Descriptors

The 0D descriptors (or constitutional descriptors) do not hold any information about atom connections. Indeed, they rely only on the chemical formula, which organizes the atom types and their occurrences within a molecule. The most usual examples are atom type, number of atoms, bonds, molecular weight (MW), and any function of atomic properties.

1D descriptors move beyond the simple atomic composition (0D) and consider fragments or features within the molecule. They reflect the occurrence of functional groups, bonds, rings, and substituents in a molecule. Despite this, the main limitation is their insensitivity to molecular configuration (Lo et al. 2018; Molecular Descriptors for Chemoinformatics 2009).

3.2 2D Descriptors

The 2D descriptors are more complex than the 1D descriptors. Topological descriptors are molecular descriptors derived from the 2D graph representation of a molecule, focusing on the connectivity of atoms. Vertices represent atoms, while edges are the bonds. They rely exclusively on the molecule's connectivity information, neglecting the 3D structure or physicochemical properties. These 2D graph representations allow the computation of topological indexes (i.e., the Balaban Index, E-state index, etc.) that give an overview of the internal atomic arrangement of compounds (Lo et al. 2018; Molecular Descriptors for Chemoinformatics 2009).

3.3 3D and 4D Descriptors

The 3D descriptors are derived from the 3D coordinates of atoms and aim to capture spatial relationships within a molecule. They incorporate a complete composition of structural data, including the three-dimensional steric shape and electron distribution in a molecule. Geometry influences electronic properties such as polarizability and dipole moment of molecules because they depend on the spatial arrangement of atoms. These descriptors have several limitations due to the need for geometry optimization, typically derived from computationally expensive optimization methods (e.g., density functional theory, Hartree–Fock, molecular mechanics). Furthermore, descriptors may depend on the chosen conformation, making their computation difficult for flexible molecules. To overcome this limitation, Hopfinger et al. proposed a 4D descriptor paradigm that is different from 3D descriptors because 3D structural properties are computed for a set of conformers (conditionally, the fourth dimension) instead of a fixed conformation (Lo et al. 2018; Molecular Descriptors for Chemoinformatics 2009).

3.4 SMILES and Graph as Descriptors

Traditional ML uses descriptors as input features to function. New deep-learning models can learn directly from chemical structures (SMILES or chemical graphs) during the training process, representing molecules as vectors in high-dimensional spaces. As a result, deep-learning models can automatically extract relevant features from the data, eliminating the need to calculate descriptors. This approach allows the model to learn which structural features are crucial as the model efficiently performs automatic feature extraction during the training process, capturing the complex relationships between atoms, bonds, their spatial arrangement, and a defined objective. This approach is powerful as a single model can perform multiple tasks simultaneously. Several levels can constitute a DL model, with a common one capturing general features and other specific levels (fine-tuning), each implemented to find the correlation between learned features and different goals. For example, a DL model can be trained to predict the efficacy and solubility of molecules toward a specific target. However, model improvement over respective single-task QSAR models is not always assured and depends on the level of correlation between activities against individual targets (Sabando et al. 2022; Tropsha et al. 2024).

4 Integration of AI in LBDD

Over time, LBDD approaches have undergone a significant paradigm shift. Initially, research developed predictive models using small sets of congeneric molecules, which shared similar structures and likely identical mechanisms of action, that is, bound to the same binding site. The fundamental limitation was that the resulting models provided poor structure–property mapping and a limited chemical space. However, the emerging concept of data mining with the advancement of computational technologies and AI techniques allows us to analyze and exploit much larger and more complex datasets containing structurally heterogeneous molecules (Fujita and Winkler 2016). This transformation has been made possible by the exponential increase in chemical/pharmaceutical data and their storage in centralized repositories that collect, manage, and archive large volumes of structured, semi-structured, and unstructured data from various sources that, in many cases, are made available online for consultation and download, especially for the field of drug discovery (ChEMBL (Gaulton et al. 2012), PubChem (Kim et al. 2016), DrugBank (Knox et al. 2024)). This huge amount of data is now preferably processed using ML (tree-based algorithms, support vector machines), DL (deep neural networks (NNs)) algorithms, which are very effective in handling complex and high-dimensional chemical data that have a non-linear relationship with the responses of chemicals. The shift from classical QSAR/pharmacophore approaches to modern combined ML-based methods has expanded the scope of drug discovery by enabling virtual screening of large chemical libraries and predicting a wide range of biological activities and properties. In LBDD models, especially QSAR, the predicted outcomes are based on the correlations identified by these complex mathematical models between the descriptors used to describe chemical structures and biological activities. This relationship is not easy to clarify. In other words, users cannot explain why the model expresses its prediction about a chemical entity, making the final model a "black box." However, while the "black box" nature of some QSAR models raises valid concerns, the field of XAI offers promising avenues to improve transparency, understand model reasoning, and unlock the full potential of AI in LBDD (Jiménez-Luna et al. 2020). For these reasons, a set of appropriate guidelines for QSAR model development will be described in the following subsections.

4.1 How to Build a QSAR Model with AI

A developed QSAR model reveals the relationship between chemical structures represented as descriptors and their endpoints. Various mathematical and statistical methods are the basis for developing predictive models, creating a general equation that describes a QSAR model: $y = f(X) + E$, where y is the predicted endpoint for a structure, f is the mathematical function used to relate the descriptors to the endpoint, and E is the error generated by a developed model that may arise due to limitations of the model, noise in the data, or inherent variability in the studied system. In short, QSAR development consists of five main steps (Fig. 3). The main one is the creation of a database containing known information, such as chemical structures and biological endpoints. In particular, the first step is the data curation process to remove any errors in the dataset. After choosing the descriptors to represent the chemicals and the mathematical method to create the function that correlates descriptors and endpoints, the database is divided into two parts: a training set and a validation/external set. A training set trains

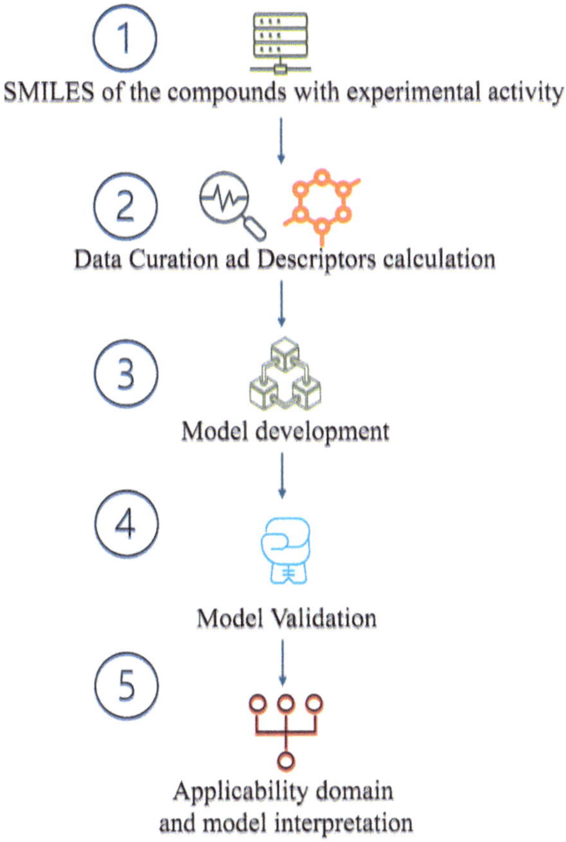

Fig. 3 Schematic representation of the QSAR model development process. (*1*) Data collection: gathering chemical structure and bioactivity data from relevant sources. (*2*) Data preprocessing: data curation and molecular descriptor calculation. (*3*) Model building: construction of the predictive model using ML or statistical techniques. (*4*) Model validation: evaluation of the model's performance through statistics. (*5*) Prediction: applicability domain determination, and interpretation of predictions

the model, modeling the mathematical function based on the available data. A validation/external set challenges the developed function to test its effectiveness on unseen data. Comparing actual data with the predicted data generated by the model on the validation set using different statistical metrics will evaluate the effectiveness of the developed model. Since these steps can generate drawbacks, the following subsection will focus on each step.

4.1.1 Dataset Curation

Researchers can easily download large datasets from different sources due to the rapid data access provided by high-speed Internet, including public repositories, while in the past, they were limited to building QSAR models based on their proprietary data. The current trend is to merge datasets from different sources to create comprehensive databases. This strategy is powerful in AI fields since the learning efficiency of these algorithms typically increases with the amount of data. Indeed, using larger datasets allows models to learn more general and robust patterns, ensuring reliable predictions for a large number of unseen data. However, the quality of these data can be dissimilar due to the different experimental protocols employed (Landrum and Riniker 2024). Furthermore, public data sources require careful evaluation due to a lack of quality control. The main problems found in public databases include structural errors when chemical structures are incorrect (i.e., structural errors) and inaccurate activity information reported (i.e., experimental errors). For example, Kramer et al. (2012), evaluating the quality of Ki data present in ChEMBL 12 within the same ligand–protein pair, found that the most common errors include unrealistically high or low activity values; multiple values for the same ligand–protein pair derived from a single publication; duplicates of the same activity for different compounds, and transcription and conversion errors. When multiple activity values exist for the same ligand–protein pairs, classical statistical data mining, such as duplicate removal, data mean aggregation, and standard deviation calculation, can mitigate these experimental errors by removing outliers. In addition to these experimental errors, structural errors are a crucial concern in the model training phase. Descriptors computed from an incorrect molecular structure can distort the final mathematical function developed and thus the final quality of the model. In summary, the presence of incorrect chemical structures in the training set will negatively affect the quality of the final model (Tropsha 2010). Initial steps in cleaning chemical records in a database include standardizing chemical structures and removing a fraction of the data that conventional cheminformatics techniques cannot handle such as inorganic and organometallic compounds, counterions, salts, and mixtures; structure validation; ring aromatization; normalization of specific chemotypes; curation of tautomeric forms; and elimination of duplicates. The importance of standardization is based on the presence of multiple SMILES that can represent the same compounds, such as aromaticity or tautomeric forms. Manual inspection of large chemical datasets is usually impossible because there are so many chemicals to analyze. For this reason, there is an effort to develop automated or semi-automated methods for data retrieval, curation, and standardization (Tropsha 2010). It is worth noting that some automated workflows, such as those created by Gadaleta et al. (2018), compare SMILES representations across different databases to identify potential inconsistencies or errors and use the frequency of occurrence. Records are evaluated based on the consistency of information retrieved from different web sources to assess the reliability of a given SMILES (Gadaleta et al. 2018). Finally, a concern arises due to the presence of activity cliffs in the database that can affect the performance of a QSAR model. Their inclusion requires special attention because the different experimental values obtained for two molecules that differ only by the presence or absence of a group or different chirality could be due to a wide range of mechanisms such as outliers of different types or even the presence of reactive functional groups (Dablander et al. 2023; Tropsha 2010).

4.1.2 Model Development and Validation

When developing a model, the main aim is to provide the scientific community with a product that meets different criteria. For a final user, the crucial aspect is understanding the final endpoint, the usability, and the most straightforward way to make a prediction. Furthermore, the comprehension of predictions, or interpretability, is a fundamental aspect. On the other hand, as with other experiments in science, other scientists must have all the elements to reproduce the model. For this reason, the Organization of Economic Cooperation and Development (OECD), an international organization of 37 countries that works to establish evidence-based international standards, proposed a set of five point seminal guidelines in 2004 for the proper development and validation of predictive QSAR (Gómez-Jiménez et al. 2018; Gramatica 2007). The principles are the following:

1. A defined endpoint.
2. An unambiguous algorithm.
3. A defined domain of applicability.
4. Appropriate measures of goodness-of-fit, robustness.
5. A mechanistic interpretation, if possible.

1. Defined Endpoint and 2. Unambiguous Algorithm

A defined endpoint is explicit in the curated dataset, which often consists of the collected experimental activity, reported as IC_{50}, EC_{50}, K_i, or other measures, and the related compound expressed as SMILE, against a specific target retrieved from the public repository mentioned in Table 1, merged or not, and curated. Developers use this dataset to train and validate models using different validation techniques, for example, splitting it into a training (typically comprising 70% or 80% of the data) and a validation set (comprising the remainder of a dataset after the training set has been generated) or employing cross-validation using the entire dataset. Cross-validation techniques, such as leave-one-out (LOO), k-fold, and

Table 1 A comprehensive list of curated databases, including their URLs, containing pharmacological information on targets and their ligands. The data includes bioactivity measurements such as IC_{50}, EC_{50}, K_i

Repository	Link
ChEMBL	https://www.ebi.ac.uk/chembl/
PubChem	https://pubchem.ncbi.nlm.nih.gov/
BindingDB	https://www.bindingdb.org/
IUPHAR/BPS	https://www.guidetopharmacology.org/
Therapeutic target database (TTD)	http://db.idrblab.net/ttd/
DrugBank	https://go.drugbank.com/
PDSP Ki database	https://pdsp.unc.edu/databases/kidb.php

validation, evaluate the performance of the model by iteratively training and testing on subsets of the original dataset. In particular, LOO training involves removing one data point at a time and predicting this point to be excluded from training. However, LOO is computationally expensive for large datasets and can lead to high variance in performance evaluation due to sensitivity to outliers. Additionally, if the training set is too large, the model's performance may be overestimated. K-fold cross-validation divides the dataset into k equal parts, then iteratively trains a model on the $k-1$ parts and predicts the remaining part as a test (Gramatica 2007). Specifying the formalism implemented during data preprocessing, dataset partitioning, descriptor computation, and the statistical method employed during model development brings transparency and makes the procedure reproducible. The dataset splitting can follow random choice or rational techniques, such as different algorithms like Kennard–Stone, minimum dissimilarity of prediction set, and sphere exclusion algorithms (Puzyn et al. 2011). These are automatic ways of dividing datasets into training and prediction sets to produce more predictive models. Based on the final endpoint, the generated model can be either classificatory or regressor. Specifically, a classification model predicts the class to which a compound belongs, judging it to be active (positive class) or inactive (negative class) toward a specific target. For this purpose, given experimental data, the user sets a discriminatory threshold to separate two or more categories of compounds. A disproportion between negative and positive instances can lead to models biased toward the majority class, failing to accurately predict the instances of the minority class. In this regard, the unbalanced ratio can measure the number of negative class data instances divided by the number of positive classes to assess the balance of the dataset (Delre et al. 2022, 2023). However, sometimes, datasets are unbalanced due to the lack of experimental data or reflect the real-world distribution. Depending on the context, some techniques to develop unbiased models employ minority or majority under sampling/oversampling approaches and specific metrics to evaluate their performance (Blagus and Lusa 2015). Regressor models predict a continuous outcome variable, in our case, the experimental activity (IC_{50}, EC_{50}, K_i). After preparing the dataset, SMILES are used to compute a set of descriptors from those described in the previous paragraphs. The correlation between the endpoint and these descriptors is pursued through statistical methods, among which ML techniques are gaining popularity due to their effectiveness. Examples of ML approaches are Random Forest, gradient boost, Support Vector Machine (SVM), and Neural Network.

- Random Forest (RF) is an ensemble learning method. The ensemble of trees constitutes the forest. After each tree training on a random subset of the data, the final prediction is an average between all the trees following a majority vote approach. Having a forest of trees reduces overfitting and improves model accuracy (Breiman 2001).
- Gradient Boosting is another method based on tree ensembles that works similarly to the random forest. However, random forest builds trees independently and in parallel, while gradient Boosting builds trees sequentially, where each new tree learns from the mistakes of the previous one. In summary, the final prediction derives from the sum of the contributions at the final prediction of each tree rather than the result of a majority vote approach (Friedman 2001). XGBoost, LightGBM, and CatBoost are different implementations of Gradient Boosting for the optimizations and variations in their tree-building process (Ahn et al. 2023).
- Support Vector Machine is a supervised ML algorithm that works by projecting data belonging to different categories of compounds in a feature space and separating them with a hyperplane developed using complex mathematical functions known as kernels (linear, polynomial, radial basis function (RBF), and sigmoid) to maximally separates the classes of data points (Cristianini and Ricci 2008).
- Neural Networks mimic the biological structure of the brain. They consist of interconnected nodes, or neurons, organized in layers and connected between them. Information flows through the network, and the connected layers cooperate to transform the input data using a mathematical function to approximate a complex function that maps inputs to outputs. During the training process, the network adjusts weights, denoting the impact of each node on the final prediction to minimize prediction errors. DL is a branch of ML that develops specific neural networks for complex tasks, emulating human behavior such as natural language, thinking, and visual reproduction (Kriegeskorte and Golan 2019).

Training these models needs careful parametrization. Knowing these parameters, the employed training and validation set, and the chosen descriptors to implement a transparent and unambiguous model (Nematzadeh et al. 2022).

3. Appropriate Measures of Goodness-of-Fit, Robustness

To evaluate the performance of a classification or regression model, several metrics can be used. For a classification model, a straightforward approach is to build a confusion matrix (Fig. 4) after comparing actual and predicted values (Ting 2010).

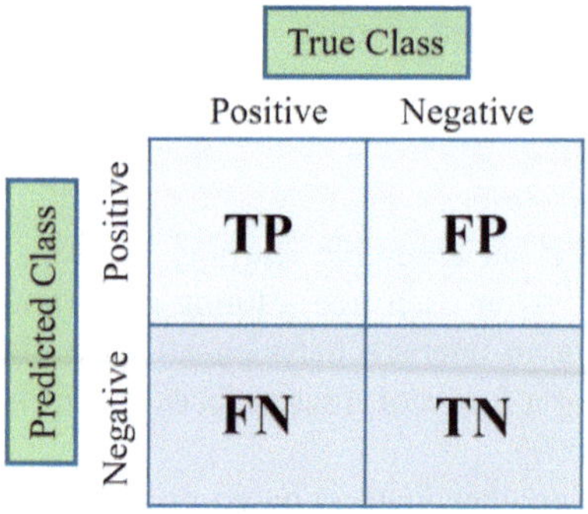

Fig. 4 Example of a confusion matrix, illustrating the four categories used to evaluate the performance of a QSAR classification model: True Positives (TP) are active compounds correctly predicted as actives; False Negatives (FN) are active compounds incorrectly classified as inactive; True Negatives (TN) are inactive compounds correctly predicted as inactive; and False Positives (FP) are inactive compounds incorrectly predicted as active

From the confusion matrix, metrics such as sensitivity, specificity, and accuracy can be calculated.

Sensitivity (SE; Eq. 2) reflects the proportion of actual positives that are correctly identified. Specificity (SP; Eq. 3) measures the proportion of actual negatives that are correctly identified. Accuracy (ACC; Eq. 4) gives an overall picture of model performance by determining the proportion of samples, active and inactive, that are correctly identified. However, accuracy leads to erroneous evaluation when the dataset is unbalanced. In this regard, the most informative and reliable metrics for assessing the performance of a model are the Balanced Accuracy (BA; Eq. 5), G-mean (Eq. 6), and the Matthews correlation coefficient (MCC; Eq. 7). BA and G-mean are the average and geometric mean of sensitivity and specificity estimated to account for the imbalance between classes by giving equal weight to both the positive and negative classes. The MCC is a correlation coefficient between the observed and predicted binary classifications, and it returns a value between -1 and $+1$. A coefficient of $+1$ signifies a perfect prediction, 0 an average random prediction, and -1 an inverse prediction. Metrics to evaluate a regressor model's performance are R^2 and Q^2 (Eq. 8), reflecting the degree of correlation between the observed and predicted activity data in external and internal validation, respectively (Chicco and Jurman 2020; Naidu et al. 2023; Thölke et al. 2023).

$$\mathrm{SE} = \frac{\mathrm{TP}}{\mathrm{TP} + \mathrm{FN}} \tag{2}$$

$$\mathrm{SP} = \frac{\mathrm{TN}}{\mathrm{TN} + \mathrm{FP}} \tag{3}$$

$$\mathrm{ACC} = \frac{\mathrm{TP} + \mathrm{TN}}{\mathrm{TP} + \mathrm{FP} + \mathrm{TN} + \mathrm{FN}} \tag{4}$$

$$\mathrm{BA} = \frac{\mathrm{SE} + \mathrm{SP}}{2} \tag{5}$$

$$G - \mathrm{mean} = \sqrt{\mathrm{SE} * \mathrm{SP}} \tag{6}$$

$$\mathrm{MCC} = \frac{\mathrm{TP} * \mathrm{TN} - \mathrm{FP} * \mathrm{FN}}{\sqrt{(\mathrm{TP} + \mathrm{FP})\,(\mathrm{TP} + \mathrm{FN})\,(\mathrm{TN} + \mathrm{FP})\,(\mathrm{TN} + \mathrm{FN})}} \tag{7}$$

Metrics to evaluate a regressor model's performance are R^2 and Q^2 (Eq. 8), reflecting the degree of correlation between the observed and predicted activity data in external and internal validation, respectively (Chicco et al. 2021).

$$R^2\left(Q^2\right) = 1 - \frac{\sum\left(y_{\mathrm{actual}} - y_{\mathrm{predicted}}\right)^2}{\sum\left(y_{\mathrm{actual}} - y_{\mathrm{average}}\right)^2} \tag{8}$$

y_{actual} and $y_{\mathrm{predicted}}$ are the actual and predicted activity values, respectively, and y_{average} indicates the average actual activity value of the training chemicals.

These metrics are suitable for all validation approaches, including cross-validation, external validation, and experimental validation.

4. A Defined Domain of Applicability

QSAR models, regardless of the statistical technique used to develop them, are trained with a given set of compounds. The relationships between chemical–physical properties and endpoints that the model learns can define its applicability domain (AD). Generally, a prediction is reliable when interpolated; thus, the system predicts a substance with physical descriptors and activity relationships similar to those it has learned. When a chemical differs from those composing the training set, the prediction extrapolated by assumption could be meaningless, random, and probably incorrect.

Methods to calculate the AD include:

- *Range-based methods* define the AD according to the range of values of each descriptor used to train the model. A model makes an unreliable prediction if one or more descriptors violate this range. However, this approach assumes that the TS uniformly covers the entire descriptor space, but some regions in the range could be empty. To address this problem, rather than the maximum and minimum values, AD could be centered on the mean value of each descriptor in the TS, considering in or out the AD, those descriptors with values upper or lower than

$\pm nz$. However, range-based methods consider each descriptor independently, without accounting for their relationships.

- *Geometric-based* methods are an evolution of interval-based methods. Computing principal component analysis (PCA) on descriptors allows all data to be projected as points into a specified space. The Convex Hull method surrounds this space, constructing the smallest convex polygon that encloses all points. An exterior compound is in AD if, by applying the previously computed PCA and plotting, the obtained point falls in the convex region.
- *Probability density-based methods:* This approach relies on the idea that the training set compounds form a distribution in the descriptor space. Methods to calculate this distribution could be parametric or nonparametric. Parametric methods assume that data follow a Gaussian or Poisson distribution. Non-parametric methods, like Kernel density estimation, grasp the distribution of descriptors, accounting for eventual discontinuity. Then, given an external compound, the algorithm computes the probability that this compound belongs to the same distribution computed previously for the TS descriptors.

For a more accurate mathematical description, the reader is referred to *Applicability Domain for QSAR Models: Where Theory Meets Reality* (Gadaleta et al. 2016). However, the estimation of the utility of AD presents a significant challenge because of the lack of a clear rationale for preferring one method over another. For this reason, there are efforts to evaluate the empirical choice to admit a compound in or out of the applicability domain through statistical analysis to reveal the reliability of the prediction in the same AD (Mora et al. 2024; Sahigara et al. 2012; Tetko et al. 2008).

5. A Mechanistic Interpretation, If Possible

AI QSAR models make mathematical relationships increasingly challenging to understand; this is an unfortunate trade-off between high model performance and transparency. Given an external dataset, we can observe the set of predictions made by the model without knowing or understanding its internal workings. As mentioned, unlike mathematical models possessing an intrinsic structure, ML models learn by developing non-linear functions. For some models, such as single decision trees, we can understand the flow of decisions that led to a final prediction. For others, such as random forests or DL models, it becomes nearly impossible to understand the internal function of these models, making them black boxes (Fig. 5). This lack of transparency makes users less confident and trusting when using the models. However, this problem is not specific to QSAR models but in all the fields involving AI approaches.

Explainable AI (XAI) is the branch that focuses on explaining internal decision-making. Therefore, the main goal is to improve the safety of social use of AI models, encouraging their wider adoption. In short, interpretability or explainability is the ability of the model to provide a prediction in a way that is easily understandable by

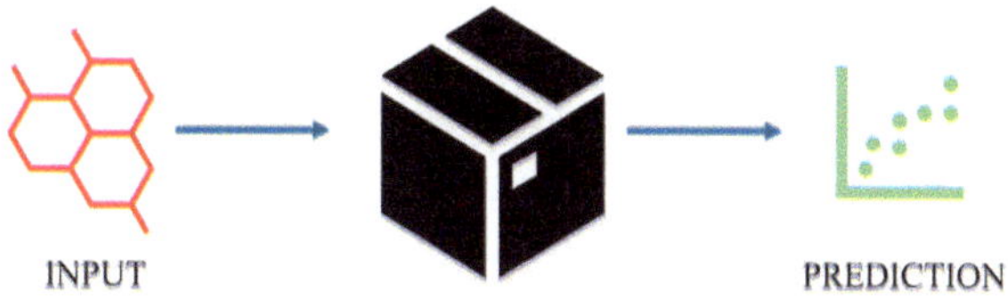

Fig. 5 Concept of a Black Box Model. The text should be positioned beneath the figure, following standard caption formatting

humans. By understanding how the model operates, one can identify if the model's actions are misaligned, misguided, or counterproductive, enabling corrective measures.

As QSARs often rely on tabular data, the most used XAI algorithms are the Shapley Additive explanations (SHAP) and Local Interpretable Model Agnostic Explanation (LIME) (Salih et al. 2024).

- SHAP is an XAI method based on game theory, considering each feature (or predictor) as a player and the model outcome as the payoff. Briefly, SHAP provides local and global explanations, meaning that it explains the role of each descriptor in predicting all instances and a specific instance. SHAP values are theoretically derived assuming feature independence and do not capture the correlation or interaction between the features. For this reason, the most important result is the descriptor ranking rather than the absolute value. Finally, it is worth noting that applying the SHAP algorithm to different ML algorithms trained on the same dataset could lead to distinct descriptor rankings.
- LIME is another XAI method that explains how the model performs locally for a specific sample in the model. LIME is model agnostic, which means it can be applied to any ML model, regardless of its type or complexity. To do this, LIME approximates the complex starting model into a simplified, linear model. To train this simplified model, LIME generates artificial data points around the sample of interest by perturbing the features of the original sample. This perturbation helps LIME understand how changes in feature values affect the model's prediction around the sample of interest. The trained simplified model is then applied to predict the instance of interest, ultimately explaining the prediction. It is crucial to highlight that LIME interpretation applies only to one subject and cannot be used or considered a general interpretation for the entire model.

Among other approaches (Belle and Papantonis 2020), SHAP and LIME described above are some examples of the most commonly used XAI in LBDD approaches (Rodríguez-Pérez and Bajorath 2020). Although these algorithms are powerful, using them requires a deep understanding of how they work to avoid misinterpreting the results provided by the XAI algorithms.

4.1.3 A Posteriori Validation

For robust external validation, after validating the model following the procedures described above, it is necessary to test model performance on compounds derived from a different distribution than those of the starting set. For this purpose, QSAR model developers use external data from sources different from those used to validate the model. This method is effective for assessing the generalizability and detecting possible overfitting. The recovery of external data involves experimental data published after validation by checking the update of the employed database since the version used to develop the model. This approach is known as time-split cross-validation (Sheridan 2013) and tests models by predicting their usefulness at a future time. Despite the name, the key factor is not time but the "design continuity" that leads to developing novel compounds after gathering knowledge from data obtained by testing compounds from the previous project. This fact leads to an overall trend of increasing potency and diversity in chemical–physical properties from the earliest to the most recent compounds. However, time-sliced public datasets do not give the order of synthesis of chemical compounds, making the time-split validation approach inapplicable. Landrum et al. developed SIMPD (Landrum et al. 2023), a multi-objective genetic algorithm for generating simulated time splits to validate ML approaches. Specifically, the SIMPD algorithm simulates time splits reproducing differences in the descriptor distributions and spatial statistics between the early compounds in a project (training set) and the later compounds (test set), predicting the performance of a predictive model when applied prospectively to project data. Another solution is to use data from a particular laboratory, but laboratories generally provide congeneric series. In conclusion, combining these strategies is nowadays the gold standard as they complement each other: external validation guarantees broad applicability, while experimental validation emphasizes biological relevance.

4.2 AI in Pharmacophore-Based LBDD

Traditional pharmacophore modeling is complex, time-consuming, and often subjective, leading to inconsistent results across different programs and researchers and generally on a subset of active compounds. Before the 2000s, Chen et al. (1999) proposed a system that analyzed a dataset of several thousand compounds and then generate suggestions for pharmacophore models based on the knowledge obtained. Finally, the model presented the obtained solutions to the researcher, who could decide on the best solution. There are multiple commercial programs providing automatic pharmacophore model generation tools such as Discovery Studio, LigandScout, MOE, PHASE, Pharmer, and PharmaGist. A recent example of how to employ ML techniques to develop regressor models having biological activity as the specific endpoint is the paper published in 2021 by Kohlbacher et al. (2021), which presented a new approach for creating quantitative pharmacophore models (QPHAR). The first step is to select a template pharmacophore generated from molecules composing the training set, which generally corresponds to the one generated by the most rigid one. Then, each pharmacophore in the training set is aligned to the template. Thus, a merged pharmacophore is generated containing the hydrophobic (H), aromatic (AR), positive/negative ionizable (PI, NI), and H-bond donor/acceptor (HBD, HBA) information in specific separate compartments. Each compartment is then clustered and refined to extract a single essential feature, creating the final pharmacophore model. Given this final model, a vector with a length corresponding to the number of features is generated for each molecule of the training set, computing the inverse Euclidean distances between the features from the reference pharmacophore and the aligned training sample features. Another concept reported in the literature focuses on the development of molecular fingerprints to describe the pharmacophoric properties of chemical structures. As mentioned earlier, fingerprints are suitable descriptors for ML training. A pharmacophore fingerprint extends the original concept and typically annotates a molecule as a unique data string. Generating pharmacophore fingerprints involves identifying potential pharmacophore locations and evaluating existing pathways within specific distance ranges. The algorithm then evaluates the presence of these pathways for each possible occurrence and encodes them in the fingerprint, setting 1 for the presence of a specific feature and 0 for its absence. Examples of pharmacophore fingerprints are the Fuzzy 3-point pharmacophore fingerprints and pharmacoprints (Bonachéra et al. 2006). These vectors or fingerprints are descriptors ready-to-use to develop a regressor model following the same principle explained for QSAR models.

4.3 Case Studies

The first case study presented addresses the critical problem of drug-resistant malaria, a disease caused by Plasmodium parasites, predominantly *P. falciparum* and *P. vivax*, which cause nearly 600,000 deaths annually. To find new drugs active on *P. falciparum* and *P. vivax*, Borba et al. (2024) attempted to build QSAR using ML algorithms (random forest, support vector machine, light gradient boosting machine) and large datasets from ChEMBL, PubChem, and the literature. They also divided the database into several subsets following the life cycle stages of Plasmodium, including asexual blood stages, gametocytes, ookinetes, and liver schizonts. Then, each subset was divided into 80% as a training set and the last 20% as an outer set by a random distribution approach, and the compounds were represented by predominantly 1D molecular descriptors, specifically different fingerprints. These were used to train different ML algorithms validated through fivefold cross-validation and an external set. These models reported internal and external BAs greater than 0.70. The application of SHAP analysis ensured the explainability of the model by marking the molecular segment responsible for the final predictions. For example, the author stated that by subjecting known compounds that share an aminoquinoline scaffold but differ in their substituents, the SHAP analysis revealed that electron-withdrawing substituents, such as chloride and bromide, at the seventh position of the quinoline ring increase antiplasmodial activity, in agreement with SAR analysis derived from experimental studies. Applying these five models, thus validated, to the ChemBridge version 2020 database, which contained 1,229,342 compounds, only 61 compounds were predicted as active by all the models. These 61 compounds were grouped by chemical similarity, resulting in 16 representative compounds. From the in vitro studies conducted on these 16 compounds, 6 showed activities in two or three stages of the *Plasmodium* life cycle, and one compound (**LDT-695**) showed more than 50% inhibition in all stages tested (liver stage, asexual blood stage, gametocytes, and ookinetes), demonstrating the effectiveness of the strategy employed.

Tsou et al. (2020), using deep neural networks (DNN) and random forest, identified several Triple-Negative breast cancer (TNBC) inhibitors as potent hits from a screening of an in-house database of 165,000 compounds. They recovered a database of 7130 molecules with previously reported MDA-MB-231 cell lines inhibitory activities collected from ChEMBL. In this study, these compounds were randomly separated into 6069 compounds (the training set) and 1061 compounds (the test set) and represented by circular topological fingerprints (ECFPs). For the regression models, a high R^2 value above 0.80 validated the system. The trained DNN and RF models were used to screen an in-house database of 165,000 compounds for potential TNBC inhibitors. The top 100 predicted compounds were experimentally tested for their ability to inhibit the MDA-MB-231 TNBC cell line. Twelve compounds (six from each model) showed inhibitory activity and, importantly, low cytotoxicity against the nonmalignant mammary epithelial cell line MCF10A. Further testing against other TNBC cell lines (BT-549 and MDA-MB-453) identified four compounds (3, 7, 8, and 10) with broader TNBC inhibitory activity.

Chang et al. (2024) aimed to discover small-molecule inhibitors targeting the epidermal growth factor receptor tyrosine kinase 858R/T790M/C797S (EGFR-TKI) triple mutation, which has emerged as a primary therapeutic option for patients in advanced stages of nonsmall cell lung cancer (NSCLC). The authors collected high-quality data on 221 compounds active against the EGFRL 858R/T790M/C797S triple mutation with high inhibitory activity ($IC_{50} < 5\ \mu M$) from 30 recent articles. Conversion of IC_{50} values to pIC_{50} provides a standardized and uniform scale. RDKit calculated the molecular descriptors from the SMILES of the compounds. Then, an integrated framework using the full quadratic effect and the Lasso model was proposed to select significant features that resulted in 12 critical descriptors. The SVR model reported a Q^2 value of 0.853 in the internal validation and an R^2 of 0.745 in the external validation. Next, the application of the model for virtual screening on a dataset containing about 50,000 compounds extracted from the CHEMBL database identified 35 potential compounds, which were further filtered using molecular docking and molecular dynamics, leading to the identification of four potentially active compounds.

Another concern in drug discovery is the design of CB2R ligands as therapeutic targets for anti-inflammatory, cancer, neurodegeneration, and other inflammatory diseases. Developing selective ligands is crucial to avoid the psychoactive side effects elicited by the CB1R activation (44% sequence homology). Delre et al. (2023) developed ALPACA, a freely accessible web platform utilizable by medicinal chemists without specific cheminformatics or programming skills, relying on ML models to predict the affinity and selectivity of CB2R ligands. The ML models were trained on bioactivity data from ChEMBL using atom pair (AP) fingerprints and five different classification algorithms. The web server makes predictions via the best-performing model (random forest) to identify molecules with high affinity for CB2R ($pK_i > 6.5$ or $pK_i > 7$) and low affinity for CB1R ($pK_i > 5.5$ or $pK_i > 6$). The authors validated the models, obtaining remarkable performances with BA and MCC in internal and external validation, gathering compounds from other sources like PubChem and BindingDB, ranging from 0.84 to 0.91 and 0.70 to 0.82, respectively.

These in-depth case studies, among others (Gonçalves et al. 2023; Lanka et al. 2023; Mamada et al. 2022; Maurya et al. 2020; Singh et al. 2023; Yadav et al. 2023; Zhang et al. 2021), offer insight into how AI and ligand-based models accelerate the drug discovery design process.

4.4 Conclusion and Future Perspective

This chapter has highlighted the pivotal role of LBDD in modern drug discovery, emphasizing its evolution from starting with simple mathematical models to advanced computational methodologies. The integration of AI marks a significant paradigm shift, enhancing the predictive power of QSAR and pharmacophore models and accelerating the identification of promising drug candidates. The increasing endeavor in molecular descriptors and the algorithm evolution process generate optimistic promise for future advancement. Furthermore, the DL method can learn the logical construction of molecules directly from SMILES or graph representation, embedding the features in a latent chemical space. However, several pitfalls can occur during a QSAR development, such as data preparation for training the model, validity assessment, and the trade-off between power and explainability. All these areas need careful and deep comprehension to develop a final model meeting the OECD principles. As shown in case studies, the synergistic combination of LBDD and AI is revolutionizing drug discovery, leading to the faster, more efficient, and more successful development of novel therapeutics.

References

Ahn JM, Kim J, Kim K. Ensemble machine learning of gradient boosting (XGBoost, LightGBM, CatBoost) and attention-based CNN-LSTM for harmful algal blooms forecasting. Toxins (Basel). 2023;15:608. https://doi.org/10.3390/toxins15100608.

Belle V, Papantonis I. Principles and practice of explainable machine learning. 2020; https://doi.org/10.48550/arXiv.2009.11698.

Blagus R, Lusa L. Joint use of over- and under-sampling techniques and cross-validation for the development and assessment of prediction models. BMC Bioinformatics. 2015;16:363. https://doi.org/10.1186/s12859-015-0784-9.

Bonachéra F, Parent B, Barbosa F, Froloff N, Horvath D. Fuzzy tricentric pharmacophore fingerprints. 1. Topological fuzzy pharmacophore triplets and adapted molecular similarity scoring schemes. J Chem Inf Model. 2006;46:2457–77. https://doi.org/10.1021/ci6002416.

Borba JVB, Salazar-Alvarez LC, Ferreira LT, Silva-Mendonça S, da Silva MFB, Sanches IH, da Costa Clementino L, Magalhães ML, Rimoldi A, Calit J, Santana S, Prudêncio M, Cravo PV, Bargieri DY, Cassiano GC, Costa FTM, Andrade CH. Innovative multistage ML-QSAR models for malaria: from data to discovery. ACS Med Chem Lett. 2024;15:1386–95. https://doi.org/10.1021/acsmedchemlett.4c00323.

Breiman L. Random forests. Mach Learn. 2001;45:5–32. https://doi.org/10.1023/A:1010933404324.

Chang H, Zhang Z, Tian J, Bai T, Xiao Z, Wang D, Qiao R, Li C. Machine learning-based virtual screening and identification of the fourth-generation EGFR inhibitors. ACS Omega. 2024;9:2314–24. https://doi.org/10.1021/acsomega.3c06225.

Chen X, Rusinko A, Tropsha A, Young SS. Automated pharmacophore identification for large chemical data sets. J Chem Inf Comput Sci. 1999;39:887–96. https://doi.org/10.1021/ci990327n.

Chicco D, Jurman G. The advantages of the Matthews correlation coefficient (MCC) over F1 score and accuracy in binary classification evaluation. BMC Genomics. 2020;21:6. https://doi.org/10.1186/s12864-019-6413-7.

Chicco D, Warrens MJ, Jurman G. The coefficient of determination R-squared is more informative than SMAPE, MAE, MAPE, MSE and RMSE in regression analysis evaluation. PeerJ Comput Sci. 2021;7:e623. https://doi.org/10.7717/peerj-cs.623.

Cristianini N, Ricci E. Support vector machines. In: Kao M-Y, editor. Encyclopedia of algorithms. Boston: Springer; 2008. p. 928–32. https://doi.org/10.1007/978-0-387-30162-4_415.

Dablander M, Hanser T, Lambiotte R, Morris GM. Exploring QSAR models for activity-cliff prediction. J Chem. 2023;15:47. https://doi.org/10.1186/s13321-023-00708-w.

Dearden JC. The history and development of quantitative structure-activity relationships (QSARs). Int J Quant Struct Prop Relat. 2016;1:1–44. https://doi.org/10.4018/IJQSPR.2016010101.

Delre P, Lavado GJ, Lamanna G, Saviano M, Roncaglioni A, Benfenati E, Mangiatordi GF, Gadaleta D. Ligand-based prediction of hERG-mediated cardiotoxicity based on the integration of different machine learning techniques. Front Pharmacol. 2022;13. https://doi.org/10.3389/fphar.2022.951083.

Delre P, Contino M, Alberga D, Saviano M, Corriero N, Mangiatordi GF. ALPACA: a machine learning platform for affinity and selectivity profiling of CAnnabinoids receptors modulators. Comput Biol Med. 2023;164:107314. https://doi.org/10.1016/j.compbiomed.2023.107314.

Friedman JH. Greedy function approximation: a gradient boosting machine. Ann Stat. 2001;29:1189–232. https://doi.org/10.1214/aos/1013203451.

Fujita T. The extrathermodynamic structure-activity correlations. In: Biological correlations—the Hansch approach, Advances in chemistry. American Chemical Society; 1974. p. 1–19. https://doi.org/10.1021/ba-1972-0114.ch001.

Fujita T, Winkler DA. Understanding the roles of the "Two QSARs". J Chem Inf Model. 2016;56:269–74. https://doi.org/10.1021/acs.jcim.5b00229.

Gadaleta D, Mangiatordi GF, Catto M, Carotti A, Nicolotti O. Applicability domain for QSAR models: where theory meets reality. Int J Quant Struct Prop Relat. 2016;1:45–63. https://doi.org/10.4018/IJQSPR.2016010102.

Gadaleta D, Lombardo A, Toma C, Benfenati E. A new semi-automated workflow for chemical data retrieval and quality checking for modeling applications. J Cheminform. 2018;10:60. https://doi.org/10.1186/s13321-018-0315-6.

Gaulton A, Bellis LJ, Bento AP, Chambers J, Davies M, Hersey A, Light Y, McGlinchey S, Michalovich D, Al-Lazikani B, Overington JP. ChEMBL: a large-scale bioactivity database for drug discovery. Nucleic Acids Res. 2012;40:D1100–7. https://doi.org/10.1093/nar/gkr777.

Gómez-Jiménez G, Gonzalez-Ponce K, Castillo-Pazos DJ, Madariaga-Mazon A, Barroso-Flores J, Cortes-Guzman F, Martinez-Mayorga K. Chapter Four – The OECD principles for (Q)SAR models in the context of knowledge discovery in databases (KDD). In: Karabencheva-Christova

TG, Christov CZ, editors. Advances in protein chemistry and structural biology, computational molecular modelling in structural biology. Academic Press; 2018. p. 85–117. https://doi.org/10.1016/bs.apcsb.2018.04.001.

Gonçalves RB, Ferraz WR, Calil RL, Scotti MT, Trossini GHG. Convergent QSAR models for the prediction of cruzain inhibitors. ACS Omega. 2023;8:38961–82. https://doi.org/10.1021/acsomega.3c03376.

Gramatica P. Principles of QSAR models validation: internal and external. QSAR & Comb Sci. 2007;26:694–701. https://doi.org/10.1002/qsar.200610151.

Güner OF, Bowen JP. Setting the record straight: the origin of the pharmacophore concept. J Chem Inf Model. 2014;54:1269–83. https://doi.org/10.1021/ci5000533.

Hansch C, Maloney PP, Fujita T, Muir RM. Correlation of biological activity of phenoxyacetic acids with Hammett substituent constants and partition coefficients. Nature. 1962;194:178–80. https://doi.org/10.1038/194178b0.

Jiménez-Luna J, Grisoni F, Schneider G. Drug discovery with explainable artificial intelligence. Nat Mach Intell. 2020;2:573–84. https://doi.org/10.1038/s42256-020-00236-4.

Kim S, Thiessen PA, Bolton EE, Chen J, Fu G, Gindulyte A, Han L, He J, He S, Shoemaker BA, Wang J, Yu B, Zhang J, Bryant SH. PubChem substance and compound databases. Nucleic Acids Res. 2016;44:D1202–13. https://doi.org/10.1093/nar/gkv951.

Knox C, Wilson M, Klinger CM, Franklin M, Oler E, Wilson A, Pon A, Cox J, Chin NEL, Strawbridge SA, Garcia-Patino M, Kruger R, Sivakumaran A, Sanford S, Doshi R, Khetarpal N, Fatokun O, Doucet D, Zubkowski A, Rayat DY, Jackson H, Harford K, Anjum A, Zakir M, Wang F, Tian S, Lee B, Liigand J, Peters H, Wang RQR, Nguyen T, So D, Sharp M, da Silva R, Gabriel C, Scantlebury J, Jasinski M, Ackerman D, Jewison T, Sajed T, Gautam V, Wishart DS. DrugBank 6.0: the DrugBank knowledgebase for 2024. Nucleic Acids Res. 2024;52:D1265–75. https://doi.org/10.1093/nar/gkad976.

Kohlbacher SM, Langer T, Seidel T. QPHAR: quantitative pharmacophore activity relationship: method and validation. J Chem. 2021;13:57. https://doi.org/10.1186/s13321-021-00537-9.

Kramer C, Kalliokoski T, Gedeck P, Vulpetti A. The experimental uncertainty of heterogeneous public K(i) data. J Med Chem. 2012;55:5165–73. https://doi.org/10.1021/jm300131x.

Kriegeskorte N, Golan T. Neural network models and deep learning. Curr Biol. 2019;29:R231–6. https://doi.org/10.1016/j.cub.2019.02.034.

Kubinyi H. Free Wilson analysis. Theory, applications and its relationship to Hansch analysis. Quant Struct Act Relat. 1988;7:121–33. https://doi.org/10.1002/qsar.19880070303.

Landrum GA, Riniker S. Combining IC50 or Ki values from different sources is a source of significant noise. J Chem Inf Model. 2024;64:1560–7. https://doi.org/10.1021/acs.jcim.4c00049.

Landrum GA, Beckers M, Lanini J, Schneider N, Stiefl N, Riniker S. SIMPD: an algorithm for generating simulated time splits for validating machine learning approaches. J Chem. 2023;15:119. https://doi.org/10.1186/s13321-023-00787-9.

Lanka G, Begum D, Banerjee S, Adhikari N, Yogeeswari P, Ghosh B. Pharmacophore-based virtual screening, 3D QSAR, docking, ADMET, and MD simulation studies: an *in silico* perspective for the identification of new potential HDAC3 inhibitors. Comput Biol Med. 2023;166:107481. https://doi.org/10.1016/j.compbiomed.2023.107481.

Lima LM, Barreiro EJ. 1.07 – beyond bioisosterism: new concepts in drug discovery. In: Chackalamannil S, Rotella D, Ward SE, editors. Comprehensive medicinal chemistry III. Oxford: Elsevier; 2017. p. 186–210. https://doi.org/10.1016/B978-0-12-409547-2.12290-5.

Lo Y-C, Rensi SE, Torng W, Altman RB. Machine learning in chemoinformatics and drug discovery. Drug Discov Today. 2018;23:1538–46. https://doi.org/10.1016/j.drudis.2018.05.010.

Mamada H, Nomura Y, Uesawa Y. Novel QSAR approach for a regression model of clearance that combines DeepSnap-Deep learning and conventional machine learning. ACS Omega. 2022;7:17055–62. https://doi.org/10.1021/acsomega.2c00261.

Maurya AK, Mulpuru V, Mishra N. Discovery of novel coumarin analogs against the α-glucosidase protein target of diabetes mellitus: pharmacophore-based QSAR, docking, and molecular dynamics simulation studies. ACS Omega. 2020;5:32234–49. https://doi.org/10.1021/acsomega.0c03871.

Molecular descriptors for chemoinformatics, 1st ed. Wiley; 2009. https://doi.org/10.1002/9783527628766.
Mora JR, Marquez EA, Pérez-Pérez N, Contreras-Torres E, Perez-Castillo Y, Agüero-Chapin G, Martinez-Rios F, Marrero-Ponce Y, Barigye SJ. Rethinking the applicability domain analysis in QSAR models. J Comput Aided Mol Des. 2024;38:9. https://doi.org/10.1007/s10822-024-00550-8.
Naidu G, Zuva T, Sibanda EM. A review of evaluation metrics in machine learning algorithms. In: Silhavy R, Silhavy P, editors. Artificial intelligence application in networks and systems. Cham: Springer; 2023. p. 15–25. https://doi.org/10.1007/978-3-031-35314-7_2.
Nematzadeh S, Kiani F, Torkamanian-Afshar M, Aydin N. Tuning hyperparameters of machine learning algorithms and deep neural networks using metaheuristics: a bioinformatics study on biomedical and biological cases. Comput Biol Chem. 2022;97:107619. https://doi.org/10.1016/j.compbiolchem.2021.107619.
Puzyn T, Mostrag-Szlichtyng A, Gajewicz A, Skrzyński M, Worth AP. Investigating the influence of data splitting on the predictive ability of QSAR/QSPR models. Struct Chem. 2011;22:795–804. https://doi.org/10.1007/s11224-011-9757-4.
Rodríguez-Pérez R, Bajorath J. Interpretation of compound activity predictions from complex machine learning models using local approximations and Shapley values. J Med Chem. 2020;63:8761–77. https://doi.org/10.1021/acs.jmedchem.9b01101.
Sabando MV, Ponzoni I, Milios EE, Soto AJ. Using molecular embeddings in QSAR modeling: does it make a difference? Brief Bioinform. 2022;23:bbab365. https://doi.org/10.1093/bib/bbab365.
Sahigara F, Mansouri K, Ballabio D, Mauri A, Consonni V, Todeschini R. Comparison of different approaches to define the applicability domain of QSAR models. Molecules. 2012;17:4791–810. https://doi.org/10.3390/molecules17054791.
Salih A, Raisi-Estabragh Z, Galazzo IB, Radeva P, Petersen SE, Menegaz G, Lekadir K. A perspective on explainable artificial intelligence methods: SHAP and LIME. 2024. https://doi.org/10.48550/arXiv.2305.02012.
Schaller D, Šribar D, Noonan T, Deng L, Nguyen TN, Pach S, Machalz D, Bermudez M, Wolber G. Next generation 3D pharmacophore modeling. WIREs Comput Mol Sci. 2020;10:e1468. https://doi.org/10.1002/wcms.1468.
Sheridan RP. Time-Split cross-validation as a method for estimating the goodness of prospective prediction. J Chem Inf Model. 2013;53:783–90. https://doi.org/10.1021/ci400084k.
Shim J, MacKerell AD. Computational ligand-based rational design: role of conformational sampling and force fields in model development. Medchemcomm. 2011;2:356–70. https://doi.org/10.1039/C1MD00044F.
Singh R, Kumar P, Sindhu J, Devi M, Kumar A, Lal S, Singh D. Parsing structural fragments of thiazolidin-4-one based α-amylase inhibitors: a combined approach employing *in vitro* colorimetric screening and GA-MLR based QSAR modelling supported by molecular docking, molecular dynamics simulation and ADMET studies. Comput Biol Med. 2023;157:106776. https://doi.org/10.1016/j.compbiomed.2023.106776.
Tetko IV, Sushko I, Pandey AK, Zhu H, Tropsha A, Papa E, Öberg T, Todeschini R, Fourches D, Varnek A. Critical assessment of QSAR models of environmental toxicity against Tetrahymena pyriformis: focusing on applicability domain and overfitting by variable selection. J Chem Inf Model. 2008;48:1733–46. https://doi.org/10.1021/ci800151m.
Thölke P, Mantilla-Ramos Y-J, Abdelhedi H, Maschke C, Dehgan A, Harel Y, Kemtur A, Mekki Berrada L, Sahraoui M, Young T, Bellemare Pépin A, El Khantour C, Landry M, Pascarella A, Hadid V, Combrisson E, O'Byrne J, Jerbi K. Class imbalance should not throw you off balance: choosing the right classifiers and performance metrics for brain decoding with imbalanced data. NeuroImage. 2023;277:120253. https://doi.org/10.1016/j.neuroimage.2023.120253.
Ting KM. Confusion matrix. In: Sammut C, Webb GI, editors. Encyclopedia of machine learning. Boston: Springer; 2010. p. 209. https://doi.org/10.1007/978-0-387-30164-8_157.
Tropsha A. Best practices for QSAR model development, validation, and exploitation. Mol Inform. 2010;29:476–88. https://doi.org/10.1002/minf.201000061.

Tropsha A, Isayev O, Varnek A, Schneider G, Cherkasov A. Integrating QSAR modelling and deep learning in drug discovery: the emergence of deep QSAR. Nat Rev Drug Discov. 2024;23:141–55. https://doi.org/10.1038/s41573-023-00832-0.

Tsou LK, Yeh S-H, Ueng S-H, Chang C-P, Song J-S, Wu M-H, Chang H-F, Chen S-R, Shih C, Chen C-T, Ke Y-Y. Comparative study between deep learning and QSAR classifications for TNBC inhibitors and novel GPCR agonist discovery. Sci Rep. 2020;10:16771. https://doi.org/10.1038/s41598-020-73681-1.

Yadav PK, Singh S, Singh AK. 3D-QSAR-based, pharmacophore modelling, virtual screening, and molecular docking studies for identification of hypoxia-inducible factor-1 inhibitor with potential bioactivity. Comput Biol Med. 2023;166:107557. https://doi.org/10.1016/j.compbiomed.2023.107557.

Zhang H, Shen C, Zhang H-R, Chen W-X, Luo Q-Q, Ding L. Discovery of novel DGAT1 inhibitors by combination of machine learning methods, pharmacophore model and 3D-QSAR model. Mol Divers. 2021;25:1481–95. https://doi.org/10.1007/s11030-021-10247-x.

Artificial Intelligence in De Novo Drug Design

Carmen Cerchia, Pietro Delre, Emanuele Falbo, and Antonio Lavecchia

Abstract Deep generative models have transformed various domains, including drug discovery and development. Building on the exponential growth in computational power and extensive public datasets, these models facilitate the rapid design of novel molecules with targeted properties, possibly expediting the hit discovery phase. In brief, generative models are trained to learn patterns from extensive datasets, providing new structures as output. This chapter explores the potential and limitations of generative models in de novo design, covering key techniques, such as recurrent neural networks (RNNs), autoencoders (AEs), generative adversarial networks (GANs), and reinforcement learning (RL), focusing on their impact in pharmaceutical research. Despite recent successes, significant obstacles remain, including the reliance on high-quality datasets and the need for robust experimental validation, in order to bridge the gap between computational predictions and practical implementation. In fact, these models tend to generate compounds that may be chemically unstable or difficult to synthesize in the laboratory. The chapter also reports benchmarks, metrics, and performance evaluation methods through the examination of diverse case studies. Finally, the chapter discusses emerging directions, such as integrating quantum computing with generative models and leveraging large language models (LLMs), also emphasizing the need for cross-disciplinary collaboration and regulatory involvement.

Keywords De novo design · Generative models · Artificial intelligence · Drug discovery

---

C. Cerchia · P. Delre · E. Falbo · A. Lavecchia (✉)
"Drug Discovery Laboratory", Dipartimento di Farmacia, Università degli Studi di Napoli "Federico II", Naples, Italy
e-mail: carmen.cerchia@unina.it; pietro.delre@unina.it; emanuele.falbo@unina.it; antonio.lavecchia@unina.it

A. Lavecchia (ed.), *Applied Artificial Intelligence for Drug Discovery*,
https://doi.org/10.1007/978-3-031-98022-0_6

1 Introduction

Tempora tempore tempera. (Latin proverb)

The discovery and development of new medicines represent one of the most complex and resource-intensive endeavors in pharmaceutical research and development. It involves a series of sequential stages, including target discovery and validation, hit identification, lead optimization, preclinical and clinical trials, regulatory approval, and post-marketing surveillance. Typically, this process spans over a decade and demands substantial financial investments, often exceeding $2.8 bn or even $4 bn per successful drug (Kiriiri et al. 2020; Henderson et al. 2023). Despite extensive efforts, the attrition rate of drug candidates remains alarmingly high. Failures frequently stem from safety concerns, insufficient efficacy, and stringent regulatory requirements (Waring et al. 2015). Traditional methodologies such as high-throughput screening (HTS) (Blay et al. 2020) and combinatorial chemistry (Furka 2022), while enabling the screening of thousands to millions of compounds, encounter significant challenges, including time, cost, and efficiency constraints.

To address these issues, computational methods have emerged as key resources in modern pharmaceutical research. These methods enhance traditional approaches by leveraging advanced computing capabilities to model, simulate, and evaluate molecular interactions and processes. Virtual screening (VS), for instance, enables the rapid analysis of extensive compound libraries, identifying potential drug candidates with desired characteristics while filtering out unsuitable ones (Lavecchia and Di Giovanni 2013). Structure-based methods, such as molecular docking and molecular dynamics simulations, evaluate receptor–ligand interactions, whereas ligand-based strategies, including pharmacophore modeling and scaffold hopping, allow the discovery of new chemotypes or optimization of known hits. These computational strategies certainly contributed to improve the efficiency of drug discovery; however, they still present inherent limitations, as exemplified by the inaccuracy of scoring functions for protein–ligand docking that do not capture subtle binding energetics, the inability of molecular mechanics to model complex protein dynamics and conformational changes, unreliable free energy calculations missing entropic contributions.

In recent years, artificial intelligence (AI) and deep learning (DL) have emerged as transformative tools in pharmaceutical research (Lavecchia 2015, 2019), particularly in de novo drug design, the process of generating novel potential drug candidates from scratch (Cerchia and Lavecchia 2023; Gangwal and Lavecchia 2024; Romanelli et al. 2024b, c).

Unlike traditional methods that rely on predefined libraries of compounds, de novo design generates from scratch novel molecular structures that (should) exhibit activity toward a specific biological target.

In brief, de novo drug design process involves three key tasks.

1. Molecule generation: This step entails designing new molecular structures, typically using two possible strategies. The first involves assembling molecules from

predefined atom groups or fragments, though this often results in compounds that are challenging to synthesize. Alternatively, virtual chemical reactions, guided by expert-coded rules, simulate laboratory conditions to produce more synthetically feasible molecules (Goldman et al. 2022; Hoffmann and Gastreich 2019). Rule-based systems are generally regarded as effective solutions for molecular generation; however, the predicted reactions may occasionally fail, and reliance on a limited set of robust reactions can unnecessarily constrain the exploration of the broader chemical space.

2. Scoring and filtering molecules: Once generated, molecules are evaluated to identify desirable candidates and eliminate unsuitable ones. This is typically achieved using different techniques, such as molecular docking or machine learning (ML) models (Lavecchia 2015, 2019). ML approaches can classify molecules as active or inactive or quantitatively predict their efficacy (quantitative structure–activity relationships, or QSAR).
3. Searching for optimized molecules: the final step is identifying molecules with the ideal balance of (predicted) binding affinity and molecular properties. De novo design is also closely related to inverse QSAR (Muratov et al. 2020; Tropsha et al. 2023), where the aim is to map desired property values back to molecular structures.

By integrating these tasks, de novo drug design aims to explore uncharted regions of chemical space, possibly uncovering novel drug candidates.

Generative AI represents a major advancement in AI, shifting from traditional tasks, like classification and prediction, to creating new, original data. By mimicking human creativity and cognitive processes, generative AI can produce diverse outputs, such as textual content or realistic images. Its versatility spans applications in creative design, language translation, summarization, and programming. At its core, generative AI relies on DL models, leading to the definition "generative models" (GMs). These latter offer a fresh approach to de novo drug design. GMs, trained on large sets of molecules, aim to learn the underlying probability distribution of training data, capturing patterns and essential features that define molecular structures and properties. By creating low-dimensional continuous representations, GMs can generate entirely new molecules by sampling from the learned data distribution (Cerchia and Lavecchia 2023; Gangwal and Lavecchia 2024; Romanelli et al. 2024b, c). The learned patterns are then leveraged for de novo molecular design, enabling the generation of novel chemical structures.

Techniques such as autoencoders (AEs), variational autoencoders (VAEs), generative adversarial networks (GANs), and transformers, to mention a few, have demonstrated remarkable potential in generating molecules with tailored properties, such as enhanced bioactivity, reduced toxicity, and improved pharmacokinetics (Romanelli et al. 2024b). Evaluation of the generated structures often leverages QSAR models, transfer learning (TL), reinforcement learning (RL), or fitness functions to optimize for desired properties. Compared to more traditional ML approaches, however, GMs feature high computational demands, reliance on large datasets, and a "black box" nature that limits interpretability. Despite these hurdles, their application in molecular design has seen exponential growth, propelled by the growing number of benchmarks and advancements in model architecture.

GMs have also accelerated early-stage drug discovery, identifying promising lead compounds in significantly shorter timeframes compared to traditional methods. For instance, a notable success story involved the discovery of kinase inhibitors, achieved through a computer-based process in just 21 days (Zhavoronkov et al. 2019).

The integration of GMs into drug discovery pipelines has sparked an era of rapid innovation, offering unparalleled capabilities in molecular design. By leveraging vast datasets (see Bonner et al. 2022 for an up-to-date overview), these models also allow exploration of the nuanced relationships between genes, proteins, and diseases, opening new possibilities for personalized medicine, analyzing genetic and clinical data to predict optimal treatments, and thereby improving patient outcomes and minimizing adverse effects.

This chapter aims to provide a concise and updated overview of the role of GMs in de novo drug design. We will explore recent advancements in the field, highlighting state-of-the-art applications. Through the recent case studies, we will examine the successes and limitations of these approaches, offering insights into their transformative potential. Finally, we will outline the challenges that remain and future directions.

2 Historical Evolution of GMs in Drug Discovery

The journey of GMs in drug discovery represents one of the most fascinating intersections of AI and pharmaceutical research. This evolution illustrates how these approaches have transformed from simple rule-based systems to increasingly sophisticated platforms.

2.1 From Early Developments to the DL Revolution

The foundations of computational molecule generation could be traced back to the late 1980s, when researchers first attempted to systematize organic molecule synthesis through computer-assisted approaches. The LHASA (Logic and Heuristics Applied to Synthetic Analysis) system, developed by E.J. Corey and colleagues, marked a crucial moment in this field (Corey 1967; Corey et al. 1985; Corey and Wipke 1969). This system, though primitive by today's standards, demonstrated the potential for computers to assist in molecular design and synthesis by applying logical rules and chemical knowledge. During this period, the focus remained largely on expert systems that encoded human knowledge rather than learning from data.

The 1990s saw the emergence of fragment-based approaches (Schneider and Fechner 2005), which represented a significant step forward in computational molecular design. These methods involved combining molecular fragments according to predefined rules to generate new compounds and contributed to fundamental principles that would later influence modern generative models. However, these early approaches were inherently limited by their reliance on human-crafted rules and predefined fragment libraries, constraining their ability to explore novel chemical space.

Molecular de novo design underwent a dramatic transformation with the advent of DL (Lecun et al. 2015). The publication of Gómez-Bombarelli's seminal paper (Gómez-Bombarelli et al. 2018) marked a watershed moment, introducing VAEs for molecular generation. This work demonstrated that DL models could learn continuous representations of molecules, enabling smooth navigation through chemical space and the generation of novel compounds with desired properties. This work marked a paradigm shift from rule-based generation to data-driven approaches.

Shortly thereafter, the field witnessed the emergence of various architectural innovations. Segler and colleagues (2018) introduced recurrent neural networks (RNNs) for generating SMILES (Simplified Molecular Input Line Entry System) strings, treating molecular generation as a language modeling task. This approach proved particularly powerful, as it allowed models to learn the implicit grammar of chemical structures directly from data. The success of these early implementations catalyzed intense interest in applying various DL architectures to molecular generation.

2.2 Architectural Diversification

The period from 2018 to 2020 marked an era of rapid diversification in generative modeling approaches. GANs emerged as a powerful framework for molecular generation, with Guimaraes and colleagues introducing ORGAN (Guimaraes et al. 2017), which combined GANs with RL to optimize molecular properties. The development of MolGAN (De Cao and Kipf 2018) represented another significant advance, as it operated directly on graph representations of molecules rather than SMILES strings. Parallel to these developments, researchers made substantial progress in improving variational approaches. The work on Junction Tree VAE (Jin et al. 2018) shifted attention toward one of the critical challenges in molecular generation: ensuring chemical validity. This approach decomposed molecules into subgraphs, enabling the generation of valid molecules while maintaining structural diversity. Kusner's Grammar VAE incorporated chemical grammar directly into the generative process, further improving the quality of generated molecules.

Further advancements were accomplished in graph-based methods. For instance, GraphRNN (You et al. 2018) demonstrated the potential of working directly with molecular graphs, while Liu's Constrained Graph VAE (Liu et al. 2018) showed the potential to incorporate chemical constraints into the generation process. As the field matured, attention increasingly turned to practical applications and validation. Perhaps most notably, recent years ushered in the first significant successes in applying GMs to real-world drug discovery challenges. Insilico Medicine's GENTRL system (Zhavoronkov et al. 2019) demonstrated the potential for end-to-end drug discovery. The development of open-source platforms like DeepChem (https://github.com/deepchem/deepchem) democratized access to these technologies, enabling broader participation in the field. The most recent period has seen a significant focus on the development of enhanced methods for multiconditional generation, allowing precise control over multiple molecular properties simultaneously.

The introduction of transformer-based models marked another significant advancement. ChemGPT (Frey et al. 2023) and subsequent models showed how large-scale pretraining could enhance molecular generation capabilities.

Finally, the emergence of large language models (LLMs) has opened further new possibilities in molecular generation and beyond (Jablonka et al. 2023). These models have demonstrated remarkable learning capabilities, potentially reducing the need for extensive task-specific training data.

3 Architectures of GMs in Drug Design

GMs leverage various DL architectures, each with unique mechanisms. Below, we present the main GM architectures (Fig. 1), their functions, advantages, and challenges (Table 1). We will not discuss the fundamental principles or the mathematical foundations of each architecture in detail, as they are thoroughly covered in other excellent works (Lavecchia 2019, 2024; Lecun et al. 2015; Tang et al. 2024).

3.1 RNNs

RNNs are designed to process sequential data (Mikolov et al. 2010), making them an intuitive choice for molecular representation through formats like SMILES. These architectures utilize feedback connections within their hidden layers, enabling them to retain information across time steps. This feature allows RNNs to capture the sequential dependencies inherent in molecular structures, such as the placement of atoms or bonds.

In practice, RNNs can be trained on large datasets of SMILES strings, enabling them to generate novel molecular sequences that adhere to chemical rules. Advanced variants such as Long Short-Term Memory (LSTM) (Hochreiter and Schmidhuber 1997) and Gated Recurrent Units (GRUs) (Cho et al. 2014) address limitations of standard RNNs, such as the vanishing gradient problem, by incorporating mechanisms that selectively retain relevant information over extended sequences.

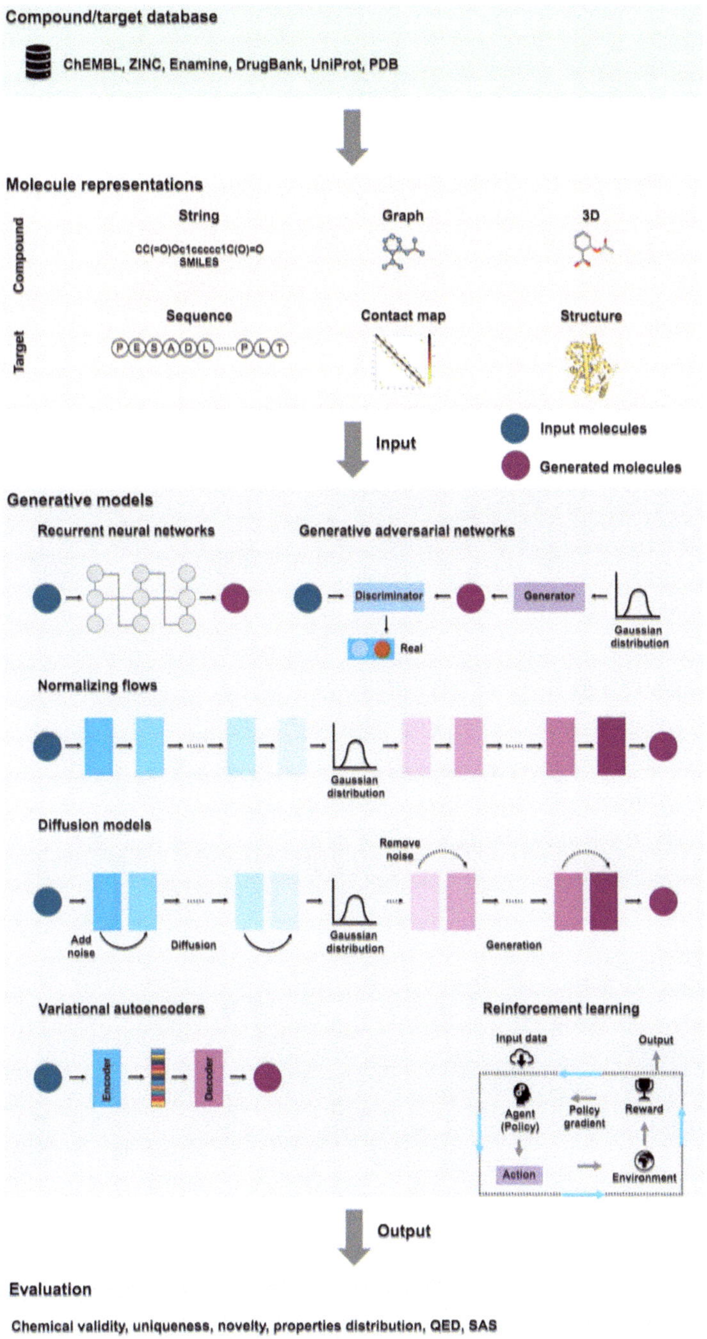

Fig. 1 Schematic overview of GM's workflow. Molecules are encoded either as strings, molecular graphs, or three-dimensional (3D) structures, whereas targets can be represented as sequence, 2D contact map, or 3D structure. The data are used to train GMs, including RNNs, GANs, normalizing flows, diffusion models, and VAEs. RL can be integrated for fine-tuning specific properties. GMs generate novel molecules, which are subsequently assessed for chemical validity, synthetic accessibility, and drug-like properties

Table 1 Comparison of the main GMs and their applications in de novo design

GM	Description	Key strengths	Key limitations	References
RNNs	Sequential models that use feedback loops to process molecular representations such as SMILES strings	Effective at capturing sequential dependencies; suitable for generating valid chemical sequences	Prone to mode collapse; struggles with long sequences; less efficient for complex molecular graphs	Blaschke et al. (2020), Lamanna et al. (2023), Loeffler et al. (2024), and Mikolov et al. (2010)
AEs and VAEs	Encode molecules into a latent space for manipulation and decode them back into chemical structures	Useful for exploring chemical space; enables smooth interpolation between molecular properties	Requires large datasets; risk of generating invalid molecules without constraints; computationally expensive	Lecun et al. (2015), Lim et al. (2018), and Romanelli et al. (2024a)
GANs	Adversarial frameworks with a generator creating new molecules and a discriminator distinguishing real from synthetic ones	Highly effective for generating novel and diverse scaffolds; adaptable for property optimization	Training instability; prone to mode collapse; high computational demand	Goodfellow et al. (2020) and Mirza and Osindero (2014)
Transformer-based models	Leverage self-attention mechanisms to capture long-range dependencies in molecular sequences and graphs	Outstanding scalability; excellent for large datasets; effective for capturing complex molecular features	High computational requirements; prone to overfitting on smaller datasets	Chithrananda et al. (2020) and Vaswani (2017)
Diffusion models	Introduce noise into molecular data and learn to reverse the process, generating realistic molecules or 3D structures	Strong at generating 3D molecular structures like protein-ligand complexes; captures complex distributions	Computationally intensive, relatively new, and requires further development for robust drug discovery applications	Hoogeboom et al. (2022)

Flow-based models	Employ invertible transformations to generate molecular graphs, enabling explicit density estimation and stepwise graph construction	Stability in training; guarantees validity at each construction step; interpretable output	Less efficient on large datasets; challenges in capturing global structural properties of complex molecules	Kobyzev et al. (2020)
RL	Optimizes molecule generation by defining and iterating reward functions for desired properties	Capable of fine-tuning models for specific properties; supports iterative learning	Requires carefully designed reward functions; computational overhead in optimization tasks	Popova et al. (2018)
TL	Utilizes pretrained models on general datasets to improve task-specific performance through fine-tuning	Reduces training time; enables effective learning in low-data scenarios	Dependent on quality and scope of pretrained datasets; less effective for novel chemical spaces	Tan et al. (2018) and Yu et al. (2022)

For instance, LSTM stands out in preserving long-term dependencies, while GRU offers computational efficiency with fewer parameters.

A prominent application of RNNs is REINVENT, developed within AstraZeneca (Blaschke et al. 2020; Loeffler et al. 2024).

However, RNNs are not without challenges. They may struggle with long sequences, particularly when the underlying molecular structure exceeds typical SMILES lengths. Moreover, these models are prone to mode collapse, where the generation process repeatedly produces a limited set of molecules. Despite these challenges, RNNs remain a cornerstone in molecular generation.

3.2 AEs and VAEs

AEs are another foundational architecture in generative modeling. These networks consist of two primary components: an encoder that compresses input data into a lower-dimensional latent space and a decoder that reconstructs the input from this representation (Lecun et al. 2015). In the context of drug design, AEs enable researchers to map molecular features into a latent space in which structurally similar molecules cluster together, facilitating local exploration of chemical space.

A significant advancement in this domain is the VAE, which introduces a probabilistic framework to the latent space. Instead of mapping molecules to fixed points, VAEs encode them as probability distributions, typically Gaussian. This approach ensures continuity in the latent space, allowing smooth interpolation between molecules and enabling the generation of diverse chemical structures. For example, sampling from regions of the latent space associated with desirable properties, such as high drug-likeness, can lead to the discovery of promising compounds.

Despite their strengths, AEs and VAEs have limitations. Standard AEs often fail to ensure meaningful organization of the latent space, which can hinder the generative process. VAEs, while addressing this issue, require careful tuning of hyperparameters and large datasets for effective training. Extensions like Conditional VAEs (CVAEs) (Lim et al. 2018) allow incorporation of specific molecular properties into the generation process, enabling the design of molecules with "tailored" characteristics (Romanelli et al. 2024a). This approach is particularly useful when optimizing for specific properties, like solubility.

3.3 GANs

GANs introduced a paradigm shift in generative modeling by employing a game-theoretic approach (Goodfellow et al. 2020). GANs consist of two frameworks: a generator, which creates synthetic molecules, and a discriminator, which distinguishes these from real molecular data. Through this adversarial process, the generator iteratively improves, producing increasingly realistic and plausible molecular structures.

A notable strength of GANs is their ability to generate highly diverse molecules, often featuring novel scaffolds not included in training data. Conditional GANs (CGANs) further enhance this capability (Mirza and Osindero 2014) by allowing the specification of desired properties during the generation process. For instance, CGANs can generate molecules with predefined physical or chemical attributes, such as lipophilicity or molecular weight.

However, GANs training is notoriously challenging due to issues like mode collapse, where the generator produces a narrow range of outputs, and instability, requiring careful balancing of the generator and discriminator networks.

3.4 *Transformer-Based Models*

Transformers, originally developed for (NLP), have rapidly gained prominence in drug design due to their ability to capture complex dependencies in data. Unlike RNNs, which process sequences step-by-step, Transformers leverage self-attention mechanisms to consider relationships across the entire sequence simultaneously (Vaswani 2017). This capability makes them particularly effective for handling molecular representations such as SMILES and molecular graphs. In molecular generation, Transformers are particularly useful for encoding and decoding molecular structures. The versatility of Transformers is matched by their computational demands. Training these models requires substantial resources, and their extensive parameterization can lead to overfitting in low-data regimes. Nevertheless, their ability to handle large and complex datasets, coupled with their scalability, positions them as a powerful tool.

3.5 *Diffusion and Flow-Based Models*

Diffusion and flow-based models represent emerging approaches in generative modeling. Diffusion models operate by gradually introducing noise into data during a forward pass and learning to reverse this process during a backward pass, ultimately generating data samples from noise (Hoogeboom et al. 2022). These models have shown promise in generating 3D molecular structures, such as protein–ligand complexes, by learning spatial relationships (Alakhdar et al. 2024).

Flow-based models (Kobyzev et al. 2020), on the other hand, use a series of invertible transformations to map data distributions, allowing explicit density estimation. This approach offers advantages in stability and interpretability compared to GANs and VAEs. For example, flow-based models can generate molecular graphs by incrementally constructing nodes and edges, ensuring validity at each step.

While diffusion and flow-based models are computationally intensive, their ability to capture complex distributions and generate diverse outputs makes them valuable additions to generative modeling.

3.6 Integrating TL and RL

TL (Tan et al. 2018; Yu et al. 2022) and RL (Popova et al. 2018) serve as complementary techniques to enhance GMs. TL allows models pretrained on large, general datasets to be fine-tuned for specific tasks, significantly reducing training time and improving performance in low-data scenarios. For instance, ChemBERTa (Chithrananda et al. 2020), a Transformer-based model, leverages TL to improve molecular representation by extracting latent features from extensive compound libraries.

RL fine-tunes generative models by optimizing for specific objectives, such as binding affinity or synthetic accessibility. By defining a reward function based on desired molecular properties, RL enables the iterative improvement of generated compounds.

4 Molecular Representations and Encoding

The choice of molecular representation plays a crucial role in the development and performance of generative models (Fig. 1). A proper representation, in both ML and DL, greatly simplifies the learning task. This principle applies to molecular representations, where capturing key structural features is essential for understanding property relationships (Wigh et al. 2022). An effective molecular representation facilitates learning by being both unambiguous and invariant to atom-numbering, while also being computable (Chuang et al. 2020). Decades of cheminformatics research have produced expert-defined molecular representations that resemble traditional feature engineering in computer vision and NLP, suggesting opportunities for cross-disciplinary insights. Similarly to how visual features of objects are hierarchically built from edges, shapes, and spatial information, molecular representations can be constructed hierarchically using DL (from local atomic environments to substructures).

4.1 String-Based Representations

String-based representations encode molecular structures as linear sequences of characters, with SMILES being the most widely adopted format. SMILES notation employs a straightforward grammar of letters and special characters to fully describe molecular structures. While a single molecule can have multiple valid SMILES strings, each SMILES string consistently corresponds to the same molecule. This many-to-one relationship has proven beneficial for data augmentation during model training, leading to significant improvements in generative model performance.

However, SMILES representations face challenges with validity constraints. To address these limitations, several alternative string-based formats have emerged.

SELFIES (Self-Referencing Embedded Strings) were designed specifically for validity-assured molecular generation (Krenn et al. 2020), as an encoding/decoding scheme for SMILES in VAEs. SELFIES represent atoms using their symbols and capture molecular connectivity by encoding branch lengths and ring sizes.

By incorporating validity constraints within its decoder, this approach aims to guarantee both syntactic and semantic validity of generated structures. However, it introduces its own complexities, as some molecules can have multiple, potentially infinite, SELFIES representations (Alberga et al. 2024), which can complicate the learning of structural and physicochemical properties (Skinnider et al. 2021).

To improve compactness and chemical motif representation, Group SELFIES were recently introduced (Cheng et al. 2023). Group SELFIES approach builds on standard SELFIES by introducing functional and chemical group tokens for molecular generative tasks. In Group SELFIES, a group is defined as a collection of atoms and bonds that represent a molecular substructure, along with its attachment points. These attachment points specify how the group can participate in bonding and include a defined maximum valency. This ensures that available valency can be accurately tracked during the decoding process. Group SELFIES could handle more complex molecular structures while maintaining robust and error-free encoding and decoding. However, a key tradeoff of Group SELFIES is slower encoding and decoding compared to standard SELFIES. Encoding is especially slow, as it requires substructure matching for each group in the set, while decoding is faster but still slower than the SELFIES decoder.

The SAFE representation offers a fresh approach to describing molecules by decomposing them into modular, interconnected fragments (Noutahi et al. 2024). Unlike traditional notations such as SMILES, which represent molecules as linear strings, SAFE encodes molecules as an unordered collection of fragments. This method retains the inherent flexibility and modularity of chemical structures while remaining fully compatible with existing SMILES parsers. SAFE proved particularly advantageous for tasks like scaffold decoration, linker design, and motif extension, as it simplifies these challenges into straightforward sequence completion problems (Lee et al. 2025). By preserving the structural integrity of molecular scaffolds and accommodating complex architectures, SAFE facilitates intuitive fragment-based molecular design without relying on the complexity of graph-based models.

4.2 Graph-Based Representations

In graph representations, atoms serve as nodes and chemical bonds as edges. This representation captures molecular topology through adjacency matrices describing atom connectivity, while node feature matrices encode atomic properties such as electronegativity, charge, and radius. Edge feature matrices can further represent bond characteristics, including order, type, and length. The implementation of graph-based representations, however, requires more sophisticated algorithms like Graph Neural Networks (GNNs). Modern approaches include Message-Passing Neural Networks (MPNNs) and Graph Convolutional Networks (GCNs), which can process complex molecular graphs and extract relevant features (Gilmer et al. 2017; Sun et al. 2020). Graph-based generative models can operate either sequentially or in single steps. Sequential approaches add atoms or fragments progressively, allowing for real-time validation of chemical constraints. However, the computational complexity of graph isomorphism, determining whether two graphs represent the same molecule, remains a significant challenge in this representation.

4.3 3D Representations

3D representations provide the most complete structural information by portraying molecules as point clouds in space, where each atom is depicted as a point in 3D space. This approach captures not only bond connectivity but also conformational preferences, offering valuable insights into molecular behavior.

For instance, the approach proposed by Gebauer and colleagues (2019) generated molecules sequentially by placing atoms in Cartesian coordinates, whereas Simm et al. (2020) adopted a similar strategy by generating molecules atom-by-atom based on internal coordinates.

3D representations face substantial computational challenges. The need to consider multiple conformers for accurate property prediction increases computational costs significantly. Recent research has addressed this limitation through the development of 3D generative models trained on conformer ensembles (Axelrod and Gómez-Bombarelli 2022), reducing the computational burden of generating training data.

SMILES-based approaches can lead to the generation of molecules that better fit target chemical spaces compared to more constrained representations like SELFIES (Skinnider et al. 2021). The performance differences between various representations often depend on the specific application and objectives. Graph-based methods provide intuitive handling of molecular structures but face computational challenges with larger molecules.

As the field of generative drug design continues to evolve, hybrid approaches combining multiple representation types may emerge to leverage their complementary strengths. The development of more efficient algorithms for handling graph isomorphism and conformational analysis could further enhance the utility of graph and 3D representations. Additionally, the integration of chemical knowledge into representation schemes may lead to more chemically aware and efficient generative models.

5 Tokenization and Encoding

The conversion of molecular representations into machine-readable formats requires tokenization strategies, vocabulary construction, and encoding. This critical pre-processing step transforms chemical notation into sequences of characters that DL models can process. Tokenization consists of the partitioning of a sequence into chunks and subsequent demarcation of such chunks (i.e., the "tokens").

Character-level tokenization breaks down the SMILES string into individual characters, usually corresponding to a given atom or bond.

Then, each token is mapped to a unique integer index, in a predefined vocabulary of all possible tokens, and then "vectorized." One-hot encoding represents the most popular approach, where each unique token is assigned a binary vector. The dimension of this vector equals the size of the complete vocabulary of possible tokens. For each token, only one entry is set to 1, while all others remain 0 (Segler et al. 2018). This encoding ensures complete orthogonality between tokens, meaning there is no implied relationship between different molecular components. The choice of tokenization strategy can also have an impact on model performance and generative capabilities. Character-level tokenization often requires longer sequences to represent molecules, potentially making it harder for models to capture relationships between tokens. To address such limitations, one-hot encoding is often replaced or complemented with learned embeddings. In this approach, each token is mapped to a dense, continuous vector in an embedding space. Embeddings capture semantic and chemical relationships between tokens, improving the model's ability to learn meaningful patterns in molecular data. The trade-off between vocabulary size and model complexity remains an active area of research. Larger vocabularies can capture more chemical information directly but increase model complexity and training requirements. On the other hand, smaller vocabularies might offer computational advantages but require the model to learn more complex combinations of tokens to represent chemical features.

6 Evaluation Metrics and Benchmarking

The assessment of generative models in drug design requires consideration of multiple, often interconnected factors. From ensuring basic chemical validity to predicting real-world utility as drug candidates, evaluation metrics serve as essential tools for gauging the performance and practical value of these models (Fig. 1).

Chemical validity stands as the most fundamental criterion, ensuring that generated molecules adhere to basic chemical rules, including proper atom valence, correct bond types, and chemically feasible structures.

Uniqueness in generated molecules quantifies the proportion of distinct molecules within the generated set, showing whether the model can produce diverse structures or tends to repeatedly generate similar molecules. The interpretation of uniqueness scores requires careful consideration of both sample size and the vastness of the chemical space being explored.

Novelty represents another important dimension of evaluation, measuring the model's capability to generate structures that are not present in the training data. While high novelty scores might initially seem desirable, they must be carefully balanced against the need to maintain drug-like properties and synthetic feasibility. A model generating entirely novel, but synthetically impossible or biologically irrelevant molecules would have limited practical utility in drug discovery.

Understanding how effectively a generative model explores chemical space requires a global analysis of molecular property distributions. Typically, a variety of properties are analyzed, including topological descriptors and physicochemical parameters, such as the octanol-water partition coefficient (logP).

The comparison of property distributions between generated molecules and reference datasets provides crucial insights into model performance. Statistical measures such as the Jensen–Shannon distance, Wasserstein distance, and Kullback–Leibler divergence offer quantitative ways to assess how well the generated molecules match the desired property profiles (Skinnider et al. 2021). These comparisons help ensure that the model produces molecules with characteristics similar to known drugs or other relevant reference compounds.

The Fréchet ChemNet Distance (FCD) (Preuer et al. 2018) offers a more advanced evaluation method by quantifying the similarity between the property distributions (chemical or biological) of generated molecules and real molecules, providing a measure of the biological relevance of the generated compounds through the ChemNet model (Mayr et al. 2018).

The practical utility of generated molecules depends heavily on their drug-likeness and synthetic feasibility. The Quantitative Estimation of Drug-likeness (QED) score (Bickerton et al. 2012) has emerged as a standard metric, providing a thorough assessment by evaluating multiple physicochemical properties against the profiles of known drugs. However, QED should be viewed as one component of a broader evaluation strategy, as it may not capture all aspects of drug-likeness relevant to specific therapeutic applications.

Synthetic accessibility, quantified through the Synthetic Accessibility Score (SAS) (Ertl and Schuffenhauer 2009), addresses the crucial question of whether generated molecules can actually be synthesized. This metric helps bridge the gap between computational design and experimental validation. Molecules with high SAS values present significant synthetic challenges, potentially limiting their practical utility regardless of their predicted properties.

Recently, the perplexity metric was applied to evaluate how well generated molecules align with specific design objectives (Moret et al. 2022). This intrinsic model score enables the identification and ranking of promising molecular designs based on the probabilities learned by the model. Furthermore, perplexity scoring helps detect undesired model biases introduced during training and supports the development of a ranking system to mitigate these biases.

The structural analysis of generated molecules provides additional insights through examination of fragments and scaffolds (Bemis and Murcko 1996; Degen et al. 2008). Various similarity metrics, including those based on molecular fingerprints and Tanimoto coefficients, help evaluate structural relationships between generated molecules and known compounds.

The introduction of standardized evaluation platforms, such as MOSES (Polykovskiy et al. 2020) and GuacaMol (Brown et al. 2019), has streamlined the assessment of generative models. These platforms incorporate diverse metrics, offer baseline models, and provide benchmark datasets, facilitating more consistent and thorough evaluations of model performance. Gao and colleagues (Gao et al. 2022) recently reported an open-source benchmark, practical molecular optimization (PMO), focusing on sample efficiency (i.e., how many molecules are required to optimize the objective) and supporting 25 molecular design algorithms.

MolScore (Thomas et al. 2024) unifies and re-implements the above-reported established benchmarks such as GuacaMol, MOSES, and PMO, while allowing users to create new benchmarks. MolScore integrates a wide range of relevant scoring functions commonly used in benchmarks, including molecular similarity, docking, predictive models, and synthesizability. It can be integrated into existing Python workflows, also including GUIs to further enhance its accessibility.

Current evaluation approaches still face important limitations. While they provide valuable insights into model performance, they should be viewed as complementary tools rather than definitive measures of success. Most metrics rely heavily on computational predictions, and the relationship between metric performance and actual success remains incompletely understood. The future development of evaluation metrics will likely focus on several key areas for improvement. Integration of experimental feedback could certainly help validate computational predictions, while the development of application-specific metrics could be more useful for specific drug discovery projects, especially in niche therapeutic areas. Improved methods for assessing scaffold novelty while maintaining drug-likeness will help balance innovation with practicality. Finally, enhanced approaches to synthetic accessibility prediction will more effectively guide the selection of promising candidates for synthesis. The ultimate validation of GMs will continue to be strongly dependent on experimental confirmation.

7 Application Cases of GMs in De Novo Drug Design

Early applications reported in the literature primarily focused on theoretical work, emphasizing model performance over experimental validation of the generated compounds (Cerchia and Lavecchia 2023; Romanelli et al. 2024b, c). The rapid growth of interest in this field has made it challenging to keep up with new publications, many of which often lack a practical, real-world application perspective. In the following section, we discuss recent success stories regarding the application of GMs, including LLMs, to solve compelling issues, such as the identification of multitarget compounds, the de novo design of peptides and proteins.

7.1 GMs Addressing the Design of Multitarget Compounds

POLYGON (POLYpharmacology Generative Optimization Network) was conceptualized to address the development of compounds targeting multiple proteins simultaneously (Munson et al. 2024). It combines a VAE for processing chemical structures and RL for optimizing compounds against multiple targets. This latter iteratively samples and optimizes molecular structures based on their ML predicted ability to inhibit multiple target proteins, as well as drug-likeness and synthetic feasibility scoring. POLYGON was trained using more than one million molecules from ChEMBL24 and achieved an accuracy of 81.9% in identifying compounds with dual-target activity at an IC_{50} threshold of 1 μM. The authors then turned their attention to pairs of synthetic-lethal proteins (Previtali et al. 2024), including various protein classes such as kinases and histone modifiers. In particular, they focused on the design of dual inhibitors of MEK1 and mTOR kinases. Of 32 synthesized compounds, the majority demonstrated IC_{50}s in the 1–10 μM range and achieved more than 50% reduction in both kinase activities. Four compounds showed particularly strong performance, maintaining significant inhibition of both targets even at 1 μM concentration. The lead compound, IDK12008, demonstrated consistent inhibitory capacity in both cell-based and cell-free assays. The current implementation of POLYGON focuses primarily on initial compound design and does not directly optimize for ADMET parameters (absorption, distribution, metabolism, excretion, and toxicity). Additionally, while off-target effects were evaluated and found to be comparable to FDA-approved kinase inhibitors, there remains room for improvement in compound selectivity.

Another recent study addressing the design of multitarget drug compounds employed a GM based on RNN, equipped with LSTM cells (Fig. 2) and employing beam search to monitor fine-tuning (Isigkeit et al. 2024). The model was pretrained with a large set of 365K molecules from ChEMBL, whereas fine-tuning was accomplished with smaller sets (molecules with $IC_{50}/EC_{50} \leq 1$ μM on the selected targets from BindingDB) by TL. The authors thus demonstrated that their model, when properly fine-tuned, can successfully generate novel molecular structures that exhibit biological activity on two intended targets. Their approach proved particularly effective for certain target pairs, with 7 out of 12 synthesized compounds showing dual activity on their intended targets.

For the FXR/sEH target pair, all three synthesized compounds (1–3) demonstrated dual activity, with compound 1 showing particularly impressive potency: an EC_{50} of 0.32 μM for FXR activation and an IC_{50} of 0.017 μM for sEH

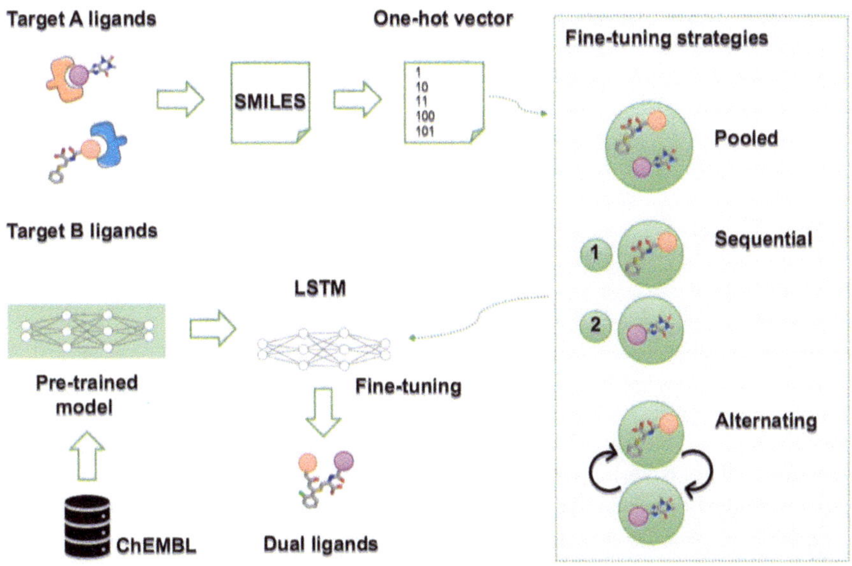

Fig. 2 Workflow of the RNN model coupled with LSTM for the design of dual target ligands. (Isigkeit et al. 2024)

inhibition. For the FXR/THRβ pair, while all three compounds (4–6) showed FXR agonism, only compound 6 demonstrated measurable THRβ activity, albeit relatively weak, with an EC_{50} greater than 10 μM. The PPARδ/sEH designs (7–9) were all confirmed as dual ligands, with compound 7 emerging as a notably potent sEH inhibitor ($IC_{50} = 0.033$ μM) while maintaining moderate PPARδ agonist activity ($EC_{50} = 3.2$ μM).

However, the study also revealed important limitations of the approach. The success of the dual ligands design appears to be heavily dependent on the chemical similarity between the known ligands of the target pairs. The method worked well for protein pairs whose known ligands occupied proximal regions in chemical space, such as FXR/sEH, FXR/THRβ, and PPARδ/sEH. In contrast, when attempting to design dual ligands for targets with more chemically diverse ligands (AT1/sEH, FFAR1/sEH, and AT1/FFAR1), the approach was less successful, generating compounds that showed activity on only one of the intended targets. Another consideration is that while the model successfully generated dual-active compounds, the potency of these compounds was often unbalanced between the two targets, with activity ranging from nanomolar to micromolar levels. This suggests that while the approach can identify promising starting points for dual-target drugs, additional optimization would likely be needed. The study also highlighted the importance of the fine-tuning strategy (Fig. 2). In fact, the pooled fine-tuning, where template molecules for both targets were used simultaneously, was more effective than sequential or alternating approaches. This insight provides valuable guidance for future applications.

7.2 LLMs Taking the Stage

The remarkable capabilities of LLMs, such as generative pre-trained transformers (GPT), bidirectional encoder representations from transformers (BERT) (Devlin 2018), and bidirectional and auto-regressive transformers (BART) (Lewis 2019), have been ultimately employed for generative tasks. These models include a combination of transformer-based architecture, large-scale training on diverse datasets, optimization techniques, and an attention mechanism, which resembles human cognition. The lifecycle of an LLM typically begins with self-supervised pretraining, where the model learns patterns from vast, unlabeled text datasets. This stage produces a "base model" capable of general tasks. Prompt engineering, by crafting specific input prompts, can guide this base model to achieve specific outcomes, but for more precise applications, the model is fine-tuned using supervised techniques. This fine-tuning results in an "instruct model," designed to follow diverse instructions without requiring task-specific adjustments. Further refinement can be achieved through alignment methods, such as instruction tuning and RL.

LLMs employ various architectures for different tasks (Ramos et al. 2025). Encoder–decoder models are optimal for sequence-to-sequence tasks, for instance, language translation or predicting chemical reactions. Encoder-only models are ideal for extracting insights, such as predicting molecular properties. Decoder-only models specialize in sequence generation or completion, enabling them to generate novel compounds.

Recent studies demonstrated that LLMs can capture and apply chemical principles embedded in their training data, enabling them to address complex problems beyond simple pattern recognition.

For an exhaustive and recent review of LLMs' application in chemistry, see the excellent work of Ramos et al. (2025). Among the notable advances, there is the integration of LLMs into autonomous "agents," i.e., systems based on a model (generally an LLM) that continuously receive observations from the environment, make decisions, and execute actions to complete a provided task. These agents may be used for data interpretation, to help chemists solve complex issues in less time, and to streamline routine lab tasks (Ramos et al. 2025).

Although LLMs lack the expert-level understanding of human chemists, studies show they can effectively utilize chemical data to perform chemistry-related tasks (Jablonka et al. 2023), including the generation of valid chemical structures, also suggesting meaningful molecular modifications, and facilitating the discovery of compounds with desired properties (Jablonka et al. 2024; Lavecchia 2024; Bran et al. 2024).

Decoder-only GPT-like architectures emerged as powerful tools for property-directed molecule generation due to their ability to capture complex patterns and relationships within molecular sequences. One of the earliest applications in this area was Adilov's 2021 work (Adilov 2021), which introduced a GPT-2-inspired causal transformer trained on SMILES strings for self-supervised learning, demonstrating the scalability and efficiency of decoder-only architectures, and surpassing encoder-only models like ChemBERTa in molecule generation and property prediction. A significant leap forward came with the development of MolGPT (Bagal et al. 2022), a six-million-parameter decoder-only model explicitly designed for molecular generation. MolGPT leveraged masked self-attention to capture long-range dependencies within SMILES strings, ensuring chemically valid outputs. Trained on benchmark datasets such as MOSES and GuacaMol, MolGPT showed outstanding results in generating novel, property-specific molecules. While its computational demands were higher than traditional VAEs, its performance on key metrics like validity, uniqueness, and diversity justified the trade-off. Advances in decoder-only models have also exploited different molecular representations, such as IUPAC nomenclature (Cho and No 2023), as well as the incorporation of 3D protein structural data to generate ligands with high binding affinity for specific protein targets (Wang et al. 2023a). In a similar fashion, cMolGPT introduced conditional generation, fine-tuning embeddings of protein–ligand pairs to produce target-specific molecules (Wang et al. 2023b).

These studies prospect an exciting future where LLMs, integrated with domain-specific knowledge, could become strategic assets in scientific research. Given the rapidly increasing number of relevant publications, we point the interested reader to a maintained and well-organized repository collecting new studies on LLM and LLM-based agents, specifically aimed at scientific applications. The repository is available at: https://github.com/ur-whitelab/LLMs-in-science.

7.3 GMs Addressing Peptide and Protein Design

Chen et al. (2024) combined GMs with molecular modeling techniques to design peptides with enhanced binding affinity for specific protein targets. The approach was successfully applied to design inhibitors for β-catenin and NF-κB essential modulator (NEMO).

The model employed a GRU-based VAE integrated with a Metropolis-Hastings (MH) sampling algorithm and trained on approximately four million peptide sequences from Uniprot, PixelDB, and THPdb databases (Fig. 3).

In the initial step, they efficiently reduced the sequence search space from billions to hundreds of potential candidates by embedding peptide sequences in a latent space and generating extensions of inhibitory peptides that are likely to enhance protein–protein interactions. The preliminary candidates were then ranked using Rosetta FlexPepDock, which evaluates binding scores after structural refinement of peptide–protein complexes. Following initial screening, top-ranked peptides underwent binding energy calculations using the molecular mechanics/generalized Born surface area (MM/GBSA) method. To further enhance specificity, iterative fine-tuning of the VAE-MH model was carried out using the binding affinity data from earlier iterations, progressively refining the latent space to focus on target-specific features.

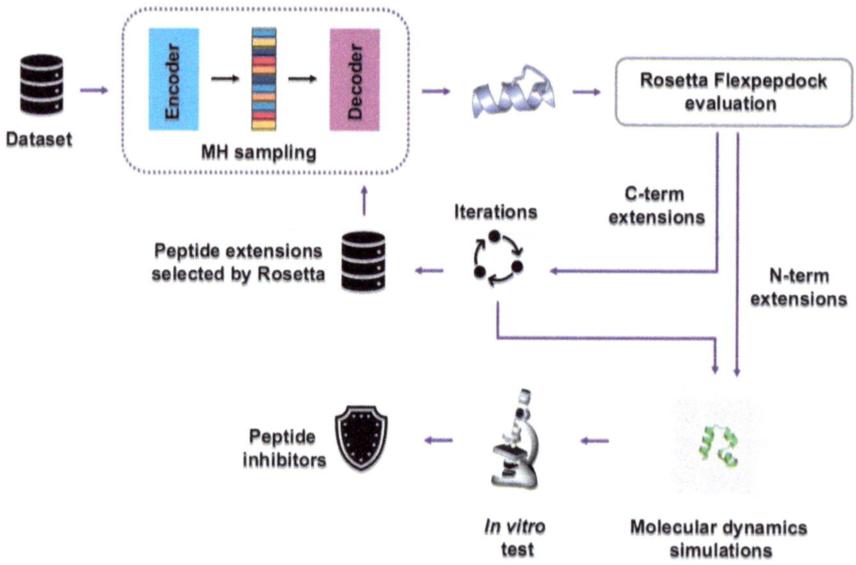

Fig. 3 Workflow of the VAE-MH model for the design of peptide inhibitors (β-catenin specific)

This approach was first applied to design β-catenin-binding peptides, specifically N- and C-terminal extensions for a previously reported peptidyl inhibitor, Peptide 9.

The experimental validation using competitive fluorescence polarization (FP)-based binding assays revealed that the best candidate (NAL-9) achieved an IC_{50} of 0.084 μM compared to Peptide 9 (0.15 μM). A second round involved the design of C-terminal extensions targeting β-catenin. Eight peptides were synthesized and tested using FP-based binding assays. Four showed enhanced binding affinity, with the most potent candidate (CAL-2) achieving an IC_{50} of 0.010 ± 0.06 μM, representing a remarkable 15-fold improvement over Peptide 9.

The authors further validated their approach using the more challenging NEMO system, characterized by an unstructured binding interface and weak initial binding ($IC_{50} \gg 100$ μM) for the 11-residue NEMO-binding domain (NBD) of IKKβ, chosen as a template for extension. After computational design and screening, four peptides were synthesized and tested. Two demonstrated substantially improved binding to NEMO, with NBD + 2 and NBD + 12 achieving IC_{50} values of approximately 50 and 75 μM, respectively. Overall, this approach efficiently narrowed down an enormous sequence space (20^5–20^9 possibilities) to a manageable number of candidates for experimental testing. On the other hand, the effectiveness varies significantly depending on (i) the direction of the peptides' elongation; (ii) the availability of structural information, and (iii) the specific characteristics of the protein–peptide interaction surface.

Ultimately, efforts in generative AI are beginning to take hold for de novo protein design, such as antibodies and synthetic binding proteins (SBP). These latter include "protein engineering" of artificial proteins, such as nanobodies and miniproteins, widely employed for research, diagnostics, and therapeutics (Li et al. 2025). SBPs are designed to bind to a particular target of interest, offering multiple advantages, like smaller size, enhanced stability, reduced immunogenicity, and overcoming the inherent limitations of traditional antibodies. De novo protein design greatly benefited from the amount of available bioinformatics databases and the powerful GMs' architectures. This generative task could be seen as the inverse of structure prediction: whereas the latter involves building an accurate 3D structure from an amino acid sequence, protein design aims to build a correct sequence that fits a given 3D structure (Tang et al. 2024).

The breakthrough in protein structure prediction achieved by DL-based approaches, particularly AlphaFold (Abramson et al. 2024), has revolutionized structural biology. Building on this success, which earned the foundational research in protein structure prediction the Nobel Prize in Chemistry (O'Leary 2024), the field is now witnessing rapid advances in protein engineering through generative AI approaches (Li et al. 2025). Unlike GMs for molecular design, GMs for protein design still lack unified metrics and robust evaluation procedures/platforms, with high variability between metrics and testing conditions of each model, making it difficult to objectively assess the quality of designed proteins (Tang et al. 2024). The tasks in protein generation are generally more complex than those in molecule generation; these include difficulties in accurately predicting protein fold categories, assigning gene function annotations, and designing crucial antibody regions. These ongoing challenges suggest considerable potential for future improvement.

A recent fascinating study applied an LLM to explore the protein design space far beyond the bounds of natural evolution, generating functional proteins that would take millions of years for evolution to discover (Hayes et al. 2025). ESM 3, a multimodal LLM trained on evolutionary data, was prompted to generate fluorescent proteins, unveiling a bright fluorescent protein at a far distance (58% sequence identity) from known fluorescent proteins. Unlike evolution, which operates within physical and functional constraints, protein language models are able to navigate the vast array of potential evolutionary pathways. Proteins can be visualized within a connected network, where each protein is linked to others by single mutational steps. Evolution traverses this network only through paths that preserve function at every step. Protein language models efficiently exploit this network, recognizing densely populated regions that evolution frequently explores and sparse areas that remain largely inaccessible. By predicting the "next step" in protein sequences, LLMs effectively simulate the movement of evolution. This study marks a significant step forward in traditional protein engineering.

8 Concluding Remarks and Future Perspectives

The integration of AI in de novo drug design represents a transformative shift in pharmaceutical research and development, though we must approach this revolution with measured optimism. The progress to date is undeniably promising; however, many challenges remain to be addressed before AI can fully realize its potential. To date, only a handful of AI-discovered or AI-designed drugs have progressed to human clinical trials, and none have yet received clinical approval (Wilczok and Zhavoronkov 2024).

These tools require careful validation: the current models rather recognize patterns in molecular representations, lacking a true conceptual understanding of chemistry. This limitation necessitates ongoing human oversight and experimental validation of AI-generated compounds (Romanelli et al. 2024b). As such, these GMs provide initial ideas but still require human expertise and several iterations of optimization before a generated molecule can be considered a true drug candidate.

Data quality and quantity continue to be limiting factors: the availability of high-quality experimental data, the bias in training datasets, and modeling of activity cliffs remain a concern, often constraining model development. Although large public databases contain millions of biological data points, the number of data points available for any specific assay remains relatively limited. Similarly, most of the data used to develop models for organic synthesis planning are stored in proprietary databases, which are not easily accessible (Walters and Barzilay 2021). Additionally, there is still a lack of standardized methods for systematically assessing the synthetic accessibility of generated compounds.

Among the most notable advancements is certainly the adoption of LLMs and LLM-based autonomous agents. These latter are reshaping multiple research fields, including drug and materials discovery, with the promise to enable the creation of smarter laboratories. The development and deployment of these agents present challenges and considerations that must be addressed to ensure trust, reliability, and ethical use (Van Dis et al. 2023). Major concerns include security vulnerabilities, privacy breaches, copyright issues, misuse, bias, lack of transparency, and the risk of generating incorrect or misleading information (Stokel-Walker and Van Noorden 2023). Additionally, ensuring robustness and scalability is crucial as these systems are applied to increasingly diverse and complex tasks. A promising direction in this area is the development of multiagent systems, where specialized agents work collaboratively to tackle specific tasks.

Success in this domain will require meaningful collaboration between data scientists and experimental chemists. This interdisciplinary approach, coupled with standardized benchmarks (not limited to model performance metrics, but clearly demonstrating the contributions of AI in augmenting the productivity and success of drug discovery programs) and transparent engagement with regulatory bodies, will be crucial. As AI-guided research advances, regulators must be engaged transparently to establish frameworks that ensure both innovation and integrity. The journey toward AI-driven drug discovery is only beginning; looking forward, several promising directions emerge.

The integration of quantum mechanics calculations with GMs could improve design capabilities; a recent study reported a hybrid quantum–classical GM applied to the design of KRAS inhibitors for cancer therapy (Ghazi Vakili et al. 2025). The generative process combined a quantum circuit Born machine (QCBM) with a 16-qubit processor, a LSTM network, and the Chemistry42 platform (Ivanenkov et al.

2023) for final validation. The QCBM utilized quantum effects to generate complex probability distributions, enabling the creation of novel molecules optimized for targeting KRAS. From one million generated compounds, 15 were selected and synthesized; two compounds, ISM061-018-2 and ISM061-022, showed significant promise, with ISM061-018-2 demonstrating strong binding affinity (1.4 μM) to KRAS-G12D in surface plasmon resonance and cell-based assays. This study highlighted the potential of quantum computing to produce experimentally validated hits that perform at least on par with more classical models (Ghazi Vakili et al. 2025).

Innovative molecular representations could better support GMs in drug design. Xiang et al. recently presented VideoMol (Xiang et al. 2024), a molecular video-based foundation model for molecular representation learning. By treating molecules as 3D dynamic videos, VideoMol captures both dynamic and physicochemical awareness using a self-supervised pretraining framework and demonstrates exceptional performances in predicting molecular targets and properties.

Ultimately, the evolution of GMs in drug discovery will be shaped by the need to balance innovation with practical utility. The field must maintain its rapid pace of technical advancement while ensuring that new developments translate into real-world impact.

Acknowledgments This work was supported by the European Union, Next Generation EU (PRIN 2022, project number: 2022P5LPHS). A.L. acknowledges funding from the PNRR Project CN000041, 'National Centre for Gene Therapy and RNA Technology-Based Drugs' (CUP UNINA E63C22000940007).

References

Abramson J, Adler J, Dunger J, Evans R, Green T, Pritzel A, Ronneberger O, Willmore L, Ballard AJ, Bambrick J, Bodenstein SW, Evans DA, Hung C-C, O'Neill M, Reiman D, Tunyasuvunakool K, Wu Z, Žemgulytė A, Arvaniti E, et al. Accurate structure prediction of biomolecular interactions with AlphaFold 3. Nature. 2024;630(8016):493–500. https://doi.org/10.1038/s41586-024-07487-w.

Adilov S. Generative pre-training from molecules; 2021. https://doi.org/10.26434/chemrxiv-2021-5fwjd.

Alakhdar A, Poczos B, Washburn N. Diffusion models in de novo drug design. J Chem Inf Model. 2024;64(19):7238–56. https://doi.org/10.1021/acs.jcim.4c01107.

Alberga D, Lamanna G, Graziano G, Delre P, Lomuscio MC, Corriero N, Ligresti A, Siliqi D, Saviano M, Contino M, Stefanachi A, Mangiatordi GF. DeLA-DrugSelf: empowering multi-objective de novo design through SELFIES molecular representation. Comput Biol Med. 2024;175:108486. https://doi.org/10.1016/j.compbiomed.2024.108486.

Axelrod S, Gómez-Bombarelli R. GEOM, energy-annotated molecular conformations for property prediction and molecular generation. Sci Data. 2022;9(1):1–14. https://doi.org/10.1038/s41597-022-01288-4.

Bagal V, Aggarwal R, Vinod PK, Priyakumar UD. MolGPT: molecular generation using a transformer-decoder model. J Chem Inf Model. 2022;62(9):2064–76. https://doi.org/10.1021/acs.jcim.1c00600.

Bemis GW, Murcko MA. The properties of known drugs. 1. Molecular frameworks. J Med Chem. 1996;39(15):2887–93. https://doi.org/10.1021/jm9602928.
Bickerton GR, Paolini GV, Besnard J, Muresan S, Hopkins AL. Quantifying the chemical beauty of drugs. Nat Chem. 2012;4(2):90–8. https://doi.org/10.1038/nchem.1243.
Blaschke T, Arús-Pous J, Chen H, Margreitter C, Tyrchan C, Engkvist O, Papadopoulos K, Patronov A. REINVENT 2.0: an AI tool for de novo drug design. J Chem Inf Model. 2020;60(12):5918–22. https://doi.org/10.1021/acs.jcim.0c00915.
Blay V, Tolani B, Ho SP, Arkin MR. High-throughput screening: today's biochemical and cell-based approaches. Drug Discov Today. 2020;25(10):1807–21. https://doi.org/10.1016/j.drudis.2020.07.024.
Bonner S, Barrett IP, Ye C, Swiers R, Engkvist O, Bender A, Hoyt CT, Hamilton WL. A review of biomedical datasets relating to drug discovery: a knowledge graph perspective. Brief Bioinform. 2022;23(6):bbac404. https://doi.org/10.1093/bib/bbac404.
Bran MA, Cox S, Schilter O, Baldassari C, White AD, Schwaller P. Augmenting large language models with chemistry tools. Nat Mach Intell. 2024;6(5):525–35. https://doi.org/10.1038/s42256-024-00832-8.
Brown N, Fiscato M, Segler MHS, Vaucher AC. GuacaMol: benchmarking models for de novo molecular design. J Chem Inf Model. 2019;59(3):1096–108. https://doi.org/10.1021/acs.jcim.8b00839.
Cerchia C, Lavecchia A. New avenues in artificial-intelligence-assisted drug discovery. Drug Discov Today. 2023;28(4):103516. https://doi.org/10.1016/j.drudis.2023.103516.
Chen S, Lin T, Basu R, Ritchey J, Wang S, Luo Y, Li X, Pei D, Kara LB, Cheng X. Design of target specific peptide inhibitors using generative deep learning and molecular dynamics simulations. Nat Commun. 2024;15(1):1611. https://doi.org/10.1038/s41467-024-45766-2.
Cheng AH, Cai A, Miret S, Malkomes G, Phielipp M, Aspuru-Guzik A. Group SELFIES: a robust fragment-based molecular string representation. Digit Discov. 2023;2(3):748–58.
Chithrananda S, Grand G, Ramsundar B Chemberta: large-scale self-supervised pretraining for molecular property prediction. ArXiv Preprint ArXiv:2010.09885. 2020.
Cho K-H, No KT. iupacGPT: IUPAC-based large-scale molecular pre-trained model for property prediction and molecule generation. 2023. https://doi.org/10.26434/chemrxiv-2023-5kjvh
Cho K, Van Merriënboer B, Gulcehre C, Bahdanau D, Bougares F, Schwenk H, Bengio Y. Learning phrase representations using RNN encoder-decoder for statistical machine translation. EMNLP 2014–conference on empirical methods in natural language processing, proceedings of the conference, p. 1724–34. 2014. https://doi.org/10.3115/v1/d14-1179.
Chuang KV, Gunsalus LM, Keiser MJ. Learning molecular representations for medicinal chemistry. J Med Chem. 2020;63(16):8705–22. https://doi.org/10.1021/acs.jmedchem.0c00385.
Corey EJ. General methods for the construction of complex molecules. Pure Appl Chem. 1967;14(1):19–38.
Corey EJ, Wipke WT. Computer-assisted design of complex organic syntheses. Science. 1969;166(3902):178–92. https://doi.org/10.1126/science.166.3902.178.
Corey EJ, Long AK, Rubenstein SD. Computer-assisted analysis in organic synthesis. Science. 1985;228(4698):408–18. https://doi.org/10.1126/science.3838594.
De Cao N, Kipf T. MolGAN: an implicit generative model for small molecular graphs. ArXiv Preprint ArXiv:1805.11973. 2018.
Degen J, Wegscheid-Gerlach C, Zaliani A, Rarey M. On the art of compiling and using "drug-like" chemical fragment spaces. ChemMedChem. 2008;3(10):1503–7. https://doi.org/10.1002/cmdc.200800178.
Devlin J. Bert: pre-training of deep bidirectional transformers for language understanding. ArXiv Preprint ArXiv:1810.04805. 2018.
Ertl P, Schuffenhauer A. Estimation of synthetic accessibility score of drug-like molecules based on molecular complexity and fragment contributions. J Chem. 2009;1:1–11.
Frey NC, Soklaski R, Axelrod S, Samsi S, Gómez-Bombarelli R, Coley CW, Gadepally V. Neural scaling of deep chemical models. Nat Mach Intell. 2023;5(11):1297–305. https://doi.org/10.1038/s42256-023-00740-3.

Furka Á. Forty years of combinatorial technology. Drug Discov Today. 2022;27(10):103308. https://doi.org/10.1016/j.drudis.2022.06.008.
Gangwal A, Lavecchia A. Unleashing the power of generative AI in drug discovery. Drug Discov Today. 2024;29(6):103992. https://doi.org/10.1016/j.drudis.2024.103992.
Gao W, Fu T, Sun J, Coley C. Sample efficiency matters: a benchmark for practical molecular optimization. Adv Neural Inf Proces Syst. 2022;35:21342–57.
Gebauer NWA, Gastegger M, Schütt KT. Symmetry-adapted generation of 3D point sets for the targeted discovery of molecules. Adv Neural Inf Proces Syst. 2019;32:7566–7578.
Ghazi Vakili M, Gorgulla C, Snider J, Nigam A, Bezrukov D, Varoli D, Aliper A, Polykovsky D, Padmanabha Das KM, Cox H III, Lyakisheva A, Hosseini Mansob A, Yao Z, Bitar L, Tahoulas D, Čerina D, Radchenko E, Ding X, Liu J, et al. Quantum-computing-enhanced algorithm unveils potential KRAS inhibitors. Nat Biotechnol. 2025. https://doi.org/10.1038/s41587-024-02526-3.
Gilmer J, Schoenholz SS, Riley PF, Vinyals O, Dahl GE. Neural message passing for quantum chemistry. International conference on machine learning; 2017. p. 1263–72.
Goldman B, Kearnes S, Kramer T, Riley P, Walters WP. Defining levels of automated chemical design. J Med Chem. 2022;65(10):7073–87.
Gómez-Bombarelli R, Wei JN, Duvenaud D, Hernández-Lobato JM, Sánchez-Lengeling B, Sheberla D, Aguilera-Iparraguirre J, Hirzel TD, Adams RP, Aspuru-Guzik A. Automatic chemical design using a data-driven continuous representation of molecules. ACS Cent Sci. 2018;4(2):268–76. https://doi.org/10.1021/acscentsci.7b00572.
Goodfellow I, Pouget-Abadie J, Mirza M, Xu B, Warde-Farley D, Ozair S, Courville A, Bengio Y. Generative adversarial networks. Commun ACM. 2020;63(11):139–44.
Guimaraes GL, Sanchez-Lengeling B, Outeiral C, Farias PLC, Aspuru-Guzik A. Objective-reinforced generative adversarial networks (organ) for sequence generation models. ArXiv Preprint ArXiv:1705.10843. 2017.
Hayes T, Rao R, Akin H, Sofroniew NJ, Oktay D, Lin Z, Verkuil R, Tran VQ, Deaton J, Wiggert M, Badkundri R, Shafkat I, Gong J, Derry A, Molina RS, Thomas N, Khan YA, Mishra C, Kim C, et al. Simulating 500 million years of evolution with a language model. Science. 2025:eads0018. https://doi.org/10.1126/science.ads0018.
Henderson RH, French D, Stewart E, Smart D, Idica A, Redmond S, Eckstein M, Clark J, Sullivan R, Keeling P. Delivering the precision oncology paradigm: reduced R&D costs and greater return on investment through a companion diagnostic informed precision oncology medicines approach. J Pharm Policy Pract. 2023;16(1):84.
Hochreiter S, Schmidhuber J. Long short-term memory. Neural Comput. 1997;9(8):1735–80. https://doi.org/10.1162/neco.1997.9.8.1735.
Hoffmann T, Gastreich M. The next level in chemical space navigation: going far beyond enumerable compound libraries. Drug Discov Today. 2019;24(5):1148–56. https://doi.org/10.1016/j.drudis.2019.02.013.
Hoogeboom E, Satorras VG, Vignac C, Welling M. Equivariant diffusion for molecule generation in 3D. International conference on machine learning; 2022. p. 8867–87.
Isigkeit L, Hörmann T, Schallmayer E, Scholz K, Lillich FF, Ehrler JHM, Hufnagel B, Büchner J, Marschner JA, Pabel J, Proschak E, Merk D. Automated design of multi-target ligands by generative deep learning. Nat Commun. 2024;15(1):7946. https://doi.org/10.1038/s41467-024-52060-8.
Ivanenkov YA, Polykovskiy D, Bezrukov D, Zagribelnyy B, Aladinskiy V, Kamya P, Aliper A, Ren F, Zhavoronkov A. Chemistry42: an AI-driven platform for molecular design and optimization. J Chem Inf Model. 2023;63(3):695–701. https://doi.org/10.1021/acs.jcim.2c01191.
Jablonka KM, Ai Q, Al-Feghali A, Badhwar S, Bocarsly JD, Bran AM, Bringuier S, Brinson LC, Choudhary K, Circi D, Cox S, de Jong WA, Evans ML, Gastellu N, Genzling J, Gil MV, Gupta AK, Hong Z, Imran A, et al. 14 examples of how LLMs can transform materials science and chemistry: a reflection on a large language model hackathon. Digi Discov. 2023;2(5):1233–50. https://doi.org/10.1039/D3DD00113J.

Jablonka KM, Schwaller P, Ortega-Guerrero A, Smit B. Leveraging large language models for predictive chemistry. Nat Mach Intell. 2024;6(2):161–9. https://doi.org/10.1038/s42256-023-00788-1.

Jin W, Barzilay R, Jaakkola T. Junction tree variational autoencoder for molecular graph generation. International conference on machine learning; 2018. p. 2323–32.

Kiriiri GK, Njogu PM, Mwangi AN. Exploring different approaches to improve the success of drug discovery and development projects: a review. Future J Pharm Sci. 2020;6(1):1–12. https://doi.org/10.1186/s43094-020-00047-9.

Kobyzev I, Prince SJD, Brubaker MA. Normalizing flows: an introduction and review of current methods. IEEE Trans Pattern Anal Mach Intell. 2020;43(11):3964–79.

Krenn M, Häse F, Nigam AK, Friederich P, Aspuru-Guzik A. Self-referencing embedded strings (SELFIES): A 100% robust molecular string representation. Mach Learn Sci Technol. 2020;1(4):45024. https://doi.org/10.1088/2632-2153/aba947.

Lamanna G, Delre P, Marcou G, Saviano M, Varnek A, Horvath D, Mangiatordi GF. GENERA: a combined genetic/deep-learning algorithm for multiobjective target-oriented de novo design. J Chem Inf Model. 2023;63(16):5107–19. https://doi.org/10.1021/acs.jcim.3c00963.

Lavecchia A. Machine-learning approaches in drug discovery: methods and applications. Drug Discov Today. 2015;20(3):318–31. https://doi.org/10.1016/j.drudis.2014.10.012.

Lavecchia A. Deep learning in drug discovery: opportunities, challenges and future prospects. Drug Discov Today. 2019;24(10):2017–32. https://doi.org/10.1016/j.drudis.2019.07.006.

Lavecchia A. Advancing drug discovery with deep attention neural networks. Drug Discov Today. 2024:104067. https://doi.org/10.1016/j.drudis.2024.104067.

Lavecchia A, Di Giovanni C. Virtual screening strategies in drug discovery: a critical review. Current medicinal chemistry. 2013;20(23):2839–60. https://doi.org/10.2174/09298673113209990001.

Lecun Y, Bengio Y, Hinton G. Deep learning. Nature. 2015;521(7553):436–44. https://doi.org/10.1038/nature14539.

Lee S, Veccham SP, Liu M. GenMol: A drug discovery generalist with discrete diffusion. ArXiv Preprint ArXiv:2501.06158. 2025. http://arxiv.org/abs/2501.06158

Lewis M. Bart: denoising sequence-to-sequence pre-training for natural language generation, translation, and comprehension. ArXiv Preprint ArXiv:1910.13461. 2019.

Li Y, Duan Z, Li Z, Xue W. Data and AI-driven synthetic binding protein discovery. Trends Pharmacol Sci. 2025;46:132. https://doi.org/10.1016/j.tips.2024.12.002.

Lim J, Ryu S, Kim JW, Kim WY. Molecular generative model based on conditional variational autoencoder for de novo molecular design. J Chem. 2018;10(1):1–9. https://doi.org/10.1186/s13321-018-0286-7.

Liu Q, Allamanis M, Brockschmidt M, Gaunt A. Constrained graph variational autoencoders for molecule design. Adv Neural Inf Proces Syst. 2018;31:7795–7804.

Loeffler HH, He J, Tibo A, Janet JP, Voronov A, Mervin LH, Engkvist O. Reinvent 4: modern AI–driven generative molecule design. J Chem. 2024;16(1):20. https://doi.org/10.1186/s13321-024-00812-5.

Mayr A, Klambauer G, Unterthiner T, Steijaert M, Wegner JK, Ceulemans H, Clevert D-A, Hochreiter S. Large-scale comparison of machine learning methods for drug target prediction on ChEMBL. Chem Sci. 2018;9(24):5441–51.

Mikolov T, Karafiát M, Burget L, Jan C, Khudanpur S. Recurrent neural network based language model. Proceedings of the 11th annual conference of the international speech communication association, INTERSPEECH 2010. 2010;2(3):1045–8. https://doi.org/10.21437/interspeech.2010-343.

Mirza M, Osindero S. Conditional generative adversarial nets. ArXiv Preprint ArXiv:1411.1784. 2014. http://arxiv.org/abs/1411.1784

Moret M, Grisoni F, Katzberger P, Schneider G. Perplexity-based molecule ranking and bias estimation of chemical language models. J Chem Inf Model. 2022;62(5):1199–206. https://doi.org/10.1021/acs.jcim.2c00079.

Munson BP, Chen M, Bogosian A, Kreisberg JF, Licon K, Abagyan R, Kuenzi BM, Ideker T. De novo generation of multi-target compounds using deep generative chemistry. Nat Commun. 2024;15(1):3636. https://doi.org/10.1038/s41467-024-47120-y.

Muratov EN, Bajorath J, Sheridan RP, Tetko IV, Filimonov D, Poroikov V, Oprea TI, Baskin II, Varnek A, Roitberg A, Isayev O, Curtalolo S, Fourches D, Cohen Y, Aspuru-Guzik A, Winkler DA, Agrafiotis D, Cherkasov A, Tropsha A. QSAR without borders. Chem Soc Rev. 2020;49(11):3525–64. https://doi.org/10.1039/d0cs00098a.

Noutahi E, Gabellini C, Craig M, Lim JSC, Tossou P. Gotta be SAFE: a new framework for molecular design. Digit Discov. 2024;3(4):796–804.

O'Leary K. AlphaFold gets an upgrade (and a Nobel). Nat Med. 2024;30(12):3393. https://doi.org/10.1038/s41591-024-03392-x.

Polykovskiy D, Zhebrak A, Sanchez-Lengeling B, Golovanov S, Tatanov O, Belyaev S, Kurbanov R, Artamonov A, Aladinskiy V, Veselov M, Kadurin A, Johansson S, Chen H, Nikolenko S, Aspuru-Guzik A, Zhavoronkov A. Molecular sets (MOSES): a benchmarking platform for molecular generation models. Front Pharmacol. 2020;11. https://doi.org/10.3389/fphar.2020.565644.

Popova M, Isayev O, Tropsha A. Deep reinforcement learning for de novo drug design. Sci Adv. 2018;4(7):eaap7885.

Preuer K, Renz P, Unterthiner T, Hochreiter S, Klambauer G. Fréchet ChemNet distance: a metric for generative models for molecules in drug discovery. J Chem Inf Model. 2018;58:1736–1741.

Previtali V, Bagnolini G, Ciamarone A, Ferrandi G, Rinaldi F, Myers SH, Roberti M, Cavalli A. New horizons of synthetic lethality in cancer: current development and future perspectives. J Med Chem. 2024;67(14):11488–521. https://doi.org/10.1021/acs.jmedchem.4c00113.

Ramos MC, Collison CJ, White AD. A review of large language models and autonomous agents in chemistry. Chem Sci. 2025;16:2514. https://doi.org/10.1039/D4SC03921A.

Romanelli V, Annunziata D, Cerchia C, Cerciello D, Piccialli F, Lavecchia A. Enhancing de novo drug design across multiple therapeutic targets with CVAE generative models. ACS Omega. 2024a;9(43):43963–76. https://doi.org/10.1021/acsomega.4c08027.

Romanelli V, Cerchia C, Lavecchia A. Deep generative models in the quest for anticancer drugs: ways forward. Front Drug Discov. 2024b;4:1362956.

Romanelli V, Cerchia C, Lavecchia A. Unlocking the potential of generative artificial intelligence in drug discovery. In: Lyu Z, editor. Applications of generative AI. Springer; 2024c. p. 37–63. https://doi.org/10.1007/978-3-031-46238-2_3.

Schneider G, Fechner U. Computer-based de novo design of drug-like molecules. Nat Rev Drug Discov. 2005;4(8):649–63. https://doi.org/10.1038/nrd1799.

Segler MHS, Kogej T, Tyrchan C, Waller MP. Generating focused molecule libraries for drug discovery with recurrent neural networks. ACS Cent Sci. 2018;4(1):120–31. https://doi.org/10.1021/acscentsci.7b00512.

Simm GNC, Pinsler R, Hernández-Lobato JM. Reinforcement learning for molecular design guided by quantum mechanics. In: Daume III H, Singh A, editors. 37th international conference on machine learning, ICML 2020, vol. PartF16814. PMLR; 2020. p. 8906–16. https://proceedings.mlr.press/v119/simm20b.html.

Skinnider MA, Stacey RG, Wishart DS, Foster LJ. Chemical language models enable navigation in sparsely populated chemical space. Nat Mach Intell. 2021;3(9):759–70.

Stokel-Walker C, Van Noorden R. What ChatGPT and generative AI mean for science. Nature. 2023;614(7947):214–6.

Sun M, Zhao S, Gilvary C, Elemento O, Zhou J, Wang F. Graph convolutional networks for computational drug development and discovery. Brief Bioinform. 2020;21(3):919–35.

Tan C, Sun F, Kong T, Zhang W, Yang C, Liu C. A survey on deep transfer learning. Artificial neural networks and machine learning–ICANN 2018: 27th international conference on artificial neural networks, Rhodes, Greece, October 4–7, 2018, proceedings, part III 27; 2018. p. 270–9.

Tang X, Dai H, Knight E, Wu F, Li Y, Li T, Gerstein M. A survey of generative AI for de novo drug design: new frontiers in molecule and protein generation. Brief Bioinform. 2024;25(4):bbae338. https://doi.org/10.1093/bib/bbae338.

Thomas M, O'Boyle NM, Bender A, De Graaf C. MolScore: a scoring, evaluation and benchmarking framework for generative models in de novo drug design. J Chem. 2024;16(1):64. https://doi.org/10.1186/s13321-024-00861-w.

Tropsha A, Isayev O, Varnek A, Schneider G, Cherkasov A. Integrating QSAR modelling and deep learning in drug discovery: the emergence of deep QSAR. Nat Rev Drug Discov. 2023:1–15. https://doi.org/10.1038/s41573-023-00832-0.

Van Dis EAM, Bollen J, Zuidema W, Van Rooij R, Bockting CL. ChatGPT: five priorities for research. Nature. 2023;614(7947):224–6.

Vaswani A. Attention is all you need. In: Advances in neural information processing systems. Cambridge: MIT Press; 2017.

Walters WP, Barzilay R. Critical assessment of AI in drug discovery. Expert Opin Drug Discov. 2021;16(9):937–47. https://doi.org/10.1080/17460441.2021.1915982.

Wang X, Gao C, Han P, Li X, Chen W, Rodríguez Patón A, Wang S, Zheng P. PETrans: de novo drug design with protein-specific encoding based on transfer learning. Int J Mol Sci. 2023a;24(2):1146.

Wang Y, Zhao H, Sciabola S, Wang W. cMolGPT: A conditional generative pre-trained transformer for target-specific de novo molecular generation. Molecules. 2023b;28(11):4430.

Waring MJ, Arrowsmith J, Leach AR, Leeson PD, Mandrell S, Owen RM, Pairaudeau G, Pennie WD, Pickett SD, Wang J, Wallace O, Weir A. An analysis of the attrition of drug candidates from four major pharmaceutical companies. Nat Rev Drug Discov. 2015;14(7):475–86. https://doi.org/10.1038/nrd4609.

Wigh DS, Goodman JM, Lapkin AA. A review of molecular representation in the age of machine learning. Wiley Interdiscip Rev Comput Mol Sci. 2022;12(5):e1603.

Wilczok D, Zhavoronkov A. Progress, pitfalls, and impact of AI-driven clinical trials. Clin Pharmacol Ther. 2024. https://doi.org/10.1002/cpt.3542.

Xiang H, Zeng L, Hou L, Li K, Fu Z, Qiu Y, Nussinov R, Hu J, Rosen-Zvi M, Zeng X. A molecular video-derived foundation model for scientific drug discovery. Nat Commun. 2024;15(1):9696.

You J, Ying R, Ren X, Hamilton W, Leskovec J. Graphrnn: generating realistic graphs with deep auto-regressive models. International conference on machine learning; 2018. p. 5708–17.

Yu F, Xiu X, Li Y. A survey on deep transfer learning and beyond. Mathematics. 2022;10(19):270–9. https://doi.org/10.3390/math10193619.

Zhavoronkov A, Ivanenkov YA, Aliper A, Veselov MS, Aladinskiy VA, Aladinskaya AV, Terentiev VA, Polykovskiy DA, Kuznetsov MD, Asadulaev A, Volkov Y, Zholus A, Shayakhmetov RR, Zhebrak A, Minaeva LI, Zagribelnyy BA, Lee LH, Soll R, Madge D, et al. Deep learning enables rapid identification of potent DDR1 kinase inhibitors. Nat Biotechnol. 2019;37(9):1038–40. https://doi.org/10.1038/s41587-019-0224-x.

Artificial Intelligence in Peptide Drug Discovery

Yajie Fu and Qifeng Bai

Abstract Peptide drugs have broad application prospects in the medical field due to their strong selectivity, strong efficacy, good biocompatibility, and low toxicity. Peptide drugs can be used in the treatment of various diseases, such as oral diseases, cardiovascular diseases, diseases related to metabolic dysfunction, autoimmune, endocrine, skin, and bone diseases. Traditional peptide drug development methods are time-consuming and labor-intensive, it is difficult to cover every potential peptide sequence combination, and expensive tests are required to evaluate the efficacy of synthesized peptides. In recent years, artificial intelligence (AI) has developed rapidly, and many traditional machine learning and deep learning algorithms have been applied to the development of peptide drugs. AI-assisted peptide drug development employs peptide databases and various models to screen or generate therapeutic peptides in a short period of time. This chapter outlines the characteristics and clinical applications of peptides, the development of traditional peptide drug development, and exemplifies the specific applications of traditional machine learning and deep learning algorithms in peptide drug development.

Keywords Artificial intelligence · AI · Machine learning · Deep learning · Peptide drug discovery

1 Introduction

Peptides are intermediate products of protein hydrolysis and are composed of amino acids connected by peptide bonds. Peptides can be divided into endogenous peptides and exogenous peptides according to their source. Endogenous peptides

Y. Fu
School of Basic Medical Sciences, Lanzhou University, Lanzhou, People's Republic of China

School of Stomatology, Lanzhou University, Lanzhou, People's Republic of China

Q. Bai (✉)
School of Basic Medical Sciences, Lanzhou University, Lanzhou, People's Republic of China
e-mail: baiqf@lzu.edu.cn

A. Lavecchia (ed.), *Applied Artificial Intelligence for Drug Discovery*,
https://doi.org/10.1007/978-3-031-98022-0_7

are peptides that naturally exist in the human body, such as growth hormone-releasing hormone, thyrotropin, and insulin. The amount of these peptides in the body is extremely low, but they have an essential effect. Exogenous peptides refer to natural peptides that exist in animals, plants, and microorganisms, as well as peptides produced after the degradation of animal and plant proteins. Peptides have many properties, including antibacterial, anticancer, antioxidant, antidiabetic, anticoagulant, etc. Due to its diverse biological properties, peptide-based therapies have been a potential treatment for many diseases for decades (Jakubczyk et al. 2020). Compared with traditional small-molecule drugs, peptide drugs have higher selectivity, target specificity, efficacy, safety, biocompatibility, and low toxicity. However, polypeptide drugs are unstable in vivo, with short half-life and low bioavailability, which limits their wide clinical application (Apostolopoulos et al. 2021).

Traditional peptide drug development is based on the modification of endogenous peptides or exogenous peptides, followed by chemical or recombinant peptide synthesis, and requires expensive and time-consuming testing to evaluate the efficacy of the peptides. Because there is a wide range of peptides, the traditional method of developing therapeutic peptides requires a lot of time, manpower, and material resources, and it is difficult to cover every potential combination of peptide sequences. With the development of AI, many traditional machine learning and deep learning algorithms have been applied to peptide drug development, especially the development of antimicrobial peptides and anticancer peptides. With the numerous databases of therapeutic peptides and accurate machine learning models, the development of peptide drugs can be significantly accelerated.

This chapter is divided into four parts. The first part outlines the properties and clinical applications of peptides. The second part introduces the development of traditional peptide drug development and the most advanced technologies currently. The third and fourth parts mainly exemplify the specific applications of traditional machine learning and deep learning algorithms in peptide drug development.

2 Properties and Applications of Peptides

Peptide drugs have higher efficacy, tolerability, and safety than traditional small-molecule drugs, and have lower production costs and relatively simple production methods than protein-based biopharmaceuticals. Therefore, peptide drugs are considered in the sweet spot between small-molecule drugs and protein-based biopharmaceuticals (Fosgerau and Hoffmann 2015). Many active substances in the human body are in the form of peptides. Peptides are involved in various fields such as hormones, nerves, cell growth, and reproduction, and can comprehensively regulate human physiological functions. Therefore, peptides are an important class of drugs that can supplement peptide hormones when endogenous peptide secretion levels are insufficient or lacking. Peptide drugs can be used in the treatment of various diseases, such as oral diseases, cardiovascular diseases, diseases related to metabolic dysfunction, autoimmune, endocrine, skin, and bone diseases. Currently, the main research focus of peptide drugs is antibacterial and anticancer treatment (Sharma et al. 2023a).

2.1 Antimicrobial Peptides

Excessive use of antibiotics leads to the emergence of drug-resistant strains. Antibiotic resistance poses a great threat to human health all over the world. An analysis in 2019 showed that antibiotic resistance poses the greatest threat to human health in sub-Saharan Africa and South Asia, but it has a significant impact on all regions (Murray et al. 2022). Therefore, it is imperative to find new antibacterial drugs. Antimicrobial peptides (AMPs) are a potential alternative to antibiotics. AMPs are endogenous peptides produced by multicellular organisms and are also defined as host defense peptides that protect the host from pathogenic microorganisms. According to different target organisms, AMPs can be divided into antibacterial peptides, antiviral peptides, antifungal peptides, and antiparasitic peptides. AMPs have broad application prospects in the medical field due to their high antibacterial activity, wide antibacterial spectrum, and wide variety. Bacteria are the main cause of various oral diseases, such as caries and periodontal disease. The discovery of AMPs provides new schemes for the treatment of common oral diseases.

The mechanism of action of AMPs is different from that of traditional antibiotics. Antibiotics kill bacteria primarily by inhibiting cytoderm synthesis, interfering with protein synthesis, and inhibiting nucleic acid replication and transcription. Most AMPs act on the bacterial cytomembrane, destroying the integrity of the membrane and causing the loss of the intracellular and extracellular barriers, thereby killing the bacteria (Yeaman and Yount 2003). Therefore, it is unlikely to induce bacterial drug resistance through AMPs, because the cytomembrane structure of bacteria rarely mutates (Zhu et al. 2022). However, high manufacturing costs, poor pharmacokinetics, and nonspecific toxicity to host cells hinder the clinical application of AMPs (Carratalá et al. 2020; Ab 2018). This urges people to design safer and more effective AMPs.

2.2 Anticancer Peptides

Cancer has become one of the most important diseases threatening human life. Common treatments include surgery, chemotherapy, radiotherapy, etc. If necessary, biological therapy, laser therapy, and other methods will be combined. Early-stage patients may be cured with timely and effective treatment, while middle- and late-stage patients are difficult to cure, and the recurrence and metastasis rates are high. In recent years, peptides have been used in the treatment of cancer due to their easy modification, low immunogenicity, and high selectivity. Anticancer peptides (ACPs) can be used as molecular targeted drugs, which can block the signal pathways needed for tumor cell proliferation and differentiation by binding with corresponding receptors (Zhao et al. 2018). In addition, ACPs can be used as drug carriers to cross the cell membrane and transport a variety of bioactive conjugates such as proteins, peptides, DNA, and siRNA into cells. The cellular uptake mechanisms can be divided into energy-independent direct osmosis and energy-dependent endocytosis (Raucher and Ryu 2015; Guidotti et al. 2017). Overall, ACPs are a promising anticancer drug to improve cancer prognosis.

2.3 Application of Peptides in Cardiovascular Diseases

Cardiovascular disease is the leading cause of death worldwide. Hypertension is one of the main causes of cardiovascular disease. The renin–angiotensin–aldosterone system (RAAS) is a pressor regulatory system produced by the kidneys in the body, which causes vascular smooth muscle contraction and water and sodium retention, producing a pressor effect. The most important enzyme in this system is the angiotensin-converting enzyme (ACE), which cleaves angiotensin I into angiotensin II, constricting blood vessels and indirectly raising blood pressure. ACE activity inhibitors are commonly used as drugs to treat hypertension (Wu et al. 2018). A study evaluated the ACE inhibitory activity of a purified peptide derived from the enzymolysis of *C. lipsoidea*. The peptide is composed of four amino acids and shows strong ACE inhibitory activity by continuous chromatography. The antihypertensive effect on spontaneously hypertensive rats (SHR) also shows that oral purified peptides significantly reduce systolic blood pressure (Ko et al. 2012). In addition, a study was conducted to isolate bioactive peptides from the *Tetradesmus obliquus* microalgae. Four peptides, WPRGYFL, GPDRPKFLGPF, WYGPDRPKFL, and SDWDRF, were selected for synthesis and specific bioactivity tests in vitro, which showed great ACE inhibitory activity (Montone et al. 2018). These studies suggest that peptides targeting RAAS may hold promise in the treatment of cardiovascular diseases.

2.4 Application of Peptides in Diabetes Mellitus

The prevalence of diabetes is increasing worldwide, becoming a serious public health problem (Ogurtsova et al. 2022). Peptide drugs, including insulin and GLP-1 receptor agonists (GLP-1RAs), have been successfully used in the treatment of type 2 diabetes mellitus (T2DM) (Wang et al. 2022). GLP-1RAs can promote insulin secretion, inhibit glucagon secretion, and delay gastric emptying (Trujillo and Nuffer 2014). Since the first GLP-1RA (exenatide) was approved by the US Food and Drug Administration (FDA) in 2005, multiple GLP-1RAs with different dosing regimens have entered the clinic, including liraglutide (2009), lixisenatide (2013), dulaglutide (2014), and semaglutide (2017) (George et al. 2018). These GLP-1RAs can promote weight loss and improve glycemic control in patients with T2DM (Nauck and Müller 2023).

3 Traditional Peptide Drug Development

In 1922, insulin extracted from animal pancreas was first used in clinics, and the development of peptide drugs began. In 1954, oxytocin and vasopressin entered the clinic as the first chemically synthesized peptide drugs (Du Vigneaud et al. 1953; Vigneaud et al. 1954). In 1963, Merrifield first proposed the solid phase peptide synthesis method (SPPS), which had the advantages of convenience and rapidity, and was a milestone in the field of peptide drug development (Merrifield 1963). The production of long peptides was limited by the synthetic methods available at the time, and insulin from animal sources often caused allergic reactions. Recombinant technology provided a simple and reliable alternative to the production of synthetic peptides, making the production of larger and more complex peptides possible. Somatostatin was the first human-derived peptide produced by recombinant technology (Itakura et al. 1977). Display technologies, including phage display, yeast display, ribosome display, mRNA display, and DNA display, are among the most advanced peptide drug discovery technologies today. Display technology can create a large peptide library and conduct multiple rounds of screening against therapeutic targets to produce high-affinity target binders (Jaroszewicz et al. 2022). Peptide drugs developed by phage display technology include romiplostim and ecallantide (Molineux and Newland 2010; Lehmann 2008). Integrated venomics is also one of the most advanced peptide drug discovery techniques. Animal venoms are a rich source of natural compounds, and many venom-derived peptides are highly potent and specific. There are still millions of neuroactive venom peptides in nature that have yet to be developed. Millions of neuroactive venom peptides remain undeveloped in nature. Integrated venomics can rapidly and accurately identify a large number of venom peptide sequences, thereby promoting the development of new peptide drugs (Vetter et al. 2011). Nevertheless, existing peptide drug development methods still rely on experimental verification techniques, which have the disadvantages of being costly and time-consuming.

4 AI-Assisted Peptide Drug Development

There are two main strategies for AI-assisted peptide drug development: identification and generation. The former refers to the screening of therapeutic peptides in naturally occurring or synthetic peptide sequences, and the latter refers to the design and synthesis of new therapeutic peptides. Some models combine the generation with identification, which is more conducive to the development of peptide drugs. The current research objects are mainly AMPs and ACPs.

4.1 *Models for Identifying Peptide Drugs*

4.1.1 Traditional Machine Learning Models

Traditional machine learning algorithms are often used for the prediction of therapeutic peptides by extracting features from known therapeutic peptides to train models and evaluate performance. These algorithms include support vector machine (SVM), random forest (RF), K-nearest neighbor (KNN), Extreme Gradient Enhancement (XGBoost), etc. SVM and RF are widely used to predict therapeutic peptides. SVM is a kind of generalized linear classifier for the binary classification of data according to supervised learning. RF is a classifier with many decision trees, which can be used for both classification and regression problems, as well as dimension reduction problems. Table 1 shows examples of the application of traditional machine learning algorithms in identifying peptide drugs.

Most of the traditional machine learning models shown in the current studies perform the binary classification task of distinguishing between therapeutic and nontherapeutic peptides. Thakur et al. created the first antiviral peptides (AVPs) database, including 604 AVPs and 452 non-AVPs, and developed the first AVP prediction method (AVPpred) based on it. The overall activity of AMP is determined by its amino acid composition, structure, and physicochemical properties. Therefore, in this study, four SVM models, namely, AVPmotif, AVPalign, AVPcompo, and AVPphyco, were trained and evaluated based on amino acid composition, sequence alignment, and physicochemical properties, respectively. These four SVM models were trained on the same datasets, and the models differ primarily in the peptide features used for training. The results showed that the AVPcompo performed best, with 90% accuracy and 0.81 correlation. In addition, a hybrid model combining the above four methods was developed. However, the performance of the hybrid model was almost similar to that of the AVPcompo (Thakur et al. 2012). Meher et al. developed iAMPpred based on SVM to predict AMP. The composition, structure, and physicochemical characteristics of the peptides were used as inputs to the SVM model for AMP prediction. Comparing iAMPpred with other AMP prediction models, such as CAMP (Thomas et al. 2009), iAMP-2 L (Xiao et al. 2013), EFC-FCBF (Veltri et al. 2017), and EFC + 307-FCBF (Veltri et al. 2017), the results showed that iAMPpred performed best. The AUC-ROC of iAMPpred was 98%, the area under precision-recall curve (AUC-PR) was 99%, and the Matthews Correlation Coefficient (MCC) was 0.91. Different from the early AMP prediction models developed based on a single dataset, iAMPpred's peptides datasets come from CAMP (Thomas et al. 2009), APD3 (Wang et al. 2016), AntiBP2 (Lata et al. 2010), LAMP (Zhao et al. 2013), and AVPred (Thakur et al. 2012). In addition, it was the first model to use normalized amino acid composition (NAAC)

Table 1 Examples of models for identifying peptide drugs

Model name	Algorithm	Role	Year	References
iAMPpred	SVM	Identify between AMPs and non-AMPs	2017	Meher et al. (2017)
AmPEPpy	RF	Identify between AMPs and non-AMPs	2021	Lawrence et al. (2021)
IAMPE	NB, KNN, SVM, RF, XGBoost	Identify between AMPs and non-AMPs	2020	Kavousi et al. (2020)
CalcAMP	Catboost, LightGBM, XGBoost, RF, extra trees	Identify between AMPs and non-AMPs	2024	Bournez et al. (2023)
SAMP	SVM	Identify between AMPs and non-AMPs	2024	Feng et al. (2024)
Deep-AmPEP30	CNN	Identify between AMPs and non-AMPs	2020	Yan et al. (2020)
AMPlify	BiLSTM	Identify between AMPs and non-AMPs	2023	Li et al. (2022)
AMP-BERT	BERT	Identify between AMPs and non-AMPs	2023	Lee et al. (2023)
sAMPpred-GAT	GAT	Identify between AMPs and non-AMPs	2023	Yan et al. (2023a)
AVPpred	SVM	Identify between AVPs and non-AVPs	2012	Thakur et al. (2012)
Deep-AVPred	CNN	Identify between AVPs and non-AVPs	2022	Sharma et al. (2022)
PGAT-ABPp	GAT	Identify between ABPs and non-ABPs	2024	Hao et al. (2024)
ACPred-fuse	RF	Identify between ACPs and non-ACPs	2020	Rao et al. (2020a)
CAPTURE	AdaBoost	Identify between ACPs and non-ACPs	2024	Ghafoor et al. (2024)
–	SVM, KNN, RF	Identify between ACPs and non-ACPs	2022	Alsanea et al. (2022)
AI4ACP	CNN	Identify between ACPs and non-ACPs	2022	Sun et al. (2022a)
ACP-GCN	GCN	Identify between ACPs and non-ACPs	2023	Rao et al. (2020b)
BertAIP	BERT	Identify between AIPs and non-AIPs	2024	Xu et al. (2024)
AIPpred	RF	Identify between AIPs and non-AIPs	2018	Manavalan et al. (2018)
PreAIP	Random Forest	Identify between AIPs and non-AIPs	2019	Khatun et al. (2019)
PreTP-stack	RF, linear discriminant analysis, XGBoost and SVM	Identify between AIPs and non-AIPs	2023	Yan et al. (2023b)
LABAMPsGCN	GCN	Predict lactic acid bacteria antimicrobial peptides	2023	Sun et al. (2022b)
UniDL4BioPep	CNN	Binary classification of peptide bioactivity	2023	Du et al. (2023)

(continued)

Table 1 (continued)

Model name	Algorithm	Role	Year	References
AMAP	SVM, XGBoost	Identify between AMPs and non-AMPs and predict functional activity against different bacteria	2019	Gull et al. (2019)
AMPFinder	RF, RNN, BiLSTM, BiGRU	Identify between AMPs and non-AMPs and predict functional activity against different bacteria	2023	Yang et al. (2023a)
AMPs-net	GCN	Identify between AMPs and non-AMPs and predict functional activity against different bacteria	2023	Ruiz Puentes et al. (2022)
DMAMP	CNN	Identify between AMPs and non-AMPs and predict functional activity against different bacteria	2024	Meng et al. (2024)
AVPIden	RF	Identify between AVPs and non-AVPs and predict the functional activity against different viruses	2021	Pang et al. (2021)
xDeep-AcPEP	CNN	Identify between ACPs and non-ACPs and predict the functional activity against six types of tumor cells	2021	Chen et al. (2021)
MLBP	CNN, BiGRU	Identify multifunctional bioactive peptides (anticancer, antidiabetic, antihypertensive, anti-inflammatory, antimicrobial, etc.)	2022	Tang et al. (2022)
DeepTPpred	CNN, LSTM	Predict multiple types of therapeutic peptides (AAP, AIP, ACP, ABP, AVP, CPP, SBP, QSP)	2023	Cui et al. (2023)
MultiPep	CNN	Multi-label classification of peptide bioactivity	2021	Grønning et al. (2021)
Deep2Pep	BiLSTM, BERT	Predict the antibacterial, antihypertensive, antioxidant, and hypoglycemic effects of peptides	2024	Chen et al. (2024a)
–	BiLSTM, BiTCN, 1D-CNN	Identify low-hemolytic peptides	2024	Sharma et al. (2023b)
AMPDeep	BERT	Predict the hemolytic activity of AMPs	2022	Salem et al. (2022)
–	BiLSTM	Predict the hemolysis, solubility, and resistance to nonspecific interactions of peptides	2023	Ansari and White (2023)
tAMPer	BiGRU, GNN	Predict the toxicity of AMPs	2024	Ebrahimikondori et al. (2024)

NB naive bayes, *Catboost* categorical boosting, *LightGBM* light gradient boosting machine, *Extra Trees* extremely randomized trees, *AdaBoost* adaptive boosting, *BiGRU* bidirectional gated recurrent unit, *ABPs* antibacterial peptides, *AAP* antiangiogenic peptides, *CPP* cell penetrating peptide, *SBP* surface binding peptide, *QSP* quorum sensing peptide characteristics to predict AMP (Meher et al. 2017). Rao et al. developed an effective ACP prediction model based on RL, called ACPredFuse. The positive dataset contained 332 experimentally verified ACPs and the negative dataset contained 2878 experimentally verified AMPs with no anticancer activity. This study adopted a multi-view feature approach to fusing the class feature vector, probability feature vector, and the best feature descriptor based on individual sequences into a new feature vector. The features were optimized by maximal relevance and maximal distance (MRMD) and sequential feature search (SFS) methods. Comparing the multi-view features with the above three single-view features, the results showed that the prediction performance of the multi-view features was better, which indicated that different view information complemented each other. ACPredFuse had the highest sensitivity (SE), accuracy (ACC), and MCC when compared to other models (AntiCP (Tyagi et al. 2013), Hajisharifi's method (Hajisharifi et al. 2014), iACP (Chen et al. 2016), ACPredFL (Wei et al. 2018), and PEPred-Suite (Wei et al. 2019; Rao et al. 2020a)).

In addition to the binary classification task of distinguishing between therapeutic and nontherapeutic peptides, some traditional machine learning models can further predict the type of bioactivity of peptides. Gull et al. developed AMAP to predict AMPs as well as types of bioactivity. The positive dataset contained 2704 peptides with 14 different types of bioactivity, such as antibacterial peptides, antifungal peptides, anticancer peptides, antiviral peptides, and anti-HIV peptides. The negative dataset contained 5156 peptides with no known antimicrobial activity. AMAP first predicted whether a peptide was AMP and then used a layered multi-label prediction approach to predict the type of bioactivity it was likely to have. In the AMP prediction task, the performance of three different machine learning models (linear AVM, nonlinear SVM, XGBoost) was evaluated, and the nonlinear SVM model showed the best performance. In the task of bioactivity type prediction, a multi-label SVM model was developed and compared with MLAMP. The results showed that MLAMP could only make predictions for five types of bioactivity, and AMAP performed significantly better (Gull et al. 2019). Pang et al. developed AVPIden based on RF to predict AVP and its functional activity against different viruses. AVPIden achieved an accuracy of 98.58 ± 0.18% in predicting AVP. Considering that AVP may act on multiple viruses simultaneously, AVPIden adopted one-against-all multi-task learning strategy to characterize targets for different viruses. The results showed that AVPIden can characterize targets for six virus families and eight viruses (Pang et al. 2021).

4.1.2 Deep Learning Models

Compared with traditional machine learning, deep learning has stronger learning capabilities and adaptability, and can process complex data, including non-linear complex problems that are difficult to handle with traditional machine learning. Deep learning algorithms for therapeutic peptide screening mainly include convolutional neural network (CNN), long short-term memory (LSTM), bi-directional long short-term memory (BiLSTM), bidirectional encoder representations from transformers (BERT), graph attention networks (GAT), graph convolutional networks (GCN), etc. Table 1 shows examples of the application of deep learning algorithms in identifying peptide drugs.

There are some models that perform binary classification tasks. Li et al. first applied the attention mechanism to AMP prediction, developing AMPlify based on BiLSTM. AMPlify was compared to three state-of-the-art models, including iAMP-2 L (Xiao et al. 2013), iAMPpred (Meher et al. 2017), and AMPScannerVr.2 (Veltri et al. 2018). AMPlify achieved the highest accuracy (93.71%), sensitivity (92.93%), specificity (94.49%), F1 score (93.66%), and OC-ROC (98.37%). In addition, to demonstrate the utility of AMPlify, AMPlify was used to search for new AMPs from the North American bullfrog genome. Finally, four AMPs with strong antimicrobial efficacy against one or more of the tested microorganisms were obtained (Li et al. 2022). Yan et al. developed sAMPpred-GAT, the first method for AMP prediction by predicting the peptide structure. sAMPpred-GAT extracted the spatial relationship and structure information between residues from the peptide sequence information and constructed a map. Then GAT was used to extract features from the map and predict whether the peptide was AMP. On XUAMP (Xu et al. 2021) dataset, the performance of sAMPpred-GAT was compared with that of 9 other models (amPEPpy (Lawrence et al. 2021), AMPfun (Chung et al. 2019), AMPEP (Bhadra et al. 2018), ADAM-HMM (Lee et al. 2015), ampir (Fingerhut et al. 2021), AMPScannerV2 (Veltri et al. 2018), AMPGram (Burdukiewicz et al. 2020), Deep-AMPEP30 (Yan et al. 2020), and CAMP-ANN (Waghu et al. 2016)). The results showed that sAMPpred-GAT achieved the highest OC-ROC (77.7%), ACC (0.715 ± 0.01), MCC (0.464 ± 0.011), and SE (0.530 ± 0.038) (Yan et al. 2023a). Sun et al. proposed a new protein-coding method, called PC6, that converted the peptide sequence into a computational matrix representing the six physicochemical properties of amino acids. On this basis, AI4ACP, an ACP prediction model, was built using CNN. After testing and evaluation, AI4ACP combined with the PC6 coding method and CNN could effectively predict ACP, with up to 0.913 accuracy, 0.925 specificity, 0.901 sensitivity, and 0.826 MCC (Sun et al. 2022a).

Deep learning algorithms can also achieve multi-label classification of peptide biological activity. Chen et al. developed a CNN-based deep learning model to predict ACPs for six types of tumor cells (breast, cervix, prostate, colon, lung, and skin). xDeep-AcPEP was trained using a multi-task learning approach. Experimental results showed that the multi-task learning model had better learning performance than the single-task learning model. On the CancerPPD (Tyagi et al. 2015) dataset, xDeep-AcPEP achieved an average mean square error of 0.1758, a Kendall correlation coefficient of 0.6156, and a Pearson correlation coefficient of 0.8086 (Chen et al. 2021). Puentes et al. developed AMPs-Net, which combined GCN with molecular dynamics simulation to predict AMPs. AMPs-Net consisted of two modules: the first module predicted whether the polypeptide was AMP, and the second module predicted the antibacterial, antiviral, antifungal, and antiparasitic activity of AMP. The experimental results proved that two novel AMPs were finally discovered (Ruiz Puentes et al. 2022).

Currently, great progress has been made in the research of AMPs and ACPs, but research on predicting other therapeutic peptides is still scarce. In addition to predicting AMPs and ACPs, some models can predict other therapeutic peptides, such as anti-inflammatory peptides (AIPs) (Xu et al. 2024; Manavalan et al. 2018; Khatun et al. 2019; Yan et al. 2023b), peptides with anti-aging properties (Kennedy et al. 2020), peptides with osteogenesis-promoting functions (Cai et al. 2022), antidiabetic peptides, and antihypertensive peptides (Tang et al. 2022; Chen et al. 2024a). Xu et al. developed the first model based on BERT that used only peptide sequence information for AIPs' prediction, called BertAIP. BertAIP was compared with other AIP prediction models, including Antiinflam (Gupta et al. 2017), AIPpred (Manavalan et al. 2018), PreAIP (Khatun et al. 2019), PreTP-EL (Guo et al. 2021), and PreTP-Stack (Yan et al. 2023b), and the results showed that BertAIP performed best (Xu et al. 2024). Kennedy et al. first applied deep learning methods to predict peptides with antiaging effects, ultimately identifying pep_RTE62G from the pea genome. In vitro experiments proved that pep_RTE62G had antiaging properties (Kennedy et al. 2020). Cai et al. developed a natural language processing (NLP) model to screen for peptides with bone-promoting functions. Finally, 28 osteogenic peptides were synthesized and tested. Among them, AIB5P had the function of promoting osteogenesis in vivo and in vitro (Cai et al. 2022). Tang et al. used a deep neural network algorithm to develop MLBP, which could simultaneously predict the antimicrobial, anticancer, anti-inflammatory, antidiabetic, antihypertensive, and other biological activities of peptides (Tang et al. 2022).

Currently, most models for identifying peptide drugs focus on the bioactivities of peptides, and few models focus on other properties of peptides. The development of peptide drugs not only focuses on the biological activity of peptides but also needs to evaluate their toxicity, hemolytic activity, solubility, and other characteristics. Ebrahimikondori et al. developed tAMPer, a multimodal model for predicting the toxicity of AMPs. tAMPer consisted of three parts: bidirectional gated recurrent units (Bi-GRU), graph neural networks (GNNs), and self-attention layers. It could jointly learn from amino acid sequence data and three-dimensional structure data. Experimental results showed that tAMPer could effectively predict the toxicity of AMPs (Ebrahimikondori et al. 2024). Salem et al. developed AMPDeep based on a pretrained model (Prot-BERT-BFD) to predict the hemolytic activity of AMPs. AMPDeep first predicted whether the AMP was secreted and then predicted its hemolytic activity (Salem et al. 2022). Sharma et al. integrated BiLSTM, BiTCN, and 1DCNN to develop a deep learning model for predicting low hemolytic peptides (Sharma et al. 2023b). Ansari et al. used BiLSTM to develop three deep learning models to predict the hemolytic activity, solubility, and resistance to nonspecific interactions of peptides. Experimental results showed that the performance of the three models was comparable to that of the most advanced models (Ansari and White 2023).

4.2 *Models for Generating Peptide Drugs*

Deep generative models (DGMs) are powerful machine learning tools that learn potential distributions from input data and generate new sample data that is similar to the training data. DGMs have become tools for peptide drug development. DGMs used in peptide drug design include generative adversarial networks (GANs), variational autoencoders (VAEs), neural language models (NLMs), and diffusion models. Table 2 shows examples of the application of DGMs.

4.2.1 Generative Adversarial Networks

The GAN model consists of a generation network module and a discrimination network module. These two modules are trained simultaneously, usually using two multilayer deep neural networks. The task of the generation network module is to generate fake data that are as close to the real data as possible, and the task of the discrimination network module is to distinguish whether the input data are real data or fake data generated by the generator. They evolve together through competition and confrontation with each other, and finally, the generator is able to

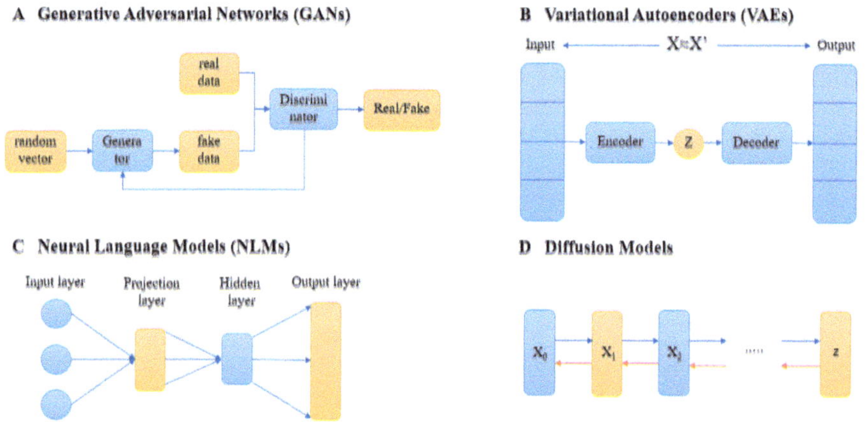

Fig. 1 Architecture of DGMs for peptide design. (**a**) The GAN model consists of a generation network module and a discrimination network module. The two modules evolve together through competition and confrontation with each other, and finally, the generator is able to generate samples that are very close to the real data. (**b**) VAE consists of an encoder and a decoder. VAE maps the input data to a low-dimensional potential space via an encoder. The decoder generates new data by reconstructing the input data using samples collected from the latent space. (**c**) NLMs utilize neural networks to learn representations of text sequences to understand, process, and generate natural language. (**d**) Diffusion models generate data by simulating the process of data gradually diffusing from random noise to target data. The forward process destroys the training data by constantly adding Gaussian noise to the original data, and ultimately reaches a pure noise state. The reverse process uses a deep network to remove noise and restore data (Fig. 1)

Table 2 Examples of models for generating peptide drugs

Model's name	Algorithm	Role	Year	References
PepGAN	GAN	Generate AMPs	2020	Tucs et al. (2020)
–	WGAN-GP	Generate AMPs	2023	Lin et al. (2023)
FBGAN	cWGAN	Generate AMPs	2018	Gupta and Zou (2019)
–	SeqGAN	Generate AMPs	2023	Cao et al. (2023)
–	WGAN, CGAN, BiGAN	Generate AMPs	2020	Ferrell et al. (2020)
AMPGAN v2	BiCGAN	Generate AMPs	2021	Van Oort et al. (2021)
Multi-CGAN	cGAN	Generate AMPs with the desired multiple properties	2023	Yu et al. (2024)
Feedback-AVPGAN	WGAN	Generate AVPs	2022	Hasegawa et al. (2022)
PandoraganGAN	LeakGAN	Generate AVPs	2023	Surana et al. (2023)
GANDALF	DCGAN	Generate ACPs	2019	Rossetto and Zhou (2019)
–	WAE	Generate AMPs	2021	Das et al. (2021)
LSSAMP	VQ-VAE	Generate AMPs	2023	Wang et al. (2023)
–	VAE	Generate AMPs	2023	Pandi et al. (2023)
HydrAMP	cVAE	Generate AMPs	2023	Szymczak et al. (2023)
–	VAE	Generate AMPs	2022	Ghorbani et al. (2022)
PepVAE	VAE	Generate AMPs and predict activity	2021	Dean et al. (2021)
–	VAE	Generate AMPs	2020	Dean and Walper (2020)
–	Binary VAE	Generate nonhemolytic antimicrobial peptides	2023	Tučs et al. (2023)
GM-pep	cVAE	Generate single bioactive peptides	2022	Chen et al. (2022)

(continued)

Table 2 (continued)

Model's name	Algorithm	Role	Year	References
–	VAE	Generate target-specific peptides and assess binding affinity	2024	Chen et al. (2024c)
–	RNN	Generate non-hemolytic antimicrobial peptides	2021	Capecchi et al. (2021)
–	LSTM	Generate AMPs	2018	Nagarajan et al. (2018)
–	LSTM	Generate AMPs	2018	Müller et al. (2018)
–	LSTM, BiLSTM	Generate AMPs specific to *E. Coli*	2021	Wang et al. (2021)
–	LSTM	Generate membrane-soluble ACPs	2018	Grisoni et al. (2018)
LSTM_Pep	LSTM, CNN	Generate therapeutic peptides with specific bioactivities	2023	Zhang et al. (2023)
AMP-diffusion	Diffusion model	Generate AMPs	2023	Chen et al. (2024b)
HYDRA	Diffusion model	De novo design peptides for various target receptors	2024	Vishva Saravanan et al. (2024)
–	Rfdiffusion	Generate peptides targeting the androgen receptor	2024	Ma et al. (2024)
–	GLAD	Generate therapeutic peptides	2024	Kong et al. (2024)
–	MMCD	Generate therapeutic peptides	2024	Wang et al. (2024)

WGAN wasserstein generative adversarial networks, *CGAN* conditional generative adversarial networks, *WGAN-GP* wasserstein, generative adversarial networks with gradient penalty, *cWGAN* conditional wasserstein generative adversarial networks, *SeqGAN* sequence generative adversarial networks, *BiGAN* bidirectional generative adversarial networks, *DCGAN* deep convolutional generative adversarial networks, *WAE* wasserstein autoencoder, *VQ-VAE* vector quantized variational autoencoder, *cVAE* conditional variational autoencoder, *GLAD* geometric latent diffusion, *MMCD* multi-modal contrastive diffusion model generate samples that are very close to the real data (Lin et al. 2022; Goodfellow et al. 2014). GAN can generate samples in parallel to greatly improve the sampling speed. In addition, the principle of GAN is the adversarial game between the generation network module and the discrimination network module, which does not approximate likelihood. Compared with other DGMs, GAN is able to generate higher-quality samples. However, the high training instability and modal collapse problems of GAN are very unfavorable for practical applications (Alqahtani et al. 2021).

Yu et al. developed the Multi-CGAN based on the conditional generative adversarial network (CGAN), which could learn from single-attribute peptide data and generate new AMPs with multiple attributes. CGAN is a variant of GAN that introduces additional conditional information based on GAN to directly control the samples generated by the model. Multi-CGAN was trained on three peptide datasets with different properties, including toxicity data, secondary structure data, and antimicrobial data. Unlike traditional CGAN, which could only generate single-attribute peptide sequences, multi-CGAN used a three-stage generation strategy to generate peptide sequences with multiple required attributes. The three-stage generation strategy included a single-label generation stage, a multi-label integration stage, and a multi-label generation stage. Experimental results showed that more than 80% of the peptide sequences generated by multi-CGAN had the required characteristics and showed low homology with the training set data (Yu et al. 2024). Hasegawa et al. combined GAN and feedback methods to propose Feedback-AVPGAN to generate novel AVPs. This study used Wasserstein GAN with a gradient penalty mechanism to improve the performance of the GAN model. Feedback methods were used to overcome the lack of experimentally proven AVPs in the datasets, allowing the model to learn from the generated data. In addition, AlphaFold2 was used to model the structures of the generated peptides, and the structures of the generated peptides were compared with the structures of known AVPs. The results showed that feedback-AVPGAN can generate new peptides with structures similar to known AVPs (Hasegawa et al. 2022). Surana et al. used LeakGAN to develop PandoraGAN to generate new AVPs. The LeakGAN model leaks advanced feature information from the discrimination network module to the generation network module during the training process. To evaluate PandoraGAN, the physicochemical properties of the generated peptides were compared to those of highly active AVPs. The results showed that the amino acid composition, net charge, alpha helix and folding tendency, instability index, Shannon entropy, and residue repetitions of the generated peptides were similar to those of highly active AVPs (Surana et al. 2023). Cao et al. combined sequence generative adversarial networks (SeqGAN), BERT, and multi-layer perceptron (MLP) to develop a model that can generate and recognize AMPs. The peptide sequences produced by SeqGAN were identified by AMP predictors based on BERT and MLP. AlphaFold2 and molecular dynamics simulation were used to screen candidate AMPs, resulting in six candidates. The results showed that one of the peptides, A-222, inhibited both Gram-positive and Gram-negative bacteria (Cao et al. 2023).

4.2.2 Variational Autoencoders

VAE is a kind of DGM developed based on a self-encoder, combining variational inference and Bayesian theory. VAE consists of an encoder and a decoder, which is able to learn from input data and generate new data similar to the training data. During training, VAE maps the input data to a low-dimensional potential space via an encoder to obtain a potential space representation and perform sampling. The decoder generates new data by reconstructing the input data using samples collected from the latent space (Wu et al. 2021). The advantages of VAEs are their native support for modeling categorical and continuous variables and their ability to provide lower bounds on the log-likelihood of data. However, VAEs provide limited support for modeling data of different dimensions, and the quality of the generated samples is not as good as other DGMs (Strokach and Kim 2022).

Szymczak et al. proposed HydrAMP to generate AMPs. HydrAMP was the first model to directly optimize both simulation generation and unconstrained generation tasks. The experimental results showed that HydrAMP outperformed Basic, PepCVAE (Das et al. 2018), and Joker (Porto et al. 2018) on the task of generating existing AMPs analogues and outperformed Basic, PepCVAE, AMP-LM (Nagarajan et al. 2018), Dean-VAE (Dean and Walper 2020), Muller-LSTM (Müller et al. 2018), and AMP-GAN (Van Oort et al. 2021) on the task of unconstrained generation. Basic is a standard conditional VAE model with potential reconstruction regularization added. Wet laboratory experiments on five bacterial strains confirmed the ability of HydrAMP to generate potent AMPs (Szymczak et al. 2023). Dean et al. developed PepVAE in combination with VAE and an AMPs activity prediction model to generate AMPs and predict antimicrobial activity. PepVAE took the peptide sequence and experimental minimum inhibitory concentration (MIC) as input values. The design of AMPs by PepVAE consisted of two stages: the first stage was to train the VAE model and the AMP activity prediction model; the second stage was to sample from the latent space, to generate new AMPs, and to predict their antimicrobial activity. Experimental results showed that PepVAE could generate AMPs against *Escherichia coli*, *Pseudomonas aeruginosa*, and *Staphylococcus aureus* (Dean et al. 2021). Tučs et al. developed a model based on the binary variational autoencoder (bVAE) and the D-Wave quantum annealer to generate nonhemolytic AMPs. The generated peptides were screened and four of them were subjected to wet laboratory verification. The results showed that three peptides had potent antimicrobial activity and two peptides had no hemolytic activity (Tučs et al. 2023).

4.2.3 Neural Language Models

NLMs utilize neural networks to learn representations of text sequences to understand, process, and generate natural language. NLMs usually adopt the structure of the recurrent neural network (RNN) and the Transformer. Compared with traditional n-gram language models, the advantage of NLMs is that they can handle long-distance dependencies. RNN is a neural network that can process sequence data and capture the temporal dependencies in the sequence through hidden states. LSTM is a variant of RNN. LSTM effectively solves the problems of gradient vanishing and gradient exploding by introducing gating mechanisms (forget gate, input gate, and output gate), and enhances the model's ability to capture long-term dependencies. Transformer introduces a self-attention mechanism, which enables the model to consider all position information in the input sequence at the same time, so as to better capture the global semantic relationship (Wan et al. 2022).

Wang et al. used LSTM and BiLSTM algorithms to build AMP generation models and prediction models to design AMPs for *Escherichia coli*. The accuracy of the AMP prediction model was 81.6 ~ 88.9%. The three-dimensional structure of the new AMPs generated by the AMP generation model was a α-helical structure with an amphiphilic surface, which was similar to the typical AMPs. This suggested that the model proposed in this study could effectively generate AMPs against *E. coli* (Wang et al. 2021). Nagarajan et al. developed an AMP generation model using the LSTM algorithm using the amino acid sequence as the input. Ten peptides were screened and synthesized from the peptide sequences generated by the LSTM model, and these ten peptides were tested for antibacterial activity. Experimental results showed that these peptides all showed broad-spectrum antimicrobial activity, among which NN2_0050 and NN2_0018 were active against multi-drug-resistant clinical isolates, including methicillin-resistant bacteria and carbapenem-resistant bacteria (Nagarajan et al. 2018). Grisoni et al. proposed an LSTM model to design film-soluble ACPs de novo. The model was trained on a dataset of alpha-helical cationic amphipathic peptides and fine-tuned on a peptide library consisting of 26 known ACPs. The anticancer activity test of the 12 peptides synthesized in this model showed that 10 of them had anticancer activity, and 6 of them could kill MCF7 cancer cells without affecting human red blood cells (Grisoni et al. 2018).

4.2.4 Diffusion Models

Diffusion models are a type of DGM. Unlike traditional generation models, diffusion models generate data by simulating the process of data gradually diffusing from random noise to target data. Diffusion models include the forward process and reverse process. The forward process destroys the training data by constantly adding Gaussian noise to the original data, and ultimately reaches a pure noise state. The reverse process uses a deep network to remove noise and restore data. Diffusion models have excellent performance in the fields of image generation, text generation, and audio generation (Cao et al. 2024; Yang et al. 2023b; Karras et al. 2024).

Chen et al. developed AMP-diffusion, which took advantage of the potential space of the most advanced protein language model ESM-2, and used the diffusion model framework to generate new AMPs. The peptides generated by AMP-diffusion had low confusion, high diversity, and similarity, and had physicochemical properties similar to those of experimentally verified AMPs (Chen et al. 2024b). Saravanan et al. proposed a hybrid deep learning approach, called HYDRA, that combined diffusion models with a binding affinity maximization algorithm for the de novo design of peptides targeting various target receptors. To demonstrate the utility of HYDRA, this study used HYDRA to generate a therapeutic peptide targeting a protein expressed by the *Plasmodium falciparum* erythrocyte membrane protein 1 (PfEMP1) gene (Vishva Saravanan et al. 2024). Wang et al. proposed a multimodal contrast diffusion model (MMCD) based on the diffusion model framework. Taking peptide sequence and structure as inputs, they constructed sequence-modal and structure-modal diffusion models, respectively to generate new therapeutic peptides. MMCD was trained on a therapeutic peptide dataset consisting of AMPs and ACPs, and the same number of nontherapeutic peptide datasets. Model evaluation results showed that MMCD was superior to other DGMs in many indicators such as antibacterial/anticancer scores, diversity, and peptide docking (Wang et al. 2024).

5 Discussion and Conclusions

This chapter provides an overview of the properties and clinical applications of peptides, the development of traditional peptide drug development, and examples of specific applications of traditional machine learning and deep learning algorithms in peptide drug development. Current research focuses on AMPs and ACPs, with less research on other therapeutic peptides. Models for identifying peptide drugs apply traditional machine learning algorithms (SVM, RF, KNN, XGBoost, etc.) and deep learning algorithms (CNN, LSTM, BiLSTM, BERT, GAT, GCN, etc.). The development of the models for identifying peptide drugs includes two stages, a binary classification task of therapeutic peptides and nontherapeutic peptides, and a multi-label classification task of peptide bioactivity. However, most models can only accomplish binary classification tasks, which limits their widespread clinical applications. Peptides have multiple bioactivities, such as antimicrobial, anticancer, antioxidant, antidiabetic, and anticoagulant, and the models for multi-label classification need further research. In addition, the development of peptide drugs should not only focus on the bioactivities of peptides, but also evaluate their toxicity, hemolytic activity, solubility, and binding ability with target proteins. Currently, models focusing on these properties are few. Models for generating peptide drugs mainly use deep learning algorithms, including GANs, VAEs, NLMs, and diffusion models. The principle of GAN is the adversarial game between the generation network module and the discrimination network module, and its implementation needs to consider various factors. First of all, GAN needs to be trained on a balanced data set among categories, otherwise the generalization ability of the model will be affected. Secondly, appropriate strategies need to be applied to characterize peptide sequences to ensure that the discrimination network module can accurately distinguish between real data and fake data. VAE consists of an encoder and a decoder, as well as is able to learn from input data and generate new data similar to the training data. The quality of samples generated by VAE is inferior to that of GAN. Considering the moonlighting nature of peptides, it is possible to discover new bioactivities when using VAEs. Therefore, validators are needed to verify the sequence of peptides generated by VAE. NLM usually adopts the structure of RNN and transformers. The advantage of NLMs is the ability to handle long distance dependencies. However, the development of NLM requires huge computational resources and needs to integrate classifiers to evaluate the performance of the model, which complicates the training process of the model. Unlike traditional generation models, diffusion models generate data by simulating the gradual diffusion of data from random noise to target data. Diffusion models have the advantages of generating high-quality data and good interpretability, but their disadvantages are long training time and high demand for computing resources. We hope that the knowledge in this chapter can be helpful to relevant researchers who are interested in the application of artificial intelligence in peptide drug discovery.

Acknowledgments We thank the Supercomputing Center of Lanzhou University and the Gansu Computation Center for their support.

References

Ab NM. The antimicrobial peptides: ready for clinical trials? Biomed J Sci Techn Res. 2018;7(4) https://doi.org/10.26717/bjstr.2018.07.001536.

Alqahtani H, Kavakli-Thorne M, Kumar G. Applications of generative adversarial networks (GANs): an updated review. Arch Comput Methods Eng. 2021;28(2):525–52. https://doi.org/10.1007/s11831-019-09388-y.

Alsanea M, Dukyil AS, Afnan RB, Alebeisat F, Islam M, et al. To assist oncologists: an efficient machine learning-based approach for anti-cancer peptides classification. Sensors (Basel, Switzerland). 2022;22(11) https://doi.org/10.3390/s22114005.

Ansari M, White AD. Serverless prediction of peptide properties with recurrent neural networks. J Chem Inf Model. 2023;63(8):2546–53. https://doi.org/10.1021/acs.jcim.2c01317.

Apostolopoulos V, Bojarska J, Chai TT, Elnagdy S, Kaczmarek K, Matsoukas J, New R, Parang K, Lopez OP, Parhiz H, Perera CO. A global review on short peptides: frontiers and perspectives. Molecules. 2021;26(2):430.

Bhadra P, Yan J, Li J, Fong S, Siu SWI. AmPEP: sequence-based prediction of antimicrobial peptides using distribution patterns of amino acid properties and random forest. Sci Rep. 2018;8(1):1697. https://doi.org/10.1038/s41598-018-19752-w.

Bournez C, Riool M, de Boer L, Cordfunke RA, de Best L, van Leeuwen R, et al. CalcAMP: a New machine learning model for the accurate prediction of antimicrobial activity of peptides. Antibiotics (Basel, Switzerland). 2023;12(4) https://doi.org/10.3390/antibiotics12040725.

Burdukiewicz M, Sidorczuk K, Rafacz D, Pietluch F, Chilimoniuk J, Rödiger S, et al. Proteomic screening for prediction and design of antimicrobial peptides with AmpGram. Int J Mol Sci. 2020;21(12) https://doi.org/10.3390/ijms21124310.

Cai M, Xiao B, Jin F, Xu X, Hua Y, Li J, et al. Generation of functional oligopeptides that promote osteogenesis based on unsupervised deep learning of protein IDRs. Bone research. 2022;10(1):23. https://doi.org/10.1038/s41413-022-00193-1.

Cao Q, Ge C, Wang X, Harvey PJ, Zhang Z, Ma Y, et al. Designing antimicrobial peptides using deep learning and molecular dynamic simulations. Brief Bioinform. 2023;24(2) https://doi.org/10.1093/bib/bbad058.

Cao H, Tan C, Gao Z, Xu Y, Chen G, Heng P-A, et al. A survey on generative diffusion models. IEEE Trans Knowl Data Eng. 2024;36:2814–30. https://doi.org/10.1109/tkde.2024.3361474.

Capecchi A, Cai X, Personne H, Köhler T, van Delden C, Reymond JL. Machine learning designs non-hemolytic antimicrobial peptides. Chem Sci. 2021;12(26):9221–32. https://doi.org/10.1039/d1sc01713f.

Carratalá JV, Serna N, Villaverde A, Vázquez E, Ferrer-Miralles N. Nanostructured antimicrobial peptides: the last push towards clinics. Biotechnol Adv. 2020;44:107603. https://doi.org/10.1016/j.biotechadv.2020.107603.

Chen W, Ding H, Feng P, Lin H, Chou KC. iACP: a sequence-based tool for identifying anticancer peptides. Oncotarget. 2016;7(13):16895–909.

Chen J, Cheong HH, Siu SWI. xDeep-AcPEP: deep learning method for anticancer peptide activity prediction based on convolutional neural network and multitask learning. J Chem Inf Model. 2021;61(8):3789–803. https://doi.org/10.1021/acs.jcim.1c00181.

Chen Q, Yang C, Xie Y, Wang Y, Li X, Wang K, et al. GM-pep: a high efficiency strategy to De Novo design functional peptide sequences. J Chem Inf Model. 2022;62(10):2617–29. https://doi.org/10.1021/acs.jcim.2c00089.

Chen L, Hu Z, Rong Y, Lou B. Deep2Pep: a deep learning method in multi-label classification of bioactive peptide. Comput Biol Chem. 2024a;109:108021. https://doi.org/10.1016/j.compbiolchem.2024.108021.

Chen T, Vure P, Pulugurta R, Chatterjee P. AMP-diffusion: integrating latent diffusion with protein language models for antimicrobial peptide generation. 2024b:2024.03.03.583201. https://doi.org/10.1101/2024.03.03.583201.

Chen S, Lin T, Basu R, Ritchey J, Wang S, Luo Y, et al. Design of target specific peptide inhibitors using generative deep learning and molecular dynamics simulations. Nat Commun. 2024c;15(1):1611. https://doi.org/10.1038/s41467-024-45766-2.

Chung CR, Kuo TR, Wu LC, Lee TY, Horng JT. Characterization and identification of antimicrobial peptides with different functional activities. Brief Bioinform. 2019;21:1098. https://doi.org/10.1093/bib/bbz043.

Cui Z, Wang SG, He Y, Chen ZH, Zhang QH. DeepTPpred: a deep learning approach with matrix factorization for predicting therapeutic peptides by integrating length information. IEEE J Biomed Health Inform. 2023;27(9):4611–22. https://doi.org/10.1109/jbhi.2023.3290014.

Das P, Wadhawan K, Chang O, Sercu T, Santos CD, Riemer M, et al. PepCVAE: semi-supervised targeted design of antimicrobial peptide sequences; 2018.

Das P, Sercu T, Wadhawan K, Padhi I, Gehrmann S, Cipcigan F, et al. Accelerated antimicrobial discovery via deep generative models and molecular dynamics simulations. Nat Biomed Eng. 2021;5(6):613–23. https://doi.org/10.1038/s41551-021-00689-x.

Dean SN, Walper SA. Variational autoencoder for generation of antimicrobial peptides. ACS Omega. 2020;5(33):20746–54. https://doi.org/10.1021/acsomega.0c00442.

Dean SN, Alvarez JAE, Zabetakis D, Walper SA, Malanoski AP. PepVAE: variational autoencoder framework for antimicrobial peptide generation and activity prediction. Front Microbiol. 2021;12:725727. https://doi.org/10.3389/fmicb.2021.725727.

Du Vigneaud V, Ressler C, Trippett S. The sequence of amino acids in oxytocin, with a proposal for the structure of oxytocin. J Biol Chem. 1953;205(2):949–57.

Du Z, Ding X, Xu Y, Li Y. UniDL4BioPep: a universal deep learning architecture for binary classification in peptide bioactivity. Brief Bioinform. 2023;24(3) https://doi.org/10.1093/bib/bbad135.

Ebrahimikondori H, Sutherland D, Yanai A, Richter A, Salehi A, Li C, et al. Structure-aware deep learning model for peptide toxicity prediction. Protein Sci. 2024;33(7):e5076. https://doi.org/10.1002/pro.5076.

Feng J, Sun M, Liu C, Zhang W, Xu C, Wang J, et al. SAMP: identifying antimicrobial peptides by an ensemble learning model based on proportionalized split amino acid composition. Brief Funct Genomics. 2024;23(6):879–90. https://doi.org/10.1093/bfgp/elae046.

Ferrell J, Remington J, Oort C, Sharafi M, Aboushousha R, Janssen-Heininger Y, et al. A generative approach toward precision antimicrobial peptide design; 2020.

Fingerhut L, Miller DJ, Strugnell JM, Daly NL, Cooke IR. Ampir: an R package for fast genome-wide prediction of antimicrobial peptides. Bioinformatics (Oxford, England). 2021;36(21):5262–3. https://doi.org/10.1093/bioinformatics/btaa653.

Fosgerau K, Hoffmann T. Peptide therapeutics: current status and future directions. Drug Discov Today. 2015;20(1):122–8. https://doi.org/10.1016/j.drudis.2014.10.003.

George C, Byun A, Howard-Thompson A. New injectable agents for the treatment of type 2 diabetes part 2-glucagon-like Peptide-1 (GLP-1) agonists. Am J Med. 2018;131(11):1304–6. https://doi.org/10.1016/j.amjmed.2018.05.043.

Ghafoor H, Asim MN, Ibrahim MA, Ahmed S, Dengel A. CAPTURE: comprehensive anti-cancer peptide predictor with a unique amino acid sequence encoder. Comput Biol Med. 2024;176:108538. https://doi.org/10.1016/j.compbiomed.2024.108538.

Ghorbani M, Prasad S, Brooks BR, Klauda JB. Deep attention based variational autoencoder for antimicrobial peptide discovery; 2022.

Goodfellow IJ, Pouget-Abadie J, Mirza M, Xu B, Warde-Farley D, Ozair S, et al. Generative adversarial nets. In: Proceedings of the 27th international conference on neural information processing systems, vol. 2. Montreal: MIT Press; 2014. p. 2672–80.

Grisoni F, Neuhaus CS, Gabernet G, Müller AT, Hiss JA, Schneider G. Designing anticancer peptides by constructive machine learning. ChemMedChem. 2018;13(13):1300–2. https://doi.org/10.1002/cmdc.201800204.

Grønning AGB, Kacprowski T, Schéele C. MultiPep: a hierarchical deep learning approach for multi-label classification of peptide bioactivities. Biol Methods Prot. 2021;6(1):bpab021. https://doi.org/10.1093/biomethods/bpab021.

Guidotti G, Brambilla L, Rossi D. Cell-penetrating peptides: from basic research to clinics. Trends Pharmacol Sci. 2017;38(4):406–24. https://doi.org/10.1016/j.tips.2017.01.003.

Gull S, Shamim N, Minhas F. AMAP: hierarchical multi-label prediction of biologically active and antimicrobial peptides. Comput Biol Med. 2019;107:172–81. https://doi.org/10.1016/j.compbiomed.2019.02.018.

Guo Y, Yan K, Lv H, Liu B. PreTP-EL: prediction of therapeutic peptides based on ensemble learning. Brief Bioinform. 2021;22(6) https://doi.org/10.1093/bib/bbab358.

Gupta A, Zou J. Feedback GAN for DNA optimizes protein functions. Nat Mach Intell. 2019;1(2):105–11. https://doi.org/10.1038/s42256-019-0017-4.

Gupta S, Sharma AK, Shastri V, Madhu MK, Sharma VK. Prediction of anti-inflammatory proteins/peptides: an insilico approach. J Transl Med. 2017;15(1):7. https://doi.org/10.1186/s12967-016-1103-6.

Hajisharifi Z, Piryaiee M, Mohammad Beigi M, Behbahani M, Mohabatkar H. Predicting anti-cancer peptides with Chou′s pseudo amino acid composition and investigating their mutagenicity via Ames test. J Theor Biol. 2014;341:34–40. https://doi.org/10.1016/j.jtbi.2013.08.037.

Hao Y, Liu X, Fu H, Shao X, Cai W. PGAT-ABPp: harnessing protein language models and graph attention networks for antibacterial peptide identification with remarkable accuracy. Bioinformatics (Oxford, England). 2024;40(8) https://doi.org/10.1093/bioinformatics/btae497.

Hasegawa K, Moriwaki Y, Terada T, Wei C, Shimizu K. Feedback-AVPGAN: feedback-guided generative adversarial network for generating antiviral peptides. J Bioinforma Comput Biol. 2022;20(6):2250026. https://doi.org/10.1142/s0219720022500263.

Itakura K, Hirose T, Crea R, Riggs AD, Heyneker HL, Bolivar F, et al. Expression in Escherichia coli of a chemically synthesized gene for the hormone somatostatin. Science (New York, NY). 1977;198(4321):1056–63. https://doi.org/10.1126/science.412251.

Jakubczyk A, Karaś M, Rybczyńska-Tkaczyk K, Zielińska E, Zieliński D. Current trends of bioactive peptides-new sources and therapeutic effect. Foods (Basel, Switzerland). 2020;9(7) https://doi.org/10.3390/foods9070846.

Jaroszewicz W, Morcinek-Orłowska J, Pierzynowska K, Gaffke L, Węgrzyn G. Phage display and other peptide display technologies. FEMS Microbiol Rev. 2022;46(2)

Karras T, Aittala M, Laine S, Aila T. Elucidating the design space of diffusion-based generative models. In: Proceedings of the 36th international conference on neural information processing systems. New Orleans: Curran Associates Inc; 2024. p. 1926.

Kavousi K, Bagheri M, Behrouzi S, Vafadar S, Atanaki FF, Lotfabadi BT, et al. IAMPE: NMR-assisted computational prediction of antimicrobial peptides. J Chem Inf Model. 2020;60(10):4691–701. https://doi.org/10.1021/acs.jcim.0c00841.

Kennedy K, Cal R, Casey R, Lopez C, Adelfio A, Molloy B, et al. The anti-ageing effects of a natural peptide discovered by artificial intelligence. Int J Cosmet Sci. 2020;42(4):388–98. https://doi.org/10.1111/ics.12635.

Khatun MS, Hasan MM, Kurata H. PreAIP: computational prediction of anti-inflammatory peptides by integrating multiple complementary features. Front Genet. 2019;10:129. https://doi.org/10.3389/fgene.2019.00129.

Ko S-C, Kang N, Kim E-A, Kang MC, Lee S-H, Kang S-M, et al. A novel angiotensin I-converting enzyme (ACE) inhibitory peptide from a marine Chlorella ellipsoidea and its antihypertensive effect in spontaneously hypertensive rats. Process Biochem. 2012;47(12):2005–11. https://doi.org/10.1016/j.procbio.2012.07.015.

Kong X, Jia Y, Huang W, Liu Y. Full-atom peptide design with geometric latent diffusion; 2024.

Lata S, Mishra NK, Raghava GP. AntiBP2: improved version of antibacterial peptide prediction. BMC Bioinformatics. 2010;11(Suppl 1):S19.

Lawrence TJ, Carper DL, Spangler MK, Carrell AA, Rush TA, Minter SJ, et al. amPEPpy 1.0: a portable and accurate antimicrobial peptide prediction tool. Bioinformatics (Oxford, England). 2021;37(14):2058–60. https://doi.org/10.1093/bioinformatics/btaa917.

Lee HT, Lee CC, Yang JR, Lai JZ, Chang KY. A large-scale structural classification of antimicrobial peptides. Biomed Res Int. 2015;2015:475062. https://doi.org/10.1155/2015/475062.

Lee H, Lee S, Lee I, Nam H. AMP-BERT: prediction of antimicrobial peptide function based on a BERT model. Protein Sci. 2023;32(1):e4529. https://doi.org/10.1002/pro.4529.

Lehmann A. Ecallantide (DX-88), a plasma kallikrein inhibitor for the treatment of hereditary angioedema and the prevention of blood loss in on-pump cardiothoracic surgery. Expert Opin Biol Ther. 2008;8(8):1187–99. https://doi.org/10.1517/14712598.8.8.1187.

Li C, Sutherland D, Hammond SA, Yang C, Taho F, Bergman L, et al. AMPlify: attentive deep learning model for discovery of novel antimicrobial peptides effective against WHO priority pathogens. BMC Genomics. 2022;23(1):77. https://doi.org/10.1186/s12864-022-08310-4.

Lin E, Lin CH, Lane HY. De Novo peptide and protein design using generative adversarial networks: an update. J Chem Inf Model. 2022;62(4):761–74. https://doi.org/10.1021/acs.jcim.1c01361.

Lin TT, Yang LY, Lin CY, Wang CT, Lai CW, Ko CF, et al. Intelligent De Novo design of novel antimicrobial peptides against antibiotic-resistant bacteria strains. Int J Mol Sci. 2023;24(7) https://doi.org/10.3390/ijms24076788.

Ma B, Liu D, Wang Z, Zhang D, Jian Y, Zhang K, et al. A top-down design approach for generating a peptide PROTAC drug targeting androgen receptor for androgenetic alopecia therapy. J Med Chem. 2024;67(12):10336–49. https://doi.org/10.1021/acs.jmedchem.4c00828.

Manavalan B, Shin TH, Kim MO, Lee G. AIPpred: sequence-based prediction of anti-inflammatory peptides using random Forest. Front Pharmacol. 2018;9:276. https://doi.org/10.3389/fphar.2018.00276.

Meher PK, Sahu TK, Saini V, Rao AR. Predicting antimicrobial peptides with improved accuracy by incorporating the compositional, physico-chemical and structural features into Chou's general PseAAC. Sci Rep. 2017;7:42362.

Meng Q, Chen G, Zheng S, Lin Y, Liu B, Tang J, et al. DMAMP: a deep-learning model for detecting antimicrobial peptides and their multi-activities. IEEE/ACM Trans Comput Biol Bioinform. 2024;21:2025. https://doi.org/10.1109/tcbb.2024.3439541.

Merrifield RB. Solid phase peptide synthesis. I. The synthesis of a Tetrapeptide. J Am Chem Soc. 1963;85(14):2149–54. https://doi.org/10.1021/ja00897a025.

Molineux G, Newland A. Development of romiplostim for the treatment of patients with chronic immune thrombocytopenia: from bench to bedside. Br J Haematol. 2010;150(1):9–20. https://doi.org/10.1111/j.1365-2141.2010.08140.x.

Montone CM, Capriotti AL, Cavaliere C, La Barbera G, Piovesana S, Zenezini Chiozzi R, et al. Peptidomic strategy for purification and identification of potential ACE-inhibitory and antioxidant peptides in Tetradesmus obliquus microalgae. Anal Bioanal Chem. 2018;410(15):3573–86. https://doi.org/10.1007/s00216-018-0925-x.

Müller AT, Hiss JA, Schneider G. Recurrent neural network model for constructive peptide design. J Chem Inf Model. 2018;58(2):472–9. https://doi.org/10.1021/acs.jcim.7b00414.

Murray CJ, Ikuta KS, Sharara F, Swetschinski L, Aguilar GR, Gray A, et al. Global burden of bacterial antimicrobial resistance in 2019: a systematic analysis. Lancet. 2022;399(10325):629–55.

Nagarajan D, Nagarajan T, Roy N, Kulkarni O, Ravichandran S, Mishra M, et al. Computational antimicrobial peptide design and evaluation against multidrug-resistant clinical isolates of bacteria. J Biol Chem. 2018;293(10):3492–509. https://doi.org/10.1074/jbc.M117.805499.

Nauck MA, Müller TD. Incretin hormones and type 2 diabetes. Diabetologia. 2023;66(10):1780–95.

Ogurtsova K, Guariguata L, Barengo NC, Ruiz PL, Sacre JW, Karuranga S, et al. IDF diabetes atlas: global estimates of undiagnosed diabetes in adults for 2021. Diabetes Res Clin Pract. 2022;183:109118. https://doi.org/10.1016/j.diabres.2021.109118.

Pandi A, Adam D, Zare A, Trinh VT, Schaefer SL, Burt M, et al. Cell-free biosynthesis combined with deep learning accelerates de novo-development of antimicrobial peptides. Nat Commun. 2023;14(1):7197. https://doi.org/10.1038/s41467-023-42434-9.

Pang Y, Yao L, Jhong JH, Wang Z, Lee TY. AVPIden: a new scheme for identification and functional prediction of antiviral peptides based on machine learning approaches. Brief Bioinform. 2021;22(6) https://doi.org/10.1093/bib/bbab263.

Porto WF, Fensterseifer ICM, Ribeiro SM, Franco OL. Joker: an algorithm to insert patterns into sequences for designing antimicrobial peptides. Biochim et Biophys Acta Gen Subj. 2018;1862(9):2043–52. https://doi.org/10.1016/j.bbagen.2018.06.011.

Rao B, Zhou C, Zhang G, Su R, Wei L. ACPred-fuse: fusing multi-view information improves the prediction of anticancer peptides. Brief Bioinform. 2020a;21(5):1846–55. https://doi.org/10.1093/bib/bbz088.

Rao B, Zhang L, Zhang GJIA. ACP-GCN: the identification of anticancer peptides based on graph convolution. Networks. 2020b;8:176005–11.

Raucher D, Ryu JS. Cell-penetrating peptides: strategies for anticancer treatment. Trends Mol Med. 2015;21(9):560–70. https://doi.org/10.1016/j.molmed.2015.06.005.

Rossetto AM, Zhou W. GANDALF: a prototype of a GAN-based peptide design method. In: Proceedings of the 10th ACM international conference on bioinformatics, computational biology and health informatics. Association for Computing Machinery; 2019. p. 61–6.

Ruiz Puentes P, Henao MC, Cifuentes J, Muñoz-Camargo C, Reyes LH, Cruz JC, et al. Rational discovery of antimicrobial peptides by means of artificial intelligence. Membranes. 2022;12(7) https://doi.org/10.3390/membranes12070708.

Salem M, Keshavarzi Arshadi A, Yuan JS. AMPDeep: hemolytic activity prediction of antimicrobial peptides using transfer learning. BMC Bioinformatics. 2022;23(1):389. https://doi.org/10.1186/s12859-022-04952-z.

Sharma R, Shrivastava S, Singh SK, Kumar A, Singh AK, Saxena S. Deep-AVPpred: artificial intelligence driven discovery of peptide drugs for viral infections. IEEE J Biomed Health Inform. 2022;26(10):5067–74. https://doi.org/10.1109/jbhi.2021.3130825.

Sharma K, Sharma KK, Sharma A, Jain R. Peptide-based drug discovery: current status and recent advances. Drug Discov Today. 2023a;28(2):103464. https://doi.org/10.1016/j.drudis.2022.103464.

Sharma R, Shrivastava S, Singh SK, Kumar A, Singh AK, Saxena S. EnDL-HemoLyt: ensemble deep learning-based tool for identifying therapeutic peptides with low hemolytic activity. IEEE J Biomed Health Inform. 2023b;28:1896. https://doi.org/10.1109/jbhi.2023.3264941.

Strokach A, Kim PM. Deep generative modeling for protein design. Curr Opin Struct Biol. 2022;72:226–36. https://doi.org/10.1016/j.sbi.2021.11.008.

Sun YY, Lin TT, Cheng WC, Lu IH, Lin CY, Chen SH. Peptide-based drug predictions for cancer therapy using deep learning. Pharmaceuticals (Basel, Switzerland). 2022a;15(4) https://doi.org/10.3390/ph15040422.

Sun TJ, Bu HL, Yan X, Sun ZH, Zha MS, Dong GF. LABAMPsGCN: a framework for identifying lactic acid bacteria antimicrobial peptides based on graph convolutional neural network. Front Genet. 2022b;13:1062576. https://doi.org/10.3389/fgene.2022.1062576.

Surana S, Arora P, Singh D, Sahasrabuddhe D, Valadi J. PandoraGAN: generating antiviral peptides using generative adversarial network. SN Comput Sci. 2023;4(5):607. https://doi.org/10.1007/s42979-023-02203-3.

Szymczak P, Możejko M, Grzegorzek T, Jurczak R, Bauer M, Neubauer D, et al. Discovering highly potent antimicrobial peptides with deep generative model HydrAMP. Nat Commun. 2023;14(1):1453. https://doi.org/10.1038/s41467-023-36994-z.

Tang W, Dai R, Yan W, Zhang W, Bin Y, Xia E, et al. Identifying multi-functional bioactive peptide functions using multi-label deep learning. Brief Bioinform. 2022;23(1) https://doi.org/10.1093/bib/bbab414.

Thakur N, Qureshi A, Kumar M. AVPpred: collection and prediction of highly effective antiviral peptides. Nucleic Acids Res. 2012;40:W199–204.

Thomas S, Karnik S, Barai RS, Jayaraman VK, Idicula-Thomas S. CAMP: a useful resource for research on antimicrobial peptides. Nucleic Acids Res. 2009;38(suppl_1):D774–D80.

Trujillo JM, Nuffer W. GLP-1 receptor agonists for type 2 diabetes mellitus: recent developments and emerging agents. Pharmacotherapy. 2014;34(11):1174–86. https://doi.org/10.1002/phar.1507.

Tucs A, Tran DP, Yumoto A, Ito Y, Uzawa T, Tsuda K. Generating ampicillin-level antimicrobial peptides with activity-aware generative adversarial networks. ACS Omega. 2020;5(36):22847–51. https://doi.org/10.1021/acsomega.0c02088.

Tučs A, Berenger F, Yumoto A, Tamura R, Uzawa T, Tsuda K. Quantum annealing designs nonhemolytic antimicrobial peptides in a discrete latent space. ACS Med Chem Lett. 2023;14(5):577–82. https://doi.org/10.1021/acsmedchemlett.2c00487.

Tyagi A, Kapoor P, Kumar R, Chaudhary K, Gautam A, Raghava GP. In silico models for designing and discovering novel anticancer peptides. Sci Rep. 2013;3:2984.

Tyagi A, Tuknait A, Anand P, Gupta S, Sharma M, Mathur D, et al. CancerPPD: a database of anticancer peptides and proteins. Nucleic Acids Res. 2015;43:D837–43. https://doi.org/10.1093/nar/gku892.

Van Oort CM, Ferrell JB, Remington JM, Wshah S, Li J. AMPGAN v2: machine learning-guided design of antimicrobial peptides. J Chem Inf Model. 2021;61(5):2198–207. https://doi.org/10.1021/acs.jcim.0c01441.

Veltri D, Kamath U, Shehu A. Improving recognition of antimicrobial peptides and target selectivity through machine learning and genetic programming. IEEE/ACM Trans Comput Biol Bioinform. 2017;14(2):300–13. https://doi.org/10.1109/tcbb.2015.2462364.

Veltri D, Kamath U, Shehu A. Deep learning improves antimicrobial peptide recognition. Bioinformatics. 2018;34(16):2740–7. https://doi.org/10.1093/bioinformatics/bty179.

Vetter I, Davis JL, Rash LD, Anangi R, Mobli M, Alewood PF, et al. Venomics: a new paradigm for natural products-based drug discovery. Amino Acids. 2011;40(1):15–28. https://doi.org/10.1007/s00726-010-0516-4.

Vigneaud V, Ressler C, Swan JM, Roberts CW. The synthesis of oxytocin1. J Am Chem Soc. 1954;76:3115–21.

Vishva Saravanan R, Choudhuri S, Ghosh B. Hybrid diffusion model for stable, affinity-driven, receptor-aware peptide generation. J Chem Inf Model. 2024;64(17):6912–25. https://doi.org/10.1021/acs.jcim.4c01020.

Waghu FH, Barai RS, Gurung P, Idicula-Thomas S. CAMPR3: a database on sequences, structures and signatures of antimicrobial peptides. Nucleic Acids Res. 2016;44(D1):D1094–7. https://doi.org/10.1093/nar/gkv1051.

Wan F, Kontogiorgos-Heintz D, de la Fuente-Nunez C. Deep generative models for peptide design. Digit Discov. 2022;1(3):195–208. https://doi.org/10.1039/d1dd00024a.

Wang G, Li X, Wang Z. APD3: the antimicrobial peptide database as a tool for research and education. Nucleic Acids Res. 2016;44(D1):D1087–93.

Wang C, Garlick S, Zloh M. Deep learning for novel antimicrobial peptide design. Biomolecules. 2021;11(3) https://doi.org/10.3390/biom11030471.

Wang L, Wang N, Zhang W, Cheng X, Yan Z, Shao G, et al. Therapeutic peptides: current applications and future directions. Signal Transduct Target Ther. 2022;7, 48(1) https://doi.org/10.1038/s41392-022-00904-4.

Wang D, Wen Z, Ye F, Li L, Zhou H. Accelerating antimicrobial peptide discovery with latent structure. In: Proceedings of the 29th ACM SIGKDD conference on knowledge discovery and data mining. Association for Computing Machinery; 2023. p. 2243–55.

Wang Y, Liu X, Huang F, Xiong Z, Zhang W. A multi-modal contrastive diffusion model for therapeutic peptide generation. Proc AAAI Conf Artif Intell. 2024;38(1):3–11.

Wei L, Zhou C, Chen H, Song J, Su R. ACPred-FL: a sequence-based predictor using effective feature representation to improve the prediction of anti-cancer peptides. Bioinformatics. 2018;34(23):4007–16.

Wei L, Zhou C, Su R, Zou Q. PEPred-suite: improved and robust prediction of therapeutic peptides using adaptive feature representation learning. Bioinformatics (Oxford, England). 2019;35(21):4272–80.

Wu CH, Mohammadmoradi S, Chen JZ, Sawada H, Daugherty A, Lu HS. Renin-angiotensin system and cardiovascular functions. Arterioscler Thromb Vasc Biol. 2018;38(7):e108–e16.

Wu Z, Johnston KE, Arnold FH, Yang KK. Protein sequence design with deep generative models. Curr Opin Chem Biol. 2021;65:18–27. https://doi.org/10.1016/j.cbpa.2021.04.004.

Xiao X, Wang P, Lin WZ, Jia JH, Chou KC. iAMP-2L: a two-level multi-label classifier for identifying antimicrobial peptides and their functional types. Anal Biochem. 2013;436(2):168–77. https://doi.org/10.1016/j.ab.2013.01.019.

Xu J, Li F, Leier A, Xiang D, Shen HH, Marquez Lago TT, et al. Comprehensive assessment of machine learning-based methods for predicting antimicrobial peptides. Brief Bioinform. 2021;22(5) https://doi.org/10.1093/bib/bbab083.

Xu T, Wang Q, Yang Z, Ying J. A BERT-based approach for identifying anti-inflammatory peptides using sequence information. Heliyon. 2024;10(12):e32951. https://doi.org/10.1016/j.heliyon.2024.e32951.

Yan J, Bhadra P, Li A, Sethiya P, Qin L, Tai HK, et al. Deep-AmPEP30: improve short antimicrobial peptides prediction with deep learning. Mol Ther Nucleic Acids. 2020;20:882–94. https://doi.org/10.1016/j.omtn.2020.05.006.

Yan K, Lv H, Guo Y, Peng W, Liu B. sAMPpred-GAT: prediction of antimicrobial peptide by graph attention network and predicted peptide structure. Bioinformatics (Oxford, England). 2023a;39(1) https://doi.org/10.1093/bioinformatics/btac715.

Yan K, Lv H, Wen J, Guo Y, Xu Y, Liu B. PreTP-stack: prediction of therapeutic peptides based on the stacked ensemble Learing. IEEE/ACM Trans Comput Biol Bioinform. 2023b;20(2):1337–44. https://doi.org/10.1109/tcbb.2022.3183018.

Yang S, Yang Z, Ni X. AMPFinder: a computational model to identify antimicrobial peptides and their functions based on sequence-derived information. Anal Biochem. 2023a;673:115196. https://doi.org/10.1016/j.ab.2023.115196.

Yang L, Zhang Z, Song Y, Hong S, Xu R, Zhao Y, et al. Diffusion models: a comprehensive survey of methods and applications. ACM Comput Surv. 2023b;56:105. https://doi.org/10.1145/3626235.

Yeaman MR, Yount NY. Mechanisms of antimicrobial peptide action and resistance. Pharmacol Rev. 2003;55(1):27–55. https://doi.org/10.1124/pr.55.1.2.

Yu H, Wang R, Qiao J, Wei L. Multi-CGAN: deep generative model-based multiproperty antimicrobial peptide design. J Chem Inf Model. 2024;64(1):316–26. https://doi.org/10.1021/acs.jcim.3c01881.

Zhang H, Saravanan KM, Wei Y, Jiao Y, Yang Y, Pan Y, et al. Deep learning-based bioactive therapeutic peptide generation and screening. J Chem Inf Model. 2023;63(3):835–45. https://doi.org/10.1021/acs.jcim.2c01485.

Zhao X, Wu H, Lu H, Li G, Huang Q. LAMP: a database linking antimicrobial peptides. PLoS One. 2013;8(6):e66557.

Zhao N, Qin Y, Liu H, Cheng Z. Tumor-targeting peptides: ligands for molecular imaging and therapy. Anti Cancer Agents Med Chem. 2018;18(1):74–86. https://doi.org/10.2174/1871520617666170419143459.

Zhu Y, Hao W, Wang X, Ouyang J, Deng X, Yu H, et al. Antimicrobial peptides, conventional antibiotics, and their synergistic utility for the treatment of drug-resistant infections. Med Res Rev. 2022;42(4):1377–422. https://doi.org/10.1002/med.21879.

Deep Learning for In Silico ADMET Prediction

Antonio Lavecchia

Abstract Deep learning (DL) is transforming in silico ADMET (absorption, distribution, metabolism, elimination, and toxicity) prediction, overcoming key limitations of traditional computational models. This chapter examines the evolution of ADMET modeling, emphasizing how DL techniques such as deep neural networks (DNNs), convolutional neural networks (CNNs), and graph neural networks (GNNs) enhance predictive accuracy by identifying complex molecular patterns. Compared to machine learning (ML) approaches like random forests (RF) and support vector machines (SVMs), DL excels in processing high-dimensional data and capturing nonlinear relationships. Advances in multitask learning, transfer learning, and molecular representations, including equivariant GNNs, have further improved model performance. Despite these advancements, challenges persist in improving in vitro–in vivo correlations, managing data heterogeneity, and enhancing interpretability. Explainable AI techniques, such as Shapley additive explanations and attention-based mechanisms, are increasing model transparency. Additionally, hybrid modeling approaches and pretraining strategies are enhancing generalization while reducing reliance on extensive datasets. In summary, this chapter outlines recent advancements, discusses limitations, and suggests future directions, ultimately enabling researchers to prioritize safer drug candidates earlier in the discovery process and accelerate regulatory approval.

Keywords Artificial intelligence · Drug discovery · Machine learning · Deep learning · Target discovery · Molecular property prediction · De novo drug design

A. Lavecchia (✉)
Drug Discovery Laboratory, Department of Pharmacy, University of Naples, Naples, Italy
e-mail: antonio.lavecchia@unina.it

A. Lavecchia (ed.), *Applied Artificial Intelligence for Drug Discovery*,
https://doi.org/10.1007/978-3-031-98022-0_8

1 Introduction

Drug discovery remains an immensely expensive, protracted, and unpredictable endeavor (Fernandez-Moure 2016; Moffat et al. 2017; Van Norman 2016a; Waring et al. 2015). Typically, the entire process of bringing a novel therapeutic agent from preclinical evaluation to regulatory approval spans 12–15 years (Fernandez-Moure 2016) and incurs costs exceeding $2.5 billion (DiMasi et al. 2016; Fernandez-Moure 2016; Kiriiri et al. 2020). This arduous journey is largely attributable to the stringent, multi-phase requirements imposed by regulatory authorities, which must meticulously validate both the safety and efficacy of new compounds during preclinical and clinical testing (Van Norman 2016b). Consequently, an inherent challenge is the extremely high attrition rate, with 80–90% of projects being terminated before human trials and nearly 95% of drugs failing during clinical testing (Fernandez-Moure 2016; Moreno and Pearson 2013). In fact, estimates indicate that only about one in 5000–10,000 identified compounds ever reaches the FDA's review stage (Petrova 2013; Sun et al. 2022).

Assessing, understanding, and refining pharmacokinetic and toxicological properties collectively encompassed under ADMET (absorption, distribution, metabolism, elimination, and toxicity) has become indispensable for reducing late-stage failures in drug development (Basavaraj and Betageri 2014; Segall and Barber 2014; Van De Waterbeemd and Gifford 2003). Over 80% of emerging drug candidates falter due to suboptimal ADMET profiles (Faqi 2016), and success rates could be markedly improved if these attributes were rigorously optimized during early phases (Di et al. 2020; Mak et al. 2022; Van Norman 2020; Xiong et al. 2021). Although in vivo and in vitro evaluations remain crucial for the initial screening of chemical leads (Chung et al. 2004), these methodologies are often prohibitively expensive, inefficient, and time-consuming even with advancements in automation (Dong et al. 2018; Van de Waterbeemd 2005; Xiong et al. 2021).

In light of these challenges, in silico ADMET prediction offers an attractive alternative, enabling high-throughput pre-screening and the prioritization of promising compounds prior to synthesis (Daina et al. 2017; Dong et al. 2018; Pires et al. 2015; de Sá et al. 2022; Wei et al. 2022; Xiong et al. 2021). The foundation of these computational approaches lies in quantitative structure–activity relationships (QSAR), first formalized by Hansch and Fujita in the 1960s (Hansch et al. 1962). Over the years, the evolution of QSAR modeling, from early linear regression models to advanced machine learning (ML) techniques such as random forest (RF), artificial neural networks, support vector machines (SVMs), decision trees (DT), and k-nearest neighbors (k-NN) has dramatically improved our ability to predict ADMET properties (Maltarollo et al. 2015), although challenges such as overfitting persist. Recent innovations in automatic feature engineering and handling large training sets have begun to address these issues.

More recently, the advent of deep learning (DL) has further advanced in silico ADMET prediction (LeCun et al. 2015; Schmidhuber 2015). DL methods have harnessed vast datasets and powerful computational resources to uncover subtle, nonlinear relationships that traditional ML techniques could not capture. Pioneering studies by Hochreiter and colleagues (Klambauer et al. 2017; Mayr et al. 2016; Unterthiner et al. 2015) have demonstrated that DL models including deep neural networks (DNNs), convolutional neural networks (CNNs), and graph neural networks (GNNs) can extract complex, abstract features from raw molecular descriptors, thereby enhancing predictive accuracy. These models excel at extracting features, effectively transforming raw input data into abstract representations that reveal intricate details of molecular structure and activity. Improved hardware, advanced techniques to mitigate overfitting, and the availability of robust training algorithms on widely accessible platforms have enabled successful DL applications in areas such as computer vision and language processing (LeCun et al. 2015). DL has also been applied in computational chemistry (Jiménez et al. 2018; Ryan et al. 2018) to generate QSAR models (Ma et al. 2015; Ramsundar et al. 2017; Xu et al. 2017a, b), with key applications in medicinal chemistry recently summarized (Gawehn et al. 2016, 2018; Ghasemi et al. 2018). Sheridan and co-workers (Ma et al. 2015; Ramsundar et al. 2017; Xu et al. 2017a, b) demonstrated that DNNs often yield models with similar or improved predictive performance compared to algorithms such as RF (Breiman 2001) on project data for relevant targets and antitargets. Another critical advantage of DL is multitask learning, which enables the simultaneous prediction of multiple ADMET endpoints by sharing hidden layers among tasks, effectively "filling gaps" in sparsely populated datasets (Kearnes et al. 2017; Korotcov et al. 2017; Ramsundar et al. 2015, 2017).

The objective of this chapter is to comprehensively review recent advances in DL methodologies for ADMET prediction, address current challenges, and propose future perspectives.

1.1 *Prediction of Physicochemical Properties*

The ADMET characteristics of a drug candidate are directly influenced by its physicochemical properties, which play a critical role in determining a drug's market success (Lipinski et al. 2001; Zhang and Wilkinson 2007). For instance, the ionization constant (pKa), a fundamental parameter that underlies properties such as the octanol–water distribution coefficient (logD) and solubility, affects a molecule's aqueous solubility and, consequently, influences the drug formulation process. Moreover, the ADMET behavior of compounds under various pH conditions is profoundly affected by their charge state (Manallack et al. 2013). Even though lead compounds with promising drug-like properties might not always be successfully commercialized, their favorable attributes continue to inspire innovative drug design. However, because physicochemical properties are not easily measured directly, precise in silico predictions facilitate the structural optimization of small-molecule drug candidates until they achieve the desired profile.

Historically, compound physicochemical properties have long been used to assess drug quality. Early efforts focused on rule-based models, with the "rule of five" emerging as a renowned indicator of compound druggability (Chagas et al. 2018). DL approaches have gained traction in this field, offering improved predictive performance. For example, Prasad et al. built a DNN model for the SAMPL6 challenge, achieving an root-mean-square error (RMSE) of 0.41 logP units (Prasad and Brooks 2020), while multitask DNN models by Wu et al. yielded similar success in predicting logP and aqueous solubility (Wu et al. 2018). Using a neural network architecture called ResNet, AI has been applied to determine key physicochemical parameters, including solubility, dissociation constant, and partition coefficient—with greater precision than non-AI models, as evidenced by enhanced yield and reduced extraction time of polysaccharides (Cui and Zhu 2021). Similarly, Cui's modified CNN model demonstrated excellent performance in predicting aqueous solubility (Cui et al. 2020).

Quantitative structure–property relationship (QSPR) models form the foundation of many in silico ADME prediction tools, correlating structural features with physicochemical properties such as solubility, lipophilicity (logP/logD), and permeability (Du et al. 2024; Wadhwa and Mittal 2022). Traditional QSPR models relied on molecular descriptors and statistical methods such as multiple linear regression and partial least squares, but recent advancements integrate ML algorithms, including DL to enhance predictive accuracy. Additionally, physiologically based pharmacokinetic (PBPK) modeling has emerged as a critical tool for simulating drug absorption and distribution in various tissues. PBPK models incorporate physicochemical properties with physiological data, improving drug disposition predictions (Balhara et al. 2022).

Graph-based approaches have also emerged as a hotspot in physicochemical property prediction. In 2013, Lusci et al. developed an undirected graph-based recurrent neural network (UG-RNN) to predict aqueous solubility, achieving performance comparable to state-of-the-art ML methods despite a small training set of only 60 compounds (Lusci et al. 2013). Later, graph CNNs utilized by Hop et al. and Coley et al. enabled dynamic fingerprint learning for predicting properties such as pKa, solubility, and melting point, wherein molecular adjacency encodes fragment similarity relevant to specific assay endpoints (Coley et al. 2017; Hop et al. 2018). As a result, DL methods for aqueous solubility prediction have outperformed traditional ML approaches, although improvements in melting point predictions remain modest. Recently, Liu et al. introduced Chemi-Net (Liu et al. 2019), a GCNN model that surpassed the commercial software Cubist used by Amgen for predicting physicochemical properties.

The integration of multi-source datasets, such as high-throughput screening data, literature-derived features, and experimental assay results, has further enhanced DL-based physicochemical property prediction (Lipkova et al. 2022). The inclusion of permeability models based on Caco-2 cell assays, which simulate intestinal drug absorption, has significantly improved the estimation of oral bioavailability (Wang and Chen 2020). Additionally, ensemble modeling approaches, which aggregate multiple predictive models (e.g., QSAR, PBPK, and DL), have demonstrated superior generalization across diverse chemical scaffolds (Yu et al. 2022). In another challenge, Francoeur and Koes integrated Natural Language Processing (NLP) to derive embedding vectors from small-molecule SMILES, which were then processed by a transformer model for aqueous solubility prediction (Francoeur and Koes 2021). Their findings indicated that simpler DL models performed better than overly complex ones, likely due to overfitting issues with limited data. To address the simultaneous prediction of multiple physicochemical properties, researchers have emphasized molecular feature learning and characterization. Examples include the use of GNN architectures (Xiong et al. 2019), the combination of traditional molecular representations with features derived from message-passing neural networks (MPNNs) (Yang et al. 2019a, b), and the development of graphical representations based on extended-connectivity circular fingerprints (ECFPs) (Rogers and Hahn 2010). Shen et al. proposed a novel molecular representation by first calculating distance matrices from molecular fingerprints and descriptors for 8 million molecules, reducing these matrices to two dimensions via uniform manifold approximation and projection (UMAP), and then mapping the data onto 2D grid maps using the Jonker–Volgenant (J–V) algorithm (Shen et al. 2021). Finally, the data were segmented into channels based on different molecular fingerprints or descriptors, and these representations were input into a CNN, achieving state-of-the-art performance across multiple datasets. Table 1 presents an overview of modern algorithms for predicting physicochemical properties, detailing their target predictions, key features, and supporting references.

1.2 *Prediction of ADMET-Related Properties*

The failure of most clinical trials is frequently attributed not to a lack of efficacy but to inadequate ADMET evaluation. In fact, the ADMET component is crucial in determining whether a drug molecule reaches its target protein in vivo, which transporter or metabolizing protein acts on it, how long it remains in the bloodstream, and when it is inactivated (Fu et al. 2022; Yin et al. 2021). Equally, the "T" (toxicity) component represents a major concern during phase I clinical trials. Thorough preliminary ADMET studies can markedly reduce the risk of clinical trial failure, thereby saving substantial time and resources (Cheng et al. 2013; Wang et al. 2015). Given that hundreds of compounds await ADMET evaluation in the early stages of drug discovery, validating each one through extensive animal studies is both time-consuming and costly. Thus, the rapid and accurate prediction of ADMET properties via AI has become increasingly widespread (Ferreira and Andricopulo 2019). Recent advances include the integration of multitask graph learning approaches, as demonstrated by Du et al., where auxiliary task selection significantly enhanced predictive performance across multiple ADMET endpoints (Du et al. 2023). For a comprehensive overview of the computational approaches used for predicting ADMET-related properties, including algorithm types, key features, applications, and corresponding references, please refer to Table 2.

Table 1 Algorithms for predicting physicochemical properties

Approach	Model/algorithm	Predicted properties	Key notes	References
DNNs	DNN (for SAMPL6 challenge)	logP	RMSE of 0.41 logP units	Prasad and Brooks (2020)
Multitask DNNs	Multitask DNN	logP, aqueous solubility	Similar predictive performance	Wu et al. (2018)
ResNet-based models	ResNet architecture	Solubility, pKa, partition coefficient	Higher precision, improved yield, and extraction time	Cui and Zhu (2021)
CNN models	Modified CNN	Aqueous solubility	Excellent performance in solubility prediction	Cui et al. (2020)
QSPR models	Traditional QSPR (MLR, PLS)	Solubility, Lipophilicity (logP/logD), permeability	Based on molecular descriptors and statistical methods	Du et al. (2024), Wadhwa and Mittal (2022)
PBPK models	Physiologically based pharmacokinetics	Absorption, distribution	Integrates physiological data for better predictions	Balhara et al. (2022)
Graph-based RNN	Undirected graph RNN (UG-RNN)	Aqueous solubility	Effective with small training sets	Lusci et al. (2013)
Graph CNNs	Graph CNN	pKa, solubility, melting point	Utilizes molecular adjacency for dynamic fingerprinting	Coley et al. (2017), Hop et al. (2018)
GCNN models	Chemi-net	Various physicochemical properties	Outperformed commercial software (cubist)	Liu et al. (2019)
Ensemble models	Ensemble (QSAR, PBPK, DL)	Multiple properties	Superior generalization across diverse scaffolds	Yu et al. (2022)
Transformer models	Transformer with SMILES Embeddings	Aqueous solubility	Simpler DL models avoid overfitting with limited data	Francoeur and Koes (2021)
GNNs	GNN architectures	Various properties	Emphasizes molecular feature learning	Xiong et al. (2019)
MPNNs	Message-passing neural networks	Physicochemical properties	Combines traditional and deep learning features	Yang et al. (2019a, b)
Fingerprint methods	ECFPs	Descriptor for property prediction	Widely used for molecular similarity	Rogers and Hahn (2010)
Novel representations	Distance matrices + UMAP + J–V + CNN	Multi-property prediction	Innovative 2D mapping; state-of-the-art performance	Shen et al. (2021)

Table 2 Summary of computational approaches for predicting ADMET-related properties

Algorithm	Type/category	Key characteristics/features	Application/usage	References
C5.0 DT	Rule-based/DT	Utilizes molecular fingerprints/descriptors to forecast CYP inhibition	Prediction of CYP inhibition for five enzymes in oxidation/hydrolysis	Su et al. (2015)
QSPR	QSPR with ML methods	Employs SVM, MLR, MARS, and RF to correlate polar surface area and logP with blood–brain partitioning	Prediction of blood–brain partition coefficient (logBB)	Zhu et al. (2018)
Adaptive ensemble classification framework (AECF)	Ensemble ML	Integrates data balancing, ensemble techniques, and optimization for high-dimensional, imbalanced data	Prediction of various ADMET properties	Yang et al. (2018a, b)
DNN-based Caco-2 permeability classifier	DNN	Utilizes molecular descriptors for Caco-2 permeability classification	Caco-2 permeability classification	Shin et al. (2018)
DNN for mouse liver microsomal stability	DNN	Predicts liver microsomal stability using molecular features	Prediction of mouse liver microsomal stability	Renn et al. (2021)
Multitask DNN	Multitask DNN	Trained on ~50,000 compounds across three species; performance depends on data overlap	Prediction of microsomal lability and passive permeability in Caco-2/TC7 cells	Wenzel et al. (2019)
DNN for Oral bioavailability	DNN	Achieves ~70% accuracy in predicting oral bioavailability in rats	Prediction of oral bioavailability in rats	Schneckener et al. (2019)

(continued)

Table 2 (continued)

Algorithm	Type/category	Key characteristics/features	Application/usage	References
DNN for CYP450 inhibition	DNN	Predicts CYP450 inhibition; ensemble XGBoost may outperform standalone DNN	Prediction of CYP450 inhibition	Li et al. (2018), Wu et al. (2019)
Multitask GCNN	Multitask graph CNNs	Utilizes 31 assay datasets; shows significant improvements over RF and single-task GCNN	Prediction of multiple ADMET endpoints (e.g., PPB, HLM, CYP inhibition)	Feinberg et al. (2020)
Multitask GCNN with transfer learning	Multitask GCNN (with transfer learning)	Integrates transfer learning to enhance predictions of pharmacokinetic properties	Prediction of oral bioavailability, plasma protein binding, Vd, and elimination half-life	Ye et al. (2018)
Hybrid framework (RNN + FCNN)	Hybrid (RNN + FCNN)	Combines RNN for SMILES extraction with FCNN for physicochemical features; merged output predicts chemical toxicity	Prediction of drug chemical toxicity	Peng et al. (2020b)
Ensemble single-task GCNN	Ensemble single-task GCNN	Yields a median RMSE improvement of 14% over RF using fingerprints	Prediction on Amgen datasets (rat PPB, solubility, RLM, human PXR activation)	Yang et al. (2019a, b)
Single and multitask GCNN	Graph CNNs	Single-task GCNNs show 10% and multitask GCNNs 38% improvement in R^2 over cubist; competitive for HLM, CYP3A4, and solubility	Prediction of multiple ADMET endpoints, including solubility and PXR activation	Liu et al. (2019)
DL for drug metabolites	DL	Outperforms rule-based SyGMa; multitask GCNN approach yields best performance among compared methods	Prediction of drug metabolites	Wang et al. (2020)
Multitask GCNN vs. fingerprint-based approaches	Multitask GCNN	Outperforms fingerprint-based methods and RF on large Bayer AG datasets; R^2 improvements of 0.26–0.28 for membrane affinity and PPB	Comprehensive ADMET endpoint prediction	Montanari et al. (2019)

ProTox-II	Webserver for toxicity prediction	Integrates molecular similarity, pharmacophores, fragment propensities, and ML; predicts multiple toxicity endpoints with confidence scores	Toxicity prediction and risk assessment	Banerjee et al. (2018)
SwissADME	Web tool for ADME and drug-likeness	Offers rapid predictive models for physicochemical properties, pharmacokinetics, and drug-likeness (e.g., BOILED-egg, iLOGP, bioavailability radar) via a user-friendly, login-free interface	Evaluation of ADME and drug-likeness in early drug discovery	Daina et al. (2017)
ADMETlab 3.0	Comprehensive ADMET platform	Employs a multitask DMPNN with molecular descriptors to predict 119 features with enhanced performance, API support, and uncertainty estimates	Comprehensive ADMET prediction	Fu et al. (2024)
pkCSM	Graph-based signature tool	Uses graph-based signatures for rapid prediction of pharmacokinetic and toxicity properties; ensures data privacy	Rapid evaluation of small-molecule ADMET properties	Pires et al. (2015)

QSAR and QSPR models have played pivotal roles in this arena. A multitude of ML techniques, when combined with QSAR/QSPR models, have yielded promising predictions of critical ADMET parameters (Tao et al. 2015). These ML approaches typically target key properties such as P-glycoprotein interactions (Smyth et al. 1998), hERG-mediated cardiotoxicity (Vandenberg et al. 2012), blood–brain barrier penetration (Kaisar et al. 2017), mutagenicity (Alves et al. 2018), cytochrome P450 enzyme inhibition (Rácz and Keserű 2020), carcinogenicity (Onakpoya et al. 2016), acute oral toxicity (Yang et al. 2018a, b), as well as respiratory toxicity and irritation/corrosion (Lei et al. 2017). New AI-powered platforms such as Interpretable-ADMET and HelixADMET leverage deep neural representations to enhance prediction accuracy and provide mechanistic insights (Wei et al. 2022; Zhang et al. 2022a, b). Furthermore, a CYP enzyme inhibition prediction model based on the C5.0 DT algorithm was developed using various molecular fingerprints or descriptors to forecast inhibition of five CYP enzymes involved in drug oxidation or hydrolysis (Su et al. 2015). Furthermore, Zhu et al. used a QSPR model to predict the blood–brain partition coefficient (logBB) for 287 compounds employing four ML methods (SVM, multivariate linear regression, multivariate adaptive regression splines, and RF) and found that polar surface area and the octanol–water partition coefficient were strongly correlated with blood–brain partitioning (Zhu et al. 2018). Additionally, DL-based ADMET predictors, such as DeepAmes (Li et al. 2023) and DeepCarc (Li et al. 2021), have shown superior performance in mutagenicity and carcinogenicity prediction, respectively.

Most ADMET datasets are imbalanced and high-dimensional, and integrated learning approaches have been applied to address these issues. By combining data balancing techniques, ensemble methods, and optimization steps, an adaptive ensemble classification framework (AECF) was established (Yang et al. 2018a, b), achieving AUROC values ranging from 0.78 to 0.91 as validated by the DrugBank database. The recently introduced multitask graph learning framework further refines this process by leveraging auxiliary tasks dynamically to optimize ADMET predictions (Du et al. 2023).

DL approaches are increasingly employed to forecast ADMET properties. For instance, Shin et al. constructed a DNN model serving as a Caco-2 permeability classifier (Shin et al. 2018), while Renn et al. developed a model to predict mouse liver microsomal stability (Renn et al. 2021). In addition, Ma et al. developed DNN models that outperform RF models for several ADMET endpoints, such as transport by P-glycoprotein and human plasma protein binding (Ma et al. 2015). More recently, Wenzel et al. successfully predicted microsomal lability across different species and passive permeability in Caco-2/TC7 cells using multitask DNN models (Wenzel et al. 2019). Specifically, they trained a multitask, fully connected DNN on large Sanofi–Aventis AG liver microsome stability datasets (~50,000 compounds) across three species, noting improved predictive performance for human liver microsomes while performance for rat and mouse liver microsomes either slightly deteriorated or remained unchanged. When overlapping molecules were excluded between the prediction set for human liver microsomes and any training molecules, no benefit was observed from multitask training. Subsequently, incorporating five additional species with smaller liver microsome datasets (200–1500 compounds each) to form an eight-species multitask model resulted in mixed performance, with macaque predictions improving (R^2 increased by 0.09 or 15%) and monkey predictions declining (R^2 decreased by 0.09 or 24%). Schneckener et al. predicted oral bioavailability in rats using a DNN model, achieving an accuracy of nearly 70% (Schneckener et al. 2019). This study suggests that DL may serve as an effective tool for forecasting in vivo pharmacokinetic behaviors. Li et al. and Wu et al. successively constructed DL models to predict CYP450 inhibition using DNNs (Li et al. 2018; Wu et al. 2019); however, the latter demonstrated that an ensemble model based on XGBoost yielded higher accuracy than the DL models. Notably, FP-ADMET has emerged as a comprehensive ML-based platform incorporating fingerprint-based modeling techniques to improve predictive power across various ADMET properties (Venkatraman 2021). Similarly, DL approaches did not exhibit a clear advantage over traditional ML methods (Liu et al. 2019), although different datasets were employed. Nonetheless, several efforts have been made to enhance DL-based ADMET prediction. In a recent study, Feinberg et al. investigated the prediction of 31 ADMET endpoints using a state-of-the-art GCNN and large, standardized assay datasets at Merck & Co., Inc. (Feinberg et al. 2020). Their single-task GCNN demonstrated strong performance improvements over RF for approximately one-third of endpoints, with R^2 enhancements ranging from 0.15–0.31 (45–133%), including plasma protein binding (PPB), HLM and RLM, CYP3A4 and 2D6 inhibition, hERG binding, and kinetic solubility in water. Notably, significant improvements were observed for rat and human PPB (R^2 improvements of 0.31 and 0.19, respectively). About one-third of endpoints showed modest improvements (R^2 improvements of 0.05–0.14), while the remaining endpoints exhibited minimal or no enhancement, particularly for in vivo pharmacokinetic parameters. The authors reported additional gains by incorporating a multitask featurization approach, most pronounced for hepatocyte stability, permeability, and P-glycoprotein efflux. Overall, their multitask GCNNs performed equivalently or better than single-task GCNNs, which in turn outperformed RF models. Using a different state-of-the-art GCNN, Yang et al. demonstrated notable performance improvements relative to RF on smaller proprietary industrial ADMET datasets as well as public datasets. On four large Amgen Inc. (CA, USA) datasets, rat PPB, solubility, RLM, and human PXR activation, their ensemble single-task GCNN models exhibited a median improvement of 14% in root-mean-square error compared with RF models using fingerprints (Yang et al. 2019a, b). Recently, Liu et al. compared another single-task GCNN model to a Cubist approach using five different Amgen Inc. datasets (Liu et al. 2019), finding that single and multitask GCNNs achieved 10% and 38% mean improvements in R^2 performance, respectively, over Cubist models. They observed that single-task GCNNs consistently outperformed Cubist in HLM, CYP3A4, and solubility endpoints, while both performed similarly for human PXR activation prediction. When applied to multiple solubility and PXR subsets, multitask GCNNs consistently outperformed Cubist models, suggesting that GCNNs can be highly competitive with RF and Cubist models, often yielding robust increases in predictivity. Ye et al. integrated transfer learning and multitask learning to markedly improve the prediction of oral bioavailability, plasma protein binding, volume of distribution at steady state, and elimination half-life (Ye et al. 2018). Peng et al. designed a hybrid framework combining an RNN and a fully connected neural network (FCNN) to extract SMILES and physicochemical properties, respectively; their combined output, fed into another fully connected neural network, successfully predicted drug chemical toxicity (Peng et al. 2020b), suggesting that integrating different types of descriptors may further enhance DL performance (Jimenez-Carretero et al. 2018). Moreover, Wang et al. developed a DL model to predict drug metabolites with an accuracy superior to the popular rule-based method systematic generation of potential metabolites (SyGMa) (Wang et al. 2020). In a comparison among a multitask graph convolutional model, a fully connected neural network, and an RF model, the multitask graph convolutional model outperformed the others. However, for more complex tasks such as predicting Caco2 permeation or in vitro metabolic stability, these multitask GCNNs were less successful, likely due to the simplicity of the constructed model, which hindered the learning of deeper features. Additionally, the multitasking model was viewed as a trial-and-error exercise, with no specific guidelines provided regarding the optimal combination of tasks. Another study by Montanari et al. compared GCNN-derived representations against two fingerprint-based approaches, simple fully connected DNNs, and RFs using eight Bayer AG proprietary ADMET datasets (39,000–236,000 compounds each) (Montanari et al. 2019). Their findings revealed that multitask GCNNs provided the most substantial improvements in predictive performance over prevailing RF models; for instance, in membrane affinity and PPB endpoints, R^2 improvements of 0.26–0.28 (65–67%) were observed. Overall, these studies indicate that DL approaches often outperform RF, with multitask models being competitive with or superior to single-task models, and GCNNs frequently surpassing fully connected DNNs trained on comparable datasets.

Other recent studies have similarly demonstrated the potential of multitask models for ADMET property prediction. Several user-friendly ADMET software tools and web servers have been developed to predict molecular ADMET properties (Banerjee et al. 2018; Daina et al. 2017; Fu et al. 2024; Pires et al. 2015; Yang et al. 2019a, b); among these, ADMETlab 2.0 (Fu et al. 2024) is widely recognized. Based on a multitask graph attention (MGA) framework, ADMETlab 2.0 predicts a wide array of ADMET properties, covering 88 parameters, including 23 ADME properties, 27 toxicity endpoints, and eight toxicophore rules. Most of its training data are sourced from open-access bioactivity databases, relevant literature, and toxicity prediction tools such as the Toxicity Estimation Software Tool (TEST). The novel architecture and diverse training datasets employed by ADMETlab 2.0 result in predictions that are often distinct from those of similar tools, making it a valuable resource for both non-expert users and medicinal chemists.

1.3 Prediction of Toxicity

The utilization of AI and ML enhances drug safety, offering a promising approach to toxicity prediction. These algorithms provide data-driven methods for safety and toxicity assessments, allowing for the detection of trends that might otherwise go unnoticed. Recent advancements in AI-based models have significantly improved the identification of critical toxicity endpoints, including lethal dose 50% (LD_{50}), drug-induced liver injury (DILI), human ether-a-go-go-related gene (hERG) inhibition, carcinogenesis, and Ames mutagenesis. Additionally, AI models now enable the prediction of systemic toxicity, subchronic and chronic effects, as well as delayed adverse reactions, contributing to a more comprehensive safety evaluation (Tran et al. 2023). Logistic regression, SVMs, and RF are examples of traditional ML approaches that can create interpretable models with moderate complexity. These approaches are particularly useful for investigating how predictors influence the risk or incidence of adverse occurrences.

In recent years, drug toxicity prediction has become an active area of DL modeling within ADMET research (Basile et al. 2019; Vo et al. 2019). DL algorithms excel at automatically detecting nonlinear patterns in data without requiring significant operator intervention. CNNs and RNNs are two commonly employed architectures in medication safety research. These models are utilized in preclinical toxicity studies to mimic patient variability, facilitate lead compound selection, and optimize trial design. Furthermore, AI-driven methods have been integrated with in vivo studies to enhance translational relevance, improving drug safety assessments by bridging the gap between computational predictions and experimental validation and reducing reliance on traditional animal testing (Feng et al. 2019).

AI-assisted medication safety and toxicity science remain in their early stages, requiring further research to evaluate their potential therapeutic impact (Kim et al. 2020). Recent advancements have emphasized the importance of high-quality data for model training. Public toxicology databases, such as CEBS (Chemical Effects in Biological Systems) (Waters et al. 2008), ToxRefDB (Feshuk et al. 2023), the Comparative Toxicogenomics Database (CTD) (Mattingly et al. 2006), and the Distributed Structure-Searchable Toxicity Database (DSSTox) (Grulke et al. 2019), have contributed significantly to AI model development by providing structured toxicity data across multiple species and experimental conditions. The integration of large-scale databases such as ToxCast and PubChem has further enhanced AI-based toxicity predictions by providing vast amounts of bioassay and chemical structure data (Mattingly et al. 2006) (Grulke et al. 2019).

In 2016, a DL model called DeepTox won the Tox21 data challenge with excellent performance (Mayr et al. 2016). Altae-Tran et al. combined iterative refinement of LSTM networks with a GCNN model for toxicity prediction, achieving better performance than traditional ML models (Altae-Tran et al. 2017). To date, multiple DL models have been reported to predict various toxicities, such as drug-induced liver injury (Matsuzaka and Uesawa 2019a), acute toxicity (Xu et al. 2017a, b), rectal toxicity (Zhen et al. 2017), hepatotoxicity (Fraser et al. 2018; Ibragimov et al. 2020; Matsuzaka et al. 2020), hERG blocking (Cai et al. 2019; Wang et al. 2020), receptor activity (Idakwo et al. 2019; Matsuzaka and Uesawa 2020), and drug–drug interactions (Deng et al. 2020; Lee et al. 2019). Recently, multitask and transfer learning strategies have enhanced prediction accuracy for multiple toxicity endpoints, improving generalization across different biological systems (Sharma et al. 2023).

A model for predicting organ-specific chemical carcinogenicity was recently developed using a Bayesian-like technique and multilevel neighborhoods of atoms (MNA) descriptors (Lagunin et al. 2018). This QSAR model identifies substances that cause cancer in specific organs. In the future, more comprehensive models could enable precise toxicity forecasting. Recent advances in carcinogenicity prediction include DeepCarc, a DL model capable of screening compounds for carcinogenic potential by analyzing molecular structure and toxicity datasets such as DrugTax and Tox21(Li et al. 2021). Additionally, CONCERTO, a hybrid graph transformer model, has outperformed previous carcinogenicity prediction models by integrating molecular fingerprint representations to improve accuracy (Fradkin et al. 2022). For the construction of predictive mutagenicity models, the Ames mutagenicity benchmark dataset is the most commonly used training dataset (Hansen et al. 2009). Created in 2009, it compiles results from Ames testing across multiple toxicology databases, including VITIC, CCRIS, and GeneTox. The dataset includes 6512 chemicals, of which 3009 are non-mutagenic and 2503 are mutagenic. The introduction of GNNs has significantly enhanced Ames mutagenicity predictions. MolGIN, based on Graph Isomorphism Networks, has shown superior performance in capturing molecular relationships compared to traditional fingerprint-based methods (Peng et al. 2020a).

In the field of cardiotoxicity prediction, several AI-based models have demonstrated high accuracy. DeepHIT, a DL model, has been developed for predicting cardiotoxic effects by leveraging high-dimensional patient data and learning temporal dependencies in drug response (Ryu et al. 2020). HergSPred, an ensemble model combining DNN, RF, and XGBoost, has been specifically designed to classify hERG blockers and non-blockers with improved interpretability (Zhang et al. 2022a, b). Similarly, CardioTox integrates multiple chemical representations to enhance the identification of compounds with potential cardiotoxic effects (Karim et al. 2021).

Drug-induced liver injury (DILI) remains one of the primary causes of drug development failure and withdrawal. Several ML-based hepatotoxicity prediction models have been developed, most relying on molecular fingerprints and Bayesian approaches. One such model was trained on a dataset of 295 compounds and validated on 237 test compounds, achieving a 60% accuracy rate (Ekins et al. 2010). Recent advancements have incorporated gene expression data and high-content imaging techniques, significantly improving DILI prediction accuracy (Chen et al. 2022). Using a regression model (deepAOT-R), molecular graph encoding CNN architecture, and a multi-classification model (deepAOT-C), acute oral toxicity prediction has been enhanced (Xu et al. 2017a, b). The test dataset includes 1673 substances; the regression model's R2 and mean absolute error (MAE) were found to be 0.864 and 0.195, respectively, while the multi-classification model's accuracy reached 95.5%. PredAOT, an ensemble learning framework, has further improved acute toxicity prediction by addressing interspecies variability (Ryu et al. 2023).

Most of the aforementioned DL-based toxicity models use 1D molecular descriptors or fingerprints as input, whereas some recent studies have designed 2D image-based neural networks to predict ADMET properties such as CYP1A2 inhibitory potency, P-glycoprotein inhibitory activity, blood-brain barrier permeability, and toxicity (Fernandez et al. 2018; Matsuzaka and Uesawa 2019b; Shi et al. 2019). Furthermore, message-passing neural networks (MPNNs) leveraging 3D molecular representations have improved toxicity assessments by capturing spatial molecular interactions (Schütt et al. 2018). Recently, Wang et al. developed a DNN model based on transcriptomic data to predict chemically induced liver toxicity (Wang et al. 2019). Similarly, high-content imaging and transcriptomic-based methods have been explored for toxicity screening, offering more biologically relevant insights into drug-induced damage (Lejal et al. 2023). Yuan et al. reported a grid-based multi-channel CNN that demonstrated improved accuracy compared to traditional DL models in toxicity prediction (Yuan et al. 2019).

Another emerging field is the prediction of adverse effects from clinical data using DL technology. For example, Men et al. predicted xerostomia due to radiation therapy for head and neck squamous cell carcinoma from a dataset of 784 patients (Men et al. 2019). AI-based models are also being optimized for personalized toxicity prediction, integrating genetic variability and metabolic pathways for enhanced patient-specific risk assessment (Rao et al. 2023). A summary of the key AI-based approaches used for toxicity prediction is provided in Table 3.

2 Advances and Future Directions in ADMET Modeling: Challenges and Opportunities

The application of DL in ADMET modeling has expanded significantly, addressing challenges such as overfitting and optimization in gradient-based learning. While DL models have demonstrated strong performance in correlating physicochemical and ADMET properties with molecular structures, a notable gap remains in their ability to establish reliable in vitro/in vivo correlations, which are crucial for assessing pharmacokinetic risks in clinical settings (Basile et al. 2019). This limitation is further compounded by the sensitivity of pharmacokinetic behaviors to small structural variations, highlighting the need for localized models tailored to specific chemical scaffolds. The success of DL-based ADMET predictions hinges on dataset size and quality, often more so than the choice of modeling technique (Van De Waterbeemd 2009). While larger datasets typically enhance model performance, multitask DL approaches have demonstrated the ability to achieve high accuracy even with limited data (Jiang et al. 2020). However, data heterogeneity remains a challenge, and models trained on inconsistent sources can suffer from reduced generalizability. The availability of high-quality proprietary data from

Table 3 Summary of AI approaches for predicting toxicity

Algorithm	Type/category	Key characteristics/features	Application/usage	References
DeepTox	DL model	Uses GCNN and LSTM for toxicity prediction; trained on Tox21 dataset	General toxicity prediction, including drug-induced toxicity	Mayr et al. (2016)
DeepCarc	DL model	Analyzes molecular structures and toxicity datasets (DrugTax, Tox21)	Carcinogenicity prediction	Li et al. (2021)
MolGIN	GNN	Uses graph isomorphism network (GIN) for mutagenicity prediction	Ames mutagenicity prediction	Peng et al. (2020a)
PredAOT	Ensemble learning model	Two-step model: Classification + regression for acute oral toxicity prediction	Acute oral toxicity prediction in mice and rats	Ryu et al. (2023)
ResNet18DNN	DL model	Uses deep residual networks and transcriptomic data for hepatotoxicity prediction	Drug-induced liver injury (DILI) prediction	Chen et al. (2022)
DeepAmes	DL model	Predicts mutagenicity using mechanistic task groupings and multitask learning	Ames test-based mutagenicity prediction	Li et al. (2023)
DeepHIT	Multitask DL model	Uses DNN and GCN for cardiotoxicity assessment	Prediction of hERG inhibition	Ryu et al. (2020)
HergSPred	Ensemble model	Combines DNN, RF, and XGBoost for hERG blocking prediction	hERG-related cardiotoxicity	Zhang et al. (2022a, b)
DeepAOT-R/C	CNN + MGECNN	Regression and classification models for acute oral toxicity	Acute toxicity prediction	Xu et al. (2017a, b)
CardioTox	DL + meta-ensemble model	Aggregates diverse chemical representations for cardiotoxicity prediction	Drug-induced cardiac toxicity assessment	Karim et al. (2021)
CONCERTO	Hybrid DNN/CNN model	Uses graph transformer and molecular fingerprint representations	Carcinogenicity prediction	Fradkin et al. (2022)

large pharmaceutical companies remains an advantage, but promoting open-access datasets and standardized benchmarks could facilitate broader advancements in DL-based ADMET modeling (Zhou et al. 2018). Moreover, the integration of multi-omics data, including transcriptomics and proteomics, has been recognized as a key strategy to improve predictive robustness and capture complex drug interactions more accurately (Piochi et al. 2023).

Although DL has progressed, it still lags behind traditional methods such as regression and recursive partitioning (e.g., RF and Cubist), though ongoing efforts aim to address these limitations (Chuang et al. 2020a; Polishchuk 2017). One of the primary obstacles remains interpretability, as understanding how a model arrives at its predictions is essential for validating structure–activity relationships and avoiding false correlations that could mislead drug discovery (Chuang et al. 2020b; Schneider et al. 2020). To enhance interpretability, techniques such as Shapley additive explanations and layer-wise relevance propagation are being explored, providing valuable insights into molecular feature importance for predictive tasks (Chuang et al. 2020a; Feinberg et al. 2020; Polishchuk 2017; Rodríguez-Pérez and Bajorath 2020; Wenzel et al. 2019). Reliable error estimation is another critical challenge, as distinguishing between aleatoric uncertainty (inherent assay errors) and epistemic uncertainty (model limitations) is essential for assessing prediction reliability (Hirschfeld et al. 2020). Hybrid modeling approaches, which integrate rule-based systems with DL architectures, are emerging as an effective strategy for improving both interpretability and predictive performance.

Molecular representation plays a crucial role in ADMET modeling (Xu et al. 2018). Unlike traditional techniques, DL-based models can process a diverse range of molecular representations, including SMILES notations, molecular graphs, and 2D structural encodings. Graph-based DL models, which treat atoms and bonds as nodes and edges, provide a more intricate molecular representation, capturing structural dependencies. However, many existing models fail to incorporate molecular fragment contributions, which can be addressed through approaches such as matched molecular pairs analysis (Fu et al. 2019; Turk et al. 2017) and response map algorithms (Wenzel et al. 2019). Structure-based ADMET prediction, particularly using protein-ligand interaction models, remains an underexplored area with significant potential (Moroy et al. 2012). Recent developments, such as equivariant GNNs, integrate geometric constraints into molecular representations, improving predictions for 3D conformations and bioactivity (Cremer et al. 2023).

Looking ahead, GCNNs are poised to become central in ADMET modeling, surpassing RF and fully connected DNNs in predictive accuracy for key endpoints. Future advancements should focus on incorporating mechanistic insights and biophysical data into DL architectures, which currently infer such properties indirectly (Schneider et al. 2020). Tools enhancing model interpretability will be essential for aligning computational predictions with biochemical mechanisms, while integrating biophysical parameters such as molecular flexibility, conformational energetics, and biomolecular interactions will further refine model accuracy. For complex pharmacokinetic endpoints, adopting systems biology approaches may be necessary to capture interdependent ADMET factors. Algorithmic innovations should prioritize enhancing generalization while reducing dependency on extensive datasets. Pretraining strategies, inspired by vision and language models, are being explored to allow DNNs trained on large datasets to perform well on domain-specific ADMET tasks (Chuang et al. 2020b; Schneider et al. 2020). Explainable AI techniques, including attention-based mechanisms and counterfactual explanations, are also being investigated to improve transparency in decision-making (Lavecchia 2024). Finally, seamlessly integrating ADMET prediction models into the iterative design-make-test cycle of medicinal chemistry will be instrumental in accelerating lead optimization and reducing late-stage attrition. Even models with moderate predictive accuracy, such as hERG toxicity classifiers, can still provide valuable insights for prioritizing candidates and identifying knowledge gaps (Sanders et al. 2017).

Acknowledgements This work was supported by the European Union, Next Generation EU (PRIN 2022, project number: 2022P5LPHS). A.L. acknowledges funding from the PNRR Project CN000041, "National Centre for Gene Therapy and RNA Technology-Based Drugs" (CUP UNINA E63C22000940007).

References

Altae-Tran H, Ramsundar B, Pappu AS, Pande V. Low data drug discovery with one-shot learning. ACS Cent Sci. 2017;3:283–93. https://doi.org/10.1021/acscentsci.6b00367.

Alves VM, Golbraikh A, Capuzzi SJ, Liu K, Lam WI, Korn DR, et al. Multi-descriptor read across (MuDRA): a simple and transparent approach for developing accurate quantitative structure–activity relationship models. J Chem Inf Model. 2018;58:1214–23. https://doi.org/10.1021/acs.jcim.8b00124.

Balhara A, Kale S, Singh S. Physiologically based pharmacokinetic (PBPK) modelling. In: Computer aided pharmaceutics and drug delivery: an application guide for students and researchers of pharmaceutical sciences. Springer; 2022. p. 255–84.

Banerjee P, Eckert AO, Schrey AK, Preissner R. ProTox-II: a webserver for the prediction of toxicity of chemicals. Nucleic Acids Res. 2018;46:W257–63. https://doi.org/10.1093/nar/gky318.

Basavaraj S, Betageri GV. Can formulation and drug delivery reduce attrition during drug discovery and development – review of feasibility, benefits and challenges. Acta Pharm Sin B. 2014;4:3–17.

Basile AO, Yahi A, Tatonetti NP. Artificial intelligence for drug toxicity and safety. Trends Pharmacol Sci. 2019;40:624–35. https://doi.org/10.1016/j.tips.2019.07.005.

Breiman L. Random forests. Mach Learn. 2001;45:5–32. https://doi.org/10.1023/A:1010933404324.

Cai C, Guo P, Zhou Y, Zhou J, Wang Q, Zhang F, et al. Deep learning-based prediction of drug-induced cardiotoxicity. J Chem Inf Model. 2019;59:1073–84.

Chagas CM, Moss S, Alisaraie L. Drug metabolites and their effects on the development of adverse reactions: revisiting Lipinski's rule of five. Int J Pharm. 2018;549:133–49. https://doi.org/10.1016/j.ijpharm.2018.07.046.

Chen Z, Jiang Y, Zhang X, Zheng R, Qiu R, Sun Y, et al. ResNet18DNN: prediction approach of drug-induced liver injury by deep neural network with ResNet18. Brief Bioinform. 2022;23:bbab503.

Cheng F, Li W, Liu G, Tang Y. In silico ADMET prediction: recent advances, current challenges and future trends. Curr Top Med Chem. 2013;13:1273–89. https://doi.org/10.2174/1568026611313990033.

Chuang KV, Gunsalus LM, Keiser MJ. Learning molecular representations for medicinal chemistry. J Med Chem. 2020a;63:8705–22. https://doi.org/10.1021/acs.jmedchem.0c00385.

Chuang KV, Gunsalus LM, Keiser MJ. Learning molecular representations for medicinal chemistry: miniperspective. J Med Chem. 2020b;63:8705–22.

Chung TDY, Terry DB, Smith LH. In vitro and in vivo assessment of ADME and PK properties during lead selection and lead optimization – guidelines, benchmarks and rules of thumb. In: Markossian S, Grossman A, Arkin M, Auld D, Austin C, Baell J, et al., editors. Assay guidance manual. Bethesda: Eli Lilly & Company and the National Center for Advancing Translational Sciences; 2004.

Coley CW, Barzilay R, Green WH, Jaakkola TS, Jensen KF. Convolutional embedding of attributed molecular graphs for physical property prediction. J Chem Inf Model. 2017;57:1757–72. https://doi.org/10.1021/acs.jcim.6b00601.

Cremer J, Medrano Sandonas L, Tkatchenko A, Clevert D-A, De Fabritiis G. Equivariant graph neural networks for toxicity prediction. Chem Res Toxicol. 2023;36:1561–73.

Cui R, Zhu F. Ultrasound modified polysaccharides: a review of structure, physicochemical properties, biological activities and food applications. Trends Food Sci Technol. 2021;107:491–508. https://doi.org/10.1016/j.tifs.2020.11.018.

Cui Q, Lu S, Ni B, Zeng X, Tan Y, Chen YD, et al. Improved prediction of aqueous solubility of novel compounds by going deeper with deep learning. Front Oncol. 2020;10 https://doi.org/10.3389/fonc.2020.00121.

Daina A, Michielin O, Zoete V. SwissADME: a free web tool to evaluate pharmacokinetics, drug-likeness and medicinal chemistry friendliness of small molecules. Sci Rep. 2017;7:42717. https://doi.org/10.1038/srep42717.

de Sá AGC, Long Y, Portelli S, Pires DEV, Ascher DB. toxCSM: comprehensive prediction of small molecule toxicity profiles. Brief Bioinform. 2022;23:bbac337. https://doi.org/10.1093/bib/bbac337.

Deng Y, Xu X, Qiu Y, Xia J, Zhang W, Liu S. A multimodal deep learning framework for predicting drug–drug interaction events. Bioinformatics. 2020;36:4316–22.

Di L, Artursson P, Avdeef A, Benet LZ, Houston JB, Kansy M, et al. The critical role of passive permeability in designing successful drugs. Chem Med Chem. 2020;15:1862–74.

DiMasi JA, Grabowski HG, Hansen RW. Innovation in the pharmaceutical industry: new estimates of R&D costs. J Health Econ. 2016;47:20–33.

Dong J, Wang N-N, Yao Z-J, Zhang L, Cheng Y, Ouyang D, et al. ADMETlab: a platform for systematic ADMET evaluation based on a comprehensively collected ADMET database. J Chem. 2018;10:29. https://doi.org/10.1186/s13321-018-0283-x.

Du B-X, Xu Y, Yiu S-M, Yu H, Shi J-Y. ADMET property prediction via multi-task graph learning under adaptive auxiliary task selection. iScience. 2023;26:108285. https://doi.org/10.1016/j.isci.2023.108285.

Du Y, Jamasb AR, Guo J, Fu T, Harris C, Wang Y, et al. Machine learning-aided generative molecular design. Nat Mach Intell. 2024;6:589–604.

Ekins S, Williams AJ, Xu JJ. A predictive ligand-based Bayesian model for human drug-induced liver injury. Drug Metab Dispos. 2010;38:2302–8.

Faqi AS. A comprehensive guide to toxicology in nonclinical drug development. Academic Press; 2016.

Feinberg EN, Joshi E, Pande VS, Cheng AC. Improvement in ADMET prediction with multitask deep featurization. J Med Chem. 2020;63:8835–48.

Feng C, Chen H, Yuan X, Sun M, Chu K, Liu H, et al. Gene expression data based deep learning model for accurate prediction of drug-induced liver injury in advance. J Chem Inf Model. 2019;59:3240–50.

Fernandez M, Ban F, Woo G, Hsing M, Yamazaki T, LeBlanc E, et al. Toxic colors: the use of deep learning for predicting toxicity of compounds merely from their graphic images. J Chem Inf Model. 2018;58:1533–43.

Fernandez-Moure JS. Lost in translation: the gap in scientific advancements and clinical application. Front Bioeng Biotechnol. 2016;4:43.

Ferreira LLG, Andricopulo AD. ADMET modeling approaches in drug discovery. Drug Discov Today. 2019;24:1157–65. https://doi.org/10.1016/j.drudis.2019.03.015.

Feshuk M, Kolaczkowski L, Watford S, Paul FK. ToxRefDB v2. 1: update to curated in vivo study data in the toxicity reference database. Frontiers. Toxicology. 2023;5:1260305.

Fradkin P, Young A, Atanackovic L, Frey B, Lee LJ, Wang B. A graph neural network approach for molecule carcinogenicity prediction. Bioinformatics. 2022;38:i84–91. https://doi.org/10.1093/bioinformatics/btac266.

Francoeur PG, Koes DR. SolTranNet–A machine learning tool for fast aqueous solubility prediction. J Chem Inf Model. 2021;61:2530–6. https://doi.org/10.1021/acs.jcim.1c00331.

Fraser K, Bruckner DM, Dordick JS. Advancing predictive hepatotoxicity at the intersection of experimental, in silico, and artificial intelligence technologies. Chem Res Toxicol. 2018;31:412–30.

Fu L, Liu L, Yang Z-J, Li P, Ding J-J, Yun Y-H, et al. Systematic modeling of log D 7.4 based on ensemble machine learning, group contribution, and matched molecular pair analysis. J Chem Inf Model. 2019;60:63–76.

Fu T, Li F, Zhang Y, Yin J, Qiu W, Li X, et al. VARIDT 2.0: structural variability of drug transporter. Nucleic Acids Res. 2022;50:D1417–31. https://doi.org/10.1093/nar/gkab1013.

Fu L, Shi S, Yi J, Wang N, He Y, Wu Z, et al. ADMETlab 3.0: an updated comprehensive online ADMET prediction platform enhanced with broader coverage, improved performance, API functionality and decision support. Nucleic Acids Res. 2024:gkae236.

Gawehn E, Hiss JA, Schneider G. Deep learning in drug discovery. Mol Inf. 2016;35:3–14. https://doi.org/10.1002/minf.201501008.

Gawehn E, Hiss JA, Brown JB, Schneider G. Advancing drug discovery via GPU-based deep learning. Expert Opin Drug Discov. 2018;13:579–82. https://doi.org/10.1080/17460441.2018.1465407.

Ghasemi F, Mehridehnavi A, Pérez-Garrido A, Pérez-Sánchez H. Neural network and deep-learning algorithms used in QSAR studies: merits and drawbacks. Drug Discov Today. 2018;23:1784–90. https://doi.org/10.1016/j.drudis.2018.06.016.

Grulke CM, Williams AJ, Thillanadarajah I, Richard AM. EPA'S DSSTox database: history of development of a curated chemistry resource supporting computational toxicology research. Comput Toxicol. 2019;12:100096.

Hansch C, Maloney PP, Fujita T, Muir RM. Correlation of biological activity of Phenoxyacetic acids with Hammett substituent constants and partition coefficients. Nature. 1962;194:178–80. https://doi.org/10.1038/194178b0.

Hansen K, Mika S, Schroeter T, Sutter A, Ter Laak A, Steger-Hartmann T, et al. Benchmark data set for in silico prediction of Ames mutagenicity. J Chem Inf Model. 2009;49:2077–81.

Hirschfeld L, Swanson K, Yang K, Barzilay R, Coley CW. Uncertainty quantification using neural networks for molecular property prediction. J Chem Inf Model. 2020;60:3770–80. https://doi.org/10.1021/acs.jcim.0c00502.

Hop P, Allgood B, Yu J. Geometric deep learning autonomously learns chemical features that outperform those engineered by domain experts. Mol Pharm. 2018;15:4371–7. https://doi.org/10.1021/acs.molpharmaceut.7b01144.

Ibragimov B, Toesca DA, Chang DT, Yuan Y, Koong AC, Xing L, et al. Deep learning for identification of critical regions associated with toxicities after liver stereotactic body radiation therapy. Med Phys. 2020;47:3721–31.

Idakwo G, Thangapandian S, Luttrell J IV, Zhou Z, Zhang C, Gong P. Deep learning-based structure-activity relationship modeling for multi-category toxicity classification: a case study of 10K Tox21 chemicals with high-throughput cell-based androgen receptor bioassay data. Front Physiol. 2019;10:1044.

Jiang J, Wang R, Wang M, Gao K, Nguyen DD, Wei G-W. Boosting tree-assisted multitask deep learning for small scientific datasets. J Chem Inf Model. 2020;60:1235–44.

Jiménez J, Škalič M, Martínez-Rosell G, De Fabritiis G. KDEEP: protein–ligand absolute binding affinity prediction via 3D-convolutional neural networks. J Chem Inf Model. 2018;58:287–96. https://doi.org/10.1021/acs.jcim.7b00650.

Jimenez-Carretero D, Abrishami V, Fernández-de-Manuel L, Palacios I, Quílez-Álvarez A, Díez-Sánchez A, et al. Tox_(R)CNN: deep learning-based nuclei profiling tool for drug toxicity screening. PLoS Comput Biol. 2018;14:e1006238. https://doi.org/10.1371/journal.pcbi.1006238.

Kaisar MA, Sajja RK, Prasad S, Abhyankar VV, Liles T, Cucullo L. New experimental models of the blood-brain barrier for CNS drug discovery. Expert Opin Drug Discov. 2017;12:89–103. https://doi.org/10.1080/17460441.2017.1253676.

Karim A, Lee M, Balle T, Sattar A. CardioTox net: a robust predictor for hERG channel blockade based on deep learning meta-feature ensembles. J Cheminform. 2021;13:60. https://doi.org/10.1186/s13321-021-00541-z.

Kearnes S, Goldman B, Pande V. Modeling Industrial ADMET Data with Multitask Networks; 2017. https://doi.org/10.48550/arXiv.1606.08793.

Kim H, Kim E, Lee I, Bae B, Park M, Nam H. Artificial intelligence in drug discovery: a comprehensive review of data-driven and machine learning approaches. Biotechnol Bioprocess Eng. 2020;25:895–930.

Kiriiri GK, Njogu PM, Mwangi AN. Exploring different approaches to improve the success of drug discovery and development projects: a review. Future J Pharm Sci. 2020;6:1–12.

Klambauer G, Unterthiner T, Mayr A, Hochreiter S. Self-normalizing neural networks; 2017. https://doi.org/10.48550/arXiv.1706.02515.

Korotcov A, Tkachenko V, Russo DP, Ekins S. Comparison of deep learning with multiple machine learning methods and metrics using diverse drug discovery data sets. Mol Pharm. 2017;14:4462–75. https://doi.org/10.1021/acs.molpharmaceut.7b00578.

Lagunin A, Rudik A, Druzhilovsky D, Filimonov D, Poroikov V. ROSC-Pred: web-service for rodent organ-specific carcinogenicity prediction. Bioinformatics. 2018;34:710–2.

Lavecchia A. Advancing drug discovery with deep attention neural networks. Drug Discov Today. 2024;104067:104067.

LeCun Y, Bengio Y, Hinton G. Deep learning. Nature. 2015;521:436–44. https://doi.org/10.1038/nature14539.

Lee G, Park C, Ahn J. Novel deep learning model for more accurate prediction of drug-drug interaction effects. BMC Bioinformatics. 2019;20:1–8.

Lei T, Chen F, Liu H, Sun H, Kang Y, Li D, et al. ADMET evaluation in drug discovery. Part 17: development of quantitative and qualitative prediction models for chemical-induced respiratory toxicity. Mol Pharm. 2017;14:2407–21. https://doi.org/10.1021/acs.molpharmaceut.7b00317.

Lejal V, Cerisier N, Rouquié D, Taboureau O. Assessment of drug-induced liver injury through cell morphology and gene expression analysis. Chem Res Toxicol. 2023;36:1456–70. https://doi.org/10.1021/acs.chemrestox.2c00381.

Li X, Xu Y, Lai L, Pei J. Prediction of human cytochrome P450 inhibition using a multitask deep autoencoder neural network. Mol Pharm. 2018;15:4336–45.

Li T, Tong W, Roberts R, Liu Z, Thakkar S. DeepCarc: deep learning-powered carcinogenicity prediction using model-level representation. Front Artif Intell. 2021;4:757780.

Li T, Liu Z, Thakkar S, Roberts R, Tong W. DeepAmes: a deep learning-powered Ames test predictive model with potential for regulatory application. Regul Toxicol Pharmacol. 2023;144:105486. https://doi.org/10.1016/j.yrtph.2023.105486.

Lipinski CA, Lombardo F, Dominy BW, Feeney PJ. Experimental and computational approaches to estimate solubility and permeability in drug discovery and development settings1. Adv Drug Deliv Rev. 2001;46:3–26. https://doi.org/10.1016/S0169-409X(00)00129-0.

Lipkova J, Chen RJ, Chen B, Lu MY, Barbieri M, Shao D, et al. Artificial intelligence for multimodal data integration in oncology. Cancer Cell. 2022;40:1095–110.

Liu K, Sun X, Jia L, Ma J, Xing H, Wu J, et al. Chemi-net: a molecular graph convolutional network for accurate drug property prediction. Int J Mol Sci. 2019;20:3389.

Lusci A, Pollastri G, Baldi P. Deep architectures and deep learning in chemoinformatics: the prediction of aqueous solubility for drug-like molecules. J Chem Inf Model. 2013;53:1563–75. https://doi.org/10.1021/ci400187y.

Ma J, Sheridan RP, Liaw A, Dahl GE, Svetnik V. Deep neural nets as a method for quantitative structure–activity relationships. J Chem Inf Model. 2015;55:263–74. https://doi.org/10.1021/ci500747n.

Mak K-K, Epemolu O, Pichika MR. The role of DMPK science in improving pharmaceutical research and development efficiency. Drug Discov Today. 2022;27:705–29. https://doi.org/10.1016/j.drudis.2021.11.005.

Maltarollo VG, Gertrudes JC, Oliveira PR, Honorio KM. Applying machine learning techniques for ADME-Tox prediction: a review. Expert Opin Drug Metab Toxicol. 2015;11:259–71. https://doi.org/10.1517/17425255.2015.980814.

Manallack DT, Prankerd RJ, Yuriev E, Oprea TI, Chalmers DK. The significance of acid/base properties in drug discovery. Chem Soc Rev. 2013;42:485–96. https://doi.org/10.1039/C2CS35348B.

Matsuzaka Y, Uesawa Y. Prediction model with high-performance constitutive androstane receptor (CAR) using DeepSnap-deep learning approach from the Tox21 10K compound library. Int J Mol Sci. 2019a;20:4855.

Matsuzaka Y, Uesawa Y. Optimization of a deep-learning method based on the classification of images generated by parameterized deep snap a novel molecular-image-input technique for quantitative structure–activity relationship (QSAR) analysis. Front Bioeng Biotechnol. 2019b;7:65.

Matsuzaka Y, Uesawa Y. DeepSnap-deep learning approach predicts progesterone receptor antagonist activity with high performance. Front Bioeng Biotechnol. 2020;7:485.

Matsuzaka Y, Hosaka T, Ogaito A, Yoshinari K, Uesawa Y. Prediction model of aryl hydrocarbon receptor activation by a novel QSAR approach, DeepSnap–deep learning. Molecules. 2020;25:1317.

Mattingly CJ, Rosenstein MC, Colby GT, Forrest JN Jr, Boyer JL. The comparative Toxicogenomics database (CTD): a resource for comparative toxicological studies. J Exp Zool A Comp Exp Biol. 2006;305:689–92.

Mayr A, Klambauer G, Unterthiner T, Hochreiter S. DeepTox: toxicity prediction using deep learning. Front Environ Sci. 2016;3:80.

Men K, Geng H, Zhong H, Fan Y, Lin A, Xiao Y. A deep learning model for predicting xerostomia due to radiation therapy for head and neck squamous cell carcinoma in the RTOG 0522 clinical trial. Int J Radiat Oncol Biol Phys. 2019;105:440–7.

Moffat JG, Vincent F, Lee JA, Eder J, Prunotto M. Opportunities and challenges in phenotypic drug discovery: an industry perspective. Nat Rev Drug Discov. 2017;16:531–43.

Montanari F, Kuhnke L, Ter Laak A, Clevert D-A. Modeling physico-chemical ADMET endpoints with multitask graph convolutional networks. Molecules. 2019;25:44.

Moreno L, Pearson AD. How can attrition rates be reduced in cancer drug discovery? vol. 8. Taylor & Francis; 2013.

Moroy G, Martiny VY, Vayer P, Villoutreix BO, Miteva MA. Toward in silico structure-based ADMET prediction in drug discovery. Drug Discov Today. 2012;17:44–55.

Onakpoya IJ, Heneghan CJ, Aronson JK. Post-marketing withdrawal of 462 medicinal products because of adverse drug reactions: a systematic review of the world literature. BMC Med. 2016;14:10. https://doi.org/10.1186/s12916-016-0553-2.

Peng Y, Lin Y, Jing X-Y, Zhang H, Huang Y, Luo GS. Enhanced graph isomorphism network for molecular ADMET properties prediction. IEEE Access. 2020a;8:168344–60.

Peng Y, Zhang Z, Jiang Q, Guan J, Zhou S. TOP: a deep mixture representation learning method for boosting molecular toxicity prediction. Methods. 2020b;179:55–64.

Petrova E. Innovation in the pharmaceutical industry: the process of drug discovery and development. In: Innovation and marketing in the pharmaceutical industry: emerging practices, research, and policies. Springer; 2013. p. 19–81.
Piochi LF, Preto AJ, Moreira IS. DELFOS—drug efficacy leveraging forked and specialized networks—benchmarking scRNA-seq data in multi-omics-based prediction of cancer sensitivity. Bioinformatics. 2023;39:btad645.
Pires DEV, Blundell TL, Ascher DB. pkCSM: predicting small-molecule pharmacokinetic and toxicity properties using graph-based signatures. J Med Chem. 2015;58:4066–72. https://doi.org/10.1021/acs.jmedchem.5b00104.
Polishchuk P. Interpretation of quantitative structure–activity relationship models: past, present, and future. J Chem Inf Model. 2017;57:2618–39. https://doi.org/10.1021/acs.jcim.7b00274.
Prasad S, Brooks BR. A deep learning approach for the blind logP prediction in SAMPL6 challenge. J Comput Aided Mol Des. 2020;34:535–42. https://doi.org/10.1007/s10822-020-00292-3.
Rácz A, Keserű GM. Large-scale evaluation of cytochrome P450 2C9 mediated drug interaction potential with machine learning-based consensus modeling. J Comput Aided Mol Des. 2020;34:831–9. https://doi.org/10.1007/s10822-020-00308-y.
Ramsundar B, Kearnes S, Riley P, Webster D, Konerding D, Pande V. Massively multitask networks for drug discovery; 2015. https://doi.org/10.48550/arXiv.1502.02072.
Ramsundar B, Liu B, Wu Z, Verras A, Tudor M, Sheridan RP, et al. Is multitask deep learning practical for pharma? J Chem Inf Model. 2017;57:2068–76. https://doi.org/10.1021/acs.jcim.7b00146.
Rao M, Nassiri V, Alhambra C, Snoeys J, Van Goethem F, Irrechukwu O, et al. AI/ML models to predict the severity of drug-induced liver injury for small molecules. Chem Res Toxicol. 2023;36:1129–39. https://doi.org/10.1021/acs.chemrestox.3c00098.
Renn A, Su B-H, Liu H, Sun J, Tseng YJ. Advances in the prediction of mouse liver microsomal studies: from machine learning to deep learning. WIREs Comput Mol Sci. 2021;11:e1479.
Rodríguez-Pérez R, Bajorath J. Interpretation of compound activity predictions from complex machine learning models using local approximations and Shapley values. J Med Chem. 2020;63:8761–77. https://doi.org/10.1021/acs.jmedchem.9b01101.
Rogers D, Hahn M. Extended-connectivity fingerprints. J Chem Inf Model. 2010;50:742–54.
Ryan K, Lengyel J, Shatruk M. Crystal structure prediction via deep learning. J Am Chem Soc. 2018;140:10158–68. https://doi.org/10.1021/jacs.8b03913.
Ryu JY, Lee MY, Lee JH, Lee BH, Oh K-S. DeepHIT: a deep learning framework for prediction of hERG-induced cardiotoxicity. Bioinformatics. 2020;36:3049–55. https://doi.org/10.1093/bioinformatics/btaa075.
Ryu JY, Jang WD, Jang J, Oh K-S. PredAOT: a computational framework for prediction of acute oral toxicity based on multiple random forest models. BMC Bioinformatics. 2023;24:66.
Sanders JM, Beshore DC, Culberson JC, Fells JI, Imbriglio JE, Gunaydin H, et al. Informing the selection of screening hit series with in silico absorption, distribution, metabolism, excretion, and toxicity profiles: miniperspective. J Med Chem. 2017;60:6771–80.
Schmidhuber J. Deep learning in neural networks: an overview. Neural Netw. 2015;61:85–117. https://doi.org/10.1016/j.neunet.2014.09.003.
Schneckener S, Grimbs S, Hey J, Menz S, Osmers M, Schaper S, et al. Prediction of oral bioavailability in rats: transferring insights from in vitro correlations to (deep) machine learning models using in silico model outputs and chemical structure parameters. J Chem Inf Model. 2019;59:4893–905.
Schneider P, Walters WP, Plowright AT, Sieroka N, Listgarten J, Goodnow RA, et al. Rethinking drug design in the artificial intelligence era. Nat Rev Drug Discov. 2020;19:353–64. https://doi.org/10.1038/s41573-019-0050-3.
Schütt KT, Sauceda HE, Kindermans P-J, Tkatchenko A, Müller K-R. SchNet–a deep learning architecture for molecules and materials. J Chem Phys. 2018;148:241722.
Segall MD, Barber C. Addressing toxicity risk when designing and selecting compounds in early drug discovery. Drug Discov Today. 2014;19:688–93.

Sharma B, Chenthamarakshan V, Dhurandhar A, Pereira S, Hendler JA, Dordick JS, et al. Accurate clinical toxicity prediction using multi-task deep neural nets and contrastive molecular explanations. Sci Rep. 2023;13:4908.

Shen WX, Zeng X, Zhu F, Wang YL, Qin C, Tan Y, et al. Out-of-the-box deep learning prediction of pharmaceutical properties by broadly learned knowledge-based molecular representations. Nat Mach Intell. 2021;3:334–43.

Shi T, Yang Y, Huang S, Chen L, Kuang Z, Heng Y, et al. Molecular image-based convolutional neural network for the prediction of ADMET properties. Chemom Intell Lab Syst. 2019;194:103853.

Shin M, Jang D, Nam H, Lee KH, Lee D. Predicting the absorption potential of chemical compounds through a deep learning approach. IEEE/ACM Trans Comput Biol Bioinform. 2018;15:432–40. https://doi.org/10.1109/TCBB.2016.2535233.

Smyth MJ, Krasovskis E, Sutton VR, Johnstone RW. The drug efflux protein, P-glycoprotein, additionally protects drug-resistant tumor cells from multiple forms of caspase-dependent apoptosis. Proc Natl Acad Sci. 1998;95:7024–9. https://doi.org/10.1073/pnas.95.12.7024.

Su B-H, Tu Y, Lin C, Shao C-Y, Lin OA, Tseng YJ. Rule-based prediction models of cytochrome P450 inhibition. J Chem Inf Model. 2015;55:1426–34. https://doi.org/10.1021/acs.jcim.5b00130.

Sun D, Gao W, Hu H, Zhou S. Why 90% of clinical drug development fails and how to improve it? Acta Pharm Sin B. 2022;12:3049–62.

Tao L, Zhang P, Qin C, Chen SY, Zhang C, Chen Z, et al. Recent progresses in the exploration of machine learning methods as in-silico ADME prediction tools. Adv Drug Deliv Rev. 2015;86:83–100. https://doi.org/10.1016/j.addr.2015.03.014.

Tran TTV, Surya Wibowo A, Tayara H, Chong KT. Artificial intelligence in drug toxicity prediction: recent advances, challenges, and future perspectives. J Chem Inf Model. 2023;63:2628–43. https://doi.org/10.1021/acs.jcim.3c00200.

Turk S, Merget B, Rippmann F, Fulle S. Coupling matched molecular pairs with machine learning for virtual compound optimization. J Chem Inf Model. 2017;57:3079–85.

Unterthiner T, Mayr A, Klambauer G, Hochreiter S. Toxicity prediction using deep learning; 2015. https://doi.org/10.48550/arXiv.1503.01445.

Van de Waterbeemd H. From in vivo to in vitro/in silico ADME: progress and challenges. Expert Opin Drug Metab Toxicol. 2005;1:1–4. https://doi.org/10.1517/17425255.1.1.1.

Van De Waterbeemd H. Improving compound quality through in vitro and in silico physicochemical profiling. Chem Biodivers. 2009;6:1760–6.

Van De Waterbeemd H, Gifford E. ADMET in silico modelling: towards prediction paradise? Nat Rev Drug Discov. 2003;2:192–204.

Van Norman GA. Drugs, devices, and the FDA: part 1: an overview of approval processes for drugs. JACC Basic Transl Sci. 2016a;1:170–9.

Van Norman GA. Drugs and devices: comparison of European and US approval processes. JACC Basic Transl Sci. 2016b;1:399–412.

Van Norman GA. Limitations of animal studies for predicting toxicity in clinical trials: part 2: potential alternatives to the use of animals in preclinical trials. Basic Transl Sci. 2020;5:387–97.

Vandenberg JI, Perry MD, Perrin MJ, Mann SA, Ke Y, Hill AP. hERG K+ channels: structure, function, and clinical significance. Physiol Rev. 2012;92:1393–478. https://doi.org/10.1152/physrev.00036.2011.

Venkatraman V. FP-ADMET: a compendium of fingerprint-based ADMET prediction models. J Cheminform. 2021;13:75. https://doi.org/10.1186/s13321-021-00557-5.

Vo AH, Van Vleet TR, Gupta RR, Liguori MJ, Rao MS. An overview of machine learning and big data for drug toxicity evaluation. Chem Res Toxicol. 2019;33:20–37.

Wadhwa P, Mittal A. Quantitative structure-property relationship (QSPR) modeling applications in formulation development. In: Computer aided pharmaceutics and drug delivery: an application guide for students and researchers of pharmaceutical sciences. Springer; 2022. p. 543–60.

Wang Y, Chen X. QSPR model for Caco-2 cell permeability prediction using a combination of HQPSO and dual-RBF neural network. RSC Adv. 2020;10:42938–52.

Wang Y, Xing J, Xu Y, Zhou N, Peng J, Xiong Z, et al. In silico ADME/T modelling for rational drug design. Q Rev Biophys. 2015;48:488–515. https://doi.org/10.1017/S0033583515000190.

Wang H, Liu R, Schyman P, Wallqvist A. Deep neural network models for predicting chemically induced liver toxicity endpoints from transcriptomic responses. Front Pharmacol. 2019;10:42.

Wang Y, Huang L, Jiang S, Yifei W, Zou J, Fu H, et al. Capsule networks showed excellent performance in the classification of hERG blockers/nonblockers. Front Pharmacol. 2020;10:1631.

Waring MJ, Arrowsmith J, Leach AR, Leeson PD, Mandrell S, Owen RM, et al. An analysis of the attrition of drug candidates from four major pharmaceutical companies. Nat Rev Drug Discov. 2015;14:475–86.

Waters M, Stasiewicz S, Alex Merrick B, Tomer K, Bushel P, Paules R, et al. CEBS – chemical effects in biological systems: a public data repository integrating study design and toxicity data with microarray and proteomics data. Nucleic Acids Res. 2008;36:D892–900. https://doi.org/10.1093/nar/gkm755.

Wei Y, Li S, Li Z, Wan Z, Lin J. Interpretable-ADMET: a web service for ADMET prediction and optimization based on deep neural representation. Bioinformatics. 2022;38:2863–71. https://doi.org/10.1093/bioinformatics/btac192.

Wenzel J, Matter H, Schmidt F. Predictive multitask deep neural network models for ADME-Tox properties: learning from large data sets. J Chem Inf Model. 2019;59:1253–68.

Wu K, Zhao Z, Wang R, Wei G-W. TopP–S: persistent homology-based multi-task deep neural networks for simultaneous predictions of partition coefficient and aqueous solubility. J Comput Chem. 2018;39:1444–54. https://doi.org/10.1002/jcc.25213.

Wu Z, Lei T, Shen C, Wang Z, Cao D, Hou T. ADMET evaluation in drug discovery. 19. Reliable prediction of human cytochrome P450 inhibition using artificial intelligence approaches. J Chem Inf Model. 2019;59:4587–601.

Xiong Z, Wang D, Liu X, Zhong F, Wan X, Li X, et al. Pushing the boundaries of molecular representation for drug discovery with the graph attention mechanism. J Med Chem. 2019;63:8749–60.

Xiong G, Wu Z, Yi J, Fu L, Yang Z, Hsieh C, et al. ADMETlab 2.0: an integrated online platform for accurate and comprehensive predictions of ADMET properties. Nucleic Acids Res. 2021;49:W5–14.

Xu Y, Ma J, Liaw A, Sheridan RP, Svetnik V. Demystifying multitask deep neural networks for quantitative structure–activity relationships. J Chem Inf Model. 2017a;57:2490–504. https://doi.org/10.1021/acs.jcim.7b00087.

Xu Y, Pei J, Lai L. Deep learning based regression and multiclass models for acute oral toxicity prediction with automatic chemical feature extraction. J Chem Inf Model. 2017b;57:2672–85.

Xu Y, Yao H, Lin K. An overview of neural networks for drug discovery and the inputs used. Expert Opin Drug Discov. 2018;13:1091–102.

Yang H, Sun L, Li W, Liu G, Tang Y. In Silico prediction of chemical toxicity for drug design using machine learning methods and structural alerts. Front Chem. 2018a;6 https://doi.org/10.3389/fchem.2018.00030.

Yang M, Chen J, Xu L, Shi X, Zhou X, Xi Z, et al. A novel adaptive ensemble classification framework for ADME prediction. RSC Adv. 2018b;8:11661–83. https://doi.org/10.1039/C8RA01206G.

Yang H, Lou C, Sun L, Li J, Cai Y, Wang Z, et al. admetSAR 2.0: web-service for prediction and optimization of chemical ADMET properties. Bioinformatics. 2019a;35:1067–9. https://doi.org/10.1093/bioinformatics/bty707.

Yang K, Swanson K, Jin W, Coley C, Eiden P, Gao H, et al. Analyzing learned molecular representations for property prediction. J Chem Inf Model. 2019b;59:3370–88.

Ye Z, Yang Y, Li X, Cao D, Ouyang D. An integrated transfer learning and multitask learning approach for pharmacokinetic parameter prediction. Mol Pharm. 2018;16:533–41.

Yin J, Li F, Zhou Y, Mou M, Lu Y, Chen K, et al. INTEDE: Interactome of drug-metabolizing enzymes. Nucleic Acids Res. 2021;49:D1233–43. https://doi.org/10.1093/nar/gkaa755.

Yu T-H, Su B-H, Battalora LC, Liu S, Tseng YJ. Ensemble modeling with machine learning and deep learning to provide interpretable generalized rules for classifying CNS drugs with high prediction power. Brief Bioinform. 2022;23:bbab377.

Yuan Q, Wei Z, Guan X, Jiang M, Wang S, Zhang S, et al. Toxicity prediction method based on multi-channel convolutional neural network. Molecules. 2019;24:3383.

Zhang M-Q, Wilkinson B. Drug discovery beyond the 'rule-of-five.'. Curr Opin Biotechnol. 2007;18:478–88. https://doi.org/10.1016/j.copbio.2007.10.005.

Zhang S, Yan Z, Huang Y, Liu L, He D, Wang W, et al. HelixADMET: a robust and endpoint extensible ADMET system incorporating self-supervised knowledge transfer. Bioinformatics. 2022a;38:3444–53. https://doi.org/10.1093/bioinformatics/btac342.

Zhang X, Mao J, Wei M, Qi Y, Zhang JZH. HergSPred: accurate classification of hERG blockers/nonblockers with machine-learning models. J Chem Inf Model. 2022b;62:1830–9. https://doi.org/10.1021/acs.jcim.2c00256.

Zhen X, Chen J, Zhong Z, Hrycushko B, Zhou L, Jiang S, et al. Deep convolutional neural network with transfer learning for rectum toxicity prediction in cervical cancer radiotherapy: a feasibility study. Phys Med Biol. 2017;62:8246.

Zhou Y, Cahya S, Combs SA, Nicolaou CA, Wang J, Desai PV, et al. Exploring tunable hyperparameters for deep neural networks with industrial ADME data sets. J Chem Inf Model. 2018;59:1005–16.

Zhu L, Zhao J, Zhang Y, Zhou W, Yin L, Wang Y, et al. ADME properties evaluation in drug discovery: in silico prediction of blood–brain partitioning. Mol Divers. 2018;22:979–90. https://doi.org/10.1007/s11030-018-9866-8.

Harnessing Artificial Intelligence to Revolutionise Molecular Modelling and Simulations

Xiaohan Lin, Yanheng Li, Xiaoxuan Yu, Xuhan Liu, Yao-Kun Lei, and Yi Isaac Yang

Abstract Molecular dynamics (MD) simulations are key computational methods for understanding physical and chemical processes, as well as predicting the thermodynamic and kinetic properties of molecular systems. Despite their widespread use, traditional MD approaches suffer from several limitations, such as sampling inefficiency, inaccuracies in force fields, and challenges related to non-scalable time and spatial scales. Recent advancements in machine learning (ML) and artificial intelligence (AI) offer new opportunities to overcome these obstacles. Beginning with molecular representations, this chapter reviews the application of various ML techniques (supervised, unsupervised, and reinforcement learning) in addressing long-standing challenges in MD. In particular, we highlight the development of machine learning-based force fields, which offer a powerful solution to enhance both accuracy and efficiency in MD simulations. AI-driven enhanced sampling methods are also emphasised, ranging from the construction of more effective collective variables to generative algorithms for efficient density estimation of the Boltzmann distribution. Furthermore, we discuss a novel perspective on the deep integration of AI and MD, emphasising their synergy at both the software and hardware levels. As this interdisciplinary field continues to evolve, it presents not only significant challenges but also a wealth of opportunities for further exploration.

Keywords Molecular dynamics · Deep molecular models · Force field · Enhanced sampling · Simulation software

Xiaohan Lin, Yanheng Li, and Xiaoxuan Yu authors contributed equally to this work.

X. Lin · Y. Li · X. Yu
College of Chemistry and Molecular Engineering, Peking University, Beijing, China

X. Liu · Y.-K. Lei · Y. I. Yang (✉)
Institute of Systems and Physical Biology, Shenzhen Bay Laboratory, Shenzhen, China
e-mail: yangyi@szbl.ac.cn

A. Lavecchia (ed.), *Applied Artificial Intelligence for Drug Discovery*,
https://doi.org/10.1007/978-3-031-98022-0_9

1 Introduction

Molecular dynamics (MD) simulates the evolution of a molecular system over time. With the correct initial 3D structure and proper force field, accurate trajectories could be generated by the MD process for calculating thermodynamic and kinetic properties. Over the years, MD has been applied to a variety of systems, including biomolecules, surfaces and materials (Suárez and Díaz 2020; Sinha et al. 2022; Wang et al. 2024a; Zhao et al. 2023; Zhou et al. 2023), providing unparalleled insights into molecular processes that are often inaccessible through experimental techniques alone.

However, traditional MD methods have a trade-off between accuracy and efficiency. MD methods require classical force fields within Newtonian mechanics to offer high-computational efficiency but suffer from low accuracy and are incapable of handling processes such as bond breaking. On the other hand, ab initio MD provides much higher computational accuracy but is prohibitively slow, limiting its application to smaller molecular systems. Moreover, researchers have developed various enhanced sampling methods to accelerate MD sampling (Torrie and Valleau 1977; Laio and Parrinello 2002; Barducci et al. 2008; Gao 2008a; Sugita and Okamoto 1999), although these methods also encounter challenges when applied to real systems with high degrees of freedom.

Recently, advancements in artificial intelligence (AI), especially deep neural networks (DNNs), have brought new perspectives to MD. DNNs are powerful nonlinear approximation tools with excellent fitting ability and simple optimisation algorithms. Moreover, AI modelling has powerful parallel computing capabilities when running on Graph Processing Units (GPUs), which have similar running processes to MD modelling, making it possible to simulate large molecular systems with AI frameworks.

In this chapter, we systematically introduce AI-driven MD methods, covering both theoretical algorithms and practical applications. In the following paragraphs, the first section will focus on AI algorithms related to MD. In the second section, we will introduce machine learning-based molecular force fields. The third section will discuss machine learning-based sampling methods. The fourth section will explore software frameworks that integrate AI with MD. Finally, we will summarise the topics covered and propose potential challenges in this rapidly evolving field.

2 Algorithms and Methods

This section will focus on AI algorithms related to MD simulations, which can be divided into molecular modelling and learning algorithms.

2.1 Molecular Modelling

Molecular modelling is a fundamental problem in computational chemistry, which aims to utilise information in molecular systems and select appropriate approaches to model interactions within the system. While applying neural networks to molecular modelling, it is necessary to consider the intrinsic properties of molecular systems, called *inductive biases* in machine learning. Incorporating inductive biases allows learning algorithms to restrict the search scope for possible solutions or prioritise one solution over another, independent of the observed data. The inductive bias of a molecular system mainly includes the symmetries within the system, which refers to the invariance and equivariance of molecular properties under transformations like rotation, translation, permutation, and sometimes reflection, and the definition of equivariance and invariance are listed below:

$$\begin{aligned} D'(g)f(\mathbf{x}) &= f(D(g)\mathbf{x}) \text{ (equivariance)} \\ f(\mathbf{x}) &= f(D(g)\mathbf{x}) \text{ (invariance)} \end{aligned} \tag{1}$$

where $\boldsymbol{x}$ stands for the molecular system, g represents the symmetric group operation, and $D(g)$ represents the group operation on the corresponding vector space.

Models that can satisfy invariance or equivariance requirements are typically classified into descriptor- and graph-based models (Zhang et al. 2020) (Fig. 1). Next, we will discuss in detail the development of invariant and equivariant models within these two categories, along with future challenges and directions.

2.1.1 Descriptor-Based Models

The "descriptors" are predefined vectors that can characterise the molecular systems. Let's assume that a molecular system X contains N atoms, and the coordinates of atoms are denoted as $\{\mathbf{x}_i\}_{i=1,\ldots,N}$. The descriptor models first transform $\mathbf{x}_i$ into a set of descriptors $\mathbf{q}(\mathbf{x}_i)$, and then exploit a simple network like multi-layer perceptron (MLP) f_θ to predict the desired property $A(X)$:

$$A(X) = \sum_{i=1}^{N} f_\theta(\mathbf{q}(\mathbf{x}_i)) \tag{2}$$

In machine learning, selecting descriptors is also known as *feature engineering*. One should find a complete set of descriptors to determine the property, and it would cause failure if one neglects even one of the essential descriptors. Due to this fact, the development of descriptor-based models mainly focused on the choice of $\mathbf{q}(\mathbf{x}_i)$. The pioneering work along this line can be dated back to the Behler-Parrinello neural network (BPNN) in 2007 (Behler and Parrinello 2007). In BPNN, the environment

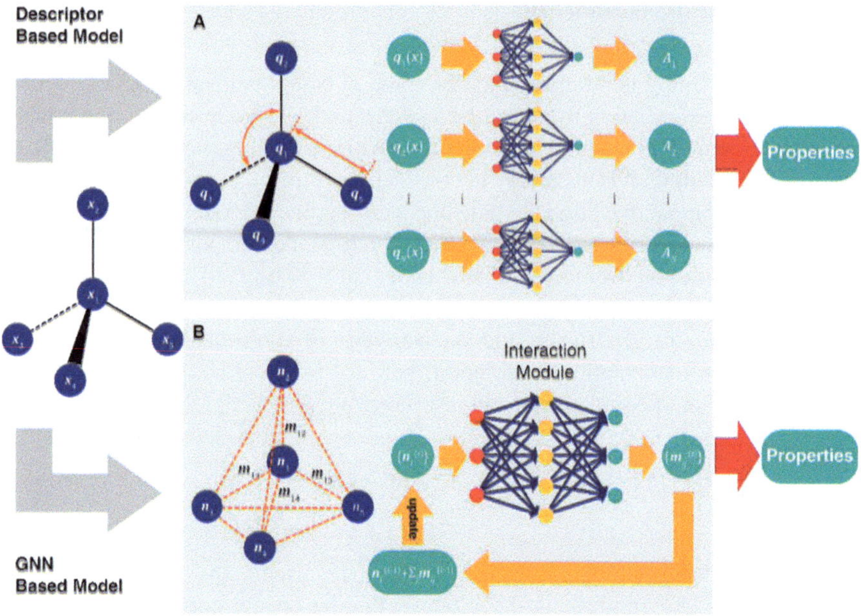

Fig. 1 Architectures of deep molecular models. (**a**) Descriptor-based models. $q(\mathbf{x}_i)$ corresponds to the i-th particle (atom) descriptors, which are further fed into a neural network. The output of each atomic neural network is then summed up to predict molecular properties. (**b**) Graph-based models. Each particle (atom) is represented by a node vector ($\mathbf{n}_i$). An interaction module collects messages ($\mathbf{m}_{ij}$) between nodes which are used to update the node vector. After several iterations of message passing, the node vectors are used to predict the molecular properties. (Reprinted with permission from *J. Phys. Chem. A* 2020, *124*, 6745–6763. Copyright 2020 American Chemical Society)

of each atom is transformed by specific symmetric functions and encoded into invariant descriptors called atomic environment vectors (AEVs):

$$G_i^1 = \sum_{j \neq i} \exp\left[-\eta\left(R_{ij} - R_s\right)^2\right] f_c\left(R_{ij}\right)$$

$$G_i^2 = 2^{1-\zeta} \sum_{j,k \neq i} \left(1 + \lambda \cos\theta_{ijk}\right)^{\zeta} \exp\left[-\eta\left(R_{ij} + R_{ik} + R_{jk}\right)^2\right] f_c\left(R_{ij}\right) f_c\left(R_{jk}\right) f_c\left(R_{ik}\right) \tag{3}$$

$$f_c\left(R_{ij}\right) = \begin{cases} 0.5 \times \left[\cos\left(\frac{\pi R_{ij}}{R_c}\right) + 1\right], & \text{for } R_{ij} \leq R_c \\ 0, & \text{for } R_{ij} > R_c \end{cases}$$

where R_{ij} is the distance between atoms i, j, and θ_{ijk} is the angle between atoms i, j, k. $f_c(R_{ij})$ is a cutoff function (Eq. 3) so that the AEVs gradually decay to zero at the cutoff radius. Based on the AEV descriptors, a set of MLPs output the atomic energies, and the summation gives overall energy. In BPNN, MLPs have shared parameters for atoms of the same types.

It is noteworthy that the AEVs introduced by BPNN, or its variants ANI and TensorMol (Smith et al. 2017; Yao et al. 2018), not only include two-body (distance) terms but also account for three-body (angular) features. Following this line, one can develop four-body descriptors if necessary (e.g. chirality descriptors). Furthermore, in a recently developed descriptor-based model, deep potential molecular dynamics (DPMD) (Zhang et al. 2018a) treats each atom in a local coordinate frame, making the descriptors invariant. The descriptors constructed using this approach are more straightforward than those in BPNN.

The descriptors we discussed above are invariant descriptors, which discard higher-order symmetry information of the system during their construction. As a result, they can only be used to predict properties that remain invariant under symmetrical operations like energy directly. While handling equivariant properties like forces, the gradients from invariant networks can be used to fit them. However, calculating gradients via backpropagation is time- and memory-consuming, significantly increasing the cost of inference. In contrast, those properties can be predicted by equivariant networks directly without relying on a backward process.

The construction of equivariant descriptors originated from atom cluster expansion (ACE) (Drautz 2019), and many molecule descriptors can be considered as exceptional cases of ACE (Behler and Parrinello 2007; Bartók et al. 2010; Shapeev 2016). By defining the following atom base A_{iv}, the atomic energy calculated by multi-body expansion can be translated into the form of Eq. 4:

$$E_i(\boldsymbol{\sigma})=\sum_{v} c_v^{(1)} A_{iv}+\sum_{v_1 v_2}^{v_1\geq v_2} c_{v_1 v_2}^{(2)} A_{iv_1} A_{iv_2}+\sum_{v_1 v_2 v_3}^{v_1\geq v_2\geq v_3} c_{v_1 v_2 v_3}^{(3)} A_{iv_1} A_{iv_2} A_{iv_3}+\cdots \tag{4}$$

$$A_{iv} = \langle \rho_i | \phi_v \rangle = \sum_j \phi_v\left(\boldsymbol{r}_{ji}\right)$$

$$\phi_v(\boldsymbol{r}) = \sqrt{4\pi} R_{nl}(r) Y_l^m\left(\hat{\boldsymbol{r}}\right)$$

where $\phi_v(\boldsymbol{r})$ is a set of basis functions. R_{nl} is a radial function depending on distance r, and Y_l^m stands for spherical harmonics depending on the unit direction vector $\hat{\boldsymbol{r}}$. The computational complexity scales linearly concerning the number of neighbour atoms.

One nicety of these descriptor-based models, especially when compared to the following graph-based models, is that the computational complexity is generally linearly scaled with the number of atoms N, when the descriptors are well formulated. However, descriptor-based models cannot scale well with the complexity of molecular systems. When the type of atoms increases, the desired descriptors will skyrocket. As a result, applications of descriptor-based models are mostly limited to relatively simple systems like material systems (Sheng et al. 2011). Worse still, since the suitable descriptors $\mathbf{q}(\mathbf{x})$ will inevitably vary for different systems, the descriptor-based models have limited transferability between different molecular systems, limiting their applications for more general purposes like force field development. Despite these shortcomings, the development of the descriptor-based models has furthered and enriched our understanding of how to represent molecular systems (Bartók et al. 2013; Behler 2016). The descriptors proposed in this field have wide applications beyond deep molecular models. For example, some descriptors can be used as suitable collective variables (CVs) or order parameters for the analysis or enhanced sampling of material systems (Piaggi and Parrinello 2017; Niu et al. 2018). Moreover, it is always defined as a machine learning problem to find good descriptors relevant to a particular property, where manual feature engineering can be replaced by automatic representation learning, which will be an essential direction for further studies.

2.1.2 Graph-Based Models

In contrast to descriptor-based models, Graph Neural Network (GNN) models deal with unstructured inputs like atom types $\{Z_i\}$ and coordinates $\{\mathbf{x}_i\}$. The inputs are vectorised by one-hot encoding and basis functions, and then learnable molecular descriptors are constructed by message passing and eventually applied to downstream tasks.

It is clear that the coordinates $\{\mathbf{x}_i\}$ will change under translation or rotation, and the symmetry of the network will be broken if we use them as the direct input. So, in invariant graph networks, the input features are predefined SE(3) invariant features. Unlike descriptor-based networks, they are usually easy to get and independent of feature engineering. For example, SchNet (Schütt et al. 2017) used the one-hot encoding of atom types as the initial node features of the network, and the distance initialised the edge features d_{ij} between atoms i, j. The following Gaussian radial basis functions (RBFs) were used to vectorise distances:

$$e_k\left(\boldsymbol{R}_i - \boldsymbol{R}_j\right) = \exp\left(-\gamma \left\|d_{ij} - \mu_k\right\|^2\right) \tag{5}$$

While γ and $\{\mu_k\}_{k=1:N}$ are hyperparameters. By vectorisation, the network will be easier to train and perform better. The initialised node and edge features were passed to a graph convolutional network (GCN). Following several iterations of message passing, features were passed to a readout module (mainly MLPs) to output the desired properties:

$$\mathbf{x}_i^0 = \mathbf{a}_{Z_i}$$

$$\mathbf{x}_i^{l+1} = \left(X^l * W^l\right)_i = \sum_{j=0}^{n_{\text{atoms}}} \mathbf{x}_j^l \circ W^l \left(\mathbf{r}_j - \mathbf{r}_i\right) \tag{6}$$

$$P = \text{Readout}\left(X^L\right) = \text{Readout}\left(\left\{\mathbf{x}_i^L\right\}\right)$$

Where $\mathbf{x}_i^l$ stands for the atom representation, which was initialised using an embedding module depending on the atom type Z_i. W^l is a weight matrix depending on the RBF-based edge features.

Following studies have attempted to replace GCNs with the recent attention algorithms to enhance the expressive ability of models. For example, MolCT (Zhang et al. 2023a) employs an attention-based message-passing approach, where the coordinate information of molecular systems is incorporated into the model through "positional embedding" module, similar to the module used in language models for encoding word positions. The network architecture of MolCT is shown in Fig. 2.

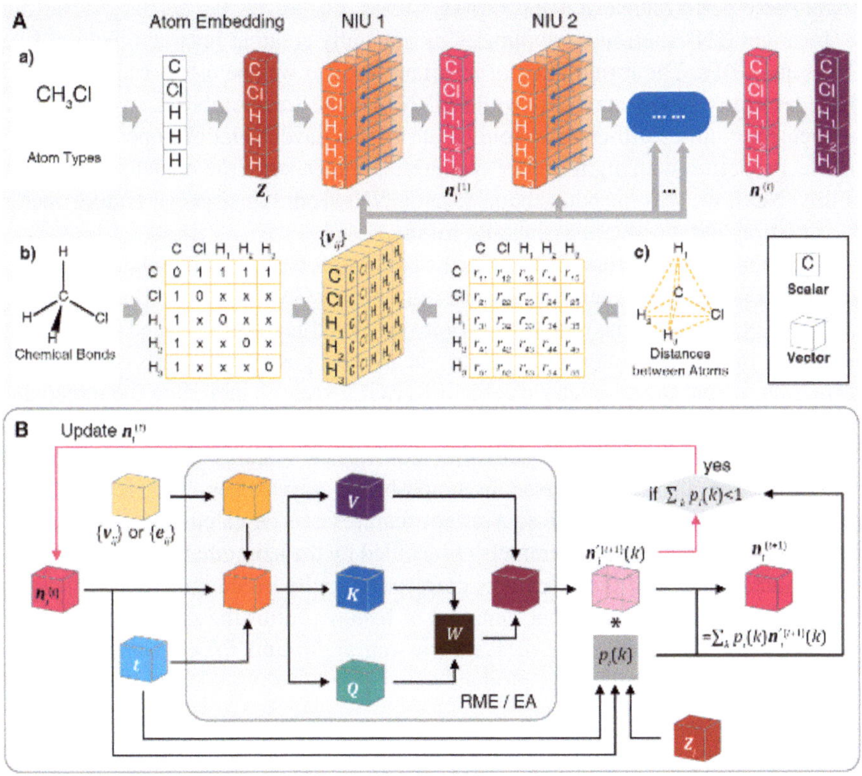

Fig. 2 Illustration of MolCT with (**a**) an architectural overview, (**b**) the ego-attention mechanism and the neural interaction unit. Adapted with permission from *arXiv:2012.11816v3*. Copyright 2023 author(s), licensed under a CC-BY-NC-SA 4.0 License

Networks based on RBFs formed the foundation model for graph-based molecular modelling. Some of the following works focus on the development of more effective RBFs. For example, MolCT (Zhang et al. 2023a) used log-Gaussian RBFs instead of Gaussian RBFs, which better aligns with the physical scenario where long-range interactions decay with the inverse of distance to the power of n. Some other works (Gasteiger et al. 2020; Gasteiger et al. 2021; Gastegger et al. 2021) introduced additional information and used other basis functions to vectorise it. For DimeNet (Gasteiger et al. 2020), in addition to distances vectorised by Bessel RBFs, the three-body angle information was also introduced and vectorised by the following Fourier-Bessel sphere basis functions (SBFs):

$$\tilde{a}_{\mathrm{SBF,ln}}(d, \alpha) = \sqrt{\frac{2}{c^3 j_{l+1}^2(z_{ln})}} j_l\left(\frac{z_{ln}}{c} d\right) Y_l^0(\alpha) \tag{7}$$

The message-passing scheme in DimeNet was also redesigned to provide angle information. According to downstream benchmarks, DimeNet could reach the desired expression ability at $N_{\mathrm{SBF}} = 6$ and $N_{\mathrm{RBF}} = 4$, while SchNet had a comparable performance at $N_{\mathrm{RBF}} = 300$.

Invariant GNNs introduce symmetry by manually creating invariant features like RBFs and SBFs. The formulation of invariant features will be more straightforward than descriptor-based methods, while GNN will be transferable between similar molecule systems. Furthermore, equivariant GNNs have been developed to directly predict properties with higher-order symmetries and enhance the model's expression ability. Next, we will introduce equivariant GNNs from two aspects: feature vector construction and message-passing algorithms.

The simplest equivariant feature vector is constructed using the relative position vector $\mathbf{r}_{ij} = \mathbf{r}_i - \mathbf{r}_j$ between atoms. For example, in EGNN and GFN (Satorras et al. 2022; Xu et al. 2022), the relative position vector is directly introduced as the input of GNNs. In VisNet (Wang et al. 2024b), the relative position vector is combined to form a set of runtime geometry calculation (RGC) vectors, including the molecular system's distance, angle, dihedral, and improper angle information.

Similar to descriptor-based networks, equivariant features based on spherical harmonics are also widely used in graph-based approaches. With the relative coordinates of atoms, a set of equivariant features can be calculated by spherical harmonic functions, and the symmetry is decided by the azimuthal quantum number l. The vector spaces of those features correspond to the irreducible representations of the SO(3) group, and the concatenation of features forms the input to GNNs.

Next, we will introduce the message-passing algorithms for equivariant GNNs. The first one will be the "tensor product":

$$\begin{bmatrix} x_1 \\ x_2 \\ x_3 \end{bmatrix} \otimes \begin{bmatrix} y_1 \\ y_2 \\ y_3 \end{bmatrix} = \begin{bmatrix} x_1 y_1 & x_1 y_2 & x_1 y_3 \\ x_2 y_1 & x_2 y_2 & x_2 y_3 \\ x_3 y_1 & x_3 y_2 & x_3 y_3 \end{bmatrix} \tag{8}$$

From the above formula, the dimension of the updated feature vector is the product of the input vectors, which will cause a dimension explosion if applied directly to the next iteration of the network. However, it is possible to decompose the vector into a direct sum of lower-dimensional SO(3) equivariant vectors, and the decomposition coefficients are Clebsch-Gordan coefficients or Wigner 3-j symbols (Khersonskii et al. 1988).

Networks based on irreducible representations include MACE, NequIP, Allegro (Batatia et al. 2022; Batzner et al. 2022; Musaelian et al. 2023), etc. Those networks have powerful expressive capabilities, which can be used to predict force or dipole moment and properties with higher-order symmetries. However, due to the high-computational complexity of tensor product and decomposition, the inference and training of such networks will be significantly slow, limiting their application in large molecule systems. Therefore, some works driven by the pursuit of efficiency have developed various new message-passing algorithms. Upon examining the commonalities among these networks, it can be observed that their equivariant operations can be classified into two types: *inner product* and *scaling*:

$$\text{Inner product}: \begin{bmatrix} x_1 \\ x_2 \\ x_3 \end{bmatrix} \cdot \begin{bmatrix} x_1 \\ x_2 \\ x_3 \end{bmatrix} = x_1^2 + x_2^2 + x_3^2 \tag{9}$$

$$\text{Scaling}: \sigma \begin{bmatrix} x_1 \\ x_2 \\ x_3 \end{bmatrix} = \begin{bmatrix} \sigma x_1 \\ \sigma x_2 \\ \sigma x_3 \end{bmatrix}$$

Networks that use the above equivariant operations include EGNN, GFN, SO3krates (Satorras et al. 2022; Xu et al. 2022; Frank et al. 2024), etc. GNNs adopt tensor products with an optional inner product and scaling operations, like the RMSNorm operation in Allegro (Musaelian et al. 2023). These operations' different selections and combinations represent a trade-off between speed and accuracy, which need to be analysed and examined based on the target molecular systems.

Graph-based models have achieved great success across various molecular systems (Wang et al. 2024a; Musaelian et al. 2023; Yuan et al. 2024). Compared to descriptor-based methods such as BPNN and DPMD (Behler and Parrinello 2007; Zhang et al. 2018a), the message-passing algorithm is a more comprehensive description of atomic environments. On the one hand, during iterations of message passing, atoms in molecular systems can receive information beyond the cutoff radius, allowing the model to capture long-range interactions. On the other hand, message passing enables the centre atom to implicitly capture multi-body information, with the order of multi-body interactions being controlled by adjusting the number of iterations. In contrast, BPNN and DPNN input the descriptors of all neighbour atoms directly into MLPs, which may lead to overly complex models and cause overfitting.

2.1.3 Learning Algorithms

After "digitalising" molecular objects using various modelling algorithms, the next step is selecting an appropriate learning algorithm to be applied to the MD process. In MD, diverse challenges arise across multiple stages, including force field preparation, sampling, thermodynamic and kinetic analysis, etc. Over the past few decades, ML algorithms have been broadly categorised into three types: supervised learning, which involves fitting a deterministic mapping based on labelled data; unsupervised learning, which focuses on summarising distributions, representations, and projective operators from unlabelled data; and reinforcement learning (RL), where policies are actively updated through system–environment interactions and on-the-fly data generation. With the vast array of ML algorithms available, careful consideration is required to match the most suitable algorithm to each specific problem in MD. In Sections 2.1.4–2.1.6, we concisely review the concepts behind standard learning algorithms and outline their potential applications in MD simulations.

2.1.4 Supervised Learning

As universal functional approximators (Hornik et al. 1989), the most direct and widely employed applications of machine learning algorithms involve constructing predictive mappings from input variables $\{x_i\}$ to target labels $\{y_i\}$. The significant success of supervised ML in solving classification and regression problems presents new opportunities for revolutionary changes at the algorithmic level in traditional MD research paradigms (Deng et al. 2009; Krizhevsky et al. 2012). In general, supervised learning algorithms learn a parameterised map $f_\theta(\cdot)$ with a supervised loss on a labelled dataset $\{(x_i, y_i)\}$ during the training process.

$$\mathcal{L}(\theta) = \frac{1}{N}\sum_{i=1}^{N} \ell\left(f_\theta\left(x_i\right), y_i\right) \tag{10}$$

The loss function ℓ represents the discrepancy between predicted values and the ground truth. For regression problems, ℓ is commonly defined as mean squared error (MSE) or mean absolute error (MAE) (Goodfellow 2016). For classification problems, where y_i corresponds to categorical or boolean labels, supervised algorithms often treat the task as a logistic regression problem (Kleinbaum et al. 2002). In this case, f_θ typically predicts "logits", which can be converted into categorical distributions through logistic or SoftMax functions. Consequently, ℓ is frequently defined as the cross-entropy between the predicted distributions and the ground truth categories. Here, we provide insights and overviews on how supervised learning can assist MD simulations in addressing challenges that are struggling to be solved by traditional algorithms.

Breaking the Boundaries of Speed and Accuracy with Machine Learning Force Fields

Classical MD simulations, based on the Born-Oppenheimer approximation and classical Hamiltonian, can simulate microscopic events at extensive timescales (ns $\sim\mu$ s) (Frenkel and Smit 2023). However, they employed an empirical force field with a simplistic functional form based on manually defined terms such as bonds, angles, and dihedrals (Wang et al. 2004; Rappé et al. 1992). This limits their ability to simulate complex chemical processes incorporating electronic degrees of freedom, such as chemical reactions involving bond formation and breakage. On the other hand, ab initio molecular dynamics (AIMD) considers the electronic freedom in each atomic configuration (Tuckerman et al. 1996; Marx and Hutter 2000). However, despite its high accuracy, it requires extensive computation for every simulation step to calculate single-point energies, thus hindering its application in large-scale simulations of real-world systems.

This dilemma is frequently encountered in physics-based methods: a trade-off between speed and accuracy. ML algorithms offer a promising solution to advance this Pareto front. Surrogate neural network models are constructed for energy and force estimation ($\mathbf{F}_\theta(\mathbf{x}) = -\nabla_{\mathbf{x}} U_\theta(\mathbf{x})$, where $\mathbf{x}$ represents atomic positions), and potential energy surfaces are fitted with matching energies and forces from physical oracles with quantum-level accuracy. This is done by minimising the following MSE loss with respect to a dataset $\{\mathbf{x}_i, U(\mathbf{x}_i), \mathbf{F}(\mathbf{x}_i)\}$:

$$\mathcal{L}(\theta) = \frac{1}{N}\sum_{i=1}^{N}\lambda_1 |U(\mathbf{x}_i) - U_\theta(\mathbf{x}_i)|^2 + \lambda_2 \|\mathbf{F}(\mathbf{x}_i) - \mathbf{F}_\theta(\mathbf{x}_i)\|_2^2 \tag{11}$$

The first term, $|U(\mathbf{x}_i) - U_\theta(\mathbf{x}_i)|^2$, aims to minimise the difference between the predicted energy and the labels, commonly referred to as "energy matching". The second term, $\|\mathbf{F}(\mathbf{x}_i) - \mathbf{F}_\theta(\mathbf{x}_i)\|_2^2$ focuses on aligning the predicted forces with the labels, which is known as "force matching".

With this approach, the simulation can proceed under the machine learning force field (MLFF) $\mathbf{F}_\theta$, which is assumed to possess ab initio accuracy and also offers improved simulation efficiency on modern AI-compatible hardware. This strategy has proven effective in simulations involving chemical species and processes with quantum effects, such as ions, metals, heterogeneous phase boundaries, chemical reactions, and even protein dynamics (Wang et al. 2024a; Yuan et al. 2024; Niu et al. 2020; Zhang et al. 2021a, 2024).

Furthermore, MD simulations serve as physical models that need to reflect the real world accurately. However, in classical MD simulations, only a few parameters of empirical force fields are optimised to match a limited amount of experimental data (Wang et al. 2004; Rappé et al. 1992). Although this approach demonstrates considerable generalisability, these parameters may be biased and inaccurate when used to acquire high-accuracy thermodynamic properties. For instance, inaccuracies in the force field could introduce systematic errors in calculating the free energies of protein–ligand binding, where results are expected to match the chemical precision of about 1 kcal/mol (Shirts 2012; Shirts and Mobley 2013).

Before the era of ML, experimental information was commonly incorporated into MD simulations using the maximum entropy principle (Cesari et al. 2018). This approach has been practical in scenarios where experimental information could be transformed into configuration constraints. However, it lacked general solutions for arbitrary experimental data, which limited its application in many valuable scenarios, such as determining thermodynamic properties like thermal stability and binding affinity. In recent years, researchers have discovered that supervising MLFFs with experimental feedback such as free energies can significantly enhance the accuracy of MD simulations (Thaler and Zavadlav 2021; Wang et al. 2023).

A detailed discussion on MLFFs can be found in Sect. 3.

Constructing Potential of Mean Force of Reaction Coordinates

The evolution of complex physical, chemical, and biological systems often exhibits characteristics of separation between fast and slow timescales. By the Mori-Zwanzig reduction, from the complete configuration $\mathbf{x}$, a free energy profile, known as the potential of mean force (PMF) $F(\mathbf{q})$ can be integrated (Zwanzig 2001; Tuckerman 2023). Here, $\mathbf{q} = \mathbf{q}(\mathbf{x})$ represents the most significant, and typically the slowest, degrees of freedom in the dynamics of $\mathbf{x}$, commonly referred to as reaction coordinates. According to statistical physics, PMF is defined formally as

$$F(\mathbf{q}) = \frac{\int d\mathbf{x} \exp(-\beta U(\mathbf{x}))\delta(\mathbf{q}(\mathbf{x}) - \mathbf{q})}{\int d\mathbf{x} \exp(-\beta U(\mathbf{x}))} \tag{12}$$

As the fluctuations of the fast-moving degrees of freedom are averaged out, $F(\mathbf{q})$ presents a smoother free energy landscape, thus enabling coarse-grained simulations to be conducted in a reduced space (the space of $\mathbf{q}$) with Langevin dynamics or Monte-Carlo sampling. This sampling strategy significantly reduces computational demand, extending MD simulations to longer timescales. However, deriving PMF from Eq. 12 requires full thermodynamic integration over the space of $\mathbf{x}$. Therefore, researchers seek the generalisable capabilities of AI, using few-shot learning on "landmark" samples $\{\mathbf{q}_i, F(\mathbf{q}_i), \nabla_{\mathbf{q}} F(\mathbf{q}_i)\}$ drawn from a predefined distribution (Schneider et al. 2017; Wang et al. 2019). A coarse-grained potential can be constructed by minimising the force matching loss:

$$\mathcal{L}(\theta) = \frac{1}{N}\sum_{i=1}^{N} \left\| \nabla_{\mathbf{q}} F(\mathbf{q}_i) - \nabla_{\mathbf{q}} F_\theta(\mathbf{q}_i) \right\|_2^2 \tag{13}$$

The accuracy of this surrogate PMF F_θ heavily depends on the sampling strategy of these "landmark" samples, which intuitively should be representative enough to allow ML algorithms to smoothly interpolate between all areas of interest in $\mathbf{q}$ space.

2.1.5 Unsupervised Learning

Although many problems in MD simulations can be modelled as "constructing a deterministic mapping" problems, there are significant challenges that well-defined "labels" are not available. Issues such as sampling inefficiency and the difficulty of extracting useful information from MD simulations cannot be directly addressed by supervised learning algorithms. Instead, researchers turn to unsupervised learning algorithms for solutions.

Generative Modelling of Molecular Systems with Statistical Physics Principles In statistical physics, MD simulations function as a "sampler" of the Boltzmann distribution based on the assumption of ergodicity (Frenkel and Smit 2023). However, two consecutive samples generated by MD simulations are correlated at a timescale of approximately femtoseconds (fs), corresponding to the bonds' vibration. On the other hand, meaningful events in complex systems (from the flipping of dihedral angles in a small molecule to large-scale mechanical movements in proteins or even phase transitions) occur at timescales much more extensive than fs due to high free energy barriers. To efficiently sample these rare events, enhanced sampling techniques such as meta-dynamics (Laio and Parrinello 2002; Barducci et al. 2008) and tempering-based algorithms (Gao 2008b; Yang and Qin Gao 2009) are adopted. While effective in some scenarios, they heavily rely on prior knowledge specific to the studied system. Researchers' ultimate goal is to develop statistically "independent" samplers of the Boltzmann distributions, where system configurations can be independently and identically sampled as long as the Hamiltonian of the system is defined.

In unsupervised learning, generative learning is a discipline dealing with distributions. Generally, generative algorithms are developed to fit and sample a distribution $p(x)$ through either an explicitly or implicitly modelled distribution $p_\theta(x) \approx p(x)$ with data $\{x_i\}$, where $x_i \sim p(x)$. With sufficient training, these algorithms can be considered independent samplers capable of generating samples from $p(x)$ in a rare-event-free manner. In a nutshell, to match the parameterised distribution $p_\theta(x)$ with the target distribution $p(x)$, which is difficult to sample. Generative learning has developed strategies such as variational autoencoders (VAEs) (Kingma 2013), generative adversarial networks (GANs) (Goodfellow et al. 2020; Arjovsky et al. 2017), and normalising flows (Rezende and Mohamed 2015; Chen et al. 2018) to efficiently minimise the divergence $D(p_\theta, p)$ between p_θ and p. $D(p_\theta, p)$ is commonly defined as a computationally feasible and sampling-efficient measure of distributional distance, including Kullback-Leibler (KL) divergence, Jensen-Shannon (JS) divergence (Cover 1999), and Wasserstein distances (Peyré and Cuturi 2019).

In fact, many generative learning algorithms are rooted in statistical physics principles. For example, energy-based models (EBMs) (Du and Mordatch 2019) mimic the equilibrium Boltzmann distribution with a parameterised energy function $p_\theta(x) \propto \exp(-U_\theta(x))$. Another example is diffusion models (Ho et al. 2020; Song and Ermon 2019; Song et al. 2020), which construct non-equilibrium stochastic differential equations (SDEs) to mimic the behaviour of particles moving according to annealed Langevin dynamics. Simply put, DMs utilise a non-equilibrium forward SDE to introduce noise into the data,

$$\mathrm{d}\mathbf{x}_t = \mathbf{f}(\mathbf{x}_t, t)\,\mathrm{d}t + \sigma_t \mathrm{d}\mathbf{w}_t \tag{14}$$

and learn a reverse SDE that is driven by a parameterised score function $\mathbf{s}_\theta(\mathbf{x}_t, t)$ (which can be conceptualised as a time-dependent force field),

$$\mathrm{d}\mathbf{x}_t = \left[\mathbf{f}(\mathbf{x}_t, t) - \sigma_t^2 \mathbf{s}_\theta(\mathbf{x}_t, t)\right]\mathrm{d}t + \sigma_t \mathrm{d}\overline{\mathbf{w}}_t \tag{15}$$

which aims to recover the data distribution from white noise. Through noise augmentation of samples and smoothing of the data distribution, DMs become powerful density learners, particularly in data-abundant regions.

Therefore, the straightforward application of generative learning in enhancing sampling of a Boltzmann distribution has demonstrated its power in predicting and sampling conformations of organic molecules, proteins, and protein-ligand complexes, serving as an alternative to computationally intensive MD simulations (Xu et al. 2022; Noé et al. 2019; Corso et al. 2023; Zheng et al. 2024). (A detailed discussion on enhanced sampling can be found in Sect. 4.) On the other hand, beyond modelling atomic-scale distributions, generative learning is employed to construct density in a projected space, which corresponds to the PMF in the study of coarse-graining (Arts et al. 2023). More interestingly, a more challenging problem, inverse coarse-graining (Zhang et al. 2023b), which seeks to reverse the coarse-graining process, is inherently a generative problem. Often, when long-timescale MD simulations are conducted in a reduced space of reaction coordinates $\mathbf{q} = \mathbf{q}(\mathbf{x})$, where $\dim(\mathbf{q}) < \dim(\mathbf{x})$, researchers aim to retrieve the fine-grained configuration $\mathbf{x}$ from the coarse-grained configuration $\mathbf{q}$ to obtain a finer structural understanding. For example, they might recover atomic-scale protein structures from their backbones or atomic-scale polymer structures from their units. In fact, to ensure consistency in statistical physics, the distribution of $\mathbf{x}$ given $\mathbf{q}$ corresponds to a conditional Boltzmann distribution,

$$p(\mathbf{x}|\mathbf{q}) = \frac{\exp(-\beta U(\mathbf{x}))\delta(\mathbf{q}(\mathbf{x}) - \mathbf{q})}{\int \exp(-\beta U(\mathbf{x}))\delta(\mathbf{q}(\mathbf{x}) - \mathbf{q})\mathrm{d}\mathbf{x}} \tag{16}$$

Given a single $\mathbf{q}$, many configurations are compatible with $\mathbf{q}$, making this problem naturally generative. Therefore, generative AI holds promise for solving coarse-graining and inverse problems simultaneously and consistently.

Dimension Reduction and Clustering on Free Energy Surfaces

A trajectory of MD simulations, $\{\mathbf{x}_t\}$ represents systems in high-dimensional space across both time and space. Utilising MD simulations to predict thermodynamic properties often treats the MD as a simple sampler of the Boltzmann distribution. However, rich kinetic dynamic information is embedded within the MD trajectory. For instance, diffusion constants of motion in certain degrees of freedom can be calculated via the time-correlation function (Zwanzig 2001). Although $\{\mathbf{x}_t\}$ lies in a formidable high-dimensional space, it contains redundant and irrelevant fluctuations of fast-moving degrees of freedom. For complex systems where fast and slow dynamics are separated, long-timescale correlations typically lie in a relatively low-dimensional space $\mathbf{q} = \mathbf{q}(\mathbf{x})$, which is also known as CVs (Yang et al. 2019). For example, $(\mathbf{q}, \dot{\mathbf{q}})$ might encompass the collective motion of several atoms in chemical reactions, important normal modes of protein dynamics, or transitions between metastable states. Therefore, clustering structures in $\{\mathbf{x}_t\}$ and constructing a dimensional reduction map $\mathbf{q} = \mathbf{q}(\mathbf{x})$ are essential for distilling kinetic models from MD trajectories.

Traditional methods such as linear discriminant analysis (LDA) (Balakrishnama and Ganapathiraju 1998; Xanthopoulos et al. 2013), time-lagged independent component analysis (TICA) (Molgedey and Schuster 1994; Pérez-Hernández et al. 2013), and Markov state models (MSMs) (Bowman et al. 2013; Chodera and Noé 2014) rely on manually specified CVs, metastable structures, or distance metrics, often based on linear or near-linear assumptions. As a result, they struggle to accurately describe projection mappings that are highly nonlinear and have limited generalisability. On the other hand, ML algorithms excel at handling complex nonlinear problems, and techniques such as dimensionality reduction, clustering, and representation learning are well-established and widely used within the ML community (Goodfellow 2016; Bishop and Nasrabadi 2006). Integrating these ideas with principles of physics can aid in extracting essential information from MD simulations. Parameterised methodologies, including auto-encoding, time-lagged analysis, and contrastive learning, have proven effective in extracting nonlinear CVs from high-dimensional configurations (Bonati et al. 2020; Trizio and Parrinello 2021; Chen et al. 2019; Wehmeyer and Noé 2018; Hernández et al. 2018; Zhang et al. 2019a). These CVs improve the system's physical understanding and feedback into MD simulations, boosting sampling efficiency through enhanced sampling techniques. (A detailed discussion of AI-enhanced CV methods can be found in Sect. 4.1.)

Unsupervised Pretraining for Molecular Modelling

In molecular sciences, labelled data, such as high-precision energy, forces, and physicochemical properties of molecules, are computationally or experimentally expensive to obtain. The limited data may not be sufficient to support accurate and generalisable supervised learning training. Therefore, efficient training strategies and computationally friendly molecular representations become crucial, as they can significantly reduce the amount of labelled data required for training. In recent years, the unsupervised pretraining-supervised fine-tuning scheme has become the mainstream in the fields of natural language processing and computer vision (OpenAI et al. 2024; Devlin et al. 2019; Radford et al. 2021), which has also inspired chemists to apply this paradigm to the training of molecular models. From the perspective of network architecture, a molecular model is composed of two parts: the predictor f and the encoder g:

$$\{\mathbf{h}_i\} = g\left(\{\mathbf{R}_i\}, \{Z_{i.}\}\right)$$

$$E = f\left(\{\mathbf{h}_i\}\right) \tag{17}$$

While the encoder takes the system's configuration $\{\mathbf{R}_i\}$, atom types $\{Z_i\}$ and sometimes other information as input and outputs the learnable atom representation vector $\{\mathbf{h}_i\}$. The predictor takes $\{\mathbf{h}_i\}$ as input and gives the prediction. In the self-supervised learning paradigm, the encoder will be pre-trained using a large dataset of configurations generated through MD simulations. The predictor is then added and fine-tuned using a small amount of data labelled by QM calculations or wet experiments.

Currently, pretraining strategies can be broadly categorised into three types: masked modelling, contrastive learning, and denoising score matching. The masked modelling approach is derived from BERT (Devlin et al. 2019), with the core idea being that part of the information from the input 3D molecular graph is masked, and then the masked part will be predicted using the encoder and another predictor (note that the predictor is not the force/energy predictor f). The masked information can include atom types, coordinates, and interatomic distances (Fang et al. 2022; Cui et al. 2024).

The contrastive learning approach is derived from the field of image processing (Radford et al. 2021; Chen et al. 2020), with the training objective of maximising the mutual information between positive sample pairs and minimising it between other sample pairs (i.e. negative sample pairs). The key point of contrastive learning lies in constructing positive sample pairs. Current methods include using the 3D molecular graph and its 2D representation as a positive pair (Liu et al. 2021; Stärk et al. 2021), or comparing the 3D graph with its noised or partly masked version (Cui et al. 2024), etc. The denoising score-matching approach is similar to the training of diffusion models. This method involves adding noise to the coordinates to get a noisy configuration, which is fed into the encoder. The network is subsequently trained to predict the direction and intensity of the noise, allowing the noisy configuration to be restored to the original one. The noise fitting process has a mathematical interpretation similar to that of training force fields (Zaidi et al. 2022). In most works, Gaussian noise is added to the configurations, while subsequent work has refined the form of the noise to incorporate physics-guided information (Feng et al. 2023; Jiao et al. 2023).

2.1.6 Reinforcement Learning

Supervised learning relies on labelled data, while most unsupervised learning algorithms depend on massive amounts of unlabelled data to achieve good performance. However, for many topics in MD simulations, the goal is to generate high-quality data. For instance, if we attempt to train a surrogate sampler using generative learning to replace MD in sampling the Boltzmann distribution, numerous data points that already follow the Boltzmann distribution are required, which presents a paradox. This observation inspires us to recognise that many applications of MD require a more flexible, adaptive, and exploratory algorithm—one that can generate its own data to learn and bootstrap itself (Liu et al. 2024).

RL is a type of machine learning algorithm where an agent or parameterised policy π_θ learns to take action by interacting with a predefined environment (Sutton and Barto 2018). Unlike other learning algorithms, RL does not explicitly rely on labelled or unlabelled data corpus but instead learns from a series of consequences of its actions, aiming to maximise cumulative rewards R_t over time. The policy explores different strategies and adjusts its actions based on feedback from the environment, which can be rewards or penalties. Therefore, RL is particularly flexible as a "data-free" learning strategy since it learns from self-generated trajectories and as a "gradient-free" learning strategy since the reward function does not necessarily need to be differentiable. Successful applications of RL in domains such as game playing and chess indicate that it is a powerful tool for solving complex problems where the searchable solution space is nearly infinitely large. Applying RL algorithms in MD simulations can be a promising solution for many challenging issues, provided that the states, actions, rewards, and policies are meticulously designed.

In fact, the potential applications of RL in MD are broader than we might imagine, as many problems in MD involve a "chicken or the egg" causality dilemma. Besides the previously discussed example of generative modelling of the Boltzmann distribution, another challenge arises in CV-based enhanced sampling. A good CV representing the system's slow degrees of freedom is required for efficient sampling. However, a comprehensive system sampling is also necessary to determine an effective CV. It is widely believed that this dilemma may be resolved within the framework of RL. Actually, widely used enhanced sampling algorithms such as meta-dynamics (Laio and Parrinello 2002; Barducci et al. 2008) already incorporate basic concepts of RL and active learning. In meta-dynamics, a repulsive bias V_{bias} is gradually added to the potential energy function U, centred at the current configuration $\mathbf{x}_t$. This repulsive bias encourages the system to escape local minima for broader exploration. From this point of view, meta-dynamics can be considered a naïve form of RL, as its "policy" (bias potential) is adaptively adjusted based on the self-sampled configurations.

3 Machine Learning Force Fields

In this section, we will focus on MLFFs. The investigation of MLFFs roughly falls into two parts: molecular modelling and training approaches. The former part has been introduced in the first section at length so that the training approaches will be the primary content of this part. Generally, the training methods of MLFFs can be divided into supervised QM-based methods and experimental data-driven methods.

3.1 QM-Based Supervised Learning

QM-based training approaches are the most commonly used in the field of MLFFs. Firstly, MD simulations are employed to sample the conformational space of molecular systems, thereby enabling researchers to generate the inputs for prepared MLFFs rapidly. After a large amount of conformational data has been generated, ab initio calculations will be used to compute the accurate energy of the given conformation, which will be used as the label. However, in more recent cases, we will calculate the QM forces of atoms as another label, even discarding the energy label. This is because using force matching (formally defined in Sect. 2.1.4) as the target accounts for higher fitting accuracy and better performance (Christensen and von Lilienfeld 2020), although the cost of computing force labels is much higher than energy. Sometimes, we will train MLFFs on large molecule systems, and it is impossible to calculate the energy and forces of the whole system using QM methods. In these cases, we must divide the interactions into short- and long-range ones. Usually, MLFFs will be used to deal with short-range many-body interactions, and long-range interactions are often regarded as classical Coulomb interactions, which PME (Particle-Mesh Ewald) or other algorithms will accelerate. Furthermore, after the training loss converges, the network may still have large prediction deviations for some conformations, which might cause the collapse of MD simulations. Returning to the first step, supplementing the data, and then continuing training will be necessary. Until the network can reach the desired force/energy prediction accuracy for the conformation space we are interested in and the MD simulation can run stably, the training process of MLFF is completed.

3.2 Experimental Data-Driven Learning

Unlike the previous two bottom-up approaches, MLFF training based on experimental data allows the network to learn continuously and progressively align with the experimentally observed results through parameter updates during the MD simulation. The observable $\langle O \rangle$ to be computed is a function of the entire MD trajectory (Thaler and Zavadlav 2021):

$$\langle O \rangle = \sum_{n=1}^{N} w_n(\theta)\, O(\mathbf{S}_n) \tag{18}$$

where $\{\mathbf{S}_n\}_{n=1,\ldots,T}$ represents the sub-trajectory sampled from the entire MD trajectory, θ denotes the network parameters, and the weight factor w_n equals $1/N$ for unbiased simulations, while in enhanced sampling, it corresponds to the respective reweighting factors. Note that the state $\mathbf{S}_T$ is linked to the previous states $\mathbf{S}_t(t < T)$ through an ordinary differential equation (ODE). Therefore, when calculating the derivative of the network parameters θ with respect to the loss function $L(\theta)$, differentiation over the entire trajectory is required to perform complete backpropagation and obtain the updated values for θ. Another key point is that when using this approach for force field parameter optimisation, the MD propagator must be differentiable, which means that irreversible MD simulations, such as Langevin dynamics, cannot be applied with this technique.

The above optimisation approach requires differentiation over the entire trajectory, which leads to high memory cost and makes it prone to gradient explosion. A reweighting-based learning approach has been proposed to overcome these drawbacks (Thaler and Zavadlav 2021; Wang et al. 2023). In this approach, MD simulation is first performed under the canonical ensemble based on the potential energy surface (PES) defined by the initial parameter $\hat{\theta}$, generating a series of reference trajectories. Subsequently, the reweighting parameter and the observables under the current force field parameter θ can be calculated using the following formula:

$$w_n = \frac{e^{-\beta\left(U_\theta(\mathbf{S}_n)-U_{\hat{\theta}}(\mathbf{S}_n)\right)}}{\sum_{j=1}^{N} e^{-\beta\left(U_\theta(\mathbf{S}_j)-U_{\hat{\theta}}(\mathbf{S}_j)\right)}} \tag{19}$$

In this case, the observable $\langle O(U_\theta)\rangle$ is still calculated through Eq. 18, but $O(\mathbf{S}_n)$ is no longer a function of θ. Therefore, only w needs to be differentiated, and the differentiation over the entire trajectory by chain rule will be omitted, leading to a significant reduction in computational cost.

The reweighting-based approach can only optimise MLFFs using the thermodynamic properties of the system but is turned off to be applied to optimise the system's dynamic properties (such as spectra), as differentiating dynamic quantities with respect to parameters inevitably requires backpropagation through the entire trajectory. To address this issue, Han and Yu (2024) proposed a trajectory differentiation method based on the adjoint method from Neural ODEs (Chen et al. 2018). Specifically, to calculate the dynamic property, we first generate a series of reference states $\{\mathbf{z}_n(0)\}$ through MD simulations under NVT or NPT ensemble, followed by evolution using an NVE integrator on these reference states to obtain the dynamic observable $O(\mathbf{z}_n)$ for each reference state. Unlike the previous reweighting method, $O(\mathbf{z}_n)$ also contains the gradient with respect to θ, which can be calculated using the adjoint method. First, we define the adjoint variable:

$$\mathbf{a}(t_i) = \left.\frac{\partial L}{\partial \mathbf{z}(t_i)}\right|_{z(t_j<t_i),\theta} \tag{20}$$

The reverse dynamics of **a** can be solved using the chain rule:

$$\mathbf{a}(t_{i-1}) = \mathbf{a}(t_i)\frac{\partial f(\mathbf{z}(t_{i-1},\theta))}{\partial \mathbf{z}(t_{i-1})} + \frac{dL}{d\mathbf{z}(t_{i-1})} \tag{21}$$

Due to the time reversibility of f, the variable $\frac{\partial f(\mathbf{z}(t_{i-1},\theta))}{\partial \mathbf{z}(t_{i-1})}$ can be computed using only $\mathbf{z}(t_i)$. Using the adjoint trajectory, the total derivative of the parameters with respect to the loss function can be obtained:

$$\frac{dL}{d\theta} = \sum_{i=K}^{1} \mathbf{a}(t_i) \frac{\partial f(\mathbf{z}(t_{i-1}), \theta)}{\partial \theta} \tag{22}$$

The computational cost by an adjoint method is comparable to regular MD simulations, making MLFF optimisation based on dynamic properties feasible.

4 Artificial Intelligence for Enhanced Sampling

Sampling of the conformational space is a challenging problem in MD simulations. As the landscape of the potential energy surface, which determines the system's dynamics, is rugged, the system is prone to getting trapped in local minima. The energy barrier between these minima leads to transitions between different conformations of rare events. Besides, the reachable timescales of the simulations are limited by the computational resources, which are always much shorter than the timescales of the critical chemical and biological processes of interest (Yang et al. 2019).

Enhanced sampling methods have been developed to overcome these limitations. Focus on the most high-profile regions of the conformational space. These methods guide the sampling process in spending more time in the regions related to the transitions of interest. Generally, enhanced sampling can be achieved by modifying the potential energy surface, increasing the temperature, or driving an adiabatically decoupled degree of freedom. Many enhanced sampling methods have been developed, which can be classified into two categories based on whether they are related to CVs, including CV-based methods and CV-free methods. CV-based methods use CVs to project the high-dimensional conformational space to a lower-dimensional space to accelerate the sampling process. Famous CV-based methods include umbrella sampling (Torrie and Valleau 1977), meta-dynamics (Laio and Parrinello 2002; Valsson et al. 2016), steered MD (Isralewitz et al. 2001), and conformational flooding (Grubmüller 1995). On the other hand, the CV-free methods do not require the definition of CVs but attempt to introduce conformations from the ensemble with a new temperature, pressure or even Hamiltonian, or smooth the whole potential energy surface. Replica exchange MD (Sugita and Okamoto 1999), Wang-Landau algorithm (Wang and Landau 2001), accelerated MD (Hamelberg et al. 2004), and integrated tempering sampling (Gao 2008a; Yang and Qin Gao 2009) are some widely used CV-free methods. The above methods are shown to be successful in improving sampling efficiency. However, selecting CVs is not trivial and often requires professional domain knowledge. Besides, most CV-based methods cannot deal with high-dimensional CVs, which limits their applications. CV-free methods always need less human intervention and prior knowledge, although they could not always achieve high efficiency.

AI techniques bring new opportunities to the field of enhanced sampling. Machine learning society has developed a large number of algorithms to deal with high-dimensional data, including dimensionality reduction, clustering and even direct distribution sampling driven by generative models (van der Maaten and Hinton 2008; Pedregosa et al. 2011; Anukrishna and Paul 2017). The field of enhanced sampling has benefited a lot from these techniques. Dimensional reduction techniques have been used to construct CVs automatically, while RL and other novel AI algorithms have strongly promoted CV-free methods. Generative models boost the efficiency of the sampling process by regarding MD as a part of a larger framework rather than the sampling framework itself. The combination of AI and enhanced sampling has proven to be a powerful tool for studying complex systems, including proteins, nucleic acids, and materials. This section will introduce the most recent advances in AI-enhanced sampling.

4.1 AI-Enhanced CV-Based Methods

The fundamental problem in the CV-based enhanced sampling method is selecting or constructing proper CVs, which could be used to distinguish different approaches in selecting and biasing them. The construction of CVs can be regarded as a feature extraction problem, in which slow degrees of freedom are extracted from the high-dimensional conformational space. Traditionally, CVs are selected by human experts based on prior knowledge. However, without simulating the rare event of interest, it is challenging for experts to select efficient CVs for a priori unknown systems (Mehdi et al. 2024). From the viewpoint of AI, the system is embedded into a latent space using molecular representation models, and CVs are then selected or constructed. In the first section of this chapter, molecular representation models are thoroughly discussed, so we will not repeat them here. We will focus on the CV construction models' targets and strategies.

Deep-LDA (Bonati et al. 2020) is a deep learning method to construct CVs based on LDA. Unbiased simulations of metastable states are used as input to derive physical descriptors. After nonlinear transformation by a DNN, the LDA method is applied to obtain the direction of maximal separation between the metastable states. The CVs are then constructed ultimately. The Deep-TDA (Trizio and Parrinello 2021) exhibits architectural similarities to Deep-LDA, with a notable distinction in its approach to CV computation. While Deep-LDA incorporates a linear transformation step, Deep-TDA circumvents this intermediary process by directly generating CVs through the neural network's output layer. The discrimination between distinct configurational basins is accomplished through a specialised loss function that constrains the projected data to conform to a predefined probability distribution.

VAMPnets (Mardt et al. 2018) (Variational Approach for Markov Processes neural networks) implements the variational approach for conformation (VAC) principle to autonomously construct Markov State Models (MSMs) from structural features. Although its success has been demonstrated across multiple applications, the architecture of VAMPnets employs a SoftMax function in the output layer for discrete metastable state detection, rendering it unsuitable for constructing continuous CVs required in enhanced sampling methods. To solve this limitation, state-free reversible VAMPnets (SRV) (Chen et al. 2019) approximate the transfer operator's eigenfunctions as nonlinear functions of the input features and avoid the usage of kernel tricks. The SRV framework maps input coordinates to an n-dimensional continuous output space, where n represents a hyperparameter corresponding to the number of slow modes to be learned. In practical implementations, SRV approximates the eigenfunctions through a linear combination $\widetilde{\psi_i}(\mathbf{x}) = \sum_j s_{ij} f_j(\mathbf{x})$, optimising the objective function $\mathcal{L} = \sum_i g\left(\widetilde{\lambda_i}\right)$, where $g\left(\widetilde{\lambda_i}\right)$ denotes any monotonically decreasing function of the machine-learned eigenvalues of the transfer operator.

Based on information bottlenecks, Wehmeyer and Noe proposed the time-lagged autoencoder (deep TAE) for CV construction (Wehmeyer and Noé 2018). By training a pair of encoder and decoder, the method attempts to generate a latent space embedding containing the most essential information in encoder training and reconstruct the input from the latent space for decoder training. The low-dimensional latent representation can be used as CVs for further sampling. In deep TAE, the coordinate at time t is used as input of the encoder and the decoder learned to predict coordinates at a later time $t + \tau$. Pande et al. extended the method and proposed variational dynamics encoder (VDE) (Hernández et al. 2018) by substituting the AE in deep TAE with VAE (Kingma and Welling 2022). In VAEs, the training target is to maximise the evidence lower bound (ELBO) composed of reconstruction loss, measuring how well the model reconstructs the input and KL divergence between the approximate posterior and the defined prior distribution for the latent space. An additional term called autocorrelation loss is used under the VDE framework, which maximises the largest dynamic eigenvalue to encourage the discovery of the slowest process in the input data.

The methods mentioned above mainly focus on the structural similarity of a given conformation set to reduce the dimension of the system. Other metrics, e.g. the time reachability of the potential energy, can also be used for CV construction. Information Distilling of Metastability (IDM) posits an essential hypothesis that hierarchical metastable states in the free energy landscape can be defined according to the temporal resolution of observation (Zhang et al. 2019a). The working flow of IDM is shown in Fig. 3. Given an observation timescale τ_{obs}, a classification neural network $\chi_\theta(\mathbf{x})$ is trained to maximise the mutual information between two conformations within a temporal neighbourhood, $\mathbf{x}_t$, $\mathbf{x}_{t+\delta t}$ ($\delta t < \tau_{\text{obs}}$), which are assumed to belong to the same metastable state under τ_{obs}. The network architecture is intentionally designed to project all configurations into the same information-rich vector space, where learnable cluster centres are optimised and selected. Consequently, IDM simultaneously performs dimensional reduction and clustering tasks end-to-end, enabling the construction of hierarchical kinetic models based on varying resolutions of timescales. The reduced representation built by ML algorithms, particularly DNNs, is highly expressive, requiring no prior knowledge except for the basic assumption of how metastable states are defined.

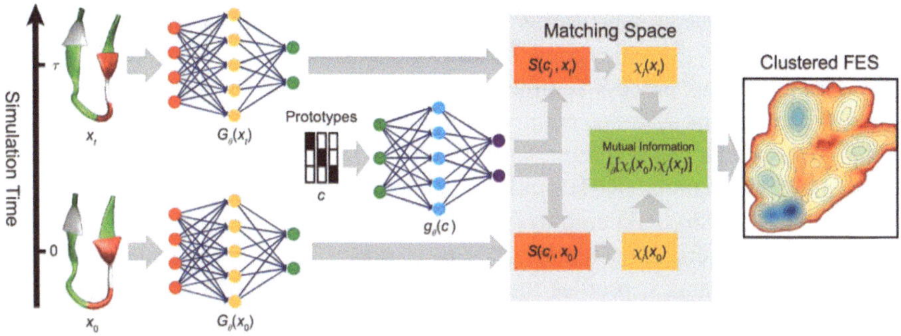

Fig. 3 Illustrative working flow of IDM. High-dimensional input vectors (**x**) and preimages or prototypes of cluster centroids (**c**) are projected into images in the matching space by G_θ and g_θ, respectively. The similarity score between **x** and **c** calculated in the matching space, S(**c**,**x**), leads to output χ(**x**), which is further used for MI (I_β) maximisation. The matching space is the reduced embedding for **x** and χ(**x**) as the soft clustering labels. Reprinted with permission from *J. Phys. Chem. Lett.* 2019, *10*, 5571–5576. Copyright 2019 American Chemical Society

Besides the construction of CVs, AI techniques can also be used in the generation of bias potential and bias force, and it is essentially an MLFF construction problem. Well-known approaches include force-biasing using neural networks (FUNN), deep variational enhanced sampling (DEEP-VES), etc. FUNN (Guo et al. 2018) combines neural networks with adaptive biasing force to accelerate the sampling process, while DEEP-VES (Bonati et al. 2019) uses the variational principle to minimise the sampled data distribution and the chosen target. These methods are efficient in enhancing the sampling process and can be used in various systems.

The construction of CVs or CV-based bias potentials and CV-driven enhanced sampling are closely interrelated techniques. CV and bias potential construction relies on simulation data, which can be efficiently obtained through enhanced sampling methods with the constructed CVs; conversely, improved CVs and bias terms can generate higher-quality simulation data. In fact, this process bears a striking resemblance to the feedback loop mechanism observed in RL. For example, the RL‡ studies the transition states of chemical reactions (Zhang et al. 2021b). A "shooting game" reward function can be formulated for transition state recognition. Specifically, this game aims to maximise the "transition path (TP) probability" probability $p(\mathrm{TP}|\,\mathbf{x})$, which is defined as the probability of successful shooting starting from conformation **x** (always represented by CVs rather than all-atom coordinates). An optimisable sampling policy is defined as a biased Boltzmann distribution $p_\theta \propto \exp\left(-\beta(U + V_\theta)\right)$ to achieve this goal. In the spirit of Actor-Critic learning, commonly adopted in RL, a critic model $p_w(\mathbf{x}) \approx p(\mathrm{TP}|\,\mathbf{x})$ is trained to estimate the expected reward of **x** sampled by the actor (sampling policy). During a training iteration, configurations **x** are generated by p_θ, then trajectories with opposite momentums are shot to evaluate the reward $p(\mathrm{TP}|\,\mathbf{x})$. These samples and their corresponding rewards are used to train the critic model $p_w(\mathbf{x})$, which in turn is used to optimise the policy p_θ. This bootstrap iteration enhances the accuracy of p_w, while increasing the "scores" in the shooting game of the sampling policy. Upon convergence, the policy can efficiently generate transition state-like structures. The entire learning process resembles game playing and is highly flexible and adaptable to various systems, requiring only knowledge of the regions of reactants and products in the phase space.

Reinforced dynamics (RiD) similarly applies RL for bias potential construction (Zhang et al. 2018b). RiD employs a neural network to construct a bias potential $V_\theta(\mathbf{q}(\mathbf{x}))$ that covers all potential CVs $\mathbf{q} = \mathbf{q}(\mathbf{x})$. This bias potential, like that used in meta-dynamics, profiles the local PMF, $V_\theta(\mathbf{q}) \approx - F(\mathbf{q})$ and serves as an exploratory policy attempting to traverse the CV space as comprehensively as possible. Furthermore, it is supervised and trained on the fly based on the structures q collected by the biased MD simulations, together with their corresponding PMF value $F(\mathbf{q}(\mathbf{x}))$, which is calculated via restrained MD. RiD surpasses meta-dynamics in capability and has been successfully applied to complex processes such as protein folding, which involve numerous CVs. However, the profiling of the PMF with V_θ in RiD, it is based on supervised learning, and as such, its accuracy highly depends on the sampling strategy. Therefore, RiD employs sophisticated protocols to estimate the uncertainty of fitting $F(\mathbf{q})$ with $V_\theta(\mathbf{q})$ and attempts to systematically improve the accuracy with extra sampling in regions with low confidence.

4.2 Generative Models for Enhanced Sampling

Learning to sample probability distributions is the realm of generative models. Generative models are trained to generate samples from a target distribution, which is usually intractable, and have achieved remarkable success in the generation of images (Betker et al. n.d.; Esser et al. 2024; Lipman et al. 2023), texts (OpenAI et al. 2024; Grattafiori et al. 2024), audio (Copet et al. 2024; Evans et al. 2024), and even complex molecule conformations (Abramson et al. 2024). The most critical challenge in this field is to match the probability distribution of the generated samples with the target distribution. Different models select different metrics and matching strategies, e.g. Wasserstein GAN (Arjovsky et al. 2017) minimises the Wasserstein distance between the generated samples and the target distribution estimated by the critic network, while normalising flows (Rezende and Mohamed 2015) enables KL divergence minimisation by constructing an invertible and normalisable mapping between the empirical distribution and a given simple prior distribution. In enhanced sampling, generative models can be used to generate samples from the target distribution, which is always the Boltzmann distribution of the system. MD simulations are used to create training data, provide losses, or act as generators with proper bias potentials applied.

Targeted adversarial learning optimised sampling (TALOS) is a strategy to sample from a predefined distribution by solving a min-max game in GAN (Zhang et al. 2019b). Transacting ideas from GANs, TALOS simultaneously trains a discriminator $D_w(\mathbf{q})$, along with a bias potential $V_\theta(\mathbf{x})$, optimising the following min-max problem (Fig. 4),

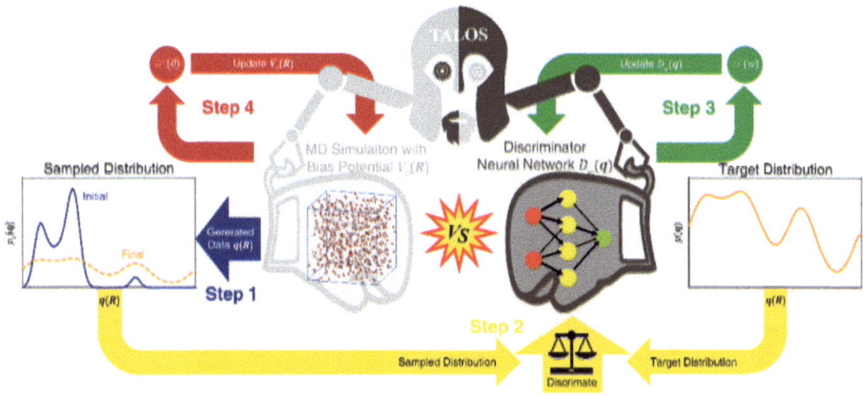

Fig. 4 Illustration and workflow of targeted adversarial learning optimised sampling (TALOS). Target samples are drawn from a target distribution defined on a descriptor space (**q**); meanwhile, a real sampler (MD engine) parallelly generates samples based on the Hamiltonian modified by bias potential $V_\theta(\mathbf{R})$. A virtual discriminator (neural network) D_w is trained to approximate the Wasserstein distance between target and biased samples and feedback to train $V_\theta(\mathbf{R})$. Adapted with permission from *J. Phys. Chem. Lett.***2019**, *10*, 5791–5797. Copyright 2019 American Chemical Society

$$\min_{V_\theta} \max_{D_w \sim \mathbb{L}1} \left[\langle D_w(\mathbf{q}) \rangle_{p(\mathbf{q})} - \langle D_w(\mathbf{q}) \rangle_{p_\theta(\mathbf{q})} \right] \tag{23}$$

$\mathbb{L}1$ is the set of 1-Lipschitz functions, and $p_\theta(\mathbf{q})$ represents the probability distribution of $\mathbf{q}(\mathbf{x})$ where $\mathbf{x}$ is sampled according to the biased Boltzmann distribution $p \propto \exp(-\beta(U + V_\theta))$. In TALOS, the "critic" D_w the network is trained to discriminate between the "fake" samples generated by p_θ and the "real" samples drawn from the target distribution p. Meanwhile, the "actor" V_θ, which is in fact a bias potential applied simulation engine (e.g. MD), is optimised to "fool" D_w by rendering p_θ resemble p. When the Nash equilibrium is reached, $p_\theta \approx p$, and thus, the PMF of $\mathbf{q}$ in biased MD simulations with V_θ aligns with the target distribution p. When p is targeted to be explorative, such as a uniform or tempered distribution commonly adopted in meta-dynamics, TALOS can serve as a general-purpose enhanced sampler.

Since the analytical form of the optimal critic D_w is unknown, thus a neural network with a Lipschitz constraint is used to approximate it following the gradient.

$$\nabla_w L(w) = \langle \nabla_w D_w(\mathbf{q}) \rangle_{p_\theta} - \langle \nabla_w D_w(\mathbf{q}) \rangle_p \tag{24}$$

Once the critic network is fixed, the bias potential is optimised by the gradient as below,

$$\nabla_\theta L(\theta) = \left\langle \left(D_w(\mathbf{q}) - \langle D_w(\mathbf{q}) \rangle_{p_\theta} \right) \beta \nabla_\theta V_\theta(\mathbf{x}) \right\rangle_{p_\theta} \tag{25}$$

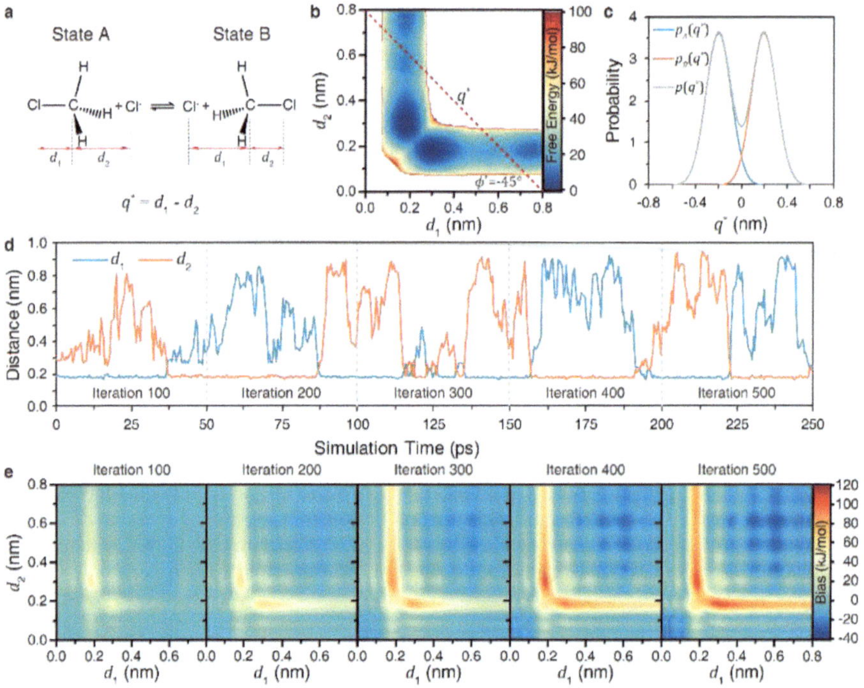

Fig. 5 Targeted adversarial learning optimised sampling for an S_N2 reaction. (**a**) The reaction under study: $Cl^- + CH_3Cl -> ClCH_3 + Cl^-$. d_1, d_2 are chosen as CVs for bias potential construction while $q^* = d_1 - d_2$ is used for target distribution definition. (**b**) 2D free energy surface over d_1, d_2 space . (**c**) The distribution of q^* for the reactant (blue), product (orange) and the interpolated target distribution (grey). (**d**) Example of the biased MD trajectories projected on d_1, d_2 at different iterations. (**e**) The optimised 2D bias potential at different training iterations. Reprinted with permission from *J. Phys. Chem. Lett.* 2019, *10*, 5791–5797. Copyright 2019 American Chemical Society

where $\beta = 1/k_BT$ is the reverse temperature. The training process of TALOS will not stop until the Nash equilibrium is reached.

Notably, TALOS does not require the bias potential and the target distribution to operate in the same space, providing flexibility in algorithm design. For example, TALOS was applied to an all-atom simulation of the S_N2 reaction of $Cl^- + CH_3Cl -> ClCH_3 + Cl^-$ (Fig. 5). Note the length of two C-Cl bonds d_1 and d_2. A two-dimensional bias potential $V_\theta(d_1, d_2)$ was applied to the MD simulation, and the difference between two bonds $q^* = d_1 - d_2$ was used as the descriptor to define the target distribution. The critic network evaluates the discrepancy between generated samples and the target distribution throughout the training process, implementing a systematic bias correction mechanism. Positive bias values serve as penalties for oversampled regions, while negative bias values are applied to the undersampled areas, facilitating improved conformational sampling. Consequently, the system was driven to a predefined target distribution, producing fast chemical reactions.

Although the target descriptor was one-dimensional, the bias potential complements the two-dimensional free energy surfaces. The flexibility in selecting both the bias potential and target distribution makes TALOS a powerful tool for sampling a variety of systems.

Based on normalising flows, the Boltzmann generator was developed to sample equilibrium distributions (Noé et al. 2019). Unlike standard generative learning, a Boltzmann generator does not learn the probability distribution estimated by the distribution of training data, but to sample $\mu(\mathbf{x}) \propto \exp(-\beta U(\mathbf{x}))$ using the energy surface $U(\mathbf{x})$ as an input. Boltzmann generators work by mapping a latent space sample $\mathbf{z}$ from a simple prior distribution, e.g. Gaussian distribution, to a sample $\mathbf{x}$ from the target Boltzmann distribution by a trainable normalising flow model $F_{\mathbf{zx}}$ (Figure 6(1)). A reweighting process is then applied to the generated samples to correct the bias between its generated distribution and the target Boltzmann distribution.

Remarkably, the Boltzmann Generator builds a bijective mapping $\mathbf{f}_\theta : \mathcal{Z} \to \mathcal{X}$, which transforms samples $\mathbf{z} \sim \mathcal{N}(\mathbf{0}, \mathbf{I})$ into samples $\mathbf{x} = \mathbf{f}(\mathbf{z})$, following the target distribution $p(\mathbf{x})$. The training objective is based on maximum likelihood estimation (MLE), where the likelihood is derived using the change of variable theorem:

$$\mathbb{E}_{\mathbf{x}\sim p(\mathbf{x})} \log p_X(\mathbf{x};\theta) = \mathbb{E}_{\mathbf{x}\sim p(\mathbf{x})} \log \left[p_{\mathcal{N}(\mathbf{0},\mathbf{I})} \left(\mathbf{f}_\theta^{-1}(\mathbf{x}) \right) \left| \frac{\partial \mathbf{f}_\theta}{\partial \mathbf{x}} \right|^{-1} \right] \tag{26}$$

After training, statistically independent samples can be generated from a one-shot inference of $\mathbf{f}_\theta$ via MLE requires ample data $\mathbf{x}_i \sim p(\mathbf{x})$, which might be costly to obtain. Thus, the Boltzmann Generator also considers another training objective under the direct supervision of the Boltzmann distribution, i.e. minimising the KL divergence between p_θ and p,

$$D_{\mathrm{KL}}\left(p_\theta \middle\| p\right) = \mathbb{E}_{\mathbf{x}\sim p_X(\mathbf{x};\theta)} \log \left[\frac{p_X(\mathbf{x};\theta)}{p(\mathbf{x})} \right] = \mathbb{E}_{\mathbf{z}\sim\mathcal{N}(\mathbf{0},\mathbf{I})} \log \left[\frac{p_X(\mathbf{f}_\theta(\mathbf{z}))}{\exp(-\beta U(\mathbf{f}_\theta(\mathbf{z})))} \right] + \text{const} \tag{27}$$

Combining both terms gives the final training objective,

$$\min_{\mathbf{f}_\theta} \mathbb{E}_{\mathbf{x}\sim p_X(\mathbf{x};\theta)} \left[\log p_X(\mathbf{f}_\theta(\mathbf{z})) + \beta U(\mathbf{x}) \right] - \mathbb{E}_{\mathbf{x}\sim \text{data}} \left[\log p_X(\mathbf{x};\theta) \right] \tag{28}$$

The invertibility of $\mathbf{f}_\theta$ is guaranteed by the network architecture, RealNVP. Multiple invertible blocks are stacked to construct a deep normalising flow model, obtaining the ability to encode a complex transformation.

The results shown in Fig. 6 (2) demonstrate the application of the Boltzmann generator to a condensed model system comprising a bistable dimer within a densely filled environment of repulsive solvent particles. The open-close transition of the dimer is a rare event for a collective rearrangement of the solvent particles is required due to the high density. A set of short MD simulations in both open and closed states is used as the initialisation. By sampling in the Gaussian latent space

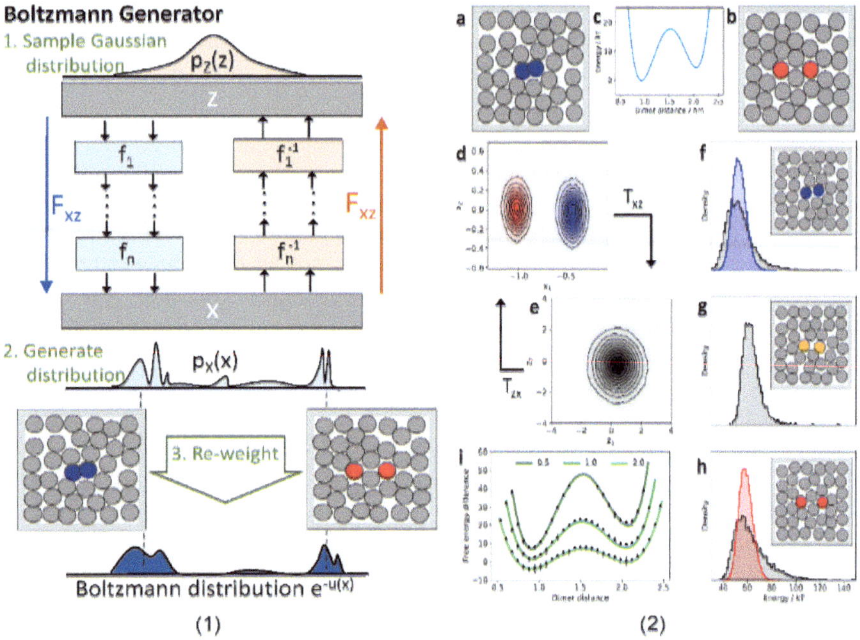

Fig. 6 Boltzmann Generators. (1) The overview of Boltzmann generators. The model is trained by minimising the difference between the generated and Boltzmann distributions. The inference proceeds by drawing latent samples $\mathbf{z}$ from a simple prior (e.g. Gaussian) and performing a transformation via a flow from the sampled latent variable $\mathbf{z}$ to configurations $\mathbf{x}$. Reweighting is required to obtain unbiased Boltzmann distribution. (2) Results on a condensed bistable system. (**a-c**) Repulsive particle system with bistable dimer. Closed (blue) and open (red) configurations are from MD simulations (used as input data). (**d-e**) Distribution of MD data on $\mathbf{x}_1$, $\mathbf{x}_2$ and in latent space coordinates $\mathbf{z}_1$, $\mathbf{z}_2$ after training. (**f-h**) Potential energy distribution from MD (coloured), and the trained Boltzmann generator (grey) for closed (blue), transition (yellow) and open (red) configurations. (**i**) Free energy differences as a function of dimer distance and relative temperatures sampled with Boltzmann generators (dots) and umbrella sampling (lines). Adapted from https://github.com/noegroup/paper_boltzmann_generators under a CC-BY-4.0 License

and mapping to the target distribution via $F_{\mathbf{zx}}$, the well-trained Boltzmann generator sampled both states and previous unobserved transition states. The generated samples have reasonable structures and close-to-equilibrium energies. The direct relationship between the temperature of the canonical ensemble and the variance of the latent space Gaussian prior allows Boltzmann generators to learn to generate thermodynamics, and physically realistic transition pathways between states can be generated through simple interpolation within the latent space.

Other generative models, including score-based diffusion models, can also enhance the sampling process. Distributional Graphormer (DiG) (Zheng et al. 2024) is an example of ensemble generation using a diffusion model. Using Graphormer models as the backbone network, DiG can be trained using the energy functions of molecular systems.

$$\begin{aligned}\sum_{i=1}^{N} \frac{1}{M} \sum_{m=1}^{M} \Bigg\| \frac{\beta_i}{2} \Bigg(\nabla \left(\boldsymbol{R}_{\mathcal{D},i}^{(m)} \cdot p\boldsymbol{s}_{\mathcal{D},i}^{\theta} \left(\boldsymbol{R}_{\mathcal{D},i}^{(m)} \right) \right) + \nabla \left\| \boldsymbol{s}_{\mathcal{D},i}^{\theta} \left(\boldsymbol{R}_{\mathcal{D},i}^{(m)} \right) \right\|^2 \\ + \nabla \left(\nabla \cdot p\boldsymbol{s}_{\mathcal{D},i}^{\theta} \left(\boldsymbol{R}_{\mathcal{D},i}^{(m)} \right) \right) \Bigg) - \frac{\partial}{\partial t} \boldsymbol{s}_{\mathcal{D},i}^{\theta} \left(\boldsymbol{R}_{\mathcal{D},i}^{(m)} \right) \Bigg\|^2 \\ + \frac{\lambda_1}{M} \sum_{m=1}^{M} \left\| \frac{1}{k_{\mathrm{B}} T} \nabla E_{\mathcal{D}} \left(\boldsymbol{R}_{\mathcal{D},1}^{(m)} \right) + \boldsymbol{s}_{\mathcal{D},1}^{\theta} \left(\boldsymbol{R}_{\mathcal{D},1}^{(m)} \right) \right\|^2\end{aligned} \tag{29}$$

The second term is weighted by λ_1 matches the score model at the final generation step to the score from the energy function, and the first term implicitly propagates the energy function supervision to intermediate time steps. The $\left\{ \mathbf{R}_{\mathcal{D},i}^{(m)} \right\}_{\{m=1\}}^{M}$ are points on a grid spanning the structure space. The high-dimensional nature of the structure space of molecular systems would lead to a massive number of points for regular grids. However, the space of actual interest is only a low-dimensional manifold of physically reasonable structures (structures with low energy) relevant to the problem, allowing the training to only use relevant structures as $\mathbf{R}_0$. Additionally, structural data can be used for score network training using the same method as regular diffusion models. Density estimation and guided structure generation can be done using standard techniques, likelihood estimation driven by probability flow ODE and classifier guidance, as with all other diffusion models.

Integrating generative models and enhanced sampling methods remain an active research field, and more advanced algorithms are expected to be developed in the future. We anticipate that the combination will promote the study of many interesting complex systems, including proteins, nucleic acids, and materials.

5 AI-Powered Molecular Simulation Software

5.1 Gaps Between AI and Molecular Simulations

As discussed in previous sections, AI has revolutionised the research paradigm in molecular simulations at a methodological level. However, a significant technical gap hinders the real-world application of AI-assisted molecular simulations, mainly due to compatibility issues between AI and molecular simulations in software and hardware. Traditionally, MD simulation packages such as CHARMM (Brooks et al. 1983), AMBER (Pearlman et al. 1995), GROMACS (Berendsen et al. 1995), NAMD (Nelson et al. 1996), LAMMPS (Plimpton 1995), OpenMM (Eastman and Pande 2010), and SPONGE (Huang et al. 2022) have been sophisticatedly developed using C/C++, FORTRAN, or hand-written CUDA kernels on conventional high-performance computing platforms, including CPUs and general-purpose GPUs (GPGPUs) (Chen et al. 2009), as well as specialised hardware like ANTON (Shaw et al. 2008, 2014, 2021). The functions required for traditional MD, including energy, forces, and dynamics calculations, have been highly optimised for speed. Therefore, the simulation packages are often closed systems, making them challenging for extension or customisation.

In contrast, as an emerging and rapidly growing field, AI has embraced top-level programming languages like Python. It operates as an open system where the underlying computational logic is abstracted and packaged, facilitating code assembly through standard interfaces like function calls, offering high flexibility and scalability. Hardware has also diverged to include AI-specialised types, optimised explicitly for AI-friendly computations such as parallelised matrix multiplications and additions, and favouring computation units designed for lower precision.

Increasingly, researchers are recognising the significant technical challenges posed by this gap. For instance, when attempting to use AI to assist in MD, integrating trained AI functions, such as neural network potentials, into MD simulation packages is often challenging and usually requires complex function calls for interfacing (Behler 2016; Bartók et al. 2017). Similarly, developing and executing codes within a unified framework becomes nearly impossible when AI algorithms aim to utilise MD's energy functions or physical engines for training (Noé et al. 2019). Consequently, the labelling, training, and inference stages are often conducted separately, which hinders the realisation of complete active learning and prevents the full integration of AI and MD into a closed-loop system.

5.2 *AI-Native Molecular Simulation with Unified Frameworks*

Despite the differences in their objectives, AI and MD simulations share many similarities in their core computational logic (Fig. 7) (Zhang et al. 2023c). For example, in AI training, the loss function $\mathcal{L}(\boldsymbol{\theta})$ for a set of parameters $\boldsymbol{\theta}$ with respect to a batch of data $\{x_i\}$, and the corresponding gradient $\nabla_{\boldsymbol{\theta}}\mathcal{L}(\boldsymbol{\theta})$ are evaluated through backpropagation. Subsequently, an optimiser such as ADAM (Kingma and Ba 2015) or SGD (Bottou 2010) is used to update $\boldsymbol{\theta}$ based on $\nabla_{\boldsymbol{\theta}}\mathcal{L}(\boldsymbol{\theta})$ and the learning rate. In parallel, in MD simulations, the energy function $U(\mathbf{x})$ for a set of coordinates $\mathbf{x}$, and the corresponding force $\mathbf{F}(\mathbf{x}) = -\nabla_{\mathbf{x}}U(\mathbf{x})$ are calculated in each time step. Similarly, an integrator such as Langevin dynamics or Nose-Hoover chains is then employed to update $\mathbf{x}$ according to $\mathbf{F}(\mathbf{x})$ and the time step $\mathrm{d}t$.

This key observation has enabled efforts to build AI-native molecular simulations, that is, rewriting MD simulation software within AI programming frameworks such as PyTorch (Paszke et al. 2019), TensorFlow (Abadi et al. 2016), JAX (Bradbury et al. 2018), and MindSpore (MindSpore Official Site | MindSpore n.d.) and leveraging highly integrated functions like tenser operators within these frameworks. Researchers have developed AI-native MD engines such as JAX-MD (Schoenholz and Cubuk 2020; Schoenholz et al. 2021), Torch-MD (Doerr et al. 2021), and MindSPONGE (Zhang et al. 2023c). Although their performance in terms of simulation speed may not match that of specialised traditional simulation software, the complete integration of AI in both software and hardware facilitates

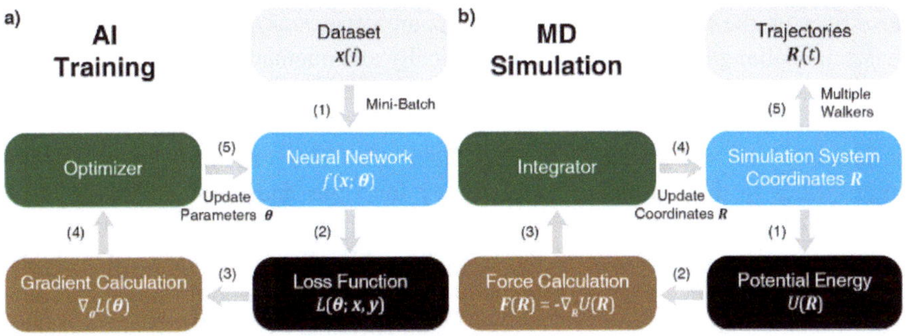

Fig. 7 Comparison between AI training and MD simulation. (**a**) The process of AI training. (**b**) The process of MD simulation. Adapted with permission from *J. Chem. Theory Comput.* **2023**, *19*, 4338–4350. Copyright 2023 American Chemical Society

a seamless union with physical engines. This integration helps MD simulations benefit from the rapid advancements in AI technology.

5.3 Fusion of AI and Molecular Simulations Brings New Opportunities

5.3.1 Auto Differentiation Makes MD "Differentiable"

In traditional MD simulations, the functional forms of energy functions are limited to simplified terms, including bonds, angles, dihedrals, and Lennard-Jones potentials, and their corresponding force functions are often analytically derived and hard-coded into the simulation software (Wang et al. 2004; Rappé et al. 1992). However, increasingly diverse energy function forms have emerged to accommodate a variety of simulation scenarios, including coarse-grained force fields (Marrink et al. 2007; Monticelli et al. 2008), Hamiltonians with floating charges, masses and virtual atoms (Groot and Warren 1997; Pearlman 1994), and MLFFs (Behler and Parrinello 2007; Batzner et al. 2022; Frank et al. 2024; Cui et al. 2024). It becomes cumbersome and sometimes challenging to derive these novel functional forms and to extend a highly optimised MD simulation package to incorporate them. Moreover, many enhanced sampling algorithms, such as meta-dynamics (Laio and Parrinello 2002; Barducci et al. 2008) and umbrella sampling (Torrie and Valleau 1977; Kästner 2011), require CVs $\mathbf{q} = \mathbf{q}(\mathbf{x})$ and the corresponding Jacobian $\nabla_{\mathbf{x}}\mathbf{q}(\mathbf{x})$ to bias the MD simulations for broader exploration. Common CVs are typically simple (often linear) combinations of internal coordinates. These CVs and their corresponding Jacobians are also hard-coded in simulation packages or plugins like PLUMED (Bonomi et al. 2009; Tribello et al. 2014). However, as discussed in previous sections, for complex systems, highly parameterised and nonlinear CVs have become the trend, and their Jacobians are often non-trivial. This complexity makes it challenging to derive them analytically and manually.

Fortunately, automatic differentiation (AD) (Paszke et al. 2017; Baydin et al. 2018), a technique natively integrated into AI frameworks, can calculate gradients analytically and automatically by the chain rule. This capability helps resolve complexities in differentiating novel energy functions and CVs. The gradient or Jacobian of these functions can be easily obtained with a simple AD call as long as their functions are adequately defined. Thus, AD offers significant extensibility for innovative advancements in MD research. Moreover, high-performance AI techniques such as Just-in-Time (JIT) compilation (Bradbury et al. 2018) can enhance the computation of customised functions by converting them into computational graphs, facilitating their rapid integration into MD production runs. Additionally, AD could automatically propagate gradients along with MD dynamics to optimise the intrinsic parameters of MD simulations based on experimental or high-precision data, conceptually rendering MD simulations "differentiable".

5.3.2 Auto Parallelisation "Vectorises" MD Simulations

Extending simulation times for a single trajectory is crucial for MD to converge the Boltzmann distribution and sample physically important metastable states and chemical events. However, there is also a demand for running multiple replicas of MD simulations with different parameters. For instance, replica exchange (Sugita and Okamoto 1999; Swendsen and Wang 1986; Liu et al. 2005) simulates multiple trajectories at different temperatures simultaneously. By allowing them to exchange positions and velocities during simulation, they can help each other overcome free energy barriers. Another critical example is free energy perturbations (FEP) (Shirts 2012; Shirts and Mobley 2013; Zwanzig 1954; Jespers et al. 2021), which require simulating multiple trajectories with different force field parameters to determine the free energy difference between two thermodynamic states. The concept and demand for "high-throughput" MD is increasingly gaining attention in research and industrial applications. However, traditional simulation packages initiate multiple processes or perform various simulations in sequence to achieve this goal.

In the realm of AI, parallelisation is a fundamental requirement. AI models are usually trained using "mini-batch" optimisation, a method where the training dataset is divided into small, manageable batches (mini-batches) of data points instead of using the entire dataset or a single data point for each training iteration. This approach reduces variance during gradient descent and balances computational efficiency with convergence stability. Mini-batch optimisation necessitates that the call to the loss function $\mathcal{L}(\mathbf{x}, \theta)$ be vectorised for multiple data points $(\mathbf{x}_1, \mathbf{x}_2, \cdots \mathbf{x}_n)$. Therefore, AI programming frameworks have focused on automatically adapting functions to accommodate vectorised inputs. Modern frameworks like JAX (Bradbury et al. 2018) even introduce a function transformation called "vmap", specifically designed to convert a function into its "batched" version that can process a batch of data in one operation. Beyond convenience, vectorised computations often offer superior speed on AI-compatible hardware such as GPUs, TPUs, and

NPUs, where tensor operations are optimised within those massive computational cores to reduce FLOPs. Therefore, an AI-native MD package can harness the power of the auto parallelisation features to run a "high-throughput" MD simulation by simply utilising the "vmap" decorator and duplicating the system into an ensemble of systems.

5.4 *Applying AI-Native Molecular Simulations to Real-World Systems*

Although AI-native molecular simulation is still a relatively young and evolving perspective, AI-native MD packages such as JAX-MD, Torch-MD, and MindSPONGE have already been applied in research. For instance, Majewski et al. utilised Torch-MD and TorchMD-Net to construct a machine learning potential for coarse-grained protein dynamics (Majewski et al. 2023). Unlike all-atom force fields, protein residues in their study are represented as amino acid-specific "beads", and complex machine learning potentials are incorporated into the force field terms. This makes traditional MD engines less practical for such simulations.

Moreover, as machine learning potentials often face challenges in terms of generalisation and smoothness, Majewski et al. attempted to integrate physics-informed prior forces—such as bond, dihedral, and nonbonded repulsive terms—into the force field to mitigate nonphysical artefacts, such as clashes. To achieve this, they reused corresponding functions from TorchMD, seamlessly integrating them with MLFFs implemented in PyTorch. This integration streamlined the training process, showcasing the advantages of the AI-native approach. After training, the coarse-grained MD simulations were also conveniently performed using Torch-MD.

Their workflow, incorporating training, coarse-grained sampling, and analysis within the PyTorch framework, successfully folded proteins in the training set into their native structures while reconstructing the associated free energy landscapes with significantly reduced computational effort. Furthermore, training on multiple proteins demonstrated generalisation to their mutants, highlighting the potential of this AI-native strategy to build coarse-grained MLFFs capable of generalising across all proteins through efficient, large-scale training.

Another study utilised JAX-MD to investigate the structure of P450-heme complexes. P450, also known as cytochrome P450 enzymes, catalyses the degradation and metabolism of xenobiotic substances. The catalytic mechanism involves heme, a small molecule located in P450's pocket, with an iron ion interacting with the heme group and the sulphur atom of a cysteine residue. This makes the P450–heme complex a crucial protein–ligand system responsible for significant biological functions in the human body. Predicting and sampling the P450–heme complex is particularly challenging. Template-based modelling and homology modelling approaches often result in steric clashes between heme and P450 residues. Additionally, conformational sampling using MD is not only computationally expensive but also prone to inaccuracies due to poorly developed force fields for iron ion interactions.

Therefore, Guo et al. developed FlexFill, a rapid and flexible docking approach that integrates data-driven and physics-based methods (Guo et al. 2024). FlexFill begins with initial structures generated through template searches and homology modelling, followed by a physical refinement step based on MD simulations. To address inaccuracies in force fields, FlexFill employs a hybrid strategy: classical force fields implemented in JAX-MD are utilised to describe atomic interactions within and between heme and P450, while a customised bond formation energy term is introduced to maintain and promote the formation of the Fe-S bond. While traditional MD engines are often inconvenient for implementing customised energy functions, JAX-MD enables simple and straightforward integration of customised force fields. Sampling is also performed using JAX-MD, leveraging an efficient local structure optimisation algorithm. The use of highly parallelised JAX operators significantly enhances the efficiency of the sampling process compared to other methods.

6 Conclusions and Prospects

Integrating AI algorithms into MD simulations has revolutionised our ability to study complex molecular systems, from surfaces and materials to large biomolecules. While these advancements have yielded promising results, several critical challenges impede the full potential of AI applications in MD. Here, we critically examine the key limitations and controversies surrounding deep learning approaches in molecular modelling and simulations while offering perspectives on potential solutions to these fundamental challenges.

6.1 More Effective Algorithms for Molecular Modelling

To conclude, networks used for molecular modelling should align with the inductive biases inherent in molecular systems, which mainly include invariance and equivariance. Descriptor-based and graph-based models were developed to satisfy the symmetrical constraint. At the same time, the former is suitable for systems with fewer atom types and more atom numbers, and the latter is ideal for smaller and more complex systems.

However, with the enrichment of GPU computing power, attempts to apply GNN to large systems have emerged (Wang et al. 2024a; Musaelian et al. 2023). Additionally, research aimed at improving descriptor-based models for complex systems has appeared (Darby et al. 2023). In addition, recent work has combined descriptors with GNNs. For example, MACE (Batatia et al. 2022) integrates ACE descriptors with message passing, significantly enhancing the network's ability to capture multi-body interactions. In the future, as our understanding of algorithms and molecular systems deepens, networks applied to molecular modelling may converge to one or a few dominant approaches, much like the unification achieved by the transformer architecture in computer vision and natural language modelling (OpenAI et al. 2024; Dosovitskiy et al. 2020; Peebles and Xie 2023; Vaswani et al. 2017).

Another direction for the future development of molecular modelling is to incorporate more physical and chemical information within the systems. For example, current molecular models do not consider the systems' response to external fields. Due to computing power limitations, it is challenging for molecular models to capture long-range interactions. Recent work (Gastegger et al. 2021) has attempted to incorporate long-range information into networks, enabling the network to model solvent effects for molecular systems.

Finally, most current molecular models are primarily used to predict properties with low-order symmetries, such as scalar properties like energy and vector properties like force and dipole moment. However, networks based on tensor products and spherical harmonics inherently have the potential to predict tensor properties. Therefore, in our view, it is one of the future challenges in this field to construct datasets and representations for properties with higher-order symmetry.

6.2 *Transferability*

A critical limitation in deep molecular models is their insufficient transferability across different systems. Despite efforts to construct deep learning models for potential energy surfaces (original and biased), most models demonstrate limited practical transferability. This fundamental challenge necessitates researchers to develop new models for each system, including the time-consuming processes of architecture selection and hyperparameter optimisation. The difficulty in leveraging pre-existing trained models further compounds this issue. Similar transferability constraints are observed in various applications, including coarse-graining, CV construction, sampling strategy design, etc.

Recent developments show promise in addressing these transferability challenges in foundation models across different domains, including molecular representation (Liu et al. 2021; Jiao et al. 2023; Zhou et al. 2022), structure embedding (Lin et al. 2023), and atomic potential (Cui et al. 2024). These models, trained on extensive datasets, have demonstrated improved generalisation capabilities across diverse systems. However, while representing a significant step forward, these foundation models still encounter considerable limitations. The continued development of large-scale foundation models and advances in transfer learning techniques may provide a pathway to enhance the transferability of deep learning models in molecular simulations.

6.3 The Challenges of MLFFs

The training of MLFFs can be divided into two approaches: the bottom-up approach based on quantum chemical (majorly DFT) labels and the top-down approach based on experimental data. Top-down approaches are more commonly applied to classical force field optimisation (Wang et al. 2023), with fewer applications in MLFFs. On the other hand, the bottom-up approach relies on expensive QM calculation, which is challenging to apply to complex systems with higher degrees, such as enzyme systems and transition metal systems. Therefore, we believe that the key direction for the future development of MLFFs is to become more cost-effective, more generalisable, and more applicable to complex systems. Exploring self-supervised pretraining methods (Cui et al. 2024) and researching separating blocks for complex systems have brought enlightenment for addressing those challenges.

6.4 Sample Efficiency and Uncertainty Management

The reliability and practical applicability of deep learning in molecular simulations face two interconnected challenges: data scarcity and uncertainty quantification. The synergy between uncertainty quantification and data efficiency becomes apparent in active learning approaches, where uncertainty estimates guide strategic data collection by identifying regions where additional training data would be most beneficial (Wang and Shang 2014; Gal et al. 2017). While these challenges are related, each presents distinct concerns that require careful consideration for the advancement of the field.

A fundamental requirement for deep molecular models is the ability to provide reliable predictions with quantified uncertainties. This is particularly crucial in molecular simulations where model predictions often guide critical decisions, e.g. drug discovery. Without robust uncertainty quantification, models may make over-confident predictions in regions where they lack sufficient knowledge, potentially leading to misleading or dangerous conclusions.

The challenge of uncertainty quantification becomes even more pronounced given the complex, high-dimensional nature of molecular systems. Current approaches primarily rely on ensemble methods, where multiple deep molecular models are trained to approximate prediction uncertainty through their collective divergence (Zhang et al. 2018b; Unke and Meuwly 2019). However, this heuristic approach lacks theoretical guarantees and incurs substantial computational costs, limiting its practical applicability. Bayesian deep learning emerges as a more promising framework for systematic uncertainty quantification (Ghahramani 2015), though it faces challenges in sampling high-dimensional parameter spaces—a problem analogous to those encountered in MD simulations.

The uncertainty quantification challenge is further complicated by the severe data limitations in molecular modelling. Unlike fields such as computer vision

(Deng et al. 2009; Krizhevsky et al. 2012) or natural language processing (Merity et al. 2016; Zhu et al. 2015) that benefit from abundant standardised datasets, molecular modelling often faces data scarcity due to computationally expensive and time-consuming data generation processes, particularly for large systems and long-timescale simulations. This challenge is compounded by the inherent over-parameterisation of deep molecular models, which typically require substantial training data to achieve reliable performance.

Several promising strategies have emerged to address these challenges. For data efficiency, transfer learning has shown considerable success by leveraging pre-trained models developed on related tasks, requiring only minimal parameter fine-tuning for specific applications (Tan et al. 2018). Meta-learning approaches optimise models for rapid adaptation to new tasks with limited data exposure, enabling few-shot or even one-shot learning capabilities (Finn et al. 2017; Santoro et al. 2016; Andrychowicz et al. 2016; Thrun and Pratt 1998). Enhanced sampling techniques from statistical physics offer potential solutions for improving the efficiency of Bayesian deep learning approaches, particularly in exploring high-dimensional parameter spaces (Welling and Teh 2011).

Looking forward, advances in both data efficiency and uncertainty quantification will be crucial for the field's development, with these challenges exhibiting a mutually reinforcing relationship. The evolution of transfer learning and meta-learning approaches enhances data efficiency, enabling better training with limited data, while improved uncertainty quantification through Bayesian deep learning methodologies helps identify where additional data would be most valuable. This virtuous cycle—where better uncertainty estimates guide more efficient data collection, and more efficient learning strategies along with more collected high-quality data enable better model performance and uncertainty quantification—represents a promising path forward. The synergistic development of these approaches is essential for the future of molecular simulations, where both efficient use of limited data and reliable uncertainty estimates are necessary for deploying these models in critical applications. As these methods mature together, they will enhance our ability to confidently apply deep learning models across the molecular sciences, while making optimal use of computationally expensive training data.

6.5 Computational Performance and Scientific Operators

A significant bottleneck in AI-driven molecular simulations stems from the limited optimisation of specialised scientific computing operators. Current AI frameworks, primarily designed for general machine learning applications, offer well-optimised implementations of standard operations like matrix multiplication and convolution. However, critical operations in scientific computing—such as Fast Fourier Transforms (FFT), matrix inversions, and eigenvalue decompositions—often lack efficient implementations. This optimisation gap substantially impacts the functionality and performance of AI-enhanced molecular simulations. Key features essential for scientific computing, such as dynamic tensor shapes in computational graphs, have historically been underserved in these frameworks. Recent developments in major AI packages have begun addressing these limitations with improved support for scientific computing operations (PyTorch Foundation 2023). As the intersection of AI and scientific computing continues to grow, particularly in molecular sciences, we anticipate significant improvements in implementing and optimising these specialised operators, leading to more efficient and capable AI-driven molecular simulation frameworks.

6.6 Continual Learning in Molecular Simulation

The practical application of AI in molecular simulations often encounters scenarios where data arrives sequentially rather than being available all at once. This sequential nature presents unique challenges, particularly in cases involving large conformational spaces, diverse chemical compounds, or varying task priorities. When dealing with complex conformational spaces or exploring chemical space across different compounds, retraining models from scratch with each new data batch becomes computationally prohibitive. The importance of different conformations varies significantly depending on the specific task, whether it's optimisation, free energy estimation, or studying dynamic properties. Continual learning, also known as lifelong learning, offers a promising framework to address these challenges by enabling models to learn incrementally from new data while maintaining performance on previously learned tasks (Wang et al. 2024c). Recent research exploring continual learning strategies in molecular simulations has achieved promising results (Eckhoff and Reiher 2023). Developing more sophisticated continual learning strategies tailored explicitly for molecular simulations could enhance AI-driven approaches in this field, including adaptive architectures that adjust their capacity based on task complexity and efficient methods for determining what information should be preserved versus updated.

6.7 Integration of Physical Laws and Experimental Data

Another critical challenge in applying deep learning to molecular simulations lies in effectively incorporating experimental data and physical laws into AI models. While the universal approximation theorem ensures that neural networks can theoretically approximate any continuous function, it can significantly enhance their efficiency and reliability if the model is constrained to respect physical principles and experimental observations. These constraints are natural regularisers, reducing overfitting tendencies and improving learning speed and generalisation capabilities. The maximum entropy principle (MaxEnt) offers one approach to incorporating such constraints through penalty terms in loss functions (Jaynes 1986;

Pressé et al. 2013). However, MaxEnt's effectiveness diminishes when handling high-dimensional, interdependent constraints, and it primarily accommodates first-order moments while neglecting higher-order statistical information. Bayesian inference presents a more comprehensive framework for incorporating experimental and physical constraints, treating them as evidence in the modelling process for energy-based models. It is straightforward to integrate experimental observations in energy-based models by the compositionality of the energy function,

$$F_\theta(x) = F_{\text{exp}}\left(q_{\text{exp}}(x)\right) + F_{\text{phys}}(x) \tag{30}$$

where F_{exp} corresponds to the experimentally measured quantities $q_{\text{exp}}(\boldsymbol{x})$, whereas F_{phys} accounts for additional physics restraints (e.g. force-field-like terms). Training of such an EBM will yield an energy function $F_\theta(\boldsymbol{x})$ whose marginal distribution over the experiment-related variables $q\text{exp}(x)$ is consistent with observations (Zhang et al. 2020). This approach is promising in bridging the gap between theoretical predictions and experimental reality. As the field advances, developing sophisticated methods for integrating physical laws and experimental constraints into deep molecular models will become increasingly crucial. This integration enlightens a key direction for future research, potentially leading to more accurate, physically meaningful, and experimentally validated simulation results.

References

Abadi M, Barham P, Chen J, Chen Z, Davis A, Dean J, Devin M, Ghemawat S, Irving G, Isard M, et al. {TensorFlow}: a system for {Large-Scale} machine learning. In: 12th USENIX symposium on operating systems design and implementation (OSDI 16); 2016. p. 265–83.

Abramson J, Adler J, Dunger J, Evans R, Green T, Pritzel A, Ronneberger O, Willmore L, Ballard AJ, Bambrick J, Bodenstein SW, Evans DA, Hung C-C, O'Neill M, Reiman D, Tunyasuvunakool K, Wu Z, Žemgulytė A, Arvaniti E, Beattie C, Bertolli O, Bridgland A, Cherepanov A, Congreve M, Cowen-Rivers AI, Cowie A, Figurnov M, Fuchs FB, Gladman H, Jain R, Khan YA, Low CMR, Perlin K, Potapenko A, Savy P, Singh S, Stecula A, Thillaisundaram A, Tong C, Yakneen S, Zhong ED, Zielinski M, Žídek A, Bapst V, Kohli P, Jaderberg M, Hassabis D, Jumper JM. Accurate structure prediction of biomolecular interactions with AlphaFold 3. Nature. 2024;630:493–500. https://doi.org/10.1038/s41586-024-07487-w.

Andrychowicz M, Denil M, Colmenarejo SG, Hoffman MW, Pfau D, Schaul T, Shillingford B, de Freitas N. Learning to learn by gradient descent by gradient descent. In: Proceedings of the 30th International conference on neural information processing systems. Red Hook: Curran Associates Inc; 2016. p. 3988–96.

Anukrishna PR, Paul V. A review on feature selection for high dimensional data. In: 2017 International conference on inventive systems and control (ICISC); 2017. p. 1–4.

Arjovsky M, Chintala S, Bottou L. Wasserstein generative adversarial networks. In: Proceedings of the 34th International conference on machine learning – volume 70. Sydney: JMLR.org; 2017. p. 214–23.

Arts M, Garcia Satorras V, Huang C-W, Zugner D, Federici M, Clementi C, Noé F, Pinsler R, van den Berg R. Two for one: diffusion models and force fields for coarse-grained molecular dynamics. J Chem Theory Comput. 2023;19:6151–9. https://doi.org/10.1021/acs.jctc.3c00702.

Balakrishnama S, Ganapathiraju A. Linear discriminant analysis-a brief tutorial. Inst Signal Inf Process. 1998;18:1–8.
Barducci A, Bussi G, Parrinello M. Well-tempered Metadynamics: a smoothly converging and tunable free-energy method. Phys Rev Lett. 2008;100:020603. https://doi.org/10.1103/PhysRevLett.100.020603.
Bartók AP, Payne MC, Kondor R, Csányi G. Gaussian approximation potentials: the accuracy of quantum mechanics, without the electrons. Phys Rev Lett. 2010;104:136403. https://doi.org/10.1103/PhysRevLett.104.136403.
Bartók AP, Kondor R, Csányi G. On representing chemical environments. Phys Rev B. 2013;87:184115. https://doi.org/10.1103/PhysRevB.87.184115.
Bartók AP, De S, Poelking C, Bernstein N, Kermode JR, Csányi G, Ceriotti M. Machine learning unifies the modeling of materials and molecules. Sci Adv. 2017;3:e1701816. https://doi.org/10.1126/sciadv.1701816.
Batatia I, Kovacs DP, Simm GNC, Ortner C, Csanyi G. MACE: higher order equivariant message passing neural networks for fast and accurate force fields. In: Presented at the Advances in Neural Information Processing Systems October 31; 2022.
Batzner S, Musaelian A, Sun L, Geiger M, Mailoa JP, Kornbluth M, Molinari N, Smidt TE, Kozinsky B. E(3)-equivariant graph neural networks for data-efficient and accurate interatomic potentials. Nat Commun. 2022;13:2453. https://doi.org/10.1038/s41467-022-29939-5.
Baydin AG, Pearlmutter BA, Radul AA, Siskind JM. Automatic differentiation in machine learning: a survey. J Mach Learn Res. 2018;18:1–43.
Behler J. Perspective: machine learning potentials for atomistic simulations. J Chem Phys. 2016;145:170901. https://doi.org/10.1063/1.4966192.
Behler J, Parrinello M. Generalized neural-network representation of high-dimensional potential-energy surfaces. Phys Rev Lett. 2007;98:146401. https://doi.org/10.1103/PhysRevLett.98.146401.
Berendsen HJ, van der Spoel D, van Drunen R. GROMACS: A message-passing parallel molecular dynamics implementation. Comput Phys Commun. 1995;91:43–56. https://doi.org/10.1016/0010-4655(95)00042-E.
Betker J, Goh G, Jing L, Brooks T, Wang J, Li L, Ouyang L, Zhuang J, Lee J, Guo Y, Manassra W, Dhariwal P, Chu C, Jiao Y, Ramesh A. Improving image generation with better captions. Comput Sci. 2023;2(3):8. https://cdn.openai.com/papers/dall-e-3.pdf.
Bishop CM, Nasrabadi NM. Pattern recognition and machine learning. Springer; 2006.
Bonati L, Zhang Y-Y, Parrinello M. Neural networks-based variationally enhanced sampling. Proc Natl Acad Sci. 2019;116:17641–7. https://doi.org/10.1073/pnas.1907975116.
Bonati L, Rizzi V, Parrinello M. Data-driven collective variables for enhanced sampling. J Phys Chem Lett. 2020;11:2998–3004. https://doi.org/10.1021/acs.jpclett.0c00535.
Bonomi M, Branduardi D, Bussi G, Camilloni C, Provasi D, Raiteri P, Donadio D, Marinelli F, Pietrucci F, Broglia RA, et al. PLUMED: a portable plugin for free-energy calculations with molecular dynamics. Comput Phys Commun. 2009;180:1961–72. https://doi.org/10.1016/j.cpc.2009.05.011.
Bottou L. Large-scale machine learning with stochastic gradient descent. In: Proceedings of COMPSTAT'2010: 19th international conference on computational StatisticsParis france, august 22–27, 2010 keynote, invited and contributed papers. Springer; 2010. p. 177–86.
Bowman GR, Pande VS, Noé F. An introduction to Markov state models and their application to long timescale molecular simulation. Springer; 2013.
Bradbury J, Frostig R, Hawkins P, Johnson MJ, Leary C, Maclaurin D, Necula G, Paszke A, VanderPlas J, Wanderman-Milne S, Zhang Q. JAX: composable transformations of Python+NumPy programs. 2018. http://github.com/jax-ml/jax.
Brooks BR, Bruccoleri RE, Olafson BD, States DJ, Swaminathan S, Karplus M. CHARMM: a program for macromolecular energy, minimization, and dynamics calculations. J Comput Chem. 1983;4:187–217. https://doi.org/10.1002/jcc.540040211.
Cesari A, Reißer S, Bussi G. Using the maximum entropy principle to combine simulations and solution experiments. Comput. 2018;6:15. https://doi.org/10.3390/computation6010015.

Chen G, Li G, Pei S, Wu B. Gpgpu supported cooperative acceleration in molecular dynamics. In: 2009 13th international conference on computer supported cooperative work in design. IEEE; 2009. p. 113–8.

Chen RT, Rubanova Y, Bettencourt J, Duvenaud DK. Neural ordinary differential equations. Adv Neural Inf Process Syst. 2018;31

Chen W, Sidky H, Ferguson AL. Nonlinear discovery of slow molecular modes using state-free reversible VAMPnets. J Chem Phys. 2019;150:214114. https://doi.org/10.1063/1.5092521.

Chen T, Kornblith S, Norouzi M, Hinton G. A simple framework for contrastive learning of visual representations. In: Proceedings of the 37th International conference on machine learning. JMLR.org; 2020. p. 1597–607.

Chodera JD, Noé F. Markov state models of biomolecular conformational dynamics. Curr Opin Struct Biol. 2014;25:135–44. https://doi.org/10.1016/j.sbi.2014.04.002.

Christensen AS, von Lilienfeld OA. On the role of gradients for machine learning of molecular energies and forces. Mach Learn Sci Technol. 2020;1:45018. https://doi.org/10.1088/2632-2153/abba6f.

Copet, J., Kreuk, F., Gat, I., Remez, T., Kant, D., Synnaeve, G., Adi, Y., Défossez, A Simple and controllable music generation., http://arxiv.org/abs/2306.05284, (2024).

Corso G, Stärk H, Jing B, Barzilay R, Jaakkola TS. DiffDock: Diffusion steps, twists, and turns for molecular docking. In: The eleventh international conference on learning representations; 2023.

Cover TM. Elements of information theory. Wiley; 1999.

Cui T, Tang C, Su M, Zhang S, Li Y, Bai L, Dong Y, Gong X, Ouyang W. Geometry-enhanced pretraining on interatomic potentials. Nat Mach Intell. 2024;6:428–36. https://doi.org/10.1038/s42256-024-00818-6.

Darby JP, Kovács DP, Batatia I, Caro MA, Hart GLW, Ortner C, Csányi G. Tensor-reduced atomic density representations. Phys Rev Lett. 2023;131:28001. https://doi.org/10.1103/PhysRevLett.131.028001.

Deng J, Dong W, Socher R, Li L-J, Li K, Fei-Fei L. Imagenet: a large-scale hierarchical image database. In: 2009 IEEE conference on computer vision and pattern recognition. Ieee; 2009. p. 248–55.

Devlin J, Chang M-W, Lee K, Toutanova K. BERT: Pre-training of Deep Bidirectional Transformers for Language Understanding. arXiv. 2019:1810.04805. http://arxiv.org/abs/1810.04805.

Doerr S, Majewski M, Pérez A, Kramer A, Clementi C, Noe F, Giorgino T, De Fabritiis G, Torch MD. A deep learning framework for molecular simulations. J. Chem Theory Comput. 2021;17:2355–63. https://doi.org/10.1021/acs.jctc.0c01343.

Dosovitskiy A, Beyer L, Kolesnikov A, Weissenborn D, Zhai X, Unterthiner T, Dehghani M, Minderer M, Heigold G, Gelly S, Uszkoreit J, Houlsby N. An image is worth 16x16 words: transformers for image recognition at scale. In: Presented at the International Conference on Learning Representations October 2; 2020.

Drautz R. Atomic cluster expansion for accurate and transferable interatomic potentials. Phys Rev B. 2019;99:14104. https://doi.org/10.1103/PhysRevB.99.014104.

Du Y, Mordatch I. Implicit generation and modeling with energy based models. Adv Neural Inf Process Syst. 2019;32

Eastman P, Pande V. OpenMM: a hardware-independent framework for molecular simulations. Comput Sci Eng. 2010;12:34–9. https://doi.org/10.1109/MCSE.2010.27.

Eckhoff M, Reiher M. Lifelong machine learning potentials. J Chem Theory Comput. 2023;19:3509–25. https://doi.org/10.1021/acs.jctc.3c00279.

Esser, P., Kulal, S., Blattmann, A., Entezari, R., Müller, J., Saini, H., Levi, Y., Lorenz, D., Sauer, A., Boesel, F., Podell, D., Dockhorn, T., English, Z., Lacey, K., Goodwin, A., Marek, Y., Rombach, R.: Scaling rectified flow transformers for high-resolution image synthesis., http://arxiv.org/abs/2403.03206, (2024).

Evans Z, Parker JD, Carr CJ, Zukowski Z, Taylor J, Pons J. Stable audio open. arXiv. 2024:2407.14358http://arxiv.org/abs/2407.14358.

Fang X, Liu L, Lei J, He D, Zhang S, Zhou J, Wang F, Wu H, Wang H. Geometry-enhanced molecular representation learning for property prediction. Nat Mach Intell. 2022;4:127–34. https://doi.org/10.1038/s42256-021-00438-4.

Feng S, Ni Y, Lan Y, Ma Z-M, Ma W-Y. Fractional denoising for 3D molecular pre-training. In: Proceedings of the 40th International conference on machine learning. PMLR; 2023. p. 9938–61.

Finn C, Abbeel P, Levine S. Model-agnostic meta-learning for fast adaptation of deep networks. In: Proceedings of the 34th International conference on machine learning. PMLR; 2017. p. 1126–35.

Frank JT, Unke OT, Müller K-R, Chmiela S. A Euclidean transformer for fast and stable machine learned force fields. Nat Commun. 2024;15:6539. https://doi.org/10.1038/s41467-024-50620-6.

Frenkel D, Smit B. Understanding molecular simulation: from algorithms to applications. Elsevier; 2023.

Gal Y, Islam R, Ghahramani Z. Deep Bayesian active learning with image data. In: Proceedings of the 34th international conference on machine learning, vol. 70. Sydney: JMLR.org; 2017. p. 1183–92.

Gao YQ. An integrate-over-temperature approach for enhanced sampling. J Chem Phys. 2008a;128:064105. https://doi.org/10.1063/1.2825614.

Gao YQ. Self-adaptive enhanced sampling in the energy and trajectory spaces: accelerated thermodynamics and kinetic calculations. J Chem Phys. 2008b;128:134111. https://doi.org/10.1063/1.2901037.

Gastegger M, Schütt KT, Müller K-R. Machine learning of solvent effects on molecular spectra and reactions. Chem Sci. 2021;12:11473–83. https://doi.org/10.1039/D1SC02742E.

Gasteiger J, Groß J, Günnemann S. Directional message passing for molecular graphs. In: International conference on learning representations (ICLR); 2020.

Gasteiger J, Becker F, Günnemann S. GemNet: Universal directional graph neural networks for molecules. In: Conference on neural information processing systems (NeurIPS); 2021.

Ghahramani Z. Probabilistic machine learning and artificial intelligence. Nature. 2015;521:452–9. https://doi.org/10.1038/nature14541.

Goodfellow I.: Deep learning, (2016).

Goodfellow I, Pouget-Abadie J, Mirza M, Xu B, Warde-Farley D, Ozair S, Courville A, Bengio Y. Generative adversarial networks. Commun ACM. 2020;63:139–44. https://doi.org/10.1145/3422622.

Grattafiori A, Dubey A, Jauhri A, Pandey A, Kadian A, Al-Dahle A, Letman A, Mathur A, Schelten A, Vaughan A, Yang A, Fan A, Goyal A, Hartshorn A, Yang A, Mitra A, Sravankumar A, Korenev A, Hinsvark A, Rao A, Zhang A, Rodriguez A, Gregerson A, Spataru A, Roziere B, Biron B, Tang B, Chern B, Caucheteux C, Nayak C, Bi C, Marra C, McConnell C, Keller C, Touret C, Wu C, Wong C, Ferrer CC, Nikolaidis C, Allonsius D, Song D, Pintz D, Livshits D, Wyatt D, Esiobu D, Choudhary D, Mahajan D, Garcia-Olano D, Perino D, Hupkes D, Lakomkin E, AlBadawy E, Lobanova E, Dinan E, Smith EM, Radenovic F, Guzmán F, Zhang F, Synnaeve G, Lee G, Anderson GL, Thattai G, Nail G, Mialon G, Pang G, Cucurell G, Nguyen H, Korevaar H, Xu H, Touvron H, Zarov I, Ibarra IA, Kloumann I, Misra I, Evtimov I, Zhang J, Copet J, Lee J, Geffert J, Vranes J, Park J, Mahadeokar J, Shah J, Linde J, van der Billock J, Hong J, Lee J, Fu J, Chi J, Huang J, Liu J, Wang J, Yu J, Bitton J, Spisak J, Park J, Rocca J, Johnstun J, Saxe J, Jia J, Alwala KV, Prasad K, Upasani K, Plawiak K, Li K, Heafield K, Stone K, El-Arini K, Iyer K, Malik K, Chiu K, Bhalla K, Lakhotia K, Rantala-Yeary L, Maaten L, van der Chen L, Tan L, Jenkins L, Martin L, Madaan L, Malo L, Blecher L, Landzaat L, Oliveira L, de Muzzi M, Pasupuleti M, Singh M, Paluri M, Kardas M, Tsimpoukelli M, Oldham M, Rita M, Pavlova M, Kambadur M, Lewis M, Si M, Singh MK, Hassan M, Goyal N, Torabi N, Bashlykov N, Bogoychev N, Chatterji N, Zhang N, Duchenne O, Çelebi O, Alrassy P, Zhang P, Li P, Vasic P, Weng P, Bhargava P, Dubal P, Krishnan P, Koura PS, Xu P, He Q, Dong Q, Srinivasan R, Ganapathy R, Calderer R, Cabral RS, Stojnic R, Raileanu R, Maheswari R, Girdhar R, Patel R, Sauvestre R, Polidoro R, Sumbaly R, Taylor R, Silva R, Hou R, Wang R,

Hosseini S, Chennabasappa S, Singh S, Bell S, Kim SS, Edunov S, Nie S, Narang S, Raparthy S, Shen S, Wan S, Bhosale S, Zhang S, Vandenhende S, Batra S, Whitman S, Sootla S, Collot S, Gururangan S, Borodinsky S, Herman T, Fowler T, Sheasha T, Georgiou T, Scialom T, Speckbacher T, Mihaylov T, Xiao T, Karn U, Goswami V, Gupta V, Ramanathan V, Kerkez V, Gonguet V, Do V, Vogeti V, Albiero V, Petrovic V, Chu W, Xiong W, Fu W, Meers W, Martinet X, Wang X, Wang X, Tan XE, Xia X, Xie X, Jia X, Wang X, Goldschlag Y, Gaur Y, Babaei Y, Wen Y, Song Y, Zhang Y, Li Y, Mao Y, Coudert ZD, Yan Z, Chen Z, Papakipos Z, Singh A, Srivastava A, Jain A, Kelsey A, Shajnfeld A, Gangidi A, Victoria A, Goldstand A, Menon A, Sharma A, Boesenberg A, Baevski A, Feinstein A, Kallet A, Sangani A, Teo A, Yunus A, Lupu A, Alvarado A, Caples A, Gu A, Ho A, Poulton A, Ryan A, Ramchandani A, Dong A, Franco A, Goyal A, Saraf A, Chowdhury A, Gabriel A, Bharambe A, Eisenman A, Yazdan A, James B, Maurer B, Leonhardi B, Huang B, Loyd B, Paola BD, Paranjape B, Liu B, Wu B, Ni B, Hancock B, Wasti B, Spence B, Stojkovic B, Gamido B, Montalvo B, Parker C, Burton C, Mejia C, Liu C, Wang C, Kim C, Zhou C, Hu C, Chu C-H, Cai C, Tindal C, Feichtenhofer C, Gao C, Civin D, Beaty D, Kreymer D, Li D, Adkins D, Xu D, Testuggine D, David D, Parikh D, Liskovich D, Foss D, Wang D, Le D, Holland D, Dowling E, Jamil E, Montgomery E, Presani E, Hahn E, Wood E, Le E-T, Brinkman E, Arcaute E, Dunbar E, Smothers E, Sun F, Kreuk F, Tian F, Kokkinos F, Ozgenel F, Caggioni F, Kanayet F, Seide F, Florez GM, Schwarz G, Badeer G, Swee G, Halpern G, Herman G, Sizov G, Guangyi Z, Lakshminarayanan G, Inan H, Shojanazeri H, Zou H, Wang H, Zha H, Habeeb H, Rudolph H, Suk H, Aspegren H, Goldman H, Zhan H, Damlaj I, Molybog I, Tufanov I, Leontiadis I, Veliche I-E, Gat I, Weissman J, Geboski J, Kohli J, Lam J, Asher J, Gaya J-B, Marcus J, Tang J, Chan J, Zhen J, Reizenstein J, Teboul J, Zhong J, Jin J, Yang J, Cummings J, Carvill J, Shepard J, McPhie J, Torres J, Ginsburg J, Wang J, Wu K, U KH, Saxena K, Khandelwal K, Zand K, Matosich K, Veeraraghavan K, Michelena K, Li K, Jagadeesh K, Huang K, Chawla K, Huang K, Chen L, Garg L, A L, Silva L, Bell L, Zhang L, Guo L, Yu L, Moshkovich L, Wehrstedt L, Khabsa M, Avalani M, Bhatt M, Mankus M, Hasson M, Lennie M, Reso M, Groshev M, Naumov M, Lathi M, Keneally M, Liu M, Seltzer ML, Valko M, Restrepo M, Patel M, Vyatskov M, Samvelyan M, Clark M, Macey M, Wang M, Hermoso MJ, Metanat M, Rastegari M, Bansal M, Santhanam N, Parks N, White N, Bawa N, Singhal N, Egebo N, Usunier N, Mehta N, Laptev NP, Dong N, Cheng N, Chernoguz O, Hart O, Salpekar O, Kalinli O, Kent P, Parekh P, Saab P, Balaji P, Rittner P, Bontrager P, Roux P, Dollar P, Zvyagina P, Ratanchandani P, Yuvraj P, Liang Q, Alao R, Rodriguez R, Ayub R, Murthy R, Nayani R, Mitra R, Parthasarathy R, Li R, Hogan R, Battey R, Wang R, Howes R, Rinott R, Mehta S, Siby S, Bondu SJ, Datta S, Chugh S, Hunt S, Dhillon S, Sidorov S, Pan S, Mahajan S, Verma S, Yamamoto S, Ramaswamy S, Lindsay S, Lindsay S, Feng S, Lin S, Zha SC, Patil S, Shankar S, Zhang S, Zhang S, Wang S, Agarwal S, Sajuyigbe S, Chintala S, Max S, Chen S, Kehoe S, Satterfield S, Govindaprasad S, Gupta S, Deng S, Cho S, Virk S, Subramanian S, Choudhury S, Goldman S, Remez T, Glaser T, Best T, Koehler T, Robinson T, Li T, Zhang T, Matthews T, Chou T, Shaked T, Vontimitta V, Ajayi V, Montanez V, Mohan V, Kumar VS, Mangla V, Ionescu V, Poenaru V, Mihailescu VT, Ivanov V, Li W, Wang W, Jiang W, Bouaziz W, Constable W, Tang X, Wu X, Wang X, Wu X, Gao X, Kleinman Y, Chen Y, Hu Y, Jia Y, Qi Y, Li Y, Zhang Y, Zhang Y, Adi Y, Nam Y, Yu W, Zhao Y, Hao Y, Qian Y, Li Y, He Y, Rait Z, DeVito Z, Rosnbrick Z, Wen Z, Yang Z, Zhao Z, Ma Z. The Llama 3 herd of models. arXiv. 2024:2407.21783. http://arxiv.org/abs/2407.21783.

Groot RD, Warren PB. Dissipative particle dynamics: Bridging the gap between atomistic and mesoscopic simulation. J Chem Phys. 1997;107:4423–35. https://doi.org/10.1063/1.474784.

Grubmüller H. Predicting slow structural transitions in macromolecular systems: conformational flooding. Phys Rev E. 1995;52:2893–906. https://doi.org/10.1103/PhysRevE.52.2893.

Guo AZ, et al. Adaptive enhanced sampling by force-biasing using neural networks. J Chem Phy. 2018; AIP Publishing, https://pubs.aip.org/aip/jcp/article/148/13/134108/75750/Adaptive-enhanced-sampling-by-force-biasing-using

Guo R, Chen C, Sang J, Cui G, Zhang F, Cui X, Li S. Advancing Template-Based Flexible Docking of P450-Heme Complexes via JAX MD. In: 2024 IEEE International Conference on Bioinformatics and Biomedicine (BIBM); 2024. p. 675–80.

Hamelberg D, Mongan J, McCammon JA. Accelerated molecular dynamics: a promising and efficient simulation method for biomolecules. J Chem Phys. 2004;120:11919–29. https://doi.org/10.1063/1.1755656.

Han B, Yu K. Refining potential energy surface through dynamical properties via differentiable molecular simulation. 2024. http://arxiv.org/abs/2406.18269.

Hernández CX, Wayment-Steele HK, Sultan MM, Husic BE, Pande VS. Variational encoding of complex dynamics. Phys Rev E. 2018;97:062412. https://doi.org/10.1103/PhysRevE.97.062412.

Ho J, Jain A, Abbeel P. Denoising diffusion probabilistic models. Adv Neural Inf Process Syst. 2020;33:6840–51.

Hornik K, Stinchcombe M, White H. Multilayer feedforward networks are universal approximators. Neural Netw. 1989;2:359–66. https://doi.org/10.1016/0893-6080(89)90020-8.

Huang Y-P, Xia Y, Yang L, Wei J, Yang YI, Gao YQ. SPONGE: A GPU-accelerated molecular dynamics package with enhanced sampling and AI-driven algorithms. Chin J Chem. 2022;40:160–8. https://doi.org/10.1002/cjoc.202100456.

Isralewitz B, Gao M, Schulten K. Steered molecular dynamics and mechanical functions of proteins. Curr Opin Struct Biol. 2001;11:224–30. https://doi.org/10.1016/S0959-440X(00)00194-9.

Jaynes ET. Bayesian methods: general background. In: Justice JH, editor. Maximum entropy and Bayesian methods in applied statistics: proceedings of the fourth maximum entropy workshop. University of Calgary, 1984. Cambridge: Cambridge University Press; 1986. p. 1–25.

Jespers W, Åqvist J, Gutiérrez-de-Terán H. Free energy calculations for protein–ligand binding prediction. Protein-Ligand Interact Drug Des. 2021:203–26. https://doi.org/10.1007/978-1-0716-1209-5_12.

Jiao R, Han J, Huang W, Rong Y, Liu Y. Energy-motivated equivariant pretraining for 3D molecular graphs. In: Proceedings of the Thirty-seventh AAAI conference on artificial intelligence and thirty-fifth conference on innovative applications of artificial intelligence and thirteenth symposium on educational advances in artificial intelligence. AAAI Press; 2023. p. 8096–104.

Kästner J. Umbrella sampling. Wiley Interdiscip Rev Comput Mol Sci. 2011;1:932–42. https://doi.org/10.1002/wcms.66.

Khersonskii VK, Moskalev AN, Varshalovich DA. Quantum theory of angular momentum. World Scientific Publishing Company; 1988.

Kingma DP. Auto-encoding variational bayes. ArXiv Prepr. 2013:ArXiv13126114.

Kingma DP, Ba J. Adam: a method for stochastic optimization. In: ICLR (poster); 2015.

Kingma, D.P., Welling, M.: Auto-encoding variational bayes. 2022. http://arxiv.org/abs/1312.6114.

Kleinbaum DG, Dietz K, Gail M, Klein M, Klein M. Logistic regression. Springer; 2002.

Krizhevsky A, Sutskever I, Hinton GE. Imagenet classification with deep convolutional neural networks. Adv Neural Inf Process Syst. 2012;25

Laio A, Parrinello M. Escaping free-energy minima. Proc Natl Acad Sci. 2002;99:12562–6. https://doi.org/10.1073/pnas.202427399.

Lin X, Chen Z, Li Y, Lu X, Fan C, Cao Z, Feng S, Gao YQ, Zhang J. ProTokens: A Machine-Learned Language for Compact and Informative Encoding of Protein 3D Structures. 2023. https://www.biorxiv.org/content/10.1101/2023.11.27.568722v1.

Lipman Y, Chen RTQ, Ben-Hamu H, Nickel M, Le M. Flow matching for generative modeling. 2023. http://arxiv.org/abs/2210.02747.

Liu P, Kim B, Friesner RA, Berne BJ. Replica exchange with solute tempering: a method for sampling biological systems in explicit water. Proc Natl Acad Sci. 2005;102:13749–54. https://doi.org/10.1073/pnas.0506346102.

Liu S, Wang H, Liu W, Lasenby J, Guo H, Tang J. Pre-training molecular graph representation with 3D g. In: Presented at the international conference on learning representations; 2021.

Liu X, Zhang J, Hou Z, Yang YI, Gao YQ. From predicting to decision making: Reinforcement learning in biomedicine. WIREs Comput Mol Sci. 2024;14:e1723. https://doi.org/10.1002/wcms.1723.

Majewski M, Pérez A, Thölke P, Doerr S, Charron NE, Giorgino T, Husic BE, Clementi C, Noé F, De Fabritiis G. Machine learning coarse-grained potentials of protein thermodynamics. Nat. Commun. 2023;14:5739. https://doi.org/10.1038/s41467-023-41343-1.

Mardt A, Pasquali L, Wu H, Noé F. VAMPnets for deep learning of molecular kinetics. Nat. Commun. 2018;9:5. https://doi.org/10.1038/s41467-017-02388-1.

Marrink SJ, Risselada HJ, Yefimov S, Tieleman DP, De Vries AH. The MARTINI force field: coarse grained model for biomolecular simulations. J Phys Chem B. 2007;111:7812–24. https://doi.org/10.1021/jp071097f.

Marx D, Hutter J. Ab initio molecular dynamics: theory and implementation. Mod Methods Algorithms Quantum Chem. 2000;1:141.

Mehdi S, Smith Z, Herron L, Zou Z, Tiwary P. Enhanced sampling with machine learning. Annu Rev Phys Chem. 2024;75:347–70. https://doi.org/10.1146/annurev-physchem-083122-125941.

Merity S, Xiong C, Bradbury J, Socher R. Pointer sentinel mixture models. 2016. http://arxiv.org/abs/1609.07843.

MindSpore Official Site | MindSpore. https://www.mindspore.cn/en

Molgedey L, Schuster HG. Separation of a mixture of independent signals using time delayed correlations. Phys Rev Lett. 1994;72:3634. https://doi.org/10.1103/PhysRevLett.72.3634.

Monticelli L, Kandasamy SK, Periole X, Larson RG, Tieleman DP, Marrink S-J. The MARTINI coarse-grained force field: extension to proteins. J. Chem. Theory Comput. 2008;4:819–34. https://doi.org/10.1021/ct700324x.

Musaelian A, Batzner S, Johansson A, Sun L, Owen CJ, Kornbluth M, Kozinsky B. Learning local equivariant representations for large-scale atomistic dynamics. Nat Commun. 2023;14:579. https://doi.org/10.1038/s41467-023-36329-y.

Nelson MT, Humphrey W, Gursoy A, Dalke A, Kalé LV, Skeel RD, Schulten K. NAMD: a parallel, object-oriented molecular dynamics program. Int J Supercomput Appl High Perform Comput. 1996;10:251–68. https://doi.org/10.1177/109434209601000401.

Niu H, Piaggi PM, Invernizzi M, Parrinello M. Molecular dynamics simulations of liquid silica crystallization. Proc Natl Acad Sci. 2018;115:5348–52. https://doi.org/10.1073/pnas.1803919115.

Niu H, Bonati L, Piaggi PM, Parrinello M. Ab initio phase diagram and nucleation of gallium. Nat Commun. 2020;11:2654. https://doi.org/10.1038/s41467-020-16372-9.

Noé F, Olsson S, Köhler J, Wu H. Boltzmann generators: sampling equilibrium states of many-body systems with deep learning. Science. 2019;365:eaaw1147. https://doi.org/10.1126/science.aaw1147.

OpenAI, Achiam J, Adler S, Agarwal S, Ahmad L, Akkaya I, Aleman FL, Almeida D, Altenschmidt J, Altman S, Anadkat S, Avila R, Babuschkin I, Balaji S, Balcom V, Baltescu P, Bao H, Bavarian M, Belgum J, Bello I, Berdine J, Bernadett-Shapiro G, Berner C, Bogdonoff L, Boiko O, Boyd M, Brakman A-L, Brockman G, Brooks T, Brundage M, Button K, Cai T, Campbell R, Cann A, Carey B, Carlson C, Carmichael R, Chan B, Chang C, Chantzis F, Chen D, Chen S, Chen R, Chen J, Chen M, Chess B, Cho C, Chu C, Chung HW, Cummings D, Currier J, Dai Y, Decareaux C, Degry T, Deutsch N, Deville D, Dhar A, Dohan D, Dowling S, Dunning S, Ecoffet A, Eleti A, Eloundou T, Farhi D, Fedus L, Felix N, Fishman SP, Forte J, Fulford I, Gao L, Georges E, Gibson C, Goel V, Gogineni T, Goh G, Gontijo-Lopes R, Gordon J, Grafstein M, Gray S, Greene R, Gross J, Gu SS, Guo Y, Hallacy C, Han J, Harris J, He Y, Heaton M, Heidecke J, Hesse C, Hickey A, Hickey W, Hoeschele P, Houghton B, Hsu K, Hu S, Hu X, Huizinga J, Jain S, Jain S, Jang J, Jiang A, Jiang R, Jin H, Jin D, Jomoto S, Jonn B, Jun H, Kaftan T, Kaiser Ł, Kamali A, Kanitscheider I, Keskar NS, Khan T, Kilpatrick L, Kim JW, Kim C, Kim Y, Kirchner JH, Kiros J, Knight M, Kokotajlo D, Kondraciuk Ł, Kondrich A, Konstantinidis A, Kosic K, Krueger G, Kuo V, Lampe M, Lan I, Lee T, Leike J, Leung J, Levy D, Li CM, Lim R, Lin M, Lin S, Litwin M, Lopez T, Lowe R, Lue P, Makanju A, Malfacini K, Manning S, Markov T, Markovski Y, Martin B, Mayer K, Mayne A, McGrew B, McKinney SM, McLeavey C, McMillan P, McNeil J, Medina D, Mehta A, Menick J, Metz L, Mishchenko A, Mishkin P, Monaco V, Morikawa E, Mossing D, Mu T, Murati M, Murk O, Mély D, Nair A, Nakano R, Nayak R, Neelakantan A, Ngo R, Noh H, Ouyang L, O'Keefe C, Pachocki J, Paino

A, Palermo J, Pantuliano A, Parascandolo G, Parish J, Parparita E, Passos A, Pavlov M, Peng A, Perelman A, Peres F, de AB, Petrov M, Pinto HP, de Michael O, Pokorny, Pokrass M, Pong VH, Powell T, Power A, Power B, Proehl E, Puri R, Radford A, Rae J, Ramesh A, Raymond C, Real F, Rimbach K, Ross C, Rotsted B, Roussez H, Ryder N, Saltarelli M, Sanders T, Santurkar S, Sastry G, Schmidt H, Schnurr D, Schulman J, Selsam D, Sheppard K, Sherbakov T, Shieh J, Shoker S, Shyam P, Sidor S, Sigler E, Simens M, Sitkin J, Slama K, Sohl I, Sokolowsky B, Song Y, Staudacher N, Such FP, Summers N, Sutskever I, Tang J, Tezak N, Thompson MB, Tillet P, Tootoonchian A, Tseng E, Tuggle P, Turley N, Tworek J, Uribe JFC, Vallone A, Vijayvergiya A, Voss C, Wainwright C, Wang JJ, Wang A, Wang B, Ward J, Wei J, Weinmann CJ, Welihinda A, Welinder P, Weng J, Weng L, Wiethoff M, Willner D, Winter C, Wolrich S, Wong H, Workman L, Wu S, Wu J, Wu M, Xiao K, Xu T, Yoo S, Yu K, Yuan Q, Zaremba W, Zellers R, Zhang C, Zhang M, Zhao S, Zheng T, Zhuang J, Zhuk W, Zoph B. GPT-4 Technical Report. arXiv. 2024:2303.08774. http://arxiv.org/abs/2303.08774

Paszke A, Gross S, Chintala S, Chanan G, Yang E, DeVito Z, Lin Z, Desmaison A, Antiga L, Lerer A. Automatic differentiation in pytorch. 2017.

Paszke A, Gross S, Massa F, Lerer A, Bradbury J, Chanan G, Killeen T, Lin Z, Gimelshein N, Antiga L, Desmaison A, Kopf A, Yang E, DeVito Z, Raison M, Tejani A, Chilamkurthy S, Steiner B, Fang L, Bai J, Chintala S. PyTorch: An imperative style, high-performance deep learning library. In: Wallach H, Larochelle H, Beygelzimer A, dAlché-Buc F, Fox E, Garnett R, editors. Advances in neural information processing systems. Curran Associates, Inc; 2019.

Pearlman DA. A comparison of alternative approaches to free energy calculations. J Phys Chem. 1994;98:1487–93. https://doi.org/10.1021/j100056a020.

Pearlman DA, Case DA, Caldwell JW, Ross WS, Cheatham TE III, DeBolt S, Ferguson D, Seibel G, Kollman P. AMBER, a package of computer programs for applying molecular mechanics, normal mode analysis, molecular dynamics and free energy calculations to simulate the structural and energetic properties of molecules. Comput Phys Commun. 1995;91:1–41. https://doi.org/10.1016/0010-4655(95)00041-D.

Pedregosa F, Varoquaux G, Gramfort A, Michel V, Thirion B, Grisel O, Blondel M, Prettenhofer P, Weiss R, Dubourg V, Vanderplas J, Passos A, Cournapeau D, Brucher M, Perrot M, Duchesnay É. Scikit-learn: machine learning in Python. J Mach Learn Res. 2011;12:2825–30.

Peebles W, Xie S. Scalable diffusion models with transformers. In: 2023 IEEE/CVF International Conference on Computer Vision (ICCV). Paris: IEEE; 2023. p. 4172–82.

Pérez-Hernández G, Paul F, Giorgino T, De Fabritiis G, Noé F. Identification of slow molecular order parameters for Markov model construction. J Chem Phys. 2013;139:015102. https://doi.org/10.1063/1.4811489.

Peyré G, Cuturi M, et al. Computational optimal transport: with applications to data science. Found Trends® Mach Learn. 2019;11:355–607. https://doi.org/10.1561/2200000073.

Piaggi PM, Parrinello M. Entropy based fingerprint for local crystalline order. J Chem Phys. 2017;147:114112. https://doi.org/10.1063/1.4998408.

Plimpton S. Fast parallel algorithms for short-range molecular dynamics. J Comput Phys. 1995;117:1–19. https://doi.org/10.1006/jcph.1995.1039.

Pressé S, Ghosh K, Lee J, Dill KA. Principles of maximum entropy and maximum caliber in statistical physics. Rev Mod Phys. 2013;85:1115–41. https://doi.org/10.1103/RevModPhys.85.1115.

PyTorch Foundation PyTorch 2.1: automatic dynamic shape compilation, distributed checkpointing. 2023. https://pytorch.org/blog/pytorch-2-1/

Radford A, Kim JW, Hallacy C, Ramesh A, Goh G, Agarwal S, Sastry G, Askell A, Mishkin P, Clark J, Krueger G, Sutskever I. Learning transferable visual models from natural language supervision. In: Proceedings of the 38th International conference on machine learning. PMLR; 2021. p. 8748–63.

Rappé AK, Casewit CJ, Colwell K, Goddard WA III, Skiff WM. UFF, a full periodic table force field for molecular mechanics and molecular dynamics simulations. J Am Chem Soc. 1992;114:10024–35. https://doi.org/10.1021/ja00051a040.

Rezende D, Mohamed S. Variational inference with normalizing flows. In: International conference on machine learning. PMLR; 2015. p. 1530–8.

Santoro A, Bartunov S, Botvinick M, Wierstra D, Lillicrap T. Meta-learning with memory-augmented neural networks. In: Proceedings of the 33rd International Conference on International Conference on Machine Learning, vol. 48. JMLR.org, New York; 2016. p. 1842–50.

Satorras VG, Hoogeboom E, Welling M. E(n) equivariant graph neural networks. 2022. http://arxiv.org/abs/2102.09844.

Schneider E, Dai L, Topper RQ, Drechsel-Grau C, Tuckerman ME. Stochastic neural network approach for learning high-dimensional free energy surfaces. Phys Rev Lett. 2017;119:150601. https://doi.org/10.1103/PhysRevLett.119.150601.

Schoenholz S, Cubuk ED. JAX MD: A Framework for Differentiable Physics. Adv Neural Inf Process Syst. 2020;33:11428–41.

Schoenholz SS, Cubuk ED, Jax MD. A framework for differentiable physics. J Stat Mech Theory Exp. 2021:124016. https://doi.org/10.1088/1742-5468/ac3ae9.

Schütt K, Kindermans P-J, Sauceda Felix HE, Chmiela S, Tkatchenko A, Müller K-R. SchNet: A continuous-filter convolutional neural network for modeling quantum interactions. In: Advances in neural information processing systems. Curran Associates, Inc; 2017.

Shapeev AV. Moment tensor potentials: a class of systematically improvable interatomic potentials. Multiscale Model Simul. 2016;14:1153–73. https://doi.org/10.1137/15M1054183.

Shaw DE, Deneroff MM, Dror RO, Kuskin JS, Larson RH, Salmon JK, Young C, Batson B, Bowers KJ, Chao JC, et al. Anton, a special-purpose machine for molecular dynamics simulation. Commun ACM. 2008;51:91–7. https://doi.org/10.1145/1364782.1364802.

Shaw DE, Grossman J, Bank JA, Batson B, Butts JA, Chao JC, Deneroff MM, Dror RO, Even A, Fenton CH, et al. Anton 2: raising the bar for performance and programmability in a special-purpose molecular dynamics supercomputer. In: SC'14: Proceedings of the international conference for high performance computing, networking, storage and analysis. IEEE; 2014. p. 41–53.

Shaw DE, Adams PJ, Azaria A, Bank JA, Batson B, Bell A, Bergdorf M, Bhatt J, Butts JA, Correia T, Dirks RM, Dror RO, Eastwood MP, Edwards B, Even A, Feldmann P, Fenn M, Fenton CH, Forte A, Gagliardo J, Gill G, Gorlatova M, Greskamp B, Grossman JP, Gullingsrud J, Harper A, Hasenplaugh W, Heily M, Heshmat BC, Hunt J, Ierardi DJ, Iserovich L, Jackson BL, Johnson NP, Kirk MM, Klepeis JL, Kuskin JS, Mackenzie KM, Mader RJ, McGowen R, McLaughlin A, Moraes MA, Nasr MH, Nociolo LJ, O'Donnell L, Parker A, Peticolas JL, Pocina G, Predescu C, Quan T, Salmon JK, Schwink C, Shim KS, Siddique N, Spengler J, Szalay T, Tabladillo R, Tartler R, Taube AG, Theobald M, Towles B, Vick W, Wang SC, Wazlowski M, Weingarten MJ, Williams JM, Yuh KA. Anton 3: Twenty microseconds of molecular dynamics simulation before lunch. In: Proceedings of the international conference for high performance computing, networking, storage and analysis. New York: Association for Computing Machinery; 2021. p. 1–11.

Sheng HW, Kramer MJ, Cadien A, Fujita T, Chen MW. Highly optimized embedded-atom-method potentials for fourteen fcc metals. Phys Rev B. 2011;83:134118. https://doi.org/10.1103/PhysRevB.83.134118.

Shirts MR. Best practices in free energy calculations for drug design. Comput Drug Discov Des. 2012;425–467 https://doi.org/10.1007/978-1-61779-465-0_26.

Shirts MR, Mobley DL. An introduction to best practices in free energy calculations. Biomol Simul Methods Protoc. 2013;271–311:271. https://doi.org/10.1007/978-1-62703-017-5_11.

Sinha S, Tam B, Wang SM. Applications of molecular dynamics simulation in protein study. Membranes. 2022;12:844. https://doi.org/10.3390/membranes12090844.

Smith JS, Isayev O, Roitberg AE. ANI-1: an extensible neural network potential with DFT accuracy at force field computational cost. Chem Sci. 2017;8:3192–203. https://doi.org/10.1039/C6SC05720A.

Song Y, Ermon S. Generative modeling by estimating gradients of the data distribution. Adv Neural Inf Process Syst. 2019;32

Song Y, Sohl-Dickstein J, Kingma DP, Kumar A, Ermon S, Poole B. Score-based generative modeling through stochastic differential equations. In: Presented at the international conference on learning representations October 2; 2020.

Stärk H, Beaini D, Corso G, Tossou P, Dallago C, Günnemann S, Lio P. 3D infomax improves GNNs for molecular property prediction. In: Presented at the NeurIPS 2021 AI for Science Workshop; 2021.

Suárez D, Díaz N. SARS-CoV-2 main protease: a molecular dynamics study. J Chem Inf Model. 2020;60:5815–31. https://doi.org/10.1021/acs.jcim.0c00575.

Sugita Y, Okamoto Y. Replica-exchange molecular dynamics method for protein folding. Chem Phys Lett. 1999;314:141–51. https://doi.org/10.1016/S0009-2614(99)01123-9.

Sutton RS, Barto AG. Reinforcement learning: an introduction. MIT press; 2018.

Swendsen RH, Wang J-S. Replica monte carlo simulation of spin-glasses. Phys Rev Lett. 1986;57:2607. https://doi.org/10.1103/PhysRevLett.57.2607.

Tan, C., Sun, F., Kong, T., Zhang, W., Yang, C., Liu, C.: A survey on deep transfer learning. In: Kůrková, V., Manolopoulos, Y., Hammer, B., Iliadis, L., and Maglogiannis, I. (eds.) Artificial neural networks and machine learning—ICANN 2018. pp. 270–279. Springer, Cham (2018).

Thaler S, Zavadlav J. Learning neural network potentials from experimental data via differentiable trajectory reweighting. Nat Commun. 2021;12:6884. https://doi.org/10.1038/s41467-021-27241-4.

Thrun S, Pratt L. Learning to learn: introduction and overview. In: Thrun S, Pratt L, editors. Learning to learn. Boston: Springer; 1998. p. 3–17.

Torrie GM, Valleau JP. Nonphysical sampling distributions in Monte Carlo free-energy estimation: umbrella sampling. J Comput Phys. 1977;23:187–99. https://doi.org/10.1016/0021-9991(77)90121-8.

Tribello GA, Bonomi M, Branduardi D, Camilloni C, Bussi G. PLUMED 2: New feathers for an old bird. Comput Phys Commun. 2014;185:604–13. https://doi.org/10.1016/j.cpc.2013.09.018.

Trizio E, Parrinello M. From enhanced sampling to reaction profiles. J Phys Chem Lett. 2021;12:8621–6. https://doi.org/10.1021/acs.jpclett.1c02317.

Tuckerman ME. Statistical mechanics: theory and molecular simulation. Oxford University Press; 2023.

Tuckerman ME, Ungar PJ, Von Rosenvinge T, Klein ML. Ab initio molecular dynamics simulations. J Phys Chem. 1996;100:12878–87. https://doi.org/10.1021/jp960480+.

Unke OT, Meuwly M. PhysNet: a neural network for predicting energies, forces, dipole moments, and partial charges. J Chem Theory Comput. 2019;15:3678–93. https://doi.org/10.1021/acs.jctc.9b00181.

Valsson O, Tiwary P, Parrinello M. Enhancing important fluctuations: rare events and metadynamics from a conceptual viewpoint. Annu Rev Phys Chem. 2016;67:159–84. https://doi.org/10.1146/annurev-physchem-040215-112229.

van der Maaten L, Hinton G. Visualizing data using t-SNE. J Mach Learn Res. 2008;9:2579–605.

Vaswani A, Shazeer N, Parmar N, Uszkoreit J, Jones L, Gomez AN, Kaiser Ł, Polosukhin I. Attention is all you need. In: Proceedings of the 31st international conference on neural information processing systems. Red Hook: Curran Associates Inc; 2017. p. 6000–10.

Wang F, Landau DP. Efficient, multiple-range random walk algorithm to calculate the density of states. Phys Rev Lett. 2001;86:2050–3. https://doi.org/10.1103/PhysRevLett.86.2050.

Wang D, Shang Y. A new active labeling method for deep learning. In: 2014 International Joint Conference on Neural Networks (IJCNN).; 2014. p. 112–9.

Wang J, Wolf RM, Caldwell JW, Kollman PA, Case DA. Development and testing of a general amber force field. J Comput Chem. 2004;25:1157–74. https://doi.org/10.1002/jcc.20035.

Wang J, Olsson S, Wehmeyer C, Pérez A, Charron NE, de Fabritiis G, Noé F, Clementi C. Machine learning of coarse-grained molecular dynamics force fields. ACS Cent Sci. 2019;5:755–67. https://doi.org/10.1021/acscentsci.8b00913.

Wang X, Li J, Yang L, Chen F, Wang Y, Chang J, Chen J, Feng W, Zhang L, Yu K. DMFF: an open-source automatic differentiable platform for molecular force field development and molecular dynamics simulation. J Chem Theory Comput. 2023;19:5897–909. https://doi.org/10.1021/acs.jctc.2c01297.

Wang T, He X, Li M, Li Y, Bi R, Wang Y, Cheng C, Shen X, Meng J, Zhang H, Liu H, Wang Z, Li S, Shao B, Liu T-Y. Ab initio characterization of protein molecular dynamics with AI2BMD. Nature. 2024a;635:1019–27. https://doi.org/10.1038/s41586-024-08127-z.

Wang Y, Wang T, Li S, He X, Li M, Wang Z, Zheng N, Shao B, Liu T-Y. Enhancing geometric representations for molecules with equivariant vector-scalar interactive message passing. Nat Commun. 2024b;15:313. https://doi.org/10.1038/s41467-023-43720-2.

Wang L, Zhang X, Su H, Zhu J. A comprehensive survey of continual learning: theory, method and application. IEEE Trans Pattern Anal Mach Intell. 2024c;46:5362–83. https://doi.org/10.1109/TPAMI.2024.3367329.

Wehmeyer C, Noé F. Time-lagged autoencoders: deep learning of slow collective variables for molecular kinetics. J Chem Phys. 2018;148:241703. https://doi.org/10.1063/1.5011399.

Welling M, Teh YW. Bayesian learning via stochastic gradient langevin dynamics. In: Proceedings of the 28th International conference on international conference on machine learning. Madison: Omnipress; 2011. p. 681–8.

Xanthopoulos P, Pardalos PM, Trafalis TB, Xanthopoulos P, Pardalos PM, Trafalis TB. Linear discriminant analysis. Robust Data Min. 2013:27–33. https://doi.org/10.1007/978-1-4419-9878-1_4.

Xu M, Yu L, Song Y, Shi C, Ermon S, Tang J. GeoDiff: a geometric diffusion model for molecular conformation generation. In: International conference on learning representations; 2022.

Yang L, Qin Gao Y. A selective integrated tempering method. J Chem Phys. 2009;131:214109. https://doi.org/10.1063/1.3266563.

Yang YI, Shao Q, Zhang J, Yang L, Gao YQ. Enhanced sampling in molecular dynamics. J Chem Phys. 2019;151:070902. https://doi.org/10.1063/1.5109531.

Yao K, Herr JE, Toth DW, Mckintyre R, Parkhill J. The TensorMol-0.1 model chemistry: a neural network augmented with long-range physics. Chem Sci. 2018;9:2261–9. https://doi.org/10.1039/C7SC04934J.

Yuan S, Han X, Zhang J, Xie Z, Fan C, Xiao Y, Gao YQ, Yang YI. Generating high-precision force fields for molecular dynamics simulations to study chemical reaction mechanisms using molecular configuration transformer. J Phys Chem A. 2024;128:4378–90. https://doi.org/10.1021/acs.jpca.4c01267.

Zaidi S, Schaarschmidt M, Martens J, Kim H, Teh YW, Sanchez-Gonzalez A, Battaglia P, Pascanu R, Godwin J. Pre-training via denoising for molecular property prediction. In: Presented at the Eleventh international conference on learning representations; 2022.

Zhang L, Han J, Wang H, Car R, E, W. Deep potential molecular dynamics: a scalable model with the accuracy of quantum mechanics. Phys Rev Lett. 2018a;120:143001. https://doi.org/10.1103/PhysRevLett.120.143001.

Zhang L, Wang H, E, W. Reinforced dynamics for enhanced sampling in large atomic and molecular systems. J Chem Phys. 2018b;148:124113. https://doi.org/10.1063/1.5019675.

Zhang J, Lei Y-K, Che X, Zhang Z, Yang YI, Gao YQ. Deep representation learning for complex free-energy landscapes. J Phys Chem Lett. 2019a;10:5571–6. https://doi.org/10.1021/acs.jpclett.9b02012.

Zhang J, Yang YI, Noé F. Targeted adversarial learning optimized sampling. J Phys Chem Lett. 2019b;10:5791–7. https://doi.org/10.1021/acs.jpclett.9b02173.

Zhang J, Lei Y-K, Zhang Z, Chang J, Li M, Han X, Yang L, Yang YI, Gao YQ. A perspective on deep learning for molecular modeling and simulations. J Phys Chem A. 2020;124:6745–63. https://doi.org/10.1021/acs.jpca.0c04473.

Zhang L, Wang H, Car R, Weinan E. Phase diagram of a deep potential water model. Phys Rev Lett. 2021a;126:236001. https://doi.org/10.1103/PhysRevLett.126.236001.

Zhang J, Lei Y-K, Zhang Z, Han X, Li M, Yang L, Isaac Yang Y, Qin Gao Y. Deep reinforcement learning of transition states. Phys Chem Chem Phys. 2021b;23:6888–95. https://doi.org/10.1039/D0CP06184K.

Zhang J, Lei Y-K, Zhou Y, Yang YI, Gao YQ. Molecular CT: unifying geometry and representation learning for molecules at different scales 2023a. http://arxiv.org/abs/2012.11816.

Zhang J, Lin X, Gao YQ, et al. Machine-learned invertible coarse graining for multiscale molecular modeling. ArXiv Prepr. 2023b:ArXiv230501243.

Zhang J, Chen D, Xia Y, Huang Y-P, Lin X, Han X, Ni N, Wang Z, Yu F, Yang L, Yang YI, Gao YQ. Artificial intelligence enhanced molecular simulations. J Chem Theory Comput. 2023c;19:4338–50. https://doi.org/10.1021/acs.jctc.3c00214.

Zhang S, Makoś MZ, Jadrich RB, Kraka E, Barros K, Nebgen BT, Tretiak S, Isayev O, Lubbers N, Messerly RA, Smith JS. Exploring the frontiers of condensed-phase chemistry with a general reactive machine learning potential. Nat Chem. 2024;16:727–34. https://doi.org/10.1038/s41557-023-01427-3.

Zhao K, Yu H, Xiong H, Lu Q, Gao YQ, Xu B. Action at a distance: organic cation induced long range organization of interfacial water enhances hydrogen evolution and oxidation kinetics. Chem Sci. 2023;14:11076–87. https://doi.org/10.1039/D3SC03300G.

Zheng S, He J, Liu C, Shi Y, Lu Z, Feng W, Ju F, Wang J, Zhu J, Min Y, Zhang H, Tang S, Hao H, Jin P, Chen C, Noé F, Liu H, Liu T-Y. Predicting equilibrium distributions for molecular systems with deep learning. Nat Mach Intell. 2024;6:558–67. https://doi.org/10.1038/s42256-024-00837-3.

Zhou G, Gao Z, Ding Q, Zheng H, Xu H, Wei Z, Zhang L, Ke G. Uni-Mol: a universal 3D molecular representation learning framework. In: Presented at the The eleventh international conference on learning representations; 2022.

Zhou S, Shi J, Liu S, Li G, Pei F, Chen Y, Deng J, Zheng Q, Li J, Zhao C, Hwang I, Sun C-J, Liu Y, Deng Y, Huang L, Qiao Y, Xu G-L, Chen J-F, Amine K, Sun S-G, Liao H-G. Visualizing interfacial collective reaction behaviour of Li–S batteries. Nature. 2023;621:75–81. https://doi.org/10.1038/s41586-023-06326-8.

Zhu, Y., Kiros, R., Zemel, R., Salakhutdinov, R., Urtasun, R., Torralba, A., Fidler, S.: Aligning books and movies: towards story-like visual explanations by watching movies and reading books. 2015. http://arxiv.org/abs/1506.06724.

Zwanzig RW. High-temperature equation of state by a perturbation method. I Nonpolar gases. J Chem Phys. 1954;22:1420–6. https://doi.org/10.1063/1.1740409.

Zwanzig R. Nonequilibrium statistical mechanics. Oxford University Press; 2001.

Drug Discovery with Quantum Machine Learning

Alexey Pyrkov, Alex Aliper, Dmitry Bezrukov, and Alex Zhavoronkov

Abstract The field of drug discovery has seen significant advancements in recent years with AI and machine learning techniques. The integration of quantum computing in the pipelines provides new opportunities. Recently, huge progress was achieved in quantum hardware, the main players announce the achievement of fault-tolerant devices by 2030 in their roadmaps. This book chapter provides an introduction into Quantum Machine Learning (QML) algorithms and an overview of how these algorithms can be used to accelerate various stages of the drug discovery pipeline. Key QML algorithms are described, along with examples of successful case studies in drug discovery. The chapter also examines the challenges and limitations of current QML approaches, as well as future directions for the field.

Keywords Quantum machine learning · Drug discovery · Quantum neural networks · Quantum de novo generation · Quantum molecular docking

1 Introduction and Historical Notes

The origins of quantum computing can be traced back to the early twentieth century, with the advent of quantum mechanics. This new branch of physics, led by pioneers like Max Planck, Albert Einstein, and Niels Bohr, revolutionized the understanding of atomic and subatomic processes. However, only in the 1980s, theoretical physicists began to explore these quantum principles for computational purposes.

In 1980, Manin (1980) and in 1981, Richard Feynman (1982), a visionary physicist and Nobel prize winner, suggested that classical computers might not

A. Pyrkov · A. Zhavoronkov (✉)
Insilico Medicine Hong Kong Ltd, Pak Shek Kok, New Territories, Hong Kong
e-mail: a.pyrkov@insilicomedicine.com; alex@insilico.com

A. Aliper · D. Bezrukov
Insilico Medicine AI Ltd, Abu Dhabi, UAE
e-mail: aliper@insilicomedicine.com; d.bezrukov@insilicomedicine.com

A. Lavecchia (ed.), *Applied Artificial Intelligence for Drug Discovery*,
https://doi.org/10.1007/978-3-031-98022-0_10

efficiently simulate quantum phenomena, proposing instead that a quantum computer would excel at such tasks. Around the same time, David Deutsch introduced the concept of a universal quantum computer (Deutsch and Penrose 1997), setting the stage for a profound shift in computational paradigms. Deutsch's work laid the groundwork for understanding how quantum mechanics could be harnessed to perform computations that are infeasible for classical computers.

The 1990s witnessed significant theoretical advancements with the development of quantum algorithms (Dalzell et al. 2023; Montanaro 2016). In 1994, Peter Shor, a mathematician at Bell Labs, proposed an algorithm (Shor 1994) that could factor large integers exponentially faster than the best-known classical algorithms. Shor's algorithm demonstrated the potential of quantum computing to solve problems considered intractable for classical machines, such as breaking widely used cryptographic codes.

In 1996, Lov Grover introduced another pivotal algorithm (Grover 1996) capable of searching an unsorted database quadratically faster than classical counterparts. Grover's algorithm further fueled interest in the field, highlighting the broad applicability of quantum computing beyond niche scientific problems.

Following the initial concept proposed by Manin and Feynman, Seth Lloyd formalized the idea of simulation of quantum systems with other more simple quantum systems in 1996 (Lloyd 1996), and over the next 25 years after that, significant advancements have been made in utilizing quantum computers to perform quantum chemistry calculations (Cao et al. 2019). His approach involves encoding quantum states that represent the electronic configurations of molecules on qubits and then simulating their time-dependent evolution under the influence of molecular Hamiltonians (the operators corresponding to the total energy of the system) with the use of Trotterization (Trotter 1959).

In 1995, the Quantum Phase Estimation (QPE) algorithm was introduced (Kitaev 1995), enabling the estimation of ground state energy with exceptional precision on fault-tolerant quantum computers. This algorithm transforms the traditional eigenvalue problem into a unitary one by calculating phases from the application of a unitary operation to an eigenfunction. The QPE achieves high-probability estimation with error ϵ, using $O(\log(1/\epsilon))$ qubits and $O(1/\epsilon)$ gates.

Parallel to the development of quantum algorithms, the quest to build hardware for quantum computers had begun on different physical platforms. The fundamental building block of a quantum computer is the qubit, which, unlike a classical bit, can exist in a superposition of states, enabling parallel information processing (Nielsen and Chuang 2000). Another necessary ingredient to overcome classical algorithms is that all these quantum algorithms should naturally handle entanglement required to model inseparable quantum states (Nielsen and Chuang 2000). Quantum entanglement is a phenomenon where two or more particles become interconnected in such a way that the state of one particle cannot be fully described without considering the state of the other(s), even when the particles are separated by very long distances. Such behavior was a subject of a famous paper of 1935 by Albert Einstein, Boris Podolsky, and Nathan Rosen, known as the EPR paradox (Einstein et al. 1935). They argued that quantum mechanics might be incomplete because it allowed for "spooky action at a distance," where measuring one particle instantaneously affected the state of another entangled particle, no matter the distance between them. Following the EPR paper, the notion of quantum entanglement was famously coined by Erwin Schrodinger (1935). The first experimental investigation in the direction of quantum computing was started from directly addressing the peculiar predictions of quantum entanglement in the 1970s by John Clauser and Stuart Freedman (1972). Using the concepts proposed by physicist John Bell in 1964 (1997), Clauser and Freedman performed experiments testing Bell's inequalities—mathematical inequalities that, if violated, would challenge the principle of local realism upheld by classical physics. Following the groundbreaking work of Clauser and Freedman, other notable physicists, such as Alain Aspect, further refined these experiments in the 1980s with increased precision and control (Aspect et al. 1982). His work provided even more robust evidence in favor of quantum entanglement reality and the violation of Bell's inequalities. In recognition of this body of work and its profound implications for our understanding of the universe, the Nobel Prize in Physics 2022 was awarded to Alain Aspect, John F. Clauser, and Anton Zeilinger (Nobel Prize 2022). All three physicists made significant contributions to experimental investigation of quantum entanglement. Anton Zeilinger, in particular, is noted for his pioneering work in quantum teleportation and the practical applications of entanglement in the fields of quantum computing and quantum communication (Bouwmeester et al. 1997).

Throughout the late 1990s and early 2000s, various groups worldwide experimented with different physical systems to realize qubits, including ions (Bruzewicz et al. 2019), photons (Slussarenko and Pryde 2019), nuclear spins (Vandersypen and Chuang 2005), and superconducting circuits (Kjaergaard et al. 2020). The very first quantum computers included trapped ion quantum computers (Monroe et al. 1995) and liquid-state nuclear magnetic resonance (NMR) computers (Cory et al. 1997; Gershenfeld and Chuang 1997). NMR quantum computing was initially a very promising approach for realizing quantum computers. It used the magnetic properties of atomic nuclei to manipulate and control qubits in molecules dissolved in a solution. Despite early successes, such as the first demonstration of Shor's algorithm, NMR quantum computing was ultimately deemed unsuccessful for achieving scalable quantum computation due to several inherent limitations. First, scalability was a major issue. NMR systems faced practical limits on the number of qubits that could be manipulated coherently. Typically, these systems could handle only a small number of qubits, beyond which the required radiofrequency pulses for manipulating nuclear spins became too complex and the signals too weak to be reliably distinguished. Second, the initialization process in NMR quantum computing was problematic. Unlike other qubit technologies, which can start from a well-defined ground state, NMR works at room temperature, where the nuclear spins are almost equally distributed across quantum states, reducing the effective polarization to a very small value. This results in a phenomenon known as the "pseudo-pure" state, where only a tiny fraction of the spins contribute to the useful quantum computation, leading to poor signal-to-noise ratios. Third, there was a theoretical realization that liquid-state NMR quantum computing, as initially conceived, might not provide any true advantage over classical computing for many tasks due to the lack of entangled states, essentially operating in a regime that classical simulations could also efficiently handle. And the rise of other more promising technologies such as trapped ions, superconducting circuits, and later developments in photonic and topological quantum computing, presented pathways with clearer scaling advantages. In 1995, Cirac and Zoller proposed using trapped ions as a practical quantum computing platform (Cirac and Zoller 1995), laying the groundwork for ion trap quantum technology. In the same year, David Wineland's group achieved the first demonstration of a quantum gate operation using trapped ions, giving the impulse to practical quantum computations (Monroe et al. 1995). In 1998, notable progress in the experimental demonstration of entanglement between multiple ions was achieved by the same group at the National Institute of Standards and Technology (NIST), a critical step for building scalable quantum systems. Another significant milestone was reached in 2004 with the demonstration of quantum teleportation between ions over a distance within a trap (Riebe et al. 2004), an experiment that emphasized the advanced manipulation capabilities of ion-trap systems. In the following years, the advancement of microfabricated trap architectures opened new possibilities for scaling ion-trap systems, such as the implementation of quantum error correction protocols was successfully demonstrated using ion traps, crucial for maintaining coherence and reliability in quantum computations (Chiaverini et al. 2004). All these defined trapped ions as one of the most prospective platforms for scalable quantum computations. Another most popular and promising platform for realization of quantum computations is superconducting qubits. A first significant experimental milestone in superconducting quantum computing was achieved in 1999 with the demonstration of the first superconducting qubit based on a Josephson junction (Nakamura et al. 1999). This breakthrough demonstrated the feasibility of using superconducting circuits to manipulate quantum states. Subsequently, the development of the phase qubit by Martinis and colleagues in 2002 (Martinis et al. 2002), marked another step toward practical quantum computation. In 2004, the implementation of the first two-qubit gate in a superconducting system was a pivotal moment, enabling entanglement between qubits and advancing towards logic operations necessary for quantum algorithms (Yamamoto et al. 2003). Another landmark was reached in 2008 with improved coherence times of superconducting qubits through innovations such as the introduction of transmon qubits (Koch et al. 2007), which reduced susceptibility to charge noise. The advent of circuit quantum electrodynamics (cQED) in the mid-2000s provided a robust framework for coupling qubits to microwave photons (Wallraff et al. 2004), further enhancing control and readout capabilities. These efforts culminated in the development of small-scale quantum devices, each demonstrating the principles underpinning quantum computation (see Fig. 1).

These small-scale quantum devices with cloud access, such as provided by IBM, attracted a lot of attention in developing a new class of algorithms, called variational quantum algorithms (VQAs) (Cerezo et al. 2021). VQAs represent a class of quantum algorithms that leverage the principles of quantum mechanics to solve

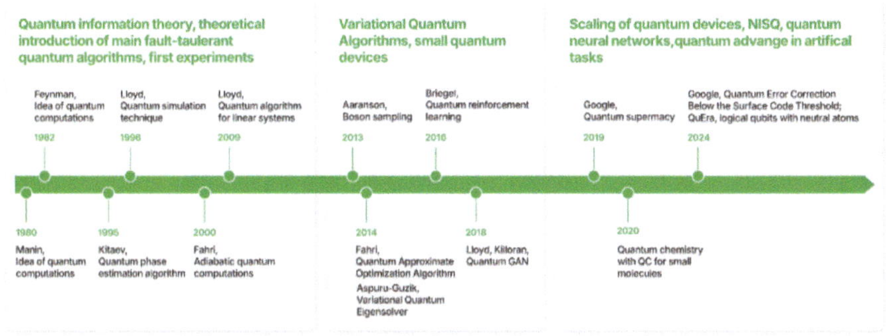

Fig. 1 Development of quantum information processing

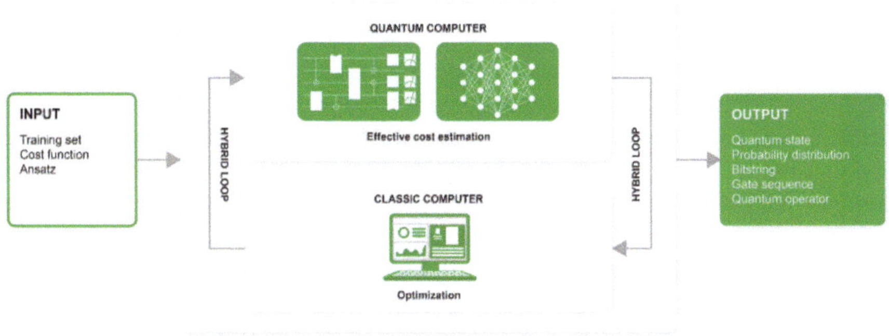

Fig. 2 Principal scheme for variational quantum algorithms. Variational quantum algorithms work in a hybrid loop between classical and quantum computing and involve trainable parameterized quantum circuits for effective cost estimation and classical optimization techniques to minimize the cost function

optimization problems and perform quantum simulations utilizing restricted quantum resources (Bharti et al. 2022). VQAs combine classical optimization techniques with quantum circuits (see Fig. 2), making them suitable for the current state of quantum hardware. A quantum circuit is used on the stage of preparation of quantum state, and the output of this circuit is measured to estimate a cost function. The quantum circuits in VQAs are parameterized, meaning that they depend on a set of adjustable parameters (often angles in quantum gates). By tuning these parameters, the algorithm can explore different quantum states and find the one that minimizes (or maximizes) the cost function. The cost function is a crucial component of VQAs. It quantifies the quality of the quantum state produced by the circuit. In many applications, this cost function is related to the energy of a quantum system, making VQAs particularly useful for quantum chemistry and materials science. This cost function is then optimized using classical optimization techniques, such as gradient descent or genetic algorithms. The iterative process continues until the optimal parameters for the quantum circuit are found. Variational Quantum Eigensolver

(VQE) (Peruzzo et al. 2014) is one of the most well-known VQAs, designed to find the ground state energy of a quantum system. It uses a parameterized quantum circuit to prepare trial states and measures the expectation value of the Hamiltonian (the energy operator) to evaluate the cost function. Classical optimization is then employed to adjust the parameters, iteratively improving the energy estimate until convergence is achieved. VQE has been successfully applied to small molecular systems, demonstrating its potential for quantum chemistry (Tilly et al. 2022). Quantum Approximate Optimization Algorithm (QAOA) (Farhi et al. 2014; Blekos et al. 2024) is another prominent VQA, aimed at solving combinatorial optimization problems. It uses a parameterized quantum circuit to encode the problem's cost function and applies a series of quantum gates to explore the solution space. The algorithm alternates between applying a problem-specific Hamiltonian and a mixing Hamiltonian, with parameters optimized classically to maximize the probability of measuring the optimal solution. QAOA has shown promise in solving problems like Max-Cut and other NP-hard problems. Variational Quantum Classifier (VQC) (Li and Deng 2021; Havlíček et al. 2019; Schuld 2019a) is a VQA designed for machine learning tasks. It employs a parameterized quantum circuit to classify data points by encoding classical data into quantum states. The circuit's parameters are optimized to minimize the classification error, allowing the algorithm to learn from the data. VQC has been explored for applications in quantum-enhanced machine learning (Haug et al. 2023; Bartkiewicz et al. 2020), where it may outperform classical classifiers in certain scenarios. The performance of VQAs heavily relies on the classical optimization process, which can be challenging, especially in high-dimensional parameter spaces. The optimization landscape may contain many local minima, making it difficult for classical optimizers to find the global minimum. Currently, researchers are exploring advanced optimization techniques (Stokes et al. 2020; Schuld 2019b), such as gradient-free methods and adaptive algorithms, to enhance the performance of VQAs (Bonet-Monroig 2023). Quantum machine learning (QML) algorithms are based on the same variational nature (Biamonte et al. 2017; Benedetti et al. 2019; Sajjan et al. 2022). There is a significant connection between the parameters of parameterized quantum circuits (PQCs) and the weights of classical neural networks. Just as neural networks consist of layers of interconnected nodes (neurons) where each connection has an associated weight, PQCs are composed of layers of quantum gates, each governed by parameters. In both cases, these parameters (weights in neural networks and angles in quantum gates) dictate how input data is transformed as it passes through the model. In classical neural networks, the learning process involves adjusting the weights through techniques like backpropagation and gradient descent to minimize a loss function. Similarly, in QML, the parameters of a PQC are optimized using classical optimization methods to minimize a cost function that quantifies the model's performance. This iterative adjustment process is fundamentally the same, as both aim to find the optimal set of parameters that yield the best predictions or classifications. However, the core idea behind QML is to leverage the unique properties of quantum mechanics such as genuine randomness, superposition, entanglement, and interference to enhance machine learning tasks. The advantages of QML include the potential for exponential expressiveness, enabling the representation of complex functions with fewer parameters than classical models (Abbas et al. 2021; Liu et al. 2021a; Du and Tao 2024; Wang et al. 2021). Additionally, the variational nature of these algorithms provides resilience to noise, making them more robust in practical applications in comparison with other quantum algorithms (Fontana et al. 2021; Sharma et al. 2020; Ilin and Arad 2024). As quantum technology continues to advance, QML is expected to scale to larger datasets and more complex problems, opening new avenues for applications in fields such as drug discovery (Pyrkov et al. 2023), finance (Doosti et al. 2024), healthcare (Pyrkov et al. 2024), and materials science (Weidman et al. 2024). Despite its promise, QML faces several challenges, including the need for efficient data encoding (Daimon and Matsushita 2024), the complexity of classical optimization, barren plateaus (McClean et al. 2018; Larocca et al. 2024), and the risk of getting trapped in local minima during training (Wierichs et al. 2020). Ongoing research aims to address these challenges by developing advanced optimization techniques and exploring new encoding strategies.

At the same time with developing VQAs, the fast progress in quantum hardware was observed. Companies such as IBM (IBM Quantum Computing n.d.), Google (Google Quantum AI n.d.), Rigetti (Quantum Computing n.d.), and D-Wave (D-Wave Systems The Practical Quantum Computing Company n.d.) have been at the forefront of developing superconducting qubit-based systems. IBM's Quantum Experience platform, launched in 2016, allowed researchers and developers to access quantum computers via the cloud, democratizing access to quantum computing resources. In 2019, Google claimed to have achieved quantum supremacy with its 53-qubit processor, Sycamore, which performed a specific computation in 200 seconds that would take the most powerful classical supercomputers thousands of years to complete (Arute et al. 2019). This announcement not only validated the potential of quantum computing but also spurred increased investment and interest in the field. Recent developments in quantum hardware have been geared towards increasing the number of qubits and improving their quality. Notable advancements in this area include IBM's Condor processor, which features 1121 qubits (IBM n.d.). Google has also made strides with its Sycamore processor, focusing on performing tasks beyond the capability of classical supercomputers, recently Google demonstrated quantum error correction below surface code threshold (Acharya et al. 2024). These processors are setting benchmarks for qubit count, but it is not just about numbers; nowadays enhancing qubit fidelity and coherence times is crucial. Scalable error correction is the main goal in this ecosystem. Trapped ion technology has been a game-changer in terms of high fidelity and precision. IonQ (n.d.), one of the leaders in this technology, is working on expanding its hardware capabilities and optimizing the laser systems used to manipulate ion qubits. The system's ability to maintain coherence over long periods of time gives it an edge in various quantum computing protocols. Quantinuum (n.d.) continues to innovate using a combination of trapped ions and other quantum technologies, pushing the envelope in terms of qubit connectivity and integration, accelerated their roadmap to achieve universal, fault-tolerant quantum computing by 2030. QuEra Computing, a leading innovator in the field of neutral atoms quantum computing, has unveiled an ambitious strategic plan to develop a series of error-corrected quantum computers. The initiative kicked off in 2024 and aims to culminate in the creation of a system featuring 100 logical error-corrected qubits by 2026 (QuEra n.d.). This announcement signals the beginning of a transformative period in quantum computing.

2 Quantum Computing Paradigms

The primary approach to quantum computation is known as gate-based quantum computing (Nielsen and Chuang 2010), where computations are represented as a series of quantum logic gates, which are unitary transformations U that operate on qubits within a circuit (see Fig. 3a). This method involves three main phases: initialization, computation, and measurement. In the initialization phase, a quantum register is set up in a specific state, typically $|0\rangle^{\otimes n}$. During computation, qubits undergo transformations using quantum gates such as the Pauli gates (X, Y, Z), the Hadamard gate (H), and controlled gates like the CNOT (C_X), collectively forming the unitary operation $U = U_m \cdot U_{m-1} \cdot \ldots \cdot U_1$ referred to as the quantum circuit. Finally, the qubits are measured, collapsing the quantum state $|\psi\rangle$ into classical bit values. This process needs repetition as measurement destroys the quantum state. Quantum gates can implement any n-qubit unitary operation by decomposing it into one- and two-qubit gates from a universal set. However, current gate-based quantum computers suffer from noise and decoherence, restricting the depth of quantum circuits to just a few dozen operations.

Adiabatic quantum computing (AQC) or quantum annealing offers an alternative approach (Albash and Lidar 2018), leveraging quantum adiabatic evolution to perform calculations (see Fig. 3b). Unlike gate-based computing, AQC does not rely on time-sequenced gates but instead focuses on solving optimization problems by evolving a system from a simple Hamiltonian H_0 to a complex one H_1. According to the adiabatic theorem, if the system starts in the ground state $|\psi_0\rangle$ of H_0 and the Hamiltonian is evolved slowly $H(t) = (1 - t/T)H_0 + (t/T)H_1$ where T is the total time, the system will remain in the ground state corresponding to the final Hamiltonian H_1, which encodes the solution to the problem. This ability to map problems, including NP-complete ones, to AQC in polynomial time highlights its versatility.

Boson sampling (Lund et al. 2017), or Gaussian boson sampling, is a quantum computing model that reveals quantum advantage for specific problems by using an optical network of beam splitters and phase shifters described by a unitary matrix U (see Fig. 3c). In this structure, a set of n indistinguishable single photons are input, evolving according to U. The photons passing through the network of m possible channels, forms an entangled state making classical computation exponentially challenging. Boson sampling optical setup can produce output probability distributions encoding solutions to complex problems. While initially theoretical, this method is gaining attention in practical applications such as determining molecular vibronic spectra and solving graph-based problems.

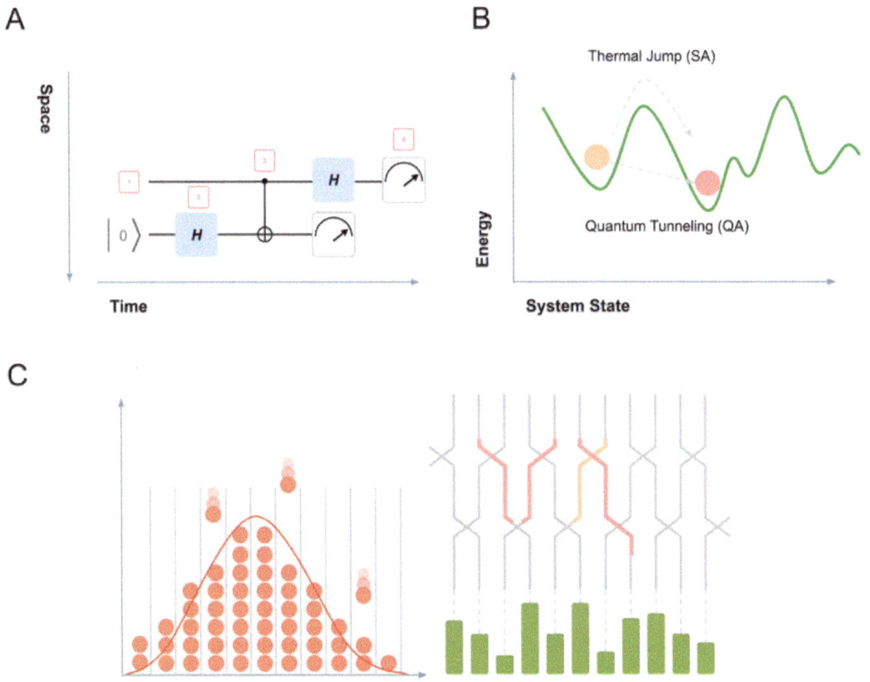

Fig. 3 Quantum computing paradigms. (**a**) Gate-based quantum computing utilizes quantum gates to manipulate qubits in a circuit model. (**b**) Adiabatic quantum computing relies on slowly evolving a quantum system from an easy-to-prepare ground state to the ground state of a problem Hamiltonian with quantum annealing. (**c**) Boson sampling is a non-universal quantum computing paradigm that focuses on sampling the output distribution of indistinguishable bosons passing through a linear optical network, demonstrating quantum advantage for specific tasks

All these quantum computation paradigms operate with quantum states as input and encoding of classical data into quantum states is necessary. The selection of a data encoding strategy is essential in quantum computations and in particular in QML (Rath and Date 2024), as it can greatly influence performance and affect the underlying computational complexity. We outline the most widely used methods below but the methods for the data encoding are under very active research.

Basis encoding is one of the most usable methods for encoding classical data into quantum states. It is a simple method for encoding binary data into a quantum system. Specifically, let D represent a classical binary dataset, where each element $x_m \in D$ is an N-bit binary string of the form $x_m = (b_{m1}, b_{m2}, \ldots, b_{mN})$, with b_{mj} being either 0 or 1. The classical dataset can then be represented by the quantum state $|D\rangle$ of N qubits, where M is the total number of basis states used for the encoding:

$$|D\rangle = \frac{1}{\sqrt{M}} \sum_{m=1}^{M} | x_m\rangle.$$

Here, x_m corresponds to the m-th element of the dataset. For example, the binary string $[1, 0, 1]$ is directly encoded into quantum state $|101\rangle$. To work with real numbers, it is needed to represent real numbers as binary strings and then transform them into a quantum state in the computational basis. The resulting quantum state is a bit-wise representation of a binary string, translated into the corresponding states of the quantum subsystems.

Amplitude encoding enables the representation of complex-valued floats as the amplitudes of a quantum state. A given complex vector x with length N can be encoded into $\log(N)$ qubits. Specifically,

$$x \to U_x|0\rangle^{\otimes \log(N)} = |x\rangle = \sum_{k=0}^{N-1} \alpha_k \mid k\rangle.$$

The primary benefit of amplitude encoding is that it requires only $n = \log NM$ qubits to encode a dataset D consisting of M inputs, each with N features. If the quantum machine learning algorithm operates in polynomial time, it exhibits a logarithmic runtime dependency on the size of the dataset. This notion of exponential speedups from qubit-efficient quantum machine learning algorithms may seem counterintuitive to machine learning practitioners, as simply loading the NM features from memory is inherently linear in NM. Indeed, these speedup claims are valid only if state preparation can also be achieved in a qubit-efficient manner. This is only feasible in certain cases that possess significant structure. For instance, to create a vector with uniform entries, we can simply apply n Hadamard gates, resulting in a qubit-efficient state preparation time. Conversely, there are subspaces within a Hilbert space that cannot be accessed from a given initial state using qubit-efficient algorithms. Thus, it remains a significant and complex open question to determine which classes of relevant states for machine learning using amplitude encoding can be prepared efficiently in terms of qubits.

Angle encoding is a method used to represent classical data within quantum states by leveraging the concept of parameterizing qubit states through rotational angles. In quantum computing, a single qubit can be visualized on the Bloch sphere, where any quantum state $|\psi\rangle$ can be described by spherical coordinates using rotation angles. Given a classical data point $x \in R$, angle encoding maps this data onto a qubit by associating it with rotation angles $\theta(x)$ around specific axes. A common approach is to define a rotation gate, such as the $R_y(\theta)$ gate, which is expressed as $R_y(\theta(x)) = \exp(-i\theta(x)Y/2)$, where Y is the Pauli-Y matrix. This operation rotates the qubit state about the y-axis on the Bloch sphere, transforming the initial state, often $|0\rangle$, into a new state: $|\psi(x)\rangle = R_y(\theta(x))|0\rangle$. Through this mechanism, multiple qubits can be used to encode a vector of classical data points $x = (x_1, x_2, \ldots, x_n)$, resulting in a quantum state that represents the entire dataset.

Another crucial part for effective realization of quantum computing algorithms is quantum compiling (Moro et al. 2021; Maronese et al. 2021; Khatri et al. 2019). It involves transforming high-level quantum algorithms into low-level instructions that can be executed on quantum hardware. The goal is to optimize the performance of quantum circuits while minimizing errors and resource usage for NISQ devices. It usually consists of several steps such as gate decomposition, scheduling, mapping to hardware, and so on. Gate decomposition involves breaking down complex quantum gates into simpler, more manageable gates that can be directly implemented on quantum hardware. This is essential because most quantum computers have a limited set of native gates. Techniques such as the Solovay–Kitaev theorem or Trotterization provide frameworks for approximating any unitary operation using a finite set of gates. Circuit optimization techniques aim to reduce the number of gates and depth of quantum circuits, which are critical for minimizing decoherence and error rates. Common methods include identifying and reusing repeated subcircuits and merging adjacent gates into a single gate when possible. Mapping quantum circuits to specific hardware architectures is essential for efficient execution on NISQ devices. This involves assigning logical qubits to physical qubits while considering connectivity constraints and error rates. Techniques such as the Quantum Routing Algorithm help in finding optimal mappings. Then, error mitigation and correction techniques are applied. High-level synthesis (HLS) for quantum computing involves generating quantum circuits from high-level descriptions, such as quantum programming languages. Tools like Qiskit (n.d.), QuTip (n.d.), and Cirq (n.d.) provide frameworks for HLS, allowing developers to focus on algorithm design rather than low-level implementation details. Readout is another fundamental step in quantum computing that transforms quantum information into classical information and may increase complexity of quantum algorithms in the meaning of time-to-solution due to the probabilistic nature of measurements in quantum mechanics.

Below we focus on high-level implementations of QML algorithms.

3 Quantum Machine Learning Algorithms

The exciting potential of quantum computing has inspired researchers to create quantum analogs for a diverse array of classical machine learning techniques. We will explore the primary quantum machine learning methods that have emerged, highlighting their unique features, advantages, and the specific classical algorithms they aim to enhance or replicate. By examining these developments, we can gain a deeper understanding of how quantum computing may revolutionize the field of machine learning and open up new avenues for data analysis and problem-solving in drug discovery.

General perspectives of QML algorithms can be estimated from the point of view of learning theory. Learning theory focuses on establishing a solid mathematical foundation for learning from data. Key questions in this field include determining the number of examples needed to learn a specific function and the computational resources required to complete a learning task. Depending on the data access model and learning goals, different learning models can be defined. Two notable models are the probably approximately correct (PAC) framework, developed by Valiant (1984), and the statistical learning theory by Vapnik (1998).

In these models, a learner aims to approximate an unknown function using a training set of input–output pairs. The training set examples are assumed to be drawn from an unknown probability distribution, and predictions are tested on points from the same distribution. The efficiency of an estimator is measured by two quantities: sample complexity and time complexity. Sample complexity refers to the minimum number of examples required to learn a function up to a certain level of approximation, which is related to the capacity of the hypotheses space and the regularity of the data distribution. Time complexity corresponds to the runtime of the best learning algorithm.

A learning algorithm is considered efficient if its runtime is polynomial in the dimension of the domain elements and inverse polynomial in the error parameters. The primary goal is to find a model that not only fits the training examples well but also guarantees good prediction performance on new observations. This ability to generalize, known as generalization capability, is a crucial aspect that distinguishes machine learning from standard optimization techniques.

Since the 1990s, researchers have been investigating how quantum resources influence learning efficiency. For a detailed overview of the current landscape of quantum learning theory, we recommend the review by Arunachalam and de Wolf (2017).

The quantum PAC model was established under the premise that learners have access to a quantum computer and an oracle that provides the training set in quantum superposition (Bshouty and Jackson 1998). Research has demonstrated that, regarding sample complexity, the quantum PAC model is comparable to the classical PAC model, differing only by constant factors (Servedio and Gortler 2004; Zhang 2010). This suggests that quantum mechanics does not significantly lessen the amount of data required for learning. However, in specific learning frameworks, such as exact learning, quantum learners can achieve polynomially greater efficiency than their classical counterparts in terms of queries to the data oracle (Angluin 1988; Servedio and Gortler 2004).

While quantum and classical models are similar in sample complexity, the quantum PAC model presents advantages in time complexity (Klivans and Servedio 2001). For example, learning disjunctive normal forms (DNFs) is a difficult problem in classical contexts, with the best-known algorithms exhibiting exponential time complexity. In contrast, the quantum PAC model allows for efficient learning of DNFs under uniform distribution (Bshouty and Jackson 1995), facilitated by an algorithm that enables exponentially faster sampling from the probability distribution defined by the coefficients of a Boolean Fourier transform (Bernstein and Vazirani 1997). Furthermore, DNF formulas can be learned efficiently even in the presence of noise.

Another instance where quantum resources are thought to be essential for efficient learning involves a class of functions identified by Kearns and Valiant (1994). This class is known to be provably difficult to learn using classical methods, yet it can be learned efficiently within the quantum PAC model, largely due to Shor's quantum factoring algorithm.

Overall, findings from quantum learning theory indicate that by harnessing quantum mechanical effects, it is possible to achieve improved generalization error or to learn functions that are challenging for classical learners, depending on the specific learning model in question.

As for the particular QML algorithms, most of the QML algorithms are naturally realizable in the gate-based paradigm. The main exclusion is the Restricted Boltzmann Machine algorithm (Zhang et al. 2018) that can be naturally realized in the adiabatic quantum computing paradigm (Albash and Lidar 2018) and we start our overview of QML algorithms with this algorithm.

3.1 *Quantum Restricted Boltzmann Machines (QRBMs)*

Restricted Boltzmann Machines (RBMs) is a class of probabilistic graphical models that are used for dimensionality reduction, classification, regression, collaborative filtering, feature learning, and topic modeling (Zhang et al. 2018; Melko et al. 2019). They are particularly well suited for recommendation systems. Here is a simplified explanation of how RBMs work on the example of a movie recommendation system:

1. An RBM has two layers: a visible layer v with n visible units and a hidden layer h with m hidden units. The visible layer represents the input data. In a movie recommendation system, this could be the user's ratings for different movies. The hidden layer learns to capture features or patterns in the visible data. These features might represent user preferences or movie genres (see Fig. 4a).
2. Every visible node is connected to every hidden node, and these connections have associated weights. However, there are no connections between the nodes within the same layer.
3. For a movie recommendation system, the input data could be a matrix where rows represent users and columns represent movies. Each visible unit v_i and hidden unit h_j are binary (they take values of 0 or 1, not watched or watched for input). The weights of the connections between visible and hidden nodes are initialized randomly.
4. The goal of training is to adjust the weights so that the model can correctly reconstruct the input data. There are two phases in training. Forward propagation: the visible layer is set to the data vector (user movie ratings), which is processed to compute the activations of the hidden units. This involves multiplying the visible vector by the weight matrix and passing it through a non-linear activation function (such as a sigmoid). This probabilistic activation attempts to capture

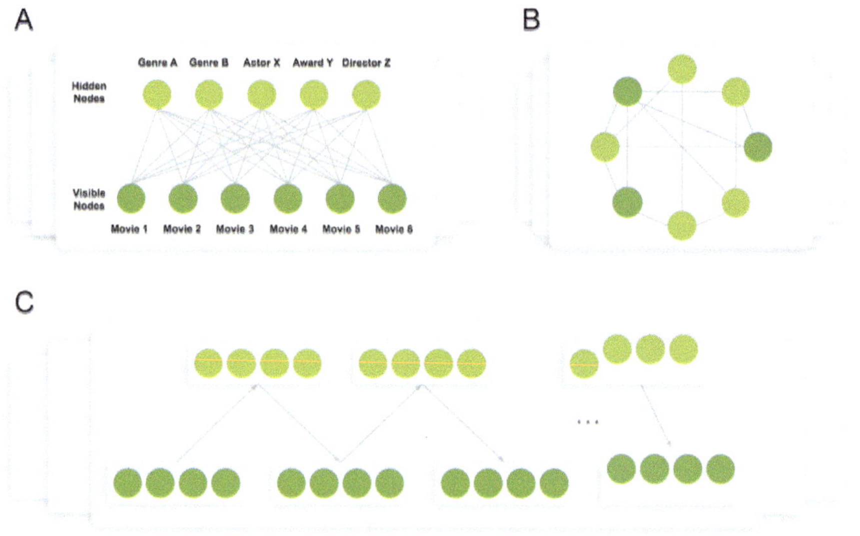

Fig. 4 Schematic representation of a Quantum Restricted Boltzmann Machine (QRBM). (**a**) The RBM consists of two layers: a visible layer representing the input data and a hidden layer capturing the latent features. The connections between the visible and hidden layers are represented by weighted edges, which are adjusted during the training process to optimize the model's performance. (**b**) Mapping a RBM to an Ising model involves interpreting the RBM's energy function in terms of the Ising Hamiltonian. In this mapping, the visible and hidden units of the RBM correspond to spins in the Ising model, where the interactions between spins are defined by the weights of the connections in the RBM. This relationship allows for the application of statistical mechanics techniques to analyze the learning dynamics and properties of RBMs. (**c**) Training a RBM evolves using contrastive divergence, a stochastic gradient descent algorithm that updates the weights by minimizing the difference between the data distribution and the model's distribution through iterative sampling and reconstruction of the visible layer. This architecture allows for efficient sampling and representation of high-dimensional data, showcasing the potential of quantum computing in machine learning applications

the features in the hidden layer. Backward propagation: from the hidden layer's activations, reconstruct the visible layer data. This is done by multiplying the hidden vector by the transpose of the weight matrix and passing it through the activation function. The contrastive divergence tries to minimize the difference between the original and reconstructed inputs, adjusting weights to reduce this reconstruction error. The weights are updated using the difference between the product of input and hidden layer activations in the positive phase and the product in the negative phase (see Fig. 4c).

Training is organized with the concepts of Energy-Based Model. The energy of a configuration (v, h) in an RBM is given by:

$$E(v,h) = -\sum_{i=1}^{n}\sum_{j=1}^{m} v_i h_j w_{i,j} - \sum_{i=1}^{n} v_i a_i - \sum_{j=1}^{m} h_j b_j$$

where $w_{i,j}$ is the weight between visible unit v_i and hidden unit h_j. a_i is the bias of visible unit v_i. b_j is the bias of hidden unit h_j.

The joint probability distribution of visible and hidden units is given by the Boltzmann distribution:

$$P(v,h) = \frac{1}{Z} e^{-E(v,h)}$$

where Z is the partition function:

$$Z = \sum_{v}\sum_{h} e^{-E(v,h)}$$

The marginal probability of a visible vector v is:

$$P(v) = \sum_{h} P(v,h)$$

Given the structure of RBMs, the conditional probability of hidden units given visible units is:

$$P(v_i = 1 \mid h) = \sigma\left(a_i + \sum_{j=1}^{m} h_j w_{i,j}\right)$$

Similarly, the conditional probability of visible units given hidden units is:

$$P(v_i = 1 \mid h) = \sigma\left(a_i + \sum_{j=1}^{m} h_j w_{i,j}\right)$$

where $\sigma(x) = \frac{1}{1+e^{-x}}$ is the sigmoid function.

The update rule for the weights is given by:

$$\Delta w_{i,j} = \eta\left(\langle v_i h_j\rangle_{\text{data}} - \langle v_i h_j\rangle_{\text{reconstruction}}\right)$$

where η is the learning rate, $\langle\cdot\rangle_{\text{data}}$ is the expectation observed in the training set, $\langle\cdot\rangle_{\text{reconstruction}}$ is the expectation computed from the reconstruction.

The biases are updated similarly:

$$\Delta a_i = \eta\left(\langle v_i\rangle_{\text{data}} - \langle v_i\rangle_{\text{reconstruction}}\right)$$

$$\Delta b_j = \eta \left(\langle h_j \rangle_{\text{data}} - \langle h_j \rangle_{\text{reconstruction}} \right)$$

To recommend movies to a user, input their existing movie ratings into the trained RBM. The RBM uses the existing ratings to determine activations of the hidden units, which represent latent features of the movie preferences. Use these hidden activations to reconstruct visible units, which in turn predict the ratings for movies that the user has not rated. These predicted ratings can be used to recommend new movies.

It can be easily reformulated to AQC problem with transforming the problem into Ising form (see Fig. 4b), which is given by:

$$E\left(s|, h|, J\right) = \sum_{i=1}^{N} h_i s_i + \sum_{i<j} J_{i,j} s_i s_j, \quad s_i \in \{-1, +1\}.$$

This represents an objective function of N variables $s = [s_1, s_2, \ldots, s_N]$ that correspond to physical Ising spins, where h_i are the biases and $J_{i,j}$ are the couplings between spins.

The weights and biases of an RBM trained using a binary dataset consisting of states 0, 1 can be converted to use states -1, 1 through the following mappings:

$$a\prime_i = \frac{a_i}{2} + \frac{\sum_j W_{i,j}}{4}$$

$$b\prime_i = \frac{b_i}{2} + \frac{\sum_j W_{i,j}}{4}$$

$$W' = \frac{W}{4}$$

These transformed weights and biases can then be used to run an RBM on AQC devices such as the D-Wave machines and other Ising solvers (Kālis et al. 2023; Dixit et al. 2021; Nomura 2021; Vieijra et al. 2020; Laydevant et al. 2024).

3.2 *Variable Depth Quantum Circuits*

A crucial part for effective realization of variational quantum computing algorithms including QML is the use of optimized parametrized quantum circuits in depth and hardware resources (Chong et al. 2017; Ge et al. 2024; Du et al. 2022). Variable Depth Quantum Circuits (VDQCs) are designed to adaptively adjust the depth of quantum circuits based on the problem's requirements and the computational constraints of available quantum hardware (Ippoliti et al. 2023; Wada et al. 2024;

Seidel et al. 2024). By varying the depth, these circuits can potentially optimize the trade-off between fidelity and circuit complexity, making them suitable for Noisy Intermediate-Scale Quantum (NISQ) devices. The depth of a quantum circuit is the maximum number of gates that act sequentially on any qubit. The depth of the circuit can be adjusted dynamically based on error rates, desired accuracy, and computational resources (Bäumer et al. 2024).

Consider a quantum state $|\psi\rangle$ represented by:

$$| \psi\rangle = \sum_{i=0}^{N-1} \alpha_i \mid i\rangle$$

where $N = 2^n$ for n qubits, and α_i are complex amplitudes satisfying $\sum_i |\alpha_i|^2 = 1$.

Represented by unitary operators U. A gate acting on $|\psi\rangle$ yields:

$$|\psi'\rangle = U\,|\psi\rangle$$

The depth D of a circuit refers to the sequential application of gates. For a sequence of gates U_1, U_2, ..., U_m, the circuit depth is the number of layers where gates act in parallel on disjoint sets of qubits. Begin with an initial set of gates $U(\theta, d)$ for a given depth d where θ are parameters. Adjust d based on feedback from the system, such as error rates or fidelity metrics. Increase depth $d \to d + 1$ if higher accuracy is required and error rates are manageable. Decrease depth $d \to d - 1$ if error accumulation exceeds a threshold or if faster computation is favored over precision. The goal is to find optimal parameters $\theta*$ and depth $d*$ that minimize a given cost function L, which may represent energy, loss, or another quantity of interest:

$$\min_{\theta, d} \{L\,(\theta, d)\}$$

Measure fidelity with respect to target state $|\varphi\rangle$:

$$F\,(\theta, d) = |\langle\varphi \mid U\,(\theta, d)|\,\psi\rangle|^2$$

Evaluate $F(\theta, d)$ and if fidelity improves, continue adjusting depth as necessary. Reevaluate based on noise and other real-world constraints (see Fig. 5).

Using simulators and hardware, dynamically updating gates and depths require real-time data processing. Advanced quantum programming environments and machine learning techniques may assist in efficiently navigating the optimal parameters and depths. Variable Depth Quantum Circuits offer an adaptable approach to quantum computation, especially useful on NISQ devices where error rates are significant. Through careful adjustment of depth, they allow experimentation with complex algorithms while managing inherent quantum noise and optimizing quantum resources.

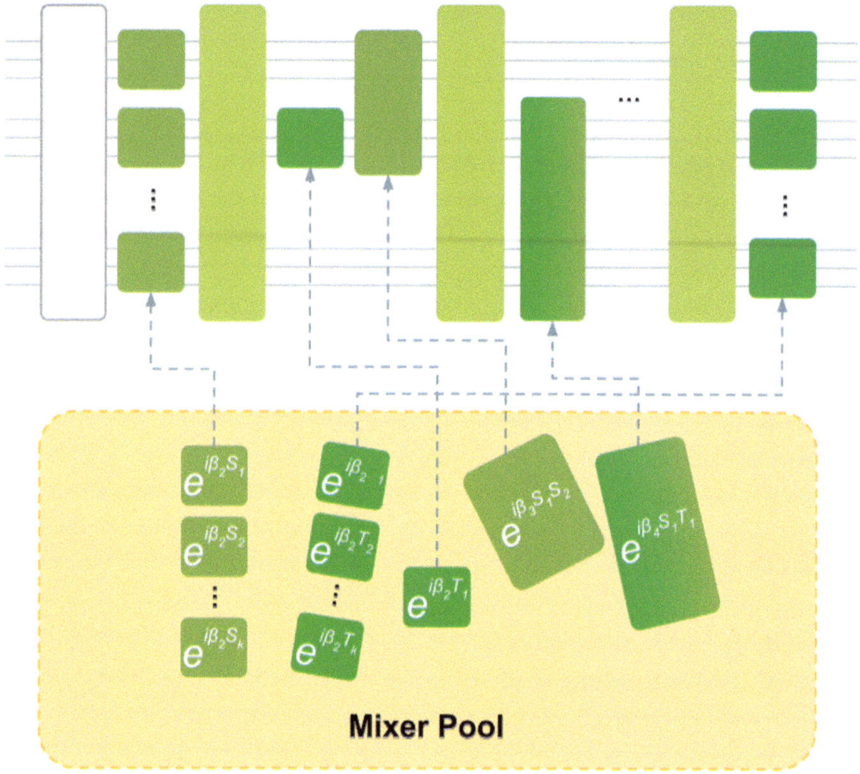

Fig. 5 Variable depth quantum circuit design allows to dynamically arrange of quantum gates across multiple layers taking into account hardware restrictions. Each layer represents a set of quantum operations applied to different qubits, with the depth of the circuit indicating the number of sequential gate applications. This design allows for the optimization of quantum algorithms by adjusting the circuit depth to balance between computational complexity and the fidelity of quantum state manipulation

3.3 Quantum Support Vector Machines

Support Vector Machines (SVM) with Quantum Kernel Estimators leverage quantum computing principles to enhance the classical SVM algorithm, particularly in tasks like clustering or classification. Quantum kernel methods take advantage of the high-dimensional Hilbert space that quantum systems inhabit, offering potentially exponential speedups over classical methods for certain problems (Rebentrost et al. 2014). In a classical SVM, the goal is to find a hyperplane that best separates data points of different classes. For two classes, the optimization problem is stated as:

$$\min_{w,b} \parallel w \parallel^2$$

subject to:

$$y_i\,(w \cdot x_i - b) \geq 1, \quad i = 1, 2, \ldots, n$$

where x_i are the data points, y_i are the class labels, and w is the normal vector to the hyperplane.

When data is not linearly separable in its original space, the kernel trick can transform the data implicitly into a higher-dimensional space where it becomes linearly separable. This is done using a kernel function $K(x_i, x_j) = \phi(x_i) \cdot \phi(x_j)$.

Quantum Kernel Estimators aim to map classical data into a quantum Hilbert space and exploit quantum states properties to compute these kernel functions more efficiently.

Given two data points x_i and x_j, the quantum kernel is defined as:

$$K_Q\left(x_i, x_j\right) = \left|\left\langle \Phi\left(x_i\right) \middle| \Phi\left(x_j\right)\right\rangle\right|^2$$

where $|\Phi(x)\rangle$ is the quantum state corresponding to classical data point x.

The quantum state can be prepared by encoding the classical data into a quantum circuit. For example, using parameterized quantum circuits, classical input features are encoded into the amplitudes or the gates applied to a quantum circuit.

Suppose you want to cluster a dataset into two classes. Harnessing the power of quantum computation, we can exploit the quantum kernel to efficiently compute distances and similarities in an exponentially large feature space: (1) Implement a quantum feature map using a parameterized quantum circuit $U(x)$ that maps classical data into a quantum state; (2) Compute the quantum kernel for all pairs of points in your dataset:

$$K_Q\left(x_i, x_j\right) = \left|\left\langle 0 \middle| U^{\dagger}\left(x_i\right) U\left(x_j\right) \middle| 0\right\rangle\right|^2$$

1. Use these quantum kernels in a classical SVM solver to find the optimal separation hyperplane in the high-dimensional space effectively created by the quantum kernel.
2. Use the solution from the SVM solver to classify new data points or to assign clusters.

The quantum kernel provides a computational advantage by performing certain operations much faster than classical algorithms, potentially allowing for the handling of more complex datasets and higher-dimensional spaces (Schuld and Petruccione 2021). Current quantum hardware is prone to errors (Li et al. 2015; Gentinetta et al. 2024), which can affect performance in this approach. Error mitigation strategies are necessary for reliable results.

3.4 Quantum k-Nearest Neighbors

Quantum k-Nearest Neighbors (kNN) algorithm (Wiebe et al. 2014) is based on the measurement of overlap between two quantum states that serves as a similarity metric analogous to Euclidean distance. This overlap can be quantified using a simple circuit called a swap test. This circuit is instrumental for measuring the "distances" or similarities between classical vectors in k-Nearest Neighbors (kNN) algorithms.

The circuit begins with input data represented in quantum states $|0\rangle$ or $|1\rangle$, formatted as $|q_1\rangle|q_2\rangle \ldots \mid q_n\rangle$. Successive controlled-SWAP gates, also known as Fredkin gates, utilize an auxiliary qubit for control. The reference state is processed by a unitary operation V which forms a test circuit in contrast to the reference circuit.

When comparing a qubit from the reference state to its counterpart in the test circuit, if the output state $|\psi\rangle$ of the test circuit matches the reference state $|\phi\rangle$, the probability of measuring the auxiliary bit in the state $|0\rangle$ is one. However, if the states differ, the probability of measuring the auxiliary bit in the zero state is given by:

$$P\left(|0\rangle_{\text{anc}}\right) = \frac{1}{2} + \frac{1}{2}\left|\left\langle\psi\middle|\phi\right\rangle\right|^2$$

This probability reflects the overlap between the two states.

Using d distinct pairs of input and reference states, a classical d-dimensional vector S is created, containing the measured probabilities for each pair of states:

$$S = [P_1, P_2, \ldots, P_d]$$

This vector S can then be employed in classical algorithms for further processing or analysis. Recently, a quantum k-nearest neighbors algorithm based on the Euclidean distance estimation was also proposed (Zardini et al. 2024).

3.5 Quantum Multi-class Classification

Multi-class classification is a fundamental task in machine learning. Theoretical insights suggest that quantum classifiers (Chalumuri et al. 2021; Giuntini et al. 2023; Pillay et al. 2024a; Du et al. 2023) can map data into high-dimensional quantum feature spaces, potentially achieving better separation of classes that are not linearly separable in classical spaces, akin to the kernel trick in classical machine learning.

Given a dataset

$$D = \{(x_i, y_i)\}_{i=1}^{M},$$

consisting of training data $\{x_i\}_{i=1}^{M}$ and their respective labels $\{y_i\}_{i=1}^{M}$, the goal of multi-class classification is to develop a model for classifying unlabeled data. This section describes the quantum multi-class classification algorithm on the basis of SWAP-test (Pillay et al. 2024a).

First step for the algorithm is to encode the test data, the training data, and their respective labels into a quantum state

$$| \Psi_i \rangle = \sum_{m=1}^{M} \sqrt{w_m} \mid 0 \rangle \mid \widetilde{x} \rangle \mid x_m \rangle \mid y_m \rangle \mid m \rangle .$$

The ancillary qubit is initialized in the ground state $|0\rangle$, the index register $\sum_{m=1}^{M} \sqrt{w_m} \mid m\rangle$ with $\sum_{m=1}^{M} w_m = 1$ is used to link the training data to their respective labels, and two different registers are used for test and training. Some unitary operator $U_{\Phi(x)}$ encodes the test and training data into quantum states as $\mid \widetilde{x} \rangle = U_{\Phi(\widetilde{x})} \mid 0 \rangle$ and $\mid x_m \rangle = U_{\phi(x_m)} \mid 0 \rangle$, respectively. This operation can be understood as applying a feature map to the classical data. The choice of operator, which corresponds to a specific encoding method, determines the type of kernel that the classifier will evaluate. The label qubit $|y_m\rangle$ is used to store the label state that each label y_m is mapped to. Each label state

$$| y_m \rangle = \cos\left(\frac{\theta_{y_m}}{2}\right) \mid 0 \rangle + e^{i\phi_{y_m}} \sin\left(\frac{\theta_{y_m}}{2}\right) \mid 1 \rangle$$

with $0 \leq \theta_{y_m} \leq \pi$ and $0 \leq \phi_{y_m} \leq 2\pi$ has an associated Bloch vector

$$y_m = \left(\cos \phi_{y_m} \sin \theta_{y_m}, \ \sin \phi_{y_m} \sin \theta_{y_m}, \ \cos \theta_{y_m} \right)$$

which will be referred to as the label vector of y_i. The label states are selected such that their Bloch vectors are maximally separated on the Bloch sphere.

A modified SWAP-test is implemented on the test and training data registers. A C-SWAP gate conditioned on the ancilla, in between two Hadamard gates applied to the ancilla, swaps the two registers. The result is

$$| \Psi_f \rangle = H_a.C - \text{SWAP}.H_a \mid \Psi_i \rangle \quad = \sum_{m=1}^{M} \frac{\sqrt{w_m}}{2} (|0\rangle \mid \psi_+ \rangle + \mid 1 \rangle | \psi_- \rangle) | y_m \rangle | m \rangle$$

with $\mid \psi_\pm \rangle = \mid \widetilde{x} \rangle \mid x_m \rangle \pm \mid x_m \rangle \mid \widetilde{x} \rangle$.

Then, state tomography in the updated basis is performed (see Pillay et al. 2024a for details) with the results of these measurements are used to construct a vector which can be referred as the predicted vector y_{pred}

$$
\begin{aligned}
y_{\text{pred}} &= \left(Tr\left(\sigma_z^{(l)} \rho_{fx}\right) \; Tr\left(\sigma_z^{(l)} \rho_{fy}\right) \; Tr\left(\sigma_z^{(l)} \rho_{fz}\right) \right), \\
&= \left(\sum_m w_m \left| \left\langle \widetilde{x} \middle| x_m \right\rangle \right|^2 \cos\phi_{y_m} \sin\theta_{y_m} \sum_m w_m \left| \left\langle \widetilde{x} \middle| x_m \right\rangle \right|^2 \sin\phi_{y_m} \sin\theta_{y_m} \right. \\
&\quad \left. \sum_m w_m \left| \left\langle \widetilde{x} \middle| x_m \right\rangle \right|^2 \cos\theta_{y_m} \right).
\end{aligned}
$$

The fidelities that are obtained from the modified SWAP-test represent a valid kernel $k\left(\widetilde{x}, x_m\right) = \left| \left\langle \widetilde{x} \middle| x_m \right\rangle \right|^2$. If $\alpha_i = \sum_{m|y_m=i} w_m k\left(\widetilde{x}, x_m\right)$, the predicted vector can be expressed as

$$
y_{\text{pred}} = \sum_{i=1}^{L} \alpha_i \left(\cos\phi_{y_i} \sin\theta_{y_i}, \; \sin\phi_{y_i} \sin\theta_{y_i}, \; \cos\theta_{y_i} \right)
$$

or equivalently,

$$
y_{\text{pred}} = \sum_{i=1}^{L} \alpha_i \, y_i .
$$

Thus, the predicted vector is a linear combination of the label vectors. The weight of each label vector, α_i, is the sum of the kernel values between the test data and the training data that have that label and high α_i indicates a high similarity between the test data and the training data belonging to class y_i. The test data can be assigned to the class y_i when the inner product between some label vector y_i and the predicted vector y_{pred} is the highest. It has been shown that the multi-class classification algorithm can handle O(R) classes, where R is the number of executions of the required circuits.

Quantum multi-class classifiers offer the potential for enhanced performance by exploiting quantum entanglement and superposition, potentially enabling processing of high-dimensional data more efficiently than classical models (Pillay et al. 2024b). However, leveraging quantum advantage depends on hardware capabilities and the problem structure.

4 Quantum Neural Networks

While quantum versions of traditional machine learning techniques show a marked theoretical quantum speedup, some of the most notable progress in a few last years has been made with artificial neural networks (ANNs) and consideration of quantum neural networks (QNNs) provides further perspectives (Mangini et al. 2021). The concept of quantum neural networks (QNNs) initially arose from efforts to understand neural processes through the lens of quantum mechanics (Schwartz et al. 2005; Adams and Petruccione 2020; Jedlicka 2017). These studies were devoted to the possibility that quantum phenomena, such as superposition, entanglement, and tunneling, may play a role in the way neural processes operate in living organisms. However, the field remains largely speculative, as experimental evidence directly supporting quantum effects in neural processes is limited. Over time, the field has changed the focus to taking advantage of the computational parallelism offered by quantum superposition and entanglement in construction of ANNs (Schuld et al. 2014). In the early stages of QNN exploration, much of the focus was on crafting quantum systems that adhered to the principles of classical ANNs. However, these endeavors largely stumbled over the challenge of reconciling the linear evolution of quantum states within circuits with the inherently non-linear behavior of classical neural networks. As a result, the focus has gradually shifted towards employing variational quantum circuits for learning data representations, instead of trying to replicate neural networks directly in a quantum context. This has led to the development of quantum adaptations of well-known classical neural network structures, such as quantum convolutional neural networks (QCNNs), quantum graph neural networks (QGNNs), quantum generative adversarial networks (QGANs), and so on, using variational quantum circuits as their foundation (Benedetti et al. 2019).

4.1 Quantum Convolutional Neural Networks

Convolutional Quantum Neural Networks (QCNNs) (Cong et al. 2019; Kerenidis et al. 2019; Hur et al. 2022) are quantum adaptations of classical convolutional neural networks (CNNs).

In a classical CNN (see Fig. 6a), the convolution layer applies a set of filters to input data to extract features. A QCNN adapts this process by utilizing quantum circuits to perform analogous operations (see Fig. 6b). The quantum convolutional layer maps input quantum states to output states, capturing feature correlations in superposition.

The input to a QCNN can be a quantum state $|\psi\rangle$, represented as a superposition of basis states. For example, in a simple 2-qubit example:

$$|\psi\rangle = \alpha\,|00\rangle + \beta\,|01\rangle + \gamma\,|10\rangle + \delta\,|11\rangle$$

Using parameterized quantum gates, QCNNs apply transformations to the input state. Consider the unitary operation $U(\theta)$ acting on qubits as a "quantum filter":

$$U(\theta) = \exp(-i\theta H)$$

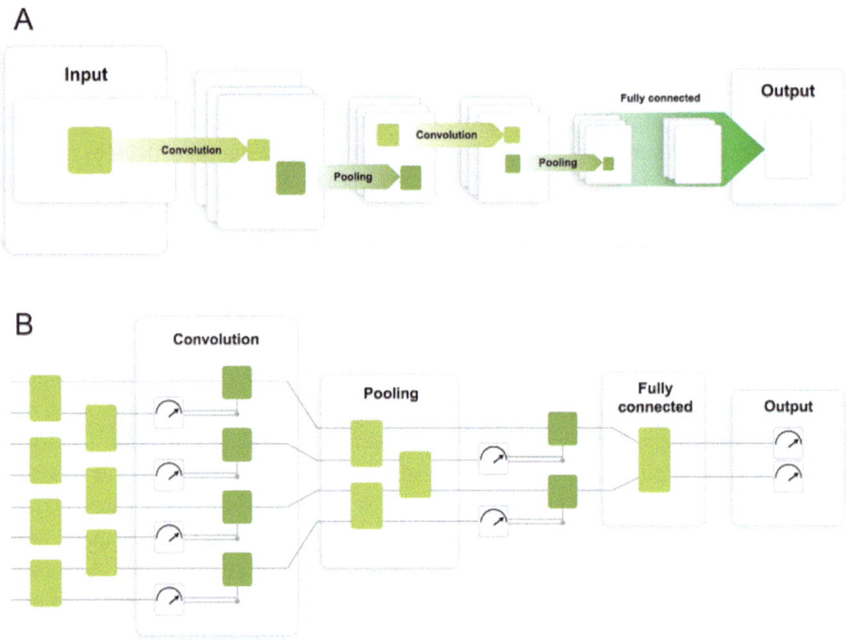

Fig. 6 (**a**) Classical CNN architecture. (**b**) Quantum Convolutional Neural Network (QCNN) is the adoption of classical CNN. QCNN utilizes quantum gates to perform convolutional operations on qubits, enabling the extraction of features from quantum states. It incorporates quantum pooling and measurement techniques to reduce dimensionality and produce classical outputs, enhancing the model's ability to learn complex patterns in quantum data. The purpose of the pooling layer is to reduce the dimensions of quantum circuit, i.e. reduce the number of qubits in circuit, while retaining as much information as possible from previously learned data

where H is a Hamiltonian operator, such as the Pauli-Z or any combination of Pauli matrices, and θ is a parameter learned during training. Entangling gates, such as Controlled-NOT (CNOT), are applied between qubits to capture dependencies between features. For instance:

$$|\psi_{\text{conv}}\rangle = CNOT_{12} \cdot (U_1(\theta_1) \otimes U_2(\theta_2)) \cdot |\psi\rangle$$

Much like classical pooling, quantum pooling reduces dimensionality, preserving essential information while discarding less important data. This can be implemented by discarding certain qubits from entangled groups or by measuring and resetting qubits. A typical operation might average or max-out over certain amplitudes, for instance:

$$|\psi_{\text{pool}}\rangle = |0\rangle\langle 0|_1 |\psi_{\text{conv}}\rangle + |1\rangle\langle 1|_1 |\psi_{\text{conv}}\rangle$$

When measuring a subset of qubits, the remaining qubits collapse to a reduced state, effectively achieving pooling.

The final layer of the QCNN often involves re-entangling the qubits and performing a series of measurements to extract the features: 1. Quantum gates similarly map the pooled state to a final output quantum state, potentially preparing data for measurement. The transformation can be denoted as:

$$|\psi_{\text{out}}\rangle = V\left(\vec{\phi}\right)|\psi_{\text{pool}}\rangle$$

where $V\left(\vec{\phi}\right)$ is another unitary transformation parameterized by a vector $\vec{\phi}$.

2. Measure the final transformed quantum state to obtain classical data, enabling comparison to known labels or further classical post-processing.

The QCNN thus iteratively applies quantum gates and measurements to its input, adapting the parameters of these operations during training to enhance performance just as classical neural networks adjust weights and biases.

In general, there are two main types of QCNNs that may provide a quantum advantage. The first type resembles the general structure depicted in Fig. 6. QCNNs with this design feature pooling layers that reduce the number of active qubits by half with each subsequent layer. This architectural decision requires only O(log(N)) parameters and effectively addresses the problem of barren plateaus. The second type can be referred to as Hybrid-QCNNs (HQCNNs) (Liu et al. 2021b). In HQCNN models, the forward pass of a convolutional filter is substituted with a quantum circuit, while pooling layers are executed classically after measurement. HQCNNs are favored because they provide greater classical control over the network, and the combination of quantum and classical elements may lead to performance improvements, albeit with increased complexity and challenges in trainability associated with the original QCNN architecture. Although QCNNs utilize only $O(\log(N))$ parameters and avoid barren plateaus, this alone does not necessarily make them superior to classical CNNs. A thorough analysis of QCNNs, which was later expanded to encompass all QML models, examined the generalization bounds of these models. This analysis serves as a framework for assessing whether a QML model can demonstrate improved performance on unseen (test) data compared to its classical equivalent. The findings indicate that if a QML model achieves a low training error on a specific task, while the classical model with the same training error is considerably more complex, the QML model is likely to outperform the classical model on unseen data. The effectiveness of classical CNNs in drug discovery has led to the investigation of QCNNs, particularly in the field of biophysics, where the relatively large input data can be divided into manageable quantum circuits through the HQCNN approach.

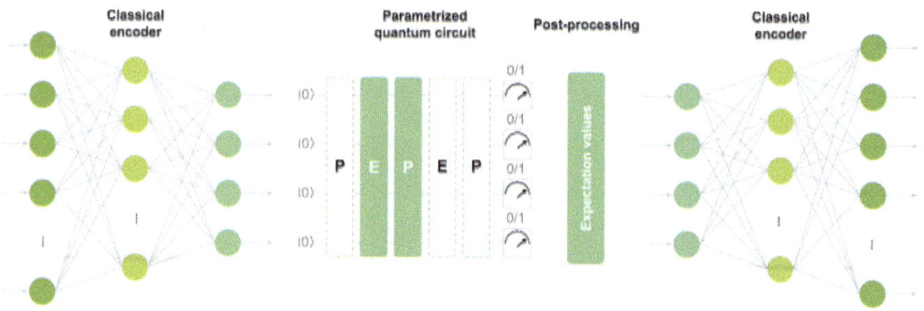

Fig. 7 Schematic representation of a hybrid classical-quantum autoencoder (Sakhnenko et al. 2022). Initially, the classical encoder compresses the data to a size of 4 units, which is then embedded into a 4-qubit quantum circuit by the encoding layer, labeled as E, and subsequently processed by the processing layer, P. The final quantum state is measured in the standard Z-basis, and the corresponding expectation values for each qubit are computed. These expectation values form the latent space. Finally, the decoder reconstructs the compressed data representation back to its original size

4.2 Quantum Autoencoder

Quantum Autoencoder (QA) (Romero et al. 2017; Bondarenko and Feldmann 2020; Sakhnenko et al. 2022) is an innovative approach that combines quantum computing techniques with classical autoencoders, aimed at efficiently encoding quantum states or classical data into a compressed or lower-dimensional representation and denoise data.

Quantum Autoencoder consists of two stages: (1) Encoding: A quantum circuit compresses the input state into a smaller number of qubits; (2) Decoding: Another quantum circuit reconstructs the original input state from the compressed state. The parameters of the quantum circuit (often gates) are optimized to minimize reconstruction error, similar to classical autoencoders (see Fig. 7).

Consider an initial pure quantum state $|\psi\rangle$ composed of a larger number of qubits:

$$|\psi\rangle = \sum_{i=1}^{N} \alpha_i \mid i\rangle$$

Here, α_i are complex amplitudes and $|i\rangle$ represents the computational basis states. For a system with n qubits, $N = 2^n$.

The initial state is passed through a parameterized quantum circuit $U(\theta)$ which is designed to reduce the number of qubits retaining significant information about $|\psi\rangle$.

$$|\psi_{\text{enc}}\rangle = U(\theta)\,|\psi\rangle$$

The unitary $U(\theta)$ is parameterized by a set of parameters θ which are to be optimized.

Part of the qubits are measured and discarded, keeping only the relevant compressed quantum information.

The encoding is inverted by another quantum circuit aiming to reconstruct the original state:

$$|\psi_{\text{dec}}\rangle = U^{\dagger}(\theta)\,|\psi_{\text{enc}}\rangle$$

To quantify the reconstruction quality, a cost or loss function is defined based on the fidelity between the original and the decoded state:

$$L(\theta) = 1 - |\langle\psi \mid \psi_{\text{dec}}\rangle|^2$$

Here, the fidelity $|\langle\psi \mid \psi_{\text{dec}}\rangle|^2$ indicates how close the reconstructed state is to the original state.

The process iteratively updates θ to minimize the loss function using optimization algorithms like gradient descent or quantum-specific algorithms like the Variational Quantum Eigensolver (VQE).

4.3 *Quantum Generative Adversarial Networks*

Generative Adversarial Networks (GANs) (Goodfellow et al. 2014; Gui et al. 2023) are machine learning models designed to produce new data samples that resemble those from a specified distribution. GANs consist of three main components: the prior distribution/noise sampling, the generator, and the discriminator. The generator produces data samples from random noise, while the discriminator assesses the authenticity of the generated samples by comparing them to real data. This adversarial training process enables the generator to improve over time, resulting in increasingly realistic samples. GANs have been applied in molecular generation and have demonstrated the ability to create novel molecular structures that meet specific desired properties.

A QGAN was introduced by Dallaire-Demers and Killoran (2018) and Lloyd and Weedbrook (2018). This research presented the idea of incorporating quantum circuits into the GAN framework, particularly utilizing quantum circuits to measure gradients. Quantum GANs provide (i) quantum speedup, enabling the learning of more complex representations of molecules thanks to the amplitude amplification property; (ii) the ability to search within exponentially large chemical spaces; and (iii) the capability to sample from distributions that are impractical for classical modeling.

In a Quantum Generative Adversarial Network (QuGAN), the noise generator plays a critical role analogous to the role of input noise in classical GANs. The noise generator serves as the source of randomness that drives the generation of diverse output states, allowing the generator to explore the space of possible outputs.

The noise generator typically starts with a simple quantum state, often the state $|0\rangle^{\otimes n}$, where n is the number of qubits. The qubits are then transformed to incorporate randomness through parameterized quantum gates.

A common method involves preparing the initial state using Hadamard gates to create a uniform superposition of all basis states. This can be represented as:

$$|\psi_{\text{initial}}\rangle = H^{\otimes n}\,|0\rangle^{\otimes n}$$

where H denotes the Hadamard gate, leading to the state:

$$|\psi_{\text{initial}}\rangle = \frac{1}{\sqrt{2^n}} \sum_{x=0}^{2^{n-1}} |\,x\rangle$$

The noise generator then processes the initial state through a series of parameterized gates, introducing additional layers of randomness and variability:

$$|\psi_Z\rangle = U(\theta)\,|\psi_{\text{initial}}\rangle$$

where $U(\theta)$ is a unitary transformation parameterized by angles θ, typically including rotations and entangling gates that vary the state amplitudes and phases in a controlled fashion.

By using a variational quantum circuit, the noise generator embeds quantum noise into the output state, which will be used as the seed for the generator's transformation of quantum data.

The output of the noise generator, $|\psi_Z\rangle$, acts as the stochastic input to the main generator circuit. The enriched randomness from this state allows the generator to produce a wide range of possible outputs, mimicking the diversity of the target data distribution.

The generator in a QuGAN is a quantum circuit designed to create a quantum state that has a probability distribution close to the target or real data distribution.

The generator applies a sequence of unitary transformations represented by a parameterized quantum circuit $U(\theta)$. The circuit manipulates the qubits to produce an output state:

$$|\psi_G\rangle = U(\theta)\,|0\rangle^{\otimes n}$$

Here, $U(\theta)$ consists of a sequence of basic quantum gates (e.g., rotations and entangling gates) where θ represents the tunable parameters (angles and phases) of the gates.

The discriminator is another quantum circuit tasked with evaluating whether a given quantum state (either real or generated) is likely to be real.

The discriminator takes an input quantum state (either $|\psi_{\text{real}}\rangle$ or $|\psi_G\rangle$) and applies its own parameterized quantum circuit $V(\phi)$:

$$|\psi_D\rangle = V(\phi)\,|\psi\rangle$$

The final output of the discriminator is obtained by performing a measurement on $|\psi_D\rangle$. The measured expectation value serves as a score for how likely the input state is real:

$$p(\text{real}|\psi) = \langle\psi|\,V^{\dagger}(\phi)\,MV(\phi)\mid\psi\rangle$$

where M is a measurement operator, often chosen to reflect the computational basis.

The objective of a QuGAN involves the adversarial interplay between the generator and the discriminator. The training aims to reach a balance where the generator produces states indistinguishable from real states.

The discriminator is trained to maximize its ability to differentiate real states from generated ones:

$$L_D(\phi|\theta) = -E_{\psi_{\text{real}}}\left[\log\, p(\text{real}|\psi_{\text{real}})\right] - E_{\psi_G}\left[\log(1 - p(\text{real}|\psi_G))\right]$$

The generator is trained to minimize the success of the discriminator in identifying fake states:

$$L_G(\theta|\phi) = -E_{\psi_G}\left[\log\, p(\text{real}|\psi_G)\right]$$

The optimization of QuGANs involves iteratively updating the parameters of both the generator and the discriminator to minimize and maximize their respective losses.

Quantum gradients are often estimated using the parameter-shift rule, which computes the derivative of an expectation value with respect to circuit parameters.

For a generic parameter θ_i:

$$\frac{\partial\langle M\rangle}{\partial\theta_i} \approx \frac{\langle M\rangle_{\theta_i+\pi/2} - \langle M\rangle_{\theta_i-\pi/2}}{2}$$

The parameters θ and ϕ are adjusted using quantum-adapted optimization algorithms, analogous to classical gradient descent:

$$\theta' \leftarrow \theta - \eta_\theta \nabla_\theta L_G$$

$$\phi' \leftarrow \phi - \eta_\phi \nabla_\phi L_D$$

where η_θ and η_ϕ are learning rates.

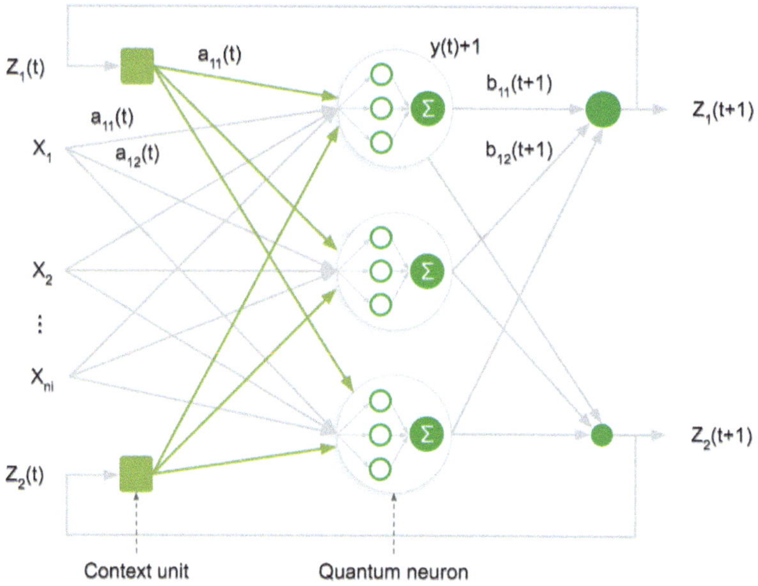

Fig. 8 Example architecture of quantum recurrent neural network for classification. (Hochreiter and Schmidhuber 1997)

4.4 Recurrent Quantum Neural Networks

Recurrent Quantum Neural Networks (RQNNs) (Bausch 2020) are at the frontier of integrating quantum computing into sequential data processing.

In Fig. 8, the architecture of RQNN on the basis of quantum neurons is presented. However, in most general case, the pipeline of RQNN is following. The first task for an RQNN is to encode classical input data into a quantum state. For a sequence $x = (x_1, x_2, \ldots, x_T)$, each time step involves a transformation from classical to quantum data. In this case, amplitude or angle encoding can be used.

The core operation of an RQNN involves recurrently processing quantum states. The current input quantum state $|\psi_t\rangle$ interacts with the previous hidden state $|\phi_{t-1}\rangle$ (acting as the quantum memory), typically using entangling operations.

The transformation at each step is represented as a unitary operation $U(\theta_t)$ applied to the combined system:

$$| \phi_t\rangle = U\ (\theta_t)\ (|\psi_t\rangle \otimes |\phi_{t-1}\rangle)$$

The operation $U(\theta_t)$ could be constructed from a series of single-qubit gates (e.g., R_x, R_z) and two-qubit gates (e.g., CNOT, CZ) to form a circuit that updates the state based on inputs and the previous hidden state.

Entangling gates in $U(\theta_t)$ establish dependencies across qubits, enabling the network to capture temporal correlations within the input sequences.

After the sequential input has been processed, the network collects outputs based on the resultant quantum states, which may be in a superposition.

The quantum states are measured to produce classical outputs. Typically, the measurement projects the quantum state onto the computational basis:

$$y_t = \text{Measure}\left(|\phi_t\rangle\right)$$

The measurement outcome y_t might represent a probability distribution over quantum state amplitudes, gleaning information about the input sequence.

Instead of direct measurement, one may compute expectation values of observables to derive quantitative outputs. If M is an observable, then:

$$\langle M \rangle_t = \langle \phi_t \,|M|\, \phi_t \rangle$$

Training involves tuning the parameters θ_t of unitary transformations to minimize the loss between the network's output and the target sequence.

Similar to the previous subsection, the parameter-shift rule can be adopted to compute gradients necessary for parameter updates.

4.5 Hybrid LSTM + QNN Architecture

In classical deep learning, Long Short-Term Memory (LSTM) networks (Hochreiter and Schmidhuber 1997) are utilized to identify both short-term and long-term dependencies across entire sequences of data.

The Long Short-Term Memory (LSTM) network is a specialized form of recurrent neural network (RNN) designed to capture long-range dependencies in sequential data. It is recognized as one of the most effective techniques for sequence modeling, achieving significant success in various applications, such as machine translation and question answering in natural language processing (NLP). LSTMs tackle the vanishing gradient problem commonly encountered in traditional RNNs by introducing an additional cell state, denoted as c_t, at each time step. This cell state enables gradients to flow without diminishing, effectively acting as the memory of the LSTM cell, which has two memory components, h_t and c_t, in contrast to the single memory component h_t found in RNNs. This feature enhances the numerical stability of LSTM training and improves prediction accuracy. The effectiveness of LSTMs has also led to their use in learning quantum evolution dynamics from experimental data that exhibit similar sequential characteristics.

The information flow in a classical LSTM cell is described by the following equations (see Fig. 9A):

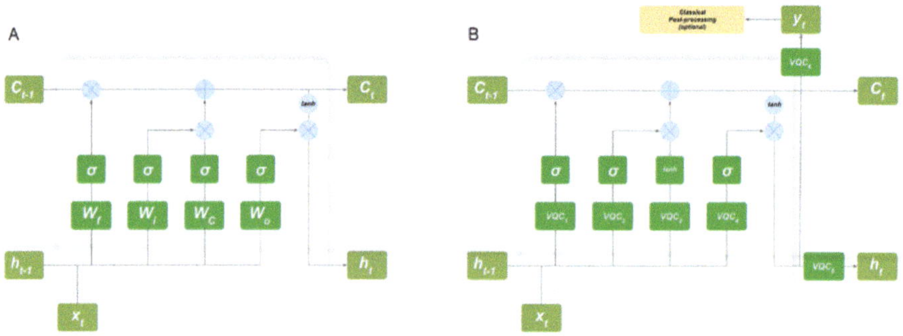

Fig. 9 Comparison of architectures of classical and quantum LSTMs. (**a**) The information flow in a classical LSTM cell allows to avoid vanishing gradients with the use of two memory components h and c, where σ denotes the sigmoid function and W_n represent the weight matrices. (**b**) The hybrid quantum-classical model improves classical architecture with the use of variational quantum circuits instead of the weight matrices enabling faster learning

$$f_t = \sigma\left(W_f \cdot v_t + b_f\right)$$

$$i_t = \sigma\left(W_i \cdot v_t + b_i\right)$$

$$\widetilde{C}_t = \tan h\left(W_C \cdot v_t + b_C\right)$$

$$c_t = f_t * c_{t-1} + i_t * \widetilde{C}_t$$

$$o_t = \sigma\left(W_o \cdot v_t + b_o\right)$$

$$h_t = o_t * \tan h\left(c_t\right)$$

where σ denotes the sigmoid function, and W_n represent the weight matrices of the classical neural networks, $n = (f, i, C, o, b_n)$ represents the bias associated with W_n, while $v_t = [h_{t-1}, x_t]$ denotes the concatenation of h_{t-1} and x_t. The symbols $*$ and $+$ indicate element-wise multiplication and addition, respectively. This will be compared with the Quantum LSTM (QLSTM) that will be introduced below.

The hybrid quantum-classical model of long short-term memory (QLSTM) that replaces the classical neural networks in LSTM cells with PQCs was proposed in (Chen et al. 2020). PQCs serve the roles of feature extraction and data compression in the QLSTM architecture. Using PQCs as the building blocks of quantum LSTM enables faster learning.

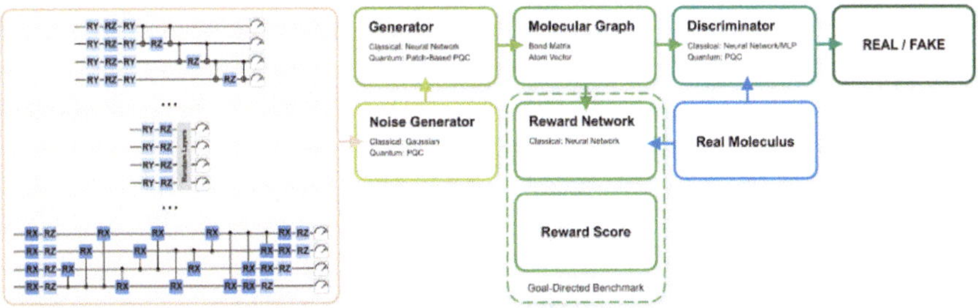

Fig. 10 Quantum MolGAN (Kao et al. 2023). The MolGAN incorporates various combinations of classical and quantum components. The classical noise generator samples from a Gaussian distribution, while the quantum generator employs a PQC. The classical generator is constructed using neural networks, whereas the quantum generator utilizes a patch-based PQC to create the molecular graph. This molecular graph is represented through a bond matrix and an atom vector. The classical discriminator is designed using either a graph-based neural network or a multilayer perceptron (MLP), while the quantum discriminator leverages a PQC

Each PQC in a QLSTM cell consists of three layers: (1) Data Encoding Layer encode classical data into the quantum state of the circuit with the angle encoding method; (2) The encoded quantum state undergoes a series of unitary operations; 3) The expectation values of each qubit are measured in the computational basis.

A QLSTM cell is constructed by stacking six PQC blocks (see Fig. 9b). These PQCs take the concatenation v_t of the hidden state h_{t-1} from the previous time step and the current input vector x_t as input. The measured values from each PQC are then passed through non-linear activation functions (sigmoid and tanh). Similar to classical LSTM, a QLSTM cell can be expressed mathematically as (see Fig. 10):

1. Forget Block

$$f_t = \sigma\left(PQC_1\left(v_t\right)\right)$$

2. Input and Update Block

$$i_t = \sigma\left(pQC_2\left(v_t\right)\right)$$

$$\widetilde{C}_t = \tan h\left(PQC_3\left(v_t\right)\right)$$

$$c_t = f_t * c_{t-1} + i_t * \widetilde{C}_t$$

3. Output Block

$$o_t = \sigma\left(PQC_4\left(v_t\right)\right)$$

$$h_t = PQC_5\left(o_t * \tan h\left(c_t\right)\right)$$

$$y_t = PQC_6\left(o_t * \tan h\left(c_t\right)\right)$$

where σ denotes the sigmoid function, tan$h$ denotes the hyperbolic tangent activation function, $v_t = [h_{t-1} x_t]$ represents the concatenation of h_{t-1} and x_t, $_*$ denotes element-wise multiplication, + denotes element-wise addition.

The QLSTM cell uses PQC_5 to transform c_t to h_t and PQC_6 to transform c_t to y_t, allowing flexibility in designing the architecture. The parameter-shift method is used to calculate the analytical gradient of the quantum circuits. This method enables the application of gradient descent optimization from classical machine learning to PQC-based models. It was shown that such QLSTM learns significantly more information than classical LSTM immediately after the first training epoch. QLSTM's loss decreases more robust and rapidly compared to classical LSTM.

4.6 Quantum Transformer

A significant portion of the AI revolution can be attributed to the transformer architecture introduced in the paper "Attention is All You Need" by Google DeepMind (Vaswani et al. 2017). This architecture was initially designed for language translation and consists of encoder and decoder components connected through a cross-attention mechanism. The encoder is particularly effective for learning a context-rich representation of a given input sequence by masking parts of the sequence and predicting the masked elements. The decoder, on the other hand, is adept at generating new sequences by predicting the subsequent parts of a sequence based on the provided context.

The core capabilities of the transformer architecture stem from the self-attention mechanism, which computes query, key, and value vectors for each input token (such as a sub-word in text). Attention scores are derived from the scaled dot product of the query and key vectors, and SoftMax is used to normalize these scores, resulting in weights that modulate the aggregation of the value vector. This process effectively captures the degree to which each token attends to every other token in the sequence. The self-attention mechanism is performed multiple times in parallel through a process known as multi-head attention.

The remarkable success of classical transformers in machine learning has understandably attracted the attention of researchers in quantum machine learning

(QML). Most implementations of quantum transformers have been adapted as Vision Transformers (ViTs) (Cherrat et al. 2024) rather than for Natural Language Processing (NLP) (Sipio et al. 2021). The main application of transformers in the realm of drug discovery has continued to focus on transformer-based generative models. In 2022, Sipio et al. discussed the emergence of quantum NLP and emphasized that the mathematical operations underlying the self-attention mechanism of transformers can all be implemented in a quantum framework (Sipio et al. 2021).

An approach to accelerating the computation of self-attention matrices by leveraging Grover's search algorithm was proposed in (Gao et al. 2023). This quantum algorithm provides a polynomial speedup compared to classical methods for specific cases and realizable with fault-tolerant computations. The attention matrix A is defined as: $A = \exp(QK^T)$, where $Q \in R^{n \times d}$ represents the matrix of n query tokens and $K \in R^{n \times d}$ represents the matrix of n key tokens. Each token is represented by a d-dimensional vector. The attention computation aims to calculate:

$$\mathrm{Att}(Q, K, V) := D^{-1}\mathrm{AV}$$

where $D \in R^{n \times n}$ is a diagonal matrix containing the self-attention scores. $V \in R^{n \times d}$ is the value matrix associated with the key tokens. The conventional classical computation of this attention matrix requires $O(n^2 d)$ time, which can be computationally expensive for large sequence lengths (n). *In the paper*, the sparsity of the attention matrix A was exploited to enhance computational efficiency. They introduce the notion of a (τ, k)-good matrix, where τ is a threshold value and k is the sparsity parameter. A matrix A is (τ, k)-good if, for each row i, the number of elements greater than τ is at most k. This assumption aligns with the observation that attention is often focused on a limited number of key tokens. The algorithm uses Grover's search to efficiently identify the k most significant entries (those greater than τ) in each row of A. This quantum algorithm provides a quadratic speedup over classical search for unstructured data, reducing the search time to $\widetilde{O}\left(\sqrt{nk}\right)$. For each row i, the algorithm defines a function $u(j)$ that returns 1 if $(QK^T)_{i,j} \geq \tau$ and 0 otherwise. Grover's search is then employed to find all indices j for which $u(j) = 1$, effectively identifying the significant entries in the row. Based on the indices returned by Grover's search, a new sparse matrix B is constructed. The elements of B corresponding to the significant entries identified by Grover's search are copied from the original matrix A. The remaining entries in B are set to zero, resulting in a sparse representation of the attention matrix. A rigorous error analysis was implemented showing that the difference between the attention computations using the original matrix A and the sparse approximation B is bounded by a small factor η. The resulting sparse matrix B allows for faster attention computation, reducing the time complexity from $O(n^2 d)$ to $\widetilde{O}\left(n^{1.5}k^{0.5}d + nkd\right)$. This quantum-enhanced approach offers a potential pathway for accelerating LLM computations, especially for tasks involving long sequences.

Recently, the Quixer model, a quantum transformer designed for language modeling tasks, was introduced (Khatri et al. 2024). This model utilizes the Linear Combination of Unitaries (LCU) and Quantum Singular Value Transform (QSVT) to process and analyze textual data.

Each word or token in the input sequence is represented by a unitary operator, a fundamental concept in quantum mechanics. This is achieved by first mapping a classical vector embedding of a token w to a set of angles using a trainable linear layer. These angles are then used to parameterize a quantum circuit U to prepare a unitary representation of the token. To capture relationships between tokens, Quixer employs the LCU technique to create a superposition of token unitaries. The goal is to represent the combined influence of multiple tokens as a weighted sum of their corresponding unitary operators:

$$M := \sum_{j=0}^{n-1} b_j U_j$$

where n is the window size, representing the number of tokens considered in the context, b_j are complex-valued coefficients, potentially trainable, that determine the weight of each unitary U_j in the superposition. The coefficients satisfy the normalization condition $\sum_{j=0}^{n-1} |b_j| = 1$. To enable more sophisticated interactions between tokens, Quixer applies a non-linear transformation to the superposition of unitaries obtained from the LCU. The QSVT is used for this purpose, allowing the application of polynomial transformations to a block-encoded matrix. Given a block encoding U_M of a matrix M, and a polynomial P_c of degree d:

$$P_c(x) = c_d \cdot x^d + c_{d-1} \cdot x^{d-1} + \cdots + c_1 \cdot x + c_0$$

with the constraints $|P_c(x)| < 1, x \in [-1, 1]$ and $\text{parity}(P_c) = d \bmod 2$

QSVT constructs a circuit to represent:

$$P_c(M) = c_d M^d + c_{d-1} M^{d-1} + \cdots + c_1 M + c_0 I$$

This polynomial transformation introduces non-linearity into the processing of token interactions. After the QSVT operation, the quantum state is measured to extract classical information. This measurement involves computing expectation values of certain observables (Hermitian operators in quantum mechanics). The obtained values are then fed into a classical feed-forward neural network (f_{out}) to produce the final output of the model. Unlike the dot product-based attention mechanism in classical transformers, Quixer models attention through the composition of unitary operators representing the tokens. Quixer demonstrates competitive performance on a language modeling task using the Penn Treebank dataset compared to classical models like LSTM and FNet and shows potential for achieving better performance with further development. The training and evaluation of Quixer are performed classically with the quantum operations simulated on conventional hardware.

Another approach with the use of Quantum Mixed-State Self-Attention Network was proposed in (Chen et al. 2024). In this case, classical input data x_s is transformed into quantum states $|x_{s,\,q}\rangle$, $|x_{s,\,k}\rangle$, and $|x_{s,\,v}\rangle$, representing queries, keys, and values, respectively, using trainable quantum embedding circuits. The core innovation lies in calculating attention weights using mixed states. The pure state query and key vectors undergo partial trace operations to extract information from the first $n/2$ qubits (subsystem A) while discarding the remaining $n/2$ qubits (subsystem B). This results in mixed-state queries (ρ_q) and keys (σ_k):

$$\rho_q = tr_B\left(|x_q\rangle\langle x_q|\right), \quad \sigma_k = tr_B\left(|x_k\rangle\langle x_k|\right).$$

The similarity between these mixed states can be directly estimated using the Hilbert–Schmidt distance, preserving quantum information and avoiding information loss:

$$D_{\mathrm{HS}}\left(\rho_q, \sigma_k\right) = tr\left[\left(\rho_q - \sigma_k\right)^2\right]$$

The model incorporates positional information using fixed quantum gates within the circuit

$$PE(s) = \frac{s-1}{S_m - 1} \times \left(PE_{\max} - PE_{\min}\right) \times 2\pi$$

This improves encoding accuracy without increasing qubit requirements. After calculating attention weights and applying them to the value states, measurements are performed to obtain classical data, which is then fed into a classical fully connected network for final prediction.

Implementation and possible acceleration of linear algebra in transformer architectures under the lens of fault-tolerant quantum computing was analyzed in (Guo et al. 2024).

In 2023, both Gao et al. (2023) and Li et al. (2024) demonstrated implementations of a quantum self-attention mechanism. More recently, Liao and Ferrie (2024) introduced complete end-to-end quantum algorithms for GPT. Notably, Guo et al.'s work includes a thorough complexity analysis and shows a theoretical quantum advantage for several normalization operations throughout the architecture. In the fields of biochemistry and drug discovery, transformer encoders have been developed to extract feature vectors from SMILES strings for use in downstream predictive tasks, while transformer decoders have been employed to generate SMILES strings with specified characteristics. Quantum-based attention mechanisms for generative pre-trained transformers are still in their early stages and many of the results reported so far have been primarily theoretical but the field is progressing rapidly.

5 Applications of QML and QNNs in Drug Discovery

5.1 Protein Folding

A groundbreaking quantum annealing approach to protein folding on a lattice was introduced by Perdomo-Ortiz et al. (2012). They examined the hydrophobic-polar model, which treats proteins as linear chains of N hydrophobic or polar beads positioned on a lattice. In this model, binary variables are used to encode the bead coordinates on the lattice, along with an additional set of auxiliary binary variables needed to formulate a quadratic Hamiltonian. Using the D-Wave quantum machine, they successfully demonstrated the protein folding of a short protein ($N = 6$) on a lattice through a growth algorithm. In 2021, a model Hamiltonian with $O(N^4)$ scaling and a corresponding quantum variational algorithm for folding a polymer chain with N monomers on a lattice was proposed (Robert et al. 2021). This model incorporates various physical and chemical properties of the protein. The variational quantum algorithm, known as CVaR-VQE, significantly reduces the number of optimization iterations required to minimize the classical cost function. As a demonstration of this algorithm, classical simulations of the folding of the 10-amino-acid protein Angiotensin and the folding of a 7-amino-acid neuropeptide using 9 qubits on an IBM 20-qubit quantum computer were conducted.

In 2022, a quantum algorithm for protein folding that combines quantum walks and deep learning was proposed (Casares et al. 2022). This algorithm, called QFold, is a scalable hybrid quantum approach that does not rely on lattice model simplifications. It utilizes the results from AlphaFold as a starting point for a quantum Metropolis–Hastings algorithm. Simulations were performed on several small peptides and compared with classical simulated annealing, demonstrating that this algorithm offers a polynomial quantum advantage. Additionally, the mapping of the protein design problem to the D-Wave quantum annealer was introduced, referred to as QPacker, which employs a large side-chain rotamer library and the Rosetta energy function (Mulligan et al. 2020). It was shown that the performance of quantum-annealer-based design is currently comparable to classical methods, and the authors believe that as quantum devices scale up, their approach will enable them to tackle larger design challenges than what is feasible with classical hardware.

Benchmarks of the QAOA protocol have been implemented to sample low-energy conformations of short peptides (Boulebnane et al. 2023). Numerical simulations on 20 qubits yielded less than promising results, indicating that deep quantum circuits are necessary and that the performance of QAOA can be similar to random sampling. This raises concerns about the feasibility of QAOA for addressing the protein folding problem in the near future.

Last achievements of quantum computing in this field in 2024 already demonstrate the potential in predicting mRNA secondary structures (Alevras et al. 2024). The feasibility of utilizing a quantum computer to address mRNA secondary structures with sequence lengths of up to 60 nucleotides, which corresponds to problems involving 10–80 qubits was investigated. To tackle the optimization challenges arising from mRNA structure prediction, a Conditional Value at Risk (CVaR)-based Variational Quantum Eigensolver (VQE) algorithm on the IBM Eagle and Heron quantum processors was employed. It was shown that even with "minimal" error mitigation and fixed-depth circuits, the hardware experiments yield accurate predictions of minimum free energy (MFE) structures that align with the results obtained from the classical solver CPLEX. Notably, Paquet et al. introduced QuantumBound (2024), a hybrid quantum neural network aimed at predicting the physicochemical properties of ligands within receptor–ligand complexes. Additionally, Jin et al. developed a QNN model to forecast potential COVID-19 variant strains using available SARS-CoV-2 RNA sequences (Jin et al. 2022). These initial successes emphasize the promise of quantum neural networks in tackling significant challenges in protein folding problems and biotherapeutics.

5.2 *Quantum De Novo Generation*

One of the most effective classical models for drug discovery is the Generative Adversarial Network (GAN) (Gui et al. 2023; Martinelli 2022), an unsupervised machine learning framework known for generating highly realistic data. GANs involve two competing neural networks: a generator and a discriminator. The generator's role is to produce new data samples that closely mimic the training data's statistical characteristics, while the discriminator's task is to distinguish between real and generated (fake) samples. Training continues until the discriminator can no longer effectively tell the difference between a sufficient number of fake and real samples. Various GAN architectures, such as the Objective-Reinforced Generative Adversarial Network (ORGAN), LatentGAN, and MolGAN, have been developed specifically for de novo drug discovery.

ORGAN merges GANs with reinforcement learning, operating on simplified molecular-input line-entry system (SMILES) representations of molecules. It maintains diversity and enhances desirable drug properties in generated samples. LatentGAN combines an autoencoder with a generative adversarial network, working with SMILES representations to optimize without worrying about syntax. MolGAN directly handles graph-structured data, creating molecules with superior chemical properties.

In 2021, QuMolGAN was proposed (Li et al. 2021), a quantum variant of MolGAN, utilizing graph representations of molecules trained on the QM9 dataset. Approximately 100 qubits are required to reconstruct these molecules, which include atom vectors and bond matrices. QuMolGAN's generator uses noise from a variational quantum circuit to create molecular graphs, while the discriminator uses Frechet distance to differentiate between fake and real molecular data.

In 2022, Insilico Medicine enhanced QuMolGAN (Kao et al. 2023) by proposing three models (see Fig. 11): (1) QuMolGAN using a variational quantum circuit as a noise generator, (2) MolGAN with a quantum generator, and (3) MolGAN with

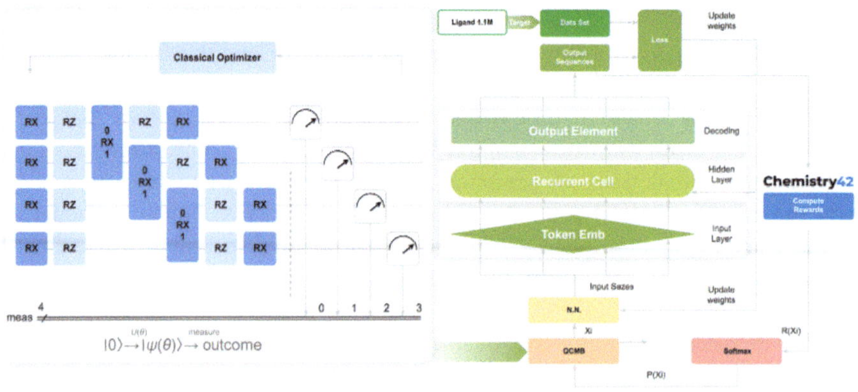

Fig. 11 Quantum LSTM. (Vakili et al. 2025)

a quantum discriminator (MolGAN-CQ). These models improved drug properties like quantitative estimation of drug-likeness (QED), solubility (logP), and synthesizability (SA). The quantum generator model employs multiple variational quantum circuits as sub-generators to form final molecules, while MolGAN-CQ utilizes a classical generator with a quantum discriminator to improve QED, logP, and SA.

Anoshin et al. introduced a hybrid quantum cycle generative adversarial network for small molecule generation, building on the cycle-consistent framework established in previous research (Anoshin et al. 2024). Their approach incorporated a hybrid generator in which a quantum circuit processed the noise vector (prior distribution) and was connected to a multi-layer perceptron (MLP) to produce molecular graphs. This method demonstrated comparable performance across various metrics, including uniqueness, validity, diversity, drug-likeness, synthesizability, and solubility, showcasing the potential of hybrid quantum-classical architectures to enhance generative models. However, the study did not provide a comprehensive comparison of the total number of parameters used, which limits the ability to make definitive claims about its expressive power.

Additionally, a scalable quantum generative autoencoder (SQ-VAE) was developed for the simultaneous reconstruction and sampling of drug molecules (Li and Ghosh 2021). SQ-VAE outperformed its classical counterpart, producing high-dimensional molecules with better drug properties. In protein design, quantum-annealer-based methods have been applied to identify amino acid side chains and conformations stabilizing protein backbones, with a discrete variational autoencoder using a Restricted Boltzmann Machine (RBM) executed on a D-Wave quantum computer for these calculations (Gircha et al. 2023).

In 2025 (Vakili et al. 2025), Insilico Medicine unveiled a LSTM+QNN architecture that effectively combines the computational capabilities of quantum algorithms, trained on a 16-qubit IBM quantum computer, with the proven reliability of classical methods for designing small molecules (see Fig. 11). This hybrid generative model was utilized to develop new KRAS inhibitors, which are critical targets in cancer therapy. The new molecules were generated using both a classical LSTM network and a Quantum Circuit Born Machine (QCBM) as the foundational generative frameworks. The LSTM network processes sequential data that encapsulates the chemical structures of ligands, while the QCBM, trained based on the quality of samples generated by the LSTM, creates complex, high-dimensional probability distributions.

In this model, Chemistry42 was used as a local filter to validate sample generation at each step, which was then applied to train the QCBM model. The QCBM, a quantum circuit model, was executed on a quantum processing unit. Subsequently, samples generated by the trained QCBM were input into the LSTM model, which produced sequences based on these samples. The reward value for each sample was calculated at every step using the local filter until epoch 20, after which we selected Chemistry42. This reward value was then utilized to train our quantum generative model.

During the first epoch, no rewards were available, so the algorithm sampled from the untrained QCBM model. From the second epoch onward, rewards were computed, enabling us to calculate the Softmax of the rewards.

The Quantum Circuit Born Machine (QCBM) model is a quantum variational generative model that requires a classical optimizer to train its parameters. The total number of these parameters is determined by the number of qubits and layers specified in the model. To optimize these parameters, the model is initially set up with randomly assigned parameters. These parameters are then updated throughout the training process. The training of the QCBM model focuses on minimizing the Exact Negative Log-Likelihood (Exact NLL) loss function.

A total of 15 promising molecules were synthesized from the generated candidates. Notably, two of these molecules, ISM061-018-2 and ISM061-22, each featuring distinct scaffolds, demonstrated effective engagement with KRAS. ISM06-018-2 was identified as a broad-spectrum KRAS inhibitor, showing a binding affinity for KRAS-G12D, while ISM061-22 exhibited specific mutant selectivity, displaying enhanced activity against KRAS G12R and Q61H mutants. This work marks the first instance of a quantum-generative model yielding experimentally confirmed biological hits, highlighting the practical potential of quantum-assisted drug discovery in producing viable therapeutics.

5.3 Quantum Molecular Docking

In 2020, a demonstration of quantum advantage in molecular docking was achieved using a photonic quantum computer known as a boson sampler (Banchi et al. 2020). The study published in Science Advances revealed that Gaussian Boson Samplers can effectively predict molecular docking configurations. In this approach, the problem was transformed into finding the maximum weighted clique in a graph, with the Gaussian Boson Sampler employed to sample large-weight cliques with high probability. The binding mode of a ligand to the tumor necrosis factor-α converting enzyme, a target associated with immune system diseases and cancer, was successfully predicted.

Additionally, it was shown that quantum annealing can address the issue of molecular unfolding (MU) (Mato et al. 2021), which is a critical phase in molecular docking. MU facilitates the removal of a molecule's initial bias by expanding it into an unfolded shape, searching for a configuration that maximizes the molecular area. This problem was reformulated as a high-order unconstrained binary optimization problem and solved using the latest D-Wave annealing hardware, with performance compared to state-of-the-art classical solvers.

Another promising application for quantum annealing is multibody docking (Pandey et al. 2022), which involves determining the optimal spatial configurations of more than two molecules in a solution space that scales exponentially with N. The multibody molecular docking problem was encoded as a quantum annealing challenge, enabling sampling with the D-Wave quantum annealer. For performance evaluation, the Rosetta REF2015 energy function was utilized, allowing for the comparison of predicted docked configurations against crystallographic data. It was found that the lowest-energy predicted configuration exhibited a root mean square deviation of 2.4A from the native crystallographic structure of a three-helix bundle.

The primary algorithms for utilizing gate-based quantum computers in molecular docking tasks include Variational Quantum Eigensolvers (VQEs) and the quantum phase estimation algorithm. These algorithms enable the construction of force fields based on precise quantum calculations for molecular fragments that exceed the capabilities of current classical quantum chemical methods (Kiss et al. 2022). As NISQ quantum devices are scaled up to thousands of qubits, the range of physical systems for which accurate ab initio quantum calculations can solve the molecular docking problem will be significantly expanded.

Furthermore, Domingo et al. (2023) and Dong et al. (2023) trained HQCNNs to predict protein–ligand binding affinities. Domingo et al. showed that their HQCNN architecture could reduce the number of parameters by 20 percent without sacrificing performance, which corresponded to a 40 percent decrease in training times, depending on the hardware used. Similarly, Dong et al. achieved competitive results compared to force field-based MM/GBSA and MM/PBSA calculations while also reducing the total number of parameters relative to classical models.

In the research conducted by Smaldone and Batista (2024), a HQCNN was trained to predict drug toxicity. This study demonstrated a method in which the weights of a convolutional layer are learned through quantum circuits, performing the underlying matrix multiplication of discrete dot products with quadratic quantum speedup. This approach matches the performance of classical models with the same number of parameters and can be transitioned to a classical CNN mid-training, facilitating noiseless training convergence.

5.4 *Molecular Dynamics and Quantum Chemistry*

The primary challenge facing current NISQ devices in calculating the interaction energy of a protein–ligand system is achieving sufficient precision in determining the total energies of individual monomers. This precision is necessary to accurately assess binding energies, which are typically around 5–10 kcal/mol, by statistically evaluating total energy expectation values. Standard Variational Quantum Eigensolver (VQE) methods require very deep circuits with thousands of parameters to reach this level of precision, which is impractical with existing quantum hardware. However, the first calculations of protein–ligand binding energies were realized on NISQ quantum computers in 2022 (Kirsopp et al. 2022). Furthermore, a quantum adaptation of symmetry-adapted perturbation theory (SAPT) was proposed to address this issue (Malone et al. 2022). This method enables the direct computation of interaction energy through expectation values rather than relying on energy differences. Its advantages include calculating interaction energy as a perturbation series in the intermolecular potential and utilizing monomer-centered basis sets, which may reduce the number of quantum gates needed. Additionally, it often facilitates favorable error cancellation for errors associated with the chosen wave function ansatz. This approach offers the potential to significantly decrease circuit depth in the VQE protocol and provides a physical rationale for breaking down interaction energy components into electrostatic, exchange, induction, and dispersion contributions, which can aid medicinal chemists in designing protein inhibitors.

The density functional theory (DFT) framework is effective for simulating quantum mechanics on classical computers. The DFT approach transforms the complex many-body problem, which has exponential complexity, into a system of noninteracting particles described by a set of single-particle states. In this context, the complexity of the problem increases linearly with the number of particles simulated. All many-body effects are encapsulated in the exchange-correlation energy, which is a functional of the density alone. Although the exact exchange-correlation functional remains unknown, approximations have significantly improved over the past 70 years and have proven surprisingly accurate for many systems. However, DFT struggles with various systems of interest, and systematically correcting errors in correlation effect approximations is generally not feasible. Recent research has focused on the intricate development of accurate exchange-correlation density functionals using matrix product states, machine learning, and quantum computing. Investigations have also been conducted on the Kohn–Sham potential inversion problem, which involves measuring the density of a time-evolved many-body system on a quantum computer. Generalizations of the Hohenberg–Kohn and Runge–Gross theorems of DFT, along with their time-dependent versions, have been proposed for qubit Hamiltonians. This opens the door to approximating many-body observables in quantum computing as single-qubit quantities that are functionals of the density. Some efforts have been made to implement mean-field approximations on quantum computers and quantum annealers, such as Hartree–

Fock, with experimental milestones achieved on a 12-qubit platform (Arute et al. 2020) and calculations of the one-particle density matrix with quantum annealing (Negre et al. 2022). It has also been demonstrated that mean-field-type approaches like DFT are expected to benefit from advancements in quantum technologies (Senjean et al. 2023). As a proof of concept, the one-dimensional Hubbard model was simulated using a quantum algorithm for site occupation functional theory, and the quantum equivalent of standard DFT was successfully applied to a hydrogen chain. The significant advantage of performing DFT on a quantum computer lies in the precision required. The numerically exact solution to the electronic structure problem aims for chemical accuracy. However, DFT on quantum computers is anticipated to be more resilient to noise compared to numerically exact wave function calculations, making it a promising near-term application for quantum computing in quantum chemistry, condensed-matter physics, and potentially biology (Pyrkov et al. 2024).

To gain a clearer insight into quantum advantage in chemical simulation, the required quantum and classical resources necessary to simulate a range of pharmaceutically significant molecules was investigated. Classical methods have demonstrated that achieving reliable simulations of these molecules demands substantial resources. Significant approximations enable quantum chemistry to handle systems of relevant size; however, these approximations are not suitable for strongly correlated systems, such as drugs interacting with the heme center of P450 proteins and adequate descriptions for these systems are computationally beyond the capabilities of classical hardware, making them strong candidates for quantum simulation. An assessment of the quantum resources needed—both in terms of total simulation time and scale—for simulating cytochrome P450 and the iron molybdenum cofactor (FeMoco) were conducted in (Goings et al. 2022; Lee et al. 2021) and (Reiher et al. 2017) respectively. FeMoco is a component of the nitrogenase enzyme, which is responsible for breaking the dinitrogen triple bond, ultimately resulting in the production of two ammonia molecules. Gaining a deeper understanding of its chemistry could provide valuable insights for designing catalysts for nitrogen fixation. On the other hand, cytochromes P450 are a family of heme proteins that play a vital role in drug metabolism. The interaction between drugs and P450 proteins can lead to adverse effects, such as drug–drug interactions or an increased rate of systemic clearance of the active compound from the body and they are frequently regarded as anti-targets in CADD. Subsequently, many studies have concentrated on determining the upper limits of the computational resources required, particularly the number of non-Clifford gates and qubits for simulating cytochrome P450 and FeMoco. With advancements in quantum algorithms and enhanced representations of chemical Hamiltonians (Lee et al. 2021; Berry et al. 2019; von Burg et al. 2021; Rocca et al. 2024), it was observed a significant reduction—by several orders of magnitude—in the quantum resources needed for sampling the eigen spectrum of molecular systems in fault-tolerant quantum computing (FTQC). Early efforts in FTQC for quantum chemistry utilizing quantum phase estimation protocol (Kitaev 1995) primarily concentrated on estimating the spectra of electronic structure Hamiltonians by Trotterizing the time evolution operator on a quantum computer (Lloyd 1996). However, more recent research has shifted towards qubitization (Low and Chuang 2019), a variant of QPE that employs a quantum walk operator. Currently, qubitization is considered the leading approach for electronic structure calculations in FTQC. The quantum algorithms currently demonstrating the best performance (in terms of lowest resource cost) either employ a Double Factorization (DF) method (von Burg et al. 2021) or utilize Tensor Hypercontraction (THC) (Lee et al. 2021). These techniques have shown promising improvements in the expected quantum computing runtime for sampling the spectrum of the electronic structure Hamiltonian, estimated to require around 10^2 hours (or approximately 10^9 non-Clifford gates) for both the FeMoco and P450 systems (Senjean et al. 2023; Goings et al. 2022) when using a specific superconducting architecture running surface code for error correction. In addition to these initial proposals, several recent studies have further enhanced by integrating DF with block-invariant symmetry-shifts (BLISS) of the Hamiltonian (Rocca et al. 2024; Loaiza and Izmaylov 2023; Oumarou et al. 2024; Berry et al. 2024). Recently, a novel Active Volume (AV) architecture has been proposed for fusion-based photonic FTQC, which promises to significantly reduce computational runtime by leveraging non-local connections among the physical components of the quantum computer (Litinski and Nickerson 2022). In traditional architectures lacking Active Volume capabilities, fault-tolerant operations necessitate the involvement of a large number of logical qubits, which could otherwise remain idle, thereby hindering the compiler's ability to utilize those qubits for other parallel tasks. Recently, it was shown that the combination of BLISS-THC, AV compilation, and certain modifications to the block encoding circuit can reduce the computational runtime of an electronic structure calculation for P450 by at least a factor of 233, in comparison to a THC calculation performed on equivalent photonic hardware lacking AV capabilities (Caesura et al. 2025).

While quantum computers require a significant number of qubits to address large active spaces, there have been a lot of demonstration experiments conducted on smaller, yet still classically manageable, systems over the past decade. Initial experiments focused on molecular hydrogen and implemented various forms of quantum phase estimation (QPE), achieving chemical accuracy through hardware-specific operations and gates. Other notable examples include calculating the bond energy of the Helium Hydride ion using phase estimation within a solid-state quantum spin register. A detailed review of recent results and techniques can be found in (Tilly et al. 2022). Beyond toy molecular systems of scientific interest, a 10-qubit Variational Quantum Eigensolver (VQE) on trapped-ion qubits was realized to tackle an industrially relevant chemical problem (Nützel et al. 2024). They computed the ground state energy of a strongly correlated metal chelate with chemical accuracy, starting from a Hartree–Fock state and employing detailed post-processing to generate the energy estimates. Recently, an End-to-End Quantum Simulation of catalytic reactions that generate chiral molecules was proposed (van Dam et al. 2024). This method combines (1) the use of high-performance computing (HPC) tools such as AutoRXN and AutoCAS to systematically identify strongly correlated chemistry within a vast chemical space, (2) highlighting the benefits of logical qubits over physical qubits during the quantum computing phase to prepare the quantum ground state of the strongly correlated active site, and (3) the implementation of optimized quantum measurements of the logical qubits through classical shadows to accurately predict various properties of the ground state, including energy levels. This hybrid workflow offers insight into how HPC, logical quantum computing, and AI can collaborate to ultimately address more complex chemical challenges such as ab initio quantum chemistry with unparalleled accuracy as reliable quantum computers continue to advance.

6 Looking Ahead

The future of QML in drug discovery is thought to be very promising, with potential to accelerate and enhance many stages of the process. We are currently entering a transformative phase of the quantum revolution, characterized by quantum hardware comprising hundreds of physical qubits with applications in practical tasks, including de novo drug design, molecular docking, protein folding, and others overviewed in this chapter.

However, realizing the full potential of QML in drug discovery requires addressing several challenges. Current hardware limitations pose one of the most significant challenges for implementing complex QML algorithms for drug discovery applications with performance higher than with classical computers. NISQ devices have a limited number of qubits and the quality of qubits is critical nowadays, as factors such as coherence time and error rates significantly influence algorithm performance allowing implementation of shallow circuits only. Additionally, the susceptibility of quantum systems to errors necessitates robust quantum error correction techniques, which require considerable overhead in terms of additional qubits, some resource estimations were presented in previous section for quantum chemistry. Other limitations include restricted qubit connectivity, which can lead to increased circuit depth. The integration of quantum and classical systems also introduces complexities that can affect performance, while the high cost and specific physical requirements of quantum hardware limit accessibility and deployment.

From the algorithmic part, quantum encoding presents one of the important challenges for QML, primarily due to the inherent complexities of representing classical data in a quantum framework. The efficient mapping of high-dimensional classical data into a quantum state, as the exponential growth of quantum state space can lead to difficulties in encoding and processing large datasets. Additionally, the design of effective quantum feature maps that preserve the relevant information while minimizing noise and errors is crucial, yet remains a complex task. The susceptibility of quantum systems to decoherence and noise further complicates the encoding process, as these factors can degrade the quality of the quantum states used for computation. Another important challenge is understanding generalization in QML models. Generalization refers to a model's ability to perform well on data it has not been trained on, a crucial aspect for ensuring the reliability and applicability of QML models in real-world scenarios. Recent research suggests that traditional methods of understanding generalization may not be sufficient for QML (Gil-Fuster et al. 2024). Studies have shown that Quantum Neural Networks (QNNs) can successfully fit random data. This unexpected behavior, termed "memorization," challenges conventional wisdom that good generalization is associated with model complexity. To address this problem, researchers are exploring alternative, non-uniform generalization measures that go beyond traditional metrics like the VC dimension and Rademacher complexity (Shalev-Shwartz and Ben-David 2014). These new measures include analyzing the speed of training convergence, examining the geometric characteristics of the minimum point in the loss function, and evaluating the model's robustness against noise. Further research into these alternative measures is crucial for gaining a deeper understanding of generalization in QML and ensuring that these models can make accurate and reliable predictions for novel molecules. Development and adaptation of new more effective QML algorithms for drug discovery is evidently necessary and on the forefront.

Nevertheless, in the coming years, it is anticipated that quantum devices with thousands of physical qubits will become available, thereby creating opportunities for early adopters to capitalize on drug discovery applications and possible practical quantum supremacy in such tasks. In addition to the advancement of quantum hardware, new quantum software solutions and algorithms are under fast development for applications of quantum computing in drug discovery. Numerous start-ups are actively working on software to enhance quantum computing algorithms for drug discovery. Classical generative chemistry algorithms may also benefit from quantum computing through the development of quantum-inspired algorithms based on tensor networks. Big pharmaceutical companies address critical questions regarding the potential impact of quantum computing on their operations, a dialogue and collaborations between pharma, AI, and quantum computing companies are already underway. Consequently, there is a growing consensus that the continued advancement of quantum hardware and software will significantly impact drug discovery and the life sciences in the near future, enabling the realization of quantum advantages in addressing real-world challenges.

References

Abbas A, Sutter D, Zoufal C, Lucchi A, Figalli A, Woerner S. The power of quantum neural networks. Nat Comput Sci. 2021;1:403–9.

Acharya R, Aghababaie-Beni L, Aleiner I, et al. Quantum error correction below the surface code threshold; 2024. https://doi.org/10.48550/arXiv.2408.13687

Adams B, Petruccione F. Quantum effects in the brain: a review. AVS Quantum Sci. 2020;2:022901.

Albash T, Lidar DA. Adiabatic quantum computation. Rev Mod Phys. 2018;90

Alevras D, Metkar M, Yamamoto T, Kumar V, Friedhoff T, Park J-E, Takeori M, LaDue M, Davis W, Galda A (2024) mRNA secondary structure prediction using utility-scale quantum computers. https://doi.org/10.48550/arXiv.2405.20328

Angluin D. Queries and concept learning. Mach Learn. 1988;2:319–42.
Anoshin M, Sagingalieva A, Mansell C, Zhiganov D, Shete V, Pflitsch M, Melnikov A. Hybrid quantum cycle generative adversarial network for small molecule generation. IEEE Trans Quantum Eng. 2024;5:1–14.
Arunachalam S, De Wolf R. Guest column: a survey of quantum learning theory. ACM SIGACT News. 2017;48:41–67.
Arute F, Arya K, Babbush R, et al. Quantum supremacy using a programmable superconducting processor. Nature. 2019;574:505–10.
Arute F, Arya K, et al. Hartree-Fock on a superconducting qubit quantum computer. Science. 2020;369:1084–9.
Aspect A, Grangier P, Roger G. Experimental realization of Einstein-Podolsky-Rosen-Bohm Gedankenexperiment: a new violation of Bell's inequalities. Phys Rev Lett. 1982;49:91–4.
Banchi L, Fingerhuth M, Babej T, Ing C, Arrazola JM. Molecular docking with Gaussian boson sampling. Science. Advances. 2020;6:eaax1950.
Bartkiewicz K, Gneiting C, Černoch A, Jirákovå K, Lemr K, Nori F. Experimental kernel-based quantum machine learning in finite feature space. Sci Rep. 2020;10:12356.
Bäumer E, Tripathi V, Wang DS, Rall P, Chen EH, Majumder S, Seif A, Minev ZK. Efficient long-range entanglement using dynamic circuits. PRX Quantum. 2024;5:030339.
Bausch J. Recurrent quantum neural networks. In: Advances in neural information processing systems. Curran Associates; 2020. p. 1368–79.
Bell JS. Speakable and unspeakable in quantum mechanics. Cambridge: Cambridge University Press; 1997.
Benedetti M, Lloyd E, Sack S, Fiorentini M. Parameterized quantum circuits as machine learning models. Quantum Sci Technol. 2019;4:043001.
Bernstein E, Vazirani U. Quantum complexity theory. SIAM J Comput. 1997;26:1411–73.
Berry DW, Gidney C, Motta M, McClean JR, Babbush R. Qubitization of arbitrary basis quantum chemistry leveraging sparsity and low rank factorization. Quantum. 2019;3:208.
Berry DW, Tong Y, Khattar T, et al. Rapid initial state preparation for the quantum simulation of strongly correlated molecules. arXiv [quant-ph]; 2024. https://doi.org/10.48550/ARXIV.2409.11748
Bharti K, Cervera-Lierta A, Kyaw TH, et al. Noisy intermediate-scale quantum algorithms. Rev Mod Phys. 2022;94 https://doi.org/10.1103/revmodphys.94.015004.
Biamonte J, Wittek P, Pancotti N, Rebentrost P, Wiebe N, Lloyd S. Quantum machine learning. Nature. 2017;549:195–202.
Blekos K, Brand D, Ceschini A, Chou C-H, Li R-H, Pandya K, Summer A. A review on quantum approximate optimization algorithm and its variants. Phys Rep. 2024;1068:1–66.
Bondarenko D, Feldmann P. Quantum autoencoders to denoise quantum data. Phys Rev Lett. 2020;124:130502.
Bonet-Monroig X. Performance comparison of optimization methods on variational quantum algorithms. Phys Rev A. 2023;107 https://doi.org/10.1103/PhysRevA.107.032407.
Boulebnane S, Lucas X, Meyder A, Adaszewski S, Montanaro A. Peptide conformational sampling using the quantum approximate optimization algorithm. NPJ Quantum Inf. 2023;9:70.
Bouwmeester D, Pan J-W, Mattle K, Eibl M, Weinfurter H, Zeilinger A. Experimental quantum teleportation. Nature. 1997;390:575–9.
Bruzewicz CD, Chiaverini J, McConnell R, Sage JM. Trapped-ion quantum computing: progress and challenges. Appl Phys Rev. 2019;6 https://doi.org/10.1063/1.5088164.
Bshouty NH, Jackson JC. Learning DNF over the uniform distribution using a quantum example oracle. In: Proceedings of the eighth annual conference on computational learning theory – COLT '95. Santa Cruz: ACM Press; 1995. p. 118–27.
Bshouty NH, Jackson JC. Learning DNF over the uniform distribution using a quantum example Oracle. SIAM J Comput. 1998;28:1136–53.
Caesura A, Cortes CL, Pol W, et al. Faster quantum chemistry simulations on a quantum computer with improved tensor factorization and active volume compilation. arXiv [quant-ph]; 2025. https://doi.org/10.48550/ARXIV.2501.06165

Cao Y, Romero J, Olson JP, et al. Quantum chemistry in the age of quantum computing. Chem Rev. 2019;119:10856–915.
Casares PAM, Campos R, Martin-Delgado MA. QFold: quantum walks and deep learning to solve protein folding. Quantum Sci Technol. 2022;7:025013.
Cerezo M, Arrasmith A, Babbush R, et al. Variational quantum algorithms. Nat Rev Phys. 2021;3:625–44.
Chalumuri A, Kune R, Manoj BS. A hybrid classical-quantum approach for multi-class classification. Quantum Inf Process. 2021;20:119.
Chen SY-C, Yoo S, Fang Y-LL. Quantum long short-term memory; 2020. https://doi.org/10.48550/arXiv.2009.01783
Chen F, Zhao Q, Feng L, Chen C, Lin Y, Lin J. Quantum mixed-state self-attention network; 2024. https://doi.org/10.48550/arXiv.2403.02871
Cherrat EA, Kerenidis I, Mathur N, Landman J, Strahm M, Li YY. Quantum vision transformers. Quantum. 2024;8:1265.
Chiaverini J, Leibfried D, Schaetz T, et al. Realization of quantum error correction. Nature. 2004;432:602–5.
Chong FT, Franklin D, Martonosi M. Programming languages and compiler design for realistic quantum hardware. Nature. 2017;549:180–7.
Cirac JI, Zoller P. Quantum computations with cold trapped ions. Phys Rev Lett. 1995;74:4091–4.
Cirq. Google Quantum AI; n.d.
Cong I, Choi S, Lukin MD. Quantum convolutional neural networks. Nat Phys. 2019;15:1273–8.
Cory DG, Fahmy AF, Havel TF. Ensemble quantum computing by NMR spectroscopy. Proc Natl Acad Sci. 1997;94:1634–9.
Daimon S, Matsushita Y. Quantum circuit generation for amplitude encoding using a transformer decoder. Phys Rev Appl. 2024;22:L041001.
Dallaire-Demers P-L, Killoran N. Quantum generative adversarial networks. Phys Rev A. 2018;98:012324.
Dalzell AM, McArdle S, Berta M, et al. Quantum algorithms: a survey of applications and end-to-end complexities; 2023. https://doi.org/10.48550/arXiv.2310.03011
Deutsch D, Penrose R. Quantum theory, the church–Turing principle and the universal quantum computer. Proc R Soc Lond A Math Phys Sci. 1997;400:97–117.
Dixit V, Selvarajan R, Alam MA, Humble TS, Kais S. Training restricted Boltzmann machines with a D-wave quantum Annealer. Front Phys. 2021;9 https://doi.org/10.3389/fphy.2021.589626.
Domingo L, Djukic M, Johnson C, Borondo F. Binding affinity predictions with hybrid quantum-classical convolutional neural networks. Sci Rep. 2023;13:17951.
Dong L, Li Y, Liu D, et al. Prediction of protein-ligand binding affinity by a hybrid quantum-classical deep learning algorithm. Adv Quantum Technol. 2023;6:2300107.
Doosti M, Wallden P, Hamill CB, Hankache R, Brown OT, Heunen C. A brief review of quantum machine learning for financial services; 2024. https://doi.org/10.48550/arXiv.2407.12618
Du Y, Tao D. On exploring the potential of quantum auto-encoder for learning quantum systems; 2024. https://doi.org/10.48550/arXiv.2106.15432
Du Y, Huang T, You S, Hsieh M-H, Tao D. Quantum circuit architecture search for variational quantum algorithms. Npj Quantum Inf. 2022;8:62.
Du Y, Yang Y, Tao D, Hsieh M-H. Problem-dependent power of quantum neural networks on multiclass classification. Phys Rev Lett. 2023;131:140601.
D-Wave Systems The Practical Quantum Computing Company. n.d.
Einstein A, Podolsky B, Rosen N. Can quantum-mechanical description of physical reality be considered complete? Phys Rev. 1935;47:777–80.
Farhi E, Goldstone J, Gutmann S. A quantum approximate optimization algorithm; 2014. https://doi.org/10.48550/arXiv.1411.4028
Feynman RP. Simulating physics with computers. Int J Theor Phys. 1982;21:467.
Fontana E, Fitzpatrick N, Ramo DM, Duncan R, Rungger I. Evaluating the noise resilience of variational quantum algorithms. Phys Rev A. 2021;104:022403.

Freedman SJ, Clauser JF. Experimental test of local hidden-variable theories. Phys Rev Lett. 1972;28:938–41.

Gao Y, Song Z, Yang X, Zhang R. Fast quantum algorithm for attention computation; 2023. https://doi.org/10.48550/arXiv.2307.08045

Ge Y, Wenjie W, Yuheng C, Kaisen P, Xudong L, Zixiang Z, Yuhan W, Ruocheng W, Junchi Y. Quantum circuit synthesis and compilation optimization: overview and prospects; 2024. https://doi.org/10.48550/arXiv.2407.00736

Gentinetta G, Thomsen A, Sutter D, Woerner S. The complexity of quantum support vector machines. Quantum. 2024;8:1225.

Gershenfeld NA, Chuang IL. Bulk spin-resonance quantum computation. Science. 1997;275:350–6.

Gil-Fuster E, Eisert J, Bravo-Prieto C. Understanding quantum machine learning also requires rethinking generalization. Nat Commun. 2024;15:2277.

Gircha AI, Boev AS, Avchaciov K, Fedichev PO, Fedorov AK. Hybrid quantum-classical machine learning for generative chemistry and drug design. Sci Rep. 2023;13:8250.

Giuntini R, Holik F, Park DK, Freytes H, Blank C, Sergioli G. Quantum-inspired algorithm for direct multi-class classification. Appl Soft Comput. 2023;134:109956.

Goings JJ, White A, Lee J, Tautermann CS, Degroote M, Gidney C, Shiozaki T, Babbush R, Rubin NC. Reliably assessing the electronic structure of cytochrome P450 on today's classical computers and tomorrow's quantum computers. Proc Natl Acad Sci USA. 2022;119:e2203533119.

Goodfellow IJ, Pouget-Abadie J, Mirza M, Xu B, Warde-Farley D, Ozair S, Courville A, Bengio Y. Generative adversarial nets. In: Proceedings of the 27th international conference on neural information processing systems, vol. 2. Cambridge: MIT Press; 2014. p. 2672–80.

Google Quantum AI. n.d.

Grover LK. A fast quantum mechanical algorithm for database search; 1996. https://doi.org/10.48550/arXiv.quant-ph/9605043

Gui J, Sun Z, Wen Y, Tao D, Ye J. A review on generative adversarial networks: algorithms, theory, and applications. IEEE Trans Knowl Data Eng. 2023;35:3313–32.

Guo N, Yu Z, Choi M, Agrawal A, Nakaji K, Aspuru-Guzik A, Rebentrost P. Quantum linear algebra is all you need for transformer architectures; 2024. https://doi.org/10.48550/arXiv.2402.16714

Haug T, Self CN, Kim MS. Quantum machine learning of large datasets using randomized measurements. Mach Learn Sci Technol. 2023;4:015005.

Havlíček V, Córcoles AD, Temme K, Harrow AW, Kandala A, Chow JM, Gambetta JM. Supervised learning with quantum-enhanced feature spaces. Nature. 2019;567:209–12.

Hochreiter S, Schmidhuber J. Long short-term memory. Neural Comput. 1997;9:1735–80.

Hur T, Kim L, Park DK. Quantum convolutional neural network for classical data classification. Quantum Mach Intell. 2022;4:3.

IBM. IBM quantum system two: the era of quantum utility is here IBM quantum computing blog; n.d.

IBM Quantum Computing. n.d.

Ilin Y, Arad I. Dissipative variational quantum algorithms for Gibbs state preparation; 2024.

IonQ. IonQ trapped ion quantum computing; n.d.

Ippoliti M, Li Y, Rakovszky T, Khemani V. Operator relaxation and the optimal depth of classical shadows. Phys Rev Lett. 2023;130:230403.

Jedlicka P. Revisiting the quantum brain hypothesis: toward quantum (neuro)biology? Front Mol Neurosci. 2017;10 https://doi.org/10.3389/fnmol.2017.00366.

Jin Y-X, Hu J-J, Li Q, Luo Z-C, Zhang F-Y, Tang H, Qian K, Jin X-M. Quantum deep learning for mutant COVID-19 strain prediction; 2022. https://doi.org/10.48550/arXiv.2203.03556

Kālis M, Locāns A, Šikovs R, Naseri H, Ambainis A. A hybrid quantum-classical approach for inference on restricted Boltzmann machines; 2023. https://doi.org/10.48550/arXiv.2304.12418

Kao P-Y, Yang Y-C, Chiang W-Y, et al. Exploring the advantages of quantum generative adversarial networks in generative chemistry. J Chem Inf Model. 2023;63:3307–18.

Kearns M, Valiant L. Cryptographic limitations on learning Boolean formulae and finite automata. J ACM. 1994;41:67–95.

Kerenidis I, Landman J, Prakash A. Quantum algorithms for deep convolutional neural networks; 2019. https://doi.org/10.48550/arXiv.1911.01117

Khatri S, LaRose R, Poremba A, Cincio L, Sornborger AT, Coles PJ. Quantum-assisted quantum compiling. Quantum. 2019;3:140.

Khatri N, Matos G, Coopmans L, Clark S. Quixer: a quantum transformer model; 2024. https://doi.org/10.48550/arXiv.2406.04305

Kirsopp JJM, Di Paola C, Manrique DZ, Krompiec M, Greene-Diniz G, Guba W, Meyder A, Wolf D, Strahm M, Muñoz Ramo D. Quantum computational quantification of protein–ligand interactions. Int J Quantum Chem. 2022;122:e26975.

Kiss O, Tacchino F, Vallecorsa S, Tavernelli I. Quantum neural networks force fields generation. Mach Learn Sci Technol. 2022;3:035004.

Kitaev AY. Quantum measurements and the Abelian Stabilizer Problem; 1995. https://doi.org/10.48550/arXiv.quant-ph/9511026

Kjaergaard M, Schwartz ME, Braumüller J, Krantz P, Wang JI-J, Gustavsson S, Oliver WD. Superconducting qubits: current state of play. Ann Rev Conden Matter Phys. 2020;11:369–95.

Klivans AR, Servedio R. Learning DNF in time. In: Proceedings of the thirty-third annual ACM symposium on theory of computing. New York: Association for Computing Machinery; 2001. p. 258–65.

Koch J, Yu TM, Gambetta J, Houck AA, Schuster DI, Majer J, Blais A, Devoret MH, Girvin SM, Schoelkopf RJ. Charge-insensitive qubit design derived from the Cooper pair box. Phys Rev A. 2007;76:042319.

Larocca M, Thanasilp S, Wang S, Sharma K, Biamonte J, Coles PJ, Cincio L, McClean JR, Holmes Z, Cerezo M. A review of barren plateaus in variational quantum computing; 2024. https://doi.org/10.48550/arXiv.2405.00781

Laydevant J, Marković D, Grollier J. Training an Ising machine with equilibrium propagation. Nat Commun. 2024;15:3671.

Lee J, Berry DW, Gidney C, Huggins WJ, McClean JR, Wiebe N, Babbush R. Even more efficient quantum computations of chemistry through tensor hypercontraction. PRX quantum. 2021;2 https://doi.org/10.1103/prxquantum.2.030305.

Li W, Deng D-L. Recent advances for quantum classifiers; 2021. https://doi.org/10.48550/arXiv.2108.13421

Li J, Ghosh S Scalable variational quantum circuits for autoencoder-based drug discovery; 2021. https://doi.org/10.48550/arXiv.2112.12563

Li Z, Liu X, Xu N, Du J. Experimental realization of a quantum support vector machine. Phys Rev Lett. 2015;114:140504.

Li J, Topaloglu RO, Ghosh S. Quantum generative models for small molecule drug discovery. IEEE Trans Quantum Eng. 2021;2:1–8.

Li G, Zhao X, Wang X. Quantum self-attention neural networks for text classification. Sci China Inf Sci. 2024;67:142501.

Liao Y, Ferrie C. GPT on a quantum computer; 2024. https://doi.org/10.48550/arXiv.2403.09418

Litinski D, Nickerson N. Active volume: An architecture for efficient fault-tolerant quantum computers with limited non-local connections. arXiv [quant-ph]; 2022. https://doi.org/10.48550/ARXIV.2211.15465

Liu Y, Arunachalam S, Temme K. A rigorous and robust quantum speed-up in supervised machine learning. Nat Phys. 2021a;17:1013–7.

Liu J, Lim KH, Wood KL, Huang W, Guo C, Huang H-L. Hybrid quantum-classical convolutional neural networks. Sci China Phys Mech Astronom. 2021b;64:290311.

Lloyd S. Universal Quantum Simulators. Science. 1996;273:1073.

Lloyd S, Weedbrook C. Quantum generative adversarial learning. Phys Rev Lett. 2018;121:040502.

Loaiza I, Izmaylov AF. Block-invariant symmetry shift: preprocessing technique for second-quantized Hamiltonians to improve their decompositions to linear combination of Unitaries. J Chem Theory Comput. 2023;19:8201–9.
Low GH, Chuang IL. Hamiltonian simulation by qubitization. Quantum. 2019;3:163.
Lund AP, Bremner MJ, Ralph TC. Quantum sampling problems, BosonSampling and quantum supremacy. NPJ Quantum Inf. 2017;3:15.
Malone FD, et al. Towards the simulation of large scale protein–ligand interactions on NISQ-era quantum computers. Chem Sci. 2022;13:3094–108.
Mangini S, Tacchino F, Gerace D, Bajoni D, Macchiavello C. Quantum computing models for artificial neural networks. Europhys Lett. 2021;134:10002.
Manin Y. Computable and uncomputable. Moscow: Sovetskoye Radio; 1980. p. 128.
Maronese M, Moro L, Rocutto L, Prati E. Quantum compiling; 2021. https://doi.org/10.48550/arXiv.2112.00187
Martinelli DD. Generative machine learning for de novo drug discovery: a systematic review. Comput Biol Med. 2022;145:105403.
Martinis JM, Nam S, Aumentado J, Urbina C. Rabi oscillations in a large Josephson-junction qubit. Phys Rev Lett. 2002;89:117901.
Mato K, Mengoni R, Ottaviani D, Palermo G. Quantum molecular unfolding; 2021. https://doi.org/10.48550/arXiv.2107.13607
McClean JR, Boixo S, Smelyanskiy VN, Babbush R, Neven H. Barren plateaus in quantum neural network training landscapes. Nat Commun. 2018;9:4812.
Melko RG, Carleo G, Carrasquilla J, Cirac JI. Restricted Boltzmann machines in quantum physics. Nat Phys. 2019;15:887–92.
Monroe C, Meekhof DM, King BE, Itano WM, Wineland DJ. Demonstration of a fundamental quantum logic gate. Phys Rev Lett. 1995;75:4714–7.
Montanaro A. Quantum algorithms: an overview. NPJ Quantum Inf. 2016;2:1–8.
Moro L, Paris MGA, Restelli M, Prati E. Quantum compiling by deep reinforcement learning. Commun Phys. 2021;4:178.
Mulligan VK, Melo H, Merritt HI, Slocum S, Weitzner BD, Watkins AM, Renfrew PD, Pelissier C, Arora PS, Bonneau R Designing peptides on a quantum computer; 2020. https://doi.org/10.1101/752485.
Nakamura Y, Pashkin YA, Tsai JS. Coherent control of macroscopic quantum states in a single-Cooper-pair box. Nature. 1999;398:786–8.
Negre CFA, Lopez-Bezanilla A, Zhang Y, Akrobotu PD, Mniszewski SM, Tretiak S, Dub PA. QUBO-based density matrix electronic structure method; 2022. https://doi.org/10.48550/arXiv.2201.04720
Nielsen MA, Chuang IL. Quantum computation and quantum information. New York: Cambridge University Press; 2000.
Nielsen MA, Chuang IL. Quantum computation and quantum information, 10th anniversary ed; 2010. https://doi.org/10.1017/CBO9780511976667
Nobel Prize. The Nobel Prize in physics; 2022. NobelPrize.org
Nomura Y. Helping restricted Boltzmann machines with quantum-state representation by restoring symmetry. J Phys Condens Matter. 2021;33:174003.
Nützel L, Gresch A, Hehn L, et al. Solving an industrially relevant quantum chemistry problem on quantum hardware. arXiv [quant-ph]; 2024. https://doi.org/10.48550/ARXIV.2408.10801
Oumarou O, Scheurer M, Parrish RM, Hohenstein EG, Gogolin C. Accelerating quantum computations of chemistry through regularized compressed double factorization. Quantum. 2024;8:1371.
Pandey M, Zaborniak T, Melo H, Galda A, Mulligan VK. Multibody molecular docking on a quantum annealer; 2022. https://doi.org/10.48550/arXiv.2210.11401
Paquet E, Soleymani F, St-Pierre-Lemieux G, Viktor HL, Michalowski W. QuantumBound – interactive protein generation with one-shot learning and hybrid quantum neural networks. Artif Intell Chem. 2024;2:100030.

Perdomo-Ortiz A, Dickson N, Drew-Brook M, Rose G, Aspuru-Guzik A. Finding low-energy conformations of lattice protein models by quantum annealing. Sci Rep. 2012;2:571.
Peruzzo A, McClean J, Shadbolt P, Yung M-H, Zhou X-Q, Love PJ, Aspuru-Guzik A, O'Brien JL. A variational eigenvalue solver on a photonic quantum processor. Nat Commun. 2014;5:4213.
Pillay SM, Sinayskiy I, Jembere E, Petruccione F. A multi-class quantum kernel-based classifier. Adv Quantum Technol. 2024a;7:202300249.
Pillay SM, Sinayskiy I, Jembere E, Petruccione F. A multi-class quantum kernel-based classifier. Adv Quantum Technol. 2024b;7:2300249.
Pyrkov A, Aliper A, Bezrukov D, Lin Y-C, Polykovskiy D, Kamya P, Ren F, Zhavoronkov A. Quantum computing for near-term applications in generative chemistry and drug discovery. Drug Discov Today. 2023;28:103675.
Pyrkov A, Aliper A, Bezrukov D, Podolskiy D, Ren F, Zhavoronkov A. Complexity of life sciences in quantum and AI era. WIREs Comput Mol Sci. 2024;14:e1701.
Qiskit. IBM quantum computing; n.d.
Quantinuum Accelerating Quantum Computing. n.d.
Quantum Computing. Rigetti computing; n.d.
QuEra. Quantum error correction QuEra; n.d.
QuTiP. Quantum toolbox in python; n.d.
Rath M, Date H. Quantum data encoding: a comparative analysis of classical-to-quantum mapping techniques and their impact on machine learning accuracy. EPJ Quantum Technol. 2024;11:1–22.
Rebentrost P, Mohseni M, Lloyd S. Quantum support vector machine for big data classification. Phys Rev Lett. 2014;113:130503.
Reiher M, Wiebe N, Svore KM, Wecker D, Troyer M. Elucidating reaction mechanisms on quantum computers. Proc Natl Acad Sci USA. 2017;114:7555–60.
Riebe M, Häffner H, Roos CF, et al. Deterministic quantum teleportation with atoms. Nature. 2004;429:734–7.
Robert A, et al. Resource-efficient quantum algorithm for protein folding. NPJ Quantum Inf. 2021;7:38.
Rocca D, Cortes CL, Gonthier JF, Ollitrault PJ, Parrish RM, Anselmetti G-L, Degroote M, Moll N, Santagati R, Streif M. Reducing the runtime of fault-tolerant quantum simulations in chemistry through symmetry-compressed double factorization. J Chem Theory Comput. 2024;20:4639–53.
Romero J, Olson JP, Aspuru-Guzik A. Quantum autoencoders for efficient compression of quantum data. Quantum Sci Technol. 2017;2:045001.
Sajjan M, Li J, Selvarajan R, Sureshbabu SH, Kale SS, Gupta R, Singh V, Kais S. Quantum machine learning for chemistry and physics. Chem Soc Rev. 2022;51:6475–573.
Sakhnenko A, O'Meara C, Ghosh KJB, Mendl CB, Cortiana G, Bernabé-Moreno J. Hybrid classical-quantum autoencoder for anomaly detection. Quantum Mach Intell. 2022;4:27.
Schrodinger E. Die gegenwortige Situation in der Quantenmechanik. Naturwissenschaften. 1935;23:807–12.
Schuld M. Quantum machine learning in feature Hilbert spaces. Phys Rev Lett. 2019a;122 https://doi.org/10.1103/PhysRevLett.122.040504.
Schuld M. Evaluating analytic gradients on quantum hardware. Phys Rev A. 2019b;99 https://doi.org/10.1103/PhysRevA.99.032331.
Schuld M, Petruccione F. Quantum models as kernel methods. In: Schuld M, Petruccione F, editors. Machine learning with quantum computers. Cham: Springer International Publishing; 2021. p. 217–45.
Schuld M, Sinayskiy I, Petruccione F. The quest for a quantum neural network. Quantum Inf Process. 2014;13:2567–86.
Schwartz JM, Stapp HP, Beauregard M. Quantum physics in neuroscience and psychology: a neurophysical model of mind–brain interaction. Philos Trans R Soc B, Biol Sci. 2005;360:1309–27.

Seidel R, Bock S, Zander R, Petrič M, Steinmann N, Tcholtchev N, Hauswirth M. Qrisp: a framework for compilable high-level programming of gate-based quantum computers; 2024. https://doi.org/10.48550/arXiv.2406.14792

Senjean B, Yalouz S, Saubanère M. Toward density functional theory on quantum computers? SciPost Phys. 2023;14:055.

Servedio RA, Gortler SJ. Equivalences and separations between quantum and classical learnability. SIAM J Comput. 2004;33:1067–92.

Shalev-Shwartz S, Ben-David S. Understanding machine learning: from theory to algorithms. Cambridge: Cambridge University Press; 2014.

Sharma K, Khatri S, Cerezo M, Coles PJ. Noise resilience of variational quantum compiling. New J Phys. 2020;22:043006.

Shor PW. Algorithms for quantum computation: discrete logarithms and factoring. In: Proceedings 35th annual symposium on foundations of computer science; 1994. p. 124–34.

Sipio RD, Huang J-H, Chen SY-C, Mangini S, Worring M. The dawn of quantum natural language processing; 2021. https://doi.org/10.48550/arXiv.2110.06510

Slussarenko S, Pryde GJ. Photonic quantum information processing: a concise review. Appl Phys Rev. 2019;6:041303.

Smaldone AM, Batista VS. Quantum-to-classical neural network transfer learning applied to drug toxicity prediction. J Chem Theory Comput. 2024;20:4901–8.

Stokes J, Izaac J, Killoran N, Carleo G. Quantum natural gradient. Quantum. 2020;4:269.

Tilly J, Chen H, Cao S, et al. The variational quantum Eigensolver: a review of methods and best practices. Phys Rep. 2022;986:1–128.

Trotter HF. On the product of semi-groups of operators. Proc Am Math Soc. 1959;10:545–51.

Vakili MG, Gorgulla C, Nigam A, et al. Quantum computing-enhanced algorithm unveils novel inhibitors for KRAS. Nat Biotechnol. 2025; https://doi.org/10.1038/s41587-024-02526-3.

Valiant LG. A theory of the learnable. Commun ACM. 1984;27:1134–42.

van Dam W, Liu H, Low GH, Paetznick A, Paz A, Silva M, Sundaram A, Svore K, Troyer M. End-to-end quantum simulation of a chemical system. arXiv [quant-ph]; 2024. https://doi.org/10.48550/ARXIV.2409.05835

Vandersypen LMK, Chuang IL. NMR techniques for quantum control and computation. Rev Mod Phys. 2005;76:1037–69.

Vapnik VN. Statistical learning theory. New York: Wiley; 1998.

Vaswani A, Shazeer N, Parmar N, Uszkoreit J, Jones L, Gomez AN, Kaiser Ł, et al. Attention is all you need. Adv Neural Inf Proces Syst. 2017;30

Vieijra T, Casert C, Nys J, De Neve W, Haegeman J, Ryckebusch J, Verstraete F. Restricted Boltzmann machines for quantum states with non-abelian or Anyonic symmetries. Phys Rev Lett. 2020;124:097201.

von Burg V, Low GH, Häner T, Steiger DS, Reiher M, Roetteler M, Troyer M. Quantum computing enhanced computational catalysis. Phys Rev Res. 2021;3 https://doi.org/10.1103/physrevresearch.3.033055.

Wada K, Raymond R, Sato Y, Watanabe HC. Sequential optimal selections of single-qubit gates in parameterized quantum circuits. Quantum Sci Technol. 2024;9:035030.

Wallraff A, Schuster DI, Blais A, Frunzio L, Huang R-S, Majer J, Kumar S, Girvin SM, Schoelkopf RJ. Strong coupling of a single photon to a superconducting qubit using circuit quantum electrodynamics. Nature. 2004;431:162–7.

Wang X, Du Y, Luo Y, Tao D. Towards understanding the power of quantum kernels in the NISQ era. Quantum. 2021;5:531.

Weidman JD, Sajjan M, Mikolas C, Stewart ZJ, Pollanen J, Kais S, Wilson AK. Quantum computing and chemistry. Cell Rep Phys Sci. 2024;5:102105.

Wiebe N, Kapoor A, Svore K. Quantum algorithms for nearest-neighbor methods for supervised and unsupervised learning; 2014. https://doi.org/10.48550/arXiv.1401.2142

Wierichs D, Gogolin C, Kastoryano M. Avoiding local minima in variational quantum eigensolvers with the natural gradient optimizer. Phys Rev Res. 2020;2:043246.

Yamamoto T, Pashkin YA, Astafiev O, Nakamura Y, Tsai JS. Demonstration of conditional gate operation using superconducting charge qubits. Nature. 2003;425:941–4.
Zardini E, Blanzieri E, Pastorello D. A quantum k-nearest neighbors algorithm based on the Euclidean distance estimation. Quantum Mach Intell. 2024;6:23.
Zhang C. An improved lower bound on query complexity for quantum PAC learning. Inf Process Lett. 2010;111:40–5.
Zhang N, Ding S, Zhang J, Xue Y. An overview on restricted Boltzmann machines. Neurocomputing. 2018;275:1186–99.

AI-Driven Discovery of microRNA Targets for Disease Therapy and Drug Development

Pietro Delre, Carmen Cerchia, and Antonio Lavecchia

Abstract Non-coding RNAs, encompassing microRNAs, long non-coding RNAs, and circular RNAs have transitioned from being viewed as genomic "dark matter" to being recognized as pivotal regulators of gene expression, controlling vast transcriptional landscapes and cellular processes. Extensive evidence now links the dysregulation of these molecules to the pathophysiology of numerous human diseases, including cancer, cardiovascular disorders, and neurodegenerative conditions, thereby offering significant therapeutic and diagnostic potential. To navigate the complexity of these interactions and harness their potential, artificial intelligence is playing an increasingly crucial role. This chapter explores this emerging synergy, reviewing Machine Learning and Deep Learning approaches applied to predict miRNA–disease associations, small molecules–miRNA, lncRNA–miRNA and circRNA–miRNA interactions. While substantial progress has been made, significant challenges remain. We will therefore delve into both the advantages and the current limitations of these AI methodologies, aiming specifically to promote the critical development of more transparent and interpretable models that can be effectively utilized by clinicians and pharmaceutical chemists in real-world applications.

Keywords microRNA · lncRNA—miRNA interactions · circRNA–miRNA interactions · Machine learning · Artificial intelligence · miRNA–disease associations

1 Introduction

Over the past two decades, our understanding of non-coding RNAs (ncRNAs) has undergone a dramatic transformation. Once considered as "junk" or genomic

P. Delre · C. Cerchia · A. Lavecchia (✉)
Dipartimento di Farmacia, Drug Discovery Laboratory, Università degli Studi di Napoli "Federico II", Naples, Italy
e-mail: pietro.delre@unina.it; carmen.cerchia@unina.it; antonio.lavecchia@unina.it

A. Lavecchia (ed.), *Applied Artificial Intelligence for Drug Discovery*,
https://doi.org/10.1007/978-3-031-98022-0_11

"dark matter," ncRNAs are now recognized as powerful regulatory molecules that control more than two-thirds of the human transcriptional output (Esteller 2011). This paradigm shift has been driven by advances in high-throughput sequencing technologies and bioinformatics, revealing that a vast portion of the human genome is transcribed into functional non-coding RNAs with critical roles in cellular homeostasis. Eukaryotic cells, in fact, produce a greater amount of ncRNAs than protein-coding mRNAs, highlighting their fundamental importance in gene regulation. Far from being inert, ncRNAs, such as microRNAs (miRNAs), long non-coding RNAs (lncRNAs), and circular RNAs (circRNAs) can silence or modulate the expression of multiple target genes simultaneously (Meng et al. 2017; O'Brien et al. 2018). This ability to coordinate and fine-tune complex signaling networks places them at the center of numerous physiological and pathological processes. A growing body of research has shown that the dysregulation of ncRNAs contributes to the onset and progression of a wide range of diseases, including cancer, cardiovascular disorders, and neurodegenerative conditions (Hussen et al. 2021; Kandettu et al. 2020; Liu et al. 2016; Meng et al. 2017). These insights have opened new avenues in biomedical research, offering alternative molecular pathways for early detection, personalized treatment strategies, and drug development. Importantly, computational biology and artificial intelligence (AI) are now playing a key role in decoding the vast and complex ncRNA landscape, enabling the discovery of novel disease associations and therapeutic targets (de Gonzalo-Calvo et al. 2024; Sheng et al. 2024). This chapter explores the emerging role of ncRNAs in disease mechanisms and therapeutic innovation, with a particular focus on microRNAs. We begin by introducing the biology of miRNAs and their potential as disease regulators. We then discuss data-driven approaches to understand miRNA-disease associations (MDAs) including key database resources and the application of predictive models. Next, we examine the potential of small molecules to modulate miRNA activity, with a special emphasis on structure-based and AI-guided computational methods. We also explore the intricate crosstalk between miRNAs, lncRNAs, and circRNAs, highlighting how these molecules form complex regulatory networks. Special attention is given to the mechanisms underlying lncRNA–miRNA (LMI) and circRNA–miRNA (CMI) interactions, alongside current computational strategies for predicting these interactions. Finally, we discuss the implications of ncRNA networks in drug development and resistance, underscoring their therapeutic potential and the remaining challenges in the field. This comprehensive overview aims to provide readers with both foundational knowledge and current advances at the intersection of ncRNA biology and computational biomedicine.

2 microRNAs: Small Regulators, Big Promise

Approximately 98% of the human genome contains ncRNAs, which were historically labeled as "junk DNA". Recent studies have demonstrated the crucial role of this noncoding portion in different biological processes. ncRNAs, transcribed

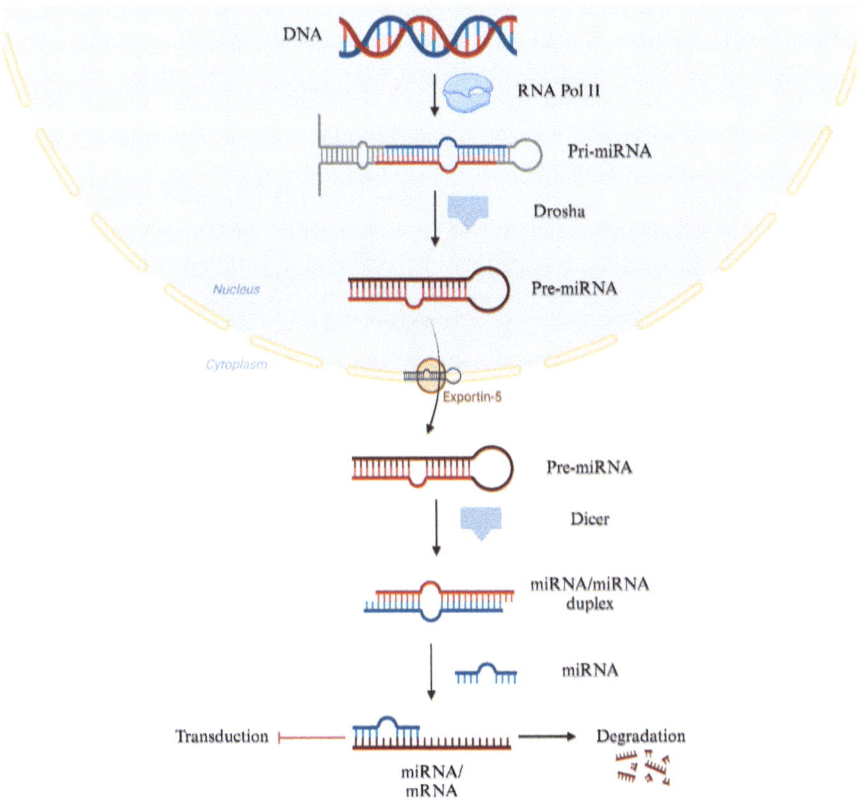

Fig. 1 Summary of the major steps in miRNA biogenesis. (Created in BioRender. Cerchia, C. 2025. https://BioRender.com/83tlscn)

from these regions, are essential regulators of gene expression and contribute to organismal development, tissue differentiation, and cell cycle regulation. Based on nucleotide length (nt), ncRNAs are classified as small (<200 nt; e.g., sncRNAs) or long (lncRNAs). miRNAs, a subclass of sncRNAs, are typically ~22 nt in length and often evolutionarily conserved (Nemeth et al. 2024). miRNA biogenesis (Fig. 1) follows a tightly regulated canonical pathway that begins with the transcription of miRNA genes, predominantly by RNA Polymerase II (Pol II), into primary miRNA transcripts (pri-miRNAs). These long transcripts typically possess a 5′ cap and a 3′ poly (A) tail and are processed in the nucleus by the microprocessor complex, generally composed of the RNase III enzyme Drosha and the double-stranded RNA-binding protein DGCR8. This cleavage generates hairpin structures of approximately 70 nucleotides known as precursor miRNAs (pre-miRNAs). Pre-miRNAs are subsequently exported from the nucleus to the cytoplasm primarily via Exportin-5. In the cytoplasm, the RNase III enzyme Dicer, often in complex with cofactors such as TRBP/PACT, performs the second processing step, cleaving the terminal loop to produce a miRNA/miRNA* duplex of approximately 22 nucleotides. This duplex is then loaded into an Argonaute (AGO) protein, the core component of the RNA-induced silencing complex (RISC). A critical step in the assembly of a functional RISC is the selection of one of the two strands of the miRNA duplex to serve as the guide strand. The other strand, the passenger strand (miRNA*), is typically expelled from the complex and subsequently degraded by cellular nucleases. The selection of the guide strand is primarily determined by the relative thermodynamic stability of the two ends of the miRNA duplex. The strand with the less stable base pairing at the 5′ end is preferentially loaded into the AGO protein. Guided by the incorporated miRNA guide strand, the AGO protein directs RISC to target messenger RNAs (mRNAs), binding usually through partial sequence complementarity in the 3′ untranslated region (3′ UTR). This targeting mainly leads to gene silencing through two main mechanisms: mRNA degradation (involving deadenylation, decapping, and sometimes AGO2-mediated cleavage) and translational repression (inhibiting initiation or post-initiation steps) (O'Brien et al. 2018). Although these repressive roles are well known, some studies have also uncovered non-canonical functions, including translational activation and transcriptional regulation in specific cellular contexts (Havens et al. 2012).

3 miRNAs in Disease Pathways: An Overview from Data Collection to AI Model Implementation

Dysregulation of miRNAs is closely linked to the onset and progression of various diseases. For instance, overexpression of miR-21 and miR-155, alongside downregulation of miR-145, is frequently observed in breast cancer patients. Recent studies underscore the therapeutic potential of miRNAs. Carolina Garcia Villanueva et al. proposed novel miRNA-targeting biologic therapies for pseudoexfoliative glaucoma (PEG), the most common secondary glaucoma characterized by fibrillar extracellular material accumulation in the anterior ocular chamber. Their work highlights the involvement of miR-122-5p, miR-146a-5p, and miR-320a-3p (Villanueva et al. 2025). Sharma et al. identified an inverse correlation between miR-502-3p levels and GABA subunit proteins, demonstrating that miR-502-3p inhibition protects GABA function. This positions miR-502-3p as a promising therapeutic target for neurological disorders like Alzheimer's disease (AD) and AD-related dementia (Sharma et al. 2024). Duan et al. revealed miR-214's suppressive role in XBP1-mediated pathways affecting HUVEC proliferation and angiogenesis, suggesting new targets for modulating vascular function and heart failure progression (Duan et al. 2015). These examples highlight the increasing role of miRNAs as therapeutic, diagnostic, and biomarker targets, driving innovation in prevention and treatment strategies (Chakraborty et al. 2021). Advances in miRNA research have enabled the creation of robust datasets, facilitating the application of AI-driven algorithms despite difficulties in matching the scale of "big data." This section examines current developments in MDAs prediction, including key databases and algorithmic methods, offering a comprehensive overview of the field's state of the art.

3.1 *Database Sources on miRNAs to Detect Their Role in Disease Onset*

Several databases including omics data have been increasing exponentially due to the advancement of systems and molecular biology. There are different types of miRNA databases. Some of these store miRNA information, containing sequences, predicted secondary structures, evolutionary conservation of miRNAs, and potential variants linking them to potential phenotypic effects. Others report experimentally validated MDAs (possible targets, protein interactions, cellular pathways and networks) or miRNA–drug association MDAs that contain experimentally validated data on how small molecules influence microRNA expression.

3.1.1 Main Databases

The onset of disease involves complex biological pathways that require an in-depth understanding of the underlying molecular mechanisms. To fully elucidate the role of miRNAs in disease development and progression, a high level of knowledge is needed across multiple levels of biological regulation. For this purpose, researchers can rely on a variety of specialized databases and bioinformatics resources that support the study of different aspects of miRNA biology (Table 1). A fundamental aspect of miRNA function is their interaction with mRNAs, which can influence entire gene regulatory networks. Dysregulation of these interactions can trigger a cascade of molecular events, ultimately disrupting

Table 1 Summary of major miRNA databases, their biological focus areas, and corresponding web resources

Database name	Focus	Web link	References
miRTarBase	miRNA–mRNA interaction	https://awi.cuhk.edu.cn/~miRTarBase/miRTarBase_2025	Cui et al. (2025)
HMDD	miRNA–disease link	http://www.cuilab.cn/hmdd	Cui et al. (2024)
MirGeneDB	miRNA phylogeny	https://www.mirgenedb.org/	Clarke et al. (2025)
miRNATissueAtlas	miRNA tissue specificity	https://ccb-compute2.cs.uni-saarland.de/mirnatissueatlas_2025	Keller et al. (2022)
miRStart	Transcription start sites of miRNAs	https://awi.cuhk.edu.cn/~miRStart2	Xu et al. (2025)
ncRNAdrug	miRNA–drug link	http://www.jianglab.cn/ncRNADrug	Cao et al. (2024)
miRNASNP	Single nucleotide polymorphisms in miRNA	https://guolab.wchscu.cn/miRNASNP//#!/	Liu et al. (2021a, b)

physiological homeostasis and contributing to disease. In this context, miRTarBase (Cui et al. 2025) is a leading resource, providing comprehensive information on experimentally validated miRNA–target interactions. Since its development in 2011, it has become a cornerstone in the field, with the latest release (v10.0) containing over 3.8 million curated MTIs extracted from approximately 13,700 scientific articles. The incorporation of the LLAMA3 large language model has significantly improved the precision of literature mining, particularly in identifying miRNA–disease associations. Additionally, the updated web interface now includes data on miRNAs involved in drug resistance, small molecules influencing miRNA expression, and miRNAs as potential biomarkers. The utility of miRTarBase is further enhanced by its integration with widely used analytical platforms such as DIANA-miRPath and g:Profiler, facilitating functional interpretation and pathway analysis. Complementing miRTarBase, the Human microRNA Disease Database (HMDD v4.0) (Cui et al. 2024) focuses specifically on experimentally supported associations between miRNAs and human diseases. The latest update includes over 53,000 entries and introduces new categories such as exosomal miRNAs, virus-encoded miRNAs, and miRNA–circRNA interactions. Standardized nomenclature, based on miRBase and Disease Ontology, ensures consistency and accuracy across the dataset, making HMDD a valuable tool for exploring the roles of miRNAs in pathophysiological contexts. From an evolutionary perspective, MirGeneDB v3.0 (Clarke et al. 2025) provides a manually curated and phylogenetically informed catalog of approximately 22,000 miRNA genes across 39 animal species. This resource facilitates the identification of evolutionarily conserved miRNA families, many of which are implicated in fundamental biological processes and common diseases. In terms of expression profiling, miRNATissueAtlas 2025 (Keller et al. 2022) offers a large-scale dataset of tissue-specific expression patterns for nine classes of non-coding RNAs, including miRNAs, in both *Homo sapiens* and *Mus musculus*. With over 61,000 uniformly annotated samples across 35 organs, this resource supports cross-species comparisons and evolutionary studies, helping researchers identify miRNAs with tissue-restricted or conserved expression relevant to disease mechanisms. To investigate transcriptional regulation, miRStart 2.0 (Xu et al. 2025) provides high-confidence annotations of transcription start sites (TSSs) for over 1700 human and 1100 mouse miRNAs, along with associated transcription factors. By integrating data from FANTOM, DBTSS, and ENCODE, and applying DL models, miRStart enables detailed exploration of the upstream regulatory networks that control miRNA expression and their potential dysregulation in disease. Finally, miRNASNP v4 (Liu et al. 2021a, b) focuses on the impact of genetic variation on miRNA biology. It contains data on more than seven million Single Nucleotide Variants (SNVs) and over 500,000 Disease-Related Variations (DRVs) affecting miRNAs and their target sites. The platform provides tools to predict structural changes in pre-miRNAs and to assess the functional consequences of gain or loss of miRNA–target interactions, which may underlie various pathological conditions. Together, these databases form a comprehensive and integrated framework that significantly enhances our understanding of miRNA biology, providing essential support for advancing translational research and the development of miRNA-based diagnostics and therapeutics.

3.2 *Application of Artificial Intelligence in miRNA-Disease Association*

The initial approach to discovering associations between miRNAs and diseases was based on calculating miRNA similarities, assuming that miRNAs with similar functions are often associated with similar diseases. Accordingly, the functional similarity of two miRNAs is estimated by measuring the semantic similarity of their associated diseases, often using structured ontologies (e.g., Disease Ontology or MeSH) (Schuurman and Leszczynski 2008) and/or phenotypic similarity between diseases, such as overlapping clinical symptoms or gene/target profiles. Despite its utility, defining and managing these metrics is challenging due to the complexity of biological systems. The initial methods were used to determine direct connections between diseases and miRNAs by analyzing the available data and predicting likelihood indices that link a given miRNA to the disease, or by using clustering methods. These similarity methods have achieved excellent predictive accuracy, but they are not effective for diseases without any known associated miRNAs (Pundhir and Gorodkin 2013; Wang et al. 2005).

3.2.1 Deep Learning Models for Predicting miRNA-Disease Associations

With the introduction of DL techniques, the methods to discover MDAs are becoming more sophisticated. Computational frameworks for predicting MDAs typically use similarity-based descriptors, obtained by mathematical transformations, such as miRNA functional similarity (FS), disease semantic similarity (SS1 and SS2), and Gaussian interaction profile (GIP) kernel similarity (Cheng et al. 2019; van Laarhoven et al. 2011; Wang et al. 2010). These descriptors are based on the notion that miRNAs with similar functions are likely to be associated with phenotypically or semantically similar diseases, and vice versa. FS is derived by analyzing the hierarchical structure of disease ontologies using directed acyclic graphs (DAGs) constructed from MeSH descriptors, where each disease node is connected through parent–child relationships and semantic contributions decay with increasing distance from the root node, regulated by a decay factor. The similarity FS between miRNAs is determined by the comparison of their associated diseases, based on the method proposed by Wang et al. The MISIM database provides the FS computed as reported by Wang et al. and stored in a dedicated .csv file (Li et al. 2019; Wang et al. 2010). SS1 is computed by summing the semantic contributions of common ancestor nodes between two disease DAGs, weighted by their hierarchical distance and normalized by the total semantic contribution of each DAG. However, SS1 does not account for the frequency of disease terms across the ontology, potentially overestimating the similarity due to common, less informative nodes. SS2 is useful to overcome this limitation, as this metric incorporates a parameter named information content (IC), assigning higher weights to less frequently occurring, more specific disease nodes, thereby providing a more accurate and discriminative comparison measure (Cheng et al. 2019). GIP kernel similarity, on the other hand, is calculated from a binary miRNA–disease association matrix, where known associations are marked as 1 and unknown associations as 0, and similarity is computed using a Gaussian kernel function applied to the interaction profiles of miRNAs (rows) (KM) and diseases (columns) (KD). The kernel parameter (σ) regulates the sensitivity of similarity estimation, with smaller values capturing local patterns and larger values reflecting the global similarities. These descriptors serve as essential inputs for DL models that aim to project miRNA and disease data into a latent space, allowing for the prediction of novel associations (Sheng et al. 2024). Depending on how these heterogeneous data sources are integrated, models are classified as either single-view or multi-view (Fig. 2). Single-view models, including MLMDA, DFELMDA, AEMDA, SMALF, and SVAEMDA, typically employ autoencoders (Ji et al. 2021a, b, 2022; Liu et al. 2021a, b, 2022; Zheng et al. 2019) or Convolutional Neural Networks (CNNs) (e.g., CNNDMP, CNNMDA, DNRLCNN (Xuan et al. 2018, 2019; Zhong et al. 2022)) to learn low-dimensional embeddings, whichare then used for downstream classification using algorithms such as Random Forest, XGBoost, or softmax layers. For instance, DFELMDA (Liu et al. 2022) uses separate autoencoders to reduce miRNA and disease features into 128-dimensional latent spaces, which are input into two parallel Random Forest classifiers whose outputs are averaged to yield the final prediction; however, such models generally lack mechanisms for adaptive fusion of heterogeneous features. In contrast, multi-view approaches, such as AMHMDA, MMGCN, NIMCGCN, and VGAMF (Ding et al. 2022; Li et al. 2020a, b; Ning et al. 2023; Tang et al. 2021a, b) treat each similarity descriptor as an independent view and integrate them using advanced techniques such as graph convolutional networks (GCNs), attention mechanisms, and tensor factorization to learn joint representations that capture cross-modal interactions. PMDFI (Tang et al. 2021a, b), for example, employs stacked autoencoders to extract high-level features from FS, KM, KD, SS1, and SS2 matrices and uses a feature interaction module to generate cross-features, which are processed by multiple Random Forest classifiers and combined via logistic regression. Similarly, MMGCN (Tang et al. 2021a, b) constructs four similarity graphs (based on miRNA sequence, function, disease semantics, and target profiles), encodes each with a GCN, uses multichannel attention to dynamically weigh the embeddings, and employs a CNN module to integrate the representations, followed by matrix factorization to predict association scores. A more recent model, SFPred (Xuan et al. 2025), constructs a heterogeneous miRNA–disease graph based on FS, SS1, SS2, and GIP similarities, extracts first-order neighborhood subgraphs for each candidate pair, transforms them into adjacency matrices, and employs a multi-layer perceptron (MLP) to iteratively update topological features. These updated matrices are then passed through a context-aware transformer (FCTransformer) to learn

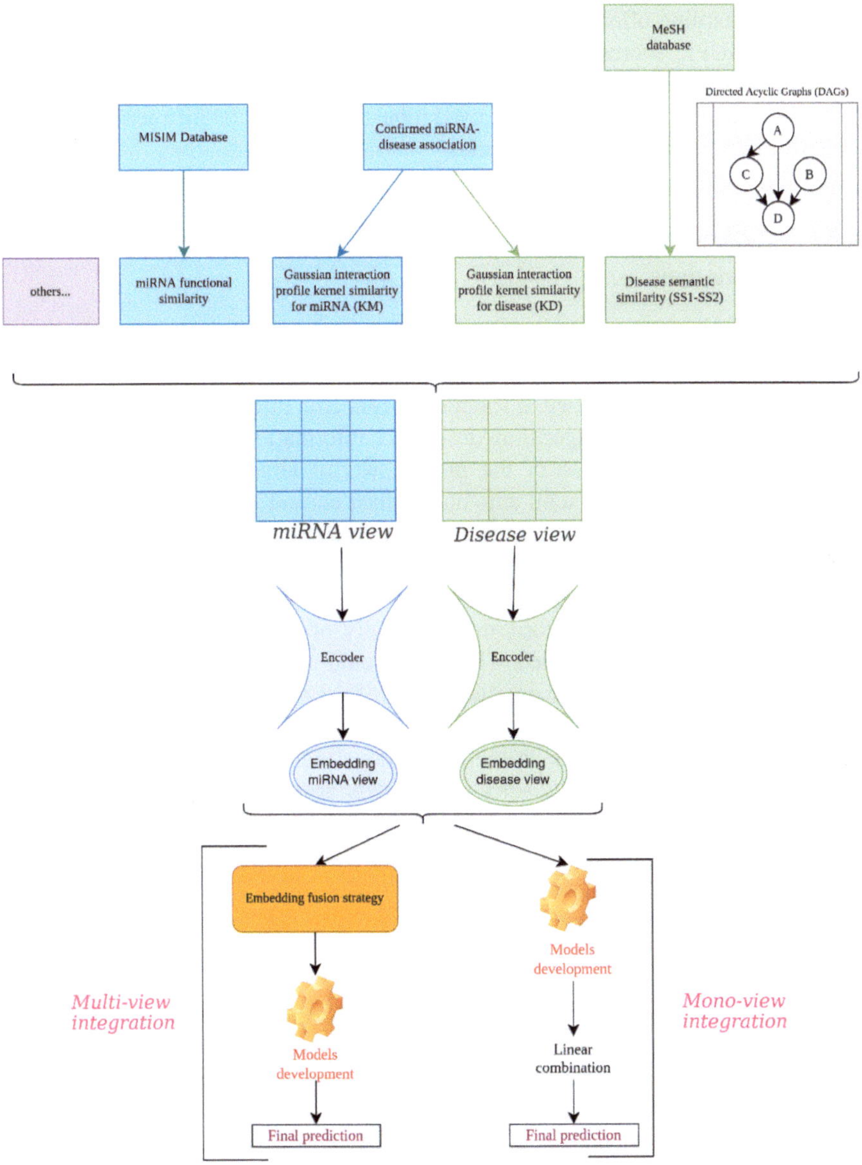

Fig. 2 Illustration of the main differences between the implementation of single-view and multi-view models for miRNA disease association prediction

high-level representations and produce final association predictions. Overall, while single-view models are computationally efficient and relatively easier to implement, they often fall short in capturing complex interdependencies among data sources and may suffer from limited generalization in cross-dataset validations. In contrast, multi-view learning offers superior predictive performance, often achieving higher AUC and AUPR scores in benchmark datasets by enabling the adaptive and context-aware integration of heterogeneous biological knowledge. Furthermore, multi-view models demonstrate improved robustness and scalability across different data distributions and disease categories.

3.2.2 Main Challenges and Future Directions in miRNA-Disease Associations

Despite the development of several models able to predict MDAs, and the encouraging results obtained, some critical issues remain that hinder their full potential. First, the quality of databases is a point of concern. In fact, they are assembled by employing automatic algorithms to extract information. However, despite the progress made, and the use of the most powerful large language models, the extraction process remains semi-automatic, since current models, despite the high accuracy reached, can make mistakes in the interpretation of complex or ambiguous contexts. Therefore, to refine the extraction process, a final manual refinement in the data curation phase is necessary, but the high quantity of data makes this final check difficult. Furthermore, the uniformity of nomenclature, as mentioned above, is still being consolidated. This issue could result in errors when data are integrated from different sources affecting the final performance of models. It is also worth noting that DL models are suitable for being trained when millions of examples are available, while in these situations the available samples are only a few thousand. Furthermore, since the models are mostly classification tasks, during training it is necessary to provide negative and positive examples, but there are only positive examples in databases. For this reason, examples are defined as negative when they are not present in the data. The main problem is that miRNAs are ubiquitous and therefore the association may be missing due to a lack of evidence yet. In this perspective, the same methods can be used to assume that the unlabeled example can belong to both the positive and negative class. Furthermore, miRNA–disease descriptors are subjected to a complex transformation process for extracting key features among the data with various methods, particularly autoencoders. As a result, the performance of the final models appears optimal, but the relationship between the models' predictions and their explainability remains obscure, making these models black boxes. However, efforts to make models increasingly interpretable are steadily increasing. Finally, an important limitation is that most models are hosted on GitHub and lack user-friendly graphical interfaces that can be easily used by non-experts in computational fields. Therefore, developers should improve the usability of these models to make them accessible to medical and pharmaceutical scientists (Sheng et al. 2024; Sun et al. 2023).

3.3 *miRNA Regulation with Small Molecules: An Emerging Role for Computational Approaches*

A direct method for targeting miRNAs involves the development of antisense oligonucleotides (ASOs), known as antagomirs, which are single-stranded RNA sequences complementary to miRNAs containing one or more modified bases (Barger and Nana-Sinkam 2015). These represent a promising solution for miRNA targeting but present various challenges when transitioning to clinical development. Indeed, despite numerous advancements, aspects requiring improvement include their stability, specific target delivery, and the knowledge of potential side effects. In essence, the refinement of their pharmacokinetic and pharmacodynamic properties is still in an early state. Small molecules, conversely, represent an advantageous approach as they are a cornerstone of therapeutics, and the pharmaceutical field possesses extensive knowledge on how to optimize them. Studies have highlighted that small molecules can modulate miRNA activity through various mechanisms. They can directly bind to specific miRNA sequences or structural motifs, thereby inhibiting their interaction with target mRNAs or influencing their processing. Alternatively, they can indirectly influence miRNA levels or function by targeting proteins involved in miRNA biogenesis or degradation, or by affecting the transcription of miRNA genes (Zhou et al. 2025).

3.3.1 Structure-Based Approaches

In this context, computational approaches are increasingly employed to accelerate the discovery of small molecules that regulate miRNAs. Specifically, structure-based methods, such as molecular docking, are utilized to predict the direct binding affinity between small molecules and target miRNAs. A primary challenge, however, lies in obtaining the initial three-dimensional miRNA structure required for these methods. Despite the general scarcity of high-resolution experimental structures, Nuclear Magnetic Resonance (NMR) studies have facilitated the computational prediction of dynamic conformational ensembles for miRNAs (Tadesse and Benhamou 2024). Furthermore, computational servers trained on such data have been developed to predict miRNA three-dimensional structures directly from their sequence. Several studies have demonstrated the effectiveness of applying these computational strategies. For instance, Duca et al. employed molecular docking to identify a molecule binding to the Dicer site of miR-372, which exhibited improved inhibitory activity compared to a known binder targeting the same site (Tran et al. 2022). In another study, Duca et al. successfully characterized the binding mode of polyamine analogues, revealing key interactions between miR-372 and one inhibitor (P3) through docking studies complemented by experimental footprinting (Becquart et al. 2018). Further studies have utilized docking to identify specific small molecule binding sites on target miRNAs. Maiti et al. performed a docking study between pre-miR-21 and aminoglycosides, which showed streptomycin exhibiting the strongest binding. This approach also enabled the identification of U-A and A-G base pairings as potential binding motifs (Bose et al. 2012). Nevertheless, miRNAs inherently exhibit conformational changes and flexibility. Accurately accounting for this dynamic nature during docking simulations remains a significant challenge, potentially impacting the reliability of binding predictions. Furthermore, more sophisticated computational approaches, such as free energy perturbation (FEP) calculations and molecular mechanics/Poisson–Boltzmann surface area (MM/PBSA) methods, can be utilized for more accurate estimation of binding free energies (Rasouli et al. 2025).

3.3.2 Artificial Intelligence Approaches

The application of DL to predict interactions between small molecules and miRNAs holds the potential to significantly accelerate the drug discovery process by enabling in silico screening and candidate prioritization. DL models possess the capability to learn from diverse data sources and correlate the chemical structures of small molecules with the primary sequences of RNAs. Various algorithms are employed for this purpose, with CNNs, Graph Neural Networks (GNNs), and Recurrent Neural Networks (RNNs) being among the most widely used. Several frameworks capable of predicting the interaction probability between small molecules and RNA have been developed, including sChemNET, SMTRI, DeepsmirUD, GCNNMMA, and RiboStrike. These models typically share a common training methodology that involves integrating structural features of both the miRNAs and the small molecules. As an illustrative example, the architecture of SMTRI is discussed; this model is readily accessible due to its implementation within a dedicated platform. SMTRI utilizes CNNs to learn the interaction patterns between small molecules and miRNAs. Data regarding RNA secondary structural motifs and bound small molecules (SMs) were obtained from databases like RNALigands [18], RPocket [21], Inforna [39], and R-Bind [40]. The initial step in the SMTRI methodology involves the identification and extraction of RNA motifs. SMTRI selectively focuses on the pocket region of an RNA sequence, specifically the RNA motifs, thereby aiming to eliminate potentially irrelevant information. These RNA motifs are recognized as unique functional loops formed during the interaction between miRNAs and their target mRNAs. The physicochemical properties of the small molecules were captured starting from their SMILES strings. From these, extended 3D fingerprints (1024-bit) and 43 additional molecular descriptors were calculated using the RDKit toolkit, forming a final 1067-bit feature vector for each molecule. Regarding RNA features, 41 statistical metrics based on nucleotide composition and combinations within the RNA motif sequence were calculated directly from the nucleotide sequence. These metrics include sequence length, GC content, nucleotide ratios, and the ratio of miRNA length within the motif. Additionally, to extract secondary structure features, a Long Short-Term Memory-based Stacked Autoencoder (LSTM-SAE) was implemented. This learns high-level latent features directly from the input RNA motif sequence, resulting in a 512-bit latent vector. The 553-bit feature vector for the RNA motif (combining the 41 statistical metrics and the 512-bit latent vector) and the 1067-bit feature vector for the small molecule were linearly concatenated to create a 1620-bit input feature vector. This vector served as the input to the CNN-based model, which comprises 17 layers, including convolutional, max-pooling, fully connected, and batch normalization layers. Trained employing positive and negative interaction examples, the model can output a probability score ranging from 0 (no interaction predicted) to 1 (interaction predicted). Intermediate scores indicate the confidence level or likelihood that the input small molecule binds to the target RNA motif. In summary, the model provides a quantitative measure of the predicted interaction, enabling users to rank and prioritize potential small molecule candidates for subsequent experimental validation. The performance of SMTRI was directly compared against several other state-of-the-art algorithms, including XGBoost, Naive Bayes (NB), RFSMMA, on independent testing sets derived from PDB, PubChem, and RPocket data. The consistently high AUC values (>0.95) achieved by SMTRI underscores its robust ability to accurately predict small molecules that target miRNA–mRNA interactions. However, a significant limitation in applying DL to predict miRNA–small molecule interactions is the scarcity of experimentally validated miRNA–small molecule interaction pairs available for model training. Databases might not be frequently updated and may contain only a small percentage of all experimentally verified interactions. This scarcity can lead to models liable to underfitting or overfitting the data. Furthermore, given the prominence of certain miRNAs in medical research compared to others, existing datasets may disproportionately represent certain types of miRNA–small molecule interactions or specific chemical classes of small molecules, leading to biased model performance. The scarcity and potential bias in training data pose significant future challenges in improving generalization to novel compounds and enhancing the interpretability of DL model predictions. Moreover, the intrinsic complexity of biological interactions involving miRNAs and SMs is another factor limiting the accuracy and reliability of these approaches.

4 Interplay Between miRNAs, lncRNAs, and circRNAs

Long non-coding RNAs (lncRNAs) are RNA molecules longer than 200 nucleotides that, in most cases, do not encode proteins (Esteller 2011). Unlike miRNAs, lncRNAs are mostly involved in regulating gene expression rather than protein synthesis. They tend to be highly specific to certain cell types or tissues and can act at different stages, namely before, during, and after transcription. lncRNAs can generate small interfering RNAs and short-ncRNAs; moreover, they can act as scaffold to create ribonucleoprotein complexes. This wide variety of functions comes from their unique ability to interact with different macromolecules within the cell, such as DNA, RNA, and proteins (Hussen et al. 2021). One of their key functions is acting as competing endogenous RNAs (ceRNAs), interfering with the biological activity of miRNAs through a process known as miRNA sponging.

By sequestering miRNAs, lncRNAs relieve repression on downstream target genes, thereby modulating gene expression. For instance, genetic sequencing of laryngeal squamous cell carcinoma has identified the lncRNA H19, which serves as a ceRNA for miR-148a-3p, upregulating DNMT1 (DNA methyltransferase enzyme) expression and promoting cancer cell proliferation, migration, and invasion. Given their regulatory significance, lncRNAs such as H19 are potential biomarkers for disease diagnosis and therapeutic targets (Wu et al. 2016). circRNAs are a distinct class of non-coding RNAs characterized by a covalently closed-loop structure lacking the 5′-cap and 3′-poly(A) tail found in linear RNAs. Based on their genetic origin, circRNAs are classified into exonic circRNAs (ecircRNAs), intronic circRNAs (iciRNAs), exon–intron circRNAs (eiciRNAs), and intergenic circRNAs (igcircRNAs) (Meng et al. 2017) (Fig. 3).

Due to its circular structure, circRNA has higher structural stability and exonuclease resistance, thus representing a strong ceRNA candidate for miRNA. While most circRNAs are non-coding, some, like circ-ZNF609, can bind ribosomes and undergo translation (Miao et al. 2021a, b) Functionally, circRNAs participate in gene regulation, neuromodulation, and tumorigenesis, often acting as miRNA sponges to counteract miRNA-mediated gene repression. Similar to lncRNAs, circRNAs harbor multiple MREs, allowing them to bind miRNAs and reduce their availability for silencing target genes (Hansen et al. 2013). This buffering effect can lead to increased expression of genes that would otherwise be downregulated by miRNAs. Their identification and characterization rely on techniques such as RNase R digestion, qPCR, and RNA sequencing. Emerging evidence highlights their role in disease regulation, particularly in cancer, where they modulate gene expression by influencing miRNA activity, thereby affecting tumor cell migration and invasion (Miao et al. 2021a, b).

4.1 Mechanisms of lncRNA–miRNA and circRNA–miRNA Interactions

Both circRNAs and lncRNAs interact with miRNAs through multiple mechanisms, playing crucial roles in post-transcriptional gene regulation. Their interactions with miRNAs occur via sponge adsorption, competition, transport, and interference, ultimately influencing gene expression. Among these mechanisms, miRNA sponging is one of the most widely studied and effective ways of regulating miRNAs. miRNA sponging refers to the ability of circRNAs and lncRNAs to sequester miRNAs, preventing them from binding to their target mRNAs. A typical miRNA sponge contains multiple miRNA recognition elements (MREs), often four to ten binding sites with mismatches at intermediate positions to evade degradation by Argonaute 2 (Ago2) (Papachristou et al. 2012). Unlike small interfering RNAs (siRNAs), which are susceptible to degradation, miRNA sponges offer prolonged inhibition and can simultaneously target multiple miRNAs with a common seed

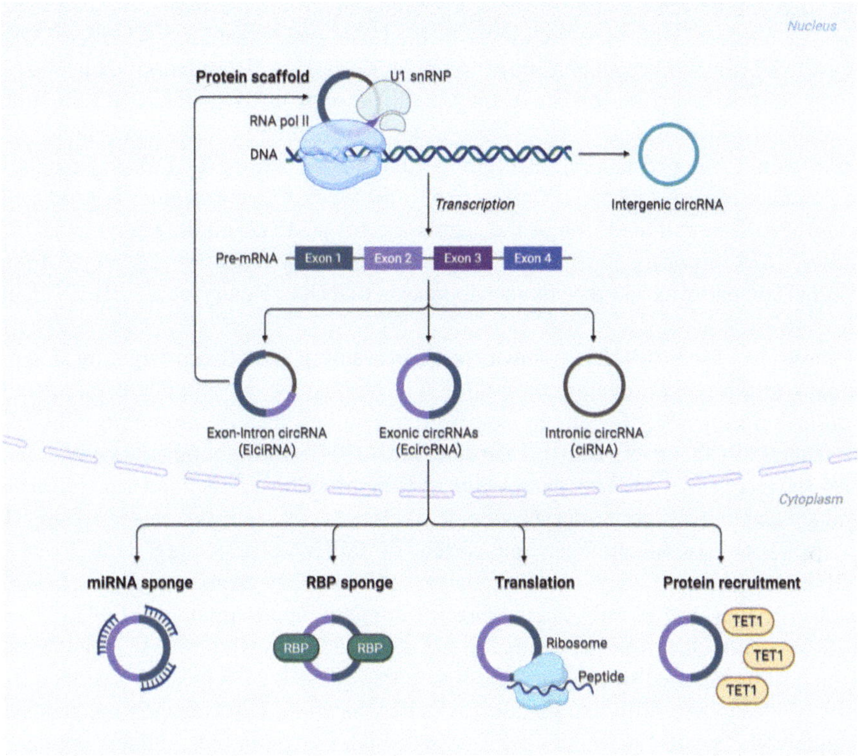

Fig. 3 Schematic representation of circRNA biogenesis and functional roles. circRNAs originate from different genomic regions through a process involving back-splicing. In the nucleus, transcription by RNA polymerase II (RNA pol II) produces pre-mRNA, which can be processed into various types of circRNAs: exon–intron circRNAs (eiciRNAs), exonic circRNAs (ecircRNAs), and intronic circRNAs (ciRNAs). Additionally, circRNAs can arise from intergenic regions. These circRNA species are then transported to the cytoplasm where they perform diverse regulatory functions, including acting as miRNA sponges, RNA-binding protein (RBP) sponges, serving as templates for translation, and recruiting proteins such as TET1. (Created in BioRender. Cerchia, C. 2025, https://BioRender.com/gwuht4a)

region (Kandettu et al. 2020). One key factor influencing the stability of miRNA sponges is their promoter type. Type II promoters, transcribed by RNA polymerase II (Pol II), play a crucial role in long transcript expression. These promoters contain regulatory elements allowing for precise transcriptional control. Type II promoters enable the addition of polyadenylation (polyA) tails. This feature prolongs transcript stability, ensuring longer stability and enhanced regulatory effects (Li et al. 2020a, b). lncRNAs can also compete with miRNAs for binding to mRNA, thereby indirectly inhibiting miRNA function. For instance, lncRNA XIST competes with miR-499a for binding to Bcl2, promoting its expression and enhancing NSCLC progression. Targeting XIST through therapeutic interventions could potentially inhibit NSCLC exacerbation (Zhang et al. 2017). On the other hand, miRNAs can also regulate lncRNA stability by incorporating them into the RISC. For example, miR-21 interacts with lncRNA GAS5 to modulate gene silencing. When ceRNAs like GAS5 are downregulated, miRNAs enhance mRNA degradation. Conversely, when ceRNA levels are elevated, they compete with mRNAs for RISC binding, weakening miRNA repression (Zhang et al. 2013). Interestingly, some lncRNAs serve as miRNA precursors, undergoing cleavage by Drosha and Dicer enzymes to generate mature miRNAs. H19-derived miRNA-675, for example, is produced under stress conditions through the regulation of the RNA-binding protein HUR. miRNA-675 inhibits gastric cancer cell proliferation by reducing activation of the Akt/mTOR pathway via the tumor suppressor RUNX1 (Keniry et al. 2012; Liu et al. 2016).

circRNAs bind miRNAs competitively, preventing them from targeting downstream mRNAs. For example, circGFRA1 inhibits the pro-apoptotic effects of miR-188-3p on NSCLC cells by competitive binding, thereby promoting NSCLC proliferation (Yao et al. 2020). Likewise, circ-0109291 competitively binds miR-188-3p, increasing ABCB1 expression and contributing to cisplatin resistance in oral squamous cell carcinoma (Gao et al. 2022).

circRNAs can function as transporters of miRNAs, relocating them to new cellular sites where they exert regulatory effects. The cerebellar degeneration-related 1 antisense (CDR1-AS) transcript contains approximately 70 MREs for miR-7. Unlike other miRNAs, miR-7 is not degraded upon binding to CDR1-AS but is instead transported to specific sites. Once transported, CDR1-AS is degraded by miR-671, releasing miR-7 to function at its new location. This transport mechanism uniquely alters miRNA availability across cellular compartments, influencing gene regulation regionally (Hansen et al. 2011).

4.2 *Unraveling lncRNA–miRNA (LMI) and circRNA–miRNA (CMI) Interactions*

lncRNAs, circRNAs, and miRNAs are intricately linked, with miRNAs acting as intermediaries between the two non-coding RNA species. Neither lncRNAs nor circRNAs function in isolation; instead, they regulate each other dynamically. The lncRNA/circRNA–miRNA–mRNA axis plays a critical role in numerous diseases, including cancer, cardiovascular disorders, and neurodegenerative conditions (Ma et al. 2023). Understanding these complex interactions opens new avenues for diagnostic and therapeutic applications targeting non-coding RNA networks.

Laboratory experiments involving lncRNAs, circRNAs, and miRNAs are undeniably expensive, complex, and time-consuming, making them particularly challenging for certain organisms, especially humans. With thousands of sequences identified in both animals and plants, large-scale experimental validation remains impractical. This underscores the urgent need for computational approaches to characterize these non-coding RNAs and optimize the design of costly and labor-intensive experiments.

Recent advances in computational techniques for predicting lncRNA–miRNA (LMI) and circRNA–miRNA (CMI) interactions are largely fueled by two primary developments. First, the explosion of high-throughput biological data has enabled the identification of previously unknown associations among non-coding RNAs. Second, there has been significant progress in the design of advanced algorithmic frameworks tailored to these prediction tasks.

With the rapid advancement of high-throughput technologies in genomics, transcriptomics, and proteomics, there is a growing potential for computational approaches to uncover novel LMI and CMI. The development of numerous publicly accessible databases and tools has significantly empowered researchers to access, integrate, and analyze a wide range of data related to lncRNAs, circRNAs, and miRNAs. These resources offer essential information such as nucleotide sequences, expression profiles, genetic variants, predicted interactions, and functional annotations, serving as inputs for computational modeling. Importantly, many of these databases include experimentally validated associations, which provide high-confidence labels for training and benchmarking prediction algorithms. Collectively, these integrated resources have laid the groundwork for the development of data-driven prediction methods. A summary of widely used databases is provided in Table 2.

4.3 Computational Methods for LMI Prediction

As more genomic datasets become available, computational strategies present new possibilities for the large-scale identification of potential LMIs. Extensive research efforts have been devoted to understanding lncRNAs and miRNAs, resulting in the development of numerous predictive models aimed at discovering their interactions. These methods can be broadly categorized into two types based on input modalities: network-based and sequence-based approaches (Sheng et al. 2023).

4.3.1 Network-Based Methods

Network-based approaches typically involve the integration of diverse data sources, such as expression profiles, sequence data, functional annotations, and known LMIs, to compute similarity measures for lncRNAs and miRNAs. Similarity metrics are then used to construct a bipartite network that incorporates known interactions alongside inferred similarities. Subsequently, ML techniques are applied to extract informative features from this network and predict new LMIs.

EPLMI was among the first computational models developed to predict novel lncRNA–miRNA interactions (LMIs), proposed by Huang et al. (2018). It is based on the key assumption that similar lncRNAs tend to share similar interaction or

Table 2 Summary of relevant lncRNA and circRNA databases useful for LMI and CMI prediction, with corresponding web resources

Database name	Focus	Web link	References
lncRNADisease	Experimentally supported and predicted ncRNA–disease associations, including lncRNAs and circRNAs, with enhanced data coverage, integration of causative links, and standardized disease terms using disease ontology and MeSH.	http://www.rnanut.net/lncrnadisease/	Lin et al. (2024)
RNADisease	An expanded database covering a wide range of RNA types (miRNA, lncRNA, circRNA, piRNA, mRNA), species, and diseases, offering RNA–disease enrichment tools and expression analysis based on thousands of cancer-related high-throughput sequencing datasets	http://www.rnadisease.org/	Chen et al. (2023)
TANRIC	An open-access platform for interactive analysis of lncRNA expression across multiple cancer types. It supports exploration of annotated or custom lncRNAs in relation to clinical and molecular features, enabling visualization, enrichment analysis, and discovery of potential biomarkers and therapeutic targets.	https://ibl.mdanderson.org/tanric/_design/basic/main.html	Li et al. (2015)
lncRNASNP	A comprehensive database of SNPs in lncRNAs across eight eukaryotic species, detailing SNP effects on lncRNA structure, expression, and miRNA binding. Includes GWAS-linked variants, TCGA/COSMIC mutations, eQTLs, and predicted or experimental associations between lncRNAs and diseases or drugs	http://gong_lab.hzau.edu.cn/lncRNASNP3/	Yang et al. (2023)
LNCipedia	A public database providing annotated sequences of 127,802 lncRNA transcripts and 56,946 genes, along with 2482 manually curated lncRNA-related research articles.	https://lncipedia.org	Volders et al. (2019)

circBank	A comprehensive circRNA database covering 10 model organisms, offering detailed information on circRNA sequences, expression profiles, regulatory elements (miRNA sites), coding potential, conservation, and experimentally validated data from public transcriptomic sources.	http://www.circbank.cn/#/home	Liu et al. (2019)
circR2Disease	An updated database of 4201 experimentally validated circRNA–disease associations involving 3077 circRNAs and 312 diseases across five species. It includes curated literature-based data and predicted associations	https://github.com/bioinforlab/CircR2Disease-v2.0	Fan et al. (2022)
circBase	A resource for exploring public circRNA datasets and analyzing circRNAs in RNA-seq data. It offers sequence-based search, custom script downloads, table export in multiple formats, and FASTA sequence retrieval	https://www.circbase.org	Glažar et al. (2014)
circRNADisease	A significantly expanded database of experimentally validated circRNA–disease associations. It also includes over 7 million mutation-related circRNAs from 30 TCGA cancer types, along with circRNA-associated genes, miRNAs, proteins, signaling pathways, phenotypes, and prognostic data	http://cgga.org.cn/circRNADisease/	Sun et al. (2024)
NONCODE	A comprehensive cross-species database of annotated non-coding RNAs, with a focus on lncRNAs in both animals and plants. It includes 644,510 lncRNAs across multiple species, updated lncRNA–disease (especially cancer) associations, tissue-specific expression and predicted functions in five plants, and conservation annotations for 23 plant species	http://www.noncode.org/	Zhao et al. (2021)

non-interaction patterns with miRNAs and vice versa. The model employs a two-way diffusion graph algorithm to infer potential LMIs. To construct the interaction network, the authors calculated three types of similarity scores for both lncRNAs and miRNAs by integrating diverse biological data sources:

1. Expression-Based Similarity, computed using expression profile data. Specifically, the Pearson correlation coefficient (PCC) was used to assess the expression similarity between pairs of lncRNAs or pairs of miRNAs.
2. Function-Based Similarity between RNAs, evaluated using biological function data. For miRNAs, the authors used known miRNA–target gene interactions, while for lncRNAs, putative functional annotations were applied. These datasets enabled the calculation of functional similarities between miRNA–miRNA and lncRNA–lncRNA pairs.
3. Sequence-Based Similarity to capture sequence-level relationships. To this aim, the model applied the Needleman–Wunsch pairwise sequence alignment algorithm, providing a quantitative measure of sequence similarity for both lncRNAs and miRNAs.

By bringing together different types of similarity data and using a two-way diffusion approach, EPLMI was able to successfully predict new lncRNA–miRNA interactions. This model laid important groundwork for future research and inspired the development of more advanced methods in the field.

More recently, Wang et al. introduced a method called GCNCRF, which combines Graph Convolutional Networks (GCNs) and Conditional Random Fields (CRFs) to predict human LMIs (Xie et al. 2024).The authors first measured how similar lncRNAs and miRNAs are to one another based on both sequence information and interaction profiles. These different similarity scores were then combined into a single similarity measure for each RNA type. Using this information, along with known LMIs, they built a network where nodes represent lncRNAs and miRNAs, and edges represent their potential interactions. Next, they used a technique called random walk with restart to extract important features from the similarity networks of lncRNAs and miRNAs. These features were then used to build a feature matrix representing the overall interaction network. The GCN part of the model was used to learn the structure of this network, identifying patterns and relationships among the RNAs by aggregating information from neighboring nodes. After that, a CRF layer with an attention mechanism was applied to refine the node representations. This step helped ensure that similar lncRNAs or miRNAs had similar feature embeddings. Finally, a decoder layer was used to calculate the likelihood of interaction between each lncRNA–miRNA pair, producing a matrix of predicted interaction scores.

The central aspect of network-based techniques is the construction of a heterogeneous lncRNA–miRNA graph, followed by the application of algorithms, such as matrix factorization or GCNs, to learn meaningful representations and estimate interaction scores. These methods are able to leverage the topological structure of the network, offering interpretable and robust performance. Moreover, they are relatively straightforward in terms of theoretical understanding and computational implementation. Nonetheless, their dependence on known interactions introduces challenges, such as potential prediction bias arising from imbalanced datasets. For instance, EPLMI relies on prior LMI data to perform message passing. Additionally, the overall prediction quality may be compromised by inconsistencies in the input data, especially since network-based models often integrate multiple data types.

4.3.2 Sequence-Based Methods

Sequence-based models rely exclusively on the nucleotide sequences of lncRNAs and miRNAs. Feature extraction in these models typically follows one of two strategies. The first involves handcrafted feature engineering, where attributes such as k-mer frequencies, composition–transition–distribution (CTD) metrics, GC content, and base-pair counts are calculated. These features are then fed into DL models, such as CNNs or LSTM networks, to learn abstract representations and predict interactions. The second strategy bypasses manual feature design by directly applying encoding techniques (e.g., one-hot encoding) to raw sequences, which are then input into DL architectures like CNNs, LSTMs, or gated recurrent units (GRUs) to automatically extract deep features and generate interaction scores.

Yang et al. developed LncMirNet, a deep learning-based model CNNs (Yang et al. 2020).

The authors extracted four different types of features from lncRNA and miRNA sequences using several techniques: k-mer (short sequences of nucleotides) analysis, composition transition distribution (CTD), and two embedding methods: doc2vec and role2vec. Each of these methods captures different aspects of the RNA sequences, from basic nucleotide patterns to more abstract representations. These extracted features were then transformed into matrix formats using a technique called Histogram-Dd, resulting in structured feature matrices for both lncRNAs and miRNAs. Each matrix had a fixed size, ensuring consistency before being processed by the CNN model. Next, these matrices were fed into separate CNN pipelines, one for lncRNAs and one for miRNAs. As the data passed through multiple CNN layers, the model learned complex and deep features from each sequence. Finally, the output representations (or "tensors") from both pipelines were combined and passed through FCLs to predict the likelihood of interaction between each lncRNA and miRNA pair. LncMirNet effectively demonstrated how DL can uncover subtle, non-linear sequence features that are useful for identifying potential LMIs.

Yu et al. developed an approach to predict potential interactions between lncRNAs and miRNAs by combining a pre-training strategy called rna2vec with deep feature extraction techniques (Kim and Lee 2023). The process started by training a set of k-mer embeddings using word2vec, a popular method originally developed for natural language processing (Mikolov et al. 2013). This allowed the model to convert RNA sequences into meaningful, low-dimensional numerical vectors that capture important sequence patterns. Next, the authors designed a hybrid neural network that combines CNNs and bi-directional gated recurrent units (Bi-GRUs). The CNN component was used to detect local patterns in the sequences, while the Bi-GRU component captured long-range dependencies and contextual relationships within the RNA sequences. To enhance the model's ability to focus on the most informative features, an attention mechanism was added. This helped the network weigh and integrate relevant features across different layers of the Bi-GRU. In the final step, the representations of lncRNAs and miRNAs were combined and passed through FCLs to generate scores that indicate the likelihood of interaction between each pair.

Sequence-based models often surpass their network-based counterparts in predictive performance by harnessing only the sequence data of lncRNAs and miRNAs. However, their success is contingent upon the deep models' capacity to learn high-level features. Despite their effectiveness, such DL models are typically resource-intensive in terms of computational time and memory usage, and the interpretability of the extracted features remains limited.

4.3.3 Key Challenges and Future Directions in LMI Prediction

Network-based approaches often integrate lncRNA and miRNA expression profiles, sequences, and functional data to build heterogeneous networks. While each data type adds valuable information, these sources can be noisy, incomplete, or outdated. For instance, databases like PMRD and microRNA.org are infrequently updated or have limited accessibility. High-quality, curated data is essential for effective model training. Strengthening collaboration between computational and biomedical researchers and promoting open data sharing are critical steps toward improving data reliability.

DL models used in sequence-based methods can capture complex, non-linear features. However, their "black-box" nature makes it difficult to interpret the biological significance of their predictions. In bioinformatics, model interpretability is crucial for understanding underlying regulatory mechanisms. Future models should incorporate interpretable architectures and visualization tools to bridge the gap between predictive accuracy and biological insight.

Effective training of LMI prediction models requires both positive (known) and negative (non-interacting) samples. However, true negative samples are rarely available. Most studies address this by randomly selecting negatives from unlabeled data, which may contain unknown positives and lead to biased results. A promising solution is Positive-Unlabeled (PU) learning, which enables models to learn from confirmed positives and unlabeled examples, improving prediction performance. There remains significant potential to refine negative sample selection using PU-learning strategies (Sheng et al. 2023).

Finally, many computational studies rely solely on performance metrics like accuracy, F1-score, AUC, and AUPR, without experimental follow-up. To advance the field, closer collaboration between computational scientists and experimental biologists is essential. This integration will help validate predictions and uncover functional roles of lncRNA–miRNA interactions.

4.4 Computational Methods for CMI Prediction

Currently, research on CMI predictions remains more limited, highlighting the need for robust and reliable prediction methods to advance the field. However, known CMIs are scattered across multiple databases with varying molecular attributes, creating challenges for computational data modeling. To address this, more recently developed models generate multiple datasets to train and optimize predictions for data with diverse characteristics. Modeling data with diverse attributes not only helps mitigate performance imbalances across different datasets, improving overall model effectiveness, but also enhances the model's adaptability to various prediction tasks and conditions. In fact, first-generation models were capable of handling only individual datasets, exhibiting significant performance disparities across different data.

To address data incompleteness, Yu et al. reported RBNE-CMI, a novel framework that models incomplete multi-attribute heterogeneous networks to enhance circRNA–miRNA interaction prediction (Fig. 4) (Yu et al. 2024). RBNE-CMI integrates two commonly used datasets, CMI-9905 (Wang et al. 2022) and CMI-753 (Wang et al. 2023) into a unified dataset (CMI-10413), comprising 10,413 interaction pairs among 1163 miRNAs and 2657 circRNAs. The CMI-9905 dataset, introduced by Wang et al. (2022) combines the Circbank and CircR2Cancer databases. This dataset contains 8905 interaction pairs involving 2346 circRNAs and 962 miRNAs. The CMI-753 was obtained in a similar fashion, and contains 753 circRNA–miRNA positive pairs between 515 circRNAs and 469 miRNAs (Wang et al. 2023). By leveraging DNABERT (Ji et al. 2021a, b) for nucleotide sequence representation and refining features with denoising autoencoders (DAE), the method preserves both molecular attribute (such as nucleotide sequence representations) and behavioral features (i.e. the interactions within the network, derived from network embeddings). The framework builds an incomplete CMI network that maps how miRNAs and circRNAs interact, even when some data is missing. It also groups similar miRNAs and circRNAs into self-similarity networks, helping to reveal hidden relationships. To refine predictions, it applies Gaussian and sigmoid similarity measures, which highlight molecular connections based on their patterns of interaction. The authors reported that RBNE-CMI achieved an AUC of 0.8792 in cross-validation and outperformed existing models even when evaluated on complete datasets, demonstrating robustness across different data conditions. A case study identified 18 out of 20 predicted interactions as potential cancer biomarkers, showcasing the method's practical utility. Some challenges remain, particularly in the construction of negative samples. Current methods often generate negative samples by randomly selecting non-interacting circRNA–miRNA pairs, but this approach risks including false negatives (pairs that may actually interact but lack experimental validation). These false negatives can mislead the model during training, reducing prediction accuracy. The authors foresee future refinements, such as exploring structured sampling strategies using balance theory-based approaches, to improve prediction accuracy.

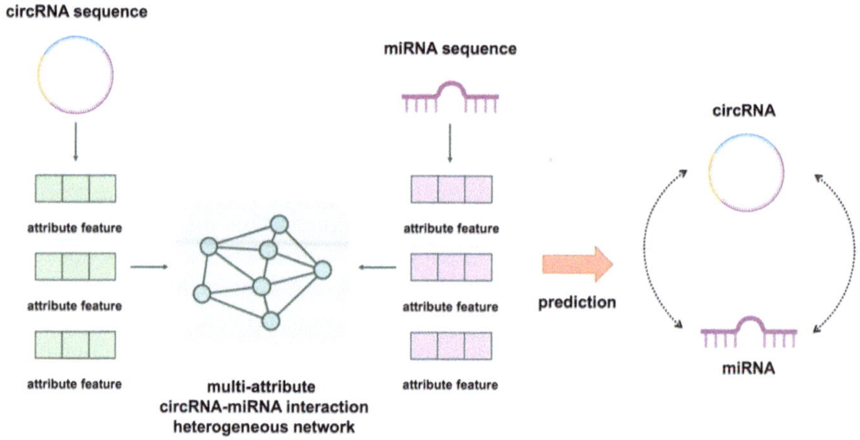

Fig. 4 Schematization of RBNE-CMI approach showing RNA molecular attribute feature extraction and multi-attribute CMI heterogeneous network embedding

Jiang et al. proposed Hither-CMI, a DL model that integrates hybrid multimodal networks and higher-order neighborhood information to enhance CMI prediction (Jiang et al. 2025). Hither-CMI constructs a hybrid multimodal network by integrating sequence, structure, and expression similarity data of circRNA and miRNA using Multiple Kernel Learning (MKL), ensuring optimal fusion of diverse biological features. This network is then combined with the CMI association network, and an enhanced Graph Convolutional Network (GCN) is applied to derive hybrid higher-order embedding representations, capturing both immediate and distant interactions. Finally, these embeddings are fed into an XGBoost classifier for training and prediction. Experimental validation demonstrates Hither-CMI's predictive performance, achieving an AUC of 0.9134. In case studies, 25 out of the top 30 predicted CMIs were confirmed by recent literature, underscoring its reliability as a preselection tool for further biological research. Additionally, Hither-CMI outperforms existing methods across multiple evaluation metrics (AUC, AUPR, accuracy), making it a robust model for CMI prediction. Despite the strengths of the Hither-CMI approach, some challenges remain: (i) balancing multimodal similarity network fusion due to inconsistent kernel function behavior, and (ii) potential loss of distant neighbor information in deep network layers. Future improvements will focus on optimizing multisource network learning and refining sensitivity to neighborhood information. Beyond CMI prediction, Hither-CMI holds promise for broader applications, including circRNA–disease and lncRNA–miRNA interaction predictions, expanding its role in biological network analysis and biomedical research.

Yin et al. introduced MRHRL, a novel computational model for predicting CMI using multirelational hypergraph representation learning (Yin et al. 2024). Simple graphs can only represent pairwise connections and thus they struggle to capture higher-order relationships among CMAs, limiting their predictive performance. Hypergraphs extend the traditional concept of edges in simple graphs to hyperedges, which can connect multiple nodes simultaneously, making them well-suited for representing sparse associations. The MRHRL approach employs hypergraph neural networks to aggregate information across multiple RNA interaction types, integrating a view attention mechanism to incorporate complementary perspectives. The model consists of two key components: (1) Multirelational Hypergraph Learning, which constructs hypergraphs under four distinct relationship types to capture complex correlations between RNA molecules, and (2) Multitask Joint Optimization, which enhances prediction accuracy and generalization by combining association prediction with a hyperedge-level reconstruction task. Experimental evaluations on three real-world datasets demonstrate that MRHRL significantly outperforms existing CMA prediction models. However, limitations remain, including the static nature of hypergraph construction and computational challenges as dataset size increases. Future research will focus on developing dynamic hypergraph structures to improve adaptability and efficiency.

4.4.1 Key Challenges and Future Directions in CMI Prediction

Computational prediction of CMIs holds significant promise for advancing biological research by offering an efficient and cost-effective means to prioritize potential candidates for experimental validation. These approaches can substantially reduce experimental burden and help formulate more targeted hypotheses. However, as the volume and complexity of biological data continue to grow, the field faces persistent challenges that hinder the development of reliable and generalizable prediction models. A major challenge lies in the variability and inconsistency of available datasets. In many cases, CMI data are derived from different experimental conditions, annotated using diverse standards, or affected by incomplete and noisy information. This lack of uniformity complicates data integration and model training. Consequently, models often perform well within specific datasets but fail to generalize across different sources due to inconsistencies in feature distributions and sample properties.

To address these issues, researchers often curate datasets by selecting samples with similar characteristics, allowing for more consistent feature representation. However, this strategy typically reduces the dataset size and introduces bias, ultimately limiting the model's predictive power. The trade-off between dataset completeness and feature quality compromises the robustness and accuracy of prediction results, especially when applied to new or heterogeneous data.

Despite these limitations, the expanding body of experimentally validated CMI data has greatly enhanced our understanding of the molecular mechanisms underlying circRNA–miRNA binding. Many existing approaches rely on extracting behavioral or structural features from circRNA–miRNA association networks, which remain central to prediction accuracy.

Improving data standardization, enhancing feature extraction techniques, and designing models with better adaptability and generalization across diverse datasets will be crucial. The integration of multiple data modalities, such as sequence, expression, and structural information, is likely to yield more biologically meaningful insights.

Computational approaches to CMI prediction have already demonstrated considerable potential; however, continued advancements in algorithm development, data integration, and interdisciplinary efforts key to enhancing the utility of computational methods in CMI.

4.5 Non-coding RNA Networks in Drug Development and Resistance: Mechanisms and Therapeutic Potential

Due to their regulatory roles in signaling pathways associated with various pathologies, these ncRNAs may represent possible targets for drug development. In fact, analyzing changes in ncRNA expression levels can help elucidate drug mechanisms. For instance, monitoring ncRNA expression offers a more accessible proxy for studying drug activity. In HCT-116 colon cancer cells, oxaliplatin upregulates miR-185-5p, which subsequently downregulates β-catenin, Cyclin D1, MMP2, and MMP9, leading to inhibited tumor progression. Rescue experiments further demonstrated the regulatory interaction between lncRNA NEAT1, miR-185-5p, and IGF2 in colon cancer (Zhuang et al. 2020). Thus, measuring NEAT1 expression post-treatment could provide insight into oxaliplatin's mechanism of action.

lncRNAs are involved in cancer development and metastasis, with their expression levels often changing based on the tumor stage and degree of metastasis. Because of these roles, along with their strong tissue specificity, lncRNAs show great potential as biomarkers for prognosis and prediction in cancer.

Multidrug Resistance (MDR) is a phenomenon where cancer cells become resistant to multiple chemotherapeutic agents with distinct structures and mechanisms, representing a major challenge in oncology. Therefore, understanding ncRNA involvement in MDR is emerging as an interesting venue. Doxorubicin is widely used against various tumors but often leads to MDR. In breast cancer, circ_0085495 is upregulated in doxorubicin-resistant cells, acting as a sponge for miR-873-5p. This sponging enhances drug resistance and promotes proliferation and metastasis (Xie and Zheng 2022). Besides circ_0085495, circ_0006528 and circ_0001667 also affect doxorubicin resistance (Cui et al. 2022).

In colorectal cancer (CRC), lncRNA BANCR sponges miR-203. Knocking down BANCR in CRC model mice reduced tumor progression and enhanced doxorubicin sensitivity via the BANCR/miR-203/CSE1L axis (Ma et al. 2018). These ncRNAs could thus serve as clinical biomarkers to assess doxorubicin resistance and guide treatment decisions.

Tamoxifen, another drug frequently used to treat breast and ovarian cancers, also encounters resistance issues. Exosomal circ_UBE2D2 contributes to tamoxifen resistance by sponging miRNAs such as miR-200a-3p. Downregulating circ_UBE2D2 enhances tamoxifen sensitivity (Hu et al. 2020). Additionally, bioinformatic studies predict that lncRNA CYTOR binds miR-125a-5p. CYTOR promotes serum response factor (SRF) expression and activates Hippo and MAPK pathways, ultimately increasing tamoxifen resistance (Liu et al. 2020). Hence, targeting specific ncRNAs such as circ_UBE2D2 and CYTOR may reduce resistance and improve therapeutic outcomes. These ncRNA–miRNA axes are valuable both as biomarkers for drug sensitivity and as therapeutic targets.

Among the RNA-targeting strategies, the use of siRNAs offers a strong rationale. siRNAs are synthetic oligonucleotides that can selectively inhibit ncRNA expression. However, they require delivery systems to overcome cell membrane barriers and enzymatic degradation. For example, in glioblastoma multiforme (GBM), tumor-targeting immunoliposomes have been used to deliver siRNAs across the blood–brain barrier to suppress lncRNA MALAT1, thereby inhibiting tumor growth (Kim et al. 2018). Indirect modulators can also affect ncRNA pathways. For instance, estradiol promotes breast cancer progression by inducing lncRNA HOTAIR expression. Therefore, estradiol antagonists or estrogen receptor inhibitors may act as indirect inhibitors of HOTAIR, offering therapeutic benefits (Bhan et al. 2013).

Overall, ncRNAs offer a promising source of novel, targeted therapies, especially in the context of personalized medicine. Expression levels of specific ncRNAs could help determine disease stage and guide treatment choices tailored to individual patients. Furthermore, the targeted modulation of regulatory networks involving ncRNAs, identified through computational approaches, holds great potential for the development of innovative therapeutic strategies.

4.6 Conclusion

Recent discoveries on ncRNA interactions have indeed sparked a profound shifting to a Copernican revolution in the scientific community's understanding of gene regulation and disease pathogenesis. Once relegated to marginal roles, these molecules are now recognized as central players orchestrating complex cellular networks, with their dysregulation implicated in a wide array of human pathologies. The subsequent accumulation of vast clinical and biological data, captured in the numerous databases cited within this chapter, has paved the way for AI to emerge as a powerful tool. AI enables the analysis of these extensive datasets, the prediction of complex interactions (including miRNA–disease associations, miRNA–small molecule interactions, lncRNA–miRNA, and circRNA–miRNA links), and the identification of potential therapeutic pathways. This chapter provided a general overview of the principles underpinning the development of state-of-the-art ML and DL model architectures employed for these tasks. Despite considerable progress yielding sophisticated predictive models, significant obstacles remain. We have highlighted how the reliability and utility of these computational approaches are often constrained by limitations in the quality, completeness, standardization, and accessibility of the foundational data used for model training. Furthermore, the intrinsic "black box" nature of many advanced algorithms hinders biological interpretability, a crucial aspect for translating in silico predictions into experimentally testable hypotheses. Encouragingly, as we have discussed, these challenges are being actively addressed through ongoing efforts, such as the pursuit of standardized nomenclature and the adoption of innovative strategies like Positive-Unlabeled (PU) learning to better manage the lack of confirmed negative samples. In conclusion, while the development of mathematically robust and complex computational models is essential, the paramount challenge, and opportunity, lies in fostering deeper collaboration between computational scientists and experimental biologists. This synergy is indispensable to ensure that in silico predictions, particularly those derived from interpretable models, are rigorously validated through in vitro and in vivo experimentation, thereby accelerating the translation of computational insights into tangible biological understanding and impactful clinical advances.

Acknowledgments A.L and P.D acknowledge funding from the PNRR Project CN000041 "National Centre for Gene Therapy and RNA Technology-based Drugs"—(CUP) UNINA E63C22000940007.

References

Barger JF, Nana-Sinkam SP. MicroRNA as tools and therapeutics in lung cancer. Respir Med. 2015;109:803–12. https://doi.org/10.1016/j.rmed.2015.02.006.

Becquart C, Le Roch M, Azoulay S, Uriac P, Di Giorgio A, Duca M. Exploring heterocycle-Spermine conjugates as modulators of oncogenic microRNAs biogenesis. ACS Omega. 2018;3:16500–8. https://doi.org/10.1021/acsomega.8b02681.

Bhan A, Hussain I, Ansari KI, Kasiri S, Bashyal A, Mandal SS. Antisense-transcript long noncoding RNA (lncRNA) HOTAIR is transcriptionally induced by estradiol. J Mol Biol. 2013;425:3707–22. https://doi.org/10.1016/j.jmb.2013.01.022.

Bose D, Jayaraj G, Suryawanshi H, Agarwala P, Pore SK, Banerjee R, Maiti S. The tuberculosis drug streptomycin as a potential cancer therapeutic: inhibition of miR-21 function by directly targeting its precursor. Angew Chem Int Ed. 2012;51:1019–23. https://doi.org/10.1002/anie.201106455.

Cao X, Zhou X, Hou F, Huang Y, Yuan M, Long M, Chen S, Lei W, Zhu J, Chen J, Zhang T, Guo A-Y, Jiang W. ncRNADrug: a database for validated and predicted ncRNAs associated with drug resistance and targeted by drugs. Nucleic Acids Res. 2024;52:D1393–9. https://doi.org/10.1093/nar/gkad1042.

Chakraborty C, Sharma AR, Sharma G, Lee S-S. Therapeutic advances of miRNAs: a preclinical and clinical update. J Adv Res. 2021;28:127–38. https://doi.org/10.1016/j.jare.2020.08.012.

Chen J, Lin J, Hu Y, Ye M, Yao L, Wu L, Zhang W, Wang M, Deng T, Guo F, Huang Y, Zhu B, Wang D. RNADisease v4.0: an updated resource of RNA-associated diseases, providing RNA-disease analysis, enrichment and prediction. Nucleic Acids Res. 2023;51:D1397–404. https://doi.org/10.1093/nar/gkac814.

Cheng L, Zhao H, Wang P, Zhou W, Luo M, Li T, Han J, Liu S, Jiang Q. Computational methods for identifying similar diseases. Mol Ther Nucleic Acids. 2019;18:590–604. https://doi.org/10.1016/j.omtn.2019.09.019.

Clarke AW, Høye E, Hembrom AA, Paynter VM, Vinther J, Wyrożemski Ł, Biryukova I, Formaggioni A, Ovchinnikov V, Herlyn H, Pierce A, Wu C, Aslanzadeh M, Cheneby J, Martinez P, Friedländer MR, Hovig E, Hackenberg M, Umu SU, Johansen M, Peterson KJ, Fromm B. MirGeneDB 3.0: improved taxonomic sampling, uniform nomenclature of novel conserved microRNA families and updated covariance models. Nucleic Acids Res. 2025;53:D116–28. https://doi.org/10.1093/nar/gkae1094.

Cui Y, Fan J, Shi W, Zhou Z. Circ_0001667 knockdown blocks cancer progression and attenuates adriamycin resistance by depleting NCOA3 via releasing miR-4458 in breast cancer. Drug Dev Res. 2022;83:75–87. https://doi.org/10.1002/ddr.21845.

Cui C, Zhong B, Fan R, Cui Q. HMDD v4.0: a database for experimentally supported human microRNA-disease associations. Nucleic Acids Res. 2024;52:D1327–32. https://doi.org/10.1093/nar/gkad717.

Cui S, Yu S, Huang H-Y, Lin Y-C-D, Huang Y, Zhang B, Xiao J, Zuo H, Wang J, Li Z, Li G, Ma J, Chen B, Zhang H, Fu J, Wang L, Huang H-D. miRTarBase 2025: updates to the collection of experimentally validated microRNA–target interactions. Nucleic Acids Res. 2025;53:D147–56. https://doi.org/10.1093/nar/gkae1072.

de Gonzalo-Calvo D, Karaduzovic-Hadziabdic K, Dalgaard LT, Dieterich C, Perez-Pons M, Hatzigeorgiou A, Devaux Y, Kararigas G. Machine learning for catalysing the integration of noncoding RNA in research and clinical practice. EBioMedicine. 2024;106:105247. https://doi.org/10.1016/j.ebiom.2024.105247.

Ding Y, Lei X, Liao B, Wu F-X. Predicting miRNA-disease associations based on multi-view variational graph auto-encoder with matrix factorization. IEEE J Biomed Health Inform. 2022;26:446–57. https://doi.org/10.1109/JBHI.2021.3088342.

Duan Q, Yang L, Gong W, Chaugai S, Wang F, Chen C, Wang P, Zou M-H, Wang DW. MicroRNA-214 is upregulated in heart failure patients and suppresses XBP1-mediated endothelial cells angiogenesis. J Cell Physiol. 2015;230:1964–73. https://doi.org/10.1002/jcp.24942.

Esteller M. Non-coding RNAs in human disease. Nat Rev Genet. 2011;12:861–74. https://doi.org/10.1038/nrg3074.

Fan C, Lei X, Tie J, Zhang Y, Wu F-X, Pan Y. CircR2Disease v2.0: an updated web server for experimentally validated circRNA-disease associations and its application. Genomics Proteomics Bioinformatics. 2022;20:435–45. https://doi.org/10.1016/j.gpb.2021.10.002.

Gao F, Han J, Wang Y, Jia L, Luo W, Zeng Y. Circ_0109291 promotes cisplatin resistance of oral squamous cell carcinoma by sponging miR-188-3p to increase ABCB1 expression. Cancer Biother Radiopharm. 2022;37:233–45. https://doi.org/10.1089/cbr.2020.3928.

Glažar P, Papavasileiou P, Rajewsky N. circBase: a database for circular RNAs. RNA. 2014;20:1666–70. https://doi.org/10.1261/rna.043687.113.

Hansen TB, Wiklund ED, Bramsen JB, Villadsen SB, Statham AL, Clark SJ, Kjems J. miRNA-dependent gene silencing involving Ago2-mediated cleavage of a circular antisense RNA. EMBO J. 2011;30:4414–22. https://doi.org/10.1038/emboj.2011.359.

Hansen TB, Jensen TI, Clausen BH, Bramsen JB, Finsen B, Damgaard CK, Kjems J. Natural RNA circles function as efficient microRNA sponges. Nature. 2013;495:384–8. https://doi.org/10.1038/nature11993.

Havens MA, Reich AA, Duelli DM, Hastings ML. Biogenesis of mammalian microRNAs by a non-canonical processing pathway. Nucleic Acids Res. 2012;40:4626–40. https://doi.org/10.1093/nar/gks026.

Hu K, Liu X, Li Y, Li Q, Xu Y, Zeng W, Zhong G, Yu C. Exosomes mediated transfer of Circ_UBE2D2 enhances the resistance of breast cancer to tamoxifen by binding to MiR-200a-3p. Med Sci Monit. 2020;26:e922253. https://doi.org/10.12659/MSM.922253.

Huang Y-A, Chan KCC, You Z-H. Constructing prediction models from expression profiles for large scale lncRNA-miRNA interaction profiling. Bioinformatics. 2018;34:812–9. https://doi.org/10.1093/bioinformatics/btx672.

Hussen BM, Azimi T, Abak A, Hidayat HJ, Taheri M, Ghafouri-Fard S. Role of lncRNA BANCR in human cancers: an updated review. Front Cell Dev Biol. 2021;9 https://doi.org/10.3389/fcell.2021.689992.

Ji C, Gao Z, Ma X, Wu Q, Ni J, Zheng C. AEMDA: inferring miRNA–disease associations based on deep autoencoder. Bioinformatics. 2021a;37:66–72. https://doi.org/10.1093/bioinformatics/btaa670.

Ji Y, Zhou Z, Liu H, Davuluri RV. DNABERT: pre-trained bidirectional encoder representations from transformers model for DNA-language in genome. Bioinformatics. 2021b;37:2112–20. https://doi.org/10.1093/bioinformatics/btab083.

Ji C, Wang Y, Gao Z, Li L, Ni J, Zheng C. A semi-supervised learning method for MiRNA-disease association prediction based on variational autoencoder. IEEE/ACM Trans Comput Biol Bioinform. 2022;19:2049–59. https://doi.org/10.1109/TCBB.2021.3067338.

Jiang C, Wang L, Yu C-Q, You Z-H, Wang X-F, Wei M-M, Shi T-L, Liang S-Z, Wang D-W. Hither-CMI: prediction of circRNA-miRNA interactions based on a hybrid multimodal network and higher-order neighborhood information via a graph convolutional network. J Chem Inf Model. 2025;65:446–59. https://doi.org/10.1021/acs.jcim.4c01991.

Kandettu A, Radhakrishnan R, Chakrabarty S, Sriharikrishnaa S, Kabekkodu SP. The emerging role of miRNA clusters in breast cancer progression. Biochim Biophys Acta Rev Cancer. 2020;1874:188413. https://doi.org/10.1016/j.bbcan.2020.188413.

Keller A, Gröger L, Tschernig T, Solomon J, Laham O, Schaum N, Wagner V, Kern F, Schmartz GP, Li Y, Borcherding A, Meier C, Wyss-Coray T, Meese E, Fehlmann T, Ludwig N. miRNATissueAtlas2: an update to the human miRNA tissue atlas. Nucleic Acids Res. 2022;50:D211–21. https://doi.org/10.1093/nar/gkab808.

Keniry A, Oxley D, Monnier P, Kyba M, Dandolo L, Smits G, Reik W. The H19 lincRNA is a developmental reservoir of miR-675 that suppresses growth and Igf1r. Nat Cell Biol. 2012;14:659–65. https://doi.org/10.1038/ncb2521.

Kim Y, Lee M. Deep learning approaches for lncRNA-mediated mechanisms: a comprehensive review of recent developments. Int J Mol Sci. 2023;24:10299. https://doi.org/10.3390/ijms241210299.

Kim S-S, Harford JB, Moghe M, Rait A, Pirollo KF, Chang EH. Targeted nanocomplex carrying siRNA against MALAT1 sensitizes glioblastoma to temozolomide. Nucleic Acids Res. 2018;46:1424–40. https://doi.org/10.1093/nar/gkx1221.

Li J, Han L, Roebuck P, Diao L, Liu L, Yuan Y, Weinstein JN, Liang H. TANRIC: an interactive open platform to explore the function of lncRNAs in cancer. Cancer Res. 2015;75:3728–37. https://doi.org/10.1158/0008-5472.CAN-15-0273.

Li J, Zhang S, Wan Y, Zhao Y, Shi J, Zhou Y, Cui Q. MISIM v2.0: a web server for inferring microRNA functional similarity based on microRNA-disease associations. Nucleic Acids Res. 2019;47:W536–41. https://doi.org/10.1093/nar/gkz328.

Li D, Zhang J, Li J. Role of miRNA sponges in hepatocellular carcinoma. Clin Chim Acta. 2020a;500:10–9. https://doi.org/10.1016/j.cca.2019.09.013.

Li J, Zhang S, Liu T, Ning C, Zhang Z, Zhou W. Neural inductive matrix completion with graph convolutional networks for miRNA-disease association prediction. Bioinformatics. 2020b;36:2538–46. https://doi.org/10.1093/bioinformatics/btz965.

Lin X, Lu Y, Zhang C, Cui Q, Tang Y-D, Ji X, Cui C. LncRNADisease v3.0: an updated database of long non-coding RNA-associated diseases. Nucleic Acids Res. 2024;52:D1365–9. https://doi.org/10.1093/nar/gkad828.

Liu G, Xiang T, Wu Q-F, Wang W-X. Long noncoding RNA H19-derived miR-675 enhances proliferation and invasion via RUNX1 in gastric cancer cells. Oncol Res. 2016;23:99–107. https://doi.org/10.3727/096504015X14496932933575.

Liu M, Wang Q, Shen J, Yang BB, Ding X. Circbank: a comprehensive database for circRNA with standard nomenclature. RNA Biol. 2019;16:899–905. https://doi.org/10.1080/15476286.2019.1600395.

Liu Y, Li M, Yu H, Piao H. lncRNA CYTOR promotes tamoxifen resistance in breast cancer cells via sponging miR-125a-5p. Int J Mol Med. 2020;45:497–509. https://doi.org/10.3892/ijmm.2019.4428.

Liu C-J, Fu X, Xia M, Zhang Q, Gu Z, Guo A-Y. miRNASNP-v3: a comprehensive database for SNPs and disease-related variations in miRNAs and miRNA targets. Nucleic Acids Res. 2021a;49:D1276–81. https://doi.org/10.1093/nar/gkaa783.

Liu D, Huang Y, Nie W, Zhang J, Deng L. SMALF: miRNA-disease associations prediction based on stacked autoencoder and XGBoost. BMC Bioinformatics. 2021b;22:219. https://doi.org/10.1186/s12859-021-04135-2.

Liu W, Lin H, Huang L, Peng L, Tang T, Zhao Q, Yang L. Identification of miRNA-disease associations via deep forest ensemble learning based on autoencoder. Brief Bioinform. 2022;23:bbac104. https://doi.org/10.1093/bib/bbac104.

Ma S, Yang D, Liu Y, Wang Y, Lin T, Li Y, Yang S, Zhang W, Zhang R. LncRNA BANCR promotes tumorigenesis and enhances adriamycin resistance in colorectal cancer. Aging (Albany NY). 2018;10:2062–78. https://doi.org/10.18632/aging.101530.

Ma B, Wang S, Wu W, Shan P, Chen Y, Meng J, Xing L, Yun J, Hao L, Wang X, Li S, Guo Y. Mechanisms of circRNA/lncRNA-miRNA interactions and applications in disease and drug research. Biomed Pharmacother. 2023;162:114672. https://doi.org/10.1016/j.biopha.2023.114672.

Meng S, Zhou H, Feng Z, Xu Z, Tang Y, Li P, Wu M. CircRNA: functions and properties of a novel potential biomarker for cancer. Mol Cancer. 2017;16:94. https://doi.org/10.1186/s12943-017-0663-2.

Miao L, Feng G, Yuan H. CircRNAs: a family number of miRNA regulatory transcriptome in laryngeal carcinoma. J Clin Lab Anal. 2021a;35:e24038. https://doi.org/10.1002/jcla.24038.

Miao Q, Ni B, Tang J. Coding potential of circRNAs: new discoveries and challenges. PeerJ. 2021b;9:e10718. https://doi.org/10.7717/peerj.10718.

Mikolov T, Chen K, Corrado G, Dean J. Efficient estimation of word representations in vector space; 2013. https://doi.org/10.48550/arXiv.1301.3781

Nemeth K, Bayraktar R, Ferracin M, Calin GA. Non-coding RNAs in disease: from mechanisms to therapeutics. Nat Rev Genet. 2024;25:211–32. https://doi.org/10.1038/s41576-023-00662-1.

Ning Q, Zhao Y, Gao J, Chen C, Li X, Li T, Yin M. AMHMDA: attention aware multi-view similarity networks and hypergraph learning for miRNA-disease associations identification. Brief Bioinform. 2023;24:bbad094. https://doi.org/10.1093/bib/bbad094.

O'Brien J, Hayder H, Zayed Y, Peng C. Overview of MicroRNA biogenesis, mechanisms of actions, and circulation. Front Endocrinol. 2018;9 https://doi.org/10.3389/fendo.2018.00402.

Papachristou DJ, Sklirou E, Corradi D, Grassani C, Kontogeorgakos V, Rao UNM. Immunohistochemical analysis of the endoribonucleases Drosha, dicer and Ago2 in smooth muscle tumours of soft tissues. Histopathology. 2012;60:E28–36. https://doi.org/10.1111/j.1365-2559.2012.04192.x.

Pundhir S, Gorodkin J. MicroRNA discovery by similarity search to a database of RNA-seq profiles. Front Genet. 2013;4:133. https://doi.org/10.3389/fgene.2013.00133.

Rasouli A, Pickard FCI, Sur S, Grossfield A, Işık Bennett M. Essential considerations for free energy calculations of RNA–small molecule complexes: lessons from the theophylline-binding RNA Aptamer. J Chem Inf Model. 2025;65:223–39. https://doi.org/10.1021/acs.jcim.4c01505.

Schuurman N, Leszczynski A. Ontologies for bioinformatics. Bioinform Biol Insights. 2008;2:187–200.

Sharma B, Torres M, Rodriguez S, Kumar S. MicroRNA-502-3p, a novel regulator of GABA function in Alzheimer's disease. Alzheimers Dement. 2024;20:e087754. https://doi.org/10.1002/alz.087754.

Sheng N, Huang L, Gao L, Cao Y, Xie X, Wang Y. A survey of computational methods and databases for lncRNA-MiRNA interaction prediction. IEEE/ACM Trans Comput Biol Bioinform. 2023;20:2810–26. https://doi.org/10.1109/TCBB.2023.3264254.

Sheng N, Xie X, Wang Y, Huang L, Zhang S, Gao L, Wang H. A survey of deep learning for detecting miRNA- disease associations: databases, computational methods, challenges, and future directions. IEEE/ACM Trans Comput Biol Bioinform. 2024;21:328–47. https://doi.org/10.1109/TCBB.2024.3351752.

Sun J, Xu M, Ru J, James-Bott A, Xiong D, Wang X, Cribbs AP. Small molecule-mediated targeting of microRNAs for drug discovery: experiments, computational techniques, and disease implications. Eur J Med Chem. 2023;257:115500. https://doi.org/10.1016/j.ejmech.2023.115500.

Sun Z-Y, Yang C-L, Huang L-J, Mo Z-C, Zhang K-N, Fan W-H, Wang K-Y, Wu F, Wang J-G, Meng F-L, Zhao Z, Jiang T. circRNADisease v2.0: an updated resource for high-quality experimentally supported circRNA-disease associations. Nucleic Acids Res. 2024;52:D1193–200. https://doi.org/10.1093/nar/gkad949.

Tadesse K, Benhamou RI. Targeting MicroRNAs with small molecules. Noncoding RNA. 2024;10:17. https://doi.org/10.3390/ncrna10020017.

Tang M, Liu C, Liu D, Liu J, Liu J, Deng L. PMDFI: predicting miRNA–disease associations based on high-order feature interaction. Front Genet. 2021a;12:656107. https://doi.org/10.3389/fgene.2021.656107.

Tang X, Luo J, Shen C, Lai Z. Multi-view multichannel attention graph convolutional network for miRNA-disease association prediction. Brief Bioinform. 2021b;22:bbab174. https://doi.org/10.1093/bib/bbab174.

Tran TPA, Poulet S, Pernak M, Rayar A, Azoulay S, Giorgio AD, Duca M. Development of 2-deoxystreptamine–nucleobase conjugates for the inhibition of oncogenic miRNA production. RSC Med Chem. 2022;13:311–9. https://doi.org/10.1039/D1MD00345C.

van Laarhoven T, Nabuurs SB, Marchiori E. Gaussian interaction profile kernels for predicting drug–target interaction. Bioinformatics. 2011;27:3036–43. https://doi.org/10.1093/bioinformatics/btr500.

Villanueva CG, Andrés-Blasco I, Martucci A, Lopez JFR, Moreno VZ, Montero MG, Nucci C, Durán MDP, Feijoo JG. Differential profile of mirnas in aqueous humour: a new potential epigenetic mechanism for pseudoexfoliation glaucoma. Acta Ophthalmol. 2025;103 https://doi.org/10.1111/aos.16970.

Volders P-J, Anckaert J, Verheggen K, Nuytens J, Martens L, Mestdagh P, Vandesompele J. LNCipedia 5: towards a reference set of human long non-coding RNAs. Nucleic Acids Res. 2019;47:D135–9. https://doi.org/10.1093/nar/gky1031.

Wang X, Zhang J, Li F, Gu J, He T, Zhang X, Li Y. MicroRNA identification based on sequence and structure alignment. Bioinformatics. 2005;21:3610–4. https://doi.org/10.1093/bioinformatics/bti562.

Wang D, Wang J, Lu M, Song F, Cui Q. Inferring the human microRNA functional similarity and functional network based on microRNA-associated diseases. Bioinformatics. 2010;26:1644–50. https://doi.org/10.1093/bioinformatics/btq241.

Wang X-F, Yu C-Q, Li L-P, You Z-H, Huang W-Z, Li Y-C, Ren Z-H, Guan Y-J. KGDCMI: a new approach for predicting circRNA–miRNA interactions from multi-source information extraction and deep learning. Front Genet. 2022;13 https://doi.org/10.3389/fgene.2022.958096.

Wang X-F, Yu C-Q, You Z-H, Qiao Y, Li Z-W, Huang W-Z, Zhou J-R, Jin H-Y. KS-CMI: a circRNA-miRNA interaction prediction method based on the signed graph neural network and denoising autoencoder. iScience. 2023;26:107478. https://doi.org/10.1016/j.isci.2023.107478.

Wu T, Qu L, He G, Tian L, Li L, Zhou H, Jin Q, Ren J, Wang Y, Wang J, Kan X, Liu M, Shen J, Guo M, Sun Y. Regulation of laryngeal squamous cell cancer progression by the lncRNA H19/miR-148a-3p/DNMT1 axis. Oncotarget. 2016;7:11553–66. https://doi.org/10.18632/oncotarget.7270.

Xie H, Zheng R. Circ_0085495 knockdown reduces adriamycin resistance in breast cancer through miR-873-5p/integrin β1 axis. Anti-Cancer Drugs. 2022;33:e166–77. https://doi.org/10.1097/CAD.0000000000001174.

Xie J, Xu P, Lin Y, Zheng M, Jia J, Tan X, Sun J, Zhao Q. LncRNA–miRNA interactions prediction based on meta-path similarity and Gaussian kernel similarity. J Cell Mol Med. 2024;28:e18590. https://doi.org/10.1111/jcmm.18590.

Xu J, Wan J, Huang H-Y, Chen Y, Huang Y, Huang J, Zhang Z, Su C, Zhou Y, Lin X, Lin Y-C-D, Huang H-D. miRStart 2.0: enhancing miRNA regulatory insights through deep learning-based TSS identification. Nucleic Acids Res. 2025;53:D138–46. https://doi.org/10.1093/nar/gkae1086.

Xuan P, Dong Y, Guo Y, Zhang T, Liu Y. Dual convolutional neural network based method for predicting disease-related miRNAs. Int J Mol Sci. 2018;19:3732. https://doi.org/10.3390/ijms19123732.

Xuan P, Sun H, Wang X, Zhang T, Pan S. Inferring the disease-associated miRNAs based on network representation learning and convolutional neural networks. Int J Mol Sci. 2019;20:3648. https://doi.org/10.3390/ijms20153648.

Xuan P, Qi X, Chen S, Gu J, Wang X, Cui H, Lu J, Zhang T. Subgraph topology and dynamic graph topology enhanced graph learning and pairwise feature context relationship integration for predicting disease-related miRNAs. J Chem Inf Model. 2025;65:1631–40. https://doi.org/10.1021/acs.jcim.4c01757.

Yang S, Wang Y, Lin Y, Shao D, He K, Huang L. LncMirNet: predicting LncRNA–miRNA interaction based on deep learning of ribonucleic acid sequences. Molecules. 2020;25:4372. https://doi.org/10.3390/molecules25194372.

Yang Y, Wang D, Miao Y-R, Wu X, Luo H, Cao W, Yang W, Yang J, Guo A-Y, Gong J. lncRNASNP v3: an updated database for functional variants in long non-coding RNAs. Nucleic Acids Res. 2023;51:D192–8. https://doi.org/10.1093/nar/gkac981.

Yao J, Xu G, Zhu L, Zheng H. circGFRA1 enhances NSCLC progression by sponging miR-188-3p. Onco Targets Ther. 2020;13:549–58. https://doi.org/10.2147/OTT.S230795.

Yin W, Wang S, Zhang Y, Qiao S, Wu W, Li H. Multirelational hypergraph representation learning for predicting circRNA-miRNA associations. J Chem Inf Model. 2024;64:8349–60. https://doi.org/10.1021/acs.jcim.4c01436.

Yu C-Q, Wang X-F, Li L-P, You Z-H, Ren Z-H, Chu P, Guo F, Wang Z-Y. RBNE-CMI: an efficient method for predicting circRNA-miRNA interactions via multiattribute incomplete heterogeneous network embedding. J Chem Inf Model. 2024;64:7163–72. https://doi.org/10.1021/acs.jcim.4c01118.

Zhang Z, Zhu Z, Watabe K, Zhang X, Bai C, Xu M, Wu F, Mo Y-Y. Negative regulation of lncRNA GAS5 by miR-21. Cell Death Differ. 2013;20:1558–68. https://doi.org/10.1038/cdd.2013.110.

Zhang Y-L, Li X-B, Hou Y-X, Fang N-Z, You J-C, Zhou Q-H. The lncRNA XIST exhibits oncogenic properties via regulation of miR-449a and Bcl-2 in human non-small cell lung cancer. Acta Pharmacol Sin. 2017;38:371–81. https://doi.org/10.1038/aps.2016.133.

Zhao L, Wang J, Li Y, Song T, Wu Y, Fang S, Bu D, Li H, Sun L, Pei D, Zheng Y, Huang J, Xu M, Chen R, Zhao Y, He S. NONCODEV6: an updated database dedicated to long non-coding RNA annotation in both animals and plants. Nucleic Acids Res. 2021;49:D165–71. https://doi.org/10.1093/nar/gkaa1046.

Zheng K, You Z-H, Wang L, Zhou Y, Li L-P, Li Z-W. MLMDA: a machine learning approach to predict and validate MicroRNA-disease associations by integrating of heterogenous information sources. J Transl Med. 2019;17:260. https://doi.org/10.1186/s12967-019-2009-x.

Zhong J, Zhou W, Kang J, Fang Z, Xie M, Xiao Q, Peng W. DNRLCNN: a CNN framework for identifying MiRNA-disease associations using latent feature matrix extraction with positive samples. Interdiscip Sci. 2022;14:607–22. https://doi.org/10.1007/s12539-022-00509-z.

Zhou M, Wang J, Peng Y, Tian X, Zhang W, Chen J, Wang Y, Wang Y, Yang Y, Zhang Y, Huo X, Wu Y, Yu Z, Xie T, Ma X. Elemene as a binding stabilizer of microRNA-145-5p suppresses the growth of non-small cell lung cancer. J Pharm Anal. 2025;15:101118. https://doi.org/10.1016/j.jpha.2024.101118.

Zhuang S-T, Cai Y-J, Liu H-P, Qin Y, Wen J-F. LncRNA NEAT1/miR-185-5p/IGF2 axis regulates the invasion and migration of colon cancer. Mol Genet Genomic Med. 2020;8:e1125. https://doi.org/10.1002/mgg3.1125.

AI in Retrosynthesis: Introduction, Methods, Evaluation, and Future Directions

Ruchi Bharti, Ajay Thakur, Uma Koul, Monika Verma, and Renu Sharma

Abstract Retrosynthesis has come a long way in employing various artificial intelligence (AI) methodologies, notably revising how synthetic routes are designed and evaluated. Among common AI-driven retrosynthesis techniques, single-step and multistep approaches utilize different algorithms such as breadth-first search, beam search, and Monte Carlo tree search, all optimizing predictive performance for complicated synthesis processes. This chapter highlights the various AI-based techniques for retrosynthetic analysis by their effort-saving, cost-saving, and accessible means for improved overall feasibility of synthetic routes. An important point of these techniques is scoring functions and datasets that serve as useful instruments for evaluating the proposed routes' synthetic efficiency, economic viability, and thermodynamic feasibility. Several examples from the pharmaceutical and biocatalysis fields demonstrate how AI will enable better drug design and biotechnological processes. By leveraging comprehensive chemical databases such as the USPTO reaction dataset, Reaxys, and Pistachio, AI systems can tap into vast amounts of reaction data, refining their predictive models. The chapter concludes on a note underlining the enormous promise of AI-retrosynthetic planning for the future of synthetic chemistry by allowing drug discovery and biotechnology innovations to become cheaper, less wasteful, and more efficient methods.

Keywords Artificial intelligence · Retrosynthesis · Synthetic routes · Machine learning · Scoring functions · Datasets

1 Introduction

Since the beginning of chemical sciences, organic synthesis has fascinated researchers because of its unique ability to produce complex molecular structures

R. Bharti (✉) · A. Thakur · U. Koul · M. Verma · R. Sharma
Department of Chemistry, University Institute of Sciences, Chandigarh University, Mohali, Punjab, India
e-mail: ruchi.uis@cumail.in

A. Lavecchia (ed.), *Applied Artificial Intelligence for Drug Discovery*,
https://doi.org/10.1007/978-3-031-98022-0_12

from simple, readily available precursors (Verma et al. 2022; Golfmann and Walker 2023; Perrone et al. 2023; Sharma et al. 2023; Javahershenas and Nikzat 2023; Javahershenas 2023; Bharti et al. 2022). However, high precision and control are essential as these often demand intricate stereochemistry, strategic positioning of functional groups, and consistent molecular architecture. The ongoing quest within the field is to design synthetic routes that enable the construction of such molecules with increased efficiency while reducing both the time and the number of steps involved compared with previously established methods (de Souza et al. 2017). Synthetic chemists often rely on the retrosynthetic strategy initially conceived by E. J. Corey to develop an optimized route toward a desired molecular target. This is based on analyzing the target molecule by decomposing it into simpler precursor structures and effectively reversing the forward synthetic process (Corey and Cheng 1995; Corey 1991; Corey 1988). By simply breaking down complex molecular architectures into manageable synthetic building blocks, the retrosynthetic analysis provides a logical framework for identifying viable synthetic routes. This method has had a profound influence on modern synthetic organic chemistry and changed how chemists approach complex molecular synthesis by allowing the identification of key disconnections, functional group transformations, and strategic intermediates that guide the construction of the target molecule through a stepwise rational sequence. The Organic Syntheses: The Disconnection Approach textbook reflected the widespread of this approach as Warren and Wyatt (2008) systematically present the principles of retrosynthetic strategies with ways to apply them in practical scenarios in this book.

1.1 Various Terms Associated with Retrosynthesis

Retrosynthesis comprises a set of key terms and concepts (Badami 2019; Mikolajczyk et al. 2023) that provide the basis for effectively understanding systematic disconnection strategies and analysis of target molecules.

1. ***Target Molecule (TM)*** The compound which is the ultimate objective of a synthetic process. It is the focal point systematically deconstructed into more straightforward, readily accessible precursors. This stepwise disconnection process generates key intermediates and guides the synthetic route design by ensuring that each step is aligned with the goal (Scheme 1).

2. ***Retrosynthetic Analysis*** (Scheme 1) involves conceptually breaking down the target molecule into successively more accessible simpler starting materials. This hypothetical decomposition is opposite to the synthetic direction where reactions are seen as proceeding backwards from the final product to the reactants. By tracing this reverse pathway, strategic disconnections, functional group interconversions, and intermediates can be identified that help simplify the synthetic challenge.

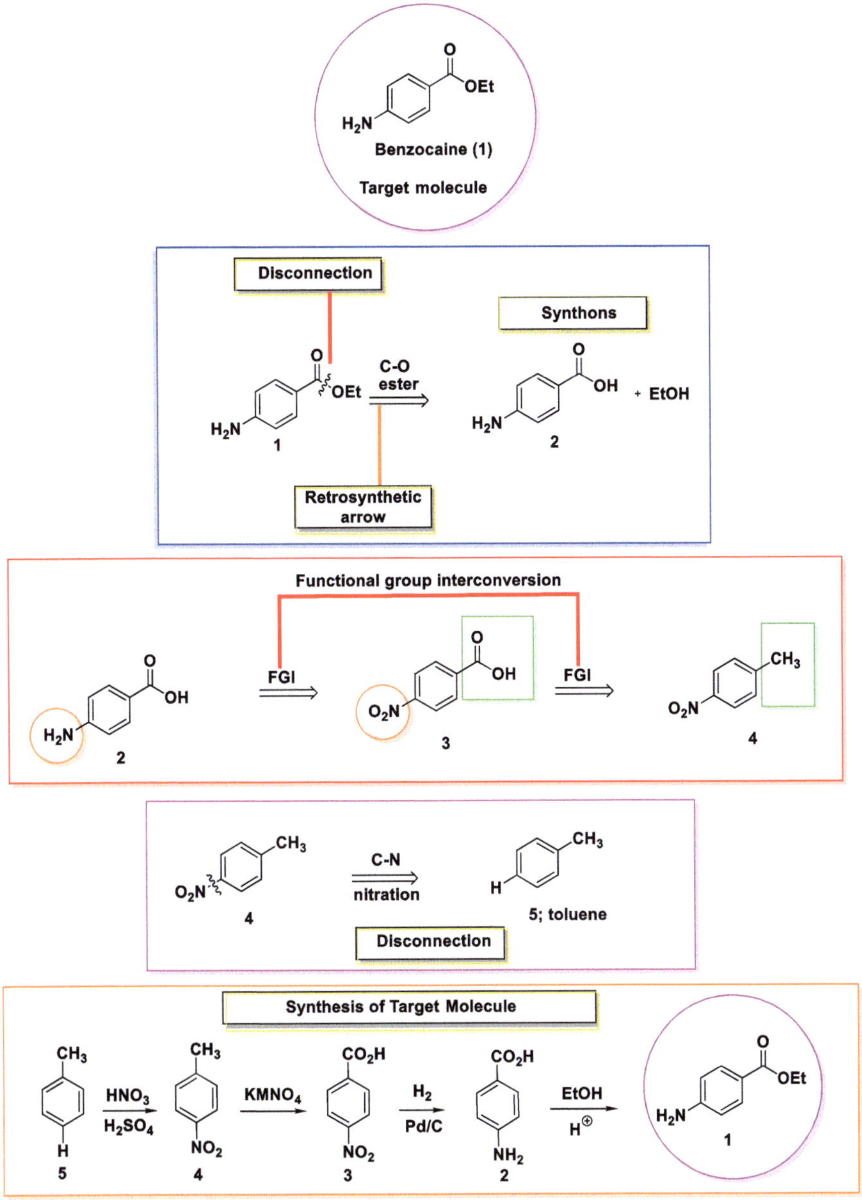

Scheme 1 Process of retrosynthesis to synthesis (Warren and Wyatt 2008)

3. ***Disconnection*** refers to the conceptual splitting of a specific bond within a molecule, which represents the opposite of a forward synthesis reaction (shown in Scheme 1). This process identifies the immediate precursor needed to build the target molecule in a forward synthesis. Often referred to as a transformation, dis-

Table 1 Synthon and their relative synthetic equivalents

Some common synthons	
Synthon	Synthetic equivalents
R^+ (alkyl cation = carbenium ion)	RCl, RBr, RI, ROTs
Ar_+ (aryl cation)	$ArN2^+X^-$
$H\overset{+}{C}=O$ (Acylium ion)	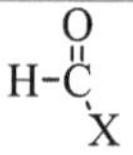 (X = Cl, NR'_2, OR')
$\overset{+}{C}H_2OH$ (oxocarbenium ion)	HCHO
$R\overset{+}{C}HOH$ (oxocarbenium ion)	RCHO
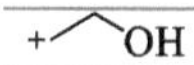	
$\overset{+}{C}H_2$-$CH_2C\equiv N$	CH_2=$CHC\equiv N$

Panda (2020)

connection highlights the logical step required to simplify the molecular structure into more accessible building blocks. In retrosynthetic schemes, disconnection is conventionally represented by a wavy line (~), which visually indicates the bond undergoing splitting during analysis.

4. The ***Retrosynthetic Arrow*** is a special notation used to represent dissociation in retrosynthetic analysis. It is represented as a double-line closed arrow (⇒) and represents the transformation of the target molecule into its immediate precursor. This arrow highlights the reverse synthetic step, guides the dissociation process, and visually indicates the direction from the product to its starting materials (shown in Scheme 1).
5. ***Synthons (S)*** (shown in Table 1) are conceptual fragments generated during retrosynthetic dissociation of a target molecule. These hypothetical units represent ideal structures, often classified as anionic or neutral fragments, based on the bond polarity at the cleavage site. Bond polarity is important in predicting these fragments' chemical reactivity and compatibility with accurate reagents during further synthesis. While S provides a theoretical framework for designing a synthetic route, they are not actual compounds or reagents. Instead, they guide the selection of appropriate chemical euivalents (reagents or intermediates) that recreate the desired bonds in a practical synthesis.
6. A ***Retron*** (Scheme 2) is a distinctive structural feature or subunit within a target molecule that indicates the potential application of a particular synthetic transformation. Each chemical reaction leaves a distinctive "signature" in the form of a specific substructure in the product, such as an enone generated via aldol condensation. The retron corresponding to that reaction must be present in the target molecule for a transformation to be applicable. Identifying the retron allows chemists to match a target molecule with the appropriate synthetic reactions, ensuring a logical and efficient synthesis route design.

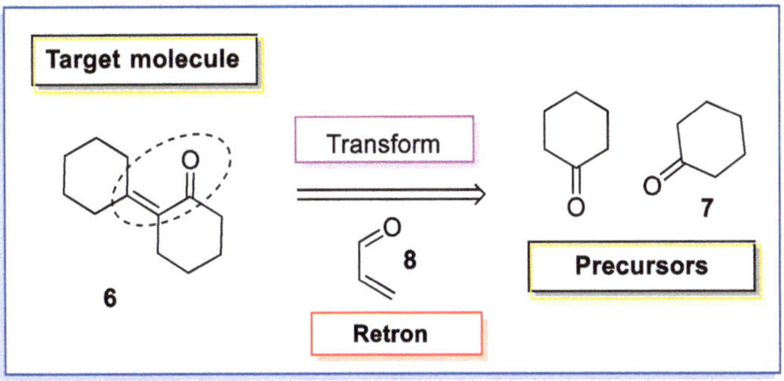

Scheme 2 Mikolajczyk et al. (2023)

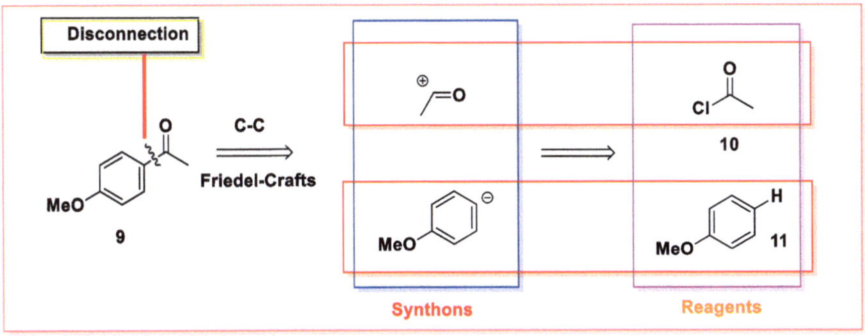

Scheme 3 Representation of synthons and reagents (Warren and Wyatt 2008)

7. A ***Reagent*** (Scheme 3) is a chemical substance used in practical synthesis to replicate the reactivity proposed by a corresponding synthon in retrosynthetic analysis. In contrast, synthons are theoretical constructs and reagents that embody their reactivity in real-world conditions. Multiple reagents can often represent a single synthon. Each can achieve the desired transformation based on its specific chemical properties. In some cases, the oxidation state of a synthon may need to be adjusted to better align with a suitable reagent and its associated chemical behavior. This alignment between synthon and reagent is critical to translating the retrosynthetic scheme into a viable and efficient forward synthesis.
8. ***Functional Group Interconversion (FGI)*** (Scheme 1) is a strategical replacement of one functional group with another in those cases where direct dissociation does not yield a suitable precursor and makes better alignment of the optimizing molecular structures with available reagents.
9. The ***Retrosynthetic Tree*** (Fig. 1) is like a map showing how a complex molecule can be broken down into simpler parts. At the top is the target molecule, which can be used to identify the suitable starting materials by stepwise analysis. Each disconnection forms a new "precursor" that is further simplified. This process

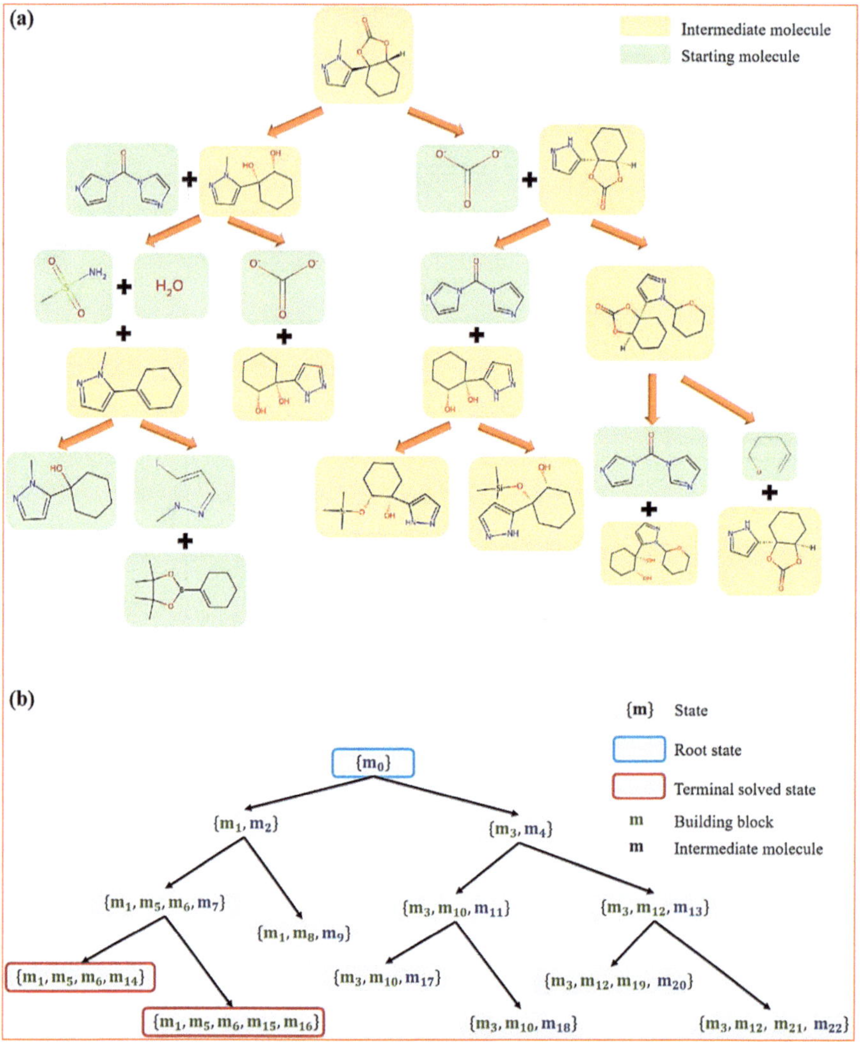

Fig. 1 (**a**) Retrosynthetic analysis and (**b**) its retrosynthetic tree representation (Zhao et al. 2024)

continues until simple and easily available materials are reached. However, breaking it down is not enough. Each time pathway is critically evaluated for its practicality by considering factors such as starting materials' availability, the synthetic sequence's length and complexity, and overall efficiency. The most important thing about this tree is its many routes to make complex molecules. The best and easiest route can then be chosen. The retrosynthetic tree forms an effective means of understanding and implementing this entire process systematically and logically.

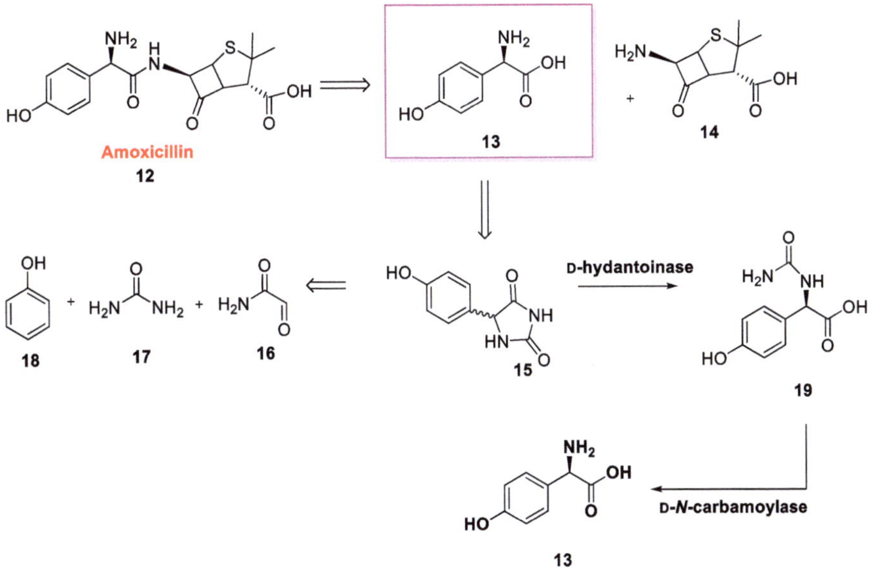

Scheme 4 Retrosynthetic synthesis of amoxicillin (Koch et al. 2019; Bucherer and Steiner 1934; Schulze and Wubbolts 1999)

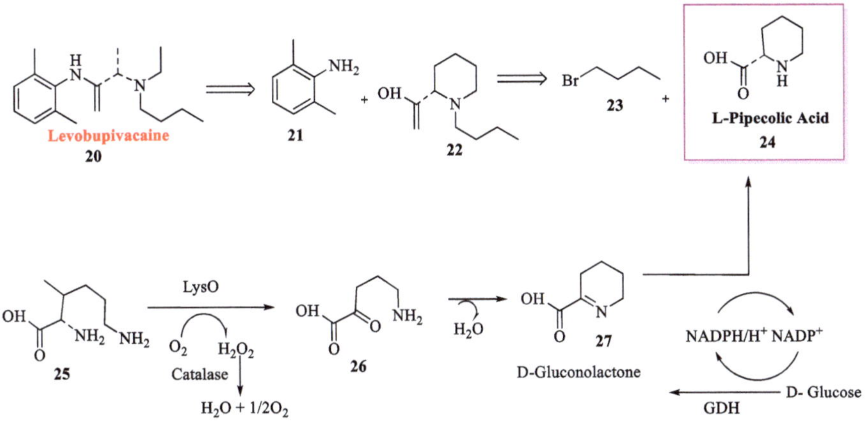

Scheme 5 Retrosynthetic analysis of levobupivacaine (May et al. 2002; Kusakabe et al. 1980; Zhu and Hua 2009; Heydari et al. 2004)

In Schemes 4-6, a series of retrosynthetic pathways previously reported in the literature is illustrated. These retrosynthetic analyses helped rationalize the design and identify possible synthetic strategies.

Retrosynthetic analysis of amoxicillin (**12**) (Scheme 4) involves the systematic fragmentation of the structure to find simple and viable precursors. It starts with a strategic disconnection of the acyclic amide bond and transforms it into some useful,

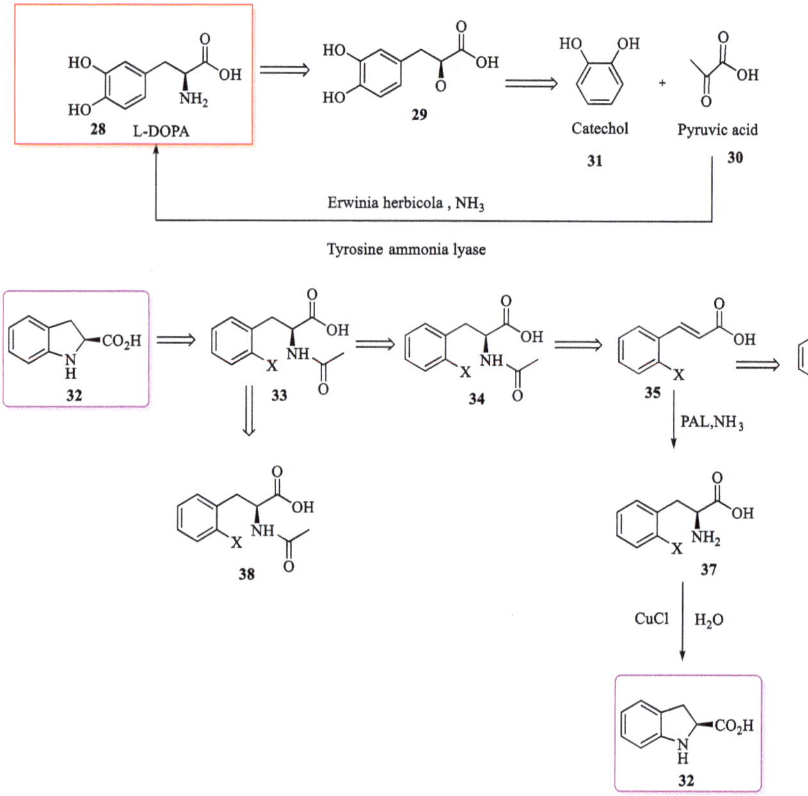

Scheme 6 Retrosynthesis of L-DOPA and indole derivative (Bürgi et al. 1966; Kumagai 2000; Tsuchida et al. 1993)

manageable intermediate. This way of fragmentation results in an α-amino acid (**13**), a key building block in the synthesis route. This intermediate is also considered by structuring it according to the structural requirements of amoxicillin's β-lactam core with certain side chains to preserve all its special chemical features (Koch et al. 2019; Bucherer and Steiner 1934; Schulze and Wubbolts 1999).

A strategic retrosynthetic analysis of levobupivacaine (**20**) (Scheme 5) focusing on the amide bond will help identify its structural components. This breakdown results in two key fragments: an aniline derivative (**21**) and a pipecolic acid derivative (**22**). A closer examination of the 22 suggests it can be further decomposed into simpler precursors, specifically bromobutane (**23**) and L-pipecolic acid (**24**). The synthesis of 24 starts from lysine (**25**); thus, lysine oxidase is the key enzyme in this reaction, which oxidizes lysine to form an intermediate **26**. The **26** then undergoes a reduction step, resulting in the efficient production of **24** (May et al. 2002; Kusakabe et al. 1980; Zhu and Hua 2009; Heydari et al. 2004).

The structure is broken down into simpler molecules in the retrosynthesis of L-DOPA (28) (Scheme 6). The first step reveals 3-(3,4-dihydroxyphenyl)-2-oxopropanoic acid (**29**), a key intermediate that further split into two main components, i.e., pyruvic acid (**30**) and catechol (**31**). 30 is preferred as a versatile starting material in this process and used to create 28 with the assistance of an enzyme such as tyrosine ammonia lyase (Bürgi et al. 1966; Kumagai 2000). This scheme also contains an associated enzymatic procedure by which an indole derivative (**32**) is synthesized from a cinnamic acid derivative (**35**). The transformation is catalyzed by phenylalanine ammonia-lyase (Tsuchida et al. 1993).

Retrosynthetic work has traditionally been based on chemists' experience, knowledge, and intuition. This approach has helped solve complex compound structures but lacks a systematic and standardized procedure. Due to this, it is often difficult to synthesize complex and new compounds. However, the process relies highly on the expertise of familiar reactions of the individual, which makes it challenging to develop appropriate strategies for unknown or highly complex molecules. Therefore, more time and effort are required, especially for new solutions or approaches if needed to be adopted. The synthesis of vitamin B12 was a notable example of the complexities of retrosynthetic methods, which take nearly 12 years (Sun and Sahinidis 2022). Thus, it was necessary to develop a process that could overcome the complexity of retrosynthesis and aid in synthesizing complex molecules.

In 1967, E. J. Corey initiated the first use of computational tools and introduced the concept of leveraging computational algorithms to systematically explore synthetic pathways (Sun and Sahinidis 2022). Since then, the field of computer-aided retrosynthesis has progressed at an accelerated pace.

2 Overview of AI-Based Methods in Retrosynthesis

In recent years, AI has changed the game for retrosynthesis. These new ideas have made the process more accurate and quicker, as seen in both one-step and multistep strategies. This has made it a lot easier and better to find good ways to make complex compounds compared to before.

2.1 Single-Step Retrosynthesis

Single-step retrosynthesis (represented in Fig. 2) is commonly viewed as predicting the structure of reactants directly from a product molecule (Zhong et al. 2023a). In this context, molecular representations are of key importance, as they provide a means to encode chemical structure representations and form the base for AI models that predict reaction paths and reactant molecules. Among the most widely used is the Simplified Molecular Input Line Entry System (SMILES). This represents

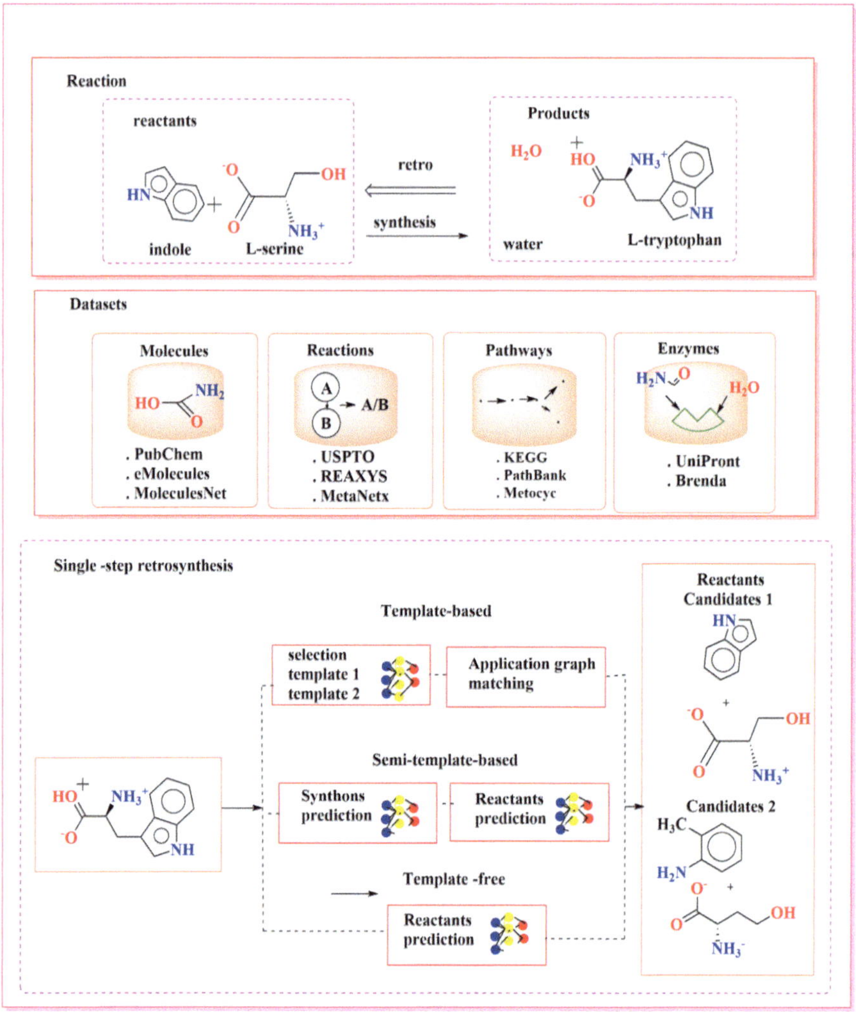

Fig. 2 Representation of single-step retrosynthesis (Adapted from Gricourt et al. 2024)

molecules as strings and exists in various formats such as canonical SMILES and rooted SMILES (Karpov et al. 2019; Wan et al. 2022a; Kim et al. 2021a; Coley et al. 2019a; Baylon et al. 2019; Zhang et al. 2023; Ucak et al. 2023; Toniato et al. 2023; Tetko et al. 2020). Another important representation involves circular fingerprints, such as hierarchical substructure and functional and extended connectivity. These methods translate local molecular features into hashed, vectorized forms, optimizing molecular comparison and database searching. These fingerprints extract molecular structure by concentrating on individual molecular parts (Baylon et al. 2019; Ucak et al. 2023; Fortunato et al. 2020; Heid et al. 2021a; Coley et al. 2017; Irwin et al. 2022;

Yang et al. 2022; Fang et al. 2023; Sankaranarayanan et al. 2022). Molecular graphs describe molecules as networks in which atoms are nodes and bonds are edges that maintain the structural connectivity of the molecule. This representation has become integral to cheminformatics and is frequently utilized in machine learning models, allowing graph-based algorithms to analyze molecular structures (Heid et al. 2021a; Wan et al. 2022b; Zhong et al. 2023b; Coley et al. 2019b; Li et al. 2023; Mao et al. 2021; Zhu et al. 2023; Yan et al. 2022; Seo et al. 2021; He et al. 2022; Yu et al. 2022; Ishida et al. 2019; Dai et al. 2019; Chen and Jung 2021). In another approach, atom environments are represented as circular, topological fragments centered on individual atoms, capturing local molecular features in a SMARTS-based format.

This technique emphasizes the immediate chemical context of atoms, which is critical for detailed structural information (Ucak et al. 2022, 2023). Taking this idea further, the signature descriptor captures all-atom environments in a molecular graph up to a set radius. These descriptors come in SMILES format and give a complete picture of how molecules are arranged (Delépine et al. 2018). Self-referencing embedded Strings (SRES) take the SMILES approach a step further by ensuring that the molecular structures are encoded more reliably and less prone to errors (Ucak et al. 2023). Although SRES enhances the robustness of molecular structures encoded in the SMILES notation, SMIRKS is a specialized extension that directly encodes chemical reactions. SMIRKS is applied to describe the conversion process from reactants to products, which is a fundamental tool for template-based retrosynthesis (Gricourt et al. 2024).

Generally, single-step retrosynthesis uses three main strategies: methods based on templates, those without templates, and approaches that combine elements of both.

2.1.1 Template-Based Retrosynthesis

These templates served as structured rules describing chemical transformations observed in historical data. This allows them to be used systematically to predict reactants. These templates are normally determined by searching for conserved reaction centers and principal structural modifications in large chemical libraries (Fortunato et al. 2020). Template reactions are typically encoded using atom-mapped SMARTS (SMiles ARbitrary Target Specification) strings, providing a versatile and powerful tool for coding chemical transformations. These strings unambiguously define the structures of reactants and products and show the correspondence (mapping) of atoms between them. A major advantage of this representation is its ability to handle stereochemical details and allows retrosynthesis models to maintain fidelity to the stereochemical requirements of target molecules (Coley et al. 2019a). Further, SMIRKS strings improve on SMARTS in that they can represent entire chemical transformations, allowing retrosynthesis models to describe reactants and reactions needed to convert one molecule to another (Gricourt et al. 2024).

Neural networks (NNs) also play an important role in retrosynthetic analysis. These models are programmed to observe patterns and chemical interactions in molecular functional groups. The NN learns similarities between the target molecule and reaction templates by observing different structural features. Such templates share similar reactivity profiles as the target molecule (Segler and Waller 2017). Another method uses the natural way of showing molecules as graphs, with atoms as points and bonds as lines, and uses special computer programs called graph neural networks (GNNs) to guess the right reaction templates. This approach works well to explain complex interactions in the molecule because it can give more details about certain parts of the molecule or how they interact, which helps a lot in picking the right template (Ishida et al. 2019; Dong et al. 2022).

This template-based approach replicates the bond changes in known chemical reactions and ensures that the predictions match the already established and understood pathways. Studying and classifying existing reaction patterns makes it more useful and easier to construct complex molecules. But this method has a serious drawback. It cannot identify or suggest completely new or unknown reaction mechanisms. Its reliance on templates limits its discovery capability.

2.1.2 Template-Free

Researchers tried several models while attempting to predict the reactions without using predefined templates. Among these, models with encoder–decoder designs that use long short-term memory cells were some of the first to be tested (Liu et al. 2017; Ucak et al. 2021). These models effectively understand how parts of molecules relate to each other in sequence, allowing them to create reaction results straight from the input structure. Also, scientists explored the Weisfeiler–Lehman networks, which use graph-based methods to study how molecules connect and to represent chemical structures (Coley et al. 2019b). In recent years, transformers like deep generative models have become popular for predicting reactions without depending on the templates. These models are complex NN and consist of two main components, i.e., an encoder and a decoder. The encoder processes the input molecular representation to capture its structural and contextual features while the decoder generates the corresponding output such as reaction prediction structures. Transformer models have advanced beyond using standard SMILES representations as inputs. They now use mixed approaches that combine molecular graphs with SMILES sequences, for example, G2GT (Lin et al. 2023), graph-truncated attention (Zhu et al. 2023), graph-enhanced transformers (Mao et al. 2021), DVMP (Zhu et al. 2023), BiG2S (Hu et al. 2023), and Graph 2SMILES (Tu and Coley 2022). In addition, transformer architectures have also been successfully used to predict the atomic environment of reactants from their product structures (Ucak et al. 2022). Using self-attention mechanisms, these models can learn and represent the intricate intra-atomic relationships crucial to correctly characterizing the local atomic environment within reactants. Apart from this, transformers have also been applied to reverse reaction modeling, facilitating the product structure conversion into the underlying reactants (Ucak et al. 2023). Such bidirectional functionality provides a strong foundation for reaction modeling, enabling synthesis prediction and retrosynthesis analysis.

2.1.3 Semi-Template-Free Retrosynthesis

This approach begins with determining the site of reaction with respect to the product's structure. Once this site is precisely located, specific bonds are strategically cleaved to generate synthons. The next step now involves predicting the corresponding reactants using advanced generative models. These models use various methods, for example, a pre-computed vocabulary approach (GraphRetro) (Somnath et al. 2021), transformer-based models (RetroSub (Fang et al. 2023), RetroPrime (Wang et al. 2021a), and RetroExpert (Yan et al. 2020)), and graph-to-graph translation models (G2Gs) (Shi et al. 2020), to map synthons back to their possible precursor molecules.

2.2 *Multistep Retrosynthesis*

Single-step retrosynthetic strategies continue to evolve but are still relatively uncommon in the context of complex molecules. Most compounds used in real life, like pharmaceuticals, agrochemicals, and advanced materials, feature complex molecular structures that are hard to produce in a single step. The synthesis of pharmaceutical substances for industrial applications is generally efficient but typically averages 8.1 steps (Carey et al. 2006). In contrast, synthesizing highly complex molecules, such as challenging academic goals or natural products, can take longer. Most of these methods exceed the industrial average, and some exceed 100 individual reaction steps (Nicolaou and Sorensen 1996). Such complexity is primarily due to the creation of highly specific structural motifs, the introduction of multiple stereocenters, and precise functional group manipulation while maintaining high levels of yield and selectivity in each step. Multistep retrosynthesis employs various search algorithms to identify feasible synthetic pathways to target molecules. This section discusses significant frameworks and methodologies in this domain, highlighting their scope, features, and significance.

2.2.1 Breadth-First Search Algorithms

RetroSynx (Wang et al. 2022) is a framework that employs a BFS algorithm. It systematically tests promising intermediate compounds to predict thermodynamically feasible pathways. The routes that deserve more attention show good theoretical and chemical feasibility (Liu et al. 2023). These template-based tools reduce the likelihood of heading down energetically infeasible pathways, but their practical efficiency is extremely low. Furthermore, as the platform's primary building blocks are provided by its proprietary sources (e.g., Aladdin Healthcare Technologies), this greatly restricts its reach and adaptability (Aladdin Healthcare Technologies 2020). Additionally, it does not have freely available source codes, which also limits its flexibility for further scientific exploration and optimization, thus restricting this tool's application in several research environments.

2.2.2 Beam Search Algorithms

Beam search has now proved to be a major strategy for retrosynthetic expansion. By combining SCScore assessment and generative log-likelihoods, this approach identifies intermediates for synthesis that are likely to require very high probabilities for feasibility (Schwaller et al. 2020). This further optimized approach has included RPScore which contributes to its effectiveness by prioritizing reactions based on their significance and practicality (Kreutter and Reymond 2023). Notable adaptations of this approach (Probst 2022) and (60) extend its applicability to synthetic biology by incorporating estimates of enzyme availability. By open-source chemical databases such as eMolecules and Enamine, this framework eliminates dependence on proprietary systems for their resources. It opens their doors for access and use towards a wider take-up in academia and industry.

2.2.3 Monte Carlo Tree Search (MCTS)

MCTS techniques apply NNs to the selection of reaction templates and the assessment of their feasibility. This approach uses a rollout process where the simulation of possible reaction pathways takes place while ensuring efficiency in computation and accuracy. Reinforcement learning is another enhancement that improves the decision process over time (Segler et al. 2018; Genheden et al. 2020). Advanced versions of MCTS include those using GNNs and increase the quality of reaction templates. This method improves reaction pathway prediction and parameter modeling, such as solvents and catalysts (Zhang et al. 2022). In addition, applying reinforcement-based structures will enable decision-making without traditional rollout processes, rendering such simulations less complex (Lin et al. 2020; Wang et al. 2020). Such techniques become successful only when supported with effective chemistry databases like ZINC and Sigma-Aldrich, which, in many ways, provide reliable molecular building blocks.

2.2.4 A*-Based and Graph-Centric Approaches

Retro* search algorithm reflects a considerable step forward in retrosynthetic planning about cost estimates on existing and potential synthetic approaches. Via a large reaction database, NNs trained on that database accurately evaluate and prioritize synthesis routes. Advanced versions of the reactors include multi-target searching capabilities with offline-trained GNNs that understand and promote the diversifying adaptiveness of routes, such as that of Retro* + (Chen et al. 2020) and RetroGraph (Kim et al. 2021b). Meanwhile, the ASICS (Xie et al. 2022) model creates a strong synergy by integrating known reaction databases and SAScores (synthetic accessibility scores), leading to optimal synthetic routing. Other frameworks, such as GNN-Retro (Jeong et al. 2022), optimize retrosynthetic analysis further through graph-based evaluation.

3 Scoring Functions and Datasets in Retrosynthetic Planning

Scoring functions and datasets form the basis for any retrosynthetic planning as they help evaluate synthetic feasibility, predict reactions, and optimize pathways. Computational tools and models assess these systems against established benchmarks and allow the selection of the most effective synthetic pathways. The following is an overview of some important scoring functions and datasets involved in retrosynthetic analysis.

3.1 Scoring Functions

The availability of structures for target molecules will be emphasized significantly in the retrosynthetic planning and benchmarks such as GASA (Yu et al. 2022) and SCScore (Han et al. 2022) have been used for synthetic feasibility evaluation. These are derived from comprehensive chemical databases, thus providing very reliable estimates of the available routes and their associated complexities. The same considerations of the availability of reactions are also given numerical values in the form of RAScore (Coley et al. 2018) and ICHO (Thakkar et al. 2021), which will decide between the synthetic pathways by qualifying the feasibility of the reaction. Regarding experimental implementation, especially in scalable systems such as continuous reactors, platforms like InFlow (Badowski et al. 2020) allow a thorough review of the process-reactive conditions. Economic factors also play a crucial role in retrosynthetic planning, and scores like CoPriNet (Plehiers et al. 2020) estimate the costs of target molecules. Many tools, such as GC-NORM-based techniques (Sanchez-Garcia et al. 2023), dGPredictor (Tang et al. 2023), and eQuilibrator (Wang et al. 2021b), help assess the way the reactions behave thermodynamically. It accurately predicts the thermodynamic optimality for proposed reactions and pathways to ensure that the synthetic process is energetically viable and assesses path complexity. Additionally, assessing path complexity is vital regarding the number of steps involved in the synthesis. CMPNN (Beber et al. 2022) is one tool that estimates this space to discover efficiencies and feasibility of the proposed synthetic routes while possibly omitting certain steps in retrosynthetic planning. In applications to pharmaceuticals, specialized tools such as RetroGNN (Li and Chen 2022) and DFRscore (Liu et al. 2022) are used to determine the suitability of retrosynthetic pathways for pharmaceutical needs. These tools allow precise evaluation of the efficiency and practicality of chemical pathways necessary for drug synthesis. Furthermore, in synthetic biology, scores such as FCNN (Kim et al. 2023) are applied to optimize pathways in biotechnology and synthetic biology applications.

3.2 Datasets

The USPTO Reaction Dataset is valuable open-source data, publicly free under the Creative Commons CC0 License. This set is based on organic reactions described in patent applications filed with the U.S. Patent and Trademark Office (USPTO). Therefore, it contains much information that can be used in numerous scientific discoveries. Its repertoire also includes sub-sets, for example, USPTO-50 k, which involves a sub-set of 50,000 reactions, and USPTO-Full, a much higher sub-set, containing a volume of nearly one million reactions. It is relatively popular in chemistry-based applications (Liu et al. 2017; Wang et al. 2021a; Correia et al. 2022; Zheng et al. 2019) and biocatalysis research (Yang et al. 2019, 2022; Zheng et al. 2022).

Reaxys is noted for being the main commercial database for an almost exhaustive collection of experimentally validated chemical data on molecular structures, chemical reactions, and their properties. It is indispensable for retrosynthetic planning to allow chemists to identify viable synthetic routes by examining historical reaction data and operating conditions in targeting molecules. Its particular utility has been with the development of complex synthetic strategies (Thakkar et al. 2023; Zhang et al. 2021; Qiao et al. 2022). Furthermore, Reaxys has been instrumental in biocatalysis work, providing valuable insights into enzymatic reactions and exploring biocatalyzed processes for sustainable chemical synthesis through its curated experimental data (Thakkar et al. 2020a).

Like Reaxys's reaction database, the Pistachio reaction database is another powerful resource, storing nearly 2.5 million unique reactions. This commercial database serves as an important apparatus for AI-powered research. It is extensively utilized to advance and improve machine learning models for reaction prediction, retrosynthesis, and formation of new synthetic pathways (Toniato et al. 2023; Aladdin Healthcare Technologies 2020; Zhang and Lapkin 2023; Thakkar et al. 2020b).

The RetroRules dataset, released under the CC-BY license, is now meant for modeling enzyme-catalyzed reactions. It is the basis of metabolic pathway discovery and metabolic engineering activities. The tools are devoted to mapping enzyme transformations. This is especially useful for biocatalysis (Latendresse et al. 2023; Heid et al. 2021b), where any reaction driven by an enzyme is identified and subsequently optimized toward green chemistry production.

MetaNetX is another CC-BY licensed and complementary source to RetroRules that integrates metabolites and biochemical pathways from more than 10 databases, such as MetaCyc, KEGG, Rhea, ChEBI, BiGG, etc. This multimodal dataset offers a complete platform for the modeling and analysis of metabolic systems. It is especially adopted in biocatalysis (Yang et al. 2022; Kreutter and Reymond 2023; Yang et al. 2019), delivering important information for enzyme selection and pathway optimization. Moreover, in synthetic biology (Delépine et al. 2018; Kim et al. 2021c; Hérisson et al. 2022), MetaNetX introduces access to a database of biochemical transformations and network reconstructions.

4 Examples of Retrosynthetic Planning Using AI

The LHASA (Logic and Inference Applied to Synthetic Analysis) system is a sophisticated computer program that engages in retrosynthetic analysis at the molecular or reaction level and has been designed specifically with complex organic compounds in mind. Its primary objective is to motivate chemists to work towards logically and methodically converting complex target molecules into simple and practical starting materials. The system thoroughly considers functional groups' compatibility, reactivity, and overall synthetic feasibility. When determining how to synthesize ethyl 4-chloro-3-methoxy-5-(methylamino)benzoate (**39**) from toluene (47), the LHASA system performs an in-depth analysis into functional group transformations and substitution patterns on the aromatic ring (Scheme 7). The first

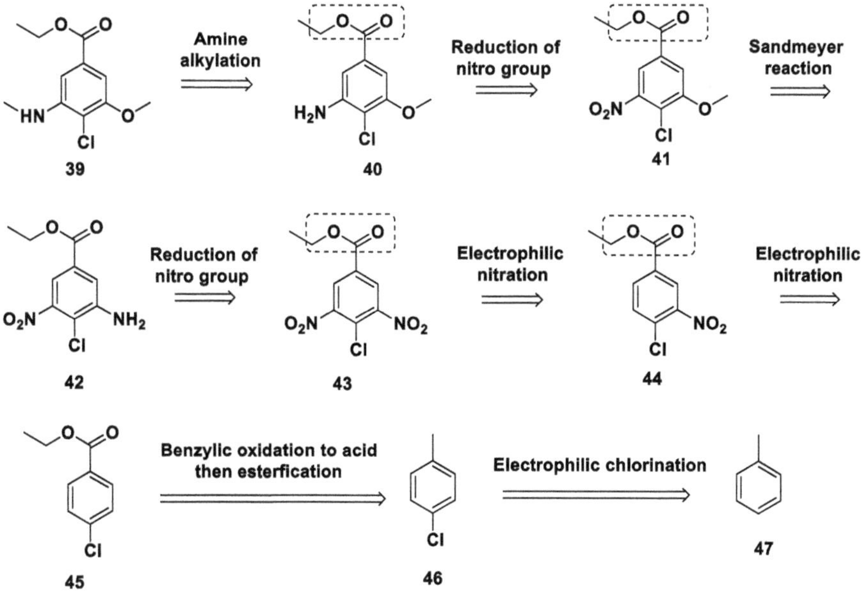

Scheme 7 Retrosynthesis of ethyl 4-chloro-3-methoxy-5-(methylamino)benzoate (Johnson et al. 1992)

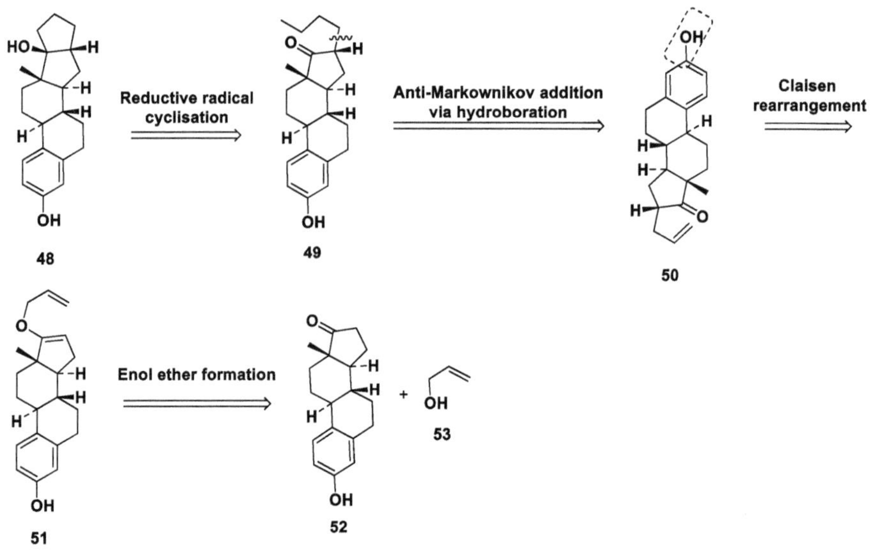

Scheme 8 Retrosynthesis of dihydroxydodecahydro-methylphenanthrene derivative (Johnson et al. 1992)

proposed suggestion is to change the ester functionality in the target molecule to a methyl group. This alteration then favors the aromatic ring's electronics, leading to increased efficiency for the following synthetic steps. Next, one will recommend removing the chlorine substituent, which indicates the ability of the system to locate the reaction feasibility and the functional group's impact. The ester group limits the electrophilic substitution by reducing the electron density on the ring, especially at para. On the other hand, if this reaction is successful, with an initial ester substituted by a methyl group, the direct electron-donating effect increases para reactivity and thus favors electrophilic substitution (Johnson et al. 1992). This perfectly indicates that the LHASA system contributes to logical clarity in retrosynthetic analysis and makes the best out of optimizing reactivity and functional group transformations in the design of synthesis routes.

The computational potential of the LHASA system is further extended to the retrosynthetic analysis of complex natural products such as the synthesis of a dihydroxydodecahydro-methylphenanthrene derivative (**48**) from estrogen (**52**) (Scheme 8 and Scheme 9) (Johnson et al. 1992).

The comprehensive Asymmetric Synthetic Knowledge Compilation and Optimization System (ASKCOS)-assisted retrosynthetic strategy for this complicated structure called branebrutinib (**57**) (Scheme 10) consists of an exemplary, systematic, computational strategy in identifying feasible synthetic routes and efficient bond disconnections and prioritizing strategic transformations. It becomes one of the major focuses concerning critical C-N bond formation. To it, ASKCOS proposes Boc-protection in the intermediate (**59**) for the N-H group. Selectivity has been

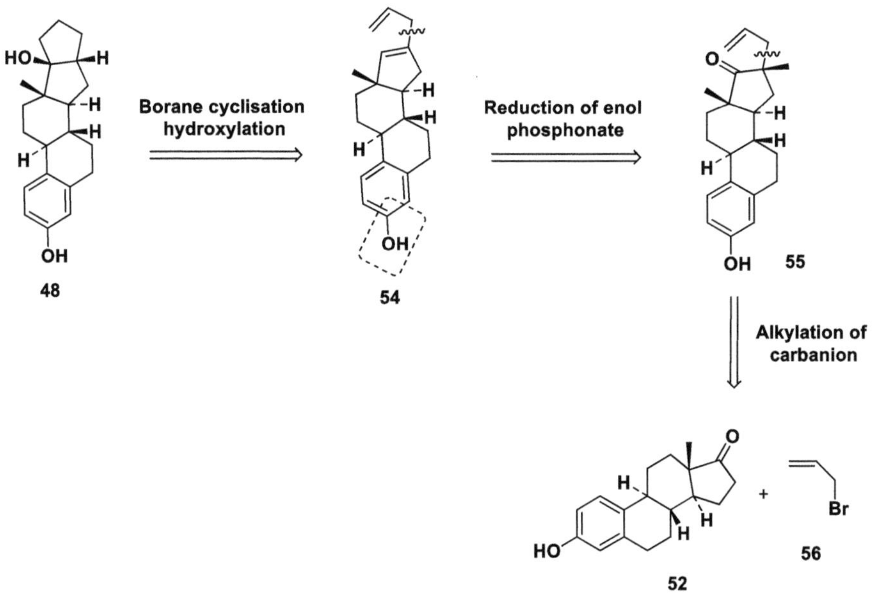

Scheme 9 Retrosynthesis of dihydroxydodecahydro-methylphenanthrene derivative (Johnson et al. 1992)

achieved by intermediate protection and avoiding side reactions, thus permitting a more.

Controlled proceeding of further reactions. At one point of synthesis reported for compound 1, the selective reaction did not proceed due to interference caused by the carboxamide moiety with the Buchwald coupling reaction, and using a nitrile in the intermediate (**60**) replacing the carboxamide allowed efficient coupling to proceed. Although ASKCOS predicted a particular C-N coupling, it has not been performed experimentally; however, convincing support for the feasibility of ASKCOS's suggested methodology has been acquired from the discovery by the researchers of a similar C-N coupling with a free carboxyamide. This indicates that the strategy under consideration is potentially practical and its model predictions derive from sound principles of synthetic chemistry (de Lange et al. 2011).

Depending on the available starting materials, ASKCOS here proposes retrosynthetic accessibility for 2,6-dichloro-N-(3-(indolin-6-yl)-1H-indazole-5-yl)benzamide derivatives (**72**) via extensive analysis of the commercial availability and synthetic feasibility of intermediates (Scheme 11). The routes suggested by the algorithm require only three steps to synthesize compound 76, in contrast to the requirement of one extra step for compounds 78 and 79; two further transformations must be accomplished to form compound 81. Such logical planning indicates the indicative ability of ASKCOS to optimize strategies based on the availability of materials and derived synthetic complexity, assuring laboratory practicality (de Lange et al. 2011).

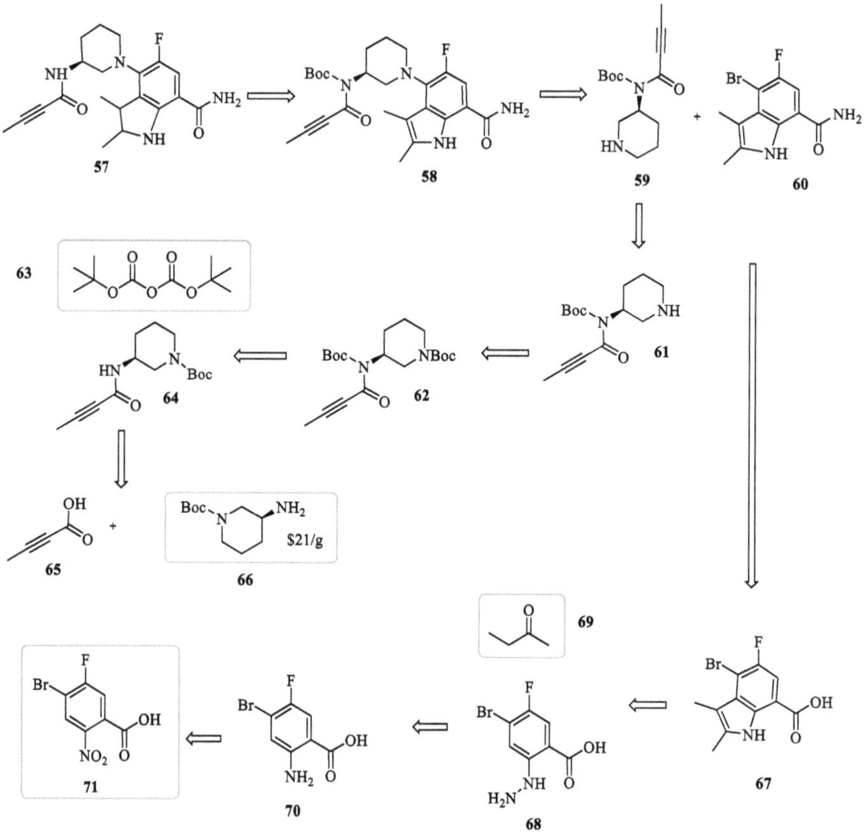

Scheme 10 (ASKCOS)-assisted retrosynthetic pathway for Branebrutinib (de Lange et al. 2011)

For the synthesis of (2E)-3-[4-({2-[2-(1,1-difluoroethyl)-4-fluorophenyl]-6-hydroxy-1-benzothiophen-3-yl}oxy)phenyl]prop-2-enoic acid, ASKCOS describes two different retrosynthetic schemes, each of which provides a one-step path to the target compound (Scheme 12). The first pathway proposes a typical Suzuki cross-coupling reaction, which produces carbon–carbon bonds by coupling an aryl halide (**84**) with an organoboron reagent (**85**). The second suggested pathway involves a palladium-catalyzed C-H insertion reaction, whereby functional group introduction may selectively occur by activating carbon–hydrogen bonds under catalytic conditions. The most interesting point concerning the protocol is the predicted uses that align closely with those found effective in experimental practices with polar solvents and high temperatures. This has truly proven the modeling power of computation in mimicking reaction conditions in vivo (de Lange et al. 2011).

EditRetro is an AI-enabled retro-synthetic analysis tool based on USPTO-50 K data. It has proved effective in projecting synthetic pathways toward complex and organic compounds. Using this approach, retrosynthetic pathways were

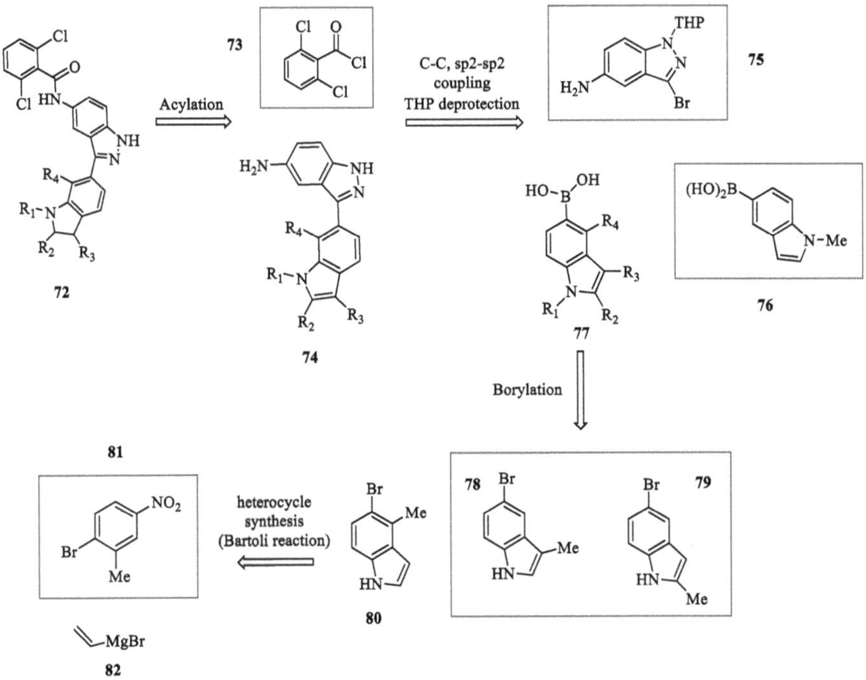

Scheme 11 Retrosynthetic accessibility for 2,6-dichloro-N-(3-(indolin-6-yl)-1H-indazole-5-yl)benzamide derivatives (de Lange et al. 2011)

successfully proposed for challenging molecules such as *N*-(3-(3-(1H-pyrrolo[2,3-b]pyridin-4-yl)imidazolidine-1-carbonyl)phenyl)-3-(trifluoromethyl)benzamide (**88,** Scheme 13), glutathione peroxidase 4 inhibitors (**95,** Scheme 13), *N*-[2-[2-(dimethylamino)ethyl-methylamino]-4-methoxy-5-[[4-(1-methylindol-3-yl)pyrimidin-2-yl]amino]phenyl]prop-2-enamide (**104,** Scheme 14), and 2-(3-cyano-4-isobutoxyphenyl)-4-methylthiazole-5-carboxylic acid (**112,** Scheme 14) (Struble et al. 2020). As shown in Scheme 13, the retrosynthetic pathway of 88 suggested by EditRetro closely matches the synthetic route documented in the methodology (Han et al. 2024). In addition, the reported synthesis of glutathione-peroxidase (**95**) (Zhang et al. 2022) employed a template-free model to investigate completely different reaction types and predict the synthetic pathway. This method, however, could identify all steps of the pathway among the top four predictions, even when classification into specific reaction types was not used (Struble et al. 2020). Additionally, the compound N-[2-[2-(dimethylamino)ethyl-methylamino]-4-methoxy-5-[[4-(1-methylindol-3-yl)pyrimidin-2-yl]amino]phenyl]prop-2-enamide (**104**) was predicted to result from the following series of transformations such as acylation, nitro group reduction, nucleophilic aromatic substitution, and finally the Suzuki cross-coupling reaction for the last step. Still, this last step moves against the literature (Zhavoronkov et al. 2019) which states

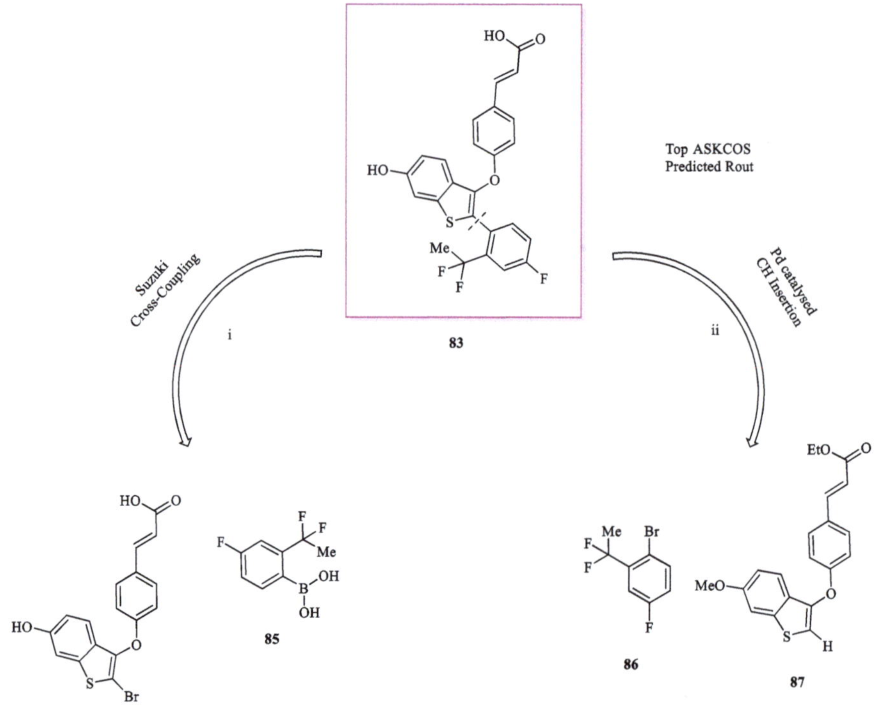

Scheme 12 ASKCOS predicted synthesis of (2E)-3-[4-({2-[2-(1,1-difluoroethyl)-4-fluorophenyl]-6-hydroxy-1-benzothiophen-3-yl}oxy)phenyl]prop-2-enoic acid (de Lange et al. 2011)

that the transformation under consideration would materially involve Friedel–Crafts arylation. However, pyrazinylindole formation is compatible with the prevalent Graph2Edits methodology (Struble et al. 2020). Furthermore, EditRetro also accurately predicted the synthesis of 2-(3-cyano-4-isobutoxyphenyl)-4-methylthiazole-5-carboxylic acid (**112**). This three-step synthesis prediction includes the hydrolysis of an ester and a Suzuki cross-coupling, which follows a similar pathway previously reported (Struble et al. 2020).

5 Conclusion and Future Aspects

Emerging AI-based retrosynthetic approaches have transformed synthetic chemistry into a versatile tool by providing excellent retrosynthetic analysis and pathway prediction for synthesizing a molecule. AI enables the use of sophisticated algorithms and scoring systems that generate output related to the efficiency, scalability, and feasibility of proposed synthetic routes. Majorly, such data as the USPTO

N-(3-(3-(1H-pyrrolo[2,3-b]pyridin-4-yl)imidazolidine-1-carbonyl)phenyl)-3-(trifluoromethyl)benzamide

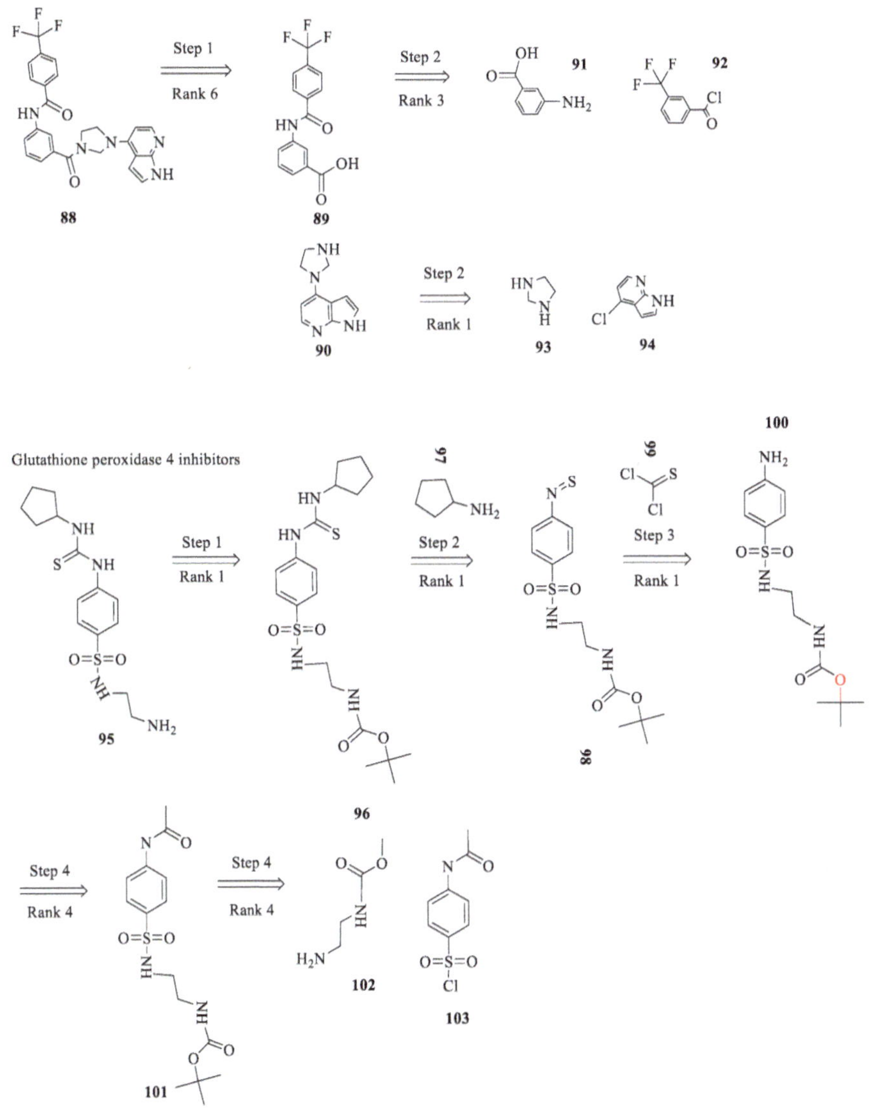

Scheme 13 EditRetro retro-synthetic analysis of *N*-(3-(3-(1H-pyrrolo[2,3-b]pyridin-4-yl)imidazolidine-1-carbonyl)phenyl)-3-(trifluoromethyl)benzamide and glutathione peroxidase 4 inhibitors (Struble et al. 2020)

Reaction Dataset and Reaxys significantly increased the exactness and pertinence of AI-based predictions, empowering chemists in the optimization process of the synthetic process with much preciseness in site-specified applications. AI-based development will inevitably be more relevant in drug discovery, biocatalysis,

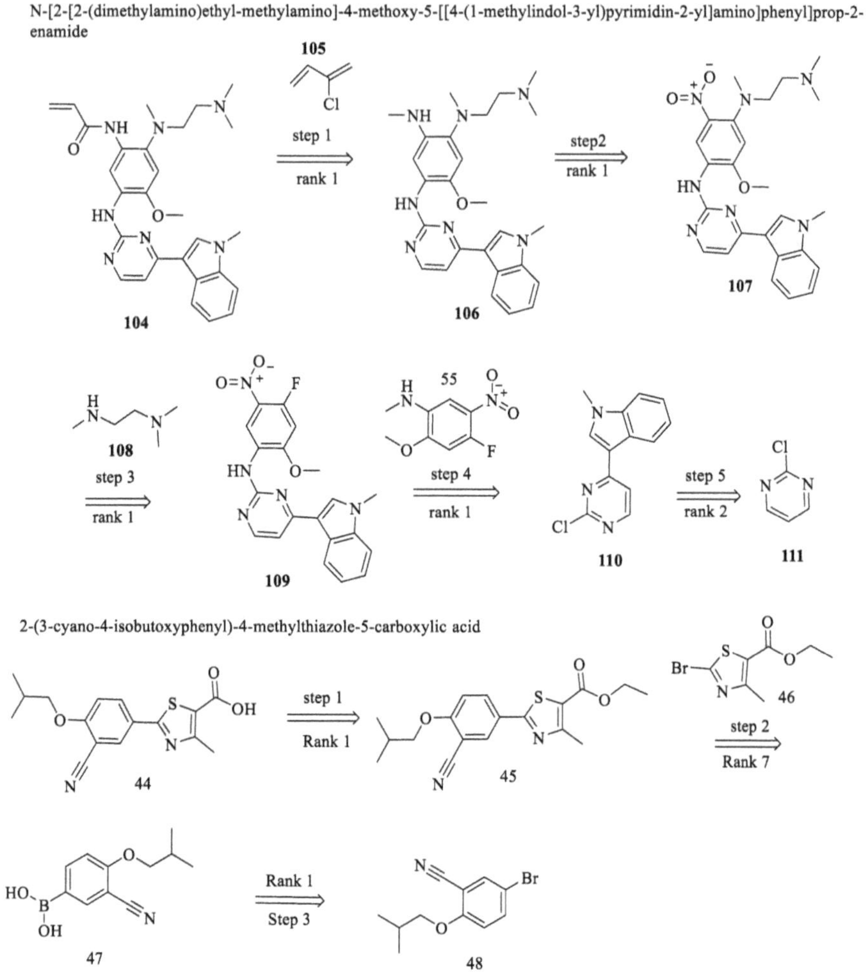

Scheme 14 EditRetro retro-synthetic analysis of *N*-[2-[2-(dimethylamino)ethyl-methylamino]-4-methoxy-5-[[4-(1-methylindol-3-yl)pyrimidin-2-yl]amino]phenyl]prop-2-enamide and 2-(3-cyano-4-isobutoxyphenyl)-4-methylthiazole-5-carboxylic acid (Struble et al. 2020)

and synthetic biology in the future. It will help answer significant sustainability challenges related to costs and efficiency. The future of retrosynthesis planning will significantly be influenced by the advent of AI capabilities, making chemical synthesis faster, safer, cleaner, and much more efficient across all industry lines. For future implementation, applying machine learning methods combined with quantum computing will enable significantly better performance in AI retrosynthesis. Combining such approaches could enhance prediction accuracy for complex reactions and molecular-level interactions by enabling more precise modeling of reaction mechanisms, energy states, and molecular dynamics. Newly emerging algorithms, reinforcement learning, and generative models are foreseen to facilitate better retrosynthetic planning by allowing systems to search and optimize reaction paths autonomously. In addition, the combination of AI with high-throughput experimental platforms, such as automated laboratories, will enable fast iteration and validation of synthetic pathways. In addition, developing real-time learning AI systems from experimental data may be used for real-time adaptive retrosynthetic planning to overcome unforeseen challenges more efficiently.

References

Aladdin Healthcare Technologies. "Deploying AI in drug discovery." Health Europa, 2020. https://www.healtheuropa.com/deploying-ai-in-drug-discovery/103477/?utm.

Badami BV. Retrosynthetic analysis: art of planning organic synthesis. Resonance. 2019;24(10):1071–86.

Badowski T, Gajewska EP, Molga K, Grzybowski BA. Synergy between expert and machine-learning approaches allows for improved retrosynthetic planning. Angew Chem Int Ed. 2020;59(2):725–30.

Baylon JL, Cilfone NA, Gulcher JR, Chittenden TW. Enhancing retrosynthetic reaction prediction with deep learning using multiscale reaction classification. J Chem Inf Model. 2019;59(2):673–88.

Beber ME, Gollub MG, Mozaffari D, Shebek KM, Flamholz AI, Milo R, Noor E. eQuilibrator 3.0: a database solution for thermodynamic constant estimation. Nucleic Acids Res. 2022;50(D1):D603–9.

Bharti R, Verma M, Thakur A, Sharma R. N-heterocyclic Carbenes (NHCs): an introduction. In: Carbene. IntechOpen; 2022.

Bucherer HT, Steiner W. Syntheses of hydantoins. I. On reactions of α-hydroxy and α-amino nitriles. J Prakt Chem. 1934;140:291–316.

Bürgi W, Richterich R, Colombo JP. L-lysine dehydrogenase deficiency in a patient with congenital lysine intolerance. Nature. 1966;211(5051):854–5.

Carey JS, Laffan D, Thomson C, Williams MT. Analysis of the reactions used for the preparation of drug candidate molecules. Org Biomol Chem. 2006;4(12):2337–47.

Chen S, Jung Y. Deep retrosynthetic reaction prediction using local reactivity and global attention. JACS Au. 2021;1(10):1612–20.

Chen B, Li C, Dai H, Song L. Retro*: learning retrosynthetic planning with neural guided a* search. In: International conference on machine learning. PMLR; 2020. p. 1608–16.

Coley CW, Rogers L, Green WH, Jensen KF. Computer-assisted retrosynthesis based on molecular similarity. ACS central science. 2017;3(12):1237–45.

Coley CW, Rogers L, Green WH, Jensen KF. SCScore: synthetic complexity learned from a reaction corpus. J Chem Inf Model. 2018;58(2):252–61.

Coley CW, Green WH, Jensen KF. RDChiral: an RDKit wrapper for handling stereochemistry in retrosynthetic template extraction and application. J Chem Inf Model. 2019a;59(6):2529–37.

Coley CW, Jin W, Rogers L, Jamison TF, Jaakkola TS, Green WH, Jensen KF. A graph-convolutional neural network model for the prediction of chemical reactivity. Chem Sci. 2019b;10(2):370–7.

Corey EJ. Robert Robinson lecture. Retrosynthetic thinking—essentials and examples. Chem Soc Rev. 1988;17:111–33.

Corey EJ. The logic of chemical synthesis: multistep synthesis of complex carbogenic molecules (nobel lecture). Angew Chem Int Ed Engl. 1991;30(5):455–65.

Corey EJ, Cheng X-M. The logic of chemical synthesis. NY: Wiley; 1995.

Correia J, Carreira R, Pereira V, Rocha M. Predicting the number of biochemical transformations needed to synthesize a compound. In: In *2022 international joint conference on neural networks (IJCNN)*. IEEE; 2022, July. p. 1–8.

Dai H, Li C, Coley C, Dai B, Song L. Retrosynthesis prediction with conditional graph logic network. Adv Neural Inf Proces Syst. 2019;32

de Lange B, Hyett DJ, Maas PJ, Mink D, van Assema FB, Sereinig N, de Vries JG. Asymmetric synthesis of (S)-2-indolinecarboxylic acid by combining biocatalysis and homogeneous catalysis. ChemCatChem. 2011;3(2):289–92.

de Souza RO, Miranda LS, Bornscheuer UT. A retrosynthesis approach for biocatalysis in organic synthesis. *Chemistry–a*. European J. 2017;23(50):12040–63.

Delépine B, Duigou T, Carbonell P, Faulon JL. RetroPath2. 0: a retrosynthesis workflow for metabolic engineers. Metab Eng. 2018;45:158–70.

Dong Z, Chen Z, Wang Q. Retrosynthesis prediction based on graph relation network. In: 2022 15th international congress on image and signal processing, BioMedical engineering and informatics (CISP-BMEI). IEEE; 2022, November. p. 1–5.

Fang L, Li J, Zhao M, Tan L, Lou JG. Single-step retrosynthesis prediction by leveraging commonly preserved substructures. Nat Commun. 2023;14(1):2446.

Finlay, M. R. V., Anderton, M., Ashton, S., Ballard, P., Bethel, P. A., Box, M. R., ... Wrigley, G. L. (2014). Discovery of a potent and selective EGFR inhibitor (AZD9291) of both sensitizing and T790M resistance mutations that spares the wild-type form of the receptor.

Fortunato ME, Coley CW, Barnes BC, Jensen KF. Machine-learned prediction of reaction template applicability for data-driven retrosynthetic predictions of energetic materials. In: AIP conference proceedings, vol. 2272, No. 1. AIP Publishing; 2020, November.

Genheden S, Thakkar A, Chadimová V, Reymond JL, Engkvist O, Bjerrum E. AiZynthFinder: a fast, robust and flexible open-source software for retrosynthetic planning. J Chem. 2020;12(1):70.

Golfmann M, Walker JC. Bicyclobutanes are unusual building blocks for complexity generation in organic synthesis. Commun Chem. 2023;6(1):9.

Gricourt G, Meyer P, Duigou T, Faulon JL. Artificial intelligence methods and models for retro-biosynthesis: a scoping review. ACS Synth Biol. 2024;13(8):2276–94.

Han P, Zhao P, Lu C, Huang J, Wu J, Shang S, et al. Gnn-retro: retrosynthetic planning with graph neural networks. In: Proceedings of the AAAI conference on artificial intelligence, vol. 36, No. 4; 2022. p. 4014–21.

Han Y, Xu X, Hsieh CY, Ding K, Xu H, Xu R, et al. Retrosynthesis prediction with an iterative string editing model. Nat Commun. 2024;15(1):6404.

He HR, Wang J, Liu Y, Wu F. Modeling diverse chemical reactions for single-step retrosynthesis via discrete latent variables. In: Proceedings of the 31st ACM international conference on Information & Knowledge Management; 2022, October. p. 717–26.

Heid E, Liu J, Aude A, Green WH. Influence of template size, canonicalization, and exclusivity for retrosynthesis and reaction prediction applications. J Chem Inf Model. 2021a;62(1):16–26.

Heid E, Goldman S, Sankaranarayanan K, Coley CW, Flamm C, Green WH. EHreact: extended Hasse diagrams for the extraction and scoring of enzymatic reaction templates. J Chem Inf Model. 2021b;61(10):4949–61.

Hérisson J, Duigou T, Du Lac M, Bazi-Kabbaj K, Sabeti Azad M, Buldum G, et al. The automated galaxy-SynBioCAD pipeline for synthetic biology design and engineering. Nat Commun. 2022;13(1):5082.

Heydari M, Ohshima T, Nunoura-Kominato N, Sakuraba H. Highly stable L-lysine 6-dehydrogenase from the thermophile Geobacillus stearothermophilus isolated from a Japanese hot spring: characterization, gene cloning and sequencing, and expression. Appl Environ Microbiol. 2004;70(2):937–42.

Hu H, Jiang Y, Yang Y, Chen JX. BiG2S: a dual task graph-to-sequence model for the end-to-end template-free reaction prediction. Appl Intell. 2023;53(24):29620–37.

Irwin R, Dimitriadis S, He J, Bjerrum EJ. Chemformer: a pre-trained transformer for computational chemistry. Mach Learn Sci Technol. 2022;3(1):015022.

Ishida S, Terayama K, Kojima R, Takasu K, Okuno Y. Prediction and interpretable visualization of retrosynthetic reactions using graph convolutional networks. J Chem Inf Model. 2019;59(12):5026–33.
Javahershenas R. Recent advances in the application of deep eutectic solvents for the synthesis of Spiro heterocyclic scaffolds via multicomponent reactions. J Mol Liq. 2023;385(122):398.
Javahershenas R, Nikzat S. Recent advances in the multicomponent synthesis of heterocycles using tetronic acid. RSC Adv. 2023;13(24):16619–29.
Jeong J, Lee N, Shin Y, Shin D. Intelligent generation of optimal synthetic pathways based on knowledge graph inference and retrosynthetic predictions using reaction big data. J Taiwan Inst Chem Eng. 2022;130(103):982.
Johnson AP, Marshall C, Judson PN. Some recent progress in the development of the LHASA computer system for organic synthesis design: starting-material-oriented retrosynthetic analysis. Recueil des Travaux Chimiques des Pays-Bas. 1992;111(6):310–6.
Karpov P, Godin G, Tetko IV. A transformer model for retrosynthesis. In: International conference on artificial neural networks. Cham: Springer; 2019. p. 817–30.
Kim E, Lee D, Kwon Y, Park MS, Choi YS. Valid, plausible, and diverse retrosynthesis using tied two-way transformers with latent variables. J Chem Inf Model. 2021a;61(1):123–33.
Kim J, Ahn S, Lee H, Shin J. Self-improved retrosynthetic planning. In: International conference on machine learning. PMLR; 2021b, July. p. 5486–95.
Kim Y, Ryu JY, Kim HU, Jang WD, Lee SY. A deep learning approach to evaluate the feasibility of enzymatic reactions generated by retrobiosynthesis. Biotechnol J. 2021c;16(5):2000605.
Kim H, Lee K, Kim C, Lim J, Kim WY. DFRscore: deep learning-based scoring of synthetic complexity with drug-focused retrosynthetic analysis for high-throughput virtual screening. J Chem Inf Model. 2023;64(7):2432–44.
Koch M, Duigou T, Faulon JL. Reinforcement learning for bioretrosynthesis. ACS Synth Biol. 2019;9(1):157–68.
Kochev N, Avramova S, Jeliazkova N. Ambit-SMIRKS: a software module for reaction representation, reaction search and structure transformation. J Chem. 2018;10(1):42.
Kreutter D, Reymond JL. Multistep retrosynthesis combining a disconnection-aware triple transformer loop with a route penalty score guided tree search. Chem Sci. 2023;14(36):9959–69.
Kumagai H. Microbial production of amino acids in Japan. History Modern Biotechnol. 2000;I:71–85.
Kusakabe H, Kodama K, Kuninaka A, Yoshino H, Misono H, Soda K. A new antitumor enzyme, L-lysine alpha-oxidase from Trichoderma viride. Purification and enzymological properties. J Biol Chem. 1980;255(3):976–81.
Latendresse M, Malerich JP, Herson J, Krummenacker M, Szeto J, Vu VA, et al. Synroute: a retrosynthetic planning software. J Chem Inf Model. 2023;63(17):5484–95.
Li B, Chen H. Prediction of compound synthesis accessibility based on reaction knowledge graph. Molecules. 2022;27(3):1039.
Li J, Fang L, Lou JG. RetroRanker: leveraging reaction changes to improve retrosynthesis prediction through re-ranking. J Chem. 2023;15(1):58.
Lin K, Xu Y, Pei J, Lai L. Automatic retrosynthetic route planning using template-free models. Chem Sci. 2020;11(12):3355–64.
Lin Z, Yin S, Shi L, Zhou W, Zhang YJ. G2GT: retrosynthesis prediction with graph-to-graph attention neural network and self-training. J Chem Inf Model. 2023;63(7):1894–905.
Liu B, Ramsundar B, Kawthekar P, Shi J, Gomes J, Luu Nguyen Q, et al. Retrosynthetic reaction prediction using neural sequence-to-sequence models. ACS Central Sci. 2017;3(10):1103–13.
Liu CH, Korablyov M, Jastrzebski S, Włodarczyk-Pruszynski P, Bengio Y, Segler M. RetroGNN: fast estimation of synthesizability for virtual screening and de novo design by learning from slow retrosynthesis software. J Chem Inf Model. 2022;62(10):2293–300.
Liu Q, Tang K, Zhang L, Du J, Meng Q. Computer-assisted synthetic planning considering reaction kinetics based on transition state automated generation method. AICHE J. 2023;69(7):e18092.
Mao K, Xiao X, Xu T, Rong Y, Huang J, Zhao P. Molecular graph enhanced transformer for retrosynthesis prediction. Neurocomputing. 2021;457:193–202.

May O, Verseck S, Bommarius A, Drauz K. Development of dynamic kinetic resolution processes for biocatalytic production of natural and nonnatural L-amino acids. Org Process Res Dev. 2002;6(4):452–7.

Mikolajczyk A, Zhdan U, Antoniotti S, Smolinski A, Jagiello K, Skurski P, Polanski J. Retrosynthesis from transforms to predictive sustainable chemistry and nanotechnology: a brief tutorial review. Green Chem. 2023;25(8):2971–91.

Nicolaou KC, Sorensen EJ. Classics in total synthesis: targets, strategies, methods. Wiley; 1996.

Panda, B. Retrosynthesis. 2020. https://www.citycollegekolkata.org/documents/online_course_materials/Retrosyntesis_Biswajit_Pandal.pdf.

Perrone S, Messa F, Troisi L, Salomone A. N-, O-and S-heterocycles synthesis in deep eutectic solvents. Molecules. 2023;28(8):3459.

Plehiers PP, Coley CW, Gao H, Vermeire FH, Dobbelaere MR, Stevens CV, et al. Artificial intelligence for computer-aided synthesis in flow: analysis and selection of reaction components. Front Chem Eng. 2020;2:5.

Probst D, Manica M, Nana Teukam YG, Castrogiovanni A, Paratore F, Laino T. Biocatalysed synthesis planning using data-driven learning. Nat Commun. 2022;13(1):964.

Qiao H, Wu Y, Zhang Y, Zhang C, Wu X, Wu Z, et al. Transformer-based multitask learning for reaction prediction under low-resource circumstances. RSC Adv. 2022;12(49):32020–6.

Sanchez-Garcia R, Havasi D, Takács G, Robinson MC, von Delft F, Deane CM. CoPriNet: graph neural networks provide accurate and rapid compound price prediction for molecule prioritisation. Digital Disco. 2023;2(1):103–11.

Sankaranarayanan K, Heid E, Coley CW, Verma D, Green WH, Jensen KF. Similarity based enzymatic retrosynthesis. Chem Sci. 2022;13(20):6039–53.

Schulze B, Wubbolts MG. Biocatalysis for industrial production of fine chemicals. Curr Opin Biotechnol. 1999;10(6):609–15.

Schwaller P, Petraglia R, Zullo V, Nair VH, Haeuselmann RA, Pisoni R, et al. Predicting retrosynthetic pathways using transformer-based models and a hyper-graph exploration strategy. Chem Sci. 2020;11(12):3316–25.

Segler MH, Waller MP. Neural-symbolic machine learning for retrosynthesis and reaction prediction. Chem Eur J. 2017;23(25):5966–71.

Segler MH, Preuss M, Waller MP. Planning chemical syntheses with deep neural networks and symbolic AI. Nature. 2018;555(7698):604–10.

Seo SW, Song YY, Yang JY, Bae S, Lee H, Shin J, et al. GTA: graph truncated attention for retrosynthesis. Proc AAAI Conf Artificial Intelligence. 2021;35(1):531–9.

Sharma S, Monga Y, Gupta A, Singh S. 2-Oxindole and related heterocycles: synthetic methodologies for their natural products and related derivatives. RSC Adv. 2023;13(21):14249–67.

Shi C, Xu M, Guo H, Zhang M, Tang J. A graph to graphs framework for retrosynthesis prediction. In: International conference on machine learning. PMLR; 2020. p. 8818–27.

Somnath VR, Bunne C, Coley C, Krause A, Barzilay R. Learning graph models for retrosynthesis prediction. Adv Neural Inf Proces Syst. 2021;34:9405–15.

Struble TJ, Alvarez JC, Brown SP, Chytil M, Cisar J, DesJarlais RL, et al. Current and future roles of artificial intelligence in medicinal chemistry synthesis. J Med Chem. 2020;63(16):8667–82.

Sun Y, Sahinidis NV. Computer-aided retrosynthetic design: fundamentals, tools, and outlook. Curr Opin Chem Eng. 2022;35(100):721.

Tang K, Zhuang Y, Wang W, Liu Q, Zhang L, Du J, Meng Q. GC-NORM-based thermodynamic framework for evaluations of organic reactions involving carbon dioxide utilization. Chem Eng Sci. 2023;278(118):913.

Tetko IV, Karpov P, Van Deursen R, Godin G. State-of-the-art augmented NLP transformer models for direct and single-step retrosynthesis. Nat Commun. 2020;11(1):5575.

Thakkar A, Selmi N, Reymond JL, Engkvist O, Bjerrum EJ. “Ring breaker”: neural network driven synthesis prediction of the ring system chemical space. J Med Chem. 2020a;63(16):8791–808.

Thakkar A, Kogej T, Reymond JL, Engkvist O, Bjerrum EJ. Datasets and their influence on the development of computer-assisted synthesis planning tools in the pharmaceutical domain. Chem Sci. 2020b;11(1):154–68.

Thakkar A, Chadimová V, Bjerrum EJ, Engkvist O, Reymond JL. Retrosynthetic accessibility score (RAscore)–rapid machine learned synthesizability classification from AI driven retrosynthetic planning. Chem Sci. 2021;12(9):3339–49.
Thakkar A, Vaucher AC, Byekwaso A, Schwaller P, Toniato A, Laino T. Unbiasing retrosynthesis language models with disconnection prompts. ACS Central Sci. 2023;9(7):1488–98.
Toniato A, Vaucher AC, Schwaller P, Laino T. Enhancing diversity in language based models for single-step retrosynthesis. Digital Discovery. 2023;2(2):489–501.
Tsuchida T, Nishimoto Y, Kotani T, Iiizumi K. Japanese patent. JP. 1993;5123177A
Tu Z, Coley CW. Permutation invariant graph-to-sequence model for template-free retrosynthesis and reaction prediction. J Chem Inf Model. 2022;62(15):3503–13.
Ucak UV, Kang T, Ko J, Lee J. Substructure-based neural machine translation for retrosynthetic prediction. J Chem. 2021;13(1):4.
Ucak UV, Ashyrmamatov I, Ko J, Lee J. Retrosynthetic reaction pathway prediction through neural machine translation of atomic environments. Nat Commun. 2022;13(1):1186.
Ucak UV, Ashyrmamatov I, Lee J. Reconstruction of lossless molecular representations from fingerprints. J Chem. 2023;15(1):26.
Verma M, Thakur A, Sharma R, Bharti R. Recent advancement in the one-pot synthesis of the tri-substituted methanes (TRSMs) and their biological applications. Curr Org Synth. 2022;19(1):86–114.
Wan Y, Liao B, Hsieh CY, Zhang S. Retroformer: pushing the limits of interpretable end-to-end retrosynthesis transformer. arXiv preprint arXiv. 2022a:*2201.12475*.
Wan Y, Liao B, Hsieh C-Y, Zhang S. Retroformer: pushing the limits of interpretable end-to-end retrosynthesis transformer. In: Proceedings of the 39th international conference on machine learning; proceedings of machine learning research, vol. 162. PMLR; 2022b. p. 22475–90.
Wang X, Qian Y, Gao H, Coley CW, Mo Y, Barzilay R, Jensen KF. Towards efficient discovery of green synthetic pathways with Monte Carlo tree search and reinforcement learning. Chem Sci. 2020;11(40):10959–72.
Wang X, Li Y, Qiu J, Chen G, Liu H, Liao B, et al. RetroPrime: a diverse, plausible and transformer-based method for single-step retrosynthesis predictions. Chem Eng J. 2021a;420(129):845.
Wang L, Upadhyay V, Maranas CD. dGPredictor: automated fragmentation method for metabolic reaction free energy prediction and de novo pathway design. PLoS Comput Biol. 2021b;17(9):e1009448.
Wang W, Liu Q, Zhang L, Dong Y, Du J. RetroSynX: a retrosynthetic analysis framework using hybrid reaction templates and group contribution-based thermodynamic models. Chem Eng Sci. 2022;248(117):208.
Warren S, Wyatt P. Organic synthesis: the disconnection approach. Wiley; 2008.
Xie S, Yan R, Han P, Xia Y, Wu L, Guo C, et al. Retrograph: retrosynthetic planning with graph search. In: Proceedings of the 28th ACM SIGKDD conference on knowledge discovery and data mining; 2022. p. 2120–9.
Yan C, Ding Q, Zhao P, Zheng S, Yang J, Yu Y, Huang J. Retroxpert: decompose retrosynthesis prediction like a chemist. Adv Neural Inf Proces Syst. 2020;33:11248–58.
Yan C, Zhao P, Lu C, Yu Y, Huang J. RetroComposer: composing templates for template-based retrosynthesis prediction. Biomolecules. 2022;12(9):1325.
Yang Q, Sresht V, Bolgar P, Hou X, Klug-McLeod JL, Butler CR. Molecular transformer unifies reaction prediction and retrosynthesis across pharma chemical space. Chem Commun. 2019;55(81):12152–5.
Yang F, Liu J, Zhang Q, Yang Z, Zhang X. CNN-based two-branch multi-scale feature extraction network for retrosynthesis prediction. BMC Bioinform. 2022;23(1):362.
Yu J, Wang J, Zhao H, Gao J, Kang Y, Cao D, et al. Organic compound synthetic accessibility prediction based on the graph attention mechanism. J Chem Inf Model. 2022;62(12):2973–86.
Zhang C, Lapkin AA. Reinforcement learning optimization of reaction routes on the basis of large, hybrid organic chemistry–synthetic biological, reaction network data. Reaction Chem Eng. 2023;8(10):2491–504.

Zhang Y, Wang L, Wang X, Zhang C, Ge J, Tang J, et al. Data augmentation and transfer learning strategies for reaction prediction in low chemical data regimes. Organic Chem Front. 2021;8(7):1415–23.

Zhang B, Zhang X, Du W, Song Z, Zhang G, Zhang G, et al. Chemistry-informed molecular graph as reaction descriptor for machine-learned retrosynthesis planning. Proc Natl Acad Sci. 2022;119(41):e2212711119.

Zhang B, Lin J, Du L, Zhang L. Harnessing data augmentation and normalization preprocessing to improve the performance of chemical reaction predictions of data-driven model. Polymers. 2023;15(9):2224.

Zhao D, Tu S, Xu L. Efficient retrosynthetic planning with MCTS exploration enhanced a* search. Commun Chem. 2024;7(1):52.

Zhavoronkov A, Ivanenkov YA, Aliper A, Veselov MS, Aladinskiy VA, Aladinskaya AV, et al. Deep learning enables rapid identification of potent DDR1 kinase inhibitors. Nat Biotechnol. 2019;37(9):1038–40.

Zheng S, Rao J, Zhang Z, Xu J, Yang Y. Predicting retrosynthetic reactions using self-corrected transformer neural networks. J Chem Inf Model. 2019;60(1):47–55.

Zheng S, Zeng T, Li C, Chen B, Coley CW, Yang Y, Wu R. Deep learning driven biosynthetic pathways navigation for natural products with BioNavi-NP. Nat Commun. 2022;13(1):3342.

Zhong Z, Song J, Feng Z, Liu T, Jia L, Yao S, Song M. Recent advances in artificial intelligence for retrosynthesis. arXiv preprint arXiv. 2023a:*2301.05864*.

Zhong W, Yang Z, Chen CYC. Retrosynthesis prediction using an end-to-end graph generative architecture for molecular graph editing. Nat Commun. 2023b;14(1):3009.

Zhu D, Hua L. Biocatalytic asymmetric amination of carbonyl functional groups–a synthetic biology approach to organic chemistry. Biotechnol J Healthcare Nut Technol. 2009;4(10):1420–31.

Zhu J, Xia Y, Wu L, Xie S, Zhou W, Qin T, Liu TY. Dual-view molecular pre-training. In: Proceedings of the 29th ACM SIGKDD conference on knowledge discovery and data mining; 2023. p. 3615–27.

Active Learning in Drug Discovery: Revolutionizing Chemical Space Exploration

Kapali Suri, Anshul Yadav, Prateek Paul, Jaspreet Kaur Dhanjal, Bapi Chatterjee, and N. Arul Murugan

Abstract Drug discovery is a complex and resource-intensive process identifying molecules capable of modulating biological targets aberrantly expressed in the disease condition. Even though machine learning (ML) advances for drug discovery, traditional approaches often struggle with the "applicability domain" problem, limiting their predictive power to identify hits and lead compounds from substantial chemical libraries. Active Learning (AL) is a type of ML technique that addresses various drawbacks, such as data quality and bias, the presence of limited experimental data and applicability domain issues by iteratively selecting the most informative samples, reducing false positives and increasing the accuracy of compound-target interactions as well the prediction of Absorption, Distribution, Metabolism, Excretion, and Toxicity (ADMET) properties, as well as binding affinities. Recently, Deep Active Learning (DeepAL), a subset of AL, has efficiently combined intelligent sample selection with deep learning models for feature extraction, thereby amplifying the benefits during the drug discovery pipeline, particularly in low-data scenarios. Hence, AL can accelerate drug discovery design by integrating advanced approaches such as transfer learning, query-based optimization, and generative models. This book chapter highlights important aspects of AL, such as fundamental architecture, applications, and challenges that will further enhance the search for innovative therapeutic candidates.

Keywords Active Learning · Machine Learning · Query Strategy · Chemical Space · Drug-like Properties · Binding Affinity

---

K. Suri · A. Yadav · P. Paul · J. K. Dhanjal · N. A. Murugan (✉)
Department of Computational Biology, Indraprastha Institute of Information Technology, New Delhi, India
e-mail: arul.murugan@iiitd.ac.in

B. Chatterjee (✉)
Department of Computer Science, Indraprastha Institute of Information Technology, New Delhi, India
e-mail: bapi@iiitd.ac.in

A. Lavecchia (ed.), *Applied Artificial Intelligence for Drug Discovery*,
https://doi.org/10.1007/978-3-031-98022-0_13

1 Introduction

Drug discovery is a complex and time-consuming process of developing new drugs against the chosen target that might be critically involved in the pathogenesis of a particular disease. Small molecules that complement the structural and electrostatic properties of a biomolecular binding site can regulate the activity of target biomolecules, either by enhancing or inhibiting their function. Apart from the above limitations, pharmaceutical research faces various limitations in discovering lead compound identification, such as vast datasets, understanding intricate biological interactions, addressing regulatory complexities, and fostering interdisciplinary collaboration (Singh et al. 2023; Hughes et al. 2011).

ML has emerged as a significant tool in modern drug discovery, assisting in various stages such as target validation (Najm et al. 2021), 3D structure prediction of biomolecule (Giangrande et al. 1975), molecular structure generation (Chen et al. 2020), virtual high throughput screening (Chen et al. 2018), and predictions for ADMET (absorption, distribution, metabolism, excretion, and toxicity) (Swanson et al. 2024). Hence, using large datasets and advanced algorithms has sped up pharmaceutical research. Despite its significant potential, ML models face various limitations, particularly when encountering chemical structures that deviate significantly from those within their training datasets. This is commonly known as the "applicability domain" problem in ML-based drug discovery, which can lead to several issues, such as false positives and false negatives and biases in the model. Another limitation is the lack of labelled data since gathering high-quality experimental data is extremely expensive and time-consuming. One approach that can be employed to avoid this generalization is incorporating experimental feedback from newly tested compounds.

Active learning (AL) in ML (Fig. 1), conceptualized over 30 years ago, has emerged as a transformative tool in drug discovery over the past 15 years (MacKay 1992; Cohn et al. 1996; Zhang and Veenker 1991; Warmuth et al. 2003). It is a specialized branch of ML designed to tackle challenges associated with large, unlabelled datasets by systematically selecting the most informative samples for expert labelling. The labelling process uses an iterative procedure given by:

$$S_t = \text{argmax}_{S \subseteq D} I\left(S, M_t\right),$$

where S_t represents the set of samples selected at iteration t, D is the unlabelled dataset, and $I(S, M_t)$ is an information gain function quantifying the expected reduction in uncertainty of the model M_t after labelling the samples in S (Bailey et al. 2024). This targeted selection enhances model performance while minimizing labelling costs, making it particularly effective for exploring extensive chemical spaces with limited labelled data.

AL employs two main strategies to guide experimental design: generating new experiments and selecting from existing options. The algorithm proposes new experiments in the first approach, often leveraging generative models such as

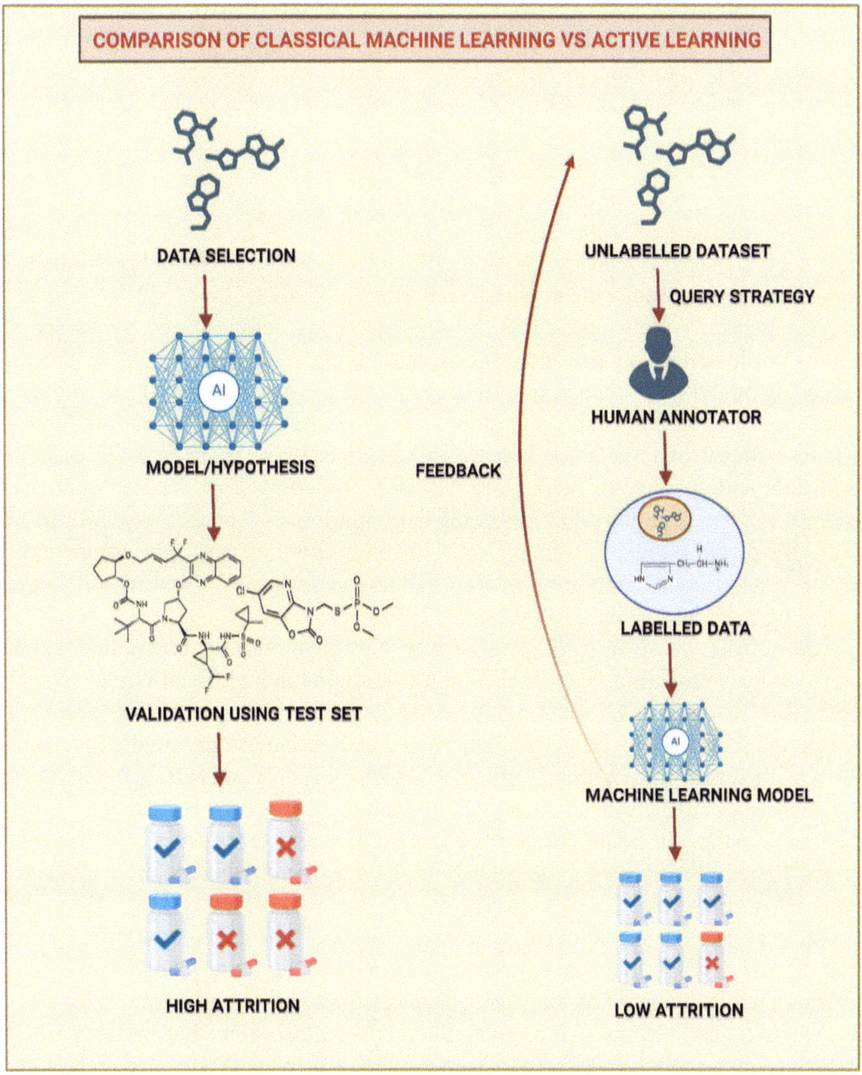

Fig. 1 Comparison of classical ML vs AL-assisted drug discovery

Variational Autoencoders (VAEs) or Generative Adversarial Networks (GANs) to design novel molecular structures. This can be mathematically represented as:

$$M_{\text{new}} = \text{Generate}\left(M_{\text{base}}, \theta\right),$$

where M_{new} denotes the newly generated molecular structure, M_{base} is the base molecular model guiding the generation, and θ is the parameter controlling the process.

In contrast, the second approach selects the most informative experiment from a predefined set of options. For instance, given that 100 molecules are available for testing, the algorithm identifies the one most likely to improve the model's predictions and advance research. This selection can be expressed as:

$$m^* = \text{argmax}_{m \in M}\, U\,(m, M)\,,$$

where m^* is the molecule chosen for testing, and M is the set of available molecules. $U(m, M)$ is a utility function estimating the expected model performance improvement if the molecule m is tested (Bailey et al. 2024).

AL has been widely utilized in drug discovery, as demonstrated by its applications in Table 1. It significantly enhances multiple stages of drug development, improving model performance across various tasks. In virtual screening, AL reduces false positives by prioritizing compounds that are most likely to exhibit strong binding affinity and meet key pharmacological criteria, such as adherence to Lipinski's Rule of Five or favourable ADMET properties. This focused approach strengthens the reliability of screening and minimizes unnecessary follow-up steps. By diversifying the training set with both common and rare compounds, AL mitigates biases, enabling models to generalize effectively and identify rare, promising molecules (Filella-Merce et al. 2023).

Across drug discovery, AL significantly aids in predicting compound–target interactions, overcoming many of the limitations found in traditional virtual screening. In drug discovery, AL holds significant potential to revolutionize drug discovery by improving molecular property predictions like binding affinity prediction, solubility, permeability prediction, further optimizing experimental design (e.g., biochemical assays, structural characterization), and reducing costs. It overcomes many challenges in the traditional virtual screening method with the most informative data points. Hence, this increases the probability of discovering therapeutic candidates with higher efficiency.

Despite so many advantages, several drawbacks limit the adoption of AL in drug discovery. First, there are infrastructure compatibility issues. High throughput screening works on predefined experimental pipelines that screen large test libraries in a single batch, whereas AL requires interactive decision-making, making small batches of compounds selected based on the previous models used for prediction. Hence, altering existing workflows to accommodate this adaptive process required substantial changes in the experimental protocol as well as proper maintenance of the data (Settles 2012; Janzen and Popa-Burke 2009). Many researchers remain hesitant to adopt AL due to past failures of computational drug discovery methods, such as in silico screening methods that over-predicted compound activity but failed in real-world validation and ML models that struggled with poor generalization, leading to high false positives in virtual screening. These past disappointments have contributed to scepticism, making researchers cautious about relying too heavily on AL-driven predictions. Even when AL models perform well, their acceptance in the scientific community is limited due to various issues like transparency. Many deep learning-based AL models function as "black boxes," making it difficult to

Table 1 Different studies on AL used in drug discovery

Purpose	Model	Strategy	Descriptor	Dataset information	Accuracy	Authors/citation
Response of compound.	Deep generative model (AL and cheminformatics tools integration)	Experimental design with algorithm updates	Heterogeneous chemical structure generation	GSK collection, eMolecules, EnamineREAL	Not directly reported as a single metric	Green et al. (2020)
Drug oral plasma exposure	DNN (phase I: Sampling informative data; phase II: Exploring chemical space)	Data-driven model refinement	Molecular features for oral drug exposure	LIFE-CHEMICALS, other sources 491 molecules and 2565 records	Accuracy of 0.856	Ding et al. (2021)
Drug response prediction	Uncertainty sampling, greedy, hybrid approaches	Exploitation via iteration-based model refinement	Molecular fingerprints, drug descriptors	CTRP, includes 494 drugs, 812 cancer cell lines, and 318,040 experiments	Statistical scores between AUC, score not given	Vasanthakumari et al. (2024)
Eye damage and irritation potential of compounds	RF, TE, GBT, RBF,	Model refinement through iterative learning cycles	Molecular descriptors and 4 ML methods	2299 compounds for serious eye damage, 5214 compounds for eye irritation.	Accuracy of 0.972 and 0.959, for eye damage model and eye irritation model, respectively	Di et al. (2022)
Binding affinity	RL with REINVENT (physics-based free energy ranking with REINVENT generative AI)	Iterative GAL cycles with varying batch sizes	Chemical space diversity, molecular descriptor.	Optimized GAL protocol with flexible batch size	Spearman ranking coefficient was above 0.7 and the R^2 coefficient was above 0.6	Loeffler et al. (2024)
Hit discovery	Exploration of chemical space using deep learning model, GCN	Feedback-based iterative learning	Molecular descriptors	LIT-PCBA collection, for VDR, ALDH1, PKM2	Not directly reported as a single metric	Van Tilborg and Grisoni (2024)

(continued)

Table 1 (continued)

Purpose	Model	Strategy	Descriptor	Dataset information	Accuracy	Authors/citation
Effect of compounds on targets	AL ML model, GM (explores categorical matrix space)	Iterative adaptive switching strategy	Categorical matrix descriptors	Synthetic, enzyme activity, protein subcellular screening	Average accuracy For IBC is again the best at 83.2%, SIB is next at 76.6%, followed By NNRLLFM and GM at 66.6% and 66.2%, respectively.	Sun and Murphy (2021)
Ligand-target interactions	RF (selection of informative ligand-target interactions)	Feedback-driven updates	Ligand-target descriptors	Bioactivity datasets, GPCR dataset,	96.2% for the SIB model, followed by the adaptive switching strategy at 94.7%, the IBC model at 93.4%, GM at 83.4%, and NNRLLFM at 82.8%	Reker et al. (2017)
Drug-likeness	XGBoost, NL, PE, AE (exploration through AL ensemble strategies	Feedback-driven updates	Graph neural network (GNN) descriptors	DEKOIS 2.0,	In vivo ability prediction: 94.8% IND ability prediction: 85.2% Market approvability prediction: 82.2%	Cai et al. (2022)
Quantum properties	GVAE (iterative feedback from generative model to quantum chemistry oracle)	Multi-target optimization via self-refinement	Ionization potential, electron affinity, and dipole moment	Subset of compounds from the ZINC15 database	Not directly reported as a single metric	Iovanac et al. (2022)

Drug-target interactions	DT, RF, SVM, BE-DTI (AL with dimensionality reduction)	Ensemble bagging-based learning	Drug-target interaction (DTI) features	KEGG, DrugBank, benchmark dataset	AUC 0.927, Sensitivity 0.886, Specificity 0.864, G-mean 0.874	Sharma and Rani (2018)
Active learning accuracy	Regression model (simulated drug-target matrices)	Stopping criteria design	Drug-target matrices	PubChem, NR, GPCR, IC, Enz	AUC for NR 93.6, GPCR 90.6, IC 86.8, Enz 90.3	Temerinac-Ott et al. (2015)
Binding affinity	Bayesian	Upper confidence bound/greedy acquisition	Structure-property models	AmpC and D4 Docking datasets	MSE 0.146 and spearman 0.919	Graff et al. (2021)
Compound's effect on multiple biological targets	Lasso regression (descriptive feature-based exploration)	AL-driven predictive modelling	Descriptive features of compounds and targets	PubChem database	AUC of 0.5	Kangas et al. (2014)
Ability of molecules to inhibit specific proteins	Generative artificial Intelligence models (AL to target unexplored chemical space)	AL to align the generative model with objectives	Molecular structures, protein-specific molecules	FDA-approved small-molecule inhibitors	C model, the percentage of molecules reaching the scoring threshold improved from 38.8% to 91.6% For the M model, this increase was from 21.7% to 80.3%.	Kyro et al. (2024)
Bioactivity and physical properties of small molecules	CNN (semi-supervised learning to improve prediction accuracy)	AL to optimize the training set	Molecular properties, bioactivity data	FreeSolv, ESOL, CatS, melting point, p450, malaria	R^2 values ranging from 0.80 to 0.95	Zhang and Lee (2019)

(continued)

Table 1 (continued)

Purpose	Model	Strategy	Descriptor	Dataset information	Accuracy	Authors/citation
Molecular properties	BAML, SOAP, DTNN, MBTR, MTM (AL for optimal training set selection)	Locality-based property prediction	Interatomic descriptors, molecular properties	QM9 dataset	Scatter plot of enthalpy predicted by machine learning	Gubaev et al. (2018)
Binding affinity	DL (AL to prioritize ligands)	Regression models to select ligands for docking	Ligand descriptors	CB2, AmpC, D4, AA2AR	Recall score was reported as roughly 74.4%	Marin et al. (2024)
Drug activity	Generative models (AL, machine vision)	AL to guide experimentation	Machine vision to extract information from imaging assays	Biological imaging data	Not directly reported as a single metric	Murphy (2011)

GBT Gradient Boosted Trees, *RBF* Radial Basis Function, *RF* Random Forest, *TE* Tree Ensemble, *DNN* Deep Neural Network, *RL* Reinforcement Learning, *GCN* Graph Neural Network, *NL* Normal Learning, *PE* Passive Ensemble, *AE* Active Ensemble, *GVAE* Grammar Variational Autoencoder, *DT* Decision Trees, *SVM* Support Vector Machines, *CNN* Convolutional Neural Network, *DL* Deep Learning, *BAML*Bonds, Angles, and Machine learning, *SOAP* Smooth Overlap of Atomic Positions, *DTNN* Deep Tensor Neural Network, *MBTR* Many-Body Tensor Representation, *MTM* Moment Tensor Model understand why certain compounds are prioritized. For instance, we do not know which features helped identify the compounds. Medicinal chemists and biologists often require clear rationales (interpretability) for why specific molecules are chosen for synthesis or testing. Another challenge is the model's generalizability, as AL models trained on limited datasets may not generalize well to structurally diverse compounds or new therapeutic targets (Dearden et al. 2009; Brown et al. 2020).

We provide a GitHub repository (Paul 2024), where we have listed the papers related to active learning in drug discovery from 2003 to 2024.

2 Fundamental Components of Active Learning

An active learning workflow has four main components: The initial training set, the machine learning algorithm, the query strategy, and the evaluation indicators to determine stopping criteria.

2.1 Initial Training Set

The initial training set is the foundation of the active learning workflow. It consists of labelled data points selected to train an ML model. A well-chosen initial training set should be diverse to prevent bias, ensuring that the model generalizes well. The number of points in this set is significantly smaller than in the unlabelled dataset.

For example, the miDruglikeness model, used in drug discovery, employs active learning to refine its predictive accuracy. Leveraging pool-based uncertainty sampling reduces reliance on large labelled datasets, accelerates training efficiency, and improves predictions for in vivo success rates and investigational new drug approvals (Cai et al. 2022).

2.2 Query Strategy

In active learning, query strategies or the acquisition functions, are crucial in selecting the most informative samples from an unlabelled dataset that significantly impacts model performance. Then, those selected samples are labelled, and the ML model is trained on them. Identifying the most informative samples effectively reduces the data required to train a model for better accuracy. By prioritizing data points that have the greatest potential to improve the model, these approaches speed up convergence, enabling the model to reach a high level of performance much faster than if it were trained on randomly selected data. The chosen samples usually denote the edge cases in which the model was poorly performed. These samples can also provide insights into the model's limitations. The selected samples by query strategies can highlight the model's limitations as they usually contain edge cases or critical data points. Query strategies help reduce inherent biases by ensuring that the chosen samples represent the wide spectrum of data distribution, that the model does not overfit a specific dataset, and that it improves its ability to generalize to unseen data. In applications where data distributions evolve (e.g. in real-time sensor systems or adaptive diagnostics), query strategies enable models to adapt dynamically by identifying new and relevant data points as they emerge. The choice of query strategy directly impacts the machine learning model's performance. A well-chosen query strategy converges faster, gives better accuracy, and reduces the labelling efforts.

In complex domains like drug discovery, where experiments and screening large chemical libraries are involved, labelling can become very expensive, so by using the query strategies, samples which have to be labelled can be reduced by a significant amount, which in turn lowers the project costs and also reduces the biases of the model towards a specific subset of data. Complex tasks like drug discovery involve multiple objectives like ADMET properties, bioactivity, and binding affinity; query strategies can select the samples that maximize the above objectives. The main categories of query strategies in active learning are uncertainty sampling, query by committee, and expected error reduction (Active Learning in Machine Learning 2024).

2.3 Uncertainty Sampling

Uncertainty sampling is one of the most used query strategies in active learning. The key idea is to select the data points where the model is most uncertain about its predictions. These are often the most informative samples, as they help the model learn more effectively. There are several methods to measure uncertainty (Fig. 2): Least Confidence is the difference between the most confident prediction and 100% confidence. The lower the confidence, the higher the uncertainty. The margin of confidence is the difference between the highest and second-highest probability. A smaller margin indicates more significant uncertainty. The ratio of confidence is the ratio of the second-highest probability to the highest probability. The higher the entropy, the higher the uncertainty. Entropy sampling is the difference between all predictions, as defined by information theory.

Nevertheless, there are various challenges with uncertainty sampling as well. It may favour noisy samples and outliers, which are not very informative. It can also select redundant samples with the same uncertainty (Monarch 2021).

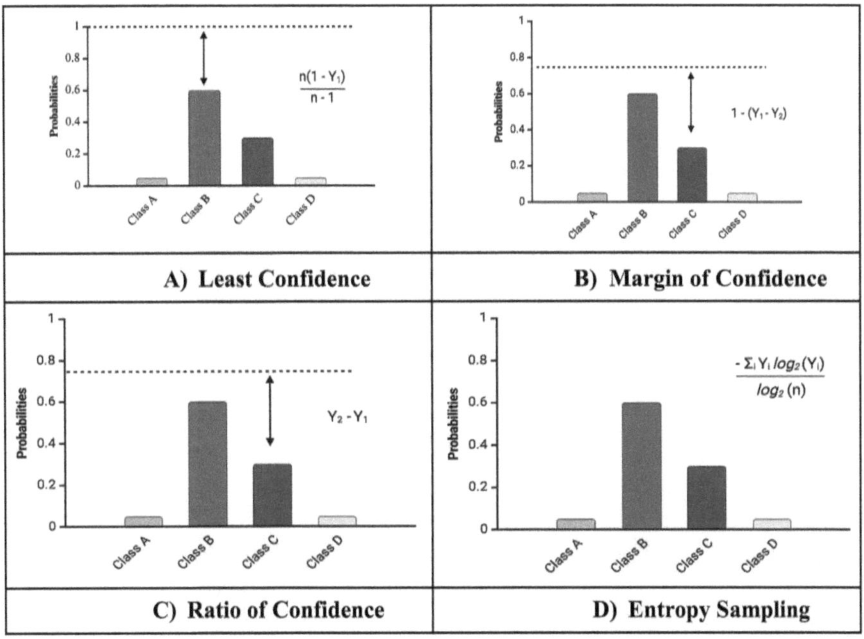

Fig. 2 Methods to measure uncertainty in uncertainty sampling

3 Diversity-Based Sampling

A diversity-based query strategy ensures that the selected data points for labelling represent a broad spectrum of the underlying data distribution. It involves measuring how distinct each candidate sample is from those already labelled. In drug discovery, compounds are selected in such a way that they cover a broad chemical space, preventing redundancy and bias toward specific molecular structures. Unlike uncertainty sampling, which focuses on model uncertainty, diversity-based approaches prioritize structural variation, enhancing the generalizability of predictive models. Common strategies include cluster-based selection, where compounds are grouped, representatives from different clusters are chosen, and maximum distance selection, which picks chemically distinct molecules based on Tanimoto similarity or Euclidean distance.

4 Query by Committee

It is another popular query strategy in active learning. This strategy requires a committee of different models to be trained on the same dataset. Then, these models vote on the samples of the unlabelled dataset, and the samples that contain the most disagreement among the committees are selected for labelling. Disagreement can be measured through Vote Entropy, calculated by applying uncertainty sampling to the distribution of model votes. Multiple models in Query by Committee methods help to capture different aspects of data distribution and reduce the bias, which can be raised from one single model. Still, this technique is computationally expensive due to the inclusion of multiple models. In addition, this method can also suffer from the problem of including redundant samples.

5 Expected Error Reduction

The intuition behind this method is that an unlabelled sample x is useful if it reduces the error on the rest of the unlabelled dataset. To find if sample x reduces the expected error, sample x is labelled using a trained model. Next, we retrain the model on this sample x, and then the model makes predictions for all other unlabelled samples in the dataset. The expected error is the difference in performance (error) before and after adding the sample x. Expected error is calculated for all the samples of the unlabelled dataset; the sample with the highest expected reduction in error is chosen to be labelled next. This method is computationally costly as it generates predictions on unlabelled datasets multiple times (Richter 2024).

Query strategies in active learning should be balanced between exploitative and explorative sampling to achieve optimal model performance. Each approach serves a unique purpose; it ensures the model's accuracy and robustness when balanced correctly.

Exploitative Strategies

The focus of exploitative strategies is on selecting data points that the model is uncertain about but are likely to provide immediate improvements to performance in critical areas. These strategies refine the model's accuracy by targeting the most impactful samples, such as those lying near decision boundaries or where the model's predictions have high uncertainty.

For example, if suppose the goal is to predict molecules with high binding affinity to a protein target, an exploitative strategy might prioritize selecting molecules that already show strong binding activity in preliminary predictions or if the goal is to optimize ADMET properties, an exploitative strategy could target compounds with favourable ADMET profiles. BRADSHAW exemplifies this by iteratively selecting compounds with high potential, significantly increasing the probability of successful drug design (Green et al. 2020).

Explorative Strategies

The focus of explorative strategies is on selecting data points that are diverse and underrepresented. This technique ensures that the model does not overfit and performs well on unseen data. In drug discovery, explorative sampling could prioritize molecules with novel chemical structures or rare functional groups, which helps the model gain information about wider chemical space and enhances its ability to predict bioactivity across structurally diverse compounds. Neither exploitative nor explorative strategies alone can fully address the challenges of active learning. Using only an exploitative strategy may lead to a highly accurate model in certain areas but fails to generalize beyond its training data, while using only an explorative strategy may end up selecting samples which contribute very little to the performance of the model.

6 Machine Learning Algorithm

The machine learning algorithm processes the initial training set and generates predictions on unlabelled data points. Training of this ML is performed initially using an initial training set, then with the data points selected using Query strategy. The ML model aims to increase its accuracy after every training iteration. ML algorithms used in active learning should perform well with small datasets and be capable of measuring confidence in their predictions, which is used in query strategy. Below are the types of machine learning models used in active learning.

Random Forest (RF) is an ensemble machine learning method that builds multiple decision trees and combines their predictions to improve accuracy. It is particularly popular in drug discovery because it easily handles large, complex datasets. One of its strengths is its ability to manage data with lots of variables, even if some of them are noisy or irrelevant (Feinberg et al. 2020; Montanari et al. 2019). For example, BRADSHAW is a molecular design system that employs RF and XGBoost for iterative compound design. It effectively filters drug candidates through computational screening, reducing development time (Green et al. 2020).

Support Vector Machines (SVM) are widely used for classification problems, and they are incredibly effective when the data have clear patterns. SVM works by finding a boundary (or hyperplane) that separates different categories, like active vs inactive compounds. In active learning, SVMs are helpful because they identify the "borderline" cases (the samples that are hardest to classify). These cases are then prioritized for labelling, significantly improving model performance over time (Wang et al. 2024).

Deep Learning algorithms, specifically neural networks, have transformed drug discovery to a great extent by efficiently detecting fine patterns in huge datasets. Deep learning methods differ from other models in the sense that they can learn complex, non-linear associations between molecular characteristics, resulting in better predictions. For example, multi-layer perceptrons (MLPs) have been utilized to predict drug–drug interactions efficiently using molecular structure data.

One of the strongest points of **Deep Learning** in active learning methods is that it can increase performance incrementally as it is provided with more data. In a process of incremental refinement in its understanding of chemical properties and interactions, the model shows a higher ability to pick drug candidates of high potential (https://www.linkedin.com/pulse/active-learning-its-applications-drug-discovery-boltzmannlabs-kb6wc). The ChemSpaceAL framework exemplifies this approach by integrating a generative pre-trained transformer (GPT) model with active learning to explore vast chemical spaces. This method has significantly improved molecular similarity and binding affinity predictions, particularly for target proteins such as c-Abl kinase and the Cas9 HNH domain, highlighting its potential for accelerating drug discovery (Kyro et al. 2024).

Naive Bayes is a powerful yet uncomplicated machine learning technique. Naive Bayes uses probabilistic techniques and carries the assumption that the features learned in the data are independent of each other. For instance, scientists use Naive Bayes to filter vast libraries of compounds in a computationally effective manner to identify candidates that are worth pursuing. Being an essential tool in speeding up the initial phases of drug discovery, Naive Bayes renders scientists more productive by having them work on leads that are worthwhile (https://www.linkedin.com/pulse/active-learning-its-applications-drug-discovery-boltzmannlabs-kb6wc).

Gaussian Processes are a desirable choice for active learning because they can simultaneously make prediction and uncertainty assessment. This is especially useful for drug discovery, where the degree of confidence in a model could have a major impact on experimental decisions. If, for instance, the model is uncertain whether a particular compound is active, that compound could be worth investigating further. Gaussian processes have feedback loops built into them, and thus point to where further data acquisition is required. By focusing on examples of higher uncertainty, scientists could potentially maximize the efficiency of active learning (Murphy 2011).

Metrics Related to Model Performance
The performance of a model can be evaluated using various metrics that assess its ability to classify data correctly are given below:

- Accuracy: (True Positives + True Negatives)/Total Predictions.
- Precision: True Positives/(True Positives + False Positives).
- Recall (Sensitivity): True Positives/(True Positives + False Negatives).
- F1-Score: 2 * (Precision * Recall)/(Precision + Recall).
- AUC (Area Under the ROC Curve): A metric that measures the model's ability to distinguish between positive and negative classes, considering all possible classification thresholds.

Metrics Related to Data Reduction
Includes data utilization rate, which is the percentage of total available data actually used in training while achieving near-optimal performance; labelling efficiency, which is the ratio of labelled data points required by the active learning algorithm

to achieve a certain level of performance compared to a random sampling strategy and query complexity which refers to the minimum number of labelled instances required to achieve a certain level of model performance.

7 Evaluation Indicators to Determine Stopping Criteria

One of the critical challenges in AL is determining when to terminate the iterative process. Over-labelling wastes resources, while early stopping may leave the model undertrained. Several evaluation indicators have been proposed to define stopping criteria. Wang et al. (2014) introduced a stability-based stopping criterion, which halts the active learning process when the change in model parameters falls below a predefined threshold. This ensures that if new data points do not significantly impact the model, further labelling is unnecessary. Another approach is the performance-based stopping criterion, which stops active learning once the desired performance is reached. However, this method requires an additional test set to evaluate the model, increasing labelling costs. The gradient-based stopping criterion suggests halting the process when additional data no longer provide meaningful information, such as when the model reaches peak performance or further reduction in uncertainty is not possible. Finally, the confidence-based stopping criterion relies on the classifier's confidence in its predictions; when the model confidently classifies most of the remaining unlabelled data, active learning is stopped. These criteria offer different perspectives on optimizing the AL process, balancing efficiency with model effectiveness (Wang et al. 2014).

For example, the moment tensor model (MTM) for molecular property prediction uses AL with a D-optimality sampling approach to achieve high accuracy with minimal data. When evaluated on the QM9 and QM7 datasets, it reached chemical accuracy with just 3500 samples, demonstrating significant efficiency improvements over conventional methods.

8 Deep Active Learning Relevance to Drug Discovery

Deep Active Learning (DeepAL) is a hybrid technique that combines active learning techniques and deep learning models to optimize the use of labelled data. It is mainly employed when obtaining labelled data is time-consuming and expensive, such as cancer detection, anomaly identification, text classification, language translation, predicting molecular properties, toxicity screening, etc. It significantly reduces the need for large datasets, a drawback in the case of traditional ML models, by strategically selecting the data points that are the most informative datasets, hence making it highly effective in low-data scenarios. It improves machine learning models by incorporating two key components: Feature Extraction and Sample Selection. Deep learning models (e.g., CNNs, Transformers, GNNs) extract complex molecular patterns from drug representations (e.g., SMILES, molecular graphs). These models learn high-dimensional feature representations, improving prediction accuracy. In Sample Selection, instead of focusing on uncertain or diverse samples, AL reduces labelling requirements by selecting the most informative samples while maximizing model performance.

A recent study by Michael Bailey et al. introduced a novel DeepAL pipeline for drug discovery, featuring two new batch selection methods, i.e. COVDROP (Coverage Dropout Method) and COVLAP (Coverage Laplace Approximation Method). COVDROP uses Monte Carlo Dropout (MC Dropout) to estimate predictive uncertainty. The model selects data samples where the model's predictions vary significantly across multiple forward passes, reducing the number of required labelled samples while improving prediction accuracy. COVLAP employs Laplace Approximation to quantify epistemic uncertainty (i.e., uncertainty due to lack of knowledge). It prioritizes samples that help reduce model uncertainty the most. The advantage of this approach is that it efficiently refines the model by choosing samples that contribute the most to learning. Both methods ensure that only the most informative data points are labelled and used for model training, minimizing redundancy and maximizing information gain.

The study demonstrated significant improvements in drug discovery tasks using COVDROP and COVLAP. COVDROP lowered RMSE (Root Mean Squared Error) across multiple datasets, improving accuracy faster. For instance, solubility and binding affinity datasets required only 450 compounds for compelling cell permeability predictions and 36 compounds to reach near-optimal RMSE for hydration-free energy predictions. This signifies that COVDROP significantly reduced the number of experiments needed to achieve optimal results. This study concludes that it significantly helps reduce redundant experiments.

Evaluation of specific protein targets, such as GSK3β and MMP-8, revealed that COVDROP consistently outperformed existing methods, achieving performance thresholds with fewer experiments. The comparative analysis further confirmed that COVDROP and COVLAP outperformed other batch selection methods like random selection, k-means sampling, BAIT, and human-based experimental ordering strategies, with statistical significance ($p < 0.05$), validating their robustness. The study highlights that these methods improve predictive performance and expand the applicability domain of predictive models, saving both time and resources in the drug discovery process.

Tilborg et al. further validated the significance of DeepAL in drug discovery by systematically evaluating various active learning strategies and deep learning architectures on large molecular libraries. Their study specifically addresses the challenges faced in low-data drug discovery scenarios, where the lack of extensive training data can hinder model performance. The authors comprehensively compare how these approaches perform in diverse chemical spaces by exploring six active learning strategies combined with two deep learning architectures. Their findings underscore that active learning, particularly when combined with deep learning models, offers a substantial advantage over traditional non-iterative methods, with DeepAL achieving up to a sixfold improvement in hit discovery efficiency. This result is particularly significant as it demonstrates how iterative, data-driven learning can vastly outperform static screening processes in identifying promising drug candidates (Van Tilborg and Grisoni 2024).

9 Applications of Active Learning in Drug Discovery

Active learning revolutionizes drug discovery by optimizing data efficiency through selective labelling of the most informative samples. This approach enhances predictive models in applications described below.

10 Preclinical Drug Development

In a recent study, an AL-based pipeline was developed to predict drug oral plasma exposure, showcasing its power in real-world applications. The model demonstrated impressive accuracy (85.6%) on an independent test set by utilizing only 30% of the training data, highlighting AL's ability to maximize data utility. Moreover, the model explored a vast chemical space of 855,000 samples, identifying 50,000 highly confident predictions. However, it is essential to note that these 50,000 highly confident predictions were not validated through in vitro or in vivo experiments within the scope of the study. They represent computational predictions made by the AL-based model. This significantly expands its applicability domain (Ding et al. 2021). Hence, AL strategically selects the most informative and fewer data points, making it particularly sound in drug discovery, leading to reduced experimental costs and optimal usage of resources.

11 Drug-Likeness Prediction

AL can be used to predict drug-likeliness, integrate AL with graph neural networks (GNNs), and enhance ensemble learning. GNNs process molecular structures like graphs, capturing complex atomic interactions. Unlike the traditional descriptors like fingerprints, GNNs models better spatial and electronic properties. This improves predictions for key drug-likeness factors such as bioavailability, binding potential, and solubility. Ensemble learning can be employed for robustness, as it combines multiple models abating individual biases and enhances predictive power. It critically ensures stability and reliability in drug-likeness assessments, even for challenging compounds. This framework is significantly helpful to pharmaceutical researchers in accelerating early-stage drug discovery and optimizing compound selection while reducing experimental costs (Cai et al. 2022).

12 Drug Response Prediction

Drug Response prediction is usually done initially on cell lines. Since it is an expensive and time-consuming process to check so many drugs, AL can be significantly helpful in selecting the most informative experiments for generating treatment response; instead of randomly selecting drugs for testing, AL selects the most informative experimental data points for developing the treatment response. Various AL approaches, including random, greedy, uncertainty, diversity, and hybrid strategies, were evaluated for 57 drugs to identify effective treatments and improve model performance. According to the results, AL significantly outperformed random and greedy sampling methods in identifying responsive cell lines and improving the accuracy of drug response predictions (Vasanthakumari et al. 2024).

13 Molecular Generation

The chemical space is vast (~10^{60} possible molecules), making randomly searching for promising drug candidates impractical. AL solves this problem by guiding the search process toward the most informative regions likely to have desired properties. For instance, in the case of c-Abl Kinase, a well-studied target with known inhibitors, the AL model learned to generate molecules similar to FDA-approved inhibitors without previous knowledge of these drugs. Moreover, it even reproduced two FDA-approved molecules exactly, demonstrating its ability to navigate chemical space efficiently. Another unexplored target with no known inhibitors, i.e. HNH Domain of Cas9, AL generated novel targets for this, showcasing AL's potential to discover drugs for targets with little prior data. Hence, integrating AL improves computational efficiency and enhances the likelihood of discovering high-value molecules. The ChemSpaceAL package's accessibility further democratizes this process, encouraging replication and broader use (Kyro et al. 2024).

14 Ligand-Binding Affinity Prediction

By allowing researchers to create small yet incredibly predictive models using only 10–25% of massive bioactivity datasets, AL in chemo genomics is revolutionizing drug development. In drug discovery, predicting how strongly a ligand can bind to a target is essential. Traditional models like deep learning and QSAR (Quantitative Structure-Activity Relationship) require more extensive labelled data for good prediction accuracy. Thereby, AL overcomes this limitation. Yuriy Khalak et al., in 2022, showed that the prediction of free energy calculations (like FEP—Free Energy Perturbation) provides highly accurate estimates of ligand-binding affinities but is computationally expensive. By using AL, one can prioritize only the use of the most relevant compounds, reducing the number of costly calculations while still efficiently identifying high-affinity binders.

Furthermore, Yuriy Khalak and colleagues (2022) showed that AL frameworks can be used with all-chemical free energy computations (Khalak et al. 2022). By prioritizing promising candidates, these iterative cycles reduce the size of vast compound libraries to find true positives using fewer experimental probes. This method demonstrates how AL-driven techniques expedite the search for potent inhibitors, including phosphodiesterase 2 (PDE2) binders while cutting expenses and time (Colliandre and Muller 2024).

15 Enhancing Screening Efficiency

Screening efficacy in drug discovery can be enhanced by employing AL with advanced matrix completion techniques. AL speeds up virtual screening using a linear regression model to predict which compounds from an extensive library will most likely bind to a target biomolecule. This method can retrieve up to 90% of the top 1% of hits after docking just 10% of the ligands, making it a highly efficient approach (Murphy 2011). Additionally, the impute-by-committee method improves matrix completion by accurately predicting missing values in compound-target interaction matrices, particularly in high-throughput screening. This method outperforms traditional imputation techniques by reducing the required experiments and increasing prediction accuracy. Furthermore, an adaptive switching strategy for active learning selects the most appropriate imputation methods, enhancing performance and ensuring more comprehensive predictions of compound effects. Together, these innovations significantly reduce computational costs and time while improving the accuracy and efficiency of virtual screening, enabling faster identification of high-affinity compounds and a better understanding of potential side effects (Cai et al. 2022).

16 Synergistic Drug Combination Screening

Drug synergy, where two drugs combined have a more significant effect than either drug alone, is especially valuable in treatments for complex diseases like cancer, infectious diseases, and neurological disorders. However, screening drug combinations experimentally is even more resource-intensive than single-drug screening because of the combinatorial explosion of possible pairs. Active learning can optimize this process by identifying the most promising drug pairs for experimental testing. In 2024, Wang et al. focused on synergistic drug combination screening, which aimed to find the combinations of drugs that, when used together, produce a more substantial therapeutic effect. This research shows that active learning can discover 60% of synergistic drug pairs while only exploring 10% of the combinatorial space (Wang et al. 2024).

17 ADMET Profiling

ADMET comprises key pharmacokinetic and safety-related properties that are critical/crucial in drug discovery. These factors determine how a drug is absorbed into the bloodstream, distributed across tissues, metabolized (mainly by the liver), excreted, and its potential toxicity. Predicting these properties accurately is crucial for reducing late-stage failures and optimizing lead compounds in the drug development pipeline. AL has emerged as a powerful approach for improving ADMET predictions by leveraging ML techniques to refine models and prioritize informative data iteratively. Several AL-based methods have shown significant potential in enhancing ADMET predictions: BERT-based ADMET Models integrating molecular graph features with descriptor-based features, offering state-of-the-art predictive performance. By capturing complex molecular relationships, they have achieved high accuracy on datasets like the Therapeutics Data Commons (TDC) (Yang et al. 2023). Multitask Deep Featurization utilizes Graph Convolutional Networks (GCNs) to achieve unprecedented accuracy. It allows the simultaneous prediction of multiple ADMET properties, enhancing efficiency by learning task-specific chemical features across various endpoints (Feinberg et al. 2020). Multitask GCN models are particularly effective in predicting physicochemical properties and can prioritize compounds with favourable ADMET profiles early in the process. These models streamline drug evaluation by addressing multiple endpoints concurrently (Montanari et al. 2019). Novel batch selection methods (COVDROP and COVLAP) for active learning were applied to optimize various properties related to ADMET in drug discovery. These methods consistently showed faster convergence and superior performance, especially with small datasets, often achieving results with fewer samples than other methods (Venkatraman 2021).

18 Comprehensive Case Studies on Active Learning Applications

BRADSHAW is an advanced molecular design system that integrates cheminformatics, experimental design, and AL to streamline drug discovery. It features an easy-to-use interface for the straightforward integration of new algorithms to enable tasks like molecular ideation, predictive modelling, and multi-parameter optimization. Utilizing AL techniques like Random Forests and XGBoost, BRADSHAW supports iterative compound design effectively. In a case study involving adenosine A2A antagonists, the system generated 10,675 potential molecules via molecular transformation strategies, filtering them using Oral Drug protocols. Impressive as it was, it replicated 72 of 176 patented compounds, demonstrating that it can generate novel as well as known successful structures.

Further optimization of MMP-12 inhibitors employed a balanced block design, selecting 100 compounds from 2500 possibilities, showcasing BRADSHAW's efficiency in lead optimization. BRADSHAW reduces computational time while fostering innovation in medicinal chemistry by enabling rapid generation, prediction, and evaluation of chemical datasets. Its modular architecture supports diverse cheminformatics tasks, proving invaluable for complex drug design challenges. This study highlights BRADSHAW's practical application in expediting drug discovery processes, from ideation to optimization, thereby enhancing the capabilities of researchers in medicinal chemistry (Green et al. 2020).

The miDruglikeness model is a significant development for predicting drug candidates as likely to pass in vivo testing, reach investigational new drug status, and achieve market approval. It employs subdivisional models designed for the specific task with graph neural networks. A novel methodological framework combining AL and ensemble learning improves predictive performance. AL, in particular, pool-based uncertainty sampling, reduces reliance on large labelled datasets by providing a focus on the most uncertain samples for labelling, thus accelerating the training of robust models. Results include high accuracy: predictions about in vivo exceed 94%, and investigational new drug and market approval predictions are higher than 80%. The active ensemble learning strategy is always better than the conventional models, such as XGBoost, in terms of accuracy, Matthews correlation coefficient, and F1-score. This highly integrated tool significantly enhances virtual screening efficiency, bettering traditional drug-likeness descriptors like the Rule of 5 and QED by offering strong predictability in workflows associated with drug discovery. It is also amenable to users via an easily accessible web server; input of compound data readily allows predictions for its usability in practice. The study outlines miDruglikeness as a critical innovation in drug discovery, providing enhanced precision and efficiency to improve clinical success rates and accelerate therapeutic development (Cai et al. 2022).

BE-DTI is an ensemble framework tailored for drug–target interaction (DTI) prediction, addressing key challenges like class imbalance and high-dimensional data through advanced machine learning techniques. The framework integrates a bagging-based ensemble learning strategy with AL and dimensionality reduction, employing decision trees as base learners and neighbourhood-balanced bagging to prioritize minority classes. BE-DTI works in three phases: Learning, where the bootstrap samples train initial models; Updation, which increases accuracy using AL to adapt instance weights dynamically; and Prediction, which uses majority voting across classifiers for robust output. Validated through ten-fold cross-validation, BE-DTI outperformed five existing methods with an AUC of 0.927, sensitivity of 0.886, specificity of 0.864, and G-mean of 0.874. Its ability to predict novel drug–target interactions makes it a potential tool in drug repositioning and discovery. BE-DTI also improves the accuracy of predictions through AL while streamlining the search for effective therapies, which makes it a very valuable tool in addressing complex biomedical challenges. Such novel methodology demonstrates what machine learning brings to pharmaceutical research: transforming efficiency in searching for therapeutic drug-target pairs towards accelerated development of effective targeted drug therapy (Sharma and Rani 2018).

ChemSpaceAL is an AL framework specifically designed for protein-specific molecular generation in drug discovery. It uses a generative pre-trained transformer (GPT) model to efficiently explore vast chemical spaces with reduced molecule evaluations. The methodology involves pretraining on large SMILES datasets, generating diverse molecules, and scoring them based on protein binding affinity. AL optimally chooses data points to update the model while trading off between accuracy and high chemical diversity. ChemSpaceAL showed a very impressive ability in case studies with c-Abl kinase, where it was able to increase the Tanimoto similarity scores from 38.8% up to 91.6% over five different AL rounds. For the Cas9 HNH domain, molecules with high scores increased from 21.3% to 52.1%. This flexible framework emphasizes the synergistic transformative nature of generative AI in combination with selective data evaluation and offers a powerful Python package that allows its inclusion into computational drug discovery pipelines (Kyro et al. 2024) .

The paper introduces an advanced machine learning (ML) algorithm for molecular property prediction, specifically addressing challenges like limited training data and outlier predictions. Using the localized interatomic interaction paradigm, the model achieves astounding predictive accuracy using far fewer training samples than other methods. Most importantly, a systematic sampling approach of training data based on a D-optimality criterion for an AL strategy has been incorporated to encourage diversity in the molecular geometries and compositions, thereby avoiding predictive errors in the outliers. Relative to benchmark collections such as QM9 and QM7, the moment tensor model (MTM) introduced here achieves unparalleled efficiency by reaching chemical accuracy using only 3500 training samples and thus outperforms the data requirements of all other methods reported to date. For instance, the MTM24 model reached a mean absolute error (MAE) of about 1 kcal/mol early in its training process. Such an improvement reduces computational costs considerably, making it highly valuable for computationally expensive applications such as materials and drug discovery (Gubaev et al. 2018).

19 Challenges and Limitations in Active Learning

AL is elemental in improving ADMET predictions by selecting instructive data points. But several limitations and issues hamper its extensive usage in the drug discovery process, like the need for high-quality and well-annotated data. Due to the scarcity of labelled data, ADMET experiments are expensive and time-consuming, restricting high AL-model accuracy. Moreover, variability in the experimental results is another factor that prevents its usage which introduces bias and noise in datasets. AL models depend on data already collected, which could be skewed towards a particular chemical. The need for high-quality, varied, and well-annotated datasets is one of the biggest obstacles to AL-based ADMET modelling (Venkatraman 2021; Tian et al. 2022; Padi et al. 2023). No AL-based model has shown complete accuracy across all ADMET endpoints due to several problems. Additionally, inter-individual variability in human metabolism due to genetic, physiological, and environmental factors makes it difficult for AL models to capture individual responses, and many models have trouble generalizing across diverse chemical scaffolds, which results in poor predictions for novel compounds. Furthermore, because adverse effects are multifaceted and entail intricate interactions at the molecular and systemic levels, predicting toxicity is extremely difficult (Yang et al. 2023).

The use of AL in biological experiments is hampered by experimental noise and stringent screening procedures. Variability in experimental parameters, such as temperature, pH, and reagent quality, makes compound annotation more difficult and lowers model reliability. Furthermore, automated screening workflows and assays are typically not flexible enough to accommodate the iterative nature of AL because they are made for standardized operations. This rigidity hinders the refinement of the model and smooth data integration. Additionally, the AL cycle is slowed considerably by the fact that experimental data frequently comes in various formats that necessitate extensive preprocessing before computer analysis.

Another challenge is that high-value compounds may be prohibitively expensive or difficult to synthesize, limiting the diversity of training data (Claus and Underwood 2002). Moreover, many AL strategies involve batch-wise acquisition of data, which can lead to redundant experiments and inefficiencies in compound selection (Murphy 2011; Schneider and Schüller 2010). Training and updating AL models can be computationally expensive, particularly for deep learning-based frameworks (Schneider and Schüller 2010).

Drug discovery involves multiple competing objectives like efficacy, toxicity, and bioavailability, making AL-based selection strategies challenging: AL strategies must balance multiple ADMET endpoints simultaneously, which requires sophisticated decision-making algorithms and selecting compounds for maximum information gain while ensuring chemical diversity is a difficult balance in AL frameworks.

Several developments, including better data-gathering techniques, are required to address these issues, enhancing training datasets with hybrid in vitro and in silico methods. Transfer learning and domain adaptation are incorporated to improve model generalization. Additionally, standardizing experimental data formats and annotation procedures and creating more adaptable high-throughput screening methodologies. Furthermore, using uncertainty quantification techniques like Bayesian optimization to enhance decision-making and using cost-aware learning to prioritize inexpensive and synthesizable molecules can be beneficial.

Furthermore, sophisticated model architectures can help with the aforementioned issues, such as integrating multi-modal learning techniques to integrate structural, bioactivity, and toxicology data and adopting self-supervised learning and generative models to alleviate the lack of labelled data. By addressing these challenges,

AL models can become more reliable and efficient, ultimately accelerating drug discovery and reducing the costs associated with experimental testing. These issues can be resolved to make AL models more dependable and effective, speeding up drug development and lowering the expense of experimental testing.

By prioritizing the most instructive data points for model training, AL is crucial in improving ADMET predictions. Several issues and restrictions hamper its extensive usage in drug development. The need for high-quality, varied, and well-annotated datasets is one of the biggest obstacles to AL-based ADMET modelling. Due to the scarcity of labelled data, ADMET experiments are costly and time-consuming, restricting the amount of high-quality labelled data available for model training. Variability in experimental outcomes is another problematic aspect. Inter-laboratory variability and inconsistent measurement methods are common problems with ADMET tests, which can introduce bias and noise into datasets.

20 Conclusion

AL has significantly transformed drug discovery by enhancing the efficiency and accuracy of predictive modelling across various applications, including ADMET profiling, ligand-binding affinity prediction, and the identification of bioactive molecules. Unlike traditional ML and high-throughput screening (HTS) methods that depend on large labelled datasets and extensive experimental validation, AL selectively chooses the most informative experiments. This targeted approach reduces the time, cost, and computational burden associated with drug discovery (Willett 2000).

In contrast to passive learning, which trains models on randomly selected data, AL utilizes query strategies like uncertainty sampling and expected error reduction to iteratively refine predictions (Han et al. 2023). This focused selection minimizes redundancy, accelerates exploration of chemical space, and enhances model convergence, making AL particularly beneficial in scenarios where data are limited or expensive to acquire (Van Tilborg and Grisoni 2024). Moreover, AL outperforms traditional docking and physics-based simulations in prioritizing high-affinity compounds. While these conventional methods provide mechanistic insights, they are often computationally intensive and constrained by scoring function accuracy. AL-driven deep learning models, such as Deep Active Learning (DeepAL), improve hit discovery efficiency by up to sixfold compared to standard virtual screening techniques (Sadybekov and Katritch 2023).

AL also addresses a critical challenge in computational drug discovery: model generalization. Conventional ML models often exhibit biases due to uneven data distribution and overfitting to well-characterized chemical subspaces. By prioritizing uncertain and diverse samples, AL systematically mitigates these biases, ensuring robust predictive performance across novel chemical entities (Murphy 2011). This advantage is particularly evident in ADMET prediction, where AL-based methods like multitask deep featurization and graph convolutional networks

(GCNs) demonstrate superior accuracy in predicting pharmacokinetic and toxicity profiles early in the drug development pipeline (Paul et al. 2020).

Despite challenges such as experimental variability and the necessity for sophisticated model architectures, integrating AL with generative models, physics-based simulations, and hybrid deep learning frameworks continues to advance drug discovery. These approaches enable researchers to systematically explore chemical space, optimize lead selection, and expedite the development of safer and more effective therapeutics. Thus, AL represents a transformative methodology that offers a data-efficient, cost-effective, and highly adaptable alternative to traditional ML and HTS-driven strategies in modern drug discovery (Wang et al. 2024; Awad Alanazi et al. 2022).

Author Contribution All the authors have made significant and equal contributions in compiling the manuscript's data, writing, and proofreading.

References

Active Learning in Machine Learning. Active learning scenarios and query strategies. In: GPTutor-Pro; 2024. https://gpttutorpro.com/active-learning-scenarios-and-query-strategies/. Accessed 28 Dec 2024.

Awad Alanazi AA, Al Fahad AI, Almorshed ASA, Alrbian AAM. Artificial intelligence in drug discovery: current applications and future directions. Int J Health Sci. 2022;6(S10):2011–40.

Bailey M, Moayedpour S, Li R, et al (2024), Deep batch active learning for drug discovery., https://elifesciences.org/reviewed-preprints/89679v2.

Brown N, Ertl P, Lewis R, et al. Artificial intelligence in chemistry and drug design. J Comput Aided Mol Des. 2020;34:709–15.

Cai C, Lin H, Wang H, et al. miDruglikeness: subdivisional drug-likeness prediction models using active ensemble learning strategies. Biomolecules. 2022;13:29.

Chen JJ, Schmucker LN, Visco DP. Pharmaceutical machine learning: virtual high-throughput screens identifying promising and economical small molecule inhibitors of complement factor C1s. Biomolecules. 2018;8:24.

Chen G, Shen Z, Iyer A, et al. Machine-learning-assisted De novo Design of Organic Molecules and Polymers: opportunities and challenges. Polymers. 2020;12:163.

Claus BL, Underwood DJ. Discovery informatics: its evolving role in drug discovery. Drug Discov Today. 2002;7:957–66.

Cohn DA, Ghahramani Z, Jordan MI. Active learning with statistical models. J Artif Intell Res. 1996;4:129–45.

Colliandre L, Muller C. Bayesian optimization in drug discovery. In: Heifetz A, editor. High performance computing for drug discovery and biomedicine. New York: Springer; 2024. p. 101–36.

Dearden JC, Cronin MTD, Kaiser KLE. How not to develop a quantitative structure–activity or structure–property relationship (QSAR/QSPR). SAR QSAR Environ Res. 2009;20:241–66.

Di P, Zheng M, Yang T, et al. Prediction of serious eye damage or eye irritation potential of compounds via consensus labelling models and active learning models based on uncertainty strategies. Food Chem Toxicol. 2022;169:113420.

Ding X, Cui R, Yu J, et al. Active learning for drug design: a case study on the plasma exposure of orally administered drugs. J Med Chem. 2021;64:16838–53.

Feinberg EN, Joshi E, Pande VS, et al. Improvement in ADMET prediction with multitask deep Featurization. J Med Chem. 2020;63:8835–48.

Filella-Merce I, Molina A, Orzechowski M, et al (2023), Optimizing drug design by merging generative AI with active learning frameworks., https://arxiv.org/abs/2305.06334.
Giangrande M, Kim YW, Mizukami H. N-terminal spin label studies of hemoglobin, ligand and pH dependence. Biochim Biophys Acta. 1975;412:187–93.
Graff DE, Shakhnovich EI, Coley CW. Accelerating high-throughput virtual screening through molecular pool-based active learning. Chem Sci. 2021;12:7866–81.
Green DVS, Pickett S, Luscombe C, et al. BRADSHAW: a system for automated molecular design. J Comput Aided Mol Des. 2020;34:747–65.
Gubaev K, Podryabinkin EV, Shapeev AV. Machine learning of molecular properties: locality and active learning. J Chem Phys. 2018;148:241727.
Han R, Yoon H, Kim G, Lee H, Lee Y. Revolutionizing medicinal chemistry: the application of artificial intelligence (AI) in early drug discovery. Pharmaceuticals (Basel). 2023;16(9):1259.
Hughes J, Rees S, Kalindjian S, et al. Principles of early drug discovery. Br J Pharmacol. 2011;162:1239–49.
Iovanac NC, MacKnight R, Savoie BM. Actively searching: inverse Design of Novel Molecules with simultaneously optimized properties. J Phys Chem A. 2022;126:333–40.
Janzen WP, Popa-Burke IG. Advances in improving the quality and flexibility of compound management. SLAS Discov. 2009;14:444–51.
Kangas JD, Naik AW, Murphy RF. Efficient discovery of responses of proteins to compounds using active learning. BMC Bioinfo. 2014;15:143.
Khalak Y, Tresadern G, Hahn DF, de Groot BL, Gapsys V. Chemical space exploration with active learning and alchemical free energies. J Chem Theory Comput. 2022;18:6259–70.
Kyro GW, Morgunov A, Brent RI, et al. ChemSpaceAL: an efficient active learning methodology applied to protein-specific molecular generation. J Chem Inf Model. 2024;64:653–65.
Loeffler HH, Wan S, Klähn M, et al. Optimal molecular design: generative active learning combining REINVENT with precise binding free energy ranking simulations. J Chem Theory Comput. acs.jctc.4c00576. 2024;20:8308.
MacKay DJC. Information-based objective functions for active data selection. Neural Comput. 1992;4:590–604.
Marin E, Kovaleva M, Kadukova M, et al. Regression-based active learning for accessible acceleration of ultra-large library docking. J Chem Inf Model. 2024;64:2612–23.
Monarch R. Uncertainty sampling cheatsheet – towards data science. In: Medium; 2021. Accessed 28 Dec 2024.
Montanari F, Kuhnke L, Ter Laak A, et al. Modeling Physico-chemical ADMET endpoints with multitask graph convolutional networks. Molecules. 2019;25:44.
Murphy RF. An active role for machine learning in drug development. Nat Chem Biol. 2011;7:327–30.
Najm M, Azencott C-A, Playe B, et al. Drug target identification with machine learning: how to choose negative examples. Int J Mol Sci. 2021;22:5118.
Padi S, Cardone A, and Sriram RD (2023), A meta-model for ADMET property prediction analysis. http://biorxiv.org/lookup/doi/10.1101/2023.12.05.570279
Paul P (2024) ActiveLearning-DrugDiscovery. GitHub repository. Available at: https://github.com/meprateekpaul/ActiveLearning-DrugDiscovery. Accessed 31 Dec 2024.
Paul D, Sanap G, Shenoy S, Kalyane D, Kalia K, Tekade RK. Artificial intelligence in drug discovery and development. Drug Discov Today. 2020;26(1):80–93.
Reker D, Schneider P, Schneider G, et al. Active learning for computational Chemogenomics. Future Med Chem. 2017;9:381–402.
Richter M. Intro to active learning. In: Inovex GmbH; 2024. https://www.inovex.de/de/blog/intro-to-active-learning/. Accessed 28 Dec 2024.
Sadybekov AV, Katritch V. Computational approaches streamlining drug discovery. Nature. 2023;616:673–85.
Schneider G, Schüller A. Adaptive combinatorial Design of Focused Compound Libraries. In: Roque ACA, editor. Ligand-macromolecular interactions in drug discovery. Totowa: Humana Press; 2010. p. 135–47.

Settles B. Active learning. Cham: Springer; 2012.
Sharma A, Rani R. BE-DTI': ensemble framework for drug target interaction prediction using dimensionality reduction and active learning. Comput Methods Prog Biomed. 2018;165:151–62.
Singh N, Vayer P, Tanwar S, et al. Drug discovery and development: introduction to the general public and patient groups. Front Drug Discov. 2023;3:1201419.
Sun H, Murphy RF. Evaluation of categorical matrix completion algorithms: toward improved active learning for drug discovery. Bioinformatics. 2021;37:3538–45.
Swanson K, Walther P, Leitz J, et al. ADMET-AI: a machine learning ADMET platform for evaluation of large-scale chemical libraries. Bioinformatics. 2024;40:btae416.
Temerinac-Ott M, Naik AW, Murphy RF. Deciding when to stop: efficient experimentation to learn to predict drug-target interactions. BMC Bioinfo. 2015;16:213.
Tian H, Ketkar R, Tao P. ADMETboost: a web server for accurate ADMET prediction. J Mol Model. 2022;28:408.
Van Tilborg D, Grisoni F. Traversing chemical space with active deep learning for low-data drug discovery. Nat Comput Sci. 2024;4:786–96.
Vasanthakumari P, Zhu Y, Brettin T, et al. A comprehensive investigation of active learning strategies for conducting anti-cancer drug screening. Cancer. 2024;16:530.
Venkatraman V. FP-ADMET: a compendium of fingerprint-based ADMET prediction models. J Cheminfo. 2021;13:75.
Wang W, Cai W, Zhang Y. Stability-based stopping criterion for active learning. IEEE; 2014. p. 1019–24.
Wang S, Allauzen A, Nghe P, Opuu V. A guide for active learning in synergistic drug discovery. bioRxiv. 2024:2024.09.13.612819.
Warmuth MK, Liao J, Rätsch G, et al. Active learning with support vector Machines in the Drug Discovery Process. J Chem Inf Comput Sci. 2003;43:667–73.
Willett P. Chemoinformatics—similarity and diversity in chemical libraries. Curr Opin Biotechnol. 2000;11:85–8.
Yang L, Jin C, Yang G, et al. Transformer-based deep learning method for optimizing ADMET properties of lead compounds. Phys Chem Chem Phys. 2023;25:2377–85.
Zhang Y, Lee AA. Bayesian semi-supervised learning for uncertainty-calibrated prediction of molecular properties and active learning. Chem Sci. 2019;10:8154–63.
Zhang B-T, Veenker G. Neural networks that teach themselves through genetic discovery of novel examples. In: In: [proceedings] 1991 IEEE international joint conference on neural networks, vol. 1. Singapore: IEEE; 1991. p. 690–5.

Large Language Models in Drug Discovery

Amit Gangwal and Antonio Lavecchia

Abstract Large language models (LLMs) like generative pre-trained Transformers and bidirectional encoder representations from Transformers have transformed natural language processing (NLP) and are increasingly applied in drug discovery. These models, trained on vast datasets, excel at text generation, comprehension, and pattern recognition, making them ideal for analyzing biomedical data, predicting drug interactions, and identifying new drug candidates. LLMs can synthesize information from multiple sources, speeding up hypothesis generation and streamlining drug development processes. They help overcome data scarcity issues by generating synthetic data, which enhances model training and prediction accuracy. In drug discovery, LLMs assist with molecule screening, target identification, and clinical trial optimization, reducing time and costs associated with traditional methods. Despite their potential, challenges remain, including data quality, interpretability, and the need for domain-specific adaptations. LLMs are also being explored for their ability to predict pharmacokinetics, toxicity, and drug–drug interactions (DDI). Future advancements will focus on integrating LLMs with other AI techniques, like Reinforcement Learning (RL) and generative models, to further enhance their capabilities in drug discovery. Collaboration between academia, industry, and regulatory bodies will be crucial for overcoming challenges and realizing the full potential of LLMs in delivering new, effective therapies.

Keywords Large language models · Natural language processing · Drug discovery · Generative AI · Transformer models · Biomedical data analysis · Machine learning · Deep learning

---

A. Gangwal
School of Pharmacy & Technology Management, SVKM NMIMS Global University, Dhule, Maharashtra, India

A. Lavecchia (✉)
"Drug Discovery" Laboratory, Department of Pharmacy, University of Naples Federico II, Naples, Italy
e-mail: antonio.lavecchia@unina.it

A. Lavecchia (ed.), *Applied Artificial Intelligence for Drug Discovery*,
https://doi.org/10.1007/978-3-031-98022-0_14

1 Introduction

Large Language Models (LLMs) have become a pivotal topic in the rise of generative artificial intelligence (AI), capturing public interest while serving as a cornerstone for organizations integrating AI into diverse business functions and applications. Beyond enterprise contexts, LLMs may appear to have surfaced abruptly alongside recent advancements in generative AI (Gangwal and Lavecchia 2024; Lavecchia 2024a; Romanelli et al. 2024). However, companies like IBM have invested years in deploying LLMs across various levels to enhance natural language understanding and natural language processing (NLP) capabilities (Boiko et al. 2023a). This evolution parallels advances in machine learning (ML) models, algorithms, neural networks, and transformer architectures, the foundational technologies driving these AI systems.

LLMs represent a distinct category of foundational models trained on massive datasets, enabling broad applicability across multiple scenarios and tasks. This approach contrasts sharply with the traditional practice of developing domain-specific models for individual use cases, a method often constrained by high costs, complex infrastructure, and suboptimal outcomes. By circumventing these limitations, LLMs provide a cost-effective and versatile solution, unlocking synergies that enhance performance. These models have advanced NLP and AI substantially and are now widely accessible to the public through tools such as OpenAI's ChatGPT-3 (Floridi and Chiriatti 2020) and GPT-4 (Achiam et al. 2023), Meta's public LLM meta-AI (LLaMA) (Touvron et al. 2023) models, and Google's bidirectional encoder representations from Transformers (BERT) (Devlin et al. 2018) and Pathways Language Model (PaLM) (Chowdhery et al. 2023). Similarly, IBM's Granite model series, part of watsonx.ai, underpins offerings like Watsonx Assistant and Watsonx Orchestrate (Oniani et al. 2024; Sejnowski 2023).

LLMs are designed to interpret and generate text in a human-like manner while extending their capabilities to other content types. Leveraging extensive training datasets, they excel in deriving meaning from context, producing coherent and relevant responses, translating across languages, summarizing text, answering questions, supporting creative writing, and even generating code. Their success stems from a vast number of parameters that enable the recognition of intricate patterns in language, facilitating diverse and complex language-related tasks. As a result, LLMs are transforming applications in various domains, such as chatbots, virtual assistants, content creation, research assistance, and language translation. As LLMs evolve, they continue to reshape our interactions with technology and redefine how we access information, positioning themselves as integral components of the modern digital landscape (He et al. 2024). Table 1 provides a detailed overview of the applications and capabilities of LLMs.

LLMs operate using deep learning (DL) methodologies alongside extensive collections of textual information. These models are typically built on a Transformer architecture, such as GPT, known for its remarkable ability to handle sequential data like text. At the core of LLMs are layered neural networks with numerous adjustable

Table 1 Major types of LLMs

Type of LLM	Architecture	Description	Accessibility	References
Generative pre-trained transformer (GPT)	Decoder-only	Pre-trained on extensive text data to generate coherent, context-aware text. Applications include translation, summarization, and chatbots (e.g., GPT-3, GPT-4)	Closed	Achiam et al. (2023), Floridi and Chiriatti (2020)
BERT	Encoder-only	Captures context bidirectionally (left-to-right and right-to-left), excelling in tasks like question answering and sentence classification (e.g., BERT, RoBERTa)	Open	Devlin et al. (2018)
T5 (text-to-text transfer transformer)	Encoder-decoder	Frames all NLP tasks as text-to-text, supporting translation, summarization, and question answering within a unified format (e.g., T5)	Open	Raffel et al. (2020)
XLNet	Decoder-only	Merges autoregressive modeling with BERT-style architecture, improving prediction for tasks involving long-term dependencies	Open	Yang et al. (2019)
BLOOM	Decoder-only	A multilingual, open-access model supporting 40+ languages for text generation, translation, and question answering with inclusivity at its core	Open	Workshop et al. (2022)
Megatron-Turing NLG	Transformer decoder	A Microsoft model specializing in fluent, coherent text generation for dialogue and content creation	Closed	Smith et al. (2022)
LaMDA (language model for dialogue applications)	Decoder-only	Google's conversational AI model designed for natural, context-aware open-ended dialogues resembling human interaction	Closed	Thoppilan et al. (2022)
OPT (open pre-trained transformer)	Decoder-only	Meta AI'S scalable alternative to GPT, optimized for text generation and machine translation tasks	Open	Zhang et al. (2022)
Flan-T5 (fine-tuned language net-T5)	Encoder-decoder	Enhanced T5 variant pre-trained across diverse tasks for better generalization in NLP challenges like reasoning, summarization, and translation	Open	Chung et al. (2024)
Codex	Decoder-only	Specialized in generating and interpreting programming code, aiding software development tasks by converting natural language into code (e.g., OpenAI codex)	Open	Chen et al. (2021)
PaLM	Decoder-only	Google's model tailored for a variety of NLP tasks, achieving superior performance in text generation and comprehension benchmarks	Closed	Chowdhery et al. (2023)

parameters fine-tuned during training. A critical component of their architecture is the attention mechanism, which hones in on relevant data segments with remarkable accuracy, enhancing the model's focus on context (Lavecchia 2024b). During training, LLMs learn to predict the next word in a sentence by analyzing the context provided by preceding words. This is achieved through probabilistic scoring of tokenized words, breaking them down into smaller sequences for processing. Tokens are then transformed into embeddings, numerical representations of contextual information. Training these models requires extensive text corpora, often comprising billions of pages, which allows them to grasp grammar, semantics, and conceptual relations through methodologies such as zero-shot and self-supervised learning. Once trained, LLMs can independently generate coherent, contextually relevant text by predicting subsequent words based on patterns learned during training. This capability enables their use in various natural language understanding and content creation tasks (Zheng et al. 2024). To enhance model reliability, techniques such as Reinforcement Learning (RL) with human feedback (RLHF) help mitigate biases, hate speech, and inaccuracies (known as "hallucinations") that can arise from training on large volumes of unstructured data. Additional methods, including prompt-tuning, fine-tuning, and prompt engineering, are essential for making enterprise-grade LLMs compliant and safe for practical applications (Boiko et al. 2023a). Figure 1 illustrates the five key functional areas of LLMs, while Table 2 provides a comparison of human intelligence and LLMs in the context of drug discovery.

In drug discovery, LLMs are transforming research by leveraging advanced NLP and data analysis capabilities. Models like GPT-4 (Achiam et al. 2023) can process extensive biomedical information, such as scientific literature, clinical trial outcomes, and genetic data, to detect patterns and generate hypotheses. This accelerates drug target identification and therapeutic development by synthesizing data from diverse sources, providing researchers with broad, holistic insights often overlooked in traditional methods. Such insights facilitate the discovery of new drug candidates and allow for the prediction of their efficacy, ultimately streamlining the discovery process. Moreover, LLMs can simulate drug interactions within biological systems, minimizing the need for lengthy lab testing and reducing development timelines. Another major contribution of LLMs in drug discovery lies in addressing data scarcity, a frequent challenge in medical research. Where datasets are limited or incomplete, LLMs can produce synthetic data that closely resemble real-world patient information, enriching the available research data. This synthetic data can be used to train additional AI models, thereby boosting their accuracy and dependability. LLMs can also optimize clinical trial designs by predicting patient responses to new treatments, helping to refine protocols and reduce costs. By embedding LLMs in the drug discovery process, researchers can make well-informed decisions, advancing the development of safer, more effective treatments. This integration of advanced AI models into drug discovery signifies a transformative step toward pioneering new and innovative therapies. Table 3 provides definitions of key technical terms referenced throughout the discussion.

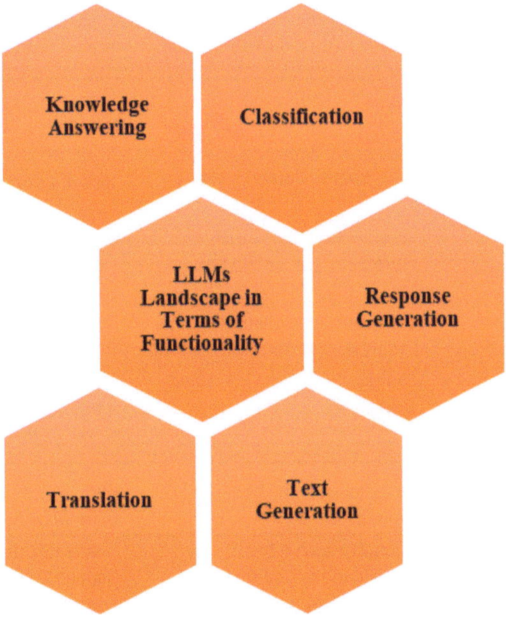

Fig. 1 LLMs demonstrate exceptional capabilities across five core areas: *Classification*, which assigns text to predefined categories; *Clustering*, which organizes similar texts into groups without predefined labels; *Response Generation*, which predicts contextually appropriate dialogue responses based on conversation history; *Text Generation*, which creates coherent text from brief prompts through techniques like few-shot learning and prompt engineering; *Translation*, which facilitates direct language conversion without intermediary steps; and *Knowledge Answering*, which leverages knowledge-intensive NLP to generate answers using an internal knowledge base, bypassing external searches

Pre-processing is essential for optimizing LLMs like GPT, BERT, and T5, ensuring smooth training and robust performance. A fundamental step in this process is tokenization, which breaks down text into smaller units (words, subwords, or characters) using a model-specific tokenizer, such as Byte-Pair Encoding (BPE) for GPT, WordPiece for BERT, or SentencePiece for multilingual models like T5. After tokenization, input sequences are standardized in length through padding and truncation: shorter sequences are extended by adding special padding tokens, whereas longer sequences are shortened (truncated) to ensure they do not exceed the model's maximum token limit. Special tokens are also added to guide the model during processing; for instance, BERT requires a [CLS] token at the beginning and [SEP] at the end of each sequence, while GPT uses <|endoftext| > to mark text boundaries. Additionally, attention masks are generated to differentiate real tokens from padding, ensuring that padding is excluded from computation. When input length exceeds model capacity, texts are divided into coherent chunks to preserve context. Post-inference, detokenization converts the model's output back to readable text. Depending on the training objective, data formatting may vary: continuous text

Table 2 Comparison of human intelligence and LLMs in drug discovery

Aspect	Human intelligence	LLMs
Data processing	Limited by cognitive capacity and time constraints	Can process vast amounts of data quickly and efficiently
Pattern recognition	Relies on experience and intuition to detect patterns	Uses advanced algorithms to identify complex patterns in large datasets
Creativity	Highly creative, capable of thinking outside the box	Generates novel hypotheses based on data, but lacks true creativity
Speed	Slower due to manual analysis and decision-making	Rapid analysis and decision-making
Accuracy	Subject to human error and biases	High accuracy in data analysis, dependent on training data quality
Adaptability	Highly adaptable to new information and changing conditions	Can adapt through retraining, but requires significant computational resources
Ethical considerations	Can evaluate ethical implications and make value-based decisions	Lacks intrinsic ethical understanding; requires human oversight
Collaboration	Effective in collaborative environments, leveraging diverse expertise	Can assist in collaborations by providing data-driven insights and recommendations
Cost	High cost due to salaries, training, and potential human error	High initial development cost, but lower operational costs over time
Scalability	Limited scalability, dependent on skilled professionals	Highly scalable, deployable across multiple projects simultaneously
Interpretability	Transparent decision-making and reasoning	Often opaque decisions; requires explainable AI techniques
Learning and improvement	Continuous learning through experience and education	Continuous improvement through retraining and AI advancements

chunks are used for unsupervised tasks like masked language modeling (BERT) or autoregressive tasks (GPT), while fine-tuning tasks (e.g., question-answering or summarization) require precise input-output pairs. Optional subword regularization can introduce tokenization variability, enhancing robustness by accounting for alternative valid tokenizations. These pre-processing techniques collectively prepare LLMs for their designated tasks, maximizing their overall effectiveness (Bahani et al. 2023). Table 4 compares the architectures and key features of GPT and BERT.

Table 3 Key concepts in LLMs and AI

Concept	Definition	Purpose	Key characteristics	Examples/models
LLMs	Models trained on large datasets to understand and generate human language, typically using DL and transformers	General-purpose models for tasks like text generation, translation, and summarization	Pre-trained on vast corpora, often fine-tuned for specific tasks; scale-driven emergent properties	GPT-3, GPT-4, LLaMA, PaLM
NLP	A branch of AI focused on enabling machines to understand, interpret, and generate human language	Encompasses tasks such as sentiment analysis, machine translation, and more	Involves techniques like tokenization, parsing, and machine learning for language tasks	Text classification, machine translation, named entity recognition (NER)
BERT	A pre-trained transformer model that understands text context bidirectionally	Specialized for tasks like question answering and sentiment analysis	Uses masked language modeling and next-sentence prediction; fine-tuned for downstream tasks	BERT, RoBERTa, ALBERT
Transformers	A neural network architecture that processes sequences in parallel using self-attention mechanisms	Revolutionized DL for NLP, enabling efficient handling of long-range dependencies	Scalable architecture; processes data non-sequentially; basis for many NLP models	BERT, GPT, T5, and various LLMs
Foundational models	Large-scale models trained on broad datasets, adaptable for various downstream tasks	Serve as a foundation for tasks across different domains through transfer learning	Versatile, multimodal, fine-tuned for specific tasks across diverse fields	GPT-4, PaLM 2, DALL·E, CLIP

1.1 LLMs

The previously dominant method for sequence-to-sequence tasks was the recurrent neural network (RNN), which leverages a form of "memory" to retain information from earlier stages in a sequence for predicting subsequent elements (Hochreiter and Schmidhuber 1997; Rumelhart et al. 1986). However, as sequence lengths increase, RNNs suffer from vanishing or exploding gradients, impairing their ability to effectively process long-range dependencies (Ribeiro et al. 2020). Consequently,

Table 4 Comparison of GPT and BERT architectures and key features

Feature	GPT	BERT
Architecture	Decoder-only transformer	Encoder-only transformer
Training objective	Autoregressive (predicts the next word)	Masked language modeling (predicts masked words)
Contextual understanding	Unidirectional (left-to-right)	Bidirectional (both left-to-right and right-to-left)
Primary use cases	Text generation, language translation, summarization	Sentiment analysis, question answering, named entity recognition
Model size	Very large (e.g., GPT-3 has 175 billion parameters)	Comparatively smaller (e.g., BERT-large has 340 million parameters)
Fine-tuning	Few-shot learning with prompts	Fine-tuning with additional layers for specific tasks
Developed by	OpenAI	Google AI
Release year	GPT-3 in 2020 (latest)	BERT in 2018

RNNs have been largely supplanted by LLM, advanced deep neural networks with billions of parameters. These models are predominantly built on the Transformer architecture, introduced by Vaswani et al. in their seminal paper *"Attention is All You Need"* (Vaswani et al. 2017), with a detailed line-by-line walkthrough available in *"The Annotated Transformer"* (Rush 2018). Transformers have revolutionized sequence modeling, becoming the backbone of nearly all state-of-the-art outcomes across various domains, including chemistry.

The Transformer architecture was the first to rely entirely on attention mechanisms for sequence-to-sequence tasks, a significant shift from earlier RNN-based methods that incorporated attention only partially (Bahdanau 2014). The "attention" mechanism identifies and emphasizes critical words in the input, enabling enhanced decision-making and improved accuracy. Technically, attention is implemented as a dot product of token embeddings combined with a learned non-linear function. The Transformer comprises two main components: the encoder and the decoder, illustrated schematically in Fig. 2a. Initially, input text is tokenized into computable integers based on the model's vocabulary, which are then transformed into numerical vectors using embedding layers (Berglund and van der Merwe 2023; Kudo 2018; Kudo and Richardson 2018; Rust et al. 2020; Song et al. 2020). The encoder stack, consisting of multiple layers (six in the original model (Vaswani et al. 2017)), processes these embeddings. Positional encodings, typically sine and cosine functions with frequencies varying by word position, allow the model to handle sequences of any length, although alternative methods can also be employed (Gehring et al. 2017). Each encoder layer integrates a multi-headed self-attention mechanism that computes attention scores for every "word" in the sequence by processing queries, keys, and values. Outputs from this mechanism are further refined through normalization and residual connections, illustrated in Fig. 2a by "bypassing" arrows, which mitigate the vanishing gradient problem (Ribeiro et

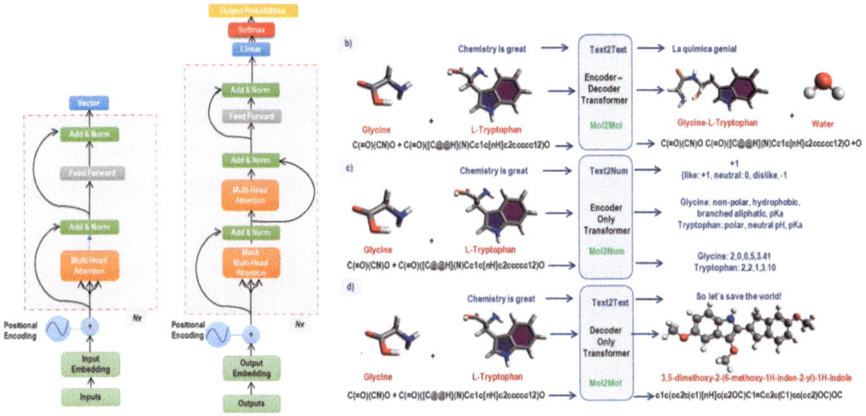

Fig. 2 Transformer architectures (Encoder-Decoder, Encoder-only, and Decoder-only) and their applications. (**a**) The encoder-decoder Transformer architecture, with an encoder (left) that encodes input into a vector and a decoder (right) that predicts the next token in sequence. (**b**) Encoder-decoder Transformers perform translation tasks, such as converting chemical reactants into products for reaction prediction. (**c**) Encoder-only Transformers generate vector outputs and are used in tasks like sentiment analysis; in chemistry, they support classification and property prediction. (**d**) Decoder-only Transformers generate likely next tokens in a sequence and are used in chemistry to create new compounds from molecular descriptions or specific instructions

al. 2020). Finally, the encoder output is passed through a pointwise feed-forward network, utilizing activation functions such as ReLU (Nair and Hinton 2010), SwiGLU (Shazeer 2020), or GELU (Hendrycks and Gimpel 2016), etc.), producing contextually rich vectors prepared for the decoder. These vectors encapsulate the input sequence's semantic and syntactic nuances, enabling sophisticated downstream tasks.

The decoder's workflow begins with the transformation of the initial token into numerical vectors, known as the output embedding step, since this input serves as the starting token for the model's final output. Following this, positional encodings are added, and the data pass through a series of six identical decoder layers. Each layer incorporates a Covered Self-Attention Mechanism, ensuring that positions do not attend to those that follow, along with an Encoder-Decoder Multi-Head Attention mechanism. It is essential that the outputs from the encoder align with those from the decoder's attention layer, as shown by the connecting square in Fig. 2a, allowing the model to focus on relevant sections of the input. A comprehensive Feed-Forward Neural Network then refines the output with greater precision. The final layer functions as a linear classifier, aligning the output with the vocabulary size, and a softmax layer transforms this output into probabilities, with the highest value indicating the predicted next word. Each sub-layer is followed by normalization and includes a residual connection. The decoder generates output sequences in an autoregressive manner, incorporating each newly predicted token until an end token signals the completion of the process. Figure 2b illustrates how the encoder-decoder model facilitates tasks like language translation or the repositioning of atoms in chemical reactions. Figure 2c demonstrates how an encoder-only model can extract additional meaning from an input sequence, represented as a numerical description, while Fig. 2d shows a simplified decoder-only structure that deduces and produces an outcome from a given input. For example, a novel peptide sequence is generated by integrating two smaller amino acids, showcasing the decoder-only model's capacity to generate new compounds.

LLMs are often tailored for specific scientific languages, such as the Simplified Molecular Input Line Entry System (SMILES) (Weininger 1988) for molecular structures and FASTA (Pearson and Lipman 1988) for biological sequences, whereas general-purpose LLMs encompass a broader range of textual data across multiple domains. In drug discovery, specialized LLMs play a pivotal role in retrosynthetic planning and predicting reaction outcomes, alongside supporting ADMET (Absorption, Distribution, Metabolism, Excretion, and Toxicity) predictions. On the other hand, general LLMs are applied in automated chemistry robotics, performing complex tasks like retrosynthetic planning. In disease research, specialized LLMs excel at extracting genomic data to identify genetic markers and networks, while general LLMs are effective at summarizing literature and generating knowledge graphs, often translating complex technical terms into accessible language. In clinical trials, while specialized LLMs are less commonly used, general LLMs are invaluable in analyzing electronic health records, facilitating patient-trial matching, and aiding trial planning. Specialized LLMs typically function as tool-like applications, where users input specific data to obtain predictions, while general LLMs act as conversational AI assistants, enabling users to ask questions and receive detailed insights. Specialized models are limited to specific training sets and languages, whereas general LLMs are increasingly being adapted to handle tasks traditionally reserved for specialized models, offering promising avenues for further research and development.

Encoder-only models, such as BERT (Devlin et al. 2018), focus on bidirectional context analysis, utilizing tasks like Masked Language Modeling (MLM) and Next Sentence Prediction (NSP) for pretraining. BERT's architecture allows for deeper understanding of input sequences, with models like RoBERTa (Liu et al. 2019) further enhancing BERT by using larger datasets, removing NSP, and refining the MLM task. In contrast, decoder-only models like GPT (Achiam et al. 2023; Floridi and Chiriatti 2020) are unidirectional, predicting the next word in a sequence based solely on preceding words, making them causal language models. This design enables them to generate sequences with a broad understanding of language. Encoder-decoder models, like BART (Lewis et al. 2019), combine the bidirectional strengths of BERT's encoder with the autoregressive decoding of GPT, excelling in sequence-to-sequence tasks by corrupting and reassembling text during pretraining. BART showcases an effective fusion of both encoder and decoder capabilities. Additionally, multi-task and multi-modal models, such as T5 (Raffel et al. 2020), approach various NLP tasks as text-to-text problems, utilizing a unified architecture with shared weights. Models like Flan-T5 (Chung et al. 2024) demonstrate the ability to generalize across tasks through instruction fine-tuning.

Furthermore, multi-modal models, like NExT-GPT (Wu et al. 2023b), enhance language models by integrating multiple input types (text, audio, and images) using specialized encoders and decoders tailored to each modality, producing coherent outputs across diverse inputs.

1.1.1 LLMs in Chemistry

ML has had a transformative impact across many scientific fields, especially chemistry. A key challenge in this domain is working with small datasets, which has led to the development of advanced ML techniques infused with specialized chemical insights, requiring substantial expertise for effective deployment. In this context, GPT-3 (Floridi and Chiriatti 2020), a powerful language model trained on vast internet text, demonstrates remarkable flexibility in tackling chemistry-related challenges. When refined for conversational chemical queries, GPT-3 can yield results that rival or surpass traditional ML approaches, particularly in data-scarce situations. Additionally, GPT-3 enables inverse design by seamlessly reversing questions, offering a potent approach to uncover new insights. Its ease of use and high performance with limited data make it a promising tool for advancing ML applications in chemistry. Beyond predictive tasks, pre-trained language models like GPT-3 also enhance literature search efficiency and help lay foundations for new research initiatives by leveraging their extensive knowledge bases (Jablonka et al. 2024).

In medicinal chemistry, a field essential to drug discovery and development, automation plays an increasing role, particularly in synthesizing novel compounds and conducting autonomous lab research. Modern approaches rely heavily on self-directed laboratory setups, where robotic manipulators execute intricate synthetic reactions. High-throughput screening demands precise and efficient compound synthesis, a key step in hit identification. Once synthesized, compounds undergo pharmacological assays to evaluate activity and selectivity. LLMs are proving invaluable in these workflows by aiding in the generation of programming code for chemical robotics. LLMs translate user specifications into experimental protocols, which are then transformed into executable instructions for robotic systems. They further assist with retrosynthetic planning and reaction prediction, accelerating molecule synthesis and drug discovery timelines (Zheng et al. 2024).

In drug discovery, specialized textual languages like SMILES (Weininger 1988) for molecular structures and FASTA (Pearson and Lipman 1988) for protein, DNA, and RNA sequences encode complex scientific information, reflecting human-developed systems for domain-specific knowledge. To navigate these languages, two main types of language models have emerged: specialized and general-purpose models. The initial framework in drug discovery relies on specialized LLMs trained in scientific languages, enabling these models to decode the statistical patterns inherent in technical texts. This allows for a deeper understanding of scientific data at their most basic level (Achiam et al. 2023; Taylor et al. 2022). Below are some examples of LLM applications in the medicinal chemistry domain:

- *Chemistry Robotics*: This field centers on the autonomous execution of experiments, where natural language instructions are translated into executable plans for robotic systems. LLMs like GPT-4 (Achiam et al. 2023) and CodeLlama (Roziere et al. 2023) have demonstrated proficiency in generating precise code to operate chemistry robots. A key example is the CLARify system (Yoshikawa et al. 2023), which uses GPT-3 (Floridi and Chiriatti 2020) to convert natural language instructions into task plans using the specialized Chemistry Description Language (XDL). These plans, optimized through PDDLStream solvers, enable safe and autonomous experiment execution with general-purpose robotic manipulators, outperforming baseline systems like SynthReader (Mehr et al. 2020). Recent studies have also shown GPT-4's capability to generate Python scripts for controlling OT-2, a liquid-handling robot, with a 95% success rate after five iterations (Inagaki et al. 2023). Additionally, emerging AI research envisions LLMs as autonomous agents that design, execute, and program experiments. Boiko et al. (2023b) showcased this potential by demonstrating LLMs using web searches for molecule synthesis information or vector searches to retrieve chemical reaction data. They also developed agents capable of coordinating multi-instrument systems, successfully performing complex experiments such as Suzuki and Sonogashira cross-coupling reactions.
- *Retrosynthetic Planning and Reaction Prediction*: Retrosynthetic planning focuses on deconstructing complex molecules into simpler precursors, while reaction prediction anticipates the outcomes of chemical reactions. Together, these tasks are fundamental for designing synthetic pathways and enabling experiments like high-throughput screening. LLMs have made significant strides in these areas. The Molecular Transformer (Schwaller et al. 2019) pioneered LLM-based retrosynthesis, using an encoder-decoder model to predict products from reactants with higher accuracy than human chemists. This model was later enhanced with hyper-graph exploration (Schwaller et al. 2020), representing molecules as nodes and reactions as hyper-arcs, where beam search identified optimal synthetic routes. Building on this foundation, Chemformer (Irwin et al. 2022), an adaptation of the BART framework (Lewis et al. 2019), introduced masked SMILES reconstruction and autoencoding to refine embeddings. Fine-tuned for retrosynthesis and reaction prediction, it achieved exceptional precision. Recently, general-purpose LLMs like Chemcrow (Bran et al. 2023) and Boiko's system (Boiko et al. 2023b) have expanded capabilities with tools for SMILES conversion, molecule pricing, patent checks, and reaction classification. Chemcrow's structured framework surpassed GPT-4 in synthesis planning, while fine-tuned GPT-3 models excelled in low-data settings, demonstrating their flexibility and transformative impact in chemical research (Jablonka et al. 2024).
- De Novo *Molecular Generation:* De novo molecule generation creates novel drug-like structures, either exploring chemical space of the training set (unconstrained) or meeting specific criteria like target affinity, off-target selectivity, physicochemical properties, synthesizability, ADME, pharmacokinetics, and toxicity (constrained). LLMs like SMILES-LSTM (Segler et al. 2018) and ORGAN (Guimaraes et al. 2018) have shown strong performance in uncon-

strained generation. MolGPT (Bagal et al. 2022), built on GPT architecture, can handle multiple constraints by recovering molecules with scaffolds and properties. General-purpose LLMs like MolT5 (Edwards et al. 2022) focus on constrained generation, using self-supervised learning on molecular data. GPT-4 (Achiam et al. 2023) has demonstrated the ability to generate molecules from text, though its performance lags behind specialized models. Multimodal approaches, such as Momu (Su et al. 2022) and GIT-Mol (Liu et al. 2024a), enhance LLM performance. Momu improves MolT5 by aligning molecular graphs with text via CLIP (Radford et al. 2021), while GIT-Mol integrates graph, image, and text data using cross-attention, boosting MolT5's molecule generation capabilities.

- De Novo *Protein Generation*: Generating new proteins can occur without restrictions or under specific constraints, such as targeting certain cellular compartments or functions. Unconstrained methods explore vast protein spaces using autoregressive LLMs, such as ProtGPT2, which produces natural-like globular sequences validated by AlphaFold2 (Ferruz et al. 2022). Constrained approaches focus on specific functions, leveraging models like ProGen (Madani et al. 2023) and PoET (Watson et al. 2023) for family-specific sequences or binders tailored to molecular targets through inverse folding. Designing protein binders is particularly challenging, requiring deep knowledge of protein folding and structural dynamics. Innovations include RFDiffusion (Watson et al. 2023), leveraging diffusion modeling for precise folding, and its extension RFdiffusion All-Atom (RFdiffusionAA) (Krishna et al. 2024), which incorporates atomic detail for designing proteins binding to small molecules or metals. ESM models (Rives et al. 2021) further validate functional proteins, achieving a 67% success rate even for sequences divergent from known proteins, with ESM 3 improving design precision (Hayes et al. 2024) (Nijkamp et al. 2022). General-purpose LLMs like ProteinDT integrate text-protein training to achieve 90% accuracy in text-guided protein design and optimization, expanding LLM utility in therapeutic development (Liu et al. 2023a).
- *Protein-Ligand Interaction Prediction.* Deciphering drug (ligand) interactions with protein targets is vital for drug discovery. Virtual screening via molecular docking and ML expedite the early identification of potential therapeutics. Specialized LLMs play pivotal roles in this process: (i) as direct predictive tools, exemplified by AlphaFold-Multimer for protein docking (Evans et al. 2021), and (ii) as integral components of advanced systems, such as combining protein LLMs with molecular fingerprints to identify binders with sub-nanomolar affinities (Singh et al. 2023). Innovative models like PSICHIC (Koh et al. 2024), trained exclusively on sequence data, surpass approaches relying on 3D structural information by accurately predicting binding residues and ligand atoms without explicit annotations. This showcases the cost-effective potential of LLMs in decoding intricate protein-ligand interactions. Meanwhile, general-purpose models like Galactica (Taylor et al. 2022) combine extensive scientific knowledge with docking score predictions, highlighting their promise in advancing molecular-level understanding and prediction of protein-ligand dynamics.

1.1.2 Main Paradigms of LLMs in Chemistry

This section explores the latest pretraining techniques for LLMs, focusing on the differences between single-modal and multi-modal objectives. It provides a comprehensive review of three key pretraining goals relevant to chemical LLMs: Masked Language Modeling (MLM), Molecule Property Prediction (MPP), and Autoregressive Token Generation (ATG).

MLM is a widely used technique in which certain tokens in input sequences are randomly hidden, and models are trained to predict these missing tokens based on the surrounding context. Chemical LLMs operate on molecular representations like SMILES or SELFIES (Self-Referencing Embedded Strings) when applied to single-domain models and on wrapped sentences in multi-domain or multi-modal setups. Single-domain models such as SMILES-BERT (Wang et al. 2019), MG-BERT (Zhang et al. 2021), ChemBERTa (Chithrananda et al. 2020), T5 Chem (Lu and Zhang 2022), Chemformer (Irwin et al. 2022), Molformer (Wu et al. 2023a), SELFormer (Yüksel et al. 2023), and BARTSmiles (Chilingaryan et al. 2024) use MLM on molecular sequences, with MG-BERT uniquely incorporating graph adjacency to refine attention to neighboring atoms. In multi-domain and multi-modal contexts, MultiTask Text+Chem T5 (Christofidellis et al. 2023), BioT5 (Pei et al. 2023), and KV-PLM (Zeng et al. 2022) apply MLM to wrapped sentences. Models like UniMap (Feng et al. 2024), MEMO (Zhu et al. 2022), and DMP (Zhu et al. 2023) further utilize MLM for text encoder pretraining, illustrating its adaptability.

The MPP objective involves pretraining chemical LLMs to predict molecular properties from sequential representations. This approach leverages cheminformatics tools to derive properties such as molecular weight, rotatable bonds, polar surface area, and SP^3-hybridized carbons, without requiring manual labeling. Studies suggest that MPP enhances model performance, as seen with models like ChemBERTa-2 (Ahmad et al. 2022) and SPT (Winter et al. 2022). However, MPP-based models are primarily designed for molecular representation learning, not for token generation.

ATG involves predicting the next token in a sequence based on the preceding tokens. This approach is central to chemical LLMs, as it directly impacts their ability to adapt and perform subsequent tasks. One clear implementation of ATG is Molecule Completion, where models are trained to complete molecular representations, as seen in models like DrugGPT (Li et al. 2023), ChemGPT (Frey et al. 2023), and MolXPT (Liu et al. 2023b). Additionally, Representation Translation focuses on generating alternative representations from specified sequences, exemplified by models such as Chemformer (Irwin et al. 2022) and X-MOL (Xue et al. 2020), which predict canonical SMILES from masked alternative representations.

In summary, these pretraining methodologies optimize the training and performance of chemical LLMs by tailoring objectives to the unique requirements of chemical data and tasks. Figure 3 provides a conceptual overview of these models.

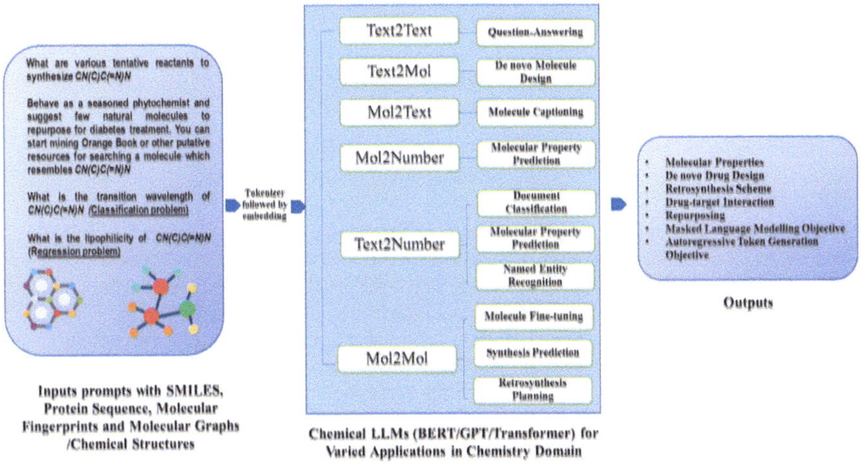

Fig. 3 Overview of LLM. This figure provides a conceptual overview of how chemical LLMs, such as BERT, GPT, and Transformer-based models, are utilized in the chemistry domain for various tasks. It highlights the different types of input data, including SMILES, protein sequences, molecular fingerprints, molecular graphs, and chemical structures, which are processed through tokenization and embedding. These processed inputs are then used to perform several tasks, such as converting text to text (Text2Text), text to molecules (Text2Mol), molecules to numbers (Mol2Number), text to numbers (Text2Number), and molecules to molecules (Mol2Mol). Each of these tasks is associated with specific applications in chemistry, including question-answering, de novo molecule design, molecular property prediction, molecular captioning, synthesis prediction, and retrosynthesis planning. The outputs generated by these models include molecular property predictions, drug-target interactions, retrosynthesis routes, repurposing suggestions, and autoregressive token generation. These diverse outputs demonstrate the broad range of applications for chemical LLMs in advancing chemistry research, particularly in areas like drug discovery and molecular design

1.1.3 Tokenization Methods of Molecules

Molecular sequence tokenization can generally be categorized into three primary approaches: (1) character-level, (2) atom-level, and (3) motif-level (Fig. 4). Character-level tokenization treats each individual character as a separate token, which may lead to incorrect segmentation of multi-character entities such as "Br." Despite this, the method has proven effective in chemical LLMs (Wang et al. 2019), demonstrating the models' remarkable ability to understand complex chemical representations. Atom-level tokenization offers a more logical approach by breaking sequences down into their fundamental components. Recent advancements in sequential representation have introduced tailored atom-level tokenizers (Heller et al. 2015) designed to handle unique characters like "[Ring1]" in SELFIES or "/h" in InChI, which are crucial for accurately representing intricate chemical structures. In contrast, SMILES lacks an inherent tokenizer, often relying on regex-based expressions for tokenization (Schwaller et al. 2018). Motif-level tokenization can be approached using either chemistry-driven or data-driven methods. Chemistry-

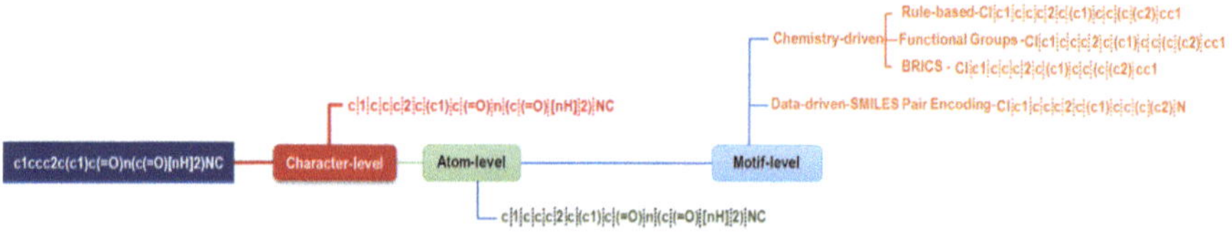

Fig. 4 Tokenization Levels in SMILES Encoding. This image illustrates three levels of SMILES encoding for the chemical structure "c1ccc2c(c1)c(=O)n(c(=O)[nH]2)NC-." The encoding is broken down into character-level, atom-level, and motif-level representations. At the character level, the molecule is represented by individual characters, including atoms and bond symbols, providing a basic textual depiction. The atom-level encoding further dissects the structure into specific atom details, focusing on each atom and its bonds. At the motif level, the structure is represented by functional groups or substructures using techniques such as rule-based methods, BRICS (Breaking of Retrosynthetically Interesting Chemical Substructures), and SMILES pair encoding, highlighting key chemical motifs or fragments. Each level captures different granularities of molecular information, supporting various computational chemistry tasks

driven techniques involve breaking down molecules into substructures of chemical significance, based on a deep understanding of chemical properties. These methods are particularly evident in graph-based studies (Jin et al. 2018). For example, Feng et al. (2024) performed motif-level tokenization on molecular representations, converting graph motifs into sequential tokens. While these methods produce chemically meaningful substructures, creating tailored rules requires specialized expertise. Moreover, transforming sequences into graphs and matching predefined patterns, such as functional groups or BRICS, can be computationally intensive.

LLMs are capable of processing a wide variety of input types, depending on their design and capabilities. The most common input is typed text, which can encompass natural language queries, technical questions, code snippets, equations, structured data (such as tables or lists), or entire documents, including articles, reports, and essays in formats like .txt, .docx, or .pdf. In addition to text, some models allow users to upload files, such as text documents, images, datasets in .csv or .json formats, or code files (.py) for analysis or debugging. Certain LLMs also support audio inputs, enabling users to interact via speech or upload audio files (.mp3 or .wav) for transcription or translation. For multimodal models, visual inputs are also supported, allowing users to upload images for analysis or description. These diverse input options enable LLMs to perform a wide range of tasks, from answering questions and generating text for analyzing data, processing images, and handling complex audio and visual requests.

In the context of drug discovery and development, LLMs can process inputs that are highly relevant to scientific research. Text inputs can include queries about drug mechanisms, target identification, or molecular interactions, as well as technical questions involving chemical formulas, reaction pathways, and pharmacokinetics or pharmacodynamics data. Researchers can input structured data such as gene target lists, molecular descriptors, protein–ligand interaction data, or high-throughput screening results in formats like .csv or .json. Attachments, including research papers in .pdf format, chemical structure files (.sdf, .mol), or clinical trial data in .csv, can be uploaded for deeper analysis or prediction tasks. Audio inputs, such as recorded discussions of experimental results or expert interviews, can be transcribed and analyzed to generate insights. For visual inputs, multimodal LLMs can process images of molecular structures, graphs, or biological pathways, and generate descriptions or interpretations. Additionally, LLMs can analyze molecular docking visualizations, protein folding images, or chemical reaction mechanisms for computational analysis or predictive modeling. These capabilities enable LLMs to assist with a variety of tasks in drug discovery, including hypothesis generation, molecule screening, drug repurposing, ADMET prediction, and more, throughout the research and development pipeline.

2 Conceptualizing LLMs in Drug Discovery

ChatGPT's rapid rise in popularity began with its launch in November 2022, when over one million users engaged with the AI model within just five days. Developed by OpenAI, ChatGPT represents a breakthrough in NLP, demonstrating an exceptional ability to generate human-like conversations by skillfully understanding context and providing relevant responses (Deng and Lin 2022). Building on this capability, researchers and practitioners have explored a wide range of medical applications for ChatGPT. Notably, in the field of drug development, AI tools like ChatGPT have shown great promise. A study published in *Nature Biotechnology* highlights how ChatGPT and other LLMs are aiding scientists in identifying novel drug targets, helping to accelerate the drug discovery process (Savage 2023). This underscores the tremendous potential of ChatGPT and LLMs across various stages of drug discovery.

Drug discovery and development is notoriously complex, multi-stage, and time-consuming, often beginning with the identification of therapeutic targets and ending with the regulatory approval of a drug candidate after extensive testing. Throughout this process, ChatGPT and LLMs play an increasingly valuable role, assisting in every phase of drug development.

Traditional drug discovery methods are slow, costly, and labor-intensive. It typically takes 15–20 years to find a viable drug candidate, with the pharmaceutical industry investing billions of dollars to bring a new drug to market (Chakraborty et al. 2014, 2010; Wouters et al. 2020). On average, the cost of developing a new drug from research to market is estimated at $2.7 billion. The challenges of high costs, inefficiency, and failure rates stem from the inability to accurately predict the efficacy and toxicity of drug candidates. This issue, coupled with the labor-intensive processes of synthesizing and testing multiple compounds, leads to low efficiency. The requirement for expensive infrastructure and skilled labor further drives up costs. Additionally, the lack of a universally accepted in vivo experimental platform capable of reliably mimicking human diseases is a key limitation in early drug development (May et al. 2022). AI-assisted ChatGPT plays a vital role across multiple phases of drug discovery, including compound multiplicity calculations, file conversion, generating input files for Gaussian software, exploring pdb files, and creating input files for docking. These contributions are crucial for the development of highly effective drug molecules (Sharma and Thakur 2023). LLMs have shown remarkable versatility, advancing through various stages from discovery to development. Their sophisticated capabilities have significantly enriched the field of drug discovery, leading to numerous beneficial outcomes. For example, LLMs excel at analyzing high-content screening data and at designing and synthesizing novel molecules. They have also enhanced our ability to predict biological activities and key ADME properties (absorption, distribution, metabolism, and excretion). As methods for molecular representation continue to evolve, the potential of LLMs for generating molecular structures opens new avenues for exploring unexplored territories in the biochemical realm (Walters and Barzilay 2021).

Understanding pharmacokinetics and pharmacodynamics is essential for successful drug development. LLMs assist in elucidating these principles for specific compounds, offering valuable insights throughout the drug discovery process. They also provide the ability to assess the toxicity profiles of drug-like molecules at various stages of development (Zhao and Wu 2023). AI has been seamlessly integrated into clinical trials, improving efficiency by shortening the duration of these trials. AI-driven tools like ChatGPT can assist in trial execution, leveraging their vast network to expedite the process. Another crucial aspect in drug discovery is understanding drug–drug interactions (DDIs), which help clarify how molecules interact with one another. Due to certain inaccuracies in ChatGPT's responses, researchers are now discussing the need for domain-specific iterations of the model. Next-generation LLMs tailored to specific domains, such as drug discovery, have the potential to provide more accurate, targeted information (Pal et al. 2024). Recent advancements have led to the creation of domain-specific models like DrugChat, which generates molecular graphs for drug discovery. DrugChat is comprised of three key components: an LLM, an adapter, and a graph neural network. Table 5 details some of the LLMs used in various stages of drug discovery.

Table 5 Leveraging the power of LLMs across drug development stages

LLM	Application	Refs
GPT-3	Genomics and transcriptomics analysis	Zheng et al. (2024)
BERT	Prediction of molecular properties and interactions	Blanchard et al. (2022)
RoBERTa	Optimization of lead compounds for improved efficacy and safety	Liu et al. (2019)
GPT-4	Assisting in patient recruitment and clinical trial design	Zheng et al. (2024)
MolecularGPT	Excels in few-shot prediction of molecular properties, outperforming baseline models	Liu et al. (2024b)
scGPT	Enhancing gene insights, cell-type annotation, and genetic predictions	Cui et al. (2024)

3 Advancements in LLMs for Drug Discovery

Recent advancements in LLMs have shown tremendous potential in reshaping various stages of the drug development pipeline. Key developments include:

Molecule Generation and Design

- *LLMs for Molecule Synthesis*: By learning patterns from extensive datasets of chemical structures and biological activity, LLMs can generate novel molecules with desired pharmacological properties. They assist in de novo drug design by proposing promising drug candidates.
- *Generative Models*: Transformer-based LLMs, such as MolGPT and ChemBERTa, are applied to generate new chemical structures by predicting the next atom or functional group in a molecule. These models are particularly focused on structure prediction and property optimization.

Target Identification

- *LLMs for Protein–Drug Interactions*: Trained on large biological datasets, including protein structures and binding affinities, LLMs can identify potential drug targets. By leveraging embeddings that capture both chemical and biological properties, they predict interactions between small molecules and complex biological targets, including novel proteins.

Drug Repurposing

- *LLMs for Predicting DDIs:* LLMs can analyze existing drug information to predict new therapeutic uses for approved drugs, accelerating the drug repurposing process. They evaluate molecular pathways and interactions to propose alternative applications for established medications.
- *Knowledge Graphs and LLMs*: When integrated with knowledge graphs, LLMs explore intricate biological networks, suggesting potential repurposing opportunities for drugs, particularly in the context of rare diseases or areas with limited datasets.

Text Mining and Literature Analysis

- *LLMs for Literature Review*: Models such as GPT-4 (Achiam et al. 2023) and BioBERT (Lee et al. 2020) are trained on vast biomedical and scientific literature, enabling rapid processing and summarization of research papers, clinical trials, and patents. This aids researchers in staying up-to-date with the latest findings and supports hypothesis generation.
- *Hypothesis Generation from Unstructured Data*: LLMs extract valuable insights from unstructured data, such as medical records and experimental data, facilitating the generation of new hypotheses related to drug mechanisms, side effects, and novel applications.

Multi-Omics Data Integration

- *LLMs for Integrating Genomic, Transcriptomic, and Proteomic Data*: LLMs are increasingly used to integrate large-scale multi-omics datasets, enhancing the understanding of disease mechanisms and aiding the discovery of novel drug targets.
- *Predictive Modeling with Multi-Modal Data*: Certain LLM architectures are designed to handle multiple data types (e.g., genomic sequences, chemical structures, clinical trial data), providing more accurate drug target predictions and biomarker identification.

AI-Driven Clinical Trials

- *Optimizing Clinical Trial Design*: LLMs optimize clinical trial design by analyzing historical trial data and predicting outcomes based on patient populations, dosages, and protocols, leading to more efficient trials and reduced failure rates.
- *Patient Stratification*: By analyzing clinical records and genomic data, LLMs identify patient subgroups most likely to respond to specific treatments, advancing precision medicine approaches.

Addressing Data Scarcity

- *LLMs for Low-Data Scenarios*: In the context of rare diseases or understudied conditions, data can be scarce. LLMs, especially those leveraging transfer learning, are valuable for orphan drug development by generalizing from limited datasets.
- *Few-Shot and Zero-Shot Learning*: LLMs are increasingly adept at few-shot and zero-shot learning, applying knowledge from similar domains to make accurate predictions, reducing the need for large labeled datasets.

Integration with Active Learning and RL

- *Active Learning*: When combined with active learning, LLMs interact with experimental data during drug development to prioritize compounds or pathways requiring further investigation.
- *RL for Drug Optimization*: LLMs integrated with RL optimize lead compounds, guiding the exploration of chemical space based on desired therapeutic outcomes.

Open-Source Models and Collaborations

- *Hugging Face Models for Chemistry and Biology*: Platforms like Hugging Face offer open-source LLMs tailored for drug discovery tasks, including protein structure prediction, molecular property prediction, and synthetic biology.
- *Collaborative AI Models*: Collaborations between large pharmaceutical companies and AI startups (e.g., NVIDIA's MegaMolBART or OpenAI's GPT) are accelerating drug discovery, creating accessible tools for the wider research community.

4 Applications of LLMs in Drug Discovery

In drug discovery, the application of LLMs for analyzing small molecules and proteins can be categorized primarily into two pre-trained model families: BERT and GPT. Both are rooted in the Transformer architecture, but they differ significantly in their training objectives and methods for processing input data. BERT, a sophisticated deep bidirectional model developed by Google, is designed to pre-train representations of language derived from unlabeled text (Devlin et al. 2018). It incorporates both left and right context in all layers to develop a deeper understanding of language. During pre-training, BERT employs a masked language modeling approach, where parts of the input tokens are masked, and the model is trained to predict the missing tokens based on their surrounding context. This technique has proven highly successful, with BERT achieving impressive results in various NLP tasks, such as named entity recognition and question answering. In drug discovery, the masking process is particularly valuable, as it enables the model to infer missing structural information from protein sequences (e.g., FASTA sequences). By masking portions of a sequence, BERT can predict the missing parts, facilitating the generation of new sequences based on these inferences. On the other hand, GPT, developed by OpenAI, is a unidirectional model that processes input data by considering only the context preceding a given word. It is trained on large text datasets for language modeling, where the model predicts the next word in a sequence based on the preceding words. In drug discovery, GPT models are often trained on vast chemical datasets, such as molecules represented in the SMILES format or large protein sequences in FASTA form. These models can quickly generate novel structures, which might be potentially useful for drug development. However, the challenge lies not in generating new structures, but in evaluating their usefulness, such as assessing toxicity, binding affinity, ease of synthesis, and drug-likeness. To address this, models like ProtGPT2 (Ferruz et al. 2022), a language model specifically designed for protein design, are trained on protein sequences. By adhering to the natural principles of protein structure, ProtGPT2 generates de novo protein sequences that align with natural amino acid propensities. Based on the GPT-2 Transformer architecture, ProtGPT2 consists of 36 layers with a model dimensionality of 1280, totaling 738 million parameters. Similarly, ProtBERT (Elnaggar et al. 2021) is a BERT-based model trained on a large corpus of protein sequences using a self-supervised approach. Pre-trained on Uniref100, a dataset of 217 million protein sequences, ProtBERT has gained an understanding of the fundamental principles of protein structure, where function is driven by structure. This makes it a valuable tool in the analysis and design of proteins for drug discovery. Figure 5 illustrates the various application areas of LLMs in the drug discovery domain.

Fine-tuning is arguably one of the most valuable applications of LLMs in drug discovery. These models have the potential to serve as the foundation for any AI-driven approach to molecular structure analysis. Just as pre-training on vast amounts of unlabeled text enables LLMs to grasp language at a deep level, training on large

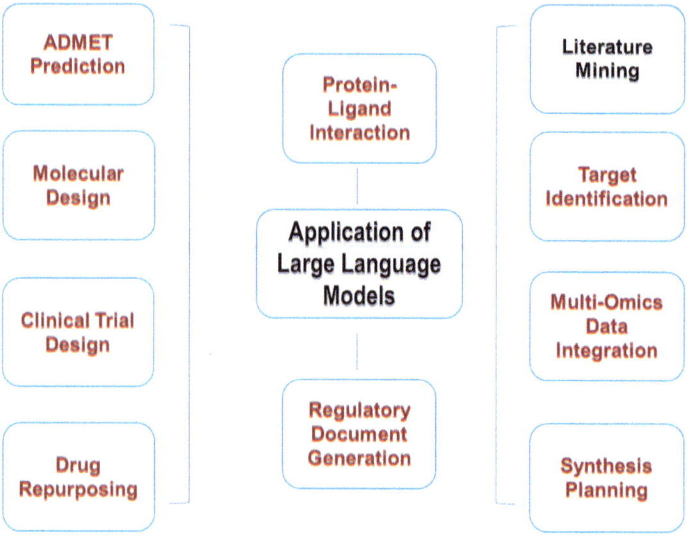

Fig. 5 The role of LLMs in various stages of drug discovery and development

datasets that cover extensive regions of molecular space allows these models to develop a profound understanding of that space. In practice, fine-tuning empowers models to specialize in solving specific tasks by leveraging smaller, labeled datasets.

Accenture Labs, for example, applies generative AI across various domains, including health and life sciences. A notable application is the use of LLMs for protein generation, particularly in the discovery of novel antimicrobial peptides. Biological drugs, or biopharmaceuticals, are derived from proteins, and in recent years, protein-based biologics have become a critical class of therapeutic molecules due to their numerous advantages. These include high specificity, low toxicity and immunogenicity, and the ability to replace or supplement the body's own proteins and hormones. GPT models trained on the protein molecular space can generate valid protein sequences, as discussed in the section on ProtGPT2 (Ferruz et al. 2022). Such models can be fine-tuned to generate antimicrobial peptides, and further customized to optimize specific properties, such as developability. For instance, with the rise of bacterial drug resistance due to the overuse of traditional antibiotics, generating novel antimicrobial peptides could provide valuable alternatives and new treatments. In addition, Accenture Labs employ LLMs as feature representations of proteins and small molecules in Autoencoder and GAN-based models, enhancing the utility of these models. The representations produced by BERT or GPT-based models can be seen as efficient data compressions of the original features, enabling more precise analyses. Table 6 outlines some example prompts that incorporate SMILES strings for various drug discovery tasks, guiding LLMs to produce relevant insights throughout the drug discovery process.

Several organizations (as shown in Table 7) are leveraging LLMs, such as ChatGPT or in-house models, to drive AI-powered drug discovery. In silico Medicine, for instance, uses ChatGPT to enhance data integration and user interaction within

Table 6 Example prompts for LLMs in drug discovery, with and without SMILES

Category	Prompt
Drug repurposing	Identify potential new uses for the drug with SMILES: CC(C)CC1 = CC=C(C=C1)C(C)C(=O)O
	Analyze the repurposing potential of CC1 = CC(=O)NC(=O)N1C2 = CC=CC=C2 for treating neurodegenerative diseases
	Suggest alternative therapeutic applications for the molecule CC(C)NCC(O)COC1 = CC=CC=C1
	Suggest existing drugs for repurposing to treat obesity
	Analyze the potential of metformin for non-diabetic conditions
	Evaluate the efficacy of anti-inflammatory drugs for obesity-related inflammation
De novo drug design	Generate novel drug-like molecules similar to the SMILES: CC(C)CC1 = CC=CC=C1
	Design new compounds based on the scaffold C1 = CC=C(C=C1)C2 = CC=CC=C2 with improved pharmacokinetic properties
	Create a library of novel molecules with the core structure C1 = CC=C(C=C1)C2 = CC=CC=C2 and optimize for binding affinity
Binding affinity prediction	Predict the binding affinity of CC(C)CC1 = CC=CC=C1 to protein target with PDB ID 1A2B
	Evaluate the docking score of CC1 = CC(=O)NC(=O)N1C2 = CC=CC=C2 against acetylcholinesterase
	Analyze the interaction strength between CC(C)NCC(O)COC1 = CC=CC=C1 and receptor GPCR
ADMET prediction	Predict the ADMET profile of CC(C)CC1 = CC=CC=C1
	Evaluate absorption and metabolism characteristics of CC1 = CC(=O)NC(=O)N1C2 = CC=CC=C2
	Analyze toxicity potential of CC(C)NCC(O)COC1 = CC=CC=C1
Natural product drug discovery	Identify natural compounds similar to SMILES: CC(C)CC1 = CC=CC=C1 and evaluate bioactivity
	Analyze the pharmacological properties of plant-derived molecules with the core structure C1 = CC=C(C=C1)C2 = CC=CC=C2
	Predict synergistic effects of combining natural product CC(C)NCC(O)COC1 = CC=CC=C1 with existing drugs
	Identify natural anti-obesity compounds from traditional medicine databases
	Analyze the bioactivity of plant extracts for treating metabolic disorders
	Predict synergistic effects of combining natural products with drugs for obesity treatment

(continued)

Table 6 (continued)

Category	Prompt
Understanding disease mechanisms	Explain the molecular mechanisms of obesity and related diseases
	Identify genetic markers associated with type 2 diabetes
	Summarize recent research on the role of gut microbiota in metabolic disorders.
Novel drug target identification	Identify novel protein targets for obesity treatment based on genomic data.
	Predict potential drug targets from transcriptomic analysis of adipose tissue.
	Analyze protein–protein interaction networks for new therapeutic targets for metabolic syndrome.
Drug design and optimization	Generate novel chemical structures for anti-obesity drugs using DL models.
	Optimize the pharmacokinetic properties of lead compounds for better bioavailability.
	Predict the ADMET profile of a new drug candidate.
Clinical trials and patient data analysis	Analyze patient data to identify biomarkers for predicting drug response in obesity treatment.
	Summarize outcomes from recent clinical trials on anti-obesity drugs.
	Predict patient adherence to new weight-loss medications using demographic and behavioral data.
Network pharmacology	Construct a network pharmacology model to study multi-target effects of a drug on obesity.
	Analyze the interaction network of bioactive compounds in herbal weight-loss formulations.
	Predict off-target effects of a drug using network pharmacology approaches.

its AI-driven target discovery platform. Sanofi has partnered with OpenAI and Formation Bio to develop AI-powered software that accelerates drug development. Additionally, Google DeepMind has launched Med-PaLM (Singhal et al. 2023), a chatbot designed to answer medical queries and support drug discovery and biomedical research.

5 Challenges and Future Directions in LLM-Driven Drug Discovery

LLMs have proven transformative in drug discovery, enabling unprecedented insights into biomedical data, molecular interaction predictions, and hypothesis generation for new drugs. However, the path to fully realize their potential is fraught with critical challenges.

Table 7 Key industry applications of LLMs in drug discovery and design

Company	Model	Application	Reference
In silico medicine	GPT-4	AI-assisted drug discovery, target identification, molecule design	Savage (2023)
Absci	GPT-4	Biologics and protein-based drug design using generative AI capabilities	Savage (2023)
Schrödinger	GPT-based models	Optimization of molecular properties and support for drug design	AI4Science and Quantum (2023)
NVIDIA	BioNeMo (GPT variant)	Molecular docking, drug-target interaction predictions	Gangwal et al. (2024a)
Valence discovery	GPT-based models	Molecular property prediction, retrosynthesis	Savage (2023)
In silico medicine	ChatPandaGPT	Molecular biology, therapeutic target discovery, pharmaceutical development	Pun et al. (2023)

One of the foremost obstacles is data quality and scarcity. Despite LLMs' capacity to process extensive datasets, high-quality, labeled data in drug discovery are limited, especially in the contexts of rare diseases or novel drug targets where experimental data are sparse or unavailable (Gangwal et al. 2024b). Additionally, existing datasets often contain noise, inconsistencies, and biases, which can impair LLM performance. Clinical trial data, for instance, are susceptible to demographic, study design, and sample size biases, which can skew LLM predictions. To overcome these limitations, researchers are increasingly turning to multi-omics data (genomics, proteomics, and metabolomics) as a means of enriching LLM-driven models. However, integrating such diverse datasets presents challenges, including harmonizing formats, scales, and granularities. Since LLMs are optimized for text data, adapting them to handle structured biological data like molecular structures or protein sequences requires substantial model modifications. Interpretability is another pressing issue. Drug discovery is a high-stakes field, and erroneous predictions can lead to costly failures in later stages. Yet, many advanced LLMs function as "black boxes," meaning their decision-making processes lack transparency. In a domain that demands mechanistic understanding of drug-target interactions, the opacity of LLM predictions undermines trust among scientists and regulatory bodies. Though explainable AI (XAI) methods for LLMs are in development, achieving a balance between model complexity and interpretability remains challenging. Generalization also poses a significant challenge. While LLMs trained on large datasets can excel at pattern recognition within familiar data, they often struggle to generalize to novel or out-of-distribution data, such as predicting a drug's efficacy for a new target or in different patient populations. This limitation is particularly relevant to drug repurposing, where the goal is to identify new applications for existing drugs. LLMs may over-rely on correlations from training data without fully understanding the underlying biological mechanisms, which can lead to spurious predictions. Addressing this issue will likely require incorporating domain-specific knowledge into models, through methods like transfer learning, fine-tuning with specialized datasets, or integrating knowledge graphs that encode biological relationships.

The future of drug discovery will likely see LLMs combined with other AI techniques, such as RL and generative models, to enhance efficiency and expand the scope of their applications. RL, which has shown success in optimizing molecular properties, could guide LLMs in exploring chemical space more effectively. Similarly, integrating LLMs with generative adversarial networks (GANs) and variational autoencoders (VAEs) could enable the generation of novel drug candidates with precisely tuned properties. LLMs are also poised to advance synthesis planning and retrosynthesis, areas traditionally challenging for AI due to the complexity of chemical reactions. By training on extensive reaction data, LLMs could assist chemists in identifying feasible synthetic routes for drug candidates. Developing models that not only predict reactions but also comprehend the subtleties of chemical reactivity remains an active area of research. Moreover, integrating LLMs into early drug discovery stages, such as target identification and hit generation, is already proving promising. However, their application in later stages, including lead optimization and preclinical testing, is still being explored. For instance, LLMs could predict pharmacokinetic and pharmacodynamic properties critical for assessing safety and efficacy in humans, though this requires high-fidelity models capable of accurately simulating complex biological systems. Another promising direction involves developing LLMs that can simulate biological systems at multiple scales, from molecular interactions to whole-organism models. This capability would allow for more comprehensive drug effect predictions, including off-target interactions and potential side effects. Achieving this level of multi-scale modeling will require advancements in both AI and computational biology. Regulatory frameworks must also evolve to address AI's growing role in drug discovery. Currently, guidelines for validating AI-generated predictions are limited. For LLMs to gain widespread acceptance, regulatory agencies like the FDA will need to establish standards around model validation, transparency, and reproducibility. This validation will likely require a combination of in silico and experimental validation, along with real-world evidence to confirm the reliability of LLM predictions.

The computational cost of training and deploying LLMs poses another challenge. Given the resource-intensive nature of drug discovery, adding LLMs, which demand significant computational power, could increase expenses, especially for smaller biotech companies and academic institutions. Interest is therefore growing in more efficient LLM architectures, such as sparse attention mechanisms or parameter-efficient fine-tuning, to ease this burden. Additionally, cloud-based platforms and AI-as-a-service (AIaaS) solutions could help democratize access to LLM-driven drug discovery tools, enabling smaller organizations to benefit from advanced models without substantial upfront investment in hardware.

Ethical considerations are equally vital. LLMs are trained on vast datasets, which may include proprietary information, patient data, or data from underserved populations. Ensuring these models uphold patient privacy, respect intellectual property rights, and adhere to ethical standards is essential. Moreover, algorithmic bias in LLMs could lead to unequal benefits across populations. Addressing these concerns will require transparency in model training, robust data governance, and collaboration among AI researchers, ethicists, and policymakers.

Finally, advancing LLM-driven drug discovery will require increased collaboration between academia, industry, and regulatory agencies. Drug discovery is inherently interdisciplinary, necessitating expertise in biology, chemistry, pharmacology, and computer science. Public-private partnerships and consortia will be critical for driving innovation, addressing data scarcity, and mitigating biases. These collaborations can also facilitate data sharing, a crucial step in overcoming the challenges of limited data and potential biases.

In conclusion, while LLMs hold vast potential for revolutionizing drug discovery, considerable challenges remain, especially regarding data quality, interpretability, generalization, and computational cost. Nevertheless, advancements in AI research, interdisciplinary collaboration, and emerging regulatory frameworks are fostering conditions for broader adoption of LLM-driven drug discovery. Promising future directions include the integration of LLMs with complementary AI techniques, the development of explainable models, and improvements in multi-scale biological simulations, all of which will enhance the effectiveness of these models. As LLMs continue to advance, they offer the promise of accelerating new drug discoveries, improving clinical trial success rates, and ultimately delivering more effective therapies to patients. Achieving this vision, however, will require concerted efforts to address the technical, ethical, and regulatory hurdles that currently stand in the way.

Acknowledgments This work was supported by the European Union, Next Generation EU (PRIN 2022, project number: 2022P5LPHS). A.L. acknowledges funding from the PNRR Project CN000041, "National Centre for Gene Therapy and RNA Technology-Based Drugs" (CUP UNINA E63C22000940007).

References

Achiam J, Adler S, Agarwal S, Ahmad L, Akkaya I, Aleman FL, et al. Gpt-4 technical report. ArXiv. 2023:Prepr ArXiv230308774.

Ahmad W, Simon E, Chithrananda S, Grand G, Ramsundar B. Chemberta-2: towards chemical foundation models. ArXiv. 2022:Prepr ArXiv220901712.

AI4Science MR, Quantum MA. The impact of large language models on scientific discovery: a preliminary study using GPT-4. ArXiv. 2023:Prepr ArXiv231107361.

Bagal V, Aggarwal R, Vinod PK, Priyakumar UD. MolGPT: molecular generation using a transformer-decoder model. J Chem Inf Model. 2022;62:2064–76. https://doi.org/10.1021/acs.jcim.1c00600.

Bahani M, El Ouaazizi A, Maalmi K. The effectiveness of T5, GPT-2, and BERT on text-to-image generation task. Pattern Recogn Lett. 2023;173:57–63.

Bahdanau D. Neural machine translation by jointly learning to align and translate. ArXiv. 2014:Prepr ArXiv14090473.
Berglund M, van der Merwe B. Formalizing BPE Tokenization. ArXiv. 2023:Prepr ArXiv230908715.
Blanchard AE, Gounley J, Bhowmik D, Chandra Shekar M, Lyngaas I, Gao S, et al. Language models for the prediction of SARS-CoV-2 inhibitors. Int J High Perform Comput Appl. 2022;36:587–602. https://doi.org/10.1177/10943420221121804.
Boiko DA, MacKnight R, Gomes G. Emergent autonomous scientific research capabilities of large language models. ArXiv. 2023a:Prepr ArXiv230405332.
Boiko DA, MacKnight R, Kline B, Gomes G. Autonomous chemical research with large language models. Nature. 2023b;624:570–8.
Bran AM, Cox S, Schilter O, Baldassari C, White AD, Schwaller P. ChemCrow: augmenting large-language models with chemistry tools 2023.
Chakraborty C, Priya Doss CG, Chen L, Zhu H. Evaluating protein-protein Interaction (PPI) networks for diseases pathway, target discovery, and drug-design using 'In silico pharmacology'. Curr Protein Pept Sci. 2014;15:561–71.
Chakraborty C, Sekhar Roy S, Hsu C-H, Wen Z-H, Lin C-S. Network building of proteins in a biochemical pathway: a computational biology related model for target discovery and drug-design. Curr Bioinforma. 2010;5:290–5.
Chen M, Tworek J, Jun H, Yuan Q, Pinto HPDO, Kaplan J, et al. Evaluating large language models trained on code. ArXiv. 2021:Prepr ArXiv210703374.
Chilingaryan G, Tamoyan H, Tevosyan A, Babayan N, Hambardzumyan K, Navoyan Z, et al. Bartsmiles: generative masked language models for molecular representations. J Chem Inf Model. 2024;64:5832–43.
Chithrananda S, Grand G, Ramsundar B. ChemBERTa: large-scale self-supervised pretraining for molecular property prediction. ArXiv. 2020:Prepr ArXiv201009885.
Chowdhery A, Narang S, Devlin J, Bosma M, Mishra G, Roberts A, et al. Palm: scaling language modeling with pathways. J Mach Learn Res. 2023;24:1–113.
Christofidellis D, Giannone G, Born J, Winther O, Laino T, Manica M. Unifying molecular and textual representations via multi-task language modelling. Int Conf Mach Learn PMLR. 2023:6140–57.
Chung HW, Hou L, Longpre S, Zoph B, Tay Y, Fedus W, et al. Scaling instruction-finetuned language models. J Mach Learn Res. 2024;25:1–53.
Cui H, Wang C, Maan H, Pang K, Luo F, Duan N, et al. scGPT: toward building a foundation model for single-cell multi-omics using generative AI. Nat Methods. 2024;21:1470–80. https://doi.org/10.1038/s41592-024-02201-0.
Deng J, Lin Y. The benefits and challenges of ChatGPT: an overview. Front Comput Intell Syst. 2022;2:81–3. https://doi.org/10.54097/fcis.v2i2.4465.
Devlin J, Chang M-W, Lee K, Toutanova K. Bert: pre-training of deep bidirectional transformers for language understanding. ArXiv. 2018:Prepr ArXiv181004805.
Edwards C, Lai T, Ros K, Honke G, Cho K, Ji H. Translation between Molecules and Natural Language 2022.
Elnaggar A, Heinzinger M, Dallago C, Rehawi G, Wang Y, Jones L, et al. Prottrans: toward understanding the language of life through self-supervised learning. IEEE Trans Pattern Anal Mach Intell. 2021;44:7112–27.
Evans R, O'Neill M, Pritzel A, Antropova N, Senior A, Green T, et al. Protein complex prediction with AlphaFold-Multimer; 2021. https://doi.org/10.1101/2021.10.04.463034.
Feng S, Yang L, Huang Y, Ni Y, Ma W, Lan Y. UniMAP: universal SMILES-graph representation learning 2024.
Ferruz N, Schmidt S, Höcker B. ProtGPT2 is a deep unsupervised language model for protein design. Nat Commun. 2022;13:4348.
Floridi L, Chiriatti M. GPT-3: its nature, scope, limits, and consequences. Minds Mach. 2020;30:681–94.

Frey NC, Soklaski R, Axelrod S, Samsi S, Gomez-Bombarelli R, Coley CW, et al. Neural scaling of deep chemical models. Nat Mach Intell. 2023;5:1297–305.
Gangwal A, Ansari A, Ahmad I, Azad AK, Kumarasamy V, Subramaniyan V, et al. Generative artificial intelligence in drug discovery: basic framework, recent advances, challenges, and opportunities. Front Pharmacol. 2024a;15 https://doi.org/10.3389/fphar.2024.1331062.
Gangwal A, Ansari A, Ahmad I, Azad AK, Sulaiman WMAW. Current strategies to address data scarcity in artificial intelligence-based drug discovery: a comprehensive review. Comput Biol Med. 2024b;179:108734.
Gangwal A, Lavecchia A. Unleashing the power of generative AI in drug discovery. Drug Discov Today. 2024;29:103992. https://doi.org/10.1016/j.drudis.2024.103992.
Gehring J, Auli M, Grangier D, Yarats D, Dauphin YN. Convolutional sequence to sequence learning. In: Int. Conf. Mach. Learn. PMLR; 2017. p. 1243–52.
Guimaraes GL, Sanchez-Lengeling B, Outeiral C, Farias PLC, Aspuru-Guzik A. Objective-Reinforced Generative Adversarial Networks (ORGAN) for sequence generation models; 2018. https://doi.org/10.48550/arXiv.1705.10843.
Hayes T, Rao R, Akin H, Sofroniew NJ, Oktay D, Lin Z, et al. Simulating 500 million years of evolution with a language model. Science 2025;387:850–8.
He Z, Liang T, Jiao W, Zhang Z, Yang Y, Wang R, et al. Exploring human-like translation strategy with large language models. Trans Assoc Comput Linguist. 2024;12:229–46.
Heller SR, McNaught A, Pletnev I, Stein S, Tchekhovskoi D. InChI, the IUPAC international chemical identifier. J Cheminfo. 2015;7:1–34.
Hendrycks D, Gimpel K. Gaussian error linear units (gelus). ArXiv Prepr. 2016:ArXiv160608415.
Hochreiter S, Schmidhuber J. Long short-term memory. Neural Comput. 1997;9:1735–80.
Inagaki T, Kato A, Takahashi K, Ozaki H, Kanda GN. LLMs can generate robotic scripts from goal-oriented instructions in biological laboratory automation. ArXiv. 2023:Prepr ArXiv230410267.
Irwin R, Dimitriadis S, He J, Bjerrum EJ. Chemformer: a pre-trained transformer for computational chemistry. Mach Learn Sci Technol. 2022;3:015022. https://doi.org/10.1088/2632-2153/ac3ffb.
Jablonka KM, Schwaller P, Ortega-Guerrero A, Smit B. Leveraging large language models for predictive chemistry. Nat Mach Intell. 2024;6:161–9.
Jin W, Barzilay R, Jaakkola T. Junction tree variational autoencoder for molecular graph generation. Proc 35th Int Conf Mach Learn PMLR. 2018:2323–32.
Koh HY, Nguyen AT, Pan S, May LT, Webb GI. Physicochemical graph neural network for learning protein–ligand interaction fingerprints from sequence data. Nat Mach Intell. 2024:1–15.
Krishna R, Wang J, Ahern W, Sturmfels P, Venkatesh P, Kalvet I, et al. Generalized biomolecular modeling and design with RoseTTAFold all-atom. Science. 2024;384:eadl2528.
Kudo T. Subword regularization: improving neural network translation models with multiple subword candidates. ArXiv Prepr. 2018:ArXiv180410959.
Kudo T, Richardson J. SentencePiece: a simple and language independent subword tokenizer and detokenizer for Neural Text Processing 2018.
Lavecchia A. Navigating the frontier of drug-like chemical space with cutting-edge generative AI models. Drug Discov Today. 2024a:104133.
Lavecchia A. Advancing drug discovery with deep attention neural networks. Drug Discov Today. 2024b;104067
Lee J, Yoon W, Sungdong K, Kim D, Sunkyu K, So CH, et al. BioBERT: a pre-trained biomedical language representation model for biomedical text mining. Bioinformatics. 2020;36:1234–40.
Lewis M, Liu Y, Goyal N, Ghazvininejad M, Mohamed A, Levy O, et al. Bart: Denoising sequence-to-sequence pre-training for natural language generation, translation, and comprehension. ArXiv. 2019:Prepr ArXiv191013461.
Li Y, Gao C, Song X, Wang X, Xu Y, Han S. DrugGPT: a GPT-based strategy for designing potential ligands targeting specific proteins. bioRxiv. 2023:2023.06. 29.543848.
Liu Y, Ott M, Goyal N, Du J, Joshi M, Chen D, et al. Roberta: a robustly optimized BERT pretraining approach. ArXiv. 2019:Prepr ArXiv190711692.
Liu S, Li Y, Li Z, Gitter A, Zhu Y, Lu J, et al. A text-guided protein design framework. ArXiv. 2023a:Prepr rXiv230204611.

Liu Z, Zhang W, Xia Y, Wu L, Xie S, Qin T, et al. Molxpt: wrapping molecules with text for generative pre-training. ArXiv. 2023b:Prepr ArXiv230510688.

Liu P, Ren Y, Tao J, Ren Z. GIT-Mol: a multi-modal large language model for molecular science with graph, image, and text. Comput Biol Med. 2024a;171:108073. https://doi.org/10.1016/j.compbiomed.2024.108073.

Liu Y, Ding S, Zhou S, Fan W, Tan Q. MolecularGPT: open large language model (LLM) for few-shot molecular property prediction. ArXiv. 2024b:Prepr ArXiv240612950.

Lu J, Zhang Y. Unified deep learning model for multitask reaction predictions with explanation. J Chem Inf Model. 2022;62:1376–87.

Madani A, Krause B, Greene ER, Subramanian S, Mohr BP, Holton JM, et al. Large language models generate functional protein sequences across diverse families. Nat Biotechnol. 2023;41:1099–106.

May LT, Bartolo BA, Harrison DG, Guzik T, Drummond GR, Figtree GA, et al. Translating atherosclerosis research from bench to bedside: navigating the barriers for effective preclinical drug discovery. Clin Sci. 2022;136:1731–58. https://doi.org/10.1042/CS20210862.

Mehr SHM, Craven M, Leonov AI, Keenan G, Cronin L. A universal system for digitization and automatic execution of the chemical synthesis literature. Science. 2020;370:101–8.

Nair V, Hinton GE. Rectified linear units improve restricted Boltzmann machines. Proc 27th Int Conf Mach Learn ICML-10. 2010:807–14.

Nijkamp E, Ruffolo J, Weinstein EN, Naik N, Madani A. ProGen2: exploring the boundaries of protein language models. Cell Syst. 2023;14:968–78. e3.

Oniani D, Hilsman J, Zang C, Wang J, Cai L, Zawala J, et al. Emerging opportunities of using large language models for translation between drug molecules and indications. Sci Rep. 2024;14:10738.

Pal S, Bhattacharya M, Lee S-S, Chakraborty C. A domain-specific next-generation large language model (LLM) or ChatGPT is required for biomedical engineering and research. Ann Biomed Eng. 2024;52:451–4. https://doi.org/10.1007/s10439-023-03306-x.

Pearson WR, Lipman DJ. Improved tools for biological sequence comparison. Proc Natl Acad Sci. 1988;85:2444–8.

Pei Q, Zhang W, Zhu J, Wu K, Gao K, Wu L, et al. Biot5: Enriching cross-modal integration in biology with chemical knowledge and natural language associations. ArXiv. 2023:Prepr ArXiv231007276.

Pun FW, Ozerov IV, Zhavoronkov A. AI-powered therapeutic target discovery. Trends Pharmacol Sci. 2023;

Radford A, Kim JW, Hallacy C, Ramesh A, Goh G, Agarwal S, et al. Learning transferable visual models from natural language supervision. Int Conf Mach Learn., PmLR. 2021:8748–63.

Raffel C, Shazeer N, Roberts A, Lee K, Narang S, Matena M, et al. Exploring the limits of transfer learning with a unified text-to-text transformer. J Mach Learn Res. 2020;21:1–67.

Ribeiro AH, Tiels K, Aguirre LA, Schön T. Beyond exploding and vanishing gradients: analysing RNN training using attractors and smoothness. Int Conf Artif Intell Stat., PMLR. 2020:2370–80.

Rives A, Meier J, Sercu T, Goyal S, Lin Z, Liu J, et al. Biological structure and function emerge from scaling unsupervised learning to 250 million protein sequences. Proc Natl Acad Sci. 2021;118:e2016239118.

Romanelli V, Cerchia C, Lavecchia A. Unlocking the potential of generative artificial intelligence in drug discovery. In: Applications of Generation AI. Springer; 2024. p. 37–63.

Roziere B, Gehring J, Gloeckle F, Sootla S, Gat I, Tan XE, et al. Code llama: open foundation models for code. ArXiv. 2023:Prepr ArXiv230812950.

Rumelhart DE, McClelland JL. Parallel Distributed Processing, Explorations in the Microstructure of Cognition I and II. MIT Press, Cambridge, 1986.

Rush A. The annotated transformer. In: Proceedings Workshop NLP Open Source Software. NLP-OSS, Melbourne: Association for Computational Linguistics; 2018. p. 52–60. https://doi.org/10.18653/v1/W18-2509.

Rust P, Pfeiffer J, Vulić I, Ruder S, Gurevych I. How good is your tokenizer? on the monolingual performance of multilingual language models. ArXiv. 2020:Prepr ArXiv201215613.

Savage N. Drug discovery companies are customizing ChatGPT: here's how. Nat Biotechnol. 2023;41:585–6. https://doi.org/10.1038/s41587-023-01788-7.

Schwaller P, Gaudin T, Lányi D, Bekas C, Laino T. "Found in translation": predicting outcomes of complex organic chemistry reactions using neural sequence-to-sequence models. Chem Sci. 2018;9:6091–8. https://doi.org/10.1039/C8SC02339E.

Schwaller P, Laino T, Gaudin T, Bolgar P, Hunter CA, Bekas C, et al. Molecular transformer: a model for uncertainty-calibrated chemical reaction prediction. ACS Cent Sci. 2019;5:1572–83.

Schwaller P, Petraglia R, Zullo VH, Nair V, Andreas Haeuselmann R, Pisoni R, et al. Predicting retrosynthetic pathways using transformer-based models and a hyper-graph exploration strategy. Chem Sci. 2020;11:3316–25. https://doi.org/10.1039/C9SC05704H.

Segler MHS, Kogej T, Tyrchan C, Waller MP. Generating focused molecule libraries for drug discovery with recurrent neural networks. ACS Cent Sci. 2018;4:120–31. https://doi.org/10.1021/acscentsci.7b00512.

Sejnowski TJ. Large language models and the reverse turing test. Neural Comput. 2023;35:309–42.

Sharma G, Thakur A. ChatGPT in Drug Discovery. 2023; https://doi.org/10.26434/chemrxiv-2023-qgs3k.

Shazeer N. Glu variants improve transformer. ArXiv. 2020:Prepr ArXiv200205202.

Singh R, Sledzieski S, Bryson B, Cowen L, Berger B. Contrastive learning in protein language space predicts interactions between drugs and protein targets. Proc Natl Acad Sci. 2023;120:e2220778120.

Singhal K, Azizi S, Tu T, Mahdavi SS, Wei J, Chung HW, et al. Large language models encode clinical knowledge. Nature. 2023;620:172–80.

Smith S, Patwary M, Norick B, LeGresley P, Rajbhandari S, Casper J, et al. Using deepspeed and megatron to train megatron-turing nlg 530b, a large-scale generative language model. ArXiv. 2022:Prepr ArXiv220111990.

Song X, Salcianu A, Song Y, Dopson D, Zhou D. Fast wordpiece tokenization. ArXiv Prepr ArXiv201215524. 2020;

Su B, Du D, Yang Z, Zhou Y, Li J, Rao A, et al. A molecular multimodal foundation model associating molecule graphs with natural language 2022.

Taylor R, Kardas M, Cucurull G, Scialom T, Hartshorn A, Saravia E, et al. Galactica: a large language model for science. ArXiv. 2022:Prepr ArXiv221109085 2023;10.

Thoppilan R, De Freitas D, Hall J, Shazeer N, Kulshreshtha A, Cheng H-T, et al. Lamda: Language models for dialog applications. ArXiv. 2022:Prepr ArXiv220108239.

Touvron H, Lavril T, Izacard G, Martinet X, Lachaux M-A, Lacroix T, et al. Llama: open and efficient foundation language models. ArXiv. 2023:Prepr ArXiv230213971.

Vaswani A, Shazeer N, Parmar N, Uszkoreit J, Jones L, Gomez AN, et al. Attention is all you need. Adv Neural Inf Process Syst. 2017;30

Walters WP, Barzilay R. Critical assessment of AI in drug discovery. Expert Opin Drug Discov. 2021;16:937–47.

Wang S, Guo Y, Wang Y, Sun H, Huang J. SMILES-BERT: large scale unsupervised pre-training for molecular property prediction. Proc 10th ACM Int Conf Bioinforma Comput Biol Health Inform. 2019:429–36.

Watson JL, Juergens D, Bennett NR, Trippe BL, Yim J, Eisenach HE, et al. De novo design of protein structure and function with RFdiffusion. Nature. 2023;620:1089–100.

Weininger D. SMILES, a chemical language and information system. 1. Introduction to methodology and encoding rules. J Chem Inf Comput Sci. 1988;28:31–6. https://doi.org/10.1021/ci00057a005.

Winter B, Winter C, Schilling J, Bardow A. A smile is all you need: predicting limiting activity coefficients from SMILES with natural language processing. Digit Discov. 2022;1:859–69.

Workshop B, Scao TL, Fan A, Akiki C, Pavlick E, Ilić S, et al. Bloom: A 176b-parameter open-access multilingual language model. ArXiv. 2022:Prepr ArXiv221105100.

Wouters OJ, McKee M, Luyten J. Estimated Research and Development investment needed to bring a new medicine to market, 2009-2018. JAMA. 2020;323:844–53. https://doi.org/10.1001/jama.2020.1166.

Wu F, Radev D, Li SZ. Molformer: motif-based transformer on 3d heterogeneous molecular graphs. Proc AAAI Conf Artif Intell. 2023a;37:5312–20.

Wu S, Fei H, Qu L, Ji W, Chua T-S. Next-gpt: any-to-any multimodal llm. ArXiv. 2023b:Prepr ArXiv230905519.

Xue D, Zhang H, Xiao D, Gong Y, Chuai G, Sun Y, et al. X-MOL: large-scale pre-training for molecular understanding and diverse molecular analysis. bioRxiv 2020:2020.12. 23.424259.

Yang Z, Dai Z, Yang Y, Carbonell J, Salakhutdinov RR, Le QV. Xlnet: Generalized autoregressive pretraining for language understanding. Adv Neural Inf Process Syst. 2019;32.

Yoshikawa N, Skreta M, Darvish K, Arellano-Rubach S, Ji Z, Bjørn Kristensen L, et al. Large language models for chemistry robotics. Auton Robot. 2023;47:1057–86.

Yüksel A, Ulusoy E, Ünlü A, Doğan T. SELFormer: molecular representation learning via SELFIES language models. Mach Learn Sci Technol. 2023;4:025035. https://doi.org/10.1088/2632-2153/acdb30.

Zeng Z, Yao Y, Liu Z, Sun M. A deep-learning system bridging molecule structure and biomedical text with comprehension comparable to human professionals. Nat Commun. 2022;13:862.

Zhang X-C, Wu C-K, Yang Z-J, Wu Z-X, Yi J-C, Hsieh C-Y, et al. MG-BERT: leveraging unsupervised atomic representation learning for molecular property prediction. Brief Bioinform. 2021;22:bbab152. https://doi.org/10.1093/bib/bbab152.

Zhang S, Roller S, Goyal N, Artetxe M, Chen M, Chen S, et al. Opt: Open pre-trained transformer language models. ArXiv. 2022:Prepr ArXiv220501068.

Zhao A, Wu Y. Future implications of ChatGPT in pharmaceutical industry: drug discovery and development. Front Pharmacol. 2023;14:1194216.

Zheng Y, Koh HY, Yang M, Li L, May LT, Webb GI, et al. Large language models in drug discovery and development: from disease mechanisms to clinical trials. ArXiv. 2024:Prepr ArXiv240904481.

Zhu Y, Chen D, Du Y, Wang Y, Liu Q, Wu S. Featurizations matter: a multiview contrastive learning approach to molecular pretraining. In: ICML 2022, vol. 2022. 2nd ed. AI Sci. Workshop.

Zhu J, Xia Y, Wu L, Xie S, Zhou W, Qin T, et al. Dual-view molecular pre-training. In: Proceedings 29th ACM SIGKDD Conference Knowledge Discover. Data Min. 2023:3615–27.

Contrastive Learning Approaches for Drug Discovery

Ana Sanchez-Fernandez and Günter Klambauer

Abstract Contrastive learning has been established as a powerful representation learning approach for unlabeled data. Models trained with contrastive learning approaches have shown to be robust and transferable across different tasks, and are currently used worldwide for applications like image generation with text prompting. In drug discovery, large amounts of unlabeled data are frequently encountered, but currently are often not used in standard machine learning methods. Therefore, contrastive learning is a suitable approach to empower drug discovery by exploiting the large amounts of unlabeled data. In this chapter, we introduce the contrastive learning framework and its most prominent applications for drug discovery. Further, we explain for which computer-aided drug discovery tasks contrastive learning methods can be used.

Keywords Contrastive learning · Drug discovery · Self-supervised learning · Representation learning · Molecular embeddings · Computational drug design

1 Introduction

During the last decade, deep learning methods have seen unprecedented flourishing, achieving state of the art in several tasks for different data modalities such as computer vision and natural language (Russakovsky et al. 2014; Lecun et al. 2015; Schmidhuber 2015). Deep learning has shown not only its potential for broad applicability but also outstanding performance in specific and well-defined tasks. Although these deep learning-based models can show exceptional and sometimes even higher than human performance for particular tasks, these models still fail to

---

A. Sanchez-Fernandez · G. Klambauer (✉)
ELLIS Unit Linz and LIT AI Lab, Institute for Machine Learning, Johannes Kepler University Linz, Linz, Austria
e-mail: sanchez@ml.jku.at; klambauer@ml.jku.at

A. Lavecchia (ed.), *Applied Artificial Intelligence for Drug Discovery*,
https://doi.org/10.1007/978-3-031-98022-0_15

generalize to new tasks and datasets (Chollet 2019). Therefore, in recent years, there has been an increased focus on training robust models whose output is transferable to unseen tasks and exhibits strong generalization.

The importance of large labeled datasets for training machine learning models has driven organizations to collect such large datasets. However, the high cost in time and resources needed for labeling and curating these datasets has put unsupervised learning methods in the spotlight. Specifically, self-supervised learning (SSL) methods have shown outstanding performance in computer vision and natural language processing. Self-supervised learning methods are those that are trained using objective functions from fully supervised methods where the "labels" are derived from the data itself. For example, randomly masking some of the pixels of an image and having the model predict them is considered a form of self-supervision (He et al. 2022). At the same time, self-supervision can be achieved in different ways. If the models learn to generate samples from the data distribution, as in the case of predicting pixels of an image, we would be talking about *generative* self-supervised learning. If, on the other hand, the models learn to distinguish between related and non-related samples within the data, we would be referring to *contrastive* self-supervised learning (Liu et al. 2020). In this work, we will focus on the latter.

In contrastive learning approaches, the objective is to learn to map inputs into a multidimensional space, i.e., an *embedding space*, in which matched samples, such as two equivalent SMILES strings, are similar to each other while unmatched samples are dissimilar. In contrast to traditional supervised methods, where a set of classes has to be defined in advance, maximizing the similarity of two sample representations allows models to generalize to unseen classes (Wu et al. 2018b). The extreme form of this would be to consider each sample as a different class, but comparing to all other samples in the dataset would be extremely computationally expensive. Therefore, contrasting against noise brings across a more computationally feasible approximation. The idea behind this principle was first introduced by Gutmann and Hyvärinen (2010), whose work presented a method to train models that are able to distinguish a certain data distribution from a noise distribution.

Contrastive learning approaches for images present a straightforward way to sample from a distribution of matched samples without having labels: using randomly augmented versions of the same image as matched samples. This concept was laid out in SimCLR (Chen et al. 2020). The authors showed that training machine learning methods in this way resulted in performance on par with fully supervised learning methods. SimCLR emphasized the importance of the choice of augmentations and large batch sizes.

Besides contrastive learning methods where representations are learned from one data modality, two samples from different data modalities that illustrate the same concept can also be regarded as matched samples to learn joint representations from. Just as two different data augmentations of the same image can be considered matched samples, an image and its corresponding text description can be regarded in the same way. The method CLIP successfully implemented this idea, training models in a self-supervised way jointly from images and text (Radford et al. 2021).

Again, although these models were not trained to predict a predefined set of classes, they showed comparable performance with fully supervised learning models that were specifically trained to do so. Moreover, the learned representations of the images were shown to be rich by showing high performance on a wide range of image classification tasks. In these, so-called downstream tasks, the contrastively learned representations were able to correctly classify images into unseen classes by pure similarity scoring, so-called zero-shot tasks. Learning a joint embedding space for images and text has had a big impact in machine learning applications, as it has made possible the retrieval of images from a given text description, and vice versa. As a result of this possibility, CLIP has also been successfully used for image generation from a given text prompt (Ramesh et al. 2022). Overall, with a multi-modal contrastive learning approach, it was possible to obtain image-encoders that produce rich, robust, and highly generalizable features.

Although contrastive learning with images and text has brought such a strong improvement to unsupervised methods, it also presents some limitations. One of them is that matching the representation of an image specifically to its text counterpart could lead to focusing only on a few features, and neglecting other relevant ones. A way of avoiding this would be to exploit the covariance structure in the data. CLOOB uses Hopfield networks to do so (Fürst et al. 2022). Instead of using the representations of each sample separately, it uses a weighted average of a given sample and the rest of the dataset. This way, the new representations have features that are more common in the most similar samples, while less common features are averaged out.

Besides computer vision and natural language processing, drug discovery is another field where usually large amounts of data are available but few of them are labeled. Correctly labeling an image by identifying the main object on it can be done by a human within a few seconds per image. However, correctly labeling a chemical structure, for example by identifying its mechanism-of-action, can take weeks and cost several thousands of dollars in wet-lab experiments. Thus, the labels for drug discovery data are by far more difficult to obtain and far less available. Nevertheless, there are databases that collect labels for molecules: large molecule databases like PubChem or ChEMBL, where structures for millions of compounds are stored, but the amount of activity labels is much more scarce (Gaulton et al. 2011; Kim et al. 2023). As the reader could infer from the explanation above, drug discovery is a domain which we could greatly benefit from pre-training our machine learning models with methods like contrastive learning. Nevertheless, the utility of machine learning for drug discovery has already been established, also in the traditional fully supervised setting.

The drug discovery pipeline aims to find compounds that revert certain adverse human conditions, such as a disease. The first challenge this poses is to assess how would it be possible to revert the given condition (Hughes et al. 2011; Emmerich et al. 2020; Tabana et al. 2023). To answer this question, a further understanding of the disease mechanism is necessary: Which pathways it affects and which genes and proteins are involved in them. Then, a *target*, which can be any druggable molecule, is selected so that small molecules can be tested against it (Knowles and Gromo

2003). In order to find candidate molecules, we would have to test a large set of molecules and check if they present the desired activity. Selecting a specific target narrows the search and simplifies the selection process.

A second challenge in the drug discovery process is to decide which molecules to test in the wet-lab, given that resources are limited and the total number of synthesizable compounds is huge (Walters et al. 1998; Melville et al. 2009). Machine learning, as a method that allows making fast predictions for a given set of data points, has brought a strong improvement in this regard (Catacutan et al. 2024). One proof of this success is the QSAR (Quantitative Structure-Activity Relationship) (Neves et al. 2018) models. The structure of a molecule reveals many properties related to drug likeness, such as solubility, lipophilicity, and many others. Therefore, the structure of a molecule can be used to predict whether it will present certain activity (Schultz et al. 2003; McKinney et al. 2000).

However, although chemical structures might be one of the most used data modalities for predicting activities, as they are easy to obtain and require low computational costs, these structures also present some caveats. One of the caveats is that even if the molecular structure of a compound contains a great amount of information regarding its properties, not having access to the structure of the target of interest might allow the model to learn irrelevant features not related to the drug-target interaction. In other words, if a model is trained using only the structure of the ligands, it might learn every feature that is correlated with the outcome of interest in the training data, even those that are not involved in the drug-target interaction. In this regard, molecular features that are informed about the structure or sequence of the target protein (e.g., Lapinsh et al. 2001; Svensson et al. 2024) might help avoid learning these spurious correlations (van Westen et al. 2011).

The target-centric approach for drug discovery explained above has been the most prominent approach during the last years (Hughes et al. 2011). Nevertheless, only focusing on finding compounds that have more affinity and bind more potently to the target in question neglects other important factors such as side effects and toxicity. In that regard, data modalities that provide a general overview of the compound effect hold a lot of promise, and one of the most resource-efficient and scalable ones is fluorescence microscopy images (Bray et al. 2016; Chandrasekaran et al. 2023).

In this book chapter, we will provide some insights about contrastive learning methods for drug discovery, with a focus on chemical structures and microscopy images.

2 Definition and Framework

In this section, we attempt to provide a definition of *contrastive learning*, both *uni-modal* and *multi-modal* contrastive learning, and at the same time also a framework, in which almost all methods can be explained (Fig. 1).

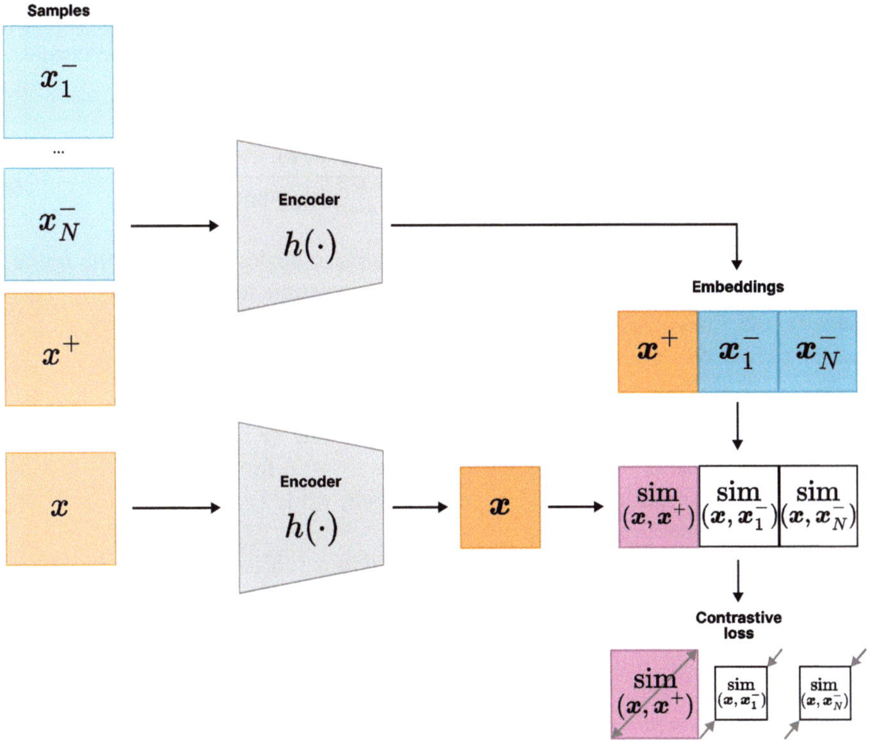

Fig. 1 General contrastive learning framework. The sample x, along with the positive pair x^+ (orange) and the negative pairs $\{x_1^-, ..., x_N^-\}$ (blue), is encoded with $h(.)$, and the pairwise similarity sim is calculated. Then, the contrastive loss is minimized with respect to the parameters of $h(.)$, increasing the similarity of the positive pair while decreasing it for the negative ones

A *uni-modal contrastive learning method* has the goal to learn representations of samples, called *embeddings*, and consists of four elements:

1. A *process to obtain matched and unmatched pairs of samples*. The process should provide $x \in \mathcal{X}$ from a domain $\mathcal{X}$ and one or more matched samples x^+ and one or more unmatched samples x^-. The set $\{x, x^+\}$ is known as matched (or: positive) pair, while $\{x, x^-\}$ is an un-matched (or: negative) pair.
2. An *encoding function* $h : \mathcal{X} \mapsto \mathbb{R}^D$, with which we obtain embeddings $\boldsymbol{x} = h(x)$. If each element of the pair is from a different data modality, a different encoder $g : \mathcal{X} \mapsto \mathbb{R}^D$ is available (see below).
3. A *similarity function* sim : $\mathbb{R}^D \times \mathbb{R}^D \mapsto \mathbb{R}$. Usually, it is the cosine similarity function $\text{sim}(\boldsymbol{x}, \boldsymbol{x}') = \boldsymbol{x}^T \boldsymbol{x} / (||\boldsymbol{x}|| \; ||\boldsymbol{x}'||)$ or the dot product $\text{sim}(\boldsymbol{x}, \boldsymbol{x}') = \boldsymbol{x}^T \boldsymbol{x}'$ for $\boldsymbol{x}, \boldsymbol{x}' \in \mathbb{R}^D$. These two similarity functions are often equivalent since the embeddings are assumed to be normalized.
4. A *contrastive loss function* encourages the embeddings of matched samples to be similar, and embeddings of un-matched samples to be dissimilar. The contrastive

loss function will be minimized with respect to the parameters of $h(.)$ to increase $\mathrm{sim}(\boldsymbol{x}, \boldsymbol{x}^+)$ and decrease $\mathrm{sim}(\boldsymbol{x}, \boldsymbol{x}^-)$. In the InfoNCE loss, we use one matched sample x^+ and N unmatched samples $\{x_1^-, ..., x_N^-\}$.

$$\ell_{\mathrm{InfoNCE}} = -\ln \frac{\exp(\mathrm{sim}(\boldsymbol{x}, \boldsymbol{x}^+)/\tau)}{\exp(\mathrm{sim}(\boldsymbol{x}, \boldsymbol{x}^+)/\tau) + \sum_{n=1}^{N} \exp(\mathrm{sim}(\boldsymbol{x}, \boldsymbol{x}_n^-)/\tau)} \tag{1}$$

While **multi-modal contrastive learning** can also be described with the framework above, we introduce a slightly different notation to make the difference between the modalities explicit. Multi-modal contrastive learning methods also consist of four elements:

1. A *process to obtain matched and unmatched pairs of samples*. The process should provide $x \in \mathcal{X}$ from a domain $\mathcal{X}$ and one or more matched samples $z^+ \in \mathcal{Z}$ and one or more unmatched samples $z^- \in \mathcal{Z}$ from a domain $\mathcal{Z}$. The set $\{x, z^+\}$ is known as matched (or: positive) pair, while $\{x, z^-\}$ is an un-matched (or: negative) pair.
2. For each domain, there is an *encoding function* $h : \mathcal{X} \mapsto \mathbb{R}^D$, with which we obtain embeddings $\boldsymbol{x} = h(x)$, and $g : \mathcal{Z} \mapsto \mathbb{R}^D$ for the domain $\mathcal{Z}$. As an example, $\mathcal{X}$ could be the space of molecular graphs, and h a graph neural network encoder. $\mathcal{Z}$ could be the space of microscopy images and g a convolutional neural network.
3. A *similarity function* $\mathrm{sim} : \mathbb{R}^D \times \mathbb{R}^D \mapsto \mathbb{R}$: see above.
4. A contrastive loss function as for uni-modal contrastive learning (see above).

3 Molecular Structures and Their Representations

Throughout this book chapter, we will be referring to different ways of representing molecules. In this section, we will briefly explain those that will be later mentioned. For a more in-depth explanation, one could refer to David et al. (2020). If already familiar with the topic, the reader can skip to the next section.

For years, chemists have been looking for descriptive representations of molecules. Moreover, the wide adoption of computers and their potential has lead to the development of machine-readable molecular representations.

- **SMILES Representation:** One of the most popular machine-readable representation of molecules are SMILES (Simplified Molecular Input Line Entry System) (Weininger 1988). SMILES are strings comprised of the sequence of atoms of a given molecule. They are obtained by selecting an atom as the initial node and then traversing the graph. The fact that we can randomly select an atom as the initial node makes SMILES non-unique, which means that several strings can represent the same molecule. At the same time, this can also be a form of data augmentation.

- **Molecular Descriptor Representation**. Another computer-readable way to represent molecules is numerical vectors. These representation are known as molecular descriptors, and they have been more precisely described in Todeschini and Consonni (2009) as:

  "The final result of a logical and mathematical procedure that transforms chemical information of a molecule, such as structural features, into useful numbers or the result of standardized experiments". The previous definition encompasses various types of descriptors, that can encode very different types of molecular information (Grisoni et al. 2018b). Although other types exist, we will explain two of the most prominent categories below.

  - **Physicochemical Descriptor Representation.** Although representations that explicitly display the set of atoms in the molecules, such as SMILES, have been extensively used, it is possible to represent them with features related to physicochemical or structural properties, such as molecular weight, log P, number of hydrogen bond donors and acceptors and number of rings, among others (Moriwaki et al. 2018; Cao et al. 2013). This kind of representations are called descriptors and they have been extensively used in the field of chem-informatics in QSAR models (Grisoni et al. 2018a; Ritchie and Macdonald 2014).
  - **Fingerprint Representation.** In this case, what is being encoded in the resulting vector is information about the structure of the molecule. One of the most widely used are ECFPs (Extended Connectivity Fingerprints) (Rogers and Hahn 2010). In essence, they provide information about increasingly larger substructures in a molecule. These substructures are defined by setting a radius from an atom of the molecule and selecting all neighboring atoms within the given radius. Then, an identifier is assigned to each substructure and the resulting set of identifiers is converted into a bit array of a certain length. They are often used jointly with the previously described physicochemical descriptors (Xue and Bajorath 2000).

- **Graph Representation.** Finally, molecules can also be encoded into graphs, defining the molecule as a set of nodes and edges. The most common way to define these graphs is setting atoms as nodes and bonds as edges. An advantage of these kind of methods is that they can also encode 3D information (Gaudelet et al. 2021).

4 Biological Images and Their Representations

Analogous to the previous section, this one is a short primer on biological images, particularly histopathology images and fluorescence microscopy images. If already familiar with these topics, the reader can skip to the next section.

4.1 Histopathology Images

The organization of living systems at a microscopical level has been a subject of study for centuries. More specifically, the study of tissues has been a scientific branch in itself since the nineteenth century, under the name of "histology" (Hussein et al. 2015). In the medical field, this kind of image results are especially useful for the diagnosis of diseases and conditions that are difficult to detect at a macroscopic level, such as cancer and infections of different kinds. The study of changes in tissue caused by disease is known as histopathology.

Histopathology samples are obtained from a biopsy sample of the patient, that is then prepared for proper visualization under the microscope. This preparation roughly consists in fixing the sample, in order to avoid its degradation, dehydrating it to harden the sample and make the later cutting easier, embedding in paraffin and finally sectioning the sample in thin slices (Dewi and Purwanto 2023). Finally, the sample is stained with dyes that allow discerning the tissue structure when light is passed through it. The most common dye is the hematoxylin and eosin. Hematoxylin is a basic stain that dyes acidic compounds such as the deoxyribonucleic acid in the cell nuclei in a purple blue color, while eosin stains the extracellular matrix and cytoplasm in pink (Fischer et al. 2008).

Traditionally, human experts have visually inspected these images to provide a diagnosis. However, over the last years there has been a digitalization process, which allows to obtain samples in an automated way and storing them in a hard disk. The direct product of this process is Whole Slide Images (WSI), which are high resolution images obtained by automatic scanning of the tissue samples (Farahani et al. 2015). This digitalization, along with the advent of machine learning methods for images has positioned diagnosis of histopathology images as a candidate task for machines to learn (Fuchs and Buhmann 2011). Therefore, in the last years, several efforts have been made to encourage the exploration of machine learning methods for histopathology images (Borovec et al. 2020; Weitz et al. 2022). To date, several methods have shown their potential to effectively classify tumors (Iizuka et al. 2020; Ahmed et al. 2022; Srinidhi et al. 2021).

4.2 Fluorescence Microscopy Images

Fluorescence microscopy is a powerful imaging technique to visualize specific structures or molecules within a sample. This method involves the use of fluorescent labels, such as dyes or proteins, which can attach to the target molecules or structures of interest. When these labels are exposed to light of a particular wavelength, they absorb energy and emit light at a longer wavelength. This emitted light is then captured by the microscope, allowing researchers to create high-contrast images that highlight the labeled components (Combs 2010; Ghiran 2011). The fluorescent labels used in this technique are selected for their ability to emit light at distinct wavelengths. This way, each employed dye can be targeted to a different specific protein, organelle, or other cellular component, which allows visualizing all the targeted cellular components simultaneously (Gustafsdottir et al. 2013).

Due to its ability to capture fine-grained changes in cells, fluorescence microscopy has been traditionally employed to study specific pathways and cellular interactions, using highly specific fluorescent labels (Singh et al. 2014). However, Carpenter et al. (2006) proposed this technique to be used to study cellular systems in a general way. For that, they proposed utilizing dyes that bind to structures whose visualization will allow to have an overview of the morphology of the cells. Usually, we can visualize the DNA in the nucleus, structures comprised of RNA, the endoplasmatic reticulum, the mitochondria and other structures as the plasma membrane, each in different "channels" of the image (Cimini et al. 2023).

At the same time, the automation of this technique allows testing a large scale of compounds without proportionally increasing the overhead. This has made possible to integrate this technique in the drug discovery pipeline, providing information about the changes that hundreds of thousands of compounds produce in the full complexity of cells (Bickle 2010). Therefore, this method is a scalable technique, that provides much more information than a potency label towards a specific target, in comparison to High Throughput Screening data.

Next we will also provide an overview on previous successes of microscopy imaging data and the history of how machine learning has been able to use these images in fully supervised learning methods.

Although, microscopy images have the potential of providing detailed information about the processes that are taking place in the cells, we need to find ways to extract this information and use it, for example, for making inference on the most likely mechanism of action of a compound. Carpenter et al. (2006) presented an open-source software that allowed extracting a large amount of features related to the morphology of the cells. These features have been the standard method to use for these kind of images since, and the concept has been adopted in industrial settings as well. Moreover, Simm et al. (2018) proved that training machine learning methods with these features increased the hit rates by a large margin in comparison to random search.

Nevertheless, machine learning methods have the potential to extract features more effectively for certain tasks than human-created software (Russakovsky et al. 2014). Following this intuition, Hofmarcher et al. (2019) leveraged the raw microscopy images to predict bioactivity and found out that Convolutional Neural Networks (CNN) trained with these images could predict assay outcomes with higher predictive performance than Fully Connected Networks (FCN) trained with morphological features.

5 Uni-modal Contrastive Learning for Drug Discovery

5.1 Uni-modal Contrastive Learning with Molecular Structures

The structure of a molecule is key for determining its physical properties and whether it presents certain biological activity (Schultz et al. 2003; McKinney et al. 2000). Therefore, finding proper ways to represent these structures is crucial to be able to capture the relevant features that determine these properties. As described in Sect. 3, several methods have been developed for representing molecules over the years. Moreover, training deep learning models using these representations has shown to be a successful strategy for predicting biological or physicochemical activities (Mayr et al. 2016; Tropsha et al. 2023). However, as it is often the case, only few labels are available for these tasks, although millions of molecular structures are publicly available. This makes contrastive learning of molecular structures an interesting pre-training method for molecules. This way, instead of training machine learning models from scratch to predict the task of interest, we will start with models that were pre-trained in a contrastive learning way and which, consequently, have already learned a rich representation of chemical structures (Fig. 2).

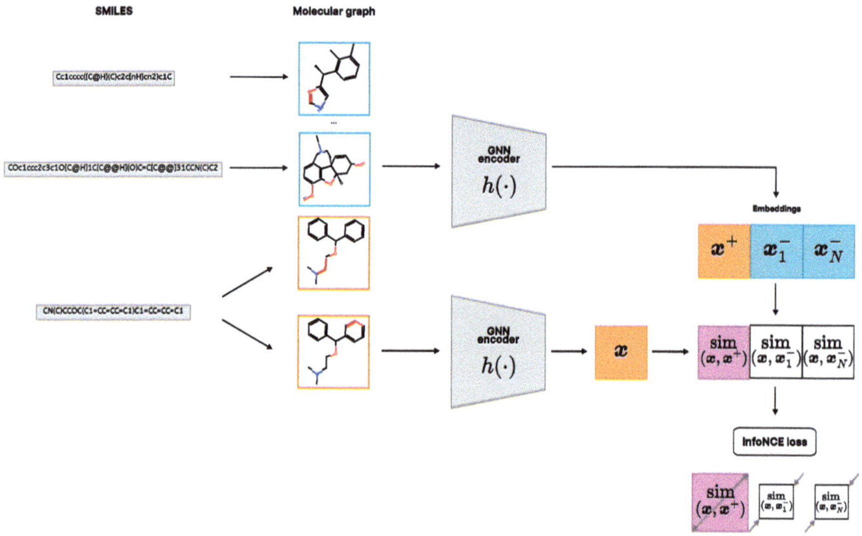

Fig. 2 Unimodal contrastive learning with molecular structures. In this case, the input samples are molecular graphs, which are constructed using the corresponding SMILES strings for each molecule. Two different random augmentations are applied to each molecular graph which produce the sample x and its positive pair x^+ (orange). Meanwhile, graphs corresponding to other molecules are the negative pairs $\{x_1^-, ..., x_N^-\}$ (blue)

One of the first methods to apply contrastive learning approaches to molecular data is MolCLR (Wang et al. 2022). In this work, the authors leverage graph neural networks to encode molecular structures and train them in a contrastive learning fashion. In this case, they perform three different data augmentation strategies to sample positive pairs: randomly masking atoms, deleting bonds, and removing parts of the graph. The authors pre-train the models with 10 million molecules collected from PubChem and evaluate their performance at different molecular property prediction tasks from MoleculeNet. The authors evaluate the performance of their method in the classification benchmark datasets BACE, HIV, BBBP (Martins et al. 2012), ClinTox (Artemov et al. 2016; Gayvert et al. 2016), SIDER (Kuhn et al. 2016), Tox21 (Mayr et al. 2016), and MUV (Wu et al. 2018a). Except for BACE and HIV, MolCLR shows improved ROC AUC with respect to the self-supervised learning methods presented in Liu et al. (2019) and Hu et al. (2019). When it comes to comparison to fully supervised methods, on-par and even improved performance was shown with respect to fully supervised methods like random forest (Breiman 2001), support vector machine (Schölkopf 1998), and several graph-based methods (Schütt et al. 2017; Kipf and Welling 2016; Xu et al. 2018; Lu et al. 2019; Yang et al. 2019). Analogously, the regression datasets ESOL (Delaney 2004), FreeSolv (Mobley and Guthrie 2014), Lipo, QM7 (Blum and Reymond 2009; Rupp et al. 2012), QM8 (Ramakrishnan et al. 2015), and QM9 (Ramakrishnan et al. 2014) were evaluated. Excluding the QM7 dataset, authors show improved performance with respect to other SSL pre-training strategies. However, for regression tasks, in some cases, fully-supervised learning methods perform significantly better as is the case for SchNet in the QM9 dataset.

However, as pointed out by Fang et al. (2021) (Molecular Contrastive Learning with Chemical Element Knowledge Graph), randomly masking atoms or bonds of a molecule could greatly transform its properties. Therefore, methods that leverage domain knowledge of molecular structures and their properties have been developed ever since.

One of them, developed by Wang et al. (2023), employs SMILES as input. In particular, they stress the importance of the methods to sample or generate positive examples in contrastive learning. They introduce a new paradigm for creating positive examples given a certain SMILES string called CLAPS (Contrastive Learning with Attention-guided Positive-sample Selection). In this paradigm, different characters of the SMILES sequences are masked based on the output weights of a trainable self-attention layer. The authors show that masking characters with a probability that is proportional to the obtained weights from self-attention yields better results than other strategies such as random masking or just masking the characters with higher weights. Once this augmentation is performed, the original and masked sequences are fed to two different transformer models that are then trained using the InfoNCE loss. For pre-training these models, the authors use the ZINC15 dataset, and then evaluated them at different downstream tasks that are relevant for drug discovery, such as physicochemical property prediction and bioactivity prediction. The authors show that this model outperforms previous state of the art methods in several benchmark datasets. For example, in the classification datasets BBBP (Martins et al. 2012) and SIDER (Kuhn et al. 2016), CLAPS shows higher ROC AUC than other molecular representations such as fingerprints, graph-based methods like MolCLR (Wang et al. 2022) or GraphCL (You et al. 2020). In the regression datasets ESOL (Delaney 2004) and FreeSolv (Mobley and Guthrie 2014), CLAPS presents the lowest root mean squared error (RMSE) out of all compared methods. Another graph-based self-supervised learning method for molecules was developed by Li et al. (2021a). In this case, the main novelty is that, to generate negative samples, the graphs are decomposed into two subgraphs and mixed with other molecular graphs.

5.2 *Uni-modal Contrastive Learning with Biological Images*

Preceded by the success of contrastive learning methods for natural images, this paradigm has also been explored for biological images (Fig. 3). Biological images, such as microscopy or histopathology images, contain information on how living systems change under different conditions.

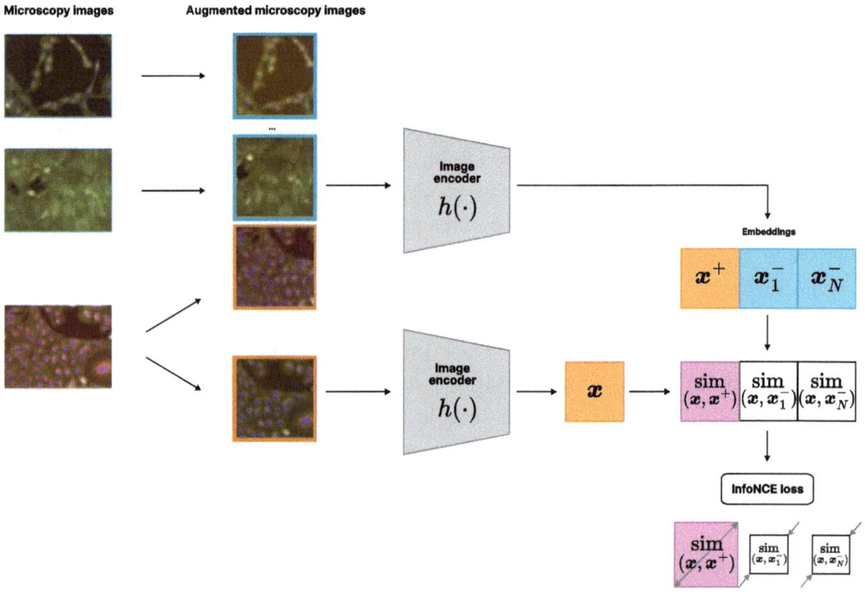

Fig. 3 Unimodal contrastive learning with biological images. The input samples are biological images, e.g., microscopy or histopathology images. Two different random augmentations (cropping, rotation, blurring) are applied to each molecular image which produce the sample x and its positive pair x^+ (orange). Meanwhile, other biological images are the negative pairs (blue)

One contrastive learning method inspired by SimCLR and applied to microscopy was developed by Perakis et al. (2021). In this work, images from the BBBC021 dataset are cropped into single cells and then two random augmentations from each image are fed into the corresponding encoders. Interestingly, the authors explore the effect of different augmentations in this data type. They show that grayscale transformation is not beneficial while cropping the images brings the highest increase in performance for the downstream task of MoA classification.

Contrastive learning has also been leveraged to perform protein subcellular localization by Tu et al. (2022). Protein subcellular localization consists of predicting in which compartment of the cells a certain protein is located. This localization then provides information on the possible functions of that protein. One of the most prominent available datasets for this task is the Human Protein Atlas, which contains 83,762 confocal microscopy images detailing the subcellular localization of 13,041 proteins across multiple cell lines (Thul and Lindskog 2018). In the manuscript, the authors show that pre-training a model with contrastive learning improves the macro and micro F1 performance metrics in protein localization with respect to other pre-trained models like SimCLR (Chen et al. 2020) or a CNN pre-trained with ImageNet (Deng et al. 2009). Furthermore, fluorescence and electron microscopy have been trained with this approach and evaluated for the downstream task of mitochondria segmentation (Li et al. 2021b).

However, experimentally obtained images usually present biases regarding confounding factors unrelated to the signal we are interested in. Usually, biological imaging data are obtained hierarchically. For example, high throughput microscopy images are usually obtained in batches, that at the same time are divided into plates, and each of the latter is comprised of a set of wells where the cells are laying. However, not only microscopy imaging datasets present this structure. Histopathology images also show a hierarchical structure as the data comprises patches that are part of a certain slide, which simultaneously belongs to a given patient. In response to this challenge, Jiang et al. (2023) developed a variant from SimCLR where the loss consists of three InfoNCE terms. In each of these terms, the positive examples are considered to be samples from the same patient, patch, and slide, respectively. This way, the authors aim to capture the hierarchical structure of the data in the learned embedding space.

6 Multi-modal Contrastive Learning for Drug Discovery

Another promising direction is to learn a representation space that brings together different data modalities. In this case, the positive pairs are two samples, one from each modality, that can be viewed as the same object.

One of the first methods to explore this possibility for life sciences was proposed by Stärk et al. (2021). In this work, the authors highlight that, although graph-based methods that encode how atoms are connected with each other in the 2-dimensional space have brought a strong improvement to property prediction tasks in comparison to other traditional methods, including the 3-dimensional coordinates of molecules' atoms can greatly enrich the resulting representations (Chen et al. 2019; Schütt et al. 2017; Gasteiger et al. 2020; Sestak et al. 2024). This might be because the spatial configuration of a molecule provides more information about whether it can fit into the binding pocket of a given target. It is also worth noting that, as the same molecule might have different conformations, the authors include all of them that are available as positive pairs. However, later work has shown that information about 3D coordinates is surprisingly ineffective for predicting biological activity (Criscuolo et al. 2024). After the contrastive pre-training, the resulting models will be informed about the relationship between two and three-dimensional molecular structures. The authors show that, for the dataset QM9 (Ramakrishnan et al. 2014), this method shows improved performance for molecular property prediction in comparison to other pre-training methods like graph contrastive learning (You et al. 2020), predicting the Gibbs free energy of the GEOM-Drugs (Axelrod and Go'mez-Bombarelli 2020) dataset, a generative model of conformers (Ganea et al. 2021) and directly predicting atoms distances.

Another rich data source that could be leveraged in the context of drug discovery consists on biomedical texts (Fig. 4). Seidl et al. (2023) leveraged assay descriptions from PubChem for activity prediction tasks. Interestingly, they optimize for the Noise Contrastive Estimation (NCE) objective function (Gutmann and Hyvärinen 2010), where the similarity between a molecule and text representation is increased if the compound is active for the given assay and decreased in case it is not. The authors show that this contrastive pre-training allows for zero-shot activity prediction, outperforming other methods such as 1-nearest-neighbor, soft-nearest-neighbor and deep learning-based models like Galactica (Taylor et al. 2022), KV-PLM (Zeng et al. 2022), or frequent hitters (Schimunek et al. 2023) at activity prediction for different splits on the PubChem (Kim et al. 2023) and on the FS-Mol (Stanley et al. 2021) datasets. Moreover, they performed linear probing with eight different benchmark datasets: BACE, BBBP (Martins et al. 2012), Clin-Tox (Artemov et al. 2016; Gayvert et al. 2016), HIV, SIDER (Kuhn et al. 2016), Tox21 (Mayr et al. 2016), ToxCast (Richard et al. 2016), Tox21-10k (Richard et al. 2021; Wu et al. 2021). In the last five mentioned datasets, CLAMP shows the highest increase in average precision. Similarly, other methods have attempted to learn from molecular structures and text using molecular structures and a text description (Liu et al. 2022; Edwards et al. 2021).

High Throughput Imaging data allow obtaining a high quantity of images in a rapid and cost-effective way. At the same time, this technology opens the possibility to scan large amounts of compounds and visualize their effect in the cells. However, obtaining labels for all of these images is not feasible in practice and activity labels are only available for a limited set of compounds. Therefore, self-supervised contrastive learning methods are suitable for leveraging unlabeled High Throughput Imaging data to pre-train models that could be useful for other downstream tasks. Moreover, jointly learning an embedding space from microscopy images and the corresponding molecular structures introduces the possibility of retrieving from a set

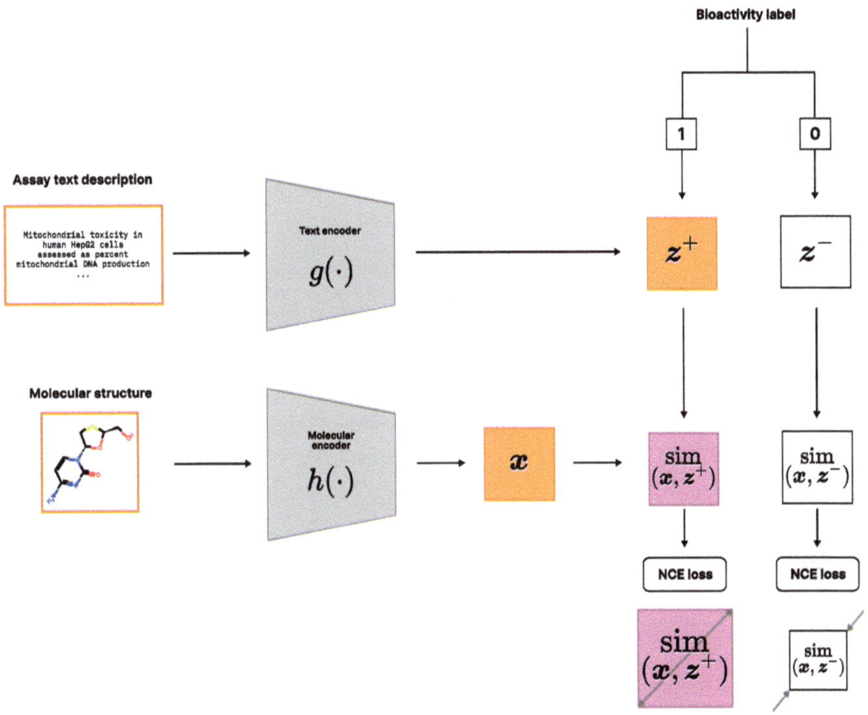

Fig. 4 Multi-modal contrastive learning with bioassay text description and molecular structures. If a molecule is active for a certain bioassay, the similarity between both embeddings will increase, while if it is inactive, it will decrease

of candidate molecules given a certain microscopy image, or vice versa (Fig. 5). This is the idea behind CLOOME, the first machine learning model trained with the goal of providing a cross-modal retrieval system of microscopy images and molecular structures (Sanchez-Fernandez et al. 2023). Learning this joint embedding space, in which the similarity of matching pairs of microscopy images and molecules is high while the similarity of unmatched pairs is low, allows using a new image as a query and finding the most similar out of a set of candidate molecules. Analogously, we could also use a molecule as a query and find the most similar images. In Sect. 7, we will provide an overview of the potential use cases of this retrieval, along with other applications.

The underlying methodology of this work consists in training a contrastive learning model with pairs of microscopy images and chemical structures. As opposed to previous approaches, where molecule encoders learned representations using activity data (Hofmarcher et al. 2019; Simm et al. 2018) or used hand-crafted representations (Carpenter et al. 2006; Bray et al. 2016), CLOOME learns representations without activity data or human expertise.

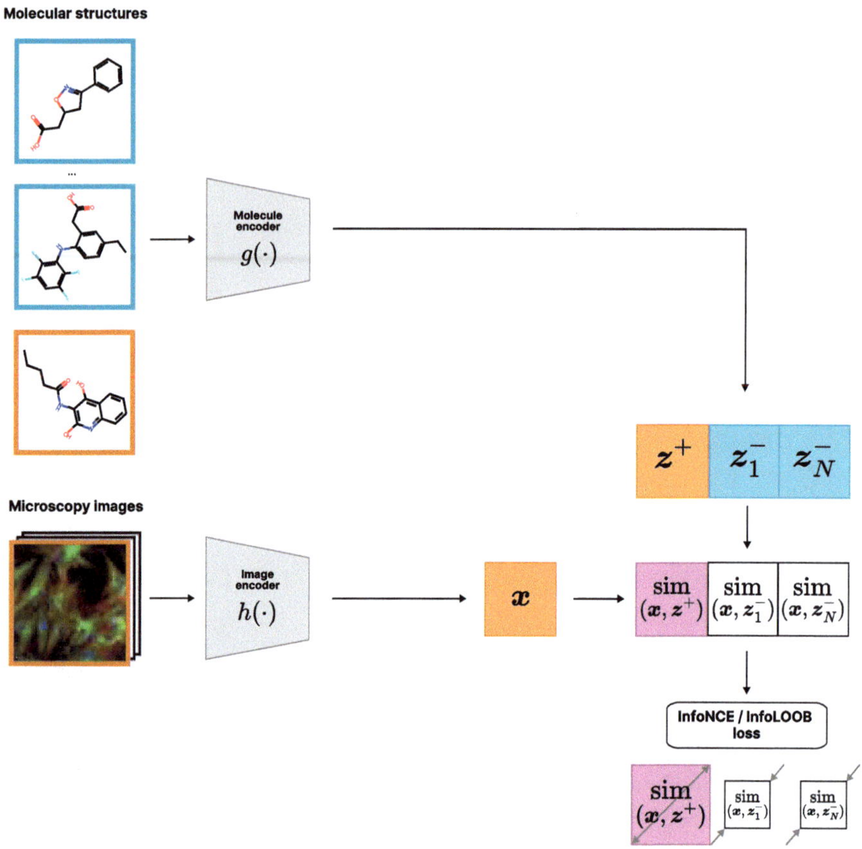

Fig. 5 Multimodal contrastive learning with microscopy images and molecular structures. The input sample x is a microscopy image of chemically perturbed cells, and its positive pair x^+ is the structure of the corresponding applied molecule (orange). The negative pairs are other molecular structures $\{x_1^-, ..., x_N^-\}$ (blue)

As hinted in Sect. 1, CLIP has the problem that a few features can be overrepresented while others are neglected (Pearl 1988; Fürst et al. 2022). This effect can be present when learning focuses only on few features and/or when the covariance structure in the data is insufficiently extracted. This problem is known as "explaining away" (Wellman and Henrion 1993). Therefore, the method CLOOB was introduced by Fürst et al. (2022) to ameliorate these drawbacks. Explaining away can be caused by saturation of the InfoNCE objective (Yeh et al. 2021; Zhang et al. 2022; Fürst et al. 2022). CLOOB introduced the InfoLOOB objective together with Hopfield networks as a promising method for contrastive learning. In CLOOME, the settings of both CLIP and CLOOB were applied to microscopy images and molecular structures, and selected the best performing model on a validation set (Sanchez-Fernandez et al. 2023).

The training dataset consists of N pairs of microscopy images and chemical structures of the corresponding applied molecules $\{(x_1, z_1), \ldots, (x_N, z_N)\}$. We assume that an adaptive image-encoder $\boldsymbol{h}^x(.)$ and an adaptive structure-encoder $\boldsymbol{h}^z(.)$ are available that map the images and chemical structures to their embeddings $\boldsymbol{x}_n = \boldsymbol{h}^x(x_n)$ and $\boldsymbol{z}_n = \boldsymbol{h}^z(z_n)$, respectively.

The stacked microscopy image embeddings are denoted as $\boldsymbol{X} = (\boldsymbol{x}_1, \ldots, \boldsymbol{x}_N)$ and the stacked structure embeddings as $\boldsymbol{Z} = (\boldsymbol{z}_1, \ldots, \boldsymbol{z}_N)$. The embeddings are normalized such that $\|\boldsymbol{x}_n\| = \|\boldsymbol{z}_n\| = 1 \; \forall n$.

For this approach, first image and structure embeddings are retrieved from stored image embeddings $\boldsymbol{U}$ and structure embeddings $\boldsymbol{V}$. $\boldsymbol{U}_{\boldsymbol{x}_i}$ denotes an image-retrieved image embedding, $\boldsymbol{U}_{\boldsymbol{z}_i}$ a structure-retrieved image embedding, $\boldsymbol{V}_{\boldsymbol{x}_i}$ an image-retrieved structure embedding, and $\boldsymbol{V}_{\boldsymbol{z}_i}$ a structure-retrieved structure embedding.

In analogy to CLOOB, these retrievals from continuous modern Hopfield networks are computed as follows:

$$\boldsymbol{U}_{\boldsymbol{x}_i} = \boldsymbol{U} \operatorname{softmax}(\beta \, \boldsymbol{U}^T \boldsymbol{x}_i) \,, \tag{2}$$

$$\boldsymbol{U}_{\boldsymbol{z}_i} = \boldsymbol{U} \operatorname{softmax}(\beta \, \boldsymbol{U}^T \boldsymbol{z}_i) \,, \tag{3}$$

$$\boldsymbol{V}_{\boldsymbol{x}_i} = \boldsymbol{V} \operatorname{softmax}(\beta \, \boldsymbol{V}^T \boldsymbol{x}_i) \,, \tag{4}$$

$$\boldsymbol{V}_{\boldsymbol{z}_i} = \boldsymbol{V} \operatorname{softmax}(\beta \, \boldsymbol{V}^T \boldsymbol{z}_i) \,, \tag{5}$$

where β is a scaling parameter of the Hopfield network, which is considered a hyperparameter. These retrieved embeddings $\boldsymbol{U}_{\boldsymbol{x}_i}, \boldsymbol{U}_{\boldsymbol{z}_i}, \boldsymbol{V}_{\boldsymbol{x}_i}$, and $\boldsymbol{V}_{\boldsymbol{z}_i}$ are also normalized to unit norm. By default, we store the current mini-batch in the modern Hopfield networks, that is, $\boldsymbol{U} = \boldsymbol{X}$ and $\boldsymbol{V} = \boldsymbol{Z}$. Note that $\boldsymbol{X}$ contains the image embeddings ($\boldsymbol{Z}$, the structure embeddings), and we use N ambiguously both as dataset size and as mini-batch size to maintain a clear notation. The choice that $\boldsymbol{U} = \boldsymbol{X}$ and $\boldsymbol{V} = \boldsymbol{Z}$ is mostly taken because of computational constraints, while $\boldsymbol{U}$ and $\boldsymbol{V}$ could hold the whole dataset.

Then, the InfoLOOB objective (Fürst et al. 2022; Poole et al. 2019) for the retrieved embeddings is used as objective function:

$$\begin{aligned} \mathrm{L}_{\mathrm{InfoLOOB}} = & -\frac{1}{N} \sum_{i=1}^{N} \ln \frac{\exp(\tau^{-1} \, \boldsymbol{U}_{\boldsymbol{x}_i}^T \boldsymbol{U}_{\boldsymbol{z}_i})}{\sum_{j \neq i}^{N} \exp(\tau^{-1} \, \boldsymbol{U}_{\boldsymbol{x}_i}^T \boldsymbol{U}_{\boldsymbol{z}_j})} \\ & -\frac{1}{N} \sum_{i=1}^{N} \ln \frac{\exp(\tau^{-1} \, \boldsymbol{V}_{\boldsymbol{x}_i}^T \boldsymbol{V}_{\boldsymbol{z}_i})}{\sum_{j \neq i}^{N} \exp(\tau^{-1} \, \boldsymbol{V}_{\boldsymbol{x}_j}^T \boldsymbol{V}_{\boldsymbol{z}_i})} \,. \end{aligned} \tag{6}$$

In this work, for the CellPainting dataset introduced by Bray et al. (2016), the following results are presented. For the downstream task of bioactivity prediction, although CLOOME did not have access to bioactivity labels during training, the resulting representations presented on-par ROC AUC with respect to fully supervised models that were particularly trained for this task (Hofmarcher et al. 2019). In the newly introduced molecule-to-image and image-to-molecule retrieval tasks, CLOOME shows higher accuracy than the available baselines at the time, i.e., a bilinear model and random retrieval. Moreover, zero-shot compound and zero-shot MoA classification tasks are introduced, where CLOOME shows higher accuracy than other representations such as software extracted features and outputs of intermediate layers of fully-supervised methods (i.e., Carpenter et al. 2006; Hofmarcher et al. 2019).

7 Contrastive Learning for Drug Discovery Tasks

As it has been shown throughout this chapter, contrastive learning methods have demonstrated considerable promise across various domains of drug discovery. Their ability to learn transferable and general-purpose representations from large, unlabeled datasets can significantly reduce the reliance on extensive and costly experimental labeling (Radford et al. 2021; Chen et al. 2020). In practice, these learned representations enable improved downstream task performance and have the potential to accelerate critical phases of the drug discovery pipeline.

A broad range of tasks stand to benefit from contrastive pre-training. Below, we discuss some of the primary areas in which contrastive learning can make a substantial impact:

1. **Bioactivity Prediction.** Predicting whether a molecule will be active against a given biological target has conventionally required large labeled datasets and careful feature engineering. Contrastive pre-training on large-scale unlabeled molecular libraries allows models to capture chemical features without using labels. Therefore, using these embeddings that already contain chemical information would lead to improved downstream activity prediction tasks, even with limited labeled data. For instance, after contrastive pre-training, graph-based or SMILES-based encoders can focus on functional groups, substructures, and other relevant features that correlate well with biological activity.
2. **Physicochemical and ADMET Property Prediction.** Many early-stage filters in drug discovery include around physicochemical and ADMET (Absorption, Distribution, Metabolism, Excretion, and Toxicity) properties (Richard et al. 2021; Hughes et al. 2011; van de Waterbeemd and Gifford 2003). Similar to bioactivity prediction, contrastive pre-training can help learners detect structure–property relationships, leveraging unlabeled molecular corpora to yield robust embeddings. These embeddings can then be adapted to predict complex properties with fewer labeled examples, reducing assay costs and speeding up early-stage screening.

3. **Mechanism-of-Action (MoA) Identification.** Identifying the MoA of novel or repurposed compounds is pivotal for understanding their therapeutic potential and possible side effects (Trapotsi et al. 2021). High-content imaging (HCI) assays offer a window into the phenotypic changes induced by drugs. Contrastive learning on these image-based assays, often combined with molecular data, can produce embeddings that emphasize phenotypically relevant features. These embeddings aid MoA classification, enabling zero-shot or few-shot inference of a compound's potential mechanism by linking it to well-studied molecular or phenotypic neighbors in the embedding space.
4. **Cross-Modality Retrieval.** Perhaps the most distinctive advantage of contrastive learning arises when multiple data modalities are available. Representations that align molecular structures with phenotypic microscopy images, protein structures, or text-based assay descriptions allow cross-modal retrieval, which can unlock the assistance of machine learning in crucial tasks for drug discovery (Sanchez-Fernandez et al. 2023). For instance:

   - **Image-to-Molecule Retrieval.** Given a microscopy image depicting cells treated with a particular compound, a model trained with contrastive learning can retrieve molecules with similar biological effects. This capability provides the possibility of *bioisosteric replacement* and *scaffold hopping* (Brown 2014), assisting medicinal chemists in identifying other compounds that produce similar cellular responses.
   - **Molecule-to-Image Retrieval.** Conversely, starting from a query molecule, one can query a reference database of images with annotated phenotypes. The retrieved images can shed light on the drug's biological effect, facilitating side-by-side comparisons to known phenotypes and guiding hypotheses about its MoA and toxicity.
   - **Text-Based Retrieval.** When paired with assay descriptions or literature-derived textual information, embeddings allow researchers to identify relevant bioassays or experimental conditions for a given molecule. This can streamline the design of follow-up experiments and improve the interpretability of model predictions.

5. **Zero-Shot and Few-Shot Learning.** Pre-trained embeddings enable fast inference on novel tasks or targets with minimal additional data. For instance, if a newly discovered protein target shares functional similarities with previously studied targets, a model leveraging contrastive embeddings may already capture relevant biochemical attributes, enabling zero-shot prediction of ligand binding or few-shot adaptation to new tasks.

Overall, contrastive learning bridges the gap between abundant unlabeled data and the increasingly complex demands of modern drug discovery. Its utility extends beyond simple classification tasks, offering a joint representation space where the relationships between different biological data modalities can be explored.

8 Discussion and Future Directions

While contrastive learning has opened new avenues in drug discovery, it is still far from a panacea. There are still some challenges that should be overcome to unleash its full potential. Below, we will elucidate these challenges and expose some possibilities to overcome them.

Data Quality and Biases Experimentally obtained datasets in drug discovery—such as microscopy images and histopathology slides—are prone to systematic biases (e.g., batch effects, plate effects, and inter-lab variability). These biases can degrade the quality of learned representations, conflating meaningful biological signals with irrelevant artifacts. Some possible strategies to address this challenge are the following. Incorporating advanced normalization strategies could help disentangle biology from technical noise, leading to more robust and generalizable embeddings. Finally, unlike natural images, biomolecular data often requires domain-specific augmentations. Therefore, carefully engineered transformations grounded in chemistry or cell biology will help ensure that positive pairs capture meaningful variations rather than trivial noise.

Scaling Up and Benchmarking The success of contrastive learning in fields like computer vision and natural language processing was catalyzed by the availability of large, high-quality datasets. In drug discovery, despite the large volumes of unlabeled data (e.g., from corporate compound archives or large-scale phenotypic screens), public datasets remain relatively sparse or heterogeneously annotated. Therefore, creating and sharing large-scale, diverse datasets that integrate structures, images, and functional annotations will enable more systematic benchmarking and broader adoption of contrastive approaches. In the domain of high throughput fluorescence microscopy, there have been increased efforts to release large datasets for biological data in recent years (Chandrasekaran et al. 2023; Fay et al. 2023). As the field matures, standardized evaluation pipelines and leaderboards for tasks such as MoA classification, property prediction, and cross-modality retrieval will encourage the development of state-of-the-art methods and promote reproducibility.

Integration with Other Paradigms Contrastive learning for biological data is often applied in isolation, but combining them with other deep learning paradigms might elevate their potential beyond retrieval. Embedding spaces derived from contrastive objectives might be integrated with generative models, reinforcement learning, or active learning strategies. One of them could be guided molecular design. Contrastively learned embeddings can serve as a foundation for generative models that propose new compounds. These models could be steered toward embedding regions that correspond to desired biological activities or safety profiles, speeding up lead optimization. Another possibility is active learning. When combined with active learning protocols, embeddings can guide the selection of the most informative compounds or assays to label, further reducing experimental costs and accelerating the iterative cycles of drug discovery.

Explainability and Interpretability A critical barrier to widespread trust and adoption of machine learning in life sciences is interpretability. Although contrastive methods provide rich representations, the reasoning behind a model's similarity scores or embedding distances remains opaque. Techniques that highlight the molecular substructures or cellular features responsible for similarity judgments can provide valuable insights, aiding medicinal chemists and biologists in decision-making. Integrating contrastive embeddings with mechanistic models or knowledge graphs might offer a more transparent framework that combines powerful representations with human-understandable domain knowledge.

Contrastive learning sets a promising direction for building versatile and robust representations in drug discovery. If we can address technical hurdles, combine them with other paradigms such as generative models, and improve their interpretability, contrastive learning has the potential to accelerate the drug discovery pipeline.

In the near future, we expect wider adoption of multi-modal contrastive learning approaches that unify heterogeneous data sources, emergence of more specialized training regimes tailored to molecular and phenotypic data and integration of contrastive learning methods for different downstream tasks in the drug discovery pipeline.

By continuing to refine these methods, we move closer to a new era in drug discovery: one where representation learning directly shapes and informs more efficient, cost-effective, and successful drug development.

References

Ahmed AA, Abouzid M, Kaczmarek E. Deep learning approaches in histopathology. Cancers. 2022;14:5264. https://doi.org/10.3390/CANCERS14215264.

Artemov AV, Putin E, Vanhaelen Q, Aliper A, Ozerov IV, Zhavoronkov A. Integrated deep learned transcriptomic and structure-based predictor of clinical trials outcomes. bioRxiv; 2016. 10.1101/095653.

Axelrod S, Go'mez-Bombarelli R. Geom: energy-annotated molecular conformations for property prediction and molecular generation. Sci Data. 2020;9 https://doi.org/10.1038/s41597-022-01288-4.

Bickle M. The beautiful cell: high-content screening in drug discovery. Anal Bioanal Chem. 2010;398:219–26. https://doi.org/10.1007/S00216-010-3788-3/METRICS.

Blum LC, Reymond JL. 970 million druglike small molecules for virtual screening in the chemical universe database GDB-13. J Am Chem Soc. 2009;131:8732. https://doi.org/10.1021/ja902302h.

Borovec J, Kybic J, Arganda-Carreras I, Sorokin DV, Bueno G, Khvostikov AV, Bakas S, Chang EI, Heldmann S, Kartasalo K, Latonen L, Lotz J, Noga M, Pati S, Punithakumar K, Ruusuvuori P, Skalski A, Tahmasebi N, Valkonen M, Venet L, Wang Y, Weiss N, Wodzinski M, Xiang Y, Xu Y, Yan Y, Yushkevich P, Zhao S, Munõz-Barrutia A. Anhir: automatic non-rigid histological image registration challenge. IEEE Trans Med Imaging. 2020;39:3042–52. https://doi.org/10.1109/TMI.2020.2986331.

Bray MA, Singh S, Han H, Davis CT, Borgeson B, Hartland C, Kost-Alimova M, Gustafsdottir SM, Gibson CC, Carpenter AE. Cell painting, a high-content image-based assay for morphological profiling using multiplexed fluorescent dyes. Nat Protoc. 2016;11:1757–74. https://doi.org/10.1038/nprot.2016.105.

Breiman L. Random forests. Mach Learn. 2001;45:5–32. https://doi.org/10.1023/A:1010933404324/METRICS.

Brown N. Bioisosteres and scaffold hopping in medicinal chemistry. Mol Inf. 2014;33:458–62. https://doi.org/10.1002/MINF.201400037.

Cao DS, Xu QS, Hu QN, Liang YZ. Chemopy: freely available python package for computational biology and chemoinformatics. Bioinformatics. 2013;29:1092–4. https://doi.org/10.1093/BIOINFORMATICS/BTT105.

Carpenter AE, Jones TR, Lamprecht MR, Clarke C, Kang IH, Friman O, Guertin DA, Chang JH, Lindquist RA, Moffat J, Golland P, Sabatini DM. Cellprofiler: image analysis software for identifying and quantifying cell phenotypes. Genome Biol. 2006;7:R100. https://doi.org/10.1186/GB-2006-7-10-R100.

Catacutan DB, Alexander J, Arnold A, Stokes JM. Machine learning in preclinical drug discovery. Nat Chem Biol. 2024;20:960–73. https://doi.org/10.1038/s41589-024-01679-1.

Chandrasekaran SN, Ackerman J, Alix E, Ando DM, Arevalo J, Bennion M, Boisseau N, Borowa A, Boyd JD, Brino L, Byrne PJ, Ceulemans H, Ch'ng C, Cimini BA, Clevert DA, Deflaux N, Doench JG, Dorval T, Doyonnas R, Dragone V, Engkvist O, Faloon PW, Fritchman B, Fuchs F, Garg S, Gilbert TJ, Glazer D, Gnutt D, Goodale A, Grignard J, Guenther J, Han Y, Hanifehlou Z, Hariharan S, Hernandez D, Horman SR, Hormel G, Huntley M, Icke I, Iida M, Jacob CB, Jaensch S, Khetan J, Kost-Alimova M, Krawiec T, Kuhn D, Lardeau CH, Lembke A, Lin F, Little KD, Lofstrom KR, Lotfi S, Logan DJ, Luo Y, Madoux F, PAM Z, Marion BA, Martin G, NJ MC, Mervin L, Miller L, Mohamed H, Monteverde T, Mouchet E, Nicke B, Ogier A, Ong AL, Osterland M, Otrocka M, Peeters PJ, Pilling J, Prechtl S, Qian C, Rataj K, Root DE, Sakata SK, Scrace S, Shimizu H, Simon D, Sommer P, Spruiell C, Sumia I, Swalley SE, Terauchi H, Thibaudeau A, Unruh A, de Waeter JV, Dyck MV, van Staden C, Warchoł M, Weisbart E, Weiss A, Wiest-Daessle N, Williams G, Yu S, Zapiec B, Żyła M, Singh S, Carpenter AE. JUMP Cell Painting dataset: morphological impact of 136,000 chemical and genetic perturbations. bioRxiv. 2023; https://doi.org/10.1101/2023.03.23.534023.

Chen T, Kornblith S, Norouzi M, Hinton G. A simple framework for contrastive learning of visual representations. In: III HD, Singh A, editors. Proceedings of the 37th international conference on machine learning, PMLR, proceedings of machine learning research, vol. 119; 2020. p. 1597–607.

Chen P, Liu W, Hsieh CY, Chen G, Zhang S. Utilizing edge features in graph neural networks via variational information maximization. arXiv preprint arXiv:1906.05488; 2019.

Chollet F. On the measure of intelligence arXiv. 1911.01547; 2019.

Cimini BA, Chandrasekaran SN, Kost-Alimova M, Miller L, Goodale A, Fritchman B, Byrne P, Garg S, Jamali N, Logan DJ, Concannon JB, Lardeau CH, Mouchet E, Singh S, Abbasi HS, Aspesi P, Boyd JD, Gilbert T, Gnutt D, Hariharan S, Hernandez D, Hormel G, Juhani K, Melanson M, Mervin LH, Monteverde T, Pilling JE, Skepner A, Swalley SE, Vrcic A, Weisbart E, Williams G, Yu S, Zapiec B, Carpenter AE. Optimizing the cell painting assay for image-based profiling. Nat Protoc. 2023;18:1981–2013. https://doi.org/10.1038/s41596-023-00840-9.

Combs CA. Fluorescence microscopy: a concise guide to current imaging methods. Curr Protoc Neurosci. 2010;50:2.1.1–2.1.14. https://doi.org/10.1002/0471142301.NS0201S50.

Criscuolo E, Özçelik R, van Tilborg D, Grisoni F. The surprising ineffectiveness of molecular dynamics coordinates for predicting bioactivity with machine learning. ChemRxiv. 2024; chemrxiv-2024-rp81v.

David L, Thakkar A, Mercado R, Engkvist O. Molecular representations in AI-driven drug discovery: a review and practical guide. J Cheminform. 2020;12(1):–22. https://doi.org/10.1186/S13321-020-00460-5.

Delaney JS. Esol: estimating aqueous solubility directly from molecular structure. J Chem Inf Comput Sci. 2004;44:1000–5. https://doi.org/10.1021/ci034243x.

Deng J, Dong W, Socher R, Li LJ, Li K, Fei-Fei L. Imagenet: a large-scale hierarchical image database. In: 2009 IEEE conference on computer vision and pattern recognition; 2009. p. 248–55. https://doi.org/10.1109/CVPR.2009.5206848.

Dewi AK, Purwanto B, Widjiati, Dewi AK, Purwanto B, Widjiati (2023) Introduction of histopathology. Molecular histopathology and cytopathology https://doi.org/10.5772/INTECHOPEN.110225.

Edwards C, Zhai C, Ji H. Text2Mol: cross-modal molecule retrieval with natural language queries. In: Moens MF, Huang X, Specia L, Yih SW, editors. Proceedings of the 2021 conference on empirical methods in natural language processing, Association for Computational Linguistics, online and Punta Cana, Dominican Republic; 2021. p. 595–607. https://doi.org/10.18653/v1/2021.emnlp-main.47.

Emmerich CH, Gamboa LM, Hofmann MC, Bonin-Andresen M, Arbach O, Schendel P, Gerlach B, Hempel K, Bespalov A, Dirnagl U, Parnham MJ. Improving target assessment in biomedical research: the got-it recommendations. Nat Rev Drug Discov. 2020;20(64):81. https://doi.org/10.1038/s41573-020-0087-3.

Fang Y, Zhang Q, Yang H, Zhuang X, Deng S, Zhang W, Qin M, Chen Z, Fan X, Chen H. Molecular contrastive learning with chemical element knowledge graph. In: Proceedings of the 36th AAAI conference on artificial intelligence, AAAI 2022, vol. 36; 2021. p. 3968–76. https://doi.org/10.1609/aaai.v36i4.20313.

Farahani N, Parwani AV, Pantanowitz L. Whole slide imaging in pathology: advantages, limitations, and emerging perspectives. Pathol Lab Med Int. 2015;7:23–33. https://doi.org/10.2147/PLMI.S59826.

Fay MM, Kraus O, Victors M, Arumugam L, Vuggumudi K, Urbanik J, Hansen K, Celik S, Cernek N, Jagannathan G, Christensen J, Earnshaw BA, Haque IS, Mabey B. RxRx3: Phenomics Map of Biology. bioRxiv; 2023. https://doi.org/10.1101/2023.02.07.527350.

Fischer AH, Jacobson KA, Rose J, Zeller R. Hematoxylin and eosin staining of tissue and cell sections, CSH protocols 2008; 2008. https://doi.org/10.1101/PDB.PROT4986.

Fuchs TJ, Buhmann JM. Computational pathology: challenges and promises for tissue analysis. Comput Med Imaging Graph. 2011;35:515–30. https://doi.org/10.1016/J.COMPMEDIMAG.2011.02.006.

Fürst A, Rumetshofer E, Lehner J, Tran VT, Tang F, Ramsauer H, Kreil D, Kopp M, Klambauer G, Bitto A, Hochreiter S. CLOOB: modern hopfield networks with InfoLOOB outperform CLIP. In: Koyejo S, Mohamed S, Agarwal A, Belgrave D, Cho K, Oh A, editors. Advances in neural information processing systems, vol. 35. Curran Associates, Inc; 2022. p. 20450–68.

Ganea O, Pattanaik L, Coley C, Barzilay R, Jensen K, Green W, Jaakkola T. GeoMol: torsional geometric generation of molecular 3D conformer ensembles. In: Ranzato M, Beygelzimer A, Dauphin Y, Liang P, Vaughan JW, editors. Advances in neural information processing systems, vol. 34. Curran Associates, Inc; 2021. p. 13757–69.

Gasteiger J, Groß J, Günnemann S. Directional message passing for molecular graphs. In: 8th international conference on learning representations (ICLR); 2020.

Gaudelet T, Day B, Jamasb AR, Soman J, Regep C, Liu G, Hayter JBR, Vickers R, Roberts C, Tang J, Roblin D, Blundell TL, Bronstein MM, Taylor-King JP (2021) Utilizing graph machine learning within drug discovery and development. Brief Bioinform 22:1–22, https://doi.org/10.1093/bib/bbab159.

Gaulton A, Bellis LJ, Bento AP, Chambers J, Davies M, Hersey A, Light Y, McGlinchey S, Michalovich D, Al-Lazikani B, Overington JP. Chembl: a large-scale bioactivity database for drug discovery. Nucleic Acids Res. 2011;40:D1100. https://doi.org/10.1093/NAR/GKR777.

Gayvert KM, Madhukar NS, Elemento O. A data-driven approach to predicting successes and failures of clinical trials. Cell Chem Biol. 2016;23:1294–301. https://doi.org/10.1016/J.CHEMBIOL.2016.07.023.

Ghiran IC. Introduction to fluorescence microscopy. Methods Mol Biol (Clifton, NJ). 2011;689:93–136. https://doi.org/10.1007/978-1-60761-950-5_7.

Grisoni F, Ballabio D, Todeschini R, Consonni V. Molecular descriptors for structure–activity applications: a hands-on approach. New York: Springer; 2018a. p. 3–53. https://doi.org/10.1007/978-1-4939-7899-1_1.

Grisoni F, Consonni V, Todeschini R. Impact of molecular descriptors on computational models. New York: Springer; 2018b. p. 171–209. https://doi.org/10.1007/978-1-4939-8639-2_5.

Gustafsdottir SM, Ljosa V, Sokolnicki KL, Anthony Wilson J, Walpita D, Kemp MM, Petri Seiler K, Carrel HA, Golub TR, Schreiber SL, Clemons PA, Carpenter AE, Shamji AF. Multiplex cytological profiling assay to measure diverse cellular states. PLoS One. 2013;8(12):null. https://doi.org/10.1371/journal.pone.0080999.

Gutmann M, Hyvärinen A. Noise-contrastive estimation: a new estimation principle for unnormalized statistical models. In: Teh YW, Titterington M, editors. Proceedings of the thirteenth international conference on artificial intelligence and statistics, PMLR, Chia Laguna Resort, Sardinia, Italy, proceedings of machine learning research, vol. 9; 2010. p. 297–304.

He K, Chen X, Xie S, Li Y, Dolla'r P, Girshick R. Masked autoencoders are scalable vision learners. In: Proceedings of the IEEE/CVF conference on Computer Vision and Pattern Recognition (CVPR); 2022. p. 16000–9.

Hofmarcher M, Rumetshofer E, Arneclevert DA, Hochreiter S, Klambauer G. Accurate prediction of biological assays with high-throughput microscopy images and convolutional networks. J Chem Inf Model. 2019;59:11631171. https://doi.org/10.1021/acs.jcim.8b00670.

Hu W, Liu B, Gomes J, Zitnik M, Liang P, Pande V, Leskovec J. Strategies for pre-training graph neural networks. In: 8th international conference on learning representations, ICLR 2020, international conference on learning representations, ICLR; 2019.

Hughes JP, Rees SS, Kalindjian SB, Philpott KL. Principles of early drug discovery. Br J Pharmacol. 2011;162:1239. https://doi.org/10.1111/J.1476-5381.2010.01127.X.

Hussein I, Raad M, Safa R, Jurjus RA, Jurjus A. Once upon a microscopic slide: the story of histology. J Cytol Histol. 2015;6 https://doi.org/10.4172/2157-7099.1000377.

Iizuka O, Kanavati F, Kato K, Rambeau M, Arihiro K, Tsuneki M. Deep learning models for histopathological classification of gastric and colonic epithelial tumours. Sci Rep. 2020;10:1–11. https://doi.org/10.1038/s41598-020-58467-9.

Jiang C, Hou X, Kondepudi A, Chowdury A, Freudiger CW, Orringer DA, Lee H, Hollon TC. Hierarchical discriminative learning improves visual representations of biomedical microscopy. In: Proceedings. IEEE computer society conference on computer vision and pattern recognition, Institute of Electrical and Electronics Engineers (IEEE), vol. 2023; 2023. p. 19798. https://doi.org/10.1109/CVPR52729.2023.01896.

Kim S, Chen J, Cheng T, Gindulyte A, He J, He S, Li Q, Shoemaker BA, Thiessen PA, Yu B, Zaslavsky L, Zhang J, Bolton EE. Pubchem 2023 update. Nucleic Acids Res. 2023;51:D1373–80. https://doi.org/10.1093/NAR/GKAC956.

Kipf TN, Welling M. Semi-supervised classification with graph convolutional networks. In: 5th international conference on Learning representations, ICLR 2017 – conference track proceedings, international conference on learning representations, ICLR; 2016.

Knowles J, Gromo G. A guide to drug discovery: target selection in drug discovery. Nat Rev Drug Discov. 2003;2:63–9. https://doi.org/10.1038/NRD986.

Kuhn M, Letunic I, Jensen LJ, Bork P. The sider database of drugs and side effects. Nucleic Acids Res. 2016;44:D1075–9. https://doi.org/10.1093/NAR/GKV1075.

Lapinsh M, Prusis P, Gutcaits A, Lundstedt T, Wikberg JE. Development of proteo-chemometrics: a novel technology for the analysis of drug-receptor interactions. Biochim Biophys Acta Gen Subj. 2001;1525:180–90. https://doi.org/10.1016/S0304-4165(00)00187-2.

Lecun Y, Bengio Y, Hinton G. Deep learning. Nature. 2015;521:436. https://doi.org/10.1038/nature14539.

Li P, Wang J, Qiao Y, Chen H, Yu Y, Yao X, Gao P, Xie G, Song S. An effective self-supervised framework for learning expressive molecular global representations to drug discovery. Brief Bioinform. 2021a;22(6):bbab109. https://doi.org/10.1093/bib/bbab109.

Li Z, Chen X, Zhao J, Xiong Z. Contrastive learning for mitochondria segmentation. In: Proceedings of the annual international conference of the IEEE engineering in medicine and biology society, EMBS; 2021b. p. 3496–500. https://doi.org/10.1109/EMBC46164.2021.9630350.

Liu S, Demirel MF, Liang Y. N-gram graph: simple unsupervised representation for graphs, with applications to molecules. In: Advances in neural information processing systems, vol. 32; 2019.

Liu X, Zhang F, Hou Z, Wang Z, Mian L, Zhang J, Tang J. Self-supervised learning: generative or contrastive. IEEE Trans Knowl Data Eng. 2020;35:857–76. https://doi.org/10.1109/TKDE.2021.3090866.

Liu S, Nie W, Wang C, Lu J, Qiao Z, Liu L, Tang J, Xiao C, Anandkumar A. Multi-modal molecule structure-text model for text-based retrieval and editing. Nat Mach Intell. 2022;5:1447–57. https://doi.org/10.1038/s42256-023-00759-6.

Lu Y, Hu Y, Li C. Drugclip: contrastive drug-disease interaction for drug repurposing. 2019; https://doi.org/10.1145/3307339.3343250.

Martins IF, Teixeira AL, Pinheiro L, Falcao AO. A Bayesian approach to in silico blood-brain barrier penetration modeling. J Chem Inf Model. 2012;52:1686–97. https://doi.org/10.1021/CI300124C.

Mayr A, Klambauer G, Unterthiner T, Hochreiter S. Deeptox: toxicity prediction using deep learning. Front Environ Sci. 2016;3:167215. https://doi.org/10.3389/FENVS.2015.00080/BIBTEX.

McKinney JD, Richard A, Waller C, Newman MC, Gerberick F. The practice of structure activity relationships (SAR) in toxicology. Toxicol Sci. 2000;56:8–17. https://doi.org/10.1093/TOXSCI/56.1.8.

Melville JL, Burke EK, Hirst JD. Machine learning in virtual screening. Comb Chem High Throughput Screen. 2009;12(4):332–43.

Mobley DL, Guthrie JP. Freesolv: a database of experimental and calculated hydration free energies, with input files. J Comput Aided Mol Des. 2014;28:711–20. https://doi.org/10.1007/S10822-014-9747-X.

Moriwaki H, Tian YS, Kawashita N, Takagi T. Mordred: a molecular descriptor calculator. J Chem. 2018;10:1–14. https://doi.org/10.1186/S13321-018-0258-Y/FIGURES/6.

Neves BJ, Braga RC, Melo-Filho CC, Moreira-Filho JT, Muratov EN, Andrade CH. Qsar-based virtual screening: advances and applications in drug discovery. Front Pharmacol. 2018;9:1275.

Pearl J. Embracing causality in default reasoning. Artif Intell. 1988;35(2):259–71.

Perakis A, Gorji A, Jain S, Chaitanya K, Rizza S, Konukoglu E. Contrastive learning of single-cell phenotypic representations for treatment classification, Lecture Notes in Computer Science (including subseries Lecture Notes in Artificial Intelligence and Lecture Notes in Bioinformatics) 12966 LNCS; 2021. p. 565–75. https://doi.org/10.1007/978-3-030-87589-3_58.

Poole B, Ozair S, van den Oord A, Alemi AA, Tucker G. On variational bounds of mutual information. In: Chaudhuri K, Salakhutdinov R, editors. Proceedings of the 36th international conference on machine learning, proceedings of machine learning research (PMLR), vol. 97; 2019. p. 5171–80.

Radford A, Kim JW, Hallacy C, Ramesh A, Goh G, Agarwal S, Sastry G, Askell A, Mishkin P, Clark J, Krueger G, Sutskever I. Learning transferable visual models from natural language supervision. In: Proceedings of the 38th international conference on machine learning, vol. 139; 2021. p. 8748–63.

Ramakrishnan R, Dral PO, Rupp M, Lilienfeld OAV. Quantum chemistry structures and properties of 134 kilo molecules. Sci Data. 2014;1:140022. https://doi.org/10.1038/sdata.2014.22.

Ramakrishnan R, Hartmann M, Tapavicza E, Lilienfeld OAV. Electronic spectra from TDDFT and machine learning in chemical space. J Chem Phys. 2015;143:084111. https://doi.org/10.1063/1.4928757.

Ramesh A, Dhariwal P, Nichol A, Chu C, Chen M. Hierarchical text-conditional image generation with clip latents. ArXiv 2204.06125; 2022.

Richard AM, Judson RS, Houck KA, Grulke CM, Volarath P, Thillainadarajah I, Yang C, Rathman J, Martin MT, Wambaugh JF, Knudsen TB, Kancherla J, Mansouri K, Patlewicz G, Williams AJ, Little SB, Crofton KM, Thomas RS. Toxcast chemical landscape: paving the road to 21st century toxicology. Chem Res Toxicol. 2016;29:1225–51. https://doi.org/10.1021/ACS.CHEMRESTOX.6B00135.

Richard AM, Huang R, Waidyanatha S, Shinn P, Collins BJ, Thillainadarajah I, Grulke CM, Williams AJ, Lougee RR, Judson RS, Houck KA, Shobair M, Yang C, Rathman JF, Yasgar A, Fitzpatrick SC, Simeonov A, Thomas RS, Crofton KM, Paules RS, Bucher JR, Austin

CP, Kavlock RJ, Tice RR. The tox21 10k compound library: collaborative chemistry advancing toxicology. Chemical research in toxicology. 2021;34:189–216. https://doi.org/10.1021/ACS.CHEMRESTOX.0C00264.

Ritchie TJ, Macdonald SJ. Physicochemical descriptors of aromatic character and their use in drug discovery. J Med Chem. 2014;57:7206–15. https://doi.org/10.1021/JM500515D.

Rogers D, Hahn M. Extended-connectivity fingerprints. J Chem Inf Model. 2010;50:742–54. https://doi.org/10.1021/ci100050t.

Rupp M, Tkatchenko A, Müller KR, von Lilienfeld OA. Fast and accurate modeling of molecular atomization energies with machine learning. Phys Rev Lett. 2012;108:058301. https://doi.org/10.1103/PhysRevLett.108.058301.

Russakovsky O, Deng J, Su H, Krause J, Satheesh S, Ma S, Huang Z, Karpathy A, Khosla A, Bernstein M, Berg AC, Fei-Fei L. Imagenet large scale visual recognition challenge. Int J Comput Vis. 2014;115:211–52. https://doi.org/10.1007/s11263-015-0816-y.

Sanchez-Fernandez A, Rumetshofer E, Hochreiter S, Klambauer G. Cloome: contrastive learning unlocks bioimaging databases for queries with chemical structures. Nat Commun. 2023;14:1–14. https://doi.org/10.1038/s41467-023-42328-w.

Schimunek J, Seidl P, Friedrich L, Kuhn D, Rippmann F, Hochreiter S, Klambauer G. Context-enriched molecule representations improve few-shot drug discovery. In: The eleventh international conference on learning representations (ICLR); 2023.

Schmidhuber J. Deep learning in neural networks: an overview. Neural Netw. 2015;61:85–117. https://doi.org/10.1016/J.NEUNET.2014.09.003.

Schölkopf B. Svms – a practical consequence of learning theory. IEEE Intell Syst Their Appl. 1998;13:18–21. https://doi.org/10.1109/5254.708428.

Schultz TW, Cronin MT, Walker JD, Aptula AO. Quantitative structure–activity relationships (QSARS) in toxicology: a historical perspective. J Mol Struct THEOCHEM. 2003;622:1–22. https://doi.org/10.1016/S0166-1280(02)00614-0.

Schütt KT, Kindermans PJ, Sauceda HE, Chmiela S, Tkatchenko A, KR M. Schnet: a continuous-filter convolutional neural network for modeling quantum interactions. Neural Inf Proces Syst. 2017;30:992–1002.

Seidl P, Vall A, Hochreiter S, Klambauer G. Enhancing activity prediction models in drug discovery with the ability to understand human language. In: Proceedings of the 40th International Conference on Machine Learning, JMLR.org, ICML'23; 2023.

Sestak F, Schneckenreiter L, Brandstetter J, Hochreiter S, Mayr A, Klambauer G. VN-EGNN: E(3)-equivariant graph neural networks with virtual nodes enhance protein binding site identification. arXiv 2404.07194; 2024.

Simm J, Klambauer G, Arany A, Steijaert M, Wegner JK, Gustin E, Chupakhin V, Chong YT, Vialard J, Buijnsters P, Ingrid V, Vapirev A, Singh S, Carpenter AE, Wuyts R, Hochreiter S, Moreau Y, Ceulemans H. Repurposing high-throughput image assays enables biological activity prediction for drug discovery HHS public access. Cell Chem Biol. 2018;25:611–8. https://doi.org/10.1016/j.chembiol.2018.01.015.

Singh S, Carpenter AE, Genovesio A. Increasing the content of high-content screening: an overview. J Biomol Screen. 2014;19:640–50. https://doi.org/10.1177/1087057114528537/ATTACHMENT/6DC3C89A-388F-49A1-84AB-CF3E7DF4CE3A/MMC1-SUPPL.PDF.

Srinidhi CL, Ciga O, Martel AL. Deep neural network models for computational histopathology: a survey. Med Image Anal. 2021;67:101813. https://doi.org/10.1016/j.media.2020.101813.

Stanley M, Bronskill J, Maziarz K, Misztela H, Lanini J, Segler M, Schneider N, Brockschmidt M. FS-Mol: a few-shot learning dataset of molecules. In: Proceedings of the neural information processing systems track on datasets and benchmarks, vol. 1; 2021.

Stärk H, Beaini D, Corso G, Tossou P, Dallago C, Günnemann S, Liò P. 3D info-max improves GNNs for molecular property prediction. In: Proceedings of machine learning research, vol. 162; 2021. p. 20479–502.

Svensson E, Hoedt PJ, Hochreiter S, Klambauer G. Hyperpcm: Robust task-conditioned modeling of drug–target interactions. J Chem Inf Model. 2024;64(7):2539–53. https://doi.org/10.1021/acs.jcim.3c01417. pMID: 38185877

Tabana Y, Babu D, Fahlman R, Siraki AG, Barakat K. Target identification of small molecules: an overview of the current applications in drug discovery. BMC Biotechnol. 2023;23:1–19. https://doi.org/10.1186/s12896-023-00815-4.

Taylor R, Kardas M, Cucurull G, Scialom T, Hartshorn A, Saravia E, Poulton A, Kerkez V, Stojnic R. Galactica: a large language model for science. arXiv 2211.09085v1; 2022.

Thul PJ, Lindskog C. The human protein atlas: a spatial map of the human proteome. Protein Sci. 2018;27:233–44. https://doi.org/10.1002/PRO.3307.

Todeschini R, Consonni V. Molecular descriptors for chemoinformatics volume I: alphabetical listing Second, Revised and Enlarged Edition. Mannheim: WILEY-VCH Verlag; 2009.

Trapotsi MA, Hosseini-Gerami L, Bender A. Computational analyses of mechanism of action (MoA): data, methods and integration. RSC Chem Biol. 2021;3:170. https://doi.org/10.1039/D1CB00069A.

Tropsha A, Isayev O, Varnek A, Schneider G, Cherkasov A. Integrating QSAR modelling and deep learning in drug discovery: the emergence of deep QSAR. Nat Rev Drug Discov. 2023;23:141–55. https://doi.org/10.1038/s41573-023-00832-0.

Tu Y, Lei H, Shen HB, Yang Y. SIFLoc: a self-supervised pre-training method for enhancing the recognition of protein subcellular localization in immunofluorescence microscopic images. Brief Bioinform. 2022;23:1–12. https://doi.org/10.1093/BIB/BBAB605.

van de Waterbeemd H, Gifford E. Admet in silico modelling: towards pre-diction paradise? Nat Rev Drug Discov. 2003;2:192–204. https://doi.org/10.1038/nrd1032.

van Westen GJP, Wegner JK, IJzerman AP, van Vlijmen HWT, Bender A. Proteochemometric modeling as a tool to design selective compounds and for extrapolating to novel targets. Med Chem Commun. 2011;2:16–30. https://doi.org/10.1039/C0MD00165A.

Walters WP, Stahl MT, Murcko MA. Virtual screening—an overview. Drug Discov Today. 1998;3(4):160–78. https://doi.org/10.1016/S1359-6446(97)01163-X.

Wang Y, Wang J, Cao Z, Farimani AB. Molecular contrastive learning of representations via graph neural networks. Nat Mach Intell. 2022;4:279–87. https://doi.org/10.1038/s42256-022-00447-x.

Wang J, Guan J, Zhou S. Molecular property prediction by contrastive learning with attention-guided positive sample selection. Bioinformatics. 2023;39 https://doi.org/10.1093/BIOINFORMATICS/BTAD258.

Weininger D. Smiles, a chemical language and information system: 1: introduction to methodology and encoding rules. J Chem Inf Comput Sci. 1988;28:31–6. https://doi.org/10.1021/CI00057A005.

Weitz P, Valkonen M, Solorzano L, Carr C, Kartasalo K, Boissin C, Koivukoski S, Kuusela A, Rasic D, Feng Y, Pouplier SKS, Sharma A, Eriksson KL, Latonen L, Laenkholm AV, Hartman J, Ruusuvuori P, Rantalainen M. Acrobat—a multi-stain breast cancer histological whole-slide-image data set from routine diagnostics for computational pathology. Sci Data. 2022;10 https://doi.org/10.1038/s41597-023-02422-6.

Wellman MP, Henrion M. Explaining 'explaining away'. IEEE Trans Pattern Anal Mach Intell. 1993;15(3):287–92. https://doi.org/10.1109/34.204911.

Wu Z, Ramsundar B, Feinberg E, Gomes J, Geniesse C, Pappu AS, Leswing K, Pande V. Moleculenet: a benchmark for molecular machine learning. Chem Sci. 2018a;9:513–30. https://doi.org/10.1039/C7SC02664A.

Wu Z, Xiong Y, Yu SX, Lin D. Unsupervised feature learning via non-parametric instance-level discrimination. In: Proceedings of the IEEE computer society conference on Computer Vision and Pattern Recognition; 2018b. p. 3733–42. https://doi.org/10.1109/CVPR.2018.00393.

Wu L, Huang R, Tetko IV, Xia Z, Xu J, Tong W. Trade off predictivity and explainability for ml-powered predictive toxicology: an in-depth investigation with tox21 datasets. Chem Res Toxicol. 2021;34:541. https://doi.org/10.1021/ACS.CHEMRESTOX.0C00373.

Xu K, Jegelka S, Hu W, Leskovec J. How powerful are graph neural networks? 2018.

Xue L, Bajorath J. Molecular descriptors in chemoinformatics, computational combinatorial chemistry, and virtual screening. Comb Chem High Throughput Screen. 2000;3:363–72. https://doi.org/10.2174/1386207003331454.

Yang K, Swanson K, Jin W, Coley C, Eiden P, Gao H, Guzman-Perez A, Hopper T, Kelley B, Mathea M, Palmer A, Settels V, Jaakkola T, Jensen K, Barzilay R. Analyzing learned molecular representations for property prediction. J Chem Inf Model. 2019;59:3370–88. https://doi.org/10.1021/ACS.JCIM.9B00237.

Yeh CH, Hong CY, Hsu YC, Liu TL, Chen Y, LeCun Y. Decoupled contrastive learning arXiv 2110.06848; 2021.

You Y, Chen T, Sui Y, Chen T, Wang Z, Shen Y. Graph contrastive learning with augmentations. Adv Neural Inf Proces Syst. 2020;33:5812–23.

Zeng Z, Yao Y, Liu Z, Sun M. A deep-learning system bridging molecule structure and biomedical text with comprehension comparable to human professionals. Nat Commun. 2022;13:1–11. https://doi.org/10.1038/s41467-022-28494-3.

Zhang C, Zhang K, Pham TX, Niu A, Qiao Z, Yoo CD, Kweon IS. Dual temperature helps contrastive learning without many negative samples: towards understanding and simplifying moco arXiv 2203.17248; 2022.

Few-Shot Learning in Drug Discovery

Johannes Schimunek and Günter Klambauer

Abstract Drug discovery is one of the research areas that is most critical for human health and well-being because of the tremendous impact of safe and efficient drugs on humans. Despite active research, the number of newly discovered drugs has been low in the last few decades because biological systems are complex and experiments are expensive. To increase the efficiency of the drug discovery process, in-silico screening has become a standard approach, using computational methods as efficient surrogates for wet lab experiments. Among these methods, AI methods have been found to be one of the most promising candidates, but they often require large amounts of data while, for many drug discovery projects, available data is scarce. Therefore, few-shot learning emerged as a subfield of AI, which refers to scenarios where the amount of data is limited and to models specifically designed for such low-data scenarios. This chapter introduces the concept of few-shot learning in the context of drug discovery and provides an overview of recent advancements aimed at overcoming the challenges posed by data scarcity.

Keywords Bioactivity prediction · QSAR modeling · Structure-activity relationships · Drug discovery · Few-shot learning · Low-data regimes

1 Introduction

Drug discovery is one of the research areas that is most critical for human health and well-being. Safe and efficient drugs (efficacy) have a tremendous impact on healthcare and society. Consequently, discovering new drugs has been a central focus for both pharmaceutical companies and academic researchers (Nicolaou 2014; Tralau-Stewart et al. 2009). However, despite substantial efforts, the number of newly approved drugs has remained disappointingly low over recent decades

J. Schimunek · G. Klambauer (✉)
ELLIS Unit Linz and LIT AI Lab, Institute for Machine Learning, Johannes Kepler University Linz, Linz, Austria
e-mail: schimunek@ml.jku.at; klambauer@ml.jku.at

A. Lavecchia (ed.), *Applied Artificial Intelligence for Drug Discovery*,
https://doi.org/10.1007/978-3-031-98022-0_16

(Dowden and Munro 2019; Hwang et al. 2016; Sun et al. 2022; Waring et al. 2015). A significant challenge lies in the fact that many drug candidates fail in late clinical phases due to issues with safety or efficacy (Bender and Cortés-Ciriano 2021a, b; Hwang et al. 2016). Such failures are particularly costly, as they follow up to a decade of work and investments amounting to billions of dollars. To address these challenges, researchers are striving to make the drug discovery process faster, less expensive, and less prone to late-stage failures.

Artificial intelligence (AI) methods, comprising computational and data-driven approaches, are among the most promising candidates to fulfill the role of making the drug discovery process faster, cheaper, and more successful. There are several reasons why AI is considered a powerful approach to drug discovery (Chen et al. 2018; Jiménez-Luna et al. 2021; Mak et al. 2024): First, its ability to learn from the available large chemical databases, such as PubChem (Kim et al. 2023), or ChEMBL (Zdrazil et al. 2024), or from large protein databases, such as PBD (Burley et al. 2023). For example, the Nobel-prize-winning method AlphaFold (Abramson et al. 2024) learns to predict 3D structures of proteins based on the amino-acid sequence, because of the availability of many of these sequence-structure pairs in PDB. Second, the fast development of new approaches over the last decade: Since Deep Learning methods have been shown to outperform classic quantitative structure-activity (QSAR) and toxicity prediction methods in benchmarks like the Tox21 data challenge (Mayr et al. 2016, 2018), there has been a surge of new methods for drug discovery, for example message-passing neural networks (Gilmer et al. 2017), generative methods (Gómez-Bombarelli et al. 2018; Segler et al. 2018), reinforcement learning methods (Olivecrona et al. 2017; Renz et al. 2019), and, more recently, large language models (Flam-Shepherd et al. 2022; Schmidinger et al. 2024; Seidl et al. 2023; Taylor et al. 2022; Zeng et al. 2022). This diverse range of AI methods, ranging from predictive models to generative approaches, forms a powerful toolkit that enhances various stages of the drug discovery process. Third, the ability to include biological knowledge and biology-informed models in the early stages of drug discovery. Traditional early-stage drug discovery is often heavily focused on chemical information about small molecules, with biological insights incorporated in later stages like lead optimization (Verbist et al. 2015). AI methods enable the integration of biological knowledge earlier in the process, e.g., through accurate bioactivity and toxicity prediction models or the analysis of high-throughput data, such as bioimaging data (Bray et al. 2016; Hofmarcher et al. 2019; Sanchez-Fernandez et al. 2023; Simm et al. 2018). Including information on biological systems early in the drug discovery process could be the key to reducing late failures (Bender and Cortés-Ciriano 2021a, b; Sun et al. 2022). These three assets, the availability of large datasets, the development of powerful approaches, and the ability to biologically inform early phases of drug discovery, make AI methods a transformative force in this field. Modern AI methods both open up completely new possibilities and often complement, augment, or even replace established in-silico methods in drug discovery.

In the early phases of drug discovery, large molecule libraries are virtually screened by in-silico methods. For many decades, virtual screening has been one of the first computational steps in drug discovery projects because the wet lab testing of large molecule libraries is too costly (Melville et al. 2009; Walters et al. 1998). Virtual screening methods rank molecules for subsequent wet-lab testing, often using QSAR-type models (Neves et al. 2018) to predict a molecule's activity or property on a drug target. QSAR models learn a mapping between a molecule's structure and its activity or property. While such models typically predict the activity of a molecule on a drug target that is the focus of a drug discovery project, additional models that predict toxic effects or ADMET properties could be used to re-rank or re-weight the ranking of the molecules (Liu et al. 2021; Unterthiner et al. 2014). However, prominently used machine learning methods, such as SVMs, Random Forests, and deep-learning-based approaches, require large amounts of training data to provide accurate predictions. This reliance on extensive data poses a challenge in early-phase drug discovery, where projects typically focus on novel targets with limited or no available data. This creates a dilemma: AI methods are highly effective when large datasets are available, which is usually the case for broadly known and thus less interesting drug targets, yet they struggle for crucial novel drug targets, where data is scarce. To address this, few-shot learning has emerged as a promising subfield of AI. Few-shot learning focuses on creating models that perform well even when only a small amount of data is available.

Few-shot learning provides strategies and models to overcome the low-data obstacle. Few-shot learning (Hochreiter et al. 2001) refers to the challenge of learning only from a few available data points. These few available data points are usually not sufficient data to train standard machine learning models like the ones mentioned above. Therefore, few-shot learning methods often incorporate meta-learning as one of the key principles to overcome the low-data obstacle. In meta-learning scenarios, a model is provided with a variety of different tasks: Being confronted with these tasks and learning useful knowledge about them, the meta-learner learns how to use the collected knowledge to build powerful models only from a few samples. Depending on the concrete strategy, the few samples can be used in different ways (see Sect. 2). Few-shot methods have shown promising performance on bioactivity benchmark datasets (Altae-Tran et al. 2017; Chen et al. 2023; Schimunek et al. 2023; Stanley et al. 2021; Wang et al. 2021) and also have been evaluated in drug discovery projects. For, Snyder et al. (2024) found that few-shot models outperform classic machine learning models and large language models in scenarios in which less than 50 data points are available.

This chapter will:

- introduce few-shot learning in drug discovery as the problem of learning from and with only a few known molecules,
- discuss gradient-based few-shot learning methods, which refine models using gradient updates on the few available data points,
- discuss similarity-based few-shot methods, which rely on comparing novel molecules to known ones, and,
- cover transductive few-shot learning methods, which assume that molecules to be predicted are presented simultaneously.

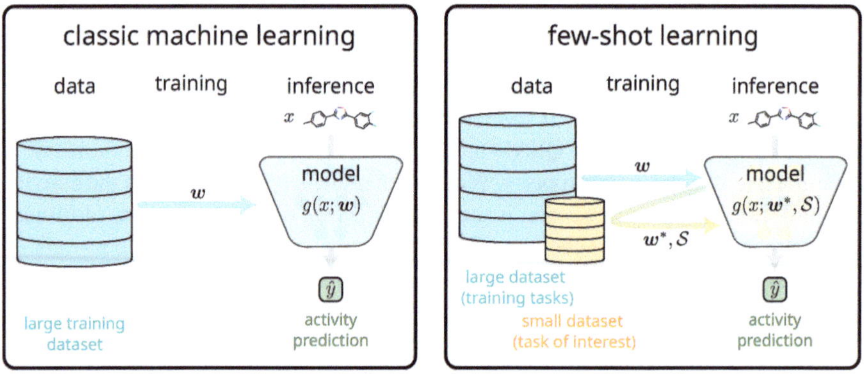

Fig. 1 Activity prediction models—comparison between classic machine learning and few-shot learning: classic machine learning models adjust learnable parameters ω on a typically large training dataset. In few-shot scenarios, the amount of available, known data S is limited. Typically, the model has access to another dataset of larger scale with training tasks

- compares few-shot model performances with other bioactivity models on a benchmark dataset and in a real-world case study, and,
- outlines on a concrete example how few-shot models can be integrated into drug discovery pipelines (Fig. 1).

2 Problem Settings

Few-shot learning refers to machine learning scenarios where the available data is limited. Few-shot models and their associated training strategies are geared to the question of how to learn from and with only a few known samples. Specifically, few-shot models are predictive models that, for a given task, use a set of known samples to make predictions for new, unseen samples. The set of known samples is referred to as the *support set*, while the set of samples for prediction is known as the *query set*. Few-shot models usually contain learnable parameters, which are trained on a training set.

Formally, let $g_{\boldsymbol{w}}(\cdot)$ be a few-shot model with learnable parameters $\boldsymbol{w}$, designed to predict some biological activity, physicochemical property, or toxicological response of untested chemical molecules from their molecular structure (Roy et al. 2015) for scenarios $\mathcal{T}_*$ where data is scarce. Typically, the model is provided with a set of training tasks $\mathcal{D}_{\text{train}} = \{\mathcal{T}_t\}_{t=1}^{T}$ during training and a set of unseen tasks during inference time (test time) $\mathcal{D}_{\text{eval}} = \{\mathcal{T}_*\} : \mathcal{D}_{\text{eval}} \cap \mathcal{D}_{\text{train}} = \varnothing$. Each task $\mathcal{T}$ comprises a set of data points, that is, pairs of molecular inputs x and associated labels y: $\mathcal{T} = \{(x_1, y_1), \ldots, (x_K, y_K)\}$. Labels are assumed to be binary or unknown: $y \in \{-1, 1, \boxminus\}$, where $\boxminus$ indicates the unknown class and molecular inputs are assumed to be symbolic or low-level representations of molecules, such as

SMILES (Mayr et al. 2018; Segler et al. 2018; Weininger 1988; Winter et al. 2019) or molecular graphs (Jiang et al. 2021; Kearnes et al. 2016; Merkwirth and Lengauer 2005; Yang et al. 2019). At inference time (test time), the amount of known data $\mathcal{S} \subset \mathcal{T}_*$ is assumed to be limited, meaning that only a small number of labels -1 or 1 are available. This labeled subset S is called the **support set** and can be used together with the training data $\mathcal{D}_{\text{train}}$ and a possibly complex training algorithm $\mathcal{A}(\cdot)$ to predict molecules $\{q_1, \dots, q_L\}$ from the query set $\mathcal{Q} \subset \mathcal{T}_*$: $\mathcal{Q} \cap \mathcal{S} = \varnothing$. Notably, samples associated with the label 1 (-1) are often considered active (inactive).

During inference, for a query molecule q, a few-shot model's prediction $\hat{y} = g\left(q; \boldsymbol{w}^*, \mathcal{S}^*\right)$ is determined by the incorporated learned weights $\boldsymbol{w}^*$ and the support set $\mathcal{S}^*$. The training algorithm $\mathcal{A} : (\boldsymbol{w}, \mathcal{S}, \mathcal{D}_{\text{train}}) \mapsto (\boldsymbol{w}^*, \mathcal{S}^*)$ in its most general form is applied to the few-shot model g during training and updates the model weights $\boldsymbol{w}$ and (or) the support set $\mathcal{S}$. Eventually, the trained model can be used to predict activity:

$$\hat{y}_l = g\left(q_l; \mathcal{A}\left(\boldsymbol{w}, \mathcal{S}, \mathcal{D}_{\text{train}}\right)\right) \quad \forall 1 \leq l \leq L. \tag{1}$$

Various training algorithms have been proposed, differing in how they utilize the training tasks and the support set. For, gradient-based few-shot methods use the support set to update the model weights (see Sect. 3.1), while similarity-based few-shot methods learn weights that enable comparison between query and support set molecules (see Sect. 3.2). Autoregressive few-shot methods (see Sect. 4.1) iteratively update the support set.

In standard few-shot scenarios (Stanley et al. 2021), the query samples are assumed to occur independently (Lazarou et al. 2021). This setting is called *inductive few-shot learning* and is described by Equation 1. Semi-supervised learning methods extend the support set $\mathcal{S} = \{(\boldsymbol{x}_1, y_1), \dots, (\boldsymbol{x}_N, y_N)\}$ with unlabeled data $\mathcal{U} = \{(\boldsymbol{x}_{N+1}, \boxminus), \dots, (\boldsymbol{x}_{N+M}, \boxminus)\}$ which frames few-shot learning as the task of learning from labeled and unlabeled data:

$$\hat{y}_l = g\left(q_l; \mathcal{A}\left(\boldsymbol{w}, \mathcal{S} \cup \mathcal{U}, \mathcal{D}_{\text{train}}\right)\right) \quad \forall 1 \leq l \leq L. \tag{2}$$

In *transductive inference*, the samples included in this additional unlabeled data set $\mathcal{U}$ are the query molecules $\{q_1, \dots, q_L\}$. This means that the query molecules are not i.i.d. and the few-shot model g has access to all query molecules simultaneously (Lazarou et al. 2021).

3 Inductive Few-Shot Methods

Inductive few-shot methods treat query molecules independently. In contrast to transductive few-shot methods, the few-shot model predicts the label for a query molecule q independently, without considering information from other query molecules.

Several attempts have been made to categorize inductive few-shot methods (Bendre et al. 2020; Bouniot et al. 2022; Duan et al. 2021; van Tilborg et al. 2024; Wang et al. 2020). These categorization efforts have been valuable for highlighting the similarities and differences between recently proposed methods. However, rather than forming clear, disjoint categories, the proposed categories often overlap or provide different perspectives on the same methods, making it difficult to assign a few-shot method to a single, distinct category. For the purpose of this chapter, we distinguish between methods that perform gradient updates on the support set and those that do not. Typically, few-shot methods in drug discovery that do not perform gradient updates rely on a similarity comparison between query and support set molecules. Therefore, we refer to them as similarity-based approaches.

3.1 Gradient-Based Few-Shot Methods

Gradient-based few-shot methods perform gradient updates on the support set. Typically, the training algorithm $\mathcal{A}$ comprises a pre-training and fine-tuning step. During pre-training, the few-shot model is provided with training tasks $\mathcal{D}_{\text{train to}}$ learn a weight configuration $\boldsymbol{w}$ which allows for efficient adaption during inference. During inference, the labeled samples from the support set are used for fine-tuning, which means that they are used to create an updated weight configuration $\boldsymbol{w}_*$ from the configuration $\boldsymbol{w}$. Depending on the specific method, this weight update during inference may affect all or only a subset of the learnable parameters.

Formally, the training algorithm $\mathcal{A}$ updates an initial weight configuration $\boldsymbol{w}^0$ in two steps: $\mathcal{A} = \mathcal{A}_{\text{fine-tune}} \circ \mathcal{A}_{\text{pretrain.}}$ In the most simple case, $\mathcal{A}_{\text{pretrain}} : \left(\boldsymbol{w}^0, \mathcal{D}_{\text{train}}\right) \mapsto \boldsymbol{w}$ takes the initial weight configuration and the training tasks as inputs. The fine-tuning step $\mathcal{A}_{\text{fine-tune}} : (\boldsymbol{w}, \mathcal{S}) \mapsto (\boldsymbol{w}^*, \cdot)$ uses the learned weights and the support set to return the updated weights, which eventually can be used to predict the bioactivity of a query molecule q:

$$\hat{y} = g_{\boldsymbol{w}^*}(q) \tag{3}$$

Notably, in this simple case, the training algorithm does not update the support set, and the support set is not directly needed as a model input to predict activity, which means that the support set returned from the training algorithm could be omitted (indicated here by "."). The choice of the training algorithm $\mathcal{A}$ is defined by the chosen few-shot method. Below, we exemplarily describe two different gradient-based few-shot methods MAML (Finn et al. 2017) and CHEF (Adler et al. 2020).

Model-agnostic meta-learning (MAML) is a strategy to learn a weight configuration from training tasks that can be adapted to novel tasks with just a few update steps. This is done by mimicking the weight optimization procedure, performed on the support set during inference, already during training on the training tasks.

Precisely, the few-shot learning problem is framed as a meta-learning problem. While sampling from the training data and mimicking inference scenarios, i.e., performing weight updates on the support set to optimize a weight configuration for a specific task and evaluating the model performance on a query set using the updated weights, the model is provided with a meta-learning scheme to optimize the weight configuration which is used as an initialization for the weight optimization procedure performed during inference. Many different variants have been proposed and applied to drug discovery, e.g. MAML and meta-learning variants have been combined with GNNs (Guo et al. 2021; Nguyen et al. 2020; Stanley et al. 2021; Wang et al. 2021). Precisely, Guo et al. (2021) refine the MAML training scheme with a module called self-attentive mechanism to control a training task's contribution to the model weight updates via measuring task importance. Nguyen et al. (2020) evaluate GNNs for two MAML versions, ANIL and FO-MAML, to reduce the computational burden of the training scheme. Wang et al. (2021) introduce Property-Aware Relation Networks (PAR) which build up on GNNs and MAML and also include two modules to update the molecule representations, a property-aware embedding function to refine molecular representations for a given task, and a relation graph learning module to allow for information-sharing between query and support-set molecules. Chen et al. (2023) propose a few-shot model ADKF-IFT, which learns deep kernel Gaussian processes with a meta-learning strategy.

Cross-domain Hebbian ensemble few-shot learning (CHEF) is a strategy that includes (a) a pre-training step in which the few-shot model is trained on provided tasks $\mathcal{D}_{\text{train}}$ (b) a fine-tuning step in which linear probing is applied to different layers of abstraction, i.e., the model's different (hidden) layers, and (c) a prediction step in which activity prediction is built by fusing the outputs from the linear probing layers. With this strategy, representations of different levels of abstraction are adapted and used for the task $\mathcal{T}_*$, which could help to address a typical obstacle applying few-shot models to novel tasks: domain shift, that is a change of the input-target distribution (Adler et al. 2020). The rationale behind this approach is that tasks associated with a large domain shift may benefit from lower-level representations, while tasks with only a small domain shift can leverage higher-level abstractions. By integrating multiple layers of representation, CHEF aims to preserve performance across tasks with varying levels of domain shift.

3.2 Similarity-Based Few-Shot Methods

Similarity-based few-shot models can be written in the following generalized form, from which different few-shot models arise as special cases (Schimunek et al. 2021):

$$\hat{y} = g_{\boldsymbol{w}}(q, S) \tag{4}$$

$$= h^{\text{assoc}} \left(\boldsymbol{h}^{\text{query}}(q), h^{\text{support}}(\mathcal{S}) \right), \tag{5}$$

where the potentially learnable model parameters $\boldsymbol{w}$ might have been trained on a training dataset $\mathcal{D}_{\text{train}}$ using a learning algorithm $\mathcal{A}$. In this formulation, similarity-based few-shot models are two place functions that take both a query molecule q and a support set $\mathcal{S}$ as inputs. The association function h^{assoc}, the query molecule encoder $\boldsymbol{h}^{\text{query}}$, and the support set encoder h^{support} can either be chemoinformatics operations, such as descriptor or similarity computations, or modules with learnable parameters, e.g., deep learning layers.

Intuitively, similarity-based few-shot models create embeddings for both the query and the support set, which are eventually associated with a predicted activity. In some models, the support set encoder h^{support} creates embeddings for each support set molecule individually. In others, h^{support} applies a pooling operation to create joint representations, e.g., class-wise embeddings.

3.2.1 Similarity Search

Similarity Search (Cereto-Massagué et al. 2015) is a classic chemoinformatic technique in which a molecule of interest q is compared with a set of known active molecules. In this context, the support set $\mathcal{S}$ consists solely of molecules from one class—the active class.

In the simplest case, activity prediction is performed by computing fingerprints or descriptor-based embeddings for both the query molecule and the support set molecules, and by applying a similarity measure $k(\cdot. \cdot)$ e.g., the Tanimoto Similarity (Tanimoto 1960).

Thus, the simple Similarity Search variant can be written as:

$$\hat{y} = \frac{1}{K} \sum_{i=1}^{K} y_i k \left(\boldsymbol{h}^{\text{desc}}(q), \boldsymbol{h}^{\text{desc}}(x_i) \right), \tag{6}$$

where $\boldsymbol{h}^{\text{desc}}$ maps a query molecule q or a support set molecule x to its chemical descriptor- or fingerprint-based embedding. Formally, we can define $\boldsymbol{h}^{\text{query}}$ to be identical to $\boldsymbol{h}^{\text{desc}}$, and h^{support} as the function which applies $\boldsymbol{h}^{\text{desc}}$ element-wise to the molecular inputs present in the support set and outputs a set of the support set molecules' embeddings along with their associated labels: $h^{\text{support}}(\mathcal{S}) = \left\{ \left(\boldsymbol{h}^{\text{des}}(x_i), y_i \right) \mid \forall i \in \{1, \ldots, K\} \right\}$. The association function h^{assoc} (a) computes pairwise similarity measures between query and support set molecules and (b) performs mean pooling across support set labels weighted by their associated similarity values.

Many different variants of Similarity Search have been proposed (Cereto-Massagué et al. 2015; Eckert and Bajorath 2007; Geppert et al. 2008; Riniker and Landrum 2013; Sheridan and Kearsley 2002; Wang et al. 2010; Willett 2014) for which some of them correspond to recently proposed few-shot methods with fixed encoders. For, Geppert et al. (2008) describe centroid-based searches (Schuffenhauer et al. 2003), in which averages (prototypes) of actives are built (compare Prototypical Networks in Sect. 3.2.3). Riniker and Landrum (2013) describe different fusion strategies for active and inactive molecules, which correspond to different fusion strategies of the support set.

3.2.2 Neural Similarity Search

The Neural Similarity Search arises as a natural variant of the Similarity Search by replacing the fixed embedding function $\boldsymbol{h}^{\text{desc}}$ with learnable embedding functions $\boldsymbol{h}_{\boldsymbol{w}_{\text{q}}}$ and $\boldsymbol{h}_{\boldsymbol{w}_{\text{s}}}$ for a query molecule q and a support set $\mathcal{S}$. Formally, this can be expressed as:

$$\boldsymbol{h}^{\text{query}} = \boldsymbol{h}_{\boldsymbol{w}_{\text{q}}} \tag{7}$$

$$h^{\text{support}}(S) = \left\{ \left(\boldsymbol{h}_{\boldsymbol{w}_{\text{S}}}(x_i), y_i \right) \mid \forall i \in \{1, \ldots, K\} \right\}. \tag{8}$$

In particular, the embedding functions $\boldsymbol{h}_{\boldsymbol{w}_{\text{q}}}$ and $\boldsymbol{h}_{\boldsymbol{w}_{\text{s}}}$ could also be hybrid encoders, combining both traditional chemoinformatics features and learnable deep learning modules (Mayr et al. 2016; Unterthiner et al. 2014).

When the query and support set encoders share weights, i.e., $\boldsymbol{w}_{\text{q}} = \boldsymbol{w}_{\text{S}} = \boldsymbol{w}$, the approach is referred to as metric learning with Siamese networks (Hertz et al. 2006; Koch et al. 2015; Ye and Guo 2018) and has been suggested for drug discovery scenarios by Torres et al. (2020), using Convolutional Neural Networks as encoders, and by Schimunek et al. (2021), using fingerprint-based MLPs as encoders. Unlike the classic Similarity Search, the association function in the Neural Similarity Search typically includes a sigmoid function σ that applies to the pooled and weighted support set labels:

$$\hat{y} = \sigma \left(\frac{1}{K} \sum_{i=1}^{K} y_i k \left(\boldsymbol{h}_{\boldsymbol{w}}(q), \boldsymbol{h}_{\boldsymbol{w}}(x_i) \right) \right). \tag{9}$$

When the similarity metric k is the dot product, Eq. 9 simplifies to:

$$\hat{y} = \sigma \left(\frac{1}{K} \sum_{i=1}^{K} y_i \boldsymbol{h}_{\boldsymbol{w}}(q)^T \boldsymbol{h}_{\boldsymbol{w}}(x_i) \right). \tag{10}$$

In particular, both Eqs. 9 and 10 may also include scaling factors or class balancing schemes (Schimunek et al. 2021), which are omitted here for simplicity.

3.2.3 Prototypical Networks (ProtoNet)

Similar to the approaches described above, prototypical networks (Snell et al. 2017) include a query encoder $\boldsymbol{h}^{\text{query}}$ and a support set encoder h^{support}. However, instead of creating separate embeddings for each support set molecule, ProtoNets computes prototypical embeddings for each class by averaging the embeddings of the support set molecules belonging to that class. Eventually, activity predictions are derived through pairwise comparisons between the query molecule and the computed prototypes.

Assuming that the query and the support set encoders share the learnable weights, as done in previous work (Snyder et al. 2024; Stanley et al. 2021; Vella and Ebejer 2022), both encoders use the same embedding function $\boldsymbol{h}_w$. The support set encoder can be defined as $h^{\text{support}} : \mathcal{S} \mapsto \left(\boldsymbol{r}^+, \boldsymbol{r}^-\right)$, where $\boldsymbol{r}^+ = \frac{1}{|\mathcal{S}^+|} \cdot \sum_{(x,y)\in\mathcal{S}^+} \boldsymbol{h}_w(x)$, and $\boldsymbol{r}^- = \frac{1}{|\mathcal{S}^-|} \cdot \sum_{(x,y)\in\mathcal{S}^-} \boldsymbol{h}_w(x)$ are the prototypical embeddings for the active and inactive classes, respectively. Here, $\mathcal{S}^+ = \{(x, y) \in \mathcal{S} \mid y = 1\}$ and $\mathcal{S}^- = \{(x, y) \in \mathcal{S} \mid y = -1\}$ are subsets of the support set containing only active and only inactive molecules. The query encoder is defined as the embedding function: $\boldsymbol{h}^{\text{query}} = \boldsymbol{h}_w$. Let $\boldsymbol{q}$ be the query embedding: $\boldsymbol{q} = \boldsymbol{h}^{\text{query}}(q)$. For a given distance metric $d(\cdot, \cdot)$, the association function h^{assoc} computes the predicted activity $\hat{y}$ as:

$$\boldsymbol{h}^{\text{assoc}} : \left(\boldsymbol{q}, \boldsymbol{r}^+, \boldsymbol{r}^-\right) \mapsto \hat{y} \tag{11}$$

$$\hat{y} = \frac{\exp\left(-d\left(\boldsymbol{q}, \boldsymbol{r}^+\right)\right)}{\exp\left(-d\left(\boldsymbol{q}, \boldsymbol{r}^+\right)\right) + \exp\left(-d\left(\boldsymbol{q}, \boldsymbol{r}^-\right)\right)} \tag{12}$$

3.2.4 Matching Networks and IterRefLSTM

Matching networks (Vinyals et al. 2016) use an embedding function to create embeddings for the query and the support set molecules, analogously to how this is done for the Neural Similarity Search discussed in Sect. 3.2.2. These embeddings are then processed by an attention-enhanced LSTM (Hochreiter and Schmidhuber 1997; Vinyals et al. 2015, 2016), which enables information sharing between the query molecule and the support set to eventually retrieve an updated query representation. As with previously discussed methods, the activity prediction is computed as a weighted sum of the support set labels, where the weights are derived from pairwise similarity comparisons between query and support set representations. Building on this concept, Altae-Tran et al. (2017) proposed the Iterative Refinement of Long Short-Term Memory (IterRefLSTM), replacing the LSTM variant used in Matching Networks. The IterRefLSTM allows for iterative updates not only to the query embedding but also to the support set embeddings.

Precisely, in the IterRefLSTM-based method, the query and support set encoders are defined as in the Neural Similarity Search variant (Eqs. 7 and 8). Let $\boldsymbol{q}$ denote the query molecule embedding, $\boldsymbol{X}$ the stacked embeddings of the support set molecules, and $\boldsymbol{y}$ or the stacked support set labels. The association function is defined as a composition of three functions, $\boldsymbol{h}^{\text{assoc}, 1}$, which enables information sharing between query and support set molecules, $\boldsymbol{h}^{\text{assoc}, 2}$, which computes similarity values using a similarity function $\boldsymbol{k}(\cdot, \cdot)$, and $\boldsymbol{h}^{\text{assoc}, 3}$, which computes the activity prediction from the weighted mean over the support set labels:

$$\boldsymbol{h}^{\text{assoc},1} : (\boldsymbol{q}, \boldsymbol{X}) \mapsto (\boldsymbol{q}', \boldsymbol{X}') \tag{13}$$

$$(\boldsymbol{q}', \boldsymbol{X}') = \text{IterRefLSTM}_J\,(\boldsymbol{q}, \boldsymbol{X}) \tag{14}$$

$$\boldsymbol{h}^{\text{assoc},2} : (\boldsymbol{q}', \boldsymbol{X}') \mapsto a \tag{15}$$

$$a = \text{softmax}\left(\boldsymbol{k}\left(\boldsymbol{q}', \boldsymbol{X}'\right)\right) \tag{16}$$

$$\boldsymbol{h}^{\text{assoc},3} : (\boldsymbol{y}, \boldsymbol{a}) \mapsto \hat{y} \tag{17}$$

$$\hat{y} = \sum_{i=1}^{N} a_i \cdot y_i \tag{18}$$

$$\boldsymbol{h}^{\text{assoc}} = \boldsymbol{h}^{\text{assoc},3} \circ \boldsymbol{h}^{\text{assoc},2} \circ \boldsymbol{h}^{\text{assoc},1}. \tag{19}$$

Here, the hyperparameter $J \in \mathbb{R}$ controls the number of iteration steps of the IterRefLSTM. We note that the input $\boldsymbol{y}$ is only introduced at $\boldsymbol{h}^{\text{assoc}, 3}$ and does not pass through $\boldsymbol{h}^{\text{assoc}, 1}$ and $\boldsymbol{h}^{\text{assoc}, 2}$ for simplicity of notation. Thus, Eq. 19 represents the composition of functions with $\boldsymbol{y}$ included only at the final stage.

3.2.5 MHNfs: Context-Enriched Molecule Representations

MHNfs (Schimunek et al. 2023) is a similarity-based few-shot method that builds on concepts of the IterRefLSTM-based method and the Neural Similarity Search variant and introduces the idea of using broader information to enrich the query and support set representations.

The query and support set encoders to share weights and include (a) a hybrid embedding function that extracts fingerprints and molecular descriptors from a molecule's low-level representation followed by an MLP, and (b) a Context Module that enriches the molecules' embeddings with context information, i.e., broader information from an external set of molecules, by associating both the query and the support set molecules with a large set of molecules present in the training data. The embedding function $\boldsymbol{h}_w$ maps a query molecule q and the support set molecules $\{x_1, \ldots, x_K\}$ to embeddings $\boldsymbol{q}$ and $\{x_1, \ldots, x_K\}$:

$$q = \boldsymbol{h}_w(q) \tag{20}$$

$$\{\boldsymbol{x}_1, \ldots, \boldsymbol{x}_K\} = \{\boldsymbol{h}_w(x_i) \mid i \in \{1, \ldots, K\}\}, \tag{21}$$

and $\boldsymbol{X}$ denotes the stacked support set molecule embeddings. The Context-Module enriches the query and support set embeddings by associating them with a large set of molecules—the context set $\boldsymbol{C}$—using a Modern Hopfield Network layer (MHN) (Ramsauer et al. 2020):

$$\boldsymbol{q}' = Hopfield(q, C) \tag{22}$$

$$\boldsymbol{X}' = Hopfield(X, C) \tag{23}$$

The association function consists of two modules, the Cross-Attention Module and the Similarity Module. The Cross-Attention Module adopts the concept of information sharing between query and support set molecules (Altae-Tran et al. 2017) but instead of the IterRefLSTM a transformer architecture (Vaswani 2017) is used:

$$\left[\boldsymbol{q}'', \boldsymbol{X}''\right] = \text{Transformer}\left(\left[\boldsymbol{q}', \boldsymbol{X}'\right]\right), \tag{24}$$

where $[\cdot, \cdot]$ indicates concatenation and the transformer performs self-attention across query and support set molecules. The Similarity Module computes predictions as a weighted sum over the support set labels, using the enriched query and support set representations to compute pairwise similarities—analogous to Eqs. 6 and 10.

3.3 In-Context Learning and Context-Enrichment

Similarity-based few-shot methods can be considered in-context learning methods if the definition of in-context learning is extended to drug discovery (Schimunek et al. 2025).

In-context learning refers to a paradigm for large language models to learn tasks solely from a few demonstrations given in the form of examples (Brown et al. 2020; Dong et al. 2022). In in-context learning scenarios, a model is provided with an optional task instruction, known data, and a query sample to generate the query sample's prediction. This is analogous to similarity-based few-shot methods (see Fig. 2), in which the few-shot model is prompted with known data (the support set) and a query sample to compute the query sample's prediction. The task instruction (i.e., "*predict activity*") is omitted because it is constant on all prompts.

In many natural language processing settings, large language models are equipped with a large context window size, which enables the model in few-shot scenarios to not only consider the few examples provided for the task but

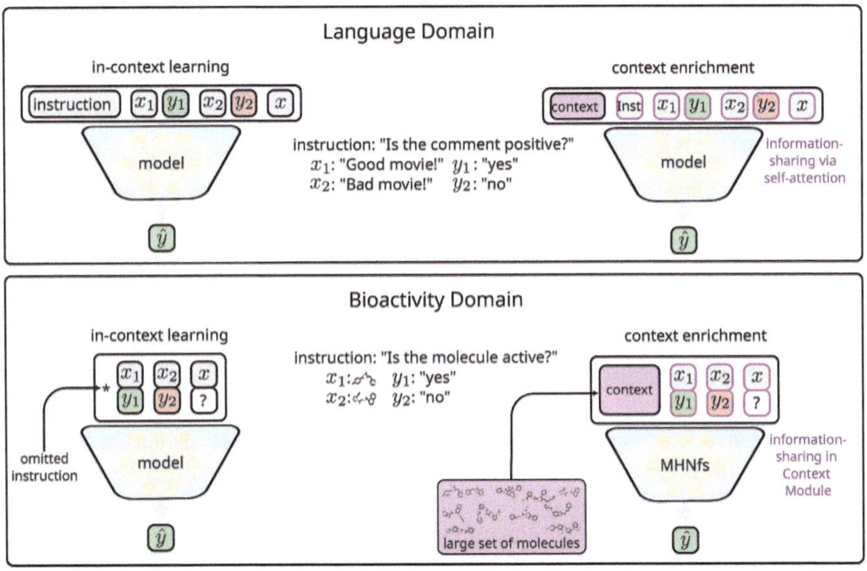

Fig. 2 In-context learning and context enrichment. Top: Language domain (adapted from Chen et al. (2021)). In in-context learning, few-shot tasks are solved by providing an instruction together with known labels and a query x to the LLM. The context window usually includes more general information and shares information via self-attention. Bottom: Bioactivity domain. Analogously, the model is provided with known data (support set) and a query to output activity. Since the instruction does not change for different tasks, it is omitted (indicated by *). Also few-shot models for drug discovery might use more general context information, e.g., a large set of training molecules (here shown exemplarily for MHNfs)

also more general information that might be additionally available, e.g., from text documents. The process of enriching the model prompt with additional, more general information— we refer to it as context enrichment (Schimunek et al. 2025)— has been shown to enhance performance in various tasks (Bai et al. 2023; Schmidinger et al. 2024; Wang et al. 2024; Zhang et al. 2023). The idea of context enrichment also has been introduced to few-shot drug discovery. Specifically, MHNfs enriches the query and supports set embeddings with context through the Context Module.

4 Transductive Few-Shot Methods

In transductive learning scenarios, a used few-shot model has access to all query molecules $\{q_1, \ldots, q_L\}$ at once (Lazarou et al. 2021), which allows the model to leverage information about the entire query set for predictions.

Section 2 and Eq. 2 describe semi-supervised learning, transductive inference, and their relationship. In semi-supervised learning, the model uses both labeled support set samples and unlabeled molecules $\mathcal{U}$ for predictions.

Recent methods used in the vision domain (Iscen et al. 2019; Lazarou et al. 2021; Liu et al. 2018; Zhu and Koniusz 2023) leverage this additional set of molecules $\mathcal{U}$ to (iteratively) augment the support set with pseudo-labeled samples. These methods rely on label propagation (Liu et al. 2018; Zhou et al. 2003; Zhu and Ghahramani 2002) which is the process of creating pseudo-labels for unlabeled samples by propagating given label information through a nearest-neighbor-based graph. In the field of drug discovery, label propagation has been used for drug-drug interaction prediction (Zhang et al. 2015), drug-target interaction prediction (Zhang et al. 2017), and for identifying drugs with specific functions in multi-drug prescriptions (Zhang et al. 2020).

Other transductive methods used in vision-based few-shot scenarios leverage information from the query set to adapt the feature space. This can be achieved, for example, by performing query-set-informed feature preprocessing (Hu et al. 2021b) or by applying embedding propagation (Hu et al. 2021a; Rodríguez et al. 2020). Embedding propagation leverages interpolations between the extracted features of a neural network based on a similarity graph. Specifically, embedding propagation works analogously to label propagation, but instead of labels, features are propagated through the graph. Embedding propagation is conceptually similar to the context-enrichment step in MHNfs, with the difference that MHNfs is not a transductive learning method because the model does not have access to all query molecules simultaneously.

4.1 Autoregressive Activity Predictions

Building on the idea of label propagation and the observation that a few-shot model's performance often correlates with the support set size (Chen et al. 2023; Schimunek et al. 2023; Stanley et al. 2021), recently an autoregressive inference strategy for inductive few-shot drug-discovery models has been suggested (Schimunek et al., 2024). In the proposed autoregressive inference mode, a given few-shot model iteratively updates the support set with pseudo-labeled samples (see Fig. 3).

A selection policy π selects the pseudo-labeled samples to be added to the support set from an unlabeled set of molecules $\mathcal{U} = \{(x_{N+1}, \boxminus), \ldots, (x_{N+M}, \boxminus)\}$ based on the model predictions:

$$\pi\left(\left\{\left(u_1, \hat{y}_1\right), \ldots, \left(u_M, \hat{y}_M\right)\right\}\right) = \left(u_i, \hat{y}_i\right) \tag{25}$$

This selection procedure is performed in an iterative fashion to gradually improve the model performance for consecutive query samples:

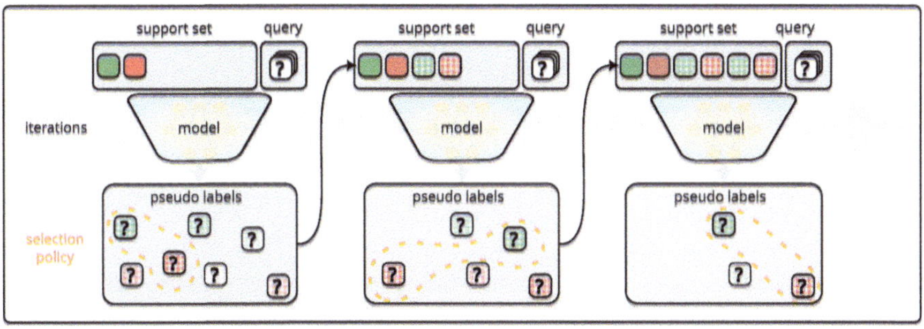

Fig. 3 Autoregressive inference mode. A few-shot model iteratively augments the support set with pseudo labeled samples

$$(u_1, \hat{y}_1) = \pi\left(\left\{(u, \hat{y}) \mid \hat{y} = g\left(U_1, \mathcal{S}\right), u \in U_1\right\}\right) \tag{26}$$

$$(u_2, \hat{y}_2) = \pi\left(\left\{(u, \hat{y}) \mid \hat{y} = g\left(U_2, \mathcal{S} \cup \left\{(u_1, \hat{y}_1)\right\}\right), u \in U_2\right\}\right) \tag{27}$$

$$\ldots \tag{28}$$

$$(u_n, \hat{y}_n) = \pi\left(\left\{(u, \hat{y}) \mid \hat{y} = g\left(U_{n-1}, \mathcal{S} \cup \left\{(u_i, \hat{y}_i)\right\}_{i=1}^{n-1}\right), u \in U_{n-1}\right\}\right) \tag{29}$$

where $U_i = (u_i, u_i + 1, \ldots, u_M)$. For simplicity, we assume that the selection policy π selects the unlabeled elements sequentially.

The autoregressive inference scheme is similar to **active learning**. In active learning, a method selects beneficial samples but, instead of using pseudo-labels, the true labels are measured and eventually samples with their true labels are added to the set of known molecules. Autoregressive strategies have been shown to improve the predictive performance of few-shot learning methods in drug discovery when the initial support set size is smaller than 4 (Schimunek et al. 2024a). Thus, autoregressive strategies should be used in these cases.

5 Few-Shot Benchmarks, Performance Metrics, and Model Comparisons

This section provides quantitative comparisons of few-shot methods in two scenarios (a) a scientifically curated few-shot benchmark dataset FS-Mol (Stanley et al. 2021) and (b) a real-world case study (Snyder et al. 2024).

Table 1 Results on FS-MOL [ΔAUC-PR] from Schimunek et al. (2023). The best method is marked bold. Error bars represent standard errors across tasks according to Stanley et al. (2021). In brackets the number of tasks per category is reported

Method	All [157]	Kin. [125]	Hydrol. [20]	Oxid.[7]
GNN-ST[a] (Stanley et al. 2021)	.029 ± .004	.027 ± .004	.040 ± .018	.020 ± .016
MAT[a] (Maziarka et al. 2020)	.052 ± .005	.043 ± .005	.095 ± .019	.062 ± .024
Random Forest[a] (Breiman 2001)	.092 ± .007	.081 ± .009	.158 ± .028	.080 ± .029
GNN-MT[a] (Stanley et al. 2021)	.093 ± .006	.093 ± .006	.108 ± .025	.053 ± .018
Similarity search	.118 ± .008	.109 ± .008	.166 ± .029	.097 ± .033
GNN-MAML[a] (Stanley et al. 2021)	.159 ± .009	.177 ± .009	.105 ± .024	.054 ± .028
PAR (Wang et al. 2021)	.164 ± .008	.182 ± .009	.109 ± .020	.039 ± .008
Frequent hitters	.182 ± .010	.207 ± .009	.098 ± .009	.041 ± .005
ProtoNet[a] (Snell et al. 2017)	.207 ± .008	.215 ± .009	.209 ± .030	.095 ± .029
Siamese Networks (Koch et al. 2015)	.223 ± .010	.241 ± .010	.178 ± .026	.082 ± .025
IterRefLSTM (Altae-Tran et al. 2017)	.234 ± .010	.251 ± .010	.199 ± .026	.098 ± .027
ADKF-IFT[b] (Chen et al. 2023)	.234 ± .009	.248 ± .020	.217 ± .017	.106 ± .008
MHNfs	.241 ± .009	.259 ± .010	.199 ± .027	.096 ± .019

[a]results from Stanley et al. (2021). [b]results from Chen et al. (2023)

5.1 *FS-Mol Few-Shot Benchmark Experiment*

Stanley et al. (2021) proposed the FS-Mol dataset to benchmark inductive few-shot learning methods in drug discovery. The dataset was extracted from ChEMBL27 (Mendez et al. 2019) and includes 489,133 measurements, 233,687 compounds, and 5120 tasks. The dataset is split into 4938 training, 40 validation, and 157 test tasks, guaranteeing disjoint task sets. Consequently, models are trained on the training tasks and eventually evaluated on the test tasks. Models are evaluated with respect to the ΔAUC-PR metric, in which the area under the precision-recall curve (AUPRC) score is compared to a random classifier.

Table 1 shows the performance values for different bioactivity models, being provided with 16 known molecules for each of the FS-Mol test tasks (Chen et al. 2023; Schimunek et al. 2023; Stanley et al. 2021). The standard error is reported across tasks. Also, performance values for different protein families, namely kinases, hydrolases, and oxidoreductases, are included. The method comparison includes few-shot methods discussed above—the Similarity Search, MAML-based models (GNN-MAML, PAR, and ADKF-IFT), Prototypical Networks, Siamese Networks, the IterRefLSTM-based model, and MHNfs—as well as traditional machine learning methods—Random Forests, Graph Neural Networks run in a single-task (GNN-ST) and multi-task (GNN-MT) setup, and Molecule Attention Transformer (MAT). The Frequent Hitter model (Schimunek et al. 2023) is a naive baseline that ignores the support set and, therefore, learns to predict a molecule's average activity.

The results indicate that few-shot models tend to outperform traditional few-shot models in scenarios in which only 16 data points are known. Interestingly, even the naive Frequent Hitter model outperforms the traditional machine learning approaches.

Notably, Snyder et al. (2024) also prepared a ChEMBL-based dataset and found that few-shot models generally outperform other machine learning approaches, namely Support Vector Machines (Mammone et al. 2009) and a transformer-based model (MolBART (Chilingaryan et al. 2023)), on scenarios with less than 50 known molecules.

5.2 *Case Study: Discovery of Novel Kinase Inhibitors for Alzheimer's Disease (AD)*

In a case study, Snyder et al. (2024) used machine learning models to find potential drug candidates for AD and showed in wet lab experiments that in particular the candidates suggested by the few-shot learning method worked well. To discover new kinase inhibitors for AD, Snyder et al. (2024) identified five different relevant kinase targets, i.e. GSK3β, ABL1, FYN, CDK5, and MARK1. Collecting known data from CheMBL to set up training sets for the machine-learning methods and known data from BindingDB as external test sets, they found that the amount of available data for one of the identified targets, MARK1, is very limited. Precisely, only five active and 13 inactive molecules have been found on ChEMBL, and on BindingDB no measurements have been available, which means no external test set could be curated for for this target to compare model performances. Therefore, Snyder et al. (2024) built a prospective test set of 13 molecules by performing a virtual screening of FDA-approved compounds. In the virtual screening step, traditional machine learning methods, namely Support Vector machines and a transformer-based model (MolBART), as well as a few-shot learning method (Prototypical Network) have been used for molecule selection. Eventually, in wet lab experiments, five molecules have been identified as novel MARK1 active inhibitors. Notably, the few-shot learning method was able to predict all of the five found inhibitors as active, outperforming the other machine learning approaches (see Table 2).

6 Applying Few-Shot Learning Methods in Drug Discovery Projects

Few-shot learning is frequently evaluated through benchmark tasks, but its true value becomes most evident in real-world drug discovery scenarios. In early-phase development, two common low-data situations often arise in practice.

Table 2 Results on the MARK1 prospective dataset in the AD case study from Snyder et al. (2024)

Method	True positives	True negatives	False positives	False negatives	p
SVM	0	8	0	5	0
MolBART	3	8	0	2	0.6
Few-shot method	5	7	1	0[a]	1

[a]Snyder et al. (2024) report a value of 1, but this appears to be a typographical error, as the main text states that the few-shot method identifies all true actives

The metric p represents the proportion of identified hits that were found by the method out of all detected hits

The first scenario relates to drug discovery projects around novel targets for which very few known ligands exist. In such projects, the target may have been identified based on emerging insights into disease biology or from fundamental research, yet only the natural ligand or a small number of active molecules have been characterized. Conventional in-silico QSAR or docking approaches often struggle to generalize from these small labeled datasets. In contrast, few-shot learning approaches are designed to work well in these settings (Schimunek et al. 2023; Stanley et al. 2021). One technique for improving predictive performance in scarce-data scenarios is the adoption of an autoregressive strategy that iteratively selects top-ranked compounds and pseudo-labels them. This procedure leads to a growing support set, from which updated predictions are derived. Wet-lab validation of selected hits can further refine the model in a closed feedback loop that capitalizes on minimal initial data while reaching substantial coverage of chemical space. This approach has already demonstrated success, such as in the discovery of new kinase inhibitors relevant to Alzheimer's disease (Snyder et al. 2024).

A second typical scenario emerges when a small set of promising compounds is known to induce beneficial phenotypic effects, although the target protein or mechanism remains unknown. Because the mechanistic details are unclear, the labeled data for these compounds often capture only a broad phenotypic endpoint. Without a known target, docking methods cannot be applied, and machine learning models tend to underperform due to the limited amount of available data. Few-shot learning can address this challenge by treating the known active compounds, and potentially a few inactive ones, as the support set. The few-shot model then ranks large numbers of query molecules, comparing them with the few reference compounds in the support set. This strategy facilitates the identification of molecules that share phenotypic profiles, advancing scaffold hopping or exploring structurally diverse chemical scaffolds that retain the desired cellular response. As additional compounds are tested in downstream experiments, newly identified active molecules can be added to the support set, incrementally enriching the model's representation of the desired effect. In both of these real-world examples, few-shot learning models not only complement well-established methods like docking but also have the potential to significantly lower experimental expenditures and accelerate iterative testing by transferring knowledge from relevant pre-training tasks and efficiently using minimal new data.

6.1 Concrete Example: Using Few-Shot Learning in Early-Stage Drug Discovery for Bioactivity Predictions

In this section, we illustrate how few-shot learning methods can be integrated into early-stage drug discovery projects. As a concrete example, we focus on the discovery and optimization of TLK2 inhibitors (Asquith et al. 2024).

Collecting Known Data Having identified TLK2 kinase as the target, we follow Asquith et al. (2024) and assume our drug discovery project begins with a literature search on oxindole inhibitors profiled by Reaction Biology Corporation, along with their associated TLK2 IC_{50} values. The identified molecules are listed in Table 3. Since only 15 known molecules were found, and Snyder et al. (2024) recommend using few-shot models when fewer than 50 molecules are available, we opt for a few-shot learning approach for virtual screening. These 15 known molecules serve as the support set. For simplicity, we make use of the MHNfs-based application (Schimunek et al. 2025), accessible at https://huggingface.co/spaces/ml-jku/mhnfs.

Data Preprocessing From the collected data, each molecule has to be converted into a suitable low-level representation, x, which depends on the specific few-shot learning method used. In our case, the MHNfs application requires molecular representations in the form of SMILES strings, which can be easily obtained from the given CAS information. Additionally, for the support set, the bioactivity labels need to be binarized using an appropriate threshold, e.g., 50 μM.

Virtual Screening The trained few-shot model, along with the support set, is then used to screen virtual libraries. These libraries may consist of in-house curated molecules or large sets of publicly accessible, unlabeled compounds, such as those from the ZINC database (Irwin et al. 2020). As an illustrative example, we compute predictions for a single query molecule: the PKR inhibitor listed in Table 3. The MHNfs application returns a prediction score of 0.926. Since the prediction scores range from 0 to 1, this suggests that the query molecule is likely to be active, making it a promising candidate for subsequent wet lab experiments.

7 Discussion and Future Directions

In drug discovery, data is often scarce because biological systems are constantly evolving as mutations emerge, pathogens adapt, and new diseases arise and we gain new insights. The COVID-19 pandemic and its associated drug discovery

Table 3 Exemplary few-shot learning study results. 15 small molecules (given first and second column) were tested by Asquith et al. (2024) and used as (top part of table)

Name	CAS	IC_{50}[μM]	Label	Prediction score	Smiles [truncated strings are indicated by "..."]
Orantinib (SU6656)	330161-87-0	>50	0		O=S(C1=CC2=C(NC(/C2=C\C(N3)=CC4=C3CCCC4)=O)C=C1) . . .
SU9516	377090-84-1	0.56	1		O=C1NC2=C(C=C(OC)C=C2)/C1=C//C3=CN=CN3
SU11652	326914-10-7	-	0		O=C(C1=C(C)NC(/C=C2C(NC3=C\2C=C(Cl)C=C3)=O)=C1C)N . . .
Syk Inhib. (GW441756)	622387-85-3	21	1		O=C1/C(C2=NC=CC=C2N1)=C\C3=CN(C)C4=C3C=CC=C4
DMBI (SU4312)	–	>50	0		O=C1NC2=C(C=CC=C2)/C1=C/C3=CC=C(N(C)C)C=C3
Cdk1 Inhib.	220749-41-7	33	1		O=C1NC2=C(C=CC=C2)/C1=C\C3=C(C1)NC4=C3C=CC=C4
GSK-3 Inhib. IX	667463-62-9	>50	0		O/N=C(C1=CC=CC=C1N2)/C2=C3C(NC4=C/3C=CC(Br)=C4)=O
GSK-3 Inhib. X	740841-15-0	>50	1		CC(O/N=C1/C(NC2=C1C=CC=C2)=C3C(NC4=C\3C=CC(Br)=C4) . . .
IC261	186611-52-9	>50	0		O=C1NC2=C(C=CC=C2)/C1=C\C3=C(OC)C=C(OC)C=C3OC
Indirubin-3/-monoxime	160807-49-8	0.44	1		O=C1NC2=C(C=CC=C2)/C1=C3NC4=C(C=CC=C4)C\3=N/O
Indirubin Derivative E804	854171-35-0	0.44	1		O=C1NC2=C(C=CC=C2)/C1=C3NC4=C(C=CC=C4)C/3=N/OCCC(O) . . .
Met Kinase Inhib. (SU11274)	658084-23-2	>50	0		O=S(C1=CC2=C(C=C1)NC(/C2=C/C3=C(C(C(N4CCN(CC4)C)=O) . . .
VEGF Receptor 2 Kinase Inhib. I	15966-93-5	>50	0		O=C(C1=C(C)NC(/C=C2C(NC3=C\2C=CC=C3)=O)=C1C)OCC
VEGF Receptor 2 Kinase Inhib. II	288144-20-7	>50	0		O=C(/C(C1=C2)=C\C3=CC4=C(CCCC4)N3)NC1=CC=C2Br
VEGF Receptor 2 Kinase Inhib. III	204005-46-9	>50	0		O=C(NC1=CC=CC=C/21)C2=C/C3=C(C)C=C(C)N3
Assumend unkown (query molecule) PKR Inhib.	608512-97-6	0.06	1	0.926	O=C1/C(C2=C(SC=N3)C3=CC=C2N1)=C\C4=CN=CN4

Molecules with IC_{50} values below 50μM were considered active, others inactive (third column). MHNfs was used to predict the activity of one molecule (CAS 608512-97-6, last row). The prediction score was 0.926, which indicates high activity and this also corresponds to the measurement, IC_{50} of 0.06μM efforts (Consortium et al. 2021; Schimunek et al. 2024b) have highlighted both the urgent need for rapid drug development pipelines and the challenges posed by low-data scenarios. Few-shot learning offers a promising approach to tackle the prevailing low-data challenges pervasive in early drug discovery, where traditional predictive modeling techniques are prone to overfitting or simply fail to generalize. By training on tasks with ample labeled data, few-shot learning models can learn robust, transferable representations, which can be quickly adapted to scenarios with only a few known samples.

Few-shot learning models are a natural fit for early-stage drug discovery, as discussed in Sect. 6.1. Beyond this, few-shot learning methods can also be valuable in later stages, such as drug repurposing or predicting off-target effects, where molecules of interest might need to be assessed in data-scarce scenarios. Additionally, few-shot learning can benefit multi-target drug design, when data for one of the targets is limited.

However, several challenges remain. The pre-training tasks need to be sufficiently relevant to the new problem domain; if tasks used for meta-training are only distantly related, the model may not yield the intended improvement in performance. Few-shot models typically experience a decline in performance under domain shift, where the targets and (or) tested molecules in the test tasks differ from those in the training set. Future research should focus on developing models and architectures that are more robust to domain shift, ensuring reliable performance across diverse biomedical applications. Another challenge is the integration of multi-modal data. Biological data often span multiple modalities, such as chemical structures, microscopy images, gene expression profiles, and proteomics data, with large domain shifts between them. Current few-shot learning methods predominantly rely on chemical structure data and lack the capability to effectively integrate these additional data types. Ensuring that a few-shot method can effectively integrate and interpret evidence from these varied modalities requires architectural and algorithmic strategies that prioritize robust and generalizable representations. A further important point involves uncertainty measures because measuring and managing uncertainty for the predictions made by the few-shot model is critical for high-stakes decision-making in pharmaceutical research.

In moving forward, one future direction lies in combining generative approaches with few-shot learning-based frameworks, thereby creating a mechanism for rapid de-novo compound design. Another prospect aims at constructing large multi-modal architectures, potentially unifying molecular structures, textual assay descriptions, and high-content imaging into a single model, which can then be adapted via few-shot principles. The role of privacy and security in data sharing may also become more pronounced, prompting interest in federated few-shot learning methods designed for collaborations across laboratories without revealing sensitive information.

Despite these challenges, few-shot learning approaches hold significant potential to transform the early stages of drug discovery by reducing the reliance on large-scale experiments and enabling more efficient identification of promising drug candidates. By bridging the gap between data scarcity and predictive performance, few-shot learning can help accelerate the development of new therapeutics.

References

Abramson J, Adler J, Dunger J, Evans R, Green T, Pritzel A, Ronneberger O, Willmore L, Ballard AJ, Bambrick J, et al. Accurate structure prediction of biomolecular interactions with alphafold 3. Nature. 2024;636(8042):E4.

Adler T, Brandstetter J, Widrich M, Mayr A, Kreil D, Kopp M, Klambauer G, Hochreiter S. Cross-domain few-shot learning by representation fusion. arXiv preprint arXiv:2010.06498. 2020.

Altae-Tran H, Ramsundar B, Pappu AS, Pande V. Low data drug discovery with one-shot learning. ACS Cent Sci. 2017;3(4):283–93.

Asquith CR, East MP, Laitinen T, Alamillo-Ferrer C, Hartikainen E, Wells CI, Axtman AD, Drewry DH, Tizzard GJ, Poso A, et al. Discovery and optimization of narrow spectrum inhibitors of tousled like kinase 2 (tlk2) using quantitative structure activity relationships. Eur J Med Chem. 2024;271:116357.

Bai Y, Lv X, Zhang J, Lyu H, Tang J, Huang Z, Du Z, Liu X, Zeng A, Hou L, et al. Longbench: a bilingual, multitask benchmark for long context understanding. arXiv preprint arXiv:2308.14508. 2023.

Bender A, Cortés-Ciriano I. Artificial intelligence in drug discovery: what is realistic, what are illusions? Part 1: Ways to make an impact, and why we are not there yet. Drug Discov Today. 2021a;26(2):511–24.

Bender A, Cortés-Ciriano I. Artificial intelligence in drug discovery: what is realistic, what are illusions? Part 2: A discussion of chemical and biological data. Drug Discov Today. 2021b;26(4):1040–52.

Bendre N, Mar'ın HT, Najafirad P. Learning from few samples: a survey. arXiv preprint arXiv:2007.15484. 2020.

Bouniot Q, Redko I, Audigier R, Loesch A, Habrard A. Improving few-shot learning through multi-task representation learning theory. In: European conference on computer vision. Springer; 2022. p. 435–52.

Bray M-A, Singh S, Han H, Davis CT, Borgeson B, Hartland C, KostAlimova M, Gustafsdottir SM, Gibson CC, Carpenter AE. Cell painting, a high-content image-based assay for morphological profiling using multiplexed fluorescent dyes. Nat Protoc. 2016;11(9):1757–74.

Breiman L. Random forests. Mach Learn. 2001;45:5–32.

Brown T, Mann B, Ryder N, Subbiah M, Kaplan JD, Dhariwal P, Neelakantan A, Shyam P, Sastry G, Askell A, et al. Language models are few-shot learners. Adv Neural Inf Proces Syst. 2020;33:1877–901.

Burley SK, Bhikadiya C, Bi C, Bittrich S, Chao H, Chen L, Craig PA, Crichlow GV, Dalenberg K, Duarte JM, et al. Rcsb protein data bank (rcsb. org): delivery of experimentally-determined pdb structures alongside one million computed structure models of proteins from artificial intelligence/machine learning. Nucleic Acids Res. 2023;51(D1):D488–508.

Cereto-Massagué A, Ojeda MJ, Valls C, Mulero M, Garcia-Vallvé S, Pujadas G. Molecular fingerprint similarity search in virtual screening. Methods. 2015;71:58–63.

Chen H, Engkvist O, Wang Y, Olivecrona M, Blaschke T. The rise of deep learning in drug discovery. Drug Discov Today. 2018;23(6):1241–50.

Chen Y, Zhong R, Zha S, Karypis G, He H. Meta-learning via language model in-context tuning. arXiv preprint arXiv:2110.07814. 2021.

Chen W, Tripp A, Hernández-Lobato JM. Meta-learning adaptive deep kernel gaussian processes for molecular property prediction. In: The eleventh international conference on learning representations; 2023.

Chilingaryan G, Tamoyan H, Tevosyan A, Babayan N, Khondkaryan L, Hambardzumyan K, Navoyan Z, Khachatrian H, Aghajanyan A. Molbart: generative masked language models for molecular representations. OpenReview; 2023.

Consortium, The COVID Moonshot, Chodera J, Lee A, London N, von Delft F. Open science discovery of oral non-covalent SARS-CoV-2 main protease inhibitors. Cambridge University Press (CUP); 2021.

Dong Q, Li L, Dai D, Zheng C, Wu Z, Chang B, Sun X, Xu J, Sui Z A survey on in-context learning. arXiv preprint arXiv:2301.00234. 2022.

Dowden H, Munro J. Trends in clinical success rates and therapeutic focus. Nat Rev Drug Discov. 2019;18(7):495–6.

Duan R, Li D, Tong Q, Yang T, Liu X, Liu X. A survey of few-shot learning: an effective method for intrusion detection. Secur Commun Netw. 2021;2021(1):4259629.

Eckert H, Bajorath J. Molecular similarity analysis in virtual screening: foundations, limitations and novel approaches. Drug Discov Today. 2007;12(5–6):225–33.

Finn C, Abbeel P, Levine S. Model-agnostic meta-learning for fast adaptation of deep networks. In: International conference on machine learning. PMLR; 2017. p. 1126–35.

Flam-Shepherd D, Zhu K, Aspuru-Guzik A. Language models can learn complex molecular distributions. Nat Commun. 2022;13(1):3293.

Geppert H, Horváth T, Gärtner T, Wrobel S, Bajorath J. Support-vector-machine-based ranking significantly improves the effectiveness of similarity searching using 2d fingerprints and multiple reference compounds. J Chem Inf Model. 2008;48(4):742–6.

Gilmer J, Schoenholz SS, Riley PF, Vinyals O, Dahl GE. Neural message passing for quantum chemistry. In: International conference on machine learning. PMLR; 2017. p. 1263–72.

Gómez-Bombarelli R, Wei JN, Duvenaud D, Hernández-Lobato JM, Sánchez-Lengeling B, Sheberla D, Aguilera-Iparraguirre J, Hirzel TD, Adams RP, Aspuru-Guzik A. Automatic chemical design using a data-driven continuous representation of molecules. ACS Cent Sci. 2018;4(2):268–76.

Guo Z, Zhang C, Yu W, Herr J, Wiest O, Jiang M, Chawla NV. Few-shot graph learning for molecular property prediction. In: Proceedings of the web conference 2021; 2021. p. 2559–67.

Hertz T, Hillel AB, Weinshall D. Learning a kernel function for classification with small training samples. In: Proceedings of the 23rd international conference on machine learning; 2006. p. 401–8.

Hochreiter S, Schmidhuber J. Long short-term memory. Neural Comput. 1997;9(8):1735–80.

Hochreiter S, Younger AS, Conwell PR. Learning to learn using gradient descent. In: Artificial neural networks—ICANN 2001: international conference Vienna, Austria, August 21–25, 2001 proceedings 11. Springer; 2001. p. 87–94.

Hofmarcher M, Rumetshofer E, Clevert D-A, Hochreiter S, Klambauer G. Accurate prediction of biological assays with high-throughput microscopy images and convolutional networks. J Chem Inf Model. 2019;59(3):1163–71.

Hu Y, Gripon V, Pateux S. Graph-based interpolation of feature vectors for accurate few-shot classification. In: 2020 25th International Conference on Pattern Recognition (ICPR). IEEE; 2021a. p. 8164–71.

Hu Y, Gripon V, Pateux S. Leveraging the feature distribution in transfer-based few-shot learning. In: International conference on artificial neural networks. Springer; 2021b. p. 487–99.

Hwang TJ, Carpenter D, Lauffenburger JC, Wang B, Franklin JM, Kesselheim AS. Failure of investigational drugs in late-stage clinical development and publication of trial results. JAMA Intern Med. 2016;176(12):1826–33.

Irwin JJ, Tang KG, Young J, Dandarchuluun C, Wong BR, Khurelbaatar M, Moroz YS, Mayfield J, Sayle RA. Zinc20—a free ultralargescale chemical database for ligand discovery. J Chem Inf Model. 2020;60(12):6065–73.

Iscen A, Tolias G, Avrithis Y, Chum O. Label propagation for deep semi-supervised learning. In: Proceedings of the IEEE/CVF conference on computer vision and pattern recognition; 2019. p. 5070–9.

Jiang D, Wu Z, Hsieh C-Y, Chen G, Liao B, Wang Z, Shen C, Cao D, Wu J, Hou T. Could graph neural networks learn better molecular representation for drug discovery? A comparison study of descriptor-based and graph-based models. J Chem. 2021;13(1):1–23.

Jiménez-Luna J, Grisoni F, Weskamp N, Schneider G. Artificial intelligence in drug discovery: recent advances and future perspectives. Expert Opin Drug Discov. 2021;16(9):949–59.

Kearnes S, McCloskey K, Berndl M, Pande V, Riley P. Molecular graph convolutions: moving beyond fingerprints. J Comput Aided Mol Des. 2016;30(8):595–608.

Kim S, Chen J, Cheng T, Gindulyte A, He J, He S, Li Q, Shoemaker BA, Thiessen PA, Yu B, et al. Pubchem 2023 update. Nucleic Acids Res. 2023;51(D1):D1373–80.

Koch G, Zemel R, Salakhutdinov R, et al. Siamese neural networks for one-shot image recognition. In: ICML deep learning workshop, volume 2. Lille; 2015.

Lazarou M, Stathaki T, Avrithis Y. Iterative label cleaning for transductive and semi-supervised few-shot learning. In: Proceedings of the ieee/cvf international conference on computer vision; 2021. p. 8751–60.

Liu Y, Lee J, Park M, Kim S, Yang E, Hwang SJ, Yang Y. Learning to propagate labels: Transductive propagation network for few-shot learning. arXiv preprint arXiv:1805.10002. 2018.

Liu Z, Ye X, Fang X, Wang F, Wu H, Wang H. Docking-based virtual screening with multi-task learning. In: 2021 IEEE international conference on bioinformatics and biomedicine (BIBM). IEEE; 2021. p. 381–5.

Mak K-K, Wong Y-H, Pichika MR. Artificial intelligence in drug discovery and development. In: Drug discovery and evaluation: safety and pharmacokinetic assays; 2024. p. 1461–98.

Mammone A, Turchi M, Cristianini N. Support vector machines. Wiley Interdiscip Rev Comput Stat. 2009;1(3):283–9.

Mayr A, Klambauer G, Unterthiner T, Hochreiter S. Deeptox: toxicity prediction using deep learning. Front Environ Sci. 2016;3:80.

Mayr A, Klambauer G, Unterthiner T, Steijaert M, Wegner JK, Ceulemans H, Clevert D-A, Hochreiter S. Large-scale comparison of machine learning methods for drug target prediction on chembl. Chem Sci. 2018;9(24):5441–51.

Maziarka L, Danel T, Mucha S, Rataj K, Tabor J, Jastrzebski S. Molecule attention transformer. arXiv preprint arXiv:2002.08264. 2020.

Melville JL, Burke EK, Hirst JD. Machine learning in virtual screening. Comb Chem High Throughput Screen. 2009;12(4):332–43.

Mendez D, Gaulton A, Bento AP, Chambers J, De Veij M, Félix E, Magariños MP, Mosquera JF, Mutowo P, Nowotka M, et al. Chembl: towards direct deposition of bioassay data. Nucleic Acids Res. 2019;47(D1):D930–40.

Merkwirth C, Lengauer T. Automatic generation of complementary descriptors with molecular graph networks. J Chem Inf Model. 2005;45(5):1159–68.

Neves BJ, Braga RC, Melo-Filho CC, Moreira-Filho JT, Muratov EN, Andrade CH. Qsar-based virtual screening: advances and applications in drug discovery. Front Pharmacol. 2018;9:1275.

Nguyen CQ, Kreatsoulas C, Branson KM. Meta-learning gnn initializations for low-resource molecular property prediction. arXiv preprint arXiv:2003.05996. 2020.

Nicolaou K. Advancing the drug discovery and development process. Angew Chem. 2014;126(35):9280–92.

Olivecrona M, Blaschke T, Engkvist O, Chen H. Molecular de-novo design through deep reinforcement learning. J Chem. 2017;9:1–14.

Ramsauer H, Schäfl B, Lehner J, Seidl P, Widrich M, Adler T, Gruber L, Holzleitner M, Pavlović M, Sandve GK, et al. Hopfield networks is all you need. arXiv preprint arXiv:2008.02217. 2020.

Renz P, Van Rompaey D, Wegner JK, Hochreiter S, Klambauer G. On failure modes in molecule generation and optimization. Drug Discov Today Technol. 2019;32:55–63.

Riniker S, Landrum GA. Open-source platform to benchmark fingerprints for ligand-based virtual screening. J Chem. 2013;5(1):1–17.

Rodríguez, P., Laradji, I., Drouin, A., and Lacoste, A. (2020). Embedding propagation: smoother manifold for few-shot classification. In Computer Vision–ECCV 2020: 16th European conference, Glasgow, UK, August 23–28, 2020, proceedings, part XXVI 16, 121–138. Springer.
Roy K, Kar S, Das RN, Roy K, Kar S, Das RN. Qsar/qspr modeling: introduction. In: A primer on QSAR/QSPR modeling: fundamental concepts; 2015. p. 1–36.
Sanchez-Fernandez A, Rumetshofer E, Hochreiter S, Klambauer G. Cloome: contrastive learning unlocks bioimaging databases for queries with chemical structures. Nat Commun. 2023;14(1):7339.
Schimunek J, Friedrich L, Kuhn D, Rippmann F, Hochreiter S, Klambauer G. A generalized framework for embedding-based few-shot learning methods in drug discovery. In: ELLIS machine learning for molecule discovery workshop; 2021.
Schimunek J, Seidl P, Friedrich L, Kuhn D, Rippmann F, Hochreiter S, Klambauer G. Context-enriched molecule representations improve few-shot drug discovery. In: The eleventh international conference on learning representations; 2023.
Schimunek J, Friedrich L, Kuhn D, Klambauer G. Autoregressive activity prediction for low-data drug discovery. In: 5th workshop on practical ML for limited/low resource settings; 2024a.
Schimunek J, Seidl P, Elez K, Hempel T, Le T, No'e F, Olsson S, Raich L, Winter R, Gokcan H, et al. A community effort in sars-cov-2 drug discovery. Mol Inf. 2024b;43(1):e202300262.
Schimunek J, Luukonen S, Klambauer G. MHNfs: prompting in-context bioactivity predictions for low data drug discovery. J Chem. 2025;65:4243.
Schmidinger N, Schneckenreiter L, Seidl P, Schimunek J, Hoedt P-J, Brandstetter J, Mayr A, Luukkonen S, Hochreiter S, Klambauer G Bio-xlstm: generative modeling, representation and in-context learning of biological and chemical sequences. arXiv preprint arXiv:2411.04165. 2024.
Schuffenhauer A, Floersheim P, Acklin P, Jacoby E. Similarity metrics for ligands reflecting the similarity of the target proteins. J Chem Inf Comput Sci. 2003;43(2):391–405.
Segler MH, Kogej T, Tyrchan C, Waller MP. Generating focused molecule libraries for drug discovery with recurrent neural networks. ACS Cent Sci. 2018;4(1):120–31.
Seidl P, Vall A, Hochreiter S, Klambauer G. Enhancing activity prediction models in drug discovery with the ability to understand human language. arXiv preprint arXiv:2303.03363. 2023.
Sheridan RP, Kearsley SK. Why do we need so many chemical similarity search methods? Drug Discov Today. 2002;7(17):903–11.
Simm J, Klambauer G, Arany A, Steijaert M, Wegner JK, Gustin E, Chupakhin V, Chong YT, Vialard J, Buijnsters P, et al. Repurposing high-throughput image assays enables biological activity prediction for drug discovery. Cell Chem Biol. 2018;25(5):611–8.
Snell J, Swersky K, Zemel RS. Prototypical networks for few-shot learning. arXiv preprint arXiv:1703.05175. 2017.
Snyder SH, Vignaux PA, Ozalp MK, Gerlach J, Puhl AC, Lane TR, Corbett J, Urbina F, Ekins S. The goldilocks paradigm: comparing classical machine learning, large language models, and few-shot learning for drug discovery applications. Commun Chem. 2024;7(1):134.
Stanley M, Bronskill JF, Maziarz K, Misztela H, Lanini J, Segler M, Schneider N, Brockschmidt M. Fs-Mol: a few-shot learning dataset of molecules. In: Thirty-fifth conference on neural information processing systems datasets and benchmarks track (round 2); 2021.
Sun D, Gao W, Hu H, Zhou S. Why 90% of clinical drug development fails and how to improve it? Acta Pharm Sin B. 2022;12(7):3049–62.
Tanimoto T. Ibm type 704 medical diagnosis program. IRE Trans Med Electron. 1960;4:280–3.
Taylor R, Kardas M, Cucurull G, Scialom T, Hartshorn A, Saravia E, Poulton A, Kerkez V, Stojnic R Galactica: a large language model for science. arXiv preprint arXiv:2211.09085. 2022.
Torres L, Monteiro N, Oliveira J, Arrais J, Ribeiro B. Exploring a siamese neural network architecture for one-shot drug discovery. In: 2020 ieee 20th international conference on bioinformatics and bioengineering (bibe); 2020. p. 168–75.
Tralau-Stewart CJ, Wyatt CA, Kleyn DE, Ayad A. Drug discovery: new models for industry–academic partnerships. Drug Discov Today. 2009;14(1–2):95–101.

Unterthiner T, Mayr A, Klambauer G, Steijaert M, Wegner JK, Ceulemans H, Hochreiter S. Deep learning as an opportunity in virtual screening. In: Proceedings of the deep learning workshop at NIPS, vol. 27. MIT Press Cambridge; 2014. p. 1–9.
van Tilborg D, Brinkmann H, Criscuolo E, Rossen L, Özçelik R, Grisoni F. Deep learning for low-data drug discovery: hurdles and opportunities. Curr Opin Struct Biol. 2024;86:102818.
Vaswani A. Attention is all you need. Adv Neural Inf Proces Syst. 2017.
Vella D, Ebejer J-P. Few-shot learning for low-data drug discovery. J Chem Inf Model. 2022;63(1):27–42.
Verbist B, Klambauer G, Vervoort L, Talloen W, Shkedy Z, Thas O, Bender A, Göhlmann HW, Hochreiter S, Consortium Q, et al. Using transcriptomics to guide lead optimization in drug discovery projects: lessons learned from the qstar project. Drug Discov Today. 2015;20(5):505–13.
Vinyals O, Bengio S, Kudlur M. Order matters: sequence to sequence for sets. arXiv preprint arXiv:1511.06391. 2015.
Vinyals O, Blundell C, Lillicrap T, Wierstra D, et al. Matching networks for one shot learning. Adv Neural Inf Proces Syst. 2016;29:3630–8.
Walters WP, Stahl MT, Murcko MA. Virtual screening—an overview. Drug Discov Today. 1998;3(4):160–78.
Wang X, Huan J, Smalter A, Lushington GH. Application of kernel functions for accurate similarity search in large chemical databases. BMC Bioinf. 2010;11:1–14. BioMed Central
Wang Y, Yao Q, Kwok JT, Ni LM. Generalizing from a few examples: a survey on few-shot learning. ACM Comput Surv. 2020;53(3):1–34.
Wang Y, Abuduweili A, Yao Q, Dou D. Property-aware relation networks for few-shot molecular property prediction. Adv Neural Inf Proces Syst. 2021;34:17441–54.
Wang Y, Wang Z, Sadeh G, Zancato L, Achille A, Karypis G, Rangwala H. Long-context protein language model. bioRxiv, 2024. p. 2024–10.
Waring MJ, Arrowsmith J, Leach AR, Leeson PD, Mandrell S, Owen RM, Pairaudeau G, Pennie WD, Pickett SD, Wang J, et al. An analysis of the attrition of drug candidates from four major pharmaceutical companies. Nat Rev Drug Discov. 2015;14(7):475–86.
Weininger D. Smiles, a chemical language and information system. 1. Introduction to methodology and encoding rules. J Chem Inf Comput Sci. 1988;28(1):31–6.
Willett P. The calculation of molecular structural similarity: principles and practice. Mol Inf. 2014;33(6–7):403–13.
Winter R, Montanari F, Noé F, Clevert D-A. Learning continuous and data-driven molecular descriptors by translating equivalent chemical representations. Chem Sci. 2019;10(6):1692–701.
Yang K, Swanson K, Jin W, Coley C, Eiden P, Gao H, Guzman-Perez A, Hopper T, Kelley B, Mathea M, et al. Analyzing learned molecular representations for property prediction. J Chem Inf Model. 2019;59(8):3370–88.
Ye M, Guo Y. Deep triplet ranking networks for one-shot recognition. arXiv preprint arXiv:1804.07275. 2018.
Zdrazil B, Felix E, Hunter F, Manners EJ, Blackshaw J, Corbett S, de Veij M, Ioannidis H, Lopez DM, Mosquera JF, et al. The chembl database in 2023: a drug discovery platform spanning multiple bioactivity data types and time periods. Nucleic Acids Res. 2024;52(D1):D1180–92.
Zeng Z, Yao Y, Liu Z, Sun M. A deep-learning system bridging molecule structure and biomedical text with comprehension comparable to human professionals. Nat Commun. 2022;13(1):862.
Zhang P, Wang F, Hu J, Sorrentino R. Label propagation prediction of drug-drug interactions based on clinical side effects. Sci Rep. 2015;5(1):12339.
Zhang W, Chen Y, Li D. Drug-target interaction prediction through label propagation with linear neighborhood information. Molecules. 2017;22(12):2056.
Zhang Y, Liu Y, Li Q, Jin R, Wen C. Lilpa: a label importance based label propagation algorithm for community detection with application to core drug discovery. Neurocomputing. 2020;413:107–33.

Zhang L, Li Y, Liu Z, Liu J, Yang M, et al.. Marathon: a race through the realm of long context with large language models. arXiv preprint arXiv:2312.09542. 2023.
Zhou D, Bousquet O, Lal T, Weston J, Schölkopf B. Learning with local and global consistency. Adv Neural Inf Proces Syst. 2003;16.
Zhu X, Ghahramani Z. Learning from labeled and unlabeled data with label propagation. ProQuest number: information to all users. 2002.
Zhu H, Koniusz P. Transductive few-shot learning with prototype-based label propagation by iterative graph refinement. In: Proceedings of the IEEE/CVF conference on computer vision and pattern recognition; 2023. p. 23996–4006.

Explainable Artificial Intelligence in Drug Discovery

Michela Proietti, Roberta Astolfi, and Alessio Ragno

Abstract This chapter explores the role of explainable artificial intelligence (XAI) in drug discovery, detailing key explainability techniques and their applications in drug design. We present a historical perspective on computational drug discovery, followed by a taxonomy of state-of-the-art XAI methods, discussing their benefits and limitations. Practical guidelines are provided to aid in selecting the most suitable explainability techniques based on the given models and data types. Additionally, we review real-world case studies where XAI enhances AI-driven drug discovery, improving model reliability and facilitating rational design. Finally, we highlight emerging research directions to advance explainability in AI-driven pharmaceutical innovation.

Keywords Explainable artificial intelligence · Post-hoc methods · Self-explainable models · Drug discovery

1 Introduction

Artificial intelligence (AI) has achieved significant results in diverse disciplines while reducing the time and costs associated with scientific research. For instance, AlphaFold (Abramson et al. 2024) and its successive variants have revolutionized the field of drug discovery with their unprecedented ability to predict the structure

M. Proietti
Department of Computer, Control, and Management Engineering A. Ruberti, Sapienza University of Rome, Rome, Italy
e-mail: mproietti@diag.uniroma1.it

R. Astolfi
Department of Drug Chemistry and Technology, Sapienza University of Rome, Rome, Italy
e-mail: roberta.astolfi@uniroma1.it

A. Ragno (✉)
LIRIS UMR 5205, INSA Lyon, CNRS, Villeurbanne, France
e-mail: alessio.ragno@insa-lyon.fr

A. Lavecchia (ed.), *Applied Artificial Intelligence for Drug Discovery*,
https://doi.org/10.1007/978-3-031-98022-0_17

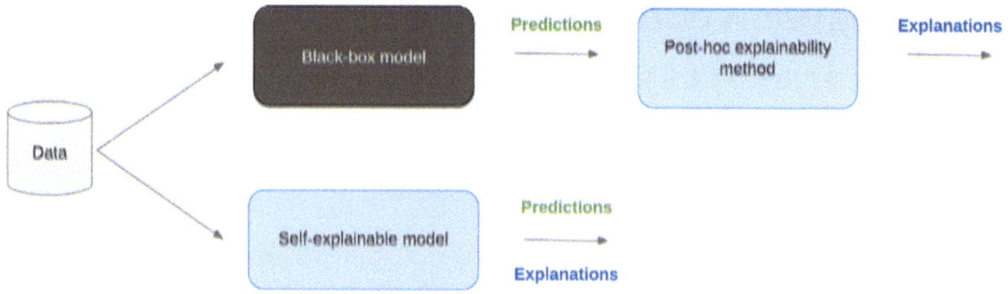

Fig. 1 Comparison between post-hoc explainability approaches and self-explainable models. Post-hoc methods produce explanations for black-box models by looking and their outputs and parameters. On the other hand, self-explainable models produce explanations alongside predictions

and interactions of all life's molecules. However, the black-box nature of these systems poses challenges in their applications to real-world scenarios, due to safety and regulatory concerns (European Union 2016). For this reason, there is increased attention towards XAI, which aims to explain the reasons behind the model's output predictions. This chapter explores the role of XAI in drug discovery, presenting existing explainability techniques and analyzing their applications in both ligand-based and structure-based drug design. We introduce a methodology-based taxonomy of state-of-the-art XAI methods, discussing their advantages and limitations and offering guidelines to select approaches based on architectures and data types. Subsequently, we review several key applications of XAI in drug discovery, analyzing real-world case studies where explainability enhances AI-based methodologies. Finally, we provide several potential future research directions to address the limitations of current explainability techniques and to encourage further collaboration between AI and pharmaceutical chemistry experts.

2 Emerging Explainability Techniques

There are several popular categorizations of XAI techniques. Figure 1 explains the main distinction between *post-hoc* and *self-explainable* methods. The former produces explanations for pre-trained models. The latter, instead, consists of architectures that simultaneously offer predictions and explanations. Another key distinction is that *local* XAI methods explain individual instances, whereas *global* methods explain overall model behavior.

To present XAI methods for drug discovery applications, we introduce a taxonomy that builds on top of the methodology-based categorizations proposed by Jiménez-Luna et al. (2020) and Ponzoni et al. (2023). Subsequently, we offer insights into what family of approaches might be more appropriate depending on the architecture and data types at hand, analyzing their advantages and drawbacks. The final purpose is to give general guidelines for researchers and medical chemists aiming to use XAI.

Table 1 Methodology-based taxonomy of explainability methods

Family	Subcategories	Popular methods
Feature attribution	Gradient-based	GradCAM (Selvaraju et al. 2020), IG (Sundararajan et al. 2017)
	Perturbation-based	Feature masking, perturbation analysis, PGExplainer (Luo et al. 2020), GraphMask (Schlichtkrull et al. 2020), SubgraphX (Yuan et al. 2021)
	Surrogate models	SHAP (Lundberg and Lee 2017), LIME (Ribeiro et al. 2016), PGMExplainer (Vu and Thai 2020), GraphLime (Huang et al. 2023), DeepLIFT (Shrikumar et al. 2017)
	Decomposition-based	LRP (Bach et al. 2015), GNN-LRP (Schnake et al. 2022), Excitation BP (Pope et al. 2019)
Instance-based	Anchors	(Ribeiro et al. 2018)
	Counterfactuals	LORE (Guidotti et al. 2019), CEODT (Carreira-Perpiñán and Hada 2021), SEDCT (Vermeire et al. 2022)
	Contrastive explanations	CEM (Prabhushankar et al. 2020)
Graph topology based	Subgraph identification	GNNExplainer (Ying et al. 2019)
	Attention-based	GAT (Veličković et al. 2017), transformers (Ying et al. 2021)
Self-explaining	Prototype-based	Prototype-based GNN (Ragno et al. 2024b)
	Concept-based	GraphCW (Proietti et al. 2024)
	Logic-based	LENs (Ciravegna et al. 2023), TELL (Ragno et al. 2024a)
Uncertainty estimation	Ensemble approaches	Bagging (Breiman 1996), Bayesian model averaging (Fragoso et al. 2018)
	Probabilistic approaches	Bayesian neural networks (Goan and Fookes 2020)

2.1 *Categorization of Explainability Techniques by Methodology*

Table 1 provides a summary of the taxonomy described by Jiménez-Luna et al. (2020) and Ponzoni et al. (2023). Given a model $M : \mathbb{R}^K \times \mathbb{R}$ and inputs $x \in \mathbb{R}^K$, each described by a set of K features $(f_1, f_2, \ldots, f_K)$, the aim of an explainability approach is to give insights into how the output $y = M(x)$ is produced.

2.1.1 Feature Attribution Methods

Feature attribution methods produce explanations in the form of attribution scores that represent the importance of each input feature f_i for the model's prediction

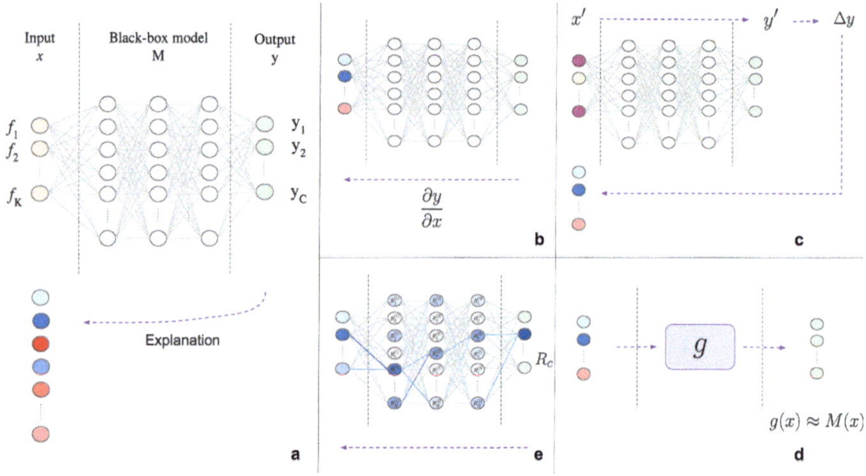

Fig. 2 Feature attribution methods. (**a**) Given a black-box model M, feature attribution methods output the importance of every input feature f_i for the prediction. (**b**) Gradient-based methods do this by computing the gradient of the output with respect to the input, indicated as ∂y/∂x. (**c**) Perturbation-based methods perturb subsets of the input features and determine how this affects the output of the model to estimate their importance for the prediction. (**d**) Surrogate methods approximate the black model with a simpler, interpretable model g such that $M(x) \approx g(x)$. (**e**) Decomposition-based approaches backpropagate the relevance of a given class c, by decomposing it among the units of each of the preceding layers, up to the input

y. This family of approaches can be divided into four subcategories, which are exemplified in Fig. 2.

Gradient-Based Methods

Gradient-based feature attribution methods quantify how input feature variations influence the model's output $M(x)$ by computing the gradient of the output with respect to the inputs x, denoted as $\frac{\partial y}{\partial x}$. The necessary derivatives can be efficiently computed using backpropagation, making gradient-based methods both computationally efficient and broadly applicable across various domains.

Several gradient-based methods, such as GradCAM and Integrated Gradients, have offered useful insights into the reasoning process of models used in diverse applications (Selvaraju et al. 2020; Sundararajan et al. 2017). Being gradients inherently tied to the model's parameters, they can reflect the underlying learned representation, allowing insights into the internal workings of the model. At the same time, this reliance on model parameters introduces sensitivity to model architecture, leading to partial information about feature contributions. Additionally, gradients can be noisy, especially when input features have minimal impact on the output. Techniques like SmoothGrad (Smilkov et al. 2017) address this by creating perturbed copies of the input and averaging the resulting gradients, thus producing sharper saliency maps and removing noisy regions.

Overall, while gradient-based feature attribution offers valuable insights into model behavior, careful application, and interpretation are essential to mitigate potential issues and ensure robust, meaningful explanations.

Perturbation-Based Methods

Perturbation-based methods rely on the actual modification or removal of subsets of the input features to measure the corresponding change in the model's predictions, thus assessing their importance. Differently from gradient-based methods, the outputs of these approaches can be directly related to a specific output variation, but their time complexity is much higher and increases with the number of input features. Additionally, the final attribution values are highly dependent on the size of the perturbation (Ancona et al. 2017).

Popular model-agnostic perturbation-based methods are feature masking, perturbation analysis (Fong and Vedaldi 2017), response randomization (Olden and Jackson 2002), and conditional multivariate models (Zintgraf et al. 2017). Other methods, such as PGExplainer (Luo et al. 2020), SubgraphX (Yuan et al. 2021), and GraphMask (Schlichtkrull et al. 2020), are specifically designed for graph neural networks (GNNs). Both PGExplainer and GraphMask use explainer networks to produce parameterized masks identifying relevant subgraphs. Specifically, PGExplainer aims to find the subgraph G_s of the input graph G_0 that maximizes the probability of obtaining the original output y_0, by learning a distribution over graph edges. GraphMask, instead, outputs a binary mask identifying all the edges that can be replaced by a learned baseline without altering the prediction. Differently from the previous approaches, SubGraphX identifies critical subgraphs by leveraging Monte Carlo tree search (Kocsis and Szepesvári 2006) to explore perturbations in graph structures.

Overall, graph-based methods are particularly valuable for highlighting important nodes, edges, or subgraphs that are crucial for a model's decision.

Surrogate Models

Given the model M, this family of methods aims at learning a surrogate model g, which approximates the behavior of M and is interpretable. In the following, given an input instance x, we identify as $x' \in \{0, 1\}^{K'}$ the corresponding binary interpretable representation, where K' is the number of simplified input features. In the case of graphs, the binary vector indicates the presence or absence of specific nodes, edges, or subgraphs. Given x', we denote by $z_i' \in \{0, 1\}^{K'}$ the perturbed samples containing a subset of the nonzero elements in x'. Given the perturbed samples z'_i, LIME (Ribeiro et al. 2016) computes the corresponding original representations $z = h_x(z')$, and gives them to the black-box model M, to get the corresponding predictions $\hat{y}_i$. Since LIME is a local method, each generated perturbation is weighted using a proximity measure, such that far-away points are ignored. Although LIME works well for structured (like tabular data) and unstructured data (like text and images), it cannot directly handle topological and relational information inherent in graph data. To fill this gap, GraphLIME (Huang et al. 2023) extends LIME to provide local explanations for predictions made by GNNs for a given node or graph. The main difference with standard LIME is that GraphLIME perturbs node features, without altering graph topology (i.e., edges).

Another popular method is DeepLIFT(Shrikumar et al. 2017), which computes importance scores by measuring the effect of setting the corresponding input feature to a reference value, which is typically an uninformative value for the feature (e.g., 0). Finally, SHAP (Lundberg and Lee 2017) is an additive feature attribution method that estimates feature scores employing Shapley values (Kuhn et al. 1960).

Among the local explanation methods, PGM-Explainer (Vu and Thai 2020) is specifically designed to explain GNNs' predictions. After creating a dataset of perturbed graphs and corresponding predictions from the black-box model, PGM-Explainer builds a probabilistic graphical model (PGM) Sucar (2021) that is used to estimate the conditional probabilities of the GNN predictions given the sampled subgraphs' features and structure. The PGM's nodes and edges represent the node and edge features in the original graph and feature dependencies, respectively.

While the methods we have described so far provide local explanations, several global surrogate explanation models aim at approximating the behavior of the black-box model employing decision trees or decision sets (Lakkaraju et al. 2017).

Decomposition-Based Methods

Decomposition-based methods provide explanations by decomposing the predictions obtained by the black-box model M into relevance scores associated with the input features.

Taylor decomposition methods (Montavon et al. 2017) and LRP (Bach et al. 2015) belong to this category. Both models backpropagate the black-box outputs layer-by-layer, following specific decomposition rules to obtain a heatmap representing the contribution of each input feature. Depending on the underlying architecture, some propagation rules might be more suited than others. Specifically, GNN-LRP (Schnake et al. 2022) extends LRP to explain the decisions made by GNNs by propagating relevance scores through the message-passing layers using rules that are tailored for graph structures. Unlike LRP, which is purely based on the network's structure and learned weights, Excitation BP (Pope et al. 2019) is a decomposition-based approach that incorporates probabilistic intuition into relevance propagation. The key idea behind Excitation BP is to allocate higher relevance scores to neurons having higher activation values relative to their layer's total activation, as more active neurons contribute more to the final decision.

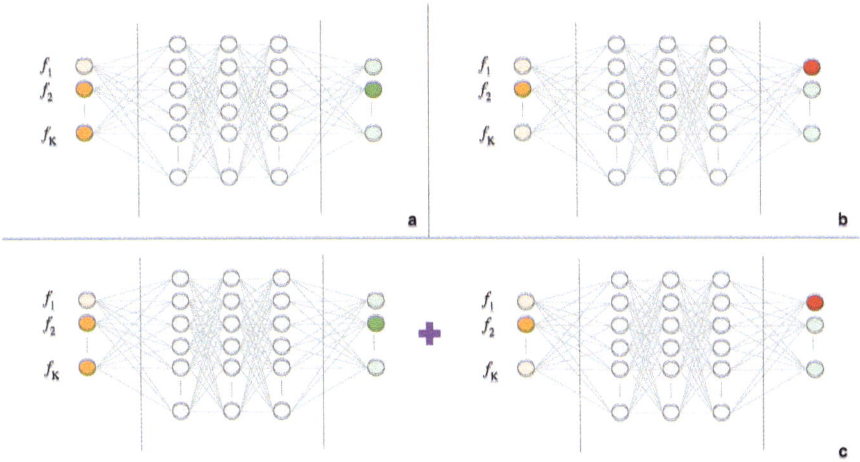

Fig. 3 Instance-based methods. Given a black-box model M, and input instance x, and the predicted class c, (**a**) an anchor represents the minimal set of features that should be present in x to keep the prediction unaltered, (**b**) a counterfactual is the smallest set of features that should be absent to obtain a different prediction by the model, (**c**) contrastive explanations consist of both pertinent positive (i.e., anchors) and pertinent negatives (i.e., counterfactuals)

Like previously described approaches, decomposition-based methods offer intuitive visual explanations that make the model's decision process more transparent.

However, these methods lack universality, as their applicability to different types of architectures might require modifications to the propagation rules.

2.1.2 Instance-Based Methods

Unlike feature attribution methods that assign importance scores to input variables, instance-based approaches explain a model's predictions through specific cases or modifications to input data. Figure 3 illustrates several existing methods.

Anchor algorithms (Ribeiro et al. 2018) belong to this category and provide rule-based, model-agnostic explanations by identifying a subset of input features such that, when these key features are fixed, the model's decision remains unchanged, regardless of variations in other variables. Therefore, in drug discovery, an anchor might highlight a particular molecular substructure consistently associated with a compound's bioactivity. Another important approach within instance-based attribution is the search for **counterfactual instances** (Guidotti 2024), or counterfactuals, which aim to answer the question: *What minimal changes would have led to a different outcome?* These explanations do not just justify why a model made a particular decision but also offer insight into how that decision could be altered. In the context of drug design, a counterfactual explanation might suggest that a particular molecule would be classified as active rather than inactive if an

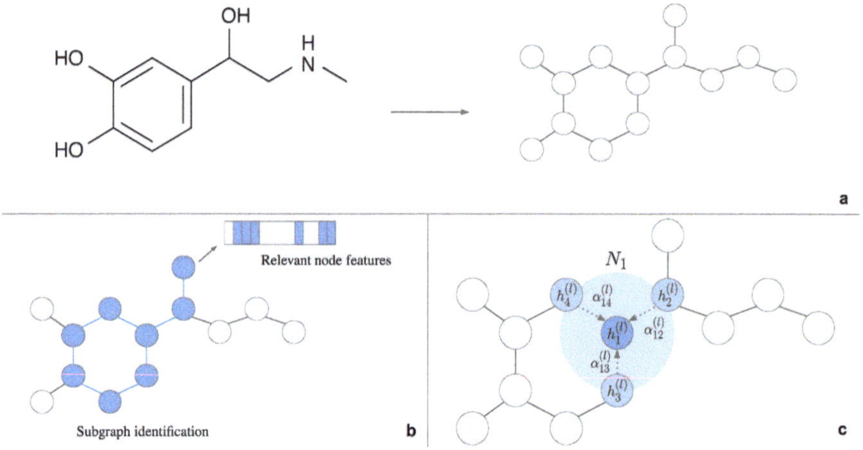

Fig. 4 Graph topology-based explanation methods. (**a**) Kekulé structure and corresponding molecular graph of adrenaline. (**b**) Subgraph identification methods identify a connected subgraph and node-level features that are relevant for a given output y of a GNN. (**c**) Attention layers in graph attention networks learn attention coefficients *α_ij^((l))* for the neighboring nodes of node *i*. (This example has been adapted from Abramson et al. 2024)

additional hydroxyl group were added at a specific position. This provides valuable guidance for medicinal chemists who want to fine-tune molecular structures for desired biological effects.

A more nuanced approach to instance-based explanations is represented by **contrastive explanations** (Prabhushankar et al. 2020), which go a step further by not only identifying what we must change to alter a prediction but also explaining why one classification was chosen over another. In drug-target prediction, for example, contrastive explanations can be used to determine why a particular compound binds to one protein but not another.

In general, instance-based approaches offer powerful ways to make AI models more transparent and actionable.

2.1.3 Graph Topology-Based Methods

Molecular graphs are a natural representation of molecules, with nodes and edges representing atoms and chemical bonds, respectively. For this reason, GNNs are widely used in drug discovery. To address their lack of interpretability, graph topology-based methods identify important structural components within the input graph that significantly influence the model's predictions. Two major approaches for explaining GNNs based on graph topology are subgraph identification methods and attention-based methods, illustrated in Fig. 4.

Subgraph Identification Methods seek to extract compact, connected subgraphs that are most responsible for a model's decision, such as functional groups in molecular graphs. One of the most influential subgraph identification methods is GNNExplainer (Ying et al. 2019), a model-agnostic approach that learns a mask over the graph structure and node features, optimizing a mutual information objective between the predicted output for the original and masked graphs. Although GNNExplainer's objective is similar to PGExplainer's, the two approaches are substantially different. The former learns an explanation for a single graph input at a time, while the latter is a probabilistic model that trains a neural network to generate explanations across multiple graph instances, thus being more scalable.

Attention-Based Methods on the other hand, focus on assigning importance scores to nodes or edges using attention mechanisms. For instance, graph attention networks (GATs) (Veličković et al. 2017) introduce an attention mechanism that computes edge-specific weights, allowing the model to prioritize certain connections over others. By implicitly highlighting the most relevant interactions, GATs offer dynamic representations of graph structures, for instance revealing which atomic interactions are most crucial for binding affinity in protein-ligand interaction prediction. Transformer-based models, such as Graphormer (Ying et al. 2021), extend this idea by utilizing global attention mechanisms, where every node attends to all other nodes in the graph. By focusing on graph structures rather than individual node features, graph topology-based methods provide insights that align more naturally with domain-specific knowledge, such as molecular substructures or network motifs. However, these methods can suffer from instability, where learned attention scores do not always correlate with meaningful feature importance. Additionally, attention mechanisms, particularly in transformers, tend to be less intuitive when applied to large, sparse graphs. Finally, subgraph identification approaches like GNNExplainer can be computationally expensive and may not always produce globally consistent explanations across different inputs.

2.1.4 Self-Explaining Methods

While most of the presented methods are post-hoc, self-explaining models integrate interpretability directly into their architecture, ensuring that their decision-making process is inherently transparent. Among the key paradigms of self-explaining models, prototype-, concept-, and logic-based approaches offer distinct methodologies to achieve interpretability while maintaining predictive accuracy.

Prototype-Based Approaches

Prototype-based models operate by learning a set of representative prototypes, which can be input samples or their subparts, summarizing key aspects of the data. In the context of graphs, Prototype-Based GNNs (Ragno et al. 2024b) are formed

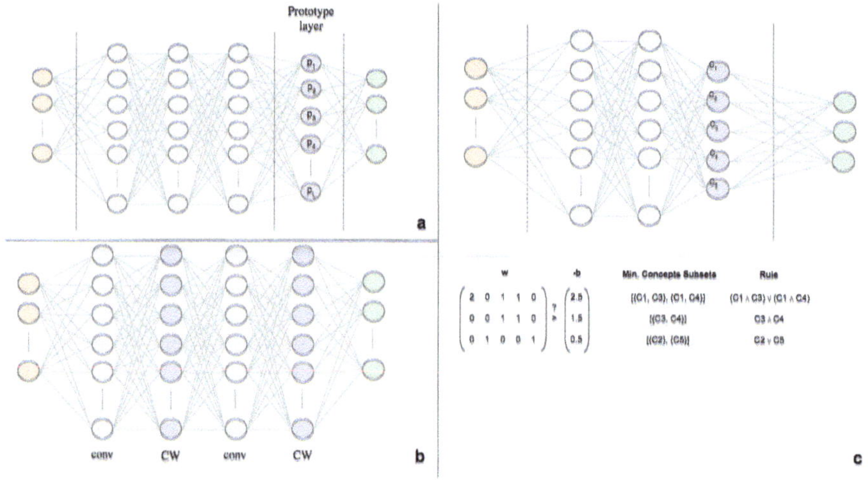

Fig. 5 Self-explaining methods. (**a**) Prototype-based networks (Ragno et al. 2024b) feature a prototype layer and produce the output based on the similarity of the input to the learned prototypes. (**b**) Concept-based methods provide explanations in the form of human-intigh-level human-interpretable concepts. For instance, concept-whitening layers (Proietti et al. 2024) are replaced to the normalization layers of a pretrained neural networks and the whole architecture is fine-tuned for a few epochs to align the axes of the latent space with the concepts of interest. (**c**) TELL (Ragno et al. 2024a) directly embeds logic constraints into a neural network layer

by three components: a GNN feature extractor made of graph convolutional layers, a prototype layer that projects the node embeddings and computes similarities between them and the learned prototypes, and a classification layer that computes output probabilities given the prototype similarities (see Fig. 5a).

While explanations provided by prototype-based approaches naturally align with human intuition, training such models is particularly challenging. If they fail to learn a representative and diverse set of prototypes, the resulting explanations may be biased or misleading.

Concept-Based Approaches

Concept-based approaches provide explanations in the form of high-level human-understandable concepts, such as chemical properties.

Graph Concept Whitening (GCW) (Proietti et al. 2024) is the first concept-based method designed to introduce interpretability into GNNs applied to drug discovery. The technique works by learning a whitening matrix to mean-center and decorrelate latent representations to ensure that different latent dimensions correspond to distinct, interpretable concepts (see Fig. 5b), thus disentangling the contribution of each concept to the model's prediction. Additionally, by using GCW in conjunction with other post-hoc methods, such as GNNExplainer, it is possible to highlight the structural parts of the molecules relevant to a certain concept that strongly influences bioactivity. One of the major advantages of GCW is that it allows models to retain high interpretability without sacrificing predictive power, as the concepts are explicitly encoded into the latent space. However, one major challenge is the dependence on predefined concepts, which requires domain expertise to define meaningful categories. Moreover, bad concept selection might prevent the model from providing meaningful explanations, thus limiting its applicability to new or less structured datasets.

Logic-Based Approaches

Logic-based approaches provide explanations by embedding structured logical rules in the learning process. Although some machine learning models, such as decision trees and Bayesian Rule Lists, naturally produce logic-based explanations, their performance is often limited when compared to deep learning models, especially in handling complex patterns and large-scale data.

To exploit the advantages of neural networks while guaranteeing interpretability, Ciravegna et al. (2023) proposed Logic-Explained Networks (LENs), a family of neural networks designed to produce logic-based explanations through a post-hoc extraction of logic rules. However, LENs face constraints in their applicability, as they are primarily designed for binary inputs, and their post-hoc explanation process does not always align with the learned model.

More recently, Ragno et al. (2024a) introduced Transparent Explainable Logic Layers (TELL), a novel approach that directly embeds logic constraints into a neural network layer. Unlike LENs, TELL ensures that explanations are perfectly aligned with the model by constraining weights to be non-negative and using a carefully designed activation function. This enables the extraction of logic rules directly from the model's parameters, extending logic-based explanations to real-valued inputs and broader applications. We provide illustrative examples in Fig. 5c.

2.1.5 Uncertainty Estimation

Uncertainty estimation methods produce explanations by quantifying how much variance or ambiguity exists in the model's predictions. Two primary approaches to uncertainty estimation include ensemble models, which aggregate multiple predictions to estimate stability, and probabilistic approaches, which explicitly model uncertainty as part of the learning process.

Ensemble Models

Ensemble models provide a natural measure of uncertainty by comparing the variation in predictions across different models trained with varying initial conditions or subsets of the data, which capture different aspects of the problem space. Intuitively, if all models agree, confidence is high, but if they diverge significantly, uncertainty is greater.

In practice, ensemble-based uncertainty estimation is often implemented through deep ensembles (Lakshminarayanan et al. 2017), where multiple deep neural networks are trained separately, for instance using different bootstrapped samples of the dataset (Breiman 1996; Freedman 1981). The predictions of these models are then aggregated, typically via averaging or majority voting or ensuring that the most informative models contribute more heavily to the final prediction (Fragoso et al. 2018).

The main limitation of ensemble models is their computational cost, as training multiple models requires more resources. Additionally, ensembles lack a structured probabilistic interpretation of uncertainty, indicating only prediction variability without explicitly modeling uncertainty sources.

Probabilistic Approaches

Unlike ensemble models, probabilistic approaches output a distribution over possible predictions, explicitly incorporating uncertainty into the model's internal structure.

One of the most established methods in this category is Bayesian neural networks (Goan and Fookes 2020), which extend traditional deep learning models by treating network weights as probability distributions rather than fixed values and inferring the posterior distribution using techniques such as Markov Chain Monte Carlo (MCMC) sampling (Spade 2020) or variational inference (Jordan et al. 1999). This is particularly useful in low-data settings, where traditional deep learning models might overfit due to insufficient training examples.

However, full Bayesian inference is often computationally expensive, requiring significant resources to approximate posterior distributions. To address this, alternative approximate Bayesian methods have been proposed (Qu et al. 2021; Gal and Ghahramani 2016; Teye et al. 2018), but they might introduce inaccuracies. Additionally, despite these approximations, the high computational cost associated with large-scale deep learning remains an active research area.

2.2 Explainability Techniques Across Different Data Modalities

Different explainability methods capture varied insights into model predictions. Figure 6 shows how different data modalities and model architectures produce distinct explanation patterns when using the same XAI approach, implying no single optimal explanation method.

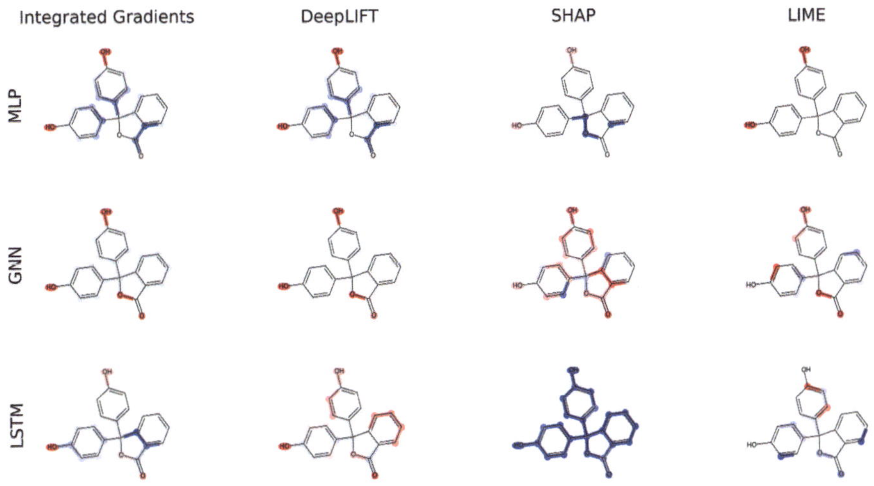

Fig. 6 Examples of explanations for gradient-based and perturbation-based methods on different models and data modalities (MLP and fingerprints, GNN and molecular graphs, LSTM and SMILES) for the prediction of water solubility. Explanations are visualized as masks highlighting positive (red) and negative (blue) contributions of the atoms to the probability of a molecule being soluble in water

Among existing modalities, graph-based representations have gained traction, especially with GNNs, which naturally capture the structural and relational properties of molecules, making them particularly powerful for tasks such as binding affinity prediction and toxicity assessment. Subgraph identification methods (e.g., GNNExplainer or GraphMask) address their lack of interpretability, pinpointing influential molecular substructures, useful for identifying pharmacophores and toxicophores. Attention-based methods provide another valuable approach by assigning importance scores to different nodes and edges in a graph, thus highlighting relevant molecular interactions, which can be particularly informative when studying drug-target interactions. However, attention scores are not always reliable indicators of causality, as they may be influenced by artifacts within the data or model training process. Perturbation-based methods, which systematically modify parts of a molecular graph and observe the effect on predictions, offer a more direct way to assess feature importance, but often at high computational cost.

In contrast, fingerprint-based molecular representations reduce molecular structures to fixed-length binary or numerical vectors, capturing essential structural features while discarding explicit spatial relationships. MLPs are often used to process fingerprints efficiently. Feature attribution methods (e.g., SHAP, LIME) can then highlight which fingerprint bits drive a model's decisions. Although intuitive, these bits may not directly map back to original chemical structures, and the presence of correlated features within fingerprints might produce misleading explanations. Surrogate models, such as decision trees, can provide more interpretable approximations of MLP behavior but introduce an additional abstraction layer. A third popular representation is SMILES, a string-based notation that encodes molecular structures in a linear form. Recurrent Neural Networks (RNNs) and Transformer-based models are particularly effective in capturing sequential dependencies between molecular subunits in SMILES strings. Explainability often involves saliency maps, where methods like Integrated Gradients and SmoothGrad highlight the most influential string segments. This approach is particularly useful for tasks such as molecular property prediction and reaction modeling, where understanding how different functional groups influence a model's decision is essential. Additionally, counterfactual explanations can reveal minimal modifications to generate different predictions. Similarly to fingerprint-based approaches, SMILES-based explanations do not always align with traditional chemical reasoning, since linear encodings do not inherently capture molecular topology. A hybrid approach that integrates graph-based methods can overcome this limitation by mapping sequence-based insights back onto the molecule's explicit structure.

Given the diverse interpretability challenges of each modality, a multi-faceted approach can be most effective. Graph-based explanations are well-suited for interpreting structural features in binding affinity or ADMET tasks, whereas fingerprint-based approaches may suffice for high-throughput screening scenarios—though caution is needed to validate feature importances. SMILES-based explainability can guide de novo molecule design, illuminating how structural tweaks affect predicted properties. In all modalities, pairing explainability with uncertainty estimation (e.g., Bayesian deep learning or ensembles) boosts interpretability by conveying confidence levels. Ultimately, there is no single best explainability technique for all drug discovery applications. The choice of method should be guided by the specific characteristics of the data, the complexity of the model, and the type of insights required. By combining different explainability approaches, researchers can achieve a more comprehensive understanding of how AI models make decisions, leading to more reliable and actionable outcomes. Table 2 compares some of these approaches, highlighting their advantages, drawbacks, and applications.

3 Applications of XAI in Drug Discovery

In this section, we explore how XAI is used in drug discovery applications to enhance interpretability and trust in AI-driven predictions. A key focus will be on different applications where explainability is crucial, such as ligand- and structure-based drug discovery and ADMET analysis. By analyzing diverse use cases, we aim to demonstrate how XAI can improve the transparency of AI models, accelerating the development of novel therapeutics. Given that XAI approaches explicitly rely on the trained models, their application should be considered only after carefully validating the accuracy and robustness of the models.

Table 2 Comparison of several explainability methods

XAI Approach	Advantages	Disadvantages	Applications
SHAP	Local and global explanations; model-agnostic.	Computationally expensive; dependent on feature correlations.	QSAR (Rodríguez-Pérez and Bajorath 2020); ADMET prediction (Wu et al. 2019)
LIME	Model-agnostic; local explanations.	Sensitive to size and type of feature perturbations.	Molecular docking (Chongjun et al. 2024)
GraphCW	Explanations in terms of human-understandable concepts.	Requires building concept datasets; concept selection impacts performance.	QSAR (Proietti et al. 2024)
GradCAM	Local explanations; model-agnostic (only differentiable models).	Requires convolutional structures; not applicable to all models.	QSAR (Zhong et al. 2021); ADMET prediction (Vangala et al. 2023)
Integrated gradients	Local explanations; model-agnostic (only differentiable models).	Computationally expensive; requires defining a baseline.	Graph classification (Preuer et al. 2019)
PGExplainer	Finds critical subgraphs relevant to model predictions.	Requires careful tuning.	Mutagenicity prediction (Luo et al. 2020)
LRP	Efficient feature estimation.	Requires architecture-specific tuning; may not generalize.	Protein-ligand interactions (Cho et al. 2020)
Prototype-based GNN	Aligns with human reasoning.	Requires careful tuning.	Molecular classification (Ragno et al. 2024b)
Bayesian neural networks	Robust against overfitting.	The inference process is complex.	Drug-likeness prediction (Beker et al. 2020)

3.1 Ligand-Based Drug Discovery: QSAR and QSPR

QSAR and QSPR modeling are widely used in drug discovery to predict molecular activity and properties based on ligands' chemical structures. These computational approaches help identify promising drug candidates, optimize molecular designs, and reduce costs by minimizing experimental efforts. In this context, XAI provides methods to interpret and validate QSAR and QSPR models.

3.1.1 SHAP-Based Interpretation for Lead Optimization

QSAR aims to identify relationships between structural motifs and molecular activity to edit a molecule's structure to maximize its potency. In a recent work, Harren et al. (2022) investigated the use of SHAP to visualize the contribution of individual atoms and molecular fragments to a compound's predicted biological activity using atom-based heatmaps. The approach is tested on two well-characterized molecular targets: Factor Xa, a serine protease critical in blood coagulation, and Renin, an aspartic protease involved in hypertension regulation. Both targets have extensive SAR data and available X-ray crystal structures, making them ideal candidates for evaluating the interpretability of SHAP.

The models used in this study include GNNs receiving data as molecular graph representations and deep neural networks taking molecular descriptors, trained to predict activity trends within lead optimization datasets. In this context, SHAP is applied to decompose the predictions, highlighting structural features such as substituents or functional groups that are critical for binding affinity or potency.

Results demonstrate that SHAP not only accurately identifies known SAR patterns but also reveals novel structural insights that align with experimental data. For example, key functional groups critical to activity against Factor Xa and Renin are highlighted, helping to validate the model's predictions and suggesting potential avenues for compound optimization. By integrating SHAP-based explanations, the study provides a framework for enhancing the trustworthiness and utility of AI-driven QSAR modeling in lead optimization.

3.1.2 QSAR Explanations and Structure-Based Techniques: A Case Study on GSK-3β

While Harren et al. (2022) show how SHAP can assist in optimizing leads by identifying critical structural features influencing biological activity, Masand et al. (2025) take a step further by coupling SHAP with structure-based techniques, such as docking and molecular dynamics, to gain deeper insights into drug-target interactions. Their work focuses on a case study involving GSK-3β modulators, a target of significant therapeutic relevance in diseases like Alzheimer's, diabetes, and cancer. By integrating explainable QSAR modeling with structural analyses, they

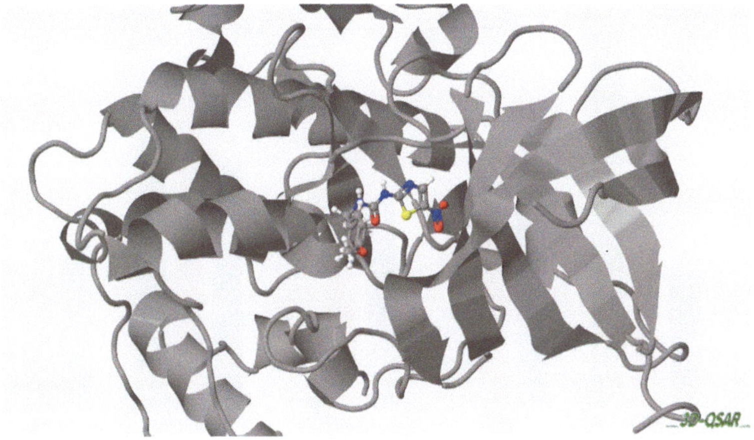

Fig. 7 3-D Visualization of the GSK-3β active site structure from PDB (1Q5K).

provide a comprehensive framework for understanding the molecular determinants of activity and binding interactions.

The study employs a dataset of 1293 molecules with experimentally measured EC50 values for GSK-3β modulation. The framework leverages Genetic Algorithms to select the most relevant molecular descriptors, while XGBoost is used to train predictive models. The authors then use SHAP to interpret these models, offering insights into how specific molecular features influenced the predictions. As a complement to these findings, molecular docking studies conducted using the GSK-3β active site structure (PDB ID: 1Q5K, Fig. 7) allow for a detailed examination of ligand-receptor interactions.

For example, fsp3CaroC5B, a descriptor representing the frequency of aromatic carbons five bonds away from sp3-carbons, emerges as a crucial feature, highlighting the importance of combining molecular rigidity from aromatic rings with the flexibility provided by sp3-carbons. Other significant descriptors include aroC_hy3, which identifies aromatic carbons with specific partial charges as vital for electronic interactions, and amdNminus_AbSA, which underscores the role of accessible polar groups, such as negatively charged amide nitrogens, in binding.

Molecular docking validates these findings by demonstrating strong alignment between the QSAR-predicted important features and experimentally validated binding interactions.

By integrating SHAP-based QSAR modeling with docking and molecular dynamics, Masand et al. (2025) present a holistic approach to analyzing drug-target interactions. This method not only provides interpretable insights into the molecular features driving GSK-3β activity but also demonstrates how structure-based techniques can be used to validate and enhance these findings. The framework serves as a valuable tool for lead optimization and rational drug design, offering a pathway to more transparent and actionable drug discovery pipelines.

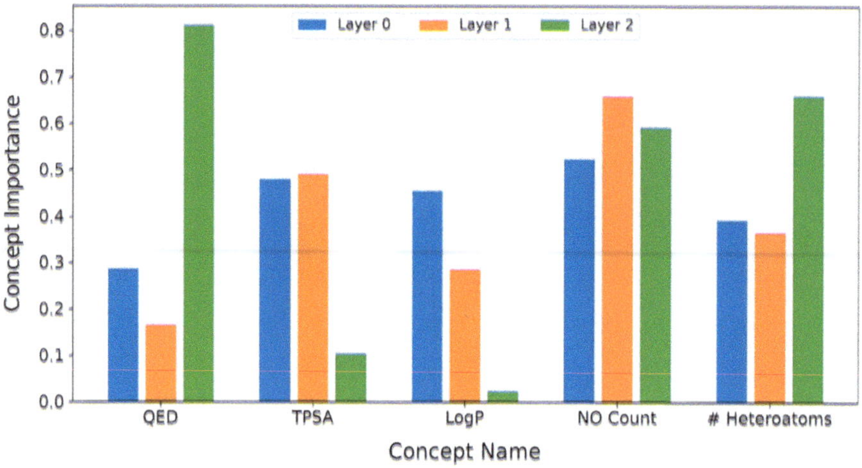

Fig. 8 Concepts importance measured at the CW layers of a three-layer GNN for the BBBP dataset. The concepts contribution is computed as the sum of the positive directional derivatives of the gradients along the channel representing that concept

3.1.3 Towards Transparent QSAR Model with Graph Concept Whitening

Post-hoc methods, such as LIME and SHAP, allow us to obtain useful explanations, but they do not inherently address the opaque nature of the underlying models, as they are applied to pre-trained black-box models. Contrarily, self-explaining models directly integrate interpretability into their architecture, providing explanations and predictions simultaneously.

For instance, as explained in Sect. 2.1.4, GCW (Proietti et al. 2024) introduces a mechanism to encode human-understandable *concepts* directly into the latent space of the model, making it possible to interpret predictions through these predefined, interpretable components.

GCW operates by partitioning the latent space of a GNN into interpretable subspaces aligned with specific human-defined concepts.

For example, when applied to QSAR tasks, GCW can map molecular features—such as the presence of aromatic rings, halogen atoms, or charged groups—to concept subspaces. During inference, the model not only predicts the activity or property of a molecule but also provides a breakdown of how each concept influenced the prediction, as shown in Fig. 8. While the five considered concepts have similar importance scores at Layer 0, they are increasingly differentiated in the subsequent layers. According to the concept scores at Layer 2, QED, NOCount, and #Heteroatoms are the most influential concepts, which confirms the findings by Sakiyama et al. (2021). This disentanglement ensures that the learned representations are both interpretable and actionable, offering insights directly at prediction time.

In QSAR modeling, GCW has been applied to datasets where the relationships between chemical structures and biological activities are well-characterized, proving its ability to provide interpretable insights while retaining high predictive accuracy.

3.1.4 Explainable QSAR for Everyone: The 3D-QSAR.com Portal

Ragno (2019) presents 3D-QSAR.com,[1] an integrated platform for computational drug discovery, providing several ligand-based, structure-based, and cheminformatics tools for molecular modeling and drug discovery workflows. Specifically, the platform provides the following services:

- Py-CoMFA, which allows researchers to construct and analyze comparative molecular field analysis models.
- Py-Align, which enables molecular alignment to standardize the spatial orientation of molecules before QSAR analysis.
- Py-ConfSearch, is a tool for conformational analysis that allows users to explore different 3D molecular conformations.
- Py-Docking, which allows users to perform molecular docking to predict how small molecules bind to target proteins.
- Py-PDB, a utility for protein structure minimization, optimizing protein-ligand interactions for docking and other simulations.
- Py-Combine, which facilitates the development of ComBinE (Comparative Binding Energy) models, integrating QSAR-like techniques with docking-derived energy terms for enhanced predictive modeling.
- Py-Graph, the most recent addition to the platform expands its capabilities by integrating GNN models for molecular classification. This represents a shift toward deep learning-based techniques, allowing users to leverage the latest advancements in AI-driven drug discovery.

Py-Graph provides an intuitive environment for training GNNs, making it accessible to researchers who may not have extensive expertise in machine learning. Py-Graph integrates DeepLIFT (Fig. 9), an XAI method that provides real-time explanations for model predictions. This technique analyzes how changes in molecular structures impact classification outcomes, offering a clear attribution of importance for each bond and atom. The results are visualized as color overlays on molecular bonds, where the color intensity reflects the influence of each structural feature on the predicted activity. This functionality allows users to interpret GNN predictions in a chemically meaningful way, facilitating the identification of key pharmacophores and guiding lead optimization efforts.

[1] https://www.3d-qsar.com

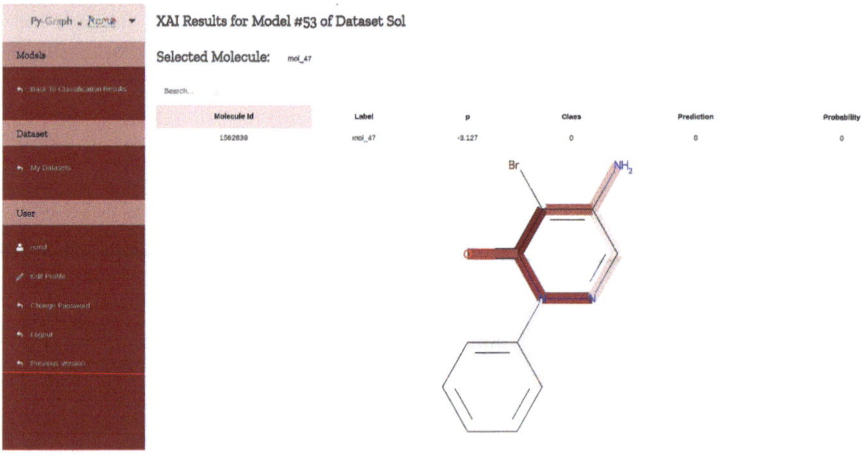

Fig. 9 Explanations provided by PyGraph for the prediction of water solubility. Explanations are visualized as masks over the most relevant atoms

By combining ligand-based QSAR methods with deep learning-based molecular modeling, 3D-QSAR.com bridges the gap between traditional computational approaches and modern AI-driven strategies.

3.1.5 Beyond QSAR: XAI Applied to Quantitative Composition-Activity Relationships on Essential Oils

Essential oils (EOs) pose a challenge for traditional QSAR models due to their complex, variable compositions. In this context, quantitative composition-activity relationship (QCAR) modeling has emerged as an alternative to QSAR, leveraging machine learning to identify the key components responsible for biological effects. Astolfi et al. (2025) apply XAI techniques to develop QCAR models to predict the antimicrobial activity of EOs against carbapenem-resistant Acinetobacter baumannii (CR-Ab), a major nosocomial pathogen with increasing antibiotic resistance.

Using a dataset of 82 EOs, built integrating experimental and literature-derived data from the AI4EssOil database,[2] the authors train and test multiple ML classifiers (SVM, Random Forest, Gradient Boosting, Decision Trees, kNN), achieving 91% accuracy in predicting EO antimicrobial efficacy. However, the key to the study is the integration of feature importance analysis using the Skater methodology to pinpoint the monoterpenes and sesquiterpenes driving activity, such as limonene, eucalyptol, alpha-pinene, linalool, beta-caryophyllene, nerol, beta-pinene, neral, and carvacrol. Experimental validation on six selected pure compounds (limonene, eucalyptol, alpha-pinene, linalool, carvacrol, and thymol) revealed inhibitory effects

[2] https://www.ai4essoil.com

against CR-Ab that are strongly correlated with the model predictions. Specifically, carvacrol and thymol demonstrate the highest antimicrobial potency, with MIC values below 0.03% v/v, supporting their essential role in EO efficacy.

Beyond antimicrobial properties, the study extends its QCAR modeling to predict cytotoxicity against 3T3-Swiss fibroblast cells. Interestingly, compounds such as limonene and eucalyptol are found to be associated with low toxicity, whereas carvacrol and thymol, despite their antimicrobial potency, exhibit strong cytotoxic effects at higher concentrations. This insight suggests that selective enrichment or blending of EO components could enhance therapeutic efficacy while reducing toxicity.

This study exemplifies how XAI can be applied beyond QSAR to more complex biological mixtures, enabling the rational design of bioactive formulations. By integrating ML with interpretability tools, the researchers demonstrated how essential oil composition can be linked directly to biological effects, providing a data-driven framework for standardizing EO-based therapeutics.

3.2 *Structure-Based Drug Discovery*

Structure-based drug discovery (SBDD) has long been a fundamental approach in drug design, relying on structural knowledge of target proteins to guide the development of bioactive compounds. While traditional simulation-based methods like docking, molecular dynamics, and free energy perturbation offer detailed insights into ligand-protein interactions, they can be computationally intensive.

In recent years, AI-driven approaches, particularly deep learning, have emerged as a faster alternative by rapidly predicting binding affinities and generating novel ligands, even for flexible proteins or complex interactions. However, despite their speed and efficiency, AI-based techniques lack the physical grounding provided by force-field-based simulations, since their predictions are purely based on statistical patterns in the training data. This limitation introduces the need for interpretable outputs. XAI meets this need by clarifying why an AI model predicts certain affinities or poses, enabling researchers to validate drug candidates more confidently. Additionally, XAI can uncover hidden patterns in ligand-target interactions, such as key hydrogen bonds and hydrophobic regions, and suggest ways to enhance binding.

In this section, we explore how XAI enhances AI-based SBDD, improving lead optimization and hit selection and accelerating therapeutic development by bridging AI's speed with the mechanistic depth of force-field methods.

3.2.1 Interpretable AI for Molecular Simulations and Virtual Screening

Computational approaches, like molecular dynamics (MD) simulations and virtual screening, are central to drug discovery, helping researchers predict molecular interactions, optimize drug candidates, and explore large chemical spaces efficiently.

However, these approaches often lack interpretability, particularly when AI-based methods are used to replace traditional physics-based simulations. Recent work in XAI tackles this issue by uncovering how AI models reach their predictions, thus improving confidence and scientific grounding. For instance, Li et al. (2022) developed an interpretable convolutional neural network (CNN) framework for analyzing MD trajectories that allows accurate classification of G-protein-coupled receptors (GPCRs), a key class of drug targets involved in various physiological processes. Additionally, the study applied local approximation methods, such as LRP, to identify key amino acid residues contributing to state classification. These explanations aligned with known activation mechanisms of GPCRs, confirming that the AI model learned biologically meaningful features rather than relying on spurious correlations. By mapping the most relevant residues onto GPCR structures, the authors demonstrated how AI-driven MD analysis could provide actionable insights into receptor function, supporting ligand design and targeted drug development.

Madushanka et al. (2024) introduced SmartCADD, which integrates AI and quantum mechanics (QM) for virtual screening and lead optimization. While traditional virtual screening predicts binding affinity by relying on molecular docking, which often underperforms due to simplified scoring functions, the SmartCADD hybrid AI-QM approach yielded more precise affinity estimates.

A key feature of SmartCADD is its explainability module, which utilizes SHAP and Integrated Gradients to interpret model predictions. When applied to HIV protease inhibitors, SHAP analysis revealed that hydrogen bond donors, aromaticity, and molecular charge distribution were the most critical features for binding affinity. These findings aligned well with QM-based interaction energy calculations, confirming that the AI model was not only accurate but also aligned with established physicochemical principles. The integration of QM and AI interpretability in SmartCADD marks a step toward reliable, physics-informed virtual screening, enhancing confidence in AI-driven predictions.

Both studies illustrate the increasing role of explainability in AI-driven molecular simulations and virtual screening. Together, these approaches highlight the potential of XAI to bridge the gap between machine learning and physics-based simulations, ensuring that AI-driven drug discovery remains not only fast and scalable but also transparent and scientifically grounded.

3.2.2 Explainable AI for Docking: From Deep Reinforcement Learning to COVID-19 Target Prediction

Molecular docking is a fundamental technique in structure-based drug discovery that predicts how small molecules bind to protein active sites, traditionally via physics-based scoring functions. While effective, these methods struggle with protein flexibility and large-scale screening. AI-driven docking offers faster alternatives by training models on experimental data but lacks the physical interpretability provided by force-field-based simulations. To address this, recent work integrates XAI into docking to enhance transparency and reveal critical ligand-target interactions.

Løver et al. (2021) used deep reinforcement learning (DRL) to automate docking, training an AI agent to iteratively optimize ligand binding conformations based on learned interaction patterns, without relying on predefined scoring functions.

To interpret the agent's decisions, the authors employed three XAI methods: Kernel SHAP, LIME, and Linear Model Trees. The obtained explanations revealed that hydrogen bonding, hydrophobic interactions, and steric fit drove the agent's decisions, aligning with standard docking principles. Additionally, LIME visualizations showed that the model can adjust ligand orientations dynamically, selecting optimal poses based on the surrounding protein environment.

This work underscored the potential of reinforcement learning for self-improving docking strategies but also emphasized the necessity of interpretability in AI-driven docking to validate predictions or refine the model for broader applicability.

Shi et al. (2023) focused on using machine learning models to predict docking scores for SARS-CoV-2 protein targets, addressing the urgent need for efficient virtual screening during the COVID-19 pandemic. To expedite screening, they trained ML models (RF, MLP, XGBoost, kNN) on molecular fingerprints and interaction features to predict docking scores for 18 different viral proteins in a unified manner. Moreover, using SHAP to interpret these scores, the authors confirmed that aromaticity, charge distribution, and functional groups contributed most to binding affinity. This matched known pharmacophoric features in viral protease inhibitors, such as key hydrogen-bonding motifs and hydrophobic interactions, reinforcing the model's biological relevance.

These studies illustrate how XAI transforms AI-driven docking from a black-box approach into a transparent tool rooted in physical principles and biological insights. By clarifying the reasoning behind AI predictions, XAI fosters trust, aids model validation, and ensures AI outputs align with established docking knowledge.

3.3 ADMET

ADMET (Absorption, Distribution, Metabolism, Excretion, and Toxicity) represents a critical set of properties that determine the pharmacokinetic and safety profiles of drug candidates. These factors highly affect a compound's efficacy, bioavailability, and safety, and poor ADMET profiles often cause late-stage failures in drug development.

The evaluation of ADMET properties is a complex task, as it involves understanding how a molecule behaves within a biological system. Experimental assessment of these properties can be time-consuming, expensive, and resource-intensive, especially when screening large compound libraries.

Machine learning models have shown great promise in predicting ADMET profiles using molecular descriptors or graph-based representations of chemical structures, allowing researchers to screen and filter drug candidates early in the development process, thus saving considerable time and costs. XAI addresses their lack of interpretability, identifying the molecular features influencing ADMET properties. For instance, post-hoc methods like SHAP and LIME can identify which structural attributes impact properties like permeability or toxicity, while self-explainable models incorporate interpretability directly into their architecture, allowing for real-time explanations of ADMET predictions.

Incorporating XAI into ADMET prediction not only fosters trust in computational tools but also guides drug optimization. By pinpointing which molecular modifications might enhance pharmacokinetics or reduce toxicity, XAI provides a rational framework for refining candidates. This section examines how XAI methods have been applied to ADMET prediction and highlights their role in informed drug design.

3.3.1 Feature Relevance in ADME Property Prediction

König and Vellido (2024) demonstrated the power of explainable machine learning for predicting ADME properties, which are critical in evaluating the pharmacokinetics of drug candidates. Their study utilized a dataset of 3521 small molecules, each annotated with six key ADME endpoints, including human and rat liver microsomal stability, plasma protein binding, solubility, and efflux ratios. The researchers applied various machine learning algorithms, with LightGBM emerging as the top-performing model due to its balance between predictive accuracy and interpretability. Using SHAP to calculate feature importance, they found logP and TPSA strongly influenced properties like metabolic stability, solubility, and plasma protein binding, reflecting lipophilicity and polarity's roles in pharmacokinetics. By visualizing SHAP values with dependence plots, the study provided interpretable insights into how these features impacted ADME outcomes. For instance, they demonstrated that molecules with higher logP values tended to exhibit reduced solubility but increased plasma protein binding, consistent with established pharmacokinetic principles. Furthermore, TPSA was shown to correlate with absorption, as highly polar molecules face challenges in crossing lipophilic biological membranes.

This approach also allowed the researchers to uncover non-linear relationships between features and ADME properties, which might have been overlooked by traditional QSAR methods. For instance, compounds with moderate logP values exhibited optimal metabolic stability, while extreme values in either direction led to diminished stability. Such nuanced insights highlight the added value of XAI in guiding drug design.

This study underscores the utility of feature relevance analysis in streamlining drug discovery workflows and improving decision-making in the early stages of development.

3.3.2 Predicting Pulmonary Toxicity with Explainable Tree-Based Models

Just as feature relevance analysis in ADME property prediction highlights the importance of molecular descriptors in understanding pharmacokinetic properties, similar explainable techniques can be extended to another critical aspect of drug discovery: toxicity prediction. Pulmonary toxicity, characterized by adverse respiratory effects such as fibrosis, bronchitis, and asthma, is a major concern in drug development, as it often leads to late-stage failures or drug withdrawals. Jaganathan et al. (2022) address this challenge by combining explainable ML models with molecular feature selection to predict pulmonary toxicity and provide interpretable insights into the structural factors contributing to toxicity.

Their study utilizes a dataset of 2527 compounds curated from toxicity-specific repositories, including PNEUMOTOX and ADReCS, encompassing both toxic and non-toxic substances.

To build the predictive models, Jaganathan et al. (2022) employ tree-based ensemble methods, including Random Forest, Extra Trees, XGBoost, and LightGBM on a selection of descriptors and fingerprints. XGBoost and LightGBM stand out for their superior accuracy, achieving 86.88% accuracy on the test set and comparable performance on an external validation set. Explainability is provided by SHAP, which quantifies the contribution of each molecular feature to the model's predictions. Key features identified include aromaticity descriptors, which are linked to the likelihood of pulmonary toxicity due to their association with reactive intermediates, and hydrophobicity-related descriptors, which influence a compound's ability to interact with cellular membranes. For instance, the study reveals that some descriptors, such as the fraction of aromatic carbons or the topological polar surface area, have non-linear impacts on toxicity. These findings align with known mechanistic insights, such as the role of aromatic compounds in oxidative stress and inflammation pathways.

The study emphasizes the growing importance of integrating interpretability into predictive models, echoing the themes observed in ADME prediction, where feature relevance serves as a foundation for actionable insights.

4 Conclusions and Future Directions

The integration of XAI in drug discovery represents a transformative shift in how computational models contribute to pharmaceutical research. The ability to interpret and rationalize AI-driven predictions is essential for ensuring reliability, reproducibility, and meaningful insights in drug design. By bridging the gap between predictive performance and human interpretability, XAI methods enhance collaboration between AI researchers and domain experts, fostering greater trust in machine learning applications.

One of the major takeaways from the current advancements in XAI for drug discovery is its potential to improve decision-making at multiple stages of the drug development pipeline. Whether in ligand-based drug design, structure-based drug discovery, or ADMET property prediction, XAI tools provide researchers with a clearer understanding of how molecular features influence biological activity and pharmacokinetics. This increased transparency helps mitigate biases, prevent erroneous model-driven assumptions, and support regulatory compliance.

Despite these advances, several challenges remain. One of the most pressing concerns is the trade-off between model complexity and interpretability. Many state-of-the-art deep learning models achieve exceptional predictive performance but lack comprehensibility, necessitating the development of novel XAI techniques that balance both aspects. Additionally, the effectiveness of interpretability methods is highly dependent on the quality and completeness of the training data. Biases and inconsistencies in molecular datasets can lead to misleading explanations, underscoring the need for rigorous data curation and validation.

Another crucial challenge is the representation of molecular structures in a manner that is both meaningful to chemists and suitable for machine learning algorithms. Traditional molecular fingerprints and descriptor-based representations offer interpretability but may not capture complex molecular interactions accurately. Graph-based learning approaches, while promising, introduce their own set of interpretability issues due to the abstraction of molecular graphs. Future research should explore hybrid representation techniques that incorporate domain knowledge to enhance explainability without sacrificing predictive power.

Beyond these representational challenges, many existing XAI methods suffer from inherent limitations that affect their usability and reliability in drug discovery applications. Post-hoc interpretability techniques, such as SHAP and LIME, while widely used, may not always provide consistent or robust explanations, particularly in high-dimensional chemical space. Perturbation-based methods can be computationally intensive and may produce unstable results depending on the level of feature masking. On the other hand, gradient-based attribution techniques, such as Integrated Gradients and DeepLIFT, are sensitive to model parameters and can struggle to provide clear mechanistic insights. At the same time, self-explainable models generally require careful training and often result in the so-called accuracy-explainability trade-off. Furthermore, concept-based approaches, such as Graph Concept Whitening, depend on predefined concepts that might not always generalize across diverse molecular datasets.

Moving forward, several key directions warrant further exploration:

1. **Development of self-explaining models**. Designing AI architectures that inherently provide explanations without the need for post-hoc interpretability methods will improve the integration of XAI in drug discovery.
2. **Standardization of XAI evaluation metrics**. Establishing benchmarks and objective criteria for assessing the quality and relevance of explanations will enhance model validation and regulatory acceptance.

3. **Improved human-AI collaboration**. Developing interactive visualization tools and user-friendly interfaces will empower medicinal chemists and pharmacologists to engage with AI models more effectively.
4. **Incorporation of causality and mechanistic understanding**. Future XAI models should move beyond correlation-based interpretations and incorporate causal inference techniques to improve hypothesis generation in drug discovery.
5. **Application to low-data scenarios**. Active learning and uncertainty estimation techniques should be further explored to optimize AI models for rare or underexplored chemical spaces.
6. **XAI-guided continual learning for knowledge transfer**. XAI can enhance continual learning strategies, allowing AI models to retain and transfer knowledge across different drug discovery tasks, such as QSAR modeling, ADMET prediction, and drug-target interaction analysis. By leveraging insights provided by explanations, models can reuse learned chemical and biological patterns, reducing the need for extensive retraining and making them particularly valuable in low-data domains where new experimental data is scarce.

Ultimately, the evolution of XAI in drug discovery will require ongoing interdisciplinary collaboration between AI experts, computational chemists, and medicinal scientists. As research continues to push the boundaries of interpretability, the integration of XAI will play a crucial role in shaping the future of data-driven drug design, accelerating innovation while maintaining scientific rigor. Addressing the limitations of existing XAI techniques will be critical in ensuring that AI models remain both powerful and interpretable, ultimately bridging the gap between computational predictions and real-world pharmaceutical applications.

References

Abramson J, Adler J, Dunger J, Evans R, Green T, Pritzel A, et al. Accurate structure prediction of biomolecular interactions with AlphaFold 3. Nature. 2024;630:493–500. https://doi.org/10.1038/s41586-024-07487-w.

Ancona M, Ceolini E, Öztireli C, Gross M. Towards better understanding of gradient-based attribution methods for Deep Neural Networks. 6th International Conference on Learning Representations (ICLR). No. 1711.06104; 2017.

Astolfi R, Oliva A, Raffo A, Sapienza F, Ragno A, Proia E, et al. Essential oils as antimicrobials against *Acinetobacter baumannii*: experimental and literature data to definite predictive quantitative composition–activity relationship models using machine learning algorithms. J Chem Inf Model. 2025;65:1378–91. https://doi.org/10.1021/acs.jcim.4c02389.

Bach S, Binder A, Montavon G, Klauschen F, Müller K-R, Samek W. On pixel-wise explanations for non-linear classifier decisions by layer-wise relevance propagation. PLoS One. 2015;10:e0130140. https://doi.org/10.1371/journal.pone.0130140.

Beker W, Wołos A, Szymkuć S, Grzybowski BA. Minimal-uncertainty prediction of general drug-likeness based on Bayesian neural networks. Nat Mach Intell. 2020;2:457–65. https://doi.org/10.1038/s42256-020-0209-y.

Breiman L. Bagging predictors. Mach Learn. 1996;24:123–40. https://doi.org/10.1007/BF00058655.

Carreira-Perpiñán MÁ, Hada SS. Counterfactual explanations for oblique decision trees: exact, efficient algorithms. Proceedings of the AAAI conference on Artificial Intelligence. 2021;35(8):2021.

Cho H, Lee EK, Choi IS. Layer-wise relevance propagation of InteractionNet explains protein–ligand interactions at the atom level. Sci Rep. 2020;10:21155. https://doi.org/10.1038/s41598-020-78169-6.

Chongjun Y, Nasr AMS, Latif MAM, Rahman MBA, Marlisah E, Tejo BA. Predicting repurposed drugs targeting the NS3 protease of dengue virus using machine learning-based QSAR, molecular docking, and molecular dynamics simulations. SAR QSAR Environ Res. 2024;35:707–28. https://doi.org/10.1080/1062936X.2024.2392677.

Ciravegna G, Barbiero P, Giannini F, Gori M, Liò P, Maggini M, et al. Logic explained networks. Artif Intell. 2023;314:103822. https://doi.org/10.1016/j.artint.2022.103822.

European Union. General data protection regulation. Regulation (EU); 2016. p. 679.

Fong R, Vedaldi A. Interpretable explanations of black boxes by meaningful perturbation. In: Proceedings of the IEEE international conference on computer vision; 2017. https://doi.org/10.1109/ICCV.2017.371.

Fragoso TM, Bertoli W, Louzada F. Bayesian model averaging: a systematic review and conceptual classification. Int Stat Rev. 2018;86:1–28. https://doi.org/10.1111/insr.12243.

Freedman DA. Bootstrapping regression models. Ann Stat. 1981:9. https://doi.org/10.1214/aos/1176345638.

Gal Y, Ghahramani Z. Dropout as a Bayesian approximation: representing model uncertainty in deep learning. In: Proceedings of the 33rd international conference on Machine Learning, PMLR, vol. 48; 2016. p. 1050–9.

Goan E, Fookes C. Bayesian neural networks: an introduction and survey. 2020:45–87. https://doi.org/10.1007/978-3-030-42553-1_3.

Guidotti R. Counterfactual explanations and how to find them: literature review and benchmarking. Data Min Knowl Discov. 2024;38:2770–824. https://doi.org/10.1007/s10618-022-00831-6.

Guidotti R, Monreale A, Ruggieri S, Turini F, Giannotti F, Pedreschi D. A survey of methods for explaining black box models. ACM Comput Surv. 2019;51:1–42. https://doi.org/10.1145/3236009.

Harren T, Matter H, Hessler G, Rarey M, Grebner C. Interpretation of structure–activity relationships in real-world drug design data sets using explainable artificial intelligence. J Chem Inf Model. 2022;62:447–62. https://doi.org/10.1021/acs.jcim.1c01263.

Huang Q, Yamada M, Tian Y, Singh D, Chang Y. GraphLIME: local interpretable model explanations for graph neural networks. IEEE Trans Knowl Data Eng. 2023;35:6968–72. https://doi.org/10.1109/TKDE.2022.3187455.

Jaganathan K, Tayara H, Chong KT. An explainable supervised machine learning model for predicting respiratory toxicity of chemicals using optimal molecular descriptors. Pharmaceutics. 2022;14:832. https://doi.org/10.3390/pharmaceutics14040832.

Jiménez-Luna J, Grisoni F, Schneider G. Drug discovery with explainable artificial intelligence. Nat Mach Intell. 2020;2:573–84. https://doi.org/10.1038/s42256-020-00236-4.

Jordan MI, Ghahramani Z, Jaakkola TS, Saul LK. An introduction to variational methods for graphical models. Mach Learn. 1999;37:183–233. https://doi.org/10.1023/A:1007665907178.

Kocsis L, Szepesvári C. Bandit based Monte-Carlo planning. 2006:282–93. https://doi.org/10.1007/11871842_29.

König C, Vellido A. Understanding predictions of drug profiles using explainable machine learning models. BioData Min. 2024;17:25. https://doi.org/10.1186/s13040-024-00378-w.

Kuhn HW, Tucker AW, Luce RD. Contributions to the theory of games. Am Math Mon. 1960;67:491. https://doi.org/10.2307/2309332.

Lakkaraju H, Kamar E, Caruana R, Leskovec J. Interpretable & explorable approximations of black box models 2017.

Lakshminarayanan B, Pritzel A, Blundell C. Simple and scalable predictive uncertainty estimation using deep ensembles. Adv Neural Inf Proces Syst. 2017;30:2017.

Li C, Liu J, Chen J, Yuan Y, Yu J, Gou Q, et al. An interpretable convolutional neural network framework for analyzing molecular dynamics trajectories: a case study on functional states for g-protein-coupled receptors. J Chem Inf Model. 2022;62:1399–410. https://doi.org/10.1021/acs.jcim.2c00085.

Løver J, Gjærum VB, Lekkas AM. Explainable AI methods on a deep reinforcement learning agent for automatic docking. IFAC-PapersOnLine. 2021;54:146–52. https://doi.org/10.1016/j.ifacol.2021.10.086.

Lundberg S, Lee S-I. A unified approach to interpreting model predictions. In: Proceedings of the 31st international conference on neural information processing systems; 2017.

Luo D, Cheng W, Xu D, Yu W, Zong B, Chen H, et al. Parameterized explainer for graph neural network. Adv Neural Inf Proces Syst. 2020;33:2020.

Madushanka A, Laird E, Clark C, Kraka E. SmartCADD: AI-QM empowered drug discovery platform with explainability. J Chem Inf Model. 2024;64:6799–813. https://doi.org/10.1021/acs.jcim.4c00720.

Masand VH, Al-Hussain S, Masand GS, Samad A, Gawali R, Jadhav S, et al. e-QSAR (Explainable AI-QSAR), molecular docking, and ADMET analysis of structurally diverse GSK3-beta modulators to identify concealed modulatory features vindicated by X-ray. Comput Biol Chem. 2025;115:108324. https://doi.org/10.1016/j.compbiolchem.2024.108324.

Montavon G, Lapuschkin S, Binder A, Samek W, Müller K-R. Explaining nonlinear classification decisions with deep Taylor decomposition. Pattern Recognit. 2017;65:211–22. https://doi.org/10.1016/j.patcog.2016.11.008.

Olden JD, Jackson DA. Illuminating the "black box": a randomization approach for understanding variable contributions in artificial neural networks. Ecol Modell. 2002;154:135–50. https://doi.org/10.1016/S0304-3800(02)00064-9.

Ponzoni I, Páez Prosper JA, Campillo NE. Explainable artificial intelligence: a taxonomy and guidelines for its application to drug discovery. WIREs Comput Mol Sci. 2023:13. https://doi.org/10.1002/wcms.1681.

Pope PE, Kolouri S, Rostami M, Martin CE, Hoffmann H. Explainability methods for graph convolutional neural networks. In: 2019 IEEE/CVF conference on Computer Vision and Pattern Recognition (CVPR). IEEE; 2019. p. 10764–73. https://doi.org/10.1109/CVPR.2019.01103.

Prabhushankar M, Kwon G, Temel D, AlRegib G. Contrastive explanations in neural networks. In: 2020 IEEE International Conference on Image Processing (ICIP). IEEE; 2020. p. 3289–93. https://doi.org/10.1109/ICIP40778.2020.9190927.

Preuer K, Klambauer G, Rippmann F, Hochreiter S, Unterthiner T. Interpretable deep learning in drug discovery. 2019:331–45. https://doi.org/10.1007/978-3-030-28954-6_18.

Proietti M, Ragno A, La Rosa B, Ragno R, Capobianco R. Explainable AI in drug discovery: self-interpretable graph neural network for molecular property prediction using concept whitening. Mach Learn. 2024:2013–44. https://doi.org/10.1007/s10994-023-06369-y.

Qu C, Liu W, Taylor CJ. Bayesian deep basis fitting for depth completion with uncertainty. In: Proceedings of the IEEE/CVF international conference on computer vision, vol. 2021; 2021.

Ragno R. www.3d-qsar.com: a web portal that brings 3-D QSAR to all electronic devices—the Py-CoMFA web application as tool to build models from pre-aligned datasets. J Comput Aided Mol Des. 2019;33:855–64. https://doi.org/10.1007/s10822-019-00231-x.

Ragno A, Plantevit M, Robardet C, Capobianco R. Transparent explainable logic layers. 2024a; https://doi.org/10.3233/FAIA240579.

Ragno A, La Rosa B, Capobianco R. Prototype-based interpretable graph neural networks. IEEE Trans Artif Intell. 2024b;5:1486–95. https://doi.org/10.1109/TAI.2022.3222618.

Ribeiro MT, Singh S, Guestrin C. Why should I trust you? In: Proceedings of the 22nd ACM SIGKDD international conference on knowledge discovery and data mining. New York: ACM; 2016. p. 1135–44. https://doi.org/10.1145/2939672.2939778.

Ribeiro MT, Singh S, Guestrin C. Anchors: high-precision model-agnostic explanations. Proc AAAI Conf Artif Intell. 2018;32 https://doi.org/10.1609/aaai.v32i1.11491.

Rodríguez-Pérez R, Bajorath J. Interpretation of machine learning models using shapley values: application to compound potency and multi-target activity predictions. J Comput Aided Mol Des. 2020;34:1013–26. https://doi.org/10.1007/s10822-020-00314-0.

Sakiyama H, Fukuda M, Okuno T. Prediction of Blood-Brain Barrier Penetration (BBBP) based on molecular descriptors of the free-form and in-blood-form datasets. Molecules. 2021;26:7428. https://doi.org/10.3390/molecules26247428.

Schlichtkrull MS, De Cao N, Titov I. Interpreting graph neural networks for NLP with differentiable edge masking. In: International conference on learning representations; 2020.

Schnake T, Eberle O, Lederer J, Nakajima S, Schutt KT, Muller K-R, et al. Higher-order explanations of graph neural networks via relevant walks. IEEE Trans Pattern Anal Mach Intell. 2022;44:7581–96. https://doi.org/10.1109/TPAMI.2021.3115452.

Selvaraju RR, Cogswell M, Das A, Vedantam R, Parikh D, Batra D. Grad-CAM: visual explanations from deep networks via gradient-based localization. Int J Comput Vis. 2020;128:336–59. https://doi.org/10.1007/s11263-019-01228-7.

Shi W, Murakoso M, Guo X, Wang MD. Development of interpretable machine learning models for COVID-19 drug target docking scores prediction. In: 2023 IEEE international conference on bioinformatics and biomedicine (BIBM). IEEE; 2023. p. 4124–31. https://doi.org/10.1109/BIBM58861.2023.10385756.

Shrikumar A, Greenside P, Kundaje A. Learning important features through propagating activation differences. In: Proceedings of the 34th international conference on machine learning; 2017.

Smilkov D, Thorat N, Kim B, Viégas F, Wattenberg M. SmoothGrad: removing noise by adding noise. 2017.

Spade DA. Markov chain Monte Carlo methods: theory and practice. 2020:1–66. https://doi.org/10.1016/bs.host.2019.06.001.

Sucar LE. Probabilistic graphical models. Cham: Springer; 2021. https://doi.org/10.1007/978-3-030-61943-5.

Sundararajan M, Taly A, Yan Q. Axiomatic attribution for deep networks. In: ICML'17: proceedings of the 34th international conference on machine learning – volume 70; 2017.

Teye M, Azizpour H, Smith K. Bayesian uncertainty estimation for batch normalized deep networks. In: Proceedings of the 35th international conference on Machine Learning, PMLR 80:4907-4916, 2018; 2018.

Vangala SR, Krishnan SR, Bung N, Srinivasan R, Roy A. pBRICS: a novel fragmentation method for explainable property prediction of drug-like small molecules. J Chem Inf Model. 2023;63:5066–76. https://doi.org/10.1021/acs.jcim.3c00689.

Veličković P, Cucurull G, Casanova A, Romero A, Liò P, Bengio Y. Graph attention networks. In: International conference on learning representations; 2017.

Vermeire T, Brughmans D, Goethals S, de Oliveira RMB, Martens D. Explainable image classification with evidence counterfactual. Pattern Anal Applic. 2022;25:315–35. https://doi.org/10.1007/s10044-021-01055-y.

Vu MN, Thai MT. PGM-explainer: probabilistic graphical model explanations for graph neural networks. In: Advances in neural information processing systems 33 (NeurIPS 2020); 2020.

Wu Z, Lei T, Shen C, Wang Z, Cao D, Hou T. ADMET evaluation in drug discovery. 19. Reliable prediction of human cytochrome P450 inhibition using artificial intelligence approaches. J Chem Inf Model. 2019;59:4587–601. https://doi.org/10.1021/acs.jcim.9b00801.

Ying R, Bourgeois D, You J, Zitnik M, Leskovec J. GNNExplainer: generating explanations for graph neural networks. Adv Neural Inf Proces Syst. 2019;32:2019.

Ying C, Cai T, Luo S, Zheng S, Ke G, He D, et al. Do transformers really perform bad for graph representation? Adv Neural Inf Proces Syst. 2021;34:2021.

Yuan H, Yu H, Wang J, Li K, Ji S. On explainability of graph neural networks via subgraph explorations. In: Proceedings of the 38th International Conference on Machine Learning, PMLR 139; 2021. p. 12241–52.

Zhong S, Hu J, Yu X, Zhang H. Molecular image-convolutional neural network (CNN) assisted QSAR models for predicting contaminant reactivity toward OH radicals: transfer learning, data augmentation and model interpretation. Chem Eng J. 2021;408:127998. https://doi.org/10.1016/j.cej.2020.127998.

Zintgraf LM, Cohen TS, Adel T, Welling M. Visualizing deep neural network decisions: prediction difference analysis. International Conference on Learning Representations; 2017.

Federated Learning in Drug Discovery: Challenges, Innovations, and Future Directions

Martina Savoia, Antonio Lavecchia, and Francesco Piccialli

Abstract The integration of federated learning (FL) in drug discovery is an approach to address challenges in data privacy and collaboration. This chapter explores the role of FL in allowing collaborative training of models among pharmaceutical companies, academic institutions, and research organizations without the need to share sensitive datasets. The work is structured to provide a comprehensive overview of FL in this domain. First, we introduce the fundamentals of FL. Next, we examine the diverse applications of FL in drug discovery, and finally, we address critical challenges such as handling non-IID data, managing large-scale datasets, and ensuring privacy. The findings underscore FL's potential in drug discovery by allowing collaboration across siloed datasets, improving model generalization, and enabling privacy-preserving innovation.

Keywords Drug discovery · Federated learning · Healthcare · Privacy-preserving · Non-IID data

1 Introduction

Drug discovery is an active research field, often requiring collaboration between pharmaceutical companies, academic institutions, and research organizations. One of the primary challenges in this domain is the need for large, high-quality datasets to train machine learning (ML) models, to accelerate the discovery and optimization

M. Savoia (✉) · F. Piccialli
Department of Mathematics and Applications "R. Caccioppoli", University of Naples Federico II, Napoli, NA, Italy
e-mail: martina.savoia@unina.it; francesco.piccialli@unina.it

A. Lavecchia
Department of Pharmacy, Drug Discovery Laboratory, University of Naples Federico II, Napoli, NA, Italy
e-mail: antonio.lavecchia@unina.it

A. Lavecchia (ed.), *Applied Artificial Intelligence for Drug Discovery*,
https://doi.org/10.1007/978-3-031-98022-0_18

of new drugs. However, due to privacy constraints, it is not always possible to share these datasets among institutions.

Federated learning (FL) (Konečný 2016) has emerged as a promising solution to address these challenges by enabling collaborative model training without requiring data to be shared among clients and server. One of the key challenges when applying ML models is the non-IID (non-independent and identically distributed) nature of the data. Unlike traditional datasets which are assumed to be homogeneous and drawn from the same distribution, real data can be distributed unevenly across different clients. Moreover, cloud environments enable easy and rapid access to data when modeling, and FL is particularly advantageous in drug discovery (Spjuth et al. 2021), where datasets held by individual institutions can often be limited in size and scope, typically focusing on specific categories of molecules or experiments. By aggregating insights across distributed datasets, FL enables the training of models on a broader and more diverse range of data than would be possible for any single institution. Healthcare Metaverses (Bashir et al. 2023) offer transformative possibilities by providing virtual spaces that simulate real-world laboratories or create entirely new experimental environments. These spaces enable collaboration among hospitals and researchers, allowing for drug design experimentation. This can be achieved either by sharing original datasets with a central medical authority or by exchanging only intermediate results to maintain data privacy. Pharmaceutical companies can leverage the metaverse to virtually simulate and test new drug designs, avoiding direct interaction with potentially dangerous chemicals and ingredients. The scalability and interoperability of an FL-based Metaverse framework make it possible for scientists worldwide to collaborate effectively. By utilizing cross-institutional and cross-national big data, this approach facilitates the development of tailored and efficient drug solutions, advancing the field of drug discovery.

A comprehensive literature review was conducted to analyze the application of FL in drug discovery. To identify relevant studies, Scopus, Google Scholar, and Web of Science sites were used. The search was conducted using "Federated learning" and "Drug discovery" keywords. A total of 19 open access documents were identified using the search criteria "federated AND learning AND drug AND discovery" on Scopus. All documents were in English and were in their final publication stage. After refining the results to focus specifically on FL and drug discovery, 18 documents were selected. Using the same search criteria on Web of Science and Google Scholar, additional documents not found on Scopus were identified, reaching a total of 40 works. Most of these are articles and conference papers. Reviews were also analyzed to identify other potentially not analyzed documents. As shown in Fig. 1, there is a growing interest in publications on FL in drug discovery, with a significant increase in 2023 and 2024.

This chapter is organized as follows. Section 2 introduces the concept of FL, its architectures, aggregation methods, and foundational principles. Section 3 reports specific applications of FL in drug discovery, covering key areas such as drug-target interaction (DTI) prediction, pharmacokinetic and pharmacodynamic

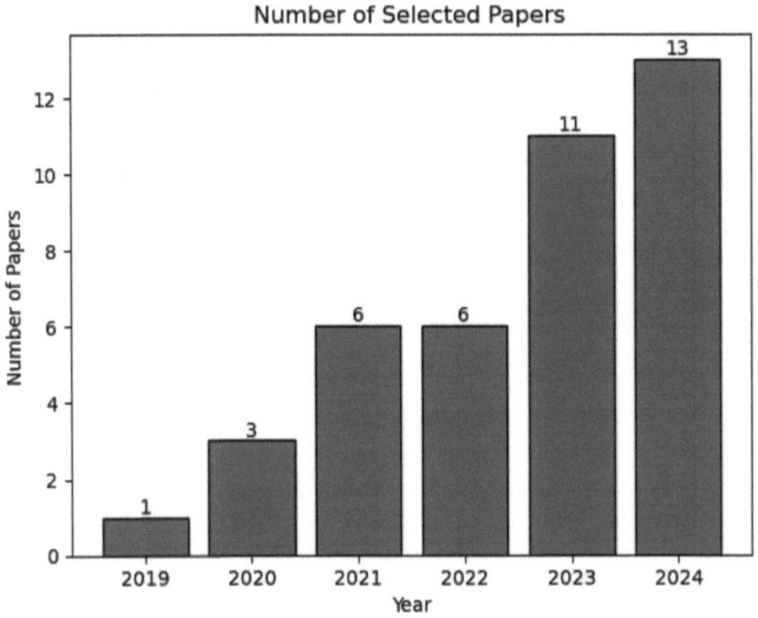

Fig. 1 Number of selected papers by year

properties prediction, and molecules generation (Fig. 2). Section 4 explores the challenges inherent to FL in drug discovery, including handling non-IID data, managing large-scale datasets, and ensuring privacy (Fig. 2). This structure provides a comprehensive overview of how FL is helpful in drug discovery, while addressing its most common challenges. Lastly, conclusions and future directions on FL in drug discovery are discussed in Sect. 5.

2 Fundamentals of Federated Learning

The centralized FL system consists of a central server and N clients $\{C_1, \ldots, C_N\}$. Each client C_i holds its own dataset D_i. The process starts with a global model $W^{(0)}$ initialized on the server, which is shared with all the participating clients as the starting point for training. Each client updates the global model using its local dataset, ensuring that sensitive data never leaves the client's device, which is crucial to maintaining privacy. After completion of the local training, clients send their model updates back to the central server. Updates from a selected group S_t of clients are averaged to create the refined global model. A commonly used approach, called FedAvg (McMahan et al. 2017), assigns importance to each client's contribution

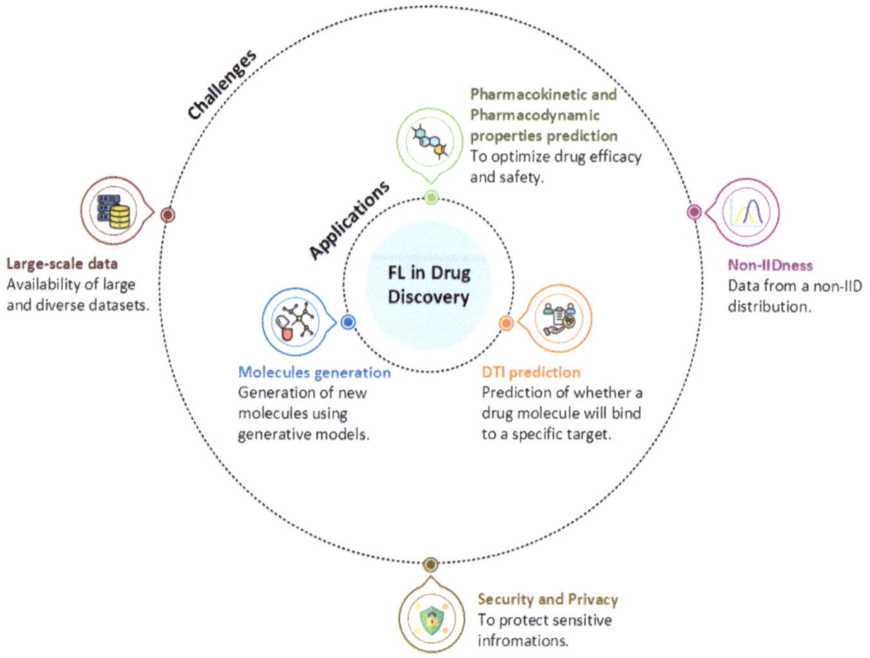

Fig. 2 Applications and challenges in drug discovery using FL

based on the size of its dataset. At the t-th round, the FedAvg aggregation can be expressed as

$$W^{(t)} = \sum_{i \in S_t} \frac{n_i}{n} W_i^{(t)}, \tag{1}$$

where $i \in S_t$ denotes the clients selected from strategy to be aggregated in the current round t, $\frac{n_i}{n}$ is the weighting factor determined by the ratio of data volume n_i of client C_i to the total data n from all selected clients, $W_i^{(t)}$ represents the model parameters of client i after local training, and $W^{(t)}$ is the aggregated global model. Each iteration is known as a "round," and the training continues until the model reaches the desired accuracy or the predetermined number of rounds is completed. In real FL scenarios, data is often distributed across clients in a non-IID manner. This non-IID nature poses challenges for model training, like heterogeneity correlated to data distributions that can vary significantly due to differences in research focus, experimental setups, or local populations. Moreover, clients often hold varying amounts of data. Combining model updates from non-IID datasets requires careful weighting and optimization to avoid bias. Aggregation techniques like FedProx (Li et al. 2020) have been developed to mitigate bias and improve the performance of

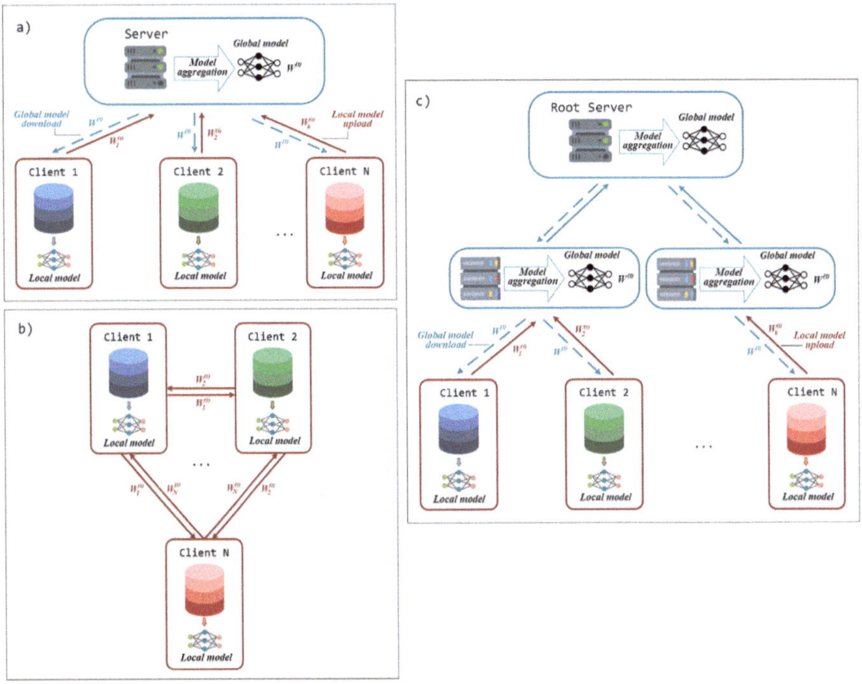

Fig. 3 FL architectures: (**a**) centralized, (**b**) decentralized, (**c**) hierarchical

the global model. Moreover, frequent exchanges of model updates between clients and the server can be resource intensive.

In centralized FL architectures, a central server coordinates the training process by aggregating updates from clients and redistributing the refined global model. However, centralized systems may face challenges related to single points of failure and bottlenecks in communication. Decentralized FL architectures address these limitations by enabling direct peer-to-peer communication between clients or employing multiple intermediate aggregation nodes. Hierarchical FL architectures combine elements of centralized and decentralized approaches, leveraging hierarchical layers of aggregation to balance scalability and robustness. For example, in drug discovery, regional servers could aggregate updates locally before sending them to a global server (Fig. 3).

In addition, FL can be categorized into vertical federated learning (VFL) and horizontal federated learning (HFL), based on how the data is distributed among participating clients. VFL is used when datasets across clients have overlapping samples but differ in the feature space. For example, different hospitals that hold data about the same patients. In HFL, datasets across clients share the same feature space but differ in the samples they contain. For example, in drug discovery, multiple pharmaceutical companies might have patient data with similar features but collected from different patients. Hence, HFL is the most common scenario in drug discovery due to the partners operating in separate chemical spaces. In addition, federated transfer learning (FTL) can be used in scenarios where the available labeled data are insufficient to train a model effectively for a specific task and can be considered as the intersection of HFL and VFL.

3 Applications of Federated Learning in Drug Discovery

FL is widely applied in addressing various challenges within drug discovery, such as de novo drug design (DNDD), which focuses on understanding and creating novel chemical entities and molecular property prediction (Shen and Nicolaou 2019). The collaborative development of the necessary models can significantly enhance experimental outcomes. To compare ML and FL and assess whether FL delivers equivalent, better, or worse results, the authors in Ali and Zafar (2024) implemented six algorithms (artificial neural networks, logistic regression, perceptron, ridge classifier, ridge classifier with cross validation, and SGD classifier) for comparative analysis in both ML and FL. For the prediction of molecules' activeness for HIV, the authors obtained slightly better results in the case of FL, thereby achieving the goal of securing data, protecting patient privacy, and eliminating data transfer costs using FL for drug discovery.

The authors in Khanna et al. (2024) provide a comprehensive overview of advancements in FL applied to drug discovery, with a particular focus on Parkinson's disease. The used datasets are the Oxford Parkinson's Disease Centre (OPDC) (Griffanti et al. 2020), Parkinson's Progression Markers Initiative (PPMI) (Kanagaraj et al. 2021), Michael J Fox Foundation (MJFF) (Vollstedt et al. 2023), the Personalized Parkinson's Project (PPP) (Bloem et al. 2019), and the Critical Path Institute for PD (CPP) global consortium (Stephenson et al. 2020). The overview highlights the advantages of leveraging FL in this domain, emphasizing its potential to accelerate the discovery of new drugs while preserving data privacy and fostering collaboration among institutions.

3.1 Multiobjective Prediction of Pharmacokinetic and Pharmacodynamic Properties

Pharmacokinetic and pharmacodynamic properties are fundamental in understanding how a drug behaves within the body and how it exerts its effects (Glassman and Muzykantov 2019). Pharmacokinetics involves the absorption, distribution, metabolism, and excretion (ADME) of a drug, while pharmacodynamics focuses on the biological effects and mechanisms through which the drug interacts with its target.

3.1.1 Molecular Property Prediction

A key challenge in federated molecular property prediction is the heterogeneity of client data, as institutions often focus on specific categories of molecules aligned with their research priorities. Heterogeneity brings difficulties to federated molecular learning (see Sect. 4.2).

What is the molecular property prediction?
Molecular property prediction is the task of predicting the properties of a molecule from its structure. Molecular properties are fundamental characteristics of molecules, including physical, chemical, and structural properties. Accurate determination of these properties is crucial for designing and utilizing chemical compounds effectively. Small molecules can be represented as graphs, and hence, graph neural networks (GNNs) have been widely applied in molecular representation learning (Deng et al. 2023).

GNNs are commonly adopted for molecular learning and property prediction due to their ability to learn state-of-the-art representations from graph-structured molecular data. In Zhu et al. (2022), the authors introduce FedChem, an FL method utilizing GNNs to address the chemical space heterogeneity, achieving accurate molecular property predictions. In this work, nine datasets from MoleculeNet (Wu et al. 2018) are used: FreeSolv, Lipophilicity, ESOL, and QM9 (regression datasets) and Tox21, SIDER, ClinTox, BBBP, and BACE (classification datasets). SpreadGNN (He et al. 2022) operates in the presence of multiple, but partial labels for each client and the absence of a central server for graph-level molecular property prediction. The results on SIDER, Tox21, MUV, QM8 datasets from MoleculeNet show that it is possible to eliminate the dependence on a central server and enable clients to learn more effectively in the presence of missing molecular properties in their respective labels. The GFedKRL (graph neural network federated knowledge relearning) framework (Ning et al. 2023) addresses challenges in FL for molecular property prediction using GNNs and knowledge distillation. GFedKRL can achieve better than baseline results for clients on eight molecule datasets (MUTAG, BZR, COX2, PTC_MR, DHFR, AIDS, NCI1, and mutagenicity) and in two scenarios for the graph classification task: In the first, all clients are from the first seven datasets in the molecular domain, with each client having a molecular graph dataset, and in the second, all clients hold from the same type of molecular dataset. FedGraphNN (He et al. 2021) is a platform that supports GNNs across seven domains, using 36 datasets. Among them, 34 are from publicly available sources such as MoleculeNet and graph kernels datasets. Moreover, the new hERG (Gaulton et al. 2017; Kim et al. 2021) and Tencent (He et al. 2019) datasets were introduced. Real examples, among others, are molecule property prediction and protein function prediction.

3.1.2 Bioactivity Prediction

Federated heterogeneous contrastive distillation (FedHCD) (Feng et al. 2024) enables joint training of global models from clients with heterogeneous data modalities, learning tasks, and molecular models. FedHCD framework views molecular FL tasks as a multimodal knowledge distillation process between clients and server. Experimentation was performed on 13 representative datasets: Tox21, ClinTox, BBBP, BACE, SIDER, MUV, ToxCast, PCBA, Lipo, ESOL, FreeSolv, QM9 from MoleculeNet, and PCQM4Mv2 from OGB Large-Scale Challenge (Hu et al. 2021). FedHCD shows superior performance across all clients, compared to baselines, demonstrating that employing contrastive learning for knowledge distillation can effectively align representation spaces across different modalities of clients by leveraging molecular characteristics and domain-specific knowledge. The authors in Xiong et al. (2022) developed a cross-silo FL framework tailored for drug discovery, utilizing a variety of drug property and activity datasets. This research covers the entire process, including federated training, model selection, and personalization. The main objective of the study is to showcase the potential of FL for predicting drug-related properties and activities while maintaining data privacy. Specifically, FL was applied to three use cases: solubility, kinase inhibitory activity, and the human Ether-à-go-go-Related Gene (hERG) liability prediction. The datasets employed for the solubility task are extracted from eChemProtal (eChemPortal 2025), EPI Suite Data website https://www.epa.gov/tsca-screening-tools/epi-suitetm-estimation-program-interface, while the Kinase dataset (https://github.com/Team-SKI/Publications) (Davis et al. 2011; Metz et al. 2011; Anastassiadis et al. 2011) was used in the kinase inhibitory activity task. Lastly, for the hERG liability task, the used datasets are NCATS (https://pubchem.ncbi.nlm.nih.gov/bioassay/588834) and JHICC (https://pubchem.ncbi.nlm.nih.gov/bioassay/1511) from Pubchem, CheMBL (Bento et al. 2014), FDA Drugs (Liu et al. 2014) and from various works Cai et al. (2019), Zhang et al. (1999), Siramshetty et al. (2020), Smalley (2017), Keserü (2003), Doddareddy et al. (2010), Sun et al. (2017). The findings reveal that federated training not only outperforms individual training but, in some instances, even exceeds the performance of centralized training on pooled datasets with inherent biases. An analysis of the neural network weight distributions suggests that FL naturally acts as a regularizer, enhancing the generalization ability of models, particularly when working with biased data. Furthermore, the study highlights the practicality of personalizing models locally, allowing users with sufficient labeled data to achieve better performance than when relying solely on the global model. HFDL-fl (horizontal federated deep learning with focal loss) (Xu et al. 2023) offers an effective solution to overcome data-sharing barriers between medical institutions, enabling privacy-preserving drug sensitivity prediction. This approach supports advancements in cancer precision medicine and other privacy-sensitive biomedical research. The authors simulate two scenarios: HFDL-Cross, involving federated training across two heterogeneous data sources, and HFDL-Within, where homogeneous data from a single source is split into two clients for federated training. These simulations demonstrate the framework's applicability in diverse data environments while maintaining data privacy and security. To predict the sensitivity of drugs in cell lines, the work includes the Genomics of Drug Sensitivity in Cancer (GDSC) (Yang et al. 2012) and Cancer Therapeutics Response Portal (CTRP) (Seashore-Ludlow et al. 2015) datasets. kMoL GitHub (2025) is an ML library for drug discovery and life sciences, with FL functions, including state-of-the-art graph-based predictive models, explainable AI components, and differential privacy. The library was benchmarked on datasets containing ADME properties, toxicity, and binding affinities values.

3.1.3 MELLODDY Project

The MELLODDY (MachinE Learning Ledger Orchestration for Drug DiscoverY) project (Oldenhof et al. 2023; Heyndrickx et al. 2023b) is an application of FL to drug discovery at an industrial scale. As the first cross-endpoint FL platform in this domain, MELLODDY enables the creation of a global federated model for drug discovery without compromising the confidentiality of proprietary datasets held by individual partners. In MELLODDY, FL is used to train QSAR models on decentralized datasets from multiple pharmaceutical companies. Lhasa Limited has developed the Effiris application (Hanser et al. 2022), leveraging a knowledge distillation framework based on the Teacher-Student paradigm to predict the on-target activity of small molecules using their SMILES (Simplified Molecular Input Line Entry System) representation. This approach involves utilizing labeled user-provided data to train a model on a defined test set. The trained model is then employed to make predictions on a larger, unlabeled dataset, enabling the transfer of knowledge from the initial model to new, unseen data points. Effiris facilitates FL within a consortium of major pharmaceutical companies, ensuring that proprietary data remains secure within the firewalls of each partner organization. The platform also features a web-based user interface to streamline collaboration among partners. By introducing a prototype platform, FL-QSAR, Chen et al. (2020) aims to enable collaborative and privacy-preserving QSAR modeling among pharmaceutical institutions. The objective of this work is to demonstrate the feasibility and effectiveness of applying HFL to QSAR analysis in drug discovery, using 15 benchmark datasets from Kaggle competition (Ma et al. 2015): 3A4, CB1, DPP4, HIVINT, HIVPROT, LOGD, METAB, NK1, OX1, OX2, PGP, PPB, RAT_F, TDI, and THROMBIN. In Huang et al. (2023), the aim is to propose a novel FL framework for collaborative drug discovery using non-IID QSAR datasets. The proposed framework is evaluated on 15 benchmark datasets (3A4, CB1, DPP4, HIVINT, HIVPROT, LOGD, METAB, NK1, OX1, OX2, PGP, PPB, RAT1_F, TDI, THROMBIN) with varying sizes, ranging from 2000 to 50,000 molecules, and offers advantages such as reduced data transmission. These datasets are divided into target activities and ADME activities. The works cited in this section are listed in Table 1.

Table 1 Summary of works cited in Sect. 3.1

Work	Datasets	Key contribution
Zhu et al. (2022)	FreeSolv, Lipophilicity, ESOL, QM9, Tox21, SIDER, ClinTox, BBBP, BACE	FedChem is an FL method utilizing GNNs to address the chemical space heterogeneity
He et al. (2022)	SIDER, Tox21, MUV, QM8	SpreadGNN operates in the presence of multiple, but partial labels for each client and the absence of a central server
Ning et al. (2023)	MUTAG, BZR, COX2, PTC_MR, DHFR, AIDS, NCI1, and mutagenicity	GFedKRL uses GNNs and knowledge distillation for the graph classification task in two scenarios
He et al. (2021)	From MoleculeNet and the new hERG and Tencent	FedGraphNN is a platform that supports GNNs across seven domains, using 36 datasets
Feng et al. (2024)	Tox21, ClinTox, BBBP, BACE, SIDER, MUV, ToxCast, PCBA, Lipo, ESOL, FreeSolv, QM9 from MoleculeNet and PCQM4Mv2 from OGB large-scale challenge	FedHCD enables joint training of global models from clients with heterogeneous data modalities, learning tasks, and molecular models. Viewing molecular FL tasks as a multimodal knowledge distillation process between clients and server
Xiong et al. (2022)	From eChemProtal and EPI Suite Data website, kinase data (https://github.com/Team-SKI/Publications) (Davis et al. 2011; Metz et al. 2011; Anastassiadis et al. 2011), NCATS, JHICC (https://pubchem.ncbi.nlm.nih.gov/bioassay/1511) from Pubchem, CheMBL, FDA-Drugs from various works	Solubility, kinase inhibitory activity, and the human Ether-à-go-go-Related Gene (hERG) liability prediction
Xu et al. (2023)	GDSC, CTRP	Sensitivity of drugs prediction in cell lines
Oldenhof et al. (2023); Heyndrickx et al. (2023b)	From multiple pharmaceutical companies	The MELLODDY project is an application of FL to drug discovery at an industrial scale. In MELLODDY, FL is used to train QSAR models on decentralized datasets
Hanser et al. (2022)	From pharmaceutical companies	The Effiris application facilitates FL within a consortium of major pharmaceutical companies
Chen et al. (2020)	3A4, CB1, DPP4, HIVINT, HIVPROT, LOGD, METAB, NK1, OX1, OX2, PGP, PPB, RAT_F, TDI, THROMBIN	FL-QSAR demonstrates the feasibility and effectiveness of applying HFL to QSAR analysis
Huang et al. (2023)	3A4, CB1, DPP4, HIVINT, HIVPROT, LOGD, METAB, NK1, OX1, OX2, PGP, PPB, RAT_F, TDI, THROMBIN	Novel FL framework for collaborative drug discovery using non-IID QSAR datasets

What are the QSAR models?
Quantitative structure-activity relationship (QSAR) models are used to predict the biological activity or pharmacological properties of chemical compounds based on their molecular structure.

3.2 Drug-Target Interaction Prediction

What is the drug-target interaction (DTI) prediction?
Drug-target interaction (DTI) prediction refers to the process of predicting an interaction score to explain how a drug interacts with a protein target in the human body (Sachdev and Gupta 2019). The interaction scores correspond to experimental measures of bioactivity, such as IC_{50}, K_i, and K_d, which indicate the strength and efficacy of the drug-target binding.

In Mittone et al. (2023), the authors delivered a benchmark privacy-preserving distributed learning implementation in the DTI domain that both meets the limiting constraints of the industry's regulatory and commercial constraints and outperforms previously available alternatives by up to 15% on the KiBA (Tang et al. 2014) dataset, focusing on measurements such as IC_{50}, K_i, and K_d, which quantify drug-target binding affinity. A trade-off is identified between the benefits of adding new data and the cost of introducing more clients. The work in Sándor and Antal (2023) presents FedMTBoost, a novel approach to model large-scale federated datasets using boosting in DTI prediction. This work explores the application of FL to multitask problems by expanding existing federated models to handle more complex datasets. The algorithm outperforms traditional baselines using the KiBA dataset, which provides affinity metrics (IC_{50}, K_i, and K_d) for kinase inhibitors, particularly in settings with limited data. FedKD-DTI (Wang et al. 2024) is a novel framework for DTI prediction that leverages federated knowledge distillation to enable multi-institutional collaboration while ensuring data privacy and security. The experimentation involves KiBA, BindingDB (Liu et al. 2007), which includes bioactivity metrics such as EC_{50} and AC_{50}, and DrugBank (Knox et al. 2024) datasets, which contains various experimental measures of bioactivity, including those found in the KiBA and BindingDB datasets, such as IC_{50}, K_i, and K_d. The works cited in this section are listed in Table 2.

Table 2 Summary of works cited in Sect. 3.2

Work	Datasets	Key contribution
Mittone et al. (2023)	KiBA	Benchmark for the DTI prediction
Sándor and Antal (2023)	KiBA	DTI prediction using FedMTBoost, a novel FL approach which uses boosting
Wang et al. (2024)	KiBA, BindingDB, DrugBank	DTI prediction using knowledge distillation for collaboration

3.3 *Molecules Generation*

The generation of new molecules is a critical task in the early stages of drug discovery, aiming to identify compounds with desired therapeutic properties. Generative adversarial networks (GANs) are largely applied in this field, enabling the de novo design of molecules and FL accelerates the discovery process.

FL-DISCO (Manu et al. 2021) is a holistic collaborative and privacy-preserving framework for drug discovery, with IID and non-IID data distributions, using the ESOL and the QM9 datasets from MoleculeNet. FL-DISCO uses GANs to generate novel molecules while preserving learned properties extracted using GNNs. FL-DISCO can generate highly novel molecular graphs that retain relevant properties within training graphs. Graph convolutional networks (GCNs) are integrated within GANs using FL to propose a novel framework for molecular discovery named GraphGANFed (Manu et al. 2024). The approach leverages FL to collaboratively train both the generator and the discriminator in a decentralized manner, ensuring data privacy, using three benchmark datasets from MoleculeNet: ESOL, QM8, and QM9. In GraphGANFed, the discriminator is implemented as a GCN, enabling it to effectively extract features from molecules represented as molecular graphs. The generated molecules exhibit high levels of novelty and diversity, demonstrating the framework's potential for advancing molecular discovery while maintaining robust privacy protections. The MolCFL (Guo et al. 2024) framework for collaborative and privacy-preserving drug discovery combines ML techniques to optimize molecular design while preserving data privacy. MolCFL employs a multilayer perceptron (MLP) as the generator and a GCN as the discriminator in a GAN, using ESOL and QM9 datasets from MoleculeNet. This enables the generation of new molecular structures by learning the graph structure of molecules. An FL framework for the discovery of potential HIV inhibitors is presented in Dutta et al. (2024). The primary focus of this work is to apply FL combined with GNNs to classify compounds as active or inactive against HIV. The used dataset is the drug therapeutics program (DTP) HIV dataset within the MoleculeNet dataset which uses representations of chemical compounds using SMILES. FL demonstrated a slight improvement in class-specific and macro-average ROC areas compared to the centralized model. The works cited in this section are listed in Table 3.

Table 3 Summary of works cited in Sect. 3.3

Work	Datasets	Key contribution
Manu et al. (2021)	ESOL, QM9	FL-DISCO uses GANs to generate novel molecules while preserving learned properties extracted using GNNs
Manu et al. (2024)	ESOL, QM8, QM9	GraphGANFed integrates GCNs into GANs and trains both the generator and discriminator in a decentralized manner for molecular discovery
Guo et al. (2024)	ESOL, QM9	MolCFL employs an MLP as the generator and a GCN as the discriminator in a GAN for the generation of new molecular structures
Dutta et al. (2024)	Drug Therapeutics Program (DTP) HIV	FL framework for the discovery of potential HIV inhibitors. FL is combined with GNNs to classify compounds as active or inactive against HIV

4 Key Challenges in Applying FL to Drug Discovery

FL has emerged as an efficient way to accelerate drug development and innovation. However, FL frameworks must address technical and methodological challenges, such as dealing with non-IID data, leveraging large datasets, enabling scalability, and maintaining privacy. Moreover, collaborative models can benefit from diverse datasets, and sharing knowledge between clients can help identify and address inconsistencies and biases present in individual datasets (Maiorescu and Semenescu 2023).

4.1 Large-Scale Data from Multiple Institutions

In drug discovery, the availability of large and diverse datasets is crucial for training robust and generalizable ML models. However, often each organization possesses datasets that are typically limited in scope and focus. FL can handle different large-scale datasets distributed across multiple institutions. In the MELLODDY project (Heyndrickx et al. 2023b), each of the ten pharmaceutical companies realized aggregated improvements on its own models through FL. In Heyndrickx et al. (2023a), authors propose conformal efficiency as a means to measure the applicability domain extension through predictive uncertainty differences. The work demonstrates that federated models can extend their applicability domain, even in the absence of labeled data. To evaluate the performance of the Effiris FL platform, Lhasa Limited organized a hackathon with participation from several industrial partners (Bassani et al. 2023). In this event, participants were tasked with uploading datasets of compounds, each labeled with positive or negative experimental activity markers for a specific biological target of interest. Once models were trained collaboratively on these datasets, the acquired knowledge was applied to predict

Table 4 Summary of works cited in Sect. 4.1

Work	Datasets	Key contribution
Heyndrickx et al. (2023b)	From multiple pharmaceutical companies	The MELLODDY project is an application of FL to drug discovery at an industrial scale
Bassani et al. (2023)	Own data	Detailed description of the participation at the hackathon organized by Lhasa Limited to evaluate the performance of the Effiris FL platform
Humbeck et al. (2021)	From six pharmaceutical companies: Amgen, Bayer, Boehringer Ingelheim, GSK, Janssen Pharmaceutica NV, Novartis	The work investigates task weighting strategies for multitask bioactivity classification models using large, real-world datasets

the labels of a pool of over 300,000 unlabeled and untested molecules. Effiris provides two prediction attributes: the "Reliability," linked to the gap in chemical space between the molecules seen by the models in the training steps and the ones being labeled, and the "Decidability," related to the confidence in the prediction. The objective of the work in Humbeck et al. (2021) is to investigate task weighting strategies for multitask bioactivity classification models in the context of federated multitask learning using large, real-world datasets from six pharmaceutical companies (Amgen, Bayer, Boehringer Ingelheim, GSK, Janssen Pharmaceutica NV, Novartis), addressing privacy-preserving drug discovery. The authors demonstrated that most weighting schemes yield comparable predictive performance (AUPR and AUROC) and convergence rates, except for those producing extreme weight distributions. The works cited in this section are listed in Table 4.

4.2 *Non-IID Data Distribution*

Non-IID data is a common challenge in FL for drug discovery. For example, pharmaceutical companies might focus on specific drug classes or can have a different amount of data. Personalized FL models address these issues by tailoring the global model to each client's specific data distribution. Non-IID data also presents an opportunity to develop models that are robust to diverse scenarios. In Zhu et al. (2022), the authors propose an FL method utilizing GNNs to address the chemical space heterogeneity (see Sect. 3.1) and achieve accurate molecular property predictions and a benchmark, FedChem, designed to simulate the distribution of chemical data across institutions. FedChem allows customization to generate scenarios with varying degrees of data heterogeneity, providing a valuable tool for evaluating FL methods in this domain. Moreover, they propose an FL with Instance reweighTing (FLIT) to alleviate the heterogeneity problem by adapting focal loss for FL. To address challenges in non-IID data distributions, the framework in Huang et al. (2023) globally shares a small subset of data among participating institutions to improve training stability and robustness. The proposed framework can achieve competitive performance to the centralized learning approach, even when the data are non-IID and decentralized across multiple clients. Results on 15 benchmark datasets demonstrate that the proposed framework outperforms local client training, indicating a significant performance improvement. In Mittone et al. (2023), the authors identified a material trade-off between the benefits of adding new data and the cost of introducing more clients. Unlike other domains where FL has been explored, in the DTI field, authors show that the input's IID nature has minimal impact, making it distinctly different from other FL applications. Choudhury et al. (2019) introduce an FL framework that can learn a global adverse drug reaction (ADR) prediction model from distributed health data held locally at different sites, using Limited IBM MarketScan Explorys Claims-EMR Dataset (LCED). The authors propose two novel methods of local model aggregation to improve the predictive capability of the global model. The first method is particularly designed for training data with imbalanced classes, assigning higher weights to sites with cases of rare events. The second approach assigns a higher weight during aggregation to sites that require fewer training samples to converge faster. In the FL context, molecular graph data among clients often exhibit non-IID characteristics and imbalances in quantity distribution (see Sect. 3.1). Experimental evaluations on three datasets demonstrate that GFedKRL (Ning et al. 2023) outperforms four public benchmark methods, showcasing its effectiveness in handling non-IID and skewed data distributions in FL for molecular property prediction. In Simm et al. (2021), the authors investigated four train-test splitting of chemical structure datasets: random splitting, locality-sensitive hashing (LSH), scaffold-based binning (scaffold network), and sphere exclusion clustering. They evaluated the effects of these methods on four large-scale pharmaceutical datasets and on a ChEMBL subset (Gaulton et al. 2017), analyzing aspects such as compound similarity within each fold, label and data imbalance, and biases in predictive performance. Both the sphere exclusion and the scaffold network achieved better grouping of similar compounds in the same fold, resulting in more realistic performance metrics than LSH. LSH and random fold splitting produced a more uniform distribution of compounds and labels across folds. This effect vanishes for tasks with 1000 compounds or more. An FL application on the cancer cell line encyclopedia (CCLE) (Barretina et al. 2012) dataset was proposed in Nolte et al. (2023). CCLE consists of responses elicited by 24 anticancer drugs screened on 504 different cancer cell lines with known genomic information, and on The National Cancer Institute (NCI) dataset (Shoemaker 2006), which consists of responses of over 55,000 chemicals screened on over 100 cancerous cell lines. To evaluate the robustness of the proposed framework, a "quasi-random allocation" of samples among clients was designed to mimic real-world scenarios where clients' databases may consist of heterogeneous population clusters. This analysis highlights the capacity of the method to perform under non-IID data. The Mechanism of Action (MoA) of compounds using fluorescence image data from cell painting was analyzed in Ju et al. (2024). The dataset (Gupta et al. 2022) employed comprises 231 drugs, each labeled with one of 10 MoA categories.

Osteosarcoma U2OS cells were exposed to these drugs for 48 hours in 384-well microplates, with six replicates per drug. The authors construct three hypothetical scenarios where organizations collaborate to train a shared model to predict 10 different labels (MoAs) but with different distributions of the available data (uniform, unbalanced, and non-IID) and with respect to the number of participants. Results show that, under the assumption that all participants possess data with homogeneous statistical distributions, a higher number of participants within the FL process leads to an improved prediction performance from the federated model. Moreover, the involvement of participants with data related only to specific MoAs not only yields federated models with notably enhanced prediction performance but also contributes novel knowledge to the federation, significantly improving prediction accuracy for their specialized MoA. A quantum FL framework is proposed in Saggi et al. (2024) and employs variational quantum circuits to tackle pill classification challenges, particularly focusing on distinguishing defective from non-defective pills under imbalanced data conditions. Furthermore, the study explores the incorporation of genomic sequencing within quantum federated systems to enhance classification tasks in healthcare. The works cited in this section are listed in Table 5.

4.3 *Privacy Concerns*

In drug discovery, data privacy is a critical concern, especially when dealing with sensitive datasets from multiple organizations. Pharmaceutical companies, research institutions, and healthcare providers often hold valuable data related to drug properties, clinical trials, and patient health, which cannot be shared due to privacy regulations and competitive reasons. FL is a decentralized approach that ensures that data privacy is preserved, as only model updates, not the raw data, are exchanged between participants. Often privacy-preserved trainings outperform centralized trainings. Patient privacy regulations, such as the General Data Protection Regulation (GDPR) and Account-ability Act (HIPAA), impose strict requirements on collecting, storing, and using personal health information (Maiorescu and Semenescu 2023). Over the years, several techniques have been developed to address these privacy concerns, including Secure Multiparty Computation (SMPC) and Differential Privacy (DP) (Smajić et al. 2023). SMPC ensures that participants can collaboratively compute results without sharing intermediate information during the process. For example, in drug discovery, this technique enables pharmaceutical companies to protect sensitive data from exposure while allowing cooperative computations among participants. DP focuses on preserving the privacy of individual data points while enabling data sharing. It has been widely applied in tasks such as drug sensitivity prediction and analysis of omics data. These privacy-preserving techniques are often integrated with FL frameworks, combining collaborative learning and advanced privacy safeguards to facilitate secure and efficient drug discovery.

Table 5 Summary of works cited in Sect. 4.2

Work	Datasets	Key contribution
Zhu et al. (2022)	FreeSolv, Lipophilicity, ESOL, QM9, Tox21, SIDER, ClinTox, BBBP, BACE	The work employed FL with Instance reweighTing (FLIT) to alleviate the heterogeneity problem by adapting focal loss for FL
Huang et al. (2023)	3A4, CB1, DPP4, HIVINT,HIVPROT, LOGD, METAB, NK1, OX1, OX2, PGP, PPB, RAT_F, TDI, THROMBIN	The authors globally share a small subset of data among participating institutions
Mittone et al. (2023)	KiBA	In the DTI field, authors show that the input's IID nature has minimal impact, making it distinctly different from other FL applications
Choudhury et al. (2019)	LCED	Two novel methods of local model aggregation. The first method is particularly designed for training data with imbalanced classes, assigning higher weights to sites with cases of rare events. The second approach assigns a higher weight to sites that require fewer training samples to converge faster
Ning et al. (2023)	MUTAG, BZR, COX2, PTC_MR, DHFR, AIDS, NCI1, and mutagenicity	GFedKRL outperforms four public benchmark methods, showcasing its effectiveness in handling non-IID and skewed data distributions
Simm et al. (2021)	Four large-scale pharmaceutical datasets and ChEMBL subset	The authors investigated four train-test splitting of chemical structure datasets: random splitting, locality-sensitive hashing (LSH), scaffold-based binning (scaffold network), and sphere exclusion clustering
Nolte et al. (2023)	CCLE	The robustness of the proposed framework is evaluated using a "quasi-random allocation" of the data
Ju et al. (2024)	Gupta et al. (2022)	MoA of compounds varying distributions of the available data (uniform, unbalanced, and non-IID) and the number of participants
Saggi et al. (2024)	–	Quantum FL for distinguish defective from non-defective pills under imbalanced data conditions

In Huang et al. (2024), a novel FL framework is proposed to advance the prediction of drug-target binding affinity (DTA) and drug-drug interactions (DDI), namely FL-DTA and FL-DDI. Datasets employed in this work are KIBA and Davis (Davis et al. 2011) for the DTA task and DrugBank for the DDI task. FL-DTA uses SMILES codes to obtain a graph-based drug representation method. FL-DDI processes tuples of two drugs and the respective interaction. Moreover, to ensure data privacy among clients, they employ secure multiparty computation (MPC)

Table 6 Summary of works cited in Sect. 4.3

Work	Datasets	Key contribution
Huang et al. (2024)	KIBA, Davis	Secure multiparty computation
Ning et al. (2023)	MUTAG, BZR, COX2, PTC_MR, DHFR, AIDS, NCI1, and mutagenicity	The GFedKRL framework uses personalized local differential privacy
Liang et al. (2022)	Inflammation of the bladder health	Blockchain-empowered coordinator-less decentralized FL platform

during the FL model aggregation phase. The GFedKRL framework (Ning et al. 2023) addresses privacy concerns by implementing personalized local differential privacy, enabling better privacy protection while controlling the noise level and enhancing model performance. Additionally, the framework employs knowledge contrast learning at the local client level to improve graph representation learning and mitigate the impact of noisy data on client models. Most of the existing FL systems deal with centralized coordinators that are vulnerable to attacks and privacy breaches. Rahasak-ML (Liang et al. 2022) is a blockchain-empowered coordinator-less decentralized FL platform to solve issues in centralized coordinator-based FL systems by providing better transparency and trust. The employed dataset is the inflammation of the bladder health dataset (Upstill-Goddard et al. 2013). It also allows small pharmaceutical companies and research institutions to build more accurate and less biased models by leveraging insights from diverse datasets across multiple sites. Additionally, it supports scalability by facilitating the connection of more participating sites and devices and offers the capability to process real-time data efficiently. The works cited in this section are listed in Table 6.

5 Conclusions ad Future Directions

The integration of FL in drug discovery is an actual research field, which allows institutions and researchers to collaboratively train machine learning models on sensitive data without compromising data security or privacy. This ability is used especially in drug-target interaction (DTI) prediction, QSAR modeling, and molecular property prediction. Most used datasets are reported in Table 7. Advancements in privacy-preserving techniques, such as differential privacy and secure multiparty computation, have enhanced the ability of FL to safeguard sensitive data. Additionally, improvements in aggregation algorithms have addressed issues of data heterogeneity and improved model performance across diverse datasets. To have more secure data, blockchain technology ensures the transparency, traceability, and trust of data and models. Furthermore, the integration of QML into FL frameworks is a recent approach to optimize molecular structures, predict binding affinities, and identify new drugs. Quantum computers can simulate complex molecular interactions and provide a broader range of potential drug candidates. QML is an

Table 7 Datasets used in works reported in this chapter

Dataset name	Works	Data type	Task
Oxford Parkinson's Disease Centre (OPDC) (Griffanti et al. 2020)	Khanna et al. (2024)	MRI	Parkinson's disease research
PPMI (Kanagaraj et al. 2021)	Khanna et al. (2024)	Clinical, imaging, omics, genetic, sensor, and biomarker data	Analysis of Parkinson's disease
MJFF (Vollstedt et al. 2023)	Khanna et al. (2024)	Demographic, clinical, genetic	Parkinson's disease-linked variant analysis
Personalized Parkinson's project (PPP) (Bloem et al. 2019)	Khanna et al. (2024)	Clinical, imaging, biospecimens, genetic, and wearable sensor data	Parkinson's disease monitoring
CPP Global Consortium (Stephenson et al. 2020)	Khanna et al. (2024)	Over 600 variables (demographics, clinical, medication usage)	Parkinson's disease cohort analysis
ESOL (Wu et al. 2018; Zhu et al. 2022)	Manu et al. (2021, 2024), Guo et al. (2024) and Feng et al. (2024)	Molecular structure and properties	Solubility prediction
QM9 (Wu et al. 2018; Zhu et al. 2022)	Manu et al. (2021, 2024), Guo et al. (2024) and Feng et al. (2024)	Geometric, energetic, electronic, and thermodynamic properties	Quantum identification of structure-property relationships
SIDER (Wu et al. 2018)	Zhu et al. (2022), He et al. (2022) and Feng et al. (2024)	Drug adverse reactions	Drug safety analysis
Tox21 (Wu et al. 2018)	Zhu et al. (2022), He et al. (2022) and Feng et al. (2024)	Chemical descriptors and substructures	Toxicity prediction
MUV (Wu et al. 2018)	He et al. (2022) and Feng et al. (2024)	High-throughput screening (HTS)	Virtual screening
QM8 (Wu et al. 2018)	He et al. (2022) and Manu et al. (2024)	Electronic spectra and excited state energy of molecules	Quantum chemistry prediction
FreeSolv (Wu et al. 2018)	Zhu et al. (2022) and Feng et al. (2024)	Hydration-free energies of small molecules in water	Solubility prediction
Lipophilicity (Wu et al. 2018)	Zhu et al. (2022)	Octanol/water distribution coefficient	Partition coefficient prediction
ClinTox (Wu et al. 2018)	Zhu et al. (2022) and Feng et al. (2024)	Drug toxicity data	Clinical toxicity prediction

(continued)

Table 7 (continued)

Dataset name	Works	Data type	Task
BACE (Wu et al. 2018)	Zhu et al. (2022) and Feng et al. (2024)	BACE-1 inhibitors data	Drug-target interaction
BBBP (Wu et al. 2018)	Zhu et al. (2022) and Feng et al. (2024)	Blood-brain barrier permeability	Permeability classification
PCBA (Wu et al. 2018)	Feng et al. (2024)	Biological activity of small molecules	Drug screening
ToxCast (Wu et al. 2018)	Feng et al. (2024)	Toxicology data	Toxicity prediction
PCQM4Mv2 (Hu et al. 2021)	Feng et al. (2024)	2D molecular graphs	HOMO-LUMO energy gap prediction
OGB (Hu et al. 2021)	Feng et al. (2024)	Graph-based molecule data	Molecular graph learning
hERG (Gaulton et al. 2017; Kim et al. 2021)	He et al. (2021)	Cardiac toxicity data	Toxicity prediction
Tencent (He et al. 2019)	He et al. (2021)	User-group bipartite graph	Social network analysis
3A4 (Ma et al. 2015)	Chen et al. (2020) and Huang et al. (2023)	CYP P450 3A4 inhibition data	ADME property prediction
CB1 (Ma et al. 2015)	Chen et al. (2020) and Huang et al. (2023)	Cannabinoid receptor binding	Drug-target interaction
DPP4 (Ma et al. 2015)	Chen et al. (2020) and Huang et al. (2023)	Dipeptidyl peptidase-4 inhibition	Drug-target interaction
HIVINT (Ma et al. 2015)	Chen et al. (2020) and Huang et al. (2023)	HIV integrase inhibition	Antiviral drug screening
LOGD (Ma et al. 2015)	Chen et al. (2020) and Huang et al. (2023)	LogD values	ADME Property prediction
METAB (Ma et al. 2015)	Chen et al. (2020) and Huang et al. (2023)	Microsomal metabolism data	Drug metabolism prediction
PGP (Ma et al. 2015)	Chen et al. (2020) and Huang et al. (2023)	P-glycoprotein transport data	Drug transport prediction
THROMBIN (Ma et al. 2015)	Chen et al. (2020) and Huang et al. (2023)	Human thrombin inhibition	Drug-target interaction
CCLE (Barretina et al. 2012)	Nolte et al. (2023)	Genomic, transcriptomic, and proteomic data	Cancer biomarker discovery
NCI (Shoemaker 2006)	Nolte et al. (2023)	Drug screening data	Drug discovery

(continued)

Table 7 (continued)

Dataset name	Works	Data type	Task
DrugBank (Knox et al. 2024)	Wang et al. (2024)	Drug and target interactions	Pharmaceutical data analysis
GDSC (Yang et al. 2012)	Xu et al. (2023)	Cancer cell line drug response	Therapeutic biomarker discovery
CTRP (Seashore-Ludlow et al. 2015)	Xu et al. (2023)	Cancer therapeutics discovery	Drug sensitivity analysis
KiBA (Tang et al. 2014)	Mittone et al. (2023), Sándor and Antal (2023) and Wang et al. (2024)	Drug-target bioactivity data	Binding affinity prediction
BindingDB (Liu et al. 2007)	Wang et al. (2024)	Protein-ligand interactions	Drug-target interaction
From other works (Liu et al. 2014; Cai et al. 2019; Zhang et al. 1999; Siramshetty et al. 2020; Smalley 2017; Keserü 2003; Doddareddy et al. 2010; Sun et al. 2017; Gupta et al. 2022)	Ju et al. (2024) and Xiong et al. (2022)	–	–

emerging field that combines the principles of quantum computing with ML and is useful for solving complex problems in various fields, including drug discovery and healthcare (Heyndrickx et al. 2023a). By leveraging quantum algorithms, molecular structures can be optimized, binding affinities can be predicted, and drug candidates can be identified more efficiently, even with sparse data. Quantum methods allow more realistic intermolecular interactions. As these technologies continue to evolve, their application will be helpful in the discovery and development of novel drugs.

Acknowledgments This work was also partially supported by the PNRR project FAIR - Future AI Research (PE00000013), Spoke 3, under the NRRP MUR program funded by the NextGenerationEU.

References

Ali K, Zafar A. Comparative study of Federated Learning and Machine Learning for Drug Discovery. In: 2024 IEEE International Conference on Computer Vision and Machine Intelligence (CVMI). IEEE; 2024. p. 1–6.

Anastassiadis T, Deacon SW, Devarajan K, Ma H, Peterson JR. Comprehensive assay of kinase catalytic activity reveals features of kinase inhibitor selectivity. Nat Biotechnol. 2011;29(11):1039–45.

Barretina J, Caponigro G, Stransky N, Venkatesan K, Margolin AA, Kim S, et al. The cancer cell line encyclopedia enables predictive modelling of anticancer drug sensitivity. Nature 2012;483(7391):603–607.

Bashir AK, Victor N, Bhattacharya S, Huynh-The T, Chengoden R, Yenduri G, et al. Federated learning for the healthcare metaverse: concepts, applications, challenges, and future directions. IEEE Int Things J 2023;10:21873–891.

Bassani D, Brigo A, Andrews-Morger A. Federated learning in computational toxicology: an industrial perspective on the Effiris Hackathon. Chem Res Toxicol. 2023;36(9):1503–17.

Bento AP, Gaulton A, Hersey A, Bellis LJ, Chambers J, Davies M, et al. The ChEMBL bioactivity database: an update. Nucleic Acids Res. 2014;42(D1):D1083–90.

Bloem BR, Marks WJ, Silva de Lima AL, Kuijf ML, Van Laar T, Jacobs BP F., et al. The personalized parkinson project: examining disease progression through broad biomarkers in early Parkinson's disease. BMC Neurol. 2019;19:1–10.

Cai C, Guo P, Zhou Y, Zhou J, Wang Q, Zhang F, et al. Deep learning-based prediction of drug-induced cardiotoxicity. J Chem Inf Model. 2019;59(3):1073–84.

Chen S, Xue D, Chuai G, Yang Q, Liu Q. FL-QSAR: a federated learning-based QSAR prototype for collaborative drug discovery. Bioinformatics 2020;36(22–23):5492–98.

Choudhury O, Park Y, Salonidis T, Gkoulalas-Divanis A, Sylla I. Predicting adverse drug reactions on distributed health data using federated learning. In: AMIA annual symposium proceedings, vol. 2019. American Medical Informatics Association; 2019. p. 313.

Davis MI, Hunt JP, Herrgard S, Ciceri P, Wodicka LM, Pallares G, et al. Comprehensive analysis of kinase inhibitor selectivity. Nat Biotechnol. 2011;29(11):1046–51.

Deng J, Yang Z, Wang H, Ojima I, Samaras D, Wang F. A systematic study of key elements underlying molecular property prediction. Nat Commun. 2023;14(1):6395.

Doddareddy MR, Klaasse EC, Shagufta, IJzerman AP, Bender A. Prospective validation of a comprehensive in silico hERG model and its applications to commercial compound and drug databases. ChemMedChem 2010;5(5):716–29.

Dutta S, de Freitas IL, Xavier PM, de Farias CM, Neira DEB. Federated learning in chemical engineering: a tutorial on a framework for privacy-preserving collaboration across distributed data sources. arXiv preprint arXiv:2411.16737. 2024.

eChemPortal. https://www.echemportal.org/echemportal/

Feng J, Wang Z, Wei Z, Li Y, Ding B, Xu H. Federated heterogeneous contrastive distillation for molecular representation learning. In: Proceedings of the 33rd ACM international conference on information and knowledge management. 2024. p. 1038–1048.

Gaulton A, Hersey A, Nowotka M, Bento AP, Chambers J, Mendez D, et al. The ChEMBL database in 2017. Nucleic Acids Res. 2017;45(D1):D945–54.

GitHub – elix-tech/kmol: kMoL is a machine learning library for drug discovery and life sciences, with federated learning capabilities. Accessed January 2025. https://github.com/elix-tech/kmol

Glassman PM, Muzykantov VR. Pharmacokinetic and pharmacodynamic properties of drug delivery systems. J Pharmacol Experim Therapeut. 2019;370(3):570–80.

Griffanti L, Klein JC, Szewczyk-Krolikowski K, Menke RA, Rolinski M, Barber TR, et al. Cohort profile: the Oxford Parkinson's disease centre discovery cohort MRI substudy (OPDC-MRI). BMJ Open 2020;10(8):e034110.

Guo Y, Gao Y, Song J. MolCFL: a personalized and privacy-preserving drug discovery framework based on generative clustered federated learning. J Biomed Inf. 2024;157:104712.

Gupta A, Harrison PJ, Wieslander H, Rietdijk J, Puigvert JC, Georgiev P, et al. Is brightfield all you need for mechanism of action prediction?. bioRxiv, 2022–10. 2022.

Hanser T, Bastogne D, Basu A, Davies R, Delaunois A, Fowkes A, et al. Using privacy-preserving federated learning to enable pre-competitive cross-industry knowledge sharing and improve QSAR models. In: Society of toxicology (SOT) annual meeting. 2022.

He C, Xie T, Rong Y, Huang W, Huang J, Ren X, Shahabi C. Cascade-BGNN: toward efficient self-supervised representation learning on large-scale bipartite graphs. arXiv preprint arXiv:1906.11994. 2019.

He C, Balasubramanian K, Ceyani E, Yang C, Xie H, Sun L, et al. FedGraphNN: a federated learning benchmark system for graph neural networks. In: ICLR 2021 workshop on distributed and private machine learning (DPML). 2021.

He C, Ceyani E, Balasubramanian K, Annavaram M, Avestimehr S. SpreadGNN: decentralized multi-task federated learning for graph neural networks on molecular data. In: Proceedings of the AAAI conference on artificial intelligence, vol. 36, no. 6. 2022. p. 6865–6873.

Heyndrickx W, Arany A, Simm J, Pentina A, Sturm N, Humbeck L, et al. Conformal efficiency as a metric for comparative model assessment befitting federated learning. Artif Intell Life Sci. 2023a;3:100070.

Heyndrickx W, Mervin L, Morawietz T, Sturm N, Friedrich L, Zalewski A, et al. Melloddy: cross-pharma federated learning at unprecedented scale unlocks benefits in QSAR without compromising proprietary information. J Chem Inf Model. 2023b;64(7):2331–44.

Hu W, Fey M, Ren H, Nakata M, Dong Y, Leskovec J. OGB-LSC: a large-scale challenge for machine learning on graphs. arXiv preprint arXiv:2103.09430. 2021.

Huang D, Ye X, Zhang Y, Sakurai T. Collaborative analysis for drug discovery by federated learning on non-IID data. Methods 2023;219:1–7.

Huang D, Ye X, Sakurai T. Multi-party collaborative drug discovery via federated learning. Comput Biol Med. 2024;171:108181.

Humbeck L, Morawietz T, Sturm N, Zalewski A, Harnqvist S, Heyndrickx W, et al. Don't overweight weights: evaluation of weighting strategies for multi-task bioactivity classification models. Molecules 2021;26(22):6959.

Ju L, Hellander A, Spjuth O. Federated learning for predicting compound mechanism of action based on image-data from cell painting. Artif Intell Life Sci. 2024;5:100098.

Kanagaraj S, Hema MS, Guptha MN. Performance analysis of classification methods for Parkinson's disease with PPMI Dataset. In: 2021 International Conference on Advancements in Electrical, Electronics, Communication, Computing and Automation (ICAECA). IEEE; 2021. p. 1–5.

Keserü GM. Prediction of hERG potassium channel affinity by traditional and hologram qSAR methods. Bioorgan Med Chem Lett 2003;13(16):2773–75.

Khanna A, Adams J, Antoniades C, Bloem BR, Carroll C, Cedarbaum J, et al. Accelerating Parkinson's disease drug development with federated learning approaches. NPJ Parkinson's Disease. 2024;10(1):225.

Kim S, Chen J, Cheng T, Gindulyte A, He J, He S, et al. PubChem in 2021: new data content and improved web interfaces. Nucleic Acids Res. 2021;49(D1):D1388–95.

Knox C, Wilson M, Klinger CM, Franklin M, Oler E, Wilson A, et al. DrugBank 6.0: the DrugBank knowledgebase for 2024. Nucleic Acids Res. 2024;52(D1):D1265–75.

Konečný J, McMahan HB, Yu FX, Richtarik P, Suresh AT, Bacon D. Federated learning: Strategies for improving communication efficiency. In: Proceedings of the 29th conference on neural information processing systems (NIPS), Barcelona. 2016. p. 5–10.

Li T, Sahu AK, Zaheer M, Sanjabi M, Talwalkar A, Smith V. Federated optimization in heterogeneous networks. Proc Mach Learn Syst. 2020;2:429–50.

Liang X, Bandara E, Zhao J, Shetty S. A blockchain-empowered federated learning system and the promising use in drug discovery. In: Blockchain in life sciences. Singapore: Springer Nature Singapore; 2022; p. 113–139.

Liu T, Lin Y, Wen X, Jorissen RN, Gilson MK. BindingDB: a web-accessible database of experimentally determined protein–ligand binding affinities. Nucleic Acids Res. 2007;35 Suppl_1:D198–201.

Liu LL, Lu J, Lu Y, Zheng MY, Luo XM, Zhu WL, et al. Novel Bayesian classification models for predicting compounds blocking hERG potassium channels. Acta Pharmacol Sinica. 2014;35(8):1093–1102.

Ma J, Sheridan RP, Liaw A, Dahl GE, Svetnik V. Deep neural nets as a method for quantitative structure–activity relationships. J Chem Inf Model. 2015;55(2):263–74.

Maiorescu R, Semenescu A. Leveraging federated learning for enhanced data management pharmaceutical industry. In: International conference interdisciplinarity in engineering. Cham: Springer Nature Switzerland; 2023. p. 399–412.

Manu D, Sheng Y, Yang J, Deng J, Geng T, Li A, et al. Fl-disco: federated generative adversarial network for graph-based molecule drug discovery: special session paper. In: 2021 IEEE/ACM international conference on computer aided design (ICCAD). IEEE; 2021. p. 1–7.

Manu D, Yao J, Liu W, Sun X. GraphGANFed: a federated generative framework for graph-structured molecules towards efficient drug discovery. IEEE/ACM Trans Comput Biol Bioinf. 2024;21:240–53.

McMahan B, Moore E, Ramage D, Hampson S, y Arcas BA. Communication-efficient learning of deep networks from decentralized data. In Artificial intelligence and statistics (pp. 1273–1282). PMLR; 2017.

Metz JT, Johnson EF, Soni NB, Merta PJ, Kifle L, Hajduk PJ. Navigating the kinome. Nat Chem Biol. 2011;7(4):200–202.

Mittone G, Svoboda F, Aldinucci M, Lane N, Lió P. A federated learning benchmark for drug-target interaction. In: Companion proceedings of the ACM web conference 2023. 2023. p. 1177–1181.

Ning Y, Wang J, Li D, Yan D, Li X. GFedKRL: graph federated knowledge re-learning for effective molecular property prediction via privacy protection. In: International conference on artificial neural networks. Cham: Springer Nature Switzerland; 2023. p. 426–438.

Nolte D, Bazgir O, Ghosh S, Pal R. Federated learning framework integrating REFINED CNN and deep regression forests. Bioinf Adv. 2023;3(1):vbad036.

Oldenhof M, Ács G, Pejó B, Schuffenhauer A, Holway N, Sturm N, et al. Industry-scale orchestrated federated learning for drug discovery. In: Proceedings of the AAAI conference on artificial intelligence, vol. 37, no. 13. 2023. p. 15576–15584.

Sándor D, Antal P. Boosting federated multitask learning: transfer effects in cross-domain drug-target interaction prediction. In: Proceedings of SAI intelligent systems conference. Cham: Springer Nature Switzerland; 2023. p. 386–399.

Sachdev K, Gupta MK. A comprehensive review of feature based methods for drug target interaction prediction. J Biomed Inf. 2019;93:103159.

Saggi MK, Bhatia AS, Kais S. Federated quantum machine learning for drug discovery and healthcare. Ann Rep Comput Chem. 2024;20:269–322.

Seashore-Ludlow B, Rees MG, Cheah JH, Cokol M, Price EV, Coletti ME, et al. Harnessing connectivity in a large-scale small-molecule sensitivity dataset. Cancer Discovery 2015;5(11):1210–23.

Shen J, Nicolaou CA. Molecular property prediction: recent trends in the era of artificial intelligence. Drug Discovery Today Technol. 2019;32:29–36.

Shoemaker RH. The NCI60 human tumour cell line anticancer drug screen. Nat Rev Cancer. 2006;6(10):813–23.

Simm J, Humbeck L, Zalewski A, Sturm N, Heyndrickx W, Moreau Y, et al. Splitting chemical structure data sets for federated privacy-preserving machine learning. J Cheminf 2021;13:1–14.

Siramshetty VB, Nguyen DT, Martinez NJ, Southall NT, Simeonov A, Zakharov AV. Critical assessment of artificial intelligence methods for prediction of hERG channel inhibition in the "big data" era. J Chem Inf Model. 2020;60(12):6007–19.

Smajić A, Grandits M, Ecker GF. Privacy-preserving techniques for decentralized and secure machine learning in drug discovery. Drug Discov Today. 2023;28:103820.

Smalley E. AI-powered drug discovery captures pharma interest. Nat Biotechnol. 2017;35(7):604–606.

Spjuth O, Frid J, Hellander A. The machine learning life cycle and the cloud: implications for drug discovery. Expert opinion on drug discovery 2021;16(9):1071–79.

Stephenson D, Alexander R, Aggarwal V, Badawy R, Bain L, Bhatnagar R, et al. Precompetitive consensus building to facilitate the use of digital health technologies to support Parkinson disease drug development through regulatory science. Digital Biomark 2020;4 Suppl 1:28–49.

Sun X, Xu B, Xue Y, Li H, Zhang H, Zhang Y, et al. Characterization and structure-activity relationship of natural flavonoids as hERG K+ channel modulators. Int Immunopharmacol. 2017;45:187–93.

Tang J, Szwajda A, Shakyawar S, Xu T, Hintsanen P, Wennerberg K, Aittokallio T. Making sense of large-scale kinase inhibitor bioactivity data sets: a comparative and integrative analysis. J Chem Inf Model. 2014;54(3):735–43.

Upstill-Goddard R, Eccles D, Fliege J, Collins A. Machine learning approaches for the discovery of gene–gene interactions in disease data. Brief Bioinf. 2013;14(2):251–60.

Vollstedt EJ, Schaake S, Lohmann K, Padmanabhan S, Brice A, Lesage S, et al. Embracing monogenic Parkinson's disease: the MJFF Global Genetic PD cohort. Mov Disord. 2023;38(2):286–303.

Wang X, Zhao Q, Wang J. FedKD-DTI: drug-target interaction prediction based on federated knowledge distillation. In: International symposium on bioinformatics research and applications. Singapore: Springer Nature Singapore; 2024. p. 95–106.

Wu Z, Ramsundar B, Feinberg EN, Gomes J, Geniesse C, Pappu AS, et al. MoleculeNet: a benchmark for molecular machine learning. Chem Sci. 2018;9(2):513–30.

Xiong Z, Cheng Z, Lin X, Xu C, Liu X, Wang D, et al. Facing small and biased data dilemma in drug discovery with enhanced federated learning approaches. Sci China Life Sci 2022;65:1–11.

Xu X, Qi Z, Han X, Xu A, Geng Z, He X, et al. Predicting anticancer drug sensitivity on distributed data sources using federated deep learning. Heliyon 2023;9(8):e18615.

Yang W, Soares J, Greninger P, Edelman EJ, Lightfoot H, Forbes S, et al. Genomics of drug sensitivity in cancer (GDSC): a resource for therapeutic biomarker discovery in cancer cells. Nucleic Acids Res. 2012;41(D1):D955–61.

Zhang S, Zhou Z, Gong Q, Makielski JC, January CT. Mechanism of block and identification of the verapamil binding domain to HERG potassium channels. Circul Res. 1999;84(9):989–98.

Zhu W, Luo J, White AD. Federated learning of molecular properties with graph neural networks in a heterogeneous setting. Patterns. 2022;3(6):100521.

Revolutionizing Drug Delivery: The Role of Artificial Intelligence in Nanomedicine and Precision Pharma

Mohammad Habeeb, Hariharan Thirumalai Vengateswaran, Huay Woon You, K. N. V. Chenchu Lakshmi, Sarita Suryabhan Pawar, and Shital Dnyaneshwar Gaikwad

Abstract Artificial intelligence (AI) is transforming drug delivery by optimizing nanomedicine design and enhancing therapeutic precision. While drug delivery systems (DDSs) have improved drug selectivity and release kinetics, their development remains challenging due to the complexity of experimental data and limitations in high-throughput analysis. Scientists struggle to efficiently acquire, manage, and interpret large datasets, slowing the rational design of smart drug carriers. Artificial intelligence and machine learning address these challenges by automating data processing, optimizing formulations, and accelerating drug discovery. Artificial intelligence-driven models analyze biological datasets, predict drug interactions, and enhance nanoparticle stability, bioavailability, and targeting accuracy. Additionally, artificial intelligence facilitates real-time monitoring and adaptive drug release, improving therapeutic efficiency, and reducing costs. By integrating AI with nanomedicine, researchers can streamline drug development and advance precision medicine. This chapter reveals artificial intelligence's role in revolutionizing DDSs, highlighting its impact on formulation design, process automation, and personalized drug delivery.

M. Habeeb (✉) · H. T. Vengateswaran
Department of Pharmaceutics, Crescent School of Pharmacy, B.S. Abdur Rahman Crescent Institute of Science and Technology, Chennai, Tamil Nadu, India

H. W. You
Pusat PERMATA@Pintar Negara, Universiti Kebangsaan Malaysia, Bangi, Selangor, Malaysia

K. N. V. C. Lakshmi
Department of Pharmaceutical Chemistry, KL College of Pharmacy, Koneru Lakshmaiah Education Foundation, Green Fields, Vaddeswaram, Andhra Pradesh, India

S. S. Pawar
Department of Pharmaceutical Chemistry, School of Pharmaceutical Sciences, Sanjivani University, Kopargaon, Ahmednagar, Maharashtra, India

S. D. Gaikwad
Department of Pharmaceutical Chemistry, Samarth Institute of Pharmacy, Pune, Maharashtra, India

A. Lavecchia (ed.), *Applied Artificial Intelligence for Drug Discovery*,
https://doi.org/10.1007/978-3-031-98022-0_19

Keywords Artificial intelligence · Drug delivery · Nanomedicine · Machine learning · Nanorobotics

1 Introduction

Industries continuously strive to improve their development to reach client requirements and demands using various approaches. One of the most critical sectors in this regard is the pharmaceutical industry, which plays a crucial role in healthcare. It functions through constant modernization and the implementation of novel methods to meet global healthcare limitations and aid in medical crises, like the latest epidemic. Nanotechnology, particularly nanomedicine, allows for the precise manufacturing of biomedical nanomedicines with anticipated characteristics and behaviors in the biological body. This comprises the ability to control pharmaceuticals' adjustable half-lives, organ and tissue accumulation, selectivity, cellular uptake, drug kinetics, and numerous factors. However, in the case of small-molecule drugs, the lack of a computable thoughtful of structure-activity relationships (QSAR) makes it nearly impossible to manually fine-tune these parameters. By integrating nanotechnology with medicine, nanomedicines have been developed for detecting, intensive care, and treating illnesses. These advanced formulations can be clinically administered to individuals/fused into biological machinery, therapeutic biodevices, and nano biosensors (He et al. 2024).

As biology, medicine, and nanotechnology have advanced, cancer treatment has quickly become more individualized and data-intensive. Individual qualities must be specifically taken into account to obtain better therapeutic impact and lower toxicity, and the development of therapeutic medications or combination therapy must be globally optimized from the outset of design. To achieve precision medicine treatment, artificial intelligence (AI) and associated knowledge have been extensively used to evaluate multi-source and multi-dimensional information (Chou et al. 2023). Artificial neural networks (ANN), deep learning (DL), and machine learning (ML) are some of the subfields of AI that is a division of computer science that proposes to achieve intricate responsibilities that need "human intelligence" through computers/computer-controlled machines. The three main categories of AI are artificial general intelligence, artificial narrow intelligence, and artificial superintelligence. AI-enabled approaches have been reported to be extensively utilized in medicine and life science, such as forecasting slight differences in the number of living species/synthetic substances from the starting point, recognizing pandemic illness outbursts initial, radiological as well as histological analysis, and muti-omics examination to develop molecular outlines (AI-driven multi-omics analysis) (Pednekar et al. 2024). AI plays a crucial role in both drug formulation and nanomedicine, optimizing nano-pharmaceutical preparations by predicting interactions between nanocarriers and biological systems. It aids in modeling drug encapsulation efficiency, release kinetics, and targeted drug delivery, ensuring improved therapeutic precision. In nanomedicine, AI enhances personalized diagnostics through biomarker analysis and imaging, enabling early cancer detection. These advancements contribute to more effective treatments, reduced toxicity, and higher survival rates in oncology.

Since multimodal mechanisms of action and the realization of their simultaneous therapeutic properties are taken into consideration, the development of nanomedicines is typically significantly more difficult. Even though the combination of AI with nanomedicines is still in its initial phases, AI can be tremendously significant in these procedures. The optimization of nano pharmaceutical preparations to advance the transporting and directing of nanomedicines can be assisted by the forecast of communications among nanocarriers and their integrated drugs, biological mediators, or membranes of cells; prediction of the medication encapsulation efficiency; and modeling of release kinetics from nanocarriers using AI algorithms. The integration of nanomedicine and AI in oncology enables personalized diagnostics through AI-driven biomarker analysis and imaging for early cancer detection. AI enhances targeted drug delivery by optimizing nano-formulations, ensuring precise, patient-specific treatment with minimal side effects. This synergy improves therapeutic outcomes, boosts survival rates, and revolutionizes cancer care (Fig. 1). For the purpose of achieving pharmacological synergism while lowering toxicity and eventually improving clinical outcomes, AI-enabled technologies also enable the choice of drug combinations along with internationally optimizing the combination treatment (Ambreen et al. 2025). This chapter aims to highlight AI-based strategies to enhance drug delivery by optimizing targeting, formulation, and treatment strategies. It highlights AI-driven advancements in predictive modeling, personalized medicine, and nanorobotics. The discussion also addresses key challenges and prospects in precision pharma.

2 Identification and Analysis of Biomarkers

In oncology, cancer biomarkers are essential because they provide crucial data for risk assessments, prognostication, therapy response, illness diagnosis, and efficient disease progression monitoring. The National Cancer Institute states that each cancer patient has distinct biomarker patterns that provide insight into the potential efficacy of particular cancer treatments. Thus, acquiring of biomarker data assists physicians in selecting the most effective cancer treatment for their patients. Additionally, patients' biomarkers vary, therefore determining a patient's unique molecular signature aids in the screening of people with cancer for cancer treatment. Additionally, these patients can be categorized based on the cancer's stage, where the tumor may be confined or have spread. Various biomarker tests have been conducted recently to assist in choosing the best cancer therapy for the patient. The majority of biomarker studies involve the identification of proteins, genetic markers, or other tumor indicators. However, multigene testing, commonly referred to as biomarker testing, comprises assessments for both single and numerous biomarkers. The Oncotype DX test, for instance, is a common cancer biomarker test used for

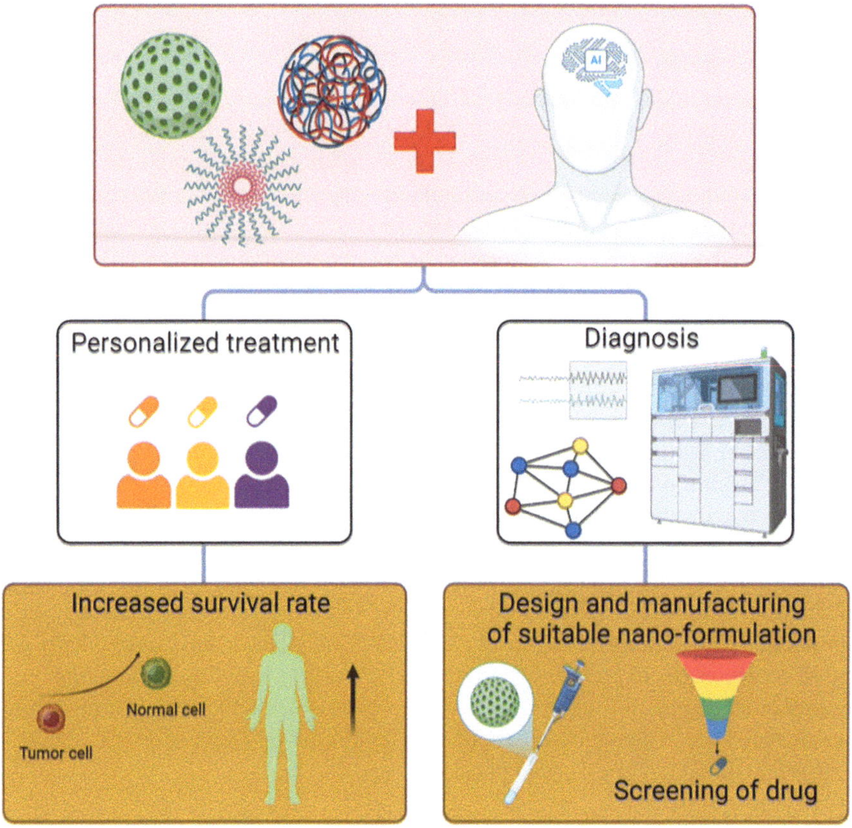

Fig. 1 Represents the integration of nanomedicine and AI for increased survival rates and optimized nano-formulation for drug development

patients with breast cancer. It determines the activity of 21 genes to determine if chemotherapy is going to be effective for the patient in question. By combining AI models for prediction and classification, these prediction requirements could be further enhanced (Gentili and Stano 2024).

Multiple nanomaterials are used in biomarker sensing applications, including carbon nanotubes, gold nanoparticles, and quantum dots. With their great sensitivity and specificity, nanoparticles are among the most promising methods for cancer biomarker detection, especially in light of recent advancements. To enhance diagnosis and treatment platforms, the potential of nanomaterials is presently being studied for both in vitro and in vivo medical applications. Furthermore, these nanomedicines offer new opportunities to obtain accurate diagnostic data that can aid in early disease detection, determining the prognosis of the condition, and customizing treatment regimens. Furthermore, ligands, antibodies, and various cellular indicators coated on drug-loaded NPs enable them to attach to the target cells, improving the response to therapy. However, obtaining molecular profiles for every patient can lead to more precise cancer treatment. These could contain biomarkers and information about the conditions. Consequently, omics-based illness profiles offer crucial genetic, proteomic, metabolomic, microbiomic, and epigenomic characteristics (Das and Chandra 2022).

3 AI for Drug Delivery

Computational methods in pharmaceutical sciences, which use multiscale simulation methods to improve the delivery of drug methods, are a result of the pharmaceutical industry's combination of AI and huge amounts of data. The various AI algorithms used in delivering drugs are presented in Table 1, with emphasis on their use in predictive modeling, drug screening, and formulation optimization. Drug-target interactions, toxicity evaluations, and nanoparticle design have been greatly improved by advanced algorithms like Artificial neural networks (ANNs) and Generative adversarial networks (GANs). These AI-powered methods accelerate the development of focused treatments, improve accuracy, and reduce the workload associated with experiments (Sumayli et al. 2023). Computational methods in pharmaceutical sciences analyze massive data and forecast drug activity using AI algorithms and machine learning approaches. Without requiring a great deal of trial-and-error testing, researchers can assess different situations and optimize drug delivery systems by simulating drug formulation as well as distribution processes. This reduces the time needed to research new drugs, lowers expenses, and boosts output. Drug delivery systems of many sizes, from interaction between molecules to macroscopic behavior, are modeled in computational pharmaceutics (Hani et al. 2023). To forecast drug behavior at every level, AI systems can examine intricate correlations between pharmacological characteristics, formulation ingredients, and physiological parameters. This enables the development of efficient drug delivery devices and enables an added thorough knowledge of the drug-delivery approach (Chen 2024). It supports in forecasting the drug's stability, in vitro medication release description, along with physicochemical features. In addition to in vivo/in vitro research on correlation, the same technique is also used to improve the evaluation of in vivo pharmacokinetic characteristics and drug distribution (Fig. 2). By using the right combination of AI tools, researchers can first identify potential risks and difficulties associated with medication delivery systems during the development stage. This enables proactive modifications and adjustments to lower hazards and enhance therapeutic efficacy. AI and computational modeling can reduce reliance on costly and time-consuming trial-and-error studies by improving predictive accuracy. However, these methods have limitations and should complement, rather than replace, empirical validation and should be correlated with the experimental findings (Vora et al. 2023).

Table 1 Represents the various algorithms utilized in drug delivery

AI algorithm	Application in drug delivery and pharma	Example cases	Practical implications	References
Artificial neural networks (ANNs)	Forecasting drug release profiles and optimizing nanoparticle formulations	Optimization of polymeric nanoparticles for controlled drug release	Reduced experimental trials by 30–50%, lowering R&D costs and enhancing formulation accuracy	Piroozmand et al. (2023)
Convolutional neural networks (CNNs)	Image-based drug screening and nanocarrier identification	High-throughput screening of nanomedicine interactions in cells	Improved nanocarrier classification accuracy by 85%, expediting drug screening processes	Aghajanpour et al. (2025)
Recurrent neural networks (RNNs)	Analyzing sequential biomedical data for drug response prediction	Predicting pharmacokinetics and pharmacodynamics	Increased prediction accuracy of drug response timelines by 40%, aiding precision medicine	Myreddy et al. (2024)
Generative adversarial networks (GANs)	Generating novel drug formulations and virtual drug libraries	Designing novel drug candidates for targeted therapy	Identified 20+ potential drug candidates in silico, reducing early-stage screening time	Das and Chandra (2022)
Support vector machines (SVMs)	Classifying drug-target interactions and predicting bioavailability	Identifying optimal ligand-receptor binding for targeted drug delivery	Enhanced ligand-receptor binding prediction by 25%, optimizing targeted therapy selection	Di Francesco et al. (2023)
Random forest (RF)	Feature selection in drug formulation and toxicity prediction	Screening excipients for nanoformulations	Increased toxicity prediction accuracy by 33%, minimizing late-stage drug failure rates	Urista et al. (2020)

Gradient boosting (e.g. XGBoost, LightGBM)	Predicting drug solubility, stability, and effectiveness	Optimizing polymer-lipid hybrid nanoparticles	Improved solubility prediction models, reducing formulation rejection by 20%	Mi et al. (2024)
Deep reinforcement learning (DRL)	Personalized dosage optimization and adaptive drug release	AI-driven precision medicine for cancer therapies	Enabled real-time dose adjustments, reducing adverse effects by 30%	Patel (2023)
Bayesian etworks	Probabilistic modeling of drug interactions and efficacy	Risk assessment of drug toxicity in polypharmacy	Identified high-risk drug interactions with 90% accuracy, improving patient safety	Furxhi et al. (2019)

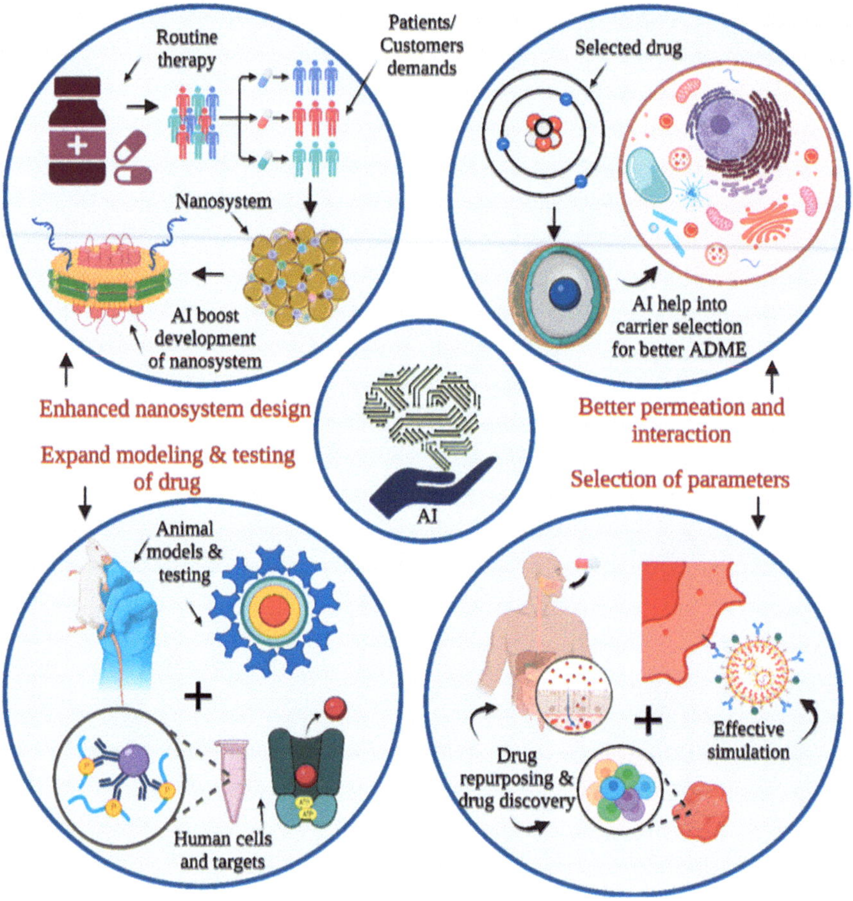

Fig. 2 Illustrates the AI approach to the problems facing the pharmaceutical industry. (Reproduced from Vora et al. 2023)

4 Nanomedicine and Computational Medicine

A biotechnology enterprise called Insilico Medicine employs AI and ML to find and develop new drugs, including optimizing drug delivery methods (Abdalla et al. 2024). An excellent illustration of how ML might improve medication delivery is its application in nanomedicine, specifically in the treatment of cancer (Eugster et al. 2024). An AI-powered platform for designing nanoparticles for tailored drug delivery in cancer therapy has been established by in silico medicine. The technology can forecast the ideal nanoparticle forms that carry drugs directly to cancer cells while causing the least amount of harm to healthy tissues by evaluating biological data and modeling drug interactions within the body (Peng et al. 2024). In one study, a new cancer medication's delivery mechanism based on nanoparticles was designed using an AI platform (Nagalakshmi et al. 2025). The nanoparticles' size, shape, and composition are optimized by the ML algorithms, ensuring that they can efficiently target tumor cells without being detected by the immune system. As a result, the medicine's therapeutic index increased, side effects decreased, and patient outcomes improved owing to a more effective drug delivery mechanism (Calmet et al. 2023). This example shows how machine learning (ML) can speed up the establishment of cutting-edge medicine delivery systems that provide more focused care with fewer side effects (Jena et al. 2024).

5 Dosage Form Designing

A scientific examination of the procedures and variables that affect the drug product's quality characteristics during the manufacturing of nanomedicines improves predictability. It lowers the possibility of failure linked to quality (Alqarni et al. 2024). In this regard, Quality by design (QbD) is a systematic approach to pharmaceutical development that emphasizes process understanding and control to ensure product quality. CQAs must be determined following the establishment of the quality target product profile (QTPP), which includes the drug product's technical qualities, formulations, and administration route. Quantitative structure–property–relationship (QSPR) simulations can be utilized to help in excipient selection. They forecast the associations between chemicals and excipients based on structure. The Vlife MDS 4.2 building module is one option. By connecting the physicochemical characteristics of polymers (substance qualities) with the formulation's CQAs, these models offer a mechanistic method for choosing compositions with desired qualities and forecast drug release mechanisms. To ensure dependable treatment performance, these parameters are constantly tracked and need to stay within predetermined bounds. Particle size, entrapment effectiveness, immunogenicity, and zeta potential are crucial factors, among others. The next stage is the definition of critical material attributes (CMAs), like the concentration of the therapeutic ingredients or the starting materials utilized to make nanomedicines. They provide the basis of the control approach together with specific CPPs like the variety of extrusion rounds in vesicle manufacturing. Operational analytical technologies are utilized to keep monitoring CPP and CMAs. The data used for process control and simulation come from a variety of sensors. A modification in the CPPs, CMAs, or CQAs is transmitted to the process control unit resulting in variations in the production process. Some of the most popular CQAs, CPPs, and CMAs in the creation of nanopharmaceuticals are covered in a recent review study.

One of the key components of Quality by design (QbD) methods is the use of Design of experiments (DoE) to systematically optimize and validate processes. It uses mathematical models to correlate the observed responses (CQAs) with the procedure variables (CPPs and CMAs). Predictions and process modifications throughout manufacturing are made easier as a result. The control strategy confirms that the final merchandise satisfies the qualities outlined in the QTPP. Fractional

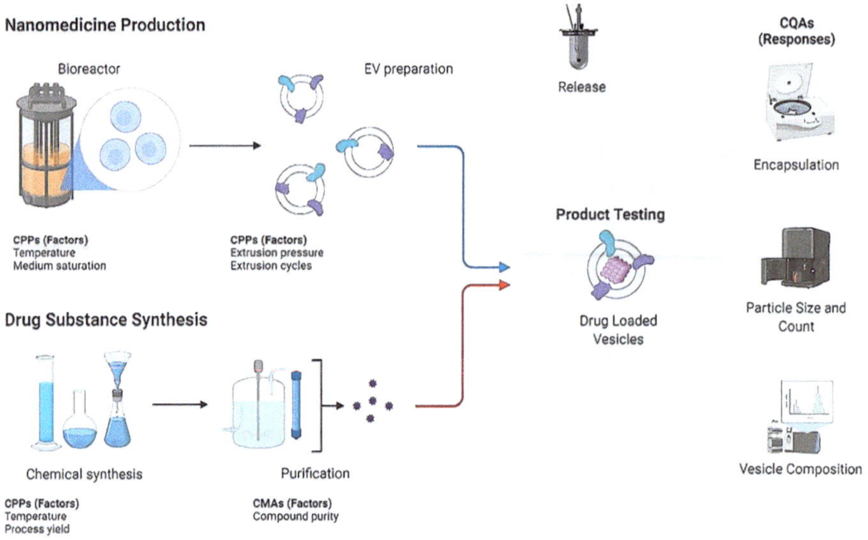

Fig. 3 Representation of the use of the design of experiments in manufacturing a medication-infused extracellular vesicle formulation for nanomedicine. (Reproduced from Villa Nova et al. 2022)

factorial and Plackett–Burman screening designs are two categories of designs used in DoE studies. These are used to simultaneously build a large number of process variables (CPPs and CMAs) that affect the product attributes. The Box–Behnken design and the central composite are included in the response area methodology for comparison. Although there are fewer process variables in these two examples, each parameter is examined at a higher resolution (Fig. 3).

6 Personalized Drug Potency Prediction Using Artificial Intelligence

Optimizing the therapy can be attained by evaluating how the patient's body reacts to drugs (Habeeb et al. 2023). The preliminary clinical report of individuals to specific medications may barely be expected grounded on omics data or unaided biomarkers because of intratumor heterogeneities, dynamic shifts in the tumor microenvironment, and the appearance of drug resistance. Drug effectiveness for precisely affected individuals has been forecast using computational models and AI algorithms, which have also been utilized to optimize treatment regimens. Computational models that inspect single-cell description, for instance, can quickly produce treatments or drug combinations that are beneficial in treating cancer. With a sensitivity of up to 100%, the clinical consequence for individuals was

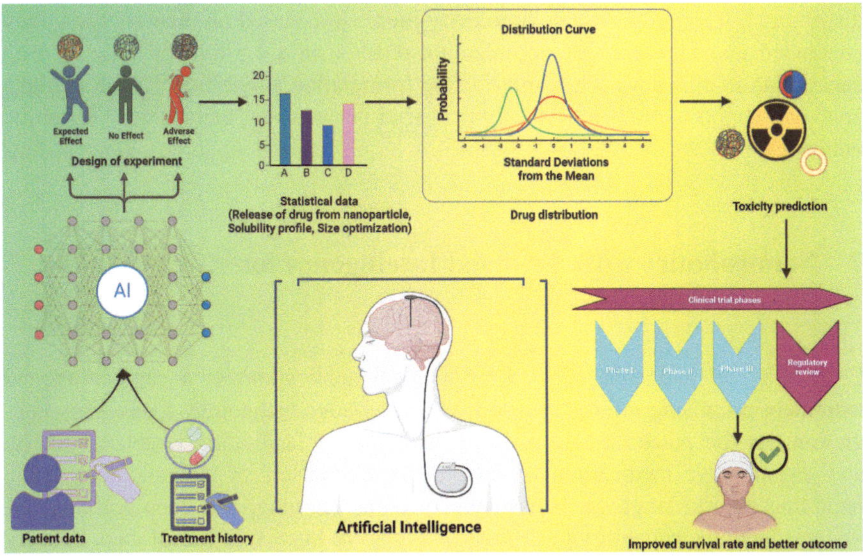

Fig. 4 AI-driven drug optimization for personalized cancer treatment, integrating patient data, computational modeling, and nanotechnology

forecast using an ML algorithm trained on gathered clinical observations regarding chemotherapy drugs and the clinical events that accompanied them. In situ diagnostics and drug prediction were also carried out using techniques derived from nanotechnology. Liposomes comprising various synthetic drugs (doxorubicin, cisplatin, gemcitabine, and caffeine) were used, and respectively liposome was identified by a distinct DNA barcode (Yaari et al. 2016). Artificial intelligence (AI) enhances nanoparticle-based drug delivery by analyzing patient data and treatment history to design optimized experiments. AI processes statistical data, including drug release profiles, solubility, and size optimization, to predict drug distribution patterns using probability models. The toxicity prediction step helps identify potential adverse effects, ensuring safer formulations before progressing to clinical trials (Phases I–III) and regulatory review. This AI-driven approach improves treatment precision, minimizes side effects, and enhances therapeutic efficacy (Fig. 4). A similar work conducted by Rampado et al. (2023) discussed the role of design of experiments (DoE) in optimizing nanoparticle-based drug delivery, which aligns with AI-driven predictive modeling (Fig. 4). Tools like Graph Neural Networks (GNN), Bayesian optimization, Support Vector Machines (SVM), and Artificial neural networks (ANN) can be used to optimize nanoparticle-based drug delivery through design of experiments (DoE). As shown, DoE helps determine expected, no, or adverse effects of drug formulations by analyzing statistical data on drug release, solubility, and size optimization. AI integrates patient data and treatment history to refine these experiments, leading to improved drug distribution patterns. The probability distribution curve in the figure represents variations in drug distribution, which further aids in toxicity prediction. Based on their analysis, they concluded that DoE-enabled optimization resulted in a 1.5-fold increase in drug encapsulation efficiency while minimizing formulation variability. This systematic approach supports drug delivery, clinical trial progression, enhancing therapeutic outcomes.

7 Nanorobotics with Artificial Intelligence for Pharmaceutical Uses

Combining AI and nanorobotics for medical applications is a captivating and extremely promising intersection of two cutting-edge technology areas. This integration has the potential to transform medicine by facilitating targeted, precise, and less invasive medication. AI encompasses a variety of technological fields, including natural language assessment, machine learning, and deep learning. AI systems have the capability to process and analyze huge volumes of data, identify patterns, and make decisions or predictions based on those findings. Nanorobots are little devices or machines with nanoscale dimensions, often less than 100 nanometers. Such nanoscale robotics can be developed for a variety of applications, such as tissue repair, surgery, and medication delivery. There are several possible medical applications for combining AI and nanorobotics. When combined, these two fields can provide precise, customized, and less invasive treatments that significantly improve patient outcomes. Addressing the related challenges, such as safety, control, and ethical concerns, will be necessary to fulfill this promise. It is anticipated that ongoing research and development in the domains of AI and nanorobotics will lead to exciting advancements in healthcare. The combination of AI with nanorobotics has the potential to transform therapeutic applications, particularly in precision medicine and tailored drug delivery. Current developments in engineering and bio-nanotechnology provide optimistic progress in integrating intelligent sensors, power, and AI in nanorobots, although challenges remain to be addressed. The integration minimizes damage to healthy tissue while offering unparalleled diagnostic and therapeutic precision. Once AI has examined an individual's genetic and medical data, nanorobots can offer these personalized treatments. Because nanorobots can perform functions without making major incisions, patients will experience less pain and heal more quickly (Ori et al. 2024) (Fig. 5).

8 Case Studies and Applications

The work by Shan He et al. in 2024 addresses the use of AI in developing nanoparticle-based drug delivery methods for neurodegenerative illnesses. They developed an AI-driven approach called IFPTML, which mixes information fusion and perturbation theory along with machine learning to forecast optimal

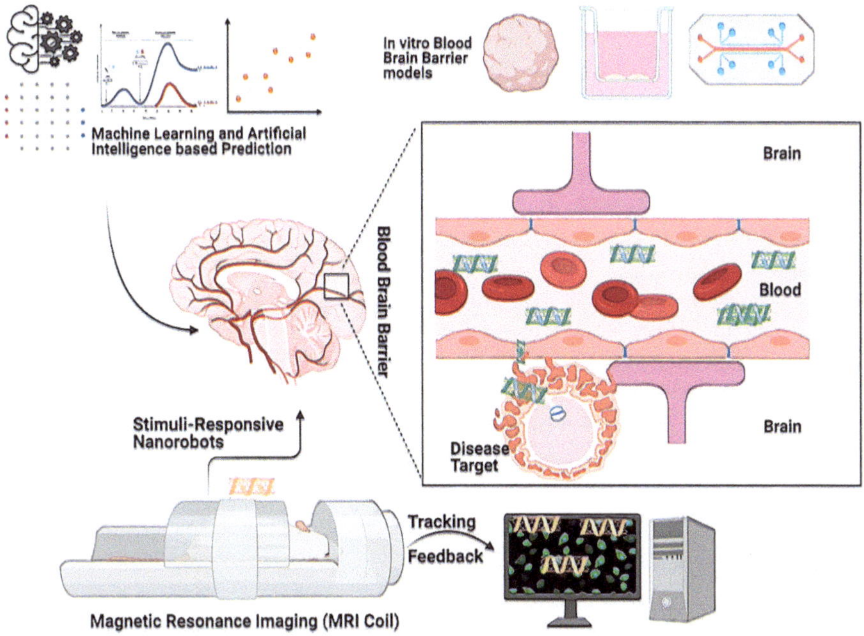

Fig. 5 AI-driven prediction and MRI-tracked nanorobots for targeted drug delivery across the blood–brain barrier. (Reproduced from Singh et al. 2021)

nanoparticle-drug pairings. They achieved great accuracy (AUROC ~0.93–0.95) by training ANN with datasets from ChEMBL and the scholarly literature. The model successfully found captivating drug-nanoparticle combinations, saving time and money on experiments. By forecasting blood–brain barrier (BBB) permeability and enhancing nanoparticle characteristics for optimal drug delivery, AI greatly enhanced drug design (He et al. 2024). Similarly, Abdulrahman Sumayli and team in 2023 evaluated the dissolution of tolfenamic acid in supercritical carbon dioxide, which was modeled and simulated in this study using AI techniques. The Dragonfly algorithm (DA) was used to optimize machine learning models, including Multi-layer Perceptron (MLP), Polynomial Regression (PR), and Extra Trees (ET), to forecast solvent density and solubility at different pressures and temperatures. With good accuracy ($R^2 = 0.98329$ for solubility and 0.99919 for density), the MLP model demonstrated its capacity to represent intricate nonlinear interactions. Additionally, PR performed exceptionally well, with an R^2 of 0.99844. These AI-powered models made precise predictions, cutting down on expenses and experimental efforts while improving the environment for producing nanomedicines using environmentally friendly procedures (Sumayli et al. 2023). Another study used AI and design of experiments (DoE) to create progesterone-incorporated solid lipid nanoparticles (PG-SLNs) for improved transdermal medication delivery. DoE was used by researchers to optimize important parameters such as medium-chain triglycerides, surfactants, and stearic acid content, while artificial neural networks

(ANN) driven by AI correctly expected the characteristics of the nanoparticles. The PG-SLNs demonstrated non-toxicity to neuronal cells and superior skin permeability, particularly when augmented with limonene. The formulation outperformed traditional suspensions in terms of performance and stability. This inventive method offers the possibility of targeted treatments for neurodegenerative diseases and uses AI to improve pharmaceutical development (Suriyaamporn et al. 2024). To maximize drug loading in polymeric nanoparticles, particularly poly(ortho ester) and poly(lactic-co-glycolic acid) (PLGA), the study investigates AI-assisted computational molecular modeling. Leila Karami and colleagues examined how granisetron, naltrexone, and risperidone interacted with these polymers using AI-powered molecular simulations. The models reduced the need for trial-and-error formulations by properly predicting controlled release qualities and drug-polymer compatibility. AI-powered simulations increased drug stability as well as bioavailability on molecular dynamics. This method accelerates the development of improved drug delivery systems with increased therapeutic efficacy (Karami et al. 2024) (Table 2).

9 Challenges and Limitations

AI-based models have many drawbacks regardless of their advantages, including the requirement for huge datasets, possible biases, and interpretability issues (Habeeb et al. 2024). In a study by Wael A. Mahdi et al., to forecast the biodistribution of drug nanoparticles for targeted cancer therapy, they used machine learning models such as Gaussian Process Regression (GPR), Extra Trees (ET), and Local Polynomial Regression (LPR). One of the biggest difficulties was data bias that resulted in biased predictions since the 534 records in the dataset had an unequal distribution of nanoparticle attributes, with some organ biodistributions being overrepresented. They addressed this by applying Z-score normalization to standardize the distribution of data and using Conditional Mutual Information (CMI) for selecting features. High-dimensional, nonlinear interactions also caused interpretability problems were reduced by choosing models that were capable of managing complex patterns and by improving predicted accuracy using the Whale optimization algorithm (WOA). These techniques advanced the design of nanoparticles for the treatment of cancer by increasing the robustness and dependability of AI-driven drug delivery predictions (Mahdi et al. 2025). Similarly, another study optimized progesterone-loaded solid lipid nanoparticles (PG-SLNs) for transdermal drug administration using Artificial neural networks (ANN) and design of experiments (DoE). Due to a lack of training data, data bias posed a significant problem that could result in overfitting and inaccurate predictions. Cross-validation and systematic DoE-generated datasets were used to prevent this and ensure balanced model training. Interpretability problems were additionally raised by the intricacy of ANN models, and these were resolved by maximizing hyperparameters such as training cycles and learning rates to attain low root mean square error (RMSE). These methods increased the model's precision and dependability, which improved PG-SLN sta-

Table 2 AI algorithms used in drug development, their applications, and outcomes in drug design

AI algorithm utilized	Application	Problem	AI-assisted solution	Key outcomes	Future directions	References
Linear discriminant analysis (LDA), artificial neural networks (ANN), multilayer perceptron (MLP), and deep learning networks (DLN)	AI-driven nanoparticle-based drug delivery for neurodegenerative disorders	Challenges in accurately predicting optimal drug-nanoparticle interactions for blood–brain barrier (BBB) penetration	Implementation of IFPTML (information fusion, perturbation theory, and machine learning) to train ANN models on ChEMBL and the literature datasets	Achieved high predictive accuracy (AUROC ~0.93–0.95) in identifying effective nanoparticle-drug conjugates, reducing experimental burden	AI-based BBB permeability models can facilitate precision-targeted neurotherapeutics and enhance computational drug screening efficiency	He et al. (2024)
Dragonfly algorithm (DA), multi-layer perceptron (MLP), polynomial regression (PR), and extra trees (ET)	AI-assisted modeling of tolfenamic acid dissolution in supercritical CO_2	Difficulty in predicting solubility and solvent density under variable pressure and temperature conditions	AI-optimized machine learning framework to model and simulate solvent density and drug solubility	High model accuracy ($R^2 = 0.98329$ for solubility, 0.99919 for density), minimized experimental iterations, and improved eco-friendly drug formulation	Expansion of AI-based solubility prediction models to optimize pharmaceutical crystallization and green chemistry processes	Sumayli et al. (2023)
Artificial neural networks (ANN), design of experiments (DoE)	Optimization of progesterone-loaded solid lipid nanoparticles (PG-SLNs) for transdermal drug delivery	Poor cutaneous permeability and physicochemical instability of conventional progesterone formulations	AI-assisted DoE approach to optimize lipid composition, surfactants, and penetration enhancers for PG-SLNs	Optimized formulation exhibited superior transdermal permeability, enhanced stability, and biocompatibility	AI-guided formulation strategies can be extended to develop personalized transdermal drug delivery systems	Suriyaamporn et al. (2024)

(continued)

Table 2 (continued)

AI algorithm utilized	Application	Problem	AI-assisted solution	Key outcomes	Future directions	References
Molecular dynamics (MD) simulations, machine learning models	Computational modeling for optimizing drug loading in polymeric nanoparticles	Limitations in empirical trial-and-error methods for drug-polymer compatibility assessment	AI-assisted MD simulations to predict molecular interactions and controlled-release kinetics of drugs within poly (ortho ester) and PLGA matrices	Improved drug-polymer interaction fidelity, optimized release kinetics, and enhanced bioavailability	Integration of AI and MD simulations for in silico design of next-generation polymeric drug delivery systems	Karami et al. (2024)

bility and drug penetration (Suriyaamporn et al. 2024). To make sure the efficacy and safety of medications, AI-based models need to be employed in conjunction with conventional experimental techniques. Because it is challenging to understand how the AI models generate their predictions, they are frequently referred to as "black boxes" that employ sophisticated algorithms. Gaining authorization from regulators for AI-based drug development devices might be difficult due to this lack of transparency since it can be difficult to prove that the model is producing reliable and precise forecasts. Additionally, a lack of transparency may cause people to lose faith in the model's predictions, especially if those projections don't match what researchers or physicians anticipate (Obrezanova 2023). High-quality, representative data are necessary for AI models to function, but when data are few, as in the case of uncommon diseases, forecasts may be inaccurate (Vora et al. 2023). Precision may be impacted by inadequate demographic representation in training data, particularly for varied patient populations. Model accuracy is further challenged by homogeneous clinical trial results and missing or erroneous information. It takes a lot of time and resources to update AI models to take into account new information, such as newly available medications or the outcomes of clinical trials. Outdated and incorrect forecasts may result from an inability to adjust to fresh facts (Tan et al. 2023). The model's ability to account for individual variability is hindered by dataset bias toward average responses, particularly in fields like cancer. It might be challenging for researchers and physicians to comprehend predictions or convert them into useful insights since complex models frequently lack interpretability. AI models' applicability in clinical research situations is further restricted by the technological know-how needed to employ them. To overcome these constraints, better data quality, continuous model improvement, and increased model explainability are expected (Jena et al. 2024).

10 Conclusion and Future Aspects

The design, development, and application of nanotechnology in healthcare are being completely transformed by the combination of AI and ML in nanomedicine. Precision medicine is made possible by these technologies that allow data-driven improvements in medication delivery, biological interactions, self-assembly, and nanomaterial production. AI makes it easier to create safer, more effective nanomedicines with regulated pharmacokinetics and therapeutic activities by utilizing intelligent platforms, real-time tuning, and predictive algorithms. The future is extremely promising, despite obstacles like restricted data availability, interpretability of models, and regulatory barriers. Drug distribution and therapy are about to be revolutionized by innovations like digital twin technology, in silico preclinical trials, and inverse nanomaterial design. A revolution in nanomedicine is about to begin as AI-driven methods stand to improve patient outcomes, tailor treatments, and combat multi-drug resistance. AI will turn healthcare into an immensely individualized and efficient field with sustained cooperation, ethical guidance, and technical advancement.

References

Abdalla Y, McCoubrey LE, Ferraro F, Sonnleitner LM, Guinet Y, Siepmann F, et al. Machine learning of Raman spectra predicts drug release from polysaccharide coatings for targeted colonic delivery. J Control Release. 2024;374:103–11.

Aghajanpour S, Amiriara H, Esfandyari-Manesh M, Ebrahimnejad P, Jeelani H, Henschel A, et al. Utilizing machine learning for predicting drug release from polymeric drug delivery systems. Comput Biol Med. 2025;188:109756.

Alqarni AA, Alqarni M, Felemban MF, Algahtani FS, Alzubaidi MA, Shukr BS. Advanced hybrid numerical-machine learning computational study on fluid flow modeling in magnetic nanocarriers for targeted drug delivery. Case Stud Therm Eng. 2024;59:104497.

Ambreen S, Umar M, Noor A, Jain H, Ali R. Advanced AI and ML frameworks for transforming drug discovery and optimization: with innovative insights in polypharmacology, drug repurposing, combination therapy and nanomedicine. Eur J Med Chem. 2025;284:117164.

Calmet H, Dosimont D, Oks D, Houzeaux G, Almirall BV, Inthavong K. Machine learning and sensitivity analysis for predicting nasal drug delivery for targeted deposition. Int J Pharm. 2023;642:123098.

Chen C. Artificial intelligence aided pharmaceutical engineering: development of hybrid machine learning models for prediction of nanomedicine solubility in supercritical solvent. J Mol Liq. 2024;397:124127.

Chou WC, Chen Q, Yuan L, Cheng YH, He C, Monteiro-Riviere NA, et al. An artificial intelligence-assisted physiologically-based pharmacokinetic model to predict nanoparticle delivery to tumors in mice. J Control Release. 2023;361:53–63.

Das KP, Chandra J. Nanoparticles and convergence of artificial intelligence for targeted drug delivery for cancer therapy: current progress and challenges. Front Med Technol. 2022;4:1067144.

Di Francesco V, Boso DP, Moore TL, Schrefler BA, Decuzzi P. Machine learning instructed microfluidic synthesis of curcumin-loaded liposomes. Biomed Microdevices. 2023;25(3):29.

Eugster R, Orsi M, Buttitta G, Serafini N, Tiboni M, Casettari L, et al. Leveraging machine learning to streamline the development of liposomal drug delivery systems. J Control Release. 2024;376:1025–38.

Furxhi I, Murphy F, Poland CA, Sheehan B, Mullins M, Mantecca P. Application of Bayesian networks in determining nanoparticle-induced cellular outcomes using transcriptomics. Nanotoxicology. 2019;13(6):827–48.

Gentili PL, Stano P. Living cells and biological mechanisms as prototypes for developing chemical artificial intelligence. Biochem Biophys Res Commun. 2024;720:150060.

Habeeb M, You WH, Aher KB, Bhavar GB, Pawar SS, Gaikwad SD. Artificial neural networks for the prediction of mechanical properties of CGNP/PLGA nanocomposites. Mater Today Proc. 2023; Article S2214785323046345. https://doi.org/10.1016/j.matpr.2023.08.354.

Habeeb M., You HW, Umapathi M, Ravikumar KK, Mishra S. Strategies of Artificial intelligence tools in the domain of nanomedicine. J Drug Deliv Sci Technol, 2024;91:105157.

Habeeb M, Vengateswaran HT, You HW, Aher KB, Bhavar GB, Gaikwad SD. Development and characterization of carboxylated copper oxide conjugated polymeric nanocomposites and correlating with computational techniques. Results Surf Interfaces. 2024;17:100323.

Hani U, et al. Computational intelligence modeling of nanomedicine preparation using advanced processing: solubility of fludrocortisone acetate in supercritical carbon dioxide. Case Stud Therm Eng. 2023;45:102968.

He S, Abarrategi JS, Bediaga H, Arrasate S, González-Díaz H. On the additive artificial intelligence-based discovery of nanoparticle neurodegenerative disease drug delivery systems. Beilstein J Nanotechnol. 2024;15:535–55.

Jena GK, Patra CN, Jammula S, Rana R, Chand S. Artificial intelligence and machine learning implemented drug delivery systems: a paradigm shift in the pharmaceutical industry. J BioX Res. 2024:7.

Karami L, Fayyazi N, Akbari M, Molaei-Baghbanan A, Maaza M. A new insight into the mechanism of loading of Granisetron, naltrexone and risperidone in poly(ortho ester) and poly

(lactic-co-glycolic acid) as the controlled release drug delivery systems using a computational molecular approach. J Drug Deliv Sci Technol. 2024;102:106377.
Mahdi WA, Alhowyan A, Obaidullah AJ. Utilization of artificial intelligence for evaluation of targeted cancer therapy via drug nanoparticles to estimate delivery efficiency to various sites. Chemom Intell Lab Syst. 2025;257:105309.
Mi K, Chou WC, Chen Q, Yuan L, Kamineni VN, Kuchimanchi Y, et al. Predicting tissue distribution and tumor delivery of nanoparticles in mice using machine learning models. J Control Release. 2024;374:219–29.
Myreddy P, Pragathi Y, Narasimham MVSP. RNN-controlled neuro-nanorobots for enhanced drug delivery. J Electr Syst. 2024;20(10s):6503–19.
Nagalakshmi R, Khan SB, Mahesh TR, Alojail M, Sangwan SR, Saraee M. Enhancing drug discovery and patient care through advanced analytics with the power of NLP and machine learning in pharmaceutical data interpretation. SLAS Technol. 2025;31:100238.
Obrezanova O. Artificial intelligence for compound pharmacokinetics prediction. Curr Opin Struct Biol. 2023;79:102546.
Ori MO, et al. Integration of artificial intelligence in Nanomedicine. Eurasian. J Sci Technol. 2024;4(2):88–104.
Patel H. Applications of Deep reinforcement learning in MEMS and nanotechnology. Qeios. 2023. Available from: https://www.qeios.com/read/POMBWN
Pednekar K, Minnee J, de Vries IJM, Prakash J. Targeted nanomedicine for reprogramming the tumor innate immune system: from bench to bedside. Eur J Pharm Biopharm. 2024;204:114510.
Peng S, Guo Y, Wang J, Wang Y, Zhang W, Zhou X, et al. The coagulation-precipitation turbidity prediction model for precision drug delivery system based on deep learning and machine vision. J Environ Chem Eng. 2024;12(2):112211.
Piroozmand F, Mohammadipanah F, Sajedi H. Artificial neural network (ANN) in drug delivery. In: A handbook of artificial intelligence in drug delivery; 2023. p. 97–122.
Rampado R, Peer D. Design of experiments in the optimization of nanoparticle-based drug delivery systems. J Control Release. 2023;358:398–419.
Singh AV, Chandrasekar V, Janapareddy P, Mathews DE, Laux P, Luch A, et al. Emerging application of nanorobotics and artificial intelligence to cross the BBB: advances in design, controlled maneuvering, and targeting of the barriers. ACS Chem Neurosci. 2021;12(11):1835–53.
Sumayli A, Mahdi WA, Alamoudi JA. Analysis of nanomedicine production via green processing: modeling and simulation of pharmaceutical solubility using artificial intelligence. Case Stud Therm Eng. 2023;51:103587.
Suriyaamporn P, Pamornpathomkul B, Wongprayoon P, Rojanarata T, Ngawhirunpat T, Opanasopit P. The artificial intelligence and design of experiment assisted in the development of progesterone-loaded solid-lipid nanoparticles for transdermal drug delivery. Pharmacia. 2024;71:1–12.
Tan P, Chen X, Zhang H, Wei Q, Luo K. Artificial intelligence aids in development of nanomedicines for cancer management. Semin Cancer Biol. 2023;89:61–75.
Urista DV, Carrué DB, Otero I, Arrasate S, Quevedo-Tumailli VF, Gestal M, et al. Prediction of antimalarial drug-decorated nanoparticle delivery systems with random forest models. Biology (Basel). 2020;9(8):198.
Villa Nova M, Lin TP, Shanehsazzadeh S, Jain K, Ng SCY, Wacker R, et al. Nanomedicine ex Machina: between model-informed development and artificial intelligence. Front Digit Health. 2022;4:799341.
Vora LK, Gholap AD, Jetha K, Thakur RRS, Solanki HK, Chavda VP. Artificial intelligence in pharmaceutical technology and drug delivery design. Pharmaceutics. 2023;15(7):1916.
Yaari Z, Da Silva D, Zinger A, Goldman E, Kajal A, Tshuva R, et al. Theranostic barcoded nanoparticles for personalized cancer medicine. Nat Commun. 2016;7(1):1–10.

Artificial Intelligence-Driven and In Silico Approaches in Health Emergencies: A Case Study on Antiviral Drug Discovery

Shuo Wang, Feiyue Ma, Brijesh Rathi, and Peng Zhan

Abstract Artificial intelligence (AI) has greatly accelerated the process of identifying and optimizing lead compounds, particularly in the face of escalating global viral threats. Researchers can use AI methods to quickly screen multimillion-compound libraries while simultaneously facilitating the de novo design of novel antiviral agents with enhanced efficacy. The success of the AI-driven approaches toward the identification of potent antiviral compounds highlights their importance in surmounting emerging viral challenges through a more cost-efficient pathway. By examining these advancements, this article provides case studies that demonstrate how AI-driven approaches can revolutionize antiviral drug development, offering a more efficient pathway to combat emerging viral challenges.

Keywords Artificial intelligence · Machine learning · Deep learning · Antiviral agents · Medicinal chemistry

1 Introduction

Identification and optimization of a lead compound to a candidate represents one of the most resource-intensive parts of the whole process of drug development. High-throughput screening (HTS) and combinatorial synthesis were once commonly used for lead compound discovery; these methods are resource-intensive and time-

S. Wang · F. Ma · P. Zhan (✉)
Department of Medicinal Chemistry, Key Laboratory of Chemical Biology (Ministry of Education), School of Pharmaceutical Sciences, Cheeloo College of Medicine, Shandong University, Shandong, People's Republic of China
e-mail: 202116762@mail.sdu.edu.cn; 202416760@mail.sdu.edu.cn; zhanpeng1982@sdu.edu.cn

B. Rathi (✉)
Laboratory for Translational Chemistry and Drug Discovery, Department of Chemistry, Hansraj College, University of Delhi, New Delhi, Delhi, India
e-mail: brijeshrathi@hrc.du.ac.in

A. Lavecchia (ed.), *Applied Artificial Intelligence for Drug Discovery*,
https://doi.org/10.1007/978-3-031-98022-0_20

consuming, but they have generated extensive datasets that represent a valuable resource (Macarron et al. 2011). However, the inherent complications of large-scale data repositories, such as noise interference and multi-source heterogeneity, make it difficult for researchers to use these data effectively. The capacity of AI to analyze vast datasets makes it an invaluable tool for developing models to identify lead compounds (LeCun et al. 2015). A notable example is the swift identification of DDR1 kinase inhibitors. Through applying deep learning (DL) techniques, a novel lead compound was successfully discovered within a remarkably short span of just 21 days (Zhavoronkov et al. 2019). This case shows that AI can speed up lead compound discovery and optimization, potentially improving the efficiency of traditional drug development pipelines.

In recent years, viral infections have continued to pose a serious threat to human health. For instance, the COVID-19 pandemic was caused by severe acute respiratory syndrome coronavirus 2 (SARS-CoV-2), which spread at an unprecedented rate globally in 2019 (Viana et al. 2022). In addition to SARS-CoV-2, other viruses such as Ebola virus (EBOV), Zika virus (ZIKV), and dengue virus (DENV) also cause incalculable losses in human beings (Ma et al. 2021; Yang et al. 2024; Gao et al. 2023). The rapid identification of lead compounds in antiviral drug development is critical during pandemics (Petersen et al. 2020). Additionally, iterative optimization of lead compounds is critical for overcoming drug resistance, frequently rendering existing therapies ineffective (De Clercq and Li 2016). AI-driven approaches can potentially discover novel chemical scaffolds with distinct resistance profiles, thereby overcoming the limitations associated with current antiviral therapies (Stokes et al. 2020).

Given these considerations, this article focuses on the application of AI in antiviral drug discovery, with particular emphasis on identifying potential lead compounds and their subsequent optimization. By leveraging AI, we would like to demonstrate its potential to revolutionize antiviral drug development, addressing both the immediate needs of emerging infections and the long-term challenges of drug resistance.

2 Application of Artificial Intelligence in Lead Discovery

In the lead discovery, screening and generative models are the two main AI approaches employed. Screening models allow for the rapid and cost-effective identification of promising lead compounds from large compound libraries, and generative models are intended to produce structurally novel compounds with significantly higher hit rates than traditional methods. These models have transformed the early stages of drug discovery, resulting in unprecedented efficiency and innovation (Bian et al. 2019).

2.1 Lead Discovery Based on Screening Models

The screening models help in the identification of potential lead compounds by analyzing interactions with specific molecular targets. The models are divided into classification and regression models based on the output. Classification models predict whether a compound would positively interact with a target, while regression models quantify such interaction (Shams et al. 2025). Most of the data for training these models was obtained from publicly available databases such as PDB, PDBbind, DrugBank, BindingDB (Wishart et al. 2018), and ChEMBL (Gaulton et al. 2017). In-depth information was included about compounds, proteins, and their interactions. Feature extraction was done on the gathered data to train a unique model for subsequent lead discovery.

2.1.1 Lead Discovery Based on Classification Models

Classification models, a type of machine learning (ML) algorithms, classify data into predefined categories. Common algorithms include decision trees, support vector machines (SVM), logistic regression, and deep learning models (Vamathevan et al. 2019; Zhu et al. 2024; Serafim et al. 2020). They have proven successful in precisely and recall-effectively identifying antiviral compounds, reducing time and cost in experimental screening.

In 2021, Huang et al. developed a biological activity-based model (BABM) to identify antiviral compounds against SARS-CoV-2 (Huang et al. 2021). The BABM approach involved several key steps. Firstly, biological activity data for compounds were collected from various sources, including internal screening libraries like the NCATS pharmaceutical collection (NPC) and the library of pharmacologically active compounds (LOPAC), as well as from the published literature. These datasets, often obtained through quantitative high-throughput screening (qHTS), provided comprehensive activity profiles for compounds across multiple assays. Secondly, the activity data were processed into activity signatures, which were binary vectors indicating the presence or absence of activity in specific assays. These signatures served as input features for ML models.

Thirdly, ML algorithms such as weighted feature significance (WFS), random forest (RF), or SVM were used to train models that could predict the activity of compounds against new targets. The models were trained on the activity signatures and optimized to recognize patterns associated with target activity. Fourthly, the trained models were validated using various methodologies. This included splitting the data into training and test sets, employing cross-validation techniques, and calculating metrics such as the area under the receiver operating characteristic curve (AUC-ROC) and positive predictive value (PPV). Experimental validation was also conducted by testing predicted compounds in live virus assays to confirm their activity. Finally, the validation process yielded significant results, with models achieving high AUC-ROC values and PPV rates, demonstrating their ability to

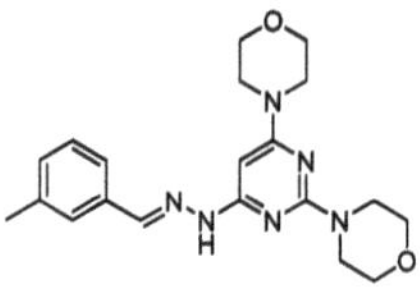

1, MLS000699212-03
EC_{50} = 500 nM

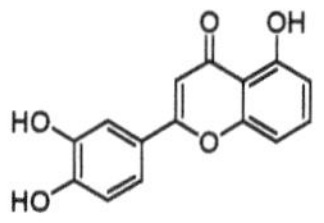

2, 5,3',4'-trihydroxyflavone
IC_{50} = 87.76 μM (M^{pro})
EC_{50} = 8.22 μM

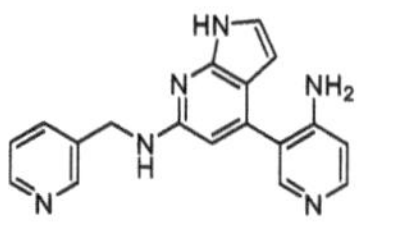

3, LabMol-301
IC_{50} = 0.8 μM (NS5 RdRp)
IC_{50} = 7.4 μM (NS2B-NS3)
EC_{50} = 6.68 μM

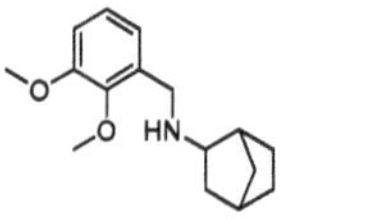

4, LabMol-212
K_d = 35 μM (NS3 helicase)
EC_{50} = 0.13 μM

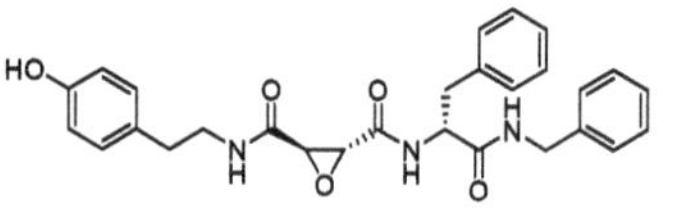

5,CAA-0225
EC_{50} = 0.20 μM

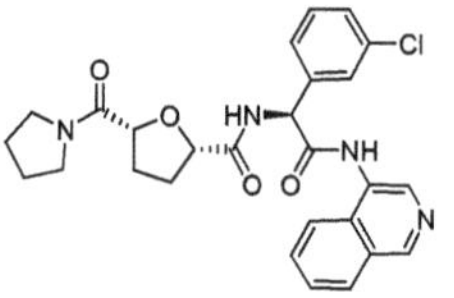

6
IC_{50} = 0.34 μM (M^{pro})
EC_{50} = 0.12 μM
CC_{50} > 10 μM

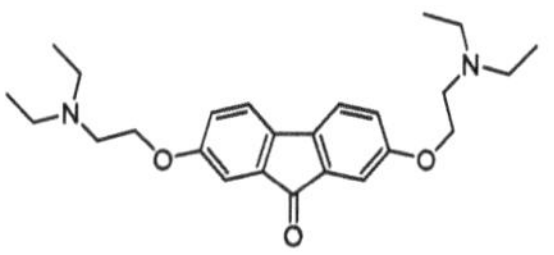

7, Tilorone
EC_{50}(predicted) = 1.95 μM

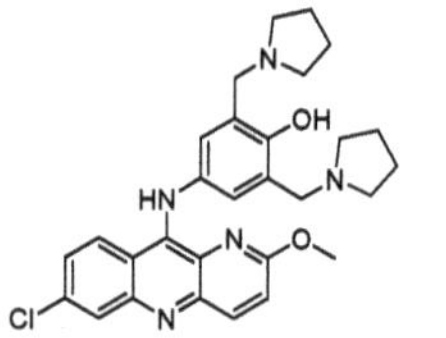

8, Pyronaridine
EC_{50}(predicted) = 0.5 μM

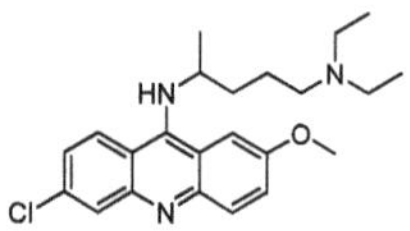

9, Quinacrine
EC_{50}(predicted) = 0.002 μM

Fig. 1 The structures of compounds **1–9**

accurately predict antiviral activity. The BABM method, which predicted compound activity against new targets by analyzing their activity patterns across multiple assays, did not rely on chemical structure information and used ML to identify potential antiviral leads.

The researchers used this model to screen a sizable commercial library of compounds (over 500,000 compounds). The model predicted 311 compounds with potential antiviral activity against SARS-CoV-2. Compounds with the highest predicted activity scores were selected for experimental validation. Benzaldehyde derivative, **MLS000699212–03** (**1**, Fig. 1), was found to be a potent inhibitor of SARS-CoV-2 with a half-maximal effective concentration (EC_{50}) of 500 nM in the cytopathic effect (CPE) assay and a dual mechanism of inhibiting viral entry and modulating autophagy. These findings highlight the potential of BABM to rapidly identify promising antiviral candidates, significantly accelerating the drug discovery pipeline. At the same time, this research showed that the screening model could significantly improve the efficiency and success rate of lead compound identification in antiviral drug discovery.

In 2021, Zhao et al. compiled training and test sets of SARS-CoV-2 M^{pro} ligands, optimized molecular descriptors, characterized molecular structures with fingerprints, analyzed chemical space diversity, and established and validated classification models using Naive Bayes (NB) and Random Projection (RP) algorithms (Zhao et al. 2021). Specifically, this study systematically compiled a comprehensive dataset of active M^{pro} ligands and generated a corresponding set of inactive compounds to form robust training and test sets. Molecular descriptors were calculated and subjected to rigorous optimization to select the most informative and relevant features. Molecular fingerprints, including ECFP_4, ECFP_6, and MACCS keys, were employed to characterize molecular structures with high precision. The chemical space diversity of the compounds was analyzed using principal component analysis (PCA) and Tanimoto similarity analysis to ensure structural diversity and coverage. Finally, classification models were constructed using NB and RP algorithms based on the optimized descriptors and fingerprints. The performance of these models was thoroughly evaluated through cross-validation and independent test set validation, ensuring their reliability and predictive power.

The researchers utilized this classification model to screen a database of 5766 natural compounds for potential SARS-CoV-2 M^{pro} inhibitors. After virtual screening, they identified a promising inhibitor named 5,3′,4′-trihydroxyflavone, which exhibited a half-maximal inhibitory concentration (IC_{50}) of 87.76 μM against M^{pro} and an EC_{50} of 8.22 μM in the CPE assay. Subsequent molecular modeling revealed that **5,3′,4′-trihydroxyflavone** (**2**, Fig. 1) forms hydrogen bonds with key residues His163, Phe140, Glu166, and Ser144 in the active site of M^{pro}, along with π-alkyl interactions with Met165 and a π–π interaction with His41. This study underlines the utility of ML in accelerating the discovery of antiviral agents and highlights the potential of natural products in combating SARS-CoV-2.

In 2022, Mottin et al. used ML models and molecular docking to identify small-molecule inhibitors targeting ZIKV (Mottin et al. 2022). The ML models developed in this study were integral to identifying compounds with cytoprotective effects against ZIKV infection. The models were trained on data from the PubChem Bioassay database, which provided 5414 molecules tested for their impact on caspase-3 activity in human progenitor cells (hNPCs). These molecules were categorized into active ($EC_{50} \leq 10$ μM) and inactive ($EC_{50} > 10$ μM) groups based on their EC_{50}. Twelve classification models were built using Python v.3.6, combining SVM and RF algorithms with various fingerprints and hybrid descriptors. The best-performing model that used the Mordred_FCFP4 hybrid descriptors and the RF algorithm, achieved an accuracy of 0.85 and an AUC of 0.80. This model was then integrated into the virtual screening workflow to prioritize compounds from the ZINC15 database that were predicted to have cytoprotective effects. The selected compounds were further evaluated for blood–brain barrier (BBB) permeability using a Bayesian ML model from Assay Central software, and were subjected to additional filters for LogP, pan-assay interference compounds (PAINS), and potential aggregators. A medicinal chemistry-based inspection of the compounds' structures and binding modes ensured that only the most promising candidates advanced to experimental validation. In addition, virtual screening through molecular docking was performed to screen over 36 million compounds.

Several promising compounds were identified, including **LabMol-301** (**3**, Fig. 1) and **LabMol-212** (**4**, Fig. 1), which possessed outstanding properties. **LabMol-301** showed dual inhibitory action against both ZIKV NS5 RdRp ($IC_{50} = 0.8\ \mu M$) and NS2B-NS3 protease ($IC_{50} = 7.4\ \mu M$). **LabMol-212** thermally destabilized the NS3 helicase with a K_d of 35 μM and was highly cytoprotective, with an EC_{50} of 0.13 μM, without significant cytotoxicity. These compounds also provided new scaffolds with unique modes of action compared to previously reported inhibitors. The approaches developed to identify discoveries such as LabMol-301 brought to light that AI was able to identify multi-target molecules with broad-spectrum antiviral activities and presented a promising method for the development of an antiviral drug. The effectiveness of the ML models in this study underscores the potential of computational methods in drug discovery for emerging viruses, particularly in identifying compounds with promising biological activity.

In 2022, Xu et al. developed classification models using two classical approaches known as quantitative structure-activity relationship (QSAR) and BABM that integrated RF and SVM algorithms as computational predictors to screen SARS-CoV-2 inhibitors (Xu et al. 2022). QSAR models involved converting compound structures into ECFP_4 and applying various ML algorithms to predict compound activity based on structural features. Feature selection and data rebalancing strategies were utilized to optimize model performance, resulting in robust predictive models for both viral entry and M^{pro} inhibitors. BABM models, on the other hand, leveraged compound activity profiles from high-throughput screening assays to predict potential inhibitors. By integrating biological activity data with structural information, BABM models demonstrated enhanced predictive power, outperforming QSAR models in certain aspects. The combination of these two approaches provided a comprehensive strategy to efficiently screen and identify promising anti-SARS-CoV-2 compounds from a large chemical library. The models achieved AUC values of 0.78 and 0.88 for viral entry and M^{pro} inhibitors, respectively.

The trained models were then used to conduct virtual screening. Among the identified compounds, **CAA-0225** (**5**, Fig. 1), an inhibitor of cathepsin L, potently inhibited SARS-CoV-2 with an EC_{50} of 0.20 μM by targeting the entry process of viruses in the CPE assay, suggesting its potential as a lead candidate against COVID-19. In conclusion, the study demonstrates that ML models can significantly enhance the speed and efficiency of anti-SARS-CoV-2 drug discovery by expanding compound screening capacities and improving hit rates. The identified compounds provide promising leads for further development of COVID-19 therapeutics.

In 2023, Saar et al. presented an ML framework to predict protein-ligand affinity using high-throughput crystallography data with a focus on SARS-CoV-2 M^{pro} (Saar et al. 2023). They utilized a classification model based on the physics-based energy descriptors, including hydrophobic, electrostatic, and hydrogen bonding interactions that describe protein-ligand complexes. The method involved collecting experimental structures of protein-ligand complexes and their biochemical measurements, extracting physics-based energy descriptors using the Open Drug Discovery

Toolkit (ODDT), and training a machine learning model to predict relative binding affinities. The model was validated using a scaffold-split approach and demonstrated superior performance compared to ligand-based and docking-based methods. The study also explored the impact of pre-training with larger datasets and the temporal robustness of the model's predictive power, revealing consistent accuracy even with limited structural data. Finally, the model guided the synthesis of a virtual library of compounds, leading to the discovery of potent inhibitors with significantly improved antiviral efficacy.

Using the model constructed for virtual screening non-covalent compounds, the most potent lead compound **6** (Fig. 1) showed an IC_{50} of 0.34 μM against M^{pro} with an EC_{50} of 120 nM in the CPE assay and no cytotoxicity at 10 μM. Compound **6** also exhibited promising in vitro ADME properties, including low plasma protein binding and excellent metabolic stability, positioning it as a promising candidate for oral antiviral development. The approach was successfully applied to design and synthesize potent inhibitors of the SARS-CoV-2 Mpro, achieving significant improvements in antiviral efficacy.

By leveraging different classification models, molecular docking, and classical QSAR methods, researchers are able to rapidly screen large compound libraries and predict their activity, thus accelerating the drug discovery process.

2.1.2 Lead Discovery Based on Regression Models

Regression models, alongside classification models, are critical tools for predicting small molecule bioactivity. Unlike classification models that predict binary target interactions, regression models quantify interaction strength, offering a nuanced understanding of compound-target engagement. This capability significantly enhances AI-driven predictions of antiviral compound efficacy by providing continuous affinity metrics. The following examples demonstrate the successful application of regression models in antiviral drug discovery.

In 2021, Rajput et al. developed an ML-based algorithm named anti-Ebola for predicting potential inhibitors against EBOV through QSAR models (Rajput and Kumar 2022). The algorithm was constructed using regression models based on 305 unique anti-Ebola compounds with experimentally determined IC_{50}/EC_{50} values sourced from the DrugRepV database. These compounds were selected for their antiviral activities expressed in terms of IC_{50}/EC_{50}, which were converted to pIC_{50} values for the development of regression-based models. The chemical descriptors and fingerprints of these compounds were calculated using the PaDEL software, which provides a comprehensive set of molecular descriptors. A feature selection process was employed to reduce the dimensionality of the data, narrowing down 17,968 descriptors to 50 key features that were most relevant for model development. Three machine learning models (SVM, RF, and artificial neural network (ANN)) were trained and validated using tenfold cross-validation to ensure robustness and generalizability. The performance of these models was quantitatively assessed using metrics such as Pearson's correlation coefficient (PCC), mean absolute error (MAE), and root mean square error (RMSE). The applicability domain of the models was further evaluated using William's plot to confirm their reliability. Additionally, the chemical diversity of the anti-Ebola compounds was analyzed using multidimensional scaling (MDS) and chemical dendrograms to explore the structural variety within the dataset.

The best-performing model was subsequently utilized for virtual screening. Among the predicted compounds, tilorone (**7**, Fig. 1), pyronaridine (**8**, Fig. 1), and quinacrine (**9**, Fig. 1) showed promising anti-EBOV activity, with predicted EC_{50} values of 1.95 μM, 0.50 μM, and 0.002 μM, respectively. Integration of these models offers a powerful tool for accelerating EBOV inhibitor discovery. The study highlights the potential of regression models in accelerating antiviral drug development, offering a promising approach to rapidly identify effective drug candidates. Although experimental validation remains essential to confirm the accuracy of these predictions, the integration of regression models provides researchers with valuable insights that can guide decision-making, especially in the face of emerging infectious diseases.

2.2 *Lead Discovery Based on Generative Models*

The discovery of lead compounds through screening models often relies on commercially available compounds, which can be limited in terms of structural novelty and may pose risks of patent infringement. In contrast, the generative models offer a significant advantage by enabling the design of novel molecular structures, thus expanding the chemical space and bypassing existing patent barriers. AI-driven models, such as variational autoencoders (VAEs), generative adversarial networks (GANs), and reinforcement learning (RL) approaches, can generate diverse and synthetically accessible molecules with desired pharmacological properties (Gangwal and Lavecchia 2024). By utilizing generative models, researchers could accelerate the development of antiviral drugs with enhanced efficacy (Zhavoronkov et al. 2019; Stokes et al. 2020). Using these advanced computational techniques, the drug discovery process can be made more efficient and robust, particularly in the search for novel antiviral agents (Gómez-Bombarelli et al. 2018; Ragoza et al. 2017).

In 2023, Mao et al. introduced TransAntivirus, a groundbreaking Transformer-based molecular generative model tailored for antiviral drug design (Mao et al. 2024). This framework employed a self-supervised pre-training strategy coupled with task-specific fine-tuning, achieving superior performance in generating novel and diverse molecular scaffolds compared to conventional methods. By integrating IUPAC and SMILES representations, the model effectively utilized the unique advantages of both systems to generate molecules with desired properties. The data preparation process was rigorous, involving meticulous filtering and standardization of molecules sourced from PubChem, which culminated in a robust training set of 30 million molecules. The model's architecture was based on the T5 Transformer, with specialized adaptations engineered to efficiently handle the intricate mappings

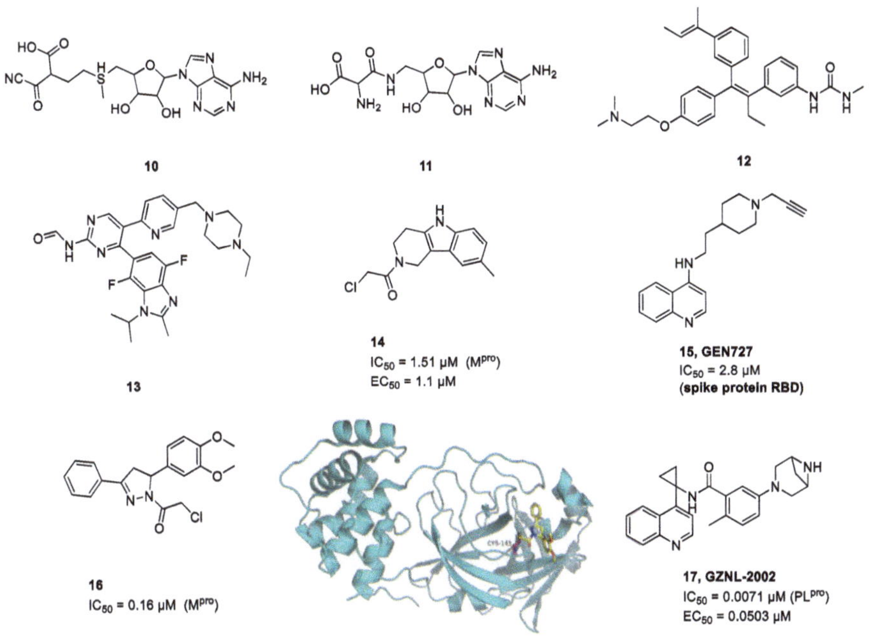

Fig. 2 Structures of compounds 9–17, and the crystal structure of the SARS-CoV-2 M^{pro} with compound 16 (PDB: 8DLB). (The figures are generated by PyMOL www.pymol.org)

between IUPAC and SMILES. TransAntivirus underwent extensive pre-training on a large dataset and was further refined through fine-tuning on specific antiviral tasks. Its performance was rigorously evaluated using key metrics such as validity, uniqueness, novelty, and diversity, where it significantly outperformed other generative models. Furthermore, comprehensive comparisons of chemical property distributions and in-depth chemical space analysis have confirmed the model's remarkable capability to generate molecules with optimized properties.

The researchers utilized this model to generate a series of compounds targeting SARS-CoV-2 RNA-dependent RNA polymerase (RdRp) and M^{pro}. Notably, two nucleoside analogues (compounds **10** and **11**) (Fig. 2) were predicted to have strong interactions with RdRp, and non-nucleoside analogues (compounds **12** and **13**) (Fig. 2) were predicted to display promising binding to M^{pro}. However, the lack of experimental validation and biological context limits its immediate applicability in antiviral drug discovery. Future work should focus on experimental validation of generated compounds, exploration of additional viral targets, and assessment of toxicity and pharmacokinetic properties.

In 2023, Saramago et al. developed an AI-driven approach to identify potential inhibitors of the SARS-CoV-2 M^{pro} using a generative model based on a recurrent neural network (Saramago et al. 2023). The model was initially trained on a dataset of over 1.6 million unique SMILES strings from the ChEMBL database, which provided a broad chemical space for learning. Subsequently, it was fine-tuned using a dataset of 289,569 compounds with known activity against SARS-CoV-1 M^{pro}, enhancing its ability to predict bioactivity against SARS-CoV-2 M^{pro}. This fine-tuned model generated 50,000 SMILES strings, which were then filtered using the "Rule of Three" criteria and predicted water solubility, resulting in 37 potential Mpro inhibitors. These compounds were further prioritized for molecular docking simulations to predict their binding modes and affinities with Mpro. Ultimately, 34 compounds were tested experimentally, five compounds showed significant inhibitory activity, with compound **14** (Fig. 2) emerging as the most potent, exhibiting an IC_{50} of 1.5 μM against M^{pro} and an EC_{50} of 1.1 μM in inhibiting SARS-CoV-2 replication in the CPE assay. Compound **14**, a reversible inhibitor with a chloromethyl amide derivative of 8-methyl-gamma-carboline, demonstrated low cytotoxicity and high binding efficiency, making it a promising lead for further development.

In 2023, Chenthamarakshan et al. made a significant advancement in the field of antiviral drug discovery by leveraging a deep generative model to expedite the identification of inhibitors targeting SARS-CoV-2 (Chenthamarakshan et al. 2023). The study utilized a VAE-based framework, CogMol, which was pre-trained on extensive datasets encompassing protein sequences, small molecules, and protein-ligand binding affinities. This foundation model enabled the generation of novel small-molecule inhibitors through an attribute-conditioned sampling approach, guided by molecular property predictors assessing drug-likeness and target binding potential. The model's utility was demonstrated by designing inhibitors for two critical SARS-CoV-2 targets: the spike protein receptor-binding domain (RBD) and the M^{pro}. The study employed a systematic approach to evaluate the synthetic feasibility of generated small molecules through retrosynthesis prediction, which utilized the IBM RXN platform to generate and optimize plausible synthetic routes for prioritized candidates, enhancing the efficiency and success rate of the drug discovery process.

Subsequently, the generated small molecules were synthesized and validated experimentally, highlighting the model's efficacy in producing promising antiviral candidates with potential broad-spectrum activity against emerging viral threats. Among these compounds, **GEN727** (**15**, Fig. 2), which targets the spike RBD, exhibited micromolar-level inhibition with an IC_{50} of 2.8 μM and demonstrated broad-spectrum antiviral activity against multiple SARS-CoV-2 variants, including Alpha, Beta, Delta, and Omicron. The ability of this compound to target a conserved lipid-binding pocket in the RBD suggests its potential as a broad-spectrum antiviral agent, particularly given its distinct mechanism compared to monoclonal antibodies. The study highlights the efficiency of AI-driven drug discovery, achieving a 50% hit rate with minimal experimental screening, and underscores the potential of generative models in addressing emerging viral threats.

In 2023, Joshi et al. introduced an AI-driven pipeline designed to discover covalent inhibitors targeting the SARS-CoV-2 M^{pro}, a crucial enzyme for viral replication (Joshi et al. 2023). This innovative approach hinges on two core components: the automated Modeling Engine (AME) and the 3D-Scaffold model. The

AME streamlines the covalent docking process by establishing bonds between electrophilic warheads and the Cys145 in the active site of M^{pro}. Utilizing AutoDockFR, it predicts binding poses and scores, subsequently ranking candidates according to these scores. Complementing this, the 3D-Scaffold model generates three-dimensional molecular structures from specified scaffolds, facilitating the creation of novel covalent candidates. Trained on drug-like molecules from the FDA and Enamine databases, this deep generative model produces various molecules with desired scaffolds and warheads.

By integrating these tools, the team achieved rapid design and high-throughput screening of covalent inhibitors, significantly streamlining the discovery process. Experimental validation of the top candidates led to the identification of four chloroacetamide-based inhibitors, compound **16** (Fig. 2) showing the most promising results, exhibiting a low IC_{50} of 160 nM. The binding mode was further confirmed by X-ray crystallography that revealed stable interactions with key residues. Despite its potent antiviral activity, compound **16** exhibited significant cytotoxicity, highlighting the need for further optimization to reduce off-target effects. The study demonstrated the efficacy of this AI-driven approach in accelerating the discovery of targeted covalent ligands for SARS-CoV-2 M^{pro}.

In 2024, Shang et al. harnessed the Tree-Invent model, a topology-constrained molecular generative model, to discover novel inhibitors of the papain-like protease (PL^{pro}) of SARS-CoV-2 (Shang et al. 2024). This model employed a topological tree structure to guide molecular generation, enabling precise control over specific substructures while allowing flexible generation of other parts. The Tree-Invent model's architecture comprised three modules: node adder, ring generator, and node connector, which collaboratively decoded molecular structures in sequence. One of the key advantages of the Tree-Invent model was its ability to exert precise control over molecular generation by constraining specific parts of the molecule, such as head, linker, and tail groups. Simultaneously, it enabled flexible exploration of chemical space through reinforcement learning. In this process, reinforcement learning was guided by docking scores that served as rewards to generate molecules with high binding affinity to the PL^{pro} active site.

In this study, the model was employed to generate molecules by converting the 2D structure of GZNL-P4 into a topological tree and employing reinforcement learning to optimize the generation process. The model successfully generated 2817 unique molecules that were subsequently filtered and optimized to identify potent PL^{pro} inhibitors with enhanced binding affinity and improved liver metabolic stability. After a series of processes including selection, synthesis, and activity validation, compound **GZNL-2002** (**17**, Fig. 2) was ultimately identified as a potent inhibitor with an IC_{50} of 7.1 nM against PL^{pro} and EC_{50} values of 0.050 μM, 0.0911 μM, and 0.1345 μM against wild-type SARS-CoV-2, Omicron BA.5, and XBB.1 variants, respectively. **GZNL-2002** demonstrated excellent liver metabolic stability, low CYP, and hERG inhibition, and favorable oral pharmacokinetic properties, making it a promising candidate for further development as a COVID-19 therapeutic.

3 Application of Artificial Intelligence in Lead Optimization

Lead optimization, a crucial stage in drug discovery, has been greatly advanced by integrating AI techniques. AI-driven approaches allow for the systematic exploration of chemical space, enabling the optimization of lead compounds by predicting essential properties such as binding affinity, selectivity, and pharmacokinetics. AI models, trained on large datasets containing molecular structures and their corresponding biological activities, can propose chemical modifications that enhance potency while minimizing off-target effects, which can not only accelerate the optimization process but also reduce the reliance on extensive experimental screening, making the drug discovery process more efficient and cost-effective (Gautam et al. 2023).

In 2024, Xiang et al. employed two open-source models named DeepFrag and EMPIRE to optimize lead compound **HCK63** (**18**, Fig. 3) targeting the SARS-CoV-2 nucleocapsid protein (N protein) (Xiang et al. 2024). The EMPIRE model demonstrated proficiency in generating structurally diverse fragments. Nevertheless, the generated fragments were limited by the lack of biological target integration. Conversely, DeepFrag could generate fragments with a focus on the atoms of both the parent ligand and the protein receptor. Researchers utilized these two models to develop a diverse library of more than 10,000 small molecules, starting from lead compound **HCK63**. Following high-throughput virtual screening, compounds exhibiting binding affinities below −6.5 kcal/mol were prioritized for synthesis. Among the 44 synthesized compounds, compound **19** (Fig. 3) emerged as the most promising candidate. Surface plasmon resonance (SPR) and microscale thermophoresis (MST) confirmed its strong binding to the N protein. In vitro assays revealed that compound **19** effectively inhibited SARS-CoV-2 replication in the CPE assay, with an EC_{50} value of 11.3 μM and low cytotoxicity ($CC_{50} > 200$ μM). Furthermore, **19** disrupted the binding of the N protein to RNA, a critical step in viral replication, demonstrating its potential as a therapeutic agent. The study underscores the power of AI in optimizing lead compounds and highlights **19** as a promising candidate for further development.

In 2024, Garnsey et al. employed a ML-driven strategy to design and optimize compound libraries for discovering potent SARS-CoV-2 PL^{pro} inhibitors (Garnsey et al. 2024). Historical bioactivity data from ChEMBL were processed to extract congeneric series, which were subsequently fragmented into core and R-group components. These components were used to train a neural network classification model capable of predicting bioisosteric replacements for specific R-groups and cores. The trained model was then applied to monomers of parallel medicinal chemistry libraries, enabling rapid exploration of chemical space and identification of promising compounds. Each R-group was meticulously checked for synthetic feasibility and potential medicinal chemistry alerts, ensuring the new compounds did not introduce problematic functionalities.

The researchers began with the known inhibitor **GRL0617** (**20**, Fig. 3), characterized by modest antiviral potency ($EC_{50} = 68.2$ μM), and subsequently employed

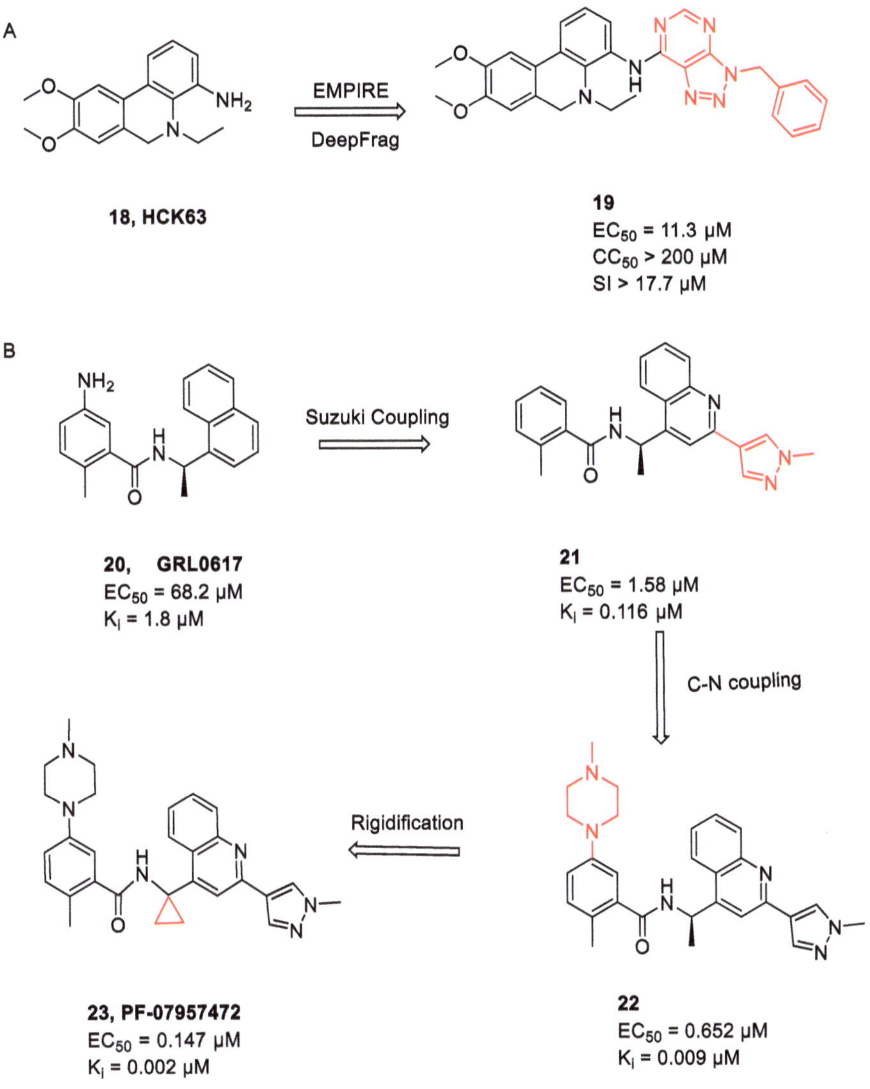

Fig. 3 The chemical revolution approaches from original molecules to promising candidates. (**a**) The discovery of compound **19**. (**b**) The discovery of **PF-07957472** (**23**)

AI-driven fragment generation based on reaction for structural optimization. They first focused on modifying the naphthalene ring system, employing a Suzuki coupling library designed with machine learning to explore diverse substituents. This led to the discovery of a quinoline-based derivative with a tenfold improvement in biochemical potency (K_i = 0.116 μM) (**21**, Fig. 3). Next, they optimized the aniline motif using C-N coupling libraries, identifying a methyl piperazine substituent that further enhanced potency (K_i = 0.009 μM) and EC_{50} = 0.652 μM

(**22**, Fig. 3). Finally, rigidification of the scaffold with a geminal-cyclopropyl group yielded the lead compound, **PF-07957472** (**23**, Fig. 3), which exhibited a threefold improvement with a K_i value of 0.002 μM and an EC_{50} of 0.147 μM in the CPE assay. The compound also showed favorable pharmacokinetic properties, including moderate oral bioavailability (41–69% across species) and low metabolic clearance and the in vivo efficacy of PF-07957472 was evaluated in a mouse-adapted SARS-CoV-2 infection model. The compound was administered orally at doses of 20, 50, and 150 mg/kg, and its antiviral activity was assessed by measuring lung viral titers and histopathology. The results demonstrated that PF-07957472 significantly reduced viral titers in the lungs, with the 50 and 150 mg/kg doses achieving reductions comparable to the positive control, Nirmatrelvir. Additionally, PF-07957472 protected mice from weight loss associated with infection, indicating its potential to mitigate infection-induced pathology. The study confirmed that PL^{pro} inhibition by PF-07957472 effectively reduced viral replication in vivo, thereby validating PL^{pro} as a viable antiviral target. These findings underscore the promise of the compound as a therapeutic agent for further development, highlighting its robust antiviral activity and potential clinical application.

This study exemplifies how AI-driven optimization can rapidly yield highly potent antiviral molecules with therapeutic potential. This systematic approach not only accelerated the drug discovery process but also ensured the synthetic accessibility and pharmacological safety of the compounds, leading to the identification of PF-07957472, a potent and selective PL^{pro} inhibitor with robust in vitro and in vivo antiviral activity.

4 Conclusion

The escalating global threat posed by viruses necessitates swift and effective responses in antiviral drug discovery. AI technologies have emerged as critical tools in addressing this challenge. This article has explored the transformative impact of AI-driven methods in identifying and optimizing antiviral agents, particularly during health emergencies (Turon et al. 2023; Zhang et al. 2025).

The ability of AI to efficiently mine vast compound datasets allows for the rapid identification of potential therapeutic agents with known safety profiles during outbreaks of emerging viral diseases (Huang et al. 2024). This is exemplified by the successful identification of ISM3312, a clinical candidate for the treatment of COVID-19 discovered using AI-driven generative models. Despite the undisclosed discovery and optimization process, the progression of ISM3312 toward clinical trials highlights the potential of AI to revolutionize antiviral drug development and offers hope for more rapid responses to emerging viral threats.

In addition to the antiviral compounds that have entered clinical trials, numerous small molecules have been discovered, particularly for emerging infectious diseases, including COVID-19, ZIKV, and DENV. The development processes of these small molecules are summarized in this paper (Table 1), with the aim of providing

Table 1 Summary of AI methodologies and case studies in antiviral drug discovery

AI model	Case example	Results	References
BABM	Huang et al developed BABM to identify antiviral compounds against SARS-CoV-2	This research identified **MLS000699212-03** as a SARS-CoV-2 entry inhibitor and autophagy modulator	Huang et al. (2021)
NB and RP models	Zhao et al. established NB and RP models to identify SARS-CoV-2 M^{pro} inhibitor	This study utilized the classification models to identify **5,3′,4′-trihydroxyflavone** as a SARS-CoV-2 M^{pro} inhibitor	Zhao et al. (2021)
SVM and RF models	Mottin et al utilized SVM and RF models to identify small molecule inhibitors targeting ZIKV	This research successfully prioritized compounds like **LabMol-301** and **LabMol-212**	Mottin et al. (2022)
QSAR and BABM models	Xu et al. identified SARS-CoV-2 entry and M^{pro} inhibitors using ML models with QSAR and BABM approaches	This study discovered **CAA-0225** as a potent SARS-CoV-2 entry inhibitor	Xu et al. (2022)
RF model	Saar et al. developed an RF model framework to predict protein-ligand affinity for SARS-CoV-2 M^{pro}	This model promoted the discovery of compound **6** as a SARS-CoV-2 M^{pro} inhibitor	Saar et al. (2023)
SVM, RF and ANN models	Rajput et al. developed SVM, RF, and ANN models for predicting potential inhibitors against EBOV	This research identified promising EBOV inhibitors like tilorone, pyronaridine, and quinacrine.	Rajput and Kumar (2022)
TransAntivirus model	Mao et al. introduced TransAntivirus model for targeting SARS-CoV-2 antiviral drug design	This model successfully generated compounds targeting SARS-CoV-2 RdRp and M^{pro}	Mao et al. (2024)
General chemical model	Saramago et al. developed a general chemical model to identify SARS-CoV-2 M^{pro} inhibitors	The model generated and identified five active inhibitors including compound **14**	Saramago et al. (2023)

(continued)

Table 1 (continued)

AI model	Case example	Results	References
CogMol model	Chenthamarakshan et al. employed CogMol to rapidly identify inhibitors targeting SARS-CoV-2 for the spike RBD and M^{pro}	This study discovered **GEN727**, which exhibited broad-spectrum antiviral activity against multiple SARS-CoV-2 variants	Chenthamarakshan et al. (2023)
AME and 3D-scaffold models	Joshi et al. developed AME and 3D-scaffold deep learning models to discover covalent inhibitors targeting SARS-CoV-2 M^{pro}	This approach facilitated rapid design and resulted in the identification of four chloroacetamide-based inhibitors	Joshi et al. (2023)
The tree-invent model	Shang et al. used the tree-invent model to discover novel inhibitors of SARS-CoV-2 PL^{pro}	This model ultimately identified **GZNL-2002** as a potent inhibitor against SARS-CoV-2 PL^{pro}	Shang et al. (2024)
DeepFrag and EMPIRE models	Xiang et al. utilized DeepFrag and EMPIRE models to optimize HCK63 for SARS-CoV-2 N protein	This study has demonstrated that compound **19** exhibits a high degree of binding to the SARS-CoV-2 N protein	Xiang et al. (2024)
DL and active learning models	Garnsey et al. used DL and active learning driven strategy to discover potent SARS-CoV-2 PL^{pro} inhibitors	This research successfully yielded **PF-07957472**, which demonstrated favorable in vitro pharmacokinetic properties and robust efficacy in a mouse-adapted model of SARS-CoV-2 infection	Garnsey et al. (2024)

insights to rapidly identify potent inhibitors and optimize lead compounds for antiviral development. Overall, AI has become an indispensable tool in antiviral drug discovery. Its integration accelerates the pipeline from compound identification to clinical application, revolutionizing the field and providing a more efficient pathway to combat emerging viral challenges.

5 Perspective

AI is fundamentally transforming antiviral drug development through advanced analysis of multi-omics biomedical data. AI systems demonstrate unprecedented capability in biomarker identification and target discovery by extracting critical insights from genomics, proteomics, and metabolomics datasets. These capabilities significantly accelerate the identification of lead compounds and lead optimization processes. AI has transformative potential in antiviral drug development, but realizing its full benefits requires overcoming data quality, model interpretability, and multi-objective balancing challenges.

The primary challenge lies in data quality and accessibility. Current ML and DL architectures require extensive, well-curated training datasets. Issues including sample bias, batch effects in multi-omics measurements, and missing data points can significantly compromise model generalizability. Addressing these limitations requires innovative approaches to data harmonization, such as federated learning frameworks for multi-institutional data integration and graph neural networks for heterogeneous data fusion (Hasin et al. 2017; Wilkinson et al. 2016).

Another significant challenge is that existing approaches often concentrate on individual data types, consequently overlooking the intricate interconnections between diverse biological systems. The development of effective multimodal fusion techniques can extract valuable insights from a wide range of sources and formats, thereby propelling drug development forward. With the emergence of big data and the advancement of GPU computing, AI can now be applied to various forms of data, including text, images, and videos. Emerging models that utilize omics data, such as deep learning-based drug classification, have shown promise in predicting drug efficacy, identifying mechanisms, and assessing toxicity. This highlights the future potential of multimodal AI in drug development (Lombardo et al. 2017; Jiménez-Luna et al. 2020).

Moreover, data scarcity, privacy concerns, limited sharing (especially for rare diseases and novel targets), underrepresentation of "negative" data, and experimental instability undermine AI's reliability in drug development. Simultaneously, centralized datasets and model interpretability are essential for collaboration and regulatory compliance. In addition, while in silico benchmarks are helpful, laboratory validation is necessary. These efforts prompt specialists to integrate both fields, accelerating the development of next-generation drugs (Catacutan et al. 2024).

Finally, as AI continues to advance, interdisciplinary collaboration between academia, industry, and regulatory bodies will be crucial for translating AI-driven discoveries into effective therapies. Regulatory agencies must develop frameworks for evaluating AI-generated compounds, ensuring their safety, efficacy, and compliance with current standards. Collaborative efforts will also be necessary to address the ethical implications of using AI in drug discovery, particularly concerns regarding transparency, bias, and data privacy (Topol 2019).

In conclusion, the integration of AI into antiviral drug discovery holds immense potential to revolutionize the field, as evidenced by the promising advancement. However, challenges such as data quality, predictive accuracy, and regulatory hurdles must be addressed to fully unlock the capabilities of AI technology. Enhancing data sharing, creating "sparse" AI algorithms, using multimodal pre-trained models, developing multimodal fusion approaches, and incorporating physical laws into AI algorithms are critical steps. Fostering cross-disciplinary collaborations, improving AI algorithms, and focusing on data integration can pave the way for a new era of antiviral drug development. AI-driven innovations will undoubtedly play a central role in combating both emerging and re-emerging viral threats, offering hope for faster and more effective treatments.

Acknowledgments The authors are supported by the Key Research and Development Program, Ministry of Science and Technology of the People's Republic of China (No. 2023YFC2606500), and the Shandong Laboratory Program (No. SYS202205).

Conflict of Interest The authors declare that they have no conflict of interest.

References

Bian Y, Wang J, Jun JJ, et al. Deep convolutional generative adversarial network (dcGAN) models for screening and design of small molecules targeting cannabinoid receptors. Mol Pharm. 2019;16(11):4451–60.

Catacutan DB, Alexander J, Arnold A, Stokes JM. Machine learning in preclinical drug discovery. Nat Chem Biol. 2024;20(8):960–73.

Chenthamarakshan V, Hoffman SC, Owen CD, et al. Accelerating drug target inhibitor discovery with a deep generative foundation model. Sci Adv. 2023;9(25):eadg7865.

De Clercq E, Li G. Approved antiviral drugs over the past 50 years. Clin Microbiol Rev. 2016;29(3):695–747.

Gangwal A, Lavecchia A. Unleashing the power of generative AI in drug discovery. Drug Discov Today. 2024;29(6):103992.

Gao S, Song L, Sylvester K, et al. Design, synthesis, and biological evaluation of Trisubstituted Piperazine derivatives as noncovalent severe acute respiratory syndrome coronavirus 2 Main protease inhibitors with improved antiviral activity and favorable Druggability. J Med Chem. 2023;66(23):16426–40.

Garnsey MR, Robinson MC, Nguyen LT, et al. Discovery of SARS-CoV-2 papain-like protease (PLpro) inhibitors with efficacy in a murine infection model. Sci Adv. 2024;10(35):eado4288.

Gaulton A, Hersey A, Nowotka M, et al. The ChEMBL database in 2017. Nucleic Acids Res. 2017;45(D1):D945–54.

Gautam S, Thakur A, Rajput A, et al. Anti-dengue: a machine learning-assisted prediction of small molecule antivirals against dengue virus and implications in drug repurposing. Viruses. 2023;16(1):45.

Gómez-Bombarelli R, Wei JN, Duvenaud D, et al. Automatic chemical design using a data-driven continuous representation of molecules. ACS Cent Sci. 2018;4(2):268–76.
Hasin Y, Seldin M, Lusis A. Multi-omics approaches to disease. Genome Biol. 2017;18(1):83.
Huang R, Xu M, Zhu H, et al. Biological activity-based modeling identifies antiviral leads against SARS-CoV-2. Nat Biotechnol. 2021;39(6):747–53.
Huang L, Xu T, Yu Y, et al. A dual diffusion model enables 3D molecule generation and lead optimization based on target pockets. Nat Commun. 2024;15(1):2657.
Jiménez-Luna J, Grisoni F, Schneider G. Drug discovery with explainable artificial intelligence. Nat Mach Intell. 2020;2(10):573–84.
Joshi RP, Schultz KJ, Wilson JW, et al. AI-accelerated design of targeted covalent inhibitors for SARS-CoV-2. J Chem Inf Model. 2023;63(5):1438–53.
LeCun Y, Bengio Y, Hinton G. Deep learning. Nature. 2015;521(7553):436–44.
Lombardo F, Desai PV, Arimoto R, et al. In Silico absorption, distribution, metabolism, excretion, and pharmacokinetics (ADME-PK): utility and best practices. An industry perspective from the international consortium for innovation through quality in pharmaceutical development. J Med Chem. 2017;60(22):9097–113.
Ma Y, Frutos-Beltrán E, Kang D, et al. Medicinal chemistry strategies for discovering antivirals effective against drug-resistant viruses. Chem Soc Rev. 2021;50(7):4514–40.
Macarron R, Banks MN, Bojanic D, et al. Impact of high-throughput screening in biomedical research. Nat Rev Drug Discov. 2011;10(3):188–95.
Mao J, Wang J, Zeb A, et al. Transformer-based molecular generative model for antiviral drug design. J Chem Inf Model. 2024;64(7):2733–45.
Mottin M, de Paula Sousa BK, et al. Discovery of new Zika protease and polymerase inhibitors through the open science collaboration project OpenZika. J Chem Inf Model. 2022;62(24):6825–43.
Petersen E, Koopmans M, Go U, et al. Comparing SARS-CoV-2 with SARS-CoV and influenza pandemics. Lancet Infect Dis. 2020;20(9):e238–44.
Ragoza M, Hochuli J, Idrobo E, et al. Protein-ligand scoring with convolutional neural networks. J Chem Inf Model. 2017;57(4):942–57.
Rajput A, Kumar M. Anti-Ebola: an initiative to predict Ebola virus inhibitors through machine learning. Mol Divers. 2022;26(3):1635–44.
Saar KL, McCorkindale W, Fearon D, et al. Turning high-throughput structural biology into predictive inhibitor design. Proc Natl Acad Sci USA. 2023;120(11):e2214168120.
Saramago LC, Santana MV, Gomes BF, et al. AI-driven discovery of SARS-CoV-2 Main protease fragment-like inhibitors with antiviral activity in vitro. J Chem Inf Model. 2023;63(9):2866–80.
Serafim MSM, Kronenberger T, Oliveira PR, et al. The application of machine learning techniques to innovative antibacterial discovery and development. Expert Opin Drug Discov. 2020;15(10):1165–80.
Shams MY, Tarek Z, Elshewey AM. A novel RFE-GRU model for diabetes classification using PIMA Indian dataset. Sci Rep. 2025;15(1):982.
Shang J, Ran T, Lu Y et al. Discovery of novel quinoline papain-like protease inhibitors for COVID-19 through topology constrained molecular generative model; 2024. bioRxiv. https://doi.org/10.1101/2024.09.07.611841.
Stokes JM, Yang K, Swanson K, et al. A deep learning approach to antibiotic discovery. Cell. 2020;180(4):688–702.
Topol EJ. High-performance medicine: the convergence of human and artificial intelligence. Nat Med. 2019;25(1):44–56.
Turon G, Hlozek J, Woodland JG, et al. First fully-automated AI/ML virtual screening cascade implemented at a drug discovery Centre in Africa. Nat Commun. 2023;14(1):5736.
Vamathevan J, Clark D, Czodrowski P, et al. Applications of machine learning in drug discovery and development. Nat Rev Drug Discov. 2019;18(6):463–77.
Viana R, Moyo S, Amoako DG, et al. Rapid epidemic expansion of the SARS-CoV-2 omicron variant in southern Africa. Nature. 2022;603(7902):679–86.

Wilkinson MD, Dumontier M, Aalbersberg IJ, et al. The FAIR guiding principles for scientific data management and stewardship. Sci Data. 2016;3:160018.
Wishart DS, Feunang YD, Guo AC, et al. DrugBank 5.0: a major update to the DrugBank database for 2018. Nucleic Acids Res. 2018;46(D1):D1074–82.
Xiang ZR, Fan SR, Ren J, et al. Utilizing artificial intelligence for precision exploration of N protein targeting phenanthridine sars-cov-2 inhibitors: a novel approach. Eur J Med Chem. 2024;279:116885.
Xu T, Xu M, Zhu W, et al. Efficient identification of anti-SARS-CoV-2 compounds using chemical structure- and biological activity-based modeling. J Med Chem. 2022;65(6):4590–9.
Yang M, Lee MK, Gao S, et al. Miniaturized modular click chemistry-enabled rapid discovery of unique SARS-CoV-2 Mpro inhibitors with robust potency and drug-like profile. Adv Sci (Weinh). 2024;11(43):e2404884.
Zhang K, Yang X, Wang Y, et al. Artificial intelligence in drug development. Nat Med. 2025;31(1):45–59.
Zhao J, Ma Q, Zhang B, et al. Exploration of SARS-CoV-2 3CLpro inhibitors by virtual screening methods, FRET detection, and CPE assay. J Chem Inf Model. 2021;61(12):5763–73.
Zhavoronkov A, Ivanenkov YA, Aliper A, et al. Deep learning enables rapid identification of potent DDR1 kinase inhibitors. Nat Biotechnol. 2019;37(9):1038–40.
Zhu TF, Qian R, Wei X, et al. PatentNetML: a novel framework for predicting key compounds in patents using network science and machine learning. J Med Chem. 2024;67(2):1347–59.

Practical and Reproducible AI-Driven Modeling Protocols in Drug Discovery

Daniela Annunziata and Francesco Piccialli

Abstract The integration of artificial intelligence (AI) into drug discovery is transforming the field by significantly increasing efficiency, reducing costs, and improving the accuracy of predictive models. This chapter presents a comprehensive and reproducible AI-driven modeling pipeline, detailing key steps such as data preprocessing, molecular descriptor computation, and machine learning-based bioactivity prediction. The workflow leverages cheminformatics and computational chemistry techniques and employs advanced supervised learning algorithms—including random forests, gradient boosting machines, and neural networks—to analyze diverse molecular representations, such as SMILES strings, molecular fingerprints, and quantum descriptors. Special emphasis is placed on Quantitative Structure-Activity Relationship (QSAR) modeling, which establishes mathematical correlations between chemical structures and biological activities to facilitate lead optimization in drug development. In addition, the chapter addresses key challenges in AI-driven drug discovery, such as data quality, overfitting, and model generalizability, by implementing robust validation strategies, including scaffold-based and time-split validation. To enhance practical applicability, an interactive Google Colab notebook is provided, allowing researchers to experiment with AI-driven methods and apply them to real-world datasets. This framework provides a powerful and reproducible approach to accelerate drug discovery while ensuring transparency and reliability in predictive modeling.

Keywords QSAR modeling · Molecular descriptors · Cheminformatics · AI-driven drug discovery · Quantum descriptors

D. Annunziata (✉) · F. Piccialli
Department of Mathematics and Applications "R.Caccioppoli", University of Naples Federico II, Naples, Italy
e-mail: daniela.annunziata@unina.it; francesco.piccialli@unina.it

A. Lavecchia (ed.), *Applied Artificial Intelligence for Drug Discovery*,
https://doi.org/10.1007/978-3-031-98022-0_21

1 Introduction

The discovery of new drugs is one of the most complex and expensive processes in the life sciences. Traditionally, this process requires years of effort and significant financial resources, with a declining probability of success as the drug advances through clinical development stages (DiMasi et al. 2016). This complex and iterative process can be widely categorized into five phases (Fig. 1). Target identification involves the discovery and validation of a biological target associated with a disease, providing the basis for subsequent screening and drug design. Once a target is identified, the next phase, hit identification, uses high-throughput screening and computational techniques to discover hits, compounds with measurable activity against the selected target. These initial hits are then advanced to lead optimization, where systematic structural modifications enhance their potency, selectivity, and pharmacokinetic properties, ultimately resulting in optimized lead compounds with improved drug-like properties. Following successful lead optimization, the most promising candidates enter clinical trials, a rigorous multiphase evaluation process in human subjects to assess safety, efficacy, dosage, and potential adverse effects. Finally, drug approval marks the final stage, where candidates who demonstrate clinical success and meet regulatory requirements receive marketing authorization, allowing for commercial distribution and clinical use. However, advanced technologies like artificial intelligence (AI) have the potential to revolutionize the field, significantly improving the speed, efficiency, and accuracy of many phases of drug discovery and development (Deng et al. 2022; Schneider et al. 2020), and it enables the analysis of vast datasets, identification of hidden patterns, and generation of accurate predictions (Ekins et al. 2019; Walters and Barzilay 2021).

Integrating AI with cheminformatics and computational chemistry addresses specific challenges in drug discovery, such as: biological activity classification (Staszak et al. 2022), solubility prediction (Chinh Nguyen et al. 2022), virtual screening (Maia et al. 2020), and toxicity prediction (Tran et al. 2023). These applications often utilize molecular representations, such as Simplified Molecular Input Line Entry System (SMILES) strings, physicochemical descriptors, and molecular

Fig. 1 The drug discovery and development pipeline, illustrating the five key stages from target identification to drug approval. This process involves the identification of a biological target, the screening and optimization of potential drug candidates, rigorous clinical evaluation, and final regulatory approval. Each phase is essential in ensuring the safety, efficacy, and viability of new therapeutic compounds before they reach the market

graphs. Machine learning and deep learning algorithms, including Random Forests, Gradient Boosting Machines, Convolutional Neural Networks, and Transformers, are frequently employed to analyze these representations. By adopting these methods, researchers can enhance the identification of promising drug candidates, reduce costs, and mitigate risks during development. One of the key areas where AI has demonstrated its transformative potential is in the development of Quantitative Structure-Activity Relationship (QSAR) models. QSAR models define a mathematical framework that relates the chemical structure of a compound to its biological or physicochemical properties (Dearden 2017; Golbraikh et al. 2017), correlating the structural features, known as descriptors, with specific responses such as activity, property, or toxicity across a range of molecules using different methodological approaches (De et al. 2022). These relationships are typically captured through molecular descriptors, encompassing simple physicochemical parameters (e.g., molecular weight, lipophilicity) as well as more intricate features derived from cheminformatics, quantum chemistry, or graph-based representations (Bahia et al. 2023; Grisoni et al. 2018). Molecular descriptors can be broadly classified into distinct categories based on the type of structural or physicochemical information they capture.

Figure 2 provides an overview of the main classes of molecular descriptors, including constitutional, geometrical, topological, electronic, and thermodynamic descriptors, each of which contributes uniquely to the characterization of molecular properties in computational drug discovery. Supervised learning algorithms, such as random forests or neural networks, enable researchers to detect patterns linking these descriptors to experimentally measured activities, thereby guiding candidate selection and lead optimization (Gupta et al. 2021; Kim et al. 2020). By analyzing these descriptors, QSAR models uncover relationships that can guide the identification of novel therapeutic compounds. When carefully validated, using metrics like the coefficient of determination (R^2) and Mean Squared Error, models can enable the rational design of novel scaffolds with enhanced drug-like properties (Cherkasov et al. 2014). Despite its potential, QSAR modeling has limitations.

The quality of the underlying data is crucial; errors or incomplete entries may introduce bias, reduce performance, and limit generalizability (Zhao et al. 2017). Similarly, the risk of overfitting is high when working with large descriptor spaces, requiring careful feature selection and hyperparameter tuning. In addition, extrapolation outside the chemical domain of the training set can lead to unreliable predictions, underscoring the importance of continuous model retraining and the incorporation of diverse, well-curated data sets. To address these challenges effectively, researchers are increasingly relying on open-source tools and reproducible workflows. Platforms like Jupyter Notebooks (Kluyver et al. 2016), integrated with libraries such as Scikit-learn (Pedregosa et al. 2011), TensorFlow (Abadi et al. 2016), and RDKit (Landrum 2016), provide accessible and customizable resources for building and deploying QSAR models. These resources empower practitioners to experiment with innovative techniques while ensuring transparency and reproducibility (Mansouri et al. 2024).

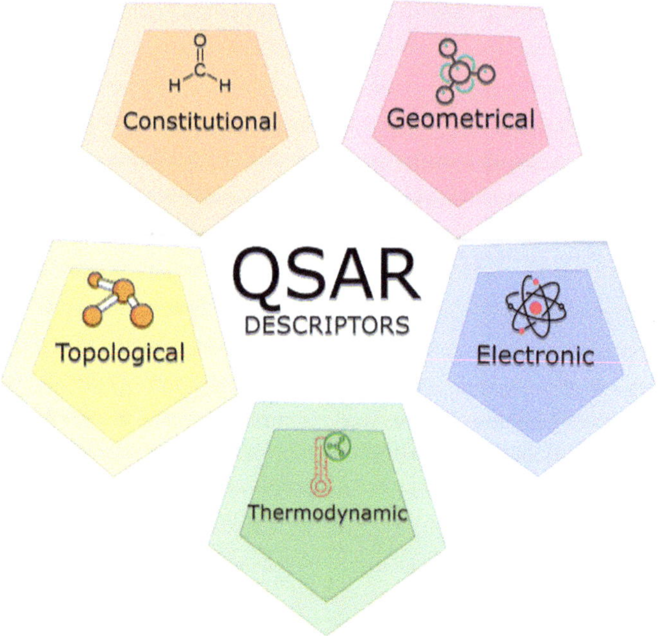

Fig. 2 The five main categories of molecular descriptors used in cheminformatics and computational chemistry. Constitutional descriptors describe basic molecular composition without considering structure, such as molecular weight or the number of hydrogen bond donors. Geometrical descriptors characterize the three-dimensional disposition of atoms, with examples including molecular volume and solvent-accessible surface area. Topological descriptors capture molecular connectivity and graph-based properties, such as the Wiener index or the Zagreb connectivity index. Electronic descriptors define electronic properties like charge distribution and orbital energies, exemplified by dipole moment and HOMO-LUMO gap. Thermodynamic descriptors relate to energy and stability, with examples including enthalpy of formation and partition coefficient (logP)

This chapter is structured to guide readers step-by-step through key AI-driven tasks in drug discovery. It begins by introducing fundamental concepts and datasets, followed by practical implementation protocols. Finally, it provides case studies and open-source notebooks to empower readers with real-world tools for addressing challenges in modern drug discovery. In the following sections, we outline a reproducible AI-driven pipeline for drug discovery, starting from data preprocessing and molecular descriptor generation to model training and evaluation. This workflow integrates theoretical insights with practical implementation and enables researchers to directly apply these methods to real-world datasets.

2 Environment Setup and Data Engineering

For researchers exploring AI-driven drug discovery, establishing a robust and reproducible computational environment is essential. The complexity of the tasks, ranging from molecular data preprocessing to deploying machine learning models, requires tools that are both accessible and versatile. In this chapter, we leverage Python (Van Rossum and Drake 2009), a widely used programming language in scientific research and its extensive ecosystem of open-source libraries.

To ensure accessibility, all examples will be implemented in Google Colab (Bisong 2019), a free cloud-based platform that eliminates the need for specialized hardware. Google Colab provides a user-friendly environment where researchers can execute Python code directly in a web browser, with access to powerful computational resources such as GPUs and TPUs. This makes Colab well-suited for cheminformatics workflows and machine learning applications in drug discovery. Furthermore, it facilitates collaboration and reproducibility, as notebooks can be easily shared and executed by other researchers. However, while Google Colab provides an immediate and accessible solution, many real-world AI-driven drug discovery workflows require a more structured and reproducible environment. To address this, we recommend using Conda to manage dependencies efficiently.

2.1 *Setting Up the Environment*

Before diving into data manipulation and model training, we need to prepare the environment by installing and importing the required libraries. These libraries are integral to our computational pipeline, powering everything from data preprocessing to advanced AI modeling. Specifically, we will rely on:

- **pandas** and **numpy**: for efficient data handling and numerical computations.
- **tensorflow**: A popular deep learning framework used for building and training neural networks,
- **matplotlib** and **seaborn**: for visualizing molecular properties and model performance.
- **rdkit**: a powerful toolkit for cheminformatics, which allows us to compute molecular descriptors, manipulate chemical structures, and transform molecular representations into machine-readable formats.
- **pyscf:** a quantum chemistry package used to compute electronic structure properties, enabling the extraction of quantum descriptors.

To ensure consistency across different executions and prevent compatibility issues, a dedicated Conda environment can be created to contain all necessary dependencies. The recommended setup consists of the following steps:

2.2 Step 1: Creating and Activating the Environment

```
conda create --name qsar_env
conda activate qsar_env
```

2.3 Step 2: Installing Required Libraries

All required dependencies can be installed using a predefined requirements file, which ensures that all users work with the same software versions.

```
matplotlib-base==3.6.3
matplotlib-inline==0.1.6
numpy==1.24.1
pandas==2.2.3
pyscf==2.7.0
rdkit==2022.09.5
scikit-learn==1.2.1
seaborn==0.12.2
tensorflow==2.11.0
xgboost==2.1.3
```

To install all dependencies at once:

```
conda install -c conda-forge --file requirements.txt
```

For users executing this workflow in Google Colab, where Conda environments are not natively supported, dependencies must be installed dynamically within the notebook:

```
!pip install rdkit pandas numpy matplotlib seaborn pyscf tensorflow
```

With the environment configured, the next step is to prepare the molecular data for AI modeling.

2.4 Datasets for AI-Driven Drug Discovery

Drug discovery workflows are data-intensive, often relying on large datasets with millions of molecular entries. One such resource is the Papyrus dataset (Béquignon et al. 2023), a curated repository that aggregates over 60 million compound-protein interactions sourced from public databases such as ChEMBL (Bento et al. 2014) and ExCAPE-DB (Sun et al. 2017), offering unparalleled breadth for cheminformatics studies. One of its key strengths lies in its standardization, including the harmonization of activity measures like pChEMBL values, which allow for meaningful comparisons across diverse datasets. Using the Papyrus dataset, researchers can access high-quality and diverse datasets that are critical for tasks such as virtual screening and Structure-Activity Relationship (SAR) modeling. Its comprehensive annotations, including molecular SMILES, target information, and activity labels, facilitate machine learning applications by providing the necessary inputs for descriptor calculation and predictive modeling. Additionally, the dataset's inclusion of quality metrics empowers practitioners to filter data for reliability, reducing noise and enhancing the robustness of downstream analyses. This makes Papyrus an ideal starting point for this chapter's pipeline, as it illustrates the type of large-scale, well-curated resource necessary to develop reproducible and impactful AI models in drug discovery. The dataset contains detailed information, including:

- **Activity_ID**: A unique identifier for each compound-protein pair, derived by concatenating the molecule's InChI (International Chemical Identifier) and the protein's unique identifier.
- **Quality**: A categorical field that assesses the reliability of the activity measurement, helping researchers focus on the highest-quality data for their analysis.
- **SMILES:** The standardized SMILES notation is the molecular structure of compounds in a textual format.
- **Connectivity**: Information about the molecule's connectivity.
- **Target_ID**: Unique identifiers for protein targets, allowing for integration with other biological datasets.
- **pchembl_value**: Log-transformed activity values derived at the highest quality level available. These values are critical for structure-activity relationship (SAR) modeling, as they provide a normalized scale for biological activity.

However, working with raw data poses significant challenges. If left unaddressed, these issues can degrade model performance or, worse, produce misleading results. Thus, data cleaning and preprocessing are indispensable steps in any AI-driven workflow.

2.5 Data Cleaning and Preprocessing

Preprocessing is a fundamental step in AI-driven drug discovery workflows because it transforms raw, unstructured datasets into clean, reliable inputs for machine learning models. Issues such as missing values, duplicate entries, or invalid molecular representations can severely compromise model accuracy and reproducibility. This section outlines a systematic approach to addressing these challenges, focusing on handling missing values, removing redundancies, and validating SMILES strings. This process begins with inspecting the dataset to identify potential issues, such as missing values, duplicate entries, or inconsistencies in molecular representations. Real-world datasets, such as those used in cheminformatics, often contain numerous challenges that can compromise the accuracy and reliability of predictive models. The first step in this process is to load the dataset into a Pandas DataFrame and perform an initial inspection to understand its structure and content.

```
import pandas as pd

#Load the dataset
df = pd.read_csv("papyrus_dataset.csv")

#Inspect the dataset
print("Dataset Information:")
print(df.info())
print("First few rows of the dataset:")
print(df.head())
```

2.5.1 Removing Missing Values

Missing values in critical fields such as pchembl_value and SMILES are common in large cheminformatics datasets. Such gaps frequently result from incomplete experimental data or inconsistencies during data curation. Rows with missing values in these fields are removed to ensure the dataset remains clean and dependable. Specifically:

- Missing pchembl_value entries are excluded because they lack the activity annotations required for predictive modeling.
- Missing SMILES strings are removed as they do not provide the molecular structure necessary for descriptor calculation.

```
df_clean = df.dropna(subset=["SMILES", "pchembl_value"])
print(f"Dataset size after removing missing values: {len(df_clean)}")
```

2.5.2 Standardization, Canonicalization, and Tautomer Normalization

Molecular representations in cheminformatics datasets vary due to multiple equivalent forms of the same compound. To ensure uniformity, a standardization pipeline is applied, which includes:

1. Removing salts and small fragments that may be present due to different data sources.
2. Canonicalizing the SMILES representation, ensuring consistent molecular representation.
3. Normalizing tautomers, selecting the most stable tautomeric form to reduce redundancy.

For this purpose, RDKit's `SaltRemover` is used to remove salts, canonical SMILES conversion is applied for consistency, and `TautomerEnumerator` is utilized to identify the most stable tautomer.

```
from rdkit import Chem
from rdkit.Chem import SaltRemover, MolStandardize
from rdkit.Chem.MolStandardize import rdMolStandardize

# Creare un'istanza del TautomerEnumerator
tautomer_enumerator = rdMolStandardize.TautomerEnumerator()

# Define the salt remover and tautomer enumerator
remover = SaltRemover.SaltRemover()
enumerator = rdMolStandardize.TautomerEnumerator()


def preprocess_molecule(smiles):
    """
    Cleans a molecule by removing salts, canonicalizing SMILES,
    and selecting the most stable tautomeric form.
    Args:
        smiles (str): Molecular representation in SMILES format.

    Returns:
        str or None: Cleaned SMILES string or None if invalid.
    """
    try:
        mol = Chem.MolFromSmiles(smiles)
        if mol is None:
            return None  # Invalid molecule
        # Remove salts
        mol = remover.StripMol(mol)
        # Normalize tautomeric forms
        mol = enumerator.Canonicalize(mol)
        # Convert back to canonical SMILES
        return Chem.MolToSmiles(mol, canonical=True)
    except Exception as e:
        print(f"Error processing {smiles}: {e}")
        return None

# Apply preprocessing to the dataset
df_clean["Processed_SMILES"] = df_clean["SMILES"].apply(
                                                    preprocess_molecule)
df_clean = df_clean.dropna(subset=["Processed_SMILES"])
```

This approach ensures that only chemically valid structures are retained, while maintaining consistency across molecular representations. Unnecessary variation due to salts, stereoisomers, or tautomeric forms is eliminated.

2.5.3 Removing Duplicates

Once molecular structures have been standardized and validated, duplicate entries are identified and removed. Redundant data can come from a variety of sources, including:

- Duplicate molecules from different databases, where the same compound appears multiple times.
- Multiple representations of the same molecule, such as stereoisomers or alternative protonation states.
- Repeated bioactivity measurements, which can lead to overrepresentation of certain compounds.

To avoid bias in model training, duplicate data sets are removed based on the canonicalized molecular structure.

```
df_clean = df_clean.drop_duplicates(subset=["Processed_SMILES"])
print(f"Dataset size after removing duplicates: {len(df_clean)}")
```

These preprocessing steps ensure the dataset is clean, standardized, and dependable for downstream modeling tasks. By addressing missing values, removing redundancies, and validating molecular representations, the dataset is optimized for generating high-quality molecular descriptors and achieving robust machine learning results.

2.6 Data Management for De Novo Design

2.6.1 Building a Vocabulary

In the domain of de novo drug discovery, the efficacy of generative models in designing novel molecules is contingent upon the representation of molecular structures. A critical step in data preprocessing is the construction of a molecular vocabulary, which defines the fundamental units, called tokens, used to describe chemical structures in a machine-readable format. The quality and granularity of this vocabulary have been demonstrated to directly influence the model's ability to generate realistic and synthesizable molecules, as they dictate the extent of the chemical space the model can explore. To construct a meaningful vocabulary for de novo design, molecular structures, typically encoded as SMILES strings, are analyzed to extract their fundamental chemical components. Two main strategies can be used for token extraction:

- Character-based vocabulary, where each individual character in a SMILES string is treated as an independent token.
- Chemically aware vocabulary, where multicharacter atomic symbols (e.g., Br, Cl, [NH + 1]) and functional groups are preserved as individual tokens, providing a more chemically meaningful representation of molecular structures.

Once the tokens have been extracted, they are assigned unique identifiers to form the vocabulary. This can be achieved using Python, specifically the `collections.Counter` library, which allows token occurrences to be counted and, if necessary, a frequency threshold applied to exclude rare tokens. Let's look at Python implementation. The first approach is to consider each character in the SMILES string as a separate token. For example, Br (bromine) would be split into 'B' and 'r', and Cl (chlorine) into 'C' and 'l', resulting in the loss of meaningful chemical information.

```
from collections import Counter
#Example molecules containing multi-character atomic symbols
smiles_list = ["CCBr", "ClC1=CC=CC=C1"]

#Character-based tokenization
tokenized_smiles = [list(smiles) for smiles in smiles_list]
#Count token frequencies
token_counts = Counter(token for tokens in tokenized_smiles for token in
tokens)
#Construct the character-based vocabulary
vocab_char = {token: idx for idx, (token, count) in
enumerate(token_counts.items())}
#Print the generated vocabulary
print("Character-based vocabulary:", vocab_char)

Output:
Character-based vocabulary: {'C': 0, 'B': 1, 'r': 2, 'l': 3, '1': 4, '=':
5}
```

To address the shortcomings of character-based tokenization, a more chemically meaningful approach is implemented that preserves atomic symbols such as Br, Cl, and functional groups as individual tokens.

The choice of vocabulary affects the accuracy and effectiveness of molecular generative models in de novo drug discovery. Chemical aware vocabulary is more aligned with real-world chemistry, providing a stronger foundation for training deep learning models in de novo drug discovery. By ensuring atomic and functional group integrity, this approach improves the model's ability to generate realistic, synthesizable molecules, leading to more effective drug design.

2.6.2 Encoding Molecular Representations

Once molecular vocabulary has been constructed, the next step in preparing data for generative models is to encode molecular structures into numerical representations that can be processed by deep learning architectures. Two common approaches to encoding tokenized SMILES strings are one-hot encoding and integer encoding. One-hot encoding is a widely used technique for converting categorical data into a machine-readable format. In this method, each unique token in the vocabulary is mapped to a binary vector, where only one position is set to 1 while all others remain 0. This ensures that tokens are treated as distinct categories without implying any hierarchical or ordinal relationships between them. While this method

```
import re
#Define chemical tokens that should be preserved
special_tokens = ["Cl", "Br", "Si", "B", "C", "N", "O", "P", "S", "F",
"I", "1","=", "#", "(", ")"]


def tokenize_smiles(smiles):
    """ Tokenizes a SMILES string while preserving chemically
   meaningful tokens. """
    pattern = "|".join(re.escape(token) for token in
   sorted(special_tokens, key=len, reverse=True))
    return re.findall(pattern, smiles)


#Tokenization with chemical token preservation
tokenized_smiles_chem = [tokenize_smiles(smiles) for smiles in
smiles_list]


#Count token frequencies
token_counts_chem = Counter(token for tokens in tokenized_smiles_chem for
token in tokens)


#Construct the chemically-aware vocabulary
vocab_chem = {token: idx for idx, (token, count) in
enumerate(token_counts_chem.items())}
#Print the generated vocabulary
print("Chemically aware vocabulary:", vocab_chem)


Output: Chemically aware vocabulary: {'C': 0, 'Br': 1, 'Cl': 2, '1': 3,
'=': 4}
```

is widely employed and effective, it can result in high-dimensional representations, particularly when dealing with extensive vocabularies. This can lead to an increase in memory usage and computational complexity. For example, given a vocabulary {'C': 0, 'Br': 1, 'Cl': 2, '1': 3, '=': 4}, the token "Cl" would be represented as [0, 0, 1, 0, 0], while "Br" would be represented as [0, 1, 0, 0, 0]. We will apply one-hot encoding in Python to the following SMILES strings:

```
from sklearn.preprocessing import OneHotEncoder
import numpy as np

#Initialize OneHotEncoder
encoder = OneHotEncoder (sparse=False,
categories=[list(vocab_chem.keys())])
#Fit encoder to known vocabulary
encoder.fit(np.array(list(vocab_chem.keys())).reshape(-1,1))
#Convert tokenized SMILES into one-hot vectors
one_hot_encoded = [encoder.transform(np.array(tokens).reshape(-1, 1)) for
tokens in tokenized_smiles_chem]
```

Integer encoding maps each unique molecular token to a corresponding integer. This approach transforms tokenized SMILES sequences into sequences of numbers that are easier for deep learning models to process and learn from. Integer encoding is particularly useful in the context of neural networks because it provides a compact and memory-efficient representation of molecular structures. Unlike one-hot encoding, which produces high-dimensional binary vectors, integer encoding produces a low-dimensional numerical format, making it more computationally efficient and scalable. This is especially beneficial when working with large molecular datasets, where reducing the input size can have a significant impact on training speed and model performance. The key advantage of integer encoding is that it preserves the categorical identity of each molecular token while allowing deep learning models to discover relationships between them. When combined with embedding layers that map discrete tokens to continuous vector spaces, integer encoding enables the model to learn chemical relationships that would be difficult to capture with other encoding methods. Table 1 provides a comparative overview of different encoding strategies used for tokenized SMILES representations.

```
def integer_encode(smiles_tokens, vocab):
    return [vocab[token] for token in smiles_tokens if token in vocab]

#Apply integer encoding
integer_encoded_smiles = [integer_encode(tokens, vocab_chem) for tokens
in tokenized_smiles_chem]
```

Table 1 Tokenization and encoding of SMILES strings into machine-readable formats

Tokenized SMILES	One-hot encoding	Integer encoding
['C', 'C', 'Br']	[[1. 0. 0. 0. 0.] [1. 0. 0. 0. 0.] [0. 1. 0. 0. 0.]]	[0, 0, 1]
['Cl', 'C', '1', '=', 'C', 'C', '=', 'C', 'C', '=', 'C', '1']	[[0. 0. 1. 0. 0.] [1. 0. 0. 0. 0.] [0. 0. 0. 1. 0.] [0. 0. 0. 0. 1.] [1. 0. 0. 0. 0.] [1. 0. 0. 0. 0.] [0. 0. 0. 0. 1.] [1. 0. 0. 0. 0.] [1. 0. 0. 0. 0.] [0. 0. 0. 0. 1.] [1. 0. 0. 0. 0.] [0. 0. 0. 1. 0.]]	[2, 0, 3, 4, 0, 0, 4, 0, 0, 4, 0, 3]

Chemically tokens are converted into one-hot binary vectors and integer sequences to serve as model inputs

3 Challenges and Best Practices in AI Model Training

Developing dependable AI models goes beyond selecting the right algorithms or preprocessing data; it also requires addressing the many challenges that arise during model training and evaluation. While earlier sections have covered data handling and feature generation, it is equally important to discuss key obstacles such as overfitting, underfitting, model generalization, and optimization strategies. These aspects, already briefly introduced, will now be explored in more detail to provide practical solutions and best practices for developing robust and reproducible AI models.

3.1 Overfitting and Underfitting

One of the most familiar challenges in AI-driven drug discovery is finding the right balance between model complexity and generalization. A model that is too complex may overfit the training data, while a model that is too simplistic may underfit and fail to capture meaningful patterns.

Overfitting occurs when a model performs exceptionally well on training data but fails to generalize to new, unseen compounds. This is often characterized by a significant gap between training and validation loss. Two main factors contribute to overfitting:

- Excessive model complexity: Deep networks with too many parameters may fit noise rather than meaningful patterns, resulting in poor generalization.
- Lack of data diversity: Training on a chemically narrow data set limits the ability of the model to make accurate predictions for novel compounds.

Several techniques can be used to mitigate these risks:

- L1/L2 Regularization: These methods penalize large weights, preventing the model from over-relying on certain features.
- Dropout layers: By randomly deactivating neurons during training, dropout forces the model to learn more generalizable representations.
- Early stopping: Monitoring validation loss and stopping training when performance begins to degrade helps prevent overfitting.
- Data Augmentation: Introducing controlled variations in molecular structures—such as conformational sampling or stereoisomer generation—enhances generalization.

However, underfitting occurs when a model is too simple to capture important patterns. This is often due to insufficient training epochs, low model complexity, or suboptimal feature selection. To improve learning, the following approaches can be used:

- Increase model complexity: Deeper networks and architectures that incorporate attention mechanisms can improve model expressiveness.
- Feature engineering: Using more informative molecular descriptors improves the quality of the data input.
- Hyperparameter optimization: Fine-tuning parameters such as learning rates, batch sizes, and activation functions through Bayesian optimization or grid search improve performance.

3.2 *Validation Strategies*

Proper validation is critical to ensuring that an AI model performs well not only on training data but also in real-world applications. Traditional validation methods, such as K-fold cross-validation, stratified sampling, and holdout validation, are widely used in machine learning, but drug discovery can require more specialized approaches. K-fold cross-validation divides the data set into K subsets (folds), trains the model on K-1 folds, and tests on the remaining fold, repeating the process K times, with each fold serving as a test set once. The final performance is averaged over all iterations, reducing the impact of data partitioning on the results. This approach provides a robust estimate of model performance while maximizing data utilization. Despite its extensive use, K-fold cross-validation may not be ideal for drug discovery, where molecular similarity between folds can lead to overestimation of performance. To mitigate this problem, alternative validation strategies such as scaffold-based splitting and time-split validation are often preferred. The scaffold-based splitting ensures that training and test sets contain structurally distinct compounds (Guo et al. 2024). Unlike random splitting, which risks inflating performance metrics by including similar molecules in both sets, scaffold-based validation better reflects real-world scenarios where AI models must predict activity for entirely new chemical scaffolds.

In addition, time-split validation provides a more realistic evaluation strategy by training models on compounds discovered in earlier years and testing them on more recent molecules (Sheridan 2013). This method simulates prospective predictions in drug discovery workflows and helps to assess how well a model can generalize to future datasets. Temporal validation has been shown to provide a more reliable estimate of model performance in real-world applications (Bosc et al. 2019).

3.3 *Data Augmentation in Drug Discovery*

Data augmentation is a powerful technique that enhances the generalization capabilities of AI models by artificially increasing dataset diversity. While commonly used in image processing and natural language processing, its role in drug discovery is uniquely tailored to molecular data. One widely used augmentation strategy is

SMILES augmentation, which takes advantage of the fact that a single molecular structure can be represented in multiple ways using SMILES notation. By randomizing these representations during training, the model becomes more robust to variations in molecular encoding. Another approach is molecular graph augmentation, where small structural modifications, such as bond perturbations, atom substitutions, or stereochemical changes, are introduced to expand the diversity of training data. This technique helps models learn a broader range of chemical relationships, improving predictions for structurally novel compounds. Beyond 2D representations, 3D conformation perturbation involves generating alternative low-energy molecular conformations through quantum chemistry simulations or molecular docking techniques. Since many molecular interactions depend on three-dimensional structure, training models with multiple conformations improves predictive performance in tasks such as binding affinity estimation.

3.4 Model Optimization

Beyond data augmentation and validation, optimizing the training process ensures model stability and efficiency. Batch normalization mitigates internal covariate shifts, leading to faster convergence and improved performance; dropout regularization prevents overreliance on specific neurons, reducing overfitting. Gradient clipping is particularly useful in deep networks, preventing abrupt weight changes that can destabilize training, and the choice of optimizer is also critical. Adjusting hyperparameters such as batch sizes, learning rate schedules, and weight decay ensures optimal performance without overfitting.

4 Predicting Biological Activity

Predicting a molecule's biological activity is one of the key challenges in drug discovery. This process aims to find whether a molecule can effectively interact with a biological target, such as an enzyme or receptor, to modulate a specific biological function. Such interactions can lead to therapeutically relevant effects, such as inhibiting a pathological enzymatic activity or activating a receptor involved in disease treatment. The goal is to develop reliable predictive models that can guide the first stages of drug discovery, reducing time and cost while improving overall success rates. Selecting a biological target is a fundamental step that defines the experimental context and guides the entire modeling pipeline. This process requires evaluating the target's biological and clinical relevance, the availability of prior knowledge, and the accessibility of experimental data. The choice of target also decides the type of data needed, which includes molecules annotated with biological activity values and structural information about the target itself. To build robust and reliable models, data collection and preprocessing are essential phases. In this study, we use the Papyrus dataset, ensuring high data quality is paramount, which includes removing errors, duplicates and missing values, and normalizing variables such as IC50 and EC50 to standard formats. In cases of unbalanced data sets, where one class of molecules is underrepresented compared to another, techniques such as oversampling or synthetic data generation can be used to balance the classes and improve the predictive performance of the model. A key element in developing accurate predictive models is how molecules are represented as input data. Traditional approaches rely on predefined molecular descriptors and fingerprints, while modern AI-driven methods leverage more flexible representations that can capture both structural and physicochemical properties. Depending on the type of machine learning model, molecular structures can be encoded in diverse ways. SMILES, for instance, require tokenization into meaningful substructures before they can be processed using natural language models. Graph Neural Networks (GNNs) operate on molecular graphs, where atoms and bonds are represented as nodes and edges, preserving connectivity patterns. Alternatively, molecular fingerprints and physicochemical descriptors transform structural information into numerical vectors suitable for conventional machine learning algorithms. Quantum descriptors, such as molecular orbital energies and dipole moments, provide an additional layer of information about electronic properties. Table 1 illustrates these different molecular representations and how they are transformed into machine-readable inputs for predictive modeling. In the next sections, we describe the key steps in detail, from molecular descriptor calculation to the implementation of classification models for bioactivity prediction. The QSAR workflow follows a structured approach, starting with data collection and molecular descriptor computation, followed by machine learning-based modeling and activity prediction. Figure 3 visually summarizes this process (Table 2).

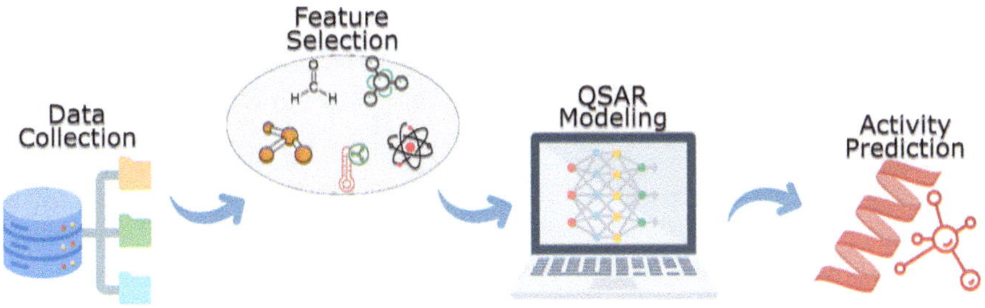

Fig. 3 Overview of the QSAR modeling pipeline. The workflow begins with data collection from molecular databases, followed by descriptor calculation to transform chemical structures into numerical representations. Machine learning models are then trained on these features to predict biological activity, guiding early-stage drug discovery

Table 2 Molecular representations and their corresponding machine learning inputs: SMILES strings can be converted into molecular graphs with adjacency and feature matrices, fingerprint-based binary vectors (e.g., ECFP4, MACCS), or descriptor-based numerical feature vectors capturing physicochemical properties such as molecular weight (MW), topological polar surface area (TPSA), and lipophilicity (LogP)

Molecular representation	Machine learning input
SMILES "C1CC2 = C(C1)C=C(C=C2)OCC"	tokenized sequence ['C', '1', 'C', 'C', '2', '=', 'C', '(', 'C', '1', ')', 'C', '=', 'C', '(', 'C', '=', 'C', '2', ')', 'O', 'C', 'C'] + vocab = {'C': 1, 'O': 2, '=': 3, '1': 4, '2': 5, '(': 6, ')': 7} ↓ numerical sequence [1, 4, 1, 1, 5, 3, 1, 6, 1, 4, 7, 1, 3, 1, 6, 1, 3, 1, 5, 7, 2, 1, 1]
Molecular graph 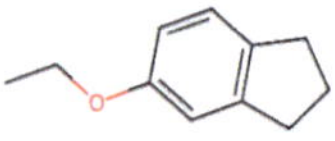	Adjacency Matrix: Binary matrix representing atom connectivity. Feature Matrix: Atomic properties like atom type, charge, number of bonds. 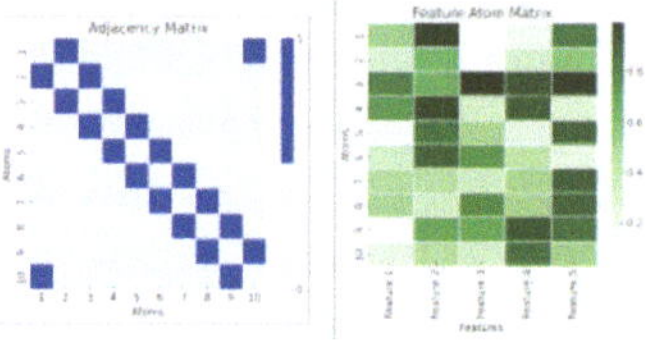
Molecular fingerprint (ECFP4, MACCS, or other fingerprint types)	Binary vector [1, 0, 1, 0, 0, 1, ..., 1, 0, 0, 1]
Molecular descriptors	Feature vector [MW: 162.23, TPSA: 9.23, LogP: 2.57, HBA: 1, HBD: 0]

4.1 Step 1: Target Selection and Data Filtering

The first step is selecting a biologically relevant target (e.g., enzyme, receptor) based on clinical significance and data availability. A well-curated dataset ensures reliable AI-driven modeling. To facilitate the extraction of protein target information, the dataset includes a dedicated annotation file, which systematically maps chemical compounds to their respective biological targets. This structured approach ensures that computational modeling relies on well-defined protein-ligand interactions, enhancing prediction accuracy. Within the dataset, each protein is uniquely identified through a target_id, which combines the protein accession number with any mutations present in its sequence.

For this study, we selected the adenosine receptor as our target, and to maintain biological relevance, we focus exclusively on Homo sapiens as the reference organism, filtering out data from animal models or other species that might introduce inconsistencies in compound behavior.

To construct a reliable and high-quality dataset, we apply the following filtering steps:

1. Select a target of interest,
2. Retain compounds with validated bioactivity,
3. Exclude low-quality data to reduce noise.

4.2 Step 2: Molecular Descriptors

Molecular descriptors quantify physicochemical, structural, and quantum properties, serving as machine-learning features to predict bioactivity. RDKit is a powerful library for cheminformatics that allows for the computation of essential descriptors directly from molecular structures encoded as SMILES strings. Key descriptors include molecular weight (MolWt), lipophilicity (LogP), total polar surface area (TPSA), the number of hydrogen bond donors (HBD) and acceptors (HBA), and the number of rotatable bonds (NumRot). Additionally, RDKit supports the calculation of molecular fingerprints, such as Extended Connectivity Fingerprints (ECFP), which encode structural information in a binary vector format suitable for machine learning. The following Python code shows how to compute these descriptors using RDKit:

This implementation computes physicochemical descriptors and ECFP fingerprints for each molecule in the dataset, producing a robust representation of molecular properties.

To better understand the dataset, visualizing the distribution of molecular descriptors can offer valuable insights. For instance, histograms can reveal the range of molecular weights or lipophilicity, while correlation heatmaps can uncover relationships between descriptors.

In addition to traditional physicochemical descriptors and molecular fingerprints, quantum descriptors provide a unique window into the electronic and molecular properties of compounds. They offer critical insights that complement conventional cheminformatics features by capturing aspects of electronic structure that classical descriptors alone cannot reveal. Quantum descriptors, such as the HOMO-LUMO gap, dipole moment, and molecular polarizability, are derived from quantum chemical calculations and play a key role in understanding molecular interactions at the electronic level. Typically, these descriptors are computed using specialized software like Gaussian, ORCA, or PySCF. The process begins by converting SMILES strings into three-dimensional molecular geometries via energy minimization techniques. These geometries serve as the foundation for quantum chemical calculations, which are often performed using density functional theory (DFT) or semiempirical methods. Although quantum mechanical simulations can be computationally intensive, strategies such as focusing on representative subsets of molecules or employing approximate semiempirical models can reduce the computational burden. Once the molecular geometries are optimized, quantum chemical calculations are conducted to determine key electronic properties, such as molecular orbital energies and dipole moments, that are then integrated with

```python
#Import necessary libraries
import pandas as pd

def select_protein_class(data, protein_data, classes=[{"15": "Adenosine
receptor"}]):
    """
    Filter the input dataset based on specific protein classifications.

    Parameters:
        data (DataFrame): A DataFrame containing biological data. It
            must include the 'target_id' column.
        protein_data (DataFrame): A DataFrame containing protein
            classification data, with required columns 'target_id' and
            'Classification'.
        classes (list of dict): A list of dictionaries specifying
            filters for different classification levels.
    Returns:
        DataFrame: A filtered DataFrame containing only the rows from
            'data' whose 'target_id' matches those in the filtered
            protein_data. The returned DataFrame is merged with the
            corresponding 'Classification' and 'Organism' columns from
            protein_data.
    """
    # If no classification filters are provided, return the original
    dataset.
    if not classes:
        return data
    # Split the 'Classification' column into separate levels using "->"
    as the delimiter.
    # Ensure that we have exactly 8 columns, filling in any missing
    levels with empty strings.
    split_classifications = (
        protein_data["Classification"]
        .str.split("->", expand=True)
        .reindex(columns=range(8), fill_value="")
    )
    protein_data[["l1", "l2", "l3", "l4", "l5", "l6", "l7", "l8"]] =
    split_classifications

    # Define a helper function to determine whether a given row matches
    all provided classification filters.
    def match_classification(row):
        for class_filter in classes:
            if all(
                str(row.get(level, "")).lower() == str(value).lower()
                for level, value in class_filter.items()
            ):
                return True
        return False
```

```
    # Apply the matching function to filter protein_data based on the
    specified classifications.
    filtered_proteins = protein_data[protein_data.apply(
                              match_classification, axis=1)]
    target_ids = filtered_proteins["target_id"].unique()
    # Return the filtered rows from 'data' that match the target_ids and
    merge in the 'Classification' and 'Organism' details.
    return data[data["target_id"].isin(target_ids)].merge(
        filtered_proteins[["target_id", "Classification", "Organism"]],
        on="target_id")

#Filter for the chosen Target ID
target = "Adenosine receptor"
adenosine = keep_protein_class(df, protein_data, classes=[{"15":
target}])
adenosine = adenosine[adenosine['Organism'] == 'Homo sapiens (Human)']
target_id = adenosine["target_id"].unique()
df_clean['target_id'] = df_clean['target_id'].fillna('')
df_target = df_clean[df_clean["target_id"].isin(target_id)]

#Display the number of selected molecules
print(f"Number of molecules with target{target_id}: {len(df_target)}")

#Exclude data with "low" quality
df_quality = df_target[df_target['Quality'] !='low']
print(f"Number of molecules with high quality: {len(df_quality)}")
```

traditional descriptors to create a comprehensive dataset. This dataset captures both macroscopic properties and quantum-level phenomena, thereby enhancing the predictive accuracy of our modeling pipeline for tasks like biological activity modeling and toxicity assessment.

This code snippet exemplifies how quantum descriptors are computed and integrated into the overall descriptor set, providing a more complete representation of each molecule. To illustrate how quantum descriptors are integrated into our workflow, the Python code demonstrates a function that converts a SMILES string into a 3D molecular structure, optimizes its geometry, and then performs a Hartree–Fock calculation using PySCF. The function calculates key quantum descriptors, including the HOMO, LUMO, their energy gap, and the dipole moment, and returns them in a structured dictionary format. To implement an effective AI-driven drug discovery workflow, it is essential to follow a structured process that stays adaptable to different experimental goals. Depending on the nature of the available data and the specific research objectives, the task of predicting biological activity can be framed as either a classification or regression problem.

- Classification Models: When bioactivity data are discretized into categorical labels (e.g., active vs. inactive, high vs. low affinity), classification models are used. These approaches are particularly useful in early-stage screening to

```python
from rdkit import Chem
from rdkit.Chem import Descriptors
from rdkit.Chem.rdFingerprintGenerator import GetMorganGenerator

# Initialize MorganGenerator
morgan_gen = GetMorganGenerator(radius=2, fpSize=2048)

#Function to calculate molecular descriptors
def calc_desc(smiles):
    mol = Chem.MolFromSmiles(smiles)
    fp_bit_vect = morgan_gen.GetFingerprint(mol)
    if mol:
        return {
            'MW': Descriptors.MolWt(mol),
            'LogP': Descriptors.MolLogP(mol),
            'TPSA': Descriptors.TPSA(mol),
            'HBD': Descriptors.NumHDonors(mol),
            'HBA': Descriptors.NumHAcceptors(mol),
            'RotB': Descriptors.NumRotatableBonds(mol),
            'FP': list(fp_bit_vect.ToBitString())
        }
    return None

#Apply the function to the dataset
df_quality['Desc'] = df_quality['SMILES'].apply(calc_desc)
df_desc = pd.json_normalize(df_quality['Desc'].dropna())
```

identify promising compounds, effectively reducing the number of candidates that proceed to costly experimental validation.

- Regression Models: When continuous bioactivity values such as IC50, EC50, or pChEMBL are available, regression models predict numerical outputs, enabling a more granular differentiation of compound potency. This is particularly valuable in QSAR modeling, where small structural variations can significantly impact molecular activity.

4.3 Step 3: Machine Learning Approaches

To build robust and dependable predictive models, various machine learning techniques can be employed, including tree-based models, artificial neural networks, and ensemble learning methods. Each approach offers unique advantages and challenges, depending on dataset characteristics and computational constraints. Random Forest (RF) is a widely used ensemble learning method that constructs multiple decision trees using bootstrap aggregation (bagging). By averaging the predictions from individual trees, RF enhances generalization and mitigates overfitting, making it particularly robust to noisy data and missing values. Due to its interpretability and efficiency, RF is often considered a strong baseline in cheminformatics. XGBoost, a

```
#Select only numeric columns for correlation calculation
numeric_cols = df_desc.drop(columns=['FP'])
correlation_matrix = numeric_cols.corr()

import seaborn as sns

#Plot the correlation heatmap
plt.figure(figsize=(12, 10))
sns.heatmap(correlation_matrix, annot=True, cmap="coolwarm", fmt=".2f")
plt.title("Heatmap Molecular Descriptors Correlations")
plt.show()

#Expand the 'FP' column into numerical columns (one column per bit in the
fingerprint)
assert len(df_desc['FP'][0]) == 2048)
fp_df = pd.DataFrame(
        df_desc['FP'].tolist(),
        columns=[f'FP_{i}' for i in range(len(df_desc['FP'][0]))]
        ).astype(int)

#Merge the expanded fingerprint columns with the rest of the dataset
(excluding the original 'FP' column)
df_desc_exp = pd.concat([df_desc.drop(columns=['FP']), fp_df], axis=1)
df_final = pd.concat([df_quality.reset_index(drop=True), df_desc_exp],
              axis=1)
```

gradient boosting algorithm, sequentially builds decision trees by learning from the errors of previous iterations. Known for its computational efficiency and predictive accuracy, XGBoost is one of the best-performing models for structured data. However, it requires careful hyperparameter tuning to prevent overfitting. Neural Networks (NNs) offer a more flexible approach, capable of capturing complex, high-dimensional relationships in molecular data. Unlike traditional machine learning models, they can autonomously learn features from raw data. This makes them particularly powerful for cheminformatics applications where intricate molecular interactions need to be modeled. However, NNs require larger datasets, substantial computational resources, and careful tuning to prevent overfitting. Choosing the most suitable predictive model depends on numerous factors, such as dataset size, molecular representation, and computational constraints. While tree-based models like RF and XGBoost provide strong baseline performances and high interpretability, neural networks offer greater flexibility in capturing nonlinear molecular relationships. A summary of the key strengths and challenges of each model is provided in Table 3.

The following sections present two distinct pipelines, explaining their differences and the rationale behind technical choices.

A. Classification Pipeline Example

This pipeline converts continuous bioactivity data into binary labels, cleans the data, partitions it into training and testing subsets, and trains a classification

```python
from rdkit import Chem
from rdkit.Chem import AllChem
from pyscf import gto, scf
import pandas as pd

# Function to calculate HOMO, LUMO, gap, and total energy from a given
SMILES string.
def calculate_hf_descriptors(index, smiles):
    try:
        # Convert the SMILES string into a 3D molecular structure.
        mol = Chem.MolFromSmiles(smiles)
        mol = Chem.AddHs(mol)  # Add hydrogen atoms.
        AllChem.EmbedMolecule(mol, AllChem.ETKDG())  # Generate a 3D
        conformation using ETKDG.
        AllChem.MMFFOptimizeMolecule(mol)  # Optimize the 3D geometry
        with the MMFF force field.

        # Extract atomic coordinates and element symbols.
        coords = mol.GetConformer().GetPositions()
        symbols = [atom.GetSymbol() for atom in mol.GetAtoms()]

        # Define the molecule for PySCF using the STO-3G basis set.
        mol_pyscf = gto.M(
            atom=[(symbols[i], tuple(coords[i]))
            for i in range(len(symbols))],
            basis="sto-3g",
        )

        # Perform a Restricted Hartree-Fock (RHF) calculation.
        mf = scf.RHF(mol_pyscf)
        mf.kernel()

        # Determine HOMO, LUMO, and the HOMO-LUMO gap.
        mo_energies = mf.mo_energy
        num_electrons = mol_pyscf.nelectron
        homo_index = num_electrons // 2 - 1
        lumo_index = homo_index + 1
        homo_energy = mo_energies[homo_index]
        lumo_energy = mo_energies[lumo_index]
        homo_lumo_gap = lumo_energy - homo_energy
        dipole = mf.dip_moment()

        # Return the calculated descriptors as a dictionary.
        return {
            "Idx": index,
            "SCF Energy": mf.e_tot,
            "HOMO": homo_energy,
            "LUMO": lumo_energy,
            "HOMO-LUMO Gap": homo_lumo_gap,
```

model. For example, we use an XGBoost classifier. The pipeline begins with careful data preprocessing, molecules are categorized as ‘active’ or ‘inactive’ based on

```
            "Dipole Moment X": dipole[0],
            "Dipole Moment Y": dipole[1],
            "Dipole Moment Z": dipole[2],
            "Dipole Moment Magnitude":
                (dipole[0] ** 2 + dipole[1] ** 2 + dipole[2] ** 2) ** 0.5,
        }
    except Exception as e:
        print(f"Error for SMILES: {smiles}, {e}")
        return None
```

Table 3 Comparison of machine learning models commonly used in drug discovery, highlighting their key strengths (e.g., robustness, accuracy) and typical challenges (e.g., overfitting, data requirements, interpretability)

Model	Advantages	Challenges
Random forest	Feature importance, robust to noise, easy to use	Less effective on sparse data, computationally expensive
XGBoost	High predictive accuracy, fast, built-in regularization	Requires careful tuning, prone to overfitting
Neural networks	Captures complex molecular relationships, adaptable to various data types	Requires large datasets, computationally intensive, tuning required

a predefined threshold, such as a pChEMBL value. Invalid or redundant entries, such as incorrectly formatted SMILES strings or duplicate records, are removed to ensure data quality. Molecular descriptors, fingerprints, and quantum descriptors together offer a comprehensive view of a molecule's properties. These descriptors enrich the dataset and improve the model's ability to predict properties influenced by electronic structure. The following function transforms raw bioactivity data into a binary classification format, categorizing compounds as either active or inactive based on a predefined threshold for the endpoint value. For instance, compounds with pchembl_value_Mean exceeding 6.5 are considered active, while those below or equal to this threshold are labeled inactive. This classification is stored in a new column, Activity_class, to ease downstream modeling.

```
def preprocess_data(data, endpoint, activity_threshold):
    active = data[data[endpoint] > activity_threshold]
    inactive = data[data[endpoint] <= activity_threshold]
    active["Activity_class"] = "A"
    inactive["Activity_class"] = "N"
    return pd.concat([active, inactive],ignore_index=True).drop(
        columns=[endpoint])
```

This step ensures a clear distinction between the classes and removes the original endpoint column to prevent data leakage during model training.

After preprocessing, the data are split into training and test subsets to evaluate model performance. Temporal splitting, which separates molecules based on their discovery timeline, reflects real-world scenarios where models are used to predict the activity of future compounds. Alternatively, random stratified splitting ensures balanced representation of activity classes, addressing the common challenge of unbalanced datasets in cheminformatics.

```
from sklearn.model_selection import train_test_split

def split_data(data, endpoint, split_by="random", split_year=None,
   test_size=0.3, random_state=42):
   if split_by.lower() == "year":
      test_set = data[data["Year"] >= split_year]
      training_set = data[data["Year"] < split_year]
   else:
      training_set, test_set = train_test_split(
         data,
         test_size=test_size,
         stratify=data[endpoint],
         random_state=random_state
      )
   X_train = training_set.drop(columns=[endpoint, "Year", "target_id"])
   y_train = training_set[endpoint]
   X_test = test_set.drop(columns=[endpoint, "Year", "target_id"])
   y_test = test_set[endpoint]
   return X_train, y_train, X_test, y_test
```

This approach ensures that both training and test sets are representative of the overall dataset while keeping the integrity of temporal splits when applicable. Before training the model, balancing the data set is an essential preprocessing step, especially when active molecules are significantly less represented than inactive ones. If ignored, this imbalance can lead to biased predictions where the model favors the majority class, reducing its ability to identify promising candidates. This is particularly problematic in drug discovery, where the proportion of bioactive compounds is typically much lower than inactive ones.

Several resampling techniques can be applied to address this problem:

- Synthetic Minority Over-Sampling Technique (SMOTE): Generates synthetic samples of the minority class by interpolating between existing instances. It helps avoid overfitting, but assumes linear relationships between data points, which may not always be the case for complex molecular data.
- Adaptive Synthetic Sampling (ADASYN): An extension of SMOTE that generates more synthetic samples in regions where the model struggles the most, thereby improving decision boundaries. However, it can introduce noise when applied to highly unbalanced datasets.
- Random Undersampling (RUS): Reduces the number of majority class samples to balance the data set. It is computationally efficient but may discard valuable information if the dataset is small.

The choice of method depends on the characteristics of the dataset and the modeling goals. Oversampling techniques such as SMOTE and ADASYN are preferred when data are limited because they help preserve minority class information, while under-sampling techniques such as RUS are useful for large datasets where computational efficiency is a priority. In addition, hybrid approaches that combine over- and under-sampling can help balance datasets more effectively while minimizing potential drawbacks.

The `train_and_evaluate` function integrates key steps for model preparation, cross-validation, and testing. Cross-validation is a technique for evaluating the reliability of machine learning models, particularly in cases where datasets are small or imbalanced. By splitting the training data into multiple folds, the model's performance can be evaluated on unseen data within each fold, providing a robust estimate of its generalizability. This process minimizes the risk of overfitting, where the model performs well on training data but does not generalize to new, unseen data. In this pipeline, stratified k-fold cross-validation ensures that the class distribution stays balanced across folds, preserving the representativeness of the training and validation sets.

To ensure feature consistency and mitigate the influence of outliers, our workflow employs a dual scaling approach: input features are scaled using `RobustScaler`, which is less sensitive to extreme values than `StandardScaler`, and for regression tasks, the output variable is scaled using `MinMaxScaler` to normalize the target values into a bounded range. This combination not only enhances the stability and convergence of gradient-based optimizers but also facilitates the integration of diverse machine learning models into the pipeline.

In classification tasks, evaluation metrics such as accuracy, precision, recall, and F1 score provide insight into performance, especially in the presence of imbalanced classes. For regression, however, metrics such as R^2, Mean Squared Error (MSE), and Mean Absolute Error (MAE) are used to assess predictive accuracy. By examining these metrics in tandem with the applied scaling techniques, researchers can make informed decisions about model optimization and selection.

By integrating preprocessing, data splitting, training, and evaluation into a unified pipeline, this workflow ensures reproducibility and flexibility. This pipeline provides not only cross-validation scores but also a detailed classification report and accuracy metrics on the test set, enabling a comprehensive evaluation of the model's performance. The pipeline is applied to the df_final dataset using the XGBoost classifier, which serves as an example of a machine learning model suitable for this type of classification task.

B. Regression Pipeline Example

While many preprocessing steps remain consistent with those in the classification pipeline, such as data cleaning, filtering, and feature scaling, the regression pipeline retains the continuous activity values (e.g., pChEMBL, IC50, or EC50) rather than converting them into categorical labels. Consequently, the data splitting function is adapted to avoid stratification, as it is not necessary when predicting continuous outcomes. Once the dataset is carefully preprocessed and appropriately

```
from sklearn.model_selection import StratifiedKFold
from sklearn.preprocessing import RobustScaler, LabelEncoder
from sklearn.metrics import classification_report, accuracy_score
import xgboost

def train_and_evaluate(X_train, y_train, X_test, y_test, model, folds=5,
    scale=True):
    if scale:
        scaler = RobustScaler ()
        X_train = pd.DataFrame(scaler.fit_transform(X_train),
            columns=X_train.columns)
        X_test = pd.DataFrame(scaler.transform(X_test),
            columns=X_test.columns)

    lblenc = LabelEncoder()
    y_train = lblenc.fit_transform(y_train)
    y_test = lblenc.transform(y_test)

    kfold = StratifiedKFold(n_splits=folds, shuffle=True,random_state=42)
    cv_scores = []
    for train_idx, val_idx in kfold.split(X_train, y_train):
        X_fold_tr, X_fold_v = X_train.iloc[train_idx],train.iloc[val_idx]
        y_fold_tr, y_fold_v = y_train[train_idx], y_train[val_idx]
        model.fit(X_fold_tr, y_fold_tr)
        cv_scores.append(model.score(X_fold_v, y_fold_v))

    model.fit(X_train, y_train)
    y_pred = model.predict(X_test)
    accuracy = accuracy_score(y_test, y_pred)
    report = classification_report(y_test, y_pred,
        target_names=lblenc.classes_)

    print(f"Cross-validation scores: {cv_scores}")
    print(f"Test accuracy: {accuracy:.2f}")
    print("Classification report:\n", report)

    return model, accuracy, report

data_class = preprocess_data(df_final, endpoint="pchembl_value_Mean",
                             activity_threshold=6.5)
X_train, y_train, X_test, y_test = split_data(
                                        data_class,
                                        endpoint="Activity_class",
                                        split_by="random",
                                        random_state=42
                                    )
model_class = xgboost.XGBClassifier(verbosity=0)


model_class, accuracy, report = train_and_evaluate(
                                    X_train, y_train,
                                    X_test, y_test,
                                    model_class
                                )
```

Table 4 Overview of classification and regression pipelines in drug discovery, outlining differences in target type, data splitting strategies, modeling approaches, and evaluation metrics

	Classification pipeline	Regression pipeline
Target	Binarized	Continuous
Splitting	Stratified	Random or temporal
Modeling	RF Classifier, XGBoost Classifier, NNs	RF Regressor, XGBoost Regressor, NNs
Evaluation	Accuracy, F1 score, Precision, Recall	R^2, MSE, MAE

partitioned into training, validation, and test sets, the next step is model selection. In this example, we employ a neural network to capture the complex, nonlinear relationships between molecular descriptors and bioactivity. The following block code demonstrates how to build and train a feedforward neural network using TensorFlow-Keras. This architecture includes batch normalization and dropout layers, which help improve stability and prevent overfitting. The Huber loss function is employed to manage outliers effectively, making it a robust choice for QSAR regression.

To provide a clear and concise comparison between the two approaches, the following table (see Table 4) summarizes the main differences between the classification and regression pipelines.

As shown in the table above, while both pipelines share common preprocessing practices and can employ similar modeling techniques, the key differences lie in how the target variable is represented and the evaluation metrics used. In classification, the focus is on balancing the classes and measuring discrete outcomes, whereas regression retains continuous values and evaluates predictive accuracy through different statistical measures.

5 Conclusion

This chapter has presented a practical and reproducible framework for applying AI-driven models, covering all essential stages, including data preprocessing, molecular descriptor generation, model training, and validation. The core challenges of QSAR modeling have been examined, with particular emphasis on the importance of data quality, rigorous validation strategies (such as scaffold-based and time-split validation), and the selection of appropriate molecular representations (fingerprints, physicochemical descriptors, and quantum parameters). These elements are critical to ensuring that AI models are not only accurate but also generalizable and applicable to real-world drug discovery workflows. To facilitate firsthand learning and practical implementation, an interactive Google Colab notebook is provided, featuring step-by-step examples on molecular feature extraction, predictive modeling, and performance evaluation. The notebook can be accessed at: https://colab.research.google.com/gist/MODAL-UNINA/2836bee25fbb6399ee29974e8d58fdb7/qsar-regression-with-deep-learning-a-

```python
import tensorflow as tf
from keras.models import Sequential
from keras.layers import Dense, Dropout, BatchNormalization, ReLU

def build_nn_regression(input_shape, initial_learning_rate=1e-4):
    """ Constructs a feedforward neural network for QSAR regression. """

    model = Sequential([
        # Input Layer + First Hidden Layer
        Dense(256, input_shape=(input_shape,)),
        BatchNormalization(),
        ReLU(),
        Dropout(0.3),

        # Second Hidden Layer
        Dense(64),
        BatchNormalization(),
        ReLU(),
        Dropout(0.2),

        # Third Hidden Layer
        Dense(16),
        BatchNormalization(),
        ReLU(),
        Dropout(0.1),

        # Fourth Hidden Layer
        Dense(4),
        BatchNormalization(),
        ReLU(),

        # Fifth Hidden Layer
        Dense(2),
        ReLU(),

        # Output Layer
        Dense(1)  # Regression output (continuous value)
    ])

    # Compile the model with RMSprop optimizer and Huber loss
    model.compile(optimizer=tf.keras.optimizers.RMSprop(
                                learning_rate=initial_learning_rate),
                  loss=tf.keras.losses.Huber(delta=1.0),
                  metrics=["mae", "mse"]
    )

    return model
```

practical-guide.ipynb. It is essential to emphasize that the architecture included in the notebook is intended solely for illustrative and demonstrative purposes. These models have been deliberately designed to be lightweight and optimized for execution in resource-constrained environments, ensuring seamless operation on Google Colab without requiring advanced computational resources. The primary objective of the notebook is to enhance practical understanding and reproducibility, rather than to develop computationally expensive or complex models. The methodologies and techniques outlined in this chapter aim to provide a solid and adaptable foundation, enabling the implementation of optimal QSAR modeling architectures. The future of AI-driven drug discovery will be shaped by more advanced methodologies, including Graph Neural Networks, Transformer architectures, and generative models. By following the approaches described, it is possible to explore and continuously refine AI applications in drug discovery, paving the way for more efficient and innovative solutions.

References

Abadi M, Barham P, Chen J, Chen Z, Davis A, Dean J, Devin M, Ghemawat S, Irving G, Isard M, et al. Tensorflow: a system for large-scale machine learning. In: 12th {USENIX} Symposium on operating systems design and implementation ({OSDI} 16). 2016. p. 265–83.

Bahia MS, Kaspi O, Touitou M, Binayev I, Dhail S, Spiegel J, Khazanov N, Yosipof A, Senderowitz H. A comparison between 2D and 3D descriptors in QSAR modeling based on bio-active conformations. Mol Inform. 2023;42(4):2200186.

Bento AP, Gaulton A, Hersey A, Bellis LJ, Chambers J, Davies M, Krüger FA, Light Y, Mak L, McGlinchey S, et al. The ChEMBL bioactivity database: an update. Nucleic Acids Res. 2014;42(D1):D1083–90.

Béquignon OJM, Bongers BJ, Jespers W, Ijzerman AP, van der Water B, van Westen GJP. Papyrus: a large-scale curated dataset aimed at bioactivity predictions. J Chem. 2023;15(1):3. https://doi.org/10.1186/s13321-022-00672-x.

Bisong E. Google colaboratory. In: Bisong E, editor. Building machine learning and deep learning models on google cloud platform: a comprehensive guide for beginners. Apress; 2019. p. 59–64. https://doi.org/10.1007/978-1-4842-4470-8_7.

Bosc N, Atkinson F, Felix E, Gaulton A, Hersey A, Leach AR. Large scale comparison of QSAR and conformal prediction methods and their applications in drug discovery. J Chem. 2019;11(1):4. https://doi.org/10.1186/s13321-018-0325-4.

Cherkasov A, Muratov EN, Fourches D, Varnek A, Baskin II, Cronin M, Dearden J, Gramatica P, Martin YC, Todeschini R, Consonni V, Kuzmin VE, Cramer R, Benigni R, Yang C, Rathman J, Terfloth L, Gasteiger J, Richard A, Tropsha A. QSAR modeling: where have you been? Where are you going to? J Med Chem. 2014;57(12):4977–5010. https://doi.org/10.1021/jm4004285.

Chinh Nguyen H, Alamray F, Kamal M, Diana T, Mohamed A, Algarni M, Su C-H. Computational prediction of drug solubility in supercritical carbon dioxide: thermodynamic and artificial intelligence modeling. J Mol Liq. 2022;354:118888. https://doi.org/10.1016/j.molliq.2022.118888.

De P, Kar S, Ambure P, Roy K. Prediction reliability of QSAR models: an overview of various validation tools. Arch Toxicol. 2022;96(5):1279–95. https://doi.org/10.1007/s00204-022-03252-y.

Dearden JC. The history and development of quantitative structure-activity relationships (QSARs). In: Oncology: breakthroughs in research and practice. IGI Global; 2017. p. 67–117.

Deng J, Yang Z, Ojima I, Samaras D, Wang F. Artificial intelligence in drug discovery: applications and techniques. Brief Bioinform. 2022;23(1):bbab430.

DiMasi JA, Grabowski HG, Hansen RW. Innovation in the pharmaceutical industry: new estimates of R&D costs. J Health Econ. 2016;47:20–33. https://doi.org/10.1016/j.jhealeco.2016.01.012.

Ekins S, Puhl AC, Zorn KM, Lane TR, Russo DP, Klein JJ, Hickey AJ, Clark AM. Exploiting machine learning for end-to-end drug discovery and development. Nat Mater. 2019;18(5):435–41. https://doi.org/10.1038/s41563-019-0338-z.

Golbraikh A, Wang XS, Zhu H, Tropsha A. Predictive QSAR modeling: methods and applications in drug discovery and chemical risk assessment. In: Handbook of computational chemistry; 2017. p. 2303–40.

Grisoni F, Ballabio D, Todeschini R, Consonni V. Molecular descriptors for structure–activity applications: a hands-on approach. In: Computational toxicology: methods and protocols. New York: Humana Press; 2018. p. 3–53.

Guo Q, Hernandez-Hernandez S, Ballester PJ. Scaffold splits overestimate virtual screening performance. In: Wand M, Malinovská K, Schmidhuber J, Tetko IV, editors. Artificial neural networks and machine Learning – ICANN 2024. Cham: Springer; 2024. p. 58–72.

Gupta R, Srivastava D, Sahu M, Tiwari S, Ambasta RK, Kumar P. Artificial intelligence to deep learning: machine intelligence approach for drug discovery. Mol Divers. 2021;25:1315–60.

Kim H, Kim E, Lee I, Bae B, Park M, Nam H. Artificial intelligence in drug discovery: a comprehensive review of data-driven and machine learning approaches. Biotechnol Bioprocess Eng. 2020;25(6):895–930. https://doi.org/10.1007/s12257-020-0049-y.

Kluyver T, Ragan-Kelley B, Pérez F, Granger B, Bussonnier M, Frederic J, Kelley K, Hamrick J, Grout J, Corlay S, Ivanov P, Avila D, Abdalla S, Willing C. Jupyter Notebooks – a publishing format for reproducible computational workflows. In: Loizides F, Schmidt B, editors. Positioning and power in Academic Publishing: players, agents and agendas. IOS Press; 2016. p. 87–90.

Landrum G. RDKit: open-source cheminformatics software. 2016. https://github.com/rdkit/rdkit/releases/tag/Release_2016_09_4

Maia EHB, Assis LC, De Oliveira TA, Da Silva AM, Taranto AG. Structure-based virtual screening: from classical to artificial intelligence. Front Chem. 2020;8:343.

Mansouri K, Moreira-Filho JT, Lowe CN, Charest N, Martin T, Tkachenko V, Judson R, Conway M, Kleinstreuer NC, Williams AJ. Free and open-source QSAR-ready workflow for automated standardization of chemical structures in support of QSAR modeling. J Chem. 2024;16(1):19. https://doi.org/10.1186/s13321-024-00814-3.

Pedregosa F, Varoquaux G, Gramfort A, Michel V, Thirion B, Grisel O, Blondel M, Prettenhofer P, Weiss R, Dubourg V, et al. Scikit-learn: machine learning in Python. J Mach Learn Res. 2011;12(Oct):2825–30.

Schneider P, Walters WP, Plowright AT, Sieroka N, Listgarten J, Goodnow RA, Fisher J, Jansen JM, Duca JS, Rush TS, Zentgraf M, Hill JE, Krutoholow E, Kohler M, Blaney J, Funatsu K, Luebkemann C, Schneider G. Rethinking drug design in the artificial intelligence era. Nat Rev Drug Discov. 2020;19(5):353–64. https://doi.org/10.1038/s41573-019-0050-3.

Sheridan RP. Time-split cross-validation as a method for estimating the goodness of prospective prediction. J Chem Inf Model. 2013;53(4):783–90. https://doi.org/10.1021/ci400084k.

Staszak M, Staszak K, Wieszczycka K, Bajek A, Roszkowski K, Tylkowski B. Machine learning in drug design: use of artificial intelligence to explore the chemical structure–biological activity relationship. Wiley Interdiscip Rev Comput Mol Sci. 2022;12(2):e1568.

Sun J, Jeliazkova N, Chupakhin V, Golib-Dzib J-F, Engkvist O, Carlsson L, Wegner J, Ceulemans H, Georgiev I, Jeliazkov V, Kochev N, Ashby TJ, Chen H. ExCAPE-DB: an integrated large scale dataset facilitating Big Data analysis in chemogenomics. J Chem. 2017;9(1):17. https://doi.org/10.1186/s13321-017-0203-5.

Van Rossum G, Drake FL. Python 3 reference manual. CreateSpace; 2009.

van Tran TT, Surya Wibowo A, Tayara H, Chong KT. Artificial intelligence in drug toxicity prediction: recent advances, challenges, and future perspectives. J Chem Inf Model. 2023;63(9):2628–43. https://doi.org/10.1021/acs.jcim.3c00200.

Walters WP, Barzilay R. Applications of deep learning in molecule generation and molecular property prediction. Acc Chem Res. 2021;54(2):263–70. https://doi.org/10.1021/acs.accounts.0c00699.

Zhao L, Wang W, Sedykh A, Zhu H. Experimental errors in QSAR modeling sets: what we can do and what we cannot do. ACS Omega. 2017;2(6):2805–12. https://doi.org/10.1021/acsomega.7b00274.

AI-Based Platforms for Drug Discovery: Current Tools and Human-Centered Design Strategies

Maria Cristina Lomuscio, Maria Giovanna Eva Papadopoulos, Miriana Calvano, Vittoria Nanna, Antonio Curci, Andrea Esposito, Nicola Corriero, Rosa Lanzilotti, Antonio Piccinno, Domenico Alberga, and Giuseppe Felice Mangiatordi

Abstract This chapter presents an overview of ten selected Artificial Intelligence (AI)-driven web platforms designed to support the early stages of the drug discovery process. Its primary aim is to serve as both a practical guide and a critical review for potential users of such platforms, including those without cheminformatics or programming expertise. Additionally, we underline the limitations of these platforms and explore potential enhancements through Human-Centered Design (HCD) strategies to help improve both existing and future web-based platforms for optimal use.

Keywords Web-platforms · Bioactivity/toxicology Prediction · Generative models · Usability · Human-centered design

Artificial Intelligence (AI)-driven methods and tools are playing an increasingly crucial role in accelerating the early phases of the Drug Discovery process, thanks to the growing availability of freely accessible repositories containing bioactivity and toxicity data. The integration of AI has created new and alternative pathways that promise to reduce both time and resources required for developing novel

M. C. Lomuscio
Department of Precision and Regenerative Medicine and Jonian Area, University of Bari Aldo Moro, Bari, Italy
e-mail: maria.lomuscio@uniba.it

M. G. E. Papadopoulos · V. Nanna · N. Corriero · D. Alberga · G. F. Mangiatordi (✉)
Institute of Crystallography, National Research Council, Bari, Italy
e-mail: mariagiovannaevapapadopoulos@cnr.it; vittoriananna@cnr.it; nicola.corriero@cnr.it; domenico.alberga@cnr.it; giuseppefelice.mangiatordi@cnr.it

M. Calvano · A. Curci · A. Esposito · R. Lanzilotti · A. Piccinno
Department of Computer Science, University of Bari Aldo Moro, Bari, Italy
e-mail: miriana.calvano@uniba.it; antonio.curci@uniba.it; andrea.esposito@uniba.it; rosa.lanzilotti@uniba.it; antonio.piccinno@uniba.it

A. Lavecchia (ed.), *Applied Artificial Intelligence for Drug Discovery*,
https://doi.org/10.1007/978-3-031-98022-0_22

therapeutics. A key advantage of machine-driven hypothesis generation is its ability to design new compounds based on multiple pharmacokinetic (Absorption, Distribution, Metabolism, Excretion, and Toxicity; ADMET) and pharmacodynamic (selectivity, bioactivity) criteria simultaneously, including potential off-target and side effects (Schneider 2018). These approaches enable a more comprehensive evaluation of compounds during the early stages of drug development, even before chemical synthesis occurs.

To ensure broad applicability of AI-based models within the scientific community, it is increasingly common for them to be made directly accessible to the public through intuitive and user-friendly web platforms. While current platforms are designed for users with varying levels of expertise in the field, to ensure intuitive and accessible interaction, they are predominantly developed by researchers with a medicinal chemistry background, particularly in cheminformatics. True advancements of these web-based platforms in this field necessitate the collaboration between medicinal chemistry and computer science, specifically through the integration of the Human-Computer Interaction (HCI) principles. To create systems that place users at the core of the process, Human-Centered Design (HCD) should be integrated to create solutions that effectively address users' needs, preferences, and behaviors while ensuring a positive experience. Regrettably, HCD concepts continue to be largely undervalued by researchers creating web platforms for drug discovery.

The aim of this chapter is to take a first step in bridging this gap by offering an in-depth analysis of AI-based platforms designed to assist experts in the early phases of drug discovery, while also emphasizing the significance of HCI basis. A selection of web-based platforms will be presented providing a practical and useful guide for potential users, including those without cheminformatics or programming skills. At the same time, AI platforms' limitations will be displayed, along with suggestions for improvements, particularly in terms of usability (cf. Designing Usable AI-Based Strategies). Finally, we will consider a practical example: a critical review of a platform, based on HCI methods, will be provided, demonstrating how usability can be enhanced to facilitate and make the interaction process more effective.

1 AI-Based Platforms for Drug Discovery: Selected Examples

In this section, we will present ten cheminformatics platforms available on the web and realized to support researchers involved in the early phases of the drug discovery process. The input required by these tools is generally represented by a two-dimensional (2D) structure or a Simplified Molecular Input Line Entry System (SMILES; Weininger 1988) representation of one or more molecules, provided by the user.

Considering their aim and the nature of the tasks they address, these platforms can be classified as either predictive or generative. The former ones are designed to predict chemical-physical, pharmacological, or toxicological properties of molecules based on Quantitative Structure-Activity Relationship (QSAR) models trained for prediction. Notably, when query compounds significantly differ from the chemicals used to train a QSAR model, the reliability of predictions is compromised. To ensure confidence in predictions, an Applicability Domain (AD) is often defined. The AD outlines the chemical space covered by the model, specifying the range within which predictions can be considered reliable (Gadaleta et al. 2016). On the other hand, generative platforms go beyond prediction, using more advanced Machine Learning (ML) methods, called Deep Learning (DL), with algorithms such as Recurrent Neural Networks (RNN), to automatically design new molecules with specific desired characteristics (Gupta et al. 2018). They could be used to (i) generate novel, potentially bioactive compounds, (ii) generate molecules that explore a new chemical space, and (iii) optimize existing molecules to improve their physicochemical properties (Lamanna et al. 2023). Notably, some cheminformatics platforms integrate both predictive and generative functionalities, enabling simultaneous molecular property estimation and de novo molecule design.

Based on a predefined set of key features, we analyze a selection of platforms that were conceived for virtual screening, binding affinity prediction, ADMET profiling, and automatic design of novel drug-like compounds. More specifically, for each platform we examine, a technical sheet specifying certain characteristics will be provided. First, the platform name will be presented including the specific version (when available), followed by general information such as the year of development and the research group behind its creation. The field *Category* will specify whether the platform is predictive or generative, detailing the type of prediction it performs. This may include tasks such as the prediction of toxicological or pharmacological activity, affinity, selectivity, ADMET properties, target fishing, bioactivity profiling, or de novo design. The *Endpoint* section will offer a detailed explanation of the platform's final aim, clarifying its intended application. Meanwhile, *Background and Significance* will feature the specific objectives and relevance of the platform for the scientific community, providing insights into its importance, particularly concerning specific protein targets. The technical aspects will be covered in *Technical information*, where the computational methods and models employed by the platform will be described. Additionally, if the *AD* is considered within the platform, it will be explicitly stated. Furthermore, *Access* will specify whether the platform is publicly available, whether registration is required, and will include a link to the webpage. The nature of the data input will be detailed in the *Input* section, explaining both the type and modality of the required information. Similarly, the *Output* section will outline the type and format of the results provided, accompanied by a visual representation (Figs. 1–10). Information about updates and engagement with the scientific community will be covered in *Community and updates*, where the year and publication of any available updates will be recorded. For user assistance, the *Technical support* section will list available resources such as contact forms, or direct contact addresses. Finally, *Further information* will include supplementary details that may be useful for users, such as guidelines, instructions, and estimated processing time, whenever these details are provided.

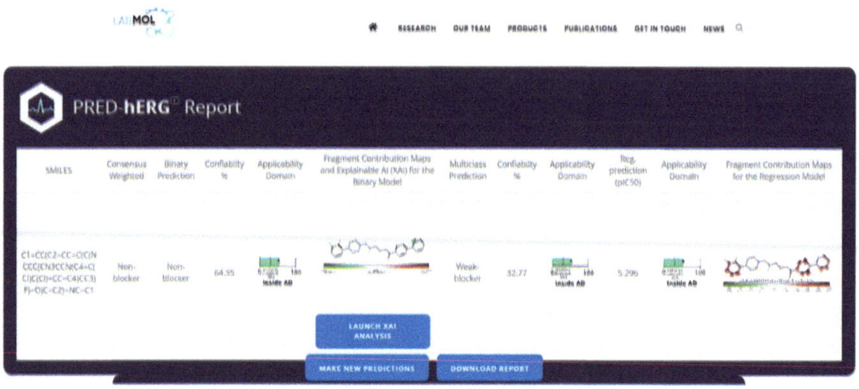

Fig. 1 Screenshot of the output generated by the PRED-hERG web platform. (http://predherg.labmol.com.br/). (Figure reproduced from Braga et al. 2014, 2015 using as input SMILES: `C1C=CN=C(C2C=CC(C(=O)NCCC(F)CN3CCN(C4C=CC=C(Cl)C = 4Cl)CC3) = CC = 2) C = 1))`

1.1 PRED-hERG (5.0)

Developed in 2014 (Braga et al. 2014, 2015) by the Laboratory for Molecular Modeling and Drug Design (LabMol, http://insightai.labmol.com.br/), which was established at the Faculty of Pharmacy of the Federal University of Goiás in Brazil.

Category: Predictive—classification and regression prediction of toxicological activity.

Endpoint: The platform performs binary and multiclass classification of chemicals that may induce cardiotoxicity by blocking the human Ether-à-go-go Related Gene (hERG) channel (Taglialatela et al. 1998). The compounds are categorized as strong blockers, moderate/weak blockers, and non-blockers. Through a regression model, it also provides a prediction of the pIC_{50} resulting from the hERG-related block.

Background and Significance: hERG is a transmembrane potassium channel involved in the regulation of cardiac action potential. Binding of structurally diverse drugs can cause hERG blockage-induced cardiotoxicity associated with ventricular arrhythmia and, in extreme cases, sudden death (Taglialatela et al. 1998; Cavalluzzi et al. 2020).

Technical information: The tool includes two classification models, namely binary and multiclass, and a regression model. All the three tasks (binary, multiclass, and regression) are integrated into a weighted consensus prediction. The platform was trained on 14,364 chemicals with hERG data extracted from ChEMBL database (Gaulton et al. 2012).

AD: The reliability of the prediction, based on the AD of the model, is reported in the platform output. Specifically, it indicates whether the input molecule falls within the AD, considering a similarity threshold of 30%.

Access: The platform is freely accessible at http://predherg.labmol.com.br/ and no registration is required.

Input: The user can either draw the 2D chemical structure of the query using the drawing tool Ketcher (Version 2.10.0; Karulin and Kozhevnikov 2011) provided in the app or paste the compound's SMILES string directly. For batch evaluations, a csv or sdf file, containing the SMILES of multiple compounds, can be uploaded. Once the compounds have been drawn or uploaded, the evaluation process begins by clicking the button 'Predict'. When the analysis is complete, the app displays the results.

Output: PRED-hERG provides (i) a classificatory and multiclass prediction of hERG blockage, with an associated percentage of confiability score, (ii) a regression prediction, with the pIC_{50} value estimation, (iii) the reliability assessment of the predictions, based on the AD of the model, (iv) a probability map, highlighting the contribution of specific chemical fragments to each prediction, and *v*) a prediction based on the consensus of the binary, multiclass, and regression models, to better assist the decision-making process of the user. An example of the platform output is available in Fig. 1. The user can also launch an eXplainable AI (XAI) analysis, enabling the visualization of SHapley Additive exPlanations (SHAP) values (Lundberg and Lee 2017) to better understand the contribution of individual features to binary classification.

Community and updates: The platform does not include user communication features. It has been updated in Alves et al. (2018a, b) and Sanches et al. (2024).

Technical support: The platform includes a 'GET IN TOUCH' button which allows users to contact the research group by completing a form to report any issues they encounter.

Further information: The platform provides detailed instructions to guide users through the input process, ensuring ease of use even for those with limited technical expertise. However, it does not allow customizing settings or parameters. Once processing is completed, the system supplies the outcomes in a few seconds. The output is presented in a clear and structured manner, with a well-organized interface that facilitates interpretation and navigation.

1.2 PRED-Skin (3.0)

Developed in 2017 (Alves et al. 2016; Braga et al. 2017) by LabMol at the Faculty of Pharmacy of the Federal University of Goiás, Brasil.

Category: Predictive—classification prediction of toxicological activity.

Endpoint: The platform performs a binary classification of chemicals based on their skin sensitization potential, categorizing compounds as sensitizer/non-sensitizer.

Background and Significance: Skin sensitization is an immunological response to certain chemicals that can manifest as an inflammatory skin reaction, mediated by delayed-type T cells causing allergic contact dermatitis (Kimber et al. 2002).

Technical information: The tool employs five QSAR models built for different skin sensitization assays, including the Direct Peptide Reactivity Assay (DPRA, *in chemico;* OECD 2024), Local Lymph Node Assay (LLNA, in vivo*;* OECD 2010), KeratinoSens (Casati et al. 2014), and the human Cell Line Activation Test (h-CLAT, in vitro*;* EU 2009) and human assays. Additionally, a Bayesian model is provided as a consensus classifier integrating predictions from all the other models.

AD: In the platform output, an indication of whether the compound is inside or outside the AD for each prediction model is provided.

Access: The platform is freely accessible at http://predskin.labmol.com.br/ and no registration is required.

Input: The user has two options for inputting its query molecule: they can either draw its 2D structure using the JSME canvas applet (Bienfait and Ertl 2013) or directly enter a SMILES string into the designated text field. Subsequently, the user can click the red 'Predict Skin Sensitization' button to launch the prediction and obtain the results.

Output: The results are displayed as a table showing the predicted skin sensitization potential. The predictions are based on an in silico integrated approach employing two models with nonanimal data, namely the *Molecular initiating event* and the *Cellular response.* Moreover, two in vivo models are being used; specifically, one built on LLNA (*Tissue/Organ response)*, and one built on human data (*Organism response).* The predictions made by these models serve as inputs for the Bayesian model, which, ultimately, generates a final ensemble prediction for the skin sensitization response in humans (Fig. 2). Alongside the predictions and their confiability, the platform assesses the compliance of the input molecule with the AD and generates maps highlighting the predicted fragment contributions to the overall response.

Community and updates: User communication features are not included. It has been updated in Alves et al. (2018a, b) and Borba et al. (2021).

Technical support: A 'GET IN TOUCH' button which allows the user to contact the research group is available.

Further information: The platform provides step-by-step instructions to guide users through the input process. Once processing is complete, the system delivers results within seconds. However, the platform does not allow customization of settings or parameters. While the output is presented in a structured format within a well-organized interface, the results page is densely packed with text-based information. This can reduce intuitiveness and increase the time required to locate and fully comprehend the output data.

1.3 CYTO-Safe

Developed in December of 2024 (Feitosa et al. 2024) by LabMol at the Faculty of Pharmacy of the Federal University of Goiás, Brasil.

Category: Predictive—classification prediction of toxicological activity.

Endpoint: The platform performs binary classification of chemicals based on their cytotoxic potential, categorizing compounds as toxic/nontoxic.

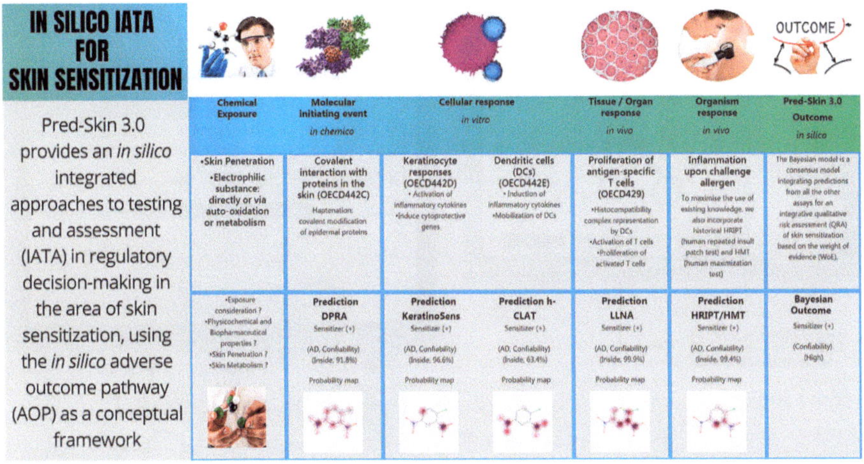

Fig. 2 Example of the output reported by the PRED-SKIN (3.0) web platform. (http://predskin.labmol.com.br/). (Figure reprinted with permission from Borba et al. (2021). Copyright 2020 American Chemical Society)

Background and Significance: Cytotoxicity is the ability of a chemical to cause damage on cellular health, often causing cell death (Khalef et al. 2024). Assessing the cytotoxic potential of molecular structures can help prioritize low-risk compounds for further validation in the early stages of the drug discovery process.

Technical information: The platform utilizes two QSAR models based on a PubChem dataset of approximately 90,000 compounds, evaluated against two cell lines, 3 T3 and HEK-293.

AD: It is considered by the platform and reported in the output specifying a similarity threshold of 10%.

Access: The platform is freely accessible at http://cytosafe.labmol.com.br/ and no registration is required.

Input: The user can draw the 2D chemical structure of the query using the drawing tool Ketcher (Version 2.10.0) provided in the app. Alternatively, it is possible to paste the compound's SMILES string directly into the drawing tool. For batch evaluations, a csv or sdf file containing the SMILES of multiple compounds can be uploaded. Once the compound(s) have been drawn or uploaded, by clicking the button 'Predict' the evaluation process begins. When the analysis is complete, the platform displays the results.

Output: The output includes a binary toxicity prediction for each cell line (toxic/nontoxic), along with a confidence percentage, and the AD assessment for the input molecule. Additionally, users can initiate an XAI analysis via a dedicated button, enabling the visualization of molecular diagrams and heatmaps and the identification of specific molecular regions that contribute to either 'cytotoxic' (red) or 'noncytotoxic' (green) outcomes (Fig. 3).

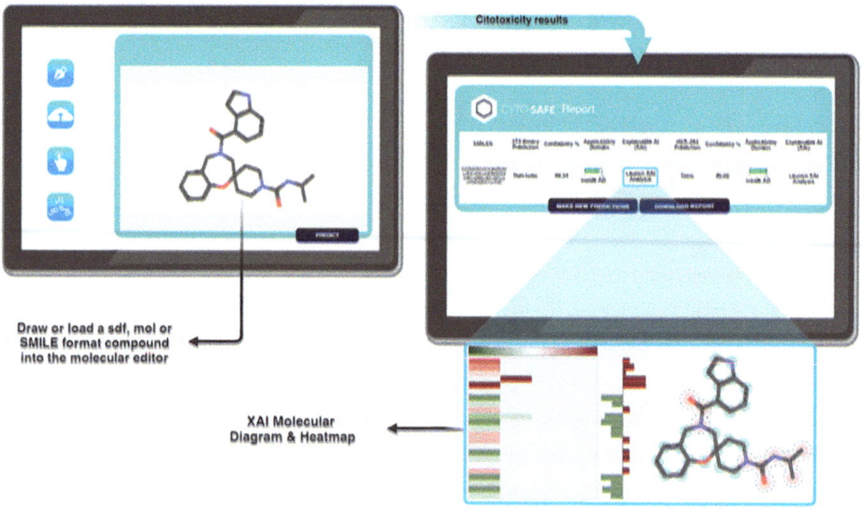

Fig. 3 General scheme of usage, outcome, and XAI of Cyto-Safe web app (http://cytosafe.labmol.com.br/. (Figure reproduced from Feitosa et al. 2024, https://creativecommons.org/licenses/by/4.0/)

Community and updates: The platform does not include user communication features and has not yet received updates.

Technical support: The platform includes a 'GET IN TOUCH' button in the top-right section, which allows users to complete a form to report any issues they encounter.

Further information: The platform provides detailed instructions to guide users through the input process. Once processing is complete, the system delivers results within seconds. However, the platform does not support customization of settings or parameters. The output is presented in a clear and structured manner, with a well-organized interface that facilitates interpretation and navigation. Publications for more information are not yet available, as the dedicated section currently displays the message 'Will be available soon'.

1.4 AMALPHI

Developed in 2024 (Lomuscio et al. 2024) by our research group as a result of the ongoing collaboration between the Institute of Crystallography, National Research Council (IC-CNR), and the Department of Pharmacy-Pharmaceutical Sciences, University of the Studies of Bari "Aldo Moro", Italy.

Category: Predictive—classification prediction of toxicological activity.

Endpoint: The platform performs binary classification of chemicals based on their Phospholipidosis (PLD) potential, categorizing compounds as either PLD inducers or non-inducers.

Background and Significance: PLD is a side effect associated with prolonged exposure to drug-like compounds, predominantly Cationic Amphiphilic Drugs (CADs), characterized by the accumulation of phospholipids in cells, particularly within lysosomes, of different organs, including liver and kidneys (Lüllmann et al. 1975).

Technical information: The tool relies on a Balanced Random Forest (BRF) algorithm used to build a classifier whose training was based on high-quality chemical collection comprising 545 curated small molecules extracted from ChEMBL version (v) 30.

AD: A binary output (yes/no) on prediction reliability, based on AD, is included.

Access: The platform is freely accessible for academic users at https://www.ba.ic.cnr.it/softwareic/amalphiportal/ and requires user registration with an email address.

Input: The user has two options for inputting its query molecule: they can either draw its 2D structure using the JSME canvas applet or directly enter a SMILES string into the designated text field. Additionally, for virtual screening applications, the user can upload a text file containing a list of SMILES strings by selecting the 'Massive' button.

Output: AMALPHI predicts the potential of each compound to act as a PLD inducer. Results are displayed as 'YES' if the BRF model predicts the compound to induce PLD, and 'NO' if it does not. The prediction is provided along with a confidence level (Fig. 4). The user can download the results as a csv file, and a link to download the predictions is also sent to the user's registered email address. Moreover, the platform 'History' page keeps a complete record of all user executions, including the uploaded input SMILES files and their corresponding outputs, ensuring easy access to past analyses.

Community and updates: The platform does not include user communication features and has not yet received updates.

Technical support: The platform offers technical support through a section accessible by clicking the 'Contact' button. Here, the user can complete a form to report any issues they encounter.

Further information: The platform provides the output in a few seconds with a well-organized interface. Settings or parameters customization are not provided.

1.5 SIGMAP

Developed in 2024 (Lomuscio et al. 2025) by our research group as a result of the ongoing collaboration between the IC-CNR, the Department of Computer

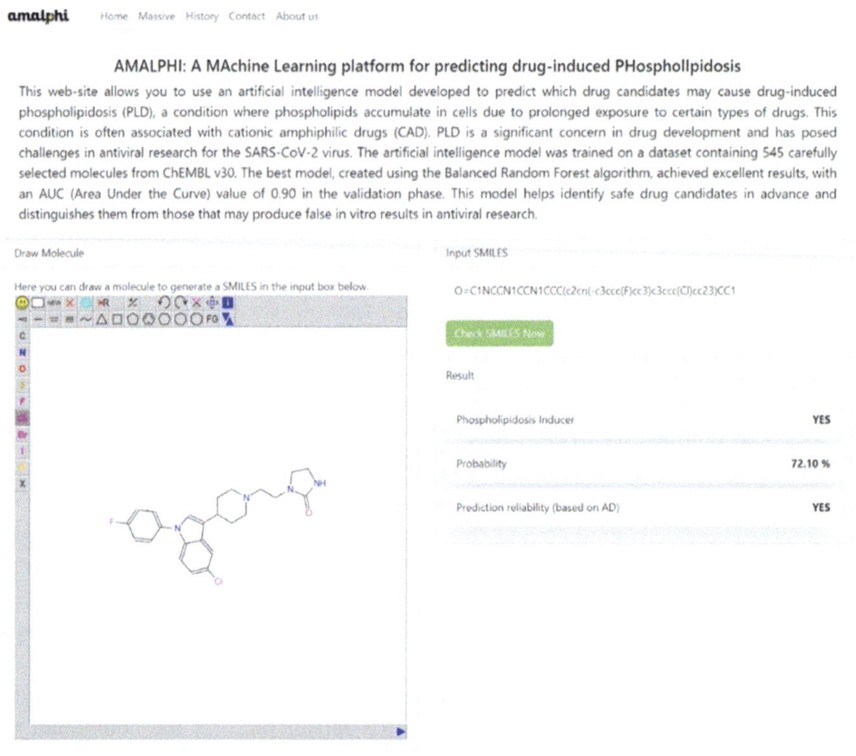

Fig. 4 Screenshot of the output page returned by the AMALPHI web platform (https://www.ba.ic.cnr.it/softwareic/amalphiportal/). (Figure reproduced from Lomuscio et al. 2024, https://creativecommons.org/licenses/by-nc-nd/4.0/)

Science, and the Department of Pharmacy-Pharmaceutical Sciences, University of the Studies of Bari "Aldo Moro", Italy.

Category: Predictive—classification prediction of pharmacological affinity.

Endpoint: The platform performs binary classification of chemicals based on their affinity potential for Sigma-1 Receptor, categorizing compounds as high affinity S1R ligands (yes/no).

Background and Significance: S1R is involved in the pathological processes of neurodegenerative diseases (Couly et al. 2023) cancer progression (Fallica et al. 2021), and viral infections, including COVID-19 (Gordon et al. 2020; Abatematteo et al. 2023). Hence, developing S1R modulators is considered a valuable therapeutic approach for addressing these diverse medical conditions.

Technical information: The system uses an ML-based model employing the support vector machine algorithm, trained on a curated dataset of high-quality bioactivity data. The dataset consists of 2,967 chemicals extracted from ChEMBL v33 and experimentally tested as potential S1R modulators.

AD: The platform includes a binary output (yes/no) for the prediction reliability, based on AD.

Access: SIGMAP is freely accessible for academic users at https://www.ba.ic.cnr.it/softwareic/sigmap/ and requires user registration with an email address.

Input: The user has two options for inputting its query molecule: they can either draw its 2D structure using the JSME canvas applet or directly enter a SMILES string into the designated text field. Additionally, the user can upload a txt file containing a list of SMILES strings by selecting the 'Massive' button.

Output: If the model identifies a molecule as S1R high-affinity binder, the result is shown as 'YES'. If low or no affinity is predicted, it is displayed as 'NO'. For predictions of high affinity, the system also provides a confidence level. When a unique query molecule is entered, SIGMAP offers additional insights through SHAP and Contrastive Explanation-based analyses (Jacovi et al. 2021), aiding in the interpretation of the results (Fig. 5). Through the Contrastive Explanation, SIGMAP can also generate 10 structural analogs of the input molecule providing the user with both similar (predicted as the input molecule) and dissimilar (counterfactual) examples. Similar examples can strengthen the classifier robustness by validating the model stability within the chemical prediction space. Additionally, counterfactual examples make the analysis actionable by bringing out small structural modifications that would alter the model prediction (Wellawatte et al. 2023). This analysis is achieved using a generative model named (Alberga et al. 2024) *DeLA-DrugSelf*, with which SIGMAP is integrated.

Community and updates: The platform does not include user communication features and has not yet received updates.

Technical support: The platform offers technical support through a section accessible by clicking the 'Contact' button. Here, the user can complete a form to report any issues they encounter.

Further information: The platform provides the output within minutes with a structured manner, facilitating interpretation. However, the platform does not support customization of settings or parameters.

1.6 Alpaca

Developed in 2023 (Delre et al. 2023) by our research group as a result of the ongoing collaboration between IC-CNR and the Department of Pharmacy-Pharmaceutical Sciences, University of the Studies of Bari "Aldo Moro", Italy.

Category: Predictive—classification prediction of pharmacological affinity and selectivity.

Endpoint: The platform performs binary classification of chemicals based on their potential affinity for Cannabinoid Receptor 1 (CB1R) and Cannabinoid Receptor 2 (CB2R) and prediction of their CB2R selectivity.

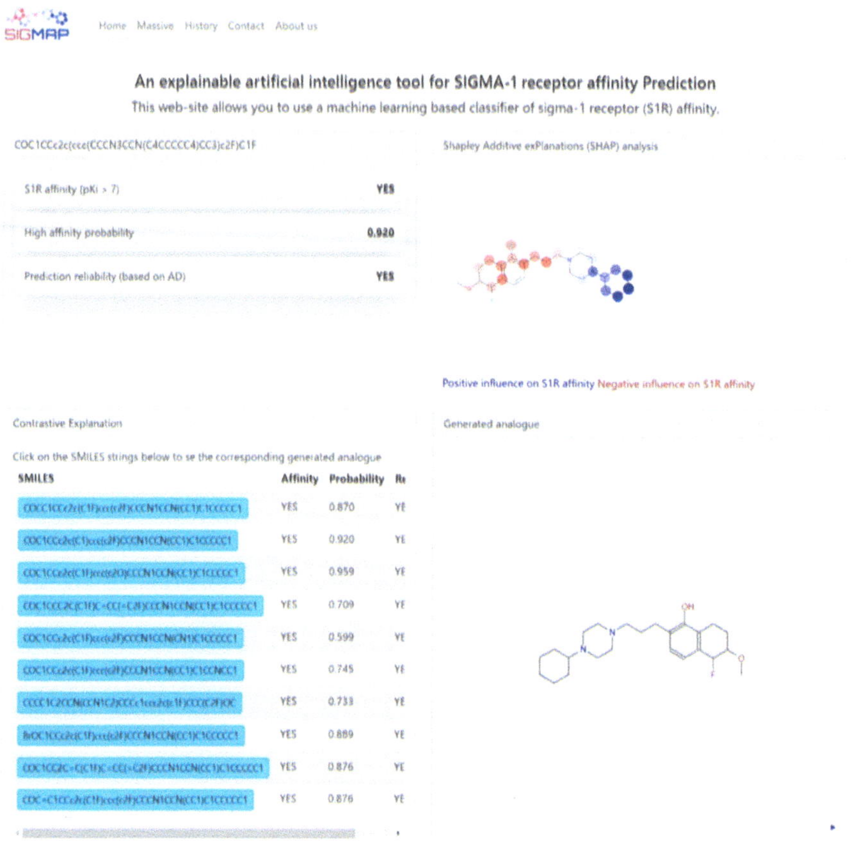

Fig. 5 Screenshot of the output page returned by the SIGMAP web platform (https://www.ba.ic.cnr.it/softwareic/sigmap/). (Figure reproduced from Lomuscio et al. 2025, https://creativecommons.org/licenses/by/3.0/)

Background and Significance: CB1R and CB2R are involved in several disorders linked to inflammation (Turcotte et al. 2016), such as neurodegenerative diseases, cancer, and neuropathic pain. CB2R is overexpressed in inflammatory state, suggesting that developing selective CB2R agents may be an effective strategy for treating neuroinflammation, while avoiding the psychotropic side effects associated with CB1R stimulation in the central nervous system (Komorowska-Müller and Schmöle 2020; Tanaka et al. 2020).

Technical information: The tool uses multiple random forest-based models, trained on three different curated datasets, comprising 3514 (for CB2R affinity models), 3846 (for CB1R affinity models), and 2183 (for CB2R/CB1R selectivity model) chemicals extracted from ChEMBL v30.

AD: The platform evaluates the input molecule and provides a binary output indicating whether it falls within or outside the AD.

Access: ALPACA is freely accessible for academic users at https://www.ba.ic.cnr.it/softwareic/alpaca/ and requires user registration with an email address.

Input: The user has two options for inputting its query molecule: they can either draw its 2D structure using the JSME canvas applet or directly enter a SMILES string into the designated text field. Additionally, the user can upload a txt file containing a list of SMILES strings by selecting the 'Massive' button.

Output: The tool predicts the affinity of the query molecule for CB2R and CB1R. For CB2R, affinity is evaluated based on pKi thresholds of 6.5 and 7, while for CB1R, the thresholds are set at pKi 5.5 and 6. If the predicted affinity exceeds the respective threshold, the result is displayed as 'YES'; otherwise, it appears as 'NO' (Fig. 6). If the query molecule is predicted to bind both CB2R and CB1R above their respective thresholds, a tertiary classifier further evaluates CB2R/CB1R selectivity. The user can download the results as a csv file, and a link is also sent to the registered email address. Additionally, the 'History' page maintains a record of all user submissions, including input SMILES and corresponding outputs.

Community and updates: The platform does not include user communication features and has not yet received updates.

Technical support: The platform offers technical support through a section accessible by clicking the 'Contact' button. Here, the user can complete a form to report any issues they encounter.

Further information: ALPACA generates the output within seconds and presents it through a well-structured interface for easy navigation. However, it does not support customization of parameters.

1.7 PLATO (r35)

Developed in 2022 (Ciriaco et al. 2022) by the Department of Pharmacy-Pharmaceutical Sciences and the Department of Chemistry of University of the Studies of Bari "Aldo Moro", Italy.

Category: Predictive—classification and regression prediction for target fishing and bioactivity profiling.

Endpoint: The platform performs classification and regression prediction of affinity on 6,297 different targets.

Background and significance: In silico target fishing can be used to predict potential side effects of a given compound, identify putative drug polypharmacology, and facilitate drug repositioning.

Technical information: PLATO is based on two multi-fingerprint similarity-based predictive approaches designed for target fishing (based on classification models) and bioactivity profiling (based on regression models; Montaruli et al. 2019; Ciriaco et al. 2021), specifically employing the Multifingerprint Similarity Search Algorithm (Mussel) (Alberga et al. 2019). The models use a set of

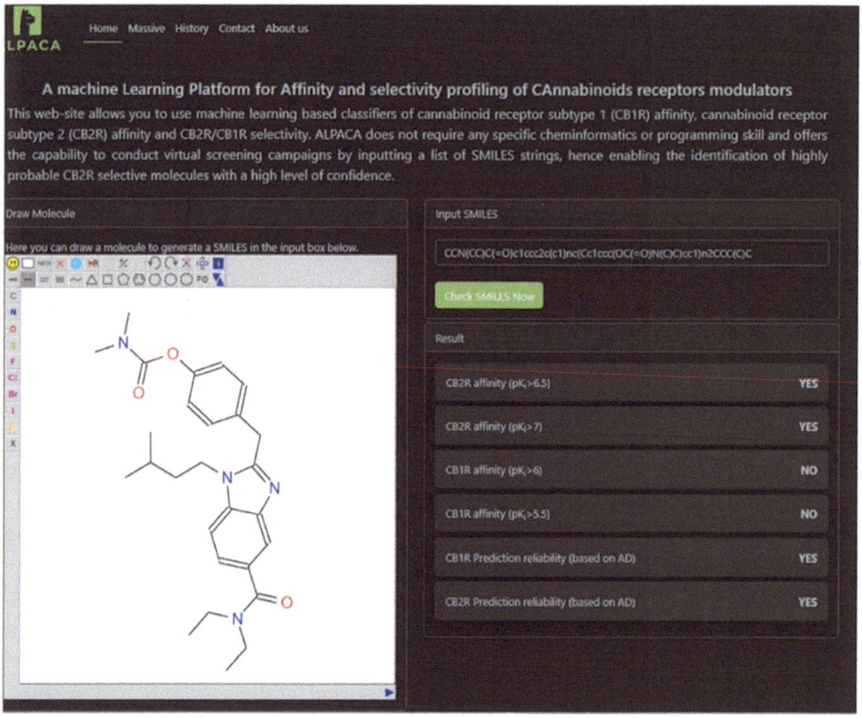

Fig. 6 Example of the output page returned by the ALPACA web platform (https://www.ba.ic.cnr.it/softwareic/alpaca/). (Figure reproduced from Delre et al. 2023, https://creativecommons.org/licenses/by/4.0/)

634,116 compounds with experimental bioactivity data for 6297 protein targets, retrieved from ChEMBL v35 (Zdrazil et al. 2024).

AD: This platform does not consider an AD and hence does not provide an AD evaluation.

Access: The platform is freely accessible for academic users at http://plato.uniba.it/plato/; no registration is required.

Input: The query molecule is entered either as SMILES or by using the JSME sketcher for opening, importing, or modifying a molecular structure. The user can then choose the output format as well as the prediction method, which can be 'Target fishing' or 'Bioactivity profiling'. The 'Get a response' button is used to launch the prediction. A batch evaluation is not possible.

Output: A downloadable pdf output is returned. In the first page, the type of prediction (target fishing or bioactivity profiling), the query chemical structure, and SMILES notation are reported (Fig. 7). The following pages contain the prediction reports. The report of the quantitative bioactivity profiling algorithm contains the target name, the score, on a scale 0–13 of the protein drug target ranking, and the 'reliable' column expressing the probability to detect a target

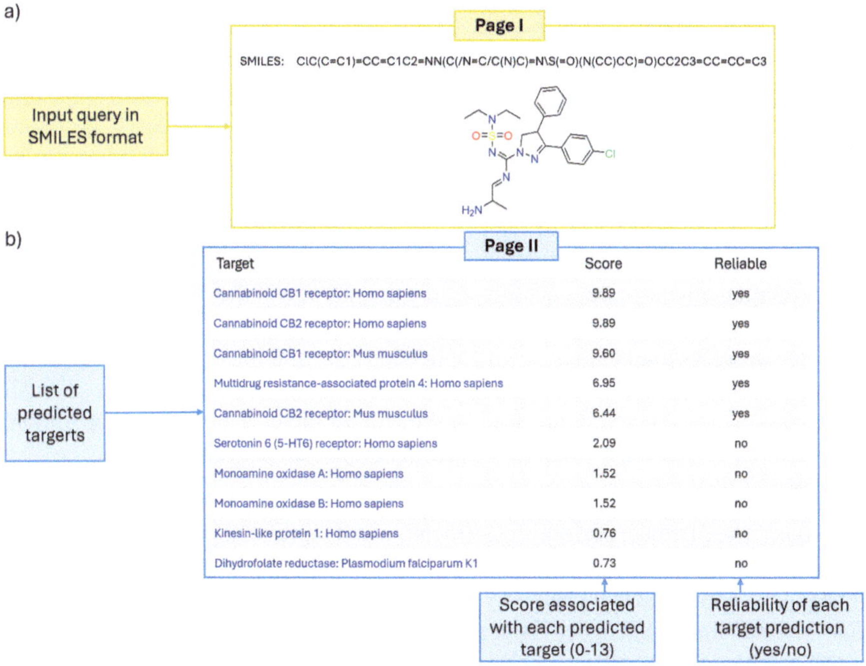

Fig. 7 Example of the target fishing output page returned by the PLATO web platform (http://plato.uniba.it/) including (**a**) the SMILES query and the (**b**) table of the predicted protein targets. (Figure adapted from Ciriaco et al. 2022, https://www.mdpi.com/openaccess)

with a degree of accuracy according to a similarity threshold. The quantitative bioactivity profiling report includes the target name, the predicted activity values expressed as IC_{50}, K_i, or EC_{50}, and the variance for the best predicted activity type.

Community and updates: The platform does not include user communication features. However, it received multiple updates, aligning with each new release of the ChEMBL database (Davies et al. 2015; Gaulton et al. 2017; Mendez et al. 2019).

Technical support: The platform provides users with a mail address for technical support, but no form to report potential issues is available.

Further information: The platform offers a downloadable user guide and generates the output within 2–20 seconds, as stated in a message that appears after the prediction process is submitted.

Once processing is complete, the output is delivered differently from other platforms. Instead of displaying a webpage, the results are provided directly to the user as a pdf or json file.

1.8 SwissADME

Developed in 2017 (Daina et al. 2017), SwissADME is an example of application within *SwissDrugDesign*, a project whose goal is to create a comprehensive web-based in silico drug design environment, freely available to researchers worldwide. The platform, providing a diverse set of tools that cover all aspects of Computer-Aided Drug Design, has been developed by the Molecular Modelling group (MMG) of the Swiss Institute of Bioinformatic (SIB) with the support of the Swiss National Science Foundation, for the scientific developments, and of the SIB, for the creation and maintenance of its web-based tools.

SwissDrugDesign includes seven applications for multiple objectives, including the prediction of molecular interactions between a target protein and a small molecule, hit identification and lead optimization, target fishing and Absorption, Distribution, Metabolism, Excretion (ADME) properties prediction.

Category: Predictive—classification and regression prediction of ADME properties.

Endpoint: The platform performs classification and regression prediction of physicochemical properties, lipophilicity, water solubility, pharmacokinetics, drug-likeness, and medicinal chemistry friendliness.

Background and Significance: These properties and parameters influence the ADME of a molecule. Assessing ADME properties is crucial in drug development to identify compounds with high biological activity and minimal toxicity, increasing their likelihood of successful clinical translation.

Technical information: The platform integrates multiple computational methods to provide a comprehensive assessment of the pharmacokinetics profile of small molecules. Open-source algorithms are at the base of some of these methods, ensuring the freedom for the global scientific community to use the results without requiring further permission. Additionally, it offers models specifically developed and tested by the MMG of SIB. Whenever possible, the platform provides multiple predictions for the same property, making possible a consensus-based evaluation of the results. The physicochemical properties, such as molecular weight, molar refractivity, and polar surface area, are computed through the OpenBabel API (O'Boyle et al. 2011). For lipophilicity prediction, four established methods, along with an in-house approach (iLOGP; Daina et al. 2014), are implemented to calculate the logP. A consensus value is then provided based on the predictions from all five methods. Three methods, ESOL model (Delaney 2004), Ali et al. (2012), and SILICOS-IT (www.silicos-it.be/software.html), are implemented to predict the water solubility. Most of the models used for pharmacokinetic prediction are machine-learning binary classifiers based on the support vector machine algorithm. Passive Gastro-Intestinal Absorption (HIA) and Blood-Brain Barrier (BBB) permeation are assessed using the BOILED-Egg classification model (Daina and Zoete 2016).

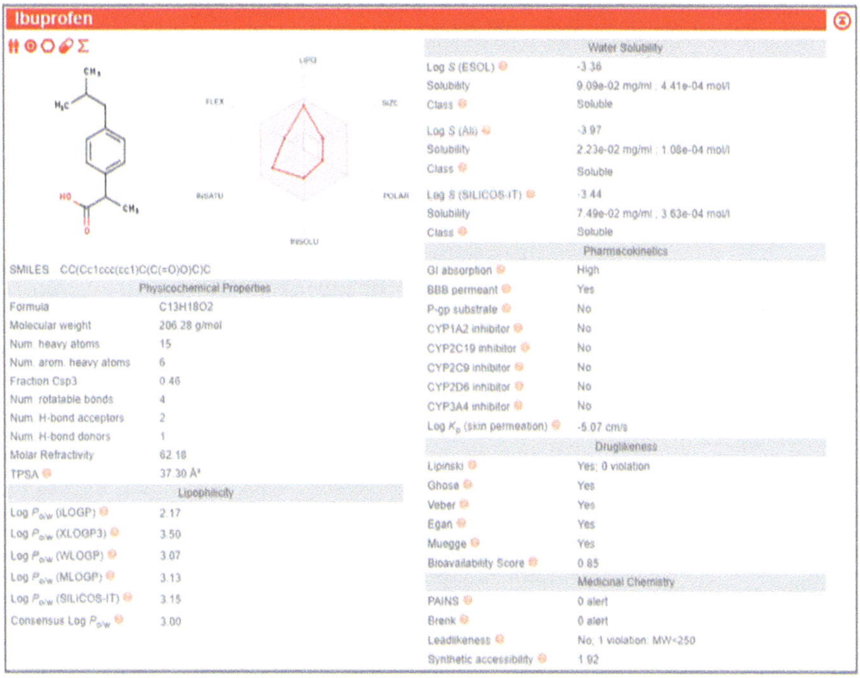

Fig. 8 Screenshot of the output generated by the SwissADME web platform. (http://www.swissadme.ch/). (Figure reproduced from Daina et al. 2017 using the following string as input: `CC(C)CC1 = CC=C(C=C1)C(C)C(=O)O Ibuprofen`)

The *drug-likeness* prediction is based on six rule-based filters, including the Lipinski rule-of-five (Lipinski 2004) and the Abbot Bioavailability Score (Martin 2005).

AD: This platform does not consider an AD and hence does not provide an AD evaluation.

Access: SwissADME is freely accessible at http://www.swissadme.ch/ and no registration is required.

Input: The input area features a molecular sketcher powered by ChemAxon's Marvin JS (Cherinka et al. 2019), enabling users to import chemical structures from files or external databases, as well as draw and edit 2D structures before adding them to a molecule list. This list, located on the right side of the submission page, serves as the actual input for computation. It can be edited as text, enabling the user to type or paste SMILES. Each entry in the list corresponds to a single molecule, represented by its SMILES notation and, optionally, a name, separated by a space.

Output: The output panels, one for each molecule submitted, are loaded immediately after calculation completion on the same web page (Fig. 8). The 2D chemical structure and the corresponding canonical SMILES are reported below the title, indicating which molecule the prediction refers to. Additionally, a bioavailability radar plot is displayed, offering a quick assessment of drug-likeness based on six physicochemical properties: lipophilicity, size, polarity, solubility, flexibility, and saturation. Then, the various models and the corresponding data are grouped in different sections of output panel ('Physicochemical Properties', 'Lipophilicity', 'Pharmacokinetics', 'Druglikeness', and 'Medicinal Chemistry'). Additional information and references appear as pop-ups when hovering over the red question mark icons next to certain entries or over the bioavailability radar plot. SwissADME output data can be exported in two ways. An option is to download a csv file, via the red icon below the SMILES list, or to copy the values to the clipboard using the adjacent red icon, allowing pasting into any text or spreadsheet application. A graphical output is also available; once all calculations are completed, the 'Show BOILED-Egg' red button appears below the sketcher, allowing users to visualize the results on the same page. The BOILED-Egg model provides an intuitive prediction of passive HIA and BBB penetration, based on the molecule's position in the WLOGP-versus-TPSA referential. The white region indicates a high probability of passive absorption in the gastrointestinal tract, while the yellow region suggests a high likelihood of brain penetration. Finally, points are blue for predicted PGP+ and red for PGP−.

Community and updates: The platform does not include user communication features. As stated in the 'For information' section, the platform received updates in look and feel while keeping the underlying technologies and parameters unchanged. As a result, the updated web tool delivers the same results as the previous version.

Technical support: The platform offers technical support through a section accessible by clicking the 'Contact' button. Here, the user can complete a form to report any issues they encounter.

Further information: The platform provides detailed instructions to guide the user through the input process. The guide is available by clicking the 'Help' button in the navigation bar. Additionally, clicking the 'Fill with an example' button provides the user with four sample SMILES, two of which are accompanied by the corresponding molecule names, offering an example of how the input should be entered. Once processing is completed, the output is provided almost immediately in a clear and structured manner, with a well-organized interface that facilitates interpretation. The platform does not allow customizing settings or parameters.

1.9 DELA-DRUGSELF

Developed in 2024 (Alberga et al. 2024) by our research group as a result of the ongoing collaboration between the IC-CNR, the Department of Pharmacy-Pharmaceutical Sciences of University of the Studies of Bari "Aldo Moro", and the Institute of Biomolecular Chemistry of CNR of Pozzuoli, Italy.

Category: Generative.

Endpoint: The platform performs automatic generation of drug-like analogues.

Background and Significance: The automatic generation of analogues supports medicinal chemists in designing derivatives of molecules already available in their laboratories, making them strong candidates for easy and cost-effective synthesis. The resulting libraries serve as valuable starting points for virtual screening procedures.

Technical information: The model consists of an R-NN composed of two layers of Long Short-Term Memory (LSTM) units. LSTM architecture is exploited in two generative approaches, namely Sampling from Scratch and Sampling with Mutations.

The Training Set of *DeLA-DrugSelf* was prepared starting from the entire ChEMBL v28 database and consists of 1,092,285 compounds. While preserving the overall architecture of its predecessor, DeLA-Drug (Creanza et al. 2022), an R-NN model with two layers of LSTM cells, DeLA-DrugSelf introduces significant advancements. The most notable improvement is the adoption of SELF-referencing Embedded Strings (SELFIES) as a molecular representation. Furthermore, it expands generation capabilities by allowing not only token substitutions but also insertions and deletions, thereby enhancing molecular diversity and design flexibility.

AD: Generative models are neither definable nor usable, as opposed to predictive models.

Access: The platform is freely available at https://www.ba.ic.cnr.it/softwareic/delaself/ and requires user registration with an email address.

Input: The user has two options for inputting its query molecule: they can either draw its 2D structure using the JSME canvas applet or directly enter a SMILES string into the designated text field. Additionally, JSME enables the direct import of mol or sdf files into the system. The user can customize the tool based on their preferences by adjusting parameters such as the number of desired compounds (ranging from 10 to 100, with a default of 10) and the number of mutations (ranging from 1 to 5, with a default of 1). Notably, when a query molecule is inserted, the web portal calculates its Quantitative Estimate of Drug-likeness (QED) score. If this score falls below a predefined threshold (<0.35), a warning is displayed, allowing the user to decide whether to proceed with analogues generation. Once the process is complete, the outcomes are displayed.

Output: The platform provides an interactive list of SELFIES, converted into SMILES format for simplicity, where the Synthetic Accessibility (SA) values do not exceed a one-unit difference compared to the query molecule (Fig. 9). The user can navigate this list in various ways. The generated compounds are ranked based on their QED score, SA value, and Tanimoto similarity to the query, allowing for an intuitive and structured exploration. Additionally, the user has the option to download the ranked list in SMILES (txt or smi format) or sdf files, making it easy to analyze the results offline.

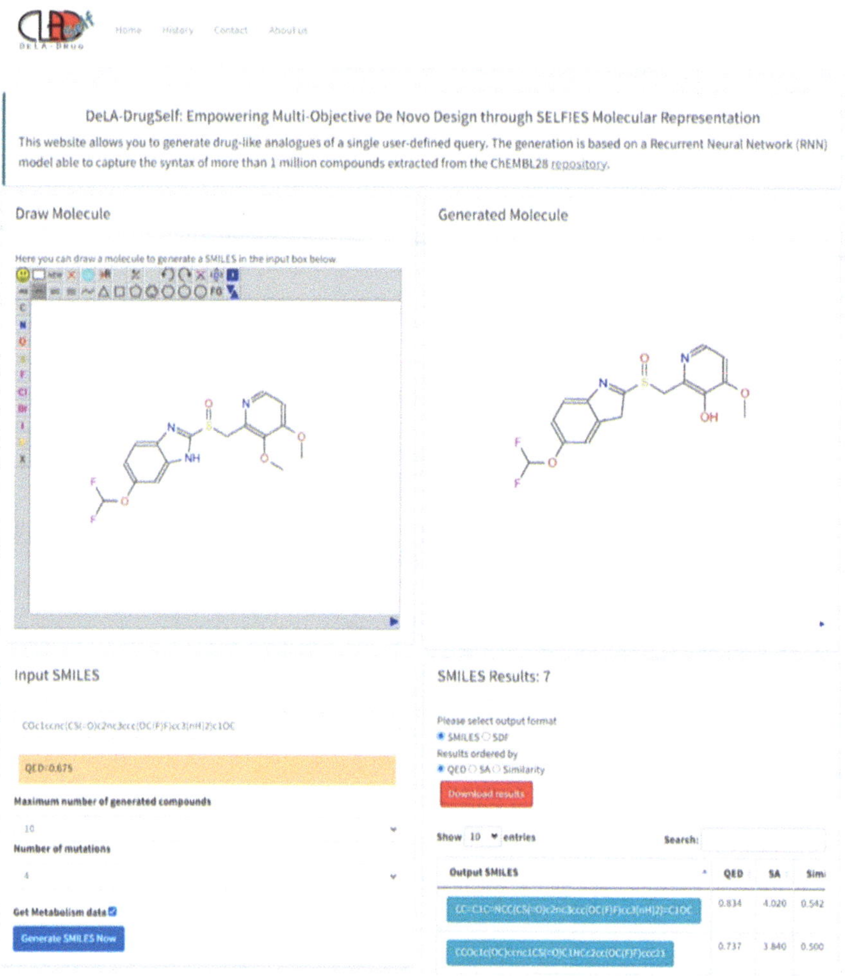

Fig. 9 Example of the output page returned by the DeLA-DrugSelf web platform (https://www.ba.ic.cnr.it/softwareic/delaself/). (Figure reproduced from Alberga et al. 2024, https://creativecommons.org/licenses/by/4.0/)

For a more detailed inspection, the 2D structures of the generated compounds are readily accessible within the JSME editor by simply clicking on the corresponding SMILES string. Moreover, if requested, the platform can predict whether the generated compounds may interact with specific cytochrome P450 isoforms, including CYP1A2, CYP2A6, CYP2B6, CYP2C8, CYP2C9, CYP2C19, CYP2D6, CYP2E1, and CYP3A4. These predictions are performed using the software CypReact (Tian et al. 2018), providing valuable insights into potential metabolic interactions.

Once the process is complete, the system sends download links for the generated data to the user's registered email. Additionally, a 'History' page keeps track of all user executions, recording details such as the input SMILES and the corresponding generated output, ensuring easy access to past results.

Community and updates: The platform does not include user communication features and has not yet received updates.

Technical support: The platform offers technical support through a section accessible by clicking the 'Contact' button. Here, the user can complete a form to report any issues they encounter.

Further information: The platform provides the output in a few seconds with a well-organized interface. Additionally, it allows customizing some parameters such as 'The maximum number of generated compounds' and 'Number of mutations'.

1.10 Deep-PK

Developed in 2024 (Myung et al. 2024) by Biosig Lab, located at St Lucia Campus in Brisbane, Australia.

Category: Predictive—classification and regression prediction of ADMET properties—and generative.

Endpoint: The platform performs classification and regression prediction of 73 endpoints—including 64 ADMET and 9 general physicochemical properties—and generation of drug-like analogues.

Background and Significance: Evaluating ADMET and general physicochemical properties is essential for identifying compounds with high biological activity and low toxicity. Examining these properties supports molecular optimization ensuring favorable pharmacokinetic and safety profiles. Moreover, the automated generation of analogues helps medicinal chemists by facilitating the design of new derivatives based on already available molecules.

Technical information: The platform incorporates multiple DL models based on Graph Neural Networks, specifically using D-MPNN of Chemprop (Yang et al. 2019), including 49 models designed for binary classification and 24 for regression tasks. Chemprop (Heid et al. 2024) was also utilized to identify key molecular substructures influencing predictions, offering insights into which molecular sites should be retained, modified, or removed.

For the optimization process in Deep-PK, the MultI-constraint MOlecule Sampling (MIMOSA) method (Fu et al. 2021) was implemented, enabling the generation of up to 100 molecular variants from a given query structure by adding, replacing, or removing substructures. Moreover, a pretrained model was employed to optimize the QED, ensuring the refinement of generated compounds based on drug-likeness criteria.

AD: Deep-PK includes binary indicators for regression tasks, specifying whether the input molecule falls or not within the AD.

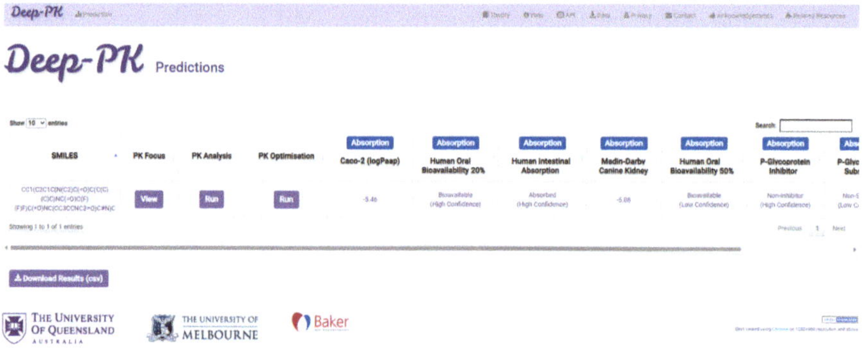

Fig. 10 Screenshot of the 'Predictions' output page generated by the Deep-PK web platform (https://biosig.lab.uq.edu.au/deeppk/). (Figure reproduced from Myung et al. 2024 using as string SMILES: `CC1(C2C1C(N(C2)C(=O)C(C(C)(C)C)NC(=O)C(F)(F)F)C(=O)NC(CC3CCNC3 = O)C#N)C)`

Access: The platform is freely accessible at https://biosig.lab.uq.edu.au/deeppk/ and no registration is required.

Input: Deep-PK supports different types of input. The user can provide a single molecule as a SMILES string or upload a file containing multiple compounds. Specifically, the platform accepts SMILES files and sdf files, each capable of containing up to 2000 molecular structures. Additionally, the user can input a molecule by drawing its structure using the JSME canvas applet, from which the corresponding SMILES representation is automatically generated.

Output: The platform shows four key prediction outputs, each presented on a dedicated page (Fig. 10). First, the 'Predictions' page displays the predicted pharmacokinetic, toxicity, and general molecular properties when all selected models have completed their analyses. Second, the 'Focus' page provides a detailed overview of the molecule, including its structure, SMILES representation, and predicted properties. Third, the 'Analysis' page allows for a deeper exploration of the molecules whose ADMET properties were predicted. The user can examine general properties, drug-likeness scores, and the importance of specific molecular substructures in the predictions. Additionally, this page features radar plots that visually represent different aspects of drug-likeness. Lastly, the 'Optimization' page is dedicated to refining the given query molecule to improve its pharmacokinetic and toxicity properties. The user can download the results from both the 'Prediction' and 'Optimization' pages as csv files, while the data from the 'Analysis' page are provided in a zip file.

Community and updates: The platform lacks user communication features and has not yet been updated.

Technical support: The platform offers technical support through a dedicated section accessible by clicking the 'Contact' button and selecting 'Deep-PK' in the 'Subject' field. The user can then write a message and send it by clicking the 'Send' button.

Table 1 Comparative overview of the platforms discussed in this chapter. Each entry is characterized by category (Predictive and/or Generative), endpoint, and relative reference

Name	Category	Endpoint	References
PRED-hERG (5.0)	Predictive	hERG-related cardiotoxicity	Braga et al. (2015), Alves et al. (2018a, b)
PRED-SKIN (3.0)	Predictive	Skin sensitization potential	Braga et al. (2017), Borba et al. (2021)
CYTO-SAFE	Predictive	Cytotoxic potential	Feitosa et al. (2024)
AMALPHI	Predictive	Phospholipidosis potential	Lomuscio et al. (2024)
SIGMAP	Predictive	Affinity potential for Sigma-1 receptor	Lomuscio et al. (2025)
ALPACA	Predictive	Affinity potential for cannabinoid receptors	Delre et al. (2023)
PLATO (r35)	Predictive	Target fishing	Ciriaco et al. (2022)
SwissADME	Predictive	ADME properties	Daina et al. (2017)
DeLA-DrugSelf	Generative	Drug-like analogues generation	Alberga et al. (2024)
Deep-PK	Predictive and generative	ADMET properties / drug-like analogues generation	Myung et al. (2024)

Further information: The platform provides detailed instructions available by clicking the 'Help' button in the navigation bar. The 'Theory' button offers detailed insights into all target predictions, including their description, interpretation, and assay type. The user can also download this information as a pdf file for offline reference. Deep-PK requires up to 15 minutes to predict all 73 properties of a single molecule. Optimizing a molecule, generating and predicting 100 analogues requires approximately 20 minutes. Once processing is completed, the output is provided with a well-organized interface.

A comprehensive overview of all the web platforms presented is provided in Table 1.

2 Human-Centered Design Strategies

AI-based web platforms have the potential to reshape the drug discovery process. These tools, however, are predominantly developed by researchers with a medicinal chemistry background, particularly in cheminformatics. We do believe that for a true advancement in the field, a closer collaboration between medicinal chemists and computer scientists specializing in HCI is highly desirable. Specifically, the HCD is the iterative design approach that places human needs, preferences, and behaviors at the core of the design and development process and focuses on understanding

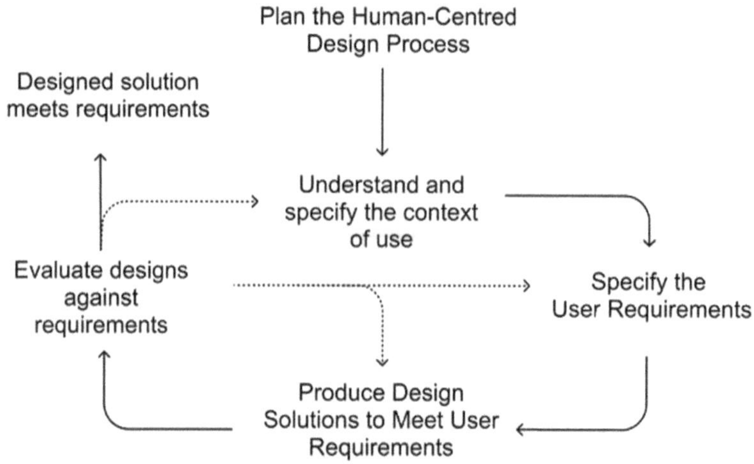

Fig. 11 The HCD model, as defined in the International Organization for Standardization (2019)

users' contexts, goals, and expectations by actively involving them throughout the design cycle (ISO 2019) (Fig. 11). Employing HCD increases the overall quality of software systems (Ardito et al. 2014) and, in particular, of their *usability*, which is defined in the ISO standard 9241:11 as *"the extent to which a system, product, or service can be used by specified users to achieve specified goals with effectiveness, efficiency, and satisfaction in a specified context of use"* (ISO 2018). HCD ensures that AI platforms are functional, useful, usable, and aligned with their values and constraints, prioritizing the needs and goals of users (Desolda et al. 2024). Through HCD, those functionalities that are considered critical for an effective, efficient, and satisfying use of the system are designed and, when possible, evaluated by users. This process ensures alignment with users' expectations while guaranteeing the stability and performance of the overall platform. A loss in these aspects could negatively affect interaction, ultimately reducing usability.

Building on the core principles of HCI, recent scientific literature introduces Human-Centered Artificial Intelligence (HCAI), as a specialization of HCI, to guide the creation of AI systems (Shneiderman 2022). HCAI systems are designed, developed, and evaluated involving users in the process; in this way, it is possible to improve the performance of the platforms and to increase the satisfaction of the users (Desolda et al. 2024).

3 Designing Usable AI-Based Tools for Drug Discovery

Designing HCAI systems is a multidisciplinary effort. Besides the two pillars of HCI and AI, other disciplines converge to provide techniques and methodologies. Particularly, ML, XAI, Software Engineering, and Ethics also contribute to HCAI

(Desolda et al. 2024). The same principles can be applied when designing usable HCAI systems for drug discovery. More specifically, in the context of HCAI, the HCD approach is fundamental when an interaction between users and system is expected. As such, two main phases are required to create HCAI systems: the actual *design* phase and the *evaluation* of the designed solution. This process for the HCAI systems requires methodologies that prioritize usability as the main principle, while ensuring ethical alignment, adaptability (i.e., the ability of a system to be customized by the user (Fischer 2023)), and adaptivity (i.e., the ability of a system to automatically change its behavior in response to usage patterns (Fischer 2023)). For simplicity, throughout the rest of this section, we will provide an overview of useful techniques to design and evaluate usable AI-based systems. If the reader is interested in exploring how researchers deal with ethical issues, transparency of the decision-making process, and other AI-related issues, we suggest reading the following books and reviews (Nam et al. 2022; Shneiderman 2022; Desolda et al. 2024).

The absence of a strong focus on usability may limit the adoption of even the most advanced AI systems, as users may struggle to use them effectively, efficiently, and with satisfaction. As previously proved by studies surveying users, a well-designed, usable interface can mitigate ambiguity, thereby reducing the risk of misinterpretation (Esposito et al. 2024). HCD strategies, often overlooked by developers of AI-based platforms, play a crucial role in ensuring that these cheminformatics platforms are truly effective. The development of an HCAI system is inherently an iterative process, where design and evaluation phases alternate in a continuous cycle. However, this process can be applied to existing systems, which require improving and refining the product via a redesign phase, as well. This way, evaluation strategies that identify potential issues and guide future updates can be implemented, although subsequent adjustments to the interaction process may require modification of the platform's back- and/or front-end, depending on the type of correction needed.

The following sections will focus on useful techniques (heavily adopted in HCI) to design and evaluate *usable* software products. We will especially focus on techniques and strategies that are considered useful for AI-based systems and may guide the further development of tools for drug discovery.

3.1 Designing for Usability

As mentioned above (cf. Human-Centered Design Strategies), one of the cornerstones of HCD is the iterative design process, which enables designers to incorporate users' feedback. However, it is recommended to involve users prior to the final stages of development. Early engagement of users in the design phase fosters solutions that better align with their needs and expertise. Certain well-established techniques have been specifically developed for this purpose in the HCI community. These techniques are reported below.

Rapid prototyping is extremely useful for obtaining quick user feedback. It consists of the creation of *prototypes*, i.e., simplified drawings which represent the interface of a system (Rogers et al. 2023). This method allows designers to freely explore ideas without allocating excessive time and resources to them, thereby facilitating the adoption of users' feedback thanks to the low costs of edits.

Participatory Design is an approach that actively involves users in the design process ensuring that their perspectives and expertise shape the system's development (Rogers et al. 2023). Leveraging rapid prototyping, designers can quickly explore ideas and get feedback from users, which in turn can support the designers by sharing their needs and requirements. This approach enables the rapid exploration of potentially innovative solutions while ensuring alignment with users' real-world experiences.

Value-Sensitive Design is an approach to design that recognizes the importance of the social, cultural, and moral environments in which software operates (Friedman and Hendry 2019). It incorporates ethical considerations and human values into design choices. Value-sensitive design is especially useful for AI-based systems, especially for highly regulated fields, such as drug discovery. It is important to note that the values are elicited by discussing them with the end-users, and it is not possible to predefine target values as 'universal truths' during the design phase.

In the context of AI-based systems for drug discovery, end-users often have a different background compared to those who designed and developed such tools (e.g., experimental vs. computational medicinal chemists). This gap must therefore be considered since the developers' understanding of the end-users' needs may not be sufficient to create *usable* software. It is, therefore, imperative that design solutions are systematically evaluated to ensure their usability.

3.2 Evaluating Usability of AI-Based Systems

Usability evaluations involve various approaches, and they usually happen when system's interaction mechanisms are modified. Thus, usually is not constantly monitored but, instead, sporadically evaluated in case of breaking changes. Two main classes of evaluation techniques can be adopted: (i) inspections exclusively by usability experts and (ii) usability studies involving end-users (Rogers et al. 2023). It is important to note that neither technique should be preferred over the other, since they reveal a complementary set of errors, particularly when inspections emphasize higher-level issues that may limit the users' interaction with the system (Nielsen 1993).

(i). *Inspections* do not require the presence of the system's end-user during the evaluation process since it is performed. The experts detect usability issues and verify the compliance system with standards. One of the most well-known methods to perform inspections is *Heuristic Evaluation* in which researchers are guided by a set of usability principles, called heuristics, to evaluate their adherence to User-Interface (UI) elements (Rogers et al. 2023). A commonly used set of heuristics is the so-called Nielsen's heuristics (Nielsen 1994).

(ii). *Usability Studies* require user participation, heterogeneous enough to gather different perspectives while still being representative of real target users of the platform. Although usability studies may be conducted in controlled environments (i.e., usability laboratories), the currently used techniques prefer informal settings, where users are directly observed while interacting with the system. This technique provides information on task efficiency, error rates, and user satisfaction. Furthermore, implementing the *thinking aloud* protocol also provides insights into the users' mental model of their tasks (Rogers et al. 2023).

In the end, the cost of conducting usability studies varies depending on the chosen method: conducting formal usability tests in controlled laboratories requires more resources than inspections that do not involve users and do not require the use of laboratories with adequate equipment. In the HCI community, the term discount usability is used to denote low-cost evaluation techniques, like inspections and informal user tests (Nielsen 1993).

4 How to Improve Usability of AI-Based Tools for Drug Discovery: An Example

This section focuses on how the inspection technique can support the evaluation phase of HCD. The ten heuristics developed by Jakob Nielsen are one of the core instruments to inspect and assess systems without involving real end-users (Nielsen 1994). The heuristics are presented and described below:

1. *Visibility of system status*: the design should consistently inform users about system status by providing timely and appropriate feedback.
2. *Match between system and the real world*: the design should use the users' language, incorporating familiar words, phrases, and concepts instead of internal jargon. It should adhere to real-world conventions, presenting information in a natural and logical order.
3. *User control and freedom*: users are prone to errors and should have clearly marked "emergency exits" that allow them to undo unwanted actions quickly and effortlessly.
4. *Consistency and standards*: users should not be left uncertain about whether different words, situations, or actions have the same meaning.
5. *Error prevention*: the design should either eliminate error-prone conditions or detect them and provide users with a confirmation option before they proceed.
6. *Recognition rather than recall*: reduce the users cognitive load by making elements, actions, and options visible. Users should not need to recall infor-

mation from one part of the interface to another, but it should be available when necessary.

7. *Flexibility and efficiency of use*: shortcuts can enhance efficiency for expert users, allowing the design to accommodate both novice and experienced users, possibly remaining hidden for beginners.
8. *Esthetic and minimalist design*: a system's interface should present only relevant information, avoiding unnecessary or extraneous content. Irrelevant elements can create confusion and hinder usability; thus, they must not be highlighted.
9. *Help users recognize, diagnose, and recover from errors*: error messages should be written in clear, simple language, enabling the user to identify the issue and offer constructive solutions.
10. *Help and documentation:* although it is ideal for the system to be intuitive enough to require no additional explanation, documentation should be available to assist users in completing their tasks.

These ten heuristics were used to propose new design solutions for an AI-based platform for drug discovery described and illustrated in the following.

4.1 Towards a More Usable AI-Based Platform for Drug Discovery

In this section, we provide a simple example of how HCI methods may improve a platform without requiring extensive resources. We considered SIGMAP as an example of an AI-based platform for drug discovery. We inspected the platform based on Nielsen's heuristics, (cf. How to Improve Usability of AI-Based Tools for Drug Discovery: An Example) and herein proposed possible solutions to the weaknesses that could negatively impact its usability. This step represents a preliminary phase aimed to identify major issues that need to be solved before we conduct a user study.

One of the main activities of SIGMAP is to predict the affinity of potential drug candidates for S1R. The section in which this output is shown is currently labeled with the SMILES corresponding to the input molecule. To ensure that the platform aligns better with the *Match between the system and the real-world* heuristic, it can be renamed 'Affinity Prediction' instead.

Another modification that could further enhance the platform's compliance with this heuristic is making it available in multiple languages. Although this rule is generally applied to the optimization of systems, scientific platforms, such as SIGMAP, are often available exclusively in English. Indeed, the scientific community adopts English as lingua franca. Since language limitations could still hinder usability and adoption, providing multilingual support would make the platform more inclusive and allow a broader range of users to interact with it more intuitively.

To foster *User Control and Freedom*, 'Cancel Prediction' button can be added to the loading page. This would ensure that users are enabled to revert their actions if needed. Currently, when an input is being processed, the UI remains active, possibly causing its involuntary interruption. To minimize errors, the interface should be kept noninteractive during this time, as suggested by *Error Prevention.*

Another change that can positively impact SIGMAP's usability consists of revising the structure of the navigation bar and footer. The navigation bar should be modified to ensure its fixed position at the top of each page allowing users to always access the basic functions of the systems, avoiding confusion and frustration. Additionally, the footer uselessly takes up excessive space on the UI and should therefore be redesigned to better align with the heuristic *Aesthetic and Minimalist Design.* Finally, the visibility and clarity of the loading process bar should be improved to comply with the *Visibility of system status* heuristic. At the moment, a throbber is shown while the system is loading, but no information on the status of the system is provided (i.e., percentage of completion). Similar changes should be implemented for other visual clues to ensure that users can distinguish them from the rest of the UI, such as enhancing contrast with their visual context.

The changes presented in this section—resulting from an evaluation which is in no way exhaustive—aim to solve issues that can generate negative feelings in users that impact the interaction process and the accomplishment of the task. To address these issues, we recommend adding new elements to the UI or modifying the structure of those that can be improved. Indeed, such high-level issues might interfere with users' feedback during a subsequent user study of the platform, hindering the recognition of deeper obstacles in the interaction (Nielsen 1993). Therefore, the heuristic rules considered in our preliminary analysis elucidate flaws and weaknesses of SIGMAP that negatively impact the usability of the platform, highlighting the areas in which it can be improved prior to a user study.

Noteworthy, these limitations are shared across the other analyzed platforms. A common issue is the lack of multilingual support, as all platforms are available only in English, contradicting the heuristic principle of *Match between the system and the real world.* Also, many platforms keep the interface interactive while processing input, violating the *Error prevention* heuristic principle. Additionally, users are often left without feedback on the system's progress, which goes against the *Visibility of system status* heuristic rule. An exception to this pattern is the Deep-PK platform, which stands out by displaying a loading page when a process is initiated. This page includes a 'Max. Waiting Time' indicator that informs users of the estimated maximum time, along with a progress bar that visually represents the process status. However, the navbar remains interactive without disrupting the ongoing process, as it enables users to open a new page. The *Aesthetic and minimalist design* of PRED-SKIN (3.0) violates this principle with an overcrowded interface, featuring dense text, excessive colors, and overwhelming images. On the other hand, this platform maintains a fixed navigation bar, improving usability, while others, like SIGMAP, feature a scrolling navigation bar, which may disorient users. Lastly, in all analyzed platforms, the footer occupies a significant portion of the page, except for Swiss-ADME and Deep-PK which present a nonintrusive design.

5 Conclusions: Towards a Symbiosis Between Humans and AI Platforms for Drug Discovery

In this chapter, we analyzed in detail a selection of AI-driven web-based platforms developed to support medicinal chemists working on early stages of the drug discovery process. These platforms are often created by cheminformatics experts, whose background does not necessarily match with that of one of the end-users, such as experimental medicinal chemists. Additionally, platform design and development are seldom assisted by the intervention of computer scientists specialized in HCD, which allows experts to iteratively refine the system. We propose that a multidisciplinary collaboration between experts in the creation of systems ensures that the end-users are placed at the core of the process resulting in *usable* platforms.

The selected web applications herein reported represent a subset of open-source models which are ideated to accelerate the drug discovery pipeline. We overviewed the platforms describing key aspects of their interface and their functional interaction (*Input*, *Output*, *Technical support*). Then, we identified strengths and weaknesses of the platforms from a cheminformatic point of view. Notably, our preliminary evaluation revealed that the analyzed platforms consistently violated heuristic principles. This underscores the importance of considering these principles when designing or updating these platforms.

However, a systematic improvement of UI requires computer scientists' expertise, implementing heuristic principles, as well as supervised user studies to ensure the system's stability under different and unexpected conditions. This synergy can promote a productive 'partnership' where human expertise and intuition complement the efficiency and scalability of AI (Desolda et al. 2024). By considering the end-users' feedback, AI platforms can be adapted and refined improving both the interaction process and the underlying AI functionalities. This reciprocal relationship defines the concept of human-AI symbiosis (Grigsby 2018), a vision where humans and AI systems work together combining the strengths and compensating for limitations in tackling complex challenges of modern drug discovery.

Acknowledgments We would like to acknowledge Davide Annicchiarico for his contribution to the preliminary evaluation of AI platforms. This work was funded by the European Union (Next Generation EU) under the program PRIN of the Ministero Dell'Università e della Ricerca (Progetti di ricerca di Rilevante Interesse Nazionale 2022—CODICE 2022Z3BBPE—CUP: B53D23015780006—Development of broad-spectrum coronavirus antiviral agents acting as allosteric modulators of the host protein sigma-1, CODICE 2022S2M7E5—CUP: B53D23008340006—Pharmacological Chaperon Therapy for Type I Mucopolysaccharidosis) and by the project "Potentiating the Italian Capacity for Structural Biology Services in Instruct Eric", acronym "ITACA.SB" (Project No. IR0000009, CUP B53C22001790006), funded by the European Union's NextGenerationEU under the MUR call 3264/2021 PNRR M4/C2/L3.1.1. The research of Miriana Calvano, Antonio Curci, and Rosa Lanzilotti is supported by the co-funding of the European Union—Next Generation EU: NRRP Initiative, Mission 4, Component 2, Investment 1.3—Partnerships extended to universities, research centers, companies and research D.D. MUR n. 341 del 15.03.2022—Next Generation EU (PE0000013—"Future Artificial Intelligence Research–FAIR"—CUP: H97G22000210007). The research of Andrea Esposito is funded by a Ph.D. fellowship within the framework of the Italian "D.M. n. 352, April 9, 2022"—under the National

Recovery and Resilience Plan, Mission 4, Component 2, Investment 3.3—h.D. Project "Human-Centered Artificial Intelligence (HCAI) techniques for supporting end users interacting with AI systems", co-supported by "Eusoft S.r.l." (CUP H91I22000410007).

References

Abatematteo FS, et al. A conformational rearrangement of the SARS-CoV-2 host protein sigma-1 is required for antiviral activity: insights from a combined in-silico/in-vitro approach. Sci Rep. 2023;13(1):12798. https://doi.org/10.1038/s41598-023-39662-w.

Alberga D, et al. A new approach for drug target and bioactivity prediction: the multifingerprint similarity search algorithm (MuSSeL). J Chem Inf Model. 2019;59(1):586–96. https://doi.org/10.1021/acs.jcim.8b00698.

Alberga D, et al. DeLA-DrugSelf: empowering multi-objective de novo design through SELFIES molecular representation. Comput Biol Med. 2024;175:108486. https://doi.org/10.1016/j.compbiomed.2024.108486.

Ali J, et al. *In Silico* prediction of aqueous solubility using simple QSPR models: the importance of phenol and phenol-like moieties. J Chem Inf Model. 2012;52(11):2950–7. https://doi.org/10.1021/ci300447c.

Alves VM, et al. QSAR models of human data can enrich or replace LLNA testing for human skin sensitization. Green Chem. 2016;18(24):6501–15. https://doi.org/10.1039/C6GC01836J.

Alves VM, et al. A perspective and a new integrated computational strategy for skin sensitization assessment. ACS Sustain Chem Eng. 2018a;6(3):2845–59. https://doi.org/10.1021/acssuschemeng.7b04220.

Alves VM, Braga RC, Andrade CH. Computational approaches for predicting hERG activity. In: Computational toxicology: risk assessment for chemicals; 2018b. p. 69–91. https://doi.org/10.1002/9781119282594.ch3.

Ardito C, et al. Investigating and promoting UX practice in industry: an experimental study. Int J Hum Comput Stud. 2014;72(6):542–51. https://doi.org/10.1016/j.ijhcs.2013.10.004.

Bienfait B, Ertl P. JSME: a free molecule editor in JavaScript. J Chem. 2013;5(1):24. https://doi.org/10.1186/1758-2946-5-24.

Borba JVB, et al. Pred-skin: a web portal for accurate prediction of human skin sensitizers. Chem Res Toxicol. 2021;34(2):258–67. https://doi.org/10.1021/acs.chemrestox.0c00186.

Braga RC, et al. Tuning hERG out: Antitarget QSAR models for drug development. Curr Top Med Chem. 2014;14(11):1399–415.

Braga RC, et al. Pred-hERG: a novel web-accessible computational tool for predicting cardiac toxicity. Mol Inform. 2015;34(10):698–701. https://doi.org/10.1002/minf.201500040.

Braga RC, et al. Pred-skin: a fast and reliable web application to assess skin sensitization effect of chemicals. J Chem Inf Model. 2017;57(5):1013–7. https://doi.org/10.1021/acs.jcim.7b00194.

Casati S et al. EURL ECVAM Recommendation on the KeratinoSensTM assay for skin sensitisation testing, JRC Publications Repository; 2014. https://doi.org/10.2788/52914.

Cavalluzzi MM, et al. Human ether-à-go-go-related potassium channel: exploring SAR to improve drug design. Drug Discov Today. 2020;25(2):344–66. https://doi.org/10.1016/j.drudis.2019.11.005.

Cherinka B, et al. Marvin: a tool kit for streamlined access and visualization of the SDSS-IV MaNGA data set. Astron J. 2019;158(2):74. https://doi.org/10.3847/1538-3881/ab2634.

Ciriaco F, et al. Quantitative polypharmacology profiling based on a multifingerprint similarity predictive approach. J Chem Inf Model. 2021;61(10):4868–76. https://doi.org/10.1021/acs.jcim.1c00498.

Ciriaco F, et al. PLATO: a predictive drug discovery web platform for efficient target fishing and bioactivity profiling of small molecules. Int J Mol Sci. 2022;23(9):5245. https://doi.org/10.3390/ijms23095245.

Couly S, Yasui Y, Su T-P. SIGMAR1 confers innate resilience against neurodegeneration. Int J Mol Sci. 2023;24(9):7767. https://doi.org/10.3390/ijms24097767.

Creanza TM, et al. DeLA-drug: a deep learning algorithm for automated Design of Druglike Analogues. J Chem Inf Model. 2022;62(6):1411–24. https://doi.org/10.1021/acs.jcim.2c00205.

Daina A, Michielin O, Zoete V. iLOGP: a simple, robust, and efficient description of *n* - Octanol/water partition coefficient for drug design using the GB/SA approach. J Chem Inf Model. 2014;54(12):3284–301. https://doi.org/10.1021/ci500467k.

Daina A, Michielin O, Zoete V. SwissADME: a free web tool to evaluate pharmacokinetics, drug-likeness and medicinal chemistry friendliness of small molecules. Sci Rep. 2017;7(1):42717. https://doi.org/10.1038/srep42717.

Daina A, Zoete V. A BOILED-egg to predict gastrointestinal absorption and brain penetration of small molecules. ChemMedChem. 2016;11(11):1117–21. https://doi.org/10.1002/cmdc.201600182.

Davies M, et al. ChEMBL web services: streamlining access to drug discovery data and utilities. Nucleic Acids Res. 2015;43(W1):W612–20. https://doi.org/10.1093/nar/gkv352.

Delaney JS. ESOL: estimating aqueous solubility directly from molecular structure. J Chem Inf Comput Sci. 2004;44(3):1000–5. https://doi.org/10.1021/ci034243x.

Delre P, et al. ALPACA: a machine learning platform for affinity and selectivity profiling of CAnnabinoids receptors modulators. Comput Biol Med. 2023;164:107314. https://doi.org/10.1016/j.compbiomed.2023.107314.

Desolda G, et al. From human-centered to symbiotic artificial intelligence: a focus on medical applications. Multimed Tools Appl. 2024; https://doi.org/10.1007/s11042-024-20414-5.

Esposito A, Desolda G, Lanzilotti R. The fine line between automation and augmentation in website usability evaluation. Sci Rep. 2024;14(1):10129. https://doi.org/10.1038/s41598-024-59616-0.

EU *EUR-Lex-02009R1223–20240424-EN-EUR-Lex*; 2009. Available at: https://eur-lex.europa.eu/eli/reg/2009/1223/2024-04-24/eng. Accessed 3 Mar 2025.

Fallica AN, et al. Recent advances in the development of sigma receptor ligands as cytotoxic agents: a medicinal chemistry perspective. J Med Chem. 2021;64(12):7926–62. https://doi.org/10.1021/acs.jmedchem.0c02265.

Feitosa FL, et al. Cyto-safe: a machine learning tool for early identification of cytotoxic compounds in drug discovery. J Chem Inf Model. 2024;64(24):9056–62. https://doi.org/10.1021/acs.jcim.4c01811.

Fischer G. Adaptive and adaptable systems: differentiating and integrating AI and EUD. In: Spano LD, et al., editors. End-user development. Cham: Springer Nature Switzerland; 2023. p. 3–18. https://doi.org/10.1007/978-3-031-34433-6_1.

Friedman B, Hendry D. Value sensitive design: shaping technology with moral imagination. Cambridge: The MIT Press; 2019.

Fu T, et al. MIMOSA: multi-constraint molecule sampling for molecule optimization. Proc AAAI Conf Artif Intell. 2021;35(1):125–33. https://doi.org/10.1609/aaai.v35i1.16085.

Gadaleta D, et al. Applicability domain for QSAR models: where theory meets reality. IJQSPR. 2016;1:45–63. https://doi.org/10.4018/IJQSPR.2016010102.

Gaulton A, et al. ChEMBL: a large-scale bioactivity database for drug discovery. Nucleic Acids Res. 2012;40(D1):D1100–7. https://doi.org/10.1093/nar/gkr777.

Gaulton A, et al. The ChEMBL database in 2017. Nucleic Acids Res. 2017;45(D1):D945–54. https://doi.org/10.1093/nar/gkw1074.

Gordon DE, et al. A SARS-CoV-2 protein interaction map reveals targets for drug repurposing. Nature. 2020;583(7816):459–68. https://doi.org/10.1038/s41586-020-2286-9.

Grigsby SS. Artificial intelligence for advanced human-machine symbiosis. In: Schmorrow DD, Fidopiastis CM, editors. Augmented cognition: intelligent technologies. Cham: Springer International Publishing. (Lecture Notes in Computer Science); 2018. p. 255–66. https://doi.org/10.1007/978-3-319-91470-1_22.

Gupta A, et al. Generative recurrent networks for *De novo* drug design. Mol Inform. 2018;37(1–2):1700111. https://doi.org/10.1002/minf.201700111.

Heid E, et al. Chemprop: a machine learning package for chemical property prediction. J Chem Inf Model. 2024;64(1):9–17. https://doi.org/10.1021/acs.jcim.3c01250.

ISO (2018) 9241-11:2018 Ergonomics of human-system interaction — Part 11: Usability: Definitions and concepts. Available at: https://www.iso.org/standard/63500.html

ISO (2019) 9241–210:2019 Ergonomics of human-system interaction – Part 210: Human-centred design for interactive systems. Available at: https://www.iso.org/standard/77520.html

Jacovi A et al. Contrastive explanations for model interpretability. In Proceedings of the 2021 conference on empirical methods in natural language processing. Proceedings of the 2021 conference on empirical methods in natural language processing, online and Punta Cana, Dominican Republic: Association for Computational Linguistics; 2021, pp. 1597–1611. https://doi.org/10.18653/v1/2021.emnlp-main.120.

Karulin B, Kozhevnikov M. Ketcher: web-based chemical structure editor. J Cheminformatics. 2011;3(S1) https://doi.org/10.1186/1758-2946-3-S1-P3.

Khalef L, et al. Cell viability and cytotoxicity assays: biochemical elements and cellular compartments. Cell Biochem Funct. 2024;42(3):e4007. https://doi.org/10.1002/cbf.4007.

Kimber I, et al. Allergic contact dermatitis. Int Immunopharmacol. 2002;2(2–3):201–11. https://doi.org/10.1016/S1567-5769(01)00173-4.

Komorowska-Müller JA, Schmöle A-C. CB2 receptor in microglia: the Guardian of Self-control. Int J Mol Sci. 2020;22(1):19. https://doi.org/10.3390/ijms22010019.

Lamanna G, et al. GENERA: a combined genetic/deep-learning algorithm for multiobjective target-oriented De novo design. J Chem Inf Model. 2023;63(16):5107–19. https://doi.org/10.1021/acs.jcim.3c00963.

Lipinski CA. Lead- and drug-like compounds: the rule-of-five revolution. Drug Discov Today Technol. 2004;1(4):337–41.

Lomuscio MC, et al. AMALPHI: a machine learning platform for predicting drug-induced PhosphoIlpidosis. Mol Pharm. 2024;21(2):864–72. https://doi.org/10.1021/acs.molpharmaceut.3c00964.

Lomuscio MC, et al. SIGMAP: an explainable artificial intelligence tool for SIGMA-1 receptor affinity prediction. RSC Med Chem. 2025;16(2):835–48. https://doi.org/10.1039/D4MD00722K.

Lüllmann H, et al. Drug-induced phospholipidoses. CRC Crit Rev Toxicol. 1975;4(2):185–218. https://doi.org/10.1080/10408447509164014.

Lundberg S, Lee S-I. A unified approach to interpreting model predictions;(2017. arXiv. 10.48550/ARXIV.1705.07874

Martin YC. A bioavailability score. J Med Chem. 2005;48(9):3164–70. https://doi.org/10.1021/jm0492002.

Mendez D, et al. ChEMBL: towards direct deposition of bioassay data. Nucleic Acids Res. 2019;47(D1):D930–40. https://doi.org/10.1093/nar/gky1075.

Montaruli M, et al. Accelerating drug discovery by early protein drug target prediction based on a multi-fingerprint similarity search †. Molecules. 2019;24(12):2233. https://doi.org/10.3390/molecules24122233.

Myung Y, de Sá AGC, Ascher DB. Deep-PK: deep learning for small molecule pharmacokinetic and toxicity prediction. Nucleic Acids Res. 2024;52(W1):W469–75. https://doi.org/10.1093/nar/gkae254.

Nam CS, Jung J-Y, Lee S, editors. Human-centered artificial intelligence: research and applications. London/San Diego/Cambridge: Academic Press; 2022. https://doi.org/10.1016/C2020-0-02460-6.

Nielsen J. Usability engineering. 3.Nachdr. Amsterdam Heidelberg: Kaufmann; 1993.

Nielsen J. Enhancing the explanatory power of usability heuristics. In: Proceedings of the SIGCHI conference on human factors in computing systems. CHI94: ACM conference on human factors in computer systems. Boston: ACM; 1994. p. 152–8. https://doi.org/10.1145/191666.191729.

O'Boyle NM, et al. Open babel: an open chemical toolbox. J Chem. 2011;3(1):33. https://doi.org/10.1186/1758-2946-3-33.

OECD. Test No. 429: Skin Sensitisation: Local Lymph Node Assay. OECD (OECD Guidelines for the Testing of Chemicals, Section 4); 2010. https://doi.org/10.1787/9789264071100-en.

OECD. Test No. 442C: In Chemico Skin Sensitisation: Direct Peptide Reactivity Assay (DPRA). OECD (OECD Guidelines for the Testing of Chemicals, Section 4); 2024. https://doi.org/10.1787/9789264229709-en.

Rogers Y, Sharp H, Preece J. Interaction design: beyond human-computer interaction. 6th ed. Hoboken: Wiley; 2023.

Sanches IH, et al. Enhancing hERG risk assessment with interpretable classificatory and regression models. Chem Res Toxicol. 2024;37(6):910–22. https://doi.org/10.1021/acs.chemrestox.3c00400.

Schneider G. Automating drug discovery. Nat Rev Drug Discov. 2018;17:97–113.

Shneiderman B. Human-centered AI. 1st ed. Oxford: Oxford University Press; 2022.

Taglialatela M, et al. Human ether-a-gogo related gene (HERG) K channels as pharmacological targets. Biochem Pharmacol. 1998;55(11):1741–6. https://doi.org/10.1016/S0006-2952(98)00002-1.

Tanaka M, Sackett S, Zhang Y. Endocannabinoid modulation of microglial phenotypes in neuropathology. Front Neurol. 2020;11:87. https://doi.org/10.3389/fneur.2020.00087.

Tian S, et al. CypReact: a software tool for in Silico reactant prediction for human cytochrome P450 enzymes. J Chem Inf Model. 2018;58(6):1282–91. https://doi.org/10.1021/acs.jcim.8b00035.

Turcotte C, et al. The CB2 receptor and its role as a regulator of inflammation. Cell Mol Life Sci. 2016;73(23):4449–70. https://doi.org/10.1007/s00018-016-2300-4.

Weininger D. SMILES, a chemical language and information system. 1. Introduction to methodology and encoding rules. J Chem Inf Comput Sci. 1988;28(1):31–6. https://doi.org/10.1021/ci00057a005.

Wellawatte GP, et al. A perspective on explanations of molecular prediction models. J Chem Theory Comput. 2023;19(8):2149–60. https://doi.org/10.1021/acs.jctc.2c01235.

Yang K, et al. Analyzing learned molecular representations for property prediction. J Chem Inf Model. 2019;59(8):3370–88. https://doi.org/10.1021/acs.jcim.9b00237.

Zdrazil B, et al. The ChEMBL database in 2023: a drug discovery platform spanning multiple bioactivity data types and time periods. Nucleic Acids Res. 2024;52(D1):D1180–92. https://doi.org/10.1093/nar/gkad1004.

AI- and ML-Driven Strategies for Drug Repurposing: Techniques, Applications, and Challenges

Yi Cong, So Nakagawa, Atsushi Ogura, Katsuhiko Mineta, Takashi Gojobori, and Toshinori Endo

Abstract This chapter provides a comprehensive overview of the role of artificial intelligence (AI) and machine learning (ML) in drug repurposing, emphasizing their potential to uncover new therapeutic applications for existing drugs. It begins by categorizing drug repurposing methodologies into data-driven and approach-driven strategies, establishing a foundational framework for subsequent discussions. The chapter then explores fundamental AI/ML techniques, including classification, regression, clustering, and dimensionality reduction, with a focus on their applications in identifying drug-disease associations and optimizing drug properties. Building on this, advanced deep learning models, such as neural networks, generative adversarial networks (GANs), and graph neural networks (GNNs), and large language models (LLMs), are explained alongside tools such as AlphaFold for protein structure prediction based on amino acid sequences, showcasing the cutting-edge integration of AI with biological data. Critical challenges, including data integration, model interpretability, and scalability, are subsequently addressed, with potential solutions such as transfer learning and federated learning highlighted. Finally, the chapter discusses future directions, focusing on predicting adverse drug reactions (ADRs) and managing CYP450-mediated metabolism, underscoring the transformative potential of AI/ML in advancing drug repurposing.

Y. Cong
Graduate School of Information Science and Technology, Hokkaido Unversity, Sapporo, Japan

S. Nakagawa
Schoolf of Medicine, Tokai University, Shibuya, Japan

A. Ogura
Nagahama Institute of Bio-Science and Technology, Nagahama, Japan

K. Mineta
Waseda University, Shinjuku, Japan

T. Gojobori
National Cheng Kung University, Tainan, Taiwan

T. Endo (✉)
Faculty of Information Science and Technology, Hokkaido Unversity, Sapporo, Japan
e-mail: endo@ist.hokudai.ac.jp

A. Lavecchia (ed.), *Applied Artificial Intelligence for Drug Discovery*,
https://doi.org/10.1007/978-3-031-98022-0_23

Keywords Drug repurposing · Machine learning · Artificial intelligence · Deep learning · Multi-omics data integration · Dimensionality and interpretability

1 Introduction

Drug repurposing (DR) is a strategy that identifies new therapeutic uses for existing drugs, initially formalized as a concept by Ashburn and Thor in 2004. This approach has gained traction as an effective alternative to traditional drug development, addressing high attrition rates and resource demands in new drug discovery. By leveraging compounds with established safety profiles, DR reduces the risk of clinical failure due to safety concerns, positioning itself as a lower-risk option in the development landscape (Ashburn and Thor 2004).

Compared to conventional drug development, which often spans 10–17 years and involves substantial financial risk, DR offers significant advantages in both time and cost. Many preclinical and safety assessments for repositioned drugs are already completed, which can dramatically shorten the development timeline and reduce early-stage investment (Pushpakom et al. 2019). Furthermore, interest in DR has grown rapidly; in 2018 alone, the number of articles on DR reached 602 (Jourdan et al. 2020). As of December 24, 2024, database search on PubMed for the keywords "(drug repurposing) OR (drug repositioning)" identified 2228 articles published in 2024 alone, reflecting the expanding research interest and investment in this field.

In recent years, DR has gained unprecedented attention due to its potential to address pressing challenges in drug development. The rapid growth of high-throughput screening technologies, the availability of large-scale biological and pharmacological datasets, and the increasing adoption of computational tools have significantly expanded the scope of DR. However, these advancements have also introduced complexities, particularly in integrating and analyzing diverse data types to uncover novel therapeutic applications. Among these advancements, AI and ML have emerged as pivotal drivers of innovation. By integrating diverse datasets and identifying hidden patterns, AI refines traditional methodologies, streamlining the identification of drug-disease associations and enabling more scalable and precise approaches.

1.1 Introduction to AI in DrugDiscovery and Repurposing

AI has become a transformative force in drug discovery and repositioning, offering data-driven tools to efficiently analyze and interpret complex biological, pharmacological, and clinical datasets. By leveraging advanced algorithms, AI can accelerate the drug development process, reduce costs, and increase the probability of identifying successful drug candidates (Choudhury et al. 2022).

ML, a major subset of AI, involves algorithms that detect patterns in large datasets, aiding in predictive modeling and classification tasks. By training on molecular, genetic, and pharmacological data, ML can reveal potential drug-disease associations and predict drug efficacy, toxicity, and pharmacokinetic properties. This data-driven approach enhances decision-making and supports hypothesis generation for repositioning existing drugs (Lavecchia 2015). Deep Learning (DL), a specialized subset of ML, uses neural networks to model complex relationships in data. DL has shown significant promise in areas like image analysis, molecule generation, and structure-activity relationship predictions. In drug repurposing, DL models can process high-dimensional data, such as omics and electronic health records, to uncover novel insights into drug mechanisms and identify new therapeutic applications(Choudhury et al. 2022; Lavecchia 2019).

AI offers several unique advantages in drug repurposing. Its ability to rapidly process and analyze multi-omics, clinical, and real-world data accelerates the identification of drug candidates and repurposing opportunities, while its capacity to reveal hidden patterns supports the discovery of therapeutic uses that may be overlooked by traditional methods. Furthermore, automated workflows powered by AI enhance scalability in drug screening and molecular modeling, streamlining repetitive tasks and conserving resources, an essential contribution to modern drug development (Perdomo-Quinteiro and Belmonte-Hernández 2024).With the advent of AI and ML, drug repurposing methodologies have evolved significantly, incorporating data-driven and approach-driven strategies to enhance efficiency and accuracy.

2 Classification of Drug Repurposing Methods

DR methodologies are broadly classified into two main frameworks: data-driven strategies and approach-driven strategies, each utilizing specific types of information or methods to identify new therapeutic applications for existing drugs.

2.1 Data-Driven Strategies

Data-driven strategies focus on leveraging specific characteristics of drugs, targets, and diseases to identify repositioning candidates. These approaches are highly efficient, particularly when high-quality data are available, as they enable large-scale screening and comprehensive exploration of existing information. By analyzing drug properties, biological targets, or disease mechanisms, data-driven strategies have demonstrated their scalability and compatibility with high-throughput computational tools. However, their success depends heavily on the quality and completeness of the datasets. Limitations such as data biases and gaps in coverage can introduce inaccuracies, potentially leading to incomplete or skewed predictions.

Despite these challenges, data-driven strategies remain indispensable in modern drug repurposing, offering a robust framework for systematic discovery.

Drug-Oriented Approaches These methods repurpose drugs based on known properties such as chemical structure, pharmacokinetics, and pharmacodynamics. By analyzing comprehensive datasets on drug properties, researchers can identify potential applications for drugs beyond their original indications. This approach is widely used in computational drug screening, where compounds are matched with potential new targets based on structural and functional similarities.

Target-Oriented Approaches Target-based repositioning centers around biological targets (proteins, genes, or pathways) that are implicated in various diseases. Using bioinformatics tools and databases, researchers link existing drugs to relevant targets, facilitating the identification of candidates by highlighting common molecular mechanisms across multiple conditions.

Disease or Therapy-Oriented Approaches This strategy prioritizes drugs that address similar pathological processes across different diseases. By comparing molecular or cellular signatures, researchers can reveal pathways or biomarkers shared among diseases, enabling the repositioning of drugs with mechanisms aligned to these shared pathways (Sahoo et al. 2021).

2.2 *Approach-Driven Strategies*

Approach-driven strategies focus on specific analytical methods or technologies that identify repositioning opportunities by analyzing network relationships, chemical structures, transcriptional responses, or chemogenomic data. These methods excel in providing deeper mechanistic insights into drug-target interactions and disease pathways. By leveraging advanced tools such as network analysis, molecular docking, and gene expression profiling, approach-driven strategies enable researchers to uncover underexplored relationships and gain a more detailed understanding of complex biological systems. However, these strategies often require significant computational resources and domain expertise, which can limit their accessibility and scalability. Additionally, the integration of heterogeneous datasets remains a challenge, as the complexity of combining different data types can introduce noise and reduce prediction accuracy. Despite these limitations, approach-driven strategies are invaluable for studies requiring detailed mechanistic insights, particularly in areas where high-quality experimental data are available. Specific examples of successful applications, such as transcriptional signature-based approaches for identifying disease-relevant drugs (Thorman et al. 2024) or ligand-based analyses using machine learning models to predict drug-target interactions (Liu et al. 2023), will be discussed in the following sections.

Network-Based Approaches Network-based approaches employ systems biology and network analysis to explore complex interactions between drugs, targets, and diseases. By constructing networks that integrate biological, pharmacological, and genomic data, researchers can identify multi-target mechanisms and repurposing opportunities. Methods like pathway analysis and protein-protein interaction (PPI) networks have demonstrated significant potential in addressing diseases involving complex interactions and polygenic factors. Additionally, knowledge graphs (KGs) further enhance network-based strategies by mapping detailed drug-target-disease relationships, allowing for data-driven predictions and visualizing potential candidates through structured knowledge integration. Embedding techniques within KGs, combined with deep learning algorithms, enables the identification of underexplored interactions, thus enriching the repositioning landscape with AI-driven insight (March-Vila et al. 2017; Perdomo-Quinteiro and Belmonte-Hernández 2024).

Ligand-Based Approaches Ligand-based repositioning relies on the principle that structurally similar molecules tend to exhibit similar biological effects. By comparing chemical structures, researchers can identify secondary targets or novel therapeutic applications for existing drugs. Molecular docking and pharmacophore modeling are especially valuable when detailed structural data on drug-receptor interactions are available, as these methods enable precise predictions of binding affinities and target interactions. For instance, recent studies have used machine learning models with molecular descriptors to predict the binding affinities of Food and Drug Administration (FDA)-approved drugs against viral targets such as the SARS-CoV-2 main protease. By integrating techniques like random forest algorithms with molecular docking, researchers can efficiently screen large compound libraries to identify high-potential candidates, accelerating the repurposing process (Liu et al. 2023; March-Vila et al. 2017).

Transcriptional Signature-Based Approaches Transcriptional signature-based methods use gene expression data to find drugs that induce similar transcriptional responses across conditions. By comparing drug-induced gene expression profiles with disease-specific signatures, researchers can identify potential repositioning candidates based on shared transcriptional effects. Resources like the Connectivity Map (CMap) offer valuable datasets for aligning gene expression profiles with disease signatures (Lamb et al. 2006). A recent study by Thorman et al. expanded on this approach by using gene knockdown transcriptional signatures alongside molecular docking to enhance drug target specificity. By analyzing over 15,000 small molecules and 4400 gene knockdowns from the LINCS database, they identified drugs whose transcriptional effects mirrored those of disease-related gene knockdowns, improving the accuracy in selecting repositioning candidates (Subramanian et al. 2017; Thorman et al. 2024).

Machine Learning (ML) Approaches Advances in ML and AI have accelerated DR research through high-throughput, data-driven analysis. ML algorithms like deep learning and matrix factorization support pattern recognition across diverse datasets, including gene expression, omics, and clinical outcomes. By integrating chemogenomic and pharmacological data, ML models achieve high accuracy in predicting drug-disease associations, offering scalable solutions for repositioning drugs across therapeutic areas. These advancements, coupled with AI-driven clustering and multi-omics data analysis, enable researchers to uncover shared molecular signatures, thus providing a robust framework for discovering new applications for existing drugs (Cong and Endo 2022).

In summary, both data-driven and approach-driven strategies play pivotal roles in drug repurposing, offering complementary strengths in leveraging existing datasets and exploring specific analytical methodologies. Recent advances in AI and ML have further enhanced these strategies, providing scalable, accurate, and efficient tools for drug-disease association prediction and mechanistic insights. However, challenges such as data heterogeneity, integration complexity, and model interpretability remain significant barriers. These limitations highlight the need for continued innovation and the development of more robust AI-driven approaches, which will be explored in subsequent sections.

3 AI and ML Techniques in Drug Repurposing

As the complexity of drug discovery continues to grow, AI and ML have become essential tools in drug repurposing. By analyzing vast and complex datasets, AI and ML provide insights that traditional methods may overlook, enabling more efficient identification of therapeutic applications for existing drugs. These technologies allow researchers to uncover hidden patterns, predict drug-disease associations, and optimize drug properties. We proposed that ML techniques in DR can be broadly categorized into four primary tasks: classification, regression, clustering, and dimensionality reduction. Each task plays a distinct role, ranging from predicting drug-disease associations and toxicity (classification and regression) to grouping drugs based on shared characteristics (clustering) and simplifying high-dimensional datasets (dimensionality reduction). These tasks collectively form the foundation of AI-driven approaches in drug discovery (Fig. 1).

Table 1 provides a quick reference, summarizing key techniques, their primary purposes, typical applications, and the main advantages and limitations associated with each. This table serves as a foundation for understanding how these techniques support different aspects of drug repurposing.

3.1 Classification, Regression, and Clustering Techniques

Classification, regression, and clustering are foundational techniques in AI and ML, playing pivotal roles in drug repurposing studies by enabling the analysis of complex datasets to uncover hidden patterns, relationships, and therapeutic mechanisms.

In drug repurposing, classification techniques assign drugs to specific categories based on shared biological, chemical, or pharmacological features, aiding in the

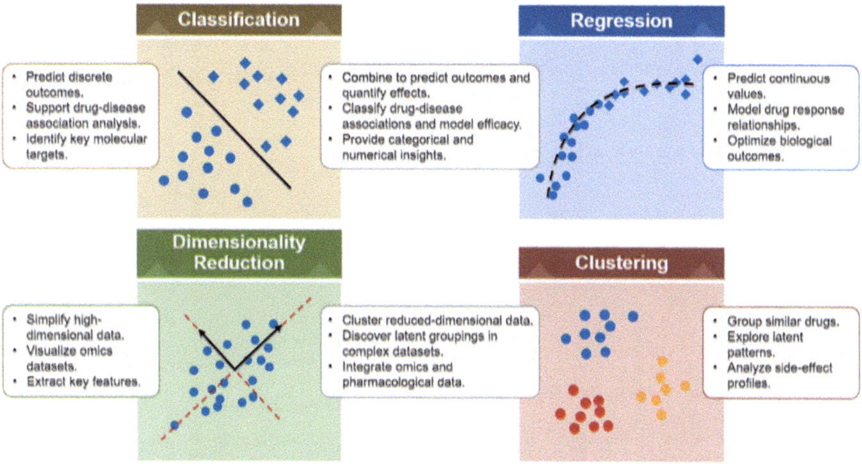

Fig. 1 Machine Learning Tasks and Applications in Drug Repurposing This figure illustrates the four primary ML tasks (classification, regression, clustering, and dimensionality reduction) and their roles in analyzing, predicting, and optimizing drug data

prediction of drug-disease associations and drug classes. Regression techniques focus on predicting quantitative variables such as drug efficacy, toxicity, and response levels, offering critical insights into therapeutic potential.

3.1.1 Logistic Regression (LR)

LR is a fundamental classification algorithm commonly used in drug repurposing studies for binary classification tasks, such as predicting the presence or absence of a specific drug effect. Although formally categorized as a regression model, LR applies the logistic (or sigmoid) function to model the probability of a binary outcome, making it suitable for classification tasks. In drug repurposing, LR can predict outcomes like drug efficacy, toxicity, or target interactions, utilizing features derived from molecular, pharmacological, or clinical data.

For instance, Bejan et al. utilized multivariable LR to examine associations between prior drug exposures and COVID-19 severity. The analysis leveraged a large-scale electronic health record (EHR) database from the Vanderbilt University Medical Center (VUMC), comprising data from 9748 patients. Drug exposure was identified through a combination of structured data and natural language processing (NLP) applied to clinical notes, enabling the extraction of relevant information from unstructured text. Exposure was defined as documented drug use within 3 months preceding the pandemic and extending to the confirmation of a COVID-19 diagnosis. To address potential confounding variables, the study applied propensity score overlap weighting to enhance the comparability of exposure groups. Logistic regression was then used to estimate associations between drug use and key

Table 1 Overview of fundamental AI and ML techniques for drug repurposing

Method	Primary purpose	Advantages	Limitations	Citations of studies
Classification & Regression				
Logistic Regression (LR)	Predicting drug efficacy, toxicity, or target interactions	Simple, interpretable, suitable for small datasets	Limited in handling nonlinear relationships	Bejan et al. (2021)
Naive Bayes (NB)	Rapid screening and initial analysis	Fast, interpretable, provides confidence scores	Assumes feature independence, struggles with dependent features	Yang and Agarwal (2011)
Support Vector Machine (SVM)	Identifying drug-disease associations and target interactions	Effective with high-dimensional data, ensures robust margins	Resource-demanding for large datasets, sensitive to kernel selection	Hsu and Si (2018) and Sengupta et al. (2024)
Random Forest (RF)	Identifying relevant molecular features	Robust to overfitting, handles high-dimensional data well	Requires significant memory and processing power for large datasets	Kitsiranuwat et al. (2022) and Liu et al. (2023)
k-Nearest Neighbor (k-NN)	Screening drugs based on chemical or pharmacological profiles	Intuitive and easy to implement, suitable for similarity-based tasks	Computationally heavy for large datasets, sensitive to distance metrics	Yang et al. (2023b) and Wang et al. (2011)
Gradient Boosting Machines (GBM) / XGBoost	Identifying key features related to drug efficacy and safety	High predictive accuracy, effective in handling complex relationships	Requires careful parameter tuning to prevent overfitting	Li et al. (2024)
Clustering				
k-Means	Grouping drugs by chemical structures or pharmacological profiles; gene expression clustering	Simple, computationally scalable for medium-sized datasets	Assumes spherical clusters, initialization-sensitive, struggles with noisy datasets	Velasco et al. (2024), Hozumi et al. (2021), Cong et al. (2022) and Madugula et al. (2021)
Hierarchical Clustering (HC)	Grouping drugs by side effect profiles; identifying functional pathways	Intuitive dendrogram visualization, no need to predefine cluster numbers	Computationally intensive for large datasets, sensitive to distance metrics	Zaza et al. (2020) and Kumari et al. (2022)

Self-Organizing Maps (SOM) / Growing Self-Organizing Maps (GSOM)	Clustering drugs by chemical and pharmacological properties; dynamically analyzing large datasets	Preserves topological relationships; GSOM adjusts resolution dynamically	SOM requires predefined grid size, GSOM involves complex parameter tuning	Hsu et al. (2003)
Markov Clustering (MCL)	Drug-drug or drug-target interaction networks	Adaptable to varying cluster sizes, efficient for large networks	Parameter tuning (inflation rate) can significantly influence results	Zhou et al. (2015)
Dimensionality reduction				
Principal Component Analysis (PCA)	Feature extraction and noise reduction; simplifies visualization of multi-omics and drug chemical data	Computationally lightweight, preserves major variance	Limited to linear patterns, unable to capture complex nonlinear relationships	Madugula et al. (2021)
t-Distributed Stochastic Neighbor Embedding (t-SNE)	Clustering analysis and visualization of high-dimensional data	Excels in capturing local patterns, suitable for clustering	Computationally demanding, parameter-sensitive, may lead to difficult interpretation	Kumari et al. (2022)
Uniform Manifold Approximation and Projection (UMAP)	Dimensionality reduction and clustering visualization; analyzes both local and global structures in multi-omics data	Efficient for large-scale datasets, preserves both local and global structures	May require parameter tuning for optimal performance, struggles with nonuniform data	Hozumi et al. (2021) and Cong et al. (2022)

outcomes, including mortality, hospitalization, ICU admission, and the need for ventilation. This analysis identified 17 drugs associated with reduced COVID-19 severity, demonstrating the utility of LR in extracting meaningful insights from high-dimensional clinical datasets (Bejan et al. 2021).

However, LR has limitations in handling complex or nonlinear relationships, which can be common in high-dimensional biological data. It is generally outperformed by more sophisticated models, such as Random Forest or Support Vector Machine, in datasets with intricate feature interactions.

3.1.2 Naive Bayes (NB)

NB, a probabilistic classification algorithm rooted in Bayes' theorem, is often applied in drug repurposing studies for its ability to handle high-dimensional data with simplicity and efficiency. Although NB assumes independence among features, which may not always hold in biological datasets, it remains effective in many practical scenarios, particularly for initial screening tasks.

For instance, Yang and Agarwal demonstrated the use of NB to identify potential drug repurposing candidates by leveraging clinical side effects as human phenotypic profiles. Their approach involved constructing a database of disease-side effect associations using data from the Side Effect Resource (SIDER) and the Pharmacogenomics Knowledgebase (PharmGKB). Metrics such as the Matthews correlation coefficient (MCC), sensitivity, and specificity were calculated to evaluate the strength of these associations, which were then used as predictive features in the model. By treating diseases as classification endpoints and side effects as feature variables, the NB model predicted the likelihood of a drug's efficacy for specific conditions. The probabilistic nature of NB allowed for the generation of confidence scores, supporting the prioritization of candidate drugs for further evaluation (Yang and Agarwal 2011).

While its assumption of feature independence can limit accuracy in cases of strong dependencies, NB's computational efficiency and ease of interpretation make it a practical choice for initial analyses in drug repurposing.

3.1.3 Support Vector Machine (SVM)

SVM is a supervised ML model designed to solve classification problems by identifying an optimal hyperplane that maximizes the margin between data classes. In drug repurposing, SVMs are particularly valuable for binary classification tasks, such as predicting whether a drug is effective for a specific disease. The ability to handle high-dimensional datasets with numerous features makes SVMs well-suited for applications such as gene expression analysis, drug-target interaction prediction, and cancer classification.

A notable example of SVM's effectiveness in high-dimensional bioinformatics applications is a study by Hsu and Si, where linear SVM was used to classify 33 types of cancer based on RNA-sequencing data from the TCGA database. The dataset included RNA expression levels for 20,531 genes across 10,471 patient samples. Preprocessing steps, such as removing features with zero variance, normalizing data, and employing feature selection methods like tree classifiers and variance thresholds, were used to reduce dimensionality. Class imbalances were addressed through a combination of under- and over-sampling methods, resulting in a balanced dataset with 300 samples per class. The model was trained and evaluated using five-fold cross-validation, achieving a high classification accuracy of 95.8% (Hsu and Si 2018). This study highlights SVM's effectiveness in processing complex biological data for classification tasks.

In addition to its success in cancer classification, SVM has demonstrated significant potential in drug repurposing for cancer treatment. For instance, Sengupta et al. developed SVM-based Quantitative Structure-Activity Relationship (QSAR) models to predict potential anticancer compounds for colorectal cancer. The study utilized pharmacological data from the Genomics of Drug Sensitivity in Cancer (GDSC) database, which included Half Maximal Inhibitory Concentration (IC50) values for 297 anticancer drugs across 46 colorectal cancer (CRC) cell lines. Chemical structures of the drugs were retrieved from PubChem, converted to 3D structures, and energy-minimized. Over 1800 chemical descriptors were calculated, followed by a feature selection process to identify relevant descriptors. QSAR models were constructed for 12 CRC cell lines using SVM with ten-fold cross-validation, and the models' performance was evaluated based on R^2 values. These models were subsequently used to predict the logIC50 values of 1627 FDA-approved drugs. Based on a logIC50 cutoff, promising drug candidates were identified and further analyzed for their target protein interactions (Sengupta et al. 2024).

By effectively processing high-dimensional biological data, SVMs have proven to be valuable tools in both drug discovery and drug repurposing. Their versatility and precision make them an indispensable approach for researchers aiming to uncover new therapeutic strategies.

3.1.4 Random Forest (RF)

RF is an ensemble learning technique that constructs multiple decision trees during training and combines their outputs to enhance classification accuracy and reduce overfitting. RF is especially effective in noisy and complex datasets, where it can pinpoint key variables contributing to model predictions. This capability allows researchers to uncover potential drug-disease associations and explore important drug-target interactions.

Kitsiranuwat et al. utilized RF to predict drug-protein interactions by integrating various protein similarity features. Their approach began with the construction of a comprehensive dataset, combining data from Search Tool for the Retrieval of Interacting Genes/Proteins (STRING), DrugBank, Kyoto Encyclopedia of Genes and Genomes (KEGG), Gene Ontology Annotation (GOA), and Comparative Toxicogenomics Database (CTD). Feature engineering yielded 13 protein similarity features, including PPI data, network topology, sequence alignment scores, and functional annotations. Using forward selection, they identified 12 key features to train the RF model, achieving a robust Area Under the Curve (AUC) greater than 93%. The predicted protein pairs were then used to enhance a drug-protein matrix, which improved the performance of existing drug repositioning algorithms (Kitsiranuwat et al. 2022). In another study, Liu et al. applied RF to predict the binding affinities of FDA-approved drugs to the SARS-CoV-2 main protease, aiming to identify potential therapeutic candidates for COVID-19. Their methodology involved curating a binding activity dataset from the Protein Data Bank (PDB) and other scientific sources. Molecular descriptors for compounds were calculated using Mold2 software, and Shannon entropy analysis reduced the feature set from 777 to 297 descriptors. Using a five-fold cross-validated RF model, they trained on 516 compounds and tested on 360, achieving robust predictive performance. The model was subsequently applied to 1188 FDA-approved drugs, with applicability domain and prediction confidence analyses identifying 10 promising candidates for COVID-19 treatment (Liu et al. 2023). These findings highlight RF's capability to process diverse pharmacological and molecular data efficiently in urgent drug repurposing contexts.

3.1.5 k-Nearest Neighbor (k-NN)

k-Nearest Neighbor (k-NN) is a simple yet powerful classification algorithm that assigns labels to data points based on the majority class of their nearest neighbors. In drug repurposing, k-NN is particularly useful for identifying drugs or compounds with similar properties, such as chemical structures or pharmacological profiles, aiding in the identification of candidates for potential repurposing applications.

Despite its simplicity, k-NN faces challenges when applied to large or high-dimensional datasets due to its high computational cost and sensitivity to distance metrics. To address these limitations, several variations of k-NN have been developed to address these challenges and enhance its applicability in complex datasets.

One such variation is *Weighted k-Nearest Neighbor (WKNN)*, introduced by Dudani in 1976, which assigns different weights to neighbors based on their distance, giving closer neighbors higher importance (Dudani 1976). Recent studies have demonstrated that WKNN has demonstrated effectiveness in drug-disease association studies by optimizing the density of drug-disease association matrices. For instance, Yang et al. applied WKNN to calculate similarity scores for drugs (based on chemical structure, ATC codes, side effects, drug-drug interactions, and target profiles) and diseases (using phenotypic and ontology similarities). By weighting contributions from the k nearest neighbors, WKNN enriched sparse drug-disease association matrices, improving prediction accuracy in identifying new drug-disease relationships (Yang et al. 2023b).

Fuzzy k-Nearest Neighbor (FKNN), introduced by Keller, Gray, and Givens in 1985, extends k-NN by incorporating fuzzy logic to assign a membership degree to each class rather than a strict label(Keller et al. 1985). This approach allows

FKNN to handle uncertain or overlapping class boundaries effectively. FKNN has been successfully applied in bioinformatics for tasks like protein classification. For instance, Wang et al. utilized FKNN to classify proteins using sequence-derived features, including amino acid composition, dipeptide composition, complexity factors, and Fourier spectrum components (Wang et al. 2011). While FKNN has not been directly used in drug repurposing, its ability to manage uncertain suggests potential for modeling complex drug-target interactions.

Additionally, *Adaptive k-Nearest Neighbor (AKNN)* dynamically adjusts the number of neighbors (k) based on local data density, making it well-suited for heterogeneous datasets. Although AKNN has not yet been extensively applied in drug repurposing, its adaptability to varying data densities indicates promise for analyzing diverse pharmacological datasets in future studies.

3.1.6 Gradient Boosting Machines (GBM)

GBM is an ensemble learning technique that combines multiple weak learners, typically decision trees, to create a powerful predictive model. GBM iteratively builds trees, with each new tree aiming to correct errors made by the previous ones, thereby reducing bias and improving overall accuracy. In drug repurposing, GBM is valuable for tasks involving classification, regression, and feature selection, enabling researchers to identify relevant molecular or pharmacological features associated with drug efficacy and safety profiles.

Extreme Gradient Boosting (XGBoost), introduced by Chen and Guestrin in 2016, is an advanced implementation of GBM that includes enhancements such as regularization, second-order optimization, and parallel processing, making it both accurate and computationally efficient (Chen and Guestrin 2016). XGBoost has demonstrated superiority over traditional models like RF in predictive accuracy and computational speed, particularly with high-dimensional datasets common in drug discovery (Sheridan et al. 2016).

In a 2024 study by Li et al. investigating sex-specific differences in ankylosing spondylitis, XGBoost was used to identify key differentially expressed genes (DEGs) associated with the disease. Gene expression datasets were sourced from the Gene Expression Omnibus (GEO) database, including five datasets containing transcriptomic data from peripheral blood mononuclear cells of male and female patients. The raw data were processed to identify differentially expressed genes (DEGs) between patients and healthy controls, as well as between male and female patients. Functional enrichment analyses, including Gene Ontology and pathway enrichment, were conducted to explore the biological relevance of these DEGs. Immune infiltration analysis provided additional insights into cell-type-specific differences, such as the proportions of IL-17-secreting immune cells across the samples. Using XGBoost, key DEGs encoding secreted proteins were identified and integrated with the Connectivity Map database to explore drugs capable of modulating the associated biological pathways (Li et al. 2024).

The application of XGBoost in this study highlights its effectiveness in processing complex gene expression data for drug repurposing. By combining accuracy, scalability, and computational efficiency, XGBoost continues to serve as a powerful tool for researchers aiming to uncover novel therapeutic candidates.

3.1.7 k-means

k-Means Clustering is one of the most widely used unsupervised learning algorithms, particularly effective for partitioning data into a predefined number (k) of clusters. The algorithm iteratively minimizes within-cluster variance by assigning data points to the nearest cluster centroid and recalculating centroids until convergence (Macqueen 1967). This approach ensures compact and well-separated clusters, making it a popular choice in various clustering tasks.

In drug repurposing, k-means is commonly used to group drugs based on their chemical structures or pharmacological profiles, facilitating the identification of compounds with similar mechanisms of action (D'haeseleer 2005). It is also been applied in gene expression analysis to classify genes with similar transcriptional patterns, uncovering pathways or targets relevant to drug discovery. The algorithm's simplicity and computational efficiency make it particularly suitable for medium-sized datasets and exploratory analyses.

For instance, a recent study on chemotherapy-induced peripheral neuropathy (CIPN) applied k-means clustering to analyze the severity of neurotoxicity symptoms by integrating patient-reported outcomes (PRO) and clinical evaluations. The analysis combined subscores from the Total Neuropathy Scale–clinical version (TNSc) and scores from the Chemotherapy-Induced Peripheral Neuropathy 20-item Quality of Life Questionnaire (QLQ-CIPN20), which assess sensory, motor, autonomic, and other neurological impairments. Using hierarchical cluster analysis to determine the number of clusters, k-means partitioned patients into three distinct groups: severely impaired, intermediately impaired, and mildly impaired. The clusters were further validated using discriminant function analysis, achieving classification accuracies of over 93% for all groups (Velasco et al. 2024).

Despite its utility, k-means has notable limitations. It assumes that clusters are spherical and evenly distributed, which may not always align with the characteristics of biological datasets. Additionally, the algorithm is sensitivity to the initial placement of centroids can lead to suboptimal results in high-dimensional or noisy data (Hozumi et al. 2021). To address these challenges, dimensionality reduction techniques are often employed before clustering, enhancing the interpretability and accuracy of results. After introducing these methods in the next section, their integration with clustering algorithms, including k-means, will be further discussed to explore their combined potential in analyzing complex biological datasets.

3.1.8 Hierarchical Clustering (HC)

HC is a foundational clustering technique introduced by Johnson, which organizes data into a hierarchical tree structure known as a dendrogram, where clusters are formed hierarchically based on the similarity or distance between data points (Johnson 1967). HC operates in two main forms: agglomerative (bottom-up) and divisive (top-down). The agglomerative approach begins with each data point as an individual cluster and iteratively merges them based on proximity, while the divisive method starts with all data points in a single cluster and recursively splits them (Everitt et al. 2011).

In drug repurposing, HC has been widely applied to analyze drug-related data and uncover hidden therapeutic potentials. For instance, Zaza et al. used HC to explore potential treatments for Raynaud's phenomenon (RP) by grouping drugs with similar adverse drug reaction (ADR) profiles. Their study utilized the World Health Organization (WHO) pharmacovigilance database to define ADR signatures, starting with 14 representative ADRs associated with erythromelalgia. Using HC, the study identified six distinct clusters of drugs based on their ADR profiles. This study exemplifies HC's utility in identifying hidden drug relationships through ADR profiling, thereby enabling innovative approaches to drug repurposing (Zaza et al. 2020).

Despite its strengths, such as providing interpretable results through dendrograms and not requiring a predefined number of clusters, HC also has limitations. It is sensitive to the choice of distance metric and linkage method, which can significantly influence clustering results. Furthermore, it is computationally intensive for large datasets, limiting its scalability in high-dimensional biological analyses.

3.1.9 Self-Organizing Maps (SOM)

SOM is an unsupervised learning algorithm that maps high-dimensional data onto a lower-dimensional grid while preserving topological relationships (Kohonen 1995). It is a powerful tool for clustering and visualizing complex datasets, particularly in biological applications such as gene expression analysis. By iteratively updating the weight vectors of grid nodes, SOM clusters similar data points, revealing patterns and relationships within the data.

One limitation of SOM is the need to predefine the grid size and structure, which can be challenging when the data distribution is unknown (Alahakoon et al. 2000). To address this, the *Growing Self-Organizing Map (GSOM)* extends SOM by dynamically adjusting grid size based on data complexity. GSOM introduces a "spread factor" to control map growth, enabling hierarchical clustering that captures coarse-to-fine patterns within the data (Alahakoon et al. 2000).

In drug repurposing, SOM and its extensions have been used to analyze complex datasets. For instance, Hsu et al. developed a Dynamic SOM Tree approach, combining GSOM and hierarchical clustering to discover cancer subtypes and identify marker genes in microarray datasets. In this method, GSOM was employed to cluster data points, with the map's resolution controlled by the spread factor (SF). Higher SF values provided finer-grained clusters, while lower SF values allowed for broader groupings. By integrating multiple GSOM layers with different resolutions, the Dynamic SOM Tree constructed a hierarchical structure that visualized relationships between clusters. The approach overcame limitations of traditional clustering methods, such as sensitivity to data order and lack of robustness to noise (Hsu et al. 2003).

3.1.10 Markov Clustering (MCL)

MCL is an unsupervised graph-based clustering method designed to identify clusters or modules within networks. It uses an iterative process of expansion and inflation to simulate stochastic flow within a graph, enhancing connections within clusters while suppressing inter-cluster edges (van Dongen 2000). These characteristics make MCL particularly effective for identifying modular structures in biological networks, such as protein-protein interaction networks (Enright 2002).

In drug repurposing, MCL has been applied to analyze drug-drug interaction (DDI) networks, uncovering shared mechanisms of action and potential new applications for existing drugs. For instance, Zhou et al. constructed a drug interaction network using data from the DrugBank database, focusing on drugs with significant DDI profiles. The study involved several steps: first, DDI information, chemical structures, ATC codes, and drug-target data were collected, resulting in a dataset of 1074 drugs and 10,835 interactions. A similarity score for drug interaction profiles was computed using the Weighted Tanimoto similarity coefficient, where only pairs with scores above a threshold (0.3) were retained to construct the network. Nodes in the network represented drugs, and edges indicated significant interaction-based similarity. Using MCL, 98 distinct clusters were identified, many of which correlated with specific drug functionalities. These clusters provided a framework for further analysis, enabling the exploration of shared properties and relationships within each group. By clustering drugs with similar interaction profiles, the method facilitated target prediction by assigning known targets of related drugs to previously uncharacterized compounds (Zhou et al. 2015).

One of MCL's strengths is its ability to handle large and complex networks efficiently, adapting to varying cluster sizes without requiring pre-specification of cluster numbers. Moreover, its probabilistic nature allows for robust identification of overlapping or interconnected clusters, which are common in biological datasets. However, MCL's dependence on parameter tuning, such as inflation rates, can influence clustering outcomes, requiring careful optimization based on the specific dataset (Enright 2002).

3.1.11 Dimensionality Reduction

Dimensionality reduction is a crucial technique in drug repurposing studies, particularly when dealing with high-dimensional biological datasets, such as gene expression or chemical structure data. By reducing the number of features, dimensionality reduction techniques help to simplify complex datasets, improve computational efficiency, and reduce noise, which enhances the model's predictive performance.

In drug repurposing, dimensionality reduction methods are rarely used in isolation. They are often combined with clustering techniques to explore relationships within datasets, such as identifying groups of similar drugs or shared biological pathways. Detailed examples of these applications will be discussed in the next section.

3.1.12 Principal Component Analysis (PCA)

PCA is a widely used linear dimensionality reduction technique that transforms data into a set of uncorrelated components, capturing variance in descending order. It leverages eigenvalue or singular value decomposition (SVD) to effectively reduce dimensionality while preserving key data structures (Jolliffe and Cadima 2016).

In drug repurposing and multi-omics data analysis, PCA aids in feature extraction and noise reduction, projecting complex biological datasets onto lower-dimensional spaces to simplify visualization and clustering tasks. This facilitates the identification of potential drug-disease associations. However, PCA's linear assumption can be limiting with highly nonlinear datasets. For such data, PCA is often used alongside nonlinear dimensionality reduction methods, such as t-SNE and UMAP, to provide more accurate representations (Hozumi et al. 2021).

3.1.13 t-Distributed Stochastic Neighbor Embedding (t-SNE)

t-SNE is a nonlinear dimensionality reduction technique developed by Laurens and Geoffrey in 2008, designed primarily for visualizing high-dimensional data in a lower-dimensional space (Laurens and Geoffrey 2008). By preserving the local structure of data, t-SNE effectively captures patterns like clusters or groupings, making it particularly useful in biological and pharmacological research.

In drug repurposing, t-SNE has been applied to visualize compound similarities, identifying clusters based on chemical or genetic features and exploring high-dimensional datasets like gene expression profiles. Despite its effectiveness, t-SNE can be computational intensity for large datasets and requires careful tuning of parameters, such as perplexity, to achieve accurate visualizations (Cai and Ma 2021).

3.1.14 Uniform Manifold Approximation and Projection (UMAP)

UMAP is a nonlinear dimensionality reduction technique designed to preserve both the local and global structure of high-dimensional data. Its foundation lies in Riemannian geometry and algebraic topology, with a mathematical basis in Laplacian eigenmaps and fuzzy simplicial sets (McInnes et al. 2018). By constructing a weighted graph representation of the data and optimizing its low-dimensional embedding using stochastic gradient descent, UMAP provides a balance between computational efficiency and high-quality visualizations (Hozumi et al. 2021; McInnes et al. 2018).

In bioinformatics and drug discovery, UMAP is particularly valuable for exploring high-dimensional datasets, such as single-cell transcriptomics and multi-omics studies, due to its scalability and ability to capture nonlinear relationships. While its performance can depend on parameter settings, UMAP remains a powerful tool for uncovering patterns in complex biological data (Hozumi et al. 2021).

3.1.15 Dimensionality Reduction-Assisted Clustering

The combination of dimensionality reduction and clustering offers a robust framework for analyzing high-dimensional biological data in drug repurposing studies. By simplifying complex datasets while preserving key patterns, dimensionality reduction methods serve as effective preprocessing steps for clustering algorithms. This approach enables the discovery of meaningful clusters, such as drug classes, molecular pathways, or disease subtypes, by addressing challenges like noise and high dimensionality in biological data.

Dimensionality reduction techniques such as t-SNE and UMAP excel at revealing local and global structures in data, while clustering algorithms like k-means and HC group similar entities based on these structures. The selection of methods depends on dataset characteristics and research goals, allowing flexibility in application.

Recent studies illustrate the utility of this hybrid approach. For instance, Hozumi et al. applied UMAP combined with k-means clustering to analyze SARS-CoV-2 genome mutation data. By first reducing the dimensionality of Single Nucleotide Polymorphism (SNP)-based profiles with UMAP, they facilitated efficient clustering of 203,344 viral sequences, identifying six distinct clusters that captured global and regional variations. This integration of dimensionality reduction and clustering improved computational efficiency and enabled the identification of evolutionary patterns (Hozumi et al. 2021). Similarly, Cong et al. leveraged UMAP and k-means clustering to identify drug-disease associations using multi-omics datasets. Gene expression data from the GEO database were first reduced to two dimensions using UMAP, followed by clustering with k-means to group diseases based on transcriptional profiles. By integrating these clusters with the Library of Integrated Network-based Cellular Signatures (LINCS) L1000 dataset, the study identified compounds capable of reversing disease-specific gene expression signatures, highlighting potential repurposing candidates(Cong et al. 2022).

In another study, Madugula et al. combined PCA with k-means clustering to analyze molecular descriptors of 1671 FDA-approved drugs. PCA reduced 1079 variables into five principal components, which were then clustered into nine groups using k-means. This process identified structurally similar drugs with shared or complementary therapeutic applications. By leveraging Prediction of Activity Spectra for Substances (PASS), the study predicted novel indications for existing drugs, such as repositioning β-lactam antibiotics for non-antimicrobial uses and discovering potential CNS applications for anesthetic drugs (Madugula et al. 2021).

Additionally, Kumari et al. employed t-SNE followed by Ordering Points to Identify the Clustering Structure (OPTICS) and HC to identify potential anti-SARS-CoV-2 compounds. After reducing dimensionality with t-SNE, clustering grouped similar compounds based on drug absorption, distribution, metabolism, and excretion (ADME) parameters. Molecular docking further validated top candidates, such as doxorubicin, which exhibited strong binding affinity to the SARS-CoV-2 main protease, supporting its potential as a repurposed therapeutic (Kumari et al. 2022). These applications demonstrate the versatility and effectiveness of combining dimensionality reduction with clustering in drug repurposing research.

As this integrated approach continues to develop, it provides a flexible solution for emerging challenges, such as integrating diverse data modalities and scaling analyses for large datasets. Future advancements are likely to enhance its efficiency and broaden its applicability in identifying novel therapeutic opportunities.

3.1.16 Deep Learning

Deep learning represents a transformative approach in drug repurposing research, leveraging its ability to discern intricate patterns within large and complex datasets. Its capacity to process high-dimensional data, such as multi-omics or real-world clinical data, has been shown to outperform classical models in predicting drug responses and uncovering novel therapeutic applications (Mohammadzadeh-Vardin et al. 2024). A detailed exploration of deep learning techniques, specific methods, and their applications in drug repurposing will be provided in the next section.

4 Deep Learning Models for Drug Repurposing

DL has revolutionized drug repurposing research by enabling the analysis of complex, high-dimensional datasets such as multi-omics, gene expression profiles, and real-world clinical data. Its ability to uncover intricate patterns and relationships within these datasets has led to significant advancements in identifying potential therapeutic applications for existing drugs. This section focuses on the most prominent deep learning architectures, including Deep Neural Networks (DNNs), Convolutional Neural Networks (CNNs), Graph Neural Networks (GNNs), Gen-

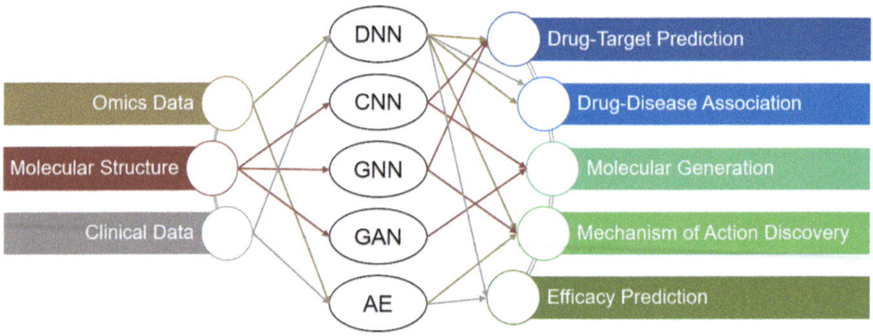

Fig. 2 Categories of deep learning models for drug repurposing This figure illustrates how diverse data types (omics, molecular structures, and clinical data) are processed by DL models to achieve key outcomes in DR. These outcomes include drug-target prediction, drug-disease association, molecular generation, mechanism of action discovery, and efficacy prediction. The relationships demonstrate the flexibility and applicability of DL methods in addressing the multifaceted challenges of DR

erative Adversarial Networks (GANs), and Autoencoders (AEs). Each of these models offers unique capabilities tailored to different aspects of drug discovery and repurposing, such as predicting drug-target interactions, generating novel chemical structures, and understanding molecular mechanisms (Fig. 2 and Table 2).

4.1 Deep Neural Networks (DNNs)

DNNs, as an extension of Artificial Neural Networks (ANNs), provide a robust framework for modeling complex nonlinear relationships in biological data. While early applications of ANNs were constrained by their limited depth and capacity to process high-dimensional datasets, DNNs, introduced by Hinton et al., overcome these challenges through multilayered architectures capable of progressively extracting abstract features (Hinton et al. 2006). This design makes DNNs particularly effective for tasks in drug repurposing, including predicting drug-disease associations, analyzing therapeutic mechanisms, and integrating multi-omics data (Mottaqi et al. 2021).

Practical applications of DNNs showcase their transformative potential. For instance, Kim et al. utilized DNNs to identify optimal drug combinations for reversing aging-related genetic signatures in acute myeloid leukemia (AML). They trained a DNN model using gene expression data from the GEO database, aligning drug-induced gene expression profiles from the Connectivity Map (CMap) with age-associated DEGs. By optimizing initial and secondary matching of drug-induced differentially expressed genes (d-DEGs) with reversed aging patterns (i-DEGs), the study identified promising drug combinations, such as vorinostat and metformin, with potential antiaging effects (Kim et al. 2021). Similarly, Amiri et

Table 2 Overview of deep learning models for drug repurposing

Method	Primary purpose	Advantages	Limitations	Citations of studies
Deep Neural Networks (DNNs)	Predicting drug-disease associations; multi-omics data integration	Handles complex, nonlinear relationships; robust across data types	Requires large datasets; interpretability issues	Kim et al. (2021) Amiri et al. (2023)
Convolutional Neural Networks (CNNs)	Predicting DTIs; molecular structure analysis	Extracts hierarchical features; good for spatial data	Limited long-range dependency handling	Zhao et al. (2022) Qian et al. (2022)
Graph Neural Networks (GNNs)	Modeling drug-target-disease relationships	Captures node and edge features in graphs; interpretable embeddings	Data sparsity; computational demand	Xu et al. (2024) Huang et al. (2024, b) Veleiro et al. (2024)
Generative Adversarial Networks (GANs)	Synthetic data generation; DTI prediction	Generates high-quality data; addresses data imbalance	Training instability; mode collapse	Sun et al. (2022) Kalemati et al. (2024)
Autoencoders (AEs)	Dimensionality reduction; feature extraction	Handles multi-omics data; reduces sparsity	Sensitivity to noise; limited global pattern capture	Mohammadzadeh-Vardin et al. (2024)

al. developed the Integration of Drug-Disease associations for drug repurposing by Deep Neural Network (IDDI-DNN) model integrating diverse drug and disease features from multiple data sources, including chemical structure, side effects, and target proteins. By fusing similarity matrices for drugs and diseases, the model effectively predicted drug-disease associations with high accuracy. For instance, it identified potential new applications for FDA-approved drugs, demonstrating robustness through metrics such as F1-score and precision-recall curve performance (Amiri et al. 2023). These studies highlight the capability of DNNs to synthesize diverse data modalities and address complex biological questions.

Despite their promise, DNNs face challenges, including their reliance on large datasets, computational demands, and limited interpretability persist. Innovations like attention mechanisms and algorithmic efficiency improvements continue to address these barriers, enabling broader adoption of DNNs in drug repurposing research. These advancements highlight the potential of DNNs to accelerate therapeutic discovery and address unmet medical needs (Kim et al. 2021; Mottaqi et al. 2021).

4.2 Convolutional Neural Networks (CNNs)

CNNs are a specialized subset of deep learning architectures that excel at capturing spatial patterns and localized features in structured data (Lecun et al. 1998). Originally developed for image recognition, CNNs have proven highly effective in biomedical applications, such as analyzing protein sequences, molecular graphs, and SMILES representations. Their ability to extract hierarchical feature representations has made them indispensable for various drug repurposing tasks (Mottaqi et al. 2021).

Building on their foundational strengths, CNNs have found critical applications in drug repurposing, particularly in predicting drug-target interactions (DTIs). One notable application is HyperAttentionDTI, an end-to-end deep learning model developed by Zhao et al. for predicting DTIs. The study utilized datasets from DrugBank, the Davis kinase binding affinity dataset (Davis), and the Kinase Inhibitor BioActivity dataset (KIBA). After filtering out inorganic compounds, very small molecules, and unrecognizable Simplified Molecular Input Line Entry System (SMILES) strings, 6655 drugs, 4294 proteins, and 17,511 positive DTI pairs were obtained. The model employed a 1D-CNN architecture to extract features from SMILES strings and protein sequences, followed by an attention mechanism that captured semantic dependencies across spatial and channel dimensions. The final output layer, comprising fully connected neural networks (FCNNs) with Leaky ReLU activation, produced interaction probability scores. Evaluations using metrics such as Area under the receiver operating characteristic curve (AUC) and Area under the precision-recall curve (AUPR) indicated that HyperAttentionDTI achieved high predictive accuracy in multiple experimental settings, including those where either drugs or proteins were excluded from the training set (Zhao et al. 2022).

Despite their promise, CNNs face challenges, including their reliance on high-quality datasets and limited capacity to capture long-range dependencies. Future innovations, such as hybrid models combining CNNs with transformers, may address these limitations and further expand their applicability in drug repurposing. For instance, Qian et al. proposed combining CNN and transformer to predict Compound-protein interaction (CAT-CPI), a hybrid model combining CNNs and transformers for compound-protein interaction prediction. Unlike traditional CNN-based models, CAT-CPI leverages CNNs to extract local geometric features from molecular images and transformers to capture long-range dependencies in protein sequences. The model was trained on balanced datasets such as Human and Celegans, as well as the imbalanced Davis dataset. CAT-CPI achieved higher predictive accuracy compared to state-of-the-art models like DeepConv-DTI and TransformerCPI, particularly on imbalanced data. Ablation studies confirmed the importance of each component, including the CNN block, transformer encoder, and k-gram-based protein feature extraction. The study concluded that using molecular images rather than SMILES or molecular graphs enhances performance, providing a novel direction for future CPI research (Qian et al. 2022). Such advancements hold promise for overcoming existing limitations and enabling more robust applications in drug discovery.

4.3 Graph Neural Networks (GNNs)

GNNs represent a powerful class of deep learning models specifically designed to process graph-structured data. Unlike traditional deep learning models, GNNs capture both the features of individual nodes and the relationships between them, making them particularly effective for tasks involving complex networks, such as DTI prediction and disease-drug association analysis. First introduced through Graph Convolutional Networks (GCNs), GNNs have since developed to include more advanced architectures, such as graph attention networks (GATs) and heterogeneous graph neural networks (Kipf and Welling 2016).

Recent advancements in GNNs highlight their utility in drug repurposing. For instance, the DTI prediction model based on Dynamic Heterogeneous Graph (DT-DHG) leverages dynamic heterogeneous graphs to represent multi-relational networks of drugs, targets, and diseases. The study utilized two key datasets: Yamanishi_08, comprising four subsets (Enzyme, Ion Channel, G Protein-coupled Receptor, and Nuclear Receptor), and the Shao dataset, known for its richer DTI data. DT-DHG integrates drug-drug similarity matrices, target-target similarity matrices, and known drug-target interactions to create a dynamic heterogeneous network. The process includes three main stages: (1) Similarity ranking to prioritize edges based on both global and local similarity measures, (2) Edge sampling using Bernoulli sampling to generate edges dynamically during training, and (3) Network fusion, where adjacency matrices are fused to form the final network. A progressive learning strategy expands receptive fields layer by layer, and Heterogeneous Graph Convolutional Networks (HGCNs) are applied to extract feature representations. Evaluation using Hits@K, AUC, and AUPR metrics showed that DT-DHG significantly outperformed baseline models on both datasets, with case studies confirming that 80% of the top-5 predicted DTIs were experimentally validated in external databases (Xu et al. 2024). TxGNN, a GNN-based framework for zero-shot drug repurposing, is capable of predicting drug-disease associations for diseases without labeled training data. The study utilized a large-scale biomedical knowledge graph (KG) comprising 17,080 diseases, 7,957 drugs, and 8,063,026 edges, and employed a heterogeneous GNN encoder to generate embeddings for nodes in the knowledge graph, along with a metric learning module to predict novel drug-disease pairs by transferring knowledge from known diseases to unexplored ones. To enhance model interpretability, the framework incorporates a TxGNN Explainer, which uses a self-explanatory approach (GraphMask) to identify key multi-hop paths between drugs and diseases, elucidating the biological rationale behind predictions. Experimental results demonstrated TxGNN's effectiveness in predicting potential treatments for rare or understudied diseases, highlighting its robustness and utility in clinical applications (Huang et al. 2024b). Another example is the Graph Embedding Neural Network Interaction Uncovering System (GeNNius), which focuses on DTI prediction by integrating GraphSAGE-based embeddings with a neural network classifier. The study constructed input graphs from datasets including DrugBank, Biomedical Network Dataset Project (BioSNAP), Binding Database (BindingDB),

Davis, and Yamanishi dataset, where nodes represented drugs or proteins, and edges represented known interactions. Drug nodes were annotated with 12 molecular descriptors, such as LogP values and molecular weight, while protein nodes used 20-dimensional amino acid composition vectors. GeNNius first generates node embeddings using a 4-layer GraphSAGE encoder, aggregating local neighborhood information. These embeddings are then fed into a FCNN classifier with sigmoid activation to predict interaction probabilities. Performance evaluation on standard benchmarks showed that GeNNius achieved superior AUROC and AUPRC scores compared to other models (Veleiro et al. 2024). These models demonstrate GNNs' capability to synthesize diverse types of biological data and extract meaningful associations critical for drug discovery.

Despite their strengths, GNNs face challenges when applied to drug repurposing. Data sparsity remains a key issue, as many potential drug-target associations are not experimentally validated, leading to limited graph structure and overfitting risks (Zhang et al. 2023). Additionally, the high computational demands of training and inference on large-scale graphs limit their practical utility, but future advancements, including hybrid models combining GNNs with transformers, may help address these challenges by capturing both local and global patterns (Wang et al. 2024). Moreover, attention mechanisms and sparsity-based pooling strategies have also shown potential to enhance model interpretability, enabling researchers to pinpoint key relationships driving predictions (Zhang et al. 2023). Such innovations are expected to expand the applicability of GNNs in drug repurposing, bridging current gaps and unlocking new therapeutic opportunities.

4.4 Generative Adversarial Networks (GANs)

GANs are a class of deep learning models originally introduced for generating realistic synthetic data by training two networks—the generator and the discriminator—in an adversarial framework (Goodfellow et al. 2014). Deep Convolutional GAN (DCGAN), a notable extension of GANs, introduced deep convolutional layers into the GAN architecture, significantly improving training stability and the quality of generated data (Radford et al. 2015). To further address limitations such as training instability and mode collapse, Wasserstein GANs (WGANs) introduced a new loss function based on Wasserstein distance (Arjovsky et al. 2017). This approach replaced the original adversarial loss with a smoother and more robust metric, enabling more stable and efficient training, especially in complex datasets. These advancements have expanded GANs' applicability, improving the generation of high-quality synthetic data and feature embeddings.

In drug repurposing, GANs have been utilized to predict DTIs and generate novel molecular structures. For instance, GANDTI is a novel method designed to predict DTIs by integrating a graph convolutional autoencoder with GANs. The datasets include DrugBank, Human Protein Reference Database (HPRD), KEGG, UniProt, and PhID, covering a wide range of drug and protein information. GANDTI constructs a drug-target heterogeneous network incorporating DDIs, target-target interactions (TTIs), and known DTIs. The network topology is represented by adjacency matrices, while similarity matrices based on drug chemical structures and target protein sequences serve as node attributes. A graph convolutional encoder with two GCN layers extracts low-dimensional embeddings of drugs and targets, which are then refined by a GAN framework to enhance robustness by aligning the embeddings with a Gaussian distribution. A multilayer perceptron (MLP) discriminator distinguishes between generated and sampled embeddings. For final DTI prediction, a LightGBM classifier is employed, leveraging ensemble learning with multiple decision trees to handle class imbalance and maximize negative sample utility (Sun et al. 2022).

Additionally, DCGAN-DTA, proposed by Kalemati et al., leverages a DCGAN framework for predicting drug-target binding affinity (DTA). This method first encodes drug SMILES strings and protein amino acid sequences, employing BLOSUM encoding to represent evolutionary features of proteins. Next, it extracts features using a customized DCGAN, where the generator synthesizes drug and protein sequences, while the discriminator learns to distinguish real from synthetic samples. The merged latent vectors are processed by an added layer, followed by a fully connected network to predict DTA. Evaluations on BindingDB and PDBBind datasets, with logarithmically transformed binding affinity values (pKd, Ki, Kd), showed that DCGAN-DTA outperformed existing methods under both warm-start and cold-start conditions. Notably, the model achieved superior concordance index (CI) scores and was validated through adversarial control experiments, confirming its robustness (Kalemati et al. 2024). These approaches highlight GANs' capacity to model nonlinear relationships and address data imbalance challenges commonly encountered in drug discovery tasks.

4.5 Autoencoders (AEs)

AEs are a class of unsupervised learning models designed to encode high-dimensional data into a compressed latent representation and reconstruct it with minimal information loss (Hinton and Salakhutdinov 2006).

In drug repurposing, AEs have demonstrated remarkable utility in uncovering novel drug-disease relationships. For instance, DeepDRA, developed by Mohammadzadeh-Vardin et al., utilizes autoencoders to predict drug response by integrating multi-omics datasets, drug descriptors, and fingerprints. The study employed datasets from GDSC, Cancer Therapeutics Response Portal (CTRP), and Cancer Cell Line Encyclopedia (CCLE), preprocessing the data by imputing missing values, extracting intersecting features, and transforming drug response values into binary categories (sensitive/resistant). The model architecture comprises two autoencoders: one for cancer cell line omics data and another for drug-related features, each with a hidden layer of 256 units. The extracted latent representations are fed into MLP classifier for drug response prediction. DeepDRA demonstrated superior performance across various metrics, including AUPRC, particularly excelling in cross-dataset validations, thus confirming its robustness and potential for drug repurposing in oncology (Mohammadzadeh-Vardin et al. 2024).

Despite their advantages, AEs face limitations, including sensitivity to noisy data and a lack of interpretability, which can hinder their clinical adoption. Additionally, conventional AEs often struggle to capture global patterns in highly heterogeneous datasets. Recent advancements, such as denoising autoencoders for noise robustness and hybrid architectures integrating AEs with GNNs or transformers, have enhanced their ability to model complex relationships and global contexts. These innovations are poised to extend AEs' applicability in drug discovery, facilitating deeper insights and more accurate predictions.

5 Integration of Natural Language Processing and AI for Biological Sequence Analysis

NLP has emerged as a transformative tool in artificial intelligence, enabling the analysis of biological sequences like DNA and proteins by treating them as a form of "language." This integration leverages NLP's ability to process unstructured and vast datasets, making it a powerful approach for tackling challenges in bioinformatics and drug repurposing.

5.1 NLP in Biological Sequence Analysis

NLP, originally developed for analyzing human language, has extended its capabilities to bioinformatics by adapting methodologies for text data to biological sequences. By representing biological sequences as vocabularies or sentences, NLP models can uncover patterns and features crucial for understanding gene functions, protein interactions, and other biological processes. This approach efficiently handles unstructured and large-scale data, addressing common bottlenecks in multi-omics data integration and drug discovery (Fig. 3).

For instance, NLP models like Tasks Assessing Protein Embeddings (TAPE) have been applied to encode protein sequences, enabling predictions of functional and interaction profiles with high accuracy (Rao et al. 2019). Similarly, tools such as DNABERT have been used to analyze genomic data, identifying disease-associated genetic variations and offering insights into the underlying molecular mechanisms (Ji et al. 2021). These models illustrate how NLP, when combined with AI, provides powerful tools for biological sequence analysis, propelling advancements in bioinformatics.

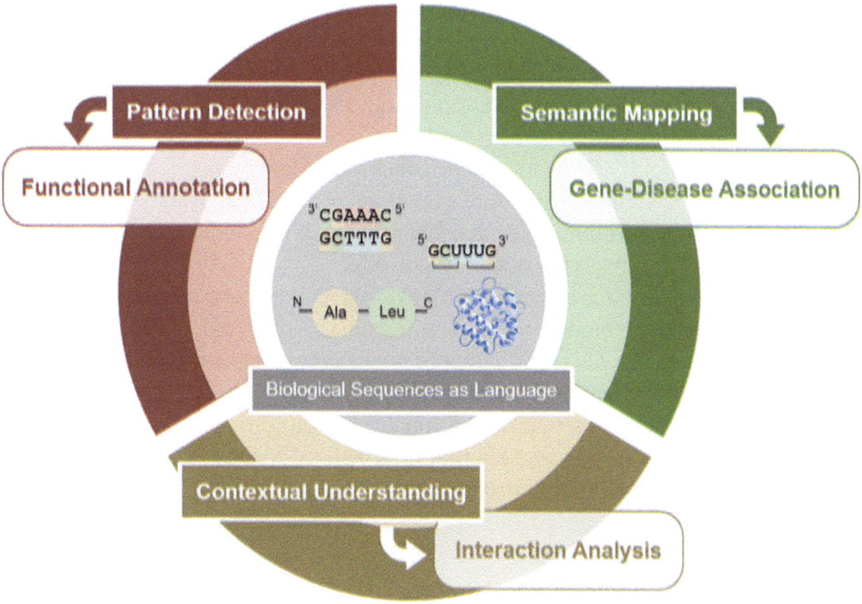

Fig. 3 Why NLP works well for biological sequence analysis Biological sequences can be viewed as "language," with patterns, contexts, and semantics similar to human language. NLP models leverage pattern detection, contextual understanding, and semantic mapping to uncover insights from DNA, RNA, and protein sequences. These abilities enable NLP to excel in tasks like functional annotation, interaction analysis, and disease association prediction

The integration of NLP and AI in this domain not only broadens the scope of NLP applications, but also equips bioinformatics with innovative methodologies for processing complex datasets, ultimately advancing the field of drug repurposing.

5.2 *AlphaFold and Its Evolution in Protein Structure Prediction*

AlphaFold has transformed structural biology and drug repurposing by providing highly accurate protein structure predictions. The release of the AlphaFold Protein Structure Database in 2022, which includes over 200 million predicted protein structures, significantly democratized access to structural data and enabled high-throughput drug screening (Varadi et al. 2022). This resource has facilitated the identification of druggable sites and the rational design of novel therapeutics, particularly for orphan drugs and complex diseases.

For instance, during the COVID-19 pandemic, Abdelkader et al. leveraged AlphaFold to construct a high-quality model of the NSP6 protein, whose structure had not yet been experimentally determined. Using this model, they performed in silico screening of 8158 compounds from DrugBank and natural product databases, ultimately identifying Venetoclax, an FDA-approved drug, as a promising therapeutic candidate with strong binding stability to NSP6. This study highlights AlphaFold's critical role in enabling rapid drug repurposing efforts during health crises (Abdelkader et al. 2022).

In 2024, AlphaFold 3 further expanded its capabilities by predicting biomolecular interactions at atomic resolution, including protein-ligand and multi-protein complex interactions (Abramson et al. 2024). These enhancements have supported multi-target drug design, improved adverse effect predictions through off-target interaction analysis, and advanced combinatorial therapy optimization. By integrating diffusion-based models with deep learning, AlphaFold 3 has streamlined drug discovery pipelines and contributed to precision medicine initiatives. Looking ahead, further advancements in AlphaFold-based models are expected to enhance drug repurposing efforts, particularly in identifying multi-target therapeutics, predicting off-target effects, and accelerating personalized treatment strategies. These innovations hold promise for addressing unmet medical needs and improving drug development efficiency.

5.3 *Large Language Models (LLMs) in Biological Sequence Analysis and Drug Repurposing*

LLMs, including GPT and BERT-style architectures, represent a breakthrough in NLP, enabling complex tasks such as language understanding, generation, and reasoning. By treating biological sequences like DNA, RNA, and protein structures as "language," LLMs can extract patterns and insights that traditional ML models might miss. Recent advancements have shown that LLMs excel at integrating and analyzing large, unstructured datasets, making them particularly valuable in bioinformatics and drug repurposing.

LLMs have demonstrated their potential in addressing key challenges in drug discovery, such as understanding DTIs, predicting side effects, and identifying novel therapeutic uses. Pretrained models, such as BERT and GPT, have been adapted to process biological data through fine-tuning, transfer learning, and integration with other computational frameworks.

One of the key applications of LLMs is their ability to identify motifs and domains within protein sequences, a task analogous to recognizing syntax and semantics in human languages. Transformer-based models, for instance, have been fine-tuned on protein datasets to predict functional sites such as binding pockets or allosteric sites, enabling deeper insights into protein interactions (Rao et al. 2019). Broader discussions on the potential of LLMs in sequence analysis further illustrate their versatility in handling complex biological data (Yang et al. 2023a).

LLMs also excel in knowledge-augmented analysis, where pretraining on biomedical literature and sequence data serves as a foundation for tasks requiring domain-specific expertise. Models like BioBERT and SciBERT have proven highly effective in integrating textual information with biological datasets. These applications have yielded valuable insights into gene-disease associations and protein interactions, further demonstrating the potential of LLMs in bridging the gap between textual and biological data (Beltagy et al. 2019; Lee et al. 2020).

Another remarkable feature of LLMs is their ability to generalize across tasks, enabling them to perform effectively even on out-of-distribution datasets—a critical capability for novel drug repurposing scenarios. Recent advancements in scaled models like GPT-4 have also revealed emergent abilities, such as reasoning through biological pathways, showcasing their potential to address complex biological questions (OpenAI et al. 2023). However, leveraging these capabilities for domain-specific applications often requires additional techniques, such as transfer learning, to adapt general-purpose models to specialized tasks in biological sequence analysis and drug discovery.

5.4 *Transfer Learning for Biological Sequence Analysis*

Transfer learning plays a pivotal role in bridging the gap between pretrained LLMs and their application in specialized domains. By fine-tuning models on smaller, domain-specific datasets, transfer learning enables LLMs to retain their general-purpose capabilities while addressing the unique challenges of biological sequence analysis. This technique involves pretraining a model on a large corpus of general data and subsequently fine-tuning it on smaller, domain-specific datasets, enabling efficient learning and reduced computational demands.

Transfer learning has also proven instrumental in integrating diverse biological datasets, including genomic, proteomic, and transcriptomic data. By fine-tuning models pretrained on large text corpora, researchers can uncover intricate relationships between genes, proteins, and drugs. This capability significantly enhances the potential of multi-omics studies in advancing drug repurposing research (Zhuang et al. 2021).

Transfer learning further enhances the utility of LLMs by leveraging knowledge from unrelated domains, significantly improving their predictive power for under-represented diseases and rare genetic conditions. For instance, this approach holds particular promise in addressing challenges associated with orphan diseases, where traditional methods often struggle due to data scarcity. By enabling generalization across domains, transfer learning provides a novel and impactful solution in such specialized contexts (Ruder et al. 2019).

In scenarios where labeled datasets are scarce, such as drug repurposing, transfer learning enables models to achieve robust performance with minimal fine-tuning. This efficiency not only lowers the barrier to AI adoption in resource-constrained domains, but also highlights the adaptability of transfer learning in addressing real-world challenges in biomedical research (Zhuang et al. 2021).

6 Challenges in AI and ML Application

AI and ML have revolutionized drug discovery and repurposing by enabling the integration and analysis of complex datasets. From predicting DTIs to identifying novel therapeutic uses, these technologies have significantly accelerated research progress. However, the application of AI and ML in this field is not without challenges. Factors such as imbalanced datasets, the integration of heterogeneous data sources, and the interpretability of deep learning models limit their full potential. Addressing these issues is critical to advancing the field and realizing the full promise of AI and ML.

This section examines the major challenges associated with the application of AI and ML in drug repurposing. These challenges can be broadly categorized into three areas: **data-related challenges**, such as imbalanced datasets and the integration of multisource data; **model performance challenges**, including dimensionality reduction, interpretability, and scalability; and **application-specific challenges**, focusing on the complexities of multi-target drug design (Fig. 4).

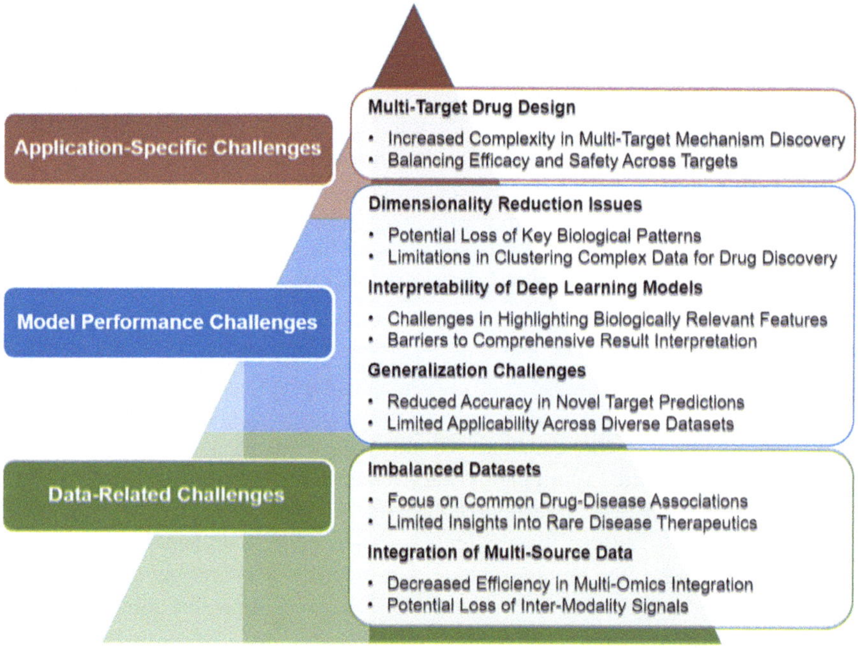

Fig. 4 Pyramid of challenges in AI applications for drug repurposing The pyramid visually categorizes challenges in AI and ML applications for drug repurposing. At the base are foundational data-related challenges like imbalanced datasets and data integration, followed by model performance issues such as interpretability and generalization. At the apex lie application-specific challenges, focusing on the complexities of multi-target drug design

6.1 Data-Related Challenges

Data are the foundation of AI and ML applications, yet the quality, quantity, and balance of available datasets remain critical challenges in drug repurposing. Imbalanced datasets, where certain classes or conditions dominate, can bias model predictions, while the integration of heterogeneous data from multiple sources often presents compatibility and standardization issues. These data-related challenges not only hinder model performance but also limit the generalizability of AI-driven solutions.

6.1.1 Handling Imbalanced Data

Imbalanced datasets represent one of the most significant challenges in applying AI and ML to drug repurposing. These datasets are often characterized by an overrepresentation of certain classes, such as common diseases or well-studied DTIs, while rare conditions and underrepresented drug classes are sparsely annotated. This imbalance can lead to biased models that favor majority classes, compromising predictive accuracy for minority classes and limiting the discovery of novel therapeutic opportunities.

To mitigate the effects of imbalanced data, various preprocessing and algorithmic techniques have been developed. Oversampling methods, such as the Synthetic Minority Oversampling Technique (SMOTE), generate synthetic examples for minority classes, enhancing data diversity during training (Chawla et al. 2002). A notable application of SMOTE in drug repurposing is the HeteroKGRep model proposed by Ceskoutsé et al. This model uses a heterogeneous knowledge graph that integrates multiple biomedical data sources—genes, diseases, drugs, and pathways—to predict novel drug-disease associations. Given the underrepresentation of rare drug-disease pairs, the authors applied SMOTE to generate synthetic samples for minority classes, improving data balance. The augmented graph, enriched with new nodes and edges, was processed by a Heterogeneous Graph Neural Network (HeteroGNN) to learn feature embeddings. These embeddings were then used by an XGBoost classifier to predict potential drug-disease associations. Experimental results showed that SMOTE significantly improved key metrics, including F1-score, AUC-ROC, and AUC-PR, highlighting the method's effectiveness in addressing class imbalance (Ceskoutsé et al. 2024).

Weighted loss functions provide an alternative by assigning higher penalties to misclassified minority samples during model training (Lin et al. 2017). This method ensures the model pays adequate attention to underrepresented classes without altering the dataset itself. Ensemble methods, such as Random Forests and Gradient Boosting, also enhance robustness against class imbalance by combining predictions from multiple models trained on resampled datasets (Dietterich 2000).

Recent advances in deep learning have introduced GANs as a powerful tool for addressing data imbalance. GANs create realistic synthetic datasets, preserving the statistical characteristics of minority classes while expanding data availability. In a study comparing SMOTE, ADASYN, and GAN techniques for class balancing in low-resource settings, García-Domínguez et al. (2023) found that models trained on GAN-enhanced data achieved the highest recall and overall performance, outperforming those using SMOTE or ADASYN (García-Domínguez et al. 2023). Additionally, Frid-Adar et al. demonstrated the effectiveness of GAN-generated medical image data in improving liver lesion classification, a concept adaptable to drug repurposing contexts (Frid-Adar et al. 2018).

Another promising technique is nearest-neighbor clustering (NNC), which groups minority class samples into denser neighborhoods, enhancing their representation in training datasets. In bioengineering, Huang et al. applied NNC to address imbalanced biomedical datasets, showcasing improved model performance and minority class recognition (Huang et al. 2024a).

Despite these advancements, several challenges persist. Techniques like SMOTE and GANs can introduce synthetic data that deviate from true biological characteristics, potentially misleading models (Chawla et al. 2002; Frid-Adar et al. 2018). Under-sampling, while effective in balancing classes, may reduce overall dataset richness (Guo et al. 2017). Additionally, imbalanced data are particularly problematic for rare diseases and orphan drugs, where real-world data scarcity limits the effectiveness of traditional methods (Johnson and Khoshgoftaar 2019; Molnar and Molnar 2023).

Collaborative approaches, such as federated learning (FL), offer a potential solution by enabling the integration of distributed datasets from multiple institutions while preserving patient privacy(Li et al. 2020). This method has shown promise in pooling underrepresented data, fostering a more balanced and diverse dataset for drug discovery applications. For instance, Mittone et al. applied FL to DTI prediction using the KIBA dataset and demonstrated up to 15% improved performance compared to the best available non-privacy preserving alternatives (Mittone et al. 2023). Their findings highlight FL's potential in overcoming data aggregation challenges caused by regulatory and commercial constraints, while maintaining high predictive accuracy.

6.1.2 Integration of Multisource Data

The integration of heterogeneous datasets, such as genomic, proteomic, and clinical data, remains a fundamental challenge in AI-driven drug repurposing. Variations in data types, formats, and scales require sophisticated methods to unify these diverse sources into a coherent analytical framework. For instance, multi-table frameworks like MFA and MDS have demonstrated efficacy in aligning multisource data into a shared coordinate space, facilitating cross-modality analyses and consensus insights (Hulot et al. 2021).

Clustering techniques play a central role in multisource data integration by revealing underlying patterns and associations across diverse datasets. For instance, CiteFuse calculates independent similarity matrices for each data modality, such as

RNA and protein expression, and employs efficient fusion algorithms to merge these matrices into a unified clustering output, effectively capturing shared structures across modalities (Kim et al. 2020). Recent studies also highlight approaches such as multi-omics pipelines, which systematically align and integrate distinct biological layers for precision medicine applications (Abdelaziz et al. 2024). For instance, a study by Tomazou et al. introduced a network-based multisource data integration methodology aimed at drug repurposing for COVID-19. By aggregating transcriptomic, proteomic, and metabolomic data from various patient-derived samples, including lung biopsies and peripheral blood mononuclear cells (PBMC), the authors constructed an integrated network that prioritized disease-associated genes and ranked potential drug candidates. This approach identified several Src tyrosine kinase inhibitors and immunomodulatory drugs as promising therapeutic options, illustrating the potential of network-based strategies in multimodal data integration for drug discovery (Tomazou et al. 2021).

In a recent review, Anokian et al. emphasized the role of GNNs in multisource data integration, highlighting their ability to model complex relationships between heterogeneous datasets. Unlike traditional matrix decomposition techniques like MFA and MDS, which reduce dimensionality to a shared latent space, GNNs leverage the inherent network structure in multisource data. By representing drugs, targets, and diseases as graph nodes and their relationships as graph edges, GNNs enable cross-modal linkages between datasets that cannot be captured by linear transformations alone (Anokian et al. 2024).

While significant progress has been made, challenges persist, such as misaligned samples and missing data caused by inconsistencies across multimodal datasets. These datasets are typically derived from different measurement platforms with varying depths and detection limits, leading to uneven sample coverage and incomplete observations. This inconsistency introduces data sparsity, which not only reduces covariance estimation accuracy (Corso et al. 2013), but also complicates the alignment of cross-modal data points. Addressing these issues requires scalable imputation techniques and domain-specific alignment algorithms. Future advancements in scalable algorithms, particularly those that integrate machine learning with domain-specific knowledge, are expected to unlock the full potential of multisource data integration for drug repurposing.

6.2 Model Performance Challenges

Model performance is a critical aspect of AI and ML applications in drug repurposing, yet several inherent challenges hinder their full potential. Issues such as dimensionality reduction trade-offs, scalability limitations, and model interpretability not only affect prediction accuracy but also constrain the broader applicability of these technologies. Addressing these challenges is essential for enhancing the reliability and acceptance of AI-driven solutions.

6.2.1 Dimensionality Reduction and Scalability Challenges

Dimensionality reduction techniques, such as t-SNE and UMAP, have shown great promise in addressing the challenges of analyzing high-dimensional data by revealing underlying structures in a more interpretable low-dimensional space (Becht et al. 2019; Kobak and Berens 2019). These methods effectively highlight local and global relationships in data, making them valuable preprocessing steps for downstream tasks such as clustering. For instance, just as discussed in this book's earlier section on **Dimensionality Reduction-Assisted Clustering**, combining dimensionality reduction techniques with clustering algorithms offers a robust strategy for handling high-dimensional datasets. Methods like t-SNE and UMAP simplify complex datasets, which are then analyzed using clustering approaches like k-means or HAC to uncover hidden patterns and associations (Cong et al. 2022; Hozumi et al. 2021; Kumari et al. 2022). Similarly, in multisource integration, dimensionality reduction can reduce noise and align data modalities, while clustering provides insights into shared structures across data types.

However, these approaches are not without challenges. A significant concern during dimensionality reduction is the potential loss of critical information, as projecting data from high-dimensional spaces to lower-dimensional ones may obscure biologically relevant patterns, such as weak correlations or rare phenotypic interactions (Kobak and Berens 2019). This issue is particularly pronounced in drug repurposing, where subtle signals in genomic or proteomic data may be critical for identifying novel therapeutic opportunities.

6.2.2 Interpretability of DL Models

The interpretability of DL models remains a critical challenge in drug repurposing, particularly in applications such as DTI prediction. While deep learning models demonstrate exceptional predictive performance, their 'black-box' nature limits transparency and hinders trust in clinical applications. This lack of interpretability complicates understanding the underlying biological mechanisms, such as the specific features driving model predictions. Chakraborty et al. extensively discuss this issue, emphasizing the critical need for model transparency to ensure reliability and applicability in real-world contexts, particularly in biomedicine (Chakraborty et al. 2017).

A notable advancement in addressing this challenge is AttentionSiteDTI, which applies self-attention mechanisms inspired by NLP models to predict DTIs. This method conceptualizes drug-target complexes as sentences, interpreting biochemical entities such as protein pockets and drug molecules through relational contexts. By leveraging attention mechanisms, the model identifies specific protein binding sites that contribute significantly to interactions, thereby offering interpretable insights while maintaining high predictive accuracy (Yazdani-Jahromi et al. 2022). This approach bridges the gap between model outputs and actionable insights, demonstrating the feasibility of interpretability in complex deep learning systems.

Building upon these general advances, frameworks specifically designed for drug repurposing have also made some progress in addressing the black-box problem. For instance, interpretable Deep learning-based Path-reasoning framework for drug repurposing (iDPath) is an interpretable deep learning-based framework developed to model drug-disease interactions with enhanced transparency. It constructs multilayer biological networks, including Graph Convolutional Networks (GCN), PPI, protein-chemical interaction (PCI), and chemical-chemical interaction (CCI) networks, to capture complex molecular relationships comprehensively. By employing attention modules at both node and path levels, iDPath identifies key biological paths contributing to drug-disease interactions, thereby elucidating the mechanism of drug action (MODA) while ensuring interpretability. Furthermore, the model's validity is supported by KEGG pathway enrichment analysis and its consistency with clinical evidence, highlighting its potential in elucidating drug mechanisms and predicting synergistic effects in drug combinations (Yang et al. 2022). By incorporating interpretability into deep learning models, methods like iDPath not only improve model transparency but also foster trust in AI-driven drug repurposing. Ensuring that model predictions are explainable remains a cornerstone for bridging computational outputs with actionable clinical insights.

Similarly, interpretability is equally critical in multi-omics data integration frameworks. Abdelaziz et al. highlight that transparent methodologies in multi-omics pipelines are essential for integrating diverse biological layers, such as genomic and proteomic data, while preserving biological meaning. While iDPath focuses on elucidating drug action mechanisms through multilayer biological networks, multi-omics pipelines aim to identify shared key features between data types, enable researchers to elucidate complex disease mechanisms, and uncover novel therapeutic targets. By emphasizing interpretability, these frameworks ensuring that computational outputs remain both biologically relevant and clinically valuable (Abdelaziz et al. 2024).

6.2.3 Generalization Issues

Generalization remains a critical challenge in the application of AI and ML to drug repurposing and multi-omics data integration. A model's ability to generalize refers to its performance on unseen data, which is essential for reliable predictions in real-world scenarios. However, many deep learning models exhibit overfitting, where they perform exceptionally well on training data but fail to maintain accuracy on test or external datasets. This issue is particularly pronounced in drug repurposing, where the complexity and variability of biological systems are not always fully captured in training datasets (Johnson and Khoshgoftaar 2019). Models trained predominantly on well-characterized drug-target interactions often struggle to predict interactions involving novel targets or rare molecular structures, a limitation that is especially detrimental in the context of rare diseases and orphan drugs (Molnar and Molnar 2023).

Beyond overfitting, data leakage due to improper data splitting strategies, such as random splits, further exacerbates generalization issues. Anokian et al. highlighted that the common practice of random splitting allows drug and cell line data to be shared across training, validation, and test sets, which introduces unintentional "information leakage." Consequently, models may achieve deceptively high-performance metrics, as they indirectly "see" parts of the test data during training. This issue underscores the need for alternative validation methods, such as leave-drug-out or leave-cell-line-out validation schemes, to ensure that unseen drug-target interactions are properly evaluated (Anokian et al. 2024).

Efforts to address generalization challenges have introduced a range of strategies aimed at improving model robustness. Regularization techniques, such as L1 and L2 regularization, are commonly employed to penalize overly complex models, thereby reducing the risk of overfitting and enhancing their ability to generalize across datasets (Goodfellow et al. 2016). Transfer learning has also emerged as a powerful tool, particularly in domains with limited labeled data. By pretraining models on large, generalized datasets and subsequently fine-tuning them on domain-specific data, transfer learning bridges the gap between data scarcity and generalization (Zhuang et al. 2021). Data augmentation methods, including synthetic data generation via GANs, further address dataset imbalances and enrich training data for underrepresented classes, contributing to improved generalization (Frid-Adar et al. 2018; Guo et al. 2017). In addition, zero-shot learning (ZSL) approaches have shown significant promise in addressing generalization issues, particularly in drug repurposing contexts. A representative example is TxGNN, a graph-based drug repurposing framework that leverages a large-scale biomedical KG to predict novel drug-disease associations. Unlike conventional models that require labeled training data for each disease, TxGNN pretrains on a KG encompassing 17,080 diseases and 7957 drugs and subsequently employs a metric learning module to transfer knowledge from well-studied diseases to underrepresented or novel diseases. By pooling information from biologically related diseases, TxGNN enhances prediction accuracy in zero-shot settings, where no labeled training data are available for specific diseases. Additionally, an explainer module identifies critical knowledge paths within the KG, providing interpretable insights into the predicted associations. Evaluations under both standard and zero-shot settings demonstrated TxGNN's superior performance, significantly expanding the scope of drug repurposing to include rare and unexplored diseases while ensuring model interpretability (Huang et al. 2024b). This capability not only expands the scope of drug repurposing efforts to include rare or unexplored diseases, but also highlights the potential of ZSL to improve model adaptability and robustness in real-world applications.

Additionally, federated learning represents a promising approach for integrating diverse datasets while preserving data privacy. By enabling decentralized training across multiple institutions, federated learning fosters collaboration and diversity in training data, which are crucial for improving model generalization without compromising individual data ownership (Li et al. 2020). The combination of these methods, along with advancements in scalable and interpretable algorithms, holds the potential to mitigate generalization issues and enhance the applicability of AI-driven models in drug repurposing and beyond.

7 Application-Specific Challenges

The application of AI and ML in drug repurposing faces domain-specific hurdles, particularly in navigating the complexities of multi-target drug design. This challenge requires tailored approaches that integrate biological expertise with advanced computational techniques, emphasizing both precision and generalizability in therapeutic discovery.

7.1 Multi-target Drug Design

The development of multi-target drugs presents a promising strategy to overcome the limitations of single-target therapies, particularly in complex diseases such as cancer and psychiatric disorders (Hopkins 2008). Unlike combination therapies, multi-target drugs offer advantages such as improved pharmacokinetics, reduced potential for resistance, and simplified administration. For instance, POLYpharmacology Generative Optimization Network (POLYGON), a generative model based on reinforcement learning, has successfully designed compounds targeting both MEK1 and mTOR, achieving dual inhibition with high accuracy and biological activity (Munson et al. 2024). This highlights the efficiency of computational approaches in accelerating multi-target drug discovery.

However, translating multi-target pharmacology into clinical success remains a significant challenge. Antolin et al. highlighted that although FDA-approved multi-target drugs underscore the potential of polypharmacology, not all drugs designed to expand their therapeutic scope have shown clinical efficacy. For instance, Dasatinib and Erlotinib, despite showing promising results in preclinical studies and early clinical trials, failed to demonstrate consistent efficacy when targeting additional pathways in larger, more rigorous trials. Dasatinib, initially approved as a second-generation breakpoint cluster region protein—tyrosine-protein kinase ABL1 (BCR-ABL) inhibitor for chronic myeloid leukemia, was repurposed to target discoidin domain-containing receptor 2 (DDR2) in squamous cell carcinoma (SCC) after DDR2 was validated as an SCC target in 2011. Although early case reports suggested clinical activity, subsequent phase II trials failed to replicate these results, highlighting the inherent unpredictability of multi-target drug development. Additionally, despite Dasatinib's potent inhibition of Src kinase, its early promise in biomarker-driven applications has not translated into clinical success. Similarly, Erlotinib, the second FDA-approved EGFR inhibitor, displayed broad polypharmacology by effectively inhibiting JAK2 V617F in preclinical polycythemia models, yet failed to show therapeutic efficacy in clinical settings. Attempts to repurpose Erlotinib for receptor tyrosine-protein kinase erbB-2 (ERBB2) and receptor tyrosine-protein kinase erbB-3 (ERBB3) mutations in pancreatic cancer also encountered significant obstacles, with phase II trials unable to validate these biomarkers as reliable predictors of efficacy (Antolin et al. 2017). These cases illustrate that while multi-target drugs offer theoretical advantages—such as overcoming resistance and addressing complex disease mechanisms—broad polypharmacology alone does not guarantee clinical success. Even with robust preclinical evidence, clinical repurposing efforts remain hampered by the intricate nature of target interactions, off-target effects, and the absence of reliable predictive biomarkers.

Additionally, many successful multi-target drugs were initially discovered serendipitously rather than through a systematic design approach, and achieving balanced efficacy across multiple targets while minimizing off-target effects requires both a comprehensive understanding of target networks and the development of optimized chemical scaffolds (Talevi 2015). Future advancements in systems pharmacology and AI-driven target prediction are expected to mitigate these challenges, improving the precision and efficacy of multi-target drug design.

8 Future Perspectives in Drug Repurposing

The future of AI and ML applications in drug repurposing hinges on addressing the challenges outlined earlier, particularly in the areas of ADRs and CYP450-mediated drug metabolism. Moving forward, significant advancements are expected in improving model accuracy, ensuring that AI and ML-driven solutions can deliver reliable predictions while also remaining interpretable and transparent.

8.1 Adverse Drug Reactions' Prediction

ADRs remain a critical yet complex task in drug development and safety evaluation. Machine learning models that integrate LINCS L1000 gene expression data with chemical structure data have shown promise in improving prediction accuracy by transforming gene expression data into enrichment vectors linked to specific biological functions (Wang et al. 2016). These integrative approaches enable the early identification of ADR risks and the detection of drug-drug interactions, addressing some of the limitations associated with traditional target-based methods.

However, predicting ADRs involves not only gene expression data but also understanding the complex relationships between drug molecules, targets, and genetic factors. Malki et al. highlight the importance of considering drug-drug-gene interactions in ADR prediction, noting that genetic variations can significantly affect drug responses and toxicity (Malki and Pearson 2020). Future advancements should focus on integrating broader biological data, such as systemic gene networks and metabolic pathways, to enhance the scope and accuracy of ADR predictions. These efforts could involve dynamic modeling approaches that better represent the interactions across different biological levels, ultimately improving the safety and efficacy of AI-driven drug development.

Interestingly, ADRs are not just limitations in clinical development but can also be leveraged for drug repurposing. By analyzing drug-induced adverse effects, machine learning models can identify new therapeutic indications for existing drugs. Lee and Chen demonstrated that drug side-effect profiles can be used to predict new drug-disease associations, revealing new therapeutic opportunities that were previously unexplored (Lee and Chen 2021). Similarly, Kaneko and Nagashima explored how understanding the mechanisms behind ADRs can provide insights into alternative uses for drugs, highlighting the potential of ADR data in drug repurposing efforts (Kaneko and Nagashima 2020).

Such improvements in ADR prediction could contribute to more reliable drug safety profiles, reduce the likelihood of adverse reactions in clinical settings, and ultimately improve therapeutic outcomes. By integrating multimodal data and advanced computational techniques, ADRs could serve as a powerful tool in the drug repurposing process.

8.2 CYP450-Mediated Drug Metabolism

Cytochrome P450 enzymes (CYP450) are a family of enzymes that play a central role in drug metabolism, particularly in the oxidation of xenobiotics, including many pharmaceuticals (El-Sherbeni and El-Kadi 2017). These enzymes are found primarily in the liver but also in other tissues, and they facilitate the biotransformation of drugs into more water-soluble metabolites that can be excreted from the body. Given their critical role in drug metabolism, CYP450 enzymes significantly impact the pharmacokinetics and toxicity profiles of drugs, influencing both their efficacy and safety (Foti 2023).

In the context of drug repurposing, CYP450 enzymes offer unique opportunities and challenges. Many drugs that are repurposed for new indications may interact with CYP450 enzymes, altering their metabolic pathways and potentially leading to unexpected side effects. This interaction is particularly important in the case of CYP450-mediated drug metabolism, where variations in enzyme activity can significantly affect the drug's therapeutic profile (Nakajima 2017).

One of the major challenges in drug repurposing is understanding and predicting how existing drugs will interact with CYP450 enzymes in different biological contexts. As CYP450 enzymes are involved in the metabolism of a wide range of substances, the prediction of their activity is crucial for assessing the safety and efficacy of repurposed drugs (Foti 2023). Recent studies have shown that by leveraging machine learning techniques, such as molecular representation models, it is possible to predict CYP450 inhibition and identify drugs that could potentially be repurposed for new therapeutic target (Ouzounis et al. 2023). These models use CYP450-related data to predict how drugs interact with the enzyme, offering insights into how drugs could be repurposed or modified to reduce adverse effects or enhance efficacy.

However, despite advancements in computational models, several challenges remain. CYP450 enzymes exhibit significant interindividual variability, which can affect the pharmacokinetics of drugs and their potential for ADRs. This variability is influenced by factors such as genetics, age, and diet, complicating efforts to predict the metabolism of drugs accurately across populations (Nakajima 2017). Furthermore, many of the current models are based on in vitro data, which may not fully capture the complexity of CYP450 activity in vivo, where factors like enzyme induction or inhibition can vary over time.

Future advancements in AI-driven models and systems pharmacology are expected to address these challenges by integrating data from multiple sources, such as CYP450 enzyme activity, genetic variations, and metabolic networks, to provide more accurate predictions of drug metabolism. Additionally, improving the mechanistic understanding of CYP450 and its role in drug interactions will be crucial for optimizing the use of repurposed drugs. Researchers are increasingly focusing on developing more predictive models that can account for both CYP450 inhibition and induction, allowing for better prediction of CYP450-related ADRs and reducing the risks associated with drug repurposing (Foti 2023).

8.3 Future Perspectives in Drug Repurposing

Looking ahead, the future of AI and ML in drug repurposing will focus on addressing the key challenges of predicting ADRs and understanding CYP450-mediated drug metabolism. Advancements in these areas are expected to significantly enhance model accuracy while ensuring that AI-driven solutions are interpretable, reliable, and clinically applicable.

A major shift in drug repurposing is the increasing emphasis on personalized medicine. By leveraging AI and ML technologies to tailor drug treatments to individual patients, we can maximize therapeutic efficacy and minimize risks. Personalized medicine involves considering genetic, environmental, and lifestyle factors that influence how drugs work and how they are metabolized in different individuals. In drug repurposing, this means identifying the most appropriate existing drugs for each patient based on their unique biological profile, reducing adverse effects and optimizing treatment outcomes.

As AI and ML models advance, they will be able to integrate multiple data sources, such as genetic information, clinical histories, and environmental factors, to predict which drugs are most likely to succeed in specific individuals. This approach will not only improve drug efficacy but also predict potential drug-drug interactions and side effects. By applying personalized treatment strategies, drug repurposing can be optimized for individual patients, aligning therapeutic options with their genetic makeup and health history.

Incorporating personalized medicine into drug repurposing through AI and ML will be key to overcoming the remaining challenges in this field. The integration of AI-driven pharmacology with precision medicine will drive the next wave of drug development, enhancing the safety, efficacy, and accessibility of repurposed drugs for diverse patient populations.

References

Abdelaziz EH, Ismail R, Mabrouk MS, Amin E. Multi-omics data integration and analysis pipeline for precision medicine: systematic review. Comput Biol Chem. 2024;113:108254. https://doi.org/10.1016/j.compbiolchem.2024.108254.

Abdelkader A, Elzemrany AA, El-Nadi M, Elsabbagh SA, Shehata MA, Eldehna WM, El-Hadidi M, Ibrahim TM. In-silico targeting of SARS-CoV-2 NSP6 for drug and natural products repurposing. Virology. 2022;573:96–110. https://doi.org/10.1016/j.virol.2022.06.008.

Abramson J, Adler J, Dunger J, Evans R, Green T, Pritzel A, Ronneberger O, Willmore L, Ballard AJ, Bambrick J, Bodenstein SW, Evans DA, Hung C-C, O'Neill M, Reiman D, Tunyasuvunakool K, Wu Z, Žemgulytė A, Arvaniti E, et al. Accurate structure prediction of biomolecular interactions with AlphaFold 3. Nature. 2024;630(8016):493–500. https://doi.org/10.1038/s41586-024-07487-w.

Alahakoon D, Halgamuge SK, Srinivasan B. Dynamic self-organizing maps with controlled growth for knowledge discovery. IEEE Trans Neural Netw. 2000;11(3):601–14. https://doi.org/10.1109/72.846732.

Amiri R, Razmara J, Parvizpour S, Izadkhah H. A novel efficient drug repurposing framework through drug-disease association data integration using convolutional neural networks. BMC Bioinf. 2023;24(1):442. https://doi.org/10.1186/s12859-023-05572-x.

Anokian E, Bernett J, Freeman A, List M, Santamaría LP, Tanoli Z, Bonnin S. Machine learning and artificial intelligence in drug repurposing—challenges and perspectives. Drug Repurposing. 2024;1(1) https://doi.org/10.58647/DRUGREPO.24.1.0004.

Antolin A, Workman P, Mestres J, Al-Lazikani B. Polypharmacology in precision oncology: current applications and future prospects. Curr Pharm Des. 2017;22(46):6935–45. https://doi.org/10.2174/1381612822666160923115828.

Arjovsky, M., Chintala, S., & Bottou, L. (2017). Wasserstein GAN.

Ashburn TT, Thor KB. Drug repositioning: identifying and developing new uses for existing drugs. Nat Rev Drug Discov. 2004;3(8):673–83. https://doi.org/10.1038/nrd1468.

Becht E, McInnes L, Healy J, Dutertre C-A, Kwok IWH, Ng LG, Ginhoux F, Newell EW. Dimensionality reduction for visualizing single-cell data using UMAP. Nat Biotechnol. 2019;37(1):38–44. https://doi.org/10.1038/nbt.4314.

Bejan CA, Cahill KN, Staso PJ, Choi L, Peterson JF, Phillips EJ. DrugWAS: drug-wide association studies for COVID-19 drug repurposing. Clin Pharmacol Ther. 2021;110(6):1537–46. https://doi.org/10.1002/cpt.2376.

Beltagy I, Lo K, Cohan A. SciBERT: a pretrained language model for scientific text. In: Proceedings of the 2019 conference on empirical methods in natural language processing and the 9th international joint conference on natural language processing (EMNLP-IJCNLP); 2019. p. 3613–8. https://doi.org/10.18653/v1/D19-1371.

Cai TT, Ma R (2021). Theoretical foundations of t-SNE for visualizing high-dimensional clustered data.

Ceskoutsé RFT, Bomgni AB, Zanfack DRG, Agany DDM, Thomas BB, Zohim EG. HeteroKGRep: heterogeneous knowledge graph based drug repositioning. Knowl-Based Syst. 2024;305:112638. https://doi.org/10.1016/j.knosys.2024.112638.

Chakraborty S, Tomsett R, Raghavendra R, Harborne D, Alzantot M, Cerutti F, Srivastava M, Preece A, Julier S, Rao RM, Kelley TD, Braines D, Sensoy M, Willis CJ, Gurram P. Interpretability of deep learning models: a survey of results. In: 2017 IEEE SmartWorld, Ubiquitous Intelligence & Computing, Advanced & Trusted Computed, Scalable Computing & Communications, Cloud & Big Data Computing, Internet of People and Smart City Innovation (SmartWorld/SCALCOM/UIC/ATC/CBDCom/IOP/SCI); 2017. p. 1–6. https://doi.org/10.1109/UIC-ATC.2017.8397411.

Chawla NV, Bowyer KW, Hall LO, Kegelmeyer WP. SMOTE: synthetic minority over-sampling technique. J Artif Intell Res. 2002;16:321–57. https://doi.org/10.1613/jair.953.

Chen T, Guestrin C. XGBoost: a scalable tree boosting system. In: Proceedings of the 22nd ACM SIGKDD international conference on knowledge discovery and data mining; 2016. p. 785–94. https://doi.org/10.1145/2939672.2939785.

Choudhury C, Arul Murugan N, Priyakumar UD. Structure-based drug repurposing: traditional and advanced AI/ML-aided methods. Drug Discov Today. 2022;27(7):1847–61. https://doi.org/10.1016/j.drudis.2022.03.006.

Cong Y, Endo T. Multi-omics and artificial intelligence-guided drug repositioning: prospects, challenges, and lessons learned from COVID-19. OMICS J Integr Biol. 2022;26(7):361–71. https://doi.org/10.1089/omi.2022.0068.

Cong Y, Shintani M, Imanari F, Osada N, Endo T. A new approach to drug repurposing with two-stage prediction, machine learning, and unsupervised clustering of gene expression. OMICS J Integr Biol. 2022;26(6):339–47. https://doi.org/10.1089/omi.2022.0026.

Corso PS, Ingels JB, Kogan SM, Foster EM, Chen Y-F, Brody GH. Economic analysis of a multi-site prevention program: assessment of program costs and characterizing site-level variability. Prev Sci. 2013;14(5):447–56. https://doi.org/10.1007/s11121-012-0316-z.

D'haeseleer P. How does gene expression clustering work? Nat Biotechnol. 2005;23(12):1499–501. https://doi.org/10.1038/nbt1205-1499.

Dietterich TG. Ensemble methods in machine learning; 2000. p. 1–15. https://doi.org/10.1007/3-540-45014-9_1.

Dudani SA. The distance-weighted k-nearest-neighbor rule. IEEE Trans Syst Man Cybern. 1976;SMC-6(4):325–7. https://doi.org/10.1109/TSMC.1976.5408784.

El-Sherbeni AA, El-Kadi AOS. Microsomal cytochrome P450 as a target for drug discovery and repurposing. Drug Metab Rev. 2017;49(1):1–17. https://doi.org/10.1080/03602532.2016.1257021.

Enright AJ. An efficient algorithm for large-scale detection of protein families. Nucleic Acids Res. 2002;30(7):1575–84. https://doi.org/10.1093/nar/30.7.1575.

Everitt BS, Landau S, Leese M, Stahl D. Hierarchical clustering. In: Shewhart WA, Wilks SS, editors. Cluster analysis. Wiley; 2011. p. 71–110. https://doi.org/10.1002/9780470977811.ch4.

Foti RS. Cytochrome P450 and other drug-metabolizing enzymes as therapeutic targets. Drug Metab Dispos. 2023;51(8):936–49. https://doi.org/10.1124/dmd.122.001011.

Frid-Adar M, Diamant I, Klang E, Amitai M, Goldberger J, Greenspan H. GAN-based synthetic medical image augmentation for increased CNN performance in liver lesion classification. Neurocomputing. 2018;321:321–31. https://doi.org/10.1016/j.neucom.2018.09.013.

García-Domínguez A, Galván-Tejada CE, Magallanes-Quintanar R, Cruz M, Gonzalez-Curiel I, Delgado-Contreras JR, Soto-Murillo MA, Celaya-Padilla JM, Galván-Tejada JI. Optimizing clinical diabetes diagnosis through generative adversarial networks: evaluation and validation. Diseases. 2023;11(4):134. https://doi.org/10.3390/diseases11040134.

Goodfellow, I. J., Pouget-Abadie, J., Mirza, M., Xu, B., Warde-Farley, D., Ozair, S., Courville, A., & Bengio, Y. (2014). Generative adversarial networks.

Goodfellow I, Bengio Y, Courville A. Deep learning. MIT Press; 2016.

Guo H, Li Y, Jennifer S, Gu M, Huang Y, Gong B. Learning from class-imbalanced data: review of methods and applications. Expert Syst Appl. 2017;73:220–39. https://doi.org/10.1016/j.eswa.2016.12.035.

Hinton GE, Salakhutdinov RR. Reducing the dimensionality of data with neural networks. Science. 2006;313(5786):504–7. https://doi.org/10.1126/science.1127647.

Hinton GE, Osindero S, Teh Y-W. A fast learning algorithm for deep belief nets. Neural Comput. 2006;18(7):1527–54. https://doi.org/10.1162/neco.2006.18.7.1527.
Hopkins AL. Network pharmacology: the next paradigm in drug discovery. Nat Chem Biol. 2008;4(11):682–90. https://doi.org/10.1038/nchembio.118.
Hozumi Y, Wang R, Yin C, Wei G-W. UMAP-assisted K-means clustering of large-scale SARS-CoV-2 mutation datasets. Comput Biol Med. 2021;131:104264. https://doi.org/10.1016/j.compbiomed.2021.104264.
Hsu Y-H, Si D. Cancer type prediction and classification based on RNA-sequencing data. In: 2018 40th annual international conference of the IEEE engineering in medicine and biology society (EMBC); 2018. p. 5374–7. https://doi.org/10.1109/EMBC.2018.8513521.
Hsu AL, Tang S-L, Halgamuge SK. An unsupervised hierarchical dynamic self-organizing approach to cancer class discovery and marker gene identification in microarray data. Bioinformatics. 2003;19(16):2131–40. https://doi.org/10.1093/bioinformatics/btg296.
Huang C-M, Lin C-H, Hung C-S, Zeng W-H, Zheng Y-C, Tsai C-M. Utilizing nearest-neighbor clustering for addressing imbalanced datasets in bioengineering. Bioengineering. 2024a;11(4):345. https://doi.org/10.3390/bioengineering11040345.
Huang K, Chandak P, Wang Q, Havaldar S, Vaid A, Leskovec J, Nadkarni GN, Glicksberg BS, Gehlenborg N, Zitnik M. A foundation model for clinician-centered drug repurposing. Nat Med. 2024b;30:3601. https://doi.org/10.1038/s41591-024-03233-x.
Hulot A, Laloë D, Jaffrézic F. A unified framework for the integration of multiple hierarchical clusterings or networks from multi-source data. BMC Bioinf. 2021;22(1):392. https://doi.org/10.1186/s12859-021-04303-4.
Ji Y, Zhou Z, Liu H, Davuluri RV. DNABERT: pre-trained bidirectional encoder representations from transformers model for DNA-language in genome. Bioinformatics. 2021;37(15):2112–20. https://doi.org/10.1093/bioinformatics/btab083.
Johnson SC. Hierarchical clustering schemes. Psychometrika. 1967;32(3):241–54. https://doi.org/10.1007/BF02289588.
Johnson JM, Khoshgoftaar TM. Survey on deep learning with class imbalance. J Big Data. 2019;6(1):27. https://doi.org/10.1186/s40537-019-0192-5.
Jolliffe IT, Cadima J. Principal component analysis: a review and recent developments. Philos Trans R Soc A Math Phys Eng Sci. 2016;374(2065):20150202. https://doi.org/10.1098/rsta.2015.0202.
Jourdan J-P, Bureau R, Rochais C, Dallemagne P. Drug repositioning: a brief overview. J Pharm Pharmacol. 2020;72(9):1145–51. https://doi.org/10.1111/jphp.13273.
Kalemati M, Zamani Emani M, Koohi S. DCGAN-DTA: predicting drug-target binding affinity with deep convolutional generative adversarial networks. BMC Genomics. 2024;25(1):411. https://doi.org/10.1186/s12864-024-10326-x.
Kaneko S, Nagashima T. Drug repositioning and target finding based on clinical evidence. Biol Pharm Bull. 2020;43(3):362–5. https://doi.org/10.1248/bpb.b19-00929.
Keller JM, Gray MR, Givens JA. A fuzzy K-nearest neighbor algorithm. IEEE Trans Syst Man Cybern. 1985;SMC-15(4):580–5. https://doi.org/10.1109/TSMC.1985.6313426.
Kim HJ, Lin Y, Geddes TA, Yang JYH, Yang P. CiteFuse enables multi-modal analysis of CITE-seq data. Bioinformatics. 2020;36(14):4137–43. https://doi.org/10.1093/bioinformatics/btaa282.
Kim SK, Goughnour PC, Lee EJ, Kim MH, Chae HJ, Yun GY, Kim YR, Choi JW. Identification of drug combinations on the basis of machine learning to maximize anti-aging effects. PLoS One. 2021;16(1):e0246106. https://doi.org/10.1371/journal.pone.0246106.
Kipf, T. N., & Welling, M. (2016). Semi-supervised classification with graph convolutional networks.
Kitsiranuwat S, Suratanee A, Plaimas K. Integration of various protein similarities using random forest technique to infer augmented drug-protein matrix for enhancing drug-disease association prediction. Sci Prog. 2022;105(3) https://doi.org/10.1177/00368504221109215.
Kobak D, Berens P. The art of using t-SNE for single-cell transcriptomics. Nat Commun. 2019;10(1):5416. https://doi.org/10.1038/s41467-019-13056-x.

Kohonen T. Self-Organizing Maps, vol. 30. Berlin/Heidelberg: Springer; 1995. https://doi.org/10.1007/978-3-642-97610-0.

Kumari P, Pradhan B, Koromina M, Patrinos GP, Van Steen K. Discovery of new drug indications for COVID-19: a drug repurposing approach. PLoS One. 2022;17(5):e0267095. https://doi.org/10.1371/journal.pone.0267095.

Lamb J, Crawford ED, Peck D, Modell JW, Blat IC, Wrobel MJ, Lerner J, Brunet J-P, Subramanian A, Ross KN, Reich M, Hieronymus H, Wei G, Armstrong SA, Haggarty SJ, Clemons PA, Wei R, Carr SA, Lander ES, Golub TR. The connectivity map: using gene-expression signatures to connect small molecules, genes, and disease. Science. 2006;313(5795):1929–35. https://doi.org/10.1126/science.1132939.

Lavecchia A. Machine-learning approaches in drug discovery: methods and applications. Drug Discov Today. 2015;20(3):318–31. https://doi.org/10.1016/j.drudis.2014.10.012.

Lavecchia A. Deep learning in drug discovery: opportunities, challenges and future prospects. Drug Discov Today. 2019;24(10):2017–32. https://doi.org/10.1016/j.drudis.2019.07.006.

Lecun Y, Bottou L, Bengio Y, Haffner P. Gradient-based learning applied to document recognition. Proc IEEE. 1998;86(11):2278–324. https://doi.org/10.1109/5.726791.

Lee CY, Chen Y-PP. Prediction of drug adverse events using deep learning in pharmaceutical discovery. Brief Bioinform. 2021;22(2):1884–901. https://doi.org/10.1093/bib/bbaa040.

Lee J, Yoon W, Kim S, Kim D, Kim S, So CH, Kang J. BioBERT: a pre-trained biomedical language representation model for biomedical text mining. Bioinformatics. 2020;36(4):1234–40. https://doi.org/10.1093/bioinformatics/btz682.

Li T, Sahu AK, Talwalkar A, Smith V. Federated learning: challenges, methods, and future directions. IEEE Signal Process Mag. 2020;37(3):50–60. https://doi.org/10.1109/MSP.2020.2975749.

Li S, Chao H, Li Z, Chen S, Zhang J, Hao W, Zhang S, Liu C, Liu H. Sex dimorphism of IL-17-secreting peripheral blood mononuclear cells in ankylosing spondylitis based on bioinformatics analysis and machine learning. BMC Musculoskelet Disord. 2024;25(1):490. https://doi.org/10.1186/s12891-024-07589-6.

Lin T-Y, Goyal P, Girshick R, He K, Dollar P. Focal loss for dense object detection. In: 2017 IEEE International Conference on Computer Vision (ICCV); 2017. p. 2999–3007. https://doi.org/10.1109/ICCV.2017.324.

Liu J, Xu L, Guo W, Li Z, Khan MKH, Ge W, Patterson TA, Hong H. Developing a SARS-CoV-2 main protease binding prediction random forest model for drug repurposing for COVID-19 treatment. Exp Biol Med. 2023; https://doi.org/10.1177/15353702231209413.

Macqueen J. Some methods for classification and analysis of multivariate observations. In: Proceedings of the 5th Berkeley symposium on mathematical statistics and probability, volume 1: statistics; 1967. p. 281–97. http://projecteuclid.org/euclid.bsmsp/1200512992.

Madugula SS, John L, Nagamani S, Gaur AS, Poroikov VV, Sastry GN. Molecular descriptor analysis of approved drugs using unsupervised learning for drug repurposing. Comput Biol Med. 2021;138:104856. https://doi.org/10.1016/j.compbiomed.2021.104856.

Malki MA, Pearson ER. Drug–drug–gene interactions and adverse drug reactions. Pharmacogenomics J. 2020;20(3):355–66. https://doi.org/10.1038/s41397-019-0122-0.

March-Vila E, Pinzi L, Sturm N, Tinivella A, Engkvist O, Chen H, Rastelli G. On the integration of in Silico drug design methods for drug repurposing. Front Pharmacol. 2017;8 https://doi.org/10.3389/fphar.2017.00298.

McInnes, L., Healy, J., & Melville, J. (2018). UMAP: uniform manifold approximation and projection for dimension reduction.. http://arxiv.org/abs/1802.03426.

Mittone G, Svoboda F, Aldinucci M, Lane N, Lió P. A federated learning benchmark for drug-target interaction. In: Companion proceedings of the ACM web conference 2023; 2023. p. 1177–81. https://doi.org/10.1145/3543873.3587687.

Mohammadzadeh-Vardin T, Ghareyazi A, Gharizadeh A, Abbasi K, Rabiee HR. DeepDRA: drug repurposing using multi-omics data integration with autoencoders. PLoS One. 2024;19(7):e0307649. https://doi.org/10.1371/journal.pone.0307649.

Molnar MJ, Molnar V. AI-based tools for the diagnosis and treatment of rare neurological disorders. Nat Rev Neurol. 2023;19(8):455–6. https://doi.org/10.1038/s41582-023-00841-y.

Mottaqi MS, Mohammadipanah F, Sajedi H. Contribution of machine learning approaches in response to SARS-CoV-2 infection. Inf Med Unlocked. 2021;23:100526. https://doi.org/10.1016/j.imu.2021.100526.

Munson BP, Chen M, Bogosian A, Kreisberg JF, Licon K, Abagyan R, Kuenzi BM, Ideker T. De novo generation of multi-target compounds using deep generative chemistry. Nat Commun. 2024;15(1):3636. https://doi.org/10.1038/s41467-024-47120-y.

Nakajima M. From the viewpoint of drug metabolism research. Yakugaku Zasshi. 2017;137(6):697–705. https://doi.org/10.1248/yakushi.16-00250-5.

OpenAI, Achiam J, Adler S, Agarwal S, Ahmad L, Akkaya I, Aleman FL, Almeida D, Altenschmidt J, Altman S, Anadkat S, Avila R, Babuschkin I, Balaji S, Balcom V, Baltescu P, Bao H, Bavarian M, Belgum J, et al. GPT-4 technical report; 2023. https://doi.org/10.48550/arXiv.2303.08774.

Ouzounis S, Panagiotopoulos V, Bafiti V, Zoumpoulakis P, Cavouras D, Kalatzis I, Matsoukas M-T, Katsila T. A robust machine learning framework built upon molecular representations predicts CYP450 inhibition: toward precision in drug repurposing. OMICS J Integr Biol. 2023;27(7):305–14. https://doi.org/10.1089/omi.2023.0075.

Perdomo-Quinteiro P, Belmonte-Hernández A. Knowledge graphs for drug repurposing: a review of databases and methods. Brief Bioinform. 2024;25(6) https://doi.org/10.1093/bib/bbae461.

Pushpakom S, Iorio F, Eyers PA, Escott KJ, Hopper S, Wells A, Doig A, Guilliams T, Latimer J, McNamee C, Norris A, Sanseau P, Cavalla D, Pirmohamed M. Drug repurposing: progress, challenges and recommendations. Nat Rev Drug Discov. 2019;18(1):41–58. https://doi.org/10.1038/nrd.2018.168.

Qian Y, Wu J, Zhang Q. CAT-CPI: combining CNN and transformer to learn compound image features for predicting compound-protein interactions. Front Mol Biosci. 2022;9 https://doi.org/10.3389/fmolb.2022.963912.

Radford, A., Metz, L., & Chintala, S. (2015). Unsupervised representation learning with deep convolutional generative adversarial networks.

Rao, R., Bhattacharya, N., Thomas, N., Duan, Y., Chen, X., Canny, J., Abbeel, P., & Song, Y. S. (2019). Evaluating protein transfer learning with TAPE.

Ruder S, Peters ME, Swayamdipta S, Wolf T. Transfer learning in natural language processing. In: Proceedings of the 2019 conference of the north, 15–18. 10.18653/v1/N19-5004; 2019.

Sahoo BM, Ravi Kumar BVV, Sruti J, Mahapatra MK, Banik BK, Borah P. Drug repurposing strategy (DRS): emerging approach to identify potential therapeutics for treatment of novel coronavirus infection. Front Mol Biosci. 2021;8 https://doi.org/10.3389/fmolb.2021.628144.

Sengupta A, Singh SK, Kumar R. Support vector machine-based prediction models for drug repurposing and designing novel drugs for colorectal cancer. ACS Omega. 2024;9(16):18584–92. https://doi.org/10.1021/acsomega.4c01195.

Sheridan RP, Wang WM, Liaw A, Ma J, Gifford EM. Extreme gradient boosting as a method for quantitative structure–activity relationships. J Chem Inf Model. 2016;56(12):2353–60. https://doi.org/10.1021/acs.jcim.6b00591.

Subramanian A, Narayan R, Corsello SM, Peck DD, Natoli TE, Lu X, Gould J, Davis JF, Tubelli AA, Asiedu JK, Lahr DL, Hirschman JE, Liu Z, Donahue M, Julian B, Khan M, Wadden D, Smith IC, Lam D, et al. A next generation connectivity map: L1000 platform and the first 1,000,000 profiles. Cell. 2017;171(6):1437–1452.e17. https://doi.org/10.1016/j.cell.2017.10.049.

Sun C, Xuan P, Zhang T, Ye Y. Graph convolutional autoencoder and generative adversarial network-based method for predicting drug-target interactions. IEEE/ACM Trans Comput Biol Bioinform. 2022;19(1):455–64. https://doi.org/10.1109/TCBB.2020.2999084.

Talevi A. Multi-target pharmacology: possibilities and limitations of the “skeleton key approach” from a medicinal chemist perspective. Front Pharmacol. 2015;6 https://doi.org/10.3389/fphar.2015.00205.

Thorman AW, Reigle J, Chutipongtanate S, Yang J, Shamsaei B, Pilarczyk M, Fazel-Najafabadi M, Adamczak R, Kouril M, Bhatnagar S, Hummel S, Niu W, Morrow AL, Czyzyk-Krzeska MF, McCullumsmith R, Seibel W, Nassar N, Zheng Y, Hildeman DA, et al. Accelerating drug discovery and repurposing by combining transcriptional signature connectivity with docking. Sci Adv. 2024;10(35) https://doi.org/10.1126/sciadv.adj3010.

Tomazou M, Bourdakou MM, Minadakis G, Zachariou M, Oulas A, Karatzas E, Loizidou EM, Kakouri AC, Christodoulou CC, Savva K, Zanti M, Onisiforou A, Afxenti S, Richter J, Christodoulou CG, Kyprianou T, Kolios G, Dietis N, Spyrou GM. Multi-omics data integration and network-based analysis drives a multiplex drug repurposing approach to a shortlist of candidate drugs against COVID-19. Brief Bioinform. 2021;22(6) https://doi.org/10.1093/bib/bbab114.

van der Laurens M, Geoffrey H. Visualizing data using t-SNE. J Mach Learn Res. 2008;9:2579–605. http://jmlr.org/papers/v9/vandermaaten08a.html.

van Dongen S. Graph clustering by flow simulation. Utrecht University; 2000.

Varadi M, Anyango S, Deshpande M, Nair S, Natassia C, Yordanova G, Yuan D, Stroe O, Wood G, Laydon A, Žídek A, Green T, Tunyasuvunakool K, Petersen S, Jumper J, Clancy E, Green R, Vora A, Lutfi M, et al. AlphaFold protein structure database: massively expanding the structural coverage of protein-sequence space with high-accuracy models. Nucleic Acids Res. 2022;50(D1):D439–44. https://doi.org/10.1093/nar/gkab1061.

Velasco R, Argyriou AA, Cornblath DR, Bruna P, Alberti P, Rossi E, Merkies ISJ, Psimaras D, Briani C, Lalisang RI, Schenone A, Cavaletti G, Bruna J. Repurposing chemotherapy-induced peripheral neuropathy grading. Eur J Neurol. 2024;31(12) https://doi.org/10.1111/ene.16457.

Veleiro U, de la Fuente J, Serrano G, Pizurica M, Casals M, Pineda-Lucena A, Vicent S, Ochoa I, Gevaert O, Hernaez M. GeNNius: an ultrafast drug-target interaction inference method based on graph neural networks. Bioinformatics. 2024;40(1) https://doi.org/10.1093/bioinformatics/btad774.

Wang P, Xiao X, Chou K-C. NR-2L: A two-level predictor for identifying nuclear receptor subfamilies based on sequence-derived features. PLoS One. 2011;6(8):e23505. https://doi.org/10.1371/journal.pone.0023505.

Wang Z, Clark NR, Ma'ayan, A. Drug-induced adverse events prediction with the LINCS L1000 data. Bioinformatics. 2016;32(15):2338–45. https://doi.org/10.1093/bioinformatics/btw168.

Wang K, Liang Y, Li X, Li G, Ghanem B, Zimmermann R, Zhou Z, Yi H, Zhang Y, Wang Y. Brave the wind and the waves: discovering robust and generalizable graph lottery tickets. IEEE Trans Pattern Anal Mach Intell. 2024;46(5):3388–405. https://doi.org/10.1109/TPAMI.2023.3342184.

Xu P, Wei Z, Li C, Yuan J, Liu Z, Liu W. Drug-target prediction based on dynamic heterogeneous graph convolutional network. IEEE J Biomed Health Inform. 2024;28(11):6997–7005. https://doi.org/10.1109/JBHI.2024.3441324.

Yang L, Agarwal P. Systematic drug repositioning based on clinical side-effects. PLoS One. 2011;6(12):e28025. https://doi.org/10.1371/journal.pone.0028025.

Yang J, Li Z, Wu W, Yu S, Chu Q, Zhang Q. Deep learning can identify explainable reasoning paths of mechanism of drug action for drug repurposing from multilayer biological network; 2022. https://doi.org/10.21203/rs.3.rs-1269212/v1.

Yang, J., Jin, H., Tang, R., Han, X., Feng, Q., Jiang, H., Yin, B., & Hu, X. (2023a). Harnessing the power of LLMs in practice: a survey on ChatGPT and beyond.

Yang M, Yang B, Duan G, Wang J. ITRPCA: a new model for computational drug repositioning based on improved tensor robust principal component analysis. Front Genet. 2023b;14 https://doi.org/10.3389/fgene.2023.1271311.

Yazdani-Jahromi M, Yousefi N, Tayebi A, Kolanthai E, Neal CJ, Seal S, Garibay OO. Attention-SiteDTI: an interpretable graph-based model for drug-target interaction prediction using NLP sentence-level relation classification. Brief Bioinform. 2022;23(4) https://doi.org/10.1093/bib/bbac272.

Zaza P, Matthieu R, Jean-Luc C, Charles K. Drug repurposing in Raynaud's phenomenon through adverse event signature matching in the World Health Organization pharmacovigilance database. Br J Clin Pharmacol. 2020;86(11):2217–22. https://doi.org/10.1111/bcp.14322.

Zhang X, Zhou Y, Lu Z, Zhai D, Luo H, Li T, Li Y. Multi-level graph neural network with sparsity pooling for recognizing Parkinson's disease. IEEE Trans Neural Syst Rehabil Eng. 2023;31:4459–69. https://doi.org/10.1109/TNSRE.2023.3330643.

Zhao Q, Zhao H, Zheng K, Wang J. HyperAttentionDTI: improving drug–protein interaction prediction by sequence-based deep learning with attention mechanism. Bioinformatics. 2022;38(3):655–62. https://doi.org/10.1093/bioinformatics/btab715.

Zhou B, Wang R, Wu P, Kong D. Drug repurposing based on drug–drug interaction. Chem Biol Drug Des. 2015;85(2):137–44. https://doi.org/10.1111/cbdd.12378.

Zhuang F, Qi Z, Duan K, Xi D, Zhu Y, Zhu H, Xiong H, He Q. A comprehensive survey on transfer learning. Proc IEEE. 2021;109(1):43–76. https://doi.org/10.1109/JPROC.2020.3004555.

Artificial Intelligence in Clinical Trials: From Protocol Design to Pharmacovigilance

Aron Schulman, Adrià Fernández-Torras, Yojana Gadiya, Khalid Saeed, Jordi Mestres, and Ziaurrehman Tanoli

Abstract Clinical trials are essential for assessing the safety and efficacy of new treatments and securing regulatory approvals. However, they are resource-intensive and often hindered by inefficiencies in protocol design, participant recruitment, and

---

Authors Aron Schulman, Adrià Fernández-Torras, and Yojana Gadiya have equally contributed to this chapter.

A. Schulman
Institute for Molecular Medicine Finland (FIMM), HiLIFE, University of Helsinki, Helsinki, Finland

iCAN Digital Precision Cancer Medicine Flagship, University of Helsinki and Helsinki University Hospital, Helsinki, Finland
e-mail: aron.schulman@helsinki.fi

A. Fernández-Torras · J. Mestres
Chemotargets SL, Parc Científic de Barcelona, Barcelona, Catalonia, Spain
e-mail: adria.fernandez@chemotargets.com; jordi.mestres@chemotargets.com

Y. Gadiya
Fraunhofer Institute for Translational Medicine and Pharmacology (ITMP), Hamburg, Germany

Bonn-Aachen International Center for Information Technology (B-IT), Bonn, Germany
e-mail: yojana.gadiya@itmp.fraunhofer.de

K. Saeed
King Abdullah International Medical Research Center (KAIMRC), King Saud Bin Abdulaziz University for Health, Riyadh, Saudi Arabia
e-mail: Khalid.saeed@helsinki.fi; saeedk@kaimrc.edu.sa

Z. Tanoli (✉)
Institute for Molecular Medicine Finland (FIMM), HiLIFE, University of Helsinki, Helsinki, Finland

iCAN Digital Precision Cancer Medicine Flagship, University of Helsinki and Helsinki University Hospital, Helsinki, Finland

Drug Discovery and Chemical Biology (DDCB) Consortium, Biocenter Finland, Helsinki, Finland

BioICAWtech, Helsinki, Finland
e-mail: ziaurrehman.tanoli@helsinki.fi

A. Lavecchia (ed.), *Applied Artificial Intelligence for Drug Discovery*,
https://doi.org/10.1007/978-3-031-98022-0_24

data management. This chapter explores how advancements in artificial intelligence (AI), particularly in large language models (LLMs), are transforming these processes. AI-powered tools streamline protocol drafting, improve eligibility criteria, address missing data issues, and enhance safety assessments. Innovations such as conversational AI for patient interaction, federated learning for secure and decentralized data analysis, and standardized metadata frameworks have advanced patient-centred methodologies and pharmacovigilance practices. These technologies provide real-time insights, broaden participant engagement and demographic diversity, and improve the generalizability of trial outcomes, while reducing costs and accelerating timelines. By examining the role of AI across the clinical trial lifecycle, this chapter highlights its current applications, innovative methodologies, and transformative future potential to create more efficient, inclusive, and impactful clinical research processes.

Keywords AI clinical trials · Protocol design · Patient recruitment · Pharmacovigilance · Large language models · Federated learning

1 Introduction

Clinical trials are indispensable to drug discovery, serving as the gold standard for evaluating the safety and efficacy of new treatments (Jager 2023). However, these trials are resource-intensive, requiring significant time, financial investment, and effort to execute (Wouters et al. 2020; Dowden and Munro 2019). Traditional clinical trial methodologies are often hindered by inefficiencies such as delays in protocol design, suboptimal participant recruitment, and challenges in data analysis (Fogel 2018; Getz et al. 2011). These limitations hinder progress, restrict the generalizability of findings, and delay the delivery of critical therapies to patients (Lai et al. 2021). Overly restrictive eligibility criteria and underrepresentation of diverse populations compromise the validity and applicability of trial outcomes (Smith et al. 2022).

A major barrier to generalizability is the underrepresentation of diverse populations in clinical trials. For example, Sauer et al (2022) reported that only 12.8% of studies disclosed race or ethnicity data, underscoring substantial underreporting and a lack of diversity. This shortfall limits the applicability of findings across demographic groups, hindering efforts to address health disparities. Additionally, challenges such as managing missing data in randomized trials can introduce bias and compromise the robustness of findings, as highlighted by Cro et al (2020). Methodological hurdles also arise when incorporating real-world evidence into clinical trials, including data heterogeneity and potential biases that can undermine the validity of results.

Recent advancements in artificial intelligence (AI) offer transformative solutions to these challenges, providing unprecedented opportunities to streamline and optimize the clinical trial process (Ghim and Ahn 2023). Machine learning (ML)

approaches, particularly large language models (LLMs), have proven highly effective in automating labour-intensive tasks such as protocol drafting and eligibility criteria optimization (Hutson 2024; Markey et al. 2024). AI-powered tools also address critical issues like missing data through advanced imputation methods (Dong et al. 2021) and enhance safety monitoring via predictive analytics (Shah and Mishra 2024). By automating these processes, AI reduces administrative burdens and improves the precision, scalability, and reproducibility of clinical research (Leviton and Loddenkemper 2023).

AI has also enabled groundbreaking innovations in patient-centric methodologies. For example, conversational AI enhances patient engagement by providing personalized guidance and real-time support throughout the trial process (Cevasco et al. 2024). Federated learning approaches alleviate long-standing concerns about data privacy and security by enabling decentralized analysis without exposing sensitive patient information (Prayitno 2021; Zhang et al. 2022a; Li et al. 2023). These advancements not only improve operational efficiency, but also ensure greater inclusivity and participant trust in clinical trials.

Beyond trial execution, AI has transformed pharmacovigilance practices by streamlining the detection and assessment of adverse drug events (ADEs) (Salas et al. 2022). Leveraging natural language processing (NLP) and predictive modelling, AI tools identify safety signals from diverse data sources such as electronic health records (EHRs) (Kompa et al. 2022) and individual case safety reports (ICSRs) (Salas et al. 2022). These advancements significantly improve drug safety monitoring, enabling timely interventions and better patient protection.

Despite its transformative potential, AI integration into clinical trials faces challenges (Askin et al. 2023). Issues such as data accessibility (Kelly et al. 2019), algorithmic transparency (Vollmer et al. 2020), and the need for representative datasets must be addressed to fully realize its benefits (Cross et al. 2024). Collaboration among researchers, policymakers, and technology developers is essential to navigate ethical, operational, and regulatory hurdles (McLennan et al. 2022). Ethical oversight is also essential to ensure that AI applications in clinical trials do not preserve biases or compromise patient trust.

This chapter explores the transformative role of AI throughout the clinical trial lifecycle, from design and execution to post-marketing surveillance. By highlighting emerging methodologies and future directions, it illustrates how AI is revolutionizing clinical research to enable more inclusive, efficient trials and accelerate the delivery of innovative therapies to patients.

2 Optimizing Clinical Trials with AI

Clinical trial data divide into five main categories: (a) trial design and registration, (b) patient enrollment, (c) study completion or termination, (d) publication, and (e) regulatory applications. Trial design and registration involve foundational details, such as study protocols and statistical analysis plans, which are critical for initiating a trial (Stevens et al. 2023). Patient enrollment collects data during clinic visits to evaluate treatment effects, often supported by automated devices like sensors and smartwatches (Triantafyllidis et al. 2024).

Advances in LLMs for text generation and processing offer significant potential to speed up clinical trial design process (Ghim and Ahn 2023). Planning trials often requires analysing past studies, which is time-consuming. LLMs simplify this process by generating automated suggestions and assisting with drafting trial documents. Tools like CliniDigest (White et al. 2023) help tackle the constant influx of new trial reports. It takes as input up to 85 user-relevant clinical trial descriptions and summarizes them into a compact, 200-word report complemented with references. Similarly, Trial2Vec (Wang and Sun 2022) converts long trial documents into embeddings, which are then used as a similarity measure between the documents. Users can find similar trials to match their use case or search for trials with an input query. The trial embedding clusters can also be visualized for better interpretability. Finally, AutoCriteria (Datta et al. 2024) improves search results by automatically extracting eligibility criteria from clinical trial protocols.

Figure 1 illustrates a further time gain for trial design provided by an LLM tool that drafts clinical trial protocols (Markey et al. 2024). The protocols are typically very lengthy and take months to prepare. With the tool, users can specify the clinical phase, indication, and the section of the report to be generated. The tool includes a retrieval-augmented generation (Lewis et al. 2020) (RAG) component that improves the quality of the generated texts and provides references. Importantly, the logic and clinical thinking in the text generated by the RAG-extended model is greatly improved when compared with a plain LLM approach. Together, these LLM tools may alleviate the sometimes-cumbersome tasks of literature review and report writing, thus allowing clinical researchers to focus on more critical aspects of the project.

2.1 AI for Missing Data Imputation

Regular clinic visits are essential for monitoring treatment effects. Missed visits can lead to participant dropout and require additional recruitment, increasing costs, and delays. During the COVID-19 pandemic, quarantines disrupted many trials, causing delays and missing data (Zame et al. 2020). To mitigate effects of missing data, deep learning (DL) methods were proposed to infer patient state from the physiological, safety, and other measurements recorded during the attended visits (Zame et al. 2020). The model predictions are then used to impute data missing due to an unattended visit (Fig. 1, middle right panel). These methods assume that missing data are independent from the unobserved state of the patient.

One competition tested the validity of machine learning (ML) and DL methods for imputing blood lab test data (Luo 2022). The competition data included 16,534 unique, real-world patient admissions, with the winning method achieving a normalized root-mean-square deviation of 0.178 on the test set. The competition

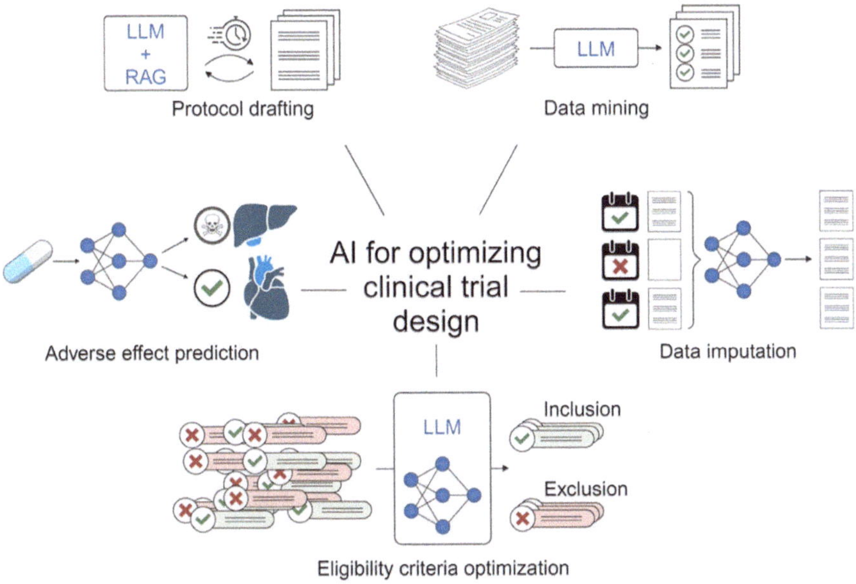

Fig. 1 AI applications for optimizing clinical trial design. Large language models automate and accelerate protocol writing and data extraction from related trial documents. Deep neural networks (DNNs) impute missing data caused by unattended clinic visits. AI designs-suitable inclusion and exclusion criteria for clinical trials. Finally, DNNs predict compound adverse effects to avert clinical trial failure

showcased how successful application of ML- and DL-based data imputation methods holds promise for increasing the robustness and reducing the time and costs of current trials, although there remains a necessity for more realistic and diverse test settings for such methods (Luo 2022).

2.2 *Eligibility Criteria Optimization*

Selecting the right participants is critical for the success of clinical trials (Kim et al. 2015). To take part, participants must meet trial-specific inclusion and exclusion criteria. The criteria aim to ensure participant safety, and that the trial results are reliable and generalizable. However, overly strict criteria often reduce patient enrollment, delaying trials and limiting representation of the target population. For example, a study found that only 41% of clinical trials involving drugs for advanced non-small-cell lung cancer (aNSCLC) allowed patients with treated brain metastases, a common condition in this group (McCoach et al. 2016). This exclusion left many potential drug recipients untested and shifted the burden to post-approval monitoring.

Recognizing these challenges, the American Society of Clinical Oncology and Friends of Cancer Research proposed modernized guidelines to make criteria more flexible (Lin et al. 2017). However, adoption of these recommendations remains slow as not all trials can apply them (Riner et al. 2023). To address this, Liu et al. developed Trial Pathfinder (Liu et al. 2021) as a data-driven solution for determining optimal eligibility criteria. They emulated trials *in silico* with real-world data on aNSCLC cases from the Flatiron database (Ma et al. 2020) and performed survival analysis to infer the effect of varying criteria on trial efficacy. In their retrospective study of 10 clinical trials, they demonstrated how using AI-defined criteria would have led to increased patient inclusion along with decreased hazard ratios. They found several redundant criteria that do not significantly affect the trial outcomes, including disease staging, blood pressure, neutrophil count, and patient age. Furthermore, they identified thresholds that can be relaxed without increasing toxicity risks, such as haemoglobin, platelet, and bilirubin levels.

Another tool, AutoTrial (Wang et al. 2023), uses LLMs with RAG components to suggest eligibility criteria tailored to specific trial needs. The input prompt allows the end users to provide detailed instructions for the model to consider when generating the eligibility criteria. Trained on data from over 70,000 trial documents, AutoTrial supports a wide range of diseases, including different cancer types, diabetes, and coronary disease, allowing for a broad range of applications. Notably, the authors report generally better performance for generating inclusion criteria than exclusion criteria. Beyond LLMs, a knowledge graph embedding-based method offers recommendations for both eligibility criteria and the overall design of clinical trials (Devarakonda et al. 2024).

2.3 *In Silico Toxicity Prediction*

Excessive drug toxicity accounts for about 30% of clinical trial failures (Sun et al. 2022). Traditional preclinical toxicity tests using animal and cell culture are costly, time-consuming, and raise ethical concerns (Hartung 2009). In silico approaches provide a faster, more ethical, and cost-efficient alternative for predicting toxicity and adverse effects.

AI-based methods are applied to predict a wide range of predefined compound toxicity properties (Tran et al. 2023). Generally, the scope of a predictive model is limited to one toxicity type at a time. For example, a well-known source of adverse effects is hERG inhibition, which may lead to cardiac arrhythmias. DeepHIT (Ryu et al. 2020) and BayeshERG (Kim et al. 2022) are deep learning approaches that represent the current state-of-the-art in hERG inhibition prediction. Both methods use the SMILES representation of compounds and convert them to molecular graphs before inputting them to a graph neural network. DeepHIT additionally trains fingerprint-based and molecular descriptor-based models to form an ensemble of predictors. Importantly, BayeshERG provides interpretability with weights from the attention mechanism to highlight the compound substructures that most contribute to the toxicity predictions.

Another example of a common reason for clinical trial failure and post-marketing withdrawal is drug-induced liver injury (DILI). To address this, DILIPredictor (Seal et al. 2024) predicts compounds with DILI risk and identifies the contributing substructures. For example, DILIPredictor correctly predicted sitaxentan as toxic and highlighted a substructure near its sulfonamide functional group associated with hepatotoxicity (Khalili et al. 2011). Notably, the tree-based model was able to differentiate between compounds hepatotoxic to mice but not humans (e.g. astaxanthin), hence, potentially informing the selection of more appropriate animal models while preventing premature animal testing.

By applying multiple toxicity prediction models sequentially, researchers can generate comprehensive safety profiles for compounds. These interpretable models guide safer drug design, lowering the risk of clinical trial failure in later stages.

2.4 AI for Clinical Trial Efficiency and Cost Reduction

AI-driven tools have optimized processes across various trial stages, from recruitment and data management to site selection and protocol design. The impact of these innovations becomes apparent when analysing specific indicators of clinical trial efficiency.

One critical metric is Time-to-First-Patient-In (TTFPI), which refers to the time required to recruit the first participant in a trial. AI tools have demonstrated the ability to expedite pre-screening, significantly reducing TTFPI. For instance, Mendel.ai can determine eligibility for breast cancer patients within minutes, compared to the traditional process that takes 19 days on average and up to 263 days for lung cancer patients (Calaprice-Whitty et al. 2009). Such reductions in prescreening time contribute to faster trial initiation and improved timelines.

Additionally, AI-powered patient matching systems have enhanced efficiency by accurately identifying suitable candidates. For example, a lung cancer trial platform achieved an exclusion accuracy of 95.7% and a qualification evaluation accuracy of 91.6%, demonstrating precision in participant selection (Alexander et al. 2020). These advancements not only improve recruitment rates but also reduce dropout risks, leading to more reliable trial outcomes while lowering costs related to extended recruitment periods and participant management.

These achievements demonstrate the potential of AI to streamline recruitment processes, improving efficiency and precision while reducing costs. However, despite these promising results, further research and validation are necessary to ensure the broad applicability of AI across diverse clinical trial settings.

3 Patient-Centric Approaches and Secure Data Analysis Using AI

Clinical trials generate diverse types of data throughout their pipeline. Study findings are typically disseminated through scientific publications and submitted to regulatory bodies, facilitating the translation of research into accessible treatments. This section focuses on the secondary use of clinical trial data, particularly during patient enrollment and study completion stages, and highlights how AI prioritizes patient-centric approaches. Recent advancements, such as federated learning for anonymized data sharing, improve the utility of clinical data by making traditional processes more scalable, efficient, and secure (Fig. 2).

Patient-reported outcomes (PROs) provide valuable insights into patients' health and treatment responses, helping clinicians optimize care (Au et al. 2010). Traditionally, PROs were collected through static surveys and primarily used for clinical decision-making and monitoring progress. However, with advancements in ML (Mello et al. 2018; Jang et al. 2024), PRO datasets now play a broader role in clinical care, healthcare management, and patient monitoring (Gilbert et al. 2015; Frost et al. 2011). By analysing large datasets of patient-reported information, researchers uncover patterns that improve treatment protocols and patient outcomes. This trend has driven new research into AI-powered methodologies for curating and extracting insights from clinical trial datasets, creating new opportunities for AI applications in medical research (Chen et al. 2024a).

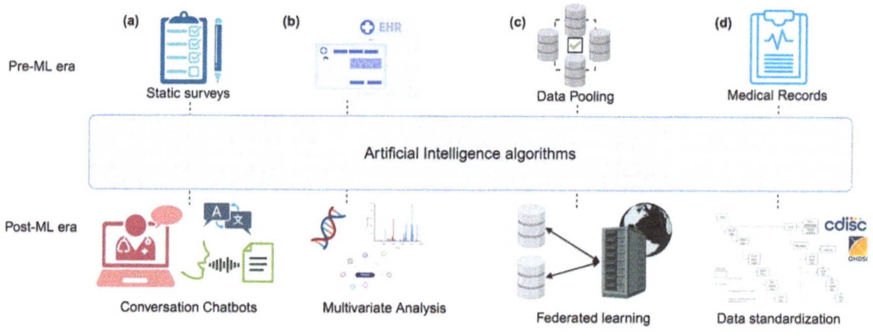

Fig. 2 The figure highlights four key areas where AI has driven significant advancements in clinical research, presented sequentially from left to right: (**a**) the evolution from static patient surveys to interactive conversational chatbots, (**b**) the application of multivariate analysis on electronic health records (EHRs) to uncover complex patterns, (**c**) the adoption of federated learning approaches for secure and decentralized data analysis, and (**d**) the standardization of medical records into metadata frameworks such as CDISC and OMOP-CDM

3.1 AI-Based Chatbots to Improve Patient Care

Recent advancements in AI, including NLP and generative pre-trained transformer (GPT) models, have led to the development of chatbots to support and monitor specific patient groups (Bibault et al. 2019; Rojc et al. 2023). Examples include kBot for asthma (Kadariya et al. 2019) and PRObot for diabetes (Pielka et al. 2024), which deliver personalized advice, reminders, and emotional support, enhancing patient engagement and adherence to treatment plans. Generalized systems like PROsBot (Fenza et al. 2023) and Chat-ePRO (Chen et al. 2024b) combine NLP with rule-based methods to provide accurate and timely information during clinical trials, highlighting AI's transformative role in healthcare.

3.2 AI Applications for Medical and Electronic Health Records (EHRs)

Medical records or EHRs, a vital source of clinical data, are anonymized to ensure patient privacy (Rodwin 2010; Nowrozy et al. 2024; Tertulino et al. 2024). Since medical records often contain unstructured text, various NLP pipelines are developed to transform them into structured formats suitable for downstream analysis (Juhn and Liu 2020; Yang et al. 2024). Standards such as CDISC and OMOP ensure uniformity and interoperability across datasets, enabling global data sharing and advancing medical research (Ramakrishnaiah et al. 2023). Notable datasets include MIMIC-IV (590,000 patients), a critical care dataset from Beth Israel Deaconess Medical Center (Johnson et al. 2023); Paediatric Intensive Care (PIC), a Chinese paediatric dataset (Zeng et al. 2020); and the Salzburg Intensive Care database (SICdb), a European dataset with records from 21,000 patients (Rodemund et al. 2024). Platforms like PhysioNet (Costa et al. 2003) democratize access to these resources, fostering innovation in medical research.

EHRs are increasingly combined with other biological data for multivariate analysis, offering deeper insights into complex conditions. For example, Sosinsky et al. combined EHR and genomic data to identify hallmark genes in cancer (Sosinsky et al. 2024). Similarly, the eMERGE Network aggregates genomic data to study genetic associations, advancing personalized medicine (Lemke et al. 2010). ML applications also extend to analysing diagnostic patterns, such as identifying markers in radiological images for Crohn's disease or examining links between sleep patterns and cognitive health (Liu et al. 2024; Wei et al. 2024).

Despite these advancements, challenges persist, particularly in accessing diverse and high-quality clinical data. Restricted availability and reliance on synthetic or simulated datasets hinder the training of robust models that generalize across varied clinical settings (Pati et al. 2024; Beaulieu-Jones et al. 2019; Youssef et al. 2023). Data centralization further complicates scalability and diversity, while privacy concerns remain a significant barrier (Kairouz et al. 2021; Malin et al. 2013; Tucker et al. 2016; Hoffman and Podgurski 2013).

3.3 Secure and Decentralized Data Analysis Mechanisms

Federated machine learning (FML) addresses data security challenges by enabling distributed model training across multiple servers without centralizing data (Kaissis et al. 2020; Yang et al. 2019). This approach adheres to key privacy principles, confidentiality, integrity, and availability (CIA), ensuring secure use of real-world clinical data under strict privacy regulations (Ham 2021). Initiatives like Canada's Collaborative Data Analysis (CODA) project (Mullie et al. 2024) and Europe's IDERHA project for lung cancer (Deist et al. 2020; Soltan et al. 2024) demonstrate the scalability of FML. Open-source tools like Google's FML toolkit further support secure and scalable implementations.

Swarm Learning (SL) offers another decentralized training approach. Unlike FML, where the final model is hosted centrally, SL develops individual models on private data at each site while sharing knowledge across models to enhance performance collectively (Saldanha et al. 2022; Schultze et al. 2022). Using blockchain technology, SL ensures secure, transparent collaboration without the need for a central coordinator. This approach mitigates risks such as data poisoning or malicious activity while maintaining the integrity of contributions through blockchain consensus.

4 Pharmacovigilance and AI

Pharmacovigilance (PV) is a critical aspect of drug safety, ensuring that safety standards observed during clinical trials are maintained when therapies reach broader and more diverse populations (Beninger 2018; Pharmacovigilance n.d.). PV involves analysing heterogeneous patient safety data from ICSRs, EHRs, and conference abstracts (Kompa et al. 2022), to identify adverse drug events (ADEs) that may be underreported or overlooked during clinical trials. Detecting new drug safety signals allows regulatory authorities to re-assess a drug's benefit-risk balance.

ICSR databases, central to PV, are extensive, complex, and often biased, making signal detection (the process of identifying adverse events potentially caused by a medicine) a challenging task. These databases currently contain over 60 million reports (FDA Adverse Events Reporting System (FAERS) Public Dashboard 2024), with numbers growing annually (Center for Drug Evaluation and Research 2019). However, many reports are poorly annotated (Durrieu et al. 2016), and event reporting is influenced by factors such as the perceived severity, underlying disease, and the reporter's background (e.g. patient vs. healthcare professionals) (Ghosh and Dewanji 2015; Meyboom et al. 1997; Castillon et al. 2019; Moore et al. 2003). Socio-demographic factors, media coverage, and global events like the COVID-19 pandemic further exacerbate these patterns (Center for Drug Evaluation and Research 2024; Whitten et al. 2008; Montes-Grajales et al. 2023), potentially leading to nocebo effects (Planès et al. 2016). A key challenge lies in identifying events that are both significantly disproportionate and drug-related, prioritizing them for clinical review (Guidance for industry good pharmacovigilance practices and pharmacoepidemiologic assessment n.d.).

4.1 AI to Detect Safety Signal in PV

Early applications of AI in PV emerged in the late 1990s to address the limitations of manual safety signal detection, offering more systematic and unbiased estimates of drug-event associations (Bate et al. 1998). Early methods, such as Bayesian Neural Networks, were particularly effective in processing large datasets, modelling complex features, and handling low-frequency events (Orre et al. 2000; Bate et al. 2002; Basile et al. 2019). Today, more than 50% of AI-related PV researches focus on signal detection (Kompa et al. 2022). While deep learning has become popular in recent years (Kompa et al. 2022), concerns about transparency and interpretability persist, particularly when performance gains over traditional methods may not fully justify these issues (Bate and Stegmann 2023).

AI applications in PV now encompass a wide range of tasks, including automatic processing of safety reports (Benin et al. 2016), language translation, term standardization (Xie et al. 2021; Berkowitz et al. 2024), report duplication detection (Norén et al. 2007; Kiguba et al. 2024), quality assurance (Ménard et al. 2019; Zou et al. 2021; Chauvet et al. 2020), causal inference (Cherkas et al. 2022; Kreimeyer et al. 2021; Wang et al. 2021), and event severity classification (Chauvet et al. 2020; Evans et al. 2020). Moreover, the importance of patient genetics in drug metabolism (Mahaux et al. 2023) has drawn attention to anticipate at-risk subpopulations (Correia Pinheiro et al. 2020) and design patient-tailored doses (Pavani et al. 2016; Miranda et al. 2024). Regulatory agencies such as the FDA launched initiatives like AI4TOX (National Center for Toxicological Research 2024a) to support AI-driven tools for regulatory science (Neyarapally et al. 2024; Gray et al. 2023; Wu et al. 2023a, b; Wang et al. 2022), while other agencies apply AI systems to monitor vaccine safety, including those for COVID-19 (Martin et al. 2022).

4.2 AI Applications to Extract Adverse Drug Events (ADEs)

Under-reporting remains a major issue in PV (Heinrich 2000), with ICSR databases estimated to cover only 10% of all ADEs (Hazell and Shakir 2006). AI methods address this gap by extracting ADEs from better-documented sources like EHRs (Linder et al. 2010; Boland and Tatonetti 2015; Casillas et al. 2016; Gaebel et al. 2015). While some information in hospital systems is structured, much of it reside in unstructured clinical narratives (Skentzos et al. 2011). NLP-based approaches have advanced significantly in extracting meaningful data from these texts (Uzuner et al. 2011; Chapman et al. 2011), particularly with the advent of large language models (LLMs) (National Center for Toxicological Research 2024b). To date, these methods are used not only to extract ADEs (Modi et al. 2024; Kovačević et al. 2013; Yang et al. 2020; Marella et al. 2017; Dandala et al. 2019; Tanaka et al. 2024) but also drug indications (Bhatt et al. 2021) and drug-drug interactions (DDI) (Hauben 2023; Bouzillé et al. 2019; Machado et al. 2023; Lim et al. 2018) that are only revealed during post-marketing, where polypharmacy becomes relevant (Pazan and Wehling 2021). Additionally, applying NLP to social media platforms, despite the challenge of noisy reports (Rees et al. 2018; Lee et al. 2021), offers a valuable and complementary source of safety information when paired with rigorous quality assessments (Comfort et al. 2018; Nikfarjam et al. 2015; Leaman et al. 2010).

4.3 Predictive Models for Adverse Drug Reactions (ADRs)

Beyond signal detection, predictive models can anticipate ADRs to prevent harm from therapies before safety concerns arise. Over hundred drugs are withdrawn from the market due to unexpected safety concerns (Craveiro et al. 2020), not detected during clinical trials, posing significant risks to patients (Clarke et al. 2006). Early efforts in ADR prediction focused on modelling safety annotations from drug label databases (Galeano and Paccanaro 2018; Zhang et al. 2018; Poleksic and Xie 2018). However, the delay in signal detection and subsequent updates to drug labels quickly rendered these databases outdated, leading to the inclusion of ICSR in modelling data (Fukuto et al. 2021; Farnoush et al. 2024; Daluwatte et al. 2020).

More recent approaches integrate drug chemistry and pharmacology with gene-disease associations and biomedical ontologies, often represented as heterogeneous networks and modelled using graph learning algorithms (Xue et al. 2020; Krix et al. 2023; Zhang et al. 2021; Bean et al. 2017; Zheng et al. 2018). These models predict not only ADR occurrences, but also event frequency (Galeano et al. 2020; Xu et al. 2022) and severity (Zhao et al. 2023). Improved transparency frameworks further enhance interpretability (Dey et al. 2018). Predictive models also extend to related areas such as DDI (Zitnik et al. 2018; Tatonetti et al. 2012; Dewulf et al. 2021), hospital events (Hobensack et al. 2023; Rajkomar et al. 2018), and inpatient risks (Bates et al. 2021), such as surgical complications (Hassan et al. 2023; Merath et al. 2020; Hashimoto et al. 2019; Hashimoto et al. 2018), falls (Bargiotas et al. 2023; Yokota et al. 2017; Fu et al. 2022; Chu et al. 2022), and healthcare-associated infections (Barchitta et al. 2021; Chen et al. 2024c; Saviauk et al. 2018; Beeler et al. 2018). Real-time safety monitoring systems now enable rapid detection and intervention for risks specific to populations, such as hypoglycaemia in diabetic patients (Chen et al. 2019), fall risk in community-dwelling older adults (Lockhart et al. 2021; Howcroft et al. 2017), and pressure ulcers in paraplegic populations (Luboz et al. 2018).

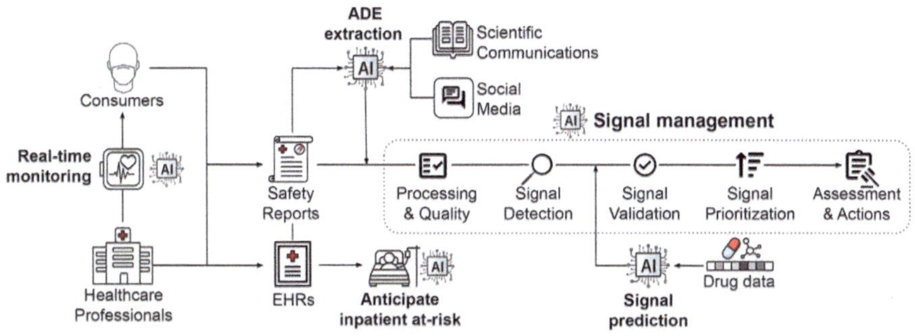

Fig. 3 Illustrative chart of AI applications in Pharmacovigilance. Signal management is central to identifying adverse drug reactions from safety reports. AI-based strategies are increasingly integrated into this process, supporting report processing and quality (e.g. duplication detection), signal detection (e.g. Bayesian modelling), signal validation (e.g. causal inference), and signal prioritization (e.g. event severity classification). To expand signal coverage, NLP-driven methods have been developed to extract adverse drug events (ADEs) from additional sources, such as scientific communications (e.g. congress abstracts, scholarly publications), social media, and Electronic Health Records (EHRs). There are also initiatives to predict safety signals directly from drug data, and AI is being applied in healthcare to anticipate inpatient risks and monitor population-specific safety in real time

4.4 Challenges in AI-Based PV

While AI is widely adopted across the PV pipeline (Fig. 3), the complexity and regulatory requirements of post-marketing data pose several challenges. First, under-reporting, data incompleteness, and the need of privacy-preserving methodologies, especially in the clinical medical domains (Li et al. 2023; Zhang et al. 2022b), limit AI-models' access to representative and high-quality data, compromising their generalizability. This is particularly prevalent in low-income regions, where the lack of reporting infrastructures hampers the collection of ADE evidence (Liang et al. 2022). Second, ADR identification requires thorough characterization and evaluation (Shakir and Layton 2002; Naidu 2013). As a result, models lacking transparency and explainability are less trusted among PV professionals (Ball et al. 2024), while fully autonomous AI approaches face scepticism, leading the PV community to advocate for human-in-the-loop solutions (Bate and Stegmann 2023; Ball and Dal Pan 2022; Tong and Baran 2024; Fia 2024; Shamim et al. 2024; Salvo et al. 2023). Third, biases inherent in PV data can lead to misinterpretation, necessitating standardized procedures and reproducibility (Fusaroli et al. 2024). Yet only 10% of studies published before 2021 have followed best practices in ML, with just 7% providing publicly available code, according to a recent review (Kompa et al. 2022).

Finally, most AI implementations lack robust post-deployment mechanisms for monitoring, evaluation, and case-report of undesired AI-system reactions. Importantly, real-time updates of AI-driven PV models are needed to accommodate the continuous flow of safety data and react to shifts in reporting patterns (Babic et al. 2019; Gerke et al. 2020). AI has as much to learn from PV as it has to offer, with PV principles currently serving as reference for AI surveillance (Balendran et al. 2024).

5 Conclusion and Future Directions

The integration of AI into clinical trials represents a paradigm shift, addressing long-standing challenges while streamlining research processes (Ghim and Ahn 2023; Askin et al. 2023). AI-driven innovations, from protocol drafting to pharmacovigilance, reduce trial time and cost, improve participant representation, and enhance generalizability (Liu et al. 2021). By automating data analysis and generating insights from large datasets, AI enables more patient-centred research approaches (Bjerring and Busch 2021).

However, the adoption of AI in clinical trials raises significant privacy and ethical challenges. Robust data privacy frameworks are critical to protecting sensitive patient information while facilitating collaborative research (Rieke et al. 2020). AI models often require access to vast amounts of personal health data, and ensuring compliance with privacy regulations remains a key challenge. Breaches of data security could lead to misuse of information, loss of trust, and legal consequences. Promising solutions such as federated learning and swarm learning enable privacy-aware data analysis without centralized data sharing, but their implementation at scale requires significant infrastructure and expertise (Rieke et al. 2020).

Ethical concerns also emerge from the potential biases inherent in AI models. Certain demographic groups are often underrepresented in training datasets, leading to models that may not perform equitably across diverse subpopulations (Cross et al. 2024). For instance, AI systems might misclassify or overlook conditions in minority ethnic or age groups, exacerbating existing health disparities (Boulton et al. 2022). Addressing these challenges demands concerted efforts and a focused commitment to develop diverse, representative datasets and conduct routine audits to identify and mitigate biases (Cross et al. 2024). Furthermore, explainability and transparency in AI decision-making are crucial for earning the trust of clinicians, regulators, and patients (Eke and Shuib 2024). Black-box models that cannot justify their predictions are less likely to be adopted in high-stakes environments such as clinical research (Amann et al. 2020).

Beyond privacy and bias, ethical discussions must also address the accountability of AI systems (García-Gómez et al. 2023). A key question is who holds responsibility when AI-generated recommendations lead to harmful outcomes. Establishing clear guidelines for accountability, combined with human oversight at critical decision points, will be essential to mitigate risks (Langer et al. 2025). Additionally, the rapid pace of AI development may outstrip regulatory frameworks, necessitating ongoing collaboration between researchers, policymakers, and regulatory agencies to update standards and ensure AI systems adhere to safety and ethical requirements (Jobin et al. 2019).

Achieving the full potential of AI in clinical trials will depend on addressing these priorities and driving innovation. Looking forward, real-time adaptive AI systems capable of learning from dynamic clinical data will improve responsiveness and accuracy (Hügle et al. 2021). Establishing global standards for data interoperability and model integration will streamline processes across institutions and geographies (Jacobsen et al. 2022). Finally, fostering collaboration among researchers, policymakers, and technology developers will be vital to driving innovation while upholding ethical and privacy standards (McLennan et al. 2022).

By prioritizing these efforts, AI has the potential to transform clinical trials into more efficient, inclusive, and impactful endeavours. This transformation will ultimately accelerate the translation of research into improved patient outcomes and promote stronger, more equitable global healthcare systems (Ghim and Ahn 2023).

Acknowledgments The work was funded by the Research Council of Finland (No. 351507 to ZT).

References

Alexander M, Solomon B, Ball DL, Sheerin M, Mullan ID, Preininger AM, Jackson GP, Herath DM. Evaluation of an artificial intelligence clinical trial matching system in Australian lung cancer patients. JAMIA Open. 2020;3:209–15.

Amann J, Blasimme A, Vayena E, et al. Explainability for artificial intelligence in healthcare: a multidisciplinary perspective. BMC Med Inform Decis Mak. 2020;20:310.

Askin S, Burkhalter D, Calado G, El Dakrouni S. Artificial Intelligence applied to clinical trials: opportunities and challenges. Health Technol. 2023;13:203–13.

Au H-J, et al. Added value of health-related quality of life measurement in cancer clinical trials: the experience of the NCIC CTG. Expert Rev Pharmacoecon Outcomes Res. 2010;10:119–28.

Babic B, Gerke S, Evgeniou T, Cohen IG. Algorithms on regulatory lockdown in medicine. Science. 2019;366:1202–4.

Balendran A, Benchoufi M, Evgeniou T, Ravaud P. Algorithmovigilance, lessons from pharmacovigilance. NPJ Digit Med. 2024;7:1–6.

Ball R, Dal Pan G. 'Artificial Intelligence' for pharmacovigilance: ready for prime time? Drug Saf. 2022;45:429–38.

Ball R, Talal AH, Dang O, Muñoz M, Markatou M. Trust but verify: lessons learned for the application of AI to case-based clinical decision-making from postmarketing drug safety assessment at the US Food and Drug Administration. J Med Internet Res. 2024;26:e50274.

Barchitta M, et al. A machine learning approach to predict healthcare-associated infections at intensive care unit admission: findings from the SPIN-UTI project. J Hosp Infect. 2021;112:77–86.

Bargiotas I, et al. Preventing falls: the use of machine learning for the prediction of future falls in individuals without history of fall. J Neurol. 2023;270:618–31.

Basile AO, Yahi A, Tatonetti NP. Artificial Intelligence for drug toxicity and safety. Trends Pharmacol Sci. 2019;40:624–35.

Bate A, Stegmann J-U. Artificial intelligence and pharmacovigilance: what is happening, what could happen and what should happen? Health Policy Technol. 2023;12:100743.

Bate A, et al. A Bayesian neural network method for adverse drug reaction signal generation. Eur J Clin Pharmacol. 1998;54:315–21.

Bate A, Lindquist M, Orre R, Edwards I, Meyboom R. Data-mining analyses of pharmacovigilance signals in relation to relevant comparison drugs. Eur J Clin Pharmacol. 2002;58:483–90.
Bates DW, et al. The potential of artificial intelligence to improve patient safety: a scoping review. NPJ Digit Med. 2021;4:1–8.
Bean DM, et al. Knowledge graph prediction of unknown adverse drug reactions and validation in electronic health records. Sci Rep. 2017;7:16416.
Beaulieu-Jones BK, et al. Privacy-preserving generative deep neural networks support clinical data sharing. Circ Cardiovasc Qual Outcomes. 2019;12:e005122.
Beeler C, et al. Assessing patient risk of central line-associated bacteremia via machine learning. Am J Infect Control. 2018;46:986–91.
Benin AL, et al. Electronic approaches to making sense of the text in the adverse event reporting system. J Healthc Risk Manag. 2016;36:10–20.
Beninger P. Pharmacovigilance: an overview. Clin Ther. 2018;40:1991–2004.
Berkowitz JS, Fatapour Y, Srinivasan A, Cortina JMA, Tatonetti NP. Biomedical text normalization through generative modeling; 2024. https://doi.org/10.1101/2024.09.30.24314663.
Bhatt A, et al. DICE: a drug indication classification and encyclopedia for AI-based indication extraction. Front Artif Intell. 2021;4:711467.
Bibault J-E, et al. Healthcare ex Machina: are conversational agents ready for prime time in oncology? Clin Transl Radiat Oncol. 2019;16:55–9.
Bjerring JC, Busch J. Artificial Intelligence and patient-centered decision-making. Philos Technol. 2021;34:349–71.
Boland MR, Tatonetti NP. Are all vaccines created equal? using electronic health records to discover vaccines associated with clinician-coded adverse events. AMIA Jt Summits Transl Sci. 2015;2015:196–200.
Boulton AJ, Del Rios M, Perkins GD. Health inequities in out-of-hospital cardiac arrest. Curr Opin Crit Care. 2022;28(3):229–36.
Bouzillé G, et al. An automated detection system of drug-drug interactions from electronic patient records using big data analytics. Stud Health Technol Inform. 2019;264:45–9.
Calaprice-Whitty D, Galil K, Salloum W, Zariv A, Jimenez B. Improving clinical trial participant prescreening with artificial intelligence (AI): a comparison of the results of AI-assisted vs standard methods in 3 oncology trials. Ther Innov Regul Sci. 2009;
Casillas A, Pérez A, Oronoz M, Gojenola K, Santiso S. Learning to extract adverse drug reaction events from electronic health records in Spanish. Expert Syst Appl. 2016;61:235–45.
Castillon G, Salvo F, Moride Y. The social impact of suspected adverse drug reactions: an analysis of the Canada vigilance spontaneous reporting database. Drug Saf. 2019;42:27–34.
Center for Drug Evaluation and Research. Reports received and reports entered into FAERS by year. FDA; 2019.
Center for Drug Evaluation and Research. Sociodemographic characteristics of adverse event reporting. FDA; 2024.
Cevasco KE, Morrison Brown RE, Woldeselassie R, Kaplan S. Patient engagement with conversational agents in health applications 2016-2022: a systematic review and meta-analysis. J Med Syst. 2024;48:40.
Chapman WW, et al. Overcoming barriers to NLP for clinical text: the role of shared tasks and the need for additional creative solutions. J Am Med Inform Assoc. 2011;18:540–3.
Chauvet R, et al. Classification of the severity of adverse drugs reactions. Stud Health Technol Inform. 2020;270:1227–8.
Chen J, et al. Detecting hypoglycemia incidents reported in patients' secure messages: using cost-sensitive learning and oversampling to reduce data imbalance. J Med Internet Res. 2019;21:e11990.
Chen J, et al. TrialBench: multi-modal artificial intelligence-ready clinical trial datasets; 2024a. https://doi.org/10.48550/ARXIV.2407.00631.
Chen Z, Wang Q, Sun Y, Cai H, Lu X. Chat-ePRO: development and pilot study of an electronic patient-reported outcomes system based on ChatGPT. J Biomed Inform. 2024b;154:104651.

Chen Y, et al. Risk assessment and prediction of nosocomial infections based on surveillance data using machine learning methods. BMC Public Health. 2024c;24:1780.
Cherkas Y, Ide J, van Stekelenborg J. Leveraging machine learning to facilitate individual case causality assessment of adverse drug reactions. Drug Saf. 2022;45:571–82.
Chu W-M, et al. A model for predicting fall risks of hospitalized elderly in Taiwan-a machine learning approach based on both electronic health records and comprehensive geriatric assessment. Front Med. 2022;9
Clarke A, Deeks JJ, Shakir SAW. An assessment of the publicly disseminated evidence of safety used in decisions to withdraw medicinal products from the UK and US markets. Drug Saf. 2006;29:175–81.
Comfort S, et al. Sorting through the safety data haystack: using machine learning to identify individual case safety reports in social-digital media. Drug Saf. 2018;41:579–90.
Correia Pinheiro L, Durand J, Dogné J. An application of machine learning in pharmacovigilance: estimating likely patient genotype from phenotypical manifestations of fluoropyrimidine toxicity. Clin Pharmacol Ther. 2020;107:944–7.
Costa M, Moody GB, Henry I, Goldberger AL. PhysioNet: an NIH research resource for complex signals. J Electrocardiol. 2003;36:139–44.
Craveiro NS, Lopes BS, Tomás L, Almeida SF. Drug withdrawal due to safety: a review of the data supporting withdrawal decision. Curr Drug Saf. 2020;15:4–12.
Cro S, Morris TP, Kahan BC, Cornelius VR, Carpenter JR. A four-step strategy for handling missing outcome data in randomised trials affected by a pandemic. BMC Med Res Methodol. 2020;20:208.
Cross JL, Choma MA, Onofrey JA. Bias in medical AI: Implications for clinical decision-making. PLOS Digit Health. 2024;3:e0000651.
Daluwatte C, Schotland P, Strauss DG, Burkhart KK, Racz R. Predicting potential adverse events using safety data from marketed drugs. BMC Bioinformatics. 2020;21:163.
Dandala B, Joopudi V, Devarakonda M. Adverse drug events detection in clinical notes by jointly modeling entities and relations using neural networks. Drug Saf. 2019;42:135–46.
Datta S, et al. AutoCriteria: a generalizable clinical trial eligibility criteria extraction system powered by large language models. J Am Med Inform Assoc. 2024;31:375–85.
Deist TM, et al. Distributed learning on 20 000+ lung cancer patients – the personal health train. Radiother Oncol. 2020;144:189–200.
Devarakonda MV, Mohanty S, Sunkishala RR, Mallampalli N, Liu X. Clinical trial recommendations using semantics-based inductive inference and knowledge graph embeddings. J Biomed Inform. 2024;154:104627.
Dewulf P, Stock M, De Baets B. Cold-start problems in data-driven prediction of drug-drug interaction effects. Pharm Basel Switz. 2021;14:429.
Dey S, Luo H, Fokoue A, Hu J, Zhang P. Predicting adverse drug reactions through interpretable deep learning framework. BMC Bioinformatics. 2018;19:476.
Dong W, et al. Generative adversarial networks for imputing missing data for big data clinical research. BMC Med Res Methodol. 2021;21:78.
Dowden H, Munro J. Trends in clinical success rates and therapeutic focus. Nat Rev Drug Discov. 2019;18:495–6.
Durrieu G, et al. Completeness of spontaneous adverse drug reaction reports sent by general practitioners to a regional pharmacovigilance centre: a descriptive study. Drug Saf. 2016;39:1189–95.
Eke CI, Shuib L. The role of explainability and transparency in fostering trust in AI healthcare systems: a systematic literature review, open issues and potential solutions. Neural Comput Applic. 2024;
Evans HP, et al. Automated classification of primary care patient safety incident report content and severity using supervised machine learning (ML) approaches. Health Informatics J. 2020;26:3123–39.
Farnoush A, Sedighi-Maman Z, Rasoolian B, Heath JJ, Fallah B. Prediction of adverse drug reactions using demographic and non-clinical drug characteristics in FAERS data. Sci Rep. 2024;14:23636.

FDA Adverse Events Reporting System (FAERS) Public Dashboard. https://fis.fda.gov/sense/app/95239e26-e0be-42d9-a960-9a5f7f1c25ee/sheet/7a47a261-d58b-4203-a8aa-6d3021737452/state/analysis (2024).
Fenza G, Orciuoli F, Peduto A, Postiglione A. Healthcare conversational agents: Chatbot for improving patient-reported outcomes. In: Barolli L, editor. Advanced information networking and applications, vol. 661. Cham: Springer; 2023. p. 137–48.
Fia W. Reflection paper on the use of Artificial Intelligence (AI) in the medicinal product lifecycle (2024).
Fogel DB. Factors associated with clinical trials that fail and opportunities for improving the likelihood of success: A review. Contemp Clin Trials Commun. 2018;11:156–64.
Frost J, Okun S, Vaughan T, Heywood J, Wicks P. Patient-reported outcomes as a source of evidence in off-label prescribing: analysis of data from patients like me. J Med Internet Res. 2011;13:e6.
Fu S, et al. A hybrid model to identify fall occurrence from electronic health records. Int J Med Inform. 2022;162:104736.
Fukuto K, Takagi T, Tian Y-S. Predicting the side effects of drugs using matrix factorization on spontaneous reporting database. Sci Rep. 2021;11:23942.
Fusaroli M, et al. The REporting of a disproportionality analysis for DrUg safety signal detection using individual case safety reports in PharmacoVigilance (READUS-PV): explanation and elaboration. Drug Saf. 2024;47:585–99.
Gaebel J, Kolter T, Arlt F, Denecke K. Extraction of adverse events from clinical documents to support decision making using semantic preprocessing. Stud Health Technol Inform. 2015;216:1030.
Galeano D, Paccanaro A. A recommender system approach for predicting drug side effects. In: 2018 International Joint Conference on Neural Networks (IJCNN); 2018. p. 1–8. https://doi.org/10.1109/IJCNN.2018.8489025.
Galeano D, Li S, Gerstein M, Paccanaro A. Predicting the frequencies of drug side effects. Nat Commun. 2020;11:4575.
García-Gómez JM, Blanes-Selva V, Cenzano JC, Cebolla-Cornejo J, Doñate-Martínez A. functional requirements to mitigate the risk of harm to patients from Artificial Intelligence in healthcare. arXiv preprint arXiv:2309.10424 (2023).
Gerke S, Babic B, Evgeniou T, Cohen IG. The need for a system view to regulate artificial intelligence/machine learning-based software as medical device. NPJ Digit Med. 2020;3:1–4.
Getz KA, Campo RA, Kaitin KI. Variability in protocol design complexity by phase and therapeutic area. Drug Inf J. 2011;45:413–20.
Ghim J-L, Ahn S. Transforming clinical trials: the emerging roles of large language models. Transl Clin Pharmacol. 2023;31:131.
Ghosh P, Dewanji A. Effect of reporting bias in the analysis of spontaneous reporting data. Pharm Stat. 2015;14:20–5.
Gilbert A, Sebag-Montefiore D, Davidson S, Velikova G. Use of patient-reported outcomes to measure symptoms and health related quality of life in the clinic. Gynecol Oncol. 2015;136:429–39.
Gray M, Xu J, Tong W, Wu L. Classifying free texts into predefined sections using AI in regulatory documents: a case study with drug labeling documents. Chem Res Toxicol. 2023;36:1290–9.
Guidance. for industry good pharmacovigilance practices and pharmacoepidemiologic assessment.
Ham JVD. Toward a better understanding of "cybersecurity". Digit Threats Res Pract. 2021;2:1–3.
Hartung T. Toxicology for the twenty-first century. Nature. 2009;460:208–12.
Hashimoto DA, Rosman G, Rus D, Meireles OR. Artificial Intelligence in surgery: promises and perils. Ann Surg. 2018;268:70–6.
Hashimoto DA, et al. Computer vision analysis of intraoperative video: automated recognition of operative steps in laparoscopic sleeve gastrectomy. Ann Surg. 2019;270:414–21.
Hassan AM, et al. Artificial Intelligence and machine learning in prediction of surgical complications: current state, applications, and implications. Am Surg. 2023;89:25–30.
Hauben M. Artificial Intelligence and data mining for the pharmacovigilance of drug–drug interactions. Clin Ther. 2023;45:117–33.

Hazell L, Shakir SAW. Under-reporting of adverse drug reactions. Drug Saf. 2006;29:385–96.
Heinrich J. Adverse events: surveillance systems for adverse events and medical errors (2000).
Hobensack M, Song J, Scharp D, Bowles KH, Topaz M. Machine learning applied to electronic health record data in home healthcare: a scoping review. Int J Med Inform. 2023;170:104978.
Hoffman S, Podgurski A. The use and misuse of biomedical data: is bigger really better? Am J Law Med. 2013;39:497–538.
Howcroft J, Kofman J, Lemaire ED. Prospective fall-risk prediction models for older adults based on wearable sensors. IEEE Trans Neural Syst Rehabil Eng. 2017;25:1812–20.
Hügle M, Kalweit G, Hügle T, Boedecker J. A dynamic deep neural network for multimodal clinical data analysis. In: Shaban-Nejad A, Michalowski M, Buckeridge DL, editors. Explainable AI in healthcare and medicine: building a culture of transparency and accountability. Cham: Springer; 2021. p. 79–92. https://doi.org/10.1007/978-3-030-53352-6_8.
Hutson M. How AI is being used to accelerate clinical trials. Nature. 2024;627:S2–5.
Jacobsen JOB, et al. The GA4GH Phenopacket schema defines a computable representation of clinical data. Nat Biotechnol. 2022;40:817–20.
Jager W. The importance of clinical trials in advancing medical research. Ann Clin Trials Vaccines Res. 2023;13:93–8.
Jang SJ, Rosenstadt J, Lee E, Kunze KN. Artificial Intelligence for clinically meaningful outcome prediction in orthopedic research: current applications and limitations. Curr Rev Musculoskelet Med. 2024;17:185–206.
Jobin A, Ienca M, Vayena E. The global landscape of AI ethics guidelines. Nat Mach Intell. 2019;1:389–99.
Johnson AEW, et al. MIMIC-IV, a freely accessible electronic health record dataset. Sci Data. 2023;10:1.
Juhn Y, Liu H. Artificial intelligence approaches using natural language processing to advance EHR-based clinical research. J Allergy Clin Immunol. 2020;145:463–9.
Kadariya D, et al. kBot: knowledge-enabled personalized Chatbot for asthma self-management. In: 2019 IEEE international conference on smart computing (SMARTCOMP). Washington, DC: IEEE; 2019. p. 138–43. https://doi.org/10.1109/SMARTCOMP.2019.00043.
Kairouz P, et al. Advances and open problems in federated learning. Found Trends® Mach Learn. 2021;14:1–210.
Kaissis GA, Makowski MR, Rückert D, Braren RF. Secure, privacy-preserving and federated machine learning in medical imaging. Nat Mach Intell. 2020;2:305–11.
Kelly CJ, Karthikesalingam A, Suleyman M, Corrado G, King D. Key challenges for delivering clinical impact with artificial intelligence. BMC Med. 2019;17:195.
Khalili H, Soudbakhsh A, Talasaz AH. Severe hepatotoxicity and probable hepatorenal syndrome associated with sulfadiazine. Am J Health Syst Pharm. 2011;68:888–92.
Kiguba R, et al. Navigating duplication in pharmacovigilance databases: a scoping review. BMJ Open. 2024;14:e081990.
Kim ES, et al. Modernizing eligibility criteria for molecularly driven trials. J Clin Oncol. 2015;33:2815–20.
Kim H, Park M, Lee I, Nam H. BayeshERG: a robust, reliable and interpretable deep learning model for predicting hERG channel blockers. Brief Bioinform. 2022;23:bbac211.
Kompa B, et al. Artificial intelligence based on machine learning in pharmacovigilance: a scoping review. Drug Saf. 2022;45:477–91.
Kovačević A, Dehghan A, Filannino M, Keane JA, Nenadic G. Combining rules and machine learning for extraction of temporal expressions and events from clinical narratives. J Am Med Inform Assoc. 2013;20:859–66.
Kreimeyer K, et al. Feature engineering and machine learning for causality assessment in pharmacovigilance: Lessons learned from application to the FDA Adverse Event Reporting System. Comput Biol Med. 2021;135:104517.
Krix S, et al. MultiGML: multimodal graph machine learning for prediction of adverse drug events. Heliyon. 2023;9:e19441.

Lai J, Forney L, Brinton DL, Simpson KN. Drivers of start-up delays in global randomized clinical trials. Ther Innov Regul Sci. 2021;55:212–27.

Langer M, Baum K, Schlicker N. Effective human oversight of AI-based systems: a signal detection perspective on the detection of inaccurate and unfair outputs. Mind Mach. 2025;35:1.

Leaman R, et al. Towards internet-age pharmacovigilance: extracting adverse drug reactions from user posts to health-related social networks. In: Proceedings of the 2010 workshop on biomedical natural language processing. Association for Computational Linguistics; 2010. p. 117–25.

Lee J-Y, et al. The use of social media in detecting drug safety-related new black box warnings, labeling changes, or withdrawals: scoping review. JMIR Public Health Surveill. 2021;7:e30137.

Lemke AA, et al. Community engagement in biobanking: experiences from the eMERGE Network. Genomics Soc Policy. 2010;6:50.

Leviton A, Loddenkemper T. Design, implementation, and inferential issues associated with clinical trials that rely on data in electronic medical records: a narrative review. BMC Med Res Methodol. 2023;23:271.

Lewis P, et al. Retrieval-augemented generation for knowledge-intensive NLP tasks. In: NeurIPS proceedings; 2020.

Li S, et al. Federated and distributed learning applications for electronic health records and structured medical data: a scoping review. J Am Med Inform Assoc. 2023;30:2041–9.

Liang L, et al. Artificial Intelligence-based pharmacovigilance in the setting of limited resources. Drug Saf. 2022;45:511–9.

Lim S, Lee K, Kang J. Drug drug interaction extraction from the literature using a recursive neural network. PLoS One. 2018;13:e0190926.

Lin NU, et al. Modernizing clinical trial eligibility criteria: recommendations of the American Society of Clinical Oncology–friends of cancer research brain metastases working group. J Clin Oncol. 2017;35:3760–73.

Linder JA, et al. Secondary use of electronic health record data: spontaneous triggered adverse drug event reporting. Pharmacoepidemiol Drug Saf. 2010;19:1211–5.

Liu R, et al. Evaluating eligibility criteria of oncology trials using real-world data and AI. Nature. 2021;592:629–33.

Liu RX, et al. Machine learning diagnosis of small-bowel crohn disease using T2-weighted MRI radiomic and clinical data. Am J Roentgenol. 2024;222:e2329812.

Lockhart TE, et al. Prediction of fall risk among community-dwelling older adults using a wearable system. Sci Rep. 2021;11:20976.

Luboz V, et al. Personalized modeling for real-time pressure ulcer prevention in sitting posture. J Tissue Viability. 2018;27:54–8.

Luo Y. Evaluating the state of the art in missing data imputation for clinical data. Brief Bioinform. 2022;23:bbab489.

Ma X, Long L, Moon S, Adamson BJS, Baxi SS. Comparison of population characteristics in real-world clinical oncology databases in the US: Flatiron Health, SEER, and NPCR; 2020. https://doi.org/10.1101/2020.03.16.20037143.

Machado J, Rodrigues C, Sousa R, Gomes LM. Drug–drug interaction extraction-based system: an natural language processing approach. Expert Syst. 2023;e13303 https://doi.org/10.1111/exsy.13303.

Mahaux O, et al. Identifying safety subgroups at risk: assessing the agreement between statistical alerting and patient subgroup risk. Drug Saf. 2023;46:601–14.

Malin BA, Emam KE, O'Keefe CM. Biomedical data privacy: problems, perspectives, and recent advances. J Am Med Inform Assoc. 2013;20:2–6.

Marella WM, Sparnon E, Finley E. Screening electronic health record-related patient safety reports using machine learning. J Patient Saf. 2017;13:31–6.

Markey N, El-Mansouri I, Rensonnet G, van Langen C, Meier C. From RAGs to riches: using large language models to write documents for clinical trials; 2024. https://doi.org/10.48550/ARXIV.2402.16406.

Martin GL, et al. Validation of Artificial Intelligence to support the automatic coding of patient adverse drug reaction reports, using nationwide pharmacovigilance data. Drug Saf. 2022;45:535–48.
McCoach CE, Berge EM, Lu X, Barón AE, Camidge DR. A brief report of the status of central nervous system metastasis enrollment criteria for advanced non–small cell lung cancer clinical trials: a review of the ClinicalTrials.gov trial registry. J Thorac Oncol. 2016;11:407–13.
McLennan S, et al. Embedded ethics: a proposal for integrating ethics into the development of medical AI. BMC Med Ethics. 2022;23:6.
Mello MM, Lieou V, Goodman SN. Clinical trial participants' views of the risks and benefits of data sharing. N Engl J Med. 2018;378:2202–11.
Ménard T, Barmaz Y, Koneswarakantha B, Bowling R, Popko L. Enabling data-driven clinical quality assurance: predicting adverse event reporting in clinical trials using machine learning. Drug Saf. 2019;42:1045–53.
Merath K, et al. Use of machine learning for prediction of patient risk of postoperative complications after liver, pancreatic, and colorectal surgery. J Gastrointest Surg. 2020;24:1843–51.
Meyboom RH, et al. Principles of signal detection in pharmacovigilance. Drug Saf. 1997;16:355–65.
Miranda O, et al. Prediction of adverse events risk in patients with comorbid post-traumatic stress disorder and alcohol use disorder using electronic medical records by deep learning models. Drug Alcohol Depend. 2024;255:111066.
Modi S, Kasmiran KA, Mohd Sharef N, Sharum MY. Extracting adverse drug events from clinical notes: a systematic review of approaches used. J Biomed Inform. 2024;151:104603.
Montes-Grajales D, Garcia-Serna R, Mestres J. Impact of the COVID-19 pandemic on the spontaneous reporting and signal detection of adverse drug events. Sci Rep. 2023;13:18817.
Moore N, et al. Biases affecting the proportional reporting ratio (PRR) in spontaneous reports pharmacovigilance databases: the example of sertindole. Pharmacoepidemiol Drug Saf. 2003;12:271–81.
Mullie L, et al. CODA: an open-source platform for federated analysis and machine learning on distributed healthcare data. J Am Med Inform Assoc. 2024;31:651–65.
Naidu RP. Causality assessment: a brief insight into practices in pharmaceutical industry. Perspect Clin Res. 2013;4:233–6.
National Center for Toxicological Research. AN FDA Artificial Intelligence (AI) program for toxicology. NCTR; 2024a.
National Center for Toxicological Research. BERTox initiative. NCTR; 2024b.
Neyarapally GA, et al. Description and validation of a novel AI tool, LabelComp, for the identification of adverse event changes in FDA labeling. Drug Saf. 2024;47:1265–74.
Nikfarjam A, Sarker A, O'Connor K, Ginn R, Gonzalez G. Pharmacovigilance from social media: mining adverse drug reaction mentions using sequence labeling with word embedding cluster features. J Am Med Inform Assoc. 2015;22:671–81.
Norén GN, Orre R, Bate A, Edwards IR. Duplicate detection in adverse drug reaction surveillance. Data Min Knowl Disc. 2007;14:305–28.
Nowrozy R, Ahmed K, Kayes ASM, Wang H, McIntosh TR. Privacy preservation of electronic health records in the modern era: a systematic survey. ACM Comput Surv. 2024;56:1–37.
Orre R, Lansner A, Bate A, Lindquist M. Bayesian neural networks with confidence estimations applied to data mining. Comput Stat Data Anal. 2000;34:473–93.
Pati S, et al. Privacy preservation for federated learning in health care. Patterns. 2024;5:100974.
Pavani A, et al. Artificial neural network-based pharmacogenomic algorithm for warfarin dose optimization. Pharmacogenomics. 2016;17:121–31.
Pazan F, Wehling M. Polypharmacy in older adults: a narrative review of definitions, epidemiology and consequences. Eur Geriatr Med. 2021;12:443–52.
Pharmacovigilance. https://www.who.int/teams/regulation-prequalification/regulation-and-safety/pharmacovigilance.

Pielka M, Schneider T, Terheyden J, Sifa R. [Vision Paper] PRObot: enhancing patient-reported outcome measures for diabetic retinopathy using Chatbots and generative AI; 2024. https://doi.org/10.48550/ARXIV.2411.02973.

Planès S, Villier C, Mallaret M. The nocebo effect of drugs. Pharmacol Res Perspect. 2016;4:e00208.

Poleksic A, Xie L. Predicting serious rare adverse reactions of novel chemicals. Bioinformatics. 2018;34:2835–42.

Prayitno, et al. A systematic review of federated learning in the healthcare area: from the perspective of data properties and applications. Appl Sci. 2021;11:11191.

Rajkomar A, et al. Scalable and accurate deep learning with electronic health records. NPJ Digit Med. 2018;1:1–10.

Ramakrishnaiah Y, Macesic N, Webb GI, Peleg AY, Tyagi S. EHR-QC: a streamlined pipeline for automated electronic health records standardisation and preprocessing to predict clinical outcomes. J Biomed Inform. 2023;147:104509.

Rees S, Mian S, Grabowski N. Using social media in safety signal management: is it reliable? Ther Adv Drug Saf. 2018;9:591–9.

Rieke N, et al. The future of digital health with federated learning. NPJ Digit Med. 2020;3:1–7.

Riner AN, et al. Call to action: overcoming enrollment disparities in cancer clinical trials with modernized eligibility criteria. JNCI Cancer Spectr. 2023;7:pkad009.

Rodemund N, Wernly B, Jung C, Cozowicz C, Koköfer A. Harnessing big data in critical care: exploring a new European dataset. Sci Data. 2024;11:320.

Rodwin MA. Patient data: property, privacy & the public interest. Am J Law Med. 2010;36:586–618.

Rojc M, Ariöz U, Šafran V, Mlakar I. Multilingual Chatbots to collect patient-reported outcomes. In: Babulak E, editor. Chatbots – the AI-driven front-line services for customers. IntechOpen; 2023. https://doi.org/10.5772/intechopen.111865.

Ryu JY, Lee MY, Lee JH, Lee BH, Oh K-S. DeepHIT: a deep learning framework for prediction of hERG-induced cardiotoxicity. Bioinformatics. 2020;36:3049–55.

Salas M, et al. the use of artificial intelligence in pharmacovigilance: a systematic review of the literature. Pharm Med. 2022;36:295–306.

Saldanha OL, et al. Swarm learning for decentralized artificial intelligence in cancer histopathology. Nat Med. 2022;28:1232–9.

Salvo F, et al. Will the future of pharmacovigilance be more automated? Expert Opin Drug Saf. 2023;22:541–8.

Sauer AB, et al. Underreporting and underrepresentation of race and ethnicity in head and neck cancer trials, 2010-2020: a systematic review. JAMA Otolaryngol Head Neck Surg. 2022;148:662–9.

Saviauk T, et al. Electronic nose in the detection of wound infection bacteria from bacterial cultures: a proof-of-principle study. Eur Surg Res. 2018;59:1–11.

Schultze JL, Büttner M, Becker M. Swarm immunology: harnessing Blockchain technology and artificial intelligence in human immunology. Nat Rev Immunol. 2022;22:401–3.

Seal S, et al. Improved detection of drug-induced liver injury by integrating predicted in vivo and in vitro data. Chem Res Toxicol. 2024;37:1290–305.

Shah IA, Mishra S. Artificial intelligence in advancing occupational health and safety: an encapsulation of developments. J Occup Health. 2024;66:uiad017.

Shakir SAW, Layton D. Causal association in pharmacovigilance and pharmacoepidemiology. Drug Saf. 2002;25:467–71.

Shamim MA, Shamim MA, Arora P, Dwivedi P. Artificial intelligence and big data for pharmacovigilance and patient safety. J Med Surg Public Health. 2024;3:100139.

Skentzos S, Shubina M, Plutzky J, Turchin A. Structured vs. unstructured: factors affecting adverse drug reaction documentation in an EMR repository. AMIA Annu Symp Proc. 2011;2011:1270–9.

Smith Z, Bilke R, Pretorius S, Getz K. Protocol design variables highly correlated with, and predictive of, clinical trial performance. Ther Innov Regul Sci. 2022;56:333–45.

Soltan AAS, et al. A scalable federated learning solution for secondary care using low-cost microcomputing: privacy-preserving development and evaluation of a COVID-19 screening test in UK hospitals. Lancet Digit Health. 2024;6:e93–e104.

Sosinsky A, et al. Insights for precision oncology from the integration of genomic and clinical data of 13,880 tumors from the 100,000 Genomes Cancer Programme. Nat Med. 2024;30:279–89.

Stevens G, Dolley S, Mogg R, Connor JT. A template for the authoring of statistical analysis plans. Contemp Clin Trials Commun. 2023;34:101100.

Sun D, Gao W, Hu H, Zhou S. Why 90% of clinical drug development fails and how to improve it? Acta Pharm Sin B. 2022;12:3049–62.

Tanaka Y, et al. OnSIDES (ON-label SIDE effectS resource) database: extracting adverse drug events from drug labels using natural language processing models; 2024. https://doi.org/10.1101/2024.03.22.24304724.

Tatonetti NP, Ye PP, Daneshjou R, Altman RB. Data-driven prediction of drug effects and interactions. Sci Transl Med. 2012;4:125ra31.

Tertulino R, Antunes N, Morais H. Privacy in electronic health records: a systematic mapping study. J Public Health. 2024;32:435–54.

Tong W, Baran SW. 50 shades of AI in regulatory science. Drug Discov Today. 2024;29:104058.

Tran TTV, Surya Wibowo A, Tayara H, Chong KT. Artificial Intelligence in drug toxicity prediction: recent advances, challenges, and future perspectives. J Chem Inf Model. 2023;63:2628–43.

Triantafyllidis A, et al. Smartwatch interventions in healthcare: a systematic review of the literature. Int J Med Inform. 2024;190:105560.

Tucker K, et al. Protecting patient privacy when sharing patient-level data from clinical trials. BMC Med Res Methodol. 2016;16:77.

Uzuner Ö, South BR, Shen S, DuVall SL. 2010 i2b2/VA challenge on concepts, assertions, and relations in clinical text. J Am Med Inform Assoc. 2011;18:552–6.

Vollmer S, et al. Machine learning and artificial intelligence research for patient benefit: 20 critical questions on transparency, replicability, ethics, and effectiveness. BMJ. 2020;l6927 https://doi.org/10.1136/bmj.l6927.

Wang Z, Sun J. Trial2Vec: zero-shot clinical trial document similarity search using self-supervision; 2022. https://doi.org/10.48550/ARXIV.2206.14719.

Wang X, Xu X, Tong W, Roberts R, Liu Z. InferBERT: a transformer-based causal inference framework for enhancing pharmacovigilance. Front Artif Intell. 2021;4

Wang X, Xu X, Tong W, Liu Q, Liu Z. DeepCausality: a general AI-powered causal inference framework for free text: a case study of LiverTox. Front Artif Intell. 2022;5

Wang Z, Xiao C, Sun J. AutoTrial: prompting language models for clinical trial design; 2023. https://doi.org/10.48550/ARXIV.2305.11366.

Wei R, et al. Linking brain structure, cognition, and sleep: insights from clinical data. Sleep. 2024;47:zsad294.

White R, Peng T, Sripitak P, Rosenberg Johansen A, Snyder M. CliniDigest: a case study in large language model based large-scale summarization of clinical trial descriptions. In: Proceedings of the 2023 ACM conference on information technology for social good. Lisbon: ACM; 2023. p. 396–402. https://doi.org/10.1145/3582515.3609559.

Whitten P, Dutta MJ, Carpenter S, Bodie GD. News media and health care providers at the crossroads of medical adverse events. In: Henriksen K, Battles JB, Keyes MA, Grady ML, editors. Advances in patient safety: new directions and alternative approaches, Vol. 1: Assessment. Rockville: Agency for Healthcare Research and Quality; 2008.

Wouters OJ, McKee M, Luyten J. Estimated research and development investment needed to bring a new medicine to market, 2009-2018. JAMA. 2020;323:844–53.

Wu L, et al. Development of benchmark datasets for text mining and sentiment analysis to accelerate regulatory literature review. Regul Toxicol Pharmacol. 2023a;137:105287.

Wu L, et al. RxBERT: enhancing drug labeling text mining and analysis with AI language modeling. Exp Biol Med. 2023b;248:1937–43.

Xie W, Ji M, Huang R, Hao T, Chow C-Y. Predicting risks of machine translations of public health resources by developing interpretable machine learning classifiers. Int J Environ Res Public Health. 2021;18:8789.
Xu X, et al. DSGAT: predicting frequencies of drug side effects by graph attention networks. Brief Bioinform. 2022;23:bbab586.
Xue R, et al. Prediction of adverse drug reactions by combining biomedical tripartite network and graph representation model. Chem Res Toxicol. 2020;33:202–10.
Yang Q, Liu Y, Chen T, Tong Y. Federated machine learning: concept and applications. ACM Trans Intell Syst Technol. 2019;10:1–19.
Yang J, et al. Development and validation of a deep learning model for detection of allergic reactions using safety event reports across hospitals. JAMA Netw Open. 2020;3:e2022836.
Yang H, et al. Enhancing psychiatric rehabilitation outcomes through a multimodal multitask learning model based on BERT and TabNet: an approach for personalized treatment and improved decision-making. Psychiatry Res. 2024;336:115896.
Yokota S, Endo M, Ohe K. Establishing a classification system for high fall-risk among inpatients using support vector machines. Comput Inform Nurs CIN. 2017;35:408–16.
Youssef A, et al. Organizational factors in clinical data sharing for artificial intelligence in health care. JAMA Netw Open. 2023;6:e2348422.
Zame WR, et al. Machine learning for clinical trials in the era of COVID-19. Stat Biopharm Res. 2020;12:506–17.
Zeng X, et al. PIC, a paediatric-specific intensive care database. Sci Data. 2020;7:14.
Zhang W, et al. Feature-derived graph regularized matrix factorization for predicting drug side effects. Neurocomputing. 2018;287:154–62.
Zhang F, Sun B, Diao X, Zhao W, Shu T. Prediction of adverse drug reactions based on knowledge graph embedding. BMC Med Inform Decis Mak. 2021;21:38.
Zhang A, Xing L, Zou J, Wu JC. Shifting machine learning for healthcare from development to deployment and from models to data. Nat Biomed Eng. 2022a;6:1330–45.
Zhang A, Xing L, Zou J, Wu JC. Shifting machine learning for healthcare from development to deployment and from models to data. Nat Biomed Eng. 2022b;6:1330–45.
Zhao H, et al. Identifying the serious clinical outcomes of adverse reactions to drugs by a multi-task deep learning framework. Commun Biol. 2023;6:1–14.
Zheng Y, et al. Predicting adverse drug reactions of combined medication from heterogeneous pharmacologic databases. BMC Bioinformatics. 2018;19:517.
Zitnik M, Agrawal M, Leskovec J. Modeling polypharmacy side effects with graph convolutional networks. Bioinformatics. 2018;34:i457–66.
Zou M, Barmaz Y, Preovolos M, Popko L, Ménard T. Using statistical modeling for enhanced and flexible pharmacovigilance audit risk assessment and planning. Ther Innov Regul Sci. 2021;55:190–6.

Leveraging Generative AI in Clinical Studies to Improve Efficiency and Quality of Drug Development

Wenlong Tang, Lorenzo Simone, Gaurav Dwivedi, Minnan Xu, Xinxin Zhu, and John W. Y. Chan

Abstract Generative artificial intelligence (AI) is emerging as a transformative technology in pharmaceutical research and development, with the potential to address longstanding challenges in trial design, execution, and documentation. As clinical development becomes increasingly complex and costly, the strategic integration of AI technologies will be crucial accelerating the delivery of new treatments to patients and maintaining a competitive advantage in the industry. We present a brief overview of current generative AI capabilities, including foundation models and emerging techniques such as retrieval-augmented generation and AI agents, and their potential applications in clinical research. The chapter examines how large language models (LLMs) can enhance various aspects of clinical trials, from patient recruitment to data analysis and reporting. Through detailed case studies, we demonstrate how LLMs can improve trial efficiency through improved site selection, enhanced patient recruitment, and accelerated documentation generation, including clinical study reports (CSRs) and patient-friendly version of investigator brochures (IBs). While precise quantification of efficiency gains and return on investment remains challenging due to the nascent state of AI adoption in clinical trials, early experiences suggest the potential for significant operational improvements. Our analysis shows that AI-driven approaches can significantly reduce manual effort while maintaining quality standards, leading to substantial time savings in clinical trial operations. We also address critical considerations for

W. Tang (✉) · G. Dwivedi · M. Xu · J. W. Y. Chan
ShinrAI Center for Artificial Intelligence/Machine Learning, Takeda Pharmaceuticals, Cambridge, MA, USA
e-mail: wenlong.tang@takeda.com; minnan.xu@takeda.com; john.chan@takeda.com

L. Simone
Department of Computer Science, University of Pisa, Pisa, Tuscany, Italy
e-mail: lorenzo.simone@di.unipi.it

X. Zhu
Center for Biomedical Data Science, Yale University, New Haven, CT, USA
e-mail: xinxin.zhu@yale.edu

A. Lavecchia (ed.), *Applied Artificial Intelligence for Drug Discovery*,
https://doi.org/10.1007/978-3-031-98022-0_25

implementing AI in clinical development, including regulatory compliance, data privacy, quality assurance, and the need for human oversight.

The chapter concludes by examining future trends in generative AI applications for clinical trials, the importance of balancing automation with human expertise to enhance, rather than replace, clinical judgment in medical research.

Keywords Generative AI · Clinical trials · Case studies · LLMs · Patient recruitment · Documentation · Multimodal models · Knowledge extraction

1 Introduction

1.1 Evolution of Clinical Trials and the Emerging Role of AI

Clinical trials have evolved into the foundation of contemporary medical research, marking a transition from basic observational techniques to the sophisticated and ethically driven methodologies we recognize today (Nellhaus and Davies 2017). This evolution has not been a linear path but rather a narrative shaped by key milestones in scientific rigor, ethics, and methodological advancements.

The advancement of clinical trials is now crucial for transforming scientific discoveries into viable healthcare solutions, guaranteeing that new treatments and interventions are safe, effective, and advantageous for patients. The origins of clinical trials can be traced back to ancient times, such as the biblical account in the Book of Daniel, which describes a simple dietary comparison experiment. While rudimentary by today's standards, this early trial introduced the idea of using control groups to evaluate interventions (Machin et al. 2006). Centuries later, in 1747, Dr. James Lind's work with sailors suffering from scurvy marked a significant shift. His systematic approach to comparing treatments was one of the first structured attempts to explore medical interventions, underscoring the importance of methodical experimentation (Trohler 2005).

As medical research entered the twentieth century, the introduction of randomization and placebo controls further shaped the field. A significant milestone was the 1943–1944 patulin trial conducted by the Medical Research Council (MRC), which marked the advent of double-blind studies and established critical principles that have since become the foundation of clinical research (Clarke 2006). However, the path has not been without ethical missteps, most infamously exemplified by the Tuskegee syphilis study, which tragically underscored the need for stringent ethical guidelines. This led to the development of the Nuremberg Code and the Declaration of Helsinki, which enshrined the principles of voluntary and informed consent, preserving participants' rights and safety. While drug development has become more systematic and ethical over the past decades, this change has led to complex technical and operational demands on drug development, making developing medicines increasingly difficult.

The complex landscape of drug development, particularly the clinical trial stages, represents a significant bottleneck in the journey to market. These phases consume a large share of research and development budgets, and the high attrition rates are concerning. Fewer than a third of Phase II compounds advance to Phase III, and more than a third of Phase III candidates fail to gain regulatory approval (Hay et al. 2014). This emphasizes the need for more efficient clinical trial processes and faster identification of drug candidates unlikely to succeed (Yamaguchi et al. 2021). These issues are intensified by a lack of robust technical infrastructure, which is essential for managing the complexities of later-phase trials. Challenges such as unreliable adherence control, insufficient patient monitoring, and ineffective endpoint selection, where the chosen endpoint may not align with the disease or trial objectives, significantly increase the costs and risks of trial failures.

AI offers opportunities to help address the pressing challenges in trials and clinical research. Through machine learning frameworks, we can grasp valuable insights from vast datasets, including electronic health records (EHR), medical literatures, and existing trial databases. This approach can aid both scientific and operational aspects of clinical trials contributing to improved decision-making and efficiency gains throughout the development life cycle.

The pharmaceutical value chain and clinical development are on the verge of a technological revolution, with the disruptive influence of AI and generative models acting as main catalysts. Leading this transformation are LLMs, sophisticated AI systems capable of processing and producing human-like text at remarkable scale and precision.

In this chapter, we will first introduce a high-level overview of their functioning to better comprehend their potential to reshape every aspect of the clinical research process, from patient recruitment to data analysis, documentation, and communication. Following this initial summary, we turn our focus to the specific applications of generative AI, exploring how these technologies can enhance site selection, improve patient engagement during recruitment, streamline the creation of essential clinical trial documentation, and facilitate efficient participant screening. We will also investigate the role of these technologies in producing clear and concise clinical study reports, as well as their ability to translate complex medical terminology into patient-friendly language within investigator brochures, thereby fostering a more inclusive and effective clinical trial environment.

1.2 Understanding Generative AI

1.2.1 Overview of Traditional and Generative AI

Machine learning is a wide-ranging area of research within artificial intelligence, grounded in statistical theory, with several subfields covering distinct types of data and challenges. Artificial intelligence itself can be broadly divided into traditional and generative AI.

Traditional AI primarily focuses on learning patterns from labeled data by mapping relationships between inputs and outputs. In this scenario, the main tasks are classification, where input data are grouped into distinct categories, and regression, which predicts continuous values. The field further includes unsupervised learning, which grasps patterns and structures in data without labels, and reinforcement learning, where systems learn through trial and error to improve decision-making.

Generative AI, on the other hand, models the statistical distribution of data, enabling it to generate synthetic data that resemble real-world samples or adapting existing data to achieve specific outcomes. These models are able to synthesize new information, integrate multimodal inputs, and produce richer context-aware information. In clinical trials, this approach can optimize various steps, such as identifying suitable candidates, improving site selection, and streamlining patient screening. It can also automate the extraction of relevant data from sources like publications and medical charts, supporting the generation of clinical trial documentation and enhancing trial design and knowledge extraction.

Figure 1 provides an overview of the key distinctions between traditional AI, focusing on classification and regression in supervised learning, and generative AI for modeling data distributions and generating new samples. It also highlights several key strengths and potential applications of each approach in the primary stages of clinical trials.

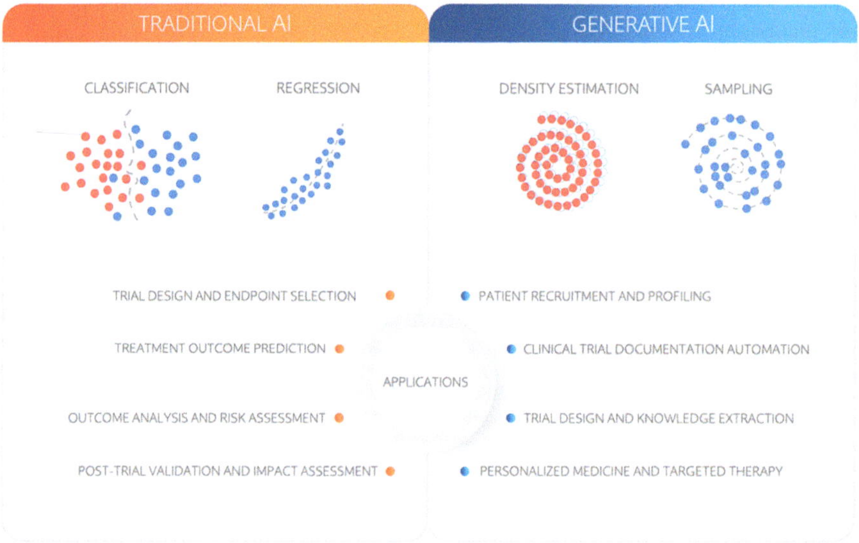

Fig. 1 A comparison of the main tasks in traditional and generative AI, where the former primarily focuses on tasks like classification and regression, and the latter enables density estimation and sampling, offering new capabilities for data generation and analysis

1.2.2 Foundation Models

Neural networks represent one of the main frameworks for traditional machine learning, typically trained on labeled datasets designed for supervised tasks. Although supervised models excel at single, specific objectives, they suffer from a well-known open problem framed as catastrophic forgetting—a phenomenon in which learning new tasks can overwrite previously acquired knowledge, leading to reduced performance on previously learned tasks (Kirkpatrick et al. 2017). The emergence of foundation models (Schneider et al. 2024) marks a significant milestone in the evolution of machine learning systems. Initially gaining traction in the field of computer vision (Radford et al. 2021), these models are characterized by several key advancements: the transformer neural network architecture, parallel scaling of training on inexpensive computing resources, availability of large datasets, and the adoption of self-supervised training methods.

The transformer architecture (Vaswani 2017), introduced in 2017, facilitates efficient processing of sequential data and allows for capturing long-range dependencies. This architecture, combined with the ability to scale training across numerous inexpensive computing units, has enabled researchers to train models on extensive corpora. The importance of parameter size also plays a critical role in the capabilities of these models; emergent properties such as chain-of-thought reasoning and responsiveness to instructions have been observed as model size exceeds two billion parameters (Wei et al. 2022).

In the domain of natural language processing (NLP), notable examples include bidirectional encoder representations from transformers (BERT) (Kenton and Toutanova 2019) and generative pretrained transformer (GPT) (Brown et al. 2020), which utilize self-supervised tasks like masked language modeling and autoregressive prediction. By predicting masked words or upcoming tokens within sentences, these models effectively capture general linguistic patterns from diverse datasets. This self-supervised approach allows foundation models to develop a broad understanding of language, which can later be fine-tuned for various downstream tasks. The fine-tuning technique, in this context, involves adjusting the model for specific tasks or examples, ranging from small tweaks to retraining with new data, which enables it to perform better on particular applications (Karabacak and Margetis 2023).

1.2.3 Evolution of Natural Language Interaction

In computer science, NLP has emerged as a core discipline with the aim of enabling machines to interact and communicate through human language. The objective of this field is to develop techniques and algorithms allowing computers to both understand human language and generate text from an underlying nonlinguistic representation of information, via natural language generation.

Symbolic NLP In the early stages of NLP, symbolic methods and decision-rule-based systems were prevalent, relying on explicitly defined rules and manually crafted knowledge representations to process and understand language. Early systems, such as SHRDLU (Johnson 1988) and ELIZA (Weizenbaum 1966), demonstrated the potential of these approaches, using predefined rules to interact with users in specific contexts, albeit with limited flexibility. During the 1970s and 1980s, symbolic NLP expanded with the development of conceptual ontologies and knowledge representations that structured real-world information for computational purposes (Moens and Steedman 1987; Nirenburg and Carbonell 1987). These rule-based systems focused on parsing, morphology, and semantics, with notable advancements in natural language understanding. Despite their early successes, symbolic approaches were ultimately limited by their inability to scale and adapt to the complexity and variability of natural language, often struggling with ambiguity, context sensitivity, and the need for vast amounts of manually crafted knowledge.

Statistical NLP Over recent decades, the way stakeholders and users interact with NLP technologies has changed significantly, along with advancements in the capabilities and goals of these systems. With the advent and rapid expansion of the internet, and unprecedented access to vast amounts of information, new challenges have arisen in effectively retrieving, interacting with, and querying data. Traditional techniques, commonly used in search engines, relied heavily on rule and keyword-based frameworks that identified specific words or patterns in a query to retrieve the most relevant documents. For instance, traditional information retrieval methods such as TF-IDF (Robertson 2004) and BM25 (Robertson et al. 2009) relied on term-matching paradigms to identify keywords based on their frequency within a document and across a larger corpus. While these methods are effective for simpler tasks like document classification and search, they fail to provide a deeper understanding of language and context. Keyword-based approaches lack a mathematical understanding of semantic meaning and the relationships between terms, which limits their effectiveness in tasks such as resolving word ambiguity, language generation, and performing question answering and summarization.

Neural NLP and Large Language Models The field has recently advanced from symbolic and statistical methods to neural models, ultimately leading to LLMs leveraging sophisticated transformer architectures, refined generative AI techniques, powerful computational resources, and vast datasets to achieve near-human levels of language comprehension and generation. This progression marks a shift in how users interact with language models, evolving from static data retrieval tools into adaptive, interactive systems. Today's LLMs have revolutionized user interaction through flexible and collaborative prompting, offering greater adaptability and supporting a broad range of applications.

Instead of following rigid rules, these models adapt to instructions, allowing users to guide their behavior for customized, context-specific applications across diverse fields. Table 1 highlights key differences across symbolic, statistical, and neural methods, illustrating how recent advancements have enabled more adaptive and context-aware language processing.

Table 1 Comparison of Symbolic, Statistical, and Neural NLP approaches across objectives, data types, flexibility, user interaction, and core methods

	Symbolic NLP	Statistical NLP	Neural NLP (LLMs)
Objective	Process human language by applying explicit rules and logical reasoning for task execution.	Model language using statistical methods, relying on features crafted from large datasets.	Learn complex language patterns with LLMs to enhance language understanding and generation.
Data	Manually crafted rules, lexicons, small, curated datasets with well-defined structures.	Large, labeled, or annotated datasets, typically structured for pattern extraction.	Large, unstructured text datasets, including scientific papers, websites, and news articles.
Flexibility	Rigid, it depends on rule adjustments to handle new tasks or data.	Flexible, it requires feature engineering and retraining for new tasks or data.	Flexible, it allows fine-tuning or prompting for new tasks with minimal retraining.
Interaction	Providing input queries or keywords; limited control over model behavior.	Interacting with predefined models by adjusting inputs or refining queries.	Entering prompts for immediate results, with minimal setup.
Methods	**Rule-based systems** (e.g., SHRDLU (Johnson 1988), ELIZA (Weizenbaum 1966)), **Ontologies and Knowledge Graphs** (e.g., WordNET (Miller 1995), SNOMED-CT (Donnelly et al. 2006), DBpedia (Auer et al. 2007))	**Statistical models** (e.g., n-gram (Brown et al. 1992), Hidden Markov Models (Eddy 1996), latent-semantic analysis (Evangelopoulos 2013)), **information retrieval methods** (e.g., TF-IDF (Robertson 2004), BM25 (Robertson et al. 2009))	**Word embeddings** (e.g., Word2Vec (Mikolov et al. 2013), GloVe (Pennington et al. 2014)) **Large Language models** (e.g., Transformers (Vaswani 2017), BERT (Kenton and Toutanova 2019), GPT (Brown et al. 2020), T5 (Raffel et al. 2020), PaLM (Chowdhery et al. 2023), LaMDA (Thoppilan et al. 2022))

1.2.4 Emerging Techniques and Capabilities of LLMs

An additional benefit of LLMs lies in their emergent abilities, which enhance their adaptability in natural language tasks. These emergent abilities develop as the model scales, with their growing complexity enabling it to generalize beyond precise instructions, produce contextually relevant responses, and even exhibit creative problem-solving. Unlike traditional models, these aptitudes develop without human intervention or the need to program specific rules; instead, they emerge autonomously from the model's exposure to extensive and diverse data.

Prompting Techniques for LLM Optimization Prompting, as a strategic technique, has recently emerged as a primary tool for assessing and leveraging the emergent capabilities of LLMs, with recent studies demonstrating their significant zero-shot reasoning skills (Kojima et al. 2022). Zero-shot prompting, which involves providing the model with task instructions without any examples, evaluates its ability to generate responses based on its existing knowledge and understanding of the task. In literature, the usage of straightforward instructional prompts has resulted in notable performance improvements across various benchmark tasks (Romera-Paredes and Torr 2015; Xian et al. 2017).

Alternatively, few-shot prompting introduces task-specific examples along with instructions, enabling in-context learning, leveraging both the model's existing knowledge and contextual examples. This approach allows the model to adapt more effectively to specific tasks with limited input, enhancing its ability to handle new scenarios. However, challenges such as instability in example variations and prompt formats necessitate careful prompt construction. Recent research (Ma et al. 2023) introduced several metrics to evaluate predictive bias, showing that better-designed prompts lead to improved performance in various downstream tasks. Few-shot prompting serves as a versatile approach, bridging the gap between zero-shot reasoning and full training for flexible task adaptation.

Retrieval-Augmented Generation (RAG) and Fine-Tuning RAG and fine-tuning are two distinct techniques for enhancing LLMs, each offering unique advantages. RAG enhances LLMs by combining parametric memory (stored within the model's parameters) with nonparametric memory (external data sources), allowing the model to better handle knowledge-intensive tasks and access relevant information beyond its initial training data.

Technically, it combines a pretrained sequence-to-sequence model with a dense vector index, like Wikipedia. This index is accessed through a neural retriever, allowing it to generate more specific, diverse, and factually accurate language (Lewis et al. 2020). Studies show that this technique excels in open-domain question-answering tasks, leveraging external data sources to support generative capabilities (Fan et al. 2024). One of the advantages of RAG is its ability to handle large-scale knowledge tasks with relatively lower initial cost for generating embeddings, making it a viable solution for many applications. While RAG effectively incorporates external data to enhance language models, it can result in larger input sizes and more verbose outputs, which may complicate information extraction in certain contexts and may be less effective than fine-tuning in tasks that require the acquisition of new, domain-specific knowledge or expertise.

Fine-tuning, on the other hand, involves various approaches for incorporating new knowledge, ranging from calibrating model parameters to introducing task-specific examples, offering flexibility in how new information is integrated into the model. This technique is particularly advantageous for teaching new skills or adapting to specialized knowledge, ensuring that the model is more aligned with specific requirements. Through this approach, LLMs can better adapt to different domains and tasks, driving specialization in areas such as coding, recommendation systems, and medical applications (Liu et al. 2024a, b; Lin et al. 2024). Techniques like multitask learning, data pruning, and parameter-efficient approaches improve accuracy, efficiency, and context relevance, optimizing fine-tuned models for complex tasks and resource constraints.

LLM Agents in the Architecture of Autonomous Systems LLM-based agents are sophisticated AI systems that integrate LLMs with essential modules such as planning, memory, and tool usage to perform complex, multistep tasks (Zhao et al. 2024). In this architecture, the LLM typically acts as the central controller, directing the sequence of operations required to accomplish a task or meet a user request. These agents extend the capabilities of LLMs beyond text generation, enabling them to execute tasks that require reasoning, decision-making, and the management of contextual information. Through the integration of memory, planning, and external tools, the agents can address complex tasks involving dynamic changes, long-term memory, and the integration of external sources of information.

In practice, LLM-based agents can enhance task execution by interacting with different modules, accessing real-time information, and dynamically adjusting their operations based on evolving inputs (Kim et al. 2024). This makes them highly adaptable for a wide range of applications, including those that require specialized knowledge or multistep workflows (Wang et al. 2023; Wang et al. 2024; Huang et al. 2022).

2 Applications of Generative AI in Clinical Trials

Clinical development programs and post-approval lifecycle management are complex and full of challenges. Generative AI can help alleviate several key pain points through applications illustrated in Fig. 2. This section describes some such challenges and provides a high-level view of how AI could help. Selected examples are described in further detail in the next section.

Documentation is a critical component of drug development for several reasons including ensuring scientific integrity, complying with regulatory requirements, facilitating effective communication, and safeguarding patient safety and rights. Its significance extends across the entire drug lifecycle. In clinical development, in particular, key documents such as the informed consent form (ICF), investigator brochure (IB), trial protocols, case report forms (CRF), clinical study reports (CSR), and briefing books for regulatory interactions, among others, are regularly created, refined, summarized, and archived. Generative AI tools powered by LLMs can greatly enhance the processes of generating, refining, and summarizing documents. Recent advancements in leveraging LLMs to automate CSR writing have started to showcase the substantial productivity gains achievable through the use of generative AI.

Monitoring ongoing clinical trials to ensure data quality and compliance with study protocols and regulatory standards is another key aspect that is essential throughout the drug life cycle. Failures in monitoring and compliance could compromise patient welfare, affect scientific integrity of studies, and could result in deviations from regulatory standards. Generative AI has the potential to improve monitoring and compliance in a few different ways. AI tools could be used to review trial protocols to flag inconsistencies in language or study design and identify

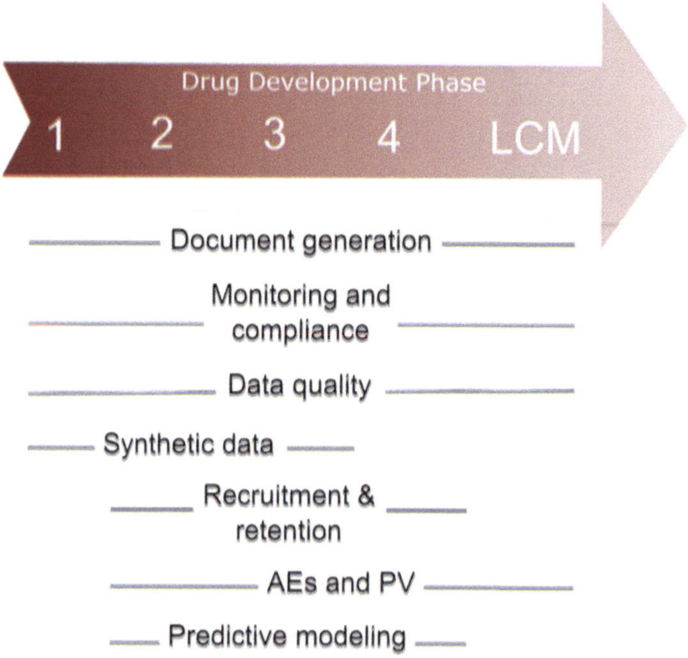

Fig. 2 Applications of generative AI at various stages of clinical development. Most applications span multiple phases, which can be used as a guide to prioritize focus areas for an organization. The applications shown here depend on LLMs as well as other types of generative models, such as time series models, generative adversarial networks, and others. *LCM* lice cycle management, *AE* adverse event, *PV* pharmacovigilance

potential deviations from regulatory standards. AI chatbots with RAG could be used to train study personnel and efficiently answer questions about study protocol. This human-AI interaction may help identify issues in the protocol, such as internal inconsistencies that should be raised with the study sponsor. Data in electronic CRFs could be monitored using AI to rapidly flag protocol deviations. For large, multicenter trials, centralized monitoring could help AI identify patterns and predict likely deviations, allowing proactive interventions such as clarifying the protocol or retraining study personnel. Patient recruitment and retention is typically a challenge for phase II, III, and IV trials. Issues with recruitment or retention can impact clinical trial timelines. Searching medical records for potential patients with suitable medical history and symptom presentation can be labor-intensive. Generative AI could help identify sites and investigators who are most likely to receive patients interested in relevant to a trial. AI-powered chatbots and assistants could answer patients' questions and help them understand the trial process, provide reminders, and share insights into the data at the end of the trial. This can boost patient engagement and interest in the trial leading to reduced dropout rates.

Generative AI can be used for synthetic data generation and predictive modeling of various aspects of clinical trials. Foundation models trained on historical patient data from clinical trials and real-world data can be used to create synthetic patient trajectories over the course of a treatment. Such models can be used for disease progression modeling as well as testing "what-if" scenarios, for example, how the disease progression is modified by a particular treatment. Patient states produced using a foundation model can be fed into another predictive model, such as a classifier to predict treatment responses or adverse events. Foundation models can be used to simulate multiple sets of inclusion/exclusion criteria to predict the impact on patient recruitment, retention, and outcomes.

While the applications discussed above span various types of generative AI models, LLMs have emerged as particularly promising tools in clinical trials. The following section examines specific case studies of LLM deployment in patient recruitment, documentation, and other aspects of clinical trials, illustrating potential pathways for integrating these technologies into existing clinical development workflows.

3 Case Studies and Real-World Examples

3.1 Enhancing Patient Recruitment Using Generative AI

3.1.1 Generative AI in Site Selection

Sentiment Analysis for Communication Between Principal Investigators and Site Engagement Team Patient recruitment is the process of finding and enrolling suitable participants for clinical trials. It is a crucial aspect of drug development and medical research, as it affects the validity, reliability, and generalizability of the results. Recruiting patients is one of the most resource-intensive steps of clinical studies. The patient recruitment industry is claimed to total $19 billion per year (Wang et al. 2003). Principal investigators (PIs) play a key role in helping to recruit and retain patients. Therefore, identifying the appropriate PIs to collaborate with is a crucial initial stage in the patient recruiting process. In the past, human reviewers were tasked with analyzing interactions between the site engagement team and PIs to categorize various sentiments—a process that was both time-consuming and labor-intensive.

Historically, this method involved extensive manual efforts to review and interpret communication logs, meeting notes, and other forms of interaction data with PIs. Each sentiment analysis by human reviewers required careful consideration of the context and nuances of language. Not only did this require a significant amount of effort, but it also introduced the possibility of human error and potential biases which could affect the accuracy of the sentiment categorization and the subsequent decisions made based on those insights (Mantyla et al. 2018).

By leveraging advanced generative AI technology, we can revolutionize this process. The AI can perform sentiment analysis of PI interactions autonomously, interpreting context and providing explanations for those sentiments, thus ensuring transparency and trust in its operations. By taking over this labor-intensive task, the technology frees up human resources for more strategic roles, enhancing overall productivity.

Moreover, generative AI can suggest future actions based on the sentiment analysis. For example, if the sentiment analysis indicates that a PI is feeling disengaged or overwhelmed, the system can recommend strategies for reengagement, such as personalized communication or additional support resources (Krugmann and Hartmann 2024). This proactive approach ensures that interactions remain positive and productive, fostering a stronger relationship between the site engagement team and the PIs.

The enhanced relationships and streamlined processes can significantly accelerate patient recruitment efforts. Studies and internal assessments estimate that this approach could lead to an estimated saving of approximately 100,000 hours across all of a large pharmaceutical company's trials each year (Hutson 2024). One practical example of how this could be operationalized is by integrating the AI system within the existing Customer Relationship Management (CRM) systems of major pharmaceutical companies. This would allow for real-time sentiment analysis and actionable insights integrated into the daily workflow of site engagement teams. The AI can flag potential issues or opportunities for engagement, notify the appropriate team members, and even automate routine follow-up communications—for instance, sending personalized thank you emails post-meetings or scheduling check-in calls based on the sentiments detected.

The data for training these AI models can be sourced from historical interactions while ensuring compliance with General Data Protection Regulation (GDPR) and other data protection regulations. Continuous learning algorithms can further refine sentiment analysis accuracy, ensuring that the generative AI model becomes more effective over time. This AI model can be deployed globally, ensuring that sentiment analysis and engagement recommendations are culturally and contextually appropriate, which is pivotal for global pharmaceutical companies. The localized AI models ensure sensitivity to cultural nuances in language and communication styles, enhancing the effectiveness of the sentiment analysis.

Aggregated Insights by Study Site or Country On top of performing sentiment analysis for each individual communication, generative AI has the powerful capability to aggregate multiple communications from a particular site, country, or study and summarize a common theme. This advanced feature offers a holistic view of site and study interactions, providing deep insights that are otherwise difficult to glean from isolated communications. The insights will enable data-driven decisions about whether we could collaborate with a site in a specific clinical trial study (Zhang et al. 2023). Generative AI can generate these summaries in a concise and readable format, which can be utilized by site engagement teams and higher management for informed decision-making. These summaries can be accompanied by visual analytics, such as graphs and charts, to illustrate trends and highlight key insights. The major component of the solution includes:

A. **Aggregated Sentiment Analysis:** The true strength of generative AI lies in its ability to synthesize data from multiple communications and derive overarching themes. This involves aggregating individual sentiment analyses to detect patterns and trends across a series of interactions. For instance, if several emails and meeting notes from a particular study show recurring expressions of concern about patient recruitment challenges, the AI can identify this as a common theme.
B. **Summarizing Common Themes:** Once the AI aggregates sentiments across multiple communications, it can summarize the findings into common themes. These summaries provide a comprehensive narrative of the site or study's overall sentiment and highlight predominant issues, strengths, and opportunities. For example, the AI might summarize that a study site is experiencing logistical issues related to patient follow-up, as evidenced by consistent negative sentiments across various communications.
C. **Actionable Insights:** Beyond summarizing the common themes, generative AI can also provide actionable insights. If a theme of delayed patient recruitment emerges, the AI might recommend specific strategies, such as enhancing community outreach efforts or providing additional training and support to the site staff. These insights can facilitate proactive measures to address issues before they escalate, improving the overall efficiency and effectiveness of the study.
D. **Customized Reports:** Generative AI can tailor these summaries and insights into customizable reports for different stakeholders. For example, a report for the site engagement team might focus on immediate operational concerns and actionable strategies, while a report for senior management might highlight broader trends and strategic implications. This level of customization ensures that the right information reaches the right audience in an optimal format.
E. **Continuous Learning:** The AI systems are continuously learning from ongoing interactions, allowing for real-time updates and progressively more accurate analyses. As new data flow in, the AI refines its understanding, adjusts its sentiment categorizations, and updates its summaries of common themes. This dynamic capability ensures that the insights remain relevant and reflective of the latest developments.

3.1.2 Generative AI in Screening Patients

Effective patient selection is another critical element in the patient recruitment process. The challenge of finding patients who meet the specific inclusion and exclusion criteria for clinical trials has traditionally involved screening large volumes of patient data manually. This process often results in delays, increased costs, and the potential for missing eligible patients who might benefit from participation (Hutson 2024). AI offers a solution to this challenge by enabling us to quickly identify the most suitable candidates based on a multitude of factors, including medical history, genetic markers, and other criteria that align with the trial's requirements. LLMs can sift through extensive EHR, patient registries, and other data sources to pinpoint potential candidates with great accuracy and speed. For example, consider a clinical trial requiring participants with specific genetic mutations linked to a rare disease—a scenario where manual screening might miss rare cases due to sheer data volume. Generative AI, trained on relevant biomedical datasets, can rapidly identify these mutations in patient records and highlight potential candidates for further screening and enrollment.

Furthermore, by analyzing social determinants of health and other patient characteristics, generative AI can also ensure diversity in clinical trials, which is critical for the generalizability of trial results (Schwartz et al. 2023). This ability helps avoid biases and ensures that trials include participants from various backgrounds and demographics. Generative AI can analyze vast datasets to find patterns and insights that human reviewers might miss, thus enhancing the precision and efficiency of the selection process. For example, generative AI can interpret clinical notes written by healthcare providers, identifying subtle mentions of symptoms or conditions that would qualify someone for a study but might not be coded in a structured format in the EHR (Lin et al. 2020).

Creating patient profiles based on these insights presents another innovative facet. These profiles can include risk scores, likelihood of adherence, and predictive outcomes, all generated through AI models. Such detailed profiling allows for a more nuanced and tailored approach to patient recruitment, ensuring trials are populated with candidates who not only meet the inclusion criteria but are also likely to remain engaged throughout the trial duration. Predictive analytics enabled by generative AI can also forecast patient retention rates, a crucial aspect of successful clinical trials. These models can analyze behavioral patterns and historical data on trial dropouts to predict which participants might be at risk of leaving the study prematurely. Armed with this information, trial coordinators can then implement targeted interventions to improve retention—such as additional follow-up, support resources, or addressing specific concerns raised by the participants.

Patient selection processes powered by generative AI also dovetail neatly with personalized medicine's future direction. As therapies become more targeted and tailored to individual genetic and phenotypic profiles, the ability to match the right patient to the right trial becomes even more critical. AI-driven patient selection ensures that trials are populated with participants most likely to benefit from the novel therapies being tested, increasing the likelihood of successful outcomes (Taherdoost and Ghofrani 2024).

Integrating generative AI into patient screening strategies will streamline patient selection processes (Askin et al. 2023). Through advanced sentiment analysis and predictive patient profiling, pharmaceutical companies stand to enhance its position as a leader in innovative clinical trial methodologies, all while fostering stronger relationships with PIs and ensuring diversity and precision in patient recruitment.

By maintaining a balanced focus on technological advancement, ethical considerations, and practical deployment strategies, pharmaceutical companies leverage generative AI to address some of the most pressing challenges in clinical trial recruitment. The future of patient recruitment has the potential to be not just faster and more efficient, but also more inclusive, personalized, and ethically sound, driven by the advanced capabilities of generative AI.

3.2 *Generative AI for Clinical Trial Documentation*

3.2.1 Leveraging Generative AI for Clinical Study Report Generation

The clinical study report (CSR) is a crucial document in the lifecycle of a clinical trial. It provides a comprehensive account of the study's methodology, analysis, results, and conclusions, serving as the official documentation required by regulatory bodies like the Food and Drug Administration (FDA) and European Medicines Agency (EMA). The traditional process of generating a CSR is typically labor-intensive, time-consuming, and riddled with complexities that entail meticulous data integration, consistency checks, and stringent adherence to regulatory guidelines. The most significant challenges include the following:

A. **Time-Consuming Process:** Creating a CSR involves the synthesis of large volumes of data from multiple sources, including clinical data, statistical analyses, adverse event reports, and patient narratives (Cobb and Haycock 2023). This process, executed largely by skilled medical writers and statisticians, spans several months.
B. **Resource-Intensive:** The labor required for CSR generation includes gathering data from various systems, cleaning and harmonizing datasets, writing narrative sections, performing statistical analyses, and ensuring adherence to regulatory formats and guidelines. This necessitates the involvement of cross-functional teams, increasing resource allocation.
C. **Consistency and Quality:** Ensuring consistency and accuracy throughout the report and other clinical and nonclinical documentation for the development program is challenging due to the sheer volume of data points and sections involved. Manual processes can lead to discrepancies, inconsistencies, and errors, jeopardizing the quality and regulatory compliance of the CSR.

Generative AI offers a transformative approach to CSR generation, significantly enhancing efficiency, consistency, and accuracy. According to the Head of Medical Writing at a major pharmaceutical company, the advanced capabilities of generative AI can streamline processes ranging from data integration to narrative creation and quality assurance (Fig. 3), cutting CSR processing time by 50% compared to manual methods. The major features of the CSR generation solution are listed as below:

A. **Automated Data Integration:** Generative AI can automate the extraction and integration of data from multiple sources such as EHRs, databases, spreadsheets, and other clinical trial management systems. The AI can collate these data

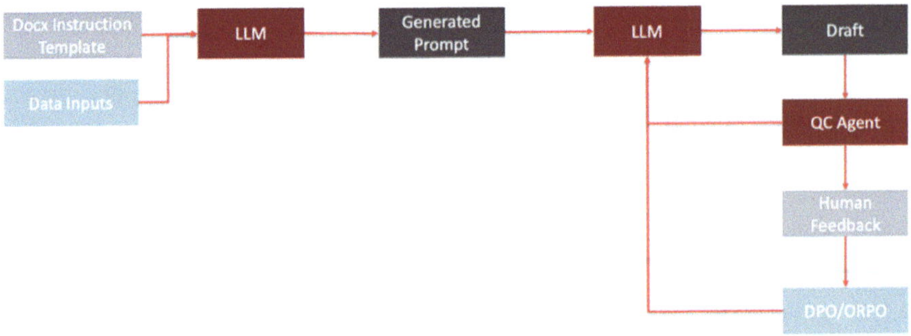

Fig. 3 A flowchart of the Clinical Study Report (CSR) generation solution powered by LLMs and AI agents. The CSR template and study data were input to Large Language Model (LLM). The agents generated prompt to make the LLM to generate a draft of CSR. The draft would go through quality control agent and human review. The feedback from the Quality Control (QC) agent and human expert would be used to fine-tune the LLM using Direct Preference Optimization (DPO) or Odds Ratio Preference Optimization (ORPO) (Hong et al. 2024)

seamlessly, ensuring that it is clean, harmonized, and ready for analysis. This step alone can drastically reduce the time and effort required to prepare data for CSR generation.

B. **Automated Data Analysis:** AI algorithms can perform automated statistical analyses on integrated data, generating immediate, accurate results. By utilizing predefined algorithms and machine learning models, generative AI can ensure that the statistical analyses adhere to regulatory standards and are performed with high precision.
C. **Automated Narrative Generation:** One of the most powerful applications of generative AI in CSR generation is automated narrative creation. The AI can draft sections of the CSR, including the study design, methodology, results, and conclusions, by interpreting and summarizing the integrated data. NLP capabilities enable the AI to write in a coherent, standardized, and regulatory-compliant format, reducing the time required for manual writing.
D. **Consistency and Quality Assurance:** Generative AI ensures consistency across all sections of the CSR by applying standardized templates and terminologies. It can cross-reference data points, verify against source data, and highlight any discrepancies. This automated quality assurance improves the overall integrity of the report, making it more reliable and regulatory-compliant.
E. **Enhanced Collaboration:** AI-powered platforms for CSR generation can facilitate enhanced collaboration among the various stakeholders involved in the process. Real-time updates, shared dashboards, and collaborative workspaces enable seamless coordination, review, and finalization of the CSR when powered by AI. Stakeholders can provide inputs, suggestions, and corrections which the AI can quickly incorporate, ensuring that the final report meets all required standards and expectations.

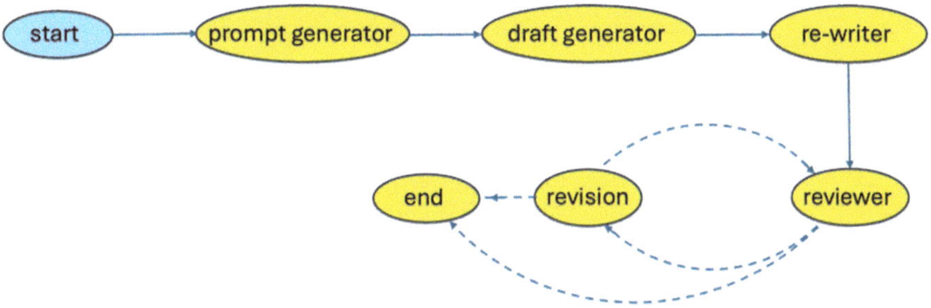

Fig. 4 An illustration of AI agents' responsibilities in CSR generation and the flow of information between agents. The AI agents generate prompts for LLMs, generate CSR drafts, review the drafts, and rewrite the CSR according to the feedback

F. **AI Agent Orchestration:** The AI agent facilitates collaborative reviews by providing real-time updates and incorporating stakeholder feedback into the final document efficiently. Using AI agents in CSR writing streamlines and enhances the drafting process by leveraging advanced language models. The agents first craft detailed prompts, identifying necessary data from both structured and unstructured sources. These prompts guide the initial draft creation. Subsequently, the agents rewrite this draft, tailoring it to match historical CSR writing styles for consistency and professionalism. They then review and suggest refinements, ensuring the draft aligns with the original prompts and addressing any discrepancies. Finally, the agents incorporate feedback from human reviewers, rechecking the document to produce a polished and accurate CSR report. The process is further illustrated in Fig. 4. This AI-assisted approach results in efficient and high-quality CSR documentation.

3.2.2 Converting Medical Jargon in Investigator Brochures to Patient-Friendly Languages

In clinical research, the Investigator's Brochure (IB) serves as a vital document, providing comprehensive safety and efficacy information about investigational products. However, the safety section of the IB, laden with technical jargons and complex medical terminologies, often becomes a barrier in effective patient communication. Transforming these sections into a patient-friendly format requires medical writers to invest substantial time and effort, a process that is both manual- and labor-intensive. This traditional approach limits efficiency and risks miscommunication, ultimately impacting patient understanding and compliance (Guttman et al. 2021). To address this challenge, the power of generative AI and Robotic Process Automation (RPA) was harnessed to automate the transformation of the IB safety section into a patient-friendly format. By leveraging LLMs, we developed an automated system that efficiently rewrites complex medical content into clear,

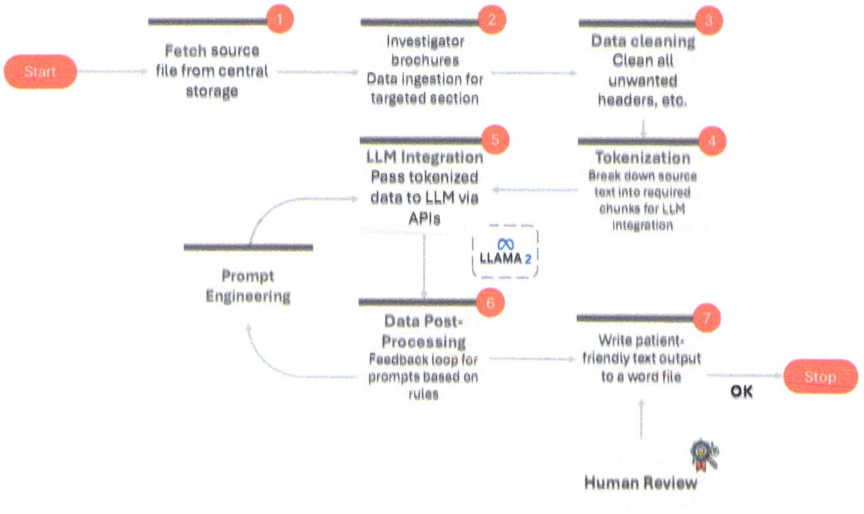

Fig. 5 A flowchart illustrating the use of an LLM to translate the safety section of an investigator's brochure into patient-friendly language

concise, and easily understandable language for patients. This innovation not only saves over 1000 h annually in a large pharmaceutical company for medical writers but also enhances the clarity of communication with patients, ensuring that crucial safety information is accessible and comprehensible. The flowchart in Fig. 5 illustrates the LLM-based method mentioned above.

The automation of transforming the IB safety sections reaps multiple benefits beyond mere timesaving. By enabling effective communication with patients, this solution fosters better understanding and retention of critical safety information, thus improving overall patient experience. Moreover, the process significantly shortens the duration for treatment handover, providing quicker access to crucial information that can potentially lead to improved health outcomes. This integration of generative AI and RPA stands as a testament to how technology can revolutionize patient care by bridging the communication gap in clinical research (Putnoki and Orosz 2024). Ultimately, this leads to a more streamlined and patient-centric approach to clinical data dissemination, enhancing the overall efficacy by saving approximately 60–90% of the transforming time and making a big impact of clinical trials (Alowais et al. 2023).

Other Use Cases in Text Generation Clinical studies require extensive documentation, including the drafting of study protocols, patient consent forms, and clinical trial reports. AI-powered text generation can streamline these tasks by generating comprehensive and consistent documents based on predefined templates and gathered data, allowing researchers to focus more on the quality and analysis of the study rather than administrative tasks.

Another vital use case is enhancing patient engagement and communication. AI text generation can create personalized patient communications, such as appointment reminders, medication instructions, and follow-up messages, tailored to individual patient needs and study requirements. Additionally, AI can assist in generating responses to patient queries about the study, ensuring consistent and accurate information dissemination. These applications not only enhance efficiency but also improve patient adherence and satisfaction (Lu et al. 2024), crucial for successful clinical trials.

3.3 Generative AI for Knowledge Extraction to Support Trial Design

Design of clinical trials requires carefully thinking through a plethora of parameters, such as patient inclusion/exclusion criteria, which efficacy endpoints to pursue, dosing of treatment, likely adverse events, etc. When designing a trial, pharmaceutical companies, among other things, gather information from past trials for the indication of interest, perform various types of quantitative analysis, then design their own trials accordingly. Trial information is derived primarily from two sources: trial databases such as ClinicalTrials.gov and publications from journals and conferences. These sources contain diverse data formats in unstructured text, tabular data, and graphs. Extracting evidence from multiple data sources in diverse data formats poses many challenges such as accurate extraction of numerical values that exist in tables and charts of publications, aligning extracted terms across multiple data sources and ontologies, and organizing output to support quantitative modeling and analysis. LLMs and multimodal models are poised to automate these tasks and drastically reduce the amount of time needed. What typically takes multiple people over weeks and months can be reduced to hours and days. Such automation will undoubtedly foster more data-driven decisions in drug-development. Prior work on leveraging LLMs to extract clinical trial information focuses on more limited aspect of trial design. AutoCriteria (Machin et al. 2006) SEETrial (Yamaguchi et al. 2021) focuses on extraction of efficacy and safety information, only found in conference abstracts. CTBench focuses on extraction of patient demographics (Neehal et al. 2024). Here, we look at the expanded use case of extraction from full text publications and handling a broader set of concepts needed for designing clinical trials.

3.3.1 Pipeline for Data Extraction from Publication

Figure 6 illustrates a pipeline needed for extraction from clinical trial literature. The first step is to find and retrieve relevant literature. These documents can be in various formats such as PDF and DOCX. Next is layout analysis tool, with popular tools

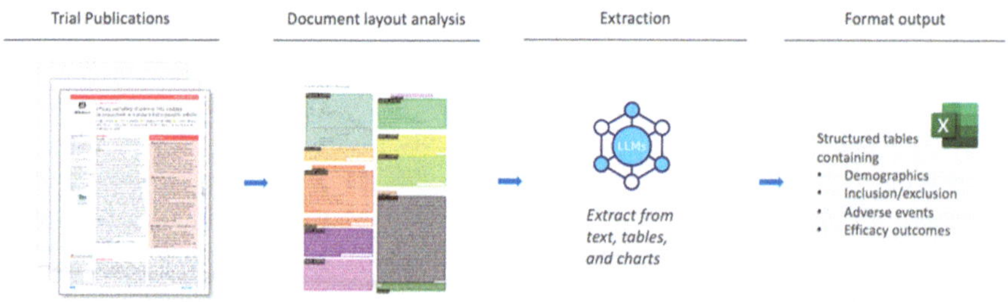

Fig. 6 Pipeline for using generative AI for knowledge extraction from clinical trial literature

being Grobid, AWS Textract, and Docling. Language models, including multimodal models, could be used for extraction of concepts from text, tables, and chart sections of publications. Finally, the extracted data can be formatted as structured data to support downstream quantitative analysis. The overall design is that of a RAG system, with relevant literature serving as the knowledge base for extraction.

3.3.2 Multimodal Models for Extraction from Charts

Clinical trial publications have abundant chart types to convey results, such as patient flow diagrams, bar plots, line graphs, scatter plots, etc. These charts present detailed numerical results not found in the narrative part of publications. These results are presented with respect to patient cohorts and subgroups and often convey how outcomes vary over time. Traditional tools such as PlotDigitizer ask users to upload charts and specify fiducial points to support downstream automated extraction. More recently, multimodal vision-language models are applied to extract information from these charts without user input (Zhang et al. 2024; Yao et al. 2024). While current multimodal models can do quite well in tasks such as chart summarization, they are not yet satisfactory for extracting accurate numerical values from graphs. Nonetheless, chart understanding is expected to improve in the coming years.

Figure 7 shows typical chart types found in clinical trial literature, and what an extraction might look like. Chart information needs to be converted to standardized tabular format with understanding of concepts important for clinical trials such as study groups, endpoints, timepoints, measurement values, etc. Current multimodal models are trained on charts and figures from general content, not clinical trial-specific charts. Hence, to achieve good performance in chart extraction from clinical trial literature, we need to build a dataset of such charts and finetune existing multimodal models on such specialized datasets.

a) Sample charts from clinical trial literature

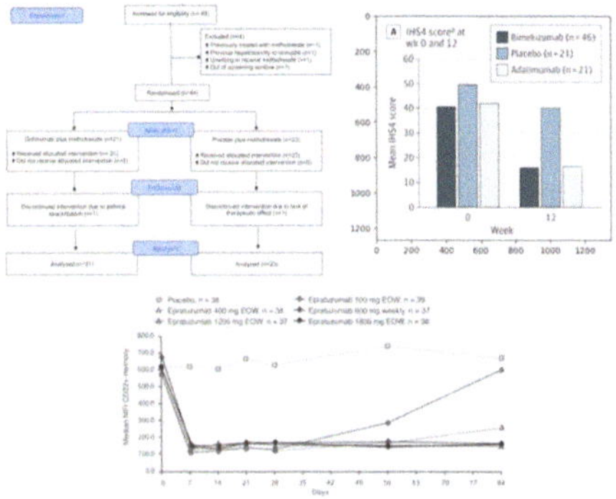

b) Sample extraction from barchart

Study_group	Endpoint	Time	Value
Placebo	IHS4 score	0	52
Bimekizumab	IHS4 score	0	41
Adalimumab	IHS4 score	0	42
Placebo	IHS4 score	12	41
Bimekizumab	IHS4 score	12	16
Adalimumab	IHS4 score	12	17

Fig. 7 (**a**) Sample charts from clinical trial literature, and (**b**) sample of extraction from bar chart using vision-language model

3.3.3 Searching, Extracting, and Integrating Information from Multiple Sources

Clinical trial information can come from multiple sources: (1) publications (Pubmed, Europe PMC, MEDLINE, Embase), (2) trial registries (ClinicalTrial.gov, EU Clinical Trials Register, Cochrane Central Register of Controlled Trials), and (3) conference proceedings, among others. To handle the diversity of sources, one can use an AI agent approach, with different LLMs specializing in extraction from each source.

An AI agent-based approach can also be suitable to handle the different phases of knowledge extraction, including (1) planner agent that takes the user request and creates a research plan which includes which concepts to extract and sources from which to extract; (2) paper and trial identification agents to find relevant clinical trials and publications; (3) data extraction and data compilation agents; and (4) reviewer agents to check on the accuracy of the extraction. As a collective, these AI agents serve as a flexible and dynamic research system that can handle knowledge search and extraction for a variety of clinical trials from a variety of sources. Such a system will be useful not only for clinical trials, but also for the broader challenge of systematic literature reviews.

As these systems continue to evolve, we can more robust and automated tools for scientific literature review and extraction, significantly reducing the manual many stages of drug development. Current literature reviews can take anywhere from two to six months, costing $100,000 and 250,000 per review, including labor costs (Michelson and Reuter 2019). Automation has the potential to shorten the process to days or weeks. However, human involvement and oversight remain essential to ensure the most relevant studies are selected and the right content is extracted.

3.4 *Generative AI's Potential Application in Post-approval Phases*

Generative AI could also play a crucial role in the post-approval phases of the pharmaceutical industry. One significant application is in drug monitoring and pharmacovigilance, where AI models can analyze vast amounts of data from various sources, including patient records and social media, to detect adverse drug reactions and ensure ongoing safety (Praveen et al. 2023). Additionally, generative AI can assist in optimizing manufacturing processes by predicting and mitigating potential issues, thus maintaining high-quality production standards (Huanbutta et al. 2024). Furthermore, AI-driven tools can enhance marketing strategies by identifying the most effective communication channels and crafting personalized messages for healthcare professionals and patients (Gupta et al. 2024). Overall, generative AI's capabilities significantly enhance efficiency, safety, and communication in the post-approval phases, thereby ensuring the success of the products.

4 Key Considerations for Using AI in Clinical Development

When implementing AI in clinical development, several critical considerations must be carefully addressed to ensure successful and responsible deployment. Foremost among these is regulatory compliance, particularly regarding data privacy and security. Clinical trial data contain sensitive patient information protected by regulations such as GDPR, HIPAA, and various national privacy laws. AI systems must be designed with robust safeguards to ensure data protection while maintaining utility.

Quality assurance and validation represent another crucial consideration. AI-generated outputs, whether documents, analyses, or predictions, must meet the same rigorous standards as traditionally produced materials. This necessitates implementing comprehensive validation processes and maintaining clear audit trails. Organizations should establish clear protocols for human review and oversight, particularly for critical decisions affecting patient safety or trial outcomes.

Integration with existing workflows presents both technical and organizational challenges. AI systems must seamlessly interface with current clinical trial management systems, electronic data capture platforms, and documentation processes. This integration requires careful consideration of system compatibility, data standardization, and user training needs. Organizations should develop clear standard operating procedures (SOPs) that define roles, responsibilities, and processes for AI-augmented workflows.

Change management and staff training are equally important considerations. Clinical development teams need appropriate training to effectively utilize AI tools while maintaining their judgment and expertise. Organizations must foster a culture that views AI as a complement to human expertise rather than a replacement.

Finally, organizations must consider the ethical implications of AI deployment in clinical development. This includes ensuring fairness and avoiding bias in patient selection, maintaining transparency in AI-driven decisions, and protecting the interests of trial participants. Regular assessment of AI system performance and impact helps ensure these tools continue to serve their intended purpose while maintaining ethical standards.

5 Conclusion

The integration of generative AI into clinical trials represents a significant advancement in pharmaceutical research and development, offering solutions to long-standing challenges while opening new avenues for innovation. This chapter has explored how generative AI, particularly LLMs and other advanced AI architectures, can enhance various aspects of clinical trials from design through execution and documentation. The evolution of AI capabilities, particularly the emergence of foundation models and LLMs, has created unprecedented opportunities in clinical research. These models' ability to process and generate human-like text, combined with their capacity for few-shot learning and chain-of-thought reasoning, makes them particularly valuable for the complex, knowledge-intensive tasks inherent in clinical trials.

As demonstrated through the case studies in this chapter, generative AI has shown practical value across multiple aspects of clinical trials. From enhancing patient recruitment through automated sentiment analysis and screening, to streamlining documentation processes in CSR generation and protocol development, to enabling sophisticated knowledge extraction for trial design, these applications are already beginning to transform clinical research operations. The technology's ability to process and analyze vast amounts of data while generating human-readable outputs has proven particularly valuable in reducing manual effort while maintaining or improving quality standards.

Despite these advantages, several important considerations must be addressed for successful implementation. Data privacy and security remain paramount, with the need to ensure compliance with regulations like GDPR and protect sensitive patient information while maintaining confidentiality in trial data. Quality control presents another crucial consideration, requiring careful validation of AI-generated content, ensuring regulatory compliance, and maintaining appropriate human oversight of critical processes. Integration challenges must also be addressed, including seamless incorporation into existing workflows, effective training of personnel, and ensuring system interoperability.

Looking to the future, several trends are likely to shape the evolution of generative AI in clinical trials. Enhanced multimodal capabilities will improve the processing of diverse data types, including images, charts, and structured data, while enabling better integration of multiple information sources and more sophisticated analysis of complex trial data. Greater automation through increased use of AI agents for complex tasks and more sophisticated orchestration of multiple AI systems will reduce the need for manual intervention in routine tasks. These advancements will lead to improved efficiency through faster trial execution, reduced costs, and better resource allocation. The integration of generative AI into clinical trials represents more than just technological advancement; it signifies a fundamental shift in how we approach clinical research. By automating routine tasks, enhancing decision-making processes, and improving communication, generative AI has the potential to accelerate the development of new treatments while maintaining or improving the quality and reliability of clinical trials.

As these technologies continue to evolve, their impact on clinical research will likely grow, leading to more efficient, cost-effective, and patient-centric trials. The key to successful implementation lies in balancing automation with human expertise, ensuring that AI serves as a tool to enhance, rather than replace, human judgment in clinical research. This balance, combined with continued technological advancement and careful attention to ethical considerations, will shape the future of clinical trials and, ultimately, the development of new therapeutic options for patients.

References

Alowais SA, Alghamdi SS, Alsuhebany N, Alqahtani T, Alshaya AI, Almohareb SN, Aldairem A, Alrashed M, Bin Saleh K, Badreldin HA, et al. Revolutionizing healthcare: the role of artificial intelligence in clinical practice. BMC Med Educ. 2023;23(1):689.

Askin S, Burkhalter D, Calado G, El Dakrouni S. Artificial intelligence applied to clinical trials: opportunities and challenges. Heal Technol. 2023;13(2):203–13.

Auer S, Bizer C, Kobilarov G, Lehmann J, Cyganiak R, Ives Z. Dbpedia: a nucleus for a web of open data. In: International semantic web conference. Springer; 2007. p. 722–35.

Brown PF, Della Pietra VJ, Desouza PV, Lai JC, Mercer RL. Class- based n-gram models of natural language. Comput Linguist. 1992;18(4):467–80.

Brown TB, Mann B, Ryder N, Subbiah M, Kaplan J, Dhariwal P, Neelakantan A, Shyam P, Sastry G, Askell A, Agarwal S, Herbert-Voss A, Krueger G, Henighan T, Child R, Ramesh A, Ziegler DM, Wu J, Winter C, Hesse C, Chen M, Sigler E, Litwin M, Gray S, Chess B, Clark J, Berner C, McCandlish S, Radford A, Sutskever I, Amodei D. Language models are few-shot learners. In: Proceedings of the 34th international conference on neural information processing systems. NIPS '20. Red Hook: Curran Associates; 2020.

Chowdhery A, Narang S, Devlin J, Bosma M, Mishra G, Roberts A, Barham P, Chung HW, Sutton C, Gehrmann S, et al. Palm: scaling language modeling with pathways. J Mach Learn Res. 2023;24(240):1–113.

Clarke M. The 1944 Patulin trial of the British medical research council. J R Soc Med. 2006;99(9):478–80.

Cobb L, Haycock N. Narratives for a clinical study report: The evolution of automation and artificial intelligence. Med Writ. 2023;32:28–31.

Datta S, Lee K, Paek H, Manion FJ, Ofoegbu N, Du J, Li Y, Huang L-C, Wang J, Lin B, et al. Autocriteria: a generalizable clinical trial eligibility criteria extraction system powered by large language models. J Am Med Inform Assoc. 2024;31(2):375–85.

Donnelly K, et al. Snomed-ct: The advanced terminology and coding system for ehealth. In: Studies in health technology and informatics, vol. 121; 2006. p. 279.

Eddy SR. Hidden Markov models. Curr Opin Struct Biol. 1996;6(3):361–5.
Evangelopoulos NE. Latent semantic analysis. Wiley Interdiscip Rev Cogn Sci. 2013;4(6):683–92.
Fan W, Ding Y, Ning L, Wang S, Li H, Yin D, Chua T-S, Li Q. A survey on rag meeting llms: Towards retrieval-augmented large language models. In: Proceedings of the 30th ACM SIGKDD conference on knowledge discovery and data mining. 2024. p. 6491–501.
Gupta R, et al. Adoption and impacts of generative artificial intelligence: theoretical underpinnings and research agenda. Int J Inf Manage Data Insights. 2024;4(1):100232.
Guttman OT, Lazzara EH, Keebler JR, Webster KL, Gisick LM, Baker AL. Dissecting communication barriers in healthcare: a path to enhancing communication resiliency, reliability, and patient safety. J Patient Saf. 2021;17(8):e1465–71.
Hay M, Thomas DW, Craighead JL, Economides C, Rosenthal J. Clinical development success rates for investigational drugs. Nat Biotechnol. 2014;32(1):40–51.
Hong J, Lee N, Thorne J. Orpo: monolithic preference optimization without reference model. arXiv preprint arXiv:2403.07691. 2024; 2(4):5.
Huanbutta K, et al. The Artificial Intelligence-driven pharmaceutical industry: a paradigm shift in drug discovery, formulation development, manufacturing, quality control, and post-market surveillance. Eur J Pharm Sci. 2024:106938.
Huang W, Abbeel P, Pathak D, Mordatch I. Language models as zero-shot planners: Extracting actionable knowledge for embodied agents. In: International conference on machine learning. PMLR; 2022. p. 9118–47.
Hutson M. How AI is being used to accelerate clinical trials. Nature. 2024;627(8003):2–5.
Johnson ML. SHRDLU, procedures, mini-world. In: Mind, language, machine: artificial intelligence in the poststructuralist age. London: Palgrave Macmillan; 1988. p. 113–22.
Karabacak M, Margetis K. Embracing large language models for medical applications: opportunities and challenges. Cureus. 2023;15(5)
Kenton JDM-WC, Toutanova LK. Bert: pre-training of deep bidirectional transformers for language understanding. In: Proceedings of naacL-HLT, vol. 1. Minneapolis; 2019. p. 2.
Kim M, Bursztyn V, Koh E, Guo S, Hwang S-W. Rada: retrieval- augmented web agent planning with llms. In: Findings of the association for computational linguistics ACL 2024, vol. 2024. p. 13511–25.
Kirkpatrick J, Pascanu R, Rabinowitz N, Veness J, Desjardins G, Rusu AA, Milan K, Quan J, Ramalho T, Grabska-Barwinska A, et al. Overcoming catastrophic forgetting in neural networks. Proc Natl Acad Sci. 2017;114(13):3521–6.
Kojima T, Gu SS, Reid M, Matsuo Y, Iwasawa Y. Large language models are zero-shot reasoners. In: Advances in neural information processing systems, vol. 35; 2022. p. 22199–213.
Krugmann JO, Hartmann J. Sentiment analysis in the age of generative AI. Cust Needs Solut. 2024;11(1):3.
Lee K, Paek H, Huang L-C, Hilton CB, Datta S, Higashi J, Ofoegbu N, Wang J, Rubinstein SM, Cowan AJ, et al. Seetrials: leveraging large language models for safety and efficacy extraction in oncology clinical trials. 2024. medRxiv, 2024–01.
Lewis P, Perez E, Piktus A, Petroni F, Karpukhin V, Goyal N, Kuttler H, Lewis M, Yih WT, Rocktaschel T, et al. Retrieval-augmented generation for knowledge-intensive NLP tasks. Adv Neural Inf Proces Syst. 2020;33:9459–74.
Lin W-C, Chen JS, Chiang MF, Hribar MR. Applications of artificial intelligence to electronic health record data in ophthalmology. Transl Vis Sci Technol. 2020;9(2):13.
Lin X, Wang W, Li Y, Yang S, Feng F, Wei Y, Chua T-S. Data- efficient fine-tuning for llm-based recommendation. In: Proceedings of the 47th International ACM SIGIR conference on research and development in information retrieval. SIGIR '24. New York: Association for Computing Machinery; 2024. p. 365–74. https://doi.org/10.1145/3626772.3657807.
Liu B, Chen C, Gong Z, Liao C, Wang H, Lei Z, Liang M, Chen D, Shen M, Zhou H, Jiang W, Yu H, Li J. Mftcoder: boosting code llms with multitask fine-tuning. In: Proceedings of the 30th ACM SIGKDD conference on knowledge discovery and data mining. KDD '24. New York: Association for Computing Machinery; 2024a. p. 5430–41. https://doi.org/10.1145/3637528.3671609.

Liu Q, Wu X, Zhao X, Zhu Y, Xu D, Tian F, Zheng Y. When moe meets llms: Parameter efficient fine-tuning for multi-task medical applications. In: Proceedings of the 47th International ACM SIGIR conference on research and development in information retrieval. SIGIR '24. New York: Association for Computing Machinery; 2024b. p. 1104–14. https://doi.org/10.1145/3626772.3657722.

Lu X, Yang C, Liangy L, Hu G, Zhong Z, Jiang Z. Artificial intelligence for optimizing recruitment and retention in clinical trials: a scoping review. J Am Med Inform Assoc. 2024;31:2749–59.

Ma H, Zhang C, Bian Y, Liu L, Zhang Z, Zhao P, Zhang S, Fu H, Hu Q, Wu B. Fairness-guided few-shot prompting for large language models. In: Thirty-seventh conference on Neural Information Processing Systems (NeurIPS). 2023.

Machin D, Day S, Green SB, editors. Textbook of clinical trials. 2nd ed. Chichester/Hoboken: Wiley; 2006.

Mantyla MV, Graziotin D, Kuutila M. The evolution of sentiment analysis – a review of research topics, venues, and top cited papers. Comput Sci Rev. 2018;27:16–32.

Michelson M, Reuter K. The significant cost of systematic reviews and meta-analyses: a call for greater involvement of machine learning to assess the promise of clinical trials. Contemp Clin Trials Commun. 2019;16:100443.

Mikolov T, Chen K, Corrado G, Dean J. Efficient estimation of word representations in vector space. In: 1st International conference on learning representations, ICLR 2013, Scottsdale, Arizona, USA, May 2–4, 2013, workshop track proceedings. 2013. http://arxiv.org/abs/1301.3781

Miller GA. Wordnet: a lexical database for English. Commun ACM. 1995;38(11):39–41.

Moens M, Steedman M. Temporal ontology in natural language. In: 25th Annual meeting of the association for computational linguistics; 1987. p. 1–7.

Neehal N, Wang B, Debopadhaya S, Dan S, Murugesan K, Anand V, Bennett KP. CTBench: a comprehensive benchmark for evaluating language model capabilities in clinical trial design. 2024. arXiv.

Nellhaus EM, Davies TH. Evolution of clinical trials throughout history. Marshall J Med. 2017;3(1):41.

Nirenburg S, Carbonell J. Integrating discourse pragmatics and propositional knowledge for multilingual natural language processing. Comput Transl. 1987;2(2):105–16.

Pennington J, Socher R, Manning CD. Glove: Global vectors for word representation. In: Proceedings of the 2014 conference on Empirical Methods in Natural Language Processing (EMNLP); 2014. p. 1532–43.

Praveen J, Kumar K, Channappa AH. Transforming pharmacovigilance using Gen AI: innovations in aggregate reporting, signal detection, and safety surveillance. J Multidiscip Res. 2023:9–16.

Putnoki AM, Orosz T. Artificial intelligence and cognitive information systems: revolutionizing business with generative artificial intelligence and robotic process automation. In: Proceedings of international conference on recent innovations in computing; 2024. p. 39–70.

Radford A, Kim JW, Hallacy C, Ramesh A, Goh G, Agarwal S, Sastry G, Askell A, Mishkin P, Clark J, et al. Learning transferable visual models from natural language supervision. In: International conference on machine learning. PMLR; 2021. p. 8748–63.

Raffel C, Shazeer N, Roberts A, Lee K, Narang S, Matena M, Zhou Y, Li W, Liu PJ. Exploring the limits of transfer learning with a unified text-to-text transformer. J Mach Learn Res. 2020;21(140):1–67.

Robertson S. Understanding inverse document frequency: on theoretical arguments for IDF. J Doc. 2004;60(5):503–20.

Robertson, S., Zaragoza, H., et al.: The probabilistic relevance framework: Bm25 and beyond. Foundations and trends® in information retrieval 3(4), 333–389 (2009).

Romera-Paredes B, Torr P. An embarrassingly simple approach to zero- shot learning. In: International conference on machine learning. PMLR; 2015. p. 2152–61.

Schneider J, Meske C, Kuss P. Foundation models: a new paradigm for artificial intelligence. Bus Inf Syst Eng. 2024;1–11

Schwartz AL, Alsan M, Morris AA, Halpern SD. Why diverse clinical trial participation matters. N Engl J Med. 2023;388(14):1252–4.
Taherdoost H, Ghofrani A. Ai's role in revolutionizing personalized medicine by reshaping pharmacogenomics and drug therapy. Intell Pharm. 2024;2(5):643–50. https://doi.org/10.1016/j.ipha.2024.08.005.
Thoppilan R, De Freitas D, Hall J, Shazeer N, Kulshreshtha A, Cheng HT, Jin A, Bos T, Baker L, Du Y, et al. Lamda: language models for dialog applications. arXiv preprint arXiv:2201.08239. 2022.
Trohler U. Lind and scurvy: 1747 to 1795. J R Soc Med. 2005;98(11):519–22.
Vaswani A. Attention is all you need. In: Advances in neural information processing systems; 2017.
Wang AE, Drayton S, Frazer HE. Delay no more: improve patient recruitment and reduce time to market in the pharmaceutical industry. Somers: IBM Institute for Business Value; 2003.
Wang Z, Zhang G, Yang K, Shi N, Zhou W, Hao S, Xiong G, Li Y, Sim MY, Chen X, et al. Interactive natural language processing. 2023. arXiv preprint arXiv:2305.13246.
Wang L, Ma C, Feng X, Zhang Z, Yang H, Zhang J, Chen Z, Tang J, Chen X, Lin Y, et al. A survey on large language model based autonomous agents. Front Comp Sci. 2024;18(6):186345.
Wei J, Tay Y, Bommasani R, Raffel C, Zoph B, Borgeaud S, Yogatama D, Bosma M, Zhou D, Metzler D, Chi EH, Hashimoto T, Vinyals O, Liang P, Dean J, Fedus W. Emergent abilities of large language models. Trans Mach Learn Res. 2022;2022
Weizenbaum J. Eliza – a computer program for the study of natural language communication between man and machine. Commun ACM. 1966;9(1):36–45.
Xian Y, Schiele B, Akata Z. Zero-shot learning-the good, the bad and the ugly. In: Proceedings of the IEEE conference on computer vision and pattern recognition. 2017. p. 4582–4591.
Yamaguchi S, Kaneko M, Narukawa M. Approval success rates of drug candidates based on target, action, modality, application, and their combinations. Clin Transl Sci. 2021;14(3):1113–22.
Yao Y, Yu T, Zhang A, Wang C, Cui J, Zhu H, Cai T, Li H, Zhao W, He Z, et al.: Minicpm-v: a gpt-4v level mllm on your phone. 2024. arXiv preprint arXiv:2408.01800.
Zhang B, Zhang L, Chen Q, Jin Z, Liu S, Zhang S. Harnessing artificial intelligence to improve clinical trial design. Commun Med. 2023;3(1):191.
Zhang L., Hu A, Xu H, Yan M, Xu Y, Jin Q, Zhang J, Huang F. Tinychart: Efficient chart understanding with visual token merging and program- of-thoughts learning. 2024. arXiv preprint arXiv:2404.16635.
Zhao A, Huang D, Xu Q, Lin M, Liu Y-J, Huang G. Expel: Llm agents are experiential learners. In: Proceedings of the AAAI conference on Artificial Intelligence, vol. 38; 2024. p. 19632–42.

AI-Driven Advances in Personalized Therapeutic Strategies for Precision Medicine

Lorenzo Simone, Y. F. Ferrari Chen, Yves A. Lussier, Peter Elkin, and Xinxin Zhu

Abstract Artificial intelligence (AI) is increasingly driving advances in precision medicine, with the potential to enable treatments that reflect the molecular and clinical profiles of each patient. This chapter examines how AI is reshaping precision therapy through methods such as predictive modeling, unsupervised learning, graph neural networks, and reinforcement learning, applied to drug response prediction, patient stratification, and adaptive treatment design. The discussion extends to generative AI, focusing on its applications in drug discovery and repurposing alongside the development of virtual patient models and digital twins for personalized health management.

Integrating multiscale biomedical data from omics, structural biology, clinical records, imaging, and population health offers a pathway to more precise therapeutic targeting and the discovery of novel clinical patterns. Case studies from recent years illustrate this potential while highlighting persistent challenges such as data

L. Simone (✉)
Department of Computer Science, University of Pisa, Pisa, Italy

School of Public Health, Yale University, New Haven, CT, USA
e-mail: lorenzo.simone@yale.edu

Y.F. Ferrari Chen
Interdisciplinary Center for Health Sciences, Scuola Superiore Sant'Anna, Pisa, Italy

Cardiology Division, Fondazione Toscana Gabriele Monasterio, Pisa, Italy
e-mail: Yu.FerrariChen@santannapisa.it

Y. A. Lussier
Department of Biomedical Informatics and Huntsman Cancer Center, University of Utah, Utah, USA
e-mail: Yves.Lussier@utah.edu

P. Elkin
University at Buffalo, The State University of New York, Buffalo, NY, USA
e-mail: elkinp@buffalo.edu

X. Zhu
Global Real-World Evidence, Boston Scientific, Marlborough, MA, USA
e-mail: katie.zhu@bsci.com

A. Lavecchia (ed.), *Applied Artificial Intelligence for Drug Discovery*,
https://doi.org/10.1007/978-3-031-98022-0_26

heterogeneity, interoperability gaps, and the ethical dimensions of large-scale data use. Looking ahead, we reflect on the transition toward integrated, multimodal decision support systems that incorporate emerging technologies such as synthetic biology and digital twins and on the steps needed to move healthcare toward a patient-centric, predictive, therapeutic, and preventive discipline.

Keywords Precision medicine · Artificial Intelligence · Machine learning · Generative AI · Personalized therapy

1 Introduction to AI-Driven Precision Therapy

The advent of precision medicine marks a transformative shift in therapeutic design away from standardized protocols toward data-informed decisions and personalization. This shift is mainly driven by two catalysts: the increasing availability of biomedical data across molecular, physiological, and behavioral levels, as well as the computational advances capable of meaningfully interpreting them. Additionally, AI has emerged as a key field integrating various disciplines and capable of distilling clinical insights from complex, multiscale, and often unstructured datasets that exceed the interpretative capacity of traditional analysis. This facilitates a highly personalized approach to diagnosis, treatment selection, and disease monitoring, promising improved patient outcomes and reduced adverse effects.

The integration of AI in clinical trials has already demonstrated promising results. AI-driven platforms enhance patient recruitment by identifying eligible participants based on genomic, demographic, and clinical criteria, thereby accelerating the trial process and improving enrolment diversity (Calaprice-Whitty et al. 2020). Beyond research applications, their use in clinical practice is spreading across three main areas: risk stratification and early disease detection through the analysis of multimodal and heterogeneous clinical data; the development of novel digital biomarkers derived from both traditional modalities (e.g., electrocardiography, echocardiography) and emerging data types (e.g., speech, patient videos); and the integration of smart devices for screening, continuous monitoring, and disease management.

The transition toward proactive and personalized healthcare has positioned AI as a transformative force, accelerating discovery and reshaping therapeutic processes—from biomarker identification to real-time tailored interventions. This convergence of data availability, computational resources, and clinical decision-making presents great opportunities, but also raises challenges from a methodological, interpretability, and implementation perspective. Effective integration of AI into clinical workflows requires addressing methodological robustness, algorithmic interpretability, and seamless implementation into healthcare settings. Beyond AI technical advances in precision medicine, successful adoption in real-world applications hinges upon rigorous clinical validation (Tikhomirov et al. 2024), ensuring equitable access to avoid exacerbating health disparities and establishing transparent and ethical use in patient care (Ong et al. 2024).

In this chapter, we provide a structured overview of the machine learning (ML) methodologies currently shaping the field of precision therapeutics, with a particular focus on widely known techniques and emerging approaches. Section 2 introduces key algorithmic frameworks, including predictive modeling, unsupervised learning, graph neural networks, reinforcement learning, and generative AI, highlighting their distinct advantages and the clinical contexts in which they have demonstrated practical utility. From cancer therapy selection to intensive care management and drug repurposing, we explore how these computational strategies are being adopted in medical decision-making, while critically addressing their translational potential and related challenges. Section 3 addresses the integration of multiscale biomedical data within omics, structural biology, clinical records, and imaging, to examine how these diverse sources can be integrated to refine therapeutic strategies or uncover new clinical insights. Finally, Sect. 4 considers the next frontiers in AI-driven precision therapy, focusing on technological advancements, ongoing challenges in real-world deployment, and the future landscape of individualized healthcare.

2 Computational Approaches in Precision Medicine: Machine Learning and Generative AI

2.1 The Transformative Role of Machine Learning in Precision Therapy

In recent years, the increasing emphasis on data acquisition, with advancements in data structures, and the growth of computational resources has been a driving force behind the success of machine learning for statistical analysis applications (Bzdok et al. 2018). In parallel, these innovations have provided a foundation for the evolution of precision medicine, requiring the collection of individual-level data, including genetic data, clinical records, environmental exposures, and lifestyle factors (MacEachern and Forkert 2021).

In this section, we explore machine learning methodologies focusing on their impact on drug response prediction, patient stratification, and adaptive treatment strategies (Fig. 1). First, we examine predictive and unsupervised learning techniques shifting the paradigm of therapeutics to personalized drug selection, building on genomic and pharmacological insights for improved outcomes. We discuss the application of graph-based models in capturing biomedical relationships, enabling more precise recommendations, referrals, and disease modeling. In the last section, we highlight reinforcement learning applications for dynamic treatment decisions, including personalized dosing, critical care interventions, and radiation therapy planning.

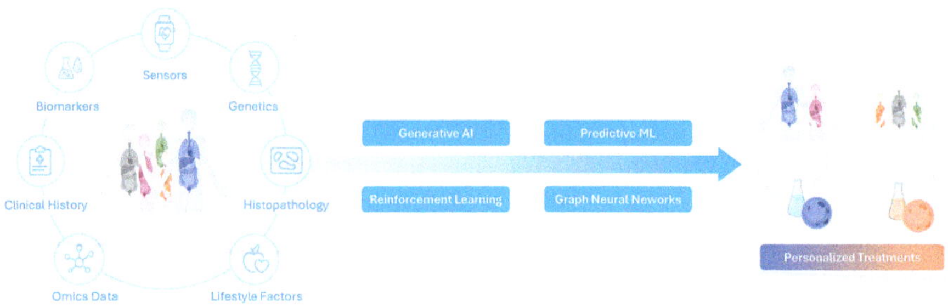

Fig. 1 Integrated patient biomedical data from biomarkers and omics to clinical history and lifestyle factors drive advanced machine learning methodologies for personalized treatment design

2.1.1 From Predictive Modeling to Clinical Pattern Discovery

Predictive machine learning relies on labeled datasets to model relationships between clinical variables and patient outcomes. Classification models categorize patients into predefined groups, such as responders and nonresponders to a specific drug, aiding in treatment selection. Regression models estimate continuous variables, such as drug dosage or disease progression, supporting precise adjustments in therapy. These techniques have been instrumental in applications such as drug response prediction, where models analyze genomic and pharmacological data to optimize cancer therapy selection. Deep learning frameworks have further enhanced predictive accuracy by integrating genetic signatures and drug molecular structures, facilitating novel drug repurposing opportunities.

While predictive models require explicit labels to correlate clinical variables and patient outcomes, unsupervised learning complements them by discovering inherent patterns within unlabeled clinical data, making it particularly valuable for patient stratification and disease subtyping. Clustering techniques group patients based on similarities in genetic, phenotypic, and clinical profiles, revealing novel subpopulations that may respond differently to treatments. The insights gained from unsupervised clustering techniques underscore the necessity of moving beyond traditional generalized treatments, motivating the shift from a one-to-many toward a many-to-one paradigm in drug response prediction.

From One-to-Many to Many-to-One: ML in Drug Response Prediction Historically, therapeutics have followed a one-to-many approach, offering a single scalable solution to a broad patient population to maximize accessibility and outcomes while minimizing side effects. Although this strategy remains regulatory and cost-efficient, it often underestimates individual variability, where genetic differences can influence treatment effectiveness (Kraus 2018). With the advent of next-generation sequencing (NGS) and machine learning, there is the potential to reverse the paradigm to many-to-one solutions allowing for revolutionary science. These tools allow for direct RNA or proteins sequencing, real-time genomic pathogen monitoring, and precision therapy based on personalized genome sequencing (Goodwin et al. 2016).

Machine learning-driven drug response prediction is redefining precision oncology by leveraging large-scale genomic and pharmacological datasets to optimize cancer therapy selection. The development of deep learning models has enabled the analysis of complex genetic signatures alongside drug molecular structures, achieving high predictive performance and uncovering novel opportunities for drug repurposing (Chang et al. 2018). Similarly, methods such as DeepDSC (Li et al. 2019) combine genetic profiles of cancer cells with chemical features of compounds to refine drug sensitivity predictions, outperforming previous approaches. More recent deep learning methodologies integrate gene interactions, expression profiles, and molecular structures to refine drug response predictions, extending to single-cell resolution (Li et al. 2024). Applied to clinical data, these models accurately identified a higher *paclitaxel* response in the pathological complete response (pCR) group compared to the residual disease group, confirming the potential in chemotherapy treatment decisions. In a related approach, (Rakaee et al. 2025) developed a deep learning model using digital pathology slides to predict immunotherapy response in advanced NSCLC. Their framework identified tumor microenvironment features linked to PD-1/PD-L1 inhibitor efficacy, demonstrating an area under the curve (AUC) of 0.66–0.75 and outperforming standard biomarkers when combined with PD-L1 status. Clinically, this tool refines patient selection for immune checkpoint blockade, underscoring how AI can guide therapy allocation for improved oncologic outcomes. To address limitations inherent in conventional machine learning approaches—particularly their assumption of independent variables—GNNs have emerged, explicitly capturing relationships within complex biomedical datasets.

Advancing Precision Therapy Through Graph Neural Networks (GNNs) Modern precision medicine is built upon large-scale heterogeneous data, ranging from patient records and genetic profiles to drug response dynamics and disease networks. A graph, as a mathematical entity, provides a solution to these complex dependencies, which are often treated as independent variables in traditional approaches. It provides a structured way to model biomedical knowledge through entities (such as diseases, medications, or genes) as nodes and their relationships as edges or connections.

Graph neural networks offer a relational understanding for clinical data, including electronic health records (EHRs), genomic profiles, and biomedical knowledge graphs, aligning predictions with how clinicians integrate diverse information when making diagnostic and treatment decisions (Chandak et al. 2023). EHRs have been effectively modelled as graphs encoding diagnoses, medication history, and procedure codes to enhance patient stratification and predict specialty care needs (Poulain and Beheshti 2024). In specialist referrals, GNNs leverage these structured representations to improve access to endocrinology and hematology care by identifying patients who would benefit from timely consultation (Fouladvand et al. 2023).

Medication optimization has also been shown to benefit from graph representations in encoding hierarchical medical knowledge to improve drug recommendations based on longitudinal patient histories (Shang et al. 2019). Representing treatment pathways as graph entities mitigates selection bias and refines personalized prescribing strategies. In oncology, this capability is practical for integrating spatially distinct biomarker data into structured intratumor networks, revealing prognostic patterns with greater detail compared to traditional models (Qiu et al. 2022).

As graph representation learning advances as an active research field, its natural adaptability to chemistry and biological entities positions it to expand beyond the above-mentioned applications in healthcare. This evolution can further enhance treatment personalization, drug discovery, and diagnostic precision, while extending to emerging areas such as drug repurposing and multi-omics integration. By translating diverse biomedical data sources into graph-structured formats, GNNs align closely with clinical reasoning processes, offering practical utility across a spectrum of healthcare applications.

2.1.2 Adaptive Decision-Making and Personalized Treatment with Reinforcement Learning

As a machine learning paradigm, reinforcement Learning (RL) draws inspiration from a multidisciplinary set of fields such as psychology, control theory, and artificial intelligence (Sutton and Barto 1999). It enables agents to learn optimal strategies, or policies, within dynamic environments through continuous exploration and interaction. In contrast to predictive ML, this approach is unaware of precomputed labels; instead refines the decision boundary based on feedback received as rewards or penalties for long-term objectives.

The unique strengths of this methodology allow for its adaptation to real-world clinical scenarios, where patient responses to treatments are often unpredictable and highly individualized. Consider, for instance, a patient as a multidimensional and dynamic environment, a prescription management system as the decision-making agent, and therapeutic interventions (drug dosages or treatment modalities) as the actions. The scalability of this strategy is applicable to various precision medicine contexts, such as preventing hypoglycemic episodes in diabetes, critical care for optimal dosing and interventions, and oncology for personalized radiation therapy planning (Jayaraman et al. 2024).

RL-Driven Strategies for Diabetes Management Building on these principles, RL's capability to learn and adapt through a balance between exploration and exploitation offers significant potential in personalized decision-making. For instance, in diabetes management, this potentially allows for the continuous optimization of insulin dosages in response to real-time glucose levels, physical activity, and diet, improving glycemic control while reducing hypoglycemic risks.

An additional key strength consists in reframing treatments from static diagnosis-based protocols to adaptive, action-oriented strategies, enabling patient-specific care. For individuals with type 1 diabetes (T1D), who constantly monitor insulin levels to maintain stable blood glucose, RL-based closed-loop systems, integrating glucose monitors and insulin pumps, have shown promise in automating this process. These systems, evaluated using an FDA-accepted T1D simulator, have demonstrated performance comparable to traditional methods, enhancing glucose regulation while reducing the risk of severe hypoglycemia, particularly in patients with low insulin sensitivity (Daskalaki et al. 2016). Further advancements include RL-driven decision support systems that personalize insulin dosing for specific scenarios, such as high-fat meals or postprandial exercise. Clinical studies have shown significant improvements in glucose outcomes and reduced hypoglycemia risk, highlighting RL's potential to tailor treatments to individual needs (Jafar et al. 2024). Beyond T1D, RL has also been applied to optimize insulin titration for type 2 diabetes. Frameworks such as *RL-DITR* have demonstrated superior glycemic control compared to both deep learning models and standard clinical methods, with clinical validation from simulation to real-world deployment (Wang et al. 2023a).

Despite advancements in clinical decision-making, the trial-and-error nature of reinforcement learning poses significant challenges to a direct implementation in real-world healthcare settings, coming from the risk of instability and unsafe action selection (Yu et al. 2021). To mitigate these risks, offline RL is often a preferred alternative, leveraging pre-collected clinical data to develop dosing policies in a controlled setting. This enables avoiding direct patient interaction during training, thus enhancing safety while preserving the ability to optimize insulin delivery for effective glucose management (Emerson et al. 2023; Nambiar et al. 2023).

Personalized Dosing and Intervention in Critical Care Clinicians in intensive care units (ICUs) manage complex cases, where treatment decisions such as medication selection and dosing, as well as the timing of critical interventions for oxygen therapy, mechanical ventilation, and sedation must be precisely personalized. In this context, reinforcement learning enables the optimization of clinical decisions based on past experiences, continuously refining treatments in response to patient-specific feedback and evolving conditions.

A leading study explored a feedback-based system for *propofol* anesthesia in human volunteers, demonstrating the effectiveness of continuously monitoring a patient's physiological response for drug delivery real-time adjustments with improved accuracy and stability (Moore et al. 2014). In a recent study (Schamberg et al. 2022), researchers introduced a continuous-action deep model based on the actor-critic framework, an RL algorithm where one network (the actor) selects actions while another (the critic) evaluates their effectiveness. This approach improved the precision of automated *propofol* infusion with an explainable pipeline to address the inherent complexity of the solution through key factors influencing the agent's decisions. More recently, an RL model was trained on data from 3000 patients (Yun et al. 2023), demonstrating the capability of maintaining stable anesthesia under dynamic conditions. Their approach adapted to factors influencing anesthetic control, such as the manual administration of *remifentanil* by anesthesiologists and variations in patient responses during surgery.

Beyond its successful application in anesthesia management, similar strategies have been applied to anticoagulation therapy in urgent care settings, where precise dosing is crucial for patient safety. Medications such as *heparin* have narrow therapeutic windows, and improper dosing can lead to severe complications, prolonged hospital stays, and increased healthcare costs. To address these challenges, a clinician-in-the-loop RL framework learning optimal dosing policies was developed from retrospective electronic medical records, demonstrating improved outcomes compared to traditional guidelines (Nemati et al. 2016). Building on this, (Lin et al. 2018) introduced a deep RL-based heparin dosing system incorporating multi-dimensional patient features, showing a significant correlation between adherence to recommendations and reduced anticoagulant-related complications. In oncology, (Blasiak et al. 2025) piloted a prospective feasibility trial (PRECISE) using CURATE.AI, an RL-like adaptive algorithm for personalizing capecitabine dosing in patients with advanced solid tumors. The platform generates individualized dose recommendations from each patient's small data on tumor response and toxicity, achieving high physician adherence and smooth clinical integration. Though not powered for efficacy endpoints, the trial demonstrated CURATE.AI's adaptability across diverse patient scenarios, underscoring the potential of reinforcement-based approaches to optimize chemotherapy dosing while balancing therapeutic impact and safety.

In critical care settings, the optimization of patient outcomes depends not only on precise dosing but also on the timely administration of interventions (Grap et al. 2012). The personal and dynamic nature of these interventions aligns with RL's ability to exploit individualized response patterns. Specifically, well-timed interventions in ICUs can help prevent ventilator-induced lung injury, enhance patient comfort, and minimize sedation-related complications.

In sepsis care, (Choi et al. 2024) applied deep reinforcement learning to 16,744 hospitalizations to derive an optimal policy for fluid administration, vasopressor use, and ventilation, which retrospectively increased survival from 76.5% to ~84.3%. This approach revealed treatment adjustments that clinicians might overlook, underlining the value of AI-based feedback. While not yet evaluated prospectively, it demonstrates RL's ability to identify data-driven protocols for complex conditions like sepsis, providing a data-driven basis for therapeutic decision support in intensive care.

Previous studies have effectively applied reinforcement techniques to develop intelligent systems that enhance patient-ventilator synchrony and comfort. For instance, (Yu et al. 2019) applied inverse RL to infer clinicians' decision-making processes, aiming to improve mechanical ventilation strategies in intensive care units. Building upon this, they introduced a supervised actor-critic approach to refine ventilatory support and sedation protocols, enabling adaptation to dynamic patient conditions (Yu et al. 2020). Similarly, (Liu et al. 2024) validated an algorithm designed to dynamically adjust mechanical ventilation settings, demonstrating improved respiratory management in critical care environments. The recent COVID-

19 pandemic highlighted the necessity for optimized oxygen delivery strategies. In response, RL-based algorithms have been developed to assist in oxygen therapy decisions for critically ill patients, aiming to maintain optimal oxygenation levels while preventing complications and reducing mortality rates compared to standard care practices (Zheng et al. 2021).

Treatment Planning in Radiation Therapy and Oncology Dynamic treatment planning plays a critical role also in advanced radiation therapy, requiring careful optimization to maximize tumor control while minimizing the exposure to surrounding organs at risk. Among recent techniques, intensity-modulated radiation therapy (IMRT) enables precise dose shaping through inverse planning, a process in which treatment objectives guide an optimization policy to determine machine parameters in order to meet multiple clinical goals.

In prostate cancer treatment, RL has been explored to automate and refine IMRTs and volumetric modulated arc therapy (VMAT) planning (Shen et al. 2021). Through the analysis of prior patient information, these models assist in determining optimal beam parameters and dose distributions, reducing the time required for manual planning while ensuring high-quality treatment delivery (Hrinivich and Lee 2020). Similarly, in lung cancer treatment, it has been applied to adaptive radiotherapy, where treatment plans evolve in response to anatomical or functional changes observed during the course of therapy (Tseng et al. 2017; Niraula et al. 2023). This enables real-time adjustments to radiation dosage, improving tumor targeting while mitigating risks such as radiation-induced lung toxicity.

Advancements in reinforcement learning offer new opportunities in precision medicine, yet present several critical challenges. Defining clinically meaningful environment representations is complex due to the high dimensionality and variability of medical data, while designing appropriate reward functions is complicated by delayed treatment outcomes and the need to balance multiple clinical objectives. Moreover, ensuring that learned policies align with established medical guidelines and safety constraints remains a fundamental concern. Overcoming these challenges requires interdisciplinary collaboration, rigorous validation through prospective clinical trials, and the development of robust evaluation frameworks to ensure the reliability and clinical utility of RL in precision therapy.

2.2 *Generative AI for Personalized Therapeutics*

In previous sections, we outlined the strengths and reliability of the predictive machine learning paradigm. Moving forward, we examine how the latest breakthroughs in Generative AI (GenAI) could redefine digital healthcare in the near future, focusing on the field of personalized therapeutics.

Differently from classification tasks, where the objective is to identify the optimal decision from a predefined set of possible outcomes, GenAI proposes creative solutions and data, building on learned patterns and contextual understanding.

This characteristic supports precision medicine applications, as the dynamic nature of biological systems requires models that can continuously adapt to evolving solutions instead of predetermined ones. In the following sections, we examine how generative models are driving advancements in personalized therapeutics, from drug development to the creation of virtual patient models, emphasizing their role in redefining the future of precision healthcare.

2.2.1 GenAI-Driven Drug Development and Repurposing

A key area where GenAI is making significant strides is drug development and repurposing. Despite recent advancements in biotechnology, drug discovery remains a highly challenging and resource-intensive process, requiring approximately 15 years and $2 billion for a small-molecule drug (Kling 1968). Preclinical research and early-stage development are responsible for a substantial share of expenses due to high attrition rates. Advancing drug discovery by improving target selection, hit identification, and optimized lead compounds with favorable pharmacokinetics could benefit clinical trial failure rates and the availability of effective treatments.

Deep generative models, recently known for their application in human language generation and interaction, are also revolutionizing molecular design with the synthesis of novel compounds tailored to specific pharmacological or chemical properties. Although the data type is structured rather than textual, resembling a graph in which atoms act as nodes and chemical bonds as edges, the underlying theoretical principle remains the same. The objective is to learn a general mathematical law, a probability distribution that governs these entities, describing how atoms arrange and bond to form compounds leveraging large molecular datasets. The prompt or textual input, instead, takes the form of constraints such as tuning binding affinity, solubility, or toxicity, guiding the development of optimized molecular structures.

Recent improvements in generative modeling allow to iteratively refine molecular structures towards desired characteristics (Morehead and Cheng 2024). This has enabled the generation of more stable and viable candidates, addressing challenges such as structural validity and property optimization. To this end, reinforcement learning is widely adopted to guide the generation of synthetic molecules through external feedback, ensuring that the resulting candidates align with therapeutic objectives and have clinical relevance (Popova et al. 2018).

While GenAI has accelerated early-stage drug discovery, its success in clinical applications remains uncertain due to several challenges. A major limitation is the difficulty of generalizing with limited data, as the chemical space far exceeds the limited experimental datasets available from past research. Differently from language models, which continuously expand their knowledge base through social media, research literature, and human feedback, molecular discovery is constrained by limited high-quality data.

The identification of truly novel compounds, beyond minor modifications of existing molecules, remains difficult. Yet, recent studies demonstrate the clinically relevant applications of optimization and repurposing of existing chemical scaffolds. Recent examples include the discovery of the antibacterial compound *abaucin* through ML screening (Liu et al. 2023), TxGNN's repurposing of existing drugs for new therapeutic uses (Huang et al. 2024), and SyntheMol's creation of novel, synthesizable antibiotics (Swanson et al. 2024).

Another significant limitation is the limited understanding of chemical principles, which impacts GenAI-based drug discovery, as it requires insights into how proteins interact with novel molecular structures, an issue absent in protein folding. To address patient-specific variability, these models need complementary advancements in chemistry, biology, and computational modeling through a multidisciplinary approach driving innovation in precision medicine.

2.2.2 Virtual Patient Modeling and Human Digital Twins

While generative AI is transforming drug discovery, another promising direction is emerging with the application of GenAI in virtual patient modeling through human digital twins. A human digital twin (DT) is a virtual and dynamic representation of an individual patient, developed within the integration of different real-time medical sources such as health records, diagnostic results, and wearable devices. A virtual model enables continuous simulation and analysis in a safe multimodal environment with applications in personalized health management, precision therapy effects, and individual risk estimation (Shen et al. 2024).

DTs that simulate specific organs, such as the living heart model (Baillargeon et al. 2014), are mainly developed using detailed imaging data, still providing spatiotemporal profiles of the electrical potential, the mechanical deformation, and the muscle fiber strain. These models are valuable tools for the prediction of mechanical devices performances, such as pacemakers and stents.

Expanding beyond organ-level simulations, disease-specific DTs integrate data across various levels, from organs to tissues and cells, with the objective of representing patient health while enabling the prediction of vital parameters and clinically relevant outcome. Generative models have been applied to clinical datasets to simulate multi-tissue expression data, facilitating the exploration of cytokine interactions within pathways like the renin-angiotensin system. This approach has been useful in identifying inflammatory cytokines, such as CXCR6 and XCL1, which influence blood pressure and are associated with diseases like SARS-CoV-2 infection (Barbiero et al. 2021).

Integrating digital twins at multiple scales in a unified framework for precision poses significant challenges due to dynamic adaptability and the stochastic nature of biological systems. While progress has been made in creating detailed virtual models, achieving mechanistic accuracy and computational feasibility requires standardization, advanced compositional models, and hypothesis-driven generative approaches (De Domenico et al. 2025). Despite these challenges, continued technological advancements promise transformative impacts, positioning digital twins as transformative innovations in individualized healthcare worldwide, capable of substantially improving patient outcomes and redefining the practice of precision therapy. To summarize the methodologies described in this section, Table 1 presents an integrated overview of the proposed paradigms, outlining their conceptual foundations, major clinical applications, and representative clinical studies that illustrate their translational impact.

3 Integrating Multiscale Data in Precision Therapy: Omics, Structural Biology, Clinical, Imaging, and Population Health

An advanced level of precision and patient-specificity can be achieved through the integration of heterogeneous biomedical data sources, each contributing different levels of detail from diverse domains. The multiscale approach not only holds the promise of unprecedented accuracy in targeted therapies, but also enables the discovery of novel clinical insights and knowledge.

In this section, we examine five key data domains, namely omics, structural biology, clinical data, imaging, and population health, and discuss their respective contributions to precision therapeutics through multimodal integration. For each domain, we introduce representative use cases from the past five years, focusing on real patient cohorts or clinical trials. We detail the AI methodologies employed, integration strategies adopted, and the resulting clinical impact, while also discussing emerging challenges and future directions, including applications in pharmacogenomics.

3.1 *Multi-omics Data for Targeted Therapies*

Advances in machine learning have significantly improved genomic variant calling and interpretation, with models like Google's DeepVariant (Poplin et al. 2018) outperforming traditional methods in detecting DNA mutations from sequencing data, highlighting the potential of AI in handling large-scale genomic data. Over the past decades, several global investments have been made in genomic research, recognizing its potential to revolutionize our understanding of diseases and human health for personalized medicine.

From a clinical application perspective, despite substantial early investments, initiatives such as The Cancer Genome Atlas (TCGA), which aimed to provide comprehensive genomic maps of common cancer types, have not met early expectations in revolutionizing targeted treatments (Tannock and Hickman 2016). Similarly, next-generation sequencing technologies of tumor genomes have proven successful

Table 1 Brief conceptual definition and overview of the machine learning methodologies and paradigms discussed in the above-mentioned studies and their translational applications in precision therapeutics

AI methodology	Brief conceptual framing	Clinical applications	Representative studies
Predictive modeling	Uses labeled datasets to model relationships between clinical variables and patient outcomes.	Classification and regression models support treatment selection, dosage adjustment, and prognosis forecasting, enabling more precise and personalized therapeutic strategies.	Cancer drug response prediction (CDRscan) (Chang et al. 2018); Drug sensitivity prediction with DeepDSC (Li et al. 2019); Deep learning integrating gene relationships for drug response (Li et al. 2024); Immunotherapy response prediction in Non-Small Cell Lung Cancer (NSCLC) (Rakaee et al. 2025).
Graph neural networks	Learn from networks where entities are represented as nodes and relationships as edges, capturing complex dependencies in structured and interconnected systems.	Integrating spatial biomarker networks for precision oncology, modeling multi-omics and biomedical knowledge graphs for personalized therapy, optimizing personalized medication recommendations and patient stratification from EHRs.	Knowledge graph for precision medicine (Chandak et al. 2023); Graph transformers for EHR modeling (Poulain and Beheshti 2024); Graph-based specialist referral prediction (Fouladvand et al. 2023); Graph-augmented transformers for recommendation (Shang et al. 2019); Intratumor biomarker networks in oncology (Qiu et al. 2022).
Reinforcement Learning	Learns optimal decision strategies through interaction with dynamic environments, refining actions based on feedback in the form of rewards or penalties to achieve long-term objectives.	Personalized insulin dosing, anesthesia and critical care optimization, anticoagulation therapy management, sepsis treatment planning, ventilator and oxygen therapy adjustment, and adaptive radiotherapy in oncology.	Insulin dosing in diabetes (Daskalaki et al. 2016; Jafar et al. 2024; Wang et al. 2023a; Emerson et al. 2023; Nambiar et al. 2023), anesthesia management (Moore et al. 2014; Schamberg et al. 2022; Yun et al. 2023), anticoagulation therapy (Nemati et al. 2016; Lin et al. 2018), chemotherapy dose personalization (Blasiak et al. 2025), sepsis treatment optimization (Choi et al. 2024), ventilator and oxygen therapy control (Yu et al. 2019; Yu et al. 2020; Liu et al. 2024; Zheng et al. 2021), and adaptive radiotherapy (Shen et al. 2021; Hrinivich and Lee 2020; Tseng et al. 2017; Niraula et al. 2023).

(continued)

Table 1 (continued)

AI methodology	Brief conceptual framing	Clinical applications	Representative studies
Generative AI	Learns underlying data distributions to generate novel, coherent outputs such as text, images, molecular structures, or simulations, enabling creative and adaptive options.	Early-stage drug discovery and repurposing, molecular structure optimization, virtual patient modeling with digital twins, and simulation of multiscale biological systems for personalized therapy development.	Molecule generation and optimization for drug discovery (Morehead and Cheng 2024; Popova et al. 2018; Liu et al. 2023; Swanson et al. 2024), drug repurposing with foundation models (Huang et al. 2024), digital twin development for precision health (Shen et al. 2024; Baillargeon et al. 2014; Barbiero et al. 2021), and multiscale modeling challenges in digital twins (De Domenico et al. 2025).

only in a small percentage of patients and in selected randomized trials for specific therapies.

To address these limitations, research interest has recently shifted toward developing a holistic view of therapeutic vulnerabilities and targets through multi-omics analysis, an area with broad applicability to cancer, neurodegenerative diseases, drug discovery, and beyond. Among the latest investments, the National Institutes of Health (NIH) launched the Multi-Omics for Health and Disease Consortium (MOHD) with a $50.3 million round, aiming to integrate diverse omics data to advance personalized healthcare. Reflecting this momentum, the global multi-omics market that was valued at $2.35 billion in 2023 is projected to expand at an annual growth rate of 15.29% through 2030, driven by advances in omics technologies, single-cell analysis, and demands from the recent COVID-19 pandemic.

In multi-omics settings, genomics provides the foundational genetic architecture, transcriptomics reveals dynamic gene expression patterns, proteomics characterizes the functional effector molecules, and metabolomics captures downstream biochemical phenotypes. The systematic integration of these layers helps construct a quantitative molecular fingerprint that enables: (1) the stratification of disease subtypes beyond histopathology, and (2) the prediction of individualized therapeutic vulnerabilities through cross-omics pathway mapping.

In oncology target discovery, noncoding genomic variants such as promoter mutations may appear undruggable when considered in isolation, yet their integration with proteomic signaling data can reveal downstream pathways with clinically actionable targets (Sammut et al. 2021). The multimodal capabilities of contemporary transformer architectures, which serve as the backbone of many large language models (LLMs), have proven highly effective in managing complex, high-dimensional omics data. These models facilitate tasks ranging from the encoding of DNA and RNA sequences (Wang et al. 2023b) to the application of graph-based approaches for modeling intricate spatial interactions.

Representation Learning in Spatial Multi-Omics Research The integration of multiple layers of molecular data requires complex encoding and alignment of diverse modalities preserving biological relevance and spatial coherence. Each omics layer brings its own information and related limitations. For instance, proteomics tends to capture fewer features but offers a closer link to functional activity, while epigenomic data reflect regulatory potential rather than direct output. As a result, when these layers are projected on a shared tissue section, they form a fragmented view and an uneven sampling of the underlying biology.

An additional challenge in aligning spatially resolved data lies in the definition of a shared distance metric for molecular relationships across various modalities. In many biological contexts, spatially adjacent cells may be functionally distinct, shaped by microenvironmental influences or infiltrating cell populations. Conversely, cells of the same type may occupy distant regions of the tissue while sharing similar molecular profiles.

To address the mismatch between spatial proximity and biological similarity, machine learning fields such as spatial representation learning offer a promising solution for encoding high-dimensional data into unified biological representations. These approaches allow for the joint modeling of molecular profiles and tissue topology, allowing to reconstruct spatial alignment through the discovery of subtle, context-dependent cell states that would otherwise remain hidden. Figure 2 illustrates the workflow behind spatial representation learning in spatial multi-omics. A single tissue slice is analyzed across multiple layers and integrated into a coherent framework. Through this process, the initial overlapping projections are transformed into distinct spatial domains, segmenting the tissue based on molecular patterns rather than arbitrary physical boundaries.

A recent trend in spatial multi-omics research builds on graph neural networks to define network topologies that capture correlated biological domains. The objective is to integrate diverse molecular layers into spatially aware representations of cellular states and their microenvironments, enabling deeper insight into tissue organization, molecular interactions, and cancer biology (Valous et al. 2024). An ambitious effort to integrate spatial multi-omics data across the epigenome, transcriptome, and proteome was introduced with SpatialGlue (Long et al. 2024), a GNN-based architecture. The method builds on a network of spatial and cross-modality relationships to reveal finer tissue structures such as cortical sublayers in the brain and demonstrates robust performance across diverse tissue types.

Beyond structural analysis, spatial multi-omics and systems pharmacology approaches also combine proteomic and transcriptomic data with virtual clinical trial models, predicting responses to treatments such as immunotherapy in hepatocellular carcinoma (Zhang et al. 2024a) . The growing interest in integrated omics representations, combined with mechanistic modeling, is a promising avenue for linking molecular profiles to clinical outcomes and ultimately advancing precise biomarker discovery and personalized therapeutic strategies.

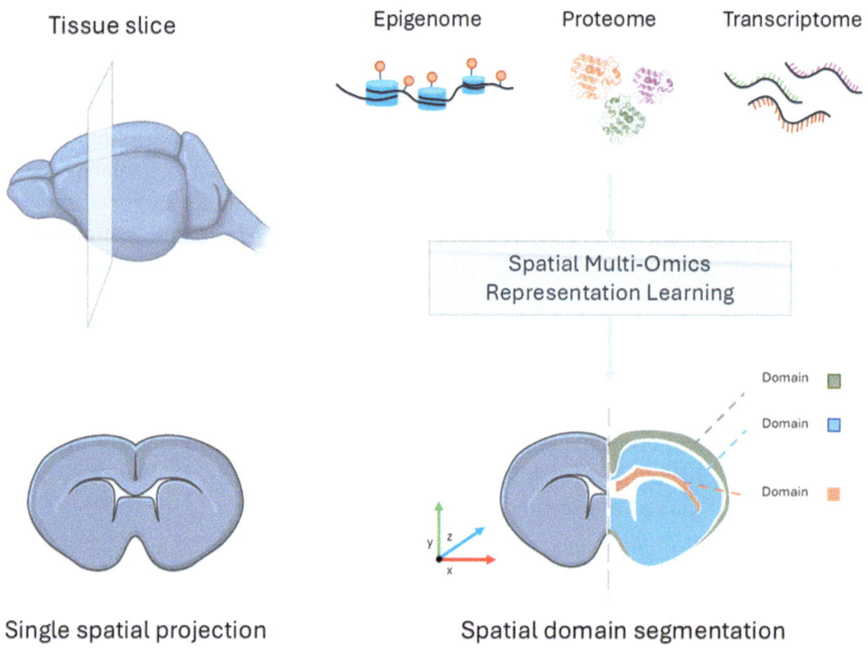

Fig. 2 Overview of spatial representation learning in multi omics. A tissue slice is analyzed across multiple omics layers including epigenomics, proteomics, and transcriptomics, and integrated using spatial multi omics representation learning. This process enables the transformation of single spatial projections into spatial domain segmentations, revealing distinct molecularly defined tissue regions

3.2 *Structural Biology Insights and the Acceleration in Pharmacogenomics*

Complementing previously described omics-based strategies, structural biology is increasingly emerging as a critical bridge between genomic variation and therapeutic intervention, offering powerful techniques to decipher the impact of mutations to molecular functions. As structural data become more deeply integrated into variant analysis, it enables a connection with machine learning tools towards a shift from descriptive genomics to mechanistic, functionally informed insights. This transition is accelerating progress in pharmacogenomics, with direct implications for rare disease diagnosis, cancer therapy, and precision drug development.

To conceptualize the broader implications of this integration, we outline the paradigms shaping the future steps of AI-augmented structural pharmacogenomics.

Biophysically Informed Drug Design At the forefront of these advances, AlphaFold 3 (Abramson et al. 2024) has revolutionized protein structure prediction, a critical aspect of drug discovery. Building on the integration of a generative model in its latest version, it directly processes raw 3D atomic coordinates to predict the structures of proteins, ligands, and antibodies with remarkable accuracy, a task that would otherwise require years using experimental methods. The technique has been validated in the Critical Assessment of Structure Prediction (CASP14) challenge, achieving performance comparable to experimental structures. This advance enables researchers to investigate the effects of genetic variants and model protein-ligand interactions by providing predicted structures for the entire human proteome and hundreds of millions of other proteins.

In the near future, the integration of AI and structural biology is expected to support personalized therapeutic strategies based on a patient's unique molecular profile, such as the structure of tumor-specific mutant proteins. The pharmaceutical industry has rapidly adopted these technologies in the drug discovery pipeline, with several new compounds entering clinical trials (Arnold 2023). Despite the appealing practical benefits, such as identifying new binding pockets for de novo drug design, guiding protein engineering, and predicting off-target effects, recent machine learning applications still require careful analysis. Generative AI models, when applied without chemically informed constraints, can produce invalid or non-synthesizable compounds, which underscores the importance of rigorous validation and expert oversight.

Mechanistic Interpretation of Genetic Variants The success of AlphaFold, which earned its inventors the 2024 Nobel Prize in Chemistry, has also driven significant incremental contributions in related fields, including the interpretation of genomic variants. Missense mutations—single nucleotide changes resulting in the substitution of one amino acid for another in a protein sequence—are often classified as variants of uncertain significance (VUS). As a result, these variants present a major interpretative challenge in clinical genomics given their unclear impact on protein function. Given recent advances in high-confidence protein conformation prediction, AlphaFold facilitates the systematic linking of genotype to phenotype, providing structural insights that help elucidate the functional consequences of these mutations.

In parallel, AlphaMissense model (Cheng et al. 2023) combines AlphaFold-derived structural context with evolutionary information to predict the pathogenicity of missense variants across the entire human proteome. Notably, this technique achieved state-of-the-art performance, accurately classifying 89% of all possible missense variants as likely pathogenic or benign. Recent advances in disease-specific variant effect prediction models, particularly in cardiovascular contexts, have further enabled more accurate and clinically meaningful interpretation of rare genetic variants (Zhan et al. 2025). From an application perspective, AI-based structure prediction is increasingly used to prioritize high-risk variants for diagnosis and therapeutic targeting. In oncology, a straightforward example consists in the identification of tumor mutations altering drug-binding sites.

3.3 *Clinical, Imaging, and Epidemiological Data for Next-Generation Therapeutics*

While molecular and structural data analyses enable therapy personalization at the target level, patient-level clinical data are essential for modeling disease progression and therapeutic outcomes. Modern healthcare repositories, including EHRs, clinical trial data, and patient registries, capture rich and diverse phenotypic profiles (diagnoses, lab results, medications, clinical notes). When effectively mined, these data sources offer a foundation for translating clinical information into actionable insights with the design of properly engineered ML architectures.

Nevertheless, clinical, imaging, and population-level datasets present unique challenges that differ from molecular data. These sources are often high in dimensionality, irregularly sampled, and predominantly unstructured, requiring tailored computational approaches. Recent advances in deep learning have made significant progress in leveraging clinical big data to inform care, from risk prediction to treatment selection (Heumos et al. 2024). This progression is visually summarized in Fig. 3, which highlights selected layers across biomedical domains such as radiomics, clinical records, digital pathology, and functional imaging, illustrating the potential integration of multiscale information to support precision therapy through machine learning techniques.

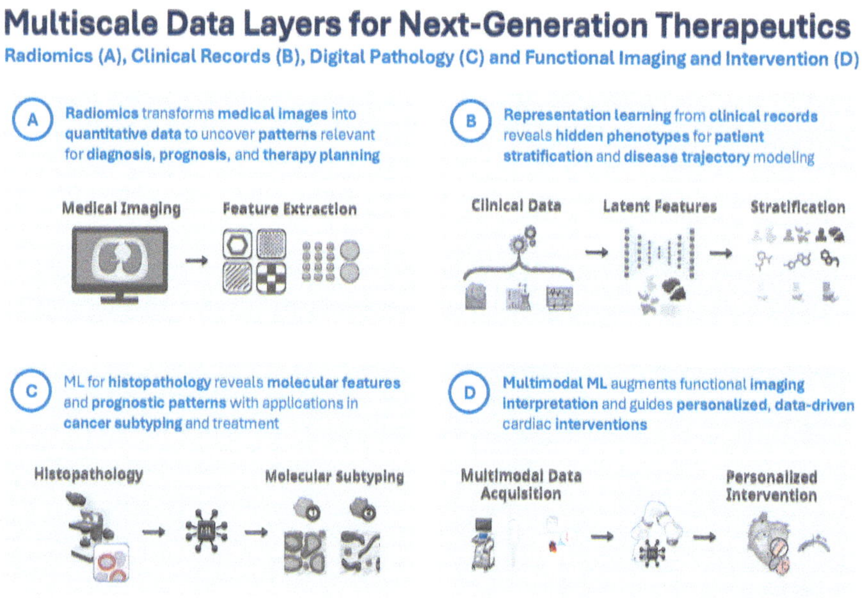

Fig. 3 Overview of machine learning applications across multiscale biomedical data domains, illustrating their role in advancing precision therapy through integrated insights from imaging, clinical records, pathology, and intervention

3.3.1 Learning from Clinical Records for Phenotyping and Personalization

The large-scale availability of routine clinical data has opened new possibilities for systematic deep learning approaches, which are designed to capture hidden patterns in patient trajectories. Instead of relying on curated variables or predefined scores, modern architectures are now capable of learning directly from raw EHRs, capturing complex temporal patterns and subtle clinical signals often overlooked in traditional frameworks.

Building on these predictive capabilities, a clinical application of this progress was presented by Google researchers training a deep neural network on de-identified hospital records (Rajkomar et al. 2018). Using clinical notes, laboratory results, and vital signs, the model predicted in-hospital mortality and readmission risk more accurately than traditional severity scoring systems.

Similarly, unsupervised representation learning can compress multiyear patient records into compressed numerical representations that summarize their clinical profiles. This methodology reduces data dimensionality by learning patterns through data reconstruction. In a recent study involving ~1.6 million patients, these patterns were used to stratify individuals into subgroups with distinct disease subtypes and progression trajectories (Landi et al. 2020). This approach uncovered clinically relevant subtypes of type II diabetes and Parkinson's disease, characterized by differing comorbidities and rates of progression. Extending this approach, (Orcutt et al. 2025) integrated large-scale EHR data from Flatiron Health with clinical trial protocols to emulate 11 landmark oncology studies. They applied multiple survival-based ML models (ensemble trees, survival forests, support vector machines, and Cox variants) to estimate metastatic mortality risk, then used unsupervised clustering to categorize patients into low-, medium-, and high-risk phenotypes. Partial synthetic data simulations helped stress-test assumptions about confounding, revealing that standard trial results may overestimate benefit for high-risk populations. This synergy of real-world records, advanced ML, and trial emulation underscores how multiscale data can refine clinical insights and guide more precise therapy.

Data-driven phenotypes have emerged as a valuable tool in therapeutics by identifying which patients are most likely to respond to specific interventions. In response to the complexity of analyzing longitudinal clinical data, ML architectures are evolving towards foundation models and LLMs capable of processing vast and heterogeneous inputs. A language model (Yang et al. 2022) with 8.9 billion parameters was trained on over 90 billion words from UF Health, PubMed, Wikipedia, and MIMIC-III. It achieved state-of-the-art performance on several clinical NLP tasks, including concept extraction, inference, and question answering. In a related development, (Heinz et al. 2025) conducted a first-of-its-kind randomized controlled trial evaluating a generative AI chatbot ("Therabot") for mental health care. Over 100 patients diagnosed with depression, anxiety, or eating disorders were randomly assigned to engage with the LLM-based chatbot or a control. After eight weeks, chatbot users showed ~51% reductions in depression severity, on par with standard in-person therapy. The authors highlight the potential for on-demand AI counseling, though they note that human clinician oversight remains essential.

Although the field is advancing rapidly, several challenges remain in integrating AI-driven clinical insights and therapeutics. A primary concern is data heterogeneity since EHRs often originate from different institutions, each with its own formats and coding standards. In addition to this, routine clinical data are rarely collected with research in mind, leading to inconsistencies and biases. As a result, without proper engineering, algorithms risk capturing confounders rather than clinically meaningful patterns. To address these issues, standards such as Fast Healthcare Interoperability Resources (FHIR (Vorisek et al. 2022)) are being adopted to improve interoperability.

Another common challenge is the interpretability of increasingly complex models, which remains a critical barrier to clinical adoption and trust. Survey data indicate that stakeholders perceive data quality, privacy, and trust as the main short-term barriers, with ethical concerns and AI dependence rising over time. Addressing these concerns through explainability techniques, collaborative engagement, and thoughtful regulatory guidance will be essential to support the safe and meaningful integration of AI into precision healthcare.

3.3.2 Medical Imaging as a Spatial and Functional Layer

Complementing clinical records, medical imaging offers a complementary layer to clinical data covering several technologies to view the human body capturing the spatial, structural, and functional characteristics of disease. Computer vision, as a machine learning field concerning image analysis, has direct applications in diagnostic and prognostic imaging, histology, pathology, therapy, and planning.

Radiomics and Imaging-Based Biomarker Discovery Among the fields in which these techniques have found early and impactful applications, there is radiomics. While traditional radiology focuses on a limited set of visible abnormalities, the latter can compute hundreds of complex descriptors including texture, shape, and intensity patterns for complex morphological analysis.

In one of the earliest large-scale studies, radiomics was applied to Computed Tomography (CT) scans from over 1000 patients with lung or head-and-neck cancers, extracting an average of 440 features per tumor (Aerts et al. 2014). Many of these imaging features had independent prognostic value for survival, often capturing clinically relevant details overlooked in traditional practice. Interestingly, the top radiomic signature, which reflected intratumor heterogeneity in the images, correlated with underlying gene expression profile in the tumors. The direct link to genomics shows that routine imaging can reveal molecular tumor traits, where high imaging heterogeneity may signal a more aggressive disease and inform decisions about combination therapy or increased surveillance.

Advances in Real-Time Imaging Analytics CT-based AI is proving its efficiency in supporting time-critical decisions and interventions with notable applications in neurology. In stroke treatment, which depends on rapid imaging to identify vessel occlusions or brain hemorrhages, a cluster-randomized trial demonstrated sped up endovascular therapy times (Martinez-Gutierrez et al. 2023). A continuous integration framework enabled the analysis of CT scans and alerted neurointerventional teams via a mobile app when a blockage was detected, reducing time-to-treatment intervals. The success of these technologies highlights the importance of an engineered integration of AI into triage and clinical practice, ensuring that patients receive the right therapy promptly. In neurodegenerative diseases, an active field is exploring AI analysis of MRI and PET images to predict which patients with mild cognitive impairment are more likely to progress to Alzheimer's disease (Xue et al. 2024) or to monitor treatment effects in multiple sclerosis lesions (Yousef et al. 2024), though such applications are still emerging in clinical practice.

Digital Pathology and Treatment Planning Further extending these capabilities, radiomics, and real-time image analysis, digital pathology represents another fertile domain for computer vision applications in precision diagnostics. Histopathology slides contain structural and molecular complexity hiding features beyond human visual interpretation. Deep foundation models for vision achieved expert-level performance in classifying non-small cell lung cancer subtypes while also predicting mutations in key oncogenic drivers, such as EGFR and STK11, directly from tissue morphology (Coudray et al. 2018). In a large-scale digital pathology study, (Amgad et al. 2024) introduced the Histomic Prognostic Signature (HiPS)—a deep learning model trained on over 1000 H&E slides of breast tumors and validated across multiple cohorts (TCGA, PLCO, CPS-II, CPS-3). By integrating histopathology features with clinical outcomes and genomic data from TCGA, the AI outperformed expert pathologists in predicting long-term patient survival. Notably, HiPS reclassified certain high-risk patients as having unexpectedly favorable prognoses, potentially sparing them unnecessary chemotherapy. This multimodal approach—combining imaging, clinical, and molecular data—underscores the transformative power of AI-driven digital pathology in precision cancer therapy.

These findings highlight the potential of deep learning for routine pathology, offering a rapid triage tool to flag likely mutation carriers and accelerate targeted interventions, even as confirmatory laboratory testing remains crucial. In therapy planning workflows, particularly in radiation oncology, tumor contouring on CT scans, a time-intensive task prone to interobserver variability, is increasingly being automated through CV systems (Hosny et al. 2022). Beyond planning, AI-powered navigation systems in interventional radiology and surgical oncology are beginning to improve real-time imaging by delineating tumor margins and guiding resections with greater spatial precision. Although still in the early stages, the integration of these methodologies points toward the inclusion of clinical copilot systems, supporting surgeons in maximizing tumor removal while preserving healthy tissue (Varghese et al. 2024).

Digital Cardiology Machine learning is transforming cardiovascular imaging by enhancing acquisition, interpretation, and decision-making, while integrating data from EHRs and pathology to enable precise diagnosis and personalized therapy. Specifically, its applications in image segmentation and automated diagnosis hold the potential to reduce healthcare costs and improve patient care value (Dey et al. 2019).

In interventional cardiology, machine learning models contribute to real-time decision support, streamlining workflows and standardizing procedures through advanced robotics. It enhances cognitive output in visual imaging interpretation, promising significant improvements in clinical practice (Sardar et al. 2019). In a recent multicenter trial, (Deisenhofer et al. 2025) demonstrated an AI-guided ablation strategy for persistent atrial fibrillation using a hybrid ensemble model (extreme gradient boosting + convolutional neural network). Their system identified regions of spatiotemporal dispersion on intracardiac electrograms, guiding more personalized ablation beyond standard pulmonary vein isolation. Patients in the AI-tailored arm achieved higher one-year freedom from AF (88% vs. 70%) with similar safety outcomes, highlighting the potential of AI to improve interventional efficacy. This example further illustrates how data-driven procedures can transform clinical decision-making in advanced cardiac electrophysiology. This technological evolution is increasingly being integrated into various aspects of Cardiac Magnetic Resonance Imaging (CMR), accelerating image reconstruction, improving efficiency, and contributing to automatic artifact correction and enhanced image quality (Zhang et al. 2024b). It has also been shown to support myocardial tissue characterization, the detection of myocardial infarction and scarring, and the automatic quantification of cardiac structures such as the volumes and masses of the left and right ventricles, thereby minimizing the need for manual correction (Lanzafame et al. 2023).

4 Future Directions: The Next Frontier in AI-Driven Precision Therapy

Looking ahead, AI in precision therapy is expected to drive a shift from a reactive model to a predictive and dynamically adaptive discipline. Future precision therapeutics will likely center around integrated, multimodal decision-support systems that combine genomic, imaging, clinical, and real-time sensor data to tailor interventions with increasing specificity. This anticipatory approach aims not only to respond to diagnosed illnesses but also to predict and mitigate disease progression before clinical symptoms arise. Rather than replacing clinicians, AI is emerging as an essential collaborative partner, augmenting human decision-making through real-time insights and personalized recommendations. As these technologies mature, the expansion of AI applications into chronic disease management and preventative care is expected to drive broader adoption beyond specialized, high-cost therapies, making precision healthcare more accessible at the population level.

As the foundations for next-generation therapeutics strengthen, new frontiers are beginning to emerge at the interface of molecular biology, computation, and clinical innovation. RNA editing technologies, capable of transiently modifying gene expression without altering the underlying genome, offer a glimpse into how therapeutic modulation could one day operate at unprecedented levels of specificity, addressing diseases such as cancer, cardiovascular, and metabolic disorders with increased flexibility (Van Norden et al. 2024). Parallel advances in patient-specific modeling, including digital twin simulations from multiscale biomedical data, promise a future where therapeutic strategies are optimized in silico before clinical application, reducing risks and personalizing care dynamically. Breakthroughs in quantum computing could further enable real-time optimization of complex treatment decisions, while AI-guided synthetic biology may accelerate the development of bespoke molecular therapies.

Despite these advancements, significant challenges remain in deploying AI-driven precision therapy in routine clinical practice. The increasing volume, velocity, and heterogeneity of health data, interoperability issues across health systems, and the need for robust validation frameworks pose barriers to widespread adoption (Bera et al. 2019; Habuza et al. 2021). Ethical considerations, including algorithmic bias and patient privacy concerns (Naik et al. 2024), require strategic policy decisions, stringent regulatory oversight, and transparent model development (Xu et al. 2019). Furthermore, the integration of AI into clinical workflows necessitates physician training and trust in AI-driven recommendations, emphasizing the need for human-AI collaboration (in particular between AI developers and clinicians) rather than full automation (Johnson et al. 2021). Bridging the gap between clinics and AI research will require a collaborative effort among stakeholders to ensure the ethical and effective implementation of AI in precision therapy. Finally, the complexity and multidimensionality of medical data pose significant technical challenges in developing AI solutions that can generalize across diverse populations and diseases.

Beyond technical innovation, the future of precision therapy will increasingly depend on how intelligently AI systems are integrated into clinical practice. LLMs, in particular, offer both transformative potential and serious caution (Clusmann et al. 2023). If their well-known limitations, such as hallucinations and data privacy vulnerabilities, can be effectively mitigated through domain-specific fine tuning, real-time validation, and regulatory oversight, they may become trusted collaborators in diagnosis, triage, and patient communication. Rather than supplanting clinical judgment, responsibly developed LLMs could strengthen decision-making at the bedside while preserving physician authority. Ultimately, the next frontier in AI-driven precision medicine will not emerge from autonomous algorithms alone, but through partnerships where human expertise and machine intelligence converge to anticipate, adapt, and transform the delivery of individualized healthcare.

References

Abramson J, Adler J, Dunger J, Evans R, Green T, Pritzel A, et al. Accurate structure prediction of biomolecular interactions with AlphaFold 3. Nature. 2024;630(8016):493–500.

Aerts HJ, Velazquez ER, Leijenaar RT, Parmar C, Grossmann P, Carvalho S, et al. Decoding tumour phenotype by noninvasive imaging using a quantitative radiomics approach. Nat Commun. 2014;5(1):4006.

AI's. potential to accelerate drug discovery needs a reality check. Nature. 2023;622(7982):217. https://doi.org/10.1038/d41586-023-03172-6.

Amgad M, Hodge JM, Elsebaie MA, Bodelon C, Puvanesarajah S, Gutman DA, et al. A population-level digital histologic biomarker for enhanced prognosis of invasive breast cancer. Nat Med. 2024;

Arnold C. Inside the nascent industry of AI-designed drugs. Nat Med. 2023;29:1292–5. https://doi.org/10.1038/s41591-023-02361-0.

Association, A. M. AMA augmented intelligence research: physician sentiments around the use of AI in heath care: motivations, opportunities, risks, and use cases. 2023.

Baillargeon B, Rebelo N, Fox DD, Taylor RL, Kuhl E. The living heart project: a robust and integrative simulator for human heart function. Eur J Mech A/Solids. 2014;48:38–47. https://doi.org/10.1016/j.euromechsol.2014.04.001.

Barbiero P, Vinas Torne R, Lió P. Graph representation forecasting of patient's medical conditions: toward a digital twin. Front Genet. 2021;12:652907. https://doi.org/10.3389/fgene.2021.652907.

Bera K, Schalper KA, Rimm DL, Velcheti V, Madabhushi A. Artificial intelligence in digital pathology–new tools for diagnosis and precision oncology. Nat Rev Clin Oncol. 2019;16(11):703–15.

Blasiak A, Truong AT, Foo N, Tan LW, Kumar KS, Tan SB, et al. Personalized dose selection platform for patients with solid tumors in the PRECISE CURATE. AI feasibility trial. npj Precis Oncol. 2025;9(1):49. https://doi.org/10.1038/s41698-025-00835-7.

Bzdok D, Altman N, Krzywinski M. Statistics versus machine learning. Nat Methods. 2018;15(4):233–4. https://doi.org/10.1038/nmeth.4642.

Calaprice-Whitty D, Galil K, Salloum W, Zariv A, Jimenez B. Improving clinical trial participant prescreening with artificial intelligence (AI): a comparison of the results of AI-assisted vs standard methods in 3 oncology trials. Ther Innov Regul Sci. 2020;54:69–74.

Chandak P, Huang K, Zitnik M. Building a knowledge graph to enable precision medicine. Sci Data. 2023;10(1):67. https://doi.org/10.1038/s41597-023-01960-3.

Chang Y, Park H, Yang HJ, Lee S, Lee KY, Kim TS, et al. Cancer drug response profile scan (CDRscan): a deep learning model that predicts drug effectiveness from cancer genomic signature. Sci Rep. 2018;8(1):8857. https://doi.org/10.1038/s41598-018-27214-6.

Cheng J, Novati G, Pan J, Bycroft C, Žemgulytė A, Applebaum T, et al. Accurate proteome-wide missense variant effect prediction with AlphaMissense. Science. 2023;381(6664)

Choi Y, Oh S, Huh JW, Joo HT, Lee H, You W, et al. Deep reinforcement learning extracts the optimal sepsis treatment policy from treatment records. Commun Med. 2024;4(1):245. https://doi.org/10.1038/s43856-024-00665-x.

Clusmann J, Kolbinger FR, Muti HS, Carrero ZI, Eckardt JN, Laleh NG, et al. The future landscape of large language models in medicine. Commun Med. 2023;3(1):141.

Coudray, N., Ocampo, P. S., Sakellaropoulos, T., Narula, N., Snuderl, M., Fenyö, D., … Tsirigos, A. (2018). Classification and mutation prediction from non–small cell lung cancer histopathology images using deep learning. Nat Med, 24(10), 1559–1567.

Daskalaki E, Diem P, Mougiakakou SG. Model-free machine learning in biomedicine: feasibility study in type 1 diabetes. PLoS One. 2016;11(7):e0158722. https://doi.org/10.1371/journal.pone.0158722.

De Domenico M, Allegri L, Caldarelli G, d'Andrea V, Di Camillo B, Rocha LM, et al. Challenges and opportunities for digital twins in precision medicine from a complex systems perspective. NPJ Digit Med. 2025;8(1):37. https://doi.org/10.1038/s41746-024-01402-3.
Deisenhofer I, Albenque JP, Busch S, Gitenay E, Mountantonakis SE, Roux A, et al. Artificial intelligence for individualized treatment of persistent atrial fibrillation: a randomized controlled trial. Nat Med. 2025:1–8.
Dey D, Slomka PJ, Leeson P, Comaniciu D, Shrestha S, Sengupta PP, Marwick TH. Artificial intelligence in cardiovascular imaging: JACC state-of-the-art review. J Am Coll Cardiol. 2019;73(11):1317–35.
Emerson H, Guy M, McConville R. Offline reinforcement learning for safer blood glucose control in people with type 1 diabetes. J Biomed Inform. 2023;142:104376. https://doi.org/10.1016/j.jbi.2023.104376.
Fouladvand S, Gomez FR, Nilforoshan H, Schwede M, Noshad M, Jee O, et al. Graph-based clinical recommender: predicting specialists procedure orders using graph representation learning. J Biomed Inform. 2023;143:104407. https://doi.org/10.1016/j.jbi.2023.104407.
Goodwin S, McPherson JD, McCombie WR. Coming of age: ten years of next-generation sequencing technologies. Nat Rev Genet. 2016;17(6):333–51. https://doi.org/10.1038/nrg.2016.49.
Grap MJ, Munro CL, Wetzel PA, Best AM, Ketchum JM, Hamilton VA, et al. Sedation in adults receiving mechanical ventilation: physiological and comfort outcomes. Am J Crit Care. 2012;21(3):e53–64. https://doi.org/10.4037/ajcc2012301.
Habuza T, Navaz AN, Hashim F, Alnajjar F, Zaki N, Serhani MA, Statsenko Y. AI applications in robotics, diagnostic image analysis and precision medicine: current limitations, future trends, guidelines on CAD systems for medicine. Inf Med Unlocked. 2021;24:100596.
Heinz MV, Mackin DM, Trudeau BM, Bhattacharya S, Wang Y, Banta HA, et al. Randomized trial of a generative AI Chatbot for mental health treatment. NEJM AI. 2025;2(4)
Heumos L, Ehmele P, Treis T, Upmeierzu Belzen J, Roellin E, May L, et al. An open-source framework for end-to-end analysis of electronic health record data. Nat Med. 2024;30(11):3369–80.
Hosny A, Bitterman DS, Guthier CV, Qian JM, Roberts H, Perni S, et al. Clinical validation of deep learning algorithms for radiotherapy targeting of non-small-cell lung cancer: an observational study. Lancet Digit Health. 2022;4(9):e657–66.
Hrinivich WT, Lee J. Artificial intelligence-based radiotherapy machine parameter optimization using reinforcement learning. Med Phys. 2020;47(12):6140–50. https://doi.org/10.1002/mp.14544.
Huang K, Chandak P, Wang Q, Havaldar S, Vaid A, Leskovec J, et al. A foundation model for clinician-centered drug repurposing. Nat Med. 2024;30(12):3601–13. https://doi.org/10.1101/2023.03.19.23287458.
Jafar A, et al. Personalized insulin dosing using reinforcement learning for high-fat meals and aerobic exercises in type 1 diabetes: a proof-of-concept trial. Nat Commun. 2024;15:1. https://doi.org/10.1038/s41467-024-50764-5.
Jayaraman P, et al. A primer on reinforcement learning in medicine for clinicians. npj Digit Med. 2024;7:1. https://doi.org/10.1038/s41746-024-01316-0.
Johnson KB, Wei WQ, Weeraratne D, Frisse ME, Misulis K, Rhee K, et al. Precision medicine, AI, and the future of personalized health care. Clin Transl Sci. 2021;14(1):86–93.
Kling RE. The government printing office. Am J Pharm Educ. 1968;32(2):211–8. https://doi.org/10.1016/s0002-9459(24)09708-0.
Kraus VB. Biomarkers as drug development tools: discovery, validation, qualification and use. Nat Rev Rheumatol. 2018;14(6):354–62. https://doi.org/10.1038/s41584-018-0005-9.
Landi I, Glicksberg BS, Lee HC, Cherng S, Landi G, Danieletto M, et al. Deep representation learning of electronic health records to unlock patient stratification at scale. NPJ Digit Med. 2020;3(1):96.
Lanzafame LR, Bucolo GM, Muscogiuri G, Sironi S, Gaeta M, Ascenti G, et al. Artificial intelligence in cardiovascular CT and MR imaging. Life. 2023;13(2):507.

Li M, Wang Y, Zheng R, Shi X, Li Y, Wu FX, Wang J. DeepDSC: a deep learning method to predict drug sensitivity of cancer cell lines. IEEE/ACM Trans Comput Biol Bioinform. 2019;18(2):575–82. https://doi.org/10.1109/tcbb.2019.2919581.

Li P, Jiang Z, Liu T, Liu X, Qiao H, Yao X. Improving drug response prediction via integrating gene relationships with deep learning. Brief Bioinform. 2024;25(3) https://doi.org/10.1093/bib/bbae153.

Lin R, Stanley MD, Ghassemi MM, Nemati S. A deep deterministic policy gradient approach to medication dosing and surveillance in the ICU. In: 2018 40th Annual international conference of the IEEE Engineering in Medicine and Biology Society (EMBC); 2018. p. 4927–31. https://doi.org/10.1109/embc.2018.8513203.

Liu G, Catacutan DB, Rathod K, Swanson K, Jin W, Mohammed JC, et al. Deep learning-guided discovery of an antibiotic targeting Acinetobacter baumannii. Nat Chem Biol. 2023;19(11):1342–50. https://doi.org/10.1038/s41589-023-01349-8.

Liu S, Xu Q, Xu Z, Liu Z, Sun X, Xie G, et al. Reinforcement learning to optimize ventilator settings for patients on invasive mechanical ventilation: retrospective study. J Med Internet Res. 2024;26:e44494. https://doi.org/10.2196/preprints.44494.

Long Y, Ang KS, Sethi R, Liao S, Heng Y, van Olst L, et al. Deciphering spatial domains from spatial multi-omics with SpatialGlue. Nat Methods. 2024;21(9):1658–67.

MacEachern SJ, Forkert ND. Machine learning for precision medicine. Genome. 2021;64(4):416–25. https://doi.org/10.1139/gen-2020-0131.

Martinez-Gutierrez JC, Kim Y, Salazar-Marioni S, Tariq MB, Abdelkhaleq R, Niktabe A, et al. Automated large vessel occlusion detection software and thrombectomy treatment times: a cluster randomized clinical trial. JAMA Neurol. 2023;80(11):1182–90.

Moore BL, Pyeatt LD, Kulkarni V, Panousis P, Padrez K, Doufas AG. Reinforcement learning for closed-loop Propofol anesthesia: a human volunteer study. In: Proceedings of the AAAI conference on Artificial Intelligence, vol. 15, issue 1; 2014. p 655–96. https://doi.org/10.1609/aaai.v24i2.18817.

Morehead A, Cheng J. Geometry-complete diffusion for 3D molecule generation and optimization. Commun Chem. 2024;7(1):150. https://doi.org/10.1038/s42004-024-01233-z.

Naik K, Goyal RK, Foschini L, Chak CW, Thielscher C, Zhu H, et al. Current status and future directions: the application of artificial intelligence/machine learning for precision medicine. Clin Pharmacol Ther. 2024;115(4):673–86.

Nambiar M, Ghosh S, Ong P, Chan YE, Bee YM, Krishnaswamy P. Deep offline reinforcement learning for real-world treatment optimization applications. In: Proceedings of the 29th ACM SIGKDD conference on knowledge discovery and data mining; 2023. p. 4673–84. https://doi.org/10.1145/3580305.3599800.

The Future of Medicine. Nat Med. 2025;31(1):1–1. https://doi.org/10.1038/s41591-024-03464-y.

Nemati S, Ghassemi MM, Clifford GD. Optimal medication dosing from suboptimal clinical examples: a deep reinforcement learning approach. In: 2016 38th Annual international conference of the IEEE Engineering in Medicine and Biology Society (EMBC); 2016. p. 2978–81. https://doi.org/10.1109/embc.2016.7591355.

Niraula D, Sun W, Jin J, Dinov ID, Cuneo K, Jamaluddin J, et al. A clinical decision support system for AI-assisted decision-making in response-adaptive radiotherapy (ARCliDS). Sci Rep. 2023;13(1):5279. https://doi.org/10.1038/s41598-023-32032-6.

Ong JCL, Chang SYH, William W, Butte AJ, Shah NH, Chew LST, et al. Medical ethics of large language models in medicine. NEJM AI. 2024;1(7) https://doi.org/10.1056/aira2400038.

Orcutt X, Chen K, Mamtani R, Long Q, Parikh RB. Evaluating generalizability of oncology trial results to real-world patients using machine learning-based trial emulations. Nat Med. 2025:1–9.

Poplin R, Chang PC, Alexander D, Schwartz S, Colthurst T, Ku A, et al. A universal SNP and small-indel variant caller using deep neural networks. Nat Biotechnol. 2018;36(10):983–7.

Popova M, Isayev O, Tropsha A. Deep reinforcement learning for de novo drug design. Sci Adv. 2018;4(7) https://doi.org/10.1126/sciadv.aap7885.

Poulain R, Beheshti R. Graph transformers on EHRs: better representation improves downstream performance. In: The twelfth international conference on learning representations. 2024.

Qiu L, Kang D, Wang C, Guo W, Fu F, Wu Q, et al. Intratumor graph neural network recovers hidden prognostic value of multi-biomarker spatial heterogeneity. Nat Commun. 2022;13(1):4250. https://doi.org/10.1038/s41467-022-31771-w.

Rajkomar A, Oren E, Chen K, Dai AM, Hajaj N, Hardt M, et al. Scalable and accurate deep learning with electronic health records. NPJ Digit Med. 2018;1(1):18.

Rakaee M, Tafavvoghi M, Ricciuti B, Alessi JV, Cortellini A, Citarella F, et al. Deep learning model for predicting immunotherapy response in advanced non– small cell lung cancer. JAMA Oncol. 2025;11(2):109–18. https://doi.org/10.1001/jamaoncol.2024.5356.

Sammut S-J, Crispin-Ortuzar M, Chin S-F, Provenzano E, Bardwell HA, Ma W, et al. Multi-omic machine learning predictor of breast cancer therapy response. Nature. 2021;601(7894):623–9. https://doi.org/10.1038/s41586-021-04278-5.

Sardar P, Abbott JD, Kundu A, Aronow HD, Granada JF, Giri J. Impact of artificial intelligence on interventional cardiology: from decision-making aid to advanced interventional procedure assistance. Cardiovasc Intervent. 2019;12(14):1293–303.

Schamberg G, Badgeley M, Meschede-Krasa B, Kwon O, Brown EN. Continuous action deep reinforcement learning for propofol dosing during general anesthesia. Artif Intell Med. 2022;123:102227. https://doi.org/10.1016/j.artmed.2021.102227.

Shang J, Ma T, Xiao C, Sun J. Pre-training of graph augmented transformers for medication recommendation. In: Proceedings of the twenty-eighth international joint conference on Artificial Intelligence, IJCAI-19; 2019. p. 5953–9. https://doi.org/10.24963/ijcai.2019/825.

Shen C, Chen L, Gonzalez Y, Jia X. Improving efficiency of training a virtual treatment planner network via knowledge-guided deep reinforcement learning for intelligent automatic treatment planning of radiotherapy. Med Phys. 2021;48(4):1909–20. https://doi.org/10.1002/mp.14712.

Shen MD, Chen SB, Ding XD. The effectiveness of digital twins in promoting precision health across the entire population: a systematic review. NPJ Digit Med. 2024;7(1):145. https://doi.org/10.1038/s41746-024-01146-0.

Sutton RS, Barto AG. Reinforcement learning. J Cogn Neurosci. 1999;11(1):126–34. https://doi.org/10.1016/s1364-6613(99)01331-5.

Swanson K, Liu G, Catacutan DB, Arnold A, Zou J, Stokes JM. Generative AI for designing and validating easily synthesizable and structurally novel antibiotics. Nat Mach Intell. 2024;6(3):338–53. https://doi.org/10.1038/s42256-024-00809-7.

Tannock IF, Hickman JA. Limits to personalized cancer medicine. N Engl J Med. 2016;375(13):1289–94.

Tikhomirov L, Semmler C, McCradden M, Searston R, Ghassemi M, Oakden-Rayner L. Medical artificial intelligence for clinicians: the lost cognitive perspective. Lancet Digit Health. 2024;6(8):e589–94. https://doi.org/10.1016/s2589-7500(24)00095-5.

Tseng HH, Luo Y, Cui S, Chien JT, Ten Haken RK, Naqa IE. Deep reinforcement learning for automated radiation adaptation in lung cancer. Med Phys. 2017;44(12):6690–705. https://doi.org/10.1002/mp.12625.

Valous NA, Popp F, Zörnig I, Jäger D, Charoentong P. Graph machine learning for integrated multi-omics analysis. Br J Cancer. 2024;131(2):205–11.

Van Norden M, Falls Z, Mandloi S, Segal BH, Baysal BE, Samudrala R, Elkin PL. The implications of APOBEC3-mediated C-to-U RNA editing for human disease. Commun Biol. 2024;7(1):529.

Varghese C, Harrison EM, O'Grady G, Topol EJ. Artificial intelligence in surgery. Nat Med. 2024;30(5):1257–68.

Vorisek CN, Lehne M, Klopfenstein SAI, Mayer PJ, Bartschke A, Haese T, Thun S. Fast healthcare interoperability resources (FHIR) for interoperability in health research: systematic review. JMIR Med Inform. 2022;10(7):e35724.

Wang G, et al. Optimized glycemic control of type 2 diabetes with reinforcement learning: a proof-of-concept trial. Nat Med. 2023a;29(10):2633–42. https://doi.org/10.1038/s41591-023-02552-9.

Wang W, Feng C, Han R, Wang Z, Ye L, Du Z, et al. trRosettaRNA: automated prediction of RNA 3D structure with transformer network. Nat Commun. 2023b;14(1):7266.

Xu J, Yang P, Xue S, Sharma B, Sanchez-Martin M, Wang F, et al. Translating cancer genomics into precision medicine with artificial intelligence: applications, challenges and future perspectives. Hum Genet. 2019;138(2):109–24.

Xue C, Kowshik SS, Lteif D, Puducheri S, Jasodanand VH, Zhou OT, et al. AI-based differential diagnosis of dementia etiologies on multimodal data. Nat Med. 2024;30(10):2977–89.

Yang X, Chen A, PourNejatian N, Shin HC, Smith KE, Parisien C, et al. A large language model for electronic health records. NPJ Digit Med. 2022;5(1):194.

Yousef H, Malagurski Tortei B, Castiglione F. Predicting multiple sclerosis disease progression and outcomes with machine learning and MRI-based biomarkers: a review. J Neurol. 2024;271(10):6543–72.

Yu C, Liu J, Zhao H. Inverse reinforcement learning for intelligent mechanical ventilation and sedative dosing in intensive care units. BMC Med Inform Decis Mak. 2019;19:111–20. https://doi.org/10.1186/s12911-019-0763-6.

Yu C, Ren G, Dong Y. Supervised-actor-critic reinforcement learning for intelligent mechanical ventilation and sedative dosing in intensive care units. BMC Med Inform Decis Mak. 2020;20:1–8. https://doi.org/10.1186/s12911-020-1120-5.

Yu C, Liu J, Nemati S, Yin G. Reinforcement learning in healthcare: a survey. ACM Comput Surv. 2021;55:1–36. https://doi.org/10.1145/3477600.

Yun WJ, Shin M, Jung S, Ko J, Lee HC, Kim J. Deep reinforcement learning-based propofol infusion control for anesthesia: A feasibility study with a 3000-subject dataset. Comput Biol Med. 2023;156:106739. https://doi.org/10.1016/j.compbiomed.2023.106739.

Zhan H, Moore JH, Zhang Z. A disease-specific language model for variant pathogenicity in cardiac and regulatory genomics. Nat Mach Intell. 2025:1–11.

Zhang S, Deshpande A, Verma BK, Wang H, Mi H, Yuan L, et al. Integration of clinical trial spatial multiomics analysis and virtual clinical trials enables immunotherapy response prediction and biomarker discovery. Cancer Res. 2024a;84(16):2734–48.

Zhang Q, Fotaki A, Ghadimi S, Wang Y, Doneva M, Wetzl J, et al. Improving the efficiency and accuracy of cardiovascular magnetic resonance with artificial intelligence–review of evidence and proposition of a roadmap to clinical translation. J Cardiovasc Magn Reson. 2024b;26(2):101051.

Zheng H, Zhu J, Xie W, Zhong J. Reinforcement learning assisted oxygen therapy for COVID-19 patients under intensive care. BMC Med Inform Decis Mak. 2021;21(1):350. https://doi.org/10.1186/s12911-021-01712-6.

Challenges and Future Directions in AI for Drug Discovery

Jinlu Zhang, Xin Shao, and Xiaohui Fan

Abstract In recent years, artificial intelligence has been widely applied in the field of drug discovery, becoming deeply integrated into all stages of drug research and development and bringing new opportunities to the advancement of biomedicine. However, the application of artificial intelligence also comes with numerous challenges that require in-depth study and response. We focus on the three main stages of drug development: target discovery, de novo drug design, and the discovery and optimization of lead compounds. By incorporating cutting-edge application examples, we explore the challenges and future directions of artificial intelligence in drug research and development.

Keywords Artificial intelligence · Drug discovery · Target discovery · De novo drug design · Lead compounds · Multi-omics data · Biomedical knowledge graphs · Virtual screening

1 Introduction

As is widely known, the development of a new drug from research to market typically requires over 10 years and substantial financial investment (Myers and Baker 2001). However, only about 10% of new drugs are approved to enter clinical research, and an even smaller proportion of drug molecules are eventually approved for market release. Especially in recent years, the research and development cycle and costs for new drugs have continuously increased, while the success rate has not seen a corresponding rise. Traditional drug development methods rely heavily on extensive preclinical candidate screening experiments and costly clinical trials to

J. Zhang · X. Shao · X. Fan (✉)
Pharmaceutical Informatics Institute, College of Pharmaceutical Sciences, Zhejiang University, Hangzhou, China

National Key Laboratory of Chinese Medicine Modernization, Innovation Center of Yangtze River Delta, Zhejiang University, Jiaxing, China
e-mail: zhangjinlu@zju.edu.cn; xin_shao@zju.edu.cn; fanxh@zju.edu.cn

A. Lavecchia (ed.), *Applied Artificial Intelligence for Drug Discovery*,
https://doi.org/10.1007/978-3-031-98022-0_27

verify their safety and efficacy, consuming significant resources and time (Hughes et al. 2011). To reduce the time and costs associated with new drug development, researchers have turned their attention to computational and big data approaches, such as virtual screening (VS) and molecular docking, but these methods have issues with poor accuracy and low efficiency (Waszkowycz et al. 2011). Artificial Intelligence (AI), including deep learning (DL) and machine learning (ML) algorithms, promises to overcome many of the problems and challenges encountered in the drug design and discovery process, offering a fresh and promising perspective for drug discovery (Blanco-Gonzalez et al. 2023).

The application of Artificial Intelligence (AI) in the field of new drug development spans the entire drug discovery process, presenting revolutionary pathways and efficiencies across various stages. At the level of target identification, AI technologies have opened new avenues for discovering novel targets. By leveraging AI, it is now possible to mine vast amounts of data to uncover disease-target-drug associations, build comprehensive knowledge bases, and establish a data-driven paradigm for target discovery and validation. This approach offers a more efficient decision-making process for initiating new drug research projects (You et al. 2022).

Moreover, AI provides more reliable methods for predicting the structures of large biomolecules, significantly aiding in the drug discovery process aimed at complex biological macromolecular complexes (Nagra et al. 2023). In the realm of candidate discovery, AI exhibits considerable potential in molecular design, virtual screening, and compound optimization. Particularly, deep generative models are capable of directly generating new molecules with drug-like properties, providing a novel tool for quickly obtaining safer and more effective drugs (Walters and Barzilay 2020). AI provides models that predict drug-related attributes, which are enhanced by integrating biochemical knowledge related to the drug and applying established scientific principles as constraints (Fang et al. 2023; Zeng et al. 2022; Vangala et al. 2023). This approach not only improves the credibility of the models (Prabhod 2023), but also enables early-stage predictions of a drug's safety and efficacy. Consequently, it reduces the associated risks and costs during the development process.

In addition, the integration of AI into drug discovery not only streamlines the technical aspects, but also brings substantial economic benefits by speeding up the development process and reducing costs. This advancement has the potential to make healthcare more accessible and affordable on a global scale (Viswa et al. 2024). Societally, the acceleration of drug discovery through AI could lead to quicker responses to pandemic situations (Keshavarzi Arshadi et al. 2020), thereby reinforcing the public health infrastructure on a global scale.

At the same time, to fully realize the potential of AI and boost its adoption in drug development, it is essential to ensure that end-users can easily engage with these tools. To achieve this, strengthening collaboration between end-users, such as medicinal chemists, pharmacologists, and clinicians, and IT professionals, including software developers, is essential. This interdisciplinary collaboration will enable end-users to interact with AI models more safely and efficiently, further advancing the development of AI technology in the field of drug discovery.

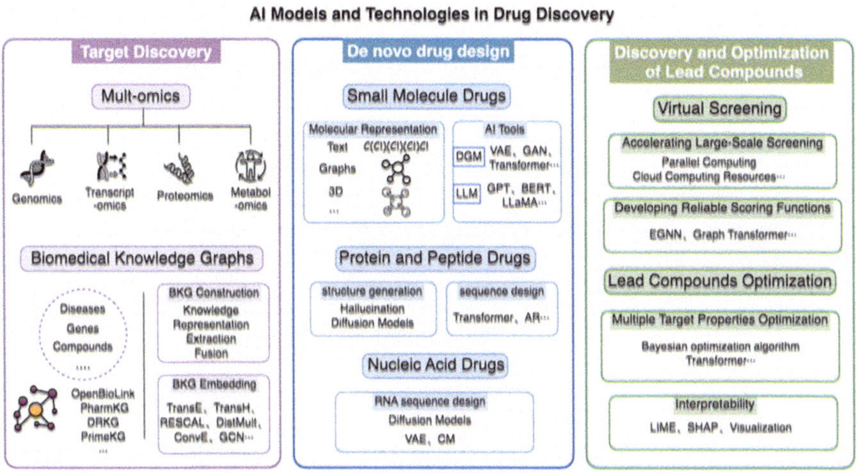

Fig. 1 AI models and technologies in drug discovery

Despite significant advancements in applying artificial intelligence to drug development, fully integrating it into the drug discovery process still faces numerous challenges. We will introduce the main challenges that artificial intelligence needs to address within the three key stages of drug discovery: target identification, de novo drug design, and the discovery and optimization of lead compounds. At the same time, by incorporating cutting-edge application examples, we will explore potential future development directions. Figure 1 shows the cutting-edge AI models and technologies related to the three key stages of drug discovery discussed in the article, further elucidating the overall structure and content of the paper. By exploring the challenges AI faces in drug discovery and its future directions, we aim to provide new insights and strategies to support the application and innovation of AI in the development of drug discovery.

This figure illustrates the three critical stages of drug discovery: target identification, de novo drug design, and the discovery and optimization of lead compounds. It also presents the relevant cutting-edge AI models and technologies associated with each stage, further detailing the structure and content of the entire article.

2 Target Discovery

A drug target refers to the site within a biological organism where a drug binds to exert its effect, including genetic loci, receptors, enzymes, ion channels, and nucleic acids. Target discovery, which encompasses the identification and early validation of targets that can alter disease, is an indispensable first step in the drug discovery process (Lindsay 2003). Despite decades of continuous innovations in experimental and omics technologies, pinpointing actionable therapeutic targets remains fraught with challenges.

Artificial Intelligence (AI) has the potential to revolutionize this field by learning from diverse biomedical data, comparing differences between diseased and non-diseased states, and analyzing RNA and/or protein expressions. AI can identify potential signaling pathways and protein interactions related to disease, thereby pinpointing candidate receptor binding sites (targets) on body cells where drugs can be effective. Therefore, AI is considered to have broad application prospects in enhancing the precision and efficiency of target discovery.

2.1 Based on Multi-omics Data

With the continuous advancement of high-throughput sequencing technologies, vast amounts of omics data, including genomics, transcriptomics, proteomics, and metabolomics, are being generated. The processing and analysis of these multi-omics datasets not only enable us to gain a deeper understanding of biological functions and mechanisms, but also have revolutionary implications for biology, medicine, and pharmacology. Omics data have led to the identification of numerous genes and proteins that play crucial roles in biological processes associated with specific diseases, providing valuable insights for drug discovery research. However, processing and analyzing these complex, high-dimensional datasets still pose significant challenges.

AI methods can efficiently extract potential knowledge from large-scale multi-omics datasets, aiding in the discovery of key genes or biological pathways and offering researchers a fresh perspective for their studies. Therefore, target identification methods that combine multi-omics data with artificial intelligence algorithms hold broad application prospects. Some researchers have already made attempts in this direction and achieved certain results. For instance, Wilson et al. (Wilson and Filipp 2018). conducted research on the regulatory networks associated with histone lysine demethylation by integrating transcriptomic and epigenomic data. They emphasized the significance of epigenetic regulators in controlling mitosis and highlighted their potential as therapeutic targets. The study revealed that epigenetic regulators, including KDM1A, KDM3A, EZH2, and DOT1L, are crucial in the development of tumors and resistance to therapy. Additionally, Pun et al. (2022). combined various bioinformatics and deep learning models trained on disease-specific multi-omics and text-based data to prioritize drug-usable genes, identifying 18 potential targets for ALS (Amyotrophic Lateral Sclerosis) treatment. Furthermore, Li et al. (2024). introduced scRank, which leverages genomic data to simulate drug perturbations computationally, sorting cell populations based on drug response. This method accurately revealed a subset of macrophages responsive to Tanshinone IIA and their potential targets in myocardial infarction.

Artificial intelligence demonstrates immense potential in integrating various omics data, such as genomics, proteomics, transcriptomics, and metabolomics.

This integration provides a comprehensive approach to uncover deeper biological insights from large datasets. By combining these data types, researchers can gain a holistic view of disease mechanisms and biological pathways, accelerating the identification of novel drug targets and uncovering patterns that traditional methods might overlook.

However, the application of multi-omics data integration in target discovery faces significant challenges. First, data heterogeneity across different sources and experimental platforms poses a major obstacle due to a variety of factors. These include intrinsic differences across data modalities, individual biological variability, variations in experimental conditions, and differences in data processing approaches. This complexity results in significant challenges in achieving consistent and reliable integration of multi-omics data. Second, datasets often contain noise or missing values, which compromise the reliability of the data and affect the accuracy of downstream analyses. Additionally, the acquisition of high-quality multi-omics data is often expensive and limited in availability, particularly for specific diseases or experimental conditions, further restricting the robustness and generalizability of predictive models.

To address these challenges, future efforts should focus on improving data processing and integration methodologies. Enhancing preprocessing workflows can significantly improve data quality. Currently, there are already established workflows for handling omics data, and the preprocessing processes for transcriptomics, metabolomics, and proteomics are generally similar. These workflows typically include steps such as noise reduction, missing data imputation, standardization, and batch effect removal. In transcriptomics, for instance, noise reduction may involve filtering low-quality reads or sequences, while missing data imputation addresses gaps in gene expression measurements. Standardization methods such as TPM, RPKM, or DESeq2's VST are commonly applied to account for sequencing depth and gene length variations. Similarly, in metabolomics, data preprocessing is equally critical, as metabolomics data are typically high-dimensional, noisy, and exhibit significant differences in magnitude across metabolites and samples. As such, data normalization is a key step in metabolomic data analysis. Common normalization methods include Probabilistic Quotient Normalization (PQN), Normalization to Internal Standards (ISN), and Quantile Normalization (QN) (Sun and Xia 2024). In proteomics, including DIA, Label-Free, and TMT-tagged proteomics technologies, the basic workflow involves log2 transformation of relative quantification values obtained from database searching, followed by data normalization and handling of missing values (Niu et al. 2022; Jayavelu et al. 2022). When removing batch effects, the Combat method is commonly used to adjust for systematic variations (Lee et al. 2019). However, these methods also present challenges, including the risk of bias from imputation techniques and data distortion from aggressive noise filtering. Furthermore, the development of advanced multi-omics data integration algorithms, capable of capturing complex cross-modal relationships within a unified framework, will be crucial. Techniques such as deep learning can play a pivotal role in extracting meaningful insights from heterogeneous datasets. Finally, the establishment of shared, high-quality multi-omics data platforms will promote broader accessibility and collaboration, providing a more reliable foundation for target discovery.

In addition to the technical challenges associated with integrating multi-omics data, significant ethical and regulatory issues must also be addressed. Omics data often originate from patient samples, raising serious concerns about privacy, consent, and data security. The development of multi-omics integration methods must consider the ethical and regulatory landscape to ensure that the resulting models are not only scientifically robust, but also ethically sound and legally compliant. For instance, advanced data anonymization techniques and secure data sharing platforms (Kendall et al. 2023) can facilitate the safe integration of different data types while respecting patient privacy. Implementing these practices not only meets ethical and regulatory requirements, but also enhances public trust in omics research, encouraging broader acceptance and participation in such studies.

2.2 Based on Biomedical Knowledge Graphs

Biomedical knowledge graphs refer to resources that integrate one or more expert information sources into a graph, where nodes represent biomedical entities and edges represent the relationships between two entities (Nicholson and Greene 2020). These graphs combine knowledge graph technology with biological information, facilitating the simplification of complex biological systems and pathological processes, thereby enabling researchers to better understand the underlying principles. Since Himmelstein et al. (2017). first created Hetionet in 2017, a comprehensive biomedical knowledge graph for drug repurposing, various other biomedical knowledge graphs such as OpenBioLink (Breit et al. 2020), PharmKG (Zheng et al. 2021), DRKG (Ioannidis et al. 2020), and PrimeKG (Chandak et al. 2023) have been developed. Each of these graphs has its unique focus and data characteristics, making significant contributions in their respective fields. Recently, in the field of target discovery, Zhou et al. (2024a). developed TarKG, a comprehensive biomedical knowledge graph specifically designed for target discovery. It integrates three highly related entities: diseases, genes, and compounds.

In recent years, machine learning (ML) models based on biomedical knowledge graphs have rapidly developed and many studies utilizing these graphs have successfully addressed numerous practical challenges in biomedical development (Nicholson and Greene 2020; Mohamed et al. 2021). Particularly in the field of target discovery, these methods have not only improved the efficiency of the process, but have also significantly enhanced its accuracy. For instance, Gogleva et al. (2022). constructed a recommendation system based on the heterogeneous biomedical knowledge graph BIKG, which integrates preclinical, clinical, and literature-based evidence. This system ranks genes by balancing various types of evidence linking them to potential resistance mechanisms of epidermal growth factor receptor inhibitors (EGFRi). Using this approach, the authors successfully identified key drivers of resistance in EGFR-mutant NSCLC. Additionally, Han et al. (2024). introduced Standigm ASK™, an innovative AI-driven drug discovery platform that integrates knowledge graphs (KG) with neural networks, aiming to uncover potential novel targets for idiopathic pulmonary fibrosis (IPF). Their study successfully identified three new genes—AMFR, MDFIC, and NR5A2—associated with IPF treatment. The effectiveness of these targets was further validated through a series of empirical experiments. These AI models based on knowledge graphs are capable of handling complex biomedical data and uncovering potential drug targets. They also assist researchers in understanding the intricate relationships between diseases and targets, as well as how these relationships impact biological pathways. This approach holds promise for accelerating drug development and enhancing the precision of therapeutic interventions.

Looking ahead, developing biomedical knowledge graphs specifically tailored for target discovery and methods for drug target identification based on knowledge graphs and artificial intelligence holds great promise. By integrating the context of specific diseases, these knowledge graphs will be capable of cross-validating information from multiple heterogeneous biomedical databases, such as proteomics, protein interaction, and drug-target relationship databases. Coupled with AI models, these systems can extract finer-grained multimodal knowledge elements and reveal intrinsic relationships between data, thus significantly accelerating the discovery of targets.

3 De Novo Drug Design

De novo drug design aims to generate molecules from scratch with specific chemical and pharmacological properties, offering structurally novel and innovative lead compounds that are crucial for originality in the drug development process. This approach includes the design of small molecules drugs, protein and peptide drugs, and nucleic acid drugs. Beyond just generating structures for evaluation, it is also essential to assess the produced molecules in terms of structural feasibility, biochemical properties, and target affinity. In recent years, the advancement of AI has further empowered de novo drug design, with a variety of generative models such as RNNs (Segler et al. 2018; Kotsias et al. 2020; Olivecrona et al. 2017), AEs (Prykhodko et al. 2019; Wang et al. 2024a), GANs (Prykhodko et al. 2019; Yu et al. n.d.), and Transformers (Fang et al. n.d.; Bagal et al. 2021; Mao et al. 2023) being utilized for molecular generation tasks.

3.1 Small Molecule Drugs

The de novo design of small molecule drugs is relatively more mature compared to that of biologics, with AI methods playing a significant role. Recent data reveal that AI-discovered small molecules constitute over 30% of the current clinical pipelines, a stark contrast to biologics such as vaccines and antibodies, which represent approximately 10% and 5%, respectively (Jayatunga et al. 2024). Additionally, the growth rate of AI-discovered small molecules is now comparable to that of classically discovered counterparts (Gangwal and Lavecchia 2024; Jayatunga et al. 2022), highlighting the advanced integration of AI in small molecule drug design.

From the perspective of molecular representation, de novo design methods can be categorized into three types: those based on one-dimensional SMILES, two-dimensional molecular graphs, and three-dimensional molecular generation. Recently, generative models for 3D drug design have gained widespread attention due to their potential to directly design ligands within protein pockets. However, current generative models for 3D drug design are faced with significant problems and challenges. These models often produce molecules with distorted conformations that fail to accurately reflect their true spatial structures (Liu et al. 2022; Lin et al. 2025). Furthermore, while these molecules may theoretically exhibit favorable binding properties, their complex structures present substantial synthetic challenges (Zhang et al. 2023). Additionally, the limited amount of data available leads to poor generalization of these models, resulting in molecules that lack innovation, stability, and affinity. To address these challenges, researchers have made numerous attempts. For instance, Zhang et al. (2024a). proposed FragGen, a novel hybrid strategy that is the first geometry-reliable, fragment-wise molecular generation method. It was created to resolve the problems of unrealistic molecular geometries and high synthetic complexity. It utilizes known molecular fragments as building blocks to generate more chemically accurate and synthesizable structures, ensuring both realistic and practically viable molecular conformations. Similarly, Zhung et al. (2024). introduced DeepICL, which enhances the generalizability of models by integrating a comprehensive understanding of protein-ligand interactions into the design process. This method uses established interaction patterns to more accurately predict molecular interactions with biological targets, thus improving the relevance and effectiveness of the molecules across different biological systems. Additionally, Jiang et al. (2024). developed PocketFlow, which tackles the issue of data scarcity by incorporating transfer learning techniques and domain-specific chemical knowledge. This approach enhances the accuracy of covalent bond modeling and the simulation of complex molecular interactions, resulting in the creation of more viable and diverse molecular structures.

Currently, the field of small molecule generation faces challenges such as the generated molecules potentially being syntactically or chemically invalid. The vastness of the chemical space, coupled with the scarcity of validated experimental data, further exacerbates these issues, resulting in molecules that lack both diversity and practicality. The rapid development of large language models offers a new approach to exploring these vast chemical spaces. By combining pretraining with fine-tuning, these large models can handle various molecular tasks, bringing new perspectives and powerful tools to the field of molecular design. For instance, the FragGPT, introduced by Yue et al. (2024)., utilizes a novel molecular representation called FU-SMILES. This representation, combined with the generative capabilities of fragment-based approaches, enhances the effectiveness of generated molecules.

It excels in various tasks such as de novo molecular design, linker design, R-group exploration, scaffold hopping, and side-chain optimization by significantly improving metrics like effectiveness, diversity, uniqueness, and novelty. Similarly, the MOLGEN model developed by Fang et al. (n.d.). employs a chemical feedback mechanism to refine the molecular generation process. This model shines in optimizing specific attributes such as penalized logP and QED scores and demonstrates superior performance in molecule docking tasks. Both approaches leverage attention mechanisms, which allow them to model long-range connections better and develop more effective language embeddings (Bagal et al. 2021; Radford 2018). This capability sets them apart from traditional neural networks, enabling more accurate predictions and simulations in molecular design tasks (Janakarajan et al. 2024; Bhattacharya et al. 2024).

3.2 *Protein and Peptide Drugs*

As researchers deepen their understanding of the metabolic pathways, pathological mechanisms, structure, and function of protein and peptide drugs, the significance of these drugs in treating complex diseases continues to grow. Compared to small molecule drugs, protein and peptide drugs offer longer half-lives, higher safety, better therapeutic effectiveness, and stronger specificity. They play an irreplaceable role in treating complex systemic diseases. Therefore, compared to the well-established field of small molecule drug development, the development of protein and peptide drugs is showing a rising trend (Nagra et al. 2023).

Protein and peptide design involves creating new amino acid sequences that correspond to proteins and peptides with desired functions. The primary strategy used in this field is directed evolution, which utilizes multiple rounds of random mutagenesis and high-throughput screening to select the most promising sequences. In recent years, a fundamental transformation in protein de novo design has occurred due to increased computational power, the accumulation of protein data, and the application of artificial intelligence (AI) technologies. Deep learning methods have particularly garnered significant attention due to their ability to utilize extensive datasets of protein and peptide sequences. These methods enable the creation of models that more precisely represent the functional properties of these sequences. Notably, breakthroughs such as AlphaFold2 (Jumper et al. 2021) and RoseTTAFold (Baek et al. 2021) have made pioneering achievements in protein structure prediction.

Currently, deep learning-based protein and peptide de novo design can be categorized into two main types: the first type is based on backbone structure design, which includes sequence optimization for fixed backbones and variable scaffolds; the second type involves direct sequence generation that does not rely on known structures, used for exploring sequence space and protein generation. Backbone-based design using deep learning models to construct energy functions that capture essential features between the scaffold structure and sequence has become an effective strategy. Compared to traditional molecular simulation methods, deep learning exhibits significant advantages in simulating van der Waals forces, hydrogen bonds, hydrophobicity, and other interaction relationships. The introduction of diffusion models has not only significantly improved prediction accuracy, but also provided a powerful tool for designing proteins with specific functions, thereby advancing the field of functional protein design, as seen in models like ProteinMPNN (Dauparas et al. 2022) and RFdiffusion (Watson et al. 2023).

In the direct sequence generation approach of protein de novo design, the rapid advancement of sequencing technology has exponentially increased the volume of protein sequence data, accumulating a vast number of unlabeled sequences. These sequences have laid a solid foundation for the application of large language models in the protein design field, allowing for a deeper understanding of the relationship between protein sequences and their functions. For instance, Ferruz et al. (2022). introduced ProtGPT2, an autoregressive Transformer model with 738 million parameters, capable of high-throughput generation of de novo protein sequences. Nijkamp et al. (2023). developed the ProGen2 model, an autoregressive language model (AR) pretrained on approximately 230 million protein sequences from the Uniref90 and BFD30 databases. Qiu et al. (2024). combined ProteinMPNN, ProGen2, and Qwen-VL to create InstructPLM, employing cross-modal alignment and instructional fine-tuning techniques. This model can generate target protein sequences based on complex biological instructions provided. This method, starting directly from sequences and not relying on predetermined structures or extensive sequence information, innovates the protein design paradigm by learning the relationships between sequences and their structural functions to explore sequence space directly.

In the future, artificial intelligence is expected to integrate multimodal biological information from fields such as genetics, transcriptomics, and proteomics to more accurately identify complex, in-depth information about proteins. This integration will enable AI to more precisely simulate biomolecular interactions and design proteins and peptides with specific functionalities. Currently, protein design models tend to be large and require substantial computational power. Future models are expected to become more lightweight as research continues to advance in areas such as knowledge distillation, weight pruning, and quantitative techniques at the sequence level. This advancement will not only improve the efficiency of protein design but also broaden the potential to create more complex and functionally diverse proteins.

3.3 Nucleic Acid Drugs

In recent years, the approval rate of nucleic acid drugs (NAD) has accelerated, covering areas such as cardiovascular and metabolic diseases, liver diseases, and a variety of rare diseases. Compared to traditional medications, nucleic acid drugs are noted for their potential to achieve high therapeutic efficacy and low drug toxicity, which makes them a prominent focus in new drug development. They are often characterized by strong specificity due to their ability to target specific genetic sequences. However, it is important to acknowledge that despite these promising features, challenges such as off-target effects still persist. Additionally, nucleic acid drugs, including mRNA and small nucleic acids, face significant hurdles related to their delivery and stability. Nucleic acid drugs are primarily divided into mRNA and small nucleic acid drugs.

mRNA drugs deliver genetic information to guide the production of therapeutic proteins within the human body, opening up new possibilities for disease treatment. Currently, mRNA technology has evolved into a versatile therapeutic option, with applications extending to vaccines, protein replacement therapies, gene editing, and cell-based therapies. Notably, during the COVID-19 pandemic, the rapid development and application of mRNA vaccines highlighted their tremendous potential. In the development of mRNA vaccines, designing RNA sequences with optimized functional properties is a critical step. Noncoding regions of RNA sequences, such as the 5′ untranslated region (5'-UTR), have a direct impact on mRNA stability and protein synthesis efficiency. However, generating RNA sequences poses significant challenges: RNA molecules are composed of various nucleotides (A, U, G, C) and can form highly complex secondary and tertiary structures. Moreover, the inherent flexibility of RNA molecules presents significant challenges for experimentally determining their three-dimensional (3D) structures. Despite recent advances in techniques such as cryo-electron microscopy, these complex and low-throughput methods still require extensive labor and specialized expertise. Deep learning (DL) aids this process by effectively capturing and extracting latent features within sequences. For example, Huang et al. (2024). have proposed a novel method based on the latent diffusion model called RNAdiffusion, which enhances RNA translation efficiency (TE) and ribosome loading rate (MRL). Sumi et al. (2024). introduced a deep generative model named RfamGen which utilizes Variational Autoencoders (VAE) and Conditional Modeling (CM) to generate artificial sequences for RNA families. This approach efficiently designs RNA family sequences by explicitly integrating sequence alignment and secondary structure information. The future integration of AI with nucleic acid sequence generation promises to not only replicate sequences similar to natural RNA, but also optimize them based on specific functional properties.

Small nucleic acid drugs, or oligonucleotide drugs, are short chains of nucleic acids composed of several nucleotides. Common among these drugs are short nucleic acid sequences designed to interact with specific target molecules within cellular pathways. The main categories include antisense oligonucleotides (ASO), small interfering RNA (siRNA), micro-RNA (miRNA), and aptamers (Miao et al. 2024). Small nucleic acid drugs are a focus of contemporary biopharmaceutical research due to their significant specificity, ease of design, short development cycles, broad target spectrum, and prolonged activity.

Currently, research and development of small nucleic acid drugs are primarily centered around siRNA. Therefore, we will use siRNA as an example to illustrate the role and future direction of AI in small nucleic acid drug development. Designing efficient and highly specific siRNA sequences has been a significant challenge for researchers. Similar to small molecules, the combinatorial space of siRNA molecules exceeds 2^42, making the number of possible molecular compositions vast. Given these challenges and characteristics, researchers are exploring the integration of AI in the development of siRNA drugs. For instance, the emerging RNAi therapy company Sylentis, based in Madrid, has developed a software called SirFinder for the rational design of siRNA (Moreno-Montañés et al. 2021). Utilizing mathematics and artificial intelligence, including neural network algorithms, SVM, and machine learning, SirFinder optimizes AI algorithms through feedback loops based on incoming experimental data. This optimization enhances the precision of experimental data interpretation, enabling the production of specific and safe siRNA molecules that selectively target desired genes without affecting others. Looking forward, in the field of small nucleic acid drug development, particularly with siRNA, AI is expected to combine multi-omics data, machine learning, and association rule mining to address issues related to gene silencing and diseases. Furthermore, AI can learn from existing databases of effective siRNA sequences and use these features to design new siRNAs. Although the application of AI in small nucleic acid drugs is not yet mature, given its potential for growth, AI is believed to significantly enhance the development of small nucleic acid drugs.

4 Discovery and Optimization of Lead Compounds

The discovery and optimization of lead compounds are critical steps in the drug development process. Lead compounds, as the starting point for developing therapeutic drugs, can be identified through various approaches, including high-throughput screening (HTS), virtual screening, and structure-based drug design. These compounds are then subjected to a series of optimization processes to enhance their pharmacokinetic and pharmacodynamic properties, such as potency, selectivity, solubility, and metabolic stability.

Currently, the discovery and optimization of lead compounds heavily rely on the expertise of medicinal chemists and require significant resource investment. This traditional approach is often slow and labor-intensive. However, with the integration of artificial intelligence, the process of lead compound discovery and optimization can achieve both speed and precision. AI-driven methods provide opportunities to improve efficiency and reduce costs, paving the way for faster and more accurate drug development.

4.1 Virtual Screening of Lead Compounds

Virtual screening refers to the prescreening of compound molecules on a computer based on preset criteria, to identify those most likely to bind to a target, thereby significantly reducing the number of compounds screened in the wet lab and enhancing the efficiency of lead compound discovery. It primarily encompasses two strategies: Ligand-Based Virtual Screening (LBVS) and Structure-Based Virtual Screening (SBVS). LBVS is particularly effective when the structure of the target is unknown but active molecules are known. It employs methods like similarity searches, quantitative structure-activity relationship (QSAR) analyses, and pharmacophore modeling to screen compounds similar to the known active molecules. On the other hand, SBVS is applicable when the target structure is known, utilizing molecular docking, pharmacophore searching, and molecular dynamics simulations to identify small molecules that can bind to the target. However, virtual screening faces challenges such as high computational costs and long processing times as the size of the virtual compound libraries expands. Additionally, for virtual screening strategies based on molecular docking, the accuracy of scoring functions needs improvement. In recent years, the application of AI algorithms in virtual screening has not only increased the number and diversity of molecules in virtual compound libraries, but has also refined screening algorithms. Developing AI-based scoring functions has made screening results more accurate.

When facing vast chemical libraries with billions of compounds, virtual screening can significantly streamline the massive dataset to just a few hundred small-molecule compounds targeted at specific biomarkers, making the discovery process of lead compounds faster and more economical. For instance, Zhou et al. (2024b). innovatively developed the RosettaVS method and integrated it into the open-source AI-accelerated virtual screening platform, OpenVS, aiming to achieve efficient and accurate virtual screening. Using the OpenVS platform, the research team successfully identified lead compounds targeting unrelated biomarkers KLHDC2 and NaV1.7 from a library of billions of compounds within seven days. These compounds demonstrated single-digit micromolar binding affinities, validating the efficiency and practicality of the method. However, it's crucial to note that despite these advancements, the quality and diversity of training data used for these AI models have significant limitations. Often, the training datasets may not fully capture the complexity of the real-world chemical spaces, which could limit the models' ability to generalize to new, unseen molecules or interactions.

Moreover, to efficiently obtain high-activity molecules, apart from requiring fast screening computational methods, developing more reliable scoring functions for molecular docking-based virtual screening is particularly crucial. For example, in the virtual screening of drugs targeting new biomarkers, there is a need for scoring functions with better generalization performance. Addressing this issue, Cao et al. (2024). employed equivariant graph neural networks to integrate physical priors related to protein-ligand interactions. They also utilized various data augmentation and data deduplication strategies to avoid potential data distribution biases due to model overfitting. Ultimately, they developed a universal protein-ligand interaction scoring method called EquiScore. In the contexts of drug virtual screening and lead compound optimization, EquiScore has demonstrated good generalization performance on new targets that were not seen during training.

Currently, virtual screening, a key technology in the discovery of lead compounds, faces several major challenges, including the escalating data processing demands due to the continuous expansion of virtual compound libraries, high computational costs, and lengthy processing times. To address these issues, future directions will focus on leveraging artificial intelligence tools to enhance the efficiency and precision of virtual screening. This involves training algorithms to recognize complex patterns between molecules and targets, enabling the rapid identification of promising candidate molecules from large-scale compound libraries. Additionally, scoring functions play a crucial role in molecular docking, yet existing functions often lack accurate predictions of binding energies. These functions typically overlook key environmental factors such as temperature, pressure, and solvent effects, which can significantly impact predictions of binding free energies. Moreover, each protein-ligand interaction is unique due to specific binding sites and recognition processes, making it challenging to develop a universal scoring function (Kitchen et al. 2004). Custom adaptations are frequently necessary to account for these unique interactions. To address these challenges, future improvements in virtual screening should focus on incorporating more comprehensive AI-driven models capable of integrating these complex variables. Such advancements will enhance the reliability of scoring functions, enabling more precise predictions across diverse chemical spaces and complex molecular interactions. Optimizing computational methods and resource management is also crucial; for instance, utilizing parallel computing and cloud computing resources can significantly accelerate the screening process while maintaining high accuracy. These strategies will help reduce the cost and time of virtual screening, thus accelerating the discovery of lead compounds and supporting new drug development.

4.2 Lead Compounds Optimization

Lead compound optimization is a crucial step in the drug discovery process, aimed at designing potential candidate drugs from biologically active compounds. During this optimization phase, active compounds undergo meticulous modifications to enhance their absorption, distribution, metabolism, excretion, and toxicity (ADMET) characteristics. Given the numerous molecular fragments in lead compounds, the combination of these fragments creates a vast chemical space, making it challenging to identify ideal molecules. To improve the efficiency of this process, researchers have developed AI-based lead compound optimization methods using technologies such as natural language processing (NLP), reinforcement learning, and generative adversarial networks (GANs).

Currently, a significant issue with AI-based lead compound optimization methods is that the optimization process should ideally address multiple dimensions, optimizing several target properties simultaneously. However, existing methods are primarily designed for optimizing single targets, which does not align with chemists' needs for optimizing multiple properties concurrently. Researchers like

Zhu et al. (2024). have explored solutions to this problem, such as the multi-objective Bayesian optimization algorithm HN-GFN based on GFlowNets, which can be used for sample-efficient multi-objective molecular optimization. Additionally, Wu et al. (2024). have introduced a multi-property molecular optimization method called Prompt-MolOpt, based on the Transformer architecture and prompt learning. This method allows for precise specification of molecular structures to be retained while flexibly optimizing the rest.

Beyond the challenges of multi-objective optimization, lead compound optimization also faces issues such as lack of sufficient interpretability. To address this, Yu et al. (2023). proposed an AI method for lead compound optimization called PBCNet (pairwise binding comparison network). This method utilizes a siamese graph convolutional neural network architecture to predict the relative binding affinity by comparing the binding modes of a set of similar ligands, balancing computational speed and accuracy effectively. PBCNet provides detailed model interpretability, including explanations at the atomic and substructural levels. This is crucial for deepening the understanding of the model's predictive mechanisms and for more effectively designing drugs.

Currently, the optimization of lead compounds faces challenges including insufficient model interpretability, suboptimal multi-objective optimization results, and weak generalization capabilities of models. As interpretable artificial intelligence models continue to evolve (Wang et al. 2024b), researchers will be able to delve deeper into how these models map molecular structures to their properties from an interpretability perspective, which is expected to enhance public trust in the optimization outcomes. To address multi-objective optimization issues, setting more precise and reasonable optimization target functions for AI models and developing new optimization strategies are anticipated to resolve these challenges, thereby achieving an effective balance among multiple objectives. Meanwhile, many scholars are investigating how to improve model generalization capabilities for unknown distributions (Chen et al. 2022). In the near future, molecular optimization models based on causal inference are expected to reduce dependency on existing data.

5 Experimental Validation

Following the identification of drug targets using artificial intelligence, rigorous experimental validation is crucial to ascertain their clinical relevance and efficacy. This validation process encompasses two main phases: target validation and pharmacological efficacy testing. For target validation, cellular models are typically used to study drug-target interactions at a molecular level, employing techniques such as Surface Plasmon Resonance (SPR) and Bio-Layer Interferometry (BLI). These methods offer high precision in measuring the dynamics and affinity of interactions but do not fully simulate the systemic and metabolic processes occurring in a living organism. Further validation is then extended to animal models and patient-derived systems, which provide more comprehensive insights into the pharmacodynamics and potential systemic effects within a whole organism.

Pharmacological efficacy studies involve both in vitro and in vivo experiments. In vitro testing, conducted at the molecular, cellular, or isolated tissue and organ levels, provides preliminary insights into a drug's action and mechanisms. Organoids are increasingly recognized for their ability to closely mimic human organ systems, which enables more accurate assessments of a drug's therapeutic effects and potential toxicities (Brooks et al. 2021). However, despite the advantages of organoid technology in efficacy testing, challenges persist due to the lack of standardized methodologies across laboratories, affecting the reproducibility and reliability of results.

6 Summary

In the field of drug development, artificial intelligence can demonstrate its capabilities in several key areas: first, by exploring data-driven patterns and integrating various biomedical information to discover new targets; second, by designing and optimizing new candidate drug molecules through generative models, guiding researchers towards ideal paths in drug optimization; and third, by using predictive models to screen lead compounds, significantly reducing the cost and time of experimental validation.

However, there are also common challenges in applying AI to the field of drug development. First, the complexity of biological data poses a challenge, as drug discovery requires the integration of chemical and biological information. Chemical data are more controllable and easier to compute; biological data involve conformational changes in receptor proteins, balancing and biased signals, which are difficult to quantify. Chemical data also tend to be more "stable" compared to biological data primarily because they are less susceptible to environmental changes and dynamic alterations. Second, current AI algorithms have incorporated only a subset of chemical indicators, leading to an incomplete set of biological evaluation metrics. This deficiency makes it difficult to comprehensively assess the biological effects of drugs. Characteristics of compounds in vivo, such as precursor compounds, metabolites, and concentration-dependent effects, are relegated to a secondary status within these models. These overlooked factors are crucial in determining whether a drug can reach its intended target, achieve therapeutic effects, and whether its side effects are within an acceptable range. Consequently, this brings greater uncertainty to AI-driven drug discovery and efficacy prediction. However, it is encouraging to note that numerous initiatives are being undertaken to address these challenges. Increasingly, platforms capable of predicting ADME (absorption, distribution, metabolism, and excretion) profiles are being developed (Fu et al. 2024; Gu et al. 2024) and integrated with virtual screening tools to prioritize compounds with favorable metabolic profiles. However, despite these advancements, the quality and quantity of training datasets remain a major challenge. Effectively training

AI models requires extensive amounts of high-quality data. However, much of the data from pharmaceutical research, derived from experimental studies, often suffer from inconsistencies in experimental conditions and data recording formats or annotations, leading to issues such as numerical discrepancies, missing data, and skewed datasets that are difficult for computers to uniformly recognize. Additionally, most high-quality and critical data are held by a few corporations and remain unpublished due to intellectual property protection concerns. Furthermore, both in academia and industry, the publicly available data predominantly consist of positive results, with a large volume of negative outcomes being difficult to access. Furthermore, AI models face a crisis of interpretability. This challenge is particularly significant because the decision-making processes of AI systems are often opaque, making it difficult for researchers and clinicians to understand how conclusions are derived. This lack of transparency can hinder trust and acceptance of AI applications in drug development.

To address the current challenges, the future of artificial intelligence in the field of drug discovery will advance in several innovative directions. First, in response to the complexity of biological data and the lack of biological evaluation metrics, future AI models are expected to learn more complex biological patterns and signals by integrating multimodal data and diverse computational methods (such as molecular dynamics simulations, molecular docking, and quantum computing (Pyrkov et al. 2023)). Additionally, incorporating more reasonable and computable biological indicators during model adjustment and optimization will provide a more comprehensive understanding of how drugs act within biological systems, thereby enhancing the accuracy and practicality of AI models in drug development. Regarding challenges in data quality and quantity, developing and establishing safer and more open data-sharing platforms is crucial. Encouraging different research institutions and enterprises to share experimental data, including both positive and negative results, can expand the data pool and improve data diversity and coverage. At the same time, developing models with greater generalization and robustness—utilizing advanced techniques such as transfer learning and data augmentation—can improve the learning ability and performance of models even when data are limited. Moreover, the generative capabilities of large language models offer new avenues for designing and optimizing novel candidate drug molecules (Zhang et al. 2024b). By using generative AI models, it is possible to explore vast chemical spaces and rapidly discover molecules with potential activity (Nakata et al. 2024). Additionally, AI systems composed of multiple AI agents represent another important development direction (Gao et al. 2024). These specialized AI agents perform their respective tasks, covering various aspects of drug development from target identification to clinical trial design. Through collaborative work, these agents can improve the efficiency and effectiveness of the entire process, promoting the intelligentization of drug research and development. In addition to focusing on technological advancements, future development strategies should also prioritize ensuring that end-users, such as pharmacists and chemists, have a comprehensive understanding of the capabilities and limitations of AI-driven tools in drug development. For developers, it's vital to clearly delineate these tools' workflows, prepare detailed user guides that emphasize both functionalities and potential drawbacks, and package these tools for ease of use by end-users. Additionally, end-users should be encouraged to acquire basic operational skills to utilize these tools efficiently. By improving the transparency regarding the capabilities and limitations of AI tools, developers can empower users to make more informed decisions. This not only facilitates quicker feedback from users to developers, but also enhances the overall effectiveness and safety of the drug development process. Considering the immense potential of artificial intelligence in transforming drug discovery, recent substantial financial investments highlight critical future trends in this field (Gangwal and Lavecchia 2024). These investments cover multiple areas of drug discovery/target discovery, molecular property prediction, new drug design, repositioning, and other areas of drug development (Gangwal and Lavecchia 2024). For example, the biotech company Genesis Therapeutics has secured hundreds of millions in funding to target areas traditionally considered "undruggable" due to scarce data, establishing several internal research and development projects. Another company, Terray Therapeutics, uses its proprietary integrated computational and experimental platform to generate vast chemical data, accelerating drug discovery (Kaufman et al. 2024). Future economic investments are expected to further enhance the use of artificial intelligence throughout various stages of drug development, from initial screening to clinical trials. In summary, with the development of AI technology and its organic integration with the biomedical field, artificial intelligence is expected to overcome current challenges, revolutionize the drug discovery process, and accelerate the development and application of new drugs.

References

Baek M, et al. Accurate prediction of protein structures and interactions using a three-track neural network. Science. 2021;373:871–6.

Bagal V, Aggarwal R, Vinod P, Priyakumar UD. MolGPT: molecular generation using a transformer-decoder model. J Chem Inf Model. 2021;62:2064–76.

Bhattacharya D, Cassady HJ, Hickner MA, Reinhart WF. Large language models as molecular design engines. J Chem Inf Model. 2024;64:7086–96.

Blanco-Gonzalez A, et al. The role of AI in drug discovery: challenges, opportunities, and strategies. Pharmaceuticals. 2023;16:891.

Breit A, Ott S, Agibetov A, Samwald M. OpenBioLink: a benchmarking framework for large-scale biomedical link prediction. Bioinformatics. 2020;36:4097–8.

Brooks A, et al. Liver organoid as a 3D in vitro model for drug validation and toxicity assessment. Pharmacol Res. 2021;169:105608.

Cao D, et al. Generic protein – ligand interaction scoring by integrating physical prior knowledge and data augmentation modelling. Nature Machine Intelligence. 2024:1–13.

Chandak P, Huang K, Zitnik M. Building a knowledge graph to enable precision medicine. Scientific Data. 2023;10:67.

Chen Y, et al. Pareto invariant risk minimization: towards mitigating the optimization dilemma in out-of-distribution generalization. 2022. arXiv preprint arXiv:2206.07766.

Dauparas J, et al. Robust deep learning – based protein sequence design using ProteinMPNN. Science. 2022;378:49–56.

Fang Y, et al. Knowledge graph-enhanced molecular contrastive learning with functional prompt. Nature Machine Intelligence. 2023;5:542–53.

Fang Y, et al. The twelfth international conference on learning representations.

Ferruz N, Schmidt S, Höcker B. ProtGPT2 is a deep unsupervised language model for protein design. Nat Commun. 2022;13:4348.

Fu L, et al. ADMETlab 3.0: an updated comprehensive online ADMET prediction platform enhanced with broader coverage, improved performance, API functionality and decision support. Nucleic Acids Res. 2024;gkae236

Gangwal A, Lavecchia A. Unlocking the potential of generative AI in drug discovery. Drug Discov Today. 2024;103992

Gao S, et al. Empowering biomedical discovery with ai agents. Cell. 2024;187:6125–51.

Gogleva A, et al. Knowledge graph-based recommendation framework identifies drivers of resistance in EGFR mutant non-small cell lung cancer. Nat Commun. 2022;13:1667.

Gu Y, et al. admetSAR3. 0: a comprehensive platform for exploration, prediction and optimization of chemical ADMET properties. Nucleic Acids Res. 2024:gkae298.

Han S, et al. Standigm ASK™: knowledge graph and artificial intelligence platform applied to target discovery in idiopathic pulmonary fibrosis. Brief Bioinform. 2024;25:bbae035.

Himmelstein DS, et al. Systematic integration of biomedical knowledge prioritizes drugs for repurposing. elife. 2017;6:e26726.

Huang K, et al. Latent diffusion models for controllable RNA sequence generation. 2024. arXiv preprint arXiv:2409.09828.

Hughes JP, Rees S, Kalindjian SB, Philpott KL. Principles of early drug discovery. Br J Pharmacol. 2011;162:1239–49.

Ioannidis VN, et al. Drkg-drug repurposing knowledge graph for covid-19. 2020. arXiv preprint arXiv:2010.09600.

Janakarajan N, Erdmann T, Swaminathan S, Laino T, Born J. Drug development supported by informatics. Springer; 2024. p. 121–41.

Jayatunga MK, Xie W, Ruder L, Schulze U, Meier C. AI in small-molecule drug discovery: a coming wave. Nat Rev Drug Discov. 2022;21:175–6.

Jayatunga MK, Ayers M, Bruens L, Jayanth D, Meier C. How successful are AI-discovered drugs in clinical trials? A first analysis and emerging lessons. Drug Discov Today. 2024:104009.

Jayavelu AK, et al. The proteogenomic subtypes of acute myeloid leukemia. Cancer Cell. 2022;40:301–17. e312.

Jiang Y, et al. Pocketflow is a data-and-knowledge-driven structure-based molecular generative model. Nature Machine Intelligence. 2024;6:326–37.

Jumper J, et al. Highly accurate protein structure prediction with AlphaFold. Nature. 2021;596:583–9.

Kaufman B, et al. Coati: Multimodal contrastive pretraining for representing and traversing chemical space. J Chem Inf Model. 2024;64:1145–57.

Kendall TJ, et al. An integrated gene-to-outcome multimodal database for metabolic dysfunction-associated steatotic liver disease. Nat Med. 2023;29:2939–53.

Keshavarzi Arshadi A, et al. Artificial intelligence for COVID-19 drug discovery and vaccine development. Frontiers in Artificial Intelligence. 2020;3:65.

Kitchen DB, Decornez H, Furr JR, Bajorath J. Docking and scoring in virtual screening for drug discovery: methods and applications. Nat Rev Drug Discov. 2004;3:935–49.

Kotsias P-C, et al. Direct steering of de novo molecular generation with descriptor conditional recurrent neural networks. Nature Machine Intelligence. 2020;2:254–65.

Lee AH, et al. Dynamic molecular changes during the first week of human life follow a robust developmental trajectory. Nat Commun. 2019;10:1092.

Li C, et al. scRank infers drug-responsive cell types from untreated scRNA-seq data using a target-perturbed gene regulatory network. Cell Reports Medicine. 2024;

Lin H, et al. Diffbp: Generative diffusion of 3D molecules for target protein binding. Chem Sci. 2025;

Lindsay MA. Target discovery. Nat Rev Drug Discov. 2003;2:831–8.

Liu M, Luo Y, Uchino K, Maruhashi K, Ji S. Generating 3d molecules for target protein binding. 2022. arXiv preprint arXiv:2204.09410.
Mao J, et al. Transformer-based molecular generative model for antiviral drug design. J Chem Inf Model. 2023;64:2733–45.
Miao Y, et al. Current status and trends in small nucleic acid drug development: leading the future. Acta Pharm Sin B. 2024;
Mohamed SK, Nounu A, Nováček V. Biological applications of knowledge graph embedding models. Brief Bioinform. 2021;22:1679–93.
Moreno-Montañés J, et al. siRNA therapeutics in ocular diseases. In: Design and delivery of SiRNA therapeutics; 2021. p. 417–42.
Myers S, Baker A. Drug discovery – an operating model for a new era. Nat Biotechnol. 2001;19:727–30.
Nagra NS, et al. The company landscape for artificial intelligence in large-molecule drug discovery. Nat Rev Drug Discov. 2023;22:949–50.
Nakata S, Mori Y, Tanaka S. Navigating ultralarge virtual chemical spaces with product-of-experts chemical language models. J Chem Inf Model. 2024;
Nicholson DN, Greene CS. Constructing knowledge graphs and their biomedical applications. Comput Struct Biotechnol J. 2020;18:1414–28.
Nijkamp E, Ruffolo JA, Weinstein EN, Naik N, Madani A. Progen2: exploring the boundaries of protein language models. Cell Systems. 2023;14:968–78. e963.
Niu L, et al. Noninvasive proteomic biomarkers for alcohol-related liver disease. Nat Med. 2022;28:1277–87.
Olivecrona M, Blaschke T, Engkvist O, Chen H. Molecular de-novo design through deep reinforcement learning. J Chem. 2017;9:1–14.
Prabhod KJ. Leveraging generative AI for personalized medicine: applications in drug discovery and development. Journal of AI-Assisted Scientific Discovery. 2023;3:392–434.
Prykhodko O, et al. A de novo molecular generation method using latent vector based generative adversarial network. J Chem. 2019;11:1–13.
Pun FW, et al. Identification of therapeutic targets for amyotrophic lateral sclerosis using PandaOmics – an AI-enabled biological target discovery platform. Front Aging Neurosci. 2022;14:914017.
Pyrkov A, et al. Quantum computing for near-term applications in generative chemistry and drug discovery. Drug Discov Today. 2023;28:103675.
Qiu J, et al. InstructPLM: aligning protein language models to follow protein structure instructions. 2024. bioRxiv, 2024.2004. 2017.589642.
Radford A. Improving language understanding by generative pre-training. (2018).
Segler MH, Kogej T, Tyrchan C, Waller MP. Generating focused molecule libraries for drug discovery with recurrent neural networks. ACS Central Science. 2018;4:120–31.
Sumi S, Hamada M, Saito H. Deep generative design of RNA family sequences. Nat Methods. 2024;21:435–43.
Sun J, Xia Y. Pretreating and normalizing metabolomics data for statistical analysis. Genes & Diseases. 2024;11:100979.
Vangala SR, Krishnan SR, Bung N, Srinivasan R, Roy A. pBRICS: a novel fragmentation method for explainable property prediction of drug-like small molecules. J Chem Inf Model. 2023;63:5066–76.
Viswa CA, Bleys J, Leydon E, Shah B, Zurkiya D. Generative AI in the pharmaceutical industry: Moving from hype to reality. McKinsey; 2024.
Walters WP, Barzilay R. Applications of deep learning in molecule generation and molecular property prediction. Acc Chem Res. 2020;54:263–70.
Wang C, Ong HH, Chiba S, Rajapakse JC. GLDM: hit molecule generation with constrained graph latent diffusion model. Brief Bioinform. 2024a;25:bbae142.
Wang Z, Huang C, Yao X. A roadmap of explainable artificial intelligence: explain to whom, when, what and how? ACM Transactions on Autonomous and Adaptive Systems. 2024b;19(4)

Waszkowycz B, Clark DE, Gancia E. Outstanding challenges in protein – ligand docking and structure-based virtual screening. Wiley Interdisciplinary Reviews: Computational Molecular Science. 2011;1:229–59.
Watson JL, et al. De novo design of protein structure and function with RFdiffusion. Nature. 2023;620:1089–100.
Wilson S, Filipp FV. A network of epigenomic and transcriptional cooperation encompassing an epigenomic master regulator in cancer. Npj Systems Biology and Applications. 2018;4:24.
Wu Z, et al. Leveraging language model for advanced multiproperty molecular optimization via prompt engineering. Nature Machine Intelligence. 2024;1-11
You Y, et al. Artificial intelligence in cancer target identification and drug discovery. Signal Transduct Target Ther. 2022;7:156.
Yu J, et al. Computing the relative binding affinity of ligands based on a pairwise binding comparison network. Nature Computational Science. 2023;3:860–72.
Yu L, Zhang W, Wang J, Yu Y.. Proceedings of the AAAI conference on artificial intelligence.
Yue J, et al. Unlocking comprehensive molecular design across all scenarios with large language model and unordered chemical language. Chem Sci. 2024;15:13727–40.
Zeng X, et al. Accurate prediction of molecular properties and drug targets using a self-supervised image representation learning framework. Nature Machine Intelligence. 2022;4:1004–16.
Zhang O, et al. ResGen is a pocket-aware 3D molecular generation model based on parallel multiscale modelling. Nature Machine Intelligence. 2023;5:1020–30.
Zhang O, et al. FragGen: towards 3D geometry reliable fragment-based molecular generation. Chem Sci. 2024a;
Zhang J, et al. The future of molecular studies through the lens of large language models. J Chem Inf Model. 2024b;64:563–6.
Zheng S, et al. PharmKG: a dedicated knowledge graph benchmark for bomedical data mining. Brief Bioinform. 2021;22:bbaa344.
Zhou C, et al. TarKG: a comprehensive biomedical knowledge graph for target discovery. Bioinformatics. 2024a;40:btae598.
Zhou G, et al. An artificial intelligence accelerated virtual screening platform for drug discovery. Nat Commun. 2024b;15:7761.
Zhu Y, et al. Sample-efficient multi-objective molecular optimization with gflownets. In: Advances in neural information processing systems, vol. 36; 2024.
Zhung W, Kim H, Kim WY. 3D molecular generative framework for interaction-guided drug design. Nat Commun. 2024;15:2688.

Index

A. Lavecchia (ed.), *Applied Artificial Intelligence for Drug Discovery*, https://doi.org/10.1007/978-3-031-98022-0

M

N

P

Q

R

S

T

U

V

W

MIX
Papier aus verantwortungsvollen Quellen
Paper from responsible sources
FSC® C105338

If you have any concerns about our products,
you can contact us on
ProductSafety@springernature.com

In case Publisher is established outside the EU,
the EU authorized representative is:
Springer Nature Customer Service Center GmbH
Europaplatz 3, 69115 Heidelberg, Germany

Printed by Libri Plureos GmbH
in Hamburg, Germany